The

AMERICAN
HERITAGE®

Science

dic·tion·ar·y

HOUGHTON MIFFLIN HARCOURT

Words included in this book that are known to have current trademark registrations are shown with initial capitals and are also identified as trademarks. No investigation has been made of common-law trademark rights in any word, because such investigation is impracticable. The inclusion of any word in this book is not, however, an expression of the Publisher's opinion as to whether or not it is subject to proprietary rights. Indeed, no word in this book is to be regarded as affecting the validity of any trademark.

Visit our website: www.houghtonmifflinbooks.com

ISBN-13: 978-0-618-88274-8
ISBN-10: 0-618-88274-X

The Library of Congress has cataloged the earlier edition as follows:

The American heritage science dictionary.– 1st ed.
p. cm.
ISBN 0-618-45504-3
1. Science–Dictionaries.
Q123.A5178 2005
503–dc22

200419696

Manufactured in the United States of America

MP 10 9 8 7 6 5 4 3 2 1

CONTENTS

ENTRIES WITH NOTES

A Closer Look

absolute zero
acid deposition
adaptation
aerodynamics
alchemy
alternation of generations
amber
antibody
artificial intelligence
aspirin
atmospheric pressure
autoimmune disease
battery
big bang
biomass
bird
black hole
blood type
Burgess Shale
cancer
capillary action
cathode-ray tube
centripetal force
circadian rhythm
color
conduction
constellation
current
cyclone
dark matter
desert
Doppler effect
dwarf star
earthquake
eclipse
electric charge
electromagnetic radiation
elementary particle
endorphin
evolution
fault
fractal
gametophyte
genetically modified organism
genomic sequencing
germination
glass
gravity
heavy water
hemoglobin
histone
hologram
hormone
infrared
iridium
laser
lightning
magnetic reversal
magnetism
meiosis
melatonin
mitosis
moon
MRI
nanotechnology
neutrino
nitric oxide
nuclear reactor
oogenesis
ozone
particle accelerator
pheromone
photosynthesis
pixel
plate tectonics
pollination
prion
protein
quantum number
radioactivity
radiocarbon dating
rainforest
red blood cell
regeneration
relativity
Schrödinger, Erwin
sex
sickle cell anemia
sodium bicarbonate
solar system
space-time
sperm
Three Age system
transpiration
turbojet
ultrasound
vaccine
vitamin
wetland
worm
zero

Biographies

Bacon, Roger
Bohr, Niels Henrik
Carver, George Washington
Copernicus, Nicolaus
Curie, Marie
Darwin, Charles
Einstein, Albert
Eratosthenes
Faraday, Michael
Fleming, Alexander
Franklin, Rosalind
Galileo Galilei
Gell-Mann, Murray
Harvey, William
Hawking, Stephen
Heisenberg, Werner
Herschel
Hutton, James
Jenner, Edward
Koch, Robert
Lavoisier, Antoine
Leakey
Leeuwenhoek, Anton van
Lyell, Charles
Maxwell, James Clerk
Mendel, Gregor
Millikan, Robert
Mitchell, Maria
Newton, Isaac
Pasteur, Louis
Pauling, Linus Carl
Priestley, Joseph
Rutherford, Ernest
Sperry, Roger Wolcott
Tesla, Nikola
Thomson, J(oseph) J(ohn)
Turing, Alan
Vesalius, Andreas
Wegener, Alfred

Usage

bacterium
berry
bug
byte
centigrade
contagious
deduction
endemic
fruit
germ
hypothesis
megabyte
metal
meteor
ocean
reduction
refraction
revolution
temperature
tidal wave
vapor
weight

EDITORIAL AND PRODUCTION STAFF

Vice President, Publisher of Dictionaries
Margery S. Berube

Vice President, Executive Editor
Joseph P. Pickett

Vice President, Managing Editor
Christopher Leonesio

Supervising Editor
Steven Kleinedler

Senior Editor
Susan Spitz

Editors
Catherine Pratt
Louise E. Robbins
Patrick Taylor

Associate Editors
Peter Chipman
Nick Durlacher
Erich Michael Groat
Uchenna Ikonné

Contributing Editors
Benjamin W. Fortson IV
Kirsten Patey Hurd
David Pritchard
Hanna Schonthal
Vali Tamm

Science Consultants
Rufus W. Burlingame III
David Hall
Allen Kropf
Rita Kropf

Contributing Writers
Agnieszka Biskup
Ann Rae Jonas
Ilene Springer

Database Production Supervisor
Christopher Granniss

Art and Production Supervisor
Margaret Anne Miles

Production Associate
Darcy Conroy

Editorial Production Associate
Brianne M. Lutfy

Administrative Coordinator
Kevin McCarthy

Intern
Ashley O'Bryan

Composition
Publication Services

Text Design
Catherine Hawkes, Cat & Mouse

FOREWORD

In college I acquired a peculiar habit. When I came across a word I didn't know, I looked it up in a dictionary, but I didn't stop there. I always took a few extra minutes to read that page of the dictionary, word for word. I was rarely disappointed during these lexicological digressions. How else would I have learned that *lexis* means "the total set of words in a language"? Or that *digression* is based on a very old Indo-European root meaning "to go or walk"?

I should be grateful that I did not have a copy of *The American Heritage Science Dictionary* in college, or I might never have graduated. Leafing through its pages is like pausing to read a chapter or two of an engaging work of nonfiction. Words, images, and ideas spill forth in profusion. The longer features headed by the phrase "A Closer Look" and the notes on word histories and usage offer a chance to reflect on the intricacies of each concept. Biographical sketches sum up the lives and work of the familiar and not-so-familiar investigators of nature. For anyone interested in the natural world and in the human ability to understand that world, this book will provide hours of pleasure.

A fellow writer once told me that everyone should read Cervantes' *Don Quixote* three times—once as an adolescent, once in middle age, and once in retirement—because each time the book conveys distinct meanings. Words, too, are understood differently at different times in life. The first time a word is encountered, the challenge is to understand what it means and to fit it into a growing body of knowledge. Later, the understanding of a word becomes deeper and richer—you come to appreciate its history and nuances. And finally, you get to see a word in broadest context, as part of the great body of knowledge that surrounds it.

Reading a dictionary is no different, really. At first, it might seem as if it would be a disjointed experience, but that's true only in one sense. On first exposure, individual words and concepts can seem remote from each other. Thus, *prime meridian* is poised between *primate* and *prime number,* which is followed by *primordial soup* and *prion.* But over time the words sort themselves into disciplines—biology, engineering, mathematics, and so on. Each word resides within its own body of knowledge, but each discipline also has connections with other bodies of knowledge. And so out of the fantastic diversity of human knowledge emerges a glimpse of unity.

In this respect, reading the dictionary reminds me of an experience that was still unknown when I was in college: surfing the World Wide Web. The Web can also seem highly disjointed. But almost anyplace on the Web can be reached in just a few clicks from another place. It is internally connected, a cohesive body of work that demonstrates how little distance separates one idea from another.

A good dictionary is a lifelong companion, continually offering new connections and new subjects to explore. The reward is not so much mastery of what a dictionary contains. It is the fun you have along the way.

Steve Olson
Author of *Mapping Human History*

PRONUNCIATION KEY

Pronunciations appear in parentheses after boldface entry words. If an entry word has a variant, and both have the same pronunciation, the pronunciation follows the variant. If the variant does not have the same pronunciation, pronunciations follow the forms to which they apply. If a word has more than one pronunciation, the first pronunciation is usually more common than the other, but often they are equally common. Pronunciations are shown within an entry where necessary.

Stress. Stress is the relative degree of emphasis with which a word's syllables are spoken. An unmarked syllable has the weakest stress in the word. The strongest, or primary, stress is indicated with a bold mark (**′**). A lighter mark (′) indicates a secondary level of stress. Words of one syllable have no stress mark, because there is no other stress level to which the syllable is compared.

Pronunciation Symbols. The pronunciation symbols used in this Dictionary are shown below. To the right of the symbols are words that show how the symbols are pronounced. The letters whose sound corresponds to the symbols are shown in boldface.

The symbol (ə) is called *schwa.* It represents a vowel with the weakest level of stress in a word. The schwa sound varies slightly according to the vowel it represents or the sounds around it:

abundant (ə-bŭn**′**dənt)
moment (mō**′**mənt)
grateful (grāt**′**fəl)
civil (sĭv**′**əl)
propose (prə-pōz**′**)

In English, the consonants *l* and *n* can be complete syllables. Examples of words with syllabic *l* and *n* are **needle** (nēd**′**l) and **sudden** (sŭd**′**n).

Foreign Symbols. Some foreign words use sounds that are not found in English. The (œ) sound is made by rounding the lips as though you were going to make the (ō) sound, but instead you make an (ā) sound. The (ü) sound is made by rounding the lips as though you were going to make the (o͞o) sound, but instead you make an (ē) sound. The (KH) sound is like a (k), but the air is forced through continuously, not stopped as with a (k). The (N) sound shows that the vowel before it is nasalized—that is, air escapes through the nose (and the mouth) when you say it.

Pronunciation Key

ă	p**a**t	ĭ	p**i**t	oi	n**oi**se	ûr	**ur**ge, t**er**m,
ā	p**ay**	ī	p**ie**, b**y**	o͝o	t**oo**k		f**ir**m, w**or**d,
âr	c**are**	îr	d**ear**, d**eer**, p**ier**	o͝or	l**ure**		h**ear**d
ä	f**a**ther	j	**j**u**dge**	o͞o	b**oo**t	v	**v**al**v**e
b	**b**i**b**	k	**k**i**ck**, **c**at, pi**que**	ou	**ou**t	w	**w**ith
ch	**ch**ur**ch**	l	**l**id, needl**e**	p	**p**o**p**	y	**y**es
d	**d**ee**d**, mill**ed**	m	**m**u**m**	r	**r**oa**r**	z	**z**ebra, **x**ylem
ĕ	p**e**t	n	**n**o, sudd**en**	s	**s**au**c**e	zh	vi**s**ion, plea-
ē	b**ee**	ng	thi**ng**	sh	**sh**ip, di**sh**		**s**ure, gara**ge**
f	**f**i**f**e, **ph**ase,	ŏ	p**o**t	t	**t**igh**t**, stopp**ed**	ə	**a**bout, it**e**m,
	rou**gh**	ō	t**oe**	th	**th**in		ed**i**ble, gall**o**p,
g	**g**a**g**	ô	c**au**ght, p**aw**	*th*	**th**is		circ**u**s
h	**h**at	ôr	c**ore**	ŭ	c**u**t	ər	butt**er**

A

A Abbreviation of **adenine, ampere, angstrom, area.**

Å Abbreviation of **angstrom.**

a– A prefix meaning "without" or "not" when forming an adjective (such as *amorphous,* without form, or *atypical,* not typical), and "absence of" when forming a noun (such as *arrhythmia,* absence of rhythm). Before a vowel or *h* it becomes *an–* (as in *anhydrous, anoxia*).

aa (ä′ä) A type of lava having a rough, jagged surface. It is relatively slow moving in its molten state, and it advances in the form of massive blocks with fissured and angular surfaces that ride on a viscous interior. The blocks range in size between the size of a football and the size of a house. See Note at **pahoehoe.**

abampere (ăb-ăm′pîr′) The unit of electromagnetic current in the centimeter-gram-second system, equal to ten amperes.

Abbevillian (ăb′ə-vĭl′ē-ən) Relating to the earliest Paleolithic archaeological sites that are found in Europe, characterized by bifacial stone hand axes.

ABC soil (ā′bē-sē′) Soil in which the A horizon, the B horizon, and the C horizon can all be seen as three distinct layers in vertical section.

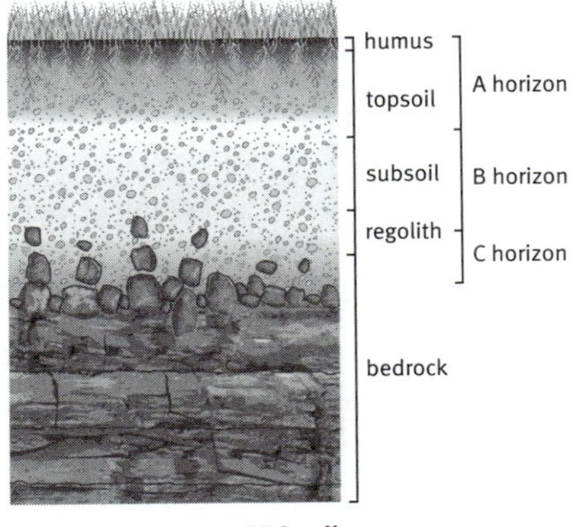

ABC soil

A soil profile showing A, B, and C horizons.

abdomen (ăb′də-mən) **1.** In vertebrates, the portion of the body between the thorax and pelvis, containing the stomach, intestines, liver, and other organs. In mammals, the abdomen is separated from the thorax by the diaphragm. **2.** In arthropods, the last, most posterior segment of the body. *—Adjective* **abdominal** (ăb-dŏm′ə-nəl).

abducent nerve (ăb-do͞o′sənt) The sixth cranial nerve, one of a pair of small motor nerves supplying the eye muscles that control lateral movement. It originates in the pons of the brain, passes through a large venous space known as the cavernous sinus, and enters the orbit of the eye.

abductor (ăb-dŭk′tər) A muscle that draws a limb or part of a limb away from the midline of the body. Compare **adductor.**

abelian group also **Abelian group** (ə-bē′lē-ən) See **commutative group.**

aberration (ăb′ə-rā′shən) **1.** A deviation in the normal structure or number of chromosomes in an organism. **2.** A defect in a lens or mirror that prevents light rays from being focused at a single point and results in a distorted or blurred image. ▸ Aberration that results in distortion of color is called **chromatic aberration.** ▸ Aberration that is caused by imperfections in the surface or shape of a spherical mirror or lens is called **spherical aberration.** See also **astigmatism, coma[2].**

abiogenesis (ā′bī-ō-jĕn′ĭ-sĭs) See **spontaneous generation.**

abiotic (ā′bī-ŏt′ĭk) Not associated with or derived from living organisms. Abiotic factors in an environment include such items as sunlight, temperature, wind patterns, and precipitation. Compare **biotic.** *—Noun* **abiosis** (ā′bī-ō′sĭs).

ablation (ă-blā′shən) **1.** The wearing away or destruction of the outer or forward surface of an object, such as a meteorite or a spacecraft, as it moves very rapidly through the atmosphere. The friction of the air striking the object heats and often melts or burns its outer layers. Spacecraft and missiles are often equipped with heat shields designed to wear away by ablation in order to prevent heat from building up in structurally important parts. **2.** The process by which snow and ice are removed from a glacier or other mass of ice. Ablation typically occurs through melting, sublimation, wind erosion, or calving. ▸ The **ablation zone** is the area of a glacier that has the lowest elevation, where annual water loss is greater than the annual accumulation of snow.

abomasum (ăb′ō-mā′səm) Plural **abomasa.** The fourth division of the stomach in ruminant animals, and the only one having glands that secrete acids and enzymes for digestion. It corresponds anatomically to the stomachs of other mammals. See more at **ruminant.**

abortion (ə-bôr′shən) **1.** Induced termination of pregnancy, involving destruction of the embryo or fetus. **2.** Any of various procedures that result in such termination. **3.** Spontaneous abortion; miscarriage. **4.** Cessation of a normal or abnormal process before completion.

ABO system (ā′bē-ō′) A classification system for human blood that identifies four major blood types based on the presence or absence of two antigens, A and B, on red blood cells. The four blood types (A, B, AB, and O, in which O designates blood that lacks both antigens) are important in determining the compatibility of blood for transfusion.

abrasion (ə-brā′zhən) **1.** The process of wearing away a surface by friction. A rock undergoes abrasion when particles of sand or small pieces of rock are carried across its surface by a glacier, stream, or the wind. **2.** A scraped area on the skin or mucous membranes.

abscess (ăb′sĕs′) A localized collection of pus surrounded by infected tissue.

abscisic acid (ăb-sĭz′ĭk) A plant hormone that maintains the water balance of plants, prevents seed embryos from germinating, and induces the dormancy of buds and seeds. *Chemical formula:* $C_{15}H_{20}O_4$.

abscissa (ăb-sĭs′ə) Plural **abscissas** or **abscissae** (ăb-sĭs′ē). The distance of a point from the y-axis on a graph in the Cartesian coordinate system. It is measured parallel to the x-axis. For example, a point having coordinates (2,3) has 2 as its abscissa. Compare **ordinate.**

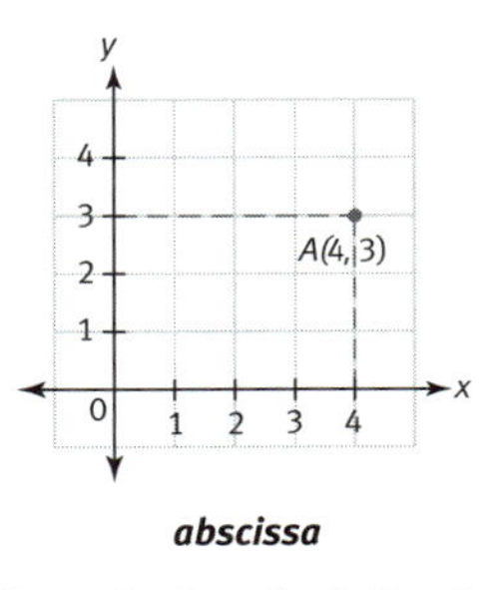

abscissa

The coordinates for A are (4,3); the abscissa is 4.

abscission (ăb-sĭzh′ən) The separation of a leaf, flower, or fruit from a plant as a result of natural structural and chemical changes. ▸ The **abscission zone** is a layer of weak, thin-walled cells that form across the base of the plant part where the break eventually occurs. A corky layer containing suberin forms beneath the abscission zone to protect the plant.

absolute convergence (ăb′sə-lo͞ot′) The mathematical property by which the sums of the absolute values of the terms in a series converge.

absolute humidity The amount of water vapor present in a unit volume of air, usually expressed in kilograms per cubic meter. Absolute humidity does not fluctuate with the temperature of the air. Compare **relative humidity.**

absolutely convergent (ăb′sə-lo͞ot′lē) Relating to or characterized by absolute convergence.

absolute magnitude See under at **magnitude.**

absolute temperature scale A temperature scale having absolute zero as the lowest temperature. Absolute temperature scales only have positive numbers. The Kelvin scale and the Rankine scale are absolute temperature scales. Compare **relative temperature scale.**

absolute value The value of a number without regard to its sign. For example, the absolute value of +3 (written $|+3|$) and the absolute value of –3 (written $|-3|$) are both 3.

absolute zero The lowest possible temperature, at which all molecules are have the least possible amount of kinetic energy. Absolute zero is equal to 0°K, –459.67°F, or –273.15°C. At temperatures approaching absolute zero, the physical characteristics of some substances change significantly. For example, some substances change from electrical insulators to conductors, while others change from conductors to insulators. Absolute zero has never been reached in laboratory experiments. See also **Bose-Einstein condensate, zero-point energy.**

A CLOSER LOOK **absolute zero**

The temperature of a substance is determined by the average velocity of its molecules: the faster they move, the warmer the substance. At *absolute zero* molecules have minimal kinetic energy (or *zero-point energy*) and heat energy cannot be extracted from them. The molecules are not motionless, however, due to the uncertainty principle of quantum mechanics, which entails that the atoms cannot have both a fixed position and zero momentum at the same time; instead, the molecules of a substance at absolute zero are always "wiggling" in some manner. Absolute zero is zero degrees Kelvin, equal to –273.15 degrees Celsius and –459.67 degrees Fahrenheit. The coldest known place in the universe is the Boomerang Nebula, where the temperature is –272° Celsius. Scientists at Massachusetts Institute of Technology have gone much lower than that by using laser traps and other techniques to cool rubidium to 2×10^{-9} degrees Kelvin.

absorptiometry (əb-sôrp′tē-ŏm′ĭ-trē) A method of chemical analysis in which a sample of a substance is exposed to electromagnetic radiation, and the amount of radiation absorbed by the sample is measured. This measurement is then used to determine the concentration or chemical composition of the substance. Absorptiometry is used in medicine to measure bone density.

absorption (əb-sôrp′shən) **1.** *Biology.* The movement of a substance, such as a liquid or solute, across a cell membrane by means of diffusion or osmosis. **2.** *Chemistry.* The process by which one substance, such as a solid or liquid, takes up another substance, such as a liquid or gas, through minute pores or spaces between its molecules. A paper towel takes up water, and water takes up carbon dioxide, by absorption. Compare **adsorption. 3.** *Physics.* The taking up and storing of energy, such as radiation, light, or sound, without it being reflected or transmitted. During absorption, the energy may change from one form into another. When radiation strikes the electrons in an atom, the electrons move to a higher orbit or state of excitement by absorption of the radiation's energy.

absorption nebula A mass of interstellar gas and dust that absorbs most or all of the light from the stars behind it. Absorption nebulae appear as irregular dark patches and collectively cover about two percent of the sky. Some absorption nebulae, especially the Coalsack near the constellations Centaurus and Southern Cross, are visible to the naked eye. Also called *dark cloud, dark nebula.* See more at **nebula.**

absorption spectrum The range of frequencies of electromagnetic radiation readily absorbed by a substance by virtue of its chemical composition. See more at **atomic spectrum.**

abyssal (ə-bĭs′əl) **1.** Relating to the greatest depths of the oceans and to the organisms that live there. The abyssal zone encompasses all depths below 4,000 m (13,120 ft). **2.** Relating to the region of the ocean bottom between the **bathyal** and **hadal zones**, from depths of approximately 2,000 to 6,000 m (6,560 to 19,680 ft). The abyssal zone includes nearly two-thirds of the Earth's surface.

abyssopelagic zone (ə-bĭs′ə-pə-lăj′ĭk) A layer of the oceanic zone lying below the **bathypelagic zone** and above the **hadopelagic zone**, at depths generally between about 4,000 and 6,000 m (13,120–19,680 ft). The abyssopelagic zone ranges in temperature from 10° to 4°C (50° to 39°F).

Ac The symbol for **actinium.**

AC Abbreviation of **alternating current.**

acanthocephalan (ə-kăn′thə-sĕf′ə-lən) Any of various, mostly small worms of the phylum Acanthocephala that live parasitically in arthropods as juveniles and in the intestines of vertebrates (especially fish) as adults. Acanthocephalans have a cylindrical, retractile proboscis that has rows of hooked spines and is used for attachment to a host. Also called *spiny-headed worm.*

acanthopterygian (ăk′ən-thŏp′tə-rĭj′ē-ən) Any of a large group of teleost fishes of the superorder Acanthopterygii, having spiny rays in the dorsal and anal fins and including the bass, perch, mackerel, and swordfish.

acarid (ăk′ə-rĭd) An arachnid of the order Acarina, which includes the mites and ticks. Acarids are small to minute, have no division between the cephalothorax and abdomen, and are often parasitic.

accelerant (ăk-sĕl′ər-ənt) A substance, such as a petroleum distillate, that is used as a catalyst, as in spreading an intentionally set fire.

acceleration (ăk-sĕl′ə-rā′shən) The rate of change of the velocity of a moving body. An increase in the magnitude of the velocity of a moving body (an increase in speed) is called a positive acceleration; a decrease in speed is called a negative acceleration. Acceleration, like velocity, is a vector quantity, so any change in the direction of a moving body is also an acceleration. A moving body that follows a curved path, even when its speed remains constant, is undergoing acceleration. See more at **gravity, relativity.**

acceleration of gravity The acceleration of a body falling freely under the influence of the Earth's gravitational pull at sea level. It is approximately equal to 9.806 m (32.16 ft) per second per second, though its measured value varies slightly with latitude and longitude. Also called *acceleration of free fall.*

acceptor (ăk-sĕp′tər) **1.** The reactant in an induced chemical reaction that has an increased rate of reaction in the presence of the inductor. **2.** An atom or molecule that receives one or more electrons from another atom or molecule, resulting in a chemical bond or flow of electric current. Compare **donor.** See also **electron carrier.**

accessory bud (ăk-sĕs′ə-rē) See under **bud.**

accessory cell See **subsidiary cell.**

accessory fruit A fruit, such as the pear or strawberry, that develops from a ripened ovary or ovaries but also has tissue derived from part of the plant outside the ovary. In pears and other pomes, the edible flesh is a swollen part of the stem, and the fruit is the seed-bearing core. In the strawberry, the receptacle is fleshy, and the fruits are achenes embedded in its surface. Also called *false fruit, pseudocarp.* Compare **aggregate fruit, multiple fruit, simple fruit.**

accessory mineral A mineral that is present in a minor amount in rocks and is not considered an essential constituent of the rock.

accommodation (ə-kŏm′ə-dā′shən) The adjustment in the focal length of the lens of the eye. Accommodation permits images at different distances to be focused on the retina.

accretion (ə-krē′shən) **1.** *Geology.* The gradual extension of land by natural forces, as in the addition of sand to a beach by ocean currents, or the extension of a floodplain through the deposition of sediments by repeated flooding. **2.** *Astronomy.* The accumulation of additional mass in a celestial object by the drawing together of interstellar gas and surrounding objects by gravity.

accretion disk A spinning disk of gas and dust surrounding a celestial object with an intense gravitational field, such as a star or a black hole. In binary star systems, the gravitational attraction of the denser star can pull matter from the other star into an accretion disk in its own orbit. The material in the accretion disk eventually spirals into the attracting star and adds to its mass. The gas in accretion disks that surround black holes becomes condensed and heated as it is sucked into the hole, emitting x-rays and other radiation that provide evidence for the presence of the black hole.

ACE inhibitor (ās) Short for *angiotensin converting enzyme inhibitor.* Any of a class of drugs that cause vasodilation by inactivating an enzyme that converts angiotensin I to the vasoconstrictor angiotensin II, used in the treatment of hypertension, congestive heart failure, and other cardiovascular disorders. See also **angiotensin.**

acellular (ā-sĕl′yə-lər) Devoid of cells. The hyphae of some fungi are acellular.

–aceous A suffix used to form adjectives meaning "made of" or "resembling" a particular substance or material, such as *silicaceous,* containing silicon.

acetaldehyde (ăs′ĭ-tăl′də-hīd′) A colorless, flammable liquid, used to manufacture acetic acid, perfumes, and drugs. *Chemical formula:* $\mathbf{C_2H_4O}$.

acetamide (ə-sĕt′ə-mīd′, ăs′ĭt-ăm′ĭd′) The crystalline amide of acetic acid, used as a solvent and wetting agent and in lacquers and explosives. *Chemical formula:* $\mathbf{CH_3CONH_2}$.

acetaminophen (ə-sē′tə-mĭn′ə-fən, ăs′ə-) A crystalline compound used in medicine to relieve pain and reduce fever. *Chemical formula:* $C_8H_9NO_2$.

acetate (ăs′ĭ-tāt′) **1.** A salt or ester of acetic acid. Salts of acetic acid contain a metal attached to the acetic acid radical CH_3COO. Esters contain another radical, such as ethyl, attached to the acetic acid radical. **2.** Cellulose acetate or a product made from it, especially fibers or film.

acetic (ə-sē′tĭk) Relating to or containing acetic acid or vinegar.

acetic acid A clear, colorless organic acid having a distinctive pungent odor. It is used as a solvent and in the manufacture of rubber, plastics, acetate fibers, pharmaceuticals, and photographic chemicals. Acetic acid is the chief acid of vinegar. *Chemical formula:* $C_2H_4O_2$.

acetone (ăs′ĭ-tōn′) A colorless, volatile, extremely flammable liquid ketone that is widely used as a solvent, for example in nail-polish remover. *Chemical formula:* C_3H_6O.

acetyl (ə-sēt′l, ăs′ĭ-tl) The radical CH_3CO, derived from acetic acid.

acetylcholine (ə-sēt′l-kō′lēn′) A substance that is released at the junction between neurons and skeletal muscle fibers, at the nerve endings of the parasympathetic nervous system, and across synapses in the central nervous system, where it acts as a neurotransmitter. *Chemical formula:* $C_7H_{16}NO_2$.

acetyl coenzyme A A compound that functions as a coenzyme in many biological reactions. It is formed as an intermediate step in the oxidation of carbohydrates, fats, and proteins. Also called *acetyl CoA. Chemical formula:* $C_{23}H_{38}N_7O_{17}P_3S$.

acetylene (ə-sĕt′l-ēn′, -ən) A colorless, highly flammable or explosive gas with a characteristic sweet odor. It is used in welding torches and in the manufacture of organic chemicals such as vinyl chloride. Acetylene is the simplest alkyne, consisting of two carbon atoms joined by a triple bond and each attached to a single hydrogen atom. Also called *ethyne. Chemical formula:* C_2H_2.

acetylene series The alkyne series. See under **alkyne.**

acetylsalicylic acid (ə-sēt′l-săl′ĭ-sĭl′ĭk) See **aspirin.**

achene also **akene** (ā-kēn′) A small, dry, one-seeded fruit in which the seed sits free inside the hollow fruit, attached only by the stem of the ovule. Achenes are indehiscent (they do not split open when ripe). The fruits of the sunflower and elm are achenes.

Acheulian also **Acheulean** (ə-sho͞o′lē-ən) Relating to a stage of tool culture of the Lower Paleolithic Period, dating from around 1.5 million to 150,000 years ago, characterized especially by flaked bifacial hand axes, cleavers, and other **core tools**. The earliest Acheulian tools have been found at African sites associated with *Homo erectus* fossils. Later examples have been found at numerous sites in the Middle East, Europe, and western and southern Asia.

achondrite (ā-kŏn′drīt′) A stony meteorite that does not contain chondrules.

achromatic (ăk′rə-măt′ĭk) Designating color perceived to have zero saturation and therefore no hue, such as neutral grays, white, or black.

acid (ăs′ĭd) Any of a class of compounds that form hydrogen ions when dissolved in water, and whose aqueous solutions react with bases and certain metals to form salts. Acids turn blue litmus paper red and have a pH of less than 7. Their aqueous solutions have a sour taste. Compare **base.** —*Adjective* **acidic.**

acid-base equilibrium The state that exists when acidic and basic ions in solution exactly neutralize each other.

acid-base indicator A substance, such as litmus paper, that indicates the degree of acidity or basicity of a solution through characteristic color changes.

acid deposition The accumulation of acids or acidic compounds on the surface of the Earth, in lakes or streams, or on objects or vegetation near the Earth's surface, as a result of their separation from the atmosphere. Acid deposition can harm the environment in a variety of ways, as by causing the acidification of lakes and streams, the leaching of minerals and other nutrients from soil, and the inhibition of nitrogen fixation and photosynthesis in plants. ► The accumulation of acids that fall to the Earth dissolved in water is known as **wet deposition**. Wet deposition includes all

A CLOSER LOOK **acid deposition**

Acid deposition—usually referred to simply as *acid rain*—actually includes two forms of pollution, wet and dry. When fossil fuels such as coal, gasoline, and oil are burned, they release the gases sulfur dioxide and nitrogen oxide. In the wet type of acid deposition, these compounds combine with water vapor in the atmosphere to form highly corrosive sulfuric and nitric acids. Prevailing winds carry the acids away from the industrial areas where they originate, and they fall to earth as rain, snow, or fog. In the dry type, the prevailing winds deposit acidic gases and particulate matter on objects in the open such as buildings, vehicles, and trees. When rain washes away this acidic matter, the runoff is even more acidic than the rainwater. Acid deposition is a serious environmental problem in parts of the world with a high density of factories, power plants, and automobiles, including much of the United States and Canada, as well as areas of Europe and Asia. It harms forests and soils and pollutes lakes and rivers, killing fish and other aquatic life. It also affects human health, contributing to respiratory diseases such as asthma and emphysema. Many scientists believe that some environmental damage caused by acid deposition could take years, even decades or centuries, to repair. Acid deposition can also damage historic buildings and monuments by corroding the stone and metal of which they are constructed. Individual and societal efforts to reduce acid deposition involve many interrelated social, economic, and political factors.

pockmarked marble gargoyle, Notre Dame Cathedral, Paris

forms of acid precipitation such as acid rain, snow, and fog. ► The accumulation of acidic particles that settle out of the atmosphere or of acidic gases that are absorbed by plant tissues or other surfaces is known as **dry deposition**.

acid precipitation Any form of precipitation, including rain, snow, hail, fog, or dew, that is high in acid pollutants, especially sulfuric and nitric acid. Acid precipitation has a pH of less than 5.6 (the normal acidity of unpolluted atmospheric water) and is often less than pH 5.0. Also called *acid rain.* See Note at **acid deposition.**

acid reflux See **heartburn.**

acne (ăk′nē) An inflammatory disease of the skin in which the sebaceous glands become clogged and infected, often causing the formation of pimples, especially on the face. It is most common during adolescence, but also occurs in infants and adults.

acoustics (ə-ko͞o′stĭks) **1.** *Used with a singular verb.* The scientific study of sound and its transmission. **2.** *Used with a plural verb.* The total effect of sound, especially as produced in an enclosed space.

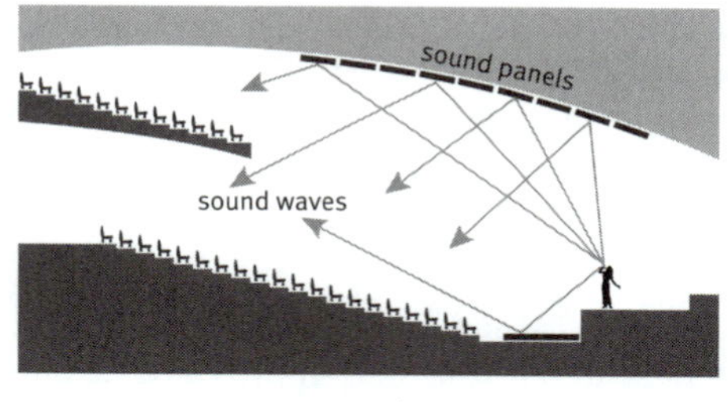

acoustics

Sound waves from a stage are deflected by sound panels and distributed throughout an auditorium.

acquired characteristic (ə-kwīrd′) A nonhereditary change of function or structure in a plant or animal made in response to the environment. Acquired characteristics include bodily changes brought about by disease or by repeated use or disuse of a body part (as in the building or atrophy of muscle tissue). The heritability of acquired characteristics was advocated by certain biological theorists like Jean-Baptiste Lamarck and rejected by Charles Darwin in his formulation of the theory of evolution.

acquired immune deficiency syndrome See **AIDS.**

acquired immunity Immunity that is not inherited. Acquired immunity can be active or passive. ▸ **Active immunity** results from the development of antibodies in response to an antigen, as from exposure to an infectious disease or through vaccination. ▸ **Passive immunity** results from the transmission of antibodies, as from mother to fetus through the placenta or by the injection of antiserum.

acre (ā′kər) A unit of area in the US Customary System, used in land and sea floor measurement and equal to 43,560 square feet or 4,047 square meters.

acrylamide (ə-krĭl′ə-mīd′) A chemical compound that is derived from acrylic acid and easily forms polymers. Acrylamide is used in synthetic fibers and sewage treatment and as a medium in electrophoresis, especially to separate macromolecules such as proteins and nucleic acids. *Chemical formula:* C_3H_5NO.

acrylic (ə-krĭl′ĭk) **1.** An acrylic resin. **2.** A paint containing acrylic resin. **3.** An acrylic fiber.

acrylic acid A colorless, corrosive liquid that readily forms polymers. It is used to make plastics, paints, synthetic rubbers, and textiles. *Chemical formula:* $C_3H_4O_2$.

acrylic fiber Any of numerous synthetic fibers polymerized from acrylonitrile.

acrylic resin Any of numerous thermoplastic or thermosetting polymers or copolymers of acrylic acid, methacrylic acid, esters of these acids, or acrylonitrile, used to produce paints, synthetic rubbers, and lightweight plastics.

acrylonitrile (ăk′rə-lō-nī′trəl, -trēl, -trīl) A colorless, poisonous, liquid organic compound having a wide variety of industrial uses, such as in the manufacture of acrylic fibers, resins, and acrylamide. *Chemical formula:* C_3H_3N.

ACTH Abbreviation of **adrenocorticotropic hormone.**

actin (ăk′tĭn) A protein found in all eukaryotic cells, forming filaments that make up a main component of the cell's supporting matrix or **cytoskeleton**. Actin and the protein myosin together make up the contractile units (called **sarcomeres**) of skeletal muscle fibers.

actinide (ăk′tə-nīd′) Any of a series of chemically similar metallic elements with atomic numbers ranging from 89 (actinium) to 103 (lawrencium). All of these elements are radioactive, and two of the elements, uranium and plutonium, are used to generate nuclear energy. See **Periodic Table.**

actinium (ăk-tĭn′ē-əm) *Symbol* **Ac** A silvery-white, highly radioactive metallic element of the actinide series that is found in uranium ores. It is about 150 times more radioactive than radium and is used as a source of alpha rays and neutrons. Its most stable isotope has a half-life of about 22 years. Atomic number 89; melting point 1,050°C (1,922°F); boiling point (estimated) 3,200°C (5,792°F); specific gravity (calculated) 10.07; valence 3. See **Periodic Table.**

actinolite (ăk-tĭn′ə-līt′) A greenish variety of amphibole. Actinolite is a monoclinic mineral, and occurs in long, slender, green needlelike crystals, or in fibrous, radiated forms in metamorphic rocks. *Chemical formula:* $Ca_2(Mg,Fe)_5Si_8O_{22}OH_2$.

actinometer (ăk′tə-nŏm′ĭ-tər) Any of several instruments used to measure radiation, such as a pyrheliometer.

actinomorphic (ăk′tə-nō-môr′fĭk) Relating to a flower that can be divided into equal halves along any diameter; radially symmetrical. The flowers of the rose and tulip are actinomorphic. Compare **zygomorphic.**

actinomycete (ăk′tə-nō-mī′sēt′) Any of various bacteria belonging to the phylum Actinobacteria that grow as branching filaments resembling fungal hyphae and are found in soil. The filaments often grow in colonies but sometimes break off into rod-shaped structures. Many species of actinomycetes produce important antibiotics such as streptomycin, while others are pathogenic in humans and other animals, especially for skin diseases. One species lives symbiotically in the roots of alders and conducts nitrogen fixation. Because of their resemblance to fungi, actinomycetes were once classified as fungi.

actinopterygian (ăk′tə-nŏp′tə-rĭj′ē-ən) See **ray-finned fish.**

action potential (ăk′shən) A momentary change in electrical potential on the surface of a neuron or muscle cell. Nerve impulses are action potentials. They either stimulate a change in polarity in another neuron or cause a muscle cell to contract.

activated charcoal (ăk′tə-vā′tĭd) Highly absorbent carbon obtained by heating granulated charcoal to expel any gases it contains, resulting in a highly porous form with a very large surface area. It is used primarily for purifying gases by adsorption, solvent recovery, or deodorization and as an antidote to certain poisons.

activation energy (ăk′tə-vā′shən) The least amount of energy needed for a chemical reaction to take place. Some elements and compounds react together naturally just by being close to each other, and their activation energy is zero. Others will react together only after a certain amount of energy is added to them. Striking a match on the side of a matchbox, for example, provides the activation energy (in the form of heat produced by friction) necessary for the chemicals in the match to ignite. Activation energy is usually expressed in terms of joules per mole of reactants.

active galactic nucleus (ăk′tĭv) See under **galactic nucleus.**

active galaxy A galaxy emitting unusually high quantities of radiation from an active galactic nucleus at its center. Some active galaxies emit more energy in radio wavelengths than they do visible light. **Seyfert galaxies** are examples of active galaxies. See more at **galactic nucleus.**

active immunity See under **acquired immunity.**

active margin See **convergent plate boundary.**

active site The part of an enzyme where the catalytic activity on the substrate takes place. See more at **enzyme.**

active transport The movement of ions or molecules across a cell membrane in the direction opposite that of diffusion, that is, from an area of lower concentration to one of higher concentration. Active transport requires the assistance of a type of protein called a carrier protein, using energy supplied by ATP.

acupuncture (ăk′yo͝o-pŭngk′chər) The practice of inserting thin needles into the body at specific points to relieve pain, treat a disease, or anesthetize a body part during surgery. Acupuncture has its origin in traditional Chinese medicine and has been in use for more than 5,000 years.

acute (ə-kyo͞ot′) **1.** Reacting readily to stimuli or impressions, as hearing or eyesight; sensitive. **2.** Relating to an illness that has a rapid onset and follows a short but severe course. Compare **chronic. 3.** Having an acute angle.

acute angle An angle whose measure is between 0° and 90°. Compare **obtuse angle.**

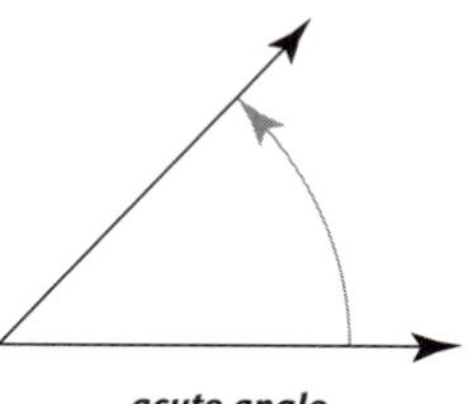

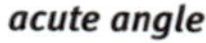

acute angle

acyclic (ā-sī′klĭk, ā-sĭk′lĭk) **1.** Not cyclic. Used especially of flowers whose parts are arranged in spirals rather than in whorls, as in magnolias. **2.** Having an open-chain molecular structure rather than a ring-shaped structure.

acyl (ăs′əl) An organic compound containing the group RCO, where R is a halogen. Acyls are formed from organic acids by replacing the hydroxyl group (OH) within the carboxyl group (COOH) with a halogen. For example, acetyl chloride (CH_3COCl) is formed by replacing the hydroxyl group of acetic acid (CH_3COOH) with a chlorine atom. The names of acyls are formed by replacing the suffix *–ic* of the acid's name with *–yl.*

Adams (ăd′əmz), **Walter Sydney** 1876–1956. American astronomer who demonstrated that the essential brightness of a star could be calculated by studying its spectrum and who introduced a method for measuring the distance of stars based on their brightness. In 1915 he discovered Sirius B, the first known white dwarf star, and his measurement of the gravitational red shift in the light leaving its surface was accepted as evidence for Albert Einstein's general theory of relativity.

adaptation (ăd′ăp-tā′shən) A change in structure, function, or behavior by which a species or individual improves its chance of survival in a specific environment. Adaptations develop as the result of natural selection operating on random genetic variations that are capable of being passed from one generation to the next. Variations that prove advantageous will tend to spread throughout the population.

A CLOSER LOOK **adaptation**

The gazelle is extremely fast, and the cheetah is even faster. These traits are *adaptations*—characteristics or behaviors that give an organism an edge in the struggle for survival. Darwinian theory holds that adaptations are the result of a two-stage process: *random variation* and *natural selection.* Random variation results from slight genetic differences. For example, one cheetah in a group may be slightly faster than the others and thus have a better chance of catching a gazelle. The faster cheetah therefore has a better chance of being well-fed and living long enough to produce offspring. Since the cheetah's young have the same genes that made this parent fast, they are more likely to be fast than the young of slower cheetahs. The process is repeated in each generation, and thereby great speed becomes an adaptation common to cheetahs. This same process of natural selection, in which the organisms best adapted to their environment tend to survive and transmit their genetic characteristics in increasing numbers to succeeding generations while those less adapted tend to be eliminated, also favors the fastest gazelles. Though evolution, in this case, may be thought of as an "arms race," animals may also adapt to their environment in a process known as *adaptive radiation,* as the so-called Darwin's finches in the Galápagos have done. On the islands, one type of finch gradually gave rise to some 13 different species of birds with differently shaped beaks, each species having adapted to its varying food niches and feeding habits. And, though we seldom think of it, humans also have an impact on an organism's adaptation to its environment. For instance, because of the misuse of antibiotics, some disease-causing bacteria have rapidly adapted to become resistant to the drugs.

adaptive radiation (ə-dăp′tĭv) The evolutionary diversification of a species or single ancestral lineage into various forms that are each adaptively specialized to a specific environmental niche. Adaptive radiation generally proceeds most rapidly in environments where there are numerous unoccupied niches or where competition for resources is minimal. See Note at **adapation.**

A/D converter Short for *analog to digital converter.* An electronic device that takes a voltage source as an input, and yields a digital numerical output indicating the strength of the voltage at the source. A/D converters are used in digital recording devices, in which an audio signal in an electronic circuit (as picked up by a microphone) is digitized and recorded on disk or tape. Compare **D/A converter.**

ADD Abbreviation of **attention deficit disorder.**

addend (ăd′ĕnd′) A number that is added to another number.

addiction (ə-dĭk′shən) **1.** A physical or psychological need for a habit-forming substance, such as a drug or alcohol. In physical addiction, the body adapts to the substance being used and gradually requires increased amounts to reproduce the effects originally produced by smaller doses. See more at **withdrawal. 2.** A habitual or compulsive involvement in an activity, such as gambling.

addition (ə-dĭsh′ən) The act, process, or operation of adding two or more numbers to compute their sum.

additive (ăd′ĭ-tĭv) *Noun.* **1.** A substance added in small amounts to something else to improve, strengthen, or otherwise alter it. Additives are used for a variety of reasons. They are added to food, for example, to enhance taste or color or to prevent spoilage. They are added to gasoline to reduce the emission of greenhouse gases, and to plastics to enhance molding capability. —*Adjective.* **2.** Relating to the production of color by the mixing of light rays of varying wavelengths. ► The **additive primaries** red, green, and blue are those colors whose wavelengths can be mixed in different proportions to produce all other spectral colors. Compare **subtractive.** See Note at **color. 3.** *Mathematics.* Marked by, produced by, or involving addition.

adductor (ə-dŭk′tər) A muscle that draws a limb or part of a limb toward the midline of the body. Compare **abductor.**

adenine (ăd′n-ēn′) A purine base that is a component of DNA and RNA, forming a base pair with thymine in DNA and with uracil in RNA. Adenine is also part of other biologically important compounds, such as ATP, NAD, and vitamin B-12, and occurs in tea. *Chemical formula:* $C_5H_5N_5$.

adenoids (ăd′n-oidz′) The small mass of lymphoid tissue located at the back of the nasal cavity. Swelling of the adenoids can block breathing through the nose.

adenosine (ə-dĕn′ə-sēn′) A compound consisting of adenine combined with ribose. Adenosine is one of the nucleotides in DNA and is also a component of ADP, AMP, and ATP. *Chemical formula:* $C_{10}H_{13}N_5O_4$.

adenosine diphosphate (dī-fŏs′fāt′) See **ADP.**

adenosine monophosphate (mŏn′ō-fŏs′fāt′) See **AMP.**

adenosine triphosphate (trī-fŏs′fāt′) See **ATP.**

adenovirus (ăd′n-ō-vī′rəs) Any of a group of DNA-containing viruses of the family *Adenoviridae* that commonly cause conjunctivitis, gastroenteritis, and respiratory infections such as colds, bronchitis, and pneumonia in humans.

ADH Abbreviation of **antidiuretic hormone.**

ADHD Abbreviation of **attention deficit hyperactivity disorder.**

adhesion (ăd-hē′zhən) **1.** The force of attraction that causes two different substances to join. Adhesion causes water to spread out over glass. Compare **cohesion. 2.** A fibrous band of abnormal tissue that binds together tissues that are normally separate. Adhesions form during the healing of some wounds, usually as a result of inflammation.

adhesive (ăd-hē′sĭv) *Noun.* **1.** A substance, such as paste or cement, that causes two surfaces to stick together. Adhesives are made of gelatin or other substances, such as epoxy, resin, or polyethylene. —*Adjective.* **2.** Relating to adhesion.

adiabatic (ăd′ē-ə-băt′ĭk) Occurring without gain or loss of heat. When a gas is compressed under adiabatic conditions, its pressure increases and its temperature rises without the gain or loss of any heat. Conversely, when a gas expands under adiabatic conditions, its pressure and temperature both decrease without the gain or loss of heat. The adiabatic cooling of air as it rises in the atmosphere is the main cause of cloud formation.

adipose (ăd′ə-pōs′) Relating to or consisting of animal fat. ► **Adipose tissue** is a type of connective tissue consisting of **adipose cells,** which are specialized to produce and store large fat globules. These globules are composed mainly of glycerol esters of oleic, palmitic, and stearic acids. Adipose tissue is the main reservoir of fat in animals.

adjacent angle (ə-jā′sənt) Either of two angles having a common side and a common vertex.

admittance (ăd-mĭt′ns) A measure of the ability of a circuit or component to allow current flow when exposed to AC voltages (its AC **conductance**). It is equal to the reciprocal of the impedance of the circuit, just as **conductivity** is equal to the reciprocal of **resistance**, and is similarly measured in mhos.

adnate (ăd′nāt′) *Botany* Joined to a part or organ of a different kind, as stamens that are joined to petals. Compare **connate.**

adolescence (ăd′l-ĕs′əns) The period of physical and psychological development from puberty to the onset of adulthood.

ADP (ā′dē′pē′) Short for *adenosine diphosphate.* An organic compound that is composed of adenosine and two phosphate groups. With the addition of another phosphate group, it is converted to ATP for the storage of energy during cell metabolism. It then forms again, from ATP, when a phosphate group is removed to release energy. *Chemical formula:* $C_{10}H_{15}N_5O_{10}P_2$.

adrenal gland (ə-drē′nəl) Either of two small endocrine glands, one located above each

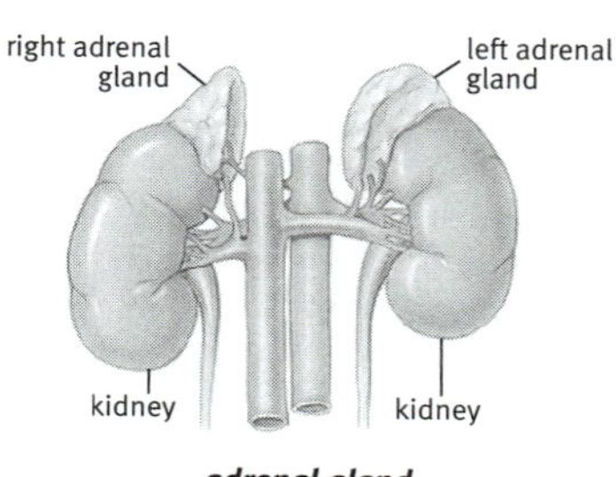

adrenal gland

placement of adrenal glands relative to kidneys

kidney. The outer portion, or cortex, secretes steroid hormones (corticosteroids). The inner portion, or medulla, secretes epinephrine and norepinephrine.

adrenaline (ə-drĕn′ə-lĭn) See **epinephrine.**

adrenergic (ăd′rə-nûr′jĭk) **1.** Relating to a neuron or axon that is activated by or capable of releasing epinephrine or an epinephrine-like substance when a nerve impulse passes. The nerve endings of the sympathetic nervous system are adrenergic. **2.** Having physiological effects similar to those of epinephrine, as certain drugs.

adrenocorticotropic hormone (ə-drē′nō-kôr′tĭ-kō-trŏp′ĭk, -trō′pĭk) A polypeptide hormone secreted by the anterior portion of the pituitary gland. It stimulates the adrenal glands to produce cortisol and other steroid hormones.

adsorption (ăd-sôrp′shən) The process by which molecules of a substance, such as a gas or a liquid, collect on the surface of another substance, such as a solid. The molecules are attracted to the surface but do not enter the solid's minute spaces as in absorption. Some drinking water filters consist of carbon cartridges that adsorb contaminants. Compare **absorption** (sense 2).

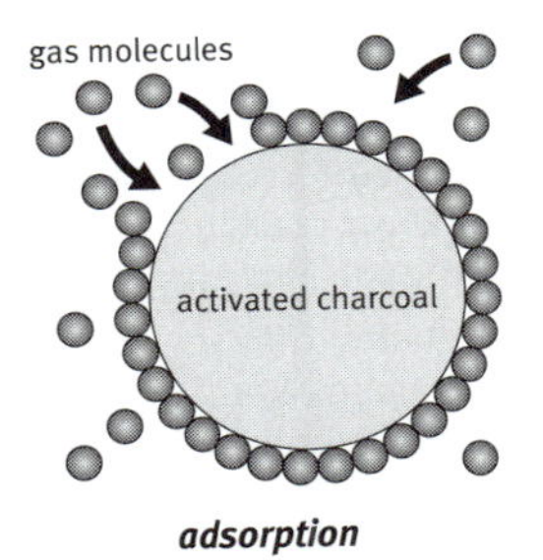

adsorption

Gas molecules are attracted to the surface of activated charcoal but are not absorbed by it.

advection (ăd-vĕk′shən) **1.** The transfer of a property of the atmosphere, such as heat, cold, or humidity, by the horizontal movement of an air mass. **2.** The rate of change of an atmospheric property caused by the horizontal movement of air. **3.** The horizontal movement of water, as in an ocean current.

adventitious root (ăd′vĕn-tĭsh′əs) A root growing from a location other than the underground, descending portion of the axis of a plant, as from a stem or leaf.

adventive (ăd-vĕn′tĭv) Not native to and not fully established in a new habitat or environment. An adventive plant may be locally or temporarily naturalized without finding conditions that allow it to spread more widely.

aeciospore (ē′sē-ə-spôr′, -shē-) A spore produced in the aecium of a rust fungus. Each spore has two nuclei and is part of a chain of spores.

aecium (ē′sē-əm, ē′shē-əm) Plural **aecia** (ē′sē-ə, ē′shē-ə). A cuplike structure of some rust fungi that contains chains of aeciospores. Aecia usually form on the bottom surface of leaves.

aeolian (ē-ō′lē-ən) See **eolian.**

aerate (âr′āt) **1.** To add a gas, such as carbon dioxide, to a liquid. **2.** To supply with oxygen. Blood is aerated in the alveoli of the lungs. **3.** To supply with air or expose to the circulation of air.

aerenchyma (â-rĕng′kə-mə) A spongy tissue with large air spaces found between the cells of the stems and leaves of aquatic plants. Aerenchyma provides buoyancy and allows the circulation of gases.

aerial root (âr′ē-əl) A root that develops from a location on a plant above the surface of the earth or water, as from a stem. For example, some orchids have aerial roots that grow from their stems and absorb water directly from the air.

aerial root

of a mangrove tree (Kandelia candel)

aeroallergen (âr′ō-ăl′ər-jən) Any of various airborne substances, such as pollen or spores, that can cause an allergic response.

aerobe (âr′ōb′) An organism, such as a bacterium, that can or must live in the presence of oxygen. Compare **anaerobe.**

aerobic (â-rō′bĭk) Occurring in the presence of oxygen or requiring oxygen to live. In aerobic respiration, which is the process used by the cells of most organisms, the production of energy from glucose metabolism requires the presence of oxygen. Compare **anaerobic.**

aerobiology (âr′ō-bī-ŏl′ə-jē) The scientific study of the sources, dispersion, and effects of airborne biological materials, such as pollen, spores, and microorganisms.

aerodynamic (âr′ō-dī-năm′ĭk) Designed to reduce or minimize the drag caused by air as an object moves though it or by wind that strikes and flows around an object. The wings and bodies of airplanes have an aerodynamic shape.

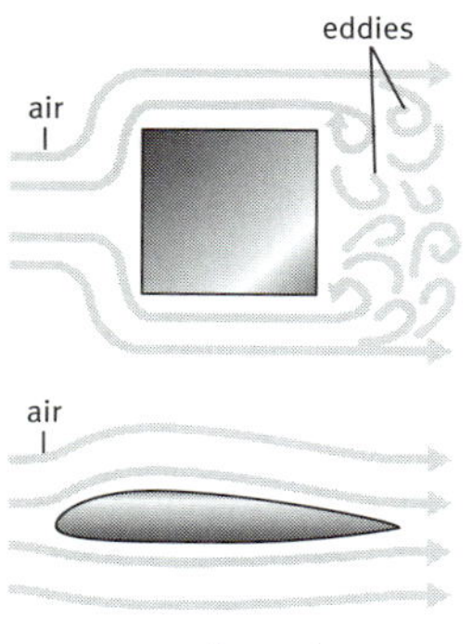

aerodynamic

top: *high drag on a less aerodynamic shape;*
bottom: *low drag on a more aerodynamic shape*

aerodynamics (âr′ō-dī-năm′ĭks) The study of the movement of air and other gases. Aerodynamics includes the study of the interactions of air with moving objects, such as airplanes, and of the effects of moving air on stationary objects, such as buildings.

A CLOSER LOOK **aerodynamics**

The two primary forces in *aerodynamics* are lift and drag. *Lift* refers to (usually upward) forces perpendicular to the direction of motion of an object traveling through the air. For example, airplane wings are designed so that their movement through the air creates an area of low pressure above the wing and an area of high pressure beneath it; the pressure difference produces the lift needed for flight. This effect is typical of *airfoil* design. *Drag* forces are parallel and opposite to the object's direction of motion and are caused largely by friction. Large wings can create a significant amount of lift, but they do so with the expense of generating a great deal of drag. Spoilers that are extended on airplane wings upon the vehicle's landing exploit this tradeoff by making the wings capable of high lift even at low speeds; low landing speeds then still provide enough lift for a gentle touchdown. Aeronautical engineers need to take into account such factors as the speed and altitude at which their designs will fly (lower air pressures at high altitudes reduce both lift and drag) in order to optimally balance lift and drag in varying conditions.

aerology (â-rŏl′ə-jē) The branch of meteorology that studies the total vertical extent of the Earth's atmosphere as opposed to the atmosphere near the Earth's surface only. The most commonly studied atmospheric factors in aerology are air temperature, atmospheric pressure, humidity, wind, and ozone levels. Radioactivity and some aspects of long-wave radiation are also studied.

aeromagnetics (âr′ō-măg-nĕt′ĭks) The scientific study of the Earth's magnetic characteristics as measured from the air.

aeronautics (âr′ə-nô′tĭks) **1.** The design, construction and operation of aircraft. **2.** The scientific study of flight through the atmosphere.

aeronomy (â-rŏn′ə-mē) The scientific study of the upper atmospheric regions of the Earth and other planets, where the ionization of gas takes place. Aeronomy is also concerned with the atmospheres around comets and satellites, or any other atmosphere where ionization, particularly of oxygen, takes place.

aerophyte (âr′ə-fīt′) See **epiphyte.**

aerosol (âr′ə-sôl′) **1.** A substance consisting of very fine particles of a liquid or solid suspended in a gas. Mist, which consists of very fine droplets of water in air, is an aerosol. Compare **emulsion, foam. 2.** A liquid substance, such as paint, an insecticide, or a hair spray, packaged under pressure for use or application as a fine spray.

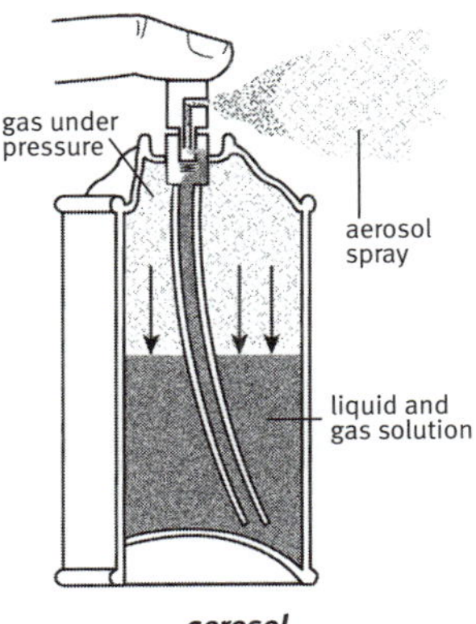

aerosol

aerospace (âr′ō-spās′) **1.** Relating to the Earth's atmosphere and the space beyond. **2.** Relating to the science and technology of flight and space travel.

aestivation (ĕs′tə-vā′shən) Another spelling of **estivation.**

affective disorder (ə-fĕk′tĭv) See **mood disorder.**

afferent (ăf′ər-ənt) Carrying sensory information toward a central organ or part, as a nerve that conducts impulses from the periphery of the body to the central nervous system. Compare **efferent.**

affinity (ə-fĭn′ĭ-tē) **1.** A relationship or resemblance in structure between species that suggests a common origin. **2.** An attraction or force between particles that causes them to combine, as the attraction between an antigen and an antibody.

afterbirth (ăf′tər-bûrth′) The placenta and fetal membranes expelled from the uterus following the birth of a mammal.

aftershock (ăf′tər-shŏk′) A less powerful earthquake that follows a more forceful one. Aftershocks usually originate at or near the focus of the main earthquakes they follow and can continue for days or months. They usually decrease in magnitude and frequency with time.

Ag The symbol for **silver.**

agar (ā′gär′, ä′gär′) A gelatinous material obtained from marine algae, especially seaweed, used as a medium for growing bacterial cultures in the laboratory and as a thickener and stabilizer in food products.

Agassiz (ăg′ə-sē), **(Jean) Louis (Rodolphe)** 1807–1873. Swiss-born American naturalist whose studies of glaciers and their movement introduced the idea of the ice age in 1840. Agassiz later revolutionized science education in the United States by emphasizing direct observation of the natural environment.

Louis Agassiz

agate (ăg′ĭt) A type of very fine-grained quartz found in various colors that are arranged in bands or in cloudy patterns. The bands form when water rich with silica enters empty spaces in rock, after which the silica comes out of solution and forms crystals, gradually filling the spaces from the outside inward. The different colors are the result of various impurities in the water.

agent (ā′jənt) A substance that can bring about a chemical reaction or a biological effect. Compare **reagent.**

Agent Orange A mixture of equal amounts of two herbicides known as 2,4-D and 2,4,5-T, and trace amounts of the toxic contaminant dioxin (a byproduct of the manufacture of 2,4,5-T). It was used in the Vietnam War to defoliate areas of forest.

agglutination (ə-glo͞ot′n-ā′shən) The clumping together of biologic material, such as red blood cells or bacteria, that is suspended in liquid, usually in response to a particular antibody.

aggregate fruit (ăg′rĭ-gĭt) A fruit, such as a raspberry or a strawberry, that consists of multiple ripened ovaries of a single flower, borne together on a common receptacle. Compare **accessory fruit, multiple fruit, simple fruit.**

aggression (ə-grĕsh′ən) Behavior that is meant to intimidate or injure an animal of the same species or of a competing species but is not predatory. Aggression may be displayed during mating rituals or to defend territory, as by the erection of fins by fish and feathers by birds.

aggressive mimicry (ə-grĕs′ĭv) A form of mimicry in which a predator (the mimic) closely resembles another organism (the model) that is attractive to a third organism (the dupe) on which the mimic preys. The anglerfish is an example of aggressive mimicry, having a modified dorsal spine that mimics a worm or small shrimp and serves as a lure to attract its prey. Compare **Batesian mimicry, Müllerian mimicry.**

agnathan (ăg′nə-thən) *Adjective.* **1.** Lacking a jaw. —*Noun.* **2.** See **jawless fish.**

Agnesi (än-yā′zē), **Maria Gaetana** 1718–1799. Italian mathematician whose major work, *Analytical Institutions* (1748), was the first comprehensive summary of the state of mathematical analysis. It brought together the work of authors writing in various languages, formulated new mathematical methods, and was widely used as a textbook for many years.

Maria Agnesi

agonic line (ā-gŏn′ĭk) An imaginary line on the Earth's surface connecting points where the magnetic declination is zero. The agonic line is a line of longitude on which a compass will show true north, since where magnetic declination is zero, magnetic north coincides with geographic north.

agonist (ăg′ə-nĭst) **1.** A muscle that actively contracts to produce a desired movement. **2.** A chemical substance, especially a drug, that can combine with a receptor on a cell to produce a physiologic response. Compare **antagonist.**

agriculture (ăg′rĭ-kŭl′chər) The science of cultivating land, producing crops, and raising livestock.

agrochemical (ăg′rə-kĕm′ĭ-kəl) **1.** A chemical, such as a hormone, fungicide, or insecticide, that improves the production of crops. **2.** A chemical or product, such as cellulose, derived from plants.

agroforestry (ăg′rō-fôr′ĭ-strē) A system of land use in which harvestable trees or shrubs are grown among or around crops or on pastureland, as a means of preserving or enhancing the productivity of the land.

agronomy (ə-grŏn′ə-mē) The scientific study of soil management and crop production, including irrigation and the use of herbicides, pesticides, and fertilizers.

A horizon In ABC soil, the uppermost, darkest zone that is rich in organic matter. The upper section of the A horizon usually contains humus along with plant and animal matter in varying stages of decay. The middle section usually contains a high concentration of quartz or other minerals that remain following the leaching away of clay, iron, and aluminum. The lower section is typically transitional in nature between the A horizon and the B horizon. Also called *zone of leaching.*

AI **1.** Abbreviation of **artificial insemination.** **2.** Abbreviation of **artificial intelligence.**

AIDS (ādz) Short for *acquired immune deficiency syndrome.* An infectious disease of the immune system caused by an human immunodeficiency virus (HIV). AIDS is characterized by a decrease in the number of helper T cells, which causes a severe immunodeficiency that leaves the body susceptible to a variety of potentially fatal infections. The virus is transmitted in infected bodily fluids such as semen and blood, as through sexual intercourse, the use of contaminated hypodermic syringes, and placental transfer between mother and fetus. Although a cure or vaccine is not yet available, a number of antiviral drugs can decrease the viral load and subsequent infections in patients with AIDS.

aiguille (ā-gwēl′) A sharply pointed mountain peak found in regions that have undergone intense glaciation. Aigulles are believed to be the remnants of the elevated areas separating two adjacent cirques.

aileron (ā′lə-rŏn′) A hinged surface that is part of the back edge of each wing on an airplane. The ailerons are moved up or down to create uneven lift on the sides of the plane to control its rolling and tilting movements.

air (âr) The colorless, odorless, tasteless mixture of gases that surrounds the Earth. Air consists of about 78 percent nitrogen and 21 percent oxygen, with the remaining part made up mainly of argon, carbon dioxide, neon, helium, methane, and krypton in decreasing order of volume. Air also contains varying amounts of water vapor, particulate matter such as dust and soot, and chemical pollutants.

air bladder 1. An air-filled sac in many fish that helps maintain buoyancy or, in some species, helps in respiration, sound production, or hearing. Also called *swim bladder.* **2.** See **float.**

airfoil (âr′foil′) A structure having a shape that provides lift, propulsion, stability, or directional control in a flying object. An aircraft wing provides lift by causing air to pass at a higher speed over the wing than below it, resulting in greater pressure below than above. Propellers are airfoils that are spun rapidly to provide propulsion. See more at **Bernoulli effect.** See Note at **Aerodynamics.**

airglow (âr′glō′) A faint photochemical luminescence in the upper atmosphere caused by the collision of x-rays and charged particles from the Sun with atoms and molecules, especially of oxygen, sodium, and the hydroxyl radical (OH). Airglow is strongest over low and middle latitudes.

air mass A widespread body of air that originates over a large area of land or ocean and assumes the temperature and humidity of that area, with characteristics distributed fairly evenly throughout the horizontal layers of the mass. Air that stands over the Caribbean Sea, for example, becomes a warm, humid maritime tropical air mass, while air that lies in the Arctic regions of northern Canada takes on the cold and dry characteristics of its surroundings and becomes a continental polar air mass. When air masses of differing properties come into contact in the middle latitudes, they frequently generate storm fronts.

air plant See **epiphyte.**

air root See **pneumatophore.**

air sac 1. An air-filled space in the body of a bird that forms a connection between the lungs and bone cavities and aids in breathing and temperature regulation. **2.** See **alveolus. 3.** A saclike enlargement in the trachea of an insect. **4.** A baglike piece of skin or tissue below the jaw of certain animals, such as the bullfrog and orangutan, that can be inflated to increase sound production.

air vesicle See **float.**

Al The symbol for **aluminum.**

alanine (ăl′ə-nēn′) A nonessential amino acid. *Chemical formula:* $C_3H_7NO_2$. See more at **amino acid.**

albedo (ăl-bē′dō) The fraction of the total light striking a surface that gets reflected from that surface. An object that has a high albedo (near 1) is very bright; an object that has a low albedo (near 0) is dark. The Earth's albedo is about 0.37. The Moon's is about 0.12.

albino (ăl-bī′nō) An organism lacking normal pigmentation or coloration. Animals that are albinos lack pigmentation due to a congenital absence of melanin. In humans and other mammals, albinos have white hair, pale skin, and usually pinkish eyes. Plants that are albinos lack normal amounts of chlorophyll or other pigments. *—Noun* **albinism** (ăl′bə-nĭz′əm).

albite (ăl′bīt′) A clear to milky white triclinic mineral of the plagioclase group. Albite is common in igneous rocks, especially granite, and in metamorphic rocks that formed at low temperatures. *Chemical formula:* $NaAlSi_3O_8$.

albumen (ăl-byo͞o′mən) The white of the egg of certain animals, especially birds and reptiles, consisting mostly of the protein albumin. The albumen supplies water to the growing embryo and also cushions it. Albumen is used commercially in making wine, vinegars, lithographs, dyes, and pharmaceuticals.

albumin (ăl-byo͞o′mĭn) A class of proteins found in egg white, milk, blood, and various other plant and animal tissues. Albumins dissolve in water and form solid or semisolid masses when heated, such as cooked egg white.

albuterol (ăl-byo͞o′tə-rôl′, -rōl′) An adrenergic stimulant used as a bronchodilator in the treatment of asthma and other obstructive lung diseases.

alchemy (ăl′kə-mē) A medieval philosophy and early form of chemistry whose aims were the transmutation of base metals into gold, the discovery of a cure for all diseases, and the preparation of a potion that gives eternal youth. The imagined substance capable of turning other metals into gold was called the philosophers' stone.

A CLOSER LOOK **alchemy**

Because their goals were so unrealistic, and because they had so little success in achieving them, the practitioners of *alchemy* in the Middle Ages got a reputation as fakers and con artists. But this reputation is not fully deserved. While they never succeeded in turning lead into gold (one of their main goals), they did make discoveries that helped to shape modern chemistry. Alchemists invented early forms of some of the laboratory equipment used today, including beakers, crucibles, filters, and stirring rods. They also discovered and purified a number of chemical elements, including mercury, sulfur, and arsenic. And the methods they developed to separate mixtures and purify compounds by distillation and extraction are still important.

alcohol (ăl′kə-hôl′) **1.** Any of a large number of colorless, flammable organic compounds that contain the hydroxyl group (OH) and that form esters with acids. Alcohols are used as solvents and for manufacturing dyes, perfumes, and pharmaceuticals. Simple alcohols, such as methanol and ethanol, are water-soluble liquids, while more complex ones, like cetyl alcohol, are waxy solids. Names of alcohols usually end in *–ol.* **2.** Ethanol.

alcoholism (ăl′kə-hô-lĭz′əm) A progressive, potentially fatal disease characterized by the excessive and compulsive consumption of alcoholic beverages and physiological and psychological dependence on alcohol. Chronic alcoholism usually results in liver and other organ damage, nutritional deficiencies and impaired social functioning.

Aldebaran (ăl-dĕb′ər-ən) A red giant star in the constellation Taurus. Aldebaran is the thirteenth brightest star in the sky, with an apparent magnitude of 0.85. *Scientific name:* Alpha Tauri.

aldehyde (ăl′də-hīd′) Any of a class of highly reactive organic compounds obtained by oxidation of certain alcohols and containing the group CHO. Aldehydes are used in manufacturing resins, dyes, and organic acids.

aldose (ăl′dōs′, -dōz′) Any of a class of simple sugars (monosaccharides) containing an aldehyde group (CHO). Galactose, glucose, and ribose are all aldoses. Compare **ketose.**

aldosterone (ăl-dŏs′tə-rōn′) A steroid hormone secreted by the adrenal cortex that regulates salt and water balance in the body. *Chemical formula:* $C_{21}H_{28}O_5$.

aldrin (ôl′drĭn) A highly poisonous white powder used as a crop pesticide and to kill termites. Because of its toxicity to animals and humans, its production has been discontinued. Aldrin is a chlorinated derivative of naphthalene closely related to dieldrin. *Chemical formula:* $C_{12}H_8Cl_6$.

aleph-null (ä′lĕf-nŭl′) The first of the transfinite cardinal numbers, corresponding to the number of elements in the set of positive integers.

Alexanderson (ăl′ĭg-zăn′dər-sən), **Ernst Frederick Werner** 1878–1975. Swedish-born American electrical engineer who in 1906 invented a high-frequency alternator that made radio communication more efficient. Alexanderson also developed the first practical television system in 1927 and the first color television receiver in 1955.

alga (ăl′gə) Plural **algae** (ăl′jē). Any of various green, red, or brown organisms that grow mostly in water, ranging in size from single cells to large spreading seaweeds. Like plants, algae manufacture their own food through photosynthesis and release large amounts of oxygen into the atmosphere. They also fix large amounts of carbon, which would otherwise exist in the atmosphere as carbon dioxide. Algae form a major component of marine plankton and are often visible as pond scum and blooms in tidal pools. Land species mostly live in moist soil and on tree trunks or rocks. Some species live in extreme environments, such as deserts, hot springs, and glaciers. Although they were once classified as plants, the algae are now considered to be protists, with the exception of the **cyanobacteria,** formerly called blue-green algae. The algae do not form a distinct phylogenetic group, but the word *alga* serves as a convenient catch-all term for various photosynthetic protist phyla, including the **green algae, brown algae,** and **red algae.**

algebra (ăl′jə-brə) A branch of mathematics in which symbols, usually letters of the alphabet, represent numbers or quantities and express general relationships that hold for all members of a specified set.

algin (ăl′jĭn) A mucilaginous polysaccharide occurring in the cell walls of brown algae. Its derivatives are widely used as thickening, stabilizing, emulsifying, or suspending agents in industrial, pharmaceutical, and food products, such as ice cream.

algorithm (ăl′gə-rĭth′əm) A finite set of unambiguous instructions performed in a prescribed sequence to achieve a goal, especially a mathematical rule or procedure used to compute a desired result. Algorithms are the basis for most computer programming.

Alhazen (ăl-hăz′ən) See **Ibn al-Haytham.**

aliasing (ā′lē-ə-sĭng) **1.** Jagged distortions in curves and diagonal lines in computer graphics caused by limited or diminished screen resolution. Compare **antialiasing. 2.** Distortion in a reproduced sound wave caused by a low sampling rate during the recording of the sound signal as digital information.

alien (ā′lē-ən) Introduced to a region deliberately or accidentally by humans. Starlings, German cockroaches, and dandelions are species that are alien to North America but have become widely naturalized in the continent. Compare **endemic, indigenous.**

alimentary canal (ăl′ə-mĕn′tə-rē) The tube or passage of the digestive system through which food passes, nutrients are absorbed, and waste is eliminated. See also **digestive tract.**

aliphatic (ăl′ə-făt′ĭk) Relating to organic compounds whose carbon atoms are linked in open chains, either straight or branched, rather than containing a benzene ring. Alkanes, alkenes, and alkynes are aliphatic compounds. Compare **aromatic.**

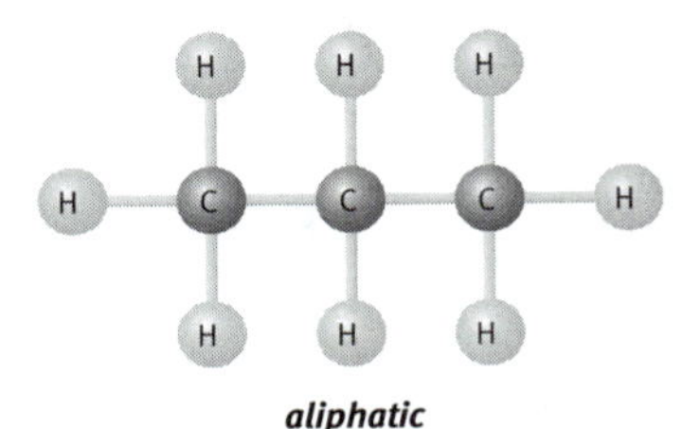

aliphatic

a straight chain of carbon atoms of a propane molecule

alkali (ăl′kə-lī′) Plural **alkalis** or **alkalies.** A hydroxide of an alkali metal. The aqueous solution of alkalis is bitter, slippery, caustic, and characteristically basic in reactions.

alkali feldspar Any of several feldspar minerals containing alkali metals and little calcium. Plagioclase, orthoclase and microcline are alkali feldspars.

alkali metal Any of a group of soft metallic elements that form alkali solutions when they combine with water. They include lithium, sodium, potassium, rubidium, cesium, and francium. Except for cesium, which has a gold sheen, alkali metals are white. The alkali metals have one electron in their outer shell, and therefore react easily with other elements and are found in nature only in compounds. See **Periodic Table.**

alkaline (ăl′kə-lĭn, -līn′) **1.** Capable of neutralizing an acid. Bases are alkaline. **2.** Relating to an alkali compound. **3a.** Having a pH greater than 7. **b.** Having a relatively low concentration of hydrogen ions.

alkaline-earth metal Any of a group of metallic elements that includes beryllium, magnesium, calcium, strontium, barium, and radium. Because the alkaline-earth metals have two electrons in their outer shell, they react easily with other elements and are found in nature only in compounds. See **Periodic Table.**

alkaloid (ăl′kə-loid′) Any of a large class of naturally occurring, complex organic compounds that contain nitrogen and have physiological effects on animals, including humans. Most alkaloids occur in plants, although some are produced by fungi and animals. Alkaloids are bases and usually form colorless crystalline solids with a bitter taste. They have a wide range of effects and are used as medicines and poisons. Morphine, quinine, strychnine, codeine, caffeine, cocaine, and nicotine are all alkaloids.

alkalosis (ăl′kə-lō′sĭs) An imbalance in the pH of body fluids, in which the blood or other body tissue is more basic than normal.

alkane (ăl′kān′) Any of a group of hydrocarbons that have carbon atoms in chains linked by single bonds and that have the general formula C_nH_{2n+2}. Alkanes can be either gaseous, liquid, or solid. They occur naturally in petroleum and natural gas, and include methane, propane and butane. Also called *paraffin.* ► The group of alkanes as a whole is called the **alkane series** or the methane or paraffin series. Its first six members are methane, ethane, propane, butane, pentane, and hexane.

alkene (ăl′kēn′) Any of a group of unsaturated hydrocarbons that have carbon atoms in chains linked by one or more double bonds and that have the general formula C_nH_{2n}. Lighter alkenes, such as ethylene, are derived from petroleum by cracking. Also called *olefin.* ► The group of alkenes as a whole is called the **alkene series** or the ethylene series. Its first five members are ethylene (or ethene), propylene (or propene), butylene (or butene), pentene, and hexene.

alkyl (ăl′kəl) A radical that has the general formula C_nH_{2n+1}, formed by removing a hydrogen atom from an alkane. Ethyl and propyl are alkyls.

alkylate (ăl′kə-lāt′) To add one or more alkyl groups to a compound.

alkyne (ăl′kīn′) Any of a group of unsaturated hydrocarbons that have carbon atoms in chains linked by one or more triple bonds and that have the general formula C_nH_{2n-2}. Alkynes can be solid, liquid, or gaseous and include acetylene. ► The group of alkynes as a whole is called the **alkyne series** or the acetylene series. The first five members of the alkyne series are acetylene (or ethyne), propyne, butyne, pentyne, and hexyne.

allantois (ə-lăn′tō-ĭs) Plural **allantoides** (ăl′ən-tō′ĭ-dēz′). A membranous sac that grows out of the lower end of the alimentary canal in embryos of reptiles, birds, and mammals. In mammals, the blood vessels of the allantois develop into the blood vessels of the umbilical cord.

allele (ə-lēl′) Any of the possible forms in which a gene for a specific trait can occur. In almost all animal cells, two alleles for each gene are inherited, one from each parent. Paired alleles (one on each of two paired chromosomes) that are the same are called **homozygous**, and those that are different are called **heterozygous**. In heterozygous pairings, one allele is usually dominant, and the other recessive. Complex traits such as height and longevity are usually caused by the interactions of numerous pairs of alleles, while simple traits such as eye color may be caused by just one pair.

allele frequency The percentage of a population of a species that carries a particular allele on a given chromosome locus.

allelopathy (ə-lē-lŏp′ə-thē, ăl′ə-) The inhibition of growth in one plant species by chemicals produced by another. For example, other plants will often not grow underneath black walnut trees, since these trees produce juglone, a chemical inhibiting plant respiration.

Allen's rule (ăl′ənz) The principle holding that in a warm-blooded animal species having distinct geographic populations, the limbs, ears, and other appendages of the animals living in cold climates tend to be shorter than in animals of the same species living in warm climates. Shorter and more compact body parts have less surface area than elongated ones and thus radiate less body heat. Allen's rule is named for the American zoologist Joel Allen (1838–1921). Compare **Bergmann's rule.**

allergen (ăl′ər-jən) A substance, such as pollen, that causes an allergic reaction.

allergy (ăl′ər-jē) An abnormally high immunologic sensitivity to certain stimuli such as drugs, foods, environmental irritants, microorganisms, or physical conditions, such as temperature extremes. These stimuli act as antigens, provoking an immunological response involving the release of inflammatory substances, such as histamine, in the body. Allergies may be innate or acquired in genetically predisposed individuals. Common symptoms include sneezing, itching, and skin rashes, though in some individuals symptoms can be severe. See also **anaphylactic shock.**

allogamy (ə-lŏg′ə-mē) See **cross-fertilization.**

allogeneic (ăl′ə-jə-nē′ĭk) also **allogenic** (ăl′ə-jĕn′ĭk) Being genetically different although belonging to or obtained from the same species, as in tissue grafts.

allograft (ăl′ə-grăft′) A graft transplanted from a donor who is of the same species as the recipient but who is genetically distinct. Compare **autograft, xenograft.**

allopatric (ăl′ə-păt′rĭk) *Ecology. Occurring* in separate, nonoverlapping geographic areas. Allopatric populations of related organisms are unable to interbreed because of geographic separation. ► The development of new species as a result of the geographic separation of populations is called **allopatric speciation**. Compare **sympatric.**

all-or-none (ôl′ər-nŭn′) Characterized by either a complete response or by a total lack of response or effect, depending on the strength of the stimulus. Neurons have an all-or-none response to impulse transmission and cannot be partially stimulated.

allosaurus (ăl′ə-sôr′əs) or **allosaur** (ăl′ə-sôr′) Any of various carnivorous dinosaurs of the genus *Allosaurus* of the late Jurassic and early Cretaceous Periods. Allosaurs were similar to but smaller than tyrannosaurs.

allotrope (ăl′ə-trōp′) Any of several crystalline forms of a chemical element. Charcoal, graphite, and diamond are all allotropes of carbon.

alloy (ăl′oi′) A metallic substance made by mixing and fusing two or more metals, or a metal and a nonmetal, to obtain desirable qualities such as hardness, lightness, and strength. Brass, bronze, and steel are all alloys.

alluvial fan (ə-lo͞o′vē-əl) A fan-shaped mass of sediment, especially silt, sand, gravel, and boulders, deposited by a river when its flow is suddenly slowed. Alluvial fans typically form where a river pours out from a steep valley through mountains onto a flat plain. Unlike deltas, they are not deposited into a body of standing water.

alluvium (ə-lo͞o′vē-əm) Plural **alluviums** or **alluvia.** Sand, silt, clay, gravel, or other matter deposited by flowing water, as in a riverbed, floodplain, delta, or alluvial fan. Alluvium is generally considered a young deposit in terms of geologic time. —*Adjective* **alluvial.**

allyl (ăl′əl) The unsaturated radical C_3H_5, derived from propene.

alopecia (ăl′ə-pē′shə) Loss of hair; baldness.

Alpha Centauri (ăl′fə sĕn-tôr′ē) A triple star system in the constellation Centaurus that is the third brightest star in the night sky, with an apparent magnitude of –0.27, and the star nearest to Earth, at a distance of about 4.2 light-years. The two brighter stars, Alpha Centauri A and B, form a binary star system, which is orbited by Proxima Centauri, a smaller, dimmer star.

alpha decay The radioactive decay of an atomic nucleus by emission of an alpha particle (two protons bound to two neutrons). When an element undergoes alpha decay, its atomic number decreases by two.

alpha-fetoprotein (ăl′fə-fē′tō-prō′tēn) An antigen produced in the liver of a fetus that can appear in certain diseases of adults, such as liver cancer. Its level in amniotic fluid can be used in the detection of certain fetal abnormalities, including Down syndrome and spina bifida.

alpha helix A common structure of proteins, characterized by a single, spiral chain of amino acids stabilized by hydrogen bonds. Compare **beta sheet, random coil.**

alpha particle A positively charged particle that consists of two protons and two neutrons bound together. It is emitted by an atomic nucleus undergoing radioactive decay and is identical to the nucleus of a helium atom. Because of their relatively large mass, alpha particles are the slowest and least penetrating forms of nuclear radiation. They can be stopped by a piece of paper. See more at **radioactive decay.**

alpha ray A stream of alpha particles.

alpha-tocopherol (ăl′fə-tō-kŏf′ə-rôl′, -rōl′) **1.** A water-insoluble alcohol that occurs in plant oils (especially wheat germ oil), egg yolks, and liver, and is also produced synthetically. Alpha-tocopherol is the principal form of vitamin E in the body. *Chemical formula:* $C_{29}H_{50}O_2$. See more at **vitamin E.**

alpine (ăl′pīn′) Resembling or characteristic of the European Alps or any other high mountain system, especially one that has been shaped by intense glacial erosion.

Altair (ăl-târ′, -tīr′) A bright white star in the constellation Aquila. Altair is a main-sequence star in the Hertzprung-Russell diagram. It is the 12th brightest star in the night sky, with an apparent magnitude of 0.77. Altair, along with Vega and Deneb, form the Summer Triangle asterism. *Scientific name:* Alpha Aquilae.

altazimuth coordinate system (ăl-tăz′ə-məth) The coordinate system in which a celestial object's position is described in terms of its **altitude** and **azimuth**. Like celestial latitude and longitude, and declination and right ascension, the altitude and azimuth coordinates are used to map objects in the sky. Unlike the other two systems, which are oriented to fixed features of the celestial sphere, the altazimuth system is based solely on an observer's location on the Earth, making it the most practical system for use in celestial navigation or surveying. Altitude can be determined using a sextant, and calculation of an observer's geographic location can then be determined using a chronometer and tables of star positions contained in a special almanac. Compare **ecliptic coordinate system, equatorial coordinate system.**

altazimuth mount A mounting for astronomical telescopes that has separate axes for horizontal and vertical rotation. Because celestial objects move in arcs across the sky, tracking their motions with an altazimuth mount requires adjustments of both axes. Compare **equatorial mount.**

alternate (ôl′tər-nĭt) **1.** Arranged singly at intervals on a stem or twig. Elms, birches, oaks, cherry trees, and hickory trees have alternate leaves. Compare **opposite. 2.** Arranged regularly between other parts, as stamens between petals on a flower.

alternate angles Two angles formed on opposite sides of a line that crosses two other lines. The angles are both exterior or both interior, but not adjacent.

alternate host 1. One of two species of host on which some pathogens, such as certain rust fungi, must develop to complete their life cycles. **2.** A species of host other than the principal host on which a parasite can survive.

alternating current (ôl′tər-nā′tĭng) An electric current that repeatedly changes its direction or strength, usually at a certain frequency or range of frequencies. The term is also used to describe alternating voltages. Power stations generate alternating current because it is easy to raise and lower the voltage of such current using transformers; thus the voltage can be raised very high for transmission (high voltages lose less power as heat than do low voltages), and lowered to safe levels for domestic and industrial use. In North America, the frequency of alternation of the direction of flow is 60 Hz, or 60 cycles per second. In other parts of the world it is 50 Hz. Compare **direct current.** See Notes at **current, Tesla.**

alternation of generations (ôl′tər-nā′shən) The regular alternation of forms or of mode of reproduction in the life cycle of an organism, especially the alternation between sexual and asexual reproductive phases in plants and some invertebrates. In plants, the alternation involves alternating generations of haploid and diploid organisms. Often one of these generations is the dominant form of the organism, and the other generation is nutritionally dependent upon it or just grows as a smaller plant. For example, in mosses and liverworts, the haploid phase is the large, familiar form of the plant. The diploid phase is smaller and grows upon the haploid phase. In angiosperms, however, the diploid phase of the organism is large and independent, while the haploid phase is reduced to the pollen grain and the eight-celled female gametophyte located in the ovule.

A CLOSER LOOK **alternation of generations**

The life cycle of fern species provides a good example of the differing roles played by the gametophyte and sporophyte in organisms that display an *alternation of generations.* The familiar large frond-bearing fern plant is the sporophyte generation of the fern. By meiosis it produces haploid spores that are dispersed and develop into gametophytes. Fern gametophytes are inconspicuous matlike plants that can make their own food by photosynthesis. The gametophytes produce both sperm and eggs. Sperm from another gametophyte reaches one of these eggs and fuses with it to form an embryo, which then grows out of the gametophyte as a new sporophyte plant. In many nonvascular plants, such as the mosses and liverworts, the sporophyte is a relatively small plant that grows in or on top of the gametophyte, which is larger. In gymnosperms and angiosperms, however, the sporophyte is the main plant form, and the gametophyte is dependent on the sporophyte.

---

alternative medicine (ôl-tûr′nə-tĭv) A variety of therapeutic or preventive health-care practices that are not typically taught or practiced in traditional medical communities and offer treatments that differ from standard medical practice. Homeopathy, herbal medicine, and acupuncture are types of alternative medicine.

alternator (ôl′tər-nā′tər) An electric generator that produces alternating current. The generator's coil is rotated (by a turbine, motor, or other power source), and its circular path causes it to cut cross a magnetic field (set up by strong magnets), first in one direction, then the other, with each cycle. The electric potential induced in the coil by this motion thus alternates between positive and negative once with each cycle, resulting in alternating current. See more at **induction** (sense 2a).

altimeter (ăl-tĭm′ĭ-tər) An instrument that measures and indicates the height above sea level at which an object, such as an airplane, is located.

altiplano (äl′tĭ-plä′nō) A high mountain plateau. The most well-known altiplano extends from Lake Titicaca, in southern Peru, to Lake Poopo in Bolivia, covering a distance of 966 km (600 mi). Its average altitude is 3,658 m (12,000 ft).

altitude (ăl′tĭ-to͞od′) **1.** The height of an object or structure above a reference level, usually above sea level or the Earth's surface. **2.** *Astronomy.* The position of a celestial object above an observer's horizon, measured in degrees along a line between the horizon (0°) and the zenith (90°). Unlike **declination** and **celestial latitude**—the corresponding points in other celestial coordinate systems—the altitude of star or other celestial object is dependent on an observer's geographic location and changes steadily as the sky passes overhead due to the rotation of the Earth. See more at **altazimuth coordinate system. 3.** *Mathematics.* The perpendicular distance from the base of a geometric figure, such as a triangle, to the opposite vertex, side, or surface.

altocumulus (ăl′tō-kyo͞o′myə-ləs) Plural **altocumuli** (ăl′tō-kyo͞om′yə-lī′). A mid-altitude cloud composed of fleecy white or gray patches or bands. Altocumulus clouds generally form between 2,000 and 6,100 m (6,560 and 20,000 ft). See illustration at **cloud.**

altostratus (ăl′tō-străt′əs) Plural **altostrati** (ăl′tō-străt′ī). A mid-altitude cloud that extends in flat, smooth sheets or layers of varying thickness. Altostratus clouds generally form between 2,000 and 6,100 m (6,560 and 20,000 ft) and often produce long, steady rain showers. See illustration at **cloud.**

altricial (ăl-trĭsh′əl) Born or hatched in a helpless condition requiring prolonged parental care, as by being naked, blind, or unable to move about. Nesting birds, monotremes, marsupials, and carnivores have altricial young. Compare **precocial.**

altruism (ăl′tro͞o-ĭz′əm) Instinctive cooperative behavior that is detrimental or without reproductive benefit to the individual but that contributes to the survival of the group to which the individual belongs. The willingness of a subordinate member of a wolf pack to forgo mating and help care for the dominant pair's pups is an example of altruistic behavior. While the individual may not reproduce, or may reproduce less often, its behavior helps ensure that a close relative does successfully reproduce, thus passing on a large share of the altruistic individual's genetic material.

alum (ăl′əm) Any of various crystalline double salts of a trivalent metal (such as aluminum, chromium, or iron) and a monovalent metal (such as potassium or sodium), especially aluminum potassium sulfate. Alum is widely used in industry as a hardener and purifier, and in medicine as an emetic and to stop bleeding.

alumina (ə-lo͞o′mə-nə) Any of several forms of aluminum oxide used in aluminum production and in abrasives, refractories, ceramics, and electrical insulation. Alumina occurs naturally as the mineral corundum and, with minor traces of chromium and cobalt, as the minerals ruby and sapphire, respectively. In its hydrated form it also occurs as the rock bauxite. Also called *aluminum oxide. Chemical formula:* **Al_2O_3.**

aluminum (ə-lo͞o′mə-nəm) *Symbol* **Al** A lightweight, silvery-white metallic element that is ductile, is found chiefly in bauxite, and is a good conductor of electricity. It is the most abundant metal in the Earth's crust and is used to make a wide variety of products from soda cans to airplane components. Atomic number 13; atomic weight 26.98; melting point 660.2°C (1,220.36°F); boiling point 2,467°C; specific gravity 2.69; valence 3. See **Periodic Table.**

aluminum oxide See **alumina.**

Alvarez (ăl′və-rĕz′), **Luis Walter** 1911–1988. American physicist who studied subatomic particles. Alvarez built a device called a hydrogen bubble chamber that made it possible to analyze the reactions occurring between atomic nuclei inside it. His observations led to the theory that protons, neutrons, and electrons are made of quarks. Alvarez won a 1968 Nobel Prize for physics for this work. With his son, geologist **Walter Alvarez** (born 1940), he later developed a theory that the extinction of dinosaurs was caused by climate changes resulting from a giant asteroid striking the Earth. See Note at **iridium.**

Luis Walter Alvarez

with his son Walter (right)

alveolus (ăl-vē′ə-ləs) Plural **alveoli** (ăl-vē′ə-lī′). Any of the tiny air-filled sacs arranged in clusters in the lungs, in which the exchange of oxygen and carbon dioxide takes place. Also called *air sac.*

Alzheimer's disease (älts′hī-mərz) A progressive, degenerative disease of the brain, commonly affecting the elderly, and associated with the development of amyloid plaques in the cerebral cortex. It is characterized by confusion, disorientation, memory failure, speech disturbances, and eventual dementia. The cause is unknown. Alzheimer's disease is named for its identifier, German psychiatrist Alois Alzheimer (1864–1915).

Am The symbol for **americium.**

AM Abbreviation of **amplitude modulation.**

amalgam (ə-măl′gəm) An alloy of mercury and another metal, especially: **1.** An alloy of mercury and silver used in dental fillings. **2.** An alloy of silver and tin used in silvering mirrors.

amber (ăm′bər) A hard, translucent, brownish-yellow substance that is the fossilized resin of ancient trees. It often contains fossil insects.

ambergris (ăm′bər-grĭs′, -grēs′) A yellow, gray, or black waxy material formed in the intestines of sperm whales that consists of a mixture of steroid derivatives. It is often found floating at sea or washed ashore, has a pleasant odor, and is added to perfumes as a fixative to slow down the rate of evaporation.

ameba (ə-mē′bə) Another spelling of **amoeba.**

amebiasis also **amoebiasis** (ăm′ə-bī′ə-sĭs) An infection or disease caused by amoebas, especially of the species *Entamoeba histolytica,* characterized by dysentery.

amenorrhea (ā-mĕn′ə-rē′ə) The absence of menstruation in a woman between puberty and menopause. Some causes include pregnancy, decreased body weight, endocrine and other medical disorders, and certain medications.

amensalism (ā-mĕn′sə-lĭz′əm) A symbiotic relationship in which one organism is harmed or inhibited and the other is unaffected. Examples of amensalism include the shading out of one plant by a taller and wider one and the inhibition of one plant by the secretions of another (known as allelopathy). Compare **commensalism, mutualism, parasitism.**

ament (ăm′ənt, ā′mənt) See **catkin.**

americium (ăm′ə-rĭsh′ē-əm) *Symbol* **Am** A synthetic, silvery-white, radioactive metallic element of the actinide series that is produced artificially by bombarding plutonium with

A CLOSER LOOK **amber**

Certain trees, especially conifers, produce a sticky substance called *resin* to protect themselves against insects. Normally, it decays in oxygen through the action of bacteria. However, if the resin happens to fall into wet mud or sand containing little oxygen, it can harden and eventually fossilize, becoming the yellowish, translucent substance known as *amber.* If any insects or other organisms are trapped in the resin before it hardens, they can be preserved, often in exquisite detail. By studying these preserved organisms, scientists are able learn key facts about life on Earth millions of years ago.

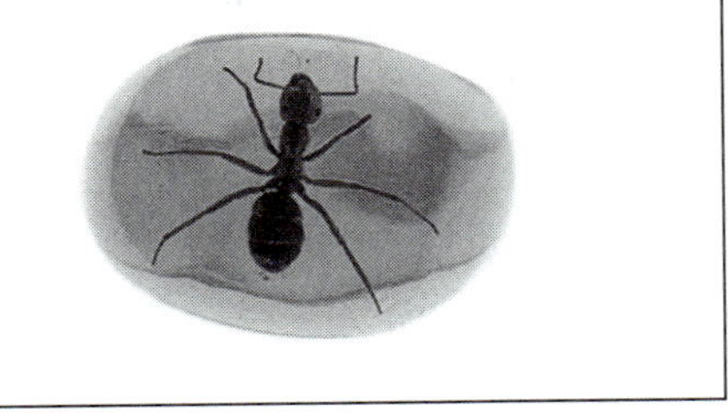

neutrons. Americium is used as a source of alpha particles for smoke detectors and gamma rays for industrial gauges. Its most stable isotope has a half-life of 7,950 years. Atomic number 95; specific gravity 11.7; valence 3, 4, 5, 6. See **Periodic Table.**

amethyst (ăm′ə-thĭst) A purple or violet, transparent form of quartz used as a gemstone. The color is caused by the presence of iron compounds in the crystal structure.

amide (ăm′īd′) Any organic compound containing the group $CONH_2$, derived from a fatty acid by replacing the hydroxyl group (OH) of the carboxyl group (COOH) with an amino group (NH_2). Amides are volatile solids.

amine (ə-mēn′, ăm′ēn) Any of a group of organic compounds that may be considered derivatives of ammonia (NH_3) in which one or more hydrogen atoms have been replaced by a hydrocarbon radical. In aniline ($C_6H_5NH_2$), for example, one hydrogen atom has been replaced by a phenyl group (C_6H_5). Amines are produced by the decay of organic matter.

amino (ə-mē′nō, ăm′ə-nō′) Relating to an amine or other chemical compound that contains the group NH_2.

amino acid Any of a large number of compounds found in living cells that contain carbon, oxygen, hydrogen, and nitrogen, and join together to form proteins. Amino acids contain a basic amino group (NH_2) and an acidic carboxyl group (COOH), both attached to the same carbon atom. Since the carboxyl group has a proton available for binding with the electrons of another atom, and the amino group has electrons available for binding with a proton from another atom, the amino acid behaves as an acid and a base simultaneously. Twenty of the naturally occurring amino acids are the building blocks of proteins, which they form by being connected to each other in chains. Eight of those twenty, called **essential amino acids,** cannot be synthesized in the cells of humans and must be consumed as part of the diet. The remaining twelve are **nonessential amino acids.**

aminobenzoic acid (ə-mē′nō-bĕn-zō′ĭk, ăm′ə-) Any of three organic acids consisting of benzoic acid with an amino group (NH_2) attached, especially PABA. *Chemical formula:* $\mathbf{C_7H_7NO_2}$.

aminoglycoside (ə-mē′nō-glī′kə-sīd′, ăm′ə-) **1.** A compound containing amino sugars in glycoside linkage. **2.** Any of a group of antibiotics, such as streptomycin, having the chemical structure of this compound, derived from species of *Streptomyces* or *Micomonospora* bacteria and usually used to treat infections caused by gram-negative bacteria.

aminophylline (ăm′ə-nŏf′ə-lĭn) A theophylline derivative that is used as a bronchodilator in the treatment of bronchial asthma, emphysema, and bronchitis. *Chemical formula:* $\mathbf{C_{16}H_{24}N_{10}O_4}$.

ammeter (ăm′mē′tər) An instrument that measures the strength of an electric current, indicating it in amperes. Ammeters typically include a galvanometer; digital ammeters typically include A/D converters as well. Compare **ohmmeter, voltmeter.**

ammonia (ə-mōn′yə) A colorless alkaline gas that is lighter than air and has a strongly pungent odor. It is used as a fertilizer and refrigerant, in medicine, and in making dyes, textiles, plastics, and explosives. *Chemical formula:* NH_3.

ammonite (ăm′ə-nīt′) Any of the ammonoids belonging to the order Ammonitida and living during the Jurassic and the Cretaceous Periods. Ammonites had a thick, very ornamental chambered shell with highly defined, wavy sutures between the chambers.

ammonite

ammonium (ə-mō′nē-əm) A positively charged ion, NH_4, derived from ammonia and found in a wide variety of organic and inorganic compounds. Compounds of ammonium chemically resemble the alkali metals.

ammonium chloride A white crystalline compound used in dry cells, as a soldering flux, and as an expectorant. Also called *sal ammoniac. Chemical formula:* NH_4Cl.

ammonium hydroxide A colorless, basic, aqueous solution of ammonia used as a household cleanser and in the manufacture of a wide variety of products, including textiles, rayon, rubber, fertilizer, and plastic. *Chemical formula:* NH_4OH.

ammonium nitrate A colorless crystalline salt used in fertilizers, explosives, and solid rocket propellants. *Chemical formula:* NH_4NO_3.

ammonium sulfate A brownish-gray to white crystalline salt used in fertilizers and water purification. *Chemical formula:* $(NH_4)_2SO_4$.

ammonoid (ăm′ə-noid′) Any of various extinct cephalopods of the subclass Ammonoidea living from the Devonian to the Cretaceous Periods. Ammonoids had a symmetrical, coiled, chambered shell with angular sutures between the chambers. They are closely related to the nautiloids, including the modern-day chambered nautilus.

amnesia (ăm-nē′zhə) Partial or total loss of memory, usually caused by brain injury or shock.

amniocentesis (ăm′nē-ō-sĕn-tē′sĭs) A procedure usually done about the sixteenth week of pregnancy in which a small sample of amniotic fluid is drawn out of the uterus through a needle inserted in the abdomen. The fluid is analyzed to determine the gender of the fetus or the presence of genetic abnormalities.

amnion (ăm′nē-ən) A thin, membranous sac filled with a watery fluid (called the **amniotic fluid**) in which the embryo or fetus of a reptile, bird, or mammal is suspended during prenatal development. Also called *amniotic sac.*

amniote (ăm′nē-ōt′) Any of the vertebrates that have an amnion during embryonic development. Reptiles, birds, and mammals are amniotes.

amniotic fluid (ăm′nē-ŏt′ĭk) The watery fluid within the amnion that surrounds the fetus. Amniotic fluid cushions the fetus from injury, allows movement, and helps to stabilize temperature. The composition of the fluid changes over the course of gestation. Initially, amniotic fluid is similar to maternal plasma. As the fetus develops, phospholipids originating from the lungs, fetal cells, lanugo, and urine are deposited in the fluid.

amniotic sac See **amnion.**

amoeba (ə-mē′bə) Plural **amoebas** or **amoebae** (ə-mē′bē). Any of various one-celled aquatic or parasitic protozoans of the genus *Amoeba* or related genera, having no definite form and consisting of a mass of protoplasm containing one or more nuclei surrounded by a flexible outer membrane. Amoebas move by means of pseudopods.

amorphous (ə-môr′fəs) **1.** Not made of crystals. Glass, amber, and plastics are amorphous substances. **2.** Lacking definite form or shape.

amoxicillin (ə-mŏk′sĭ-sĭl′ĭn) An antibiotic derived from penicillin, having an antibacterial spectrum of action similar to that of ampicillin. *Chemical formula:* $C_{16}H_{19}N_3O_5$.

AMP (ā′ĕm-pē′) Short for *adenosine monophosphate.* An organic compound that is composed of adenosine and one phosphate group. It is one of the nucleotides present in DNA and RNA, and is also the fundamental component of ATP and ADP. During certain cellular metabolic processes, AMP forms from ADP when the latter loses a phosphate group, and AMP forms ADP by acquiring a phosphate group. *Chemical formula:* $\mathbf{C_{10}H_{14}N_5O_7P}$.

ampere (ăm′pîr′) The SI unit used to measure electric current. Electric current through any given cross-section (such as a cross-section of a wire) may be measured as the amount of electrical charge moving through that cross-section in one second. One ampere is equal to a flow of one coulomb per second, or a flow of 6.28×10^{18} electrons per second.

Ampère (ăm′pîr′, äm-pĕr′), **André Marie** 1775–1836. French mathematician and physicist who is best known for his analysis of the relationship between magnetic force and electric current. He formulated Ampère's law, which describes the strength of the magnetic field produced by the flow of energy through a conductor. The ampere unit of electric current is named for him.

amphetamine (ăm-fĕt′ə-mēn′) Any of a group of drugs that stimulate the central nervous system, resulting in elevated blood pressure, heart rate, and other metabolic functions. Amphetamines are used in the treatment of certain neurological conditions, such as attention deficit hyperactivity disorder and narcolepsy. The drugs are highly addictive and are sometimes abused.

amphibian (ăm-fĭb′ē-ən) A cold-blooded, smooth-skinned vertebrate of the class Amphibia. Amphibians hatch as aquatic larvae with gills and, in most species, then undergo metamorphosis into four-legged terrestrial adults with lungs for breathing air. The eggs of amphibians are fertilized externally and lack an amnion. Amphibians evolved from lobe-finned fish during the late Devonian Period and include frogs, toads, newts, salamanders, and caecilians.

WORD HISTORY **amphibian**

Amphibians, not quite fish and not quite reptiles, were the first vertebrates to live on land. These cold-blooded animals spend their larval stage in water, breathing through their gills. In adulthood they usually live on land, using their lungs to breath air. This double life is also at the root of their name, *amphibian,* which, like many scientific words, derives from Greek. The Greek prefix *amphi–* means "both," or "double," and the Greek word *bios* means "life." Both these elements are widely used in English scientific terminology: *bios,* for example, is seen in such words as *biology, antibiotic,* and *symbiotic.*

amphibole (ăm′fə-bōl′) Any of a large group of usually dark minerals composed of a silicate joined to various metals, such as magnesium, iron, calcium or sodium. Amphiboles occur as columnar or fibrous prismatic crystals in igneous and metamorphic rocks. Most are monoclinic, but some are orthorhombic. Hornblende, actinolite and glaucophane are amphiboles. *Chemical formula:* $\mathbf{(Mg,Fe,Ca,Na)_{2\text{-}3}(Mg,Fe,Al)_5(Si,Al)_8O_{22}OH_2}$.

amphibolite (ăm-fĭb′ə-līt′) A metamorphic rock composed chiefly of amphibole and plagioclase and having little or no quartz.

amphioxus (ăm′fē-ŏk′səs) See **lancelet.**

amphoteric (ăm′fə-tĕr′ĭk) Capable of reacting chemically as either an acid or a base. Water, ammonia, and the hydroxides of certain metals are amphoteric.

ampicillin (ăm′pĭ-sĭl′ĭn) An antibiotic derived from penicillin that has a broad antibacterial spectrum of action. It is effective against gram-negative and gram-positive bacteria and is used primarily to treat gonorrhea and infections of the respiratory, urinary, and intestinal tracts.

amplification (ăm′plə-fĭ-kā′shən) An increase in the magnitude or strength of an electric current, a force, or another physical quantity, such as a radio signal.

amplitude (ăm′plĭ-to͞od′) **1.** *Physics.* One half the full extent of a vibration, oscillation, or wave. The amplitude of an ocean wave is the maximum height of the wave crest above the level of calm water, or the maximum depth of the wave trough below the level of calm water. The amplitude of a pendulum swinging through an angle of 90° is 45°. Compare **frequency. 2.** *Electronics.* The amount by which a voltage or current changes from zero or an average value.

amplitude modulation A method of transmitting signals, such as sound or digital information, in which the value of the signal is given by the amplitude of a high frequency carrier wave. In AM radio transmission, for example, the signal to be carried is a sound wave, and its increasing and decreasing value is reflected in the increasing and decreasing amplitude of the radio frequency carrier wave. In fiber optics, the carrier wave is a beam of light. Compare **frequency modulation.**

amygdala (ə-mĭg′də-lə) Plural **amygdalae** (ə-mĭg′də-lē). An almond-shaped mass of gray matter in the front part of the temporal lobe of the cerebrum that is part of the **limbic system** and is involved in the processing and expression of emotions, especially anger and fear.

amyl (ăm′əl) The radical C_5H_{11}, derived from pentane. Amyl occurs in eight isomeric forms. Also called *pentyl.*

amyl alcohol (ăm′əl, ā′məl) Any of various colorless oils derived from pentane, one of which is the main constituent of fusel oil. *Chemical formula:* C_5H_{12}.

amylase (ăm′ə-lās′) Any of various enzymes that cause starches to break down into smaller sugars, especially maltose, by hydrolysis. There are two types of amylases, *alpha-amylases* and *beta-amylases.* In humans, an alpha-amylase known as **ptyalin** is present in saliva and is also produced by the pancreas for secretion into the small intestine. Beta-amylases are found in bacteria, molds, yeasts, and the seeds of plants.

amyloid (ăm′ə-loid′) *Noun.* **1.** A hard waxy substance consisting of protein and polysaccharides that results from the degeneration of tissue and is deposited in organs or tissues of the body in various chronic diseases. —*Adjective.* **2.** Starchlike.

amyotrophic lateral sclerosis (ā′mī-ə-trō′fĭk, -ə-trŏf′ĭk, ā-mī′-) A chronic, progressive neurologic disease marked by gradual degeneration of the neurons in the spinal cord that control voluntary muscle movement. The disorder causes muscle weakness and atrophy and usually results in death. Also called *Lou Gehrig's disease,* after the American baseball player (1903–41) who was the first public figure to suffer from the disease.

anabatic (ăn′ə-băt′ĭk) Relating to warm, rising wind currents, especially those that are driven up the slopes of hills, mountains, and peaks. When air comes in contact with the warm ground surface, the air heats up, becomes less dense, and rises upward. Anabatic winds are especially common during the daytime in fair weather conditions. Compare **katabatic.**

anabolic steroid (ăn′ə-bŏl′ĭk) A group of synthetic hormones that promote the storage of protein and the growth of tissue, sometimes used by athletes to increase muscle size and strength.

anabolism (ə-năb′ə-lĭz′əm) The phase of metabolism in which complex molecules, such as the proteins and fats that make up body tissue, are formed from simpler ones. Compare **catabolism.** —*Adjective* **anabolic.**

anadromous (ə-năd′rə-məs) Relating to fish, such as salmon or shad, that migrate up rivers from the sea to breed in fresh water.

anaerobe (ăn′ə-rōb′) An organism, such as a bacterium, that can or must live in the absence of oxygen. Compare **aerobe.**

anaerobic (ăn′ə-rō′bĭk) Occurring in the absence of oxygen or not requiring oxygen to live. Anaerobic bacteria produce energy from food molecules without the presence of oxygen. Compare **aerobic.**

analgesic (ăn′əl-jē′zĭk) A drug used to eliminate pain; a painkiller. Aspirin and acetaminophen are analgesics.

analog or **analogue** (ăn′ə-lôg′) *Adjective.* **1.** Measuring or representing data by means of one or more physical properties that can express any value along a continuous scale. For example, the position of the hands of a clock is an analog representation of time. Compare **digital.** —*Noun.* **2.** An organ or structure that is similar in function to one in another kind of organism but is of dissimilar evolutionary origin. The wings of birds and the wings of insects are analogs. **3.** A chemical compound that has a similar structure and similar chemical properties to those of another compound, but differs from it by a single element or group. The antibiotic amoxicillin, for example, is an analog of penicillin, differing from the latter by the addition of an amino group. Compare **homologue.**

analogous (ə-năl′ə-gəs) **1.** Similar in function but having different evolutionary origins, as the wings of a butterfly and the wings of a bird. **2.** Similar in chemical properties and differing in chemical structure only with respect to one element or group.

analysis (ə-năl′ĭ-sĭs) **1.** The separation of a substance into its constituent elements, usually by chemical means, for the study and identification of each component. ► **Qualitative analysis** determines what substances are present in a compound. ► **Quantitative analysis** determines how much of each substance is present in a compound. **2.** A branch of mathematics concerned with limits and convergence and principally involving differential calculus, integral calculus, sequences, and series.

analytic geometry (ăn′ə-lĭt′ĭk) The use of algebra to solve problems in geometry, whereby geometric figures are represented by algebraic equations and plotted using coordinates.

anaphase (ăn′ə-fāz′) The stage of cell division in mitosis or meiosis in which the doubled set of chromosomes separates into two identical groups that move to opposite ends of the cell. Anaphase is preceded by metaphase and followed by telophase. See more at **meiosis, mitosis.**

anaphylactic shock (ăn′ə-fə-lăk′tĭk) A sudden, life-threatening allergic reaction, characterized by dilation of blood vessels with a sharp drop in blood pressure and bronchial spasm with shortness of breath. Anaphylactic shock is caused by exposure to a foreign substance, such as a drug or bee venom, to which the individual has been previously exposed. The substances act as antigens, provoking a preliminary immune response during the first exposure that results in a full-blown, immediate response during secondary exposure, called an *immediate hypersensitivity reaction.* Emergency treatment, including epinephrine injections, must be administered to prevent death. Also called *anaphylaxis.*

anaphylaxis (ăn′ə-fə-lăk′sĭs) See **anaphylactic shock.**

anapsid (ə-năp′sĭd) A reptile having a skull with no temporal openings. The earliest reptiles, the cotylosaurs, were anapsids, as are modern turtles. Anapsids probably gave rise to the diapsids and synapsids. Compare **diapsid, synapsid, therapsid.**

anatomy (ə-năt′ə-mē) **1.** The structure of an organism or any of its parts. **2.** The scientific study of the shape and structure of organisms and their parts. *—Adjective* **anatomical** (ăn′ə-tŏm′ĭ-kəl)

Anaxagoras (ăn′ăk-săg′ər-əs), 500?–428 BCE. Greek philosopher and astronomer who was the first to explain eclipses correctly. He also stated that all matter was composed of infinitesimally small particles, that the Sun and stars were glowing stones, and that the Moon took its light from the Sun.

andalusite (ăn′də-lo͞o′sīt′) A hard, grayish white to pinkish brown orthorhombic mineral. Andalusite occurs as nearly square prisms, often with cross-shaped cross sections, in metamorphic rocks. It is a polymorph of kyanite and sillimanite, but forms at shallower depths than they do, and at higher temperatures than kyanite and lower temperatures than sillimanite. *Chemical formula:* $\mathbf{Al_2SiO_5}$.

andesine (ăn′də-zēn′) A white, gray, green or yellow mineral of the plagioclase feldspar group. Andesine occurs in igneous rocks such as andesite and diorite. *Chemical formula:* $\mathbf{(Na,Ca)(Si,Al)_4O_8}$.

andesite (ăn′dĭ-zīt′) A gray, fine-grained volcanic rock. Andesite consists mainly of sodium-rich plagioclase and one or more mafic minerals such as biotite, hornblende, or pyroxene. It often contains small, visible crystals (phenocrysts) of plagioclase. It is the fine-grained equivalent of diorite.

androecium (ăn-drē′shē-əm, -shəm) Plural **androecia** (ăn-drē′shē-ə, -shə). The male reproductive organs of a flower considered as a group; the stamens. Compare **gynoecium.**

androgen (ăn′drə-jən) Any of several steroid hormones, especially testosterone, that regulate the growth, development, and function of the male reproductive system. The main source of androgens in the body are the testes.

androgynous (ăn-drŏj′ə-nəs) Having both female and male characteristics. *—Noun* **androgyny** (ăn-drŏj′ə-nē).

Andromeda (ăn-drŏm′ĭ-də) A constellation in the Northern Hemisphere near Perseus and Pegasus. It contains a spiral-shaped galaxy that, at a distance of 2.2 million light-years, is the farthest celestial object visible to the naked eye.

-ane A suffix used to form the names of saturated hydrocarbons, such as *ethane*. The suffix is shortened to *-an-* before other suffixes to indicate compounds derived from saturated hydrocarbons, as in *ethanol*.

anemia (ə-nē′mē-ə) A deficiency in the oxygen-carrying component of the blood, as in the amount of hemoglobin or the number or volume of red blood cells. Iron deficiency, often caused by inadequate dietary consumption of iron, and blood loss are common causes of anemia. See also **aplastic anemia, hemolytic anemia,** and **sickle cell anemia.** *—Adjective* **anemic.**

anemograph (ə-nĕm′ə-grăf′) An anemometer that uses a recording device to document wind speed and direction continuously.

anemometer (ăn′ə-mŏm′ĭ-tər) An instrument that measures the speed of the wind or of another flowing fluid. The most basic type of anemometer consists of a series of cups mounted at the end of arms that rotate in the wind. The speed with which the cups rotate indicates the wind speed. In this form, the anemometer also indicates the direction of the wind. ▸ Other anemometers include the **pressure-tube anemometer**, which uses the pressure generated by the wind to measure its speed, and the **hot-wire anemometer**, which uses the rate at which heat from a hot wire is transferred to the surrounding air to measure wind speed.

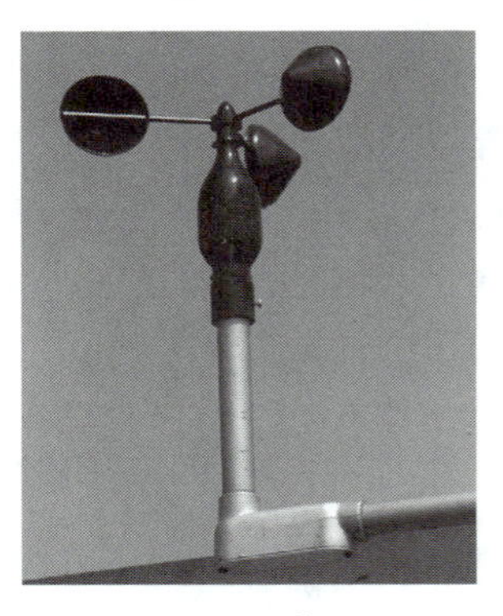

anemometer

anemone (ə-nĕm′ə-nē) See **sea anemone.**

anemophilous (ăn′ə-mŏf′ə-ləs) Pollinated by the wind.

aneroid barometer (ăn′ə-roid′) A barometer consisting of a thin elastic disk covering a chamber that contains a partial vacuum. High atmospheric pressure pushes against the disk and causes it to bulge inward, while low pressure does not push as hard, allowing the disk to bulge outward. An aneroid barometer is smaller and more portable than a mercury barometer and, when used with a barograph, can record up to a week's worth of data. Aneroid barometers are used extensively in aviation as part of altimeters.

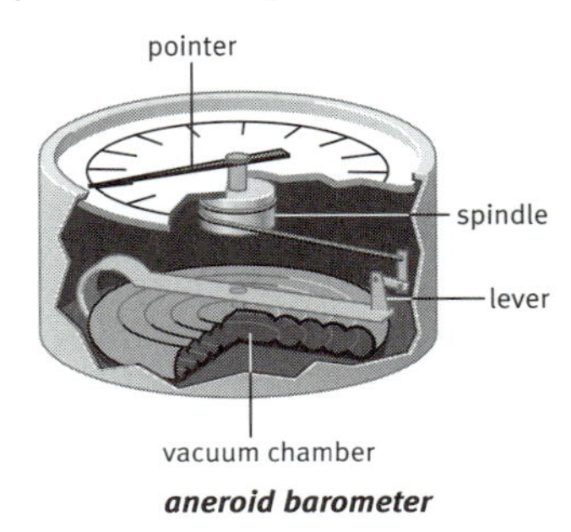

aneroid barometer

anesthesia (ăn′ĭs-thē′zhə) Total or partial loss of sensation to touch or pain, caused by nerve injury or disease, or induced intentionally, especially by the administration of anesthetic drugs, to provide medical treatment. The first public use of ether to anesthetize a patient in Boston in 1846 initiated widespread acceptance of anesthetics in the Western world for surgical procedures and obstetrics. *General anesthesia,* administered as inhalation or intravenous agents, acts primarily on the brain, resulting in a temporary loss of consciousness. *Regional* or *local anesthesia* affects sensation in a specific anatomic area, and includes topical application of local anesthetics, blocking of peripheral nerves, spinal anesthesia, and epidural anesthesia, which is used commonly during childbirth.

anesthesiology (ăn′ĭs-thē′zē-ŏl′ə-jē) The branch of medicine that deals with the study and application of anesthetics.

anesthetic (ăn′ĭs-thĕt′ĭk) A drug that temporarily depresses neuronal function, producing total or partial loss of sensation with or without the loss of consciousness.

aneuploid (ăn′yə-ploid′) Having a chromosome number that is not a multiple of the haploid number for the species. Many types of tumor cells are aneuploid. Compare **diploid, haploid.**

aneuploidy (ăn′yə-ploi′dē) The state or condition of being aneuploid.

aneurysm (ăn′yə-rĭz′əm) A localized, blood-filled dilation of a blood vessel or cardiac chamber caused by disease, such as arteriosclerosis, or weakening of the vessel or chamber wall. A ruptured aneurysm results in hemorrhage and is often fatal.

angina pectoris (ăn-jī′nə pĕk′tər-ĭs) Episodic constricting chest pain, often radiating to the left shoulder and down the left arm, caused by an insufficient supply of blood to the heart. Coronary artery disease is a common cause of angina pectoris.

angiogenesis (ăn′jē-ō-jĕn′ĭ-sĭs) The formation of new blood vessels, especially blood vessels that supply oxygen and nutrients to cancerous tissues.

angiogram (ăn′jē-ə-grăm′) An x-ray of one or more blood vessels produced by angiography and used in diagnosing pathology in the cardiovascular system, such as arteriosclerosis.

angiography (ăn′jē-ŏg′rə-fē) Examination of the blood vessels using x-rays following the injection of a radiopaque substance.

angioplasty (ăn′jē-ə-plăs′tē) The surgical repair of a blood vessel, such as an obstructed coronary artery, usually by inflating a small balloon at the end of a catheter.

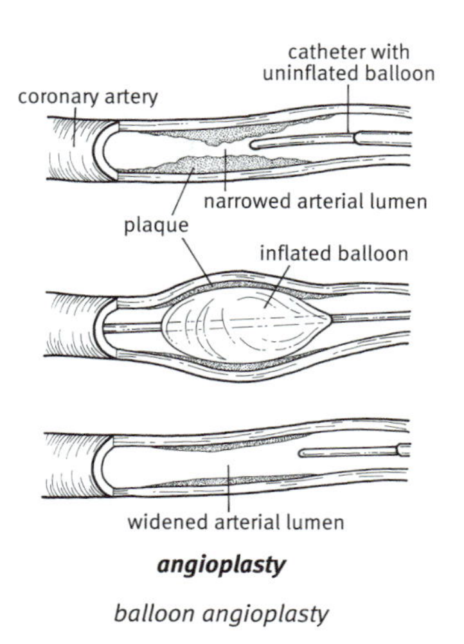

angioplasty

balloon angioplasty

angiosperm (ăn′jē-ə-spûrm′) Any of a large group of plants that produce flowers. They develop seeds from ovules contained in ovaries, and the seeds are enclosed by fruits which develop from carpels. They are also distinguished by the process of **double fertilization.** The majority of angiosperms belong to two large classes: **monocotyledons** and **eudicotyledons.** The angiosperms are the largest phylum of living plants, existing in some 235,000 species. They range from small floating plants only one millimeter (0.04 inch) in length to towering trees that are over 100 meters (328 ft) tall. Compare **gymnosperm.**

angiotensin (ăn′jē-ō-tĕn′sĭn) Any of three polypeptide hormones that function in the body in controlling arterial pressure. The most important is known as angiotensin II, a powerful vasoconstrictor that stimulates steroid production by the adrenal glands, reduces fluid loss from the kidneys, and also functions as a neurotransmitter. Angiotensin II is formed from inactive angiotensin I by the action of *angiotensin-converting enzyme* (or *ACE*). See also **ACE inhibitor, renin.**

angle (ăng′gəl) **1.** A geometric figure formed by two lines that begin at a common point or by two planes that begin at a common line. **2.** The space between such lines or planes, measured in degrees. See also **acute angle, obtuse angle, right angle.**

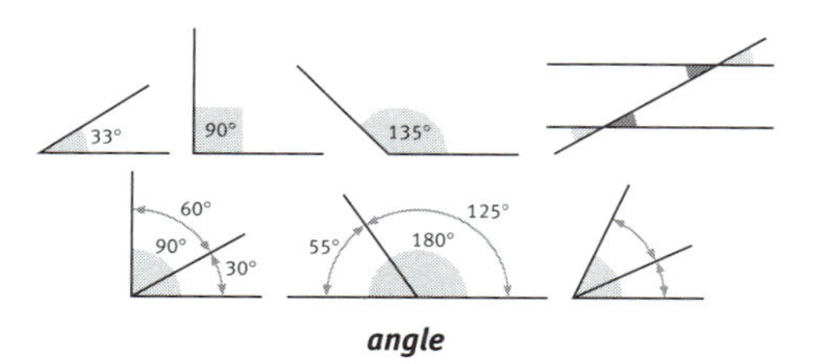

angle

top left to right: *acute, right, and obtuse angles; two pairs of alternating angles (the exterior alternating pair is light gray, the interior alternating pair is dark gray);* bottom left to right: *complementary angles, supplementary angles, and adjacent angles*

angle of incidence The angle formed by a ray or wave, as of light or sound, striking a surface and a line perpendicular to the surface at the point of impact.

angle of reflection The angle formed by a ray or wave reflected from a surface and a line perpendicular to the surface at the point of reflection.

angle of refraction The angle formed by the path of refracted light or other radiation and a line drawn perpendicular to the refracting surface at the point where the refraction occurred.

angstrom (ăng′strəm) A unit of length equal to one hundred-millionth (10^{-10}) of a meter. It was once used to measure wavelengths of light and the diameters of atoms, but has now been mostly replaced by the nanometer.

Ångström (ăng′strəm), **Anders Jonas** 1814–1874. Swedish physicist and astronomer who pioneered the use of the spectroscope in the analysis of radiation. By studying the spectrum of visible light given off by the Sun, Ångström discovered that there is hydrogen in the Sun's atmosphere. The angstrom unit of measurement is named for him.

angular acceleration (ăng′gyə-lər) The rate of change of angular velocity with respect to time. Angular acceleration is measured in revolutions per minute squared or in radians per second squared.

angular deformation A change in the shape of a body, generally due to **sheer stress**, such that a straight line connecting two points within the body before the deformation is not parallel with a straight line connecting them after the deformation.

angular displacement The distance an object moves when following a circular path. It is represented by the length of the arc of a circle drawn to represent the motion of the object about a fixed point.

angular momentum (ăng′gyə-lər) A measure of the momentum of a body in rotational motion. The angular momentum of rigid bodies is conserved; thus, a spinning sphere will continue to spin unless acted on by an outside force. Changes in angular momentum are equivalent to **torque**. In classical mechanics, angular momentum is equal to the product of the angular velocity of the body and its moment of inertia around the axis of rotation. It is a vector quantity; the vector points up along the axis of counterclockwise rotation. In quantum mechanics, the angular momentum of a physical system is quantized and can only take on discrete values. See also **Planck's constant, spin.**

angular velocity The rate of change of an angle associated with an object with respect to a reference point. For example, the average angular velocity of an object moving around a central reference point once every second is 2π radians per second; the Earth spins around its axis with an angular velocity of 2π radians per day. Also called **angular frequency.**

anhydride (ăn-hī′drīd′) A chemical compound formed from another, especially an acid, by the removal of water.

anhydrous (ăn-hī′drəs) Not containing water, especially water of crystallization.

aniline (ăn′ə-lĭn) A colorless, oily, poisonous compound used in the manufacture of rubber, dyes, resins, pharmaceuticals, and varnishes. Aniline is an amine of benzene. *Chemical formula:* C_6H_7N.

animal (ăn′ə-məl) Any of the multicellular organisms belonging to the kingdom Animalia. All animals are eukaryotes, with each of their cells having a nucleus containing DNA. Most animals develop from a blastula and have a digestive tract, nervous system, the ability to move voluntarily, and specialized sensory organs for recognizing and responding to stimuli in the environment. Animals are heterotrophs, feeding on plants, other animals, or organic matter. The first animals probably evolved from protists and appeared during the Precambrian Era.

animal husbandry The branch of agriculture concerned with the care and breeding of domestic animals such as cattle, hogs, sheep, and horses.

anion (ăn′ī′ən) An ion with net negative charge, having more electrons than protons. In electrolysis, anions migrate to a positively charged anode. Compare **cation.**

anion exchange See **ion exchange.**

anisotropic (ăn-ī′sə-trō′pĭk, -trŏp′ĭk, ăn′ī-) Differing according to orientation, as light scattered by a liquid crystal; light striking the liquid crystal's surface at a 90° angle might not be reflected (so the surface appears dark when viewed head-on), while light striking it at shallower angles is reflected (so the surface appears illuminated when viewed from a shallow angle). Compare **isotropic.**

ankylosaurus (ăng′kə-lō-sôr′əs) or **ankylosaur** (ăng′kə-lō-sôr′) A large, herbivorous dinosaur of the genus *Ankylosaurus* of the Cretaceous Period. Ankylosaurs had a squat, heavily armored body and a clubbed tail.

annelid (ăn′ə-lĭd) Any of various worms or wormlike animals of the phylum Annelida, characterized by an elongated, cylindrical body divided into ringlike segments. Most annelids have movable bristles called setae, and include earthworms, leeches, and polychetes (marine worms).

annual (ăn′yo͞o-əl) *Adjective.* **1.** Completing a life cycle in one growing season. —*Noun.* **2.** An annual plant. Annuals germinate, blossom, produce seed, and die in one growing season. They are common in environments with short growing seasons. Most desert plants are annuals, germinating and flowering after rainfall. Many common weeds, wild flowers, garden flowers, and vegetables are annuals. Examples of annuals include tomatoes, corn, wheat, sunflowers, petunias, and zinnias. Compare **biennial, perennial.**

annual ring See under **growth ring.**

annular (ăn′yə-lər) Forming or shaped like a ring.

annular eclipse See under **eclipse.**

annulus (ăn′yə-ləs) Plural **annuluses** or **annuli** (ăn′yə-lī′). **1.** A ringlike figure, part, structure, or marking, such as a growth ring on the scale of a fish. **2.** A ring or group of specialized cells around the sporangia of many ferns. By changing shape in response to variations in humidity, it breaks open the sporangium and then releases the spores with a whipping motion. **3.** The ringlike remains of a membrane (called a **veil**), found around the stipes of certain basidiomycete mushrooms. The presence or absence of an annulus is often used to identify the species of an individual mushroom. **4.** The figure bounded by and containing the area between two concentric circles.

anode (ăn′ōd′) **1.** The positive electrode in an electrolytic cell, toward which negatively charged particles are attracted. The anode has a positive charge because it is connected to the positively charged end of an external power supply. **2.** The positively charged element of an electrical device, such as a vacuum tube or a diode, to which electrons are attracted. **3.** The negative electrode of a voltaic cell, such as a battery. The anode gets its negative charge from the chemical reaction that happens inside the battery, not from an external source. Compare **cathode.**

anorexia (ăn′ə-rĕk′sē-ə) **1.** Loss of appetite, especially as a result of disease. **2.** Anorexia nervosa.

anorexia nervosa (nûr-vō′sə) An eating disorder characterized by a distorted body image, fear of becoming obese, persistent aversion to food, and severe weight loss and malnutrition. It most commonly affects teenage girls and young women, who often develop amenorrhea, osteoporosis and other abnormalities.

anorthite (ăn-ôr′thīt) A white to gray triclinic mineral of the plagioclase feldspar group. Anorthite is the plagioclase mineral that is richest in calcium and occurs in alkaline igneous rocks such as gabbro. *Chemical formula:* $\mathbf{CaAl_2Si_2O_8}$.

anorthoclase (ăn-ôr′thə-klās′) A white to colorless triclinic mineral of the alkali feldspar group. It is a common constituent in the matrices of slightly alkaline volcanic rocks. *Chemical formula:* $\mathbf{(Na,K)AlSi_3O_8}$.

anorthosite (ăn-ôr′thə-sīt′) An igneous rock consisting almost entirely of plagioclase feldspar, especially the labradorite variety. Anorthosites have been identified among rock samples collected from the Moon.

antagonist (ăn-tăg′ə-nĭst) **1.** A muscle that opposes the action of another muscle, as by relaxing while the other one contracts, thereby producing smooth, coordinated movement. **2.** A chemical substance, such as a drug, that interferes with the physiological action of another substance, especially by combining with and blocking its nerve receptor. Compare **agonist.**

Antarctic Circle (ănt-ärk′tĭk) The parallel of latitude approximately 66°33' south. It forms the boundary between the South Temperate and South Frigid zones.

antenna (ăn-tĕn′ə) **1.** Plural **antennae** (ăn-tĕn′ē). One of a pair of long, slender, segmented appendages on the heads of insects, centipedes, millipedes, and crustaceans. Most antennae are organs of touch, but some are sensitive to odors and other stimuli. **2.** Plural **antennas.** A metallic device for sending or receiving electromagnetic waves, such as radio waves. Some antennas can send waves in or receive waves from all directions; others are designed to work only in a range of directions.

anthelion (ănt-hē′lē-ən, ăn-thē′-) Plural **anthelia.** A bright white or colored spot appearing at times in the sky opposite the Sun. White anthelia are believed to form from light that is reflected off atmospheric ice crystals; colored anthelia are believed to form from light that is refracted by atmospheric ice crystals. Compare **parhelion.**

anther (ăn′thər) The pollen-bearing part at the upper end of the stamen of a flower. Most anthers occur at the tip of a slender, stemlike filament and have two lobes. Each lobe contains two pollen sacs. When pollen matures in the pollen sacs, the lobes of the anthers burst open in the process known as dehiscence to release the pollen. See more at **flower.**

antheridiophore (ăn′thə-rĭd′ē-ə-fôr′) A structure that bears the antheridia in some liverworts. See more at **liverwort.**

antheridium (ăn′thə-rĭd′ē-əm) Plural **antheridia** (ăn′thə-rĭd′ē-ə). An organ in certain organisms that produces male gametes. Antheridia are found in many groups of organisms, including the bryophytes, ferns, ascomycete fungi, and some algae. Most gymnosperms and all angiosperms, however, have lost the antheridium, and its role is filled by the pollen grain. Compare **archegonium.**

antherozoid (ăn′thər-ə-zō′ĭd) See **spermatozoid.**

anthesis (ăn-thē′sĭs) The period during which a flower is fully open and functional. Also called *efflorescence.*

anthracene (ăn′thrə-sēn′) A crystalline hydrocarbon that consists of three benzene rings fused together. It is extracted from coal tar and is used to make dyes and organic chemicals. *Chemical formula:* $C_{14}H_{10}$.

anthracite (ăn′thrə-sīt′) A hard, shiny coal that has a high carbon content. It is valued as a fuel because it burns with a clean flame and without smoke or odor, but it is much less abundant than bituminous coal. Compare **bituminous coal, lignite.**

anthraquinone (ăn′thrə-kwĭ-nōn′, -kwĭn′ōn′) A yellow crystalline powder that is insoluble in water and used chiefly in the manufacture of dyes. *Chemical formula:* $C_{14}H_8O_2$.

anthrax (ăn′thrăks′) An infectious, usually fatal disease of mammals, especially cattle and sheep, caused by the bacterium *Bacillus anthracis.* The disease is transmitted to humans through cutaneous contact, ingestion, or inhalation. Cutaneous anthrax is marked by the formation of a necrotic skin ulcer, high fever, and toxemia. Inhalation anthrax leads to severe pneumonia that is usually fatal.

anthropogenesis (ăn′thrə-pə-jĕn′ĭ-sĭs) The scientific study of the origin and development of humans.

anthropogenic (ăn′thrə-pə-jĕn′ĭk) Caused or influenced by humans. Anthropogenic carbon dioxide is that portion of carbon dioxide in the atmosphere that is produced directly by human activities, such as the burning of fossil fuels, rather than by such processes as respiration and decay.

anthropography (ăn′thrə-pŏg′rə-fē) The branch of anthropology that deals with the geographical distribution of specific human cultures.

anthropoid ape (ăn′thrə-poid′) A primate belonging to the family Pongidae, which includes the chimpanzee, bonobo, gorilla, and orangutan. Orangutans are arboreal whereas the other three species are terrestrial or semiarboreal. Anthropoid apes move in trees mainly by arm-swinging and on the ground by quadrupedal walking in which the upper body weight is borne on the knuckles. Also called *great ape, pongid.* Compare **hominid.**

anthropology (ăn′thrə-pŏl′ə-jē) The scientific study of humans, especially of their origin, their behavior, and their physical, social, and cultural development.

anthropometry (ăn′thrə-pŏm′ĭ-trē) The study of human body measurement for use in anthropological classification and comparison. The use of such data as skull dimensions and body proportions in the attempt to classify human beings into racial, ethnic, and national groups has been largely discredited, but anthropometric techniques are still used in physical anthropology and paleoanthropology, especially to study evolutionary change in fossil hominid remains.

anti– A prefix whose basic meaning is "against." It is used to form adjectives that mean "counteracting" (such as *antiseptic,* preventing infection). It is also used to form nouns referring to substances that counteract other substances (such as *antihistamine,* a substance counteracting histamine), and nouns meaning "something that displays opposite, reverse, or inverse characteristics of something else" (such as *anticyclone,* a storm that circulates in the opposite direction from a cyclone). Before a vowel it becomes *ant–,* as in *antacid.*

antialiasing (ăn′tē-ā′lē-ə-sĭng, ăn′tī-) In computer graphics, a software process for removing or reducing the jagged distortions in curves and diagonal lines so that the lines appear smooth or smoother.

antiatom (ăn′tē-ăt′əm, ăn′tī-) An atom composed of antiparticles. An antiatom consists of positrons, antiprotons, and antineutrons. It has the same mass and spin as an ordinary atom, and the same amount of charge and magnetic moment, but the charge and magnetic moment are the opposite of those of an ordinary atom.

antibaryon (ăn′tē-băr′ē-ŏn′, ăn′tī-) The antiparticle that corresponds to a baryon.

antibiosis (ăn′tē-bī-ō′sĭs, ăn′tī-) **1.** An association between two or more organisms that is detrimental to at least one of them. Allelopathy (the production of chemicals by one plant species that inhibit the growth of another) is an example of antibiosis. **2.** The antagonistic association between an organism and the metabolic substances produced by another.

antibiotic (ăn′tĭ-bī-ŏt′ĭk) *Noun.* **1.** A substance, such as penicillin, that is capable of destroying or weakening certain microorganisms, especially bacteria or fungi, that cause infections or infectious diseases. Antibiotics are usually produced by or synthesized from other microorganisms, such as molds. They inhibit pathogens by interfering with essential intracellular processes, including the synthesis of bacterial proteins. Antibiotics do not kill viruses and are not effective in treating viral infections. *–Adjective.* **2.** Relating to antibiotics. **3.** Relating to antibiosis.

antibody (ăn′tĭ-bŏd′ē) Any of numerous proteins produced by **B lymphocytes** in response

A CLOSER LOOK **antibodies**

Like other vertebrates, humans possess an effective immune system that uses *antibodies* to fight bacteria, viruses, and cancer cells. Antibodies are complex, Y-shaped protein molecules. The immune system's B lymphocytes, which are produced by the bone marrow, develop into plasma cells that can generate a huge variety of antibodies, each one capable of combining with and destroying an *antigen,* a foreign molecule. Antibodies react to very specific characteristics of different antigens, binding them to the top ends of their Y formation. Once the antibody and antigen combine, the antibodies deactivate the antigen or lead it to *macrophages* (a kind of white blood cell) that ingest and destroy it. High numbers of a particular antibody may persist for months after an invasion, eventually diminishing. However, the B cells can quickly manufacture more of the same antibody if exposure to the antigen recurs. Vaccines work by "training" B cells to recognize and react quickly to potential disease molecules.

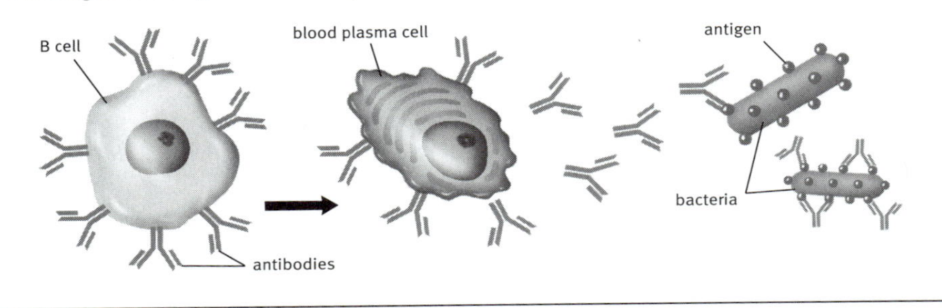

to the presence of specific foreign antigens, including microorganisms and toxins. Antibodies consist of two pairs of polypeptide chains, called **heavy chains** and **light chains**, that are arranged in a Y-shape. The two tips of the Y are the regions that bind to antigens and deactivate them. Also called *immunoglobulin.*

anticline (ăn′tĭ-klīn′) A fold of rock layers that slope downward on both sides of a common crest. Anticlines form when rocks are compressed by plate-tectonic forces. They can be as small as a hill or as large as a mountain range. Compare **syncline.**

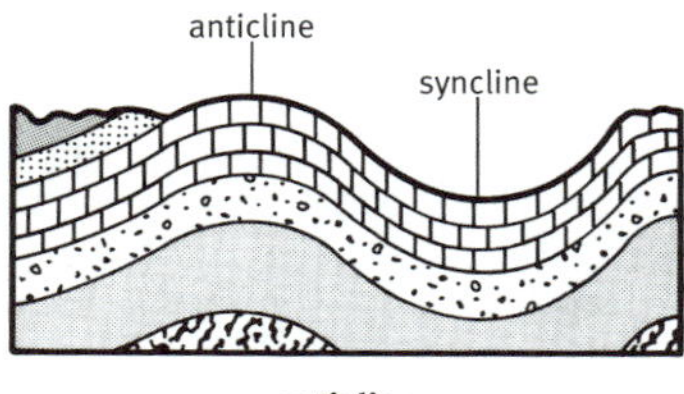

anticline

anticoagulant (ăn′tē-kō-ăg′yə-lənt, ăn′tī-) A substance that prevents the clotting of blood.

anticodon (ăn′tē-kō′dŏn, ăn′tī-) A sequence of three adjacent nucleotides in transfer RNA that binds to a corresponding codon in messenger RNA and designates a specific amino acid during protein synthesis.

anticonvulsant (ăn′tē-kən-vŭl′sənt, ăn′tī-) A drug that prevents or treats seizures.

anticyclone (ăn′tē-sī′klōn′, ăn′tī-) A large-scale system of winds that spiral outward around a region of high atmospheric pressure. In the Northern Hemisphere the wind in an anticyclone moves clockwise, and in the Southern Hemisphere it moves counterclockwise. Compare **cyclone.**

antidepressant (ăn′tē-dĭ-prĕs′ənt, ăn′tī-) A drug used to prevent or treat depression.

antideuteron (ăn′tē-do͞o′tə-rŏn′, -dyo͞o′-, ăn′tī-) The antiparticle that corresponds to a deuteron.

antidiuretic hormone (ăn′tē-dī′ə-rĕt′ĭk, ăn′tī-) A polypeptide hormone secreted by the posterior portion of the pituitary gland that constricts blood vessels, raises blood pressure, and reduces excretion of urine. Also called *vasopressin.*

antidote (ăn′tĭ-dōt′) A substance that counteracts the effects of a poison.

antielectron (ăn′tē-ĭ-lĕk′trŏn′, ăn′tī-) See **positron.**

antigen (ăn′tĭ-jən) A substance that stimulates the production of an antibody when introduced into the body. Antigens include toxins, bacteria, viruses, and other foreign substances. Compare **antibody.** See Note at **blood type.**

antihelium (ăn′tē-hē′lē-əm, ăn′tī-) The antimatter that corresponds to helium.

antihistamine (ăn′tē-hĭs′tə-mēn′) Any of various drugs that relieve cold or allergy symptoms by blocking the action of histamine.

antihydrogen (ăn′tē-hī′drə-jən, ăn′tī-) The antimatter that corresponds to hydrogen. Antihydrogen has been useful in studies of the relationship between matter and antimatter, because its matter equivalent (hydrogen) is one of the most studied and most well understood forms of matter.

anti-inflammatory (ăn′tē-ĭn-flăm′ə-tôr′ē, ăn′tī-) Preventing or reducing inflammation. Aspirin and ibuprofen are anti-inflammatory drugs.

antilepton (ăn′tē-lĕp′tŏn, ăn′tī-) The antiparticle that corresponds to a lepton.

antilogarithm (ăn′tē-lô′gə-rĭth′əm, ăn′tī) The number whose logarithm is a given number. For example, the logarithm of 1,000 (10^3) is 3, so the antilogarithm of 3 is 1,000. In algebraic notation, if $\log x = y$, then $\text{antilog } y = x$.

antimatter (ăn′tĭ-măt′ər) A form of matter that consists of antiparticles.

antimony (ăn′tə-mō′nē) *Symbol* **Sb** A metalloid element having many forms, the most common of which is a hard, very brittle, shiny, blue-white crystal. It is used in a wide variety of alloys, especially with lead in car batteries, and in the manufacture of flameproofing compounds. Atomic number 51; atomic weight 121.76; melting point 630.5°C (1,167°F); boiling point 1,380°C (2,516°F); specific gravity 6.691; valence 3, 5. See **Periodic Table.**

antineutrino (ăn′tē-no͞o-trē′nō, ăn′tī-) The antiparticle that corresponds to the neutrino.

antineutron (ăn′tē-no͞o′trŏn′, ăn′tī-) The antiparticle that corresponds to the neutron.

antinode (ăn′tĭ-nōd′) In a standing wave, the region or point of maximum amplitude between two adjacent nodes. Compare **node** (sense 3).

antinucleon (ăn′tē-no͞o′klē-ŏn′, ăn′tī-) The antiparticle that corresponds to a nucleon.

antioxidant (ăn′tē-ŏk′sĭ-dənt, ăn′tī-) A chemical compound or substance that inhibits oxidation. Certain vitamins, such as vitamin E, are antioxidants and may protect body cells from damage caused by the oxidative effects of free radicals.

antiparticle (ăn′tē-pär′tĭ-kəl, ăn′tī-) A subatomic particle, such as an antiproton, having the same mass as its corresponding particle, but opposite values of other properties such as charge, parity, spin, and direction of magnetic moment. For example, the antiparticle of the electron is the positron, which has a charge that is equal in magnitude to that of the electron but opposite in sign. Some particles, such as photons, are nondistinct from their antiparticles. When a particle and its antiparticle collide, they may annihilate one other and produce other particles.

antipodal (ăn-tĭp′ə-dəl) *Noun.* **1.** In angiosperms, any of three nuclei located at the end of the embryo sac opposite the micropyle. The antipodals degenerate at or shortly after fertilization. See more at **embryo sac.** —*Adjective.* **2.** Situated on the opposite side or sides of the Earth.

antipodes (ăn-tĭp′ə-dēz′) Two places on directly opposite sides of the Earth, such as the North Pole and the South Pole.

antiproton (ăn′tē-prō′tŏn′, ăn′tī-) The antiparticle that corresponds to the proton.

antiquark (ăn′tē-kwôrk′, ăn′tī-) The antiparticle that corresponds to a quark.

antirejection (ăn′tē-rĭ-jĕk′shən, ăn′tī-) Preventing rejection of a transplanted tissue or organ, as a drug or treatment.

antisense (ăn′tē-sĕns′, ăn′tī-) Relating to a nucleotide sequence that is complementary to a sequence of messenger RNA. When antisense DNA or RNA is added to a cell, it binds to a specific messenger RNA molecule and inactivates it.

antisepsis (ăn′tĭ-sĕp′sĭs) The destruction of pathogenic microorganisms in order to prevent infection.

antiseptic (ăn′tĭ-sĕp′tĭk) A substance that inhibits the proliferation of infectious microorganisms.

antiserum (ăn′tĭ-sîr′əm) Plural **antiserums** or **antisera.** Human or animal serum containing one or more antibodies that are specific for one or more antigens and are administered to confer immunity. The antibodies in an antiserum result from previous immunization or exposure to an agent of disease. See also **acquired immunity.**

antisocial personality disorder (ăn′tē-sō′shəl, ăn′tī-) See under **personality disorder.**

antitoxin (ăn′tē-tŏk′sĭn, ăn′tī-) **1.** An antibody formed in response to and capable of neutralizing a specific toxin of biological origin. Compare **toxin. 2.** An animal or human serum containing antitoxins, used to prevent or treat diseases caused by biological toxins, such as tetanus, botulism, and diphtheria.

antitrades (ăn′tĭ-trādz′) Winds blowing steadily from west to east in the upper levels of the troposphere, above and in a direction counter to the surface trade winds of the tropics. In the middle latitudes of the North and South Temperate Zones, the antitrades merge with the prevailing westerly surface winds. Compare **trade winds.**

antivenin (ăn′tē-vĕn′ĭn, ăn′tī-) **1.** An antitoxin active against the venom of a snake, spider, or other venomous animal or insect. **2.** An animal serum containing antivenins, used in medicine to treat poisoning caused by animal or insect venom.

anus (ā′nəs) The opening at the lower end of the digestive tract through which solid waste is excreted. —*Adjective* **anal.**

anxiety (ăng-zī′ĭ-tē) A state of apprehension and fear resulting from the anticipation of a threatening event or situation. ▸ In psychiatry, a patient has an **anxiety disorder** if normal psychological functioning is disrupted or if anxiety persists without an identifiable cause.

anxiolytic (ăng′zē-ō-lĭt′ĭk) A drug used to treat acute or chronic anxiety.

aorta (ā-ôr′tə) Plural **aortas** or **aortae** (ā-ôr′tē). The main artery of the circulatory system, arising from the left ventricle of the heart in mammals and birds and carrying blood with high levels of oxygen to all the arteries of the body except those of the lungs.

aortic valve (ā-ôr′tĭk) A valve located in the aorta that prevents the backflow of blood to the left ventricle.

apastron (ə-păs′trən) The point at which an object, such as a planet or comet, is farthest from the center of mass of the star it is orbiting. Compare **periastron.**

apatite (ăp′ə-tīt′) Any of several usually green, transparent, hexagonal minerals consisting of calcium phosphate with either fluorine, hydroxyl, chlorine, or carbonate. Apatite occurs in igneous, metamorphic, and sedimentary rocks, and is used as a source of phosphate for making fertilizers. *Chemical formula:* $Ca_5(PO_4CO_3)_3(F,OH,Cl)$.

apatosaurus (ə-păt′ə-sôr′əs) or **apatosaur** (ə-păt′ə-sôr′) A very large sauropod dinosaur of the genus *Apatosaurus* (or *Brontosaurus*) of the late Jurassic Period. Apatosaurs had a long neck and tail and a relatively small head. See Note at **brontosaurus.**

apex (ā′pĕks) The highest point, especially the vertex of a triangle, cone, or pyramid.

Apgar score (ăp′gär) A score that assesses the general physical condition of a newborn infant by assigning a value of 0, 1, or 2 to each of five criteria: heart rate, respiratory effort, muscle tone, skin color, and response to stimuli. The five scores are added together, with a perfect score being 10. Apgar scores are usually evaluated at one minute and five minutes after birth. The Apgar score is named for the system's deviser, American physician Virginia Apgar (1909–1974).

aphanitic (ăf′ə-nĭt′ĭk) Of or relating to an igneous rock in which the crystals are so fine that individual minerals cannot be distinguished with the naked eye. Aphanitic rocks are extrusive rocks that cooled so quickly that crystal growth was inhibited. Compare **phaneritic.**

aphasia (ə-fā′zhə) Partial or total loss of the ability to articulate ideas or comprehend spoken or written language, resulting from damage to the brain that is caused by injury or disease.

aphelion (ə-fē′lē-ən) The point at which an orbiting body, such as a planet or comet, is farthest away from the Sun. Compare **apogee, perihelion.**

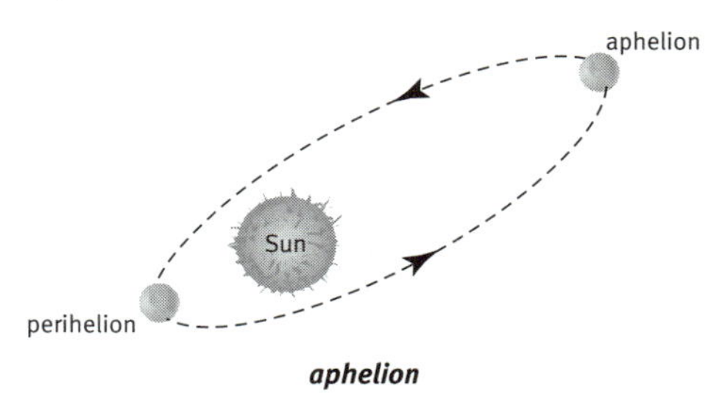

aphelion

aphotic (ā-fō′tĭk) **1.** Having no light. **2.** Relating to the region of a body of water that is not reached by sunlight and in which photosynthesis cannot occur. Most ocean water lies in the aphotic zone. Compare **photic.**

aphyllous (ā-fĭl′əs) Bearing no leaves; leafless. Aphyllous plants conduct photosynthesis in their stems and branches.

apical dominance (ā′pĭ-kəl, ăp′ĭ-) Inhibition of the growth of lateral buds by the terminal bud of a plant shoot. In most plants, apical dominance results from the release of **auxin** by the apical meristem.

apical meristem A meristem at the tip of a plant shoot or root that produces auxin and causes the shoot or root to increase in length. Growth that originates in the apical meristem is called **primary growth.** In vascular plants, the apical meristem produces three kinds of primary meristems: the protoderm, ground meristem, and procambium. These in turn produce primary tissues. See more at **meristem.**

apicomplexan (ăp′ĭ-kəm-plĕk′sən) Any of a phylum (Apicomplexa) of protozoans that are parasitic in animals, especially animal bloodstreams, and are distinguished by a variety of organelles, including fibrils and microtubules, located at one end (called the apical end) of the cell. These organelles help the apicomplexan invade an animal cell. Apicomplexans form spores and reproduce sexually in an **alternation of generations**; many have complex life cycles and are transmitted to animals hosts by bloodsucking insects. Apicomplexans include the organisms that were once classified as **sporozoans**, including the protozoans that cause malaria and toxoplasmosis.

aplastic anemia (ā-plăs′tĭk) A form of anemia in which the capacity of the bone marrow to generate red blood cells is defective. Aplastic anemia may be caused by bone marrow disease or exposure to toxic agents, such as radiation, chemicals, or drugs.

apnea (ăp′nē-ə, ăp-nē′ə) The temporary absence or cessation of breathing.

apoapsis (ăp′ō-ăp′sĭs) Plural **apoapsides** (ăp′ō-ăp′sĭ-dēz′). The point at which an orbiting object is farthest away from the body it is orbiting. This point is sometimes given a name that is specific to the body being orbited. For example, the apoapsis of an object orbiting Earth is its **apogee** (from *gaia,* the Greek word for Earth), and the apoapsis of an object orbiting the Sun is its **aphelion** (from *hēlios,* the Greek word for Sun). According to **Kepler's laws of planetary motion**, an object is at its lowest velocity at the apoapsis. Compare **periapsis.**

apoenzyme (ăp′ō-ĕn′zīm) The protein component of an enzyme, to which the coenzyme attaches to form an active enzyme.

apogee (ăp′ə-jē) **1.** The point farthest from Earth's center in the orbit of the Moon or an artificial satellite. **2.** The point in an orbit that is most distant from the body being orbited. Compare **aphelion, perigee.**

apolipoprotein (ăp′ə-lĭp′ō-prō′tēn′, -lī′pō-) Any of various proteins that combine with a lipid to form a lipoprotein, such as high density lipoprotein (HDL) and low density lipoprotein (LDL). Apolipoproteins are important in the transport of cholesterol in the body and the regulation of the level of cholesterol in cells and blood.

apolune (ăp′ə-lo͞on′) The point at which an object orbiting the Moon, such as an artificial satellite, is farthest from the Moon's center.

apomixis (ăp′ə-mĭk′sĭs) The development of an embryo without the occurrence of fertilization. **Parthenogenesis** is one form of apomixis. In plants, another form of apomixis also occurs, in which the embryo develops from the somatic cells of the ovule surrounding the embryo sac, not from the egg cell within the embryo sac itself. Such embryos are clones of the parent plant, and valuable cultivars of plants such as the fig are propagated using seeds produced through this kind of apomixis.

apoptosis (ăp′əp-tō′sĭs, ăp′ə-tō′-) A natural process of self-destruction in certain cells, such as epithelial cells and erythrocytes, that are genetically programmed to have a limited life span or are damaged. Apoptosis can be induced either by a stimulus, such as irradiation or toxic drugs, or by removal of a repressor agent. The cells disintegrate into membrane-bound particles that are then eliminated by phagocytosis. Also called *programmed cell death.*

aposematic coloration (ăp′ə-sə-măt′ĭk) See **warning coloration.**

apothecaries' weight (ə-pŏth′ĭ-kĕr′ēz) A system of weights used in pharmacy and based on an ounce equal to 480 grains and a pound equal to 12 ounces. It has been largely replaced by measures of the metric system.

apothecium (ăp′ə-thē′sē-əm, -shē-) Plural **apothecia** (ăp′ə-thē′sē-ə, -shē-). A disk-shaped or cup-shaped ascocarp of some lichens and the fungi Ascomycetes.

apparent horizon (ə-pâr′ənt) See **horizon** (sense 1a).

apparent magnitude See under **magnitude.**

apparent temperature See **heat index.**

apparent time Solar time as measured by the Sun's actual position with respect to an observer's local meridian and equal for any observer to the **hour angle** of the Sun at that location. Sundials register apparent time, which differs continuously with the observer's longitude and is not standardized across a time zone. Each degree of longitude represents a difference of four minutes in apparent time. Because of irregularities in Earth's orbit and the tilt of its axis, a day as measured by apparent time varies somewhat in length throughout the year. Also called *apparent solar time.* ▸ The moment when the Sun crosses an observer's local meridian is called the **apparent noon**. Compare **local mean time.** See more at **mean time, solar time.**

appendicitis (ə-pĕn′dĭ-sī′tĭs) Inflammation of the appendix, usually caused by a blockage or infection.

appendix (ə-pĕn′dĭks) Plural **appendixes** or **appendices** (ə-pĕn′-dĭ-sēz′). A tubular projection attached to the cecum of the large intestine and located on the lower right side of the abdomen. Also called *vermiform appendix.*

applet (ăp′lĭt) A small computer program that has limited features, requires limited memory resources, and is designed to be downloaded from the Internet to run on a webpage. An applet cannot read or write data on the user's machine.

application (ăp′lĭ-kā′shən) A computer program with an interface, enabling people to use the computer as a tool to accomplish a specific task. Word processing, spreadsheet, and communications software are all examples of applications.

apron (ā′prən) An area covered by a blanketlike deposit of glacial, eolian, marine, or alluvial sediments, especially an area at the foot of a mountain or in front of a glacier.

apsis (ăp′sĭs) Plural **apsides** (ăp′sĭ-dēz′). In the path of an orbiting body, either of the two points at which it is closest to or farthest away from the body it is orbiting. See also **apoapsis, periapsis.**

aquaculture (ăk′wə-kŭl′chər, ä′kwə-) **1.** The science of cultivating marine or freshwater food fish, such as salmon and trout, or shellfish, such as oysters and clams, under controlled conditions. **2.** See **hydroponics.**

aqua regia (rē′jē-ə, rē′jə) A corrosive, fuming, volatile mixture of hydrochloric and nitric acids. Aqua regia is used for testing metals and dissolving platinum and gold.

Aquarius (ə-kwâr′ē-əs) A constellation in the Southern Hemisphere near Pisces and Aquila. Aquarius (the Water Bearer) is the 11th sign of the zodiac.

aquatic (ə-kwăt′ĭk) Relating to, living in, or growing in water.

aqueous (ā′kwē-əs) **1.** *Chemistry.* Relating to or dissolved in water. **2.** *Geology.* Formed from matter deposited by water. Certain sedimentary rocks, such as limestone, are aqueous.

aqueous humor The clear, watery fluid that fills the chamber of the eye between the cornea and the lens.

aquiclude (ăk′wĭ-klo͞od′) An impermeable body of rock or stratum of sediment that acts as a barrier to the flow of groundwater.

aquifer (ăk′wə-fər) An underground layer of permeable rock, sediment (usually sand or gravel), or soil that yields water. The pore spaces in aquifers are filled with water and are interconnected, so that water flows through them. Sandstones, unconsolidated gravels, and porous limestones make the best aquifers. They can range from a few square kilometers to thousands of square kilometers in size.

Aquila (ăk′wə-lə) A constellation in the Northern Hemisphere near Aquarius and Hercules. Aquila (the Eagle) contains the bright star Altair.

Ar The symbol for **argon.**

Arabic numeral (ăr′ə-bĭk) One of the numerical symbols 1, 2, 3, 4, 5, 6, 7, 8, 9, or 0. They are called Arabic numerals because they were introduced into western Europe from sources of Arabic scholarship.

arachnid (ə-răk′nĭd) Any of various arthropods of the class Arachnida, such as spiders, scorpions, mites, and ticks. Arthropods are characterized by four pairs of segmented legs and a body that is divided into two regions, the cephalothorax and the abdomen.

arachnoid (ə-răk′noid′) The delicate membrane that encloses the spinal cord and brain and lies between the pia mater and dura mater.

aragonite (ə-răg′ə-nīt′, ăr′ə-gə-) A usually white, yellowish, or pink orthorhombic mineral that can occur in many different colors. Aragonite occurs as acicular (needlelike) or tabular crystals, or as fibrous aggregates. It is found in gypsum deposits, at the tips of calcite crystals, in mollusk shells and pearls, and in living reef structures. It is a polymorph of calcite. *Chemical formula:* **$CaCO_3$.**

Arber (är′bər), **Werner** Born 1929. Swiss microbiologist who postulated the existence of *restriction enzymes,* selective enzymes that break down molecules of DNA into pieces small enough to be separated for individual study but large enough to retain bits of the original substance's genetic information. These enzymes (later isolated by Hamilton Smith) laid the foundation for the science of genetic engineering, and for this work Arber shared the 1978 Nobel Prize for physiology or medicine with Smith and Daniel Nathans.

arboreal (är-bôr′ē-əl) Relating to or living in trees.

arbovirus (är′bə-vī′rəs) Any of a large group of RNA viruses that are transmitted primarily by arthropods, such as mosquitoes and ticks. The more than 400 species were originally considered to be a single group, but are now divided among four families: Togaviridae, Flaviviridae, Bunyaviridae, and Arenaviridae. These viruses cause a variety of infectious diseases in humans, including rubella, yellow fever, and dengue.

arc (ärk) **1.** A segment of a circle. **2.** See **electric arc.**

arc cosecant The inverse of the cosecant function.

arc cosine The inverse of the cosine function.

arc cotangent The inverse of the cotangent function.

Archaean (är-kē′ən) Another spelling of **Archean.**

archaebacterium (är′kē-băk-tîr′ē-əm) Plural **archaebacteria.** See **archaeon.**

archaeology or **archeology** (är′kē-ŏl′ə-jē) The scientific study of past human life and culture by the examination of physical remains, such as graves, tools, and pottery.

archaeon (är′kē-ŏn′) Plural **archaea.** Any of a group of microorganisms that resemble bacteria but are different from them in certain aspects of their chemical structure, such as the composition of their cell walls. Archaea usually live in extreme, often very hot or salty environments, such as hot mineral springs or deep-sea hydrothermal vents, but some are also found in animal digestive systems. The archaea are considered a separate kingdom in some classifications, but a division of the prokaryotes (Monera) in others. Some scientists believe that archaea were the earliest forms of cellular life. Also called *archaebacterium.*

archaeopteryx (är′kē-ŏp′tər-ĭks) An extinct primitive bird of the genus *Archaeopteryx* of the Jurassic Period, having characteristics of both birds and dinosaurs. Like dinosaurs, it had a long, bony tail, claws at the end of its fingers, and teeth. Like birds it had wings and feathers. Many scientists regard it as evidence that birds evolved from small carnivorous dinosaurs. See Note at **bird.**

Archaeozoic (är′kē-ə-zō′ĭk) Another spelling of **Archeozoic.**

archaic Homo sapiens (är-kā′ĭk) Relating to or being an early form or subspecies of *Homo sapiens,* anatomically distinct from modern humans. Neanderthals in Europe and Solo man in Asia are usually classed as archaic humans. Though archaic humans belong to the same species as modern humans, not all archaic groups or populations are necessarily ancestral to *Homo sapiens sapiens.* According to certain models of human evolution, modern humans replaced archaic populations throughout Asia and Europe after migrating out of Africa in comparatively recent times. In other models, widely separated but interbreeding archaic groups in different parts of the world evolved independently into today's physiologically distinct geographic populations.

Archean (är-kē′ən) The earlier of the two divisions of the Precambrian Eon, from about 3.8 to 2.5 billion years ago. During this time the Earth had a reducing atmosphere consisting primarily of methane, ammonia, and other gases that would be toxic to most modern life forms. There was little free oxygen. Rocks from the earliest part of the Archean are predominantly volcanic and are similar to pillow basalts, suggesting that they formed underwater. Rocks from the later part of the Archean appear to have formed on continents. It is believed that about 70% of the continental masses formed during this time. Fossils preserved in rocks from this period of time include remains of cyanobacteria, the first single-celled forms of life. These organisms are preserved in the form of stromatolites and oncolites. See Chart at **geologic time.**

archegoniophore (är′kĭ-gō′nē-ə-fôr′) A structure that bears the archegonia in some liverworts. See more at **liverwort.**

archegonium (är′kĭ-gō′nē-əm) Plural **archegonia.** The egg-producing organ occurring in bryophytes (such as mosses and liverworts), ferns, and most gymnosperms. The archegonium is a multicellular, often flask-shaped structure that contains a single egg. Compare **antheridium.**

Archeozoic also **Archaeozoic** (är′kē-ə-zō′ĭk) See **Archean.**

Archimedean solid (är′kə-mē′dē-ən, -mĭ-dē′-) A polyhedron whose faces are regular polygons and whose angles are all congruent. The faces may all be of the same type, in which case the solid is a regular polyhedron, or they may be of different types. There are only thirteen Archimedean solids. See more under **polyhedron.**

Archimedes (är′kə-mē′dēz), 287–212 BCE. Greek mathematician, engineer, and inventor. He made numerous mathematical discoveries, including the ratio of the radius of a circle to its circumference as well as formulas for the areas and volumes of various geometric figures. Archimedes created the science of mechanics, devising the first general theory of levers and finding methods for determining the center of gravity of a variety of bodies. He also invented an early type of pump called the Archimedian screw.

archipelago (är′kə-pĕl′ə-gō′) **1.** A large group of islands. **2.** A sea, such as the Aegean, or an area in a sea containing a large number of scattered islands.

archosaur (är′kə-sôr′) Any of various mostly reptilian animals of the subclass Archosauria. Archosaurs are diapsids that began to evolve in the late Permian Period, and are characterized by skulls with long, narrow snouts and teeth set in sockets. Archosaurs include the extinct dinosaurs and pterosaurs and the modern crocodilians and birds.

arc lamp An electric light in which a current produces light when an arc traverses the gap between two incandescent electrodes in a container filled with a gas, such as xenon. The two electrodes are usually made of carbon and are eventually vaporized by the heat they generate. Arc lamps are used to produce intense light (as in spotlights) and to produce heat for welding.

arc secant The inverse of the secant function.

arc sine The inverse of the sine function.

arc tangent The inverse of the tangent function.

Arctic Circle (ärk′tĭk) The parallel of latitude approximately 66°33' north. It forms the boundary between the North Temperate and North Frigid zones.

Arctic Stream See **Labrador Current.**

Arcturus (ärk-to͝or′əs) A giant star in the constellation Boötes. It is the brightest star in the Northern Hemisphere and the fourth brightest star in the sky, with an apparent magnitude of 0.00. *Scientific name: Alpha Boötes.*

area (âr′ē-ə) The extent of a surface or plane figure as measured in square units.

arenaceous (ăr′ə-nā′shəs) **1.** Resembling, derived from, or containing sand. **2.** Growing in sandy areas.

arête (ə-rāt′) A sharp, narrow ridge or spur commonly found above the snow line in mountainous areas that have been sculpted by glaciers. Arêtes form as the result of the continued backward erosion of adjoining cirques.

argillite (är′jə-līt′) A highly compacted sedimentary or slightly metamorphic rock consisting primarily of particles of clay or silt. Argillite differs from mudstone in that it does not have the same fine laminations, and from shale and slate in that it is not fissile.

arginine (är′jə-nēn′) An amino acid that is essential for children but not for adults. *Chemical formula:* $C_6H_{14}N_4O_2$.

argon (är′gŏn′) *Symbol* **Ar** A colorless, odorless element in the noble gas group. Argon makes up about one percent of the atmosphere. It is used in electric light bulbs, fluorescent tubes, and radio vacuum tubes. Atomic number 18; atomic weight 39.948; melting point –189.2°C; boiling point –185.7°C. See **Periodic Table.**

arid (ăr′ĭd) Very dry, especially having less precipitation than is needed to support most trees or woody plants. Deserts have arid climates.

Aries (âr′ēz) A constellation in the Northern Hemisphere near Taurus and Pisces. Aries (the Ram) is the first sign of the zodiac.

aril (ăr′əl) A fleshy seed cover which arises from the funiculus (the stalk of the ovule). Arils, such the red berry-like arils of the yew, are often brightly colored to attract animals who eat them and disperse the seeds. The spice mace is the aril of the nutmeg seed.

Aristotle (ăr′ĭ-stŏt′l), 384–322 BCE. Greek philosopher and scientist who wrote about virtually every area of knowledge, including most of the sciences. Throughout his life he made careful observations, collected specimens, and summarized all the existing knowledge of the natural world. He pioneered the study of zoology, developing a classification system for all animals and making extensive taxonomic studies. His systematic approach later evolved into the basic scientific method in the Western world.

arithmetic (ə-rĭth′mĭ-tĭk) The mathematics of integers, rational numbers, real numbers, or complex numbers under the operations of addition, subtraction, multiplication, and division.

arithmetic mean (ăr′ĭth-mĕt′ĭk) The value obtained by dividing the sum of a set of quantities by the number of quantities in the set. For example, if there are three test scores 70, 83, and 90, the arithmetic mean of the scores is their sum (243) divided by the number of scores (3), or 81. See more at **mean.** Compare **average, median, mode.**

arithmetic progression (ăr′ĭth-mĕt′ĭk) A sequence of numbers such as 1, 3, 5, 7, 9 ..., in which each term after the first is formed by adding a constant (in this case, 2) to the preceding number. Compare **geometric progression.**

arkose (är′kōs) A usually pinkish or red sandstone consisting primarily of quartz and feldspar. Arkose usually forms as the result of the rapid disintegration of granite in areas of vigorous erosion. Its grains are usually angular and poorly sorted (mixed randomly in differing sizes).

armature (är′mə-chər) **1.** The part of an electric motor or generator that consists of wire wound around an iron core and carries an electric current. In motors and generators using direct current, the armature rotates within a magnetic field; in motors and generators using alternating current a magnetic field is rotated about the armature. **2.** A piece of soft iron connecting the poles of a magnet. **3.** The part of an electromagnetic device, such as a relay or loudspeaker, that moves or vibrates.

armillary sphere (är′mə-lĕr′ē, är-mĭl′ə-rē) An early astronomical device made of fixed and movable rings representing circles of the celestial sphere, such as the ecliptic and the celestial equator. It was used as early as the third century BCE as both a teaching instrument and an observational tool.

aromatic (ăr′ə-măt′ĭk) Relating to an organic compound containing at least one benzene ring or similar ring-shaped component. Naphthalene and TNT are aromatic compounds. Compare **aliphatic.**

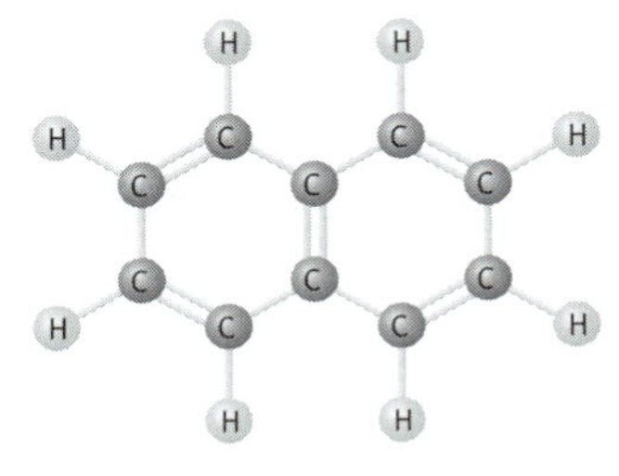

aromatic

The organic compound naphthalene is composed of two benzene rings.

ARPANET (är′pə-nĕt) A computer network developed by the Advanced Research Project Agency (now the Defense Advance Research Projects Agency) in the 1960s and 1970s as a means of communication between research laboratories and universities. ARPANET was the predecessor to the Internet.

array (ə-rā′) **1.** *Mathematics.* A rectangular arrangement of quantities in rows and columns, as in a matrix. **2.** Numerical data ordered in a linear fashion, by magnitude.

Arrhenius (ə-rē′nē-əs), **Svante August** 1859–1927. Swedish physicist and chemist who developed the theory of electrolytic dissociation, which explained the process by which ions are formed or separated. For this work he was awarded the Nobel Prize for chemistry in 1903. He also investigated osmosis, toxins, and antitoxins.

arrhythmia (ə-rĭ*th*′mē-ə) An abnormal rhythm of the heart, often detectable on an electrocardiogram. Electrical impulses in the heart normally originate in the sinoatrial node of the right atrium during diastole and are transmitted through the atrioventricular node to the ventricles, causing the muscle contraction that usually occurs during systole. However, abnormalities of electrical conduction during diastole or systole can result in various alterations of the heartbeat, such as changes in heart rate, skipped or irregular beats, and fibrillation of the heart muscle, which can be life-threatening. These electrical disturbances can be caused by metabolic abnormalities, inadequate blood supply (as in coronary artery disease), drug effects, chronic disease, and other factors. Arrhythmias are sometimes treated with the implantation of a pacemaker.

arroyo (ə-roi′ō) A small, deep gully or channel of an ephemeral stream. Arroyos usually have relatively flat floors and are flanked by steep sides consisting of unconsolidated sediments. They are usually dry except after heavy rainfall.

arsenate (är′sə-nĭt, är′sə-nāt′) A salt containing the radical AsO_4.

arsenic (är′sə-nĭk) *Symbol* **As** A metalloid element most commonly occurring as a gray crystal, but also found as a yellow crystal and in other forms. Arsenic and its compounds are highly poisonous and are used to make insecticides, weed killers, and various alloys. Atomic number 33; atomic weight 74.922; valence 3, 5. Gray arsenic melts at 817°C (at 28 atm pressure), sublimes at 613°C, and has a specific gravity of 5.73. See **Periodic Table.**

arteriole (är-tîr′ē-ōl′) Any of the smaller branches of an artery, especially one that connects with a capillary.

arteriosclerosis (är-tîr′ē-ō-sklə-rō′sĭs) A thickening, hardening, and loss of elasticity of the arterial walls that results in impaired blood circulation. See also **atherosclerosis.**

artery (är′tə-rē) Any of the blood vessels that carry oxygenated blood away from the heart to the body's cells, tissues, and organs. Arteries are flexible, elastic tubes with muscular walls that expand and contract to pump blood through the body. —*Adjective* **arterial** (är-tîr′ē-əl)

artesian well (är-tē′zhən) A deep well that passes through impermeable rock or sediment and reaches water that is held under pressure in a confined aquifer. In aquifers of this type, the water in the lower regions is trapped between two layers of impermeable rock and cannot rise to the level of the water table in the upper, unconfined regions. When a well penetrates the confined region, the pressure forces the water to rise within the well until it reaches the elevation of the water table in the unconfined region (a level known as the potentiometric surface). ▸ In a **flowing artesian well** the water is under enough pressure to rise all the way to the surface without being pumped and must be capped to control the flow.

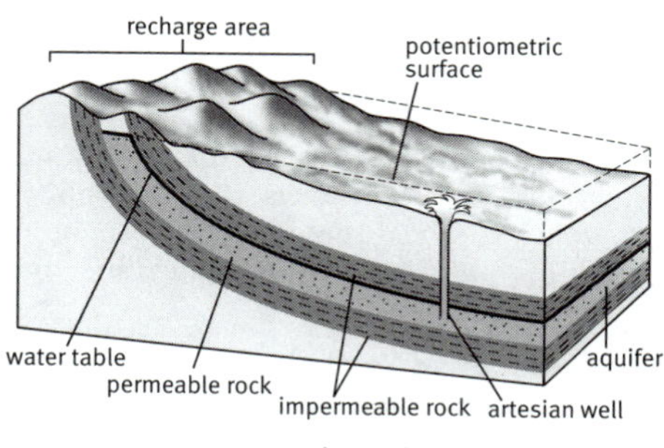

artesian well

arthritis (är-thrī′tĭs) Acute or chronic inflammation of one or more joints, usually accompanied by pain and stiffness, resulting from infection, trauma, degenerative changes, autoimmune disease, or other causes. See also **osteoarthritis, rheumatoid arthritis.**

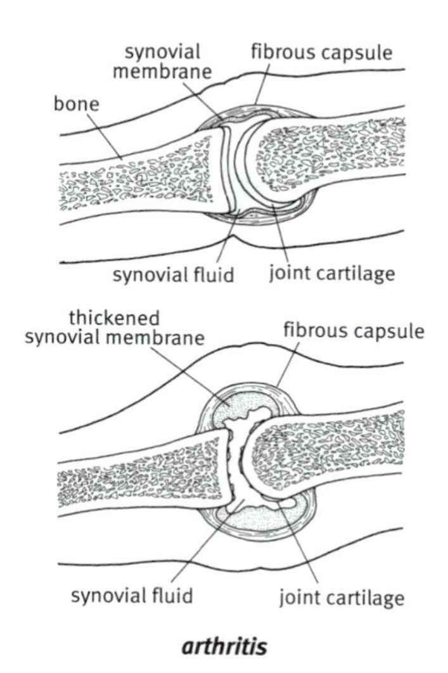

arthritis

top: *healthy finger joint;* bottom: *osteoarthritis in a finger joint showing inflammation of the synovial membrane and deterioration of the articular cartilage*

arthropod (är′thrə-pŏd′) Any of numerous invertebrate animals of the phylum Arthopoda, characterized by an exoskeleton made of chitin and a segmented body with pairs of jointed appendages. Arthropods share many features with annelids and may have evolved from them in the Precambrian Era. Arthropods include the insects, crustaceans, arachnids, myriapods, and extinct trilobites, and are the largest phylum in the animal kingdom.

arthroscopy (är-thrŏs′kə-pē) Examination of the interior of a joint, such as the knee, using a type of endoscope that is inserted into the joint through a small incision.

articulation (är-tĭk′yə-lā′shən) **1.** The arrangement of parts connected by joints. **2.** A fixed or movable joint between bones. **3.** A movable joint between inflexible parts of the body of an animal, as the divisions of an appendage in arthropods. **4.** A joint between two separable parts, as a leaf and a stem.

artifact also **artefact** (är′tə-făkt′) **1.** An object produced or shaped by human craft, especially a tool, weapon, or ornament of archaeological or historical interest. **2.** An artificial product or effect observed in a natural system, especially one introduced by the technology used in scientific investigation or by experimental error.

artificial insemination (är′tə-fĭsh′əl) The introduction of semen into the vagina or uterus without sexual contact.

artificial intelligence The ability of a computer or other machine to perform actions thought to require intelligence. Among these actions are logical deduction and inference, creativity, the ability to make decisions based on past experience or insufficient or conflicting information, and the ability to understand spoken language.

A CLOSER LOOK **artificial intelligence**

The goal of research on *artificial intelligence* is to understand the nature of thought and intelligent behavior and to design intelligent systems. A computer is not really intelligent; it just follows directions very quickly. At the same time, it is the speed and memory of modern computers that allows researchers to manage the huge quantities of data necessary to model human thought and behavior. An intelligent machine would be more flexible than a computer and would engage in the kind of "thinking" that people actually do. An example is vision. In theory, a network of sensors combined with systems for interpreting the data could produce the kind of pattern recognition that we take for granted as seeing and understanding what we see. In fact, developing software that can recognize subtle differences in objects (such as those we use to recognize human faces) is very difficult. The recognition of differences that we can perceive without deliberate effort would require massive amounts of data and elaborate guidelines to be recognized by an artificial intelligence system. According to the famous Turing Test, proposed in 1950 by British mathematician and logician Alan Turing, a machine would be considered intelligent if it could convince human observers that another human, rather than a machine, was answering their questions in conversation.

artificial life The simulation of biological phenomena through the use of computer models and robotics.

artificial selection Modification of a species by human intervention so that certain desirable traits are represented in successive generations. The different breeds of domestic dogs and the large ears of maize corn are products of artificial selection. Compare **natural selection.**

artiodactyl (är′tē-ō-dăk′təl) Any of various hoofed mammals of the order Artiodactyla, having an even number of toes on each foot. Artiodactyls include the pig, sheep, ox, deer, giraffe, and hippopotamus. Also called *even-toed ungulate.*

As The symbol for **arsenic.**

asbestos (ăs-bĕs′təs) Any of several fibrous mineral forms of magnesium silicate. Asbestos is resistant to heat, flames, and chemical action. Some forms have been shown to cause lung diseases. For this reason, asbestos is no longer used to make insulation, fireproofing material, and brake linings.

asbestosis (ăs′bĕs-tō′sĭs) A chronic, progressive lung disease caused by prolonged inhalation of asbestos particles.

aschelminth (ăsk′hĕl-mĭnth′) Any of various, mostly microscopic wormlike invertebrates of the group Aschelminthes, including the nematodes, rotifers, and gastrotrichs. Aschelminths typically inhabit water or moist soil, are often parasitic, and have complex internal structures. The status of the Aschelminthes as a single taxonomic group is controversial, and many researchers consider its members to be separate phyla.

ASCII (ăs′kē) A code that assigns the numbers 0 through 127 to the letters of the alphabet, the digits 0 through 9, punctuation marks, and certain other characters. For example, the capital letter A is coded as 65 (binary 1000001). By standardizing the values used to represent written text, ASCII enables computers to exchange information. Basic, or standard, ASCII uses seven bits for each character code, giving it 2^7, or 128, unique symbols. Various larger character sets, called extended ASCII, use eight bits for each character, yielding 128 additional codes numbered 128 to 255. Compare **Unicode.**

ascocarp (ăs′kə-kärp′) An ascus-bearing structure found in the fungi known as ascomycetes. Ascocarps are composed of interwoven hyphae, and in many species they are visible, forming the most prominent part of the fungus. Ascocarps may be cup-shaped, spherical, or flask-shaped.

ascomycete (ăs′kō-mī′sēt′) Any of various fungi belonging to the phylum Ascomycota, characterized by the presence of sexually produced spores formed within an **ascus.** Like most fungi, ascomycetes also reproduce asexually by the formation of nonsexual spores called conidia at the ends of filaments known as hyphae. Yeasts, many molds that cause food spoilage, and the edible fungi known as morels and truffles, are ascomycetes. A number of serious plant diseases, including ergot, the powdery mildews that attack fruit, and Dutch elm disease, are also caused by ascomycetes.

ascorbate (ə-skôr′bāt, -bĭt) A salt or ester of ascorbic acid (vitamin C).

ascorbic acid (ə-skôr′bĭk) See **vitamin C.**

ascospore (ăs′kə-spôr′) A sexually produced fungal spore formed within an ascus of ascomycetes. Ascospores have a haploid number of chromosomes and are formed by meiosis of the diploid zygote that results when the nuclei of sexually compatible hyphae fuse together. When an ascospore is released and lands in a place that is rich in nutrients, it germinates and sends out hyphae of its own.

ascus (ăs′kəs) Plural **asci** (ăs′ī′, -kī′). A membranous, often club-shaped structure inside which ascospores are formed through sexual reproduction in species of the fungi known as ascomycetes. The ascus is unique to ascomycetes and distinguishes them from other kinds of fungi. Asci are formed when two hyphae that are sexually compatible conjugate. Each ascus typically develops eight ascospores. Asci swell at maturity until they burst, shooting the ascospores into the air.

-ase A suffix used to form the names of enzymes. It is often added to the name of the compound that the enzyme breaks down, as in *lactase,* which breaks down lactose. It is also added to a word describing the enzyme's activity, as in *transferase,* which catalyzes the transfer of a chemical group from one molecule to another, or *reductase,* which catalyzes the reduction of an organic compound.

asepsis (ə-sĕp′sĭs, ā-sĕp′sĭs) **1.** The state of being free of pathogenic microorganisms. **2.** The process of removing microorganisms that cause infection.

aseptic (ə-sĕp′tĭk, ā-sĕp′tĭk) Free of microorganisms that cause disease.

asexual reproduction (ā-sĕk′sho͞o-əl) See under **reproduction.**

asparagine (ə-spăr′ə-jēn′) A nonessential amino acid. *Chemical formula:* $\mathbf{C_4H_8N_2O_3}$. See more at **amino acid.**

aspartame (ăs′pər-tām′, ə-spär′-) An artificial sweetener formed from aspartic acid. *Chemical formula:* $\mathbf{C_{14}H_{18}N_2O_5}$.

aspartic acid (ə-spär′tĭk) A nonessential amino acid. *Chemical formula:* $\mathbf{C_4H_7NO_4}$. See more at **amino acid.**

Asperger's syndrome (ăs′pər-gərz) A developmental disorder characterized by impairment in social interactions and repetitive behavior patterns. It is named after its identifier, Austrian psychiatrist Hans Asperger (1906–1980).

asphalt (ăs′fôlt′) A thick, sticky, dark-brown mixture of petroleum tars used in paving, roofing, and waterproofing. Asphalt is produced as a byproduct in refining petroleum or is found in natural beds.

asphyxia (ăs-fĭk′sē-ə) A condition characterized by an extreme decrease in the amount of oxygen in the body accompanied by an increase of carbon dioxide, caused by an an inability to breathe. Asphyxia usually results in loss of consciousness and sometimes death.

aspirin (ăs′pər-ĭn, ăs′prĭn) A white crystalline compound derived from salicylic acid and used in medicine to relieve fever and pain and as an anticoagulant. Also called *acetylsalicylic acid. Chemical formula:* $C_9H_8O_4$.

A CLOSER LOOK **aspirin**

Ninety percent of the population experiences at least one headache each year. The most common type is a tension headache, which is caused by stress and is characterized by tightening of the muscles in the base of the neck and along the scalp. *Aspirin* alleviates headaches by blocking the body's production of *prostaglandins,* hormones that contribute to pain by stimulating muscle contraction and blood vessel dilation. For thousands of years, people chewed the bark of willow trees to control headache and other pain. The study of the properties of this medicinal plant led German chemist Hermann Kolbe to synthesize *acetylsalicylic acid* (ASA), a building block of aspirin, in 1859. A pure form of ASA wasn't prepared until 1897, by Felix Hoffman, a chemist in the Bayer chemical factory in Germany. After publication of successful clinical trials, aspirin was distributed in powder form in 1899 and as a tablet in 1900. Aspirin possesses a number of properties that make it one of the most recommended drugs. Besides being an *analgesic,* or pain reliever, it also reduces inflammation that often accompanies injuries or diseases, such as arthritis. It is also an *antipyretic compound,* or fever reducer. Aspirin is the only over-the-counter analgesic approved for prevention of cardiovascular disease. New research suggests that aspirin may also decrease the risk of some forms of stroke. Additional studies indicate that aspirin may play a role in reducing the risks of ovarian cancer.

assay (ăs′ā, ə-sā′) **1.** A quantitative determination of the amount of a given substance in a particular sample. Assays are regularly used to determine the purity of precious metals. They can be performed by wet methods or dry methods. In the wet method, the sample is dissolved in a reagent, like an acid, until the purified metal is separated out. In the dry method, the sample is mixed with a flux (a substance such as borax or silica that helps lower the melting temperature) and then the sample is heated to the point where impurities in the metal fuse with the flux, leaving the purified metal as a residue. **2.** A bioassay.

assemblage (ə-sĕm′blĭj) A collection of artifacts from a single datable component of an archaeological site. Depending on the site and culture, an assemblage may be associated with a single limited activity, as with stone tools found at a butchering site, or may reflect a broad range of cultural life, as with artifacts that are found in a communal living site.

assimilation (ə-sĭm′ə-lā′shən) The conversion of nutrients into living tissue; constructive metabolism.

association (ə-sō′sē-ā′shən, -shē-) A large number of organisms in a specific geographic area constituting a community with one or two dominant species.

associative (ə-sō′shə-tĭv) Of or relating to the property of an operation, such as addition or multiplication, which states that the grouping of numbers undergoing the operation does not change the result. For example, $3 + (4 + 5)$ is equal to $(3 + 4) + 5$. See also **commutative, distributive.**

astatine (ăs′tə-tēn′) *Symbol* **At** A highly unstable, rare, radioactive element that is the heaviest of the halogen elements. Its most stable isotope has a half-life of 8.3 hours. Atomic number 85; melting point 302°C; boiling point 337°C; valence probably 1, 3, 5, 7. See **Periodic Table.**

asterism (ăs′tə-rĭz′əm) A pattern of stars that is not one of the traditionally established, named constellations. Asterisms may constitute a part of a larger constellation, as in the case of the seven stars in Ursa Major that make up the Big Dipper, or they may be formed of individual stars in several different constellations, as in the case of the Summer Triangle, made up of Deneb (in Cygnus), Altair (in Aquila), and Vega (in Lyra).

asteroid (ăs′tə-roid′) Any of numerous **small solar system bodies** that orbit the Sun primarily in the **asteroid belt**, a region between the orbits of Mars and Jupiter. Asteroids are intermediate in size between planets and meteoroids with diameters that measure between approximately one hundred and several hundred kilometers. While more than 1,800 asteroids have been cataloged, and as many as a million or more smaller ones may exist, their total mass has been estimated to be less than three percent of the Moon's. Asteroids are thought to be left over from the early formation of the solar system, when planetesimals in a **protoplanetary disk** were scattered after coming under Jupiter's gravitational influence. The continuing collision of planetesimals that remained between Jupiter and Mars caused many of them to fragment, creating the asteroids that exist today. Also called *minor planet, planetoid.*

asthenosphere (ăs-thĕn′ə-sfîr′) The upper part of the Earth's mantle, extending from a depth of about 75 km (46.5 mi) to about 200 km (124 mi). The asthenosphere lies beneath the lithosphere and consists of partially molten rock. Seismic waves passing through this layer are significantly slowed. Isostatic adjustments (the depression or uplift of continents by buoyancy) take place in the asthenosphere, and magma is believed to be generated there. Compare **atmosphere, hydrosphere, lithosphere.**

asthma (ăz′mə) A common inflammatory disease of the lungs characterized by episodic airway obstruction caused by extensive narrowing of the bronchi and bronchioles. The narrowing is caused by spasm of smooth muscle, edema of the mucosa, and the presence of mucus in the airway resulting from an immunologic reaction that can be induced by allergies, irritants, infection, stress, and other factors in a genetically predisposed individual. Common symptoms of asthma include wheezing, coughing, and shortness of breath.

astigmatism (ə-stĭg′mə-tĭz′əm) A visual defect in which the unequal curvature of one or more refractive surfaces of the eye, usually the cornea, prevents light rays from focusing clearly at a single point on the retina, resulting in blurred vision.

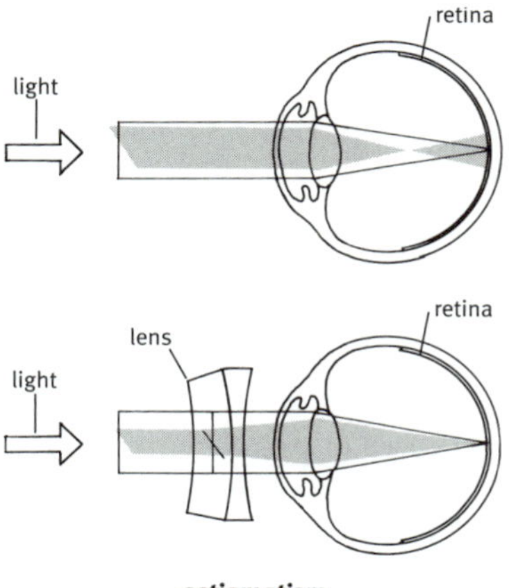

astigmatism

top: *astigmatic vision;* bottom: *corrected vision*

astral (ăs′trəl) Relating to or coming from the stars; stellar.

astringent (ə-strĭn′jənt) A substance or preparation, such as alum, that draws together or constricts body tissues, resulting in decreased flow of blood or other secretions.

astro– A prefix that means "star" (as in *astrophysics*), "celestial body" (as in *astronomy*), or "outer space" (as in *astronaut*).

astrobiology (ăs′trō-bī-ŏl′ə-jē) See **exobiology.**

astrocyte (ăs′trə-sīt′) A star-shaped cell, especially one of the glial cells that support the tissue of the central nervous system.

astrodynamics (ăs′trō-dī-năm′ĭks) The dynamics of natural and artificial bodies in outer space.

astrolabe (ăs′trə-lāb′) An ancient instrument used widely in medieval times by navigators and astronomers to determine latitude, longitude, and time of day. The device employed a disk with 360 degrees marked on its circumference. Users took readings from an indicator that pivoted around the center of the suspended device like the hand of a clock. The astrolabe was replaced by the sextant in the 18th century.

astrometric binary (ăs′trə-mĕt′rĭk) See under **binary star.**

astrometry (ə-strŏm′ĭ-trē) The scientific measurement of the positions and motions of celestial bodies.

astronomical unit (ăs′trə-nŏm′ĭ-kəl) A unit of length equal to the average distance from Earth to the Sun, approximately 149.6 million km (92.8 million mi). It is used especially to measure distances within the solar system. Compare **light year, parsec.**

astronomical year A solar year. See under **solar time.**

astronomy (ə-strŏn′ə-mē) The scientific study of the universe and the objects in it, including stars, planets, nebulae, and galaxies. Astronomy deals with the position, size, motion, composition, energy, and evolution of celestial objects. Astronomers analyze not only visible light but also radio waves, x-rays, and other ranges of radiation that come from sources outside the Earth's atmosphere.

astrophysics (ăs′trō-fĭz′ĭks) The branch of astronomy that deals with the physical and chemical processes that occur in stars, galaxies, and interstellar space. Astrophysics deals with the structure and evolution of stars, the properties of interstellar space and its interactions with systems of stars, and with the structure and dynamics of clusters of stars such as galaxies.

asymptote (ăs′ĭm-tōt′) A line whose distance to a given curve tends to zero. An asymptote may or may not intersect its associated curve.

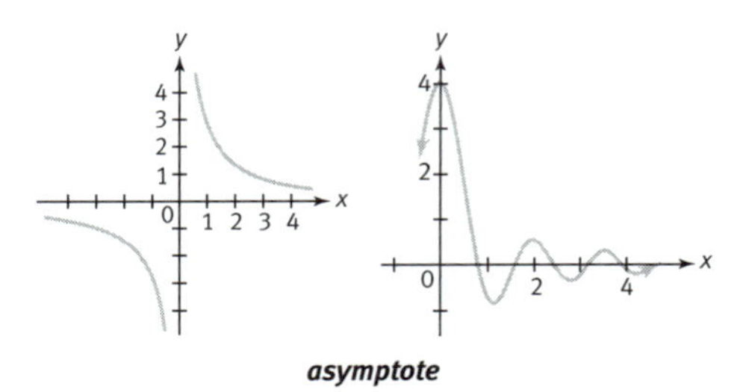

asymptote

left: *The x and y axes are asymptotes of the hypberola xy = 1.* right: *The x axis is the asymptote of the curve* $y = \frac{\sin x}{x}$.

At The symbol for **astatine.**

atavistic (ăt′ə-vĭs′tĭk) Relating to an inherited trait that reappears in an individual after being absent from a strain of organism for several generations. Atavistic traits were formerly thought to be throwbacks to ancestral types but are now known to be due to the inheritance of a pair of recessive genes.

ataxia (ə-tăk′sē-ə) Loss of muscular coordination as a result of damage to the central nervous system.

-ate A suffix used to form the name of a salt or ester of an acid whose name ends in *-ic,* such as *acetate,* a salt or ester of acetic acid. Such salts or esters have one oxygen atom more than corresponding salts or esters with names ending in *-ite.* For example, a *sulfate* is a salt of sulfuric acid and contains the group SO_4, while a sulfite contains SO_3. Compare **-ite** (sense 2).

atherosclerosis (ăth′ə-rō-sklə-rō′sĭs) A form of arteriosclerosis characterized by the deposition of plaques containing cholesterol and lipids on the innermost layer of the walls of large- and medium-sized arteries. Individuals with atherosclerosis have a higher risk of coronary artery disease and stroke. Smoking, high blood pressure, diabetes mellitus, and elevated levels of fat in the blood contribute to the development of atherosclerosis.

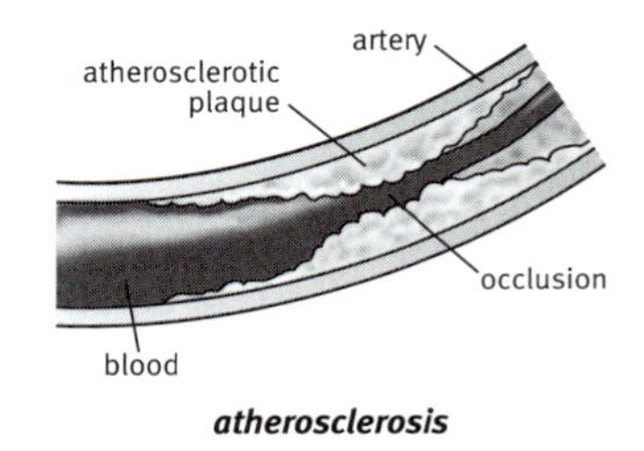

atherosclerosis

an artery narrowed by the buildup of plaque

atmometer (ăt-mŏm′ĭ-tər) An instrument that measures the rate of evaporation of water into the atmosphere. Atmometers usually measure the evaporation of water from a free water surface (such as a pan of water set into the ground so that the water's surface is even with the ground's surface) or from a porous, water-saturated surface (such as filter paper placed over a graduated cylinder of water). In the first type, the difference between the ground surface and the water level is used to calculate the volume of water that has evaporated in a given period of time. In the second type, the water volume is read directly from the graduated cylinder. Also called *evaporimeter.*

atmosphere (ăt′mə-sfîr′) **1.** The mixture of gases surrounding the Earth or other celestial body, held in place by gravity. It forms distinct layers at different heights. The Earth's atmosphere consists, in ascending order, of the troposphere (containing 90% of the atmosphere's mass), the stratosphere, the mesosphere, the thermosphere, and the exosphere. The atmosphere is composed primarily of nitrogen (78%) and oxygen (21%) and plays a major role in the **water cycle**, the **nitrogen cycle**, and the **carbon cycle**. See more at **exosphere, mesosphere, stratosphere, thermosphere, troposphere. 2.** A unit of pressure equal to the pressure of the air at sea level, about 14.7 pounds per square inch, or 1,013 millibars.

exosphere
thermosphere
mesosphere
stratosphere
troposphere

atmosphere

atmospheric pressure (ăt′mə-sfîr′ĭk) The pressure at any location on the Earth, caused by the weight of the column of air above it. At sea level, atmospheric pressure has an average value of one atmosphere and gradually decreases as altitude increases. Also called *barometric pressure.*

A CLOSER LOOK **atmospheric pressure**

The weight of the air mass, or atmosphere, that envelopes Earth exerts pressure on all points of the planet's surface. Meteorologists use barometers to measure this *atmospheric pressure* (also called *barometric pressure*). At sea level the atmospheric pressure is approximately 1 kilogram per square centimeter (14.7 pounds per square inch), which will cause a column of mercury in a mercury barometer to rise 760 millimeters (30.4 inches). The pressure is frequently expressed in pascals, after the French mathematician and philosopher Blaise Pascal, who studied the transmission of pressure in confined fluids. Subtle variations in atmospheric pressure greatly affect the weather. Low pressure generally brings rain. In areas of low air pressure, the air is less dense and relatively warm, which causes it to rise. The expanding and rising air naturally cools, and the water vapor in the air condenses, forming clouds and the drops that fall as rain. In high pressure areas, conversely, the air is dense and relatively cool, which causes it to sink. The water vapor in the sinking air does not condense, leaving the skies sunny and clear.

---

atmospherics (ăt′mə-sfîr′ĭks) **1.** Electromagnetic radiation that is in the same range as radio frequencies and is produced by natural phenomena (such as lightning) and interferes with radio communications. **2.** The radio interference produced by this electromagnetic radiation.

atoll (ăt′ôl′, ā′tôl′) A coral island or series of coral islands forming a ring that nearly or entirely encloses a shallow lagoon. Atolls are surrounded by deep ocean water and range in diameter from about 1 km (0.62 mi) to over 100 km (62 mi). They are especially common in the western and central Pacific Ocean and are believed to form along the fringes of underwater volcanoes. Compare **barrier reef, fringing reef.**

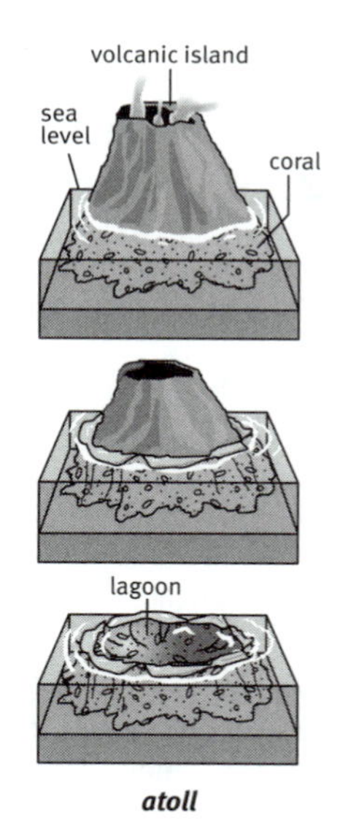

atoll

An atoll forms as a coral reef builds up around an eroding volcanic island.

atom (ăt′əm) The smallest unit of an element, consisting of at least one proton and (for all elements except hydrogen) one or more neutrons in a dense central nucleus, surrounded by one or more shells of electrons. In electrically neutral atoms, the number of protons equals the number of electrons. Atoms remain intact in chemical reactions except for the removal, transfer, or exchange of certain electrons. Compare **compound.** See also **ion, isotope, orbital.**

atomic (ə-tŏm′ĭk) **1.** Relating to an atom or to atoms. **2.** Employing nuclear energy.

atomic bomb A very destructive bomb that derives its explosive power from the fission of atomic nuclei. Atomic bombs usually have plutonium 239 or uranium 235 as their fissionable material. Also called *atom bomb.*

atomic clock An extremely precise clock whose rate is controlled by a periodic process (such as vibration, or the absorption or emission of electromagnetic radiation) that occurs at a steady rate in atoms or molecules. The standard atomic clock is based on the vibrations of cesium atoms and is so accurate that it would gain or lose less than one second in three million years. Atomic clocks are used to help track satellites, run navigation systems, and study movements of the Earth's crust.

atomic energy See **nuclear energy.**

atomic force microscope A microscope that uses a tiny probe mounted on a cantilever to scan the surface of an object. The probe is extremely close to—but does not touch—the surface. As the probe traverses the surface, attractive and repulsive forces arising between it and the atoms on the surface induce forces on the probe that bend the cantilever. The amount of bending is measured and recorded, providing a map of the atoms on the surface. Atomic force microscopes can achieve magnification of a factor of 5×10^6, with a resolution of 2 angstroms, sufficient to resolve individual carbon atoms. Also called *scanning force microscope.*

atomic mass The mass of a given atom or molecule, expressed in atomic mass units. Compare **atomic weight.** See also **mass number.**

atomic mass unit A unit of mass equal to $\frac{1}{12}$ the mass of an atom of the most common isotope of carbon (carbon 12), which is assigned a mass of 12 and has a value of 1.660×10^{-24} grams. A hydrogen atom has a mass of 1 atomic mass unit since its mass is $\frac{1}{12}$ the mass of carbon 12. Also called *dalton.*

atomic number The number of protons in the nucleus of an atom. In electrically neutral atoms, this number is also equal to the number of electrons orbiting about the atom's nucleus. The atomic number of an element determines its position in the Periodic Table, and is usually denoted by the letter Z and written as a subscript before an element's symbol, as in $_{92}U$.

atomic pile An early type of nuclear reactor whose core consisted of layers of graphite block interspersed with uranium, designed to create a sustained fission reaction.

atomic spectrum The range of characteristic frequencies of electromagnetic radiation that are readily absorbed and emitted by an atom. The atomic spectrum is an effect of the quantized orbits of electrons around the atom. An electron can jump from one fixed orbital to another: if the orbital it jumps to has a higher energy, the electron must absorb a photon of a certain frequency; if it is of a lower energy, it must give off a photon of a certain frequency. The frequency depends on the difference in energy between the orbitals. Explaining this phenomenon was crucial to the development of quantum mechanics. The atomic spectrum of each chemical element is unique and is largely responsible for the color of matter. Atomic spectra can also be analyzed to determine the composition of objects, such as stars, that are far away. See more at **orbital.** See also **spectrum.**

atomic weight The mass of an atom expressed in atomic mass units. The atomic weight of an element having more than one principal isotope is calculated both from the atomic masses of the isotopes and from the relative abundance of each isotope in nature. For example, the atomic weight of the element chlorine is 35.453, determined by averaging the atomic masses and relative abundances of its two main naturally occurring isotopes, which have atomic masses of about 35 and 37. Also called *relative atomic mass.* Compare **atomic mass.** See also **mass number.**

atom smasher See **particle accelerator.**

ATP (ā′tē′pē′) Short for *adenosine triphosphate.* An organic compound, $C_{10}H_{16}N_5O_{13}P_3$, that is composed of adenosine and three phosphate groups. It serves as a source of energy for many metabolic processes. ATP releases energy when it is broken down into ADP by hydrolysis during cell metabolism.

atrioventricular node (ā′trē-ō-vĕn-trĭk′yə-lər) A small area of specialized cardiac muscle cells that conducts impulses from the sinoatrial node to the ventricles and initiates the contraction of the ventricles. It is located in the right atrium.

atrium (ā′trē-əm) Plural **atria** or **atriums.** A chamber of the heart that receives blood from the veins and forces it by muscular contraction into a ventricle. Mammals, birds, reptiles, and amphibians have two atria; fish have one.

atrophy (ăt′rə-fē) A wasting or decrease in the size of an organ or tissue, as from death and reabsorption of cells, diminished proliferation of cells, pressure, lack of oxygen, malnutrition, decreased function, or hormonal changes.

atropine (ăt′rə-pēn′, -pĭn) A poisonous, bitter, crystalline alkaloid derived from deadly nightshade and related plants. It is used as a drug to dilate the pupils of the eye and to inhibit muscle spasms. *Chemical formula:* $\mathbf{C_{17}H_{23}NO_3}$.

attention deficit hyperactivity disorder A syndrome of disordered behavior, usually diagnosed in childhood, characterized by a persistent pattern of impulsiveness, inattentiveness, and sometimes hyperactivity that interferes with academic, occupational, or social performance. Also called *attention deficit disorder.*

attractor (ə-trăk′tər) A set of states of a dynamic physical system toward which that system tends to evolve, regardless of the starting conditions of the system. ▸ A **point attractor** is an attractor consisting of a single state. For example, a marble rolling in a smooth, rounded bowl will always come to rest at the lowest point, in the bottom center of the bowl; the final state of position and motionlessness is a point attractor. ▸ A **periodic attractor** is an attractor consisting of a finite or infinite set of states, where the evolution of the system results in moving cyclically through each state. The ideal orbit of a planet around a star is a periodic attractor, as are periodic **oscillations**. A periodic attractor is also called a *limit-cycle.* ▸ A **strange attractor** is an attractor for which the evolution through the set of possible physical states is nonperiodic (chaotic), resulting in an evolution through a set of states defining a fractal set. Most real physical systems (including the actual orbits of planets) involve strange attractors.

Au The symbol for **gold.**

AU Abbreviation of **astronomical unit.**

audiology (ô′dē-ŏl′ə-jē) The scientific study of hearing, especially the diagnosis and treatment of hearing disorders.

auditory (ô′dĭ-tôr′ē) Relating to or involving the organs or sense of hearing.

auditory nerve Either of the eighth pair of cranial nerves that carries sensory impulses from the ear to the brain. The auditory nerve transmits information related to sound and balance.

Audubon (ô′də-bŏn′), **John James** 1785–1851. American ornithologist and artist. His effort to catalog every species of bird in the United States resulted in the publication of *The Birds of America* (1827–1838), a collection of 1,065 life-size engravings of birds found in eastern North America. It is considered a classic work in ornithology and in American art.

augen (ou′gən) A large mineral grain or grain cluster having the shape of an eye in cross-section and occurring in foliated metamorphic rocks such as schist and gneiss. Augens form when large mineral crystals are sheared and deformed during the process of metamorphism.

augend (ô′jĕnd′) A quantity to which the addend is added.

augite (ô′jīt′) A glassy, dark-green to black variety of pyroxene. *Chemical formula:* $\mathbf{(Ca,Na)(Mg,Fe,Al)(Si,Al)_2O_6}$.

aureole (ôr′ē-ōl′) **1.** A band of metamorphic rock surrounding a body of cooled magma. Aureoles form through the process of contact metamorphism. See more at **contact metamorphism. 2.** See **corona** (sense 2).

auricle (ôr′ĭ-kəl) **1.** The visible part of the outer ear. **2.** An atrium of the heart.

Auriga (ô-rī′gə) A constellation in the Northern Hemisphere near Gemini and Perseus. Auriga (the Charioteer) contains the bright star Capella.

Aurignacian (ôr′ĭg-nā′shən, ôr′ēn-yā′-) Relating to an Upper Paleolithic culture in Europe between the Mousterian and Solutrean cultures, dating from around 32,000 to 25,000 years ago and characterized by flaked stone, bone, and antler tools such as scrapers, awls, and burins (engraving tools). Aurignacian culture is associated with Cro-Magnon populations and is especially noted for its well-developed art tradition, including engraved and sculpted animal forms and female figurines thought to be fertility objects. The earliest fully developed cave art, such as the painted animals in the Lascaux cave in southwest France, dates from this period.

aurora (ə-rôr′ə) Plural **auroras** or **aurorae** (ə-rôr′ē). A brilliant display of bands or folds of variously colored light in the sky at night, especially in polar regions. Charged particles from the **solar wind** are channeled through the Earth's magnetic field into the polar regions. There the particles collide with atoms and molecules in the upper atmosphere, ionizing them and making them glow. Auroras are of greatest intensity and extent during periods of increased sunspot activity, when they often interfere with telecommunications on Earth. ▸ An aurora that occurs in southern latitudes is called an **aurora australis** (ô-strā′lĭs) or **southern lights**. When it occurs in northern latitudes it is called an **aurora borealis** (bôr′ē-ăl′ĭs) or **northern lights**. See also **magnetic storm.**

austral (ô′strəl) Relating to the south or to southern regions of the globe.

australopithecine (ô-strā′lō-pĭth′ĭ-sēn′) Any of several early hominids of the genus *Australopithecus* of eastern and southern Africa, known from fossils dating from about 4 million to about 1 million years ago. The most complete australopithecine skeleton found so far, named Lucy by its discoverers, is estimated to be just over 3 million years old. While many scientists believe that australopithecines are ancestors of modern humans, not enough fossils have yet been found to establish any direct descent.

autecology (ô′tĭ-kŏl′ə-jē) The branch of ecology that deals with the biological relationship between an individual organism or an individual species and its environment. Compare **synecology.**

autism (ô′tĭz′əm) A developmental disorder characterized by severe deficits in social interaction and communication and by abnormal behavior patterns, such as the repetition of specific movements or a tendency to focus on certain objects. Autism is evident in the first years of life. Its cause is unknown. —*Adjective* **autistic.**

auto– A prefix meaning "oneself," as in *autoimmune,* producing antibodies or immunity against oneself. It also means "by itself, automatic," as in *autonomic,* governing by itself.

autoantibody (ô′tō-ăn′tĭ-bŏd′ē) An antibody that reacts with the cells, tissues, or native proteins of the individual in which it is produced. Autoimmune diseases are caused by the presence of autoantibodies.

autoclave (ô′tō-klāv′) An airtight steel vessel used to heat substances and objects under very high pressures. Autoclaves are used in laboratory experiments and for sterilization.

autoecious (ô-tē′shəs) Relating to a parasite that spends all its life on a single host. The term is used especially of certain types of rust fungi, but can also be applied to other parasites, such as aphids. Compare **heteroecious.**

autogamy (ô-tŏg′ə-mē) **1.** See **self-fertilization.** **2.** The union of nuclei within and arising from a single cell, as in certain protozoans and fungi.

autograft (ô′tō-grăft′) A graft transferred from one position to another in or on the body of an individual. Compare **allograft, xenograft.**

autoimmune disease (ô′tō-ĭ-myo͞on′) A disease in which impaired function and the destruction of tissue are caused by an immune reaction in which abnormal antibodies are produced and attack the body's own cells and tissues. Autoimmune diseases include a wide variety of disorders, including many disorders of connective tissue, such as systemic lupus erythematosus and rheumatoid arthritis.

A CLOSER LOOK **autoimmune disease**

A wide variety of disorders are classified as *autoimmune diseases,* ranging from systemic lupus erythematosus to type I diabetes, and many other disorders are suspected of having an autoimmune component. Autoimmune diseases can thus affect a wide variety of bodily tissues and processes, such as the skin, liver, kidneys, or other organs, or the chemical reactions essential to metabolism. Each disease has a characteristic set of *autoantibodies* (antibodies that attack normal cells or structures in the body itself). In some of these diseases, the autoantibodies that are produced actually cause the tissue and organ damage. In other cases, the antibodies are considered to be characteristic markers of the disease but do not cause disease themselves. It is thought that the autoantibodies are generated by an immunologic reaction with bodily proteins, but the reasons that a specific set of bodily proteins should provoke an immune response that results in disease remain obscure. The genetic makeup of the individual, environmental influences, and infectious disease organisms may all contribute to a person's susceptibility to autoimmune disease. For reasons that are not clear, the prevalence of many autoimmune diseases is much higher in women than in men. Recently there have been dramatic improvements in the diagnosis and treatment of autoimmune disorders. New tests for diagnostically important autoantibodies have been discovered. Corticosteroids are used to reduce inflammation, and anticancer drugs that kill rapidly dividing cells are used to deplete activated cells in the immune system. The most promising new drugs consist of genetically engineered monoclonal antibodies that block just one part of the immune system. By selectively shutting down the part of the immune system involved in the autoimmune response, the drugs allow some people to see dramatic improvement in their symptoms with minimal side effects.

---

autologous (ô-tŏl′ə-gəs) Derived or transplanted from the same individual's body.

autonomic nervous system (ô′tə-nŏm′ĭk) The part of the vertebrate nervous system that regulates involuntary activity in the body by transmitting motor impulses to cardiac muscle, smooth muscle, and the glands. The muscular activity of the heart and of the circulatory, digestive, respiratory, and urogenital systems is controlled by the autonomic nervous system. The autonomic nervous system is divided into two parts: the **sympathetic nervous system** and the **parasympathetic nervous system**.

autopsy (ô′tŏp′sē) A medical examination of a dead body to determine the cause of death or to study pathologic changes.

autotroph (ô′tə-trŏf′) An organism that manufactures its own food from inorganic substances, such as carbon dioxide and ammonia. Most autotrophs, such as green plants, certain algae, and photosynthetic bacteria, use light for energy. Some autotrophs, such as chemosynthetic bacteria, obtain their energy from inorganic compounds such as hydrogen sulfide by combining them with oxygen. Compare **heterotroph.** —*Adjective* **autotrophic** (ô′tə-trŏf′ĭk, -trō′fĭk)

autumnal equinox (ô-tŭm′nəl) See under **equinox.**

auxin (ôk′sĭn) Any of various hormones or similar substances that promote and regulate the growth and development of plants. Auxins are produced in the meristem of shoot tips and move down the plant, causing various effects. Auxins cause the cells below the shoot apex to expand or elongate, and this (rather than cell division) is what causes the plant to increase in height. In woody plants, auxins also stimulate cell division in the cambium, which produces vascular tissue. Auxins inhibit the growth of lateral buds so that the plant grows upwards more than outwards. They can be produced artificially in laboratories for such purposes as speeding plant growth and regulating how fast fruit will ripen.

avalanche (ăv′ə-lănch′) **1.** The sudden fall or slide of a large mass of material down the side of a mountain. Avalanches may contain snow, ice, rock, soil, or a mixture of these materials. Avalanches can be triggered by changes in temperature, by sound vibrations, or by vibrations in the earth itself. **2.** A process resulting in the production of large numbers of ionized particles, in which electrons or ions collide with molecules, with each collision itself producing an additional electron or ion that in turn collides with other molecules. Avalanches are what generate the pulses of electric current that are registered by Geiger counters.

avascular (ā-văs′kyə-lər) Not associated with or supplied by blood vessels.

average (ăv′ər-ĭj) A number, especially the arithmetic mean, that is derived from and considered typical or representative of a set of numbers. Compare **arithmetic mean, median, mode.**

Avery (ā′və-rē), **Oswald Theodore** 1877–1955. Canadian-born American bacteriologist who demonstrated in 1944 that DNA was the material that caused genetic changes in bacteria. His work was vital to scientists who later established that DNA is the carrier of genetic information in all living organisms.

avian (ā′vē-ən) Relating to birds.

Avicenna (ăv′ĭ-sĕn′ə) *See* **Ibn Sina, Hakim.**

AV node (ā′vē′) See **atrioventricular node.**

Avogadro (ä′və-gä′drō), **Amedeo** 1776–1856. Italian chemist and physicist who formulated the hypothesis known as Avogadro's law in 1811.

Avogadro's law (ä′və-gä′drōz) The principle that equal volumes of all gases under identical conditions of pressure and temperature contain the same number of molecules. Avogadro's law is true only for ideal gases (gases in which there is no interaction between the individual molecules).

Avogadro's number The number of atoms or molecules in a mole of a substance, approximately 6.0225×10^{23}. It is based on the number of carbon atoms in 12 grams of carbon 12.

avoirdupois weight (ăv′ər-də-poiz′) A system of weights and measures based on a pound containing 16 ounces or 7,000 grains, and equal to 453.59 grams. Avoirdupois weight is used in the United States to weigh everything except gems, precious metals, and drugs. Compare **troy weight.**

awn (ôn) A slender, bristlelike appendage found on the spikelets of many grasses.

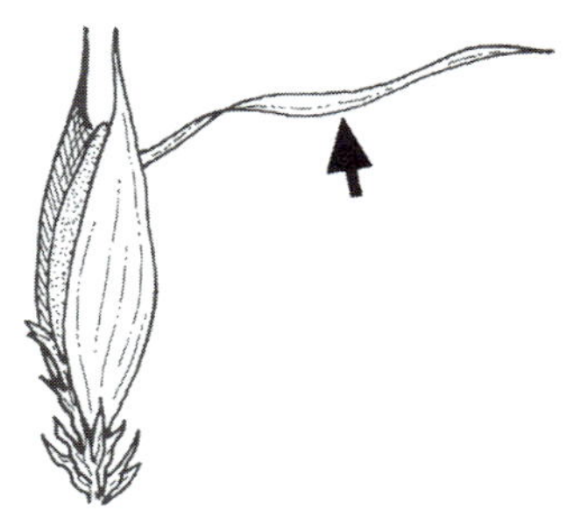

awn

The awn of California oatgrass (Danthonia californica) *extending from the bracts enclosing the spikelet.*

axial strain (ăk′sē-əl) See under **strain.**

axial stress A **stress** that tends to change the length of a body. ► **Compressive stress** is axial stress that tends to cause a body to become shorter along the direction of applied force. **Tensile stress** is axial stress that tends to cause a body to become longer along the direction of applied force. Compare **shear stress, strain.** See also **torsion.**

axil (ăk′sĭl) The angle between the upper side of a leaf or stem and the stem or branch that supports it. A bud is usually found in the axil.

axillary bud (ăk′sə-lĕr′ē) See under **bud.**

axiom (ăk′sē-əm) A principle that is accepted as true without proof. The statement "For every two points *P* and *Q* there is a unique line that contains both *P* and *Q*" is an axiom because no other information is given about points or lines, and therefore it cannot be proven. Also called *postulate.*

axis (ăk′sĭs) Plural **axes** (ăk′sēz′). **1.** An imaginary line around which an object rotates. In a rotating sphere, such as the Earth and other planets, the two ends of the axis are called poles. The 23.45° tilt of the Earth's axis with respect to the plane of its orbit around the Sun causes the Northern and Southern Hemispheres to point toward and away from the Sun at different times of the year, creating seasonal patterns of weather and climate. Other planets in the solar system have widely varying tilts to their axes, ranging from near 0° for Mercury to 177° for Venus. **2.** *Mathematics.* **a.** A line, ray, or line segment with respect to which a figure or object is symmetrical. **b.** A reference line from which distances or angles are measured in a coordinate system, such as the *x*-axis and *y*-axis in the Cartesian coordinate system. **3.** *Anatomy.* The second cervical vertebra, which serves as a pivot for the head. **4.** *Botany.* The main stem or central part of a plant or plant part, about which other plant parts, such as branches or leaflets, are arranged. —*Adjective* **axial.**

axon (ăk′sŏn′) The long portion of a neuron that conducts impulses away from the body of the cell. Also called *nerve fiber.*

azimuth (ăz′ə-məth) The position of a celestial object along an observer's horizon. Azimuth is a horizontal angle measured clockwise in degrees from a reference direction, usually the north or south point of the horizon, to the point on the horizon intersected by the object's line of **altitude** (a line from the observer's zenith through the object to the horizon). If north is the reference point (0°), then east has an azimuth of 90°, south is 180°, and so forth through 360°. See more at **altazimuth coordinate system.**

azimuthal equidistant projection (ăz′ə-mŭth′əl) An azimuthal map projection designed so that a straight line from the central point on the map to any other point gives the shortest distance between the two points. Azimuthal equidistant maps are used mainly for plotting direction and distance from the map's central point, but measurements originating from other points on the map can be greatly distorted.

azimuthal projection A map projection in which a globe, as of the Earth, is assumed to rest on a flat surface onto which its features are projected. An azimuthal projection produces a circular map with a chosen point—the point on the globe that is tangent to the flat surface—at its center. When the central point is either of Earth's poles, parallels appear as concentric circles on the map and meridians as straight lines radiating from the center. Directions from the central point to any other point on the map are accurate, although distances and shapes in some azimuthal projections are distorted away from the center. Compare **conic projection, cylindrical projection.**

azimuthal quantum number See under **quantum number**.

azine (ăz′ēn′, ā′zēn′) Any of various organic compounds, such as pyridine or pyrimidine, that have a ring structure like that of benzene but with one or more carbon atoms replaced by a nitrogen atom. Azines are heterocyclic compounds.

azo (ăz′ō, ā′zō) Containing two nitrogen atoms joined by a double bond, with each nitrogen atom attached to another group.

azo– or **az–** A prefix that means "containing an azo group attached at both ends to other groups," as in the compound *azobenzene.*

azole (ăz′ōl′, ā′zōl′) **1.** Any of various compounds having a ring structure made of five atoms, one of which is always nitrogen and another of which is either a second nitrogen or an atom of oxygen or sulfur. **2.** See **pyrrole.**

AZT (ā′zē-tē′) A nucleoside analogue antiviral drug that inhibits the replication of retroviruses such as HIV by interfering with the enzyme reverse transcriptase.

azurite (ăzh′ə-rīt′) A dark-blue monoclinic mineral occurring as a mass of crystals (an aggregate) or in the form of blades with wedge-shaped tips. It is often found together with the mineral malachite in copper deposits. Azurite is used as a source of copper, as a gemstone, and as a dye for paints and fabrics. *Chemical formula:* $Cu_3(CO_3)_2(OH)_2$.

B

B **1.** The symbol for **boron.** **2.** The symbol for **magnetic field.**

Ba The symbol for **barium.**

Babbage (băb′ĭj), **Charles** 1792–1871. British mathematician who is considered a pioneer of computer science. In 1837 Babbage described an idea for the analytical engine, a machine that could be programmed with punched cards to perform complex calculations. Although Babbage never finished building the analytical engine, his idea is recognized as the forerunner of the modern computer.

bacillus (bə-sĭl′əs) Plural **bacilli** (bə-sĭl′ī′). Any of various pathogenic bacteria, especially one that is rod-shaped.

backbone (băk′bōn′) See **vertebral column.**

backcross (băk′krôs′, -krŏs′) *Verb.* **1.** To cross a hybrid with one of its parents or with an individual genetically identical to one of its parents. Backcrossing is used in research to isolate genetic characteristics found in one of the parents. *—Noun.* **2.** The act of making such a cross. **3.** An individual resulting from such a cross.

background extinction (băk′ground′) The ongoing extinction of individual species due to environmental or ecological factors such as climate change, disease, loss of habitat, or competitive disadvantage in relation to other species. Background extinction occurs at a fairly steady rate over geological time and is the result of normal evolutionary processes, with only a limited number of species in an ecosystem being affected at any one time. Compare **mass extinction.**

backscatter (băk′skăt′ər) **1.** The deflection of radiation or particles by electromagnetic or nuclear forces through angles greater than 90° to the initial direction of travel. **2.** The radiation or particles so deflected.

backshore (băk′shôr′) The area of a shore that lies between the average high tide mark and the vegetation. The backshore is affected by waves only during severe storms. Compare **foreshore.**

Bacon (bā′kən), **Roger** 1214?–1292. English scientist and philosopher who is noted for the wide range of his knowledge and writing on scientific topics. Bacon pioneered the idea that mathematics is fundamental to science and that experimentation is essential to test scientific theories.

BIOGRAPHY **Roger Bacon**

Roger Bacon was something of a Renaissance man before there was a Renaissance. Over the course of his long life, his energetic research would lead him to study everything from languages to mathematics to optics. He is most remembered for his insistence on the importance of pursuing fruitful lines of scientific research through experimentation. His writings describe countless experiments; while the majority were probably never performed by him, the profusion alone of experimental ideas is nothing short of astounding. His own laboratory work dealt primarily with alchemy, optics, and mechanics. He was among the first to apply geometric and mathematical principles to problems in optics and the behavior of light, allowing him to make important observations on reflection and refraction. His interest in mechanics led him to describe flying machines and other devices that had not yet been invented. He was the first person in the West to come up with a recipe for gunpowder, and he suggested reforms to the calendar, which would ultimately be implemented hundreds of years later. His novel ways of pursuing knowledge were sometimes viewed with suspicion, resulting at one time in imprisonment; but he bravely resisted all strictures on his intellectual life, even when that meant having to write and work in secret.

bacteria (băk-tîr′ē-ə) Plural of **bacterium.**

bacteriology (băk-tîr′ē-ŏl′ə-jē) The scientific study of bacteria, especially bacteria that cause disease.

bacteriophage (băk-tîr′ē-ə-fāj′) A virus that infects and destroys bacterial cells.

bacterium (băk-tîr′ē-əm) Plural **bacteria.** Any of a large group of one-celled organisms that lack a cell nucleus, reproduce by fission or by forming spores, and in some cases cause disease. They are the most abundant lifeforms on Earth, and are found in all living things and in all of the Earth's environments. Bacteria usually live off other organisms. Bacteria make up most of the kingdom of prokaryotes (Monera or Prokaryota), with one group (the archaea or archaebacteria) often classified as a separate kingdom. See also **archaeon, prokaryote.** —*Adjective* **bacterial.**

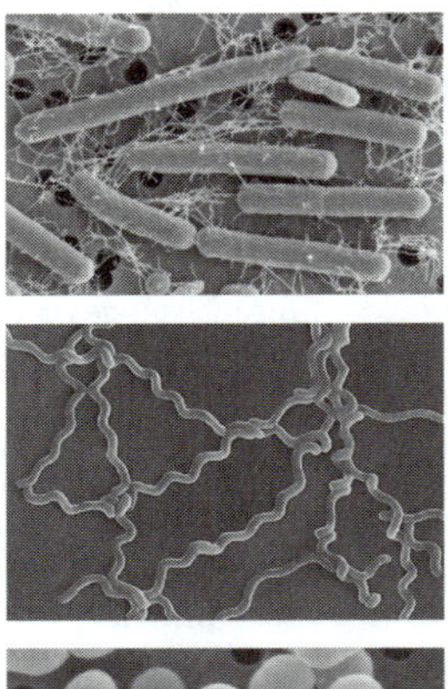

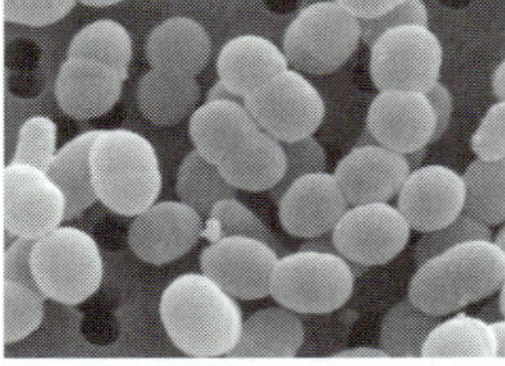

bacterium

top: *rod-shaped* Lactobacillus acidophilus; center: *spiral-shaped* Leptospira interrogans; bottom: *sphere-shaped* Leuconostoc citreum

USAGE **bacterium/bacteria**

It is important to remember that *bacteria* is the plural of *bacterium,* and that saying *a bacteria* is incorrect. It is correct to say *The soil sample contains millions of bacteria,* and *Tetanus is caused by a bacterium.*

Baekeland (bāk′lănd′), **Leo Hendrik** 1863–1944. Belgian-born American chemist who in 1907 developed Bakelite, the first plastic to harden permanently after heating. Originally used as an insulator, his invention proved to be a versatile and inexpensive material for manufacturing products such as telephones, cameras, and furniture.

Baily's beads (bā′lēz) A discontinuous, beadlike pattern of sunlight visible along the edge of the darkened Moon's disk in the seconds before and after totality during a full solar eclipse. The pattern is caused by light shining through the uneven lunar topography silhouetted along the curved edges of the disk. Baily's beads are named after British astronomer Francis Baily (1774–1844), who first observed them in 1836.

baking powder (bā′kĭng) A mixture of baking soda, a nonreactive filler (such as starch), and at least one slightly acidic compound (such as cream of tartar). Baking powder works as a leavening agent in baking by releasing carbon dioxide when mixed with a liquid, such as milk or water.

baking soda See **sodium bicarbonate.**

balance (băl′əns) To adjust a chemical equation so that the number of each type of atom and the total charge on the reactant (left-hand) side of the equation matches the number and charge on the product (right-hand) side of the equation.

baleen (bə-lēn′) A flexible horny substance hanging in fringed plates from the upper jaw of baleen whales. It is used to strain plankton from seawater when feeding. Also called *whalebone.*

baleen whale Any of several usually large whales of the suborder Mysticeti, having a symmetrical skull with two blowholes and plates of baleen instead of teeth. Baleen whales include the humpback, blue, fin, minke, and right whales, and the rorquals. Compare **toothed whale.**

ball-and-socket joint (bôl′ən-sŏk′ĭt) **1.** A joint, such as the shoulder or hip joint, in which a spherical knob or knoblike part of one bone fits into a cavity or socket of another, so that some degree of rotary motion is possible in every direction. **2.** A mechanical device consisting of a spherical knob at the end of a shaft that fits securely into a socket. Ball-and-socket joints are used to connect parts of a machine that require rotary movement in nearly all directions. Ball-and-socket joints allow the front wheels of a car to be turned by the steering mechanism.

ball bearing **1.** A ring-shaped track containing hard metal balls that roll freely, used to reduce friction where a rotating element (such as an axle) is attached to a fixed point. **2.** A hard ball used in such a track.

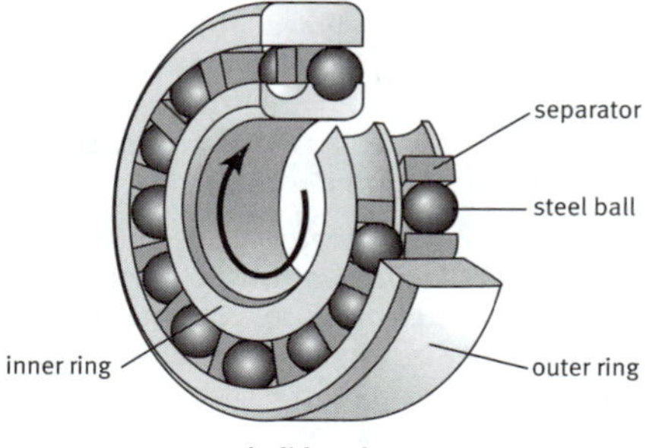

ball bearing

ballistics (bə-lĭs′tĭks) The scientific study of the characteristics of projectiles, such as bullets or missiles, and the way they move in flight.

ball lightning A rare form of lightning in the shape of a glowing red ball, associated with thunderstorms and sometimes accompanied by a loud noise. Ball lightning is thought to consist of ionized gas.

balsam (bôl′səm) Any of several aromatic resins that flow from certain plants and that contain considerable amounts of benzoic acid, cinnamic acid, or both, or their esters. Balsams are used in perfumes and medicines.

Baltimore (bôl′tə-môr′), **David** Born 1938. American microbiologist who discovered the enzyme reverse transcriptase, which is capable of passing information from RNA to DNA. Prior to this discovery, it was assumed that information could flow only from DNA to RNA. He won a 1975 Nobel Prize for his research into the connection between viruses and cancer.

band (bănd) A specific range of electromagnetic wavelengths or frequencies, as those used in radio broadcasting.

bandgap (bănd′găp′) The difference in energy in a substance between electron orbitals in which the electrons are not free to move (the **valence band**) and orbitals in which they are relatively free and will carry a current (the **conduction band**). In semiconductors, some electrons are sustained in the conduction band by thermal energy. Energy released when an electron in the conduction band falls into a hole in the valence band is called **bandgap radiation**. See also **hole, semiconductor laser.**

bandwidth (bănd′wĭdth′, -wĭth′) **1.** The numerical difference between the upper and lower frequencies of a band of electromagnetic radiation, especially an assigned range of radio frequencies. **2.** The amount of data that can be passed along a communications channel in a given period of time. For analog devices, such as standard telephones, bandwith is the range of frequencies that can be transmitted and is expressed in hertz (cycles per second). For digital devices, bandwidth is measured in bits per second. The wider the bandwidth, the faster data can be sent.

Banks (băngks), Sir **Joseph** 1743–1820. British botanist who took part in Captain James Cook's voyage around the world (1768–1771), during which he discovered and cataloged many species of plant and animal life.

Banneker (băn′ĭ-kər), **Benjamin** 1731–1806. American mathematician and astronomer who correctly predicted a solar eclipse in 1789. Although he had little formal education, Banneker published an almanac (1791–1802) containing his calculations of tidal cycles and future eclipses.

Benjamin Banneker

Banting (băn′tĭng), Sir **Frederick Grant** 1891–1941. Canadian physician who with the Scottish physiologist John Macleod won a 1923 Nobel Prize for the discovery of the hormone insulin. Banting and his assistant Charles Best experimented on diabetic dogs, demonstrating that insulin lowered their blood sugar. Insulin was tested and proven effective on humans within months of the first experiments with dogs. In acknowledgment of Best's work, Banting gave him a share of his portion of the Nobel Prize.

bar (bär) **1.** A unit used to measure atmospheric pressure. It is equal to a force of 100,000 newtons per square meter of surface area, or 0.987 atmosphere. **2.** An elongated, offshore ridge of sand, gravel, or other unconsolidated sediment, formed by the action of waves or longshore currents and submerged at least during high tide. Bars are especially common near the mouths of rivers or estuaries. **3.** A ridgelike mound of sand, gravel or silt formed within a stream, along its banks, or at its mouth. Bars form where the stream's current slows down, causing sediment to be deposited.

barb (bärb) **1.** A sharp point projecting backward, as on the stinger of a bee. **2.** One of the hairlike branches on the shaft of a feather.

barbel (bär′bəl) A slender, whiskerlike feeler extending from the head of certain fish, such as the catfish. It is used for touch and taste.

barbiturate (bär-bĭch′ər-ĭt) Any of a group of drugs that act as depressants of the central nervous system, are highly addictive, and are used primarily as sedatives and anticonvulsants. Phenobarbital and pentobarbital are examples of barbiturates.

barbule (bär′byo͞ol) A small barb or pointed projection, especially one that fringes the edges of the barbs of feathers.

barchan dune (bär-kän′) A large, crescent-shaped dune lying at right angles to the prevailing wind and having a steep, concave leeward side with the crescent tips pointing downwind. Barchan dunes form on flat, hard surfaces where the sand supply is limited. They can reach heights of up to 30 m (98 ft) and widths of 350 m (1,148 ft). See illustration at **dune.**

Bardeen (bär-dēn′), **John** 1908–1991. American physicist who, with William Brattain and William Shockley, invented the transistor in 1947. For this work all three shared a 1956 Nobel Prize for physics. In 1972 Bardeen shared another Nobel Prize for physics with American physicists Leon Neil Cooper and John Robert Schrieffer for their development of the theory of superconductivity.

bariatrics (băr′ē-ăt′rĭks) The branch of medicine that deals with the causes, prevention, and treatment of obesity.

barite (bâr′īt) A usually white, clear, or yellow orthorhombic mineral. Barite occurs as flattened blades or in a circular pattern of crystals that looks like a flower and, when colored red by iron stains, is called a desert rose. It is found in limestone, in clay-rich rocks, and in sandstones. Barite is used as a source of barium. *Chemical formula:* $BaSO_4$.

barium (bâr′ē-əm) *Symbol* **Ba** A soft, silvery-white metallic element of the alkaline-earth group. It occurs only in combination with other elements, especially in barite. Barium compounds are used in x-raying the digestive system and in making fireworks and white pigments. Atomic number 56; atomic weight 137.33; melting point 725°C; boiling point 1,140°C; specific gravity 3.50; valence 2. See **Periodic Table.**

barium sulfate A fine white powder used in making textiles, rubber, and plastic. It is also used in diagnostic imaging of the digestive tract. Barium sulfate occurs in nature as the mineral barite. *Chemical formula:* $BaSO_4$.

bark (bärk) The protective outer covering of the trunk, branches, and roots of trees and other woody plants. Bark includes all tissues outside the vascular cambium. In older trees, bark is usually divided into inner bark, consisting of living phloem, and outer bark, consisting of the periderm (the phelloderm, cork cambium, and cork) and all the tissues outside it. The outer bark is mainly dead tissue that protects the tree from heat, cold, insects, and other dangers. The appearance of bark varies according to the manner in which the periderm forms, as in broken layers or smoother rings. Bark also has lenticels, porous corky areas that allow for the exchange of water vapor and gases with the interior living tissues.

barnacle (bär′nə-kəl) Any of various small marine crustaceans of the subclass Cirripedia that form a hard shell in the adult stage and attach themselves to underwater surfaces, such as rocks, the bottoms of ships, and the skin of whales.

Barnard (bär′nərd), **Christiaan Neethling** 1923–2001. South African surgeon who performed the first successful human heart transplant in 1967.

Barnard's star (bär′nərdz) A dim, main-sequence red dwarf in the constellation Ophiuchus that is the second nearest star to Earth after the **Alpha-Centauri** system. Although it is only 5.98 light-years from our solar system, it is too faint to be seen with the unaided eye. Barnard's star has a greater proper motion (movement with respect to the background stars that is caused by an object's own motion rather than by how it is viewed from Earth) than any other star. Barnard's star is named for its identifier, American astronomer Edward Emerson Barnard (1857–1923).

barogram (băr′ə-grăm′) The continuous record of atmospheric pressure made by a barograph.

barograph (băr′ə-grăf′) An instrument that continuously records changes in atmospheric pressure. A barograph typically consists of an aneroid barometer connected to a pen; the pen is in contact with a piece of paper mounted on a cylinder that rotates once on a daily or weekly basis. As the atmospheric pressure changes, the pen is displaced in proportion to the change, thus a record of the pressure is traced onto the rotating sheet of paper.

barometer (bə-rŏm′ĭ-tər) An instrument for measuring atmospheric pressure. Barometers are used in determining height above sea level and in forecasting the weather. The two primary types of barometers are the aneroid and the mercury barometer.

barometric pressure (băr′ə-mĕt′rĭk) See **atmospheric pressure.**

baroreceptor (băr′ə-rĭ-sĕp′tər) A cell or sense organ found in the walls of the body's major arteries and stimulated by changes in blood pressure. Signals from baroceptors lead to a reduction in arterial blood pressure.

barred spiral galaxy (bärd) A spiral galaxy with a barlike bulge in the center, extending between the core and the spiral arms. About a third of spiral galaxies have this straight bar of stars, gas, and dust extending out from the nucleus. As with other spiral galaxies, barred spiral galaxies rotate, and new stars form from the dust and gas in their arms. Astronomers believe that some elliptical galaxies containing hints of a bar and spiral might once have been barred spiral galaxies.

barrier island (băr′ē-ər) A long, narrow sand island that is parallel to the mainland and serves to protect the coast from erosion. Barrier islands typically have dunes along the exposed outer side, zones of vegetation in the interior, and swampy areas along the inner lagoon.

barrier reef A long, narrow ridge of coral that runs parallel to the mainland and is separated from it by a deep lagoon. Compare **atoll, fringing reef.**

barycenter (băr′ĭ-sĕn′tər) The **center of mass** of two or more bodies, usually bodies orbiting around each other, such as the Earth and the Moon.

baryon (băr′ē-ŏn′) Any of a family of subatomic particles composed of three quarks or three antiquarks. They are generally more massive than mesons, and interact with each other via the strong force. Baryons form a subclass of hadrons and are subdivided into nucleons and hyperons. Protons and neutrons are baryons. See Table at **subatomic particle.**

baryon number A quantum number equal to the number of baryons in a system of subatomic particles minus the number of antibaryons. Baryons have a baryon number of +1, while antibaryons have a baryon number of –1. Quarks and antiquarks have baryon numbers of $+\frac{1}{3}$ and $-\frac{1}{3}$, respectively (baryons consists of three quarks). Mesons, bosons, and leptons all have baryon numbers of 0. Although the baryon number has always remained unchanged in reactions observed in experiments, it is postulated that in interactions that take place under conditions of very high energies (as during the formation of the universe, for example), proton decay may take place, and baryon number conservation may be violated. See also **isospin, strangeness.**

basal cell carcinoma (bā′səl, -zəl) A slow-growing neoplasm that is locally invasive but rarely metastasizes. It is derived from **basal cells**, the deepest layer of epithelial cells of the epidermis or hair follicles.

basal metabolic rate The rate at which energy is used by an organism at complete rest, measured in humans by the heat given off per unit time, and expressed as the calories released per kilogram of body weight or per square meter of body surface per hour.

basalt (bə-sôlt′, bā′sôlt′) A dark, fine-grained, igneous rock consisting mostly of plagioclase feldspar and pyroxene, and sometimes olivine. Basalt makes up most of the ocean floor and is the most common type of lava. It sometimes cools into characteristic hexagonal columns, as in the Giant's Causeway in Anterim, Northern Island. It is the fine-grained equivalent of gabbro.

base (bās) **1.** *Chemistry.* **a.** Any of a class of compounds that form hydroxyl ions (OH) when dissolved in water, and whose aqueous solutions react with acids to form salts. Bases turn red litmus paper blue and have a pH greater than 7. Their aqueous solutions have a bitter taste. Compare **acid. b.** See **nitrogen base. 2.** *Mathematics.* **a.** The side or face of a geometric figure to which an altitude is or is thought to be drawn. The base can be, but is not always, the bottom part of the figure. **b.** The number that is raised to various powers to generate the principal counting units of a number system. The base of the decimal system, for example, is 10. **c.** The number that is raised to a particular power in a given mathematical expression. In the expression a^n, a is the base.

base pair Any of the pairs of nucleotides connecting the complementary strands of a molecule of DNA or RNA and consisting of a purine linked to a pyrimidine by hydrogen bonds. The base pairs are adenine-thymine and guanine-cytosine in DNA, and adenine-uracil and guanine-cytosine in RNA or in hybrid DNA-RNA pairing. Base pairs may be thought of as the rungs of the DNA ladder.

base-pairing The hydrogen bonding of complementary nitrogen bases, one purine and one pyrimidine. Base-pairing occurs between the complementary strands of a DNA molecule or a DNA/RNA hybrid, and in the complementary pairing of codons and anticodons during the process of translation.

BASIC (bā′sĭk) A simple programming language developed in the 1960s that is widely taught to students as a first programming language.

basidiocarp (bə-sĭd′ē-ə-kärp′) A basidioma. This term is no longer in scientific use.

basidioma (bə-sĭd′ē-ō′mə) Plural **basidiomata** (bə-sĭd′ē-ō′mə-tə). A club-shaped, fleshy, spore-producing structure characteristic of many species of basidiomycete fungi. The basidioma grows out of the mass of hyphae known as a **mycelium** and bears the spore-dispersing structures called **basidia**. Mushrooms, toadstools, stinkhorns, and puffballs are basidiomata.

basidiomycete (bə-sĭd′ē-ō-mī′sēt′) Any of various fungi belonging to the phylum Basidomycota, bearing sexually produced spores on a **basidium.** All hyphae of basidiomycetes are divided into segments by septa and go through three stages of development. In the final stage, each segment has two nuclei, and the hyphae grow to produce basidia and disperse basidiospores. The basidiomycetes are the most familiar forms of fungi and include mushrooms, puffballs, shelf fungi, rusts, and smuts.

basidiospore (bə-sĭd′ē-ə-spôr′) A sexually produced fungal spore borne on a basidium in the fungi known as basidiomycetes. Basidiospores are produced by the union of the nuclei at the tip of a binucleated segment of a hypha. The resulting zygote then divides by meiosis into four haploid nuclei, each of which migrates to the very tip to be released as a basidiospore. A typical mushroom produces billions of basidiospores.

basidium (bə-sĭd′ē-əm) Plural **basidia.** A small, specialized, club-shaped structure typically bearing four basidiospores at the tips of minute projections in the fungi known as basidiomycetes. The basidium is unique to basidiomycetes and distinguishes them from other kinds of fungi.

basin (bā′sĭn) **1.** A region drained by a river and its tributaries. **2.** A low-lying area on the Earth's surface in which thick layers of sediment have accumulated. Some basins are bowl-shaped while others are elongate. Basins form through tectonic processes, especially in fault-bordered intermontane areas or in areas where the Earth's crust has warped downwards. They are often a source of valuable oil. **3.** An artificially enclosed area of a river or harbor designed so that the water level remains unaffected by tidal changes.

basipetal (bā-sĭp′ĭ-tl) Relating to the movement of substances, such as hormones, or the progressive development of tissues in a direction away from the tip and towards the base of a root or shoot in a plant.

basis (bā′sĭs) Plural **bases** (bā′sēz′). A set of independent vectors whose linear combinations define a vector space, such as a reference frame used to establish a coordinate system.

bast fiber (băst) Any of various durable fibers found in the phloem of certain eudicot plants, especially flax, hemp, and jute, used in making rope and baskets.

Batesian mimicry (bāt′sē-ən) A form of protective mimicry in which an unprotected species (the mimic) closely resembles an unpalatable or harmful species (the model), and therefore is similarly avoided by predators. The close resemblance between certain harmless flies and stinging bees, and the similarity between the colored stripes of the nonpoisonous king snake and those of the highly venomous coral snake, are examples of Batesian mimicry. Batesian mimicry is named after the British naturalist Henry Walter Bates (1825–92). Compare **aggressive mimicry, Müllerian mimicry.**

batholith (băth′ə-lĭth′) A large mass of igneous rock that has intruded and melted surrounding strata at great depths. Batholiths usually have a surface area of over 100 km^2 (38 mi^2).

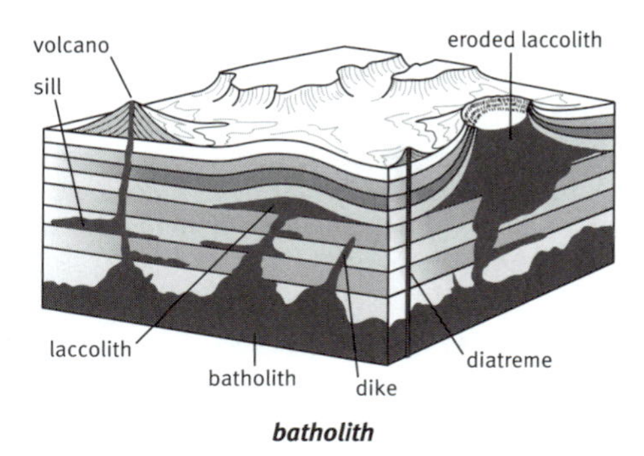

batholith

bathometer (bə-thŏm′ĭ-tər) An instrument that is used to measure water depth without the use of a sounding line. The bathometer does not require a line to extend to the bottom because it measures the difference in the gravitational effect of the water surface and of solid ground.

bathyal (băth′ē-əl) Relating to the region of the ocean bottom between the sublittoral and abyssal zones, from depths of approximately 200 to 2,000 m (656 to 6,560 ft).

bathymetry (bə-thĭm′ĭ-trē) The measurement of the depth of bodies of water, particularly of oceans and seas.

bathypelagic zone (băth′ə-pə-lăj′ĭk) A layer of the oceanic zone lying below the **mesopelagic zone** and above the **abyssopelagic zone**, at depths generally between about 1,000 and 4,000 m (3,280–13,120 ft). The bathypelagic zone receives no sunlight and water pressure is considerable. The abundance and diversity of marine life decreases with depth through this and the lower zones.

bathyscaphe (băth′ĭ-skăf′, -skāf′) A free-diving vessel used to explore the ocean at great depths. The original bathyscaphe, constructed in 1948, was made of a cylindrical metal float and a suspended steel ball that could hold two people. The float contained gasoline used to lift the vessel, and heavy iron material used for ballast. Design improvements allowed the second bathyscaphe in 1960 to descend to a record 10,912 m (35,791 ft) in the Marianas Trench in the western Pacific Ocean, almost to the deepest level ever sounded on Earth.

bathysphere (băth′ĭ-sfîr′) A hollow, spherical steel diving chamber in which people are lowered by cable from a surface vessel to explore the ocean depths. In 1934 a bathysphere carrying William Beebe and an associate reached a record depth of over 923 m (3,028 ft). Because space in the bathysphere is cramped, dives longer than three-and-a-half hours are intolerable, and it was eventually supplanted by the **bathyscaphe**.

bathythermograph (băth′ĭ-thûr′mə-grăf′) A device that records water temperature in relation to ocean depth.

batrachian (bə-trā′kē-ən) Relating to tailless amphibians, such as frogs and toads.

battery (băt′ə-rē) A device containing an electric cell or a series of electric cells storing energy that can be converted into electrical power (usually in the form of direct current). Common household batteries, such as those used in a flashlight, are usually made of dry cells (the chemicals producing the current are made into a paste). In other batteries, such as car batteries, these chemicals are in liquid form.

A CLOSER LOOK battery

A *battery* stores chemical energy, which it converts to electrical energy. A typical battery, such as a car battery, is composed of an arrangement of *galvanic cells.* Each cell contains two metal electrodes, separate from each other, immersed within an electrolyte containing both positive and negative ions. A chemical reaction between the electrodes and the electrolyte, similar to that found in electroplating, takes place, and the metals dissolve in the electrolyte, leaving electrons behind on the electrodes. However, the metals dissolve at different rates, so a greater number of electrons accumulate at one electrode (creating the negative electrode) than at the other electrode (which becomes the positive electrode). This gives rise to an electric potential between the electrodes, which are typically linked together in series and parallel to one another in order to provide the desired voltage at the battery terminals (12 volts, for example, for a car battery). The buildup of charge on the electrodes prevents the metals from dissolving further, but if the battery is hooked up to an electric circuit through which current may flow, electrons are drawn out of the negative electrodes and into the positive ones, reducing their charge and allowing further chemical reactions.

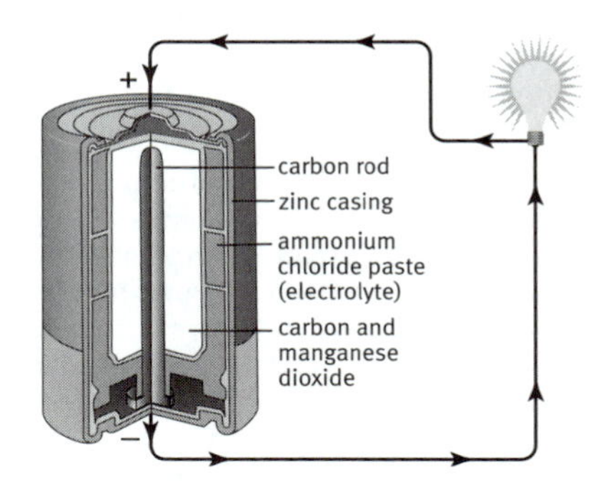

Electrons flow from the zinc casing to the carbon rod, lighting the bulb in the process. The zinc casing acts as a negative electrode; the carbon rod acts as a positive electrode. The ammonium chloride paste acts as the electrolyte and the carbon and the manganese dioxide mixture around the carbon rod extends the battery's life.

bauxite (bôk′sīt′) A soft, whitish to reddish-brown rock consisting mainly of hydrous aluminum oxides and aluminum hydroxides along with silica, silt, iron hydroxides, and clay minerals. Bauxite forms from the breakdown of clays and is a major source of aluminum.

bay (bā) **1.** A body of water partially enclosed by land but having a wide outlet to the sea. A bay is usually smaller than a gulf. **2.** A space in the cabinet of a personal computer where a storage device, such as a disk drive or CD-ROM drive, can be installed.

bayou (bī′o͞o) A sluggish, marshy stream connected with a river, lake, or gulf. Bayous are common in the southern United States.

BBS Abbreviation for **bulletin board system.**

BCE or BCE Abbreviation for **before the Common Era.**

B cell Any of the lymphocytes that develop into plasma cells in the presence of a specific antigen. The plasma cells produce antibodies that attack or neutralize the antigen in what is called the **humoral immune response.** B cells mature in the bone marrow before being released into the blood. Also called *B lymphocyte.* Compare **T cell.**

BCG vaccine (bē′sē-jē′) A preparation used to immunize individuals against tuberculosis, consisting of attenuated live human tubercle bacilli (*Mycobacterium tuberculosis*), the bacteria that cause tuberculosis.

B complex See **vitamin B complex.**

Be The symbol for **beryllium.**

beach (bēch) The area of accumulated sand, stone, or gravel deposited along a shore by the action of waves and tides. Beaches usually slope gently toward the body of water they border and have a concave shape. They extend landward from the low water line to the point where there is a distinct change in material (as in a line of vegetation) or in land features (as in a cliff).

beaker (bē′kər) A wide, cylindrical glass container with a pouring lip, used especially in laboratories.

beard (bîrd) A tuft or group of hairs or bristles on certain plants, such as barley and wheat. The individual strands of a beard are attached to a sepal or petal.

beard (on wheat)

beat (bēt) A fluctuation or pulsation, usually repeated, in the amplitude of a signal. Beats are generally produced by the superposition of two waves of different frequencies; if the signals are audible, this results in fluctuations between louder and quieter sound.

beat wave A wave whose amplitude varies periodically, produced by the superposition of two waves of different frequencies.

Beaufort scale (bō′fərt) A scale for classifying the force of the wind, ranging from 0 (*calm*) to 12 (*hurricane*). A wind classified as 0 has a velocity of less than 1.6 km (1 mi) per hour; a wind classified as 12 has a velocity of over 119 km (74 mi) per hour. Other categories include *light air*, five levels of *breeze*, four levels of *gale*, and *storm*. The scale was devised in 1805 as a means of describing the effect of different wind velocities on ships at sea. It is named after an admiral in the British navy, Sir Francis Beaufort (1774–1857).

beauty quark (byo͞o′tē) See **bottom quark.**

becquerel (bĕ-krĕl′, bĕk′ə-rĕl′) The SI derived unit used to measure the rate of radioactive decay. When the nucleus of an atom emits nucleons (protons and/or neutrons) and is thereby transformed into a different nucleus, decay has occurred. A decay rate of one becquerel for a given quantity means there is one such atomic transformation per second.

Becquerel Family of French physicists, including **Antoine César** (1788–1878), a founder of electrochemistry; his son **Alexandre Edmond** (1820–91), noted for his research on phosphorescence, magnetism, electricity, and optics; and his grandson **Antoine Henri** (1852–1908), who discovered spontaneous radioactivity in uranium. Antoine Henri Becquerel's work led to the discovery of radium by Marie and Pierre Curie, with whom he shared the 1903 Nobel Prize for physics.

bed (bĕd) **1.** A layer of sediments or rock, such as coal, that extends under a large area and has a distinct set of characteristics that distinguish it from other layers below and above it. **2.** The bottom of a body of water, such as a lake, stream, or ocean.

bedrock (bĕd′rŏk′) The solid rock that lies beneath the soil and other loose material on the Earth's surface.

Beebe (bē′bē), **(Charles) William** 1877–1962. American naturalist, explorer, and author whose numerous expeditions include an oceanic descent of 923 m (3,028 ft) in a bathysphere he helped design, a record for this type of submersible.

behavior (bĭ-hāv′yər) **1.** The actions displayed by an organism in response to its environment. **2.** One of these actions. Certain animal behaviors (such as nest building) result from **instinct**, while others (such as hunting) must be learned. **3.** The manner in which a physical system, such as a gas, subatomic particle, or ecosystem, acts or functions, especially under specified conditions.

behavioral science (bĭ-hāv′yə-rəl) Any of various scientific disciplines, such as sociology, anthropology, or psychology, in which the actions and reactions of humans and animals are studied through observational and experimental methods.

behavior modification The alteration of human behavior using basic learning techniques such as biofeedback and positive or negative reinforcement (as in rewards and punishments).

bel (bĕl) A unit of measurement equal to ten decibels.

belemnite (bĕl′əm-nīt′) **1.** Any of various extinct cephalopod mollusks of the order Belemnoidea that lived from the Triassic into the Tertiary Period. Belemnites had a large, cone-shaped internal shell with a complex structure that served as a support for muscles and as a hydrostatic device. Belemnites were closely related to the present-day squids and cuttlefishes. **2.** The fossilized internal shell of one of these cephalopods. Belemnites are used as **index fossils**.

Bell (bĕl), **Alexander Graham** 1847–1922. Scottish-born American scientist and inventor whose lifelong interest in the education of deaf people led him to conceive the idea of transmitting speech by electric waves. In 1876 his experiments with a telegraph resulted in his invention of the telephone. He later produced the first successful sound recorder, an early hearing aid, and many other devices.

belladonna (bĕl′ə-dŏn′ə) A preparation of the dried leaves or roots of deadly nightshade or related plants in the genus Belladonna, once used as a medicine. Belladonna contains several alkaloids that affect the nervous system by blocking the effects of acetylcholine.

Bell Burnell (bĕl′ bûr′nĕl′), **Susan Jocelyn** Born 1943. British astronomer. In 1967, working with astronomer Antony Hewish, she discovered the first pulsar.

bell curve A symmetrical bell-shaped curve that represents the distribution of values, frequencies, or probabilities of a set of data. It slopes downward from a point in the middle corresponding to the mean value, or the maximum probability. Data that reflect the aggregate outcome of large numbers of unrelated events tend to result in bell curve distributions. ▸ The **Gaussian** or **normal distribution** is a mathematically well-defined bell curve used in statistics and in science generally.

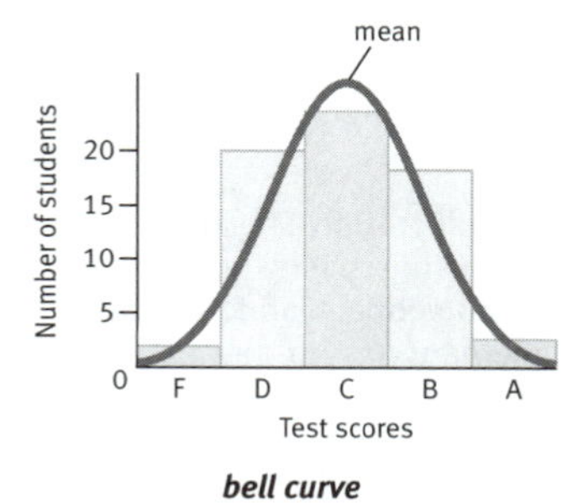

bell curve

graph showing the distribution of a set of test scores where the average grade is a C

belt (bĕlt) A geographic region that is distinctive in a specific respect.

benign (bĭ-nīn′) Not life-threatening or severe, and likely to respond to treatment, as a tumor that is not malignant. Compare **malignant.**

benign prostatic hyperplasia (hī′pər-plā′zhə) Benign enlargement of the prostate gland commonly occurring in men after the age of 50, and sometimes causing compression of the urethra and obstruction of urine flow.

benthic (bĕn′thĭk) Relating to the bottom of a sea or lake or to the organisms that live there.

benthos (bĕn′thŏs′) **1.** The bottom of a sea or lake. **2.** The organisms living on sea or lake bottoms. The benthos are divided into sessile organisms (those that are attached to the bottom or to objects on or near the bottom) and vagrant organisms (those that crawl or swim along the bottom). Compare **nekton, plankton.** See more at **epifauna, infauna.**

benzaldehyde (bĕn-zăl′də-hīd′) A colorless aromatic oil that smells like almonds. It is obtained naturally from certain nuts and plant leaves, or made synthetically. It is used in perfumes and as a solvent and flavoring. *Chemical formula:* C_7H_6O.

benzalkonium chloride (bĕn′zăl-kō′nē-əm) A yellow-white powder prepared in an aqueous solution and used as a detergent, fungicide, bactericide, and spermicide. Benzalkonium chloride is a mixture of the chlorides of various organic compounds having a benzene ring attached to an ammoniated alkane.

benzene (bĕn′zēn′) A colorless flammable liquid that is derived from petroleum. Benzene is used to make detergents, insecticides, motor fuels, and many other chemical products. *Chemical formula:* C_6H_6. See more at **benzene ring.** —*Adjective* **benzyl.**

benzene hexachloride (hĕk′sə-klôr′īd′) A musty-smelling crystalline substance that is used as an insecticide. It is prepared by adding chlorine to benzene. Also called *hexachlorobenzene. Chemical formula:* $C_6H_6Cl_6$.

benzene ring A hexagonal arrangement of six carbon atoms, each atom bonded to its adjacent atoms by a single covalent bond, and by an unusual ring bond of electrons shared by all six carbon atoms. The benzene ring is a basic component of many organic compounds,

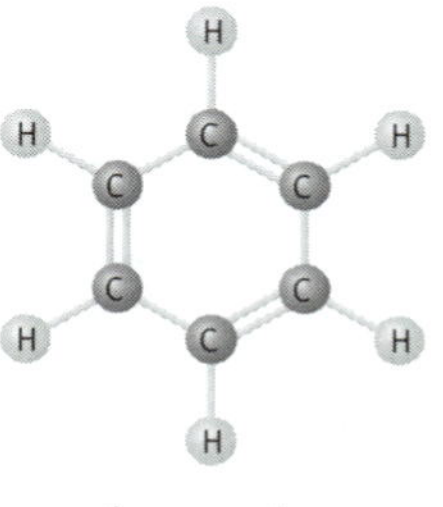

benzene ring

especially the aromatic hydrocarbons. In benzene itself, each carbon atom is also bonded to a hydrogen atom; in derivatives of benzene, one or more of the hydrogen atoms is replaced by other atoms or groups of atoms.

benzidine (bĕn′zĭ-dēn′) A yellowish, white, or reddish-gray crystalline powder that is produced synthetically and is carcinogenic. It is used in making dyes, as a reagent, and in detecting blood stains. *Chemical formula:* $\mathbf{C_{12}H_{12}N_2}$.

benzoate (bĕn′zō-āt′) A salt or ester of benzoic acid, having the general formula C_6H_5COOR, where R is an element or group that has replaced the hydrogen in the carboxyl group (COOH) of benzoic acid. In salts of benzoic acid, R is a metal, while in esters of benzoic acid, R is another radical, such as methyl.

benzofuran (bĕn′zō-fyo͝or′ăn′) A colorless, sweet-smelling, oily liquid found in coal tar and used to make thermoplastic resins for paints and varnishes. *Chemical formula:* $\mathbf{C_8H_6O}$.

benzoic acid (bĕn-zō′ĭk) A white, crystalline acid used in preserving food, as a cosmetic, and in medicine. Benzoic acid consists of a benzene ring with a carboxyl group (COOH) attached. It occurs naturally in some plants. *Chemical formula:* $\mathbf{C_7H_6O_2}$.

benzoin (bĕn′zō-ĭn, -zoin′) **1.** A resin obtained from the bark of certain tropical Asian trees of the genus *Styrax* and used in making perfumes and in medicine as an antiseptic. **2.** A very toxic white or yellowish crystalline compound derived from benzaldehyde. It oxidizes easily and is used as a reducing agent. *Chemical formula:* $\mathbf{C_{14}H_{12}O_2}$.

benzophenone (bĕn′zō-fĭ-nōn′, -fē′nōn) A white crystalline compound used in perfumes to prevent evaporation and in the manufacture of insecticides. Benzophenone is a ketone. *Chemical formula:* $\mathbf{C_{13}H_{10}O}$.

benzoyl (bĕn′zō-ĭl′, -zoil′) The radical C_6H_5CO, derived from benzoic acid.

benzoyl peroxide A flammable white granular solid used as a bleaching agent for flour, fats, waxes, and oils, as a polymerization catalyst, and in pharmaceuticals, especially to treat acne. *Chemical formula:* $\mathbf{C_{14}H_{10}O_4}$.

benzyl (bĕn′zĭl) The radical $C_6H_5CH_2$, derived from toluene.

Bergmann's rule (bûrg′mənz) The principle holding that in a warm-blooded animal species having distinct geographic populations, the body size of animals living in cold climates tends to be larger than in animals of the same species living in warm climates. Animals with larger bodies are generally more massive and thus produce more body heat. They also have smaller surface areas relative to their mass, resulting in a relatively lower rate of heat radiation. The Bergmann rule is named for the German biologist Karl Bergmann (1814–65). Compare **Allen's rule.**

beriberi (bĕr′ē-bĕr′ē) A disease caused by a deficiency of thiamine, endemic in eastern and southern Asia, and characterized by neurological symptoms, cardiovascular abnormalities, and edema.

berkelium (bər-kē′lē-əm, bûrk′lē-əm) *Symbol* **Bk** A synthetic, radioactive metallic element of the actinide series that is produced from americium, curium, or plutonium. Its most stable isotope has a half-life of about 1,400 years. Atomic number 97; melting point 986°C; valence 3, 4. See **Periodic Table.**

berm (bûrm) **1.** A nearly horizontal or landward-sloping portion of a beach formed by the deposition of sediment by storm waves. A beach may have no berm at all, or it may have more than one berm. **2.** A narrow man-made ledge or shelf, as along the top or bottom of a slope.

Bernard (bĕr-när′), **Claude** 1813–1878. French physiologist noted for his study of the chemical reactions involved in the digestive system and of the connection between the liver and the nervous system. His work laid the foundation for experimental medicine.

Bernoulli (bər-no͞o′lē) Family of Swiss mathematicians. **Jacques** (or **Jakob**) (1654–1705) was a major developer of calculus and made an important contribution to probability theory. His brother **Jean** (or **Johann**) (1667–1748) also developed calculus and contributed to the study of complex numbers and trigonometry. Jean's son **Daniel** (1700–1782) pioneered the modern field of hydrodynamics and anticipated the kinetic theory of gases, indicating that gas pressure would increase with increasing energy. He was also one of the first scientists to understand the concept of conservation of energy.

Bernoulli distribution See **binomial distribution.**

Bernoulli effect The phenomenon of falling pressure in a fluid as the velocity of the fluid stream increases. The Bernoulli effect accounts for the lift generated by the airfoil shape of an airplane wing. See also **Bernoulli's law, lift.** See Note at **aerodynamics.**

Bernoulli's law (bər-nōō′lēz) **1.** A law of fluid mechanics stating the relationship between the velocity, density, and pressure of a fluid. Mathematically, the law states that $P + \frac{1}{2}\rho v^2 =$ constant, where P is the pressure (in newtons per square meter), ρ is the density of the fluid (in kilograms per square meter), and v is the velocity (in meters per second). If no energy is added to the system, an increase in velocity is accompanied by a decrease in density and/or pressure. The law is directly related to the principle of conservation of energy. See also **Bernoulli effect. 2.** See **law of large numbers.**

Bernoulli trial A random event in which one of two possible outcomes can occur (usually denoted *success* or *failure*), with the properties that the probability of each outcome is the same in each trial and that the outcome of each trial is independent of the outcomes of the other trials. If the probability of success is *p*, the probability of failure is $1 - p$. The flip of a coin is a Bernoulli trial (where the probability of both success and failure is 0.5), as is the roll of a die (where success might be arbitrarily defined as rolling a six and failure as rolling any other number, with the probability of success being 0.167 and the probability of failure 0.833).

berry (bĕr′ē) **1.** A simple fruit that has many seeds in a fleshy pulp. Grapes, bananas, tomatoes, and blueberries are berries. Compare **drupe, pome.** See more at **simple fruit. 2.** A seed or dried kernel of certain kinds of grain or other plants such as wheat, barley, or coffee.

USAGE **berry**

Cucumbers and tomatoes aren't usually thought of as berries, but to a botanist they are in fact berries, while strawberries and raspberries are not. In botany, a berry is a fleshy kind of *simple fruit* consisting of a single ovary that has multiple seeds. Other true berries besides cucumbers and tomatoes are bananas, oranges, grapes, and blueberries. Many fruits that are popularly called berries have a different structure and thus are not true berries. For example, strawberries and raspberries are *aggregate fruits*, developed from multiple ovaries of a single flower. The mulberry is not a true berry either. It is a *multiple fruit*, like the pineapple, and is made up of the ovaries of several individual flowers.

beryl (bĕr′əl) A usually green or bluish-green hexagonal mineral occurring as transparent to translucent prisms in igneous and metamorphic rocks. Transparent varieties, such as emeralds and aquamarine, are valued as gems. Beryl is the main source of the element beryllium. *Chemical formula:* $\mathbf{Be_3Al_2Si_6O_{18}}$.

beryllium (bə-rĭl′ē-əm) *Symbol* **Be** A hard, lightweight, steel-gray metallic element of the alkaline-earth group, found in various minerals, especially beryl. It has a high melting point and is corrosion-resistant. Beryllium is used to make sturdy, lightweight alloys and aerospace structural materials. It is also used as a neutron moderator in nuclear reactors. Atomic number 4; atomic weight 9.0122; melting point 1,278°C; boiling point 2,970°C; specific gravity 1.848; valence 2. See **Periodic Table.**

Berzelius (bər-zē′lē-əs), Baron **Jöns Jakob** 1779–1848. Swedish chemist who is regarded as one of the founders of modern chemistry. Berzelius developed the concepts of the ion and of ionic compounds and made extensive determinations of atomic weights. In 1811 he introduced the classical system of chemical symbols, in which the names of elements are identified by one or two letters.

Bessemer process (bĕs′ə-mər) A method for making steel by forcing compressed air through molten iron to burn out carbon and other impurities.

Best (bĕst), **Charles Herbert** 1899–1978. American-born Canadian physiologist who assisted Frederick Banting in the discovery of the hormone insulin. In acknowledgment of his work, Banting shared his portion of the 1923 Nobel Prize with Best. In addition to further refining the use of insulin, Best later discovered the vitamin choline and the enzyme histaminase, which breaks down histamine.

beta-blocker (bā′tə-blŏk′ər) A drug that blocks the excitatory effects of epinephrine on the cardiovascular system by binding to cell-surface receptors (called beta-receptors). Beta-blockers are used to treat high blood pressure, angina, and certain abnormal heart rhythms.

beta-carotene (bā′tə-kăr′ə-tēn′) A phytochemical that is an isomer of carotene found in dark green and yellow fruits and vegetables and that is converted to vitamin A, primarily in the liver. See more at **carotene.**

beta decay A form of radioactive decay caused by the weak nuclear force, in which a beta particle (electron or positron) is emitted. ► In **beta-minus decay**, a neutron in an atomic nucleus decays into a proton, an electron, and an antineutrino. The electron and antineutrino are emitted from the nucleus, while the proton remains. The atomic number of the atom is thereby increased by 1. The decay of Carbon-14 into Nitrogen-14, a phenomenon useful in carbon dating, is an example of beta-minus decay. ► In **beta-plus decay**, a proton in an atomic nucleus decays into a neutron, a positron, and a neutrino. The positron and neutrino are emitted from the nucleus, while the neutron remains. The atomic number of the atom is thereby reduced by 1. The decay of Carbon-10 to Boron-10 is an example of beta-plus decay. See also **W boson.**

betaine (bē′tə-ēn′, -ĭn) **1.** Any of a class of organic salts that are derived from amino acids and have a cationic (positively charged) component that consists of a nitrogen atom attached to three methyl (CH_3) groups. **2.** A salt of this class that is a sweet crystalline alkaloid first found in sugar beets but also widely occurring in other plants and in animals. Betaine is used in the treatment of muscular weakness and degeneration. *Chemical formula:* $C_5H_{11}NO_2$.

beta particle A high-speed electron or positron, usually emitted by an atomic nucleus undergoing radioactive decay. Beta particles are given off naturally by decaying neutrons in radioactive atoms and can be created in particle accelerators. Beta particles have greater speed and penetrating power than alpha particles but can be stopped by a sheet of aluminum that is 2 to 3 mm thick. See more at **radioactive decay.**

beta ray A stream of beta particles.

beta sheet A secondary structure that occurs in many proteins and consists of two or more parallel adjacent polypeptide chains arranged in such a way that hydrogen bonds can form between the chains. Unlike in an alpha helix where the amino acids form a coil, in beta sheets the amino acids are arranged in a zigzag pattern that forms a straight chain. The proteins in silk have a beta-sheet structure. Compare **alpha helix, random coil.**

betatron (bā′tə-trŏn′) A type of particle accelerator that uses changing magnetic fields to accelerate electrons. Energies of several hundred million electron volts can be achieved in a betatron. See also **particle accelerator.**

beta version The version of a software product that is used in the final stage of testing before it is commercially released. Testing of a beta version is generally conducted by an independent tester outside of the company developing the product.

Betelgeuse (bēt′l-jo͞oz′) A red supergiant star in the constellation Orion. It is a variable star with a brightest apparent magnitude of 0.5. *Scientific name:* Alpha Orionis. See Note at **Rigel.**

Bethe (bā′tə), **Hans Albrecht** 1906–2005. German-born American physicist who was instrumental in the development of quantum physics. Bethe also played an important role in the development of the atomic bomb, later working to educate the public about the threat of nuclear weapons. In 1967 he received a Nobel Prize for explaining that the Sun and other stars derive their energy from a series of nuclear reactions which came to be known as the **carbon cycle**, or Bethe cycle.

Bethe cycle See **carbon cycle** (sense 2).

Bh The symbol for **bohrium.**

Bhaskara (bäs′kə-rə), 1114–1185? Indian mathematician who wrote the first work showing a systematic use of the decimal system.

B horizon In ABC soil, the second and middle zone, characterized by an accumulation of soluble or suspended organic material, clay, iron, or aluminum. These materials originate in the A horizon and are transported to the B horizon through the process of eluviation. B horizons often have a blocky structure and are reddish in color due to the presence of oxidized metals. Also called *zone of accumulation, zone of illuviation.*

Bi The symbol for **bismuth.**

bicarbonate (bī-kär′bə-nāt′) The group HCO_3 or a compound containing it, such as sodium bicarbonate. When heated, bicarbonates give off carbon dioxide.

bicarbonate of soda See **sodium bicarbonate.**

biceps (bī′sĕps′) Either of two muscles, *biceps brachii* of the arm or *biceps femoris* of the leg, each with two points of origin. The biceps of the arm bends the elbow, while the biceps of the leg helps to bend the knee as part of the hamstring.

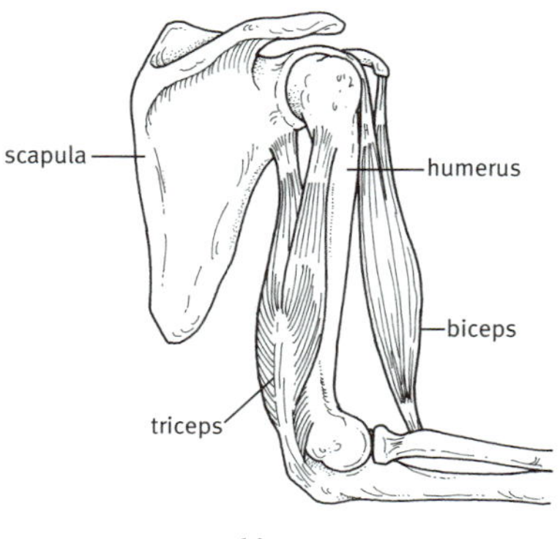

biceps

biconcave (bī′kŏn-kāv′) Concave on both sides or surfaces.

biconvex (bī′kŏn-vĕks′) Convex on both sides or surfaces.

bicuspid (bī-kŭs′pĭd) *Adjective.* **1.** Having two points or cusps. —*Noun.* **2.** A bicuspid tooth, especially a premolar.

bicyclic (bī-sī′klĭk, -sĭk′lĭk) **1.** Relating to a compound having two closed rings. Naphthalene is a bicyclic compound. **2.** Composed of or arranged in two distinct whorls, as the petals of a flower.

biennial (bī-ĕn′ē-əl) *Adjective.* **1.** Completing a life cycle normally in two growing seasons. —*Noun.* **2.** A biennial plant. In the first year, biennials normally produce a short stem, a rosette of leaves, and a fleshy root that acts as food supply. In the second season, biennials blossom, produce seed, use up their food supply, and die. Carrots, parsnips, and sugar beets are examples of biennials. Compare **annual, perennial.**

bifacial (bī-fā′shəl) Flaked in such a way as to produce a cutting edge that is sharp on both sides. Used of a stone tool. ► Bifacial tools are known as a **bifaces** and include such early **core tools** as hand axes and cleavers as well as later **flake tools** such as blades and spear or arrow points. Compare **unifacial.**

bifurcate (bī′fər-kāt′, bī-fûr′-) Forked or divided into two parts or branches, as the Y-shaped styles of certain flowers or the tongues of snakes.

big bang (bĭg) The explosion of an extremely small, hot, and dense body of matter that, according to some cosmological theories, gave rise to the universe between 12 and 20 billion years ago. Compare **big crunch, steady state theory.** See also **open universe.**

A CLOSER LOOK **big bang**

In the 1920s astronomer Edwin Hubble discovered that wherever one looked in space, distant galaxies were rapidly moving away from Earth, and the more distant the galaxy the greater its speed. Through this observation he determined that the universe was becoming larger. Hubble also found that the ratio between a galaxy's distance and velocity (speed and direction of travel) was constant; this value is called the *Hubble constant.* By calculating the distance and velocity of various galaxies and working backward, astronomers could determine how long ago the expansion began—in other words, the age of the universe. The figure, which scientists are constantly refining, is currently thought to be between 12 and 20 billion years. According to the widely accepted theory of the *big bang,* the universe was originally smaller than a dime and almost infinitely dense. A massive explosion, which kicked off the expansion, was the origin of all known space, matter, energy, and time. Scientists are also attempting to calculate how much mass the universe contains in order to predict its future. If there is enough mass, the gravity attracting all its pieces to each other will eventually stop the expansion and pull the universe back together in a *big crunch.* There may not be enough mass, however, to result in an eventual collapse. If that is the case, then the universe will expand forever, and all galaxies and matter will drift apart, eventually becoming dark and cold.

big crunch The convergence of all matter, energy, and space into a single, minute point. This convergence is hypothesized to be the final event in the universe in some cosmological theories. Compare **big bang.** See also **closed universe.**

Big Dipper An asterism composed of seven stars in the constellation Ursa Major. Four stars form the bowl and three form the handle in the outline of a dipper.

bight (bīt) A long, gradual bend or curve in a shoreline. A bight can be larger than a bay, or it can be a segment of a bay.

bijection (bī-jĕk′shən) *Mathematics* A function that is both an injection and a surjection. In a bijection, each member of the range corresponds to an element of the domain that is mapped onto it, and there is a one-to-one correspondence between the members of the domain and the range. All linear functions, such as $y = x + 3$, are bijections. Compare **injection, surjection.**

bilateral symmetry (bī-lăt′ər-əl) Symmetrical arrangement of an organism or part of an organism along a central axis, so that the organism or part can be divided into two equal halves. Bilateral symmetry is a characteristic of animals that are capable of moving freely through their environments. Compare **radial symmetry.**

bile (bīl) A bitter, alkaline, brownish-yellow or greenish-yellow fluid that is secreted by the liver, concentrated and stored in the gallbladder, and discharged into the duodenum of the small intestine. It helps in the digestion of fats and the neutralization of acids, such as the hydrochloric acid secreted by the stomach. Bile consists of salts, acids, cholesterol, lipids, pigments, and water. ▸ **Bile salts** help in the emulsification, digestion, and absorption of fats. ▸ **Bile pigments** are waste products formed by the breakdown of hemoglobin from old red blood cells.

bile duct Any of the passages that carry bile from the liver or gallbladder to the duodenum.

bilirubin (bĭl′ĭ-ro͞o′bĭn) A reddish-yellow pigment that is a constituent of bile and gives it its color. Bilirubin is a porphyrin derived from the degradation of heme. It is often a constituent of gallstones, and also causes the skin discoloration seen in jaundice. *Chemical formula:* $C_{33}H_{36}N_4O_6$.

binary (bī′nə-rē) **1.** Having two parts. **2.** *Mathematics.* Based on the number 2 or the binary number system.

binary coded decimal *Computer Science.* A code in which a string of four binary digits represents each decimal number 0 through 9 as a means of preventing calculation errors due to rounding and conversion. For example, since the binary equivalent of 3 is 0011 and the binary equivalent of 6 is 0110, 36 is represented as 0011 0110.

binary digit Either of the digits 0 or 1, used in the binary number system. See more at **bit.**

binary fission See **fission** (sense 2).

binary number system A method of representing numbers that has 2 as its base and uses only the digits 0 and 1. Each successive digit represents a power of 2. For example, 10011 represents $(1 \times 2^4) + (0 \times 2^3) + (0 \times 2^2) + (1 \times 2^1) + (1 \times 2^0)$, or $16 + 0 + 0 + 2 + 1$, or 19.

binary star A system of two stars that orbit a common center of mass, appearing as a single star when visible to the unaided eye. The orbital periods of binary stars range from several hours to several centuries. By some estimates, at least half of the stars in the Milky Way galaxy are members of binary star systems. Also called *double star.* ▸ Binary stars are divided into four main classes based on how their dual nature is detected. A **visual binary** can be resolved telescopically into its two components. Only one star of an **astrometric binary** is visible, but the unseen component can be identified from its gravitational effect on the visible star, causing it to oscillate slightly, or wobble, against the background of more distant stars. The two components of a **spectroscopic binary** are identified based on their varying orbital velocities toward or away from Earth as revealed by periodic Doppler shifts in their spectral lines. In an **eclipsing binary**, the two components orbit each other in such a way that they periodically obscure or eclipse each other as viewed from Earth, causing changes in their observed brightness. Eclipsing binaries are also considered a kind of variable star. ▸ Two stars that lie very close to each other along an observer's line of sight but that are not associated with each other in a gravitational system are known as **optical binaries.** Although they appear close to each other in the sky, such stars are actually very distant from each other in space. See also **multiple star, variable star.**

bind (bīnd) To combine with, form a bond with, or be taken up by a chemical or chemical structure. An enzyme, for example, is structured in such a way as to be able to bind with its substrate.

binocular (bə-nŏk′yə-lər) *Adjective.* **1.** Relating to or involving both eyes at once, as in binocular vision. —*Noun.* **2.** An optical device, such as a pair of field glasses, consisting of two small telescopes, designed for use by both eyes at once. Often used in the plural as *binoculars.*

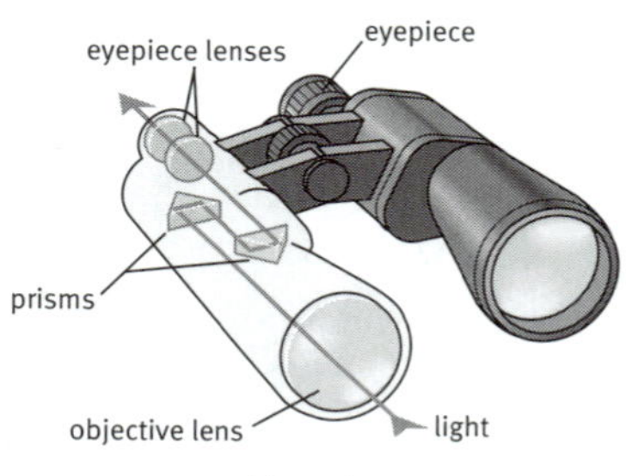

binocular

binocular vision Vision that incorporates images from two eyes simultaneously. The slight differences between the two images—seen from slightly different positions—make it possible to perceive distances between objects in what is known as depth perception. Also called *stereoscopic vision.*

binomial (bī-nō′mē-əl) A mathematical expression that is the sum of two monomials, such as $3a + 2b$.

binomial distribution The frequency distribution of the probability of a specified number of successes in an arbitrary number of repeated independent Bernoulli trials. Also called *Bernoulli distribution.*

binomial nomenclature The scientific system of naming an organism using two terms, the first being the genus and the second the species. The terms are usually Greek or Latin. For example, the scientific name of the narrow-leaf firethorn is *Pyracantha angustifolia,* where the genus is Greek for "fire thorn" and the species is Latin for "having narrow leaves."

binomial theorem *Mathematics* The theorem that specifies the expansion of any power of a binomial, that is, $(a + b)^m$. According to the binomial theorem, the first term of the expansion is x^m, the second term is $mx^{m-1}y$, and for each additional term the power of x decreases by 1 while the power of y increases by 1, until the last term y^m is reached. The coefficient of x^{m-r} is $m![r!(m-r)!]$. Thus the expansion of $(a + b)^3$ is $a^3 + 3a^2b + 3ab^2 + b^3$.

bioaccumulation (bī′ō-ə-kyo͞om′yə-lā′shən) The accumulation of a substance, such as a toxic chemical, in various tissues of a living organism. Bioaccumulation takes place within an organism when the rate of intake of a substance is greater than the rate of excretion or metabolic transformation of that substance. Compare **biomagnification.**

bioactive (bī′ō-ăk′tĭv) Relating to a substance that has an effect on living tissue.

bioassay (bī′ō-ăs′ā′, -ă-sā′) **1.** Determination of the relative purity of a substance, such as a drug or hormone, by comparing its effects with those of a standard preparation on a culture of living cells or a test organism. **2.** A test used to determine such purity.

biocenology (bī′ō-sə-nŏl′ə-jē) The study of ecological communities and of interactions among their members.

biocenosis (bī′ō-sĭ-nō′sĭs) Plural **biocenoses** (bī′ō-sĭ-nō′sēz). A group of interacting organisms that live in a particular habitat and form a self-regulating ecological community.

biochemical oxygen demand (bī′ō-kĕm′ĭ-kəl) The amount of oxygen required by aerobic microorganisms to decompose the organic matter in a sample of water, such as one polluted by sewage. It is used as a measure of the degree of water pollution.

biochemistry (bī′ō-kĕm′ĭ-strē) The scientific study of the chemical composition of living matter and of the chemical processes that go on in living organisms.

biocide (bī′ə-sīd′) A chemical agent, such as a pesticide or herbicide, that is capable of destroying living organisms.

biocontrol (bī′ō-kən-trōl′) See **biological control.**

bioconversion (bī′ō-kən-vûr′zhən) The conversion of organic materials, such as plant or animal waste, into usable products or energy sources by biological processes or agents, such as certain microorganisms.

biodegradable (bī′ō-dĭ-grā′də-bəl) Capable of being decomposed by the action of biological agents, especially bacteria.

biodiesel (bī′ō-dē′zəl, -səl) See under **biofuel.**

biodiversity (bī′ō-dĭ-vûr′sĭ-tē) The number, variety, and genetic variation of different organisms found within a specified geographic region.

bioenergetics (bī′ō-ĕn′ər-jĕt′ĭks) The scientific study of the flow and transformation of energy in and between living organisms and between living organisms and their environment.

bioengineering (bī′ō-ĕn′jə-nîr′ĭng) See **biomedical engineering.**

bioethanol (bī′ō-ĕth′ə-nôl′) See under **biofuel.**

bioethics (bī′ō-ĕth′ĭks) The study of the ethical and moral implications of medical research and practice.

biofeedback (bī′ō-fēd′băk′) The technique of using monitoring devices to obtain information about an involuntary function of the central or autonomic nervous system, such as body temperature or blood pressure, in order to gain some voluntary control over the function. Using biofeedback, individuals can be trained to respond to abnormal measurements in involuntary function with specific therapeutic actions, such as muscle relaxation, meditation, or changing breathing patterns. Biofeedback has been used to treat medical conditions such as hypertension and chronic anxiety.

biofilm (bī′ō-fĭlm′) A complex structure adhering to surfaces that are regularly in contact with water, consisting of colonies of bacteria and usually other microorganisms such as yeasts, fungi, and protozoa that secrete a mucilaginous protective coating in which they are encased. Biofilms can form on solid or liquid surfaces as well as on soft tissue in living organisms, and are typically resistant to conventional methods of disinfection. Dental plaque, the slimy coating that fouls pipes and tanks, and algal mats on bodies of water are examples of biofilms. While biofilms are generally pathogenic in the body, causing such diseases as cystic fibrosis and otitis media, they can be used beneficially in treating sewage, industrial waste, and contaminated soil.

bioflavonoid (bī′ō-flā′və-noid′) See **flavonoid.**

biofouling (bī′ō-fou′lĭng) The impairment or degradation of underwater surfaces or equipment as a result of the growth of living organisms. Organisms such as bacteria, protozoans, algae, and crustaceans can accumulate in large numbers on surfaces like pipes, tanks, and ships' hulls, resulting in corrosion, clogging, contamination, or a decrease in the efficiency of moving parts.

biofuel (bī′ō-fyo͞o′əl) Fuel produced from renewable resources, especially plant biomass, vegetable oils, and treated municipal and industrial wastes. Biofuels are considered neutral with respect to the emission of carbon dioxide because the carbon dioxide given off by burning them is balanced by the carbon dioxide absorbed by the plants that are grown to produce them. The use of biofuels as an additive to petroleum-based fuels can also result in cleaner burning with less emission of carbon monoxide and particulates. ▸ Ethanol produced by fermenting the sugars in biomass materials such as corn and agricultural residues is known as **bioethanol**. Bioethanol is used in internal-combustion engines either in pure form or more often as a gasoline additive. ▸ **Biodiesel** is made by processing vegetable oils and other fats and is also used either in pure form or as an additive to petroleum-based diesel fuel. ▸ **Biogas** is a mixture of methane and carbon dioxide produced by the anaerobic decomposition of organic matter such as sewage and municipal wastes by bacteria. It is used especially in the generation of hot water and electricity.

biogas (bī′ō-găs′) See under **biofuel.**

biogenesis (bī′ō-jĕn′ĭ-sĭs) Generation of living organisms from other living organisms.

biogeochemical cycle (bī′ō-jē′ō-kĕm′ĭ-kəl) The flow of chemical elements and compounds between living organisms and the physical environment. Chemicals absorbed or ingested by organisms are passed through the food chain and returned to the soil, air, and water by such mechanisms as respiration, excretion, and decomposition. As an element moves through this cycle, it often forms compounds with other elements as a result of metabolic processes in living tissues and of natural reactions in the atmosphere, hydrosphere, or lithosphere. See more at **carbon cycle, nitrogen cycle.**

biogeochemistry (bī′ō-jē′ō-kĕm′ĭ-strē) The scientific study of the relationship between the geochemistry of a region and the animal and plant life in that region.

biogeography (bī′ō-jē-ŏg′rə-fē) The scientific study of the geographic distribution of plant and animal life. Factors affecting distribution include the geologic history of a region, its climate and soil composition, and the presence or absence of natural barriers like deserts, oceans, and mountains. Biotic factors such as interactions among competing species, coevolutionary influences, and the reproductive and nutritional requirements of populations and species are also studied. ► A **biogeographic region** is a large, generally continuous division of the Earth's surface having a distinctive biotic community. Biogeographic regions are usually defined separately for floral and faunal communities and are largely restricted to the terrestrial areas of the Earth.

biohazard (bī′ō-hăz′ərd) A biological agent, such as an infectious microorganism, that constitutes a threat to humans or to the environment, especially one produced in biological research or experimentation.

bioinformatics (bī′ō-ĭn′fər-măt′ĭks) *Used with a singular verb.* Information technology as applied to the life sciences, especially the technology used for the collection and analysis of genomic data.

biological clock (bī′ə-lŏj′ĭ-kəl) An internal system that controls an organism's circadian rhythms, the cycles of behavior that occur regularly in a day. In mammals, the biological clock is located near the point in the brain where the two optic nerves cross. In many birds, the biological clock is located in the pineal gland. In protists and fungi, the individual cells themselves regulate circadian rhythms.

biological control Control of pests by disrupting their ecological status, as through the use of organisms that are natural predators, parasites, or pathogens. Examples of biocontrol include the use of ladybugs to prey on aphids and scale insects and the treatment of turf with spores of the bacterium *Bacillus popilliae,* which cause milky disease in Japanese beetle larvae.

biology (bī-ŏl′ə-jē) The scientific study of life and of living organisms. Botany, zoology, and ecology are all branches of biology.

bioluminescence (bī′ō-lo͞o′mə-nĕs′əns) The emission of light by living organisms, such as fireflies, glowworms, and certain fish, jellyfish, plankton, fungi, and bacteria. It occurs when a pigment (usually luciferin) is oxidized without giving off heat. Although it is believed that bioluminescence is involved in animal communication, its function in many organisms has yet to be understood. Bioluminescence is a form of chemiluminescence. Compare **chemiluminescence.**

biomagnification (bī′ō-măg′nə-fĭ-kā′shən) The increasing concentration of a substance, such as a toxic chemical, in the tissues of organisms at successively higher levels in a food chain. As a result of biomagnification, organisms at the top of the food chain generally suffer greater harm from a persistent toxin or pollutant than those at lower levels. Compare **bioaccumulation.**

biomass (bī′ō-măs′) **1.** The total amount of living material in a given habitat, population, or sample. Specific measures of biomass are generally expressed in dry weight (after removal of all water from the sample) per unit area of land or unit volume of water. **2.** Renewable organic materials, such as wood, agricultural crops or wastes, and municipal wastes, especially when used as a source of fuel or energy. Biomass can be burned directly or processed into biofuels such as ethanol and methane. See more at **biofuel.**

A CLOSER LOOK **biomass**

When biologist J.B.S. Haldane was once asked if the study of life on Earth gave him any insights into God, he replied jokingly that his research revealed that God must have "an inordinate fondness for beetles." Haldane's comment is based on the fact that there are more beetle species—almost 400,000 now known—than any other animal species. Beetles are just a fragment of the Earth's *biomass,* the matter that makes up the Earth's living organisms. Insects alone—which comprise almost one million known species and perhaps millions yet to be discovered—create an amazing amount of biomass. The number of individual insects is about 10 quintillion (10,000,000,000,000,000,000). Insects probably have more biomass than any other type of land animal. In comparison, if the weight of the Earth's human population were added up, the biomass of the insect population would be 300 times as great. Biomass also refers to the organic material on Earth that has stored sunlight in the form of chemical energy. Biomass fuels, including wood, wood waste, straw, manure, sugar cane, and many other byproducts from a variety of agricultural processes, continue to be a major source of energy in much of the developing world. There are many who advocate the use of biomass for energy as it is readily available, whereas fossil fuels, such as petroleum, coal, or natural gas, take millions of years to form in the Earth and are finite and subject to depletion as they are consumed.

---

biome (bī′ōm′) A large community of plants and animals that occupies a distinct region. Terrestrial biomes, typically defined by their climate and dominant vegetation, include grassland, tundra, desert, tropical rainforest, and deciduous and coniferous forests. There are two basic aquatic biomes, freshwater and marine, which are sometimes further broken down into categories such as lakes and rivers or pelagic, benthic, and intertidal zones.

biomechanics (bī′ō-mĭ-kăn′ĭks) The scientific study of the role of mechanics in biological systems. The study of biomechanics includes the analysis of motion in animals, the fluid dynamics of blood, and the role of mechanical processes in the development of disease.

biomedical engineering (bī′ō-mĕd′ĭ-kəl) The application of engineering techniques to the understanding of biological systems and to the development of therapeutic technologies and devices. Kidney dialysis, pacemakers, synthetic skin, artificial joints, and protheses are some products of biomedical engineering. Also called *bioengineering*.

biometeorology (bī′ō-mē′tē-ə-rŏl′ə-jē) The study of the relationship between atmospheric conditions, such as temperature and humidity, and living organisms. Biometeorology encompasses several areas of study, including terrestrial and aquatic biology, mortality, urban design, and architecture.

bionics (bī-ŏn′ĭks) The use of a system or design found in nature, such as the ability of plants to store solar energy or the aerodynamic design of bird wings, as a model for designing machines and other artificial systems.

bionomics (bī′ə-nŏm′ĭks) See **ecology.**

biophysics (bī′ō-fĭz′ĭks) The scientific study of biological processes in terms of the laws of physics. Phenomena such as echolocation in bats and the stresses and strains in skeletal and muscular structures are analyzed and explained in biophysics.

biopsy (bī′ŏp′sē) A sample of tissue removed from a living body by a medical provider for diagnostic purposes.

bioreactor (bī′ō-rē-ăk′tər) An apparatus, such as a large fermentation chamber, for growing organisms such as bacteria or yeast under controlled conditions. Bioreactors are used in the biotechnological production of substances such as pharmaceuticals, antibodies, or vaccines, or for the bioconversion of organic waste.

bioregion (bī′ō-rē′jən) An area constituting a natural ecological community with characteristic flora, fauna, and environmental conditions and bounded by natural rather than artificial borders.

bioremediation (bī′ō-rĭ-mē′dē-ā′shən) The use of biological agents, such as bacteria, fungi, or green plants, to remove or neutralize contaminants, as in polluted soil or water. Bacteria and fungi generally work by breaking down contaminants such as petroleum into less harmful substances. Plants can be used to aerate polluted soil and stimulate microbial action. They can also absorb contaminants such as salts and metals into their tissues, which are then harvested and disposed of. ► The use of green plants to decontaminate polluted soil or water is called **phytoremediation**.

biorhythm (bī′ō-rĭ*th*′əm) A recurring biological process, such as sleep, that is controlled by the circadian rhythms of an organism. See more at **circadian rhythm.**

biosolids (bī′ō-sŏl′ĭdz) Solid or semisolid organic material obtained from treated wastewater, often used as a fertilizer or soil amendment.

biosphere (bī′ə-sfîr′) **1.** The parts of the land, sea, and atmosphere in which organisms are able to live. The biosphere is an irregularly shaped, relatively thin zone in which life is concentrated on or near the Earth's surface and throughout its waters. **2.** All the Earth's ecosystems considered as a single, self-sustaining unit.

biostratigraphy (bī′ō-strə-tĭg′rə-fē) The study and categorization of rock strata based on their fossil content and distribution. Biostratigraphic data are often considered together with radiometric and paleoenvironmental data as a means of dating rock strata. Compare **lithostratigraphy.**

biota (bī-ō′tə) The organisms of a specific region or period considered as a group.

biotechnology (bī′ō-tĕk-nŏl′ə-jē) **1.** The use of a living organism to solve an engineering problem or perform an industrial task. Using bacteria that feed on hydrocarbons to clean up an oil spill is one example of biotechnology. **2.** The use of biological substances or techniques to engineer or manufacture a product or substance, as when cells that produce antibodies are cloned in order to study their effects on cancer cells. See more at **genetic engineering.**

biotic (bī-ŏt′ĭk) **1.** Consisting of living organisms. An ecosystem is made up of a biotic community (all of the naturally occurring organisms within the system) together with the physical environment. **2.** Associated with or derived from living organisms. The biotic factors in an environment include the organisms themselves as well as such items as predation, competition for food resources, and symbiotic relationships. Compare **abiotic.**

biotic potential The maximum capacity of an individual or population to reproduce under optimal environmental conditions. Populations rarely reproduce at their biotic potential because of limiting factors such as disease, predation, and restricted food resources.

biotin (bī′ə-tĭn) A water-soluble organic acid belonging to the vitamin B complex that is important in the metabolism of carbohydrates and fatty acids. It is also a cofactor for some coenzymes that catalyze the synthesis of organic acids in the body. Biotin is found in liver, egg yolks, milk, yeast, and some vegetables. *Chemical formula:* $C_{10}H_{16}N_2O_3S$.

biotite (bī′ə-tīt′) A dark-brown or dark-green to black mica. Biotite is monoclinic and is found in igneous, metamorphic, and sedimentary rocks. *Chemical formula:* $K(Mg,Fe)_3(Al,Fe)Si_3O_{10}(OH)_2$.

biotope (bī′ə-tōp′) A usually small or well-defined area that is uniform in environmental conditions and in its distribution of animal and plant life.

bioturbation (bī′ō-tər-bā′shən) The stirring or mixing of sediment or soil by organisms, especially by burrowing, boring, or ingestion.

biotype (bī′ə-tīp′) A group of organisms having the same or nearly the same genotype, such as a particular strain of an insect species.

biped (bī′pĕd′) An animal having two feet, such as a bird or human. —*Adjective* **bipedal.**

bipinnate (bī-pĭn′āt′) Relating to compound leaves that grow opposite each other on a larger stem; twice-compound or twice-pinnate. Bipinnate leaves have a feathery appearance. The acacia, coffeetree, and silktree have bipinnate leaves.

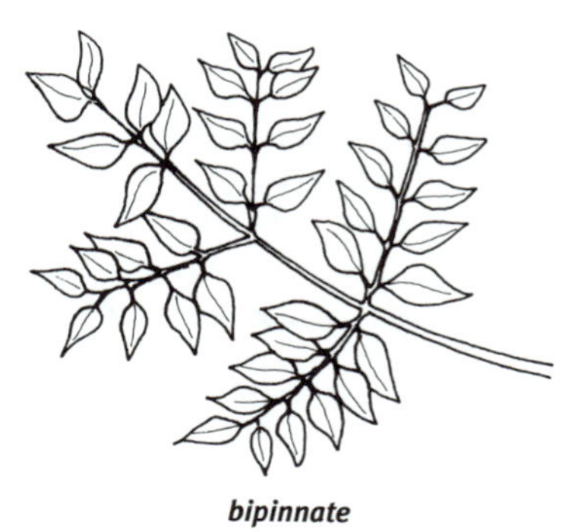

bipinnate

bipolar (bī-pō′lər) **1.** Relating to or having two poles or charges. **2.** Relating to a semiconductor device, such as a transistor, that exploits the electrical characteristics of contact between two substances, one with an inherent positive charge, the other with an inherent negative charge. **3.** Relating to or involving both of the Earth's polar regions. **4.** Relating to a neuron that has two processes or extremities. **5.** Relating to bipolar disorder.

bipolar disorder Any of several mood disorders usually characterized by periods of depression that alternate with mania.

bipolar transistor A type of transistor made of three layers of semiconductors. Each layer has been treated so that the layer in the middle (called the base) has an inherent electric charge, while the layers around it (the emitter and collector) have an inherent opposite charge. A bipolar transistor with a negative base is designated PNP, and one with a positive base is designated NPN. When subjected to current flow, the base acts like a gate, enhancing or inhibiting the current flow from the emitter to the collector. Bipolar transistors are used primarily to amplify or switch current signals, and are common in audio equipment. Compare **field effect transistor.**

bird (bûrd) Any of numerous warm-blooded, egg-laying vertebrate animals of the class Aves. Birds have wings for forelimbs, a body covered with feathers, a hard bill covering the jaw, and a four-chambered heart.

A CLOSER LOOK **birds and dinosaurs**

It is generally believed that birds are descended from dinosaurs and probably evolved from them during the Jurassic Period. While most paleontologists believe that birds evolved from a small dinosaur called the theropod, which in turn evolved from the thecodont, a reptile from the Triassic Period, other paleontologists believe that birds and dinosaurs both evolved from the thecodont. There are some who even consider the bird to be an actual dinosaur. According to this view, the bird is an avian dinosaur, and the older dinosaur a nonavian dinosaur. Although there are variations of thought on the exact evolution of birds, the similarities between birds and dinosaurs are striking and undeniable. Small meat-eating dinosaurs and primitive birds share about twenty characteristics that neither group shares with any other kind of animal; these include tubular bones, the position of the pelvis, the shape of the shoulder blades, a wishbone-shaped collarbone, and the structure of the eggs. Dinosaurs had scales, and birds have modified scales—their feathers—and scaly feet. Some dinosaurs also may have had feathers; a recently discovered fossil of a small dinosaur indicates that it had a featherlike covering. In fact, some primitive fossil birds and small meat-eating dinosaurs are so similar that it is difficult to tell them apart based on their skeletons alone.

---

birefringence (bī′rĭ-frĭn′jəns) **1.** The property or capacity of splitting a beam of light into two beams, each refracted at a different angle, and each polarized at a right angle to the other. Certain crystals such as calcite and quartz have this property. **2.** The difference in the index of refraction between two beams passing through a substance that has this property. See also **refraction, index of refraction.**

birth (bûrth) *Noun.* **1.** The emergence and separation of offspring from the body of its mother, seen in all mammals except monotremes. *—Adjective.* **2.** Present at birth, as a defect in a bodily structure.

birth control Planned interference with conception in order to control the number of offspring born. Birth control techniques include drugs containing hormones, the diaphragm, and the intrauterine device.

birth rate The ratio of total live births to total population in a specified community or area over a specified period of time. The birth rate is often expressed as the number of live births per 1,000 of the population per year.

bisect (bī′sĕkt′, bī-sĕkt′) To cut or divide into two parts, especially two equal parts.

Bishop (bĭsh′əp), **(John) Michael** Born 1936. American molecular biologist who, working with Harold Varmus, discovered oncogenes. For this work, Bishop and Varmus shared the 1989 Nobel Prize for physiology or medicine.

bismuth (bĭz′məth) *Symbol* **Bi** A brittle, pinkish-white, crystalline metallic element that occurs in nature as a free metal and in various ores. Bismuth is the most strongly diamagnetic element and has the highest atomic number of all stable elements. It is used to make low-melting alloys for fire-safety devices. Atomic number 83; atomic weight 208.98; melting point 271.3°C; boiling point 1,560°C; specific gravity 9.747; valence 3, 5. See **Periodic Table.**

bit (bĭt) The smallest unit of computer memory. A bit holds one of two possible values, either of the binary digits 0 or 1. The term comes from the phrase *binary digit*. See Note at **byte.**

bitmap (bĭt′măp′) *Computer Science.* A set of bits that represents a graphic image. Each bit or group of bits corresponds to a pixel in the image. Optical scanners and fax machines convert text or pictures into bitmaps.

bit stream or **bitstream** (bĭt′strēm′) The transmission of binary digits as a simple, unstructured sequence of bits.

bitumen (bĭ-to͞o′mən) Any of various flammable mixtures of hydrocarbons and other substances found in asphalt and tar. Bitumens occur naturally or are produced from petroleum and coal.

bituminous coal (bĭ-to͞o′mə-nəs) A soft type of coal that burns with a smoky, yellow flame. Bituminous coal is the most abundant form of coal. It has a high sulfur content, and when burned, gives off sulfurous compounds that contribute to air pollution and acid rain. Compare **anthracite, lignite.**

bivalent (bī-vā′lənt) *Chemistry.* Having a valence of 2.

bivalve (bī′vălv′) Any of various mollusks of the class Bivalvia, having a shell consisting of two halves hinged together. Clams, oysters, scallops, and mussels are bivalves. The class Bivalvia is also called Pelecypoda, and was formerly called Lamellibranchia. Compare **univalve.**

Bk The symbol for **berkelium.**

Black (blăk), Sir **James Whyte** Born 1924. British pharmacologist who discovered the first beta-blocker, which led to the development of safer and more effective drugs to treat high blood pressure and heart disease. Black also developed a blocker for gastric acid production that revolutionized the treatment of stomach ulcers. He shared with Gertrude Elion and George Hitchings the 1988 Nobel Prize for physiology or medicine.

Black, Joseph 1728–1799. British chemist who in 1756 discovered carbon dioxide, which he called "fixed air." In addition to further studies of carbon dioxide, Black formulated the concepts of latent heat and heat capacity.

blackbody (blăk′bŏd′ē) A theoretically perfect absorber and emitter of every frequency of electromagnetic radiation. The radiation emitted by a blackbody is a function only of its temperature.

blackbody radiation The electromagnetic radiation that a perfect blackbody would give off at a given temperature. A warm blackbody would emit radiation with a higher average frequency than a cooler one.

Black Death A widespread epidemic of bubonic plague that occurred in several outbreaks between 1347 and 1400. It originated in Asia and then swept through Europe, where it killed about a third of the population. See more at **bubonic plague.**

black dwarf The theoretical celestial object that remains after a white dwarf has used up all of its fuel and cooled off completely to a solid mass of extremely dense, cold carbon. A white dwarf will eventually become a black dwarf unless it has a companion star from which it can take sufficient mass to pass the **Chandrasekhar limit** and collapse into a **neutron star** or **black hole**. No black dwarf has ever been observed. Because the estimated cooling time for a white dwarf is in the trillions of years, it is unlikely that there are many, if any, black dwarfs in our universe, which is only 12 to 18 billion years old. See Note at **dwarf star.**

black hole An extremely dense celestial object whose gravitational field is so strong that not even light can escape from its vicinity. Black holes are believed to form in the aftermath of a supernova with the collapse of the star's core. See also **event horizon.** See more at **star.**

black light Invisible ultraviolet radiation. Black light causes certain fluorescent materials to emit visible light.

black lung An occupational disease of coal miners caused by the long-term inhalation of coal dust, characterized by the presence of coal around the bronchioles on x-ray images. Most patients develop no symptoms, but a few advance to progressive obstruction of the airways and destruction of lung tissue with accompanying shortness of breath and eventual respiratory disability.

black smoker See under **hydrothermal vent.**

blackwater (blăk′wô′tər) Wastewater containing bodily or other biological wastes, as from toilets, dishwashers, or kitchen drains. Compare **graywater.**

A CLOSER LOOK **black hole**

When a very massive star ends its life in a supernova explosion, the remaining matter collapses in upon itself. If there is enough mass in this collapsed star, it becomes a *black hole.* A black hole is so dense that its gravitational forces are strong enough to prevent anything that comes close enough to the region known as the *event horizon* from escaping. Even light cannot escape, since the *escape velocity* (the velocity needed for an object to escape some larger object's gravitational field) necessary to escape a black hole is greater than the speed of light. Black holes are extremely dense: for the Sun, which has a diameter of about 1,390,000 kilometers (862,000 miles), to be as dense as a black hole, its entire mass would have to be squeezed down to a ball fewer than 3 kilometers (5 miles) across. Some theorists postulate that the material in a black hole may be compressed to a single point of infinite density called a *singularity.* Because astronomers cannot directly observe a black hole, they infer its existence from the effects of its gravitational pull. For example, when a black hole results from the collapse of one star in a binary star system, it attracts material from the remaining star. This material forms an *accretion disk,* which compresses and heats up until it emits detectable x-rays. Black holes are thought to reside in the centers of many galaxies, including our own Milky Way.

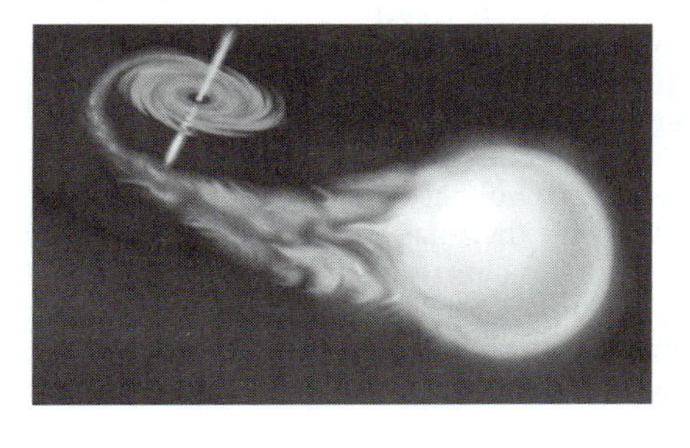

Blackwell (blăck′wĕl′), **Elizabeth** 1821–1910. British-born American physician who was the first woman doctor in the United States. In 1851 she founded an infirmary for women and children in New York City that her sister **Emily Blackwell** (1826–1910), also a physician, directed. Emily Blackwell was the first woman doctor to perform major surgeries on a regular basis.

Elizabeth Blackwell

bladder (blăd′ər) **1.** A sac-shaped muscular organ that stores the urine secreted by the kidneys, found in all vertebrates except birds and the monotremes. In mammals, urine is carried from the kidneys to the bladder by the ureters and is later discharged from the body through the urethra. **2.** An air bladder.

blade (blād) **1a.** The expanded part of a leaf or petal. Also called *lamina.* See more at **leaf. b.** The leaf of grasses and similar plants. **2.** A stone tool consisting of a slender, sharp-edged, unserrated flake that is at least twice as long as it is wide. Blade tools were developed late in the stone tool tradition, after core and flake tools, and were probably used especially as knives.

blastocoel or **blastocoele** (blăs′tə-sēl′) The fluid-filled, central cavity of a blastula.

blastocyst (blăs′tə-sĭst′) The modified blastula that is characteristic of placental mammals. It has an outer layer, known as a **trophoblast**, that participates in the development of the placenta. The inner layer of cells develops into the embryo.

blastomere (blăs′tə-mîr′) Any of the cells resulting from the cleavage of a fertilized ovum. In the initial stage of development, the blastomeres adhere to each other in a mass called a **morula**.

blastomycete (blăs′tə-mī′sēt) Any of various yeastlike deuteromycetes of the class Blastomycetes that are pathogenic in humans and animals.

blastula (blăs′chə-lə) Plural **blastulas** or **blastulae** (blăs′chə-lē′). An animal embryo at the stage immediately following the division of the fertilized egg cell, consisting of a ball-shaped layer of cells around a fluid-filled cavity known as a blastocoel. Compare **gastrula.** See also **blastocyst.**

blazar (blā′zär′) An extremely bright, starlike object characterized by rapid changes in luminosity and a flat spectrum. Originally thought to be ordinary irregular **variable stars**, their spectral properties now lead astronomers to consider blazars as a class of **active galactic nuclei**. Blazars emit radiation over a very wide range of frequencies, from radio to gamma rays, with their jets pointed at the observer. This orientation accounts for their peculiar properties, specifically the variability and intensity of their brightness, and it also distinguishes blazars from another class of active galactic nucleus, **quasars.**

bleach (blēch) A chemical agent used to whiten or remove color from textiles, paper, food, and other substances and materials. Chlorine, sodium hypochlorite, and hydrogen peroxide are bleaches. Bleaches remove color by oxidation or reduction.

blight (blīt) **1.** Any of numerous plant diseases that cause leaves, stems, fruits, and tissues to wither and die. Rust, mildew, and smut are blights. **2.** The bacterium, fungus, or virus that causes such a disease.

blindness (blīnd′nĭs) A lack or impairment of vision in which maximal visual acuity after correction by refractive lenses is one-tenth normal vision or less in the better eye. Blindness can be genetic but is usually acquired as a result of injury, cataracts, or diseases such as glaucoma or diabetes. In Asia and Africa, trachoma is a common infectious cause of blindness.

blind spot (blīnd) The small region of the retina where fibers of the optic nerve emerge from the eyeball. The blind spot has no rods or cones, so no light or visual image can be transmitted.

blizzard (blĭz′ərd) A violent snowstorm with winds blowing at a minimum speed of 56 km (35 mi) per hour and visibility of less 400 m (0.25 mi) for three hours.

BL Lac object (bē′ĕl′ läk′) A distant radio galaxy with an intensely bright active galactic nucleus that emits a jet of material directly toward the Earth. BL Lac objects may be a class of blazar, though there is evidence that they are an optical effect caused by gravitational lensing of some other bright object by an intervening galaxy. Also known as *BL Lacertae objects.*

block and tackle (blŏk) An arrangement of pulleys and ropes used to reduce the amount of force needed to move heavy loads. One pulley is attached to the load, and rope or chains connect this pulley to a fixed pulley. Each pulley may have multiple grooves or wheels for the rope to pass over numerous times. Pulling the rope or chain slowly draws the load-bearing pulley toward the fixed one with high **mechanical advantage**.

blood (blŭd) **1.** The fluid tissue that circulates through the body of a vertebrate animal by the pumping action of the heart. Blood is the transport medium by which oxygen and nutrients are carried to body cells and waste products are picked up for excretion. Blood consists of plasma in which red blood cells, white blood cells, and platelets are suspended. **2.** A fluid that is similar in function in many invertebrate animals.

blood-brain barrier A physiological mechanism that alters the permeability of capillaries in the brain, so that some substances, such as certain drugs, are prevented from entering brain tissue, while other substances are allowed to enter freely.

blood cell A red blood cell or white blood cell that is contained in the blood.

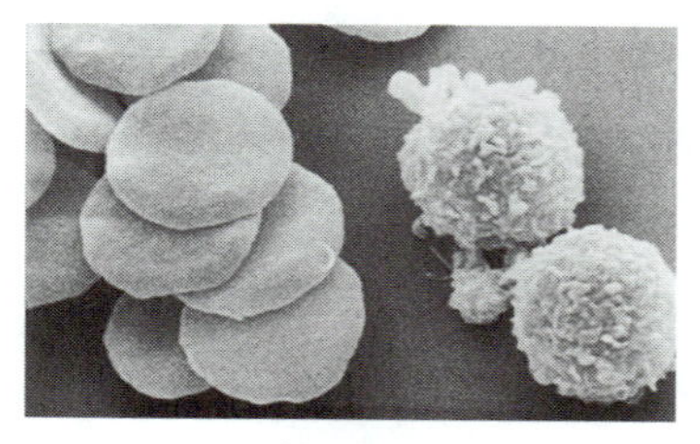

blood cell

photomicrograph of red (left) *and white blood cells*

blood clot 1. See **embolism.** 2. See **thrombus.**

blood count See **complete blood count.**

blood gas Any of the dissolved gases in blood plasma, especially oxygen and carbon dioxide.

blood group See **blood type.**

blood plasma The clear, liquid part of the blood, composed mainly of water and proteins, in which the blood cells are suspended. The blood plasma of mammals also contains platelets.

blood poisoning See **toxemia.**

blood pressure The pressure of the blood in the vessels, especially the arteries, as it circulates through the body. Blood pressure varies with the strength of the heartbeat, the volume of blood being pumped, and the elasticity of the blood vessels. Arterial blood pressure is usually measured by means of a sphygmomanometer and reported in millimeters of mercury as a fraction, with the numerator equal to the blood pressure during systole and the denominator equal to the blood pressure during diastole. See more at **hypertension, hypotension.**

blood serum Blood plasma from which the protein fibrinogen, which causes clotting of the blood, has been removed.

blood sugar Sugar in the form of glucose in the blood. The concentration of blood sugar is measured in milligrams of glucose per 100 milliliters of blood.

blood type Any of the four main types into which human blood is divided: A, B, AB, and O. Blood types are based on the presence or absence of specific antigens on red blood cells. Also called *blood group.*

A CLOSER LOOK **blood type**

Blood transfusions used to be the treatment of last resort, since they often caused death. But in the 1890s Austrian scientist Karl Landsteiner began to solve the transfusion puzzle when he found that all human red blood cells belonged to one of four groups which he named A, B, AB, and O. The types refer to *antigens* found on the surface of these cells. Antibodies circulating in a person's blood normally recognize the antigens in that same person's blood cells and don't react with them. However, if a person with one blood type is transfused with blood of another type, the antibodies bind to the foreign antigens, causing dangerous clumping of the blood. Thus the key to a successful transfusion is to give a person blood that has matching antigens. In the first half of the twentieth century, the study of Rhesus monkeys, which share many biological characteristics with humans, gave rise to the recognition of a human blood protein called the *Rh factor.* People who have this blood protein are considered Rh positive, while individuals who lack the protein are referred to as Rh negative. The Rh factor is connected to an individual's blood type. If a person has type AB blood with a positive Rh factor, his or her blood type is referred to as AB positive. The Rh factor causes a problem in a fetus whose blood is Rh positive and whose mother is Rh negative because the mother's negative blood attacks the positive blood of the fetus. In this instance, a blood transfusion to the fetus can save its life.

---

blood vessel An elastic tube or passage in the body through which blood circulates; an artery, a vein, or a capillary.

blotch (blŏch) Any of several plant diseases caused by fungi and resulting in brown or black dead areas on leaves or fruit.

blubber (blŭb′ər) The thick layer of fat between the skin and the muscle layers of whales and other marine mammals. It insulates the animal from heat loss and serves as a food reserve.

blue-green alga (blo͞o′grēn′) See under **cyanobacterium.**

blue jet A conical discharge of blue light that starts from the top of active thunderstorm clouds and proceeds upward. Blue jets can last up to several hundred milliseconds and are believed to connect the top of a thundercloud with the ionosphere.

blueschist (blo͞o′shĭst′) A bluish schist that gets its color from the presence of a sodic amphibolite, glaucophane, or a variety of glaucophane called crossite. Blueschist forms under conditions of high pressure but low temperature, as at the boundaries of converging tectonic plates.

blue shift See under **Doppler effect.**

blue vitriol A blue, crystalline compound of copper sulfate. Blue vitriol is soluble in water, and is a very important industrial salt of copper. It is used in insecticides and germicides, in electrolytes for batteries, and in electroplating baths. *Chemical formula:* $CuSO_4 \cdot 5H_2O$.

B lymphocyte also **B-lymphocyte** (bē′lĭm′fə-sīt′) See **B cell.**

Boas (bō′ăz), **Franz** 1858–1942 German-born American anthropologist who emphasized the systematic analysis of culture and language structures.

body mass index (bŏd′ē) A measurement of the relative percentages of fat and muscle mass in the human body, in which weight in kilograms is divided by height in meters squared. The result is used as an index of obesity.

body wave A seismic wave that travels through the Earth rather than across its surface. Body waves usually have smaller amplitudes and shorter wavelengths than surface waves and travel at higher speeds. Primary waves and secondary waves are body waves. Compare **surface wave.**

bog (bôg) An area of wet, spongy ground consisting mainly of decayed or decaying peat moss (sphagnum) and other vegetation. Bogs form as the dead vegetation sinks to the bottom of a lake or pond, where it decays slowly to form peat. Peat bogs are important to global ecology, since the undecayed peat moss stores large amounts of carbon that would otherwise be released back into the atmosphere. Global warming may accelerate decay in peat bogs and release more carbon dioxide, which in turn may cause further warming.

Niels Bohr

Bohr (bôr), **Niels Henrik David** 1885–1962. Danish physicist who investigated atomic structure and radiation. Bohr discovered that electrons orbit the nucleus of an atom at set distances, changing levels only when energy is lost or gained and emitting or absorbing radiation in the process. His concepts were fundamental to the later development of quantum mechanics.

BIOGRAPHY **Niels Bohr**

In 1922 Danish physicist Niels Bohr was awarded the Nobel Prize for physics for his ability to build upon the findings of Ernest Rutherford and develop a theory of atomic structure that would contribute significantly to the development of quantum mechanics. At the beginning of the twentieth century, before Bohr's discovery, scientists thought that atoms were a loosely combined mixture of electrons, protons, and neutrons. In 1911 Ernest Rutherford discovered that atoms had an extremely small, positively charged nucleus that contained no electrons, and he developed an atomic model that resembled the solar system, with negatively charged electrons orbiting a central nucleus. Rutherford's model was considered puzzling because it predicted that atoms should be unstable: since the electrons were orbiting the nucleus, they were undergoing acceleration, but accelerating electric charges give off electromagnetic energy, so the orbiting electrons should have been constantly giving off energy, and ultimately spiraling into the nucleus. But electrons did not do this. To explain the atom's apparent stability, Bohr postulated that electrons travel only in discrete orbits of different sizes and energy levels around the nucleus, and that increases or decreases in an electron's energy cause it to jump to a higher or lower orbit, absorbing or emitting energy in the form of electromagnetic radiation. Bohr's model explained why hydrogen, the simplest atom, emits and absorbs light only of certain frequencies depending on the difference in energy levels of the orbits between which the electron moves. Later in his career, Bohr developed the concept of *complementarity* to encompass wave-particle duality, the phenomenon that under some conditions light exhibits wavelike behavior and under other conditions particlelike behavior.

bohrium (bôr′ē-əm) *Symbol* **Bh** A synthetic, radioactive element that is produced by bombarding bismuth with chromium ions. Its most long-lived isotopes have mass numbers of 261, 262, and 264 with half-lives of 11.8 milliseconds, 0.1 second, and 0.44 second, respectively. Atomic number 107. See **Periodic Table.**

Bohr magneton See under **magneton.**

Bohr theory An early model of atomic structure, in which electrons circulate around the nucleus in discrete, stable orbits with different energy levels. This model was the first to predict and explain the atomic spectrum of the hydrogen atom, which arises as the electron jumps from one orbit to another orbit of lower energy, giving off electromagnetic radiation of predictable frequencies. Later models of atomic structure abandoned the idea of circular orbits, and explained the stable orbits as standing waves. See also **atomic spectrum, orbital.**

boil (boil) To change from a liquid to a gaseous state by being heated to the boiling point and being provided with sufficient energy. Boiling is an example of a phase transition.

boiling point (boi′lĭng) The temperature at which a liquid changes to a vapor or gas. This temperature stays the same until all the liquid has vaporized. As the temperature of a liquid rises, the pressure of escaping vapor also rises, and at the boiling point the pressure of the escaping vapor is equal to that exerted on the liquid by the surrounding air, causing bubbles to form. Typically boiling points are measured at sea level. At higher altitudes, where atmospheric pressure is lower, boiling points are lower. The boiling point of water at sea level is 100°C (212°F), while at the top of Mount Everest it is 71°C (159.8°F).

boll (bōl) The seed-bearing capsule of certain plants, especially cotton and flax.

bolometric magnitude (bō′lə-mĕt′rĭk) See under **magnitude.**

Boltzmann (bôlts′män′), **Ludwig** 1844–1906. Austrian physicist who developed statistical mechanics, the branch of physics that explains how the properties of atoms (such as mass and structure) determine the visible properties of matter (such as viscosity and heat conduction). Through his investigations of thermodynamics, Boltzmann developed numerous theories about the laws governing atomic motion and energy.

bond (bŏnd) A force of attraction that holds atoms or ions together in a molecule or crystal. Bonds are usually created by a transfer or sharing of one or more electrons. There are single, double, and triple bonds. See also **coordinate bond, covalent bond, ionic bond, metallic bond, polar bond.**

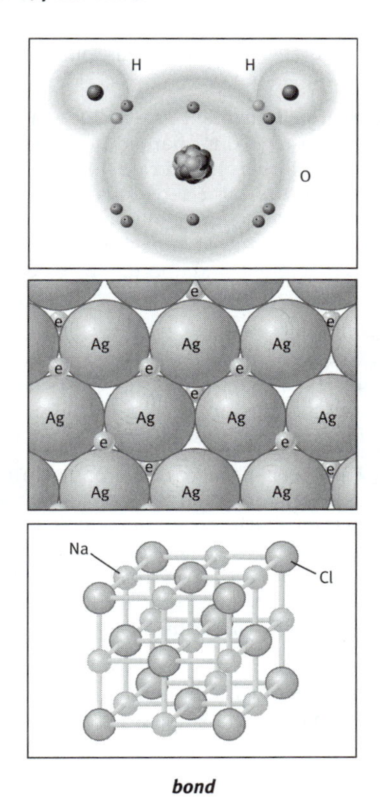

bond

top to bottom: *covalent, metallic, and ionic bonds*

bone (bōn) **1.** The hard, dense, calcified tissue that forms the skeleton of most vertebrates, consisting of a matrix made up of collagen fibers and mineral salts. There are two main types of bone structure: compact, which is solid and hard, and cancellous, which is spongy in appearance. Bone serves as a framework for the attachment of muscles and protects vital organs, such as the brain, heart, and lungs. See more at **osteoblast, osteocyte. 2.** Any of the structures made of bone that constitute a skeleton, such as the femur. The human skeleton consists of 206 bones.

bone density A measurement corresponding to the mineral density of bone, expressed as mineral mass per unit volume of bone and usually assessed by a radiologic scan of the lower spine or hip. Bone density measurements are used to diagnose and monitor osteoporosis and other bone diseases.

bone marrow The spongy, red tissue that fills the bone cavities of mammals. Bone marrow is the source of red blood cells, platelets, and most white blood cells.

bone marrow transplant A technique in which bone marrow is transplanted from one individual to another, or removed from and transplanted to the same individual, in order to stimulate production of blood cells. It is used in the treatment of malignancies, certain forms of anemia, and immunologic deficiencies.

Bonnet (bô-nĕ′), **Charles** 1720–1793. Swiss naturalist who discovered parthenogenesis when he observed that aphid eggs could develop without fertilization. Bonnet was also one of the first scientists to study photosynthesis.

bony fish (bō′nē) Any of numerous ray-finned fishes belonging to the infraclass Teleostei or Teleostomi, having a skeleton that is completely made of bone, rather than partially or completely made of cartilage. Most living species of fish are bony fish. Also called *teleost.* Compare **cartilaginous fish, jawless fish.**

book lung (bo͝ok) The breathing organ of scorpions, spiders, and some other arachnids. It consists of membranes arranged in several parallel folds like the pages of a book.

Boole (bo͞ol), **George** 1815–1864. British mathematician who wrote important works in various areas of mathematics. He developed a system of mathematical symbolism to express logical relations that is now known as Boolean algebra.

Boolean algebra (bo͞o′lē-ən) A form of symbolic logic, in which variables, which stand for propositions, have only the values "true" (or "1") and "false" (or "0"). Relationships between these values are expressed by the **Boolean operators** AND, OR, and NOT. For example, "a + b" means "a OR b", and its value is true as long as either a is true or b is true (or both). Boolean logic can be used to solve logical problems, and provides the mathematical tools fundamental to the design of digital computers. It is named after the mathematician George Boole. Also called *Boolean logic.* See also **logic gate.**

boost (bo͞ost) A linear map from one reference frame to another in which each coordinate is increased or decreased by an independent constant or linear function. A boost corresponds to a shift of the entire coordinate system without any rotation of its axes.

booster (bo͞o′stər) An additional dose of an immunizing agent, such as a vaccine or toxoid, given at a time period of weeks to years after the initial dose to sustain the immune response elicited by the first dose. Tetanus, diphtheria, and measles vaccines are commonly given in booster doses.

Boötes (bō-ō′tēz) A constellation in the Northern Hemisphere near Virgo and Corona Borealis. It contains the bright star Arcturus. Boötes (the Plowman or Herdsman) is one of the earliest recorded constellations.

borate (bôr′āt′) A salt or ester of boric acid, containing the radical BO_3.

borax (bôr′ăks′) A white, crystalline powder and mineral used as an antiseptic, as a cleansing agent, and in fusing metals and making heat-resistant glass. The mineral is an ore of boron and also occurs in yellowish, blue, or green varieties. *Chemical formula:* $\mathbf{Na_2B_4O_7 \cdot 10H_2O}$.

bore (bôr) **1.** In fluid mechanics, a jump in the level of moving water, generally propagating in the opposite direction to the current. Strong ocean tides can cause bores to propagate up rivers. **2a.** The white, shallow portion of a wave after it breaks. The bore carries ocean water onto the beach. **b.** A tidal wave caused by the surge of a flood tide upstream in a narrowing estuary or by colliding tidal currents.

boreal (bôr′ē-əl) **1.** Relating to the north or to northern areas. **2.** Relating to the north wind. **3.** Relating to the forest areas of the Northern Temperate Zone that are dominated by coniferous trees such as spruce, fir, and pine.

boric acid (bôr′ĭk) A white or colorless crystalline compound that occurs naturally or is produced artificially from borax. It is used as an antiseptic and preservative, and in cements, enamels, and cosmetics. *Chemical formula:* H_3BO_3.

boron (bôr′ŏn′) *Symbol* **B** A shiny, brittle, black metalloid element extracted chiefly from borax. It is a good electrical conductor at high temperatures and a poor conductor at low temperatures. Boron is necessary for the growth of land plants and is used in the preparation of soaps, abrasives, and hard alloys. It is also used in the control rods of nuclear reactors as a neutron absorber. Atomic number 5; atomic weight 10.811; melting point 2,300°C; sublimation point 2,550°C; specific gravity (crystal) 2.34; valence 3. See **Periodic Table.**

Bose (bōs), **Satyendra Nath** 1894–1974. Indian physicist who derived the quantum statistics of photons by assuming that photons with the same energy were indistinguishable particles. His idea inspired Albert Einstein to apply the same concept to other particles, including atoms, predicting the possibility of Bose-Einstein condensates. Particles behaving in accordance with Bose's statistics are today called bosons.

Bose-Einstein condensate A state of matter that forms at low temperatures or high densities in which the wave functions of the bosons that make up the matter overlap. The bosons all fall into the same ground level quantum state. Bose-Einstein condensates were predicted by Einstein in 1925 but not observed experimentally until 1995. Also called *superatom.* See also **state of matter.**

Bose-Einstein statistics See under **statistical mechanics.**

boson (bō′sŏn) Any of a class of elementary or composite particles, including the photon, pion, and gluon, that are not subject to the Pauli exclusion principle (that is, any two bosons can potentially be in the same **quantum state**). The value of the spin of a boson is always an integer. Mesons are bosons, as are the gauge bosons (the particles that mediate the fundamental forces). They are named after the physicist Satyendra Nath Bose. Compare **fermion.** See Note at **elementary particle.** See Table at **subatomic particle.**

bot (bŏt) A software program that imitates the behavior of a human, as by querying search engines or participating in chatroom discussions.

botany (bŏt′n-ē) **1.** The scientific study of plants, including their growth, structure, physiology, reproduction, and pathology, as well as their economic use and cultivation by humans. **2.** The plant life of a particular area.

botryoidal (bŏt′rē-oid′l) Shaped like a bunch of grapes. Certain minerals and parts of organisms can be botryoidal.

bottleneck (bŏt′l-nĕk′) An abrupt and severe reduction in the number of individuals during the history of a species, resulting in the loss of diversity from the gene pool. The generations following the bottleneck are more genetically homogenous than would otherwise be expected. Bottlenecks often occur in consequence of a catastrophic event.

bottom quark (bŏt′əm) A quark with a charge of $-\frac{1}{3}$. Its mass is smaller than that of the top quark, but larger than that of all the other quarks. The bottom quark is sometimes referred to as the *beauty quark,* though this term is falling out of use. See Table at **subatomic particle.**

botulinum toxin (bŏch′ə-lī′nəm) Any of several enzymes produced by the bacterium *Clostridium botulinum* that are extremely potent neurotoxins. Botulinum toxin interferes with the ability of neurons to release acetylcholine at nerve-muscle junctures, thereby inducing the paralysis of botulism. Botulinum toxin is resistant to enzymatic digestion in the body and is used as an antispasmodic and a treatment for wrinkles by paralyzing facial muscles. Also called *botulin.*

botulism (bŏch′ə-lĭz′əm) A severe, sometimes fatal food poisoning caused by eating food infected with the bacterium *Clostridium botulinum,* which produces **botulinum toxin.** The bacterium grows in food that has been improperly preserved.

Bouguer anomaly (bo͞o-zhâr′) The difference between the expected value of gravity at a given location (taking into account factors such as latitude, longitude, altitude, and the rotation of the Earth) and its actual value. Bouguer anomalies suggest the existence of locally dense or light regions of the Earth, where a large meteorite or an oil field might be buried beneath the surface. Bouguer anomalies can be measured in several ways depending on whether the density and shape of the terrain between the measuring point and sea level is calculated, estimated, or ignored.

Bouguer correction An approximation of the Earth's gravitational pull as a function of elevation. The additional mass beneath an object standing on the ground at a high elevation (such as on a mountain) leads to a higher amount of gravitational force. If the strength of gravity at sea level is known, then its value at higher elevations can be approximated using the Bougeur correction. Standardly, the correction equals 0.4186 ρh, where ρ is the assumed average density of the Earth, and h is the difference in altitude between the place where the strength of gravity is known and where it is being approximated. This value is added to the value of gravity at the lower location to yield an approximation of its value at the higher one. Compare **free-air correction.**

boundary condition (boun**′**də-rē) *Mathematics.* The set of conditions specified for the behavior of the solution to a set of differential equations at the boundary of its domain. Boundary conditions are important in determining the mathematical solutions to many physical problems.

bovid (bō**′**vĭd) Any of various hoofed, horned ruminant mammals of the family Bovidae, which includes cattle, sheep, goats, buffaloes, bisons, antelopes, and yaks.

bovine (bō**′**vīn′) Characteristic of or resembling cows or cattle.

bovine growth hormone A naturally occurring hormone of cattle that regulates growth and milk production. It may also be produced by genetic engineering and administered to cows to increase milk production. Also called *bovine somatotropin.*

bovine spongiform encephalopathy (spŭn**′**jĭ-fôrm′) See **mad cow disease.**

bowel (bou**′**əl) The intestine.

Bowen's reaction series (bō**′**ənz) A schematic description of the order in which minerals form during the cooling and solidification of magma and of the way the newly formed minerals react with the remaining magma to form yet another series of minerals. The series is named after American geologist Norman L. Bowen (1887–1956), who first described the scheme.

Bowman's capsule (bō**′**mənz) A cup-shaped structure around the glomerulus of each nephron of the vertebrate kidney. It serves as a filter to remove organic wastes, excess inorganic salts, and water. Bowman's capsule is named after its identifier, English physician and physiologist, Sir William Bowman (1816–1892).

Boyle (boil), **Robert** 1627–1691. English physicist and chemist who is regarded as a founder of modern chemistry. Boyle rejected the traditional theory that all matter was composed of four elements and defined an element as a substance that cannot be reduced to other, simpler substances or produced by combining simpler substances. Boyle also conducted important physics experiments with Robert Hooke that led to the development of Boyle's law.

Boyle's law (boilz) The principle that the volume of a given mass of an **ideal gas** is inversely proportional to its pressure, as long as temperature remains constant. Boyle's law is a subcase of the **ideal gas law**. Compare **Charles's law.**

B particle Either of two subatomic particles in the meson family, one neutral and one positively charged, both of which have masses 10,331 times that of an electron and average lifetimes of 1.6×10^{-12} seconds. B particles contain **bottom quarks**. See Table at **subatomic particle.**

Br The symbol for **bromine.**

brachial (brā**′**kē-əl) **1.** Relating to or involving the arm. **2.** Relating to the forelimb or wing of a vertebrate.

brachiation (brā′kē-ā**′**shən, brăk′ē-) Movement in which the suspended body swings by the arms from one hold to another, as in gibbons and arboreal primates. Adaptations used in brachiation, such as relatively long arms and a freely rotating shoulder joint, may have contributed to the development of bipedalism in protohumans.

brachiopod (brā**′**kē-ə-pŏd′) Any of various marine invertebrate animals of the phylum Brachiopoda that resemble clams. Brachiopods have paired upper and lower shells attached to a usually stationary stalk and hollow tentacles covered with cilia that sweep food particles into the mouth. Brachiopods are probably related to the phoronids and bryozoans, and were extremely abundant throughout the Paleozoic Era.

brachiosaurus (brā′kē-ə-sôr′əs) or **brachiosaur** (brā′kē-ə-sôr′). A very large sauropod dinosaur of the genus *Brachiosaurus* of the Jurassic and Cretaceous Periods. It had forelegs that were longer than its hind legs, and nostrils on top of its head.

bracket fungus (brăk′ĭt) See **shelf fungus.**

brackish (brăk′ĭsh) Containing a mixture of seawater and fresh water. Brackish water is somewhat salty.

bract (brăkt) A modified leaf growing just below a flower or flower stalk. Bracts are generally small and inconspicuous, but some are showy and petallike, as the brightly colored bracts of bougainvillaea or the white or pink bracts of flowering dogwoods.

bracteole (brăk′tē-ōl′) A small bract.

Bragg (brăg), Sir **William Henry** 1862–1942. British physicist who invented the x-ray spectrometer, a device used to measure x-ray wavelengths. With his son, the physicist Sir **William Lawrence Bragg** (1890–1971), he developed the technique of x-ray crystallography, used to determine the atomic structure of crystals. Both the father and son were awarded a joint Nobel Prize for physics in 1915 for this work.

Brahe (brä, brä′hē), **Tycho** 1546–1601. Danish astronomer who made the most accurate and extensive observations of the planets and stars before the telescope was invented. Brahe determined the position of 777 stars, demonstrated that comets follow regular paths, and observed the supernova of 1572, which became known as Tycho's star. Although Brahe did not accept the Copernican heliocentric model of the solar system, his careful observations allowed Johannes Kepler to prove that Copernicus was essentially correct.

braided stream (brā′dĭd) A stream consisting of multiple small, shallow channels that divide and recombine numerous times forming a pattern resembling the strands of a braid. Braided streams form where the sediment load is so heavy that some of the sediments are deposited as shifting islands or bars between the channels.

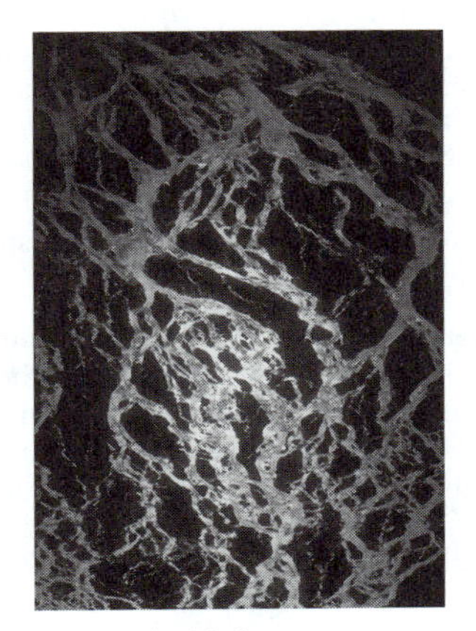

braided stream

aerial photograph of the Firth River, Yukon Territory, Canada

brain (brān) **1.** The part of the nervous system in vertebrates that is enclosed within the skull, is connected with the spinal cord, and is composed of gray matter and white matter. It is the control center of the central nervous system, receiving sensory impulses from the rest of the body and transmitting motor impulses for the regulation of voluntary movement. The brain also contains the centers of consciousness, thought, language, memory, and emotion. See more at **brainstem, cerebellum, cerebrum. 2.** A bundle of nerves in many invertebrate animals that is similar to the vertebrate brain in function and position.

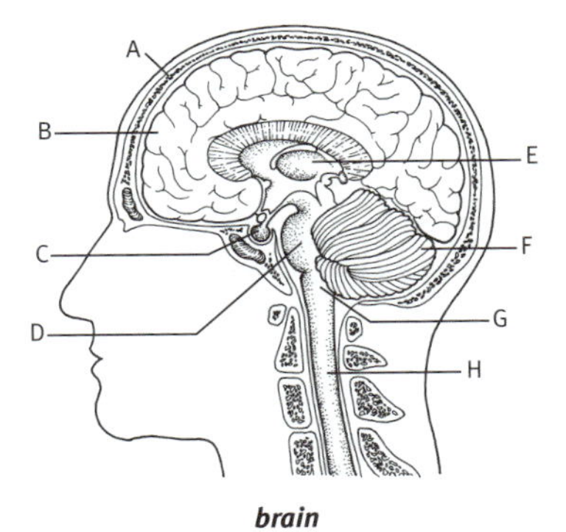

brain

A. *skull*, B. *cerebrum*, C. *pituitary gland*, D. *pons*, E. *thalamus*, F. *cerebellum*, G. *medulla oblongata*, H. *spinal cord*

brain death Permanent brain damage resulting in loss of brain function, manifested by cessation of breathing and other vital reflexes, unconsciousness with unresponsiveness to stimuli, absence of muscle activity, and a flat electroencephalogram for a predetermined length of time. Patients who are brain-dead may still exhibit normal function of the heart, lungs, and other vital organs if they are receiving artificial life support.

brainstem (brān′stĕm′) The part of the vertebrate brain located at the base of the brain and made up of the medulla oblongata, pons, and midbrain. The brainstem controls and regulates vital body functions, including respiration, heart rate, and blood pressure. See also **reticular formation.**

branchiopod (brăng′kē-ə-pŏd′) Any of various primitive aquatic crustaceans of the class Branchiopoda, such as the fairy shrimp and water flea. Branchiopods are characterized by flattened, leaflike thoracic appendages.

brass (brăs) A yellowish alloy of copper and zinc, usually 67 percent copper and 33 percent zinc. It sometimes includes small amounts of other metals. Brass is strong, ductile, and resistant to many forms of corrosion.

Brattain (brăt′n), **Walter Houser** 1902–1987. American physicist who, with John Bardeen and William Shockley, invented the transistor in 1947. For this work all three shared the 1956 Nobel Prize for physics.

BRCA (bē′är′sē′ā′) One of two genes (designated BRCA1 and BRCA2) that help repair damage to DNA, but when inherited in a defective state increase the risk of breast and ovarian cancer.

breaker (brā′kər) **1.** A wave that crests or breaks into foam, as against a shoreline. **2.** A circuit breaker.

breaker zone The nearshore zone between the outermost breakers and the bore area where wave water rushes onto the beach. Because the water in the breaker zone is shallow (usually between five and ten meters deep), most waves there are unstable. Also called **surf zone.**

breakwater (brāk′wô′tər) An offshore barrier, such as a jetty, that protects a harbor or shore from the full impact of waves.

breastbone (brĕst′bōn′) See **sternum.**

breccia (brĕch′ē-ə, brĕch′ə, brĕsh′-) A rock composed of angular fragments embedded in a fine-grained matrix. Breccias form from explosive volcanic ejections, the compaction of talus, or plate tectonic processes. Breccias differ from conglomerates in that their fragments are angular instead of rounded.

breech delivery (brēch) Delivery of a fetus with the buttocks or feet appearing first.

breed (brēd) *Verb.* **1.** To produce or reproduce by giving birth or hatching. **2.** To raise animals or plants, often to produce new or improved types. —*Noun.* **3.** A group of organisms having common ancestors and sharing certain traits that are not shared with other members of the same species. Breeds are usually produced by mating selected parents.

breeder reactor (brē′dər) A nuclear reactor that is used to create fissionable material (such as plutonium-239) by exposing nonfissionable material (such as uranium-238) to radiation. The source of the radiation is usually other fissionable material. Breeder reactors produce more fissionable material than they use up.

bridge (brĭj) A structure spanning and providing passage over a gap or barrier, such as a river or roadway.

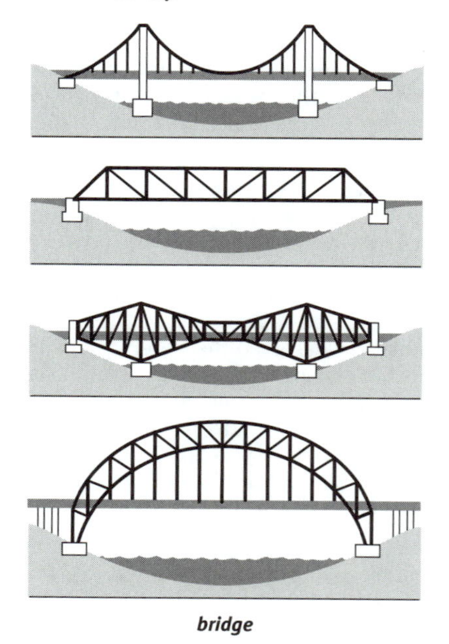

bridge

top to bottom: *suspension, through truss, cantilever, and steel arch bridge profiles*

brine (brīn) **1.** Water saturated with or containing large amounts of a salt, especially sodium chloride. The high salt content is usually due to evaporation or freezing. **2.** The water of a sea or ocean.

British thermal unit (brĭt′ĭsh) A unit used mainly to measure heat but also applied to other forms of energy. One British thermal unit is equal to the amount of heat needed to raise the temperature of one pound of water by one degree Fahrenheit, or 251.997 calories.

brittle (brĭt′l) Having a tendency to break when subject to high stress. Brittle materials have undergone very little strain when they reach their elastic limit, and tend to break at that limit. Compare **ductile.**

broad-leaved (brôd′lēvd′) also **broad-leafed** (brôd′lēft′) Having broad leaves rather than needlelike or scalelike leaves. Broad leaves are adapted to maximizing photosynthesis by capturing large amounts of sunlight. Since the gases that are exchanged with the atmosphere in photosynthesis must be dissolved in water, most broad-leaved plants grow in regions with dependable rainfall. See more at **leaf, transpiration.**

Broca's area (brō′kəz) An area located in the frontal lobe of the brain, usually in the left cerebral hemisphere. It is associated with the motor control of speech. Broca's area is named for French surgeon and anthropologist Pierre Paul Broca (1824–80), who first located the area.

Broglie (brô-glē′), Prince **Louis-Victor de** 1892–1987. French physicist who, influenced by Albert Einstein's concept that waves can behave as particles, proposed that the opposite was also true: that electrons, for example, can behave as waves. His work developed the study of wave mechanics, which was important in the development of quantum physics, and he was awarded the Nobel Prize for physics in 1929.

bromate (brō′māt′) A salt or ester containing the group BrO_3.

bromeliad (brō-mē′lē-ăd′) Any of various tropical American plants of the family Bromeliaceae, most of which are epiphytes. They usually have long stiff leaves, colorful flowers, and showy bracts. The bromeliads include the pineapple, the Spanish moss, and numerous ornamental plants.

bromide (brō′mīd′) A compound, such as potassium bromide, containing bromine and another element or radical.

bromine (brō′mēn) *Symbol* **Br** A reddish-brown volatile element of the halogen group found in compounds occurring in ocean water. The pure form is a nonmetallic liquid that gives off a highly irritating vapor. It is used to make dyes, sedatives, and photographic film. Atomic weight 79.904; atomic number 35; melting point 7.2°C; boiling point 58.78°C; specific gravity 3.12; valence 1, 3, 5, 7. See **Periodic Table.**

bronchial tube (brŏng′kē-əl) A bronchus or any of the tubes branching from a bronchus. The bronchial tubes decrease in size as they descend into the lungs.

bronchiole (brŏng′kē-ōl′) Any of the small, thin-walled tubes that branch from a bronchus and end in the alveolar sacs of the lung.

bronchiolitis (brŏng′kē-ə-lī′tĭs) An acute viral infection of the bronchioles seen mostly in infants and young children, caused most frequently by the respiratory syncytial virus.

bronchitis (brŏng-kī′tĭs) Inflammation of the mucous membrane of the bronchial tubes, often as a result of a cold or other viral infection. Smoking is also a common cause of chronic bronchitis.

bronchodilator (brŏng′kō-dī-lā′tər, -dī′lā-) A drug that widens the air passages of the lungs and eases breathing by relaxing bronchial smooth muscle.

bronchopulmonary (brŏng′kō-po͝ol′mə-nĕr′ē, -pŭl′-) Relating to or involving the bronchial tubes and the lungs.

bronchus (brŏng′kəs) Plural **bronchi** (brŏng′kī′, brŏng′kē′). Either of the two main branches of the trachea that lead to the lungs, where they divide into smaller branches.

brontosaurus (brŏn′tə-sôr′əs) or **brontosaur** (brŏn′tə-sôr′) An earlier name for **apatosaurus.**

WORD HISTORY **brontosaurus**

Take a little deception, add a little excitement, stir them with a century-long mistake, and you have the mystery of the brontosaurus. Specifically, you have the mystery of its name.

For 100 years this 70-foot-long, 30-ton vegetarian giant had two names. This case of double identity began in 1877, when bones of a large dinosaur were discovered. The creature was dubbed *apatosaurus,* a name that meant "deceptive lizard" or "unreal lizard." Two years later, bones of a larger dinosaur were found, and in all the excitement, scientists named it *brontosaurus* or "thunder lizard." This name stuck until scientists decided it was all a mistake—the two sets of bones actually belonged to the same type of dinosaur. Since it is a rule in taxonomy that the first name given to a newly discovered organism is the one that must be used, scientists have had to use the term *apatosaurus.* But "thunder lizard" had found a lot of popular appeal, and many people still prefer to call the beast *brontosaurus.*

---

bronze (brŏnz) **1.** A yellow or brown alloy of copper and tin, sometimes with small amounts of other metals such as lead or zinc. Bronze is harder than brass and is used both in industry and in art. **2.** An alloy of copper and certain metals other than tin, such as aluminum.

Bronze Age A period of human culture between the Stone Age and the Iron Age, characterized by the use of weapons and implements made of cast bronze. The beginning of the Bronze Age is generally dated before 3000 BCE in parts of Mediterranean Europe, the Middle East, and China. See Note at **Three Age system.**

brown alga (broun ăl′gə) Any of various photosynthetic protists belonging to the phylum Phaeophyta, almost all of which live in marine environments. Brown algae have chlorophylls a and c as well large quantities of the pigment fucoxanthin, which gives the group their characteristic brownish colors. This pigment absorbs the range of blue light frequencies available to brown algae when they are submerged and allows them to live in deeper waters than green algae. The brown algae store their food in a compound called laminarin and transport it throughout their bodies in a compound called mannitol (unlike plants and green algae). The cell walls of brown algae are made of cellulose and algin. Their bodies vary from small filaments to the immense leaflike thalli of **kelp.** Species of brown algae dominate shoreline and coastal ecosystems in cooler waters, and huge masses of the brown alga *Sargassum* cover the warm Sargasso sea in the Atlantic. See more at **alga.**

brown dwarf A celestial body with insufficient mass to sustain the nuclear fusion that produces radiant energy in normal stars. It is believed that a brown dwarf is formed with enough mass to start nuclear fusion in its core, but without enough for the fusion to become self-sustaining. Theory suggests that a body with about one percent of the mass of the Sun—or ten times the mass of Jupiter—can generate this initial fusion, but that it needs at least eight percent of the Sun's mass to sustain the fusion. After the fusion ends, the dwarf still glows for a period from radiating heat, with a surface temperature of about 2,500°K (4,532°F) or less. See Note at **dwarf star.**

brownfield (broun′fēld′) A piece of industrial or commercial property that is abandoned or underused and often environmentally contaminated, especially one considered as a potential site for redevelopment. Compare **greenfield.**

Brownian motion (brou′nē-ən) The random movement of microscopic particles suspended in a liquid or gas, caused by collisions between these particles and the molecules of the liquid or gas. This movement is named for its identifier, Scottish botanist Robert Brown (1773–1858). See also **kinetic theory.**

brown rot Any of several plant diseases, especially a disease of peach, plum, cherry, and related plants, caused by an ascomycete fungus and characterized by wilting and browning of the flowers and leaves and rotting of the fruits.

browser (brou′zər) A program that accesses and displays files and other data available on the Internet and other networks. Entering a website's URL in the address window of a browser will bring up that website in the browser's main window.

brucellosis (bro͞o′sə-lō′sĭs) An infectious disease caused by bacteria of the genus *Brucella,* transmitted to humans by contact with infected domestic animals such as cattle, sheep, goats, and dogs. In humans, brucellosis is marked by fever, malaise, and headache. It can also occur in some forms of wildlife, such as bison, and can cause spontaneous abortions in infected animals.

bruxism (brŭk′sĭz′əm) The habitual, involuntary grinding or clenching of the teeth, usually during sleep and sometimes associated wth stress.

bryophyte (brī′ə-fīt′) A member of a large group of seedless green plants including the mosses, liverworts, and hornworts. Bryophytes lack the specialized tissues xylem and phloem that circulate water and dissolved nutrients in the vascular plants. Bryophytes generally live on land but are mostly found in moist environments, for they have free-swimming sperm that require water for transport. In contrast to the vascular plants, the gametophyte (haploid) generation of bryophytes constitutes the larger plant form, while the small sporophyte (diploid) generation grows on or within the gametophyte and depends upon it for nutrition.

bryozoan (brī′ə-zō′ən) Any of various small aquatic invertebrate animals of the phylum Bryozoa that are capable of forming vast mosslike or branching colonies attached to seaweed or hard surfaces. Bryozoans reproduce by budding and feed on minute particles of plankton that they capture with tentacles. They are probably related to the phoronids and are the only animal phylum that does not appear in the fossil record until the early Ordovician Period.

B-s particle (bē′ĕs′) An electrically neutral meson having a mass 10,507 times that of the electron and a mean lifetime of approximately 1.6×10^{-12} seconds. See Table at **subatomic particle.**

Btu Abbreviation of **British thermal unit.**

bubble chamber (bŭb′əl) A device used to observe the paths of charged subatomic particles. A bubble chamber consists of a container filled with very dense fluid that is close to boiling. The moving particles create tracks of bubbles in the fluid that can be photographed and analyzed. Bubble chambers have been largely supplanted in laboratories by more sensitive particle detectors that do not rely on the human eye. Compare **cloud chamber.**

bubo (bo͞o′bō) Plural **buboes.** A swelling of a lymph node, especially of the armpit or groin, that is characteristic of bubonic plague.

bubonic plague (bo͞o-bŏn′ĭk) See under **plague** (sense 2).

buckminsterfullerene (bŭk′mĭn-stər-fo͝ol′ə-rēn′) An extremely stable, ball-shaped carbon molecule whose structure looks like a geodesic dome. It is believed to occur naturally in soot, and was the first fullerene to be discovered. Also called *buckyball. Chemical formula:* C_{60}.

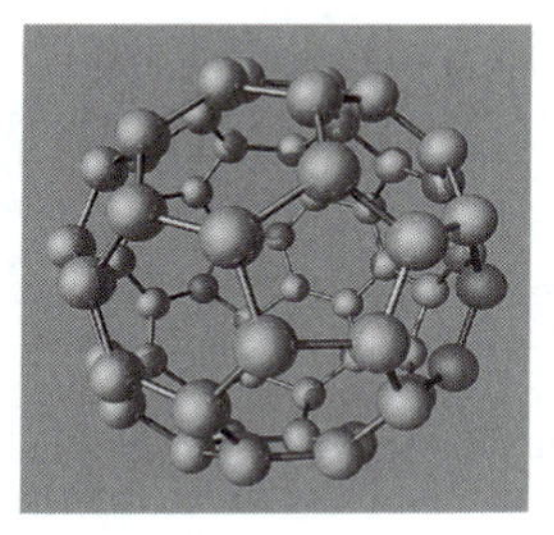

buckminsterfullerene

three-dimensional model

buckyball (bŭk′ē-bôl′) See **buckminsterfullerene.**

bud (bŭd) *Noun.* **1.** A small swelling on a branch or stem, containing an undeveloped shoot, leaf, or flower. Some species have mixed buds containing two of these structures, or even all three. ► **Terminal buds** occur at the end of a stem, twig, or branch. ► **Axillary buds,** also known as **lateral buds**, occur in the axils of leaves (in the upper angle of where the leaf grows from the stem). ► **Accessory buds** often occur clustered around terminal buds or above and on either side of axillary buds. Accessory buds are usually smaller than terminal and axillary buds. **2.** A small rounded outgrowth on an asexually reproducing organism, such as a yeast or hydra, that is capable of developing into a new individual. See more at **budding. 3.** A tiny part or structure, such as a taste bud, that is shaped like a plant bud. —*Verb.* **4.** To form or produce a bud or buds.

budding (bŭd′ĭng) A form of asexual reproduction in living organisms in which new individuals form from outgrowths (buds) on the bodies of mature organisms. These outgrowths grow by means of mitotic cell division. Many simple multicellular animals such as hydras and unicellular organisms such as yeasts reproduce by budding.

buffer (bŭf′ər) **1.** *Chemistry.* A substance that prevents change in the acidity of a solution when an acid or base is added to the solution or when the solution is diluted. Buffers are used to make solutions of known pH, especially for instrument calibration purposes. Natural buffers also exist in living organisms, where biochemical reactions are very sensitive to changes in pH. **2.** *Computer Science.* A device or an area of a computer that temporarily stores data that is being transferred between two machines that process data at different rates, such as a computer and a printer.

Buffon (bo͞o-fôɴ′), Comte **Georges Louis Leclerc de** 1707–1788. French naturalist who spent his life compiling the *Histoire naturelle,* in which he attempted to discuss all of the facts about the natural world known at that time. It eventually reached 44 volumes and laid the foundation for later studies in biology, zoology, and anatomy.

bug (bŭg) **1.** An insect belonging to the suborder Heteroptera. See more at **true bug. 2.** An insect, spider, or similar organism. Not in scientific use.

USAGE **bug**

The word *bug* is often used to refer to tiny creatures that crawl along, such as insects and even small animals that are not insects, such as spiders and millipedes. But for scientists the word has a much narrower meaning. In the strictest terms bugs are those insects that have mouthparts adapted for piercing and sucking. The mouthparts of these bugs are contained in a beak-shaped structure. Thus scientists would classify a louse but not a beetle or a cockroach as a bug. In fact, scientists often call lice and their relatives *true bugs* to distinguish them better from what everyone else calls "bugs."

bulb (bŭlb) A rounded underground storage organ that contains the shoot of a new plant. A bulb consists of a short stem surrounded by fleshy scales (modified leaves) that store nourishment for the new plant. Tulips, lilies, and onions grow from bulbs. Compare **corm, rhizome, runner, tuber.**

bulimia (bo͞o-lē′mē-ə) An eating disorder characterized by uncontrolled rapid ingestion of large quantities of food over a short period of time, followed by self-induced vomiting, fasting, and other measures to prevent weight gain. It is most common among young women and teenage girls.

bulk modulus (bŭlk) See under **modulus of elasticity.**

bulletin board (bo͝ol′ĭ-tn) An electronic communication system that allows users to send or read electronic messages, files, and other data that are of general interest and addressed to no particular person. Bulletin boards were widely used before the Internet became popular, and many of their functions are now served by websites and newsgroups for specific topics or groups.

bundle branch block (bŭn′dl) A block in electrical conduction within the ventricles of the heart due to interruption of conduction in one of the two main branches of the bundle of His. It is typically not accompanied by symptoms, but can indicate heart disease or damage.

bundle of His (hĭs) The bundle of cardiac muscle fibers that passes from the atrioventricular node to the interventricular septum and then to the ventricles. It conducts the electrical impulses that regulate the heartbeat from the right atrium to the left and right ventricles. The bundle of His is named after its discoverer, German cardiologist Wilhelm His (1836–1934).

bundle scar A small mark on a leaf scar indicating a point where a vein from the leaf was once connected with the stem. Species often differ in the number and placement of the bundle scars within a single leaf scar, and bundle scars are used to identify trees in winter or from twigs without leaves.

bundle sheath A layer or region of compactly arranged cells surrounding a vascular bundle in a plant. The bundle sheaths regulate the movement of substances between the vascular tissue and the parenchyma and, in leaves, protect the vascular tissue from exposure to air.

Bunsen (bŭn′sən), **Robert Wilhelm** 1811–1899. German chemist who with Gustav Kirchhoff developed the technique of spectroscopic analysis, leading to their discovery of the elements cesium and rubidium. Bunsen also invented various kinds of laboratory equipment, although the Bunsen burner itself was probably constructed on an earlier design by Michael Faraday.

Bunsen burner A small gas burner used in laboratories. It consists of a vertical metal tube connected to a gas fuel source, with adjustable holes at its base. These holes allow air to enter the tube and mix with the gas in order to make a very hot flame.

buoyancy (boi′ən-sē) The upward force that a fluid exerts on an object that is less dense than itself. Buoyancy allows a boat to float on water and provides lift for balloons.

bur also **burr** (bûr) A type of pseudocarp in which the outer surface possesses hooks or barbs. Burs become caught in the feathers or hair of animals, which then carry them away to disperse the seeds.

burette (byo͝o-rĕt′) A graduated glass tube having a tapered bottom with a valve. It is used especially in laboratories to pour a measured amount of liquid from one container into another.

Burgess Shale (bûr′jĭs) A rock formation in the western Canadian Rockies that contains numerous fossilized invertebrates from the early Cambrian Period.

A CLOSER LOOK **Burgess Shale**

Animals in the period known as the Cambrian Explosion sported bizarre combinations of legs, spines, segments, and heads found in no present-day animals. Many of these animals became extinct, leaving no descendants, whereas others may have evolved into groups that are familiar to us today. Most of our knowledge about these early life forms comes from the *Burgess Shale*, a 540-million-year-old formation of black shale discovered in 1909 by Charles Walcott in the Rocky Mountains of British Columbia. The unique process of fossilization that occurred in the Burgess Shale allowed exquisite preservation of these early animals. While in most cases a reaction to oxygen causes the soft parts of animals to rot away prior to fossilization, the Burgess Shale animals were killed instantly by a mudslide deep in the ocean, where there is a lack of oxygen. After burying the animals, the mud hardened into shale, preserving the soft animal parts. At the time of his discovery, Walcott was able to classify the fossils as ancestors of modern animals. The Burgess Shale was reexamined in the mid-1960s, and many new, unknown fossils were found. When Harry Whittington, Derek Briggs, and Simon Conway Morris studied these new fossils in the 1970s and 1980s, they realized that many of them did not fit into the modern classification system. The implication that there were more basic animal forms in the Cambrian Period than there are today shook up traditional ideas about evolution. In 1989 Stephen Jay Gould brought the Burgess Shale to wide public attention with the publication of his book *Wonderful Life*.

---

burl (bûrl) A large, rounded outgrowth on the trunk or branch of a tree. Burls develop from one or more twig buds whose cells continue to multiply but never differentiate so that the twig can elongate into a limb. Burls do not usually cause harm to trees.

burl

burn (bûrn) *Verb.* **1.** To be on fire; undergo combustion. A substance burns if it is heated up enough to react chemically with oxygen. **2.** To cause a burn to a bodily tissue. —*Noun.* **3.** Tissue injury caused by fire, heat, radiation (such as sun exposure), electricity, or a caustic chemical agent. Burns are classified according to the degree of tissue damage, which can include redness, blisters, skin edema and loss of sensation. Bacterial infection is a serious and sometimes fatal complication of severe burns.

bursa (bûr′sə) Plural **bursae** (bûr′sē) or **bursas.** A flattened sac containing a lubricating fluid that reduces friction between two moving structures in the body, as a tendon and a bone.

bursitis (bər-sī′tĭs) Inflammation of a bursa, most commonly of the shoulder. Symptoms include pain, tenderness, and limitation of motion.

butadiene (byo͞o′tə-dī′ēn′) A colorless, highly flammable hydrocarbon obtained from petroleum and used to make synthetic rubber. *Chemical formula:* C_4H_6.

butane (byo͞o′tān′) An organic compound found in natural gas and produced from petroleum. Butane is used as a household fuel, refrigerant, and propellant in aerosol cans. It is the fourth member of the alkane series. *Chemical formula:* C_4H_{10}.

butanol (byo͞o′tə-nôl′, -nōl′, nŏl′) Either of the two butyl alcohols that are derived from butane and have a straight chain of carbon atoms. Butanols are used as solvents and in organic synthesis. *Chemical formula:* $C_4H_{10}O$.

butanone (byo͞o′tə-nōn′) A colorless, flammable compound used as a solvent and in cleaning fluids and celluloid. Butanone is a ketone. Also called *methyl ethyl ketone. Chemical formula:* C_4H_8O.

butene (byo͞o′tēn′) See **butylene.**

butte (byo͞ot) A steep-sided hill with a flat top, often standing alone in an otherwise flat area. A butte is smaller than a mesa.

butterfly effect (bŭt′ər-flī′) A phenomenon in which a small perturbation in the initial condition of a system results in large changes in later conditions. Such phenomena are common in complex dynamical systems and are studied in chaos theory.

butyl (byo͞ot′l) The radical C_4H_9, derived from butane.

butyl alcohol Any of four isomeric compounds that are alcohols and have four carbon atoms. Butyl alcohols are used as solvents and in organic synthesis. The two straight-chained isomers are called **butanol.** *Chemical formula:* $C_4H_{10}O$.

butylate (byo͞ot′l-āt′) To bring a butyl group into a compound.

butylene (byo͞ot′l-ēn′) Any of three gaseous hydrocarbons that consist of four carbon atoms in a chain with a double bond between two of the carbons. They are part of the alkene series. Butylenes are used to make synthetic rubber. *Chemical formula:* C_4H_8.

butyraldehyde (byo͞o′tə-răl′də-hīd′) A transparent, highly flammable liquid used in making resins. *Chemical formula:* C_4H_8O.

butyrate (byo͞o′tə-rāt′) A salt or ester of butyric acid, containing the radical C_3H_7CO.

butyric acid (byo͞o-tîr′ĭk) Either of two colorless fatty acids found in butter and certain plant oils. It has an unpleasant odor and is used in emulsifying agents, disinfectants and drugs. *Chemical formula:* $C_4H_8O_2$.

bypass (bī′păs′) A passage created surgically to divert the flow of blood or other bodily fluid or to circumvent an obstructed or diseased organ.

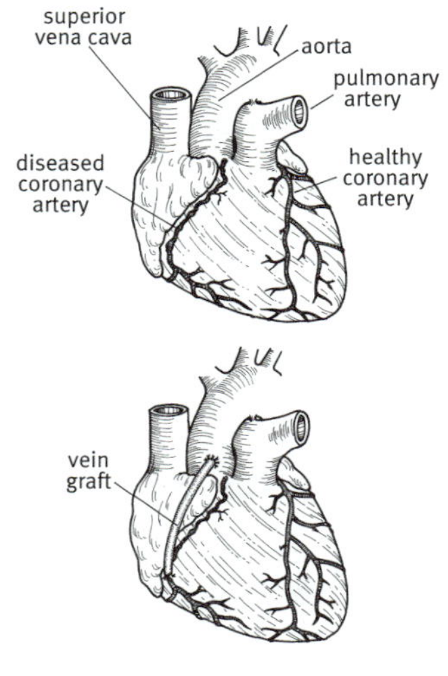

bypass

In this example of coronary bypass surgery, a vein graft bypasses a diseased right coronary artery.

by-product (bī′prŏd′əkt) Something that is produced in the process of making something else. When plants produce carbohydrates by photosynthesis, oxygen is released as a by-product. Asphalt and paraffin are by-products of the process of refining crude oil into gasoline.

Byron (bī′rən), **Augusta Ada.** Countess of Lovelace. 1815–1852. British mathematician who collaborated with Charles Babbage in the development of the analytical engine, an early computer. Byron's most important contribution was the compilation of detailed notations about how the machine could be programmed.

byte (bīt) A sequence of adjacent bits operated on as a unit by a computer. A byte usually consists of eight bits. Amounts of computer memory are often expressed in terms of megabytes (1,048,576 bytes) or gigabytes (1,073,741,824 bytes).

USAGE **byte/bit**

The word *bit* is short for *bi*nary digi*t*. A bit consists of one of two values, usually 0 or 1. Computers use bits because their system of counting is based on two options: switches on a microchip that are either *on* or *off*. Thus, a computer counts to seven in bits as follows: 0, 1, 10 [2], 11 [3], 100 [4], 101 [5], 110 [6], 111 [7]. Notice that the higher the count, the more adjacent bits are needed to represent the number. For example, it requires two adjacent bits to count from 0 to 3, and it takes three adjacent bits to count from 0 to 7. A sequence of bits can represent not just numbers but other kinds of data, such as the letters and symbols on a keyboard. The sequence of 0s and 1s that make up data are usually counted in groups of 8, and these groups of 8 bits are called *bytes*. In origin, *byte* is simply a respelling of *bite*, a byte being the number of bits that a computer can take at one bite, so to speak. The spelling change was intended to prevent confusion in written documents, since *bite* becomes identical to *bit* if the *e* at the end of *bite* is accidentally dropped. To transmit one keystroke on a typical keyboard requires one byte of information (or 8 bits). To transmit a three-letter word requires three bytes of information (or 24 bits).

bytownite (bī-tou′nīt) A bluish to dark gray triclinic mineral of the plagioclase feldspar group that contains a higher proportion of calcium than sodium. Bytownite occurs in alkaline igneous rocks. *Chemical formula:* **$(Ca,Na)(Si,Al)_4O_8$.**

C

c The symbol for the speed of light in a vacuum.

C 1. The symbol for **carbon. 2.** Abbreviation of **capacitance, capacitor, capacity, Celsius, charge conjugation, coulomb, cytosine. 3.** A programming language developed in 1972 and commonly used for writing professional software. With only a small number of built-in functions, it requires less memory than other languages, and because most if its functions are not specific to particular computers, it can be used on many different kinds of machines. The Unix operating system was written in C.

Ca The symbol for **calcium.**

cache (kăsh) An area of computer memory devoted to the high-speed retrieval of frequently used or requested data.

cachexia (kə-kĕk**′**sē-ə) Severe weight loss, anorexia, and general debility that occur as a result of chronic disease. Cachetic patients exhibit signs of malnutrition, including muscle wasting.

cadmium (kăd**′**mē-əm) *Symbol* **Cd** A rare, soft, bluish-white metallic element that occurs mainly in zinc, copper, and lead ores. Cadmium is plated onto other metals and alloys to prevent corrosion, and it is used in rechargeable batteries and in nuclear control rods as a neutron absorber. Atomic number 48; atomic weight 112.41; melting point 320.9°C; boiling point 765°C; specific gravity 8.65; valence 2. See **Periodic Table.**

caducous (kə-do͞o**′**kəs) Detaching or dropping off at an early stage of development. The gills of most amphibians and the sepals or stipules of certain plants are caducous.

caffeine (kă-fēn**′**) A bitter white alkaloid found in tea leaves, coffee beans, and various other plant parts. It is a mild stimulant. Caffeine is a xanthine and similar in structure to theobromine and theophylline. *Chemical formula:* $C_8H_{10}N_4O_2$.

calcar (kăl**′**kär′) A spur or spurlike projection, such as one found on the base of a petal or on the wing or leg of a bird.

calcareous (kăl-kâr**′**ē-əs) Composed of or containing calcium or calcium carbonate. Calcareous rocks contain as much as 50 percent calcium carbonate.

calciferol (kăl-sĭf**′**ə-rôl′, -rōl′) **1.** Any of several sterols that are forms of vitamin D, especially ergocalciferol (vitamin D_2) and cholecalciferol (vitamin D_3). See more at **vitamin D.**

calcification (kăl′sə-fĭ-kā**′**shən) **1.** *Medicine.* **a.** The accumulation of calcium or calcium salts in a body tissue. Calcification normally occurs in the formation of bone, but can be deposited abnormally, as in the lungs. **b.** A structure that has undergone calcification. **2.** *Geology.* **a.** The replacement of organic material, especially original hard material such as bone, with calcium carbonate during the process of fossilization. **b.** The accumulation of calcium in certain soils, especially soils of cool temperate regions where leaching takes place very slowly.

calcination (kăl′sə-nā**′**shən) The process of heating a substance to a high temperature but below the melting or fusing point, causing loss of moisture, reduction or oxidation, and dissociation into simpler substances. The term was originally applied to the method of driving off carbon dioxide from limestone to obtain lime (calcium oxide). Calcination is also used to extract metals from ores.

calcite (kăl**′**sīt′) A usually white, clear, pale-yellow or blue orthorhombic mineral. Calcite occurs in many different forms and is the main component of chalk, limestone, and marble. It is a polymorph of aragonite. *Chemical formula:* $CaCO_3$.

calcitonin (kăl′sĭ-tō**′**nĭn) A peptide hormone secreted by the thyroid gland that stimulates bone formation and lowers blood calcium and phosphate levels.

calcium (kăl**′**sē-əm) *Symbol* **Ca** A silvery-white, moderately hard metallic element of the alkaline-earth group that occurs in limestone and gypsum. It is a basic component of leaves, bones, teeth, and shells, and is essential for the normal growth and development of most animals and plants. Calcium is used to make plaster, cement, and alloys. Atomic number 20; atomic weight 40.08; melting point 842 to 848°C; boiling point 1,487°C; specific gravity 1.55; valence 2. See **Periodic Table.**

calcium carbonate A white or colorless crystalline compound occurring naturally in chalk, limestone, and marble and in the minerals calcite and aragonite. It is used to make toothpaste, white paint, and cleaning powder. *Chemical formula:* $CaCO_3$.

calcium chloride A white crystalline salt that attracts water very strongly. It is used in refrigeration and is spread on roads to melt ice and control dust. *Chemical formula:* $CaCl_2$.

calcium oxide A white solid compound that has a very high melting point and reacts with water to form calcium hydroxide. It is often prepared commercially by heating limestone. Calcium oxide is used as an alkali for treating acidic soils and in manufacturing steel, paper, and glass, and it is the main component of lime. *Chemical formula:* **CaO.**

calcium phosphate Any of three powdery phosphates of calcium: **1.** A colorless powder used in baking powders, as a plant food, as a plastic stabilizer, and in glass. Calcium phosphate is deliquescent, and will dissolve in the water it absorbs from the atmosphere if it is not kept in a closed container. *Chemical formula:* $Ca(H_2PO_4)_2$. **2.** A white crystalline powder used as an animal food, as a plastic stabilizer, and in glass and toothpaste. *Chemical formula:* $CaHPO_4$. **3.** A white powder that is used in ceramics, rubber, fertilizers, and for various purposes in the food industry. *Chemical formula:* $Ca_3(PO_4)_2$.

calculus (kăl′kyə-ləs) Plural **calculi** (kăl′kyə-lī′) or **calculuses. 1.** The branch of mathematics that deals with limits and the differentiation and integration of functions of one or more variables. See more at **calculus of variations, differential calculus, integral calculus. 2.** A solid mass, usually composed of inorganic material, formed in a cavity or tissue of the body. Calculi are most commonly found in the gallbladder, kidney, or urinary bladder. Also called *stone.*

calculus of variations Mathematical analysis of the maxima and minima of definite integrals, the integrands of which are functions of independent variables, dependent variables, and the derivatives of one or more dependent variables. Compare **differential calculus, integral calculus.**

caldera (kăl-dâr′ə, -dîr′ə, käl-) A large, roughly circular crater left after a volcanic explosion or the collapse of a volcanic cone. Calderas are typically much wider in diameter than the openings of the vents from which they were formed.

calibrate (kăl′ə-brāt′) **1.** To check, adjust, or standardize a measuring instrument, usually by comparing it with an accepted model. **2.** To measure the diameter of the inside of a tube.

caliche (kə-lē′chē) See **hardpan.**

californium (kăl′ə-fôr′nē-əm) *Symbol* **Cf** A synthetic, radioactive metallic element of the actinide series that is produced from curium or berkelium and is used in chemical analyses. Its most stable isotope, Cf 251, has a half-life of 800 years. Atomic number 98. See **Periodic Table.**

caliper (kăl′ə-pər) An instrument consisting of two curved legs connected at a hinge, used to measure thickness and distance. Often used in the plural as *calipers.*

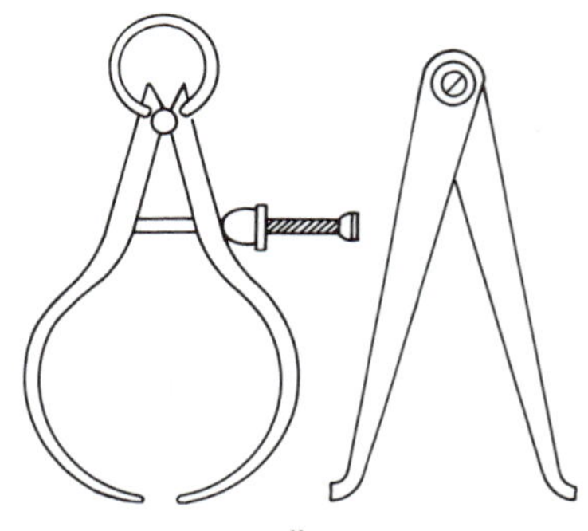

caliper

left: *outside spring caliper;* right: *inside firm-joint caliper*

Callisto (kə-lĭs′tō) One of the four brightest satellites of Jupiter and the eighth in distance from the planet. Originally sighted by Galileo, it is the largest planetary satellite.

callus (kăl′əs) **1.** An area of the skin that has become hardened and thick, usually because of prolonged pressure or rubbing. **2.** The hard bony tissue that develops around the ends of a fractured bone during healing.

calorie (kăl′ə-rē) **1.** A unit of energy equal to the amount of heat needed to raise the temperature of one gram of water by one degree

Celsius. One calorie is equivalent to 4.1868 joules. Also called *small calorie.* **2. Calorie** A unit of heat equal to the amount of heat needed to raise the temperature of 1,000 grams of water by one degree Celsius. This unit is used as a measure of the energy released by food as it is digested by the human body. Also called *kilocalorie, large calorie.*

Calvin (kăl′vĭn), **Melvin** 1911–1997. American chemist who won a Nobel Prize in 1961 for determining the chemical reactions that occur during photosynthesis. This series of reactions is now known as the Calvin cycle.

Calvin cycle A series of chemical reactions that occurs as part of the dark reactions of photosynthesis, in which carbon is broken away from gaseous carbon dioxide and fixed as organic carbon in compounds that are ultimately used to make sugars and starch as food. The Calvin cycle starts with a five-carbon sugar molecule, to which the carbon of carbon dioxide is attached by a covalent bond. This unstable molecule breaks apart into two three-carbon molecules, which are reduced by the electron-carriers ATP and NADPH (which were created by the earlier light reactions) into three-carbon molecules that are available for the synthesis of sugar and starch. It takes three carbon dioxide molecules to produce enough carbon for the synthesis of one of these three-carbon molecules and to regenerate the five-carbon sugar so the cycle can begin again. See more at **photosynthesis.**

calyptra (kə-lĭp′trə) **1.** In some bryophyte plants, a structure that covers the young sporophyte as it develops within the tissues of its gametophyte parent. The calyptra, which consists of a thickening of the archegonium walls, eventually breaks open as the spore capsule grows. **2.** See **root cap.**

calyx (kā′lĭks, kăl′ĭks) The sepals of a flower considered as a group. The calyx is the outermost whorl of a flower. See more at **sepal.**

cambium (kăm′bē-əm) Plural **cambiums** or **cambia.** A cylindrical layer of tissue in the stems and roots of many seed-bearing plants, consisting of cells that divide rapidly to form new layers of tissue. Cambium is a kind of **meristem** and is most active in woody plants, where it lies between the bark and wood of the stem. It is usually missing from monocotyledons, such as the grasses. ► The **vascular cambium** forms tissues that carry water and nutrients throughout the plant. On its outer surface, the vascular cambium forms new layers of phloem, and on its inner surface, new layers of xylem. The growth of these new tissues causes the diameter of the stem to increase. ► The **cork cambium** creates cells that eventually become bark on the outside and cells that add to the cortex on the inside. In woody plants, the cork cambium is part of the **periderm.** See also **secondary growth.**

Cambrian (kăm′brē-ən, kām′-) The first period of the Paleozoic Era, from about 540 to 505 million years ago. During this time warm seas and desert land areas were widespread, and animal life diversified rapidly during what is known as the **Cambrian Explosion**. See Chart at **geologic time.**

Cambrian Explosion The rapid diversification of multicellular animal life that took place around the beginning of the Cambrian Period. It resulted in the appearance of almost all modern animal phyla. See Note at **Burgess Shale.**

camouflage (kăm′ə-fläzh′) Protective coloring or another feature that conceals an animal and enables it to blend into its surroundings. Compare **warning coloration.**

camphor (kăm′fər) A white, gumlike, crystalline compound that has a strong odor. Camphor is volatile and is used as an insect repellent and in making plastics and explosives. *Chemical formula:* $C_{10}H_{16}O$.

camphor oil The oil that is obtained from the wood of the camphor tree and is used to make camphor. It is extracted from the wood by steam distillation.

cancer (kăn′sər) **1.** A disease characterized by any of various malignant **neoplasms** composed of abnormal cells that tend to proliferate rapidly and invade surrounding tissue. Without treatment such as chemotherapy or radiation, cancer cells can metastasize to other body sites and cause organ failure and death. **2.** A malignant tumor.

A CLOSER LOOK **cancer**

The human immune system often fights off stray cancer cells just as it does bacteria and viruses. However, when cancer cells establish themselves in the body with their own blood supply and begin replicating out of control, cancer becomes a threatening *neoplasm,* or tumor. It takes a minimum of one billion cancer cells for a neoplasm to be detectable by conventional radiology and physical examinations. *Cancer,* which represents more than 100 separate diseases, destroys tissues and organs through invasive growth in a particular part of the body and by metastasizing to distant tissues and organs through the bloodstream or lymph system. Heredity, lifestyle habits (such as smoking), and a person's exposure to certain viruses, toxic chemicals, and excessive radiation can trigger genetic changes that affect cell growth. The altered genes, or *oncogenes,* direct cells to multiply abnormally, thereby taking on the aggressive and destructive characteristics of cancer. Treatments such as surgery, chemotherapy, and radiation are effective with many cancers, but they also end up killing healthy cells. Gene therapy attempts to correct the faulty DNA that causes the uncontrolled growth of cancer cells. Researchers are investigating other treatments, such as immunotherapy (the stimulation of the body's natural defenses), vectorization (aiming chemicals specifically at cancer cells), and nanotechnology (targeting cancer cells with minute objects the size of atoms).

Cancer A faint constellation in the Northern Hemisphere near Leo and Gemini. Cancer (the Crab) is the fourth sign of the zodiac.

candela (kăn-dĕl′ə) The SI unit used to measure the brightness of a source of light (its luminous intensity). By definition, one square centimeter of a blackbody at the freezing point of platinum emits one-sixtieth of a candela of radiation. See Table at **measurement.** See also **lumen, luminous flux.**

candida (kăn′dĭ-də) Any of the yeastlike deuteromycete fungi of the genus *Candida* that are normally present on the skin and in the mucous membranes of the mouth, intestinal tract, and vagina. Certain species may become pathogenic, especially *C. albicans,* which causes thrush and other infections.

candidate species (kăn′dĭ-dāt′) A plant or animal species that is classified by a government agency as a candidate for possible listing as an endangered or threatened species.

candlepower (kăn′dl-pou′ər) Luminous intensity, expressed in candelas.

canid (kăn′ĭd, kā′nĭd) Any of various carnivorous mammals of the family Canidae, which includes the dogs, wolves, foxes, coyotes, and jackals.

canine (kā′nīn) *Adjective.* **1.** Characteristic of or resembling dogs, wolves, or related animals. **2.** Relating to any of the four pointed teeth that are located behind the incisors in most mammals. In carnivores, the canine teeth are adapted for cutting and tearing meat. —*Noun.* **3.** A canine tooth.

Canis Major (kā′nĭs) A constellation in the Southern Hemisphere near Orion. Canis Major (the Greater Dog) contains Sirius, the brightest star in the night sky.

Canis Minor A constellation in the Northern Hemisphere near Hydra and the celestial equator. Canis Minor (the Lesser Dog) contains the bright star Procyon.

Cannon (kăn′ən), **Annie Jump** 1863–1941. American astronomer noted for her work on classifying stellar spectra. Cannon classified the spectra of 225,300 stars brighter than magnitude 8.5, as well as 130,000 fainter stars.

cantilever (kăn′tl-ē′vər, -ĕv′ər) A projecting structure, such as a beam, that is supported at one end and that carries a load at the other end or along its length. Cantilevers are important structures in the design of bridges and cranes.

canyon (kăn′yən) A long, deep, narrow valley with steep cliff walls, cut into the Earth by running water and often having a stream at the bottom.

capacitance (kə-păs′ĭ-təns) A measure of the ability of a configuration of materials to store electric charge. In a **capacitor,** capacitance depends on the size of the plates, the type of insulator, and the amount of space between the plates. Most electrical components display capacitance to some degree; even the spaces between components of a circuit have a natural capacitance. Capacitance is measured in farads. Compare **inductance.**

capacitor (kə-păs′ĭ-tər) An electrical device consisting of two conducting plates separated by an electrical insulator (the **dielectric**), designed to hold an electric charge. Charge builds up when a voltage is applied across the plates, creating an electric field between them. Current can flow through a capacitor only as the voltage across it is changing, not when it is constant. Capacitors are used in power supplies, amplifiers, signal processors, oscillators, and logic gates. Compare **induction coil, resistor.**

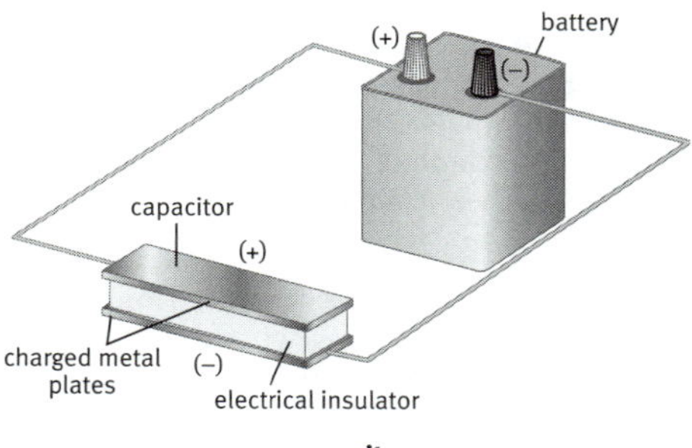

capacitor

A capacitor is charged when electrons from a power source, such as a battery, flow to one of the two plates. Because the electrons cannot pass through the insulating layer, they build up on the first plate, giving it a negative charge. Electrons on the other plate are attracted to the positive terminal of the battery, causing that plate to become positively charged.

cape (kāp) A point or head of land projecting into a body of water.

Capella (kə-pĕl′ə) A binary star in the constellation Auriga. It is one of the brightest stars in the night sky, with an apparent magnitude of 0.08. *Scientifc name:* Alpha Auriga.

capillary (kăp′ə-lĕr′ē) Any of the tiny blood vessels that connect the smallest arteries (arterioles) to the smallest veins (venules). Capillaries form a network throughout the body for the exchange of oxygen, metabolic waste products, and carbon dioxide between blood and tissue cells.

capillary action The movement of a liquid along the surface of a solid caused by the attraction of molecules of the liquid to the molecules of the solid.

A CLOSER LOOK **capillary action**

The paper towel industry owes its existence to *capillary action,* both for the way paper towels soak up liquids and for the trees out of which the towels are made. Molecules of water are naturally attracted to each other and form temporary hydrogen bonds with each other; their attraction for each other on the surface of a liquid, for example, gives rise to *surface tension.* But they are also attracted in a similar way to other molecules, called *hydrophilic* molecules, such as those in the sides of a narrow glass tube inserted into a cup of water, in the fibers of a towel, or in the cells of tree tissue known as *xylem.* These attractive forces can draw water upward against the force of gravity to a certain degree. However, they are not strong enough to draw water from the roots of a tree to its highest leaves. An additional, related force, referred to as *transpiration pull,* is required to do that. As water evaporates from the tiny pores, or *stomata,* of leaves, water from adjacent cells is drawn in to replace it by *osmosis.* Again, intermolecular attractive forces cause other water molecules to follow along, ultimately drawing water up from the roots of the tree.

---

capitate (kăp′ĭ-tāt′) *Noun.* **1.** The largest of the carpal bones. —*Adjective.* **2.** Forming a headlike mass or dense cluster, as the flowers of plants in the composite family.

capitulum (kə-pĭch′ə-ləm) Plural **capitula. 1.** A small knob or head-shaped part, such as a protuberance of a bone or the tip of an insect's antenna. **2.** An inflorescence consisting of a compact mass of small stalkless flowers, as in the English daisy. The yellow central portion of the capitulum of a daisy consists of **disk flowers,** while the outer white, petallike structures are actually **ray flowers.** The capitulum is the characteristic inflorescence of the composite family (Asteraceae) of flowering plants.

Capricorn (kăp′rĭ-kôrn′) or **Capricornus** (kăp′rĭ-kôr′nəs) A constellation in the Southern Hemisphere near Aquarius and Sagittarius. Capricorn (the Goat) is the tenth sign of the zodiac.

capsaicin (kăp-sā′ĭ-sĭn) A colorless, extremely pungent, crystalline compound that is the primary active principle producing the heat of red peppers. It is a strong irritant to skin and mucous membranes and is used in medicine as a topical analgesic. Capsaicin is highly stable, retaining its potency for long periods and despite cooking or freezing. *Chemical formula:* **$C_{18}H_{27}NO_3$.**

capsid (kăp′sĭd) The protein shell that surrounds a virus particle (known as a **virion**). See more at **virus.**

capsule (kăp′səl, -so͞ol) **1.** A dry dehiscent fruit that develops from two or more carpels, as in the poppy and the cottonwood tree. **2.** The sporangium (the hollow spore-producing structure) of mosses and other bryophytes. **3.** The outer layer of viscous polysaccharide or polypeptide slime with which some bacteria cover their cell walls. Capsules provide defense against phagocytes and prevent the bacteria from drying out.

captured rotation (kăp′chərd) See **synchronous rotation.**

carapace (kăr′ə-pās′) A hard outer covering or shell made of bone or chitin on the back of animals such as turtles, armadillos, lobsters, and crabs.

carbamate (kär′bə-māt′, kär-băm′āt′) A salt or ester containing the radical NH_2COO. Carbamates are often used as insecticides.

carbamoyl (kär-băm′ō-ĭl′) The radical NH_2CO, derived from carbamic acid.

carbide (kär′bīd′) A chemical compound consisting of carbon and a more electropositive element, such as calcium or tungsten. Many carbides, especially those made of carbon and a metal, are very hard and are used to make cutting tools and abrasives.

carbohydrate (kär′bō-hī′drāt′) Any of a large class of organic compounds consisting of carbon, hydrogen, and oxygen, usually with twice as many hydrogen atoms as carbon or oxygen atoms. Carbohydrates are produced in green plants by photosynthesis and serve as a major energy source in animal diets. Sugars, starches, and cellulose are all carbohydrates.

carbolic acid (kär-bŏl′ĭk) See **phenol.**

carbon (kär′bən) *Symbol* **C** A naturally abundant, nonmetallic element that occurs in all organic compounds and can be found in all known forms of life. Diamonds and graphite are pure forms, and carbon is a major constituent of coal, petroleum, and natural gas. Carbon generally forms four covalent bonds with other atoms in larger molecules. Atomic number 6; atomic weight 12.011; sublimation point above 3,500°C; boiling point 4,827°C; specific gravity of amorphous carbon 1.8 to 2.1, of diamond 3.15 to 3.53, of graphite 1.9 to 2.3; valence 2, 3, 4. See **Periodic Table.** —*Adjective* **carbonaceous.**

carbon 12 A stable isotope of carbon, having six protons and six neutrons in the nucleus. Carbon 12 makes up most naturally occurring carbon.

carbon 14 A naturally occurring radioactive isotope of carbon having six protons and eight neutrons in the nucleus. Carbon 14 is important in dating archaeological and biological remains by **radiocarbon dating**.

carbon-14 dating See **radiocarbon dating.**

carbonate (kär′bə-nāt′) *Noun.* **1.** A salt or ester of carbonic acid, containing the group CO_3. The reaction of carbonic acid with a metal results in a salt (such as sodium carbonate), and the reaction of carbonic acid with an organic compound results in an ester (such as diethyl carbonate). **2.** Any other compound containing the group CO_3. Carbonates include minerals such as calcite and aragonite. **3.** Sediment or a sedimentary rock formed by the precipitation of organic or inorganic carbon from an aqueous solution of carbonates of calcium, magnesium, or iron. Limestone is a carbonate rock. —*Verb.* **4.** To add carbon dioxide to a substance, such as a beverage.

carbon cycle **1.** The continuous process by which carbon is exchanged between organisms and the environment. Carbon dioxide is absorbed from the atmosphere by plants and algae and converted to carbohydrates by photosynthesis. Carbon is then passed into the food chain and returned to the atmosphere by the respiration and decay of animals, plants, and other organisms. The burning of fossil fuels also releases carbon dioxide into the atmosphere. **2.** A cycle of thermonuclear reactions caused by the absorption of protons by the nucleus of a carbon-12 atom, in which helium and isotopes of nitrogen, carbon, and oxygen are produced, and resulting in the regeneration of a carbon-12 atom so that the process can begin again. The carbon cycle is thought to be the source of significant amounts of energy in the Sun and other stars. Also called *Bethe cycle, carbon-nitrogen cycle.*

See also **proton-proton chain, triple alpha process.**

carbon dating See **radiocarbon dating.**

carbon dioxide A colorless, odorless gas that is present in the atmosphere and is formed when any fuel containing carbon is burned. It is breathed out of an animal's lungs during respiration, is produced by the decay of organic matter, and is used by plants in photosynthesis. Carbon dioxide is also used in refrigeration, fire extinguishers, and carbonated drinks. *Chemical formula:* CO_2.

carbon fiber An extremely strong, thin fiber, consisting of long, chainlike molecules of pure carbon that are made by charring synthetic fibers such as rayon in the absence of oxygen. Carbon fibers are used in high-strength composite materials in aircraft, automobiles, architectural structures, and in other applications where light materials capable of withstanding high stress are required.

carbon fixation The process in plants and algae by which atmospheric carbon dioxide is converted into organic carbon compounds, such as carbohydrates, usually by photosynthesis. See more at **carbon cycle.**

carbonic acid (kär-bŏn′ĭk) A weak, unstable acid present in solutions of carbon dioxide in water. It gives carbonated beverages their sharp taste. *Chemical formula:* H_2CO_3.

Carboniferous (kär′bə-nĭf′ər-əs) The period of geologic time from about 360 to 286 million years ago. The term is used throughout the world, although this period of time has been separated into the Mississippian (lower Carboniferous) and Pennsylvanian (upper Carboniferous) in the United States. During this time, widespread swamps formed in which plant remains accumulated and later hardened into coal. See Chart at **geologic time.**

carbon monoxide A colorless, odorless gas formed when a compound containing carbon burns incompletely because there is not enough oxygen. It is present in the exhaust gases of automobile engines and is very poisonous. *Chemical formula:* CO.

carbon nanotube The most common form of nanotube, composed entirely of carbon atoms. See more at **nanotube.**

carbon-nitrogen cycle See **carbon cycle** (sense 2).

carbon tetrachloride A colorless, nonflammable, poisonous liquid having a strong odor. It is used to make refrigerants, aerosol propellants, and pharmaceuticals. It is also used in petroleum refining and as a solvent. Until the mid-1960s, it was used as a cleaning fluid and in fire extinguishers. *Chemical formula:* CCl_4.

carbonyl (kär′bə-nĭl′) The radical CO, found in a wide range of chemical compounds, especially in aldehydes and ketones.

carbonyl chloride See **phosgene.**

carboxyl (kär-bŏk′səl) The radical COOH, characteristic of all carboxylic acids.

carboxylate (kär-bŏk′sə-lāt′) A salt or ester of an organic acid, containing the radical COO. Soaps, which are usually the sodium or potassium salts of fatty acids, are carboxylates.

carboxylic acid (kär′bŏk-sĭl′ĭk) An organic acid containing one or more carboxyl groups. Carboxylic acids often have names ending in *–oic acid,* such as *benzoic acid.* Amino acids, fatty acids, and many other important organic compounds are carboxylic acids.

carcinogen (kär-sĭn′ə-jən) A substance or agent that can cause cells to become cancerous by altering their genetic structure so that they multiply continuously and become malignant. Asbestos, DDT, and tobacco smoke are examples of carcinogens.

carcinoma (kär′sə-nō′mə) Plural **carcinomas** or **carcinomata** (kär′sə-nō′mə-tə). Any of various cancerous tumors that are derived from epithelial tissue of the skin, blood vessels, or other organs and that tend to metastasize to other parts of the body. See also **basal cell carcinoma, squamous cell carcinoma.**

cardiac (kär′dē-ăk′) Relating to or involving the heart.

cardiac cycle The rhythmic contraction (systole) and relaxation (diastole) of the chambers of the heart that corresponds to one heartbeat. During the cardiac cycle, blood is pumped out of the heart into the aorta and the pulmonary artery, and blood reenters the heart from the venae cavae and the pulmonary veins. The heart's valves open and close in response to pressure differences between the chambers.

cardiac output The volume of blood pumped per minute by each ventricle of the heart. Cardiac output is equal to the *stroke volume* (the amount of blood pumped from a ventricle in a single heartbeat) times the heart rate. It is used as a measure of the overall health of the heart.

cardinal number (kär′dn-əl) A number, such as 3, 11, or 412, used in counting to indicate quantity but not order. Compare **ordinal number.**

cardinal point One of the four principal directions on a compass (north, south, east, or west).

cardioid (kär′dē-oid′) A heart-shaped plane curve, the locus of a fixed point on a circle that rolls on the circumference of another circle with the same radius.

cardiology (kär′dē-ŏl′ə-jē) The branch of medicine that deals with diagnosis and treatment of disorders of the heart.

cardiomyopathy (kär′dē-ō-mī-ŏp′ə-thē) Any of various structural or functional abnormalities of the cardiac muscle, usually characterized by loss of muscle efficiency and sometimes heart failure. Cardiomyopathy can result from numerous causes, including congenital defects, acute or chronic infections, coronary artery disease, drugs and toxins, metabolic disorders, connective tissue disorders, or nutritional deficiencies. In some patients, the cause is unknown.

cardiopulmonary (kär′dē-ō-po͝ol′mə-nĕr′ē, -pŭl′-) Relating to or involving the heart and the lungs.

cardiopulmonary bypass A procedure that circulates and oxygenates the blood while surgery is performed on the heart. With this process, blood is diverted from the heart and lungs through a **heart-lung machine** and oxygenated blood is returned to the aorta.

cardiopulmonary resuscitation See **CPR.**

cardiovascular (kär′dē-ō-văs′kyə-lər) Relating to or involving the heart and blood vessels.

cardiovascular system See **circulatory system**

caries (kâr′ēz) Plural **caries.** Decay of a bone or tooth. Dental plaque formed by bacteria initiates a progressive process of decay that, if left unchecked, leads to tooth loss.

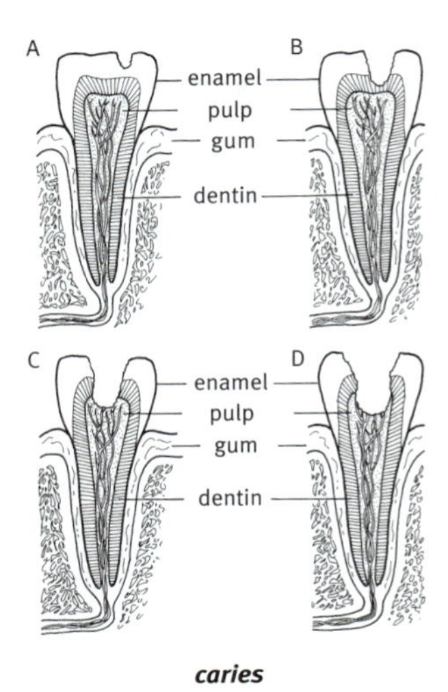

caries

four stages of progressive dental disease: A: *bacterial acid destroying enamel and forming a cavity;* B: *unchecked decay spreading to the dentin;* C: *enlarged cavity allowing bacteria to attack exposed pulp at the center of the tooth;* D: *untreated infected pulp causing eventual death of the pulp and the tooth*

carnitine (kär′nĭ-tēn′) A betaine commonly occurring in the liver and in skeletal muscle that is essential for fatty acid transport across mitochondrial membranes. *Chemical formula:* $C_7H_{15}NO_3$.

carnivore (kär′nə-vôr′) **1a.** An animal that feeds chiefly on the flesh of other animals. Carnivores include predators such as lions and alligators, and scavengers such as hyenas and vultures. In a food chain, carnivores are either secondary or tertiary consumers. Compare **detritivore, herbivore. b.** Any of various generally meat-eating mammals of the order Carnivora. Carnivores have large, sharp canine teeth and large brains, and the musculoskeletal structure of their forelimbs permits great flexibility for springing at prey. Many carnivores remain in and defend a single territory. Dogs, cats, bears, weasels, raccoons, hyenas, and (according to some classifications) seals and walruses are all carnivores. **2.** A plant that eats insects, such as a Venus flytrap.

Carnot (kär-nō′), **Nicolas Léonard Sadi** 1796–1832. French physicist and engineer who founded the science of thermodynamics. He was the first to analyze the working cycle and efficiency of the steam engine according to scientific principles. Through his experiments Carnot developed what would become the second law of thermodynamics and laid the foundation for work by Kelvin, Joule, and others.

carotene (kăr′ə-tēn′) Any of various organic compounds that occur as orange-yellow to red pigments in many plants and in animal tissue. In plant leaves, carotenes aid in the absorption of light energy by transferring the energy to chlorophyll and act as antioxidants protecting chlorophyll from damage by oxidation. In animals, carotenes are converted to vitamin A primarily in the liver. They are members of the carotenoid family of compounds and give plants such as carrots, pumpkins, and dandelions their characteristic color. *Chemical formula:* $C_{40}H_{56}$. See also **xanthophyll.**

carotenoid (kə-rŏt′n-oid′) Any of a class of yellow to red pigments found especially in plants, algae, and photosynthetic bacteria. Carotenoids generally consist of conjoined units of the hydrocarbon isoprene, with alternating single and double bonds. The carotenoids absorb light energy of certain frequencies and transfer it to chlorophyll for use in photosynthesis. They also act as antioxidants for chlorophyll, protecting it from damage by oxidation in the presence of sunlight. Carotenoids are nutritionally important for many animals, giving flamingoes their color, for example, and also have antioxidant properties. There are many types of carotenoids, including carotenes and xanthophylls. See more at **photosynthesis.**

Carothers (kə-rŭ*th*′ərz), **Wallace Hume** 1896–1937. American chemist who developed the synthetic material nylon, which was patented in 1937.

carotid artery (kə-rŏt′ĭd) Either of the two major arteries, one on each side of the neck, that carry blood to the head.

carotid body A mass of tissue near the carotid sinus that contains chemical receptors sensitive to oxygen and pH levels in the blood and that is involved in the regulation of respiratory activity.

carotid sinus A dilated region that is located at the point of bifurcation of the carotid artery and contains pressure receptors (called baroreceptors), which upon stimulation cause slowing of the heart rate, vasodilation, and a decrease in blood pressure.

carpal (kär′pəl) *Adjective.* **1.** Relating to or involving the wrist. *—Noun.* **2.** Any of the bones of the human wrist or the joint corresponding to the wrist in some other vertebrates, such as dinosaurs.

carpal tunnel syndrome Chronic pain, numbness, or tingling in the hand, caused by compression of the median nerve in the wrist. It can be caused by repetitive bending and extension of the wrist, as in keyboarding, or by medical conditions such as rheumatoid arthritis and diabetes.

carpel (kär′pəl) One of the individual female reproductive organs in a flower. A carpel is composed of an ovary, a style, and a stigma, although some flowers have carpels without a distinct style. In origin, carpels are leaves (megasporophylls) that have evolved to enclose the ovules. The term *pistil* is sometimes used to refer to a single carpel or to several carpels fused together. See more at **flower.**

carpellate (kär′pə-lāt′, -lĭt) Having carpels but no stamens. Female flowers are carpellate.

carpophore (kär′pə-fôr′) **1.** A fleshy, spore-producing body of basidiomycetes and ascomycetes. In common usage, the term *mushroom* is applied to carpophores that have a distinctive stipe and cap. **2.** A slender stalk that supports each half of a dehisced fruit in many members of the parsley family.

carpus (kär′pəs) Plural **carpi** (kär′pī′). **1.** The group of eight bones lying between the forearm and the metacarpals and forming the wrist in humans. **2.** The group of bones making up the joint corresponding to the wrist in some vertebrates, such as dinosaurs.

carrageenan (kăr′ə-gē′nən) A gelatinous material derived from Irish moss (*Chondrus crispus*) and other species of red algae. It is widely used as a thickening, stabilizing, emulsifying, or suspending agent in industrial, pharmaceutical, and food products.

carrier (kăr′ē-ər) **1.** A person, animal, or plant that serves as a host for a pathogen and can transmit it to others, but is immune to it. Mosquitoes are carriers of malaria. **2.** An organism that carries a gene for a trait but does not show the trait itself. Carriers can produce offspring that express the trait by mating with another carrier of the same gene. See more at **recessive.**

carrying capacity (kăr′ē-ĭng) The maximum population of a particular organism that a given environment can support without detrimental effects.

Carson (kär′sən), **Rachel Louise** 1907–1964. American marine biologist and writer whose best-known book, *Silent Spring* (1962), was an influential study of the dangerous effects of synthetic pesticides on food chains. Public reaction to the book resulted in stricter controls on pesticide use and shaped the ideas of the modern environmental movement.

Rachel Carson

Cartesian coordinate system (kär-tē′zhən) A system in which the location of a point is given by coordinates that represent its distances from perpendicular lines that intersect at a point called the origin. A Cartesian coordinate system in a plane has two perpendicular lines (the *x*-axis and *y*-axis); in three-dimensional space, it has three (the *x*-axis, *y*-axis, and *z*-axis). Compare **polar coordinate system.**

cartilage (kär′tl-ĭj) A strong, flexible connective tissue that is found in various parts of the body, including the joints, the outer ear, and the larynx. During the embryonic development of most vertebrates, the skeleton forms as cartilage before most of it hardens into bone. In cartilaginous fish, the mature fish retains a skeleton made of cartilage.

cartilaginous fish (kär′tl-ăj′ə-nəs) Any of various fishes of the class Selachii (or Chondrichthyes), having a skeleton that is made of cartilage. Cartilaginous fishes breathe through gill slits, of which there are usually five, and their toothlike or platelike scales (called denticles) are made of dentine and enamel. Sharks, rays, skates, sawfish, and chimaeras are cartilaginous fishes. Compare **bony fish, jawless fish.**

cartography (kär-tŏg′rə-fē) The art or technique of making maps or charts.

Carver (kär′vər), **George Washington** 1864?–1943. American botanist and educator whose work was instrumental in improving the agricultural efficiency of the United States.

George Washington Carver

BIOGRAPHY **George Washington Carver**

George Washington Carver played a central role in revitalizing Southern agriculture after the Civil War, when Southern farms produced ever smaller cotton crops. His promotion of crop rotation methods helped to restore Southern farmlands, which had been depleted by the exclusive cultivation of cotton. Carver also introduced two new crops, peanuts and sweet potatoes, that would produce well in Alabama soil. To make them economically beneficial to farmers, he developed 325 products from peanuts, including peanut butter, plastics, synthetic rubber, shaving cream, and paper. He also developed hundreds of other products from sweet potatoes and from dozens of other native plants, including soybeans and cotton. During his forty-seven years as head of the agriculture department at Tuskegee Institute in Alabama, he taught the importance of crop diversification and soil conservation. Carver also introduced movable schools that brought practical agricultural knowledge directly to farmers.

caryopsis (kăr′ē-ŏp′sĭs) Plural **caryopses** (kăr′ē-ŏp′sēz′) or **caryopsides** (kăr′ē-ŏp′sĭ-dēz′). A small, dry, one-seeded fruit of a cereal grass, having the fruit and the seed walls united. Also called *grain.*

cascade (kăs-kād′) A series of chemical or physiological processes that occur in successive stages, each of which is dependent on the preceding one, to produce a culminating effect. The steps involved in the clotting of blood occur as a cascade.

casein (kā′sēn′, -sē-ĭn) A white, tasteless, odorless mixture of related phosphoproteins precipitated from milk by rennin. Casein is very nutritious, as it contains all of the essential amino acids as well as all of the common nonessential ones. It is the basis of cheese and is used to make plastics, adhesives, paints, and foods.

Casimir effect (kăz′ə-mîr′) The effect of a net attractive force between objects in a vacuum, caused by quantum mechanical **vacuum fluctuations** creating **radiation pressure.** The radiation can be thought of as an atmosphere of **virtual particles**. The amount of radiation pressure on the objects is decreased in the gap between them, due to limits on the wavelength of the radiation in the gap. The gap is thus an area of lower radiation pressure, drawing the objects toward it. This force is strong enough to be of great importance at scales encountered in nanotechnology. The Casimir effect is named after Dutch physicist Hendrik Casimir (1909–2000).

Cassini division (kə-sē′nē) The large gap between Saturn's two most prominent rings (the outer A ring and middle B ring), appearing as a dark void but actually containing small amounts of opaque material. It is formed as a result of particles being removed from the area by the gravitational pull of Mimas, one of the smaller of Saturn's moons. The division is named after the Italian astronomer, Giovanni D. Cassini (1625–1712), who discovered it in 1675.

Cassiopeia (kăs′ē-ə-pē′ə) A W-shaped constellation in the Northern Hemisphere near Andromeda and Cepheus.

caste (kăst) A specialized group carrying out a specific function within a colony of social insects. For example, in an ant colony, members of the caste of workers forage for food outside the colony or tend eggs and larvae, while the members of the caste of soldiers, often larger with stronger jaws, are responsible for defense of the colony.

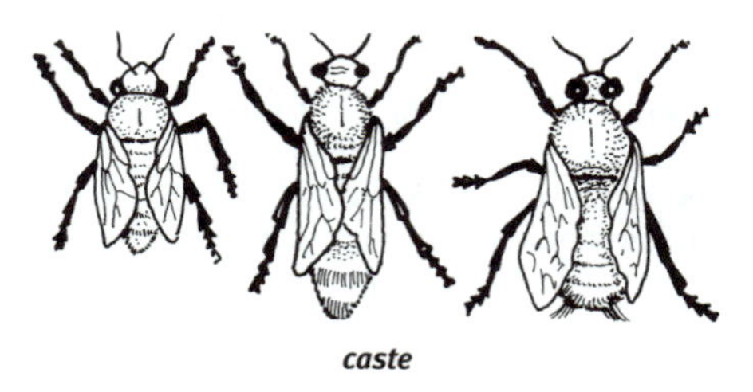

caste

left to right: *worker, queen, and drone honey bees*

Castor (kăs′tər) A bright multiple star in the constellation Gemini, with a combined apparent magnitude of 0.08. *Scientific name:* Alpha Geminorum.

catabolism (kə-tăb′ə-lĭz′əm) The phase of metabolism in which energy, in the form of ATP, is produced by the breakdown of complex molecules, such as starches, proteins and fats, into simpler ones. Compare **anabolism.** —*Adjective* **catabolic.**

cataclastic (kăt′ə-klăs′tĭk) Relating to rocks consisting of cemented fragments that originate from the mechanical breakdown of rock associated with plate tectonic processes. Cataclastic rocks form in regions that have undergone intense metamorphism and are associated with other metamorphic features such as folds and faults. They typically contain bent, broken, and granular minerals.

catadioptric (kăt′ə-dī-ŏp′trĭk) Relating to both the reflection and refraction of light, especially by a combination of mirrors and lenses or by a prism. Catadioptric systems are used in Fresnel lenses, optical calibration equipment, and some telescopes. Compare **catoptric, dioptric.**

catalyst (kăt′l-ĭst) A substance that starts or speeds up a chemical reaction while undergoing no permanent change itself. The enzymes in saliva, for example, are catalysts in digestion. —*Adjective* **catalytic** (kăt′l-ĭt′ĭk)

catalyze (kăt′l-īz′) To modify, especially to increase, the rate of a chemical reaction through the action of a catalyst.

cataract (kăt′ə-răkt′) **1.** An opacity of the lens of the eye or the membrane that covers it, causing impairment of vision or blindness. **2.** A waterfall in which a large volume of water flows over a steep precipice.

catchment area (kăch′mənt, kĕch′-) **1.** The area drained by a river or body of water. **2.** The area that absorbs water that contributes to a specific region's groundwater supply.

catechol (kăt′ĭ-kôl′, -kōl′) A biologically important organic phenol occurring naturally in lignins and resins. It has two hydroxyl groups attached to a benzene ring. Catechol is very caustic and is used in dyeing and as a photographic developer and an antiseptic. *Chemical formula:* $C_6H_6O_2$.

catecholamine (kăt′ĭ-kō′lə-mēn′, -kô′-) Any of a group of amines derived from catechol that have important physiological effects as neurotransmitters and hormones and include epinephrine, norepinephrine, and dopamine.

caterpillar (kăt′ər-pĭl′ər) The wormlike larva of a butterfly or moth. Caterpillars have thirteen body segments, with three pairs of stubby legs on the thorax and several on the abdomen, six eyes on each side of the head, and short antennae. Caterpillars feed mostly on foliage and are usually brightly colored. Many have poisonous spines.

catheter (kăth′ĭ-tər) A hollow, flexible tube inserted into a body cavity, duct, or vessel to allow the passage of fluids or distend a passageway.

cathode (kăth′ōd′) **1.** The negative electrode in an electrolytic cell, toward which positively charged particles are attracted. The cathode has a negative charge because it is connected to the negatively charged end of an external power supply. **2.** The source of electrons in an electrical device, such as a vacuum tube or diode. **3.** The positive electrode of a voltaic cell, such as a battery. The cathode gets its positive charge from the chemical reaction that happens inside the battery, not from an external source. Compare **anode.**

cathode ray A beam of electrons streaming from the negatively charged end of a vacuum tube (the cathode) toward a positively charged plate (the anode).

cathode-ray tube A sealed tube in which electrons are emitted by a heated, negatively charged element (the cathode), and travel in a beam toward a positively charged plate (the

A CLOSER LOOK cathode-ray tube

Cathode-ray tubes (CRTs), also called *electron-ray tubes,* provide the visual display in such devices as conventional television sets, computer monitors, hospital heart monitors, and laboratory oscilloscopes. CRTs are generally made of funnel-shaped glass vacuum tubes. At the larger end of the tube is a phosphor-coated screen, and at the other end is an *electron gun.* The gun consists of a heated cathode, or negative electrode, which emits electrons, and a control grid, which controls the intensity of the beam of electrons to vary the brightness of the image. The gun directs the electron beam, or *cathode ray,* toward the screen, where a positively charged anode attracts the electrons. Outside the tube, coils creating a magnetic field or plates creating an electric field both focus and steer the beam. Wherever the beam strikes the screen, it causes the phosphors to glow. Shapes and images can be formed by manipulating the beam so that its focal point on the screen sweeps across it in various paths and with different brightness. In most CRTs, the beam follows a zigzag path that covers the entire screen many times per second. Color screens use three separate beams that strike three individually colored phosphor cells (having the three primary colors red, blue, and green) that are very close together. The color combinations appear to the eye (at a distance to the screen) as one point of a single color.

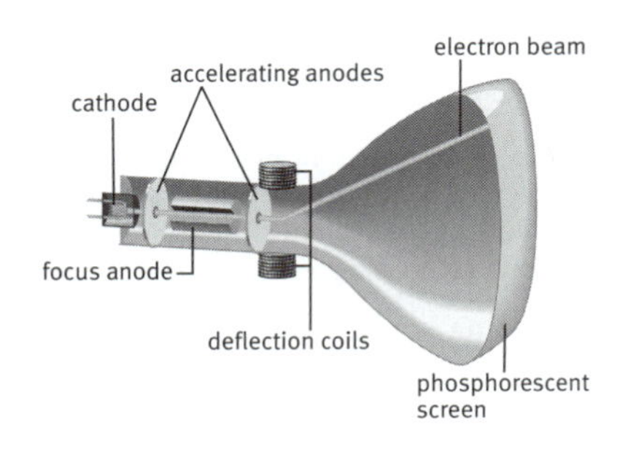

anode). Depending on the properties of the plate and the speed of the electrons, cathode-ray tubes can generate x-rays, visible light, and other frequencies of electromagnetic radiation. They are central to most television screens, in which the electron beams form images on a phosphor-coated screen.

cation (kăt′ī′ən) An ion with net positive charge, having more protons than electrons. In electrolysis, cations migrate to a negatively charged cathode. Compare **anion.**

cation exchange See **ion exchange.**

catkin (kăt′kĭn) A long, thin, indeterminate inflorescence of tiny, petalless flowers growing on willows, birches, oaks, poplars, and certain other trees. The flowers on a catkin are either all male or all female. The female flowers are usually pollinated by the wind. Also called *ament.* See illustration at **inflorescence.**

catoptric (kə-tŏp′trĭk) Relating to the reflection of light, especially by a mirror. Catoptric lenses are used in Fresnel lenses and many telescopes. Compare **catadioptric, dioptric.**

CAT scan (kăt) Short for *computerized axial tomography scan.* An image of a body structure produced by computerized axial tomography.

caudal (kôd′l) Relating to or near the tail or hind parts of an animal.

caudex (kô′dĕks′) **1.** The thickened, usually underground base of the stem of many perennial herbaceous plants, from which new leaves and flowering stems arise. **2.** The trunk of a palm or tree fern.

causality paradox (kô-zăl′ĭ-tē) A paradox resulting from hypothetical time travel, in which an individual travels back in time and performs actions that would ultimately have made the time travel impossible (as by killing one's parents at a time preceding one's birth).

cauterize (kô′tə-rīz′) To burn or sear with a cautery, as in surgical procedures.

cautery (kô′tə-rē) An agent or instrument used to destroy tissue, as in surgery, by burning, searing, cutting, or scarring, including caustic substances, electric currents, and lasers.

cave (kāv) A naturally occurring underground hollow or passage, especially one with an opening to the surface of the Earth. Caves can form through a variety of processes, including the dissolution of limestone by flowing water, the differential cooling of volcanic magma (which occurs when the outside surface of the lava cools, but the inside continues to flow downwards, forming a hollow tube), or the action of wind and waves along a rocky coast.

Cavendish (kăv′ən-dĭsh), **Henry** 1731–1810. British chemist and physicist who in 1766 discovered hydrogen, which he called "inflammable air." He also demonstrated that it is the lightest of all the gases and established that water is a compound of hydrogen and oxygen. In 1798, Cavendish estimated with great accuracy the mean density of the Earth.

cavern (kăv′ərn) A large cave.

cavitation (kăv′ĭ-tā′shən) The formation of bubblelike gaps in a liquid. Mechanical forces, such as the moving blades of a ship's propeller or sudden negative changes in pressure, can cause cavitation.

cavity (kăv′ĭ-tē) **1.** A hollow; a hole. **2.** A hollow area within the body. **3.** A pitted area in a tooth caused by caries.

cay (kē, kā) A small, low island composed largely of coral or sand. Also called *key.*

CBC Abbreviation of **complete blood count.**

cc Abbreviation of **cubic centimeter.**

Cd The symbol for **cadmium.**

cDNA Abbreviation of **complementary DNA.**

CD-ROM (sē′dē′rŏm′) A compact disk containing permanently stored data that cannot be altered.

Ce The symbol for **cerium.**

CE or CE Abbreviation for **Common Era.**

cecum (sē′kəm) Plural **ceca.** A large pouch forming the beginning of the large intestine. The appendix and the ileum of the small intestine both connect to the cecum.

ceilometer (sē-lŏm′ĭ-tər) An instrument for calculating the altitude of the lowest cloud layer in the sky. A ceilometer consists of a projector that shines an intense beam of light (usually laser light) at the cloud layer and a photoelectric cell that detects the light as it is reflected back. The distance between the projector and the photoelectric cell and the angle at which the light is detected are used to calculate the altitude of the cloud layer by triangulation;

alternatively, the time required for the light to be reflected back to the detector may be used to measure the distance, as in radar.

C. elegans (sē′ ĕl′ə-gănz) A nematode *(Caenorhabditis elegans)* that lives in soil, feeds on bacteria, and reaches lengths of about 1 mm (0.04 inch). It was the first animal whose genome was completely sequenced, and is widely used as a "model organism" by researchers in genetics and developmental biology because it has a small genome and transparent skin.

celestial (sə-lĕs′chəl) **1.** Relating to the sky or the heavens. Stars and planets are celestial bodies. **2.** Relating to the celestial sphere or to any of the coordinate systems by which the position of an object, such as a star or planet, is represented on it.

celestial equator A great circle separating the northern and southern hemispheres of the celestial sphere and lying in the same plane as the Earth's equator. The celestial equator is a projection of the Earth's equator outward onto the celestial sphere and is used as the reference point in determining a celestial body's **declination**.

celestial horizon A great circle on the celestial sphere having a plane that passes through the center of the Earth at a right angle to the line formed by an observer's zenith and nadir. The celestial horizon divides the celestial sphere into two equal hemispheres based on the observer's location, with one hemisphere representing the half of the sky visible to the observer at that location and the other representing the half that is hidden from the observer below the Earth's horizon. The celestial horizon is used as the reference point in determining a celestial body's **altitude**. Also called *rational horizon*. Compare **sensible horizon.**

celestial latitude On the celestial sphere, the position of a celestial object north or south of the **ecliptic**. Celestial latitude is measured in degrees along a great circle drawn through the object being measured and the poles of the ecliptic, with positive values north of the ecliptic and negative values south of it, so that the ecliptic itself is 0° and the north and south ecliptic poles are +90° and –90° respectively. Compare **celestial longitude.** See more at **ecliptic coordinate system.**

celestial longitude The position of a celestial object east of the vernal equinox along the **ecliptic**. Celestial longitude is measured in degrees eastward from the vernal equinox (0°) to the point where a great circle drawn through the object and the poles of the ecliptic intersects the ecliptic. Compare **celestial latitude.** See more at **ecliptic coordinate system.**

celestial mechanics The science of the motion of celestial bodies under the influence of gravitational forces. See more at **Kepler's laws of planetary motion, relativity.**

celestial meridian 1. A great circle on the celestial sphere passing through the celestial poles and an observer's **zenith**. An observer's celestial meridian is not a fixed reference on the celestial sphere but rather changes with the observer's location on Earth. Stars transit an observer's celestial meridian (that is, cross directly overhead in the sky at that location) once every 24 hours. Also called *local meridian, meridian.* **2.** In the **equatorial coordinate system**, a great circle on the celestial sphere passing through the celestial poles and the **vernal equinox**. It represents the zero point for the horizontal coordinate in this system, having a right ascension of 0 hours.

celestial navigation Navigation of a ship or aircraft based on the observed positions of celestial bodies. See more at **altazimuth coordinate system.**

celestial pole Either of the two points at which a northward or southward projection of the Earth's axis intersects the celestial sphere. The north and south celestial poles are analogous to Earth's geographic poles and are used in determining right ascension in the **equatorial coordinate system**. Depending on which hemisphere an observer is in, the stars and other celestial objects appear to revolve once around the north or south celestial pole every 24 hours, an effect produced by the rotation of the Earth on its axis. Because of the **precession** of Earth's axis, the celestial poles gradually shift position in the sky over a nearly 26,000-year cycle.

celestial sphere An imaginary sphere with Earth at its center. The stars, planets, Sun, Moon, and other celestial bodies appear to be located on this sphere, and the sphere appears to rotate around the Earth's extended axis once every 24 hours, carrying the celestial bodies with it overhead and giving them their diurnal motions. The celestial sphere is essentially a spherical map of the sky that provides the basis for the coordinate systems used in celestial navigation and in specifying the positions and motions of celestial objects. See more at **altazimuth coordinate system, ecliptic coordinate system, equatorial coordinate system.**

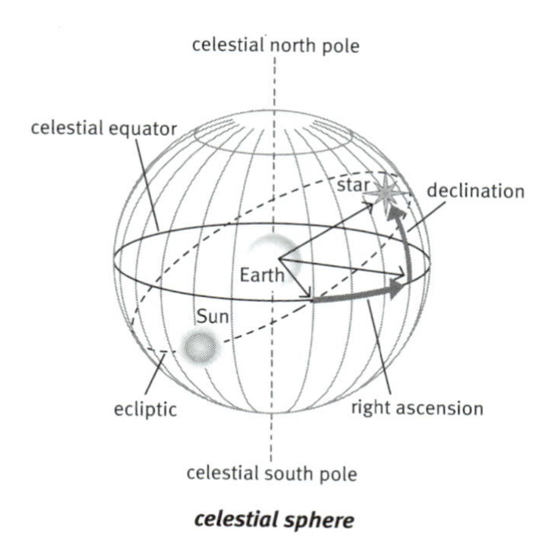

celestial sphere

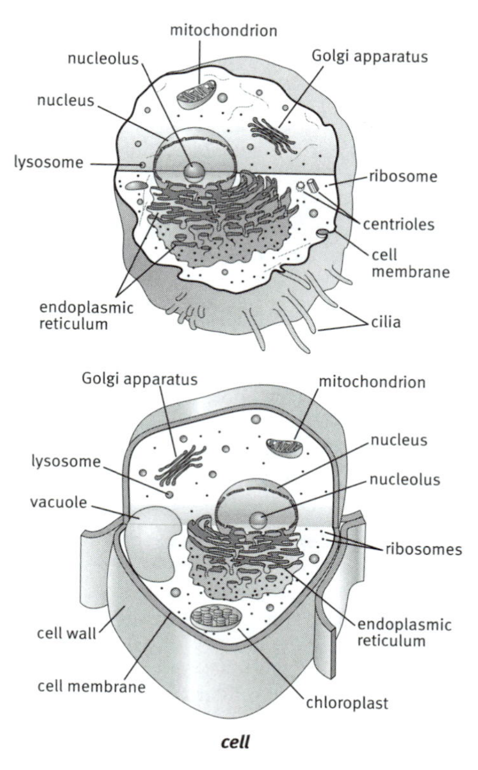

cell

top: *animal cell;* bottom: *plant cell*

celiac disease (sē′lē-ăk′) A gastrointestinal disease characterized by an inability to absorb the protein gluten, resulting in diarrhea, the passage of stools having a high fat content, and nutritional and vitamin deficiencies. Individuals with celiac disease must avoid ingesting products made from grains containing gluten, including wheat, rye, barley, and oats.

cell (sĕl) **1.** The basic unit of living matter in all organisms, consisting of protoplasm enclosed within a cell membrane. All cells except bacterial cells have a distinct nucleus that contains the cell's DNA as well as other structures (called organelles) that include mitochondria, the endoplasmic reticulum, and vacuoles. The main source of energy for all of a cell's biological processes is ATP. See more at **eukaryote, prokaryote. 2.** Any of various devices, or units within such devices, that are capable of converting some form of energy into electricity. Cells contain two electrodes and an electrolyte. See more at **electrolytic cell, solar cell, voltaic cell.** —*Adjective* **cellular.**

cell body The portion of a neuron that contains the nucleus but does not incorporate the dendrites or axon.

cell division The process by which a cell divides into two or more cells. Among prokaryotes, cell division occurs by simple **fission.** Among eukaryotes, the cell nucleus divides first, and then a new cell membrane is formed between the nuclei to form the new cell. Cell division is used as a means of reproduction in organisms that reproduce asexually, as by fission or spore formation, and sexually reproducing organisms form gametes through cell division. Cell division is also the source of tissue growth and repair in multicellular organisms. The two types of cell division in eukaryotic organisms are **mitosis** and **meiosis.**

cell-mediated immune response An immune response produced when T cells, especially cytotoxic T cells, that are sensitized to foreign antigens attack and lyse target cells. In addition to direct cytotoxicity, T cells can stimulate the production of lymphokines that activate macrophages. Cell-mediated immune responses are important in defense against pathogens, autoimmune diseases, some acquired allergies, and other immune reactions.

cell-mediated immunity Immunity resulting from a **cell-mediated immune response.** Also called *cellular immunity.* Compare **humoral immunity.**

cell membrane The thin membrane that forms the outer surface of the protoplasm of a cell and regulates the passage of materials in and out of the cell. It is made up of proteins and lipids and often contains molecular receptors. The membranes of organelles within the cell are made of the same basic material as the cell membrane. In plant cells, the cell membrane is surrounded by a rigid cell wall. Also called *plasma membrane.* Compare **cell wall.** See more at **cell.**

cell plate A partition that is formed during telophase of cell division in plants and some algae and that separates the two newly formed daughter cells.

cell sap The liquid contained within a vacuole of a plant cell. See more at **vacuole.**

cellular immunity (sĕl′yə-lər) See **cell-mediated immunity.**

cellular respiration The process of cell catabolism in which cells turn food into usable energy in the form of ATP. In this process glucose is broken down in the presence of molecular oxygen into six molecules of carbon dioxide, and much of the energy released is preserved by turning ADP and free phosphate into ATP. Cellular respiration occurs as a series of chemical reactions catalyzed by enzymes, the first of which is **glycolysis**, a series of anaerobic reactions in which glucose (a 6-carbon molecule) is split into two molecules of lactate (a 3-carbon molecule), producing a net gain of two ATP molecules. In a series of aerobic reactions, lactate is converted to pyruvate, which enters the mitochondrion and combines with oxygen to form an acetyl group, releasing carbon dioxide. The acetyl group (CH_3CO) is then combined with coenzyme A as acetyl coenzyme A, and enters the **Krebs cycle**. During this series of reactions, each acetyl group is oxidized to form two molecules of carbon dioxide, and the energy released is transferred to four electron carrier molecules. The electron carrier molecules then release their energy in a process that results in the pumping of protons (hydrogen ions) out across the inner membrane of the mitochondrion. The potential energy of the protons generated by one acetyl group is later released when they recross the membrane and are used to form three molecules of ATP from ADP and phosphate in the process of oxidative **phosphorylation**. The pyruvate from one molecule of glucose drives two turns of the Krebs cycle. Thus, during cellular respiration one molecule of glucose, as well as oxygen, ADP, and free phosphate are catabolized to yield six molecules of carbon dioxide and an increase in usable energy in the form of eight molecules of ATP.

cellular slime mold See under **slime mold.**

cellulase (sĕl′yə-lās′) Any of several enzymes produced chiefly by fungi, bacteria, and protozoans that catalyze the breakdown of cellulose. Protozoans in the guts of termites produce the cellulase needed for the termites to digest wood.

cellulose (sĕl′yə-lōs′) **1.** A carbohydrate that is a polymer composed of glucose units and that is the main component of the cell walls of most plants. It is insoluble in water and is used to make paper, cellophane, textiles, explosives, and other products. **2.** See **cellulose acetate.**

cellulose acetate Any of several compounds obtained by treating cellulose with acetic anhydride. Cellulose acetate is used in lacquers, photographic film, transparent sheeting, and cigarette filters. Also called *acetate, cellulose.*

cell wall The outermost layer of cells in plants, bacteria, fungi, and many algae that gives shape to the cell and protects it from infection. In plants, the cell wall is made up mostly of cellulose, determines tissue texture, and often is crucial to cell function. Compare **cell membrane.**

Celsius (sĕl′sē-əs) Relating to a temperature scale on which the freezing point of water is 0° and the boiling point of water is 100° under normal atmospheric pressure. See Note at **centigrade.**

Celsius, Anders 1701–1744. Swedish astronomer who invented the centigrade thermometer in 1742.

cementation (sē′mĕn-tā′shən) A metallurgical coating process in which a metal or alloy such as iron or steel is immersed in a powder of another metal, such as zinc, chromium, or aluminum, and heated to a temperature below the melting point of either. Cementation is often employed to increase resistance to oxidation.

Cenozoic (sĕn′ə-zō′ĭk) The most recent era of geologic time, from about 65 million years ago to the present. The Cenozoic Era is characterized by the formation of modern continents and the diversification of mammals and plants. Grasses also evolved during the Cenozoic. The climate was warm and tropical toward the beginning of the era and cooled significantly in the second half, leading to several ice ages. Humans first appeared near the end of this era. See Chart at **geologic time.**

Centaur (sĕn′tôr′) Any of a group of icy bodies similar to both asteroids and comets, orbiting the Sun in elliptical paths mostly in the region between Saturn and Neptune. Centaurs range in diameter from around 100 to 400 km (62 to 248 mi) and are believed to be Kuiper belt objects that have escaped into the vicinity of the gas-giant planets. Centaurs are considered to have unstable orbits, and gravitational encounters with the large outer planets could send them into the inner solar system or alternatively could eject them from the solar system into interstellar space. **Chiron**, the first such body to be classified as a Centaur, was discovered in 1977.

Centaurus (sĕn-tôr′əs) A constellation in the Southern Hemisphere near the Southern Cross and Libra. Centaurus (the Centaur) contains Alpha Centauri, the star nearest Earth.

center of gravity (sĕn′tər) The center of mass of an object in the presence of a uniform gravitational field.

center of mass A point in space determined by a distribution of mass (such as a solid object, a collection of objects, or a gas), such that a uniform force acting on the whole distribution acts as if the distribution were located at just that point. For example, the gravitational force on Earth attracts small objects as if it were pulling them from their center of mass (in this case, the center of mass is also called the **center of gravity**). Two stars in orbit around each other revolve around their collective center of mass. See also **barycenter.**

centi– A prefix meaning "a hundredth," as in *centigram,* a hundredth of a gram.

centigrade (sĕn′tĭ-grād′) See **Celsius.**

USAGE **centigrade**

Because of confusion over the prefix *centi–,* which originally meant 100 but developed the meaning $\frac{1}{100}$, scientists agreed to stop using the term *centigrade* in 1948. The term *Celsius* is now standard.

centigram (sĕn′tĭ-grăm′) A unit of weight in the metric system equal to 0.01 gram. See Table at **measurement.**

centiliter (sĕn′tə-lē′tər) A unit of volume in the metric system equal to 0.01 liter. See Table at **measurement.**

centimeter (sĕn′tə-mē′tər) A unit of length in the metric system equal to 0.01 meter. See Table at **measurement.**

centimeter-gram-second system A system of measurement in which the basic units of length, mass, and time are the centimeter, gram, and second. It is used in the United States chiefly in science, medicine, and engineering. Also called *cgs system.*

centipede (sĕn′tə-pēd′) Any of various flattened, wormlike arthropods of the class Chilopoda, whose bodies are divided into many segments, each with one pair of legs. The front legs are modified into venomous pincers used to catch prey. Compare **millipede.**

centipoise (sĕn′tə-poiz′) A centimeter-gram-second unit of dynamic viscosity equal to 0.01 poise.

central angle (sĕn′trəl) An angle formed by two rays from the center of a circle, with the center forming the vertex.

central nervous system The part of the nervous system in vertebrate animals that consists of the brain and spinal cord. Compare **peripheral nervous system.**

central processing unit The part of a computer that interprets and carries out instructions provided by the software. It tests and manipulates data, and transfers information to and from other components, such as the working memory, disk drive, monitor, and keyboard. The central processing units of personal computers are generally implemented on a single chip, called a **microprocessor**.

centrifugal (sĕn-trĭf′yə-gəl, -trĭf′ə-) **1.** Moving or directed away from a center or axis, usually as a result of being spun around the center or axis. **2.** Operated in the manner of a centrifuge. **3.** Transmitting nerve impulses away from the brain or spinal cord; efferent. **4.** Developing or progressing outward from a center or axis, as in the growth of plant structures. For example, in a centrifugal inflorescence such as a cyme, the flowers in the center or tip open first while those on the edge open last. Compare **centripetal.**

centrifugal force An effect that seems to cause an object moving in a curve to be pushed away from the curve's center. Centrifugal force is not a true force but is actually the effect of inertia, in that the moving object's natural tendency is to move in a straight line. See Note at **centripetal force.**

centrifuge (sĕn′trə-fyo͞oj′) A machine that separates substances of different densities in a sample by rotating the sample at very high speed, causing the substance to be displaced outward, sometimes through a series of filters or gratings. Denser substances tend to be displaced from the center more than ones that are less dense.

centriole (sĕn′trē-ōl′) Either of a pair of cylinder-shaped bodies found in the centrosome of most eukaryotic organisms other than plants. During cell division (both mitosis and meiosis), the centrioles move apart to help form the spindle, which then distributes the chromosomes in the dividing cell. See more at **cell, meiosis, mitosis.**

centripetal (sĕn-trĭp′ĭ-tl) **1.** Moving or directed toward a center or axis, particularly one around which an object is spinning. **2.** Transmitting nerve impulses toward the brain or spinal cord; afferent. **3.** Developing or progressing inward toward a center or axis, as in the growth of plant structures. For example, in the disk of the inflorescence of a sunflower, the florets near the edge open first, and the ones in the center last. Compare **centrifugal.**

centripetal force A force acting on a moving body at an angle to the direction of motion, tending to make the body follow a circular or curved path. The force of gravity acting on a satellite in orbit is an example of a centripetal force; the friction of the tires of a car making a turn similarly provides centripetal force on the car.

A CLOSER LOOK centripetal force

In a popular carnival ride, people stand with their backs against the wall of a cylindrical chamber. The chamber spins rapidly, the floor drops out, but the riders remain pressed against the wall without falling. Although the riders may insist they stay aboard because of an outward force pushing them against the wall, the reality is the opposite: the riders are subject to an inward, or *centripetal,* force. As the ride spins, it forces the riders to travel in a circle. According to Isaac Newton's law of inertia, objects in motion tend to travel in a straight line at constant speed unless acted on by an external force. To make an object travel along a curved path, some force—the centripetal force—must push the object toward the *center of curvature* of that path. In the case of the circular path, the direction of the force is toward the center of rotation. The wall of the ride's cylindrical chamber accomplishes this by pushing the riders toward the center (with the friction between the riders and the wall holding the riders up). The force of the Earth's gravity acts as a centripetal force on orbiting objects, such as the Earth's Moon, which is constantly being accelerated toward the center of the Earth, as in free fall. The Moon has enough inertia not to plummet into the Earth but not so much that it can escape the Earth's pull, and thus it will orbit almost indefinitely.

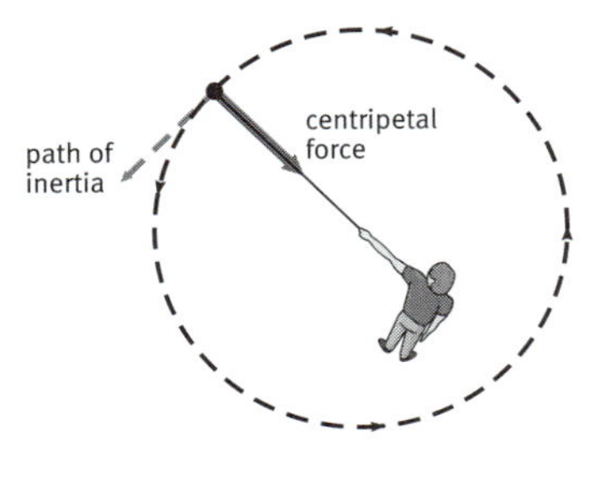

centromere (sĕn′trə-mîr′) The region of the chromosome to which the spindle fiber is attached during cell division (both mitosis and meiosis). The centromere is the constricted point at which the two chromatids forming the chromosome are joined together. See more at **meiosis, mitosis.**

centrosome (sĕn′trə-sōm′) A specialized region of the cytoplasm that is located next to the nucleus of a cell and contains the centrioles. The cells of most eukaryotes except plants have centrosomes.

cephalic (sə-făl′ĭk) Located on or near the head.

cephalochordate (sĕf′ə-lə-kôr′dāt′) A member of the subphylum Cephalochordata of the phylum Chordata, whose only living species are the lancelets. See more at **lancelet.** Compare **hemichordate.**

cephalopod (sĕf′ə-lə-pŏd′) Any of various marine mollusks of the class Cephalopoda, having long tentacles around the mouth, a large head, a pair of large eyes, and a sharp beak. Cephalopods have the most highly developed nervous system of all invertebrates. Many cephalopods squirt a cloud of dark inky liquid to confuse predators. Cephalopods include the octopus, squid, cuttlefish, and nautilus, and the extinct ammonites, belemnites, and other nautiloids.

cephalothorax (sĕf′ə-lə-thôr′ăks′) The combined head and thorax of arachnids, such as spiders, and of many crustaceans, such as crabs.

Cepheid (sē′fē-ĭd, sĕf′ē-) Any of a class of variable stars whose luminosity fluctuates with an extremely regular period. There is a strong correlation between the **absolute magnitude** of a Cepheid's luminosity and its period. By comparing the **apparent magnitude** of a Cepheid to the absolute magnitude corresponding to its period, it is possible to determine fairly accurately how distant the Cepheid is from Earth. Also called *Cepheid variable.*

Cepheus (sē′fyo͞os′, -fē-əs) A constellation in the Northern Hemisphere near Cassiopeia and Draco.

ceramic (sə-răm′ĭk) Any of various hard, brittle, heat- and corrosion-resistant materials made typically of metallic elements combined with oxygen or with carbon, nitrogen, or sulfur. Most ceramics are crystalline and are poor conductors of electricity, though some recently discovered copper-oxide ceramics are superconductors at low temperatures.

ceratopsian (sĕr′ə-tŏp′sē-ən) Any of various dinosaurs of the group Ceratopsia of the late Jurassic and Cretaceous Periods. Ceratopsians were ornithischians characterized by skulls with a parrotlike beak, a broad bony frill extending back over the neck, and often one or more horns. Most ceratopsians walked on all fours and grew to medium or large size. Triceratops was a ceratopsian.

cereal (sîr′ē-əl) A grass, such as corn, rice, sorghum, or wheat, whose starchy grains are used as food. Cereals are annual plants, and cereal crops must be reseeded for each growing season. Cereal grasses were domesticated during the Neolithic Period and formed the basis of early agriculture.

cerebellum (sĕr′ə-bĕl′əm) Plural **cerebellums** or **cerebella.** The part of the vertebrate brain that is located below the cerebrum at the rear of the skull and that coordinates balance and muscle activity. In mammals, the cerebellum is made up of two connecting hemispheres that consist of a core of white matter surrounded by gray matter.

cerebral (sĕr′ə-brəl, sə-rē′brəl) Relating to or involving the brain or cerebrum.

cerebral cortex The layer of gray matter in vertebrates that covers the cerebral hemispheres and is composed of folds of neurons and axons. The cerebral cortex is responsible for higher functions of the nervous system, including voluntary muscle activity and learning, language, and memory.

cerebral hemisphere Either of the two symmetrical halves of the cerebrum, designated right and left. In mammals, the cerebral hemispheres are connected by the corpus callosum, a transverse band of nerve fibers.

cerebral palsy A motor disorder often caused by brain injury occurring at or before birth, characterized by muscular impairment and symptoms such as poor coordination, spasm, abnormal stiffness, speech difficulties, and sometimes paralysis. Some children with cerebral palsy have accompanying neurologic conditions such as epilepsy, learning disorders, and mental retardation.

cerebrospinal fluid (sĕr′ə-brō-spī′nəl, sə-rē′brō-) The clear fluid that fills the cavities of the brain and covers the surfaces of the brain and spinal cord. It lubricates the tissues and cushions them from shock and injury.

cerebrovascular accident (sĕr′ə-brō-văs′kyə-lər, sə-rē′brō-) See **stroke.**

cerebrum (sĕr′ə-brəm, sə-rē′brəm) Plural **cerebrums** or **cerebra.** The largest part of the vertebrate brain, filling most of the skull and consisting of two cerebral hemispheres divided by a deep groove and joined by the corpus callosum, a transverse band of nerve fibers. The cerebrum processes complex sensory information and controls voluntary muscle activity. In humans it is the center of thought, learning, memory, language, and emotion.

Cerenkov effect or **Cherenkov effect** (chə-rĕng′kôf, -kəf) The emission of electromagnetic radiation by charged particles traveling through a medium at a speed faster than the speed of light in that medium. This radiation is called **Cerenkov radiation.** When it results from cosmic rays penetrating the atmosphere at high speeds, it generally takes the form of bluish light. This effect is named after its discoverer, Russian physicist Pavel Alekseyevich Cerenkov (1904–1990).

Ceres (sîr′ēz) The closest dwarf planet to the Sun, with an orbit in the asteroid belt. Ceres was the first object in the asteroid belt to be discovered (1801). Initially considered a planet, it was reclassified as an asteroid in the mid-1800s and as a dwarf planet in 2006. It has a diameter of about 960 km (595 mi). See more at **dwarf planet.**

cerium (sîr′ē-əm) *Symbol* **Ce** A shiny, gray metallic element of the lanthanide series. It is ductile and malleable and is used in electronic components, alloys, and lighter flints. It is also used in glass polishing and as a catalyst in self-cleaning ovens. Atomic number 58; atomic weight 140.12; melting point 795°C; boiling point 3,468°C; specific gravity 6.67 to 8.23; valence 3, 4. See **Periodic Table.**

cervical (sûr′vĭ-kəl) **1.** Relating to or involving the cervix of the uterus. **2.** Relating to or located in or near the neck.

cervid (sûr′vĭd) Any of various hoofed mammals of the family Cervidae, which includes the deer and elk. Male cervids typically grow antlers that are shed yearly.

cervix (sûr′vĭks) A neck-shaped anatomical structure, especially the narrowed, lower end of the uterus that extends into the vagina.

cesarean section also **caesarean section** (sĭ-zâr′ē-ən) A surgical incision through the abdominal wall and uterus, performed to deliver a fetus.

cesium (sē′zē-əm) *Symbol* **Cs** A soft, ductile, silvery-white element of the alkali group. It is liquid at room temperature and is the most reactive of all metals. Cesium is used to make photoelectric cells, electron tubes, and atomic clocks. Atomic number 55; atomic weight 132.905; melting point 28.5°C; boiling point 690°C; specific gravity 1.87; valence 1. See **Periodic Table.**

cestode (sĕs′tōd′) Any of various parasitic flatworms of the class Cestoda, having a long flat body that usually has a specialized organ of attachment at one end (the scolex). Cestodes may consist of a single segment or be divided into numerous identical rectangular segments. Food is absorbed through the outer covering of the body. Cestodes inhabit the liver and digestive tract of many vertebrate animals and also affect some invertebrates. They can attain a length of over 15 m (49 ft). Also called *tapeworm.*

cetacean (sĭ-tā′shən) Any of various, often very large aquatic mammals of the order Cetacea, having a hairless body that resembles that of a fish. Cetaceans have an elongated skull, a flat, horizontal tail, forelimbs modified into broad flippers, and no hind limbs. They breathe through blowholes located usually at the top of the skull. Whales, dolphins, and porpoises are cetaceans. See more at **baleen whale, toothed whale.**

cetane (sē′tān′) A colorless liquid hydrocarbon derived from petroleum. It is used as a solvent and a component of diesel fuels. *Chemical formula:* $C_{16}H_{34}$.

cetyl alcohol (sēt′l) A waxy alcohol used in lubricants, detergents, cosmetics, and emulsifiers. *Chemical formula:* $C_{16}H_{34}O$.

Cf The symbol for **californium.**

CFC Abbreviation of **chlorofluorocarbon.**

cg Abbreviation of **centigram.**

cgs system (sē′gē′ĕs′) See **centimeter-gram-second system.**

Chadwick (chăd′wĭk), Sir **James** 1891–1974. British physicist who in 1932 discovered the neutron. For this work, he received the 1935 Nobel Prize for chemistry.

chaetognath (kē′tŏg-năth′) Any of various small, wormlike marine invertebrates of the phylum Chaetognatha, having often transparent or translucent bodies that are pointed at one end. The head has small hooks for grasping prey, and it contains a large cerebral ganglion of sensory nerves. Chaetognaths are popularly called *arrowworms,* and are thought to be related to the echinoderms and chordates.

chain (chān) A group of atoms, often of the same element, bound together in a line, branched line, or ring to form a molecule. ▸ In a **straight chain**, each constituent atom is attached to other single atoms, not to groups of atoms. ▸ In a **branched chain**, side groups are attached to the chain. ▸ In a **closed chain**, the atoms are arranged in the shape of a ring.

Chain, Sir **Ernst Boris** 1906–1979. German-born British bacteriologist who, with Howard Florey, developed and purified penicillin in 1939. For this work, they shared a 1945 Nobel Prize with Alexander Fleming, who first discovered the antibiotic in 1928.

chain reaction A process in which the result of one event triggers another event, usually of the same kind, which in turn triggers yet another event, so that the overall reaction tends to be self-sustaining. Nuclear fission reactions are chain reactions, in which the splitting of an atomic nucleus releases neutrons that penetrate other nuclei, causing them to split. The spread of heat through a substance is also a chain reaction, as fast-moving molecules in a hot part of the substance collide with neighboring molecules, passing on their kinetic energy to them, thereby making more of the substance warmer. See more at **fission.** See Note at **nuclear reactor.** See also **kinetic theory.**

chalaza (kə-lā′zə) Plural **chalazae** (kə-lā′zē) or **chalazas. 1.** One of two spiral bands of tissue in an egg that connect the yolk to the lining membrane at either end of the shell. **2.** The region of a plant ovule that is opposite the micropyle, where the integuments and nucellus are joined.

chalcedony (kăl-sĕd′n-ē) A type of quartz that has a waxy luster and varies from transparent to translucent. It is usually white, pale-blue, gray, brown, or black and is often found as a lining in cavities. Agate, flint, and onyx are forms of chalcedony. *Chemical formula:* **SiO_2.**

Chalcolithic (kăl′kə-lĭth′ĭk) The period of human culture preliminary to the Bronze Age, characterized by the use of copper and stone tools. The Chalcolithic Period is generally recognized only for Europe and central and western Asia. Also called *Copper and Stone Age.* See Note at **Three Age system.**

chalcopyrite (kăl′kə-pī′rīt′) A brassy yellow, metallic, tetragonal mineral, usually occurring as shapeless masses of grains. Chalcopyrite is found in igneous rocks and copper-rich shales, and it is an important ore of copper. Because of its shiny look and often yellow color, it is sometimes mistaken for gold, and for this reason it is also called *fool's gold. Chemical formula:* **$CuFeS_2$.**

chalicothere (kăl′ĭ-kə-thîr′) Any of various large extinct mammals of the family Chalicotheriidae of the Eocene to the Pleistocene Epochs. Chalicotheres were odd-toed ungulates related to horses, rhinos, and tapirs, and had three toes on each foot ending in distinctive curved claws rather than hooves. Some species had front claws so long that they walked on their knuckles.

chalk (chôk) A soft, white, gray, or yellow limestone consisting mainly of calcium carbonate and formed primarily from the accumulation of fossil microorganisms such as foraminifera and calcareous algae. Chalk is used in making lime, cement, and fertilizers, and as a whitening pigment in ceramics, paints, and cosmetics. The chalk used in classrooms is usually artificial.

Chandrasekhar (chăn′drə-sā′kär), **Subrahmanyan** 1910–1995. Indian-born American astrophysicist who studied the physical processes surrounding the structure and evolution of stars, and determined the limit (now named for him) regarding the growth of white dwarfs. For this work, he received (jointly with

American physicist William Fowler) the 1983 Nobel Prize for physics.

Chandrasekhar limit The maximum size of a stable white dwarf, approximately 3×10^{30} kg (about 1.4 times the mass of the Sun). Stars with mass higher than the Chandrasekhar limit ultimately collapse under their own weight and become neutron stars or black holes. Stars with a mass below this limit are prevented from collapsing by the degeneracy pressure of their electrons. See more at **degeneracy pressure.**

channel (chăn′əl) **1.** A specified frequency band for the transmission and reception of electromagnetic signals, as for television signals. **2.** The part of a field effect transistor, usually U-shaped, through which current flows from the source to the drain. See more at **field effect transistor. 3.** A pathway through a protein molecule in a cell membrane that modulates the electrical potential across the membrane by controlling the passage of small inorganic ions into and out of the cell. **4.** The bed or deepest part of a river or harbor. **5.** A large strait, especially one that connects two seas.

chaos (kā′ŏs′) The behavior of systems that follow deterministic laws but appear random and unpredictable. Chaotic systems very are sensitive to initial conditions; small changes in those conditions can lead to quite different outcomes. One example of chaotic behavior is the flow of air in conditions of turbulence. See more at **fractal.**

character (kăr′ək-tər) **1.** *Genetics.* A structure, function, or attribute determined by a gene or a group of genes. **2.** *Computer Science.* A symbol, such as a letter, number, or punctuation mark, that occupies one byte of memory. See more at **ASCII.**

characteristic (kăr′ək-tə-rĭs′tĭk) The part of a logarithm to the base 10 that is to the left of the decimal point. For example, if 2.749 is a logarithm, 2 is the characteristic. Compare **mantissa.**

charcoal (chär′kōl′) A black porous form of carbon produced by heating wood or bone in little or no air. Charcoal is used as a fuel, for drawing, and in air and water filters.

charge (chärj) **1.** A fundamental property of the elementary particles of which matter is made that gives rise to attractive and repulsive forces. There are two kinds of charge: color charge and electric charge. See more at **color charge, electric charge. 2.** The amount of electric charge contained in an object, particle, or region of space.

charge conjugation A reversal of the sign of the charge, the magnetic moment, and other internal quantum numbers of every particle in the system to which it is applied, effectively turning every particle into its antiparticle. In the mathematics of quantum mechanics, this operation is denoted by the operator "C". It had long been assumed that charge conjugation, together with **parity conjugation**, should leave the behavior of a physical system unchanged, but observations of kaon decay suggest this is not always the case. See also **conservation law.**

charge-coupled device A device made up of semiconductors arranged in such a way that the electric charge output of one semiconductor charges an adjacent one.

charge density The electric charge per unit area or per unit volume of a body or of a region of space.

charge parity invariance See **CP invariance.**

Charles (chärlz), **Jacques Alexandre César** 1746–1823. French physicist and inventor who formulated Charles's law in 1787. In 1783 he became the first person to use hydrogen in balloons for flight.

Charles's law (chärl′zĭz) The principle that the volume of a given mass of an ideal gas is proportional to its temperature as long as its pressure remains constant. Charles's law is a subcase of the **ideal gas law**. Compare **Boyle's law.**

charm (chärm) One of the flavors of quarks, contributing to the charm number—a quantum number—for hadrons. ► A **charmed particle** is a particle that contains at least one charmed quark or charmed antiquark. The charmed quark was hypothesized to account for the longevity of the **J/psi particle** and to explain differences in the behavior of leptons and hadrons. See more at **flavor.**

charm quark A quark with a charge of $+\frac{2}{3}$ and a charm of +1. Its mass is greater than that of the electron and greater than that of the up quark, down quark, and strange quark, but less than that of the bottom quark and top quark. Also called *charmed quark*. See Table at **subatomic particle.**

Charon (kâr′ən) The largest of Pluto's three moons. Charon is just over half Pluto's size and orbits it so closely that it was not discovered as a distinct body until 1978. Although Pluto is covered with methane ice, Charon appears to be covered with water ice.

chelate (kē′lāt′) A chemical compound in the form of a ring that contains a metal ion attached by coordinate bonds to at least two nonmetal ions. Many commercial dyes as well as important biological substances, such as chlorophyll and the heme of hemoglobin, are chelates.

chelation (kĭ-lā′shən) The combination of a metal ion with a chemical compound to form a ring. Chelation is used in the industrial separation and extraction of metals and to treat metal poisoning.

chelicerate (kə-lĭs′ər-āt′, -ĭt) Any of various, mostly terrestrial arthropods of the subphylum Chelicerata, having a body divided into two main parts: a cephalothorax and an abdomen. Chelicerates have specialized feeding appendages (chelicerae) and lack antennae. Chelicerates include the arachnids and merostomes (horseshoe crabs) and are closely related to the extinct trilobites.

chelonian (kĭ-lō′nē-ən) Any of various reptiles of the order Chelonia (or Testudines), which includes the turtles and tortoises. Chelonians lack teeth and usually have a hard shell that protects the body and consists of bony plates fused to the vertebrae and ribs. Unlike all other living reptiles, the skulls of chelonians lack temporal openings, which is characteristic of the earliest known reptiles (called anapsids). Chelonians evolved during the late Permian or Triassic Period and have changed little since.

chemical (kĕm′ĭ-kəl) *Adjective.* **1.** Relating to or produced by means of chemistry. *—Noun.* **2.** A substance having a specific molecular composition, obtained by or used in a chemical process.

chemical engineering The branch of engineering that deals with the manufacture of products through chemical processes. These products include, among others, pharmaceuticals, pulp and paper, petrochemicals, microelectronic devices, polymers, and products used in food processing and in biotechnology.

chemical name The name of a chemical compound that shows the names of each of its elements or subcompounds. For example, the chemical name of aspirin is acetylsalicylic acid. Compare **trivial name.**

chemiluminescence (kĕm′ə-lo͞o′mə-nĕs′əns) The emission of light by a substance as a result of undergoing a chemical reaction that does not involve an increase in its temperature. Chemiluminescence usually occurs when a highly oxidized molecule, such as a peroxide, reacts with another molecule. The bond between the two oxygen atoms in a peroxide is relatively weak, and when it breaks the atoms must reorganize themselves, releasing energy in the form of light. Compare **bioluminescence.**

chemistry (kĕm′ĭ-strē) **1.** The scientific study of the structure, properties, and reactions of the chemical elements and the compounds they form. **2.** The composition, structure, properties, and reactions of a substance.

chemoautotroph (kē′mō-ô′tə-trŏf′) See **chemotroph.**

chemokine (kē′mō-kīn′) See **cytokine.**

chemosynthesis (kē′mō-sĭn′thĭ-sĭs) The formation of organic compounds using the energy released from chemical reactions instead of the energy of sunlight. Bacteria living in aphotic areas of the ocean are able to survive by chemosynthesis. They use energy derived from the oxidation of inorganic chemicals, such as sulfur released from deep hydrothermal vents, to produce their food. Compare **photosynthesis.**

chemosynthetic bacteria (kē′mō-sĭn-thĕt′ĭk) Bacteria that make food by chemosynthesis. Nitrifying bacteria are a type of chemosynthetic bacteria, as are the bacteria that live around vents in the bottom of the ocean.

chemotaxis (kē′mō-tăk′sĭs, kĕm′ō-) The characteristic movement or orientation of an organism or cell along a chemical concentration gradient either toward or away from the chemical stimulus. Bacteria exhibit chemotaxis when they move toward a source of nutrients.

chemotherapy (kē′mō-thĕr′ə-pē) **1.** The treatment of disease, especially cancer, using drugs that are destructive to malignant cells and tissues. **2.** The treatment of disease using chemical agents or drugs that are selectively toxic to the causative agent of the disease, such as a microorganism.

chemotroph (kē′mō-trŏf′) An organism that manufactures its own food through chemosynthesis (the oxidation of inorganic chemical compounds) as opposed to photosynthesis. The sulfur-oxidizing bacteria found at deep-sea hydrothermal vents and nitrifying bacteria in the soil are chemotrophs. Also called *chemoautotroph*. Compare **phototroph.**

Cherenkov effect (chə-rĕng′kôf, -kəf) Variant of **Cerenkov effect.**

Cherenkov radiation Variant of **Cerenkov radiation.**

chert (chûrt) A hard, brittle sedimentary rock consisting of microcrystalline quartz. It is often reddish-brown to green but can also occur in a variety of other colors, especially white, pink, brown, or black. Chert often contains impurities such as calcium, iron-oxide, or the remains of silica-rich organisms. It usually occurs as nodules in limestone and dolomite and has curved fractures.

chickenpox (chĭk′ən-pŏks′) A highly contagious infectious disease, usually of children, caused by the varicella-zoster virus of the genus *Varicellavirus.* The infection is characterized by fever, and itching skin blisters that start on the trunk of the body and spread to the extremities. Also called *varicella.*

chill factor (chĭl) See **wind-chill factor.**

chimney (chĭm′nē) **1.** An elongated opening in a volcano through which magma reaches the Earth's surface. **2.** A stack of minerals that have precipitated out of a hydrothermal vent on the floor of a sea or ocean. See more at **hydrothermal vent. 3.** An isolated column of rock along a coastline, formed by the erosion of a sea cliff by waves. Chimneys are smaller than stacks.

chinook (shĭ-no͝ok′, chĭ-) **1.** A moist, warm wind blowing from the sea in coastal regions of the Pacific Northwest. **2.** A warm, dry wind descending from the eastern slopes of the Rocky Mountains, causing a rapid rise in temperature. These winds often melt snow quite rapidly, at times at a rate of up to a foot per hour. See also **foehn.**

chip (chĭp) See **integrated circuit.**

chi particle (kī, kē) An electrically neutral meson having a mass 6,687 times that of the electron and a mean lifetime of approximately 1.5×10^{-20} seconds. See Table at **subatomic particle.**

chirality (kī-răl′ĭ-tē) The characteristic of a structure (usually a molecule) that makes it impossible to superimpose it on its mirror image. Also called *handedness.* See also **helicity, invariance, optical isomer.**

Chiron (kī′rŏn′) A large cometlike body with an orbit mostly between Saturn and Uranus. Discovered in 1977, Chiron was originally identified as an asteroid, but it has since been reclassified as a Centaur. Like a comet, Chiron has been observed to display a nebulous coma in its closest approach to the Sun, but at approximately 200 km (124 mi) in diameter it is far larger than any other known comet. See more at **Centaur.**

chiropractic medicine (kī′rə-prăk′tĭk) A system for treating disorders of the body, especially those of the bones, muscles, and joints, by manipulating the vertebrae of the spine and related structures.

chi-square (kī′skwâr′) A test statistic that is calculated as the sum of the squares of observed values minus expected values divided by the expected values.

chitin (kīt′n) A tough, semitransparent substance that is the main component of the exoskeletons of arthropods, such as the shells of crustaceans and the outer coverings of insects. Chitin is also found in the cell walls of certain fungi and algae. Chemically, it is a nitrogenous polysaccharide (a carbohydrate).

chlamydia (klə-mĭd′ē-ə) Plural **chlamydiae** (klə-mĭd′ē-ē′). Any of various bacteria of the genus *Chlamydia,* several species of which cause common infections in humans and animals, including neonatal conjunctivitis, pneumonia, bronchitis, pharyngitis, and sexually transmitted infections of the pelvis and urethra.

chloramine (klôr′ə-mēn′) **1.** One of three bactericidal compounds that form when chlorine and ammonia react in water. Chloramines are used to purify drinking water, since they are more stable than chlorine and produce fewer harmful by-products. **2.** Any of various organic compounds containing a chlorine atom attached to a nitrogen atom, especially one of three sodium salts that are used as antiseptics and germicides. The most widely used is called *chloramine-T.*

chloramphenicol (klôr′ăm-fĕn′ĭ-kôl′, -kōl′) An antibiotic derived from the soil bacterium *Streptomyces venezuelae* or produced synthetically, and effective against a broad spectrum of microorganisms. *Chemical formula:* $C_{11}H_{12}Cl_2N_2O_5$.

chlorate (klôr′āt′) A chemical compound containing the group ClO_3.

chlordane (klôr′dān′) also **chlordan** (klôr′dăn′) A colorless, odorless, viscous liquid that occurs in several isomers and was formerly used as an insecticide. Because it can damage the liver and nervous system and remains as a toxin in the environment for many years, chlordane was banned in 1988. *Chemical formula:* $C_{10}H_6Cl_8$.

chlorenchyma (klə-rĕng′kə-mə) Plant tissue consisting of parenchyma cells that contain chloroplasts and forming the basic green tissue of plant leaves and stems.

chloride (klôr′īd′) A compound, such as ammonium chloride, containing chlorine and another element or radical.

chlorinate (klôr′ə-nāt′) To add chlorine or one of its compounds to a substance. Water and sewage are chlorinated to be disinfected, and paper pulp is chlorinated to be bleached.

chlorine (klôr′ēn′) *Symbol* **Cl** A greenish-yellow, gaseous element of the halogen group that can combine with most other elements and is found chiefly in combination with the alkali metals as chlorates and chlorides. Chlorine is highly irritating and poisonous. It is used in purifying water, as a disinfectant and bleach, and in the manufacture of numerous chemical compounds. Atomic number 17; atomic weight 35.453; freezing point −100.98°C; boiling point −34.6°C; specific gravity 1.56 (−33.6°C); valence 1, 3, 5, 7. See **Periodic Table.** See Note at **chlorophyll.**

chlorinity (klôr-ĭn′ĭ-tē) A measure of the amount of chlorine and other halides in water, especially seawater. Chlorinity is measured in terms of the mass of silver required to precipitate completely the halogens in a 0.3285 kg (0.7227 lb) sample of seawater. It used to be measured in terms of the total mass, in grams, of chlorine and other halides in 1 g of seawater.

chlorite[1] (klôr′īt′) A usually green or black, flaky mineral that looks like mica. Chlorite is either monoclinic or triclinic and occurs in low-grade metamorphic rocks (rocks that have undergone little metamorphism). It often forms by the alteration of dark minerals (often rich in iron and magnesium) during metamorphism. *Chemical formula:* $(Mg,Fe,Al)_6(Si,Al)_4O_{10}(OH)_8$.

chlorite[2] (klôr′īt′) A salt containing the group ClO_2.

chlorofluorocarbon (klôr′ō-flo͝or′ō-kär′bən) A fluorocarbon containing chlorine. Chlorofluorocarbons are destructive to the Earth's ozone layer. For this reason, the production and use of chlorofluorocarbons has been sharply reduced in recent years.

chloroform (klôr′ə-fôrm′) A colorless, toxic, sweet-tasting liquid formed by combining methane with chlorine. It is used as a solvent and was once widely used as an anesthetic. *Chemical formula:* $CHCl_3$.

chlorophyll (klôr′ə-fĭl) Any of several green pigments found in photosynthetic organisms, such as plants, algae, and cyanobacteria. At its molecular core, chlorophyll has a porphyrin structure but contains a magnesium atom at its center and a long carbon side chain. Chlorophyll absorbs red and blue wavelengths of light, but reflects green. When it absorbs light energy, a chlorophyll molecule enters a higher energy state in which it easily gives up an electron to the first available electron-accepting molecule nearby. This electron moves through a chain of acceptors and is ultimately used in the synthesis of ATP, which provides chemical energy for plant metabolism. Plants rely on two forms of chlorophyll, chlorophyll a (*chemical formula:* $C_{66}H_{72}MgN_4O_5$) and chlorophyll b (*chemical formula:* $C_{66}H_{70}MgN_4O_6$), which have slightly different light absorbing properties. All plants, algae, and cyanobacteria have chlorophyll a, since only this compound can pass an electron to acceptors in oxygen-producing photosynthetic reactions. Chlorophyll b absorbs light energy that is then transferred to chlorophyll a. Several protist groups such as brown algae and diatoms lack chlorophyll b but have another pigment, chlorophyll c, instead. Other closely related pigments are used by various bacteria in photosynthetic reactions that do not produce oxygen. See more at **photosynthesis.**

WORD HISTORY **chlorophyll**

From its name, one might think that chlorophyll has chlorine in it, but it doesn't. The *chloro–* of *chlorophyll* comes from the Greek word for "green"; chlorophyll in fact is the chemical compound that gives green plants their characteristic color. The name of the chemical element *chlorine* comes from the same root as the prefix *chloro–*, and is so called because it is a greenish-colored gas.

chlorophyte (klôr′ə-fīt′) See **green algae.**

chloroplast (klôr′ə-plăst′) A plastid in the cells of green plants and green algae that contains chlorophylls and carotenoid pigments and creates glucose through photosynthesis. In plants, chloroplasts are usually disk-shaped and can reorient themselves in the cell to vary their exposure to sunlight. Chloroplasts contain the saclike membranes known as **thylakoids**, which contain the chlorophyll and are arranged in stacklike structures known as **grana.** Besides conducting photosynthesis, plant chloroplasts store starch and are involved in amino acid synthesis. Like mitochondria, chloroplasts have their own DNA that is different from the DNA in the nucleus, and chloroplasts are therefore believed to have evolved from symbiont bacteria, their DNA being a remnant of their past existence as independent organisms. See more at **cell, photosynthesis.**

chlorosis (klə-rō′sĭs) The yellowing or whitening of normally green plant tissue because of loss or decreased production of chlorophyll, often as a result of disease, insufficient light, or inadequate sources of iron and magnesium.

cholecalciferol (kō′lĭ-kăl-sĭf′ə-rôl′, -rōl′) See **vitamin D_3.**

cholecystectomy (kō′lĭ-sĭ-stĕk′tə-mē) Surgical removal of the gallbladder.

cholelithiasis (kō′lə-lĭ-thī′ə-sĭs) The presence or formation of gallstones in the gallbladder or bile ducts.

cholera (kŏl′ər-ə) An infectious, sometimes fatal disease of the small intestine caused by the bacterium *Vibrio cholerae.* It is spread from contaminated water and food and causes severe diarrhea, vomiting, and dehydration.

cholesterol (kə-lĕs′tə-rôl′) A sterol found widely in animal and plant tissues. It is a main component of blood plasma and cell membranes, and it is an important precursor of many steroid hormones (such as the estrogens, testosterone, and cortisol), vitamin D_2, and bile acids. In vertebrates, cholesterol is manufactured by the liver or absorbed from food in the intestine. Higher than normal amounts of cholesterol in the blood are associated with higher risk for developing coronary artery disease and atherosclerosis. *Chemical formula:* $C_{27}H_{46}O$. See also **high-density lipoprotein, low-density lipoprotein.**

choline (kō′lēn′) A natural amine often classed in the vitamin B complex. It is incorporated into the structure of many other biologically important molecules, such as acetylcholine and lecithin. *Chemical formula:* $C_5H_{15}NO_2$.

cholinergic (kō′lə-nûr′jĭk) **1.** Relating to a neuron or axon that is activated by or is capable of releasing acetylcholine when a nerve impulse passes. The nerve endings of the parasympathetic nervous system are cholinergic. **2.** Having physiological effects similar to those of acetylcholine, as certain drugs.

chondrite (kŏn′drīt′) A stony meteorite that contains chondrules embedded in a fine matrix of the silicate minerals olivine and pyroxene. About 85 percent of all meteorites are chondrites.

chondroitin sulfate (kŏn-drō′ĭ-tĭn) One of several classes of sulfated glycosaminoglycans that are a major constituent of various connective tissues, especially blood vessels, bone, and cartilage. Chondroitin sulfate is used as an over-the-counter dietary supplement by some people with symptoms of arthritis.

chondrule (kŏn′dro͞ol) A small round granule of olivine or pyroxene occurring in many stony meteorites. Chondrules are thought to have formed from the condensation of hot gases in the solar system.

chopper (chŏp′ər) A crudely flaked, unifacial **core tool**, especially one associated with the Oldowan stone culture of the early Paleolithic Period.

chord (kôrd) **1.** A line segment that joins two points on a curve. **2.** A straight line connecting the leading and trailing edges of an airfoil.

chordate (kôr′dāt′) Any of a large group of animals of the phylum Chordata, having at some stage of development a notochord (flexible spinal column) and nerve cord running along the back, a tail stretching above and behind the anus, and gill slits. Chordates probably evolved before the Cambrian Period and are related to the hemichordates, echinoderms, and chaetognaths. The vertebrates, tunicates, and cephalochordates are the three main groups of chordates.

chorion (kôr′ē-ŏn′) The outer membrane that encloses the embryo of a reptile, bird, or mammal. In mammals, the chorion contributes to the development of the placenta.

chorionic villus sampling (kôr′ē-ŏn′ĭk) A prenatal procedure performed in the first trimester of pregnancy to detect chromosomal defects, consisting of retrieval and examination of tissue from the chorionic villi (projections on the surface of the chorion). The tissue is obtained by passing a catheter through the cervix into the placenta.

C horizon In ABC soil, the lowermost zone, consisting mainly of unconsolidated, weathered rock fragments. By comparison with the A and B horizons, the C horizon is relatively unaffected by the processes of soil formation and therefore does not exhibit as much internal layering.

choroid (kôr′oid′) *Noun.* **1.** The dark-brown vascular coat of the eye between the sclera and the retina. —*Adjective.* **2.** Resembling the chorion; membranous. **3.** Relating to or involving the choroid of the eye.

chromate (krō′māt′) A chemical compound containing the group CrO_4.

chromatic (krō-măt′ĭk) Relating to color or colors.

chromatic aberration See under **aberration.**

chromatid (krō′mə-tĭd) Either of the two strands formed when a chromosome duplicates itself as part of the early stages of cell division. The chromatids are joined together by a single centromere and later separate to become individual chromosomes. See more at **meiosis, mitosis.**

chromatin (krō′mə-tĭn) The substance distributed in the nucleus of a cell that condenses to form chromosomes during cell division. It consists mainly of DNA and proteins called **histones**.

chromatography (krō′mə-tŏg′rə-fē) A technique used to separate the components of a chemical mixture by moving the mixture along a stationary material, such as gelatin. Different components of the mixture are caught by the material at different rates and form isolated bands that can then be analyzed.

chromium (krō′mē-əm) *Symbol* **Cr** A hard, shiny, steel-gray metallic element that is rust-resistant and does not tarnish easily. It is used to plate other metals, to harden steel, and to make stainless steel and other alloys. Atomic number 24; atomic weight 51.996; melting point 1,890°C; boiling point 2,482°C; specific gravity 7.18; valence 2, 3, 6. See **Periodic Table.**

chromodynamics (krō′mō-dī-năm′ĭks) See **quantum chromodynamics.**

chromoplast (krō′mə-plăst′) A plastid in plant cells that synthesizes and contains pigments other than chlorophyll, usually yellow or orange carotenoids. See Note at **photosynthesis.**

chromosome (krō′mə-sōm′) A structure in all living cells that consists of a single molecule of DNA bonded to various proteins and that carries the genes determining heredity. In all eukaryotic cells, the chromosomes occur as threadlike strands in the nucleus. During cell reproduction, these strands coil up and condense into much thicker structures that are easily viewed under a microscope. Chromosomes occur in pairs in all of the cells of eukaryotes except the reproductive cells, which have one of each chromosome, and some red blood cells (such as those of mammals) that expel their nuclei. In bacterial cells and other prokaryotes, which have no nucleus, the chromosome is a circular strand of DNA located in the cytoplasm.

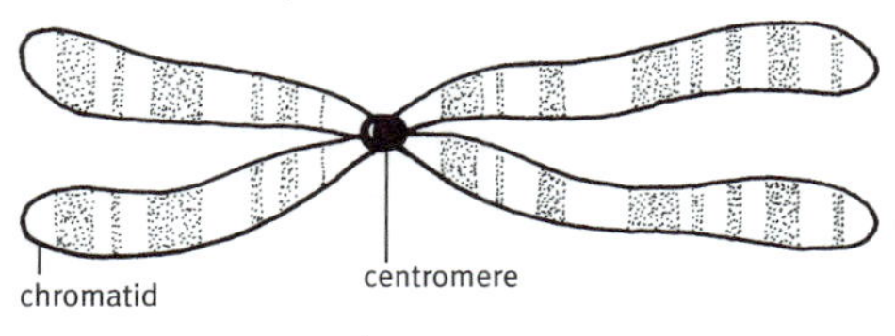

chromosome

chromosphere (krō′mə-sfîr′) A glowing, transparent layer of gas surrounding the photosphere of a star. The Sun's chromosphere is several thousand kilometers thick, is composed mainly of hydrogen at temperatures of 6,000° to 20,000°K, and gives off reddish light.

chronic (krŏn′ĭk) Relating to an illness or medical condition that is characterized by long duration or frequent recurrence. Diabetes and hypertension are chronic diseases. Compare **acute.**

chronic fatigue syndrome A syndrome characterized by debilitating fatigue and a combination of flulike symptoms such as sore throat, swollen lymph glands, low-grade fever, headache, and muscle pain or weakness. The cause is unknown.

chronic obstructive pulmonary disease Any of several pulmonary diseases, especially emphysema, in which chronic airway obstruction causes respiratory abnormalities, including shortness of breath.

chronometer (krə-nŏm′ĭ-tər) An extremely accurate clock or other timepiece. Chronometers are used in scientific experiments, navigation, and astronomical observations. It was the invention of a chronometer capable of being used aboard ship, in 1762, that allowed navigators for the first time to accurately determine their longitude at sea.

chrysalis (krĭs′ə-lĭs) **1.** The pupa of certain kinds of insects, especially of moths and butterflies, that is inactive and enclosed in a firm case or cocoon from which the adult eventually emerges. **2.** The case or cocoon of a chrysalis.

chyme (kīm) The thick semifluid mass of partly digested food that is passed from the stomach to the duodenum.

chymosin (kī′mə-sĭn) See **rennin.**

-cide A suffix that means "a killer of." It is used to form the names of chemicals that kill a specified organism, such as *pesticide,* a chemical that kills pests.

cilium (sĭl′ē-əm) Plural **cilia.** A tiny hairlike projection on the surface of some cells and microscopic organisms, especially protozoans. Cilia are capable of whipping motions and are used by some microorganisms, such as paramecia, for movement. Cilia lining the human respiratory tract act to remove foreign matter from air before it reaches the lungs.

cinder cone (sĭn′dər) A steep, conical hill consisting of glassy volcanic fragments that accumulate around and downwind from a volcanic vent. Cinder cones range in size from tens to hundreds of meters tall.

cinnamic acid (sə-năm′ĭk) A white crystalline organic acid obtained from cinnamon or from balsams, or made synthetically. It is used to manufacture perfumery compounds. *Chemical formula:* $C_9H_8O_2$.

ciprofloxacin (sĭp′rō-flŏk′sə-sĭn) A synthetic antibiotic with a broad spectrum of antibacterial activity, used mostly to treat skin, urinary tract, and respiratory tract infections.

circadian rhythm (sər-kā′dē-ən) A daily cycle of biological activity based on a 24-hour period and influenced by regular variations in the environment, such as the alternation of night and day. Circadian rhythms include sleeping and waking in animals, flower closing and opening in angiosperms, and tissue growth and differentiation in fungi. See also **biological clock.**

A CLOSER LOOK **circadian rhythm**

The *circadian rhythm,* present in humans and most other animals, is generated by an internal clock that is synchronized to light-dark cycles and other cues in an organism's environment. This internal clock accounts for waking up at the same time every day even without an alarm clock. It also causes nocturnal animals to function at night when diurnal creatures are at rest. Circadian rhythms can be disrupted by changes in daily schedule. Biologists have observed that birds exposed to artificial light for a long time sometimes build nests in the fall instead of the spring. While the process underlying circadian rhythm is still being investigated, it is known to be controlled mainly by the release of hormones. In humans, the internal clock is located within the brain's hypothalamus and pineal gland, which releases melatonin in response to the information it receives from photoreceptors in the retina. Nighttime causes melatonin secretion to rise, while daylight inhibits it. Even when light cues are absent, melatonin is still released in a cyclical manner.

---

circinate (sûr′sə-nāt′) Rolled up in the form of a coil with the tip in the center, as an unexpanded fern frond. See more at **vernation.**

circle (sûr′kəl) A closed curve whose points are all on the same plane and at the same distance from a fixed point (the center).

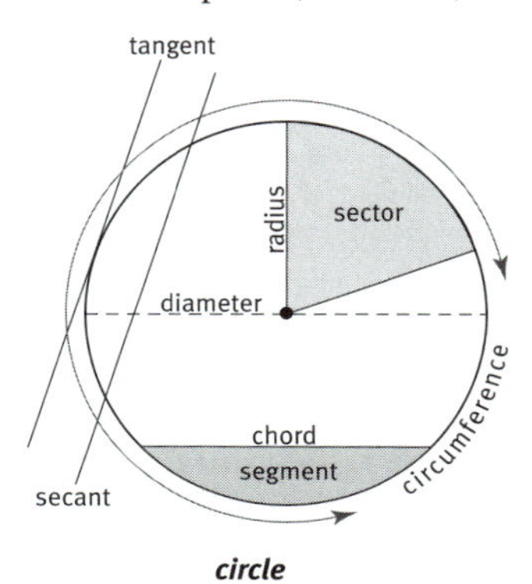

circle

circuit (sûr′kĭt) **1.** A closed path through which an electric current flows or may flow. ► Circuits in which a power source is connected to two or more components (such as light bulbs, or logic gates in a computer circuit), one after the other, are called **series circuits.** If the circuit is broken, none of the components receives a current. Circuits in which a power source is directly connected to two or more components are called **parallel circuits.** If a break occurs in the circuit, only the component along whose path the break occurs stops receiving a current. **2.** A system of electrically connected parts or devices.

circuit

Bulbs in a simple circuit (top) *and parallel circuit* (bottom left) *emit bright light since each is directly connected in its own circuit to a power source. Bulbs in a series circuit* (bottom right) *emit dim light since each consumes a portion of the battery's power coming through a single circuit.*

circuit board See **printed circuit board.**

circuit breaker A switch that automatically interrupts the flow of electric current if the current exceeds a preset limit, measured in amperes. Circuit breakers are used most often as a safety precaution where excessive current through a circuit could be hazardous. Unlike fuses, circuit breakers can usually be reset and reused.

circulate (sûr′kyə-lāt′) To move in or flow through a circle or a circuit. Blood circulates through the body as it flows out from the heart to the tissues and back again.

circulation (sûr′kyə-lā′shən) The flow of fluid, especially blood, through the tissues of an organism to allow for the transport and exchange of blood gases, nutrients, and waste products. In vertebrates, the circulation of blood to the tissues and back to the heart is caused by the pumping action of the heart. Oxygen-rich blood is carried away from the heart by the arteries, and oxygen-poor blood is returned to the heart by the veins. The circulation of lymph occurs in a separate system of vessels (the lymphatic system). Lymph is pumped back to the heart by the contraction of skeletal muscles.

circulatory system (sûr′kyə-lə-tôr′ē) The system that circulates blood through the body, consisting of the heart and blood vessels. In all vertebrates and certain invertebrates, the circulatory system is completely contained within a network of vessels (known as a **closed circulatory system**). In arthropods and many other invertebrates, a substance analogous to blood (known as hemolymph) is pumped through vessels that open into the intercellular spaces (in what is known as an **open circulatory system**). In vertebrates, the lymphatic system is also considered part of the circulatory system.

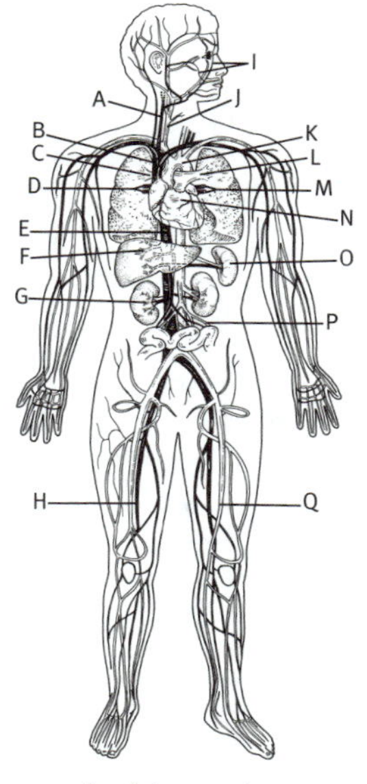

circulatory system

Anterior view of circulatory system with veins in black and arteries in white. A. *jugular vein,* B. *superior vena cava,* C. *right pulmonary artery (to lung),* D. *right pulmonary vein (from lung),* E. *inferior vena cava,* F. *blood vessels in liver,* G. *blood vessels in kidney,* H. *femoral vein,* I. *facial veins,* J. *carotid artery,* K. *aorta,* L. *left pulmonary artery (to lung),* M. *left pulmonary vein (from lung),* N. *heart,* O. *spleen,* P. *blood vessels feeding intestines,* Q. *femoral artery*

circum– A prefix meaning "around", as in *circumscribe,* to draw a figure around another figure.

circumference (sər-kŭm′fər-əns) **1a.** The boundary line of a circle. **b.** The boundary line of a figure, area, or object. **2.** The length of such a boundary. The circumference of a circle is computed by multiplying the diameter by pi.

circumpolar (sûr′kəm-pō′lər) **1.** Located or found in one of Earth's polar regions. **2.** Denoting a star that from a given observer's latitude does not go below the horizon during its diurnal motion. The closer an observer is to one of the poles, the greater the portion of the sky that contains circumpolar stars. At the pole itself, all stars are circumpolar.

circumscribe (sûr′kəm-skrīb′) To draw a figure around another figure so as to touch as many points as possible. A circle that is circumscribed around a triangle touches it at each of the triangle's three vertices.

cirque (sûrk) A steep, amphitheatre-shaped hollow occurring at the upper end of a mountain valley, especially one forming the head of a glacier or stream. Cirques are formed by the erosive activity of glaciers and often contain a small lake.

cirrhosis (sĭ-rō′sĭs) A chronic disease of the liver characterized by the replacement of normal tissue with scar tissue and the loss of functional liver cells. It is most commonly caused by chronic alcohol abuse, but can also result from nutritional deprivation or infection, especially by the hepatitis virus.

cirripede (sîr′ə-pēd′) also **cirriped** (sîr′ə-pĕd′) Any of various small marine crustaceans of the subclass Cirripedia, which includes the barnacles. Some cirripedes become internal parasites of other marine invertebrates in the adult stage, while others (the barnacles) attach themselves to objects and grow a hard shell. Cirripedes are related to copepods.

cirrocumulus (sîr′ō-kyo͞om′yə-ləs) Plural **cirrocumuli** (sîr′ō-kyo͞om′yə-lī′). A high-altitude cloud composed of a series of small, regularly arranged cloudlets in the form of ripples or grains. Cirrocumulus clouds generally form between 6,100 and 12,200 m (20,000 and 40,000 ft) and are composed exclusively of ice crystals. See illustration at **cloud.**

cirrostratus (sîr′ō-străt′əs) Plural **cirrostrati** (sîr′ō-străt′ī′). A thin, hazy, high-altitude cloud composed of ice crystals, often covering the sky in sheets and producing a halo effect around the sun. Cirrostratus clouds generally form between 6,100 and 12,200 m (20,000 and 40,000 ft). See illustration at **cloud.**

cirrus (sîr′əs) Plural **cirri** (sîr′ī′). A high-altitude cloud composed of feathery white patches or bands of ice crystals. Cirrus clouds generally form between 6,100 and 12,200 m (20,000 and 40,000 ft). See illustration at **cloud.**

cisplatin (sĭs-plā′tn) A platinum-containing chemotherapeutic drug used in the treatment of metastatic ovarian or testicular cancers and advanced bladder cancer. *Chemical formula:* $Cl_2H_6N_2Pt$.

cistron (sĭs′trŏn′) A section of DNA that contains the genetic code for a single polypeptide and functions as a hereditary unit.

citrate (sĭt′rāt′) A salt or ester of citric acid.

citric acid (sĭt′rĭk) A white, odorless acid that has a sour taste and occurs widely in plants, especially in citrus fruit, and is formed during the Krebs cycle. It is used in medicine and as a flavoring. Ions of citric acid are a by-product of the metabolism of carbohydrates during the Krebs cycle.*Chemical formula:* $C_6H_8O_7$.

citric acid cycle See **Krebs cycle.**

citrine (sĭ-trēn′, sĭt′rēn′) A pale-yellow variety of crystalline quartz resembling topaz. The coloring is caused by the presence of a small amount of iron in the crystal structure.

citronella (sĭt′rə-nĕl′ə) The pale-yellow, lemon-scented oil obtained from the leaves of a tropical Asian grass (*Cymbopogon nardus*), used in insect repellents and perfumes. Citronella consists primarily of an aldehyde of octane.

citrulline (sĭt′rə-lēn′) An amino acid originally isolated from watermelon, occurring mostly in the liver as an intermediate in the conversion of ornithine to arginine during urea formation. *Chemical formula:* $C_6H_{13}N_3O_3$.

citrus (sĭt′rəs) **1.** Any of various evergreen trees or shrubs bearing fruit with juicy flesh and a thick rind. Citrus trees are native to southern and southeast Asia but are grown in warm climates around the world. Many species have spines. The orange, lemon, lime, and grapefruit are citrus trees. **2.** The usually edible fruit of one of these trees or shrubs.

civil engineering (sĭv′əl) The branch of engineering that specializes in the design and construction of structures such as bridges, roads, and dams.

cl Abbreviation of **centiliter.**

Cl The symbol for **chlorine.**

Clactonian (klăk-tō′nē-ən) Relating to an early Paleolithic tool culture of northwest Europe, characterized by simple core and flake tools. Traditionally considered to predate the Acheulian culture, it is now thought by some to be contemporaneous.

clade (klād) A grouping of organisms made on the basis of phylogenetic relationship, rather than purely on shared features. Clades consist of a common ancestor and all its descendants. The class Aves (birds) is a clade, whereas the class Reptilia (reptiles) is not, since it does not include birds, which are descended from the dinosaurs, a kind of reptile. Many modern taxonomists prefer to use clades in classification, and not all clades correspond to traditional groups like classes, orders, and phyla. Compare **grade.**

cladistics (klə-dĭs′tĭks) A system of classification based on the phylogenetic relationships and evolutionary history of groups of organisms, rather than purely on shared features. Many modern taxonomists prefer cladistics to the traditional hierarchies of Linnean classification systems. Compare **Linnean.**

cladogram (klăd′ə-grăm′, klā′də-) A branching treelike diagram used to illustrate evolutionary (phylogenetic) relationships among organisms. Each node, or point of divergence, has two branching lines of descendance, indicating evolutionary divergence from a common ancestor. The endpoints of the tree represent individual species, and any node together with its descendant branches and subbranches constitutes a clade.

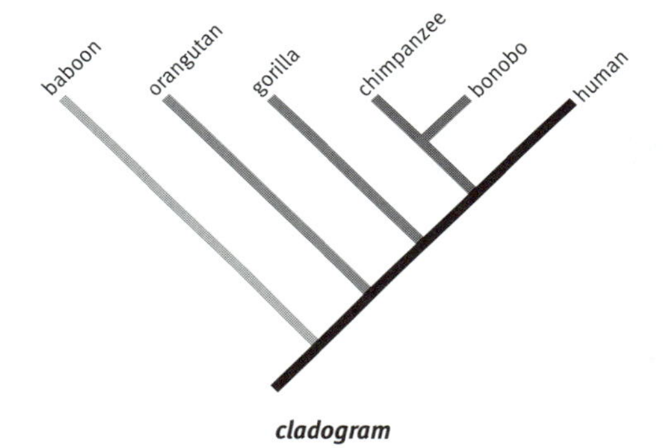

cladogram

Evolutionary relationship among primates, showing that chimpanzees and bonobos are more closely related to each other than to humans, and that all three are more closely related to each other than they are to gorillas, orangutans, and baboons.

cladophyll (klăd′ə-fĭl′) A photosynthetic branch or portion of a stem that resembles and functions as a leaf, as in the asparagus. Also called *cladode.*

class (klăs) A taxonomic category of organisms ranking above an order and below a phylum or division. In modern taxonomic schemes, the names of classes end in *–phyceae* for the various groups of algae, *–mycetes* for fungi, and *–opsida* for plants (as in *Liliopsida,* the class of plants also termed monocotyledons). The names of classes belonging to phyla of the animal kingdom, however, are formed in various ways, as *Osteichthyes* the bony fishes, *Aves,* the birds, and *Mammalia,* the mammals, all of which are classes belonging to the subphylum Vertebrata (the vertebrates) in the phylum Chordata. See Table at **taxonomy.**

classical conditioning (klăs′ĭ-kəl) A process of behavior modification in which a subject learns to respond in a desired manner such that a neutral stimulus (the *conditioned stimulus*) is repeatedly presented in association with a stimulus (the *unconditioned stimulus*) that elicits a natural response (the *unconditioned response*) until the neutral stimulus alone elicits the same response (now called the *conditioned response*). For example, in the experiments of Pavlov, food is the unconditioned stimulus that produces salivation, a reflex or unconditioned response. The bell is the conditioned stimulus, which eventually produces salivation in the absence of food. This salivation is the conditioned response.

classical physics Physics that does not make use of quantum mechanics or the theory of relativity. Newtonian mechanics, thermodynamics, and Maxwell's theory of electromagnetism are all examples of classical physics. Many theories in classical physics break down when applied to extremely small objects such as atoms or to objects moving near the speed of light. ▸ **Classical mechanics** refers to the classical physics of bodies and forces, especially Newton's laws of motion and the principles of mechanics based on them. Compare **quantum mechanics.**

classification (klăs′ə-fĭ-kā′shən) The systematic grouping of organisms according to the structural or evolutionary relationships among them. Organisms are normally classified by observed similarities in their body and cell structure or by evolutionary relationships based on the analysis of sequences of their DNA. See more at **cladistics, Linnean.** See Table at **taxonomy.**

clast (klăst) A rock fragment or grain resulting from the breakdown of larger rocks. —*Adjective* **clastic** (klăs′tĭk)

clavicle (klăv′ĭ-kəl) Either of two slender bones that extend from the upper part of the sternum (breastbone) to the shoulder. Also called *collarbone.*

claw (klô) **1.** A sharp, curved nail at the end of a toe of a mammal, reptile, or bird. **2.** A pincer, as of a lobster or crab, used for grasping.

clay (klā) A stiff, sticky sedimentary material that is soft and pliable when wet and consists mainly of various silicates of aluminum. Clay particles are smaller than silt, having a diameter less than 0.0039 mm. Clay is widely used to make bricks, pottery, and tiles.

claystone (klā′stōn′) A fine-grained, dark gray to pink sedimentary rock consisting primarily of clay that is compacted and hardened. Claystone is similar to shale but without laminations.

clean room (klēn) A room that is maintained free of contaminants, such as dust or bacteria. Clean rooms are used in laboratory work and in the production of precision parts for electronic or aerospace equipment. Also called *white room.*

clear-air turbulence (klîr′âr′) Atmospheric turbulence that occurs under tranquil and cloudless conditions at high altitudes near jet streams, mountain ranges, and developing storm systems. It subjects aircraft to strong updrafts and downdrafts.

cleavage (klē′vĭj) **1.** *Geology.* The breaking of certain minerals along specific planes, making smooth surfaces. These surfaces are parallel to the faces of the molecular crystals that make up the minerals. A mineral that exhibits cleavage breaks into smooth pieces with the same pattern of parallel surfaces regardless of how many times it is broken. Some minerals, like quartz, do not have a cleavage and break into uneven pieces with rough surfaces. **2a.** *Biology.* The series of mitotic cell divisions by which a single fertilized egg cell becomes a many-celled blastula. Each division produces cells half the size of the parent cell. **b.** Any of the single cell divisions in such a series.

cleaver (klē′vər) A bifacial stone tool flaked to produce a straight, sharp, relatively wide edge at one end. Cleavers are early core tools associated primarily with the Acheulian tool culture.

cleft lip (klĕft) A congenital deformity characterized by a vertical cleft or pair of clefts in the upper lip, with or without involvement of the palate.

cleft palate Incomplete closure of the palate during development of an embryo, resulting in a split along part or all of the roof of the mouth.

cleistogamous (klī-stŏg′ə-məs) Of or relating to a flower that does not open and is self-pollinated in the bud. The fertile flowers of the violet are inconspicuous and cleistogamous, while the plant's more familiar showy flowers are usually infertile.

cleistothecium (klī′stə-thē′sē-əm) Plural **cleistothecia.** A closed spherical ascocarp, such as the truffle.

clickstream (klĭk′strēm′) The sequence of links that are clicked on while browsing a website or series of websites.

client (klī′ənt) A program that runs on a personal computer or workstation connected to a computer network and requests information from a file server.

client/server network A computer network in which one centralized, powerful computer (called the **server**) is a hub to which many less powerful personal computers or workstations (called **clients**) are connected. The clients run programs and access data that are stored on the server. Compare **peer-to-peer network.**

climate (klī′mĭt) The general or average weather conditions of a certain region, including temperature, rainfall, and wind. On Earth, climate is most affected by latitude, the tilt of the Earth's axis, the movements of the Earth's wind belts, the difference in temperatures of land and sea, and topography. Human activity, especially relating to actions relating to the depletion of the ozone layer, is also an important factor.

climatology (klī′mə-tŏl′ə-jē) The scientific study of climates, including the causes and long-term effects of variation in regional and global climates. Climatology also studies how climate changes over time and is affected by human actions.

climax community (klī′măks′) An ecological community in which populations of plants or animals remain stable and exist in balance with each other and their environment. A climax community is the final stage of succession, remaining relatively unchanged until destroyed by an event such as fire or human interference. See more at **succession.**

cline (klīn) A gradual change in an inherited characteristic across the geographic range of a species, usually correlated with an environmental transition such as altitude, temperature, or moisture. For example, the body size in a species of warm-blooded animals tends to be larger in cooler climates (a latitudinal cline), while the flowering time of a plant may tend to be later at higher altitudes (an altitudinal cline). In species in which the gene flow between adjacent populations is high, the cline is typically smooth, whereas in populations with restricted gene flow the cline usually occurs as a series of relatively abrupt changes from one group to the next.

clinical psychology (klĭn′ĭ-kəl) See under **psychology.**

clinopyroxene (klī′nə-pîr′-ŏk-sēn′) Any variety of the mineral pyroxene that crystallizes in the monoclinic system. Diopside and augite are clinopyroxenes.

clitoris (klĭt′ər-ĭs, klĭ-tôr′ĭs) A sensitive external organ of the reproductive system in female mammals and some other animals that is capable of becoming erect. It is located above or in front of the urethra.

cloaca (klō-ā′kə) Plural **cloacae** (klō-ā′sē′). **1.** The body cavity into which the intestinal, urinary, and genital canals empty in birds, reptiles, amphibians, most fish, and monotremes. The cloaca has an opening for expelling its contents from the body, and in females it serves as the depository for sperm. Also called *vent.* **2.** See **vent** (sense 2a).

clone (klōn) *Noun.* **1.** A cell, group of cells, or organism that is produced asexually from and is genetically identical to a single ancestor. The cells of an individual plant or animal, except for gametes and some cells of the immune system, are clones because they all descend from a single fertilized cell and are genetically identical. A clone may be produced by fission, in the case of single-celled organisms, by budding, as in the hydra, or in the laboratory by putting the nucleus of a diploid cell into an egg that has had its nucleus removed. Some plants can produce clones from horizontal stems, such as runners. Clones of other cells and some plants and animals can also be produced in a laboratory. See also **therapeutic cloning.** **2.** A copy of a sequence of DNA, as from a gene, that is produced by genetic engineering. The clone is then transplanted into the nucleus of a cell from which genetic material has been removed. —*Verb.* **3.** To produce or grow a cell, a group of cells, or an organism from a single original cell. **4.** To make identical copies of a DNA sequence. See more at **genetic engineering.**

closed chain (klōsd) An arrangement of atoms of the same type that forms a ring. Benzene and naphthalene are closed-chain compounds.

closed circuit **1.** An electric circuit through which current can flow in an uninterrupted path. Compare **open circuit.** **2.** A television system in which the signal is usually sent by cable to a limited number of receivers.

closed interval A set of numbers consisting of all the numbers between a pair of given numbers and including the endpoints.

closed system A physical system that does not interact with other systems. A closed system obeys the **conservation laws** in its physical description. Also called *isolated system.* Compare **open system.**

closed universe A model of the universe in which the curvature of space is roughly spherical, entailing that the universe has finite size. An object moving in a straight line in a closed universe would eventually return to its starting point. According to most current cosmological theories, the universe is closed if it is sufficiently dense, and therefore possesses enough gravitational force to stop or reverse the expansion started by the big bang (resulting in what is called the **big crunch**). Compare **open universe.** See Note at **big bang.**

clot (klŏt) A soft insoluble mass that is formed when blood or lymph gels. During blood clotting, white blood cells, red blood cells, platelets, and various clotting factors interact in a cascade of chemical reactions that are initiated by a wound. When a body tissue is injured, calcium ions and platelets act on prothrombin to produce the enzyme thrombin. Thrombin then catalyzes the conversion of the protein fibrinogen into fibrin, a fibrous protein that holds the clot together. An abnormal clot inside the blood vessels or the heart (a **thrombus** or an **embolus**) can obstruct blood flow.

clotting factor (klŏt′ĭng) Any of various components of plasma involved in the coagulation of blood, including fibrinogen, prothrombin, and calcium ions. Hereditary deficiency of clotting factors can cause coagulation disorders such as **hemophilia**.

cloud (kloud) **1.** A visible body of very fine water droplets or ice particles suspended in the atmosphere at altitudes ranging up to several miles above sea level. Clouds are formed when air that contains water vapor cools below the dew point. **2.** A distinguishable mass of particles or gas, such as the collection of gases and dust in a nebula.

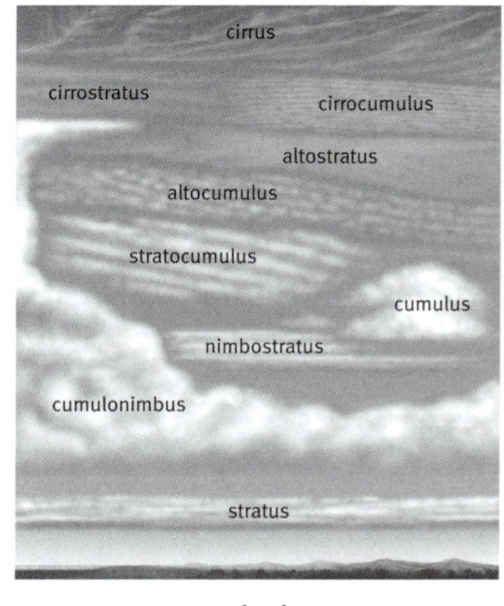

cloud

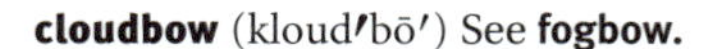
cloudbow (kloud′bō′) See **fogbow.**

cloud chamber A device used to observe the movements of charged atomic and subatomic particles, such as ions, electrons, or muons. Cloud chambers consist of a closed container filled with a gas that is on the verge of condensing. Charged particles passing through the gas ionize the atoms in their path, forming visible lines of condensation. Compare **bubble chamber.**

cloud forest A tropical forest, often near peaks of coastal mountains, that usually has constant cloud cover throughout the year.

cloud forest

Peñas Blancas Valley, Monteverde Biological Cloud Forest Preserve, Costa Rica

cloud seeding A method of making a cloud give up its moisture as rain, especially by releasing particles of dry ice or silver iodide into cold clouds. Dry ice freezes water droplets in the cloud, turning them into nuclei for the formation of raindrops. Silver iodide particles are used because they have a crystal structure similar to ice and can also serve as nuclei for raindrop formation.

cluster headache (klŭs′tər) A recurring headache, most common in males, characterized by severe pain in one eye or the adjacent temple, watering of the eye, and runny nose.

cm Abbreviation of **centimeter.**

Cm The symbol for **curium.**

cnidarian (nī-dâr′ē-ən) Any of various invertebrate animals of the phylum Cnidaria (or Coelenterata), having a body with radial symmetry and tentacles that bear microscopic stinging capsules called nematocysts. The tentacles surround a mouth that opens into a saclike internal cavity and that is used both for ingesting food and for eliminating wastes. Cnidarians evolved in the Precambrian Era, but it is not known from what type of organism. Cnidarians include the jellyfishes, hydras, sea anemones, and corals. Also called *coelenterate.*

Co The symbol for **cobalt.**

coadaptation (kō′ăd′ăp-tā′shən) The reciprocal adaptation of two or more genetically determined features through natural selection. Coadaptation can occur between interacting genes or structures within an organism or between two or more interacting species.

coagulation (kō-ăg′yə-lā′shən) The process of changing from a liquid to a gel or solid state by a series of chemical reactions, especially the process that results in the formation of a blood clot. See more at **clot.**

coal (kōl) A dark-brown to black solid substance formed from the compaction and hardening of fossilized plant parts in the presence of water and in the absence of air. Carbonaceous material accounts for more than 50 percent of coal's weight and more than 70 percent of its volume. Coal is widely used as a fuel, and its combustion products are used as raw material for a variety of products including cement, asphalt, wallboard and plastics. See more at **anthracite, bituminous coal, lignite.**

coal tar A thick, sticky, black liquid obtained through the destructive distillation (heating in the absence of air) of coal. It is used as a source of many organic compounds, such as benzene, naphthalene, and phenols, which are used in dyes, drugs, and other compounds.

coaxial cable (kō-ăk′sē-əl) A cable consisting of an electrically conductive wire surrounded by a layer of insulating material, a layer of shielding material, and an outer layer of insulating material, usually plastic or rubber. The purpose of the shielding layer is to reduce external electrical interference. Coaxial cables are used for transmission of high-frequency audio, video, computer network and other signals.

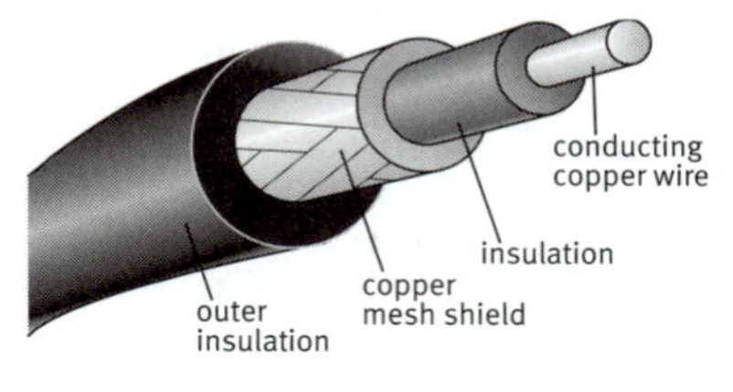

coaxial cable

cobalamin (kō-băl′ə-mĭn) See **vitamin B_{12}.**

cobalt (kō′bôlt′) *Symbol* **Co** A silvery-white, hard, brittle metallic element that occurs widely in metal ores. It is used to make magnetic alloys, heat-resistant alloys, and blue pigment for ceramics and glass. Atomic number 27; atomic weight 58.9332; melting point 1,495°C; boiling point 2,900°C; specific gravity 8.9; valence 2, 3. See **Periodic Table.**

cobble (kŏb′əl) A rock fragment larger than a pebble and smaller than a boulder. Pebbles have a diameter between 64 and 256 mm (2.56 and 10.24 inches) and are often rounded.

COBOL (kō′bôl′) A programming language developed in the late 1950s and early 1960s and used especially for business applications. It is closer to English than many other high-level languages, making it easier to learn.

cocaine (kō-kān′) A colorless or white crystalline alkaloid extracted from coca leaves. Cocaine is sometimes used in medicine as a local anesthetic, especially for the eyes, nose, or throat. It is also widely used as an illicit drug for its euphoric and stimulating effects. *Chemical formula:* $C_{17}H_{21}NO_4$.

coccus (kŏk′əs) Plural **cocci** (kŏk′sī, kŏk′ī). Any of various bacteria having a round or ovoid form such as **streptococcus** or **staphylococcus**, usually grouped in chains.

coccyx (kŏk′sĭks) Plural **coccyges** (kŏk-sī′jēz, kŏk′sĭ-jēz′). A small triangular bone at the base of the spine in humans and apes. It is composed of several fused vertebrae. Also called *tailbone.*

cochlea (kŏk′lē-ə) Plural **cochleae** (kŏk′lē-ē′, -lē-ī′) or **cochleas.** A spiral-shaped cavity of the inner ear and the main organ of hearing. The cochlea contains the nerve endings that transmit sound vibrations from the middle ear to the auditory nerve.

cochlear implant (kŏk′lē-ər) A surgically implanted electronic device that allows people with severe hearing loss to recognize some sounds. It consists chiefly of a microphone and receiver, a processor that converts speech into electronic signals, and an array of electrodes that transmit the signals to the auditory nerve in the inner ear.

Cockcroft (kŏk′krôft′), Sir **John Douglas** 1897–1967. British physicist who, with Ernest Walton, was the first to successfully split an atom using a particle accelerator in 1932. For this work they shared the 1951 Nobel Prize for physics.

cocoon (kə-ko͞on′) **1.** A case or covering of silky strands spun by an insect larva and inhabited for protection during its pupal stage. **2.** A similar protective structure, such as the egg cases made by spiders or earthworms.

code (kōd) **1.** A system of signals used to represent letters or numbers in transmitting messages. **2.** The instructions in a computer program. Instructions written by a programmer in a programming language are often called *source code.* Instructions that have been converted into machine language that the computer understands are called *machine code* or *executable code.* See also **programming language.**

codeine (kō′dēn′) An alkaloid narcotic derived from opium or morphine and used primarily as an analgesic and a cough suppressant. *Chemical formula:* $C_{18}H_{21}NO_3$.

codominant (kō-dŏm′ə-nənt) **1.** Relating to two alleles of a gene pair in a heterozygote that are both fully expressed. When alleles for both white and red are present in a carnation, for example, the result is a pink carnation since both alleles are codominant. **2.** Being one of two or more of the most common or important species in an ecological community. Like a dominant species, codominant species often influence the presence and type of other species in the community.

codon (kō′dŏn′) A sequence of three adjacent nucleotides on a strand of a nucleic acid (such as DNA) that constitutes the genetic code for a specific amino acid that is to be added to a polypeptide chain during protein synthesis. Some amino acids are coded for by more than one codon, and some codons do not signal a particular amino acid but rather signal a stop to protein synthesis.

coefficient (kō′ə-fĭsh′ənt) **1.** A number or symbol multiplied with a variable or an unknown quantity in an algebraic term. For example, 4 is the coefficient in the term $4x$, and x is the coefficient in $x(a + b)$. **2.** A numerical measure of a physical or chemical property that is constant for a system under specified conditions. The speed of light in a vacuum, for example, is a constant.

coefficient of friction A measure of the amount of resistance that a surface exerts on or substances moving over it, equal to the ratio between the maximal frictional force that the surface exerts and the force pushing the object toward the surface. The coefficient of friction is not always the same for objects that are motionless and objects that are in motion; motionless objects often experience more friction than moving ones, requiring more force to put them in motion than to sustain them in motion. ► The **static coefficient of friction** is the coefficient of friction that applies to objects that are motionless. ► The **kinetic** or **sliding coefficient of friction** is the coefficient of friction that applies to objects that are in motion. See also **drag, friction.**

coelacanth (sē′lə-kănth′) Any of various fishes of the order Coelacanthiformes, having lobed, fleshy fins. Coelacanths are crossopterygians, the ancient group of lobe-finned fishes that gave rise to land vertebrates. They were known only from Paleozoic and Mesozoic fossils until a living species (*Latimeria chalumnae*) was found in the Indian Ocean in 1938. At least one other (*Malania anjouanae*) has been discovered since then.

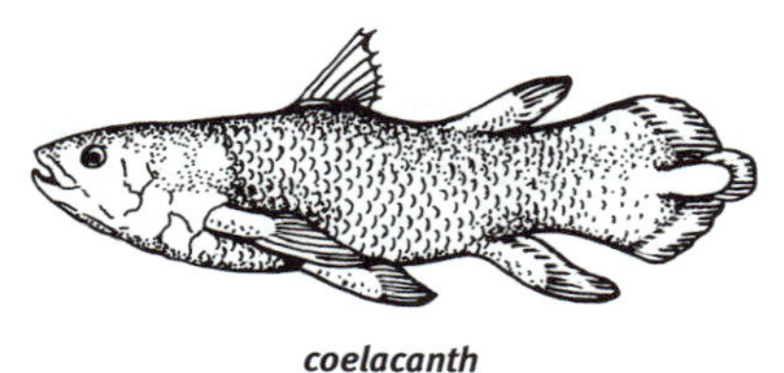

coelacanth

coelenterate (sĭ-lĕn′tə-rĭt) See **cnidarian.**

coelom (sē′ləm) The body cavity that forms from the mesoderm during the embryonic development of more complex animals. The coelom suspends the gut in fluid in the middle of the body, protecting it from gravity and allowing great increases in body size. The presence or absence of a coelom is important for the classification of animal phyla. See more at **deuterostome, protostome.**

coenzyme (kō-ĕn′zīm′) A nonprotein organic substance that usually contains a vitamin or mineral and combines with a specific protein, called an **apoenzyme**, to form an active enzyme system.

coenzyme A A coenzyme that consists of a nucleotide linked to pantothenic acid (part of the vitamin B complex), is present in all living cells, and functions as an acyl group carrier. Coenzyme A is necessary for fatty acid synthesis and oxidation, pyruvate oxidation, and other acetylation reactions. In **cellular respiration**, each of two acetyl groups derived from the original glucose molecule attaches itself to coenzyme A as *acetyl coenzyme A* and then enters the **Krebs cycle.**

coenzyme Q See **ubiquinone.**

coercivity (kō′ər-sĭv′ĭ-tē) The magnetic flux density needed to reduce the magnetization of a material (especially a **ferromagnetic** material) from complete saturation to zero. Coercivity is measured in teslas. Compare **remanance.**

coevolution (kō′ĕv-ə-lo͞o′shən) The evolution of two or more species that interact closely with one another, with each species adapting to changes in the other. The mutually beneficial development of flowering plants and insects such as bees and butterflies that pollinate them is an example of coevolution, as is the mutually antagonistic development of prey and predator species in which defensive adaptations in the one are matched by counteradaptations in the other aimed at neutralizing or overcoming them.

cofactor (kō′făk′tər) A substance, such as a metallic ion or a coenzyme, that must be associated with an enzyme for the enzyme to function. Cofactors work by changing the shape of an enzyme or by actually participating in the enzymatic reaction.

cofunction (kō′fŭngk′shən) The trigonometric function of the complement of an angle. The tangent, for example, is the cofunction of the cotangent.

cogeneration (kō-jĕn′ə-rā′shən) A process in which an industrial facility uses its waste energy to produce heat or electricity. Compare **trigeneration.**

cognition (kŏg-nĭsh′ən) The mental process of knowing, including awareness, perception, reasoning, and judgment.

coherence (kō-hîr′əns, -hĕr′-) A property holding for two or more waves or fields when each individual wave or field is in phase with every other one. Lasers, for example, emit almost perfectly coherent light; all the photons emitted by a laser have the same frequency and are in phase. Since quantum states can be described by a **wave equation**, coherence can hold for quantum states in general, though only among **bosons**. Coherence is generally possible in physical systems that may undergo **superposition**. Maintaining coherence of light is important in fiber optic communications. See also **Bose-Einstein condensate.**

cohesion (kō-hē′zhən) The force of attraction that holds molecules of a given substance together. It is strongest in solids, less strong in liquids, and least strong in gases. Cohesion of molecules causes drops to form in liquids (as when liquid mercury is poured on a piece of glass), and causes condensing water vapor to form the droplets that make clouds. Compare **adhesion.**

cold-blooded (kōld′blŭd′ĭd) Having a body temperature that changes according to the temperature of the surroundings. Fish, amphibians, and reptiles are cold-blooded.

cold front The forward edge of an advancing mass of cold air that pushes under a mass of warm air. Cold fronts often cause precipitation; water vapor in the rising warm air condenses and forms clouds, often resulting in heavy rain, thunderstorms, hail, or snow. Winter cold fronts can cause temperatures to drop significantly. Summer cold fronts reduce humidity as drier, cooler air displaces the humid, warmer air. On a weather map, a cold front is depicted as a blue line with triangles that point in the direction in which the cold air is moving. Compare **occluded front, warm front.** See illustration at **front.**

coleopteran (kō′lē-ŏp′tər-ən, kŏl′ē-) Any of numerous insects of the order Coleoptera, characterized by forewings modified to form tough protective covers for the membranous hind wings. Coleopterans include the beetles, weevils, and fireflies.

coleoptile (kō′lē-ŏp′tĭl, kŏl′ē-) A protective sheath enclosing the shoot tip and embryonic leaves of grasses.

colic (kŏl′ĭk) **1.** Severe abdominal pain, often caused by spasm, obstruction, or distention of any of the hollow viscera, such as the intestines. **2.** A condition seen in infants less than three months old, marked by periods of inconsolable crying lasting for hours at a time for at least three weeks. The cause is unknown.

colitis (kə-lī′tĭs) Inflammation of the colon.

collagen (kŏl′ə-jən) Any of various tough, fibrous proteins found in bone, cartilage, skin, and other connective tissue. Collagens have great tensile strength, and provide these body structures with the ability to withstand forces that stretch them. Collagens consist of three polypeptide chains arranged in a triple helix, and are bundled together in fibers. When boiled in water, collagen is converted into gelatin.

collarbone (kŏl′ər-bōn′) See **clavicle.**

collenchyma (kə-lĕng′kə-mə) A supportive tissue of plants, consisting of elongated living cells with unevenly thickened, nonlignified walls. Collenchyma cells remain alive at maturity. Compare **parenchyma, sclerenchyma.**

collimator (kŏl′ə-mā′tər) A device that turns incoming radiation, such as light, into parallel beams. Simple collimators consists of a tube having a narrow, variable slit at one end and a convex lens at the other. Radiation entering the tube through the slit exits the lens in the form of parallel beams. Collimators are used to establish focal lengths of lenses and to measure the distance of distant objects whose position is known. See illustration at **spectroscope.**

collinear (kə-lĭn′ē-ər) **1.** Sharing a common line, such as two intersecting planes. **2.** Lying on the same line, such as a set of points.

collision zone (kə-lĭzh′ən) See **convergent plate boundary.**

colloid (kŏl′oid′) A mixture in which very small particles of one substance are distributed evenly throughout another substance. The particles are generally larger than those in a solution, and smaller than those in a suspension. Paints, milk, and fog are colloids. Compare **solution, suspension.**

colluvium (kə-lo͞o′vē-əm) Plural **colluviums** or **colluvia.** A loose deposit of rock debris accumulated through the action of rainwash or gravity at the base of a gently sloping cliff or slope.

colon (kō′lən) The longest part of the large intestine, extending from the cecum to the rectum. Water and electrolytes are absorbed, solidified, and prepared for elimination as feces in the colon. The colon also contains bacteria that help in the body's absorption of nutrients from digested material.

colonization (kŏl′ə-nĭ-zā**′**shən) The spreading of a species into a new habitat. For example, flying insects and birds are often the first animal species to initiate colonization of barren islands newly formed by vulcanism or falling water levels. The first plant species to colonize such islands are often transported there as airborne seeds or through the droppings of birds.

colonoscopy (kō′lə-nŏs**′**kə-pē) Inspection of the interior surface of the colon with a flexible endoscope that is equipped to obtain tissue samples and inserted through the rectum.

colony (kŏl**′**ə-nē) A group of the same kind of animals, plants, or one-celled organisms living or growing together. Organisms live in colonies for their mutual benefit, and especially their protection. Multicellular organisms may have evolved out of colonies of unicellular organisms.

colony stimulating factor Any of various glycoproteins that act to promote the production and differentiation of specific blood cell types. Colony stimulating factors circulate in the blood, acting as hormones, and are also secreted locally. An example is erythropoietin, which is produced in the kidney and regulates the formation of red blood cells from progenitor cells in the bone marrow.

color (kŭl**′**ər) **1.** The sensation produced by the effect of light waves striking the retina of the eye. The color of something depends mainly on which wavelengths of light it emits, reflects, or transmits. **2.** Color charge. See also **hadron.**

A CLOSER LOOK **color**

When beams of colored light are mixed, or added, their wavelengths combine to form other *colors.* All spectral colors can be formed by mixing wavelengths corresponding to the *additive primaries* red, green, and blue. When two of the additive primaries are mixed in equal proportion, they form the *complement* of the third. Thus cyan (a mixture of green and blue) is the complement of red; magenta (a mixture of blue and red) is the complement of green; and yellow (a mixture of red and green) is the complement of blue. Mixing the three additive primaries in equal proportions reconstitutes white light. When light passes through a color filter, certain wavelengths are absorbed, or subtracted, while others are transmitted. The *subtractive primaries* cyan, magenta, and yellow can be combined using overlapping filters to form all other colors. When two of the subtractive primaries are combined in equal proportion, they form the additive primary whose wavelength they share. Thus overlapping filters of cyan (blue and green) and magenta (blue and red) filter out all wavelengths except blue; magenta (blue and red) and yellow (red and green) transmit only red; and yellow (red and green) and cyan (blue and green) transmit only green. Combining all three subtractive primaries in equal proportions filters out all wavelengths, producing black. Light striking a colored surface behaves similarly to light passing through a filter, with certain wavelengths being absorbed and others reflected. Pigments are combined to form different colors by a process of subtractive absorption of various wavelengths.

---

colorblind (kŭl**′**ər-blīnd′) Unable to distinguish certain colors. Humans who are colorblind usually cannot distinguish red from green. Many animals, including cats and dogs, are colorblind and unable to distinguish more than a few colors.

color charge A property of quarks and gluons that determines their strong force interaction with each other (especially the attractive and repulsive forces between them). Color charge is considered to be a form of charge, much like electric charge. Although there are three basic color charges (designated blue, green, and red) along with associated anticolors, color charge has nothing to do with the more familiar notion of colors seen by the eye.

color force See **strong force.**

colostomy (kə-lŏs**′**tə-mē) Surgical construction of an opening from the colon through the abdominal wall to the outside of the body for the purpose of excretion.

colposcope (kŏl**′**pə-skōp′) An instrument used for inspection of epithelial cells of the cervix and vagina. —*Noun* **colposcopy** (kŏl-pŏs**′**kə-pē)

coma[1] (kō**′**mə) Plural **comas.** A state of deep unconsciousness, usually resulting from brain trauma or metabolic disease, in which an individual is incapable of sensing or responding to external stimuli.

coma² (kō′mə) Plural **comae** (kō′mē). **1.** *Astronomy.* The brightly shining cloud of gas that encircles the nucleus and makes up the major portion of the head of a comet near the Sun. As a comet moves along its orbit away from the Sun, the gas and dust of the coma dissipate, leaving only the nucleus. A coma can have a diameter of up to 100,000 km (62,000 mi.). See more at **comet. 2.** *Physics.* A diffuse, comet-shaped image of a point source of light or radiation caused by aberration in a lens or mirror. The image appears progressively elongated with distance from the center of the field of view.

comb jelly (kōm) See **ctenophore.**

combustion (kəm-bŭs′chən) **1.** The process of burning. **2.** A chemical change, especially through the rapid combination of a substance with oxygen, producing heat and, usually, light. See also **spontaneous combustion.**

combustion chamber An enclosure in which combustion, especially of a fuel or propellant, is initiated and controlled. See also **internal-combustion engine.**

comet (kŏm′ĭt) A celestial object that orbits the Sun along an elongated path. A comet that is not near the Sun consists only of a nucleus—a solid core of frozen water, frozen gases, and dust. When a comet comes close to the Sun, its nucleus heats up and releases a gaseous coma that surrounds the nucleus. A comet forms a tail when solar heat or wind forces dust or gas off its coma, with the tail always streaming away from the Sun. ▸ **Short-period comets** have orbital periods of less than 200 years and come from the region known as the Kuiper belt. **Long-period comets** have periods greater than 200 years and come from the Oort cloud. See more at **Kuiper belt, Oort cloud.** See Note at **solar system.**

comet

Comet Kohoutek, visible from November 1973 until January 1974

commensalism (kə-mĕn′sə-lĭz′əm) A symbiotic relationship in which one organism derives benefit while causing little or no harm to the other. Examples of commensalism include epiphytic plants, which depend on a larger host plant for support but which do not derive any nourishment from it, and remoras, which attach themselves to sharks and feed on their leavings without appreciably hindering their hosts. Compare **amensalism, mutualism, parasitism.**

comminuted fracture (kŏm′ə-no͞o′tĭd) See under **fracture.**

common cold (kŏm′ən) A respiratory infection caused by any of several viruses, such as adenovirus or rhinovirus, in which the mucous membranes of the mouth, nose, and throat become inflamed. Common-cold symptoms include fever, nasal discharge, sneezing, and coughing.

common denominator A quantity into which all the denominators of a set of fractions may be divided without a remainder. For example, the fractions $\frac{1}{3}$ and $\frac{2}{5}$ have a common denominator of 15.

common divisor A number that is a factor of two or more numbers. For example, 3 is a common divisor of both 9 and 15. Also called *common factor.*

Common Era The period beginning with the year traditionally thought to have been birth of Jesus.

common logarithm A logarithm having 10 as its base. Compare **natural logarithm.**

common multiple A number divisible by each of two or more numbers without a remainder. For example, 12 is a common multiple of 2, 3, 4, and 6.

common salt See **salt** (sense 2).

communicable (kə-myo͞o′nĭ-kə-bəl) Capable of being transmitted from a person or animal to another person or animal, either through direct or indirect transmission, including insect or other vectors. Chickenpox is a communicable disease.

community (kə-myo͞o′nĭ-tē) A group of organisms or populations living and interacting with one another in a particular environment. The organisms in a community affect each other's abundance, distribution, and evolutionary adaptation. Depending on how broadly one views the interaction between organisms, a community can be small and local, as in a pond or tree, or regional or global, as in a biome.

commutative (kə-myo͞oʹtə-tĭv, kŏmʹyə-tāʹtĭv) Of or relating to binary operations for which changing the order of the inputs does not change the result of the operation. For example, addition is commutative, since $a + b = b + a$ for any two numbers a and b, while subtraction is not commutative, since $a - b \neq a - b$ unless both a and b are zero. See also **associative, distributive.**

commutative group A mathematical group in which the result of multiplying one member by another is independent of the order of multiplication. Also called *abelian group.*

commutator (kŏmʹyə-tāʹtər) **1.** The arrangement of contact points in an electric motor connecting an external direct current power supply and the rotating electric coils that use the power, used to generate the AC voltages needed by the coils. The commutator is located at the rotating shaft of the motor, where two power contacts are swept underneath two metal brushes, supplying positive and negative voltage to the coils. When the motor has rotated 180 degrees, the power contacts are each moving under the opposite brush, reversing the polarity of the voltage supplied to the coils. **2.** In a group or an algebra, an element of the form $ghg^{-1}h^{-1}$ where g and h are elements of the group or algebra. If g and h **commute**, the commutator is the identity element. The commutator is often written $[g, h]$.

commute (kə-myo͞otʹ) To yield the same result regardless of order. For example, numbers commute under addition, which is a **commutative** operation. Generally, any two operators H and G commute if their commutator is zero, i.e. $HG - GH = 0$.

compact disk or **compact disc** (kŏmʹpăktʹ) A small optical disk on which data such as music, text, or graphic images is digitally encoded.

compaction (kəm-păkʹshən) The process by which the porosity of a given form of sediment is decreased as a result of its mineral grains being squeezed together by the weight of overlying sediment or by mechanical means.

companion cell (kəm-pănʹyən) A specialized parenchyma cell, located in the phloem of flowering plants and closely associated with the development and function of a sieve-tube element. Companion cells probably provide ATP, proteins, and other substances to the sieve-tube elements, whose cytoplasm lacks many structures necessary for cell maintenance.

compass (kŭmʹpəs) **1.** A device used to determine geographical direction, usually consisting of a magnetic needle mounted on a pivot, aligning itself naturally with the Earth's magnetic field so that it points to the Earth's geomagnetic north or south pole. **2.** A device used for drawing circles and arcs and for measuring distances on maps, consisting of two legs hinged together at one end.

competence (kŏmʹpĭ-təns) **1.** The ability of bacteria to be undergo genetic **transformation**. **2.** The ability to respond immunologically to an antigen, as in an immune cell responding to a virus. **3.** The ability to function normally because of structural integrity, as in a heart valve.

competition (kŏmʹpĭ-tĭshʹən) The simultaneous demand by two or more organisms for limited environmental resources, such as nutrients, living space, or light.

competitive exclusion principle (kəm-pĕtʹĭ-tĭv) The principle that when two species compete for the same critical resources within an environment, one of them will eventually outcompete and displace the other. The displaced species may become locally extinct, by either migration or death, or it may adapt to a sufficiently distinct niche within the environment so that it continues to coexist noncompetitively with the displacing species. Also called *Gause's law.*

compiled language (kəm-pīldʹ) See under **programming language.**

compiler (kəm-pīʹlər) A computer program associated with certain programming languages that converts the instructions written in those languages into **machine code** that can later be executed directly by a computer. See more at **programming language.**

complement (kŏm′plə-mənt) **1.** A group of proteins in blood serum that interact systematically as part of the body's immune response to destroy disease-causing antigens, especially bacteria. Complement proteins interact with antibodies and other chemical substances to cause the disintegration of foreign cells and enhance other immune functions such as phagocytosis. **2.** A complementary color.

complementarity (kŏm′plə-mən-târ′ĭ-tē) The concept that the underlying properties of entities (especially subatomic particles) may manifest themselves in contradictory forms at different times, depending on the conditions of observation; thus, any physical model of an entity exclusively in terms of one form or the other will be necessarily incomplete. For example, although a unified quantum mechanical understanding of such phenomena as light has been developed, light sometimes exhibits properties of waves and sometimes properties of particles (an example of wave-particle duality). See also **uncertainty principle.**

complementary angles (kŏm′plə-mĕn′tə-rē) Two angles whose sum is 90°.

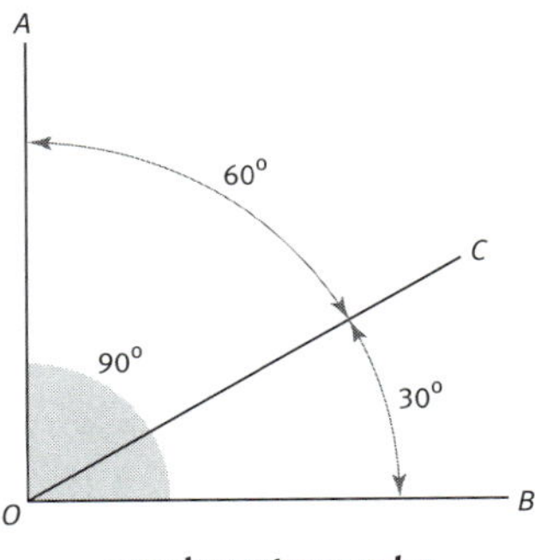

complementary angles

angles AOC and COB are complementary

complementary color A secondary color that, when combined with the primary color whose wavelength it does not contain, produces white light. Thus yellow, which is produced by mixing the primary colors red and green, is the complementary color of blue. See Note at **color.**

complementary DNA Single-stranded DNA synthesized in the laboratory using messenger RNA as a template and the enzyme reverse transcriptase. Complementary DNA is used for many purposes such as mapping chromosomes, creating clones, and sequencing genes.

complementary medicine A method of delivering of health care that combines the therapies and philosophies of conventional medicine with those of alternative medicine, such as acupuncture, herbal medicine, and biofeedback.

complete blood count (kəm-plēt′) A count performed as a diagnostic laboratory test, indicating the red blood cell count and the white blood cell count in one microliter of whole blood and other quantitative information about blood composition, such as cell volume, hematocrit, and hemoglobin content. This information is used in the diagnosis of anemia, infections, and other medical disorders.

complete flower A flower having all four floral parts: sepals, petals, stamens, and carpels. Compare **incomplete flower.** See also **perfect flower.**

complexity theory (kəm-plĕk′sĭ-tē) **1.** See **computational complexity. 2.** Any of various branches of mathematics, physics, computer science, and other fields, concerned with the emergence of order and structure in complex and apparently chaotic systems. See also **chaos.**

complex number (kŏm′plĕks′) A number that can be expressed in terms of i (the square root of -1). Mathematically, such a number can be written $a + bi$, where a and b are real numbers. An example is $4 + 5i$.

complex salt A salt that contains two different types of metal atoms, one of which does not ionize when in solution. Mercury iodide (HgI_2) is a complex salt. Compare **double salt, simple salt.**

composite family A very large family of flowering plants, Asteraceae (or Compositae), comprising about 1,100 genera and more than 20,000 species, including the daisy, lettuce, and marigold. The composite plants are eudicots and are considered to be the most highly evolved plants. Their inflorescences are characterized by many small flowers arranged in a head that resembles a single flower and arises from an involucre of bracts. A head may consist of both ray flowers and disk flowers, as in the sunflower, of disk flowers only, as in the burdock, or of ray flowers only, as in the dandelion.

composite number (kəm-pŏz′ĭt) An integer that can be divided by at least one other integer besides itself and 1 without leaving a remainder. 24 is a composite number since it can be divided by 2, 3, 4, 6, 8, and 12. No prime numbers are composite numbers. Compare **prime number.**

composite particle A subatomic particle that is composed of two or more elementary particles. The protons and neutrons in the nucleus of an atom are composite particles, as they are composed of quarks; electrons orbiting the nucleus are not composite particles. See also **elementary particle, subatomic particle.**

composite volcano See under **volcano.**

compost (kŏm′pōst′) A mixture of decayed or decaying organic matter used to fertilize soil. Compost is usually made by gathering plant material, such as leaves, grass clippings, and vegetable peels, into a pile or bin and letting it decompose as a result of the action of aerobic bacteria, fungi, and other organisms.

compound (kŏm′pound′) **1.** A substance consisting of atoms or ions of two or more different elements in definite proportions joined by chemical bonds into a molecule. The elements cannot be separated physically. Water, for example, is a compound having two hydrogen atoms and one oxygen atom per molecule. —*Adjective.* **2.** Composed of more than one part, as a compound eye or leaf.

compound eye An eye consisting of hundreds or thousands of tiny light-sensitive parts (called ommatidia), with each part serving to focus light on the retina to create a portion of an image. Most insects and some crustaceans have compound eyes.

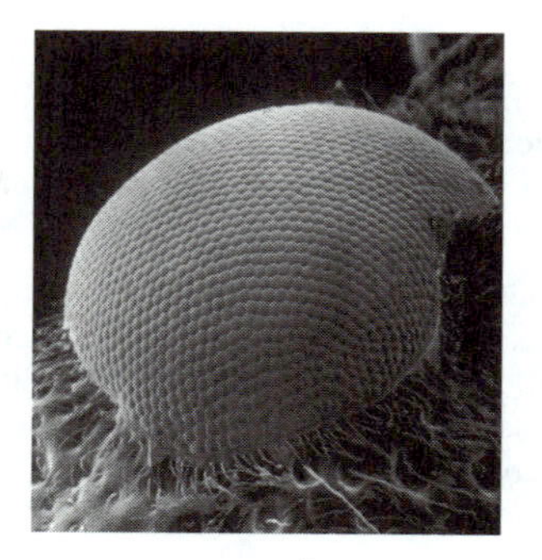

compound eye

compound eye of a beetle

compound interest Interest computed on the original principal plus any accrued interest. Thus if 5% is the rate of interest per year and the principal is $1000, the compound amount after one year will be $1050, after two years it will be $1050 × 0.05 = $1102.50, after three years it will be $1102.50 × 0.05 = $1157.63, and so forth. Mathematically, if P is the original principal and I the rate of interest expressed as a decimal, the compound amount at the end of the nth year will be $P(1 + I)^n$. The growth of the compound amount is exponential and not linear. Compare **simple interest.**

compound leaf A leaf that is composed of two or more leaflets on a common stalk. Clover, roses, sumac, and walnut trees have compound leaves.

compound lens See **lens** (sense 2b).

compression (kəm-prĕsh′ən) **1a.** A force that tends to shorten or squeeze something, decreasing its volume. **b.** The degree to which a substance has decreased in size (in volume, length, or some other dimension) after being or while being subject to stress. See also **strain.** **2.** The re-encoding of data (usually the binary data used by computers) into a form that uses fewer bits of information than the original data. Compression is often used to speed the transmission of data such as text or visual images, or to minimize the memory resources needed to store such data.

Compton (kŏmp′tən), **Arthur Holly** 1892–1962. American physicist who showed that when particles of light (called photons) collide with other particles, such as electrons, they lose energy and momentum and the light's wavelength increases. For his discovery of this phenomenon (which became known as the Compton effect) he shared the 1927 Nobel Prize for physics with Charles Wilson. He also discovered the electrical nature of cosmic rays.

Compton effect An increase in the wavelength of electromagnetic radiation, especially of x-rays or gamma-rays, when the photons constituting the radiation collide with free electrons. As a result of the Compton effect, the photons transfer some of their energy to the electrons. It is mainly through the Compton effect that matter absorbs radiant energy.

computational complexity (kŏm′pyo͝o-tā′shə-nəl) A mathematical characterization of the difficulty of a mathematical problem which describes the resources required by a computing machine to solve the problem. The mathematical study of such characterizations is called computational complexity theory and is important in many branches of theoretical computer science, especially cryptography.

computer (kəm-pyo͞o′tər) A programmable machine that performs high-speed processing of numbers, as well as of text, graphics, symbols, and sound. All computers contain a central processing unit that interprets and executes instructions; input devices, such as a keyboard and a mouse, through which data and commands enter the computer; memory that enables the computer to store programs and data; and output devices, such as printers and display screens, that show the results after the computer has processed data.

computerized axial tomography (kəm-pyo͞o′tə-rīzd′) Tomography in which computer analysis of a series of cross-sectional x-ray images made along a single axis of a bodily structure or tissue is used to construct a three-dimensional image of that structure. The technique is used in diagnostic studies of internal bodily structures, as in the detection of tumors or brain aneurysms.

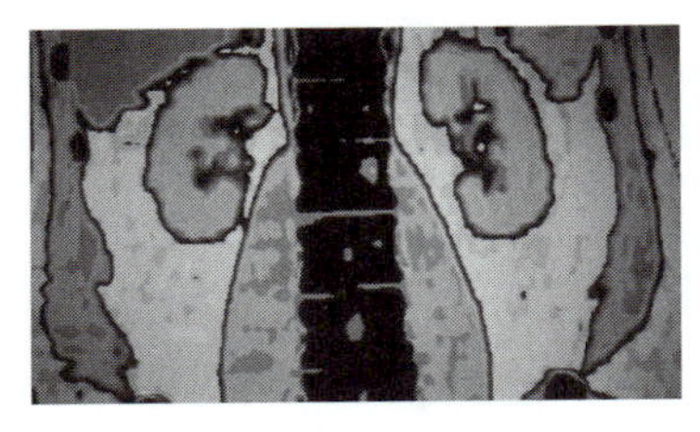

computerized axial tomography

scan of a healthy abdomen and kidneys

computer science The study of the design and operation of computer hardware and software, and of the application of computer technology to science, business, and the arts.

concave (kŏn′kāv′) Curved inward, like the inside of a circle or sphere.

concentration (kŏn′sən-trā′shən) The amount of a particular substance in a given amount of another substance, especially a solution or mixture.

conception (kən-sĕp′shən) The formation of a zygote resulting from the union of a sperm and egg cell; fertilization.

conchoidal (kŏng-koid′l) Of or relating to a mineral or rock surface that is characterized by smooth, shell-like curves. Obsidian and quartz often have conchoidal fractures.

concussion (kən-kŭsh′ən) An injury to a soft structure, especially the brain, produced by a violent blow or impact and followed by a temporary, sometimes prolonged, loss of function. A concussion of the brain results in transient loss of consciousness or memory.

condensation (kŏn′dən-sā′shən) The change of a gas or vapor to a liquid, either by cooling or by being subjected to increased pressure. When water vapor cools in the atmosphere, for example, it condenses into tiny drops of water, which form clouds.

condensed matter physics (kən-dĕnst′) The scientific study of the properties of solids, liquids, and other forms of matter in which atoms or particles adhere to each other or are otherwise highly concentrated. Solid-state physics is a branch of condensed matter physics. See also **state of matter.**

condenser (kən-dĕn′sər) **1.** An apparatus used to condense vapor, usually using cooling or pressurization. **2.** See **capacitor. 3.** A mirror, lens, or combination of lenses used to gather light and direct it upon an object or through a projection lens.

conditioning (kən-dĭsh′ə-nĭng) See **classical conditioning.**

conductance (kən-dŭk′təns) **1.** A measure of the ability of a material to carry electric current. For direct current, conductance is called **conductivity** and is equal to 1/R, where R is the resistance of the material. For alternating current, conductance is called **admittance**. Conductance is measured in mhos. See more at **admittance. 2.** See **thermal conductance.**

conduction (kən-dŭk′shən) The transfer of energy, such as heat or an electric charge, through a substance. In heat conduction, energy is transferred from molecule to molecule by direct contact; the molecules themselves do not necessarily change position, but simply vibrate more or less quickly against each other. In electrical conduction, energy is transferred by the movement of electrons or ions. Compare **convection.** See also **radiation.**

A CLOSER LOOK **conduction, convection, and radiation**

Heat is a form of energy that manifests itself in the motion of molecules and atoms, as well as subatomic particles. Heat energy can be transferred by conduction, convection, or radiation. In *conduction* heat spreads through a substance when faster atoms and molecules collide with neighboring slower ones, transferring some of their kinetic energy to them. This is how the handle of a teaspoon sticking out of a cup of hot tea eventually gets hot, though it is not in direct contact with the hot liquid. When a fluid is heated, portions of the fluid near the source of the heat tend to become less dense and expand outward, causing currents in the fluid. When these less dense regions rise, cooler portions flow in to take their place, which are then themselves subject to heating. This current flow is called *convection.* Many ocean currents are convection currents caused by the uneven heating of the ocean waters by the Sun. *Radiation* transmits heat in the form of electromagnetic waves, especially *infrared* waves, which have a lower frequency than visible light but a higher frequency than microwaves. Atoms and molecules in a substance struck by such radiation readily absorb the energy from these waves, thereby increasing their own kinetic energy and thus the temperature of the substance.

conduction band The electron orbital or orbitals, generally the outermost orbitals, in atoms in a conductor or semiconductor, in which the electrons are free enough to move and thereby carry an electric current. Compare **valence band.** See also **bandgap.**

conductivity (kŏn′dŭk-tĭv**′**ĭ-tē) **1.** The ability to transfer heat, electricity, or sound by conduction. **2.** See **conductance.**

conductor (kən-dŭk**′**tər) A material or an object that conducts heat, electricity, light, or sound. Electrical conductors contain electric charges (usually electrons) that are relatively free to move through the material; a voltage applied across the conductor therefore creates an electric current. Insulators (electrical nonconductors) contain no charges that move when subject to a voltage. Compare **insulator.** See also **resistance, superconductivity.**

condyle (kŏn**′**dīl′) A round, protruding part at the end of a bone, especially one that forms part of a joint.

cone (kōn) **1.** A three-dimensional surface or solid object in which the base is a circle and upper surface narrows to form a point. The surface of a cone is formed mathematically by moving a line that passes through a fixed point (the vertex) along a circle. **2.** A rounded or elongated reproductive structure that consists of sporophylls or scales arranged spirally or in an overlapping fashion along a central stem, as in conifers and cycads. For example, the familiar woody pinecone is actually the female cone, made up of ovule-bearing scales. The smaller male cones of the pine consist of thin overlapping microsporophylls. These produce pollen that is carried by the wind to fertilize ovules in the female cones. When the seeds in the female cones mature, the cones of many pine species expand to release them. In some pine species, cones release seeds only in response to the presence of fire. See also **strobilus. 3.** One of the cone-shaped cells in the retina of the eye of many vertebrate animals. Cones are extremely sensitive to light and can distinguish among different wavelengths. Cones are responsible for vision during daylight and for the ability to see colors. Compare **rod.**

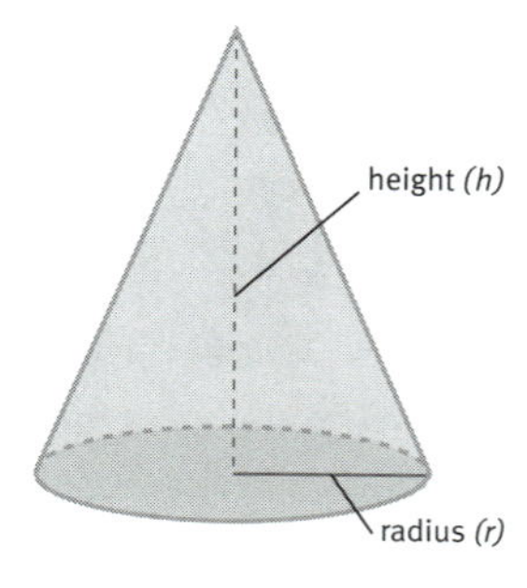

cone

The equation for determining the volume (V) of a cone is $V = \frac{1}{3}\pi r^2 h$.

confluence (kŏn**′**flo͞o-əns) **1.** A flowing together of two or more streams or two or more glaciers. **2.** The point of juncture of such streams or glaciers. **3.** The combined stream or glacier formed by this juncture.

conformal (kən-fôr′məl) **1.** Relating to the mapping of a surface or region onto another surface so that all angles between intersecting curves remain unchanged. **2.** Relating to a map projection in which small areas are rendered with true shape.

congenital (kən-jĕn′ĭ-tl) Existing at or before birth, as a defect or medical condition.

congestive heart failure (kən-jĕs′tĭv) See **heart failure.**

conglomerate (kən-glŏm′ə-rāt′) A coarse-grained sedimentary rock consisting of round rock fragments cemented together by hardened silt, clay, calcium carbonate, or a similar material. The fragments (known as clasts) have a diameter of at least 2 mm (0.08 inches), vary in composition and origin, and may include pebbles, cobbles, boulders, or fossilized seashells. Conglomerates often form through the transportation and deposition of sediments by streams, alluvial fans, and glaciers.

congruent (kŏng′gro͞o-ənt, kən-gro͞o′ənt) Relating to geometric figures that have the same size and shape. Two triangles are congruent, for example, if their sides are of the same length and their internal angles are of the same measure.

conic projection (kŏn′ĭk) A map projection in which the surface features of a globe are depicted as if projected onto a cone typically positioned so as to rest on the globe along a parallel (a line of equal latitude). In flattened form a conic projection produces a roughly semicircular map with the area below the apex of the cone at its center. When the central point is either of Earth's poles, parallels appear as concentric arcs and meridians as straight lines radiating from the center. Distances along the meridians remain true to scale, while the distortion along the parallels is progressively greater moving away from the parallel on which the cone is assumed to rest. Conic projections centered over a pole are often used in regional or national maps of temperate zones, where the distortion in the middle latitudes (the resting point of the cone) is minimal. Compare **azimuthal projection, cylindrical projection.**

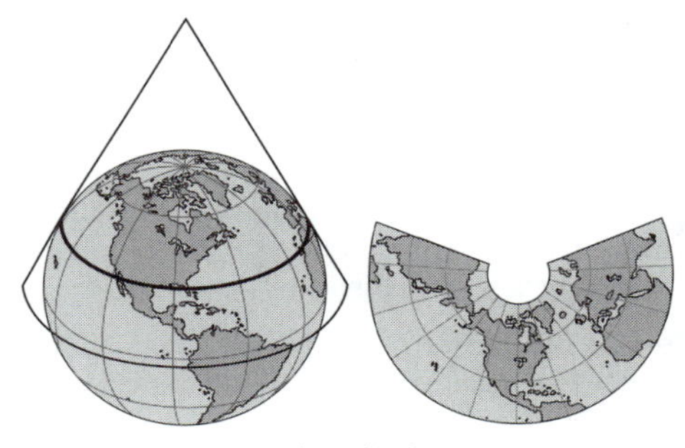

conic projection

conic section A curve formed by the intersection of a plane with a cone. Conic sections can appear as circles, ellipses, hyperbolas, or parabolas, depending on the angle of the intersecting plane relative to the cone's base.

conidiophore (kə-nĭd′ē-ə-fôr′) An asexual reproductive structure that develops at the tip of a fungal hypha and produces conidia.

conidium (kə-nĭd′ē-əm) Plural **conidia.** An asexually produced fungal spore, formed on a conidiophore. Most conidia are dispersed by the wind and can endure extremes of cold, heat, and dryness. When conditions are favorable, they germinate and grow into hyphae.

conifer (kŏn′ə-fər) Any of various gymnosperms that bear their reproductive structures in cones and belong to the phylum Coniferophyta. Conifers evolved around 300 million years ago and, as a group, show many adaptations to drier and cooler environments. They are usually evergreen and often have drought-resistant leaves that are needle-shaped or scalelike. They depend on the wind to blow pollen produced by male cones to female cones, where fertilization takes place and seeds develop. Conifers are widely distributed, but conifer species dominate the northern forest biome known as the taiga. There are some 550 species of conifers, including the pines, firs, spruces, hemlocks, cypresses, junipers, yews, and redwoods. See more at **pollination, seed-bearing plant.** —*Adjective* **coniferous.**

conjugate angles (kŏn′jə-gĭt) Two angles whose sum is 360°.

conjugated protein (kŏn′jə-gā′tĭd) A compound, such as hemoglobin, made up of a protein molecule and a nonprotein prosthetic group.

conjugation (kŏn′jə-gā′shən) **1.** A type of sexual reproduction in single-celled organisms, such as bacteria and some algae and fungi, in which two organisms or cells from the same species join together to exchange genetic material before undergoing cell division. **2.** The fusion of two gametes to form a zygote, as in some algae and fungi.

conjugation tube A slender tube in certain bacteria, algae, and fungi that connects two individuals during conjugation and through which the transfer of genetic material occurs.

conjunction (kən-jŭngk′shən) The position of two celestial bodies when they have the same celestial longitude, especially a configuration in which a planet or the Moon lies on a straight line from Earth to or through the Sun. Planets in this position are not visible to the naked eye because they are in line with the Sun and obscured by its glare; the Moon in this position is new. ► The inner planets Mercury and Venus have two conjunction points with Earth. Either planet is at **inferior conjunction** when it lies directly between the Earth and the Sun, and is at **superior conjunction** when it lies directly opposite Earth on the far side of the Sun. The outer planets have only one conjunction point with Earth, when they lie opposite Earth on the far side of the Sun. Compare **opposition.** See more at **elongation.**

conjunctiva (kŏn′jŭngk-tī′və) Plural **conjunctivas** or **conjunctivae** (kŏn′jŭngk-tī′vē). The mucous membrane that lines the inside of the eyelid and covers the surface of the eyeball.

conjunctivitis (kən-jŭngk′tə-vī′tĭs) Inflammation of the conjunctiva, characterized by redness, itching, and often accompanied by a discharge. Bacterial and viral infection and allergies are common causes of conjunctivitis.

connate (kŏn′āt′, kŏ-nāt′) *Botany* Joined with a part or organ of the same kind, as leaves that are joined at the base. Compare **adnate.**

connective tissue (kə-nĕk′tĭv) Tissue that connects, supports, binds, or encloses the structures of the body. Connective tissues are made up of cells embedded in an extracellular matrix and include bones, cartilage, mucous membranes, fat, and blood.

conodont (kō′nə-dŏnt′, kŏn′ə-) Any of various minute, toothlike or bladelike fossils made of the mineral apatite and dating from the Cambrian to the late Triassic Period. They are virtually the only preserved parts of extinct eellike animals that are now thought to have been primitive vertebrates similar to the modern hagfishes. Conodonts grew in paired assemblages in the head region of the animal and probably formed part of the feeding apparatus. They are the most widespread microfossils of the Paleozoic Era and are very important for determining the age of rock strata.

conservation (kŏn′sûr-vā′shən) The protection, preservation, management, or restoration of natural environments and the ecological communities that inhabit them. Conservation is generally held to include the management of human use of natural resources for current public benefit and sustainable social and economic utilization.

conservation biology The scientific study of the conservation of biological diversity and the effects of humans on the environment.

conservation law Any of various principles, such as the conservation of charge and the conservation of energy, that require some measurable property of a closed system to remain constant as the system changes. Conservation laws can be directly related to principles of symmetry. See also **invariance, Noether's theorem.**

conservation of charge A conservation law stating that the total electric charge of a closed system remains constant over time, regardless of other possible changes within the system.

conservation of energy A principle stating that the total energy of a closed system remains constant over time, regardless of other possible changes within the system. It is related to the symmetry of time invariance. See also **invariance, thermodynamics.**

conservation of mass A principle of classical physics stating that the total mass of a closed system is unchanged by interaction of its parts. The principle does not hold under Special Relativity, since mass and energy can be converted into one another.

conservation of momentum A conservation law stating that the total linear momentum of a closed system remains constant through time, regardless of other possible changes within the system.

constant (kŏn′stənt) **1.** A quantity that is unknown but assumed to have a fixed value in a specified mathematical context. **2.** A theoretical or experimental quantity, condition, or factor that does not vary in specified circumstances. Avogadro's number and Planck's constant are examples of constants.

constant region The portion of the amino acid sequence of an antibody's heavy or light chains that determines the class of the antibody and does not vary within a given class. The constant region terminates in a free carboxyl group (COOH). Unlike the variable region, the constant region does not interact with antigens. Compare **variable region.**

constellation (kŏn′stə-lā′shən) **1.** A group of stars seen as forming a figure or design in the sky, especially one of 88 officially recognized groups, many of which are based on mythological traditions from ancient Greek and Middle Eastern civilizations. **2.** An area of the sky occupied by one of the 88 recognized constellations. These irregularly defined areas completely fill the celestial sphere and divide it into nonoverlapping sections used in describing the location of celestial objects.

consumer (kən-so͞o′mər) A heterotrophic organism that feeds on other organisms in a food chain. ► Herbivores that feed on green plants and detritivores that feed on decaying matter are called **primary consumers.** Carnivores that feed on herbivores or detritivores are called **secondary consumers,** while those that feed on other carnivores are called **tertiary consumers.** Compare **producer.**

contact (kŏn′tăkt′) **1.** *Electricity.* **a.** A connection between two conductors that allows an electric current to flow. **b.** A part or device that makes or breaks a connection in an electrical circuit. **2.** *Geology.* The place where two different types of rock, or rocks of different ages, come together.

contact metamorphism A type of metamorphism in which the mineralogy and texture of a body of rock are changed by exposure to the

A CLOSER LOOK constellation

Various cultures throughout history have chosen different groups of stars in the night sky to form different *constellations.* While it was once thought that the Greeks were responsible for determining many of the constellations known today, it is now believed that the mythological origins of the 48 ancient constellations predate the Greeks and originate instead from ancient Middle Eastern civilizations. In the seventeenth and eighteenth centuries another 40 constellations were invented by Europeans for navigational purposes. The boundaries of the 88 constellations currently recognized were defined in the 1920s by the International Astronomical Union. There is no scientific reason why there are exactly 88; the modern constellations are only a convenient way to break up the sky to locate the position of celestial objects or track satellites. Although the stars in any given constellation may look like they're neighbors, they can actually be many light-years apart, and if seen from another part of the galaxy they would form different groups and shapes altogether. Constellation names are usually given in Latin, such as Ursa Major (Great Bear) or Centaurus (Centaur), and individual stars in constellations are named in order of brightness, using the Greek alphabet, with the genitive case of the constellation following. Therefore, Alpha Centauri is the brightest star in the constellation Centaurus, Beta Centauri is the second brightest star, and so on. The stars within our galaxy are rushing through space in various directions, and as the millennia pass, the arrangements of the star groups as seen from Earth will change, inevitably altering the constellations as we know them.

pressure and extreme temperature associated with a body of intruding magma. Contact metamorphism often results in the formation of valuable minerals, such as garnet and emery, through the interaction of the hot magma with adjacent rock. See also **aureole.** Compare **regional metamorphism.**

contagion (kən-tā′jən) **1.** The transmission of an infectious disease resulting from direct or indirect contact between individuals or animals. **2.** A disease that is transmitted in this way. **3.** The agent that causes a contagious disease, such as a bacterium or a virus.

contagious (kən-tā′jəs) **1.** Capable of being transmitted by direct or indirect contact, as an infectious disease. **2.** Bearing contagion, as a person or animal with an infectious disease that is contagious.

USAGE **contagious/infectious**

A *contagious* disease is one that can be transmitted from one living being to another through direct or indirect contact. Thus the flu, which can be transmitted by coughing, and cholera, which is often acquired by drinking contaminated water, are contagious diseases. Although *infectious* is also used to refer to such diseases, it has a slightly different meaning in that it refers to diseases caused by *infectious agents*—agents such as viruses and bacteria that are not normally present in the body and can cause an infection. While the notion of contagiousness goes back to ancient times, the idea of infectious diseases is more modern, coming from the germ theory of disease, which was not proposed until the later nineteenth century. *Contagious* and *infectious* are also used to refer to people who have *communicable* diseases at a stage at which transmission to others is likely.

continent (kŏn′tə-nənt) One of the seven great landmasses of the Earth. The continents are Africa, Antarctica, Asia, Australia, Europe, North America, and South America.

continental crust (kŏn′tə-nĕn′tl) See under **crust.**

continental divide A region of high ground, from each side of which the river systems of a continent flow into different continental-scale drainage basins. ► In North America, the **Continental Divide** is a series of mountain ridges stretching from Alaska to Mexico, marking the separation of drainage basins that empty into the Pacific Ocean or Bering Sea from those that empty into the Arctic or Atlantic Oceans or the Gulf of Mexico.

continental drift A theory stating that the Earth's continents have been joined together and have moved away from each other at different times in the Earth's history. The theory was first proposed by Alfred Wegener in 1912. While his general idea of continental movement eventually became widely accepted, his explanation for the mechanism of the movement has been supplanted by the theory of plant tectonics. See more at **plate tectonics.**

continental rift A long, narrow fissure in the Earth marking a zone of the lithosphere that has become thinner due to extensional forces associated with plate tectonics. Continental rifts are thousands of kilometers in length and hundreds of kilometers in width, and they are associated with normal faults and with grabens.

continental rise A wide, gentle incline from an ocean bottom to a continental slope. A continental rise consists mainly of silts, muds, and sand, and can be several hundreds of miles wide. Although it usually has a smooth surface, it is sometimes crosscut by submarine canyons.

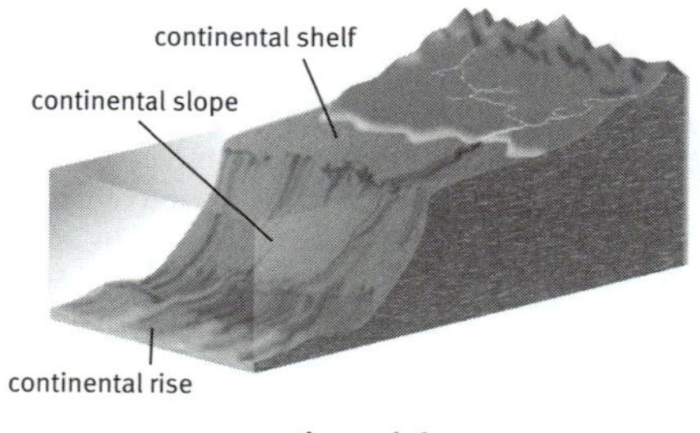

continental rise

continental shelf The part of the edge of a continent between the shoreline and the continental slope. It is covered by shallow ocean waters and has a very gentle slope.

continental slope The sloping region between a continental shelf and a continental rise. A continental slope is typically about 20 km (12.4 mi) wide, consists of muds and silts, and is often crosscut by submarine canyons.

continuous (kən-tĭn′yo͞o-əs) **1.** Relating to a line or curve that extends without a break or irregularity. **2.** A function in which changes, however small, to any x-value result in small changes to the corresponding y-value, without sudden jumps. Technically, a function is continuous at the point c if it meets the following condition: for any positive number ε, however small, there exists a positive number δ such that for all x within the distance δ from c, the value of $f(x)$ will be within the distance ε from $f(c)$. Polynomials, exponential functions, and trigonometric functions are examples of continuous functions.

continuous variation Variation within a population in which a graded series of intermediate phenotypes falls between the extremes. Height in human beings, for example, exists in continuous variation.

contour line (kŏn′to͝or′) A line on a map joining points of equal elevation above a given level, usually mean sea level. ► The change in elevation between one contour line and the next is the **contour interval.**

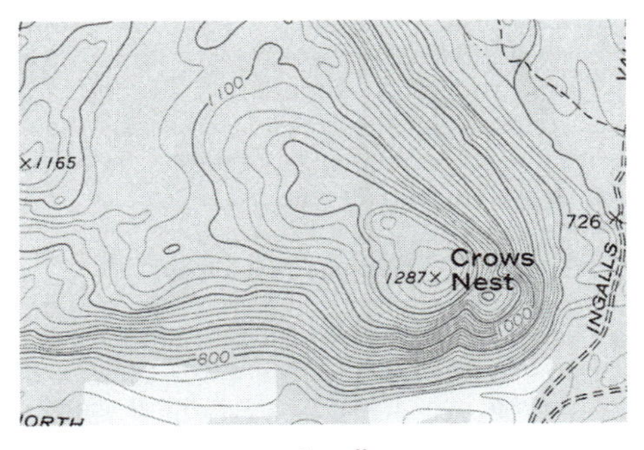

contour line

Closely spaced contour lines on the right indicate a steeper grade than the more loosely spaced lines on the left.

contour map A map that shows elevations above sea level and surface features of the land by means of contour lines.

contraceptive (kŏn′trə-sĕp′tĭv) A substance or device capable of preventing pregnancy.

contraction (kən-trăk′shən) The shortening and thickening of a muscle for the purpose of exerting force on or causing movement of a body part. See more at **muscle.**

contrast medium (kŏn′trăst′) A substance, such as barium or air, used in radiologic studies to increase the contrast of an image. In x-ray imaging, a positive contrast medium absorbs x-rays more strongly than the tissue or structure being examined; a negative contrast medium, less strongly.

control (kən-trōl′) A standard of comparison for checking or verifying the results of an experiment. In an experiment to test the effectiveness of a new drug, for example, one group of subjects (the control group) receives an inactive substance or **placebo**, while a comparison group receives the drug being tested.

control grid An electrode used in **triodes**, **tetrodes**, and **pentodes**, placed between the cathode and the plate (usually very close to the cathode). Electrons emitted by the cathode must pass through the control grid to reach the positively charged plate and create current flow through the tube, but a negative voltage on the intervening control grid impedes that flow; the more negative the control grid, the smaller the flow. Small variations in the voltage of the grid thus have a great effect on the current flow.

convection (kən-vĕk′shən) Current in a fluid caused by uneven distribution of heat. For example, air on a part of the Earth's surface warmed by strong sunlight will be heated by contact with the ground and will expand and flow upward, creating a region of low pressure below it; cooler surrounding air will then flow in to this low pressure region. The air thus circulates by convection, creating winds. See Note at **conduction.**

convection zone A region of turbulent plasma between a star's core and its visible photosphere at the surface, through which energy is transferred by convection. In the convection zone, hot plasma rises, cools as it nears the surface, and falls to be heated and rise again.

converge (kən-vûrj′) **1.** To tend toward or approach an intersecting point. **2.** In calculus, to approach a limit.

convergence (kən-vûr′jəns) **1.** *Mathematics.* The property or manner of approaching a limit, such as a point, line, or value. **2.** *Biology.* The evolution of superficially similar structures in unrelated species as they adapt to similar environments. Examples of convergence are the development of fins independently in both fish and whales and of wings in insects, birds, and bats. Also called *convergent evolution.* Compare **divergence.**

convergent plate boundary (kən-vûr′jənt) A tectonic boundary where two plates are moving toward each other. If the two plates are of equal density, they usually push up against each other, forming a mountain chain. If they are of unequal density, one plate usually sinks beneath the other in a subduction zone. The western coast of South America and the Himalayan Mountains are convergent plate boundaries. Also called *active margin, collision zone.* See more at **tectonic boundary.** Compare **divergent plate boundary.**

converter (kən-vûr′tər) **1.** An electrical device that changes the form of an electric signal or power source, as by converting alternating current to direct current, or an analog signal to a digital signal. Compare **rectifier, transformer. 2.** An electronic device that changes the frequency of a radio or other electromagnetic signal.

convex (kŏn′vĕks′) Curving outward, like the outer boundary of a circle or sphere.

cookie (ko͝ok′ē) A collection of information, usually including a username and the current date and time, stored on the local computer of a person using the World Wide Web, used chiefly by websites to identify users who have previously registered or visited the site. Cookies are used to relate one computer transaction to a later one.

Cooper pair (ko͞o′pər) A pair of weakly bound electrons in a superconductor. Cooper pairs have the property of bosons and can thus reside together in a ground state. They are named after their discoverer, American physicist Leon N. Cooper (born 1930). See also **Bose-Einstein condensate.**

coordinate (kō-ôr′dn-ĭt) One of a set of numbers that determines the position of a point. Only one coordinate is needed if the point is on a line, two if the point is in a plane, and three if it is in space.

coordinate bond A type of covalent bond in which both the shared electrons are contributed by one of the two atoms. Also called *dative bond.* See more at **covalent bond.**

coordinated universal time (kō-ôr′dn-ā′tĭd) A highly accurate time scale based on time measured by an atomic clock. Because coordinated universal time is unaffected by the gradual slowing of Earth's rotation, leap seconds are added as needed to synchronize it with the sidereal (rotation-based) universal time. Coordinated universal time is used internationally for scientific and technical purposes. Compare **universal time.**

coordinate system A system of representing points in a space of given dimensions by coordinates, such as the Cartesian coordinate system or the system of celestial longitude and latitude.

copepod (kō′pə-pŏd′) Any of various very small crustaceans of the subclass Copepoda, having an elongated body and a forked tail. Unlike most crustaceans, copepods lack a carapace over the back and do not have compound eyes. They are abundant in both salt and fresh water, and are an important food source for many water animals. Copepods include the water fleas.

Copernicus (kō-pûr′nə-kəs), **Nicolaus** 1473–1543. Polish astronomer whose theory that Earth and other planets revolve around the Sun provided the foundation for modern astronomy. His model displaced earlier theories that positioned Earth at the center of the solar system with all objects orbiting it.

BIOGRAPHY **Nicolaus Copernicus**

Nicolaus Copernicus originally studied canon law and medicine in Italy in preparation for a career in the Catholic Church. While in Italy he became interested in astronomy, which he then pursued in his spare time while working as a church administrator in Frauenberg, Poland. In a brief essay, *Commentariolis (Little Commentary),* he introduced his *heliocentric* system, in which the Sun is at the center of the universe, with all the planets and stars revolving around it in circular orbits. Copernicus was trying to account for the movements of the planets that, in the days prior to the invention of telescopes, were visible to the unaided eye and that did not fit the older Earth-centered, or *geocentric,* model of the universe of the Greek astronomer Ptolemy. Copernicus published a longer, more complete account of his theory, *De revolutionibus orbium coelestium (On the Revolutions of the Heavenly Spheres),* in 1543, just before he died. His heliocentric model of the universe was disputed by most astronomers of the time, as well as by the Church. After Copernicus's death, the few defenders of his ideas included Johannes Kepler and Galileo Galilei. *De revolutionibus* was on the Catholic Church's *Index of Forbidden Books* from 1616 to 1835. Theoretical support for Copernicus's system was provided almost 150 years after the publication of *De revolutionibus* by Sir Isaac Newton's theory of universal gravitation. His other great accomplishment was his proposal that the Earth rotates once daily on its own axis.

---

copolymer (kō-pŏl′ə-mər) A polymer of two or more different monomers. The synthetic rubber used to make tire treads and shoe soles, for example, is a copolymer made of the monomers butadiene and styrene.

copper (kŏp′ər) *Symbol* **Cu** A reddish-brown, ductile, malleable metallic element that is an excellent conductor of heat and electricity. It is widely used for electrical wires, water pipes, and rust-resistant parts, either in its pure form or in alloys such as brass and bronze. Atomic number 29; atomic weight 63.546; melting point 1,083°C; boiling point 2,595°C; specific gravity 8.96; valence 1, 2. See **Periodic Table.** See Note at **element.**

Copper and Stone Age See **Chalcolithic.**

copper sulfate A crystalline salt that is soluble in water and slightly soluble in alcohol. In its anhydrous form it is white, and in its hydrous form it is blue. Copper sulfate is used in agriculture, textile dyeing, leather treatment, electroplating, and the manufacture of germicides. In its blue form it is also called *blue vitriol. Chemical formula:* $CuSO_4 \cdot 5H_2O$.

coprocessor (kō′prŏs′ĕs-ər) A microprocessor distinct from the central processing unit. A coprocessor performs specialized functions that the central processing unit cannot perform or cannot perform as well and as quickly.

coprolite (kŏp′rə-līt′) Fossilized excrement. Analysis of the fossilized animal and plant remains within coprolites provides important information about the diet and environment of ancient biota.

coquina (kō-kē′nə) A soft porous limestone, composed of shells and fragments of shell and coral that are partially cemented by material that is high in calcium carbonate and has not completely hardened.

coral (kôr′əl) **1.** Any of numerous small, sedentary cnidarians (coelenterates) of the class Anthozoa. Corals often form massive colonies in shallow sea water and secrete a cup-shaped skeleton of calcium carbonate, which they can retreat into when in danger. Corals are related to the sea anemones and have stinging tentacles around the mouth opening that are used to catch prey. **2.** A hard, stony substance consisting of the skeletons of these animals. It is typically white, pink, or reddish and can form large reefs that support an abundance of ocean fish.

coral reef A mound or ridge of living coral, coral skeletons, and calcium carbonate deposits from other organisms such as calcareous algae, mollusks, and protozoans. Most coral reefs form in warm, shallow sea waters and rise to or near the surface, generally in the form of a **barrier reef**, **fringing reef**, or **atoll**. Coral reefs grow upward from the sea floor as the polyps of new corals cement themselves to the skeletons of those below and in turn provide support for algae and other organisms whose secretions serve to bind the skeletons together. The resulting structure provides a critical habitat for a wide variety of fish and marine invertebrates. Coral reefs also protect shores against erosion by causing large waves to break and lose some of their force before reaching land. The Great Barrier Reef off the northeastern coast of Australia extends for some 2,000 km (1,240 mi), making it the world's largest coral reef.

cordate (kôr′dāt′) Having a heart-shaped outline. Often used of leaves, such as those of the morning glory or linden.

cordierite (kôr′dē-ə-rīt′) A light-blue to dark-blue or gray orthorhombic mineral. Cordierite is a silicate of magnesium, aluminum, and sometimes iron, and is found in granites and in metamorphic rocks that form under relatively low-pressure conditions. *Chemical formula:* $(Mg,Fe)_2Al_4Si_5O_{18}$.

cordillera (kôr′dl-yâr′ə) A long and wide chain of mountains, especially the main mountain range of a large landmass. Cordilleras can include the valleys, basins, rivers, lakes, plains, and plateaus between parallel chains of a single mountain system, or they can consist solely of a string of connected mountain peaks.

cordite (kôr′dīt′) An explosive powder consisting of nitrocellulose, nitroglycerin, and petroleum jelly, used as a propellant for guns. It does not generate smoke and is shaped into cords.

core (kôr) **1.** The central or innermost portion of the Earth, lying below the mantle and probably consisting of iron and nickel. It is divided into a liquid outer core, which begins at a depth of 2,898 km (1,800 mi), and a solid inner core, which begins at a depth of 4,983 km (3,090 mi). **2.** A piece of magnetizable material, such as a rod of soft iron, that is placed inside an electrical coil or transformer to intensify and provide a path for the magnetic field produced by the current running through the wire windings. **3.** The central part of a nuclear reactor where atomic fission occurs. The core contains the fuel, the coolant, and the moderator. **4.** A long, cylindrical sample of soil, rock, or ice collected with a drill to study the strata of material that are not visible from the surface. **5.** A stone from which one or more flakes have been removed, serving as a tool in itself or as a source of flakes from which other tools could be fashioned. Stones used as cores include flint, chert, and obsidian. See more at **core tool.**

core tool A stone tool consisting of a core that is flaked (struck with another rock or similar material) to produce a cutting edge or edges. Core tools date at least to the beginning of the Oldowan tool industry and are the earliest stone tools known to have been deliberately fashioned by humans. Core tools include choppers, cleavers, and hand axes. Compare **flake tool.**

Coriolis effect (kôr′ē-ō′lĭs) The observed effect of the Coriolis force, especially the deflection of objects or substances (such as air) moving along the surface of the Earth, rightward in the Northern Hemisphere and leftward in the Southern Hemisphere. The Coriolis effect is named after the French engineer Gustave Gaspard Coriolis (1792–1843).

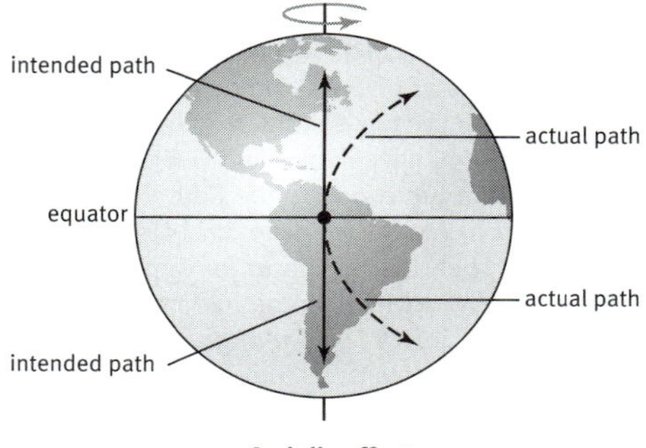

Coriolis effect

Coriolis force A velocity-dependent pseudo force used mathematically to describe the motion of bodies in rotating reference frames such as the Earth's surface. Bodies moving on the plane of rotation appear to experience a force, leftward if the rotation of the reference frame is clockwise, rightward if counterclockwise. Such motion gives rise to the Coriolis effect.

cork (kôrk) **1.** The outermost layer of tissue in woody plants that is resistant to the passage of water vapor and gases and that becomes the bark. Cork is secondary tissue, formed on the outside of the tissue layer known as **cork cambium**. The cell walls of cork cells contain suberin. Once they mature, cork cells die. Also called *phellem.* **2.** The lightweight, elastic outer bark of the cork oak, which grows near the Mediterranean Sea. Cork is used for bottle stoppers, insulation, and other products.

cork cambium A layer of cambium near the surface of the stems of woody plants that produces cork to the outside and phelloderm to the inside. It forms the middle layer of the periderm. Also called *phellogen.* See more at **cambium.**

corm (kôrm) A fleshy underground stem that is similar to a bulb but stores its food as stem tissue and has fewer and thinner leaflike scales. The crocus and gladiolus produce new shoots from corms. Compare **bulb, rhizome, runner, tuber.**

cornea (kôr′nē-ə) The tough transparent membrane of the outer layer of the eyeball that covers the iris and the pupil.

corolla (kə-rŏl′ə, kə-rō′lə) The petals of a flower considered as a group or unit. See more at **flower.**

corollary (kôr′ə-lĕr′ē) A statement that follows with little or no proof required from an already proven statement. For example, it is a theorem in geometry that the angles opposite two congruent sides of a triangle are also congruent. A corollary to that statement is that an equilateral triangle is also equiangular.

corona (kə-rō′nə) Plural **coronas** or **coronae** (kə-rō′nē). **1.** The luminous, irregular envelope of gas outside the chromosphere of a star. The Sun's corona is composed of ionized gas between approximately 1,000,000°K and 2,000,000°K and has an extremely low density. This phenomenon is visible only during a solar eclipse. **2.** A faintly colored luminous ring appearing to surround a celestial body (such as the Moon or Sun) that is visible through a haze or thin cloud, caused by diffraction of light from suspended matter in the intervening medium. Also called *aureole.* **3.** A faint glow of the air in the region of very strong electric fields, caused by ionization of the air molecules and flow of current in that region in corona discharge. **4.** The crownlike upper portion of a bodily part or structure, such as the top of the head. **5.** A crown-shaped structure on the inner side of the petals of some flowers, such as the daffodil.

Corona Borealis (bôr′ē-ăl′ĭs) A constellation (the Northern Crown) in the Northern Hemisphere between Hercules and Boötes.

corona discharge An electrical discharge characterized by a corona, occurring when one of two conducting surfaces (such as electrodes) of differing voltages has a pointed shape, resulting in a highly concentrated electric field at its tip that ionizes the air (or other gas) around it. Corona discharge can result in power loss in the transmission of electric power, and is used in photocopying machines and air-purification devices. See also **electric arc.**

coronal mass ejection (kôr′ə-nəl, kŏr′-, kə-rō′nəl) A massive, bubble-shaped burst of plasma expanding outward from the Sun's corona, in which large amounts of superheated particles are emitted at nearly the speed of light. The emissions can cause disturbances in the solar wind that disrupt satellites and create powerful magnetic storms on Earth. They were first observed in the early 1970s, when photographs taken from satellites revealed coronal activity that could not be seen in images taken from Earth.

coronary (kôr′ə-nĕr′ē) Relating to or involving the heart.

coronary artery Either of two arteries that originate in the aorta and supply blood to the muscular tissue of the heart.

coronary artery disease Atherosclerosis of the coronary arteries, which can cause angina pectoris or heart attack. A positive family history, hypertension, smoking, diabetes mellitus, and elevated blood lipids increase the risk of developing coronary artery disease.

corpus callosum (kə-lō′səm) Plural **corpora callosa.** The transverse band of nerve fibers that connects the right and left cerebral hemispheres.

corpuscle (kôr′pə-səl) **1.** Any of various cellular or small multicellular structures in the body, especially a red or white blood cell. **2.** See **particle** (sense 2).

corpus luteum (lo͞o′tē-əm) Plural **corpora lutea.** A yellow mass of cells that forms from a mature ovarian follicle after ovulation and that secretes progesterone. If fertilization of the egg occurs, the corpus luteum persists for the first few months of pregnancy.

corrosion (kə-rō′zhən) The breaking down or destruction of a material, especially a metal, through chemical reactions. The most common form of corrosion is rusting, which occurs when iron combines with oxygen and water.

cortex (kôr′tĕks′) **1.** The outer layer of an organ or body part, such as the cerebrum or the adrenal glands. **2.** The region of tissue lying between the epidermis (the outermost layer) and the vascular tissue in the roots and stems of plants. It is composed of collenchyma, parenchyma, and sclerenchyma. In roots the cortex transfers water and minerals from the epidermis to the vascular tissue, which distributes them to other parts of the plant. The cortex also provides structural support and stores food manufactured in the leaves. See illustration at **xylem.**

corticosteroid (kôr′tĭ-kō-stîr′oid′, -stĕr′-) Any of the steroid hormones, such as cortisol or aldosterone, produced by the cortex of the adrenal gland. Corticosteroids are also produced synthetically for medicinal purposes.

cortisol (kôr′tĭ-sôl′, -sōl′) The principal steroid hormone produced by the adrenal cortex. It regulates carbohydrate metabolism and the immune system and maintains blood pressure. When natural or synthetic cortisol is used as a pharmaceutical, it is known as **hydrocortisone.**

cortisone (kôr′tĭ-sōn′) A steroid hormone that is easily formed from or converted to cortisol in the blood and is also produced synthetically for use as a pharmaceutical. The effects of cortisone on body tissues are similar to those of naturally or synthetically produced cortisol.

corundum (kə-rŭn′dəm) An extremely hard mineral occurring in many colors, either as shapeless grains or as rhombohedral crystals. It also occurs in gem varieties such as ruby and sapphire and in a dark-colored variety that is used for polishing and scraping. Corundum is found in igneous and carbonate rocks. *Chemical formula:* Al_2O_3.

corymb (kôr′ĭmb, -ĭm) An indeterminate inflorescence whose outer flowers have longer stalks than the inner flowers, so that together they form a round cluster that is rather flat on top. The outer flowers open before the inner ones. Yarrow and the hawthorn have corymbs. See illustration at **inflorescence.**

cos Abbreviation of **cosine.**

cosecant (kō-sē′kănt′) **1.** The ratio of the length of the hypotenuse in a right triangle to the length of the side opposite an acute angle. The cosecant is the inverse of the sine. **2.** The reciprocal of the ordinate of the endpoint of an arc of a unit circle centered at the origin of a Cartesian coordinate system, the arc being of length x and measured counterclockwise from the point (1, 0) if x is positive or clockwise if x is negative. **3.** A function of a number x, equal to the cosecant of an angle whose measure in radians is equal to x.

cosh Abbreviation of **hyperbolic cosine.**

cosine (kō′sīn′) **1.** The ratio of the length of the side adjacent to an acute angle of a right triangle to the length of the hypotenuse. **2.** The abscissa of the endpoint of an arc of a unit circle centered at the origin of a Cartesian coordinate system, the arc being of length x and measured counterclockwise from the point (1, 0) if x is positive or clockwise if x is negative. **3.** A function of a number x, equal to the cosine of an angle whose measure in radians is equal to x.

cosmic (kŏz′mĭk) Relating to the universe or the objects in it.

cosmic background radiation Weak radio-frequency radiation traveling through outer space with equal intensity in every direction. It is thought to be residual radiation from a period shortly following the **big bang**, when the universe was very hot. Also called *microwave background.*

cosmic ray A ray of radiation of extraterrestrial origin, consisting of one or more charged particles such as protons, alpha particles, and larger atomic nuclei. Cosmic rays entering the atmosphere collide with atoms, producing secondary radiation, such as pions, muons, electrons, and gamma rays. Cosmic rays (and secondary radiation) can be easily seen in a cloud chamber.

cosmid (kŏz′mĭd) A hybrid **vector** that has been spliced with plasmid DNA for cloning large pieces of DNA.

cosmogony (kŏz-mŏg′ə-nē) The branch of cosmology that studies the origin of the universe and the larger objects found within it, such as the solar system.

cosmography (kŏz-mŏg′rə-fē) The study of the visible universe that includes the measurement and cataloging of its objects and structures.

cosmology (kŏz-mŏl′ə-jē) **1.** The scientific study of the origin, evolution, and structure of the universe. **2.** A specific theory or model of the origin and evolution of the universe.

cosmos (kŏz′məs, kŏz′mōs′) The universe, especially when considered as an orderly and harmonious whole.

costa (kŏs′tə) Plural **costae** (kŏs′tē). A rib or a riblike part, such as the midrib of a leaf or a thickened anterior vein or margin of an insect's wing.

cot Abbreviation of **cotangent.**

cotangent (kō-tăn′jənt) **1.** The ratio of the length of the adjacent side of an acute angle in a right triangle to the length of the opposite side. The cotangent is the inverse of the tangent. **2.** The ratio of the ordinate to the abscissa of the endpoint of an arc of a unit circle centered at the origin of a Cartesian coordinate system, the arc being of length *x* and measured counterclockwise from the point (1, 0) if *x* is positive or clockwise if *x* is negative. **3.** A function of a number *x*, equal to the cotangent of an angle whose measure in radians is equal to *x*.

coth Abbreviation of **hyperbolic cotangent.**

cotidal (kō-tīd′l) **1.** Indicating coincidence of high tides or low tides. **2.** Relating to a line that passes through each location on a coastal map where tides occur at the same time of day. Cotidal maps show variation in the height of the tides and indicate the time of high tide occurrence. Cotidal charts can be made using tide gauge data from gauges at regular intervals along the coast; however, computer modeling is increasingly used, especially where the tides vary greatly over a short distance.

cotyledon (kŏt′l-ēd′n) A leaf of the embryo of a seed-bearing plant. Most cotyledons emerge, enlarge, and become green after the seed has germinated. Cotyledons either store food for the growing embryo (as in monocotyledons) or absorb food that has been stored in the endosperm (as in other angiosperms) for eventual distribution to the growing parts of the embryo. Also called *seed leaf.* See more at **eudicotyledon, monocotyledon.**

cotylosaur (kŏt′l-ə-sôr′) Any of a number of extinct reptiles of the group Cotylosauria or Captorhinida of the Carboniferous and Permian Periods that include the ancestors of all terrestrial vertebrates except for amphibians. Cotylosaurs evolved from early amphibians, had four sprawling legs and a long tail, and a skull without temporal openings. Also called *stem reptile.*

cough (kôf, kŏf) The act of expelling air from the lungs suddenly and noisily, often to keep the respiratory passages free of irritating material.

coulomb (ko͞o′lŏm′, ko͞o′lōm′) The SI derived unit used to measure electric charge. One coulomb is equal to the quantity of charge that passes through a cross-section of a conductor in one second, given a current of one ampere.

Coulomb, Charles Augustin de 1736–1806. French physicist who was a pioneer in the study of magnetism and electricity. He is best known for the formulation of **Coulomb's law,** which he developed as a result of his investigations of Joseph Priestley's work on electrical repulsion. Coulomb also established a law governing the attraction and repulsion of magnetic poles. The coulomb unit of electric charge is named for him.

Coulomb force The force exerted by stationary objects bearing electric charge on other stationary objects bearing electric charge. If the charges are of the same sign, then the force is repulsive; if they are of opposite signs, the force is attractive. The strength of the force is described by Coulomb's law. Also called *electrostatic force.*

Coulomb's law A law stating that the strength of the force exerted by one point charge on another depends on the strength of the charges and on the distance between them. Since Coulomb's law is an **inverse square law,** higher charges entail stronger force, while greater distances entail weaker force. The force is understood as arising from the electric field that surrounds the charges. The force is repulsive if the charges have the same sign, and attractive if they have opposite sign.

coulometer (ko͞o-lŏm′ĭ-tər, ko͞o′lə-mē′tər) A device for determining the amount of a substance released during electrolysis by measuring the electrical charge that results from the electrolysis. Coulometers can be used to detect and measure trace amounts of substances such as water.

coumarin (ko͞o′mər-ĭn) A fragrant crystalline compound extracted from several plants, such as tonka beans and sweet clover, or produced synthetically. Coumarin and its derivatives are widely used in perfumes, as anticoagulants, and as rodenticides. *Chemical formula:* $C_9H_6O_2$.

countercurrent (koun′tər-kûr′ənt) A current that flows in an opposite direction to the flow of another current.

Cousteau (ko͞o-stō′), **Jacques Yves** 1910–1997. French underwater explorer who helped develop the Aqua-Lung (an underwater breathing apparatus) and designed the bathyscaphe and a small, deep-sea submarine known as a diving saucer. Cousteau also devised an underwater television system and produced several films that widely popularized marine biology.

Jacques Cousteau

covalent bond (kō-vā′lənt) A chemical bond formed when electrons are shared between two atoms. Usually each atom contributes one electron to form a pair of electrons that are shared by both atoms. See more at **coordinate bond, double bond, polar bond.**

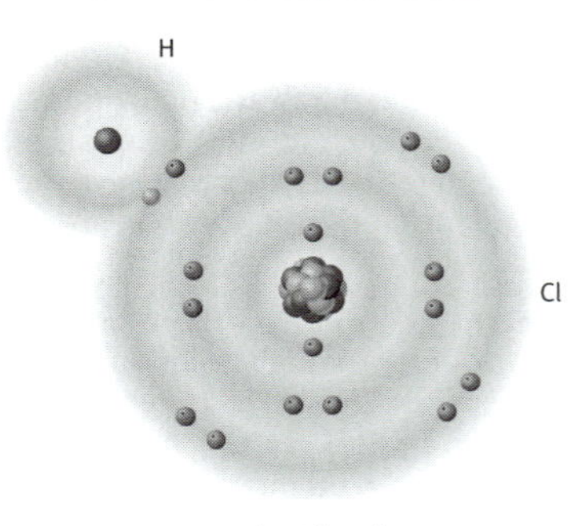

covalent bond

covalent bonding in a hydrogen chloride molecule

coxsackievirus also **Coxsackie virus** (ko͝ok-sä′kē-vī′rəs, kŏk-săk′ē-) Any of several species of enteroviruses that are associated with a variety of acute infections primarily affecting infants and young children, including meningitis, infections of the heart muscle and pericardium, and pleurisy.

CP invariance (sē′pē′) Short for *charge parity invariance.* A form of **invariance** in which reversing the sign of all charges and all spatial dimensions in the description of the physical system results in an indistinguishable physical system. CP invariance does not appear to hold in nature, as evidenced by the decay of **kaons**. See also **charge conjugation, parity conjugation.**

CPR (sē′pē-är′) Short for *cardiopulmonary resuscitation.* An emergency procedure in which the heart and lungs are made to work by manually compressing the chest overlying the heart and forcing air into the lungs. CPR is used to maintain circulation when the heart stops pumping, usually because of disease, drugs, or trauma.

CPU Abbreviation of **central processing unit.**

Cr The symbol for **chromium.**

cracking (krăk′ĭng) The process of breaking down complex chemical compounds by heating them. Sometimes a catalyst is added to lower the amount of heat needed for the reaction. Cracking is used especially for breaking petroleum molecules into shorter molecules and to extract low-boiling fractions, such as gasoline, from petroleum. See also **hydrocracking.**

cranial (krā′nē-əl) Located in or involving the skull or cranium.

cranial nerve Any of the 12 pairs of nerves in humans and other mammals that connect the muscles and sensory organs of the head and upper chest directly to the brain through openings in the skull. The cranial nerves include the optic nerve and the auditory nerve.

craniotomy (krā′nē-ŏt′ə-mē) Surgical incision into the skull.

cranium (krā′nē-əm) Plural **craniums** or **crania.** The vertebrate skull, especially the part that encloses and protects the brain.

crater (krā′tər) **1.** A bowl-shaped depression at the top of a volcano or at the mouth of a geyser. Volcanic craters can form because of magma explosions in which a large amount of lava is thrown out from a volcano, leaving a hole, or because the roof of rock over an underground magma pool collapses after the

crater

Barringer Meteor Crater near Winslow, Arizona

magma has flowed away. **2.** A shallow, bowl-shaped depression in a surface, formed by an explosion or by the impact of a body, such as a meteorite.

craton (krā′tŏn′) A large portion of a continental plate that has been relatively undisturbed since the Precambrian era and includes both shield and platform layers.

C-reactive protein (sē′rē-ăk′tĭv) An antibody found in the blood in certain acute and chronic conditions including infections and cancers. It is a nonspecific indicator of inflammation and therefore not diagnostic of any one disease.

creatine (krē′ə-tēn′, -tĭn) A nitrogenous organic acid that is found in the muscle tissue of vertebrates, mainly combined with phosphorus, and that supplies energy for muscle contraction. *Chemical formula:* $C_4H_9N_3O_2$.

creatinine (krē-ăt′n-ēn′, -ĭn) A compound formed by the metabolism of creatine, found in muscle tissue and blood and normally excreted in the urine as a metabolic waste. Measurement of creatinine levels in the blood is used to evaluate kidney function. *Chemical formula:* $C_4H_7N_3O$.

creodont (krē′ə-dŏnt′) Any of various extinct carnivorous mammals of the order Creodonta of the Paleocene to the Pliocene Epochs. Creodonts had long, low skulls with crests to which chewing muscles were attached. They were the dominant carnivorous mammals for millions of years, and were once believed to be ancestral to modern carnivores.

creosote (krē′ə-sōt′) **1.** A yellow or brown oily liquid obtained from coal tar and used as a wood preservative and disinfectant. **2.** A colorless to yellowish oily liquid containing phenols, obtained by the destructive distillation of wood tar, especially from the wood of a beech, and formerly used as an expectorant in treating chronic bronchitis.

crescent (krĕs′ənt) Partly but less than half illuminated. Used to describe the Moon or a planet. Compare **gibbous.**

crest (krĕst) The part of a wave with greatest magnitude; the highest part of a wave. Compare **trough.** See more at **wave.**

Cretaceous (krĭ-tā′shəs) The third and last period of the Mesozoic Era, from about 144 to 65 million years ago. During this time the supercontinent Pangaea continued to split up, with modern-day South America and Africa splitting apart, the Atlantic Ocean widening, and India disconnecting itself entirely from the other landmasses to which it was attached. Dinosaurs continued to be the dominant terrestrial animals, but many insect groups, modern mammals and birds, and the angiosperms (flowering plants) also first appeared. The Cretaceous Period ended with a mass extinction event in which about 75 percent of all species, including marine, freshwater, and terrestrial organisms, became extinct. See Chart at **geologic time.**

Creutzfeldt-Jakob disease (kroits′fĕlt-yä′kôp) A rare, usually fatal encephalopathy that occurs most often in middle age and is likely caused by a prion. It is characterized by progressive dementia and gradual loss of muscle control. The disease is named after its discoverers, German pathologist Hans Gerhard Creutzfeldt (1885–1964) and German neurologist Alfons Maria Jakob (1884–1931).

crevasse (krĭ-văs′) **1.** A deep fissure in a glacier or other body of ice. Crevasses are usually caused by differential movement of parts of the ice over an uneven topography. **2.** A large, deep fissure in the Earth caused by an earthquake. **3.** A wide crack or breach in the bank of a river. Crevasses usually form during floods. ► The sediments that spill out through the crevasse and fan out along the external margin of the river's bank form a **crevasse splay** deposit.

crib death (krĭb) See **sudden infant death syndrome.**

Crick (krĭk), **Francis Harry Compton** 1916–2004. British biologist who with James D. Watson identified the structure of DNA in 1953. By analyzing the patterns cast by x-rays striking DNA molecules, they found that DNA has the structure of a double helix, consisting of two spirals linked together at the base, forming ladderlike rungs. For this work they shared the 1962 Nobel Prize for physiology or medicine with Maurice Wilkins. See Note at **Rosalind Franklin.**

crinoid (krī′noid′) Any of various marine echinoderms of the class Crinoidea. Crinoids have a cup-shaped body with five or more feathery arms and sometimes a stalk for attachment to a surface. The arms contain reproductive organs and sensory tube feet. Crinoids were common during the Paleozoic Era and are important index fossils. Sea lilies and feather stars are types of crinoids.

crista (krĭs′tə) Plural **cristae** (krĭs′tē). Any of the folds of the inner membrane of a mitochondrion. See more at **mitochondrion.**

critical angle (krĭt′ĭ-kəl) The smallest angle of incidence at which radiation, such as light, is completely reflected from the boundary between two media. At angles smaller than the critical angle, some of the radiation enters the second material and is refracted.

critical mass 1. The smallest mass of a fissionable material that will sustain a nuclear chain reaction. **2.** The amount of matter needed to generate sufficient gravitational force to halt the current expansion of the universe. See also **big bang, big crunch.**

critical point 1. *Physics.* The temperature and pressure at which the liquid and gaseous phases of a pure substance become unstable and fluctuate locally within the substance. The critical point of water is at a temperature of 374°C (705.2°F) and a pressure of 218 atmospheres, at which point it becomes opaque. Compare **triple point. 2.** *Mathematics.* **a.** A maximum, minimum, or point of inflection of a curve. **b.** A point at which the derivative of a function is zero, infinite, or undefined.

critical pressure The pressure of a substance at its critical point.

critical state The state of matter of a substance at its critical point.

critical temperature 1. The temperature of a substance at its critical point. **2.** The temperature at which a material becomes a superconductor. **3.** The temperature at which a property of a material, such as its magnetism, changes.

crocodilian (krŏk′ə-dĭl′ē-ən) Any of various semiaquatic reptiles of the order Crocodilia, including the alligators, crocodiles, caimans, and gavials. Crocodilians are squat, massive, and lizardlike, with long, powerful jaws, long, heavy tails, short legs, and thick, plated skin. Like dinosaurs, crocodilians are archosaurs, and their closest modern relatives are the birds.

Crohn's disease (krōnz) A gastrointestinal disease characterized by inflammation of the ileum, resulting in abdominal cramping, diarrhea, and weight loss. It is named after American physician Burrill Bernard Crohn (1884–1983), who first described it.

Cro-Magnon (krō-măg′nən, -măn′yən) An early form of modern human *(Homo sapiens)* inhabiting Europe in the late Paleolithic Period, from about 40,000 to 10,000 years ago, characterized by a broad face and tall stature. It is known from skeletal remains first found in the Cro-Magnon cave in southern France. Cro-Magnons coexisted with European Neanderthal populations for several thousand years, although there is little evidence of interbreeding. See more at **Aurignacian.**

Crookes (kro͝oks), Sir **William** 1832–1919. British chemist and physicist who discovered thallium in 1861 and invented the radiometer (1873–76). He also developed the Crookes tube, a modified vacuum tube that was later used by W.C. Roentgen and J.J. Thomson in experiments that led to the discovery of x-rays and the electron, respectively.

cross (krôs) *Noun.* **1.** A plant or animal produced by crossbreeding; a hybrid. —*Verb.* **2.** To crossbreed or cross-fertilize plants or animals.

crossbreed (krôs′brēd′) *Verb.* **1.** To produce a hybrid animal or plant by breeding two animals or two plants of different species or varieties. For example, crossbreeding a male donkey with a female horse will produce a mule. —*Noun.* **2.** An animal or a plant produced by breeding two animals or plants of different species or varieties; a hybrid.

cross-fertilization The fertilization that occurs when the nucleus of a male sex cell from one individual joins with the nucleus of a female sex cell from another individual. In plants, cross-pollination is an example of cross-fertilization. Also called *allogamy.* Compare **self-fertilization.** —*Verb* **cross-fertilize.**

crossing over or **crossing-over** (krô′sĭng-ō′vər) The exchange of genetic material between homologous chromosomes that occurs during meiosis and contributes to genetic variability.

cross-multiply To multiply the numerator of one of a pair of fractions by the denominator of the other.

crossopterygian (krŏ-sŏp′tə-rĭj′ē-ən) Any of a mostly extinct group of lobe-finned fishes of the order Crossopterygii, whose only living member is the coelacanth. One group of crossopterygians is thought to have evolved into terrestrial vertebrates beginning in the Devonian Period. See more at **coelacanth.**

cross-pollination The transfer of pollen from the male reproductive organ (an anther or a male cone) of one plant to the female reproductive organ (a stigma or a female cone) of another plant. Insects and wind are the main agents of cross-pollination. Most plants reproduce by cross-pollination, which increases the genetic diversity of a population (increases the number of heterozygous individuals). Mechanisms that promote cross-pollination include having male flowers on one plant and female flowers on another, having pollen mature before the stigmas on the same plant are chemically receptive to being pollinated, and having anatomical arrangements (such as stigmas that are taller than anthers) that make self-pollination less likely.

cross product See **vector product.**

cross section In particle physics, an expression of the probability of the occurrence of an event, typically the scattering of subatomic particles, over a given area.

cross-stratification An arrangement of sediment strata deposited at an angle to the main stratification. The most common type of cross-stratification is produced by the migration of sand dunes or ripples. ▸ Cross-stratification in which the individual strata are greater than 1 cm (0.39 inch) in thickness is called **cross-bedding.** ▸ Cross-stratification in which the individual strata are less than 1 cm (0.39 inch) in thickness is called **cross-lamination.**

croup (kro͞op) An acute infection that affects the upper and lower respiratory tracts, especially the larynx, trachea, and bronchi, and is caused most commonly by viruses of the genus *Paramyxovirus.* It is characterized by labored breathing and obstruction below the glottis, accompanied by a barking cough.

crucible (kro͞o′sə-bəl) A heat-resistant container used to melt ores, metals, and other materials.

crucifer (kro͞o′sə-fər) Any of various plants in the mustard family (Cruciferae or Brassicaceae), including many important food plants, such as bok choy, cabbage, and radishes, as well as certain ornamental flowers.

crude oil (kro͞od) Unrefined petroleum.

crust (krŭst) The solid, outermost layer of the Earth, lying above the mantle. ▸ The crust that includes continents is called **continental crust** and is about 35.4 to 70 km (22 to 43.4 mi) thick. It consists mostly of rocks, such as granites and granodiorites, that are rich in silica and aluminum, with minor amounts of iron, magnesium, calcium, sodium, and potassium. ▸ The crust that includes ocean floors is called **oceanic crust** and is about 4.8 to 9.7 km (3 to 6 mi) thick. It has a similar composition to that of continental crust, but has higher concentrations of iron, magnesium, and calcium and is denser than continental crust. The predominant type of rock in oceanic crust is basalt.

crustacean (krŭ-stā′shən) Any of various widespread arthropods of the class Crustacea that live mostly in water and have a hard shell, a segmented body, and jointed appendages. Crustaceans include crabs, lobsters, shrimp, barnacles, and copepods.

cryobiology (krī′ō-bī-ŏl′ə-jē) The scientific study of the effects of very low temperatures on living organisms.

cryogenics (krī′ə-jĕn′ĭks) The scientific study of how matter behaves at very low temperatures, sometimes approaching absolute zero, and how such temperatures can be achieved and maintained. See also **superconductivity, superfluid.**

cryptocrystalline (krĭp′tō-krĭs′tə-lĭn, -līn′) Having a microscopic crystalline structure, as the mineral chalcedony does.

crystal (krĭs′təl) **1.** A homogenous solid formed by a repeating, three-dimensional pattern of atoms, ions, or molecules and having smooth external surfaces with characteristic angles between them. Crystals can occur in many sizes and shapes. ▸ The particular arrangement in space of these atoms, molecules, or ions, and the way in which they are joined, is called a **crystal lattice.** There are seven crystal groups or systems. Each is defined on the basis of the geometrical arrangement of the crystal lattice. **2a.** A natural or synthetic material, such as quartz or ceramic, that consists of such crystals. When subjected to mechanical stresses, crystalline materials can generate an electric charge or, when subjected to an electric field, they can generate mechanical vibrations in what is known as the **piezoelectric effect**. **b.** An electrical device, such as an oscillator or a diode used for detecting radio signals, made of such a material. *—Adjective* **crystalline.**

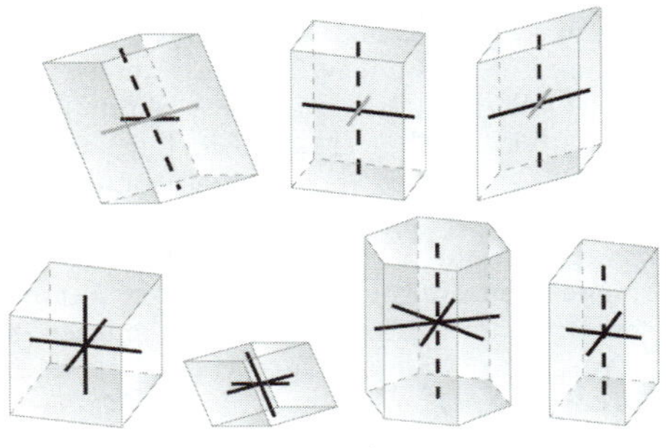

crystal

Crystal lattice systems (clockwise from top left)*: triclinic, orthorhombic, monoclinic, tetragonal, hexagonal, trigonal, and cubic.*

Cs The symbol for **cesium.**

csc Abbreviation of **cosecant.**

csch Abbreviation of **hyperbolic cosecant.**

ctenophore (tĕn′ə-fôr′) Any of various, mostly small marine invertebrates of the phylum Ctenophora, having transparent, gelatinous bodies bearing eight rows of comblike cilia. Ctenophores have a branched digestive tract that also has circulatory function. Most ctenophores feed on plankton and are bioluminescent, producing brilliant displays of blue or green light at night. Ctenophores are related to cnidarians but are more highly evolved because their bodies have a true mesoderm. Also called *comb jelly.*

CT scan (sē′tē′) See **CAT scan.**

Cu The symbol for **copper.**

cube (kyo͞ob) *Verb.* **1.** To multiply a number or a quantity by itself three times; raise to the third power. For example, five cubed is $5 \times 5 \times 5$. *—Noun.* **2.** The product that results when a number or quantity is cubed. For example, the cube of 5 is 125. **3.** A solid having six equal square faces or sides.

cube root The number whose cube is equal to a given number. For example, the cube root of 125 is 5, since $5^3 = 125$.

cubic (kyo͞o′bĭk) **1.** Referring to a volume unit of measurement. **2.** Involving a number or a variable that has been raised to the third power. **3.** Relating to a crystal having three axes of equal length intersecting at right angles. The mineral pyrite has cubic crystals. Also called *isometric.* See illustration at **crystal.**

cud (kŭd) Food that has been partly digested and brought up from the first stomach to the mouth again for further chewing by ruminants, such as cattle and sheep.

cuesta (kwĕs′tə) A ridge with a gentle slope on one side and a cliff or escarpment on the other. The gentler slope is formed by the differential erosion of underlying rock, and the cliff consists of an outcrop of harder, more resistant rock.

culm (kŭlm) The stem of a grass or similar plant.

cultigen (kŭl′tə-jən) An organism, especially a cultivated plant, such as the banana, not known to have a wild or uncultivated counterpart.

cultivar (kŭl′tə-vär′, -vâr′) A variety of a plant that has been created or selected intentionally and maintained through cultivation.

cultural anthropology (kŭl′chər-əl) The branch of anthropology that studies the development of human cultures based on ethnographic, linguistic, social, and psychological data. Compare **physical anthropology.**

culture (kŭl′chər) *Noun.* **1.** A growth of microorganisms, viruses, or tissue cells in a specially prepared nutrient medium under supervised conditions. **2.** The totality of socially transmitted behavior patterns, arts, beliefs, institutions, and all other products of human work and thought. Culture is learned and shared within social groups and is transmitted by nongenetic means. *—Verb.* **3.** To grow microorganisms, viruses, or tissue cells in a nutrient medium.

cumulonimbus (kyo͞om′yə-lō-nĭm′bəs) Plural **cumulonimbi** (kyo͞om′yə-lō-nĭm′bī). An extremely dense, vertically developed cloud with a low, dark base and fluffy masses that tower to great heights. Cumulonimbus clouds usually produce heavy rains, thunderstorms, or hailstorms. Also called *thundercloud.* See illustration at **cloud.**

cumulus (kyo͞om′yə-ləs) Plural **cumuli** (kyo͞om′yə-lī′). A dense, white, fluffy cloud with a flat base, a multiple rounded top, and a well-defined outline. The bases of cumulus clouds form primarily in altitudes below 2,000 m (6,560 ft), but their tops can reach much higher. Cumulus clouds are generally associated with fair weather but can also bring rain when they expand to higher levels. The clouds' edges are well-defined when they are composed of water droplets and fuzzy when made up of ice crystals. See illustration at **cloud.**

cupric (ko͞o′prĭk, kyo͞o′-) Containing copper, especially copper with a valence of 2. Compare **cuprous.**

cuprous (ko͞o′prəs, kyo͞o′-) Containing copper, especially copper with a valence of 1. Compare **cupric.**

curare (ko͞o-rä′rē, kyo͞o-) **1.** A dark, resinous extract obtained from several tropical American woody plants, especially *Chondrodendron tomentosum* or certain species of *Strychnos,* used as an arrow poison by some Indian peoples of South America. **2.** A purified preparation of an alkaloid obtained from *Chondrodendron tomentosum,* used in medicine and surgery to relax skeletal muscles.

curie (kyo͝or′ē, kyo͝o-rē′) A unit used to measure the rate of radioactive decay. Radioactive decay is measured by the rate at which the atoms making up a radioactive substance are transformed into different atoms. One curie is equal to 37 billion (3.7×10^{10}) of these transformations per second. Many scientists now measure radioactive decay in becquerels rather than curies.

Curie, Marie 1867–1934. Polish-born French chemist who pioneered research into radioactivity. Following Antoine Henri Becquerel's discovery of radioactivity, she investigated uranium with her husband, **Pierre Curie** (1859–1906). Together they discovered the elements radium and polonium. Marie Curie later isolated pure radium and developed the use of radioactivity in medicine.

Marie Curie

BIOGRAPHY Marie Curie

The study of radioactivity owes much of its start and early development to Marie Curie, born Maria Skłodowska in Poland in 1867. She was exposed to science early by her father, a mathematician and physicist, and in her young adulthood she moved to Paris, where she soon met many prominent physicists, including Pierre Curie, whom she married in 1895. In 1896 Henri Becquerel discovered a new phenomenon that Curie would soon name *radioactivity,* and together with Pierre she discovered two new elements, polonium and radium, in 1898. For their discovery of radioactivity, the three won the 1903 Nobel Prize for physics. In 1906, after her husband died unexpectedly, she filled his vacant professorship at the Sorbonne, becoming the first woman to teach there. In 1911 she became the first person to win a second Nobel Prize (for chemistry), which she received for the isolation of pure radium. This was an important feat because, before the invention of particle accelerators, radioactivity could only be effectively studied if one had an abundant and concentrated supply of highly radioactive sources; much of her work was spent developing techniques to create such stockpiles. Curie also saw the need for such supplies in medicine. Her frequent exposure to radioactivity apparently precipitated the leukemia that took her life in 1934, but her work was continued by her daughter Irène (1897–1956), already an important nuclear physicist in her own right.

Curie point The temperature at which a phase change in the magnetic or ferroelectric properties of a substance occurs, especially the change from ferromagnetism to paramagnetism that occurs with increasing temperature. Also called *Curie temperature.* See also **magnetic reversal, state of matter.**

curium (kyo͝or′ē-əm) *Symbol* **Cm** A synthetic, silvery-white, radioactive metallic element of the actinide series that is produced artificially from plutonium or americium. Curium isotopes are used to provide electricity for satellites and space probes. Its most stable isotope has a half-life of 16.4 million years. Atomic number 96; melting point (estimated) 1,350°C; valence 3. See **Periodic Table.**

current (kûr′ənt) **1.** A flowing movement in a liquid, gas, plasma, or other form of matter, especially one that follows a recognizable course. **2.** A flow of positive electric charge. The strength of current flow in any medium is related to voltage differences in that medium, as well as the electrical properties of the medium, and is measured in amperes. Since electrons are stipulated to have a negative charge, current in an electrical circuit actually flows in the opposite direction of the movement of electrons. See also **electromagnetism, Ohm's law.** See Note at **electric charge.**

A CLOSER LOOK **current**

Electric *current* is the phenomenon most often experienced in the form of *electricity.* Any time an object with a net electric charge is in motion, such as an electron in a wire or a positively charged ion jetting into the atmosphere from a solar flare, there is an electric current; the total current moving through some cross-sectional area in a given direction is simply the amount of positive charge moving through that cross-section. Current is sometimes confused with *electric potential* or *voltage,* but a voltage difference between two points (such as the two terminals of a battery) means only that current can *potentially* flow between them; how much does in fact flow depends on the resistance of the material between the two points. Electrical signals transmitted through a wire generally propagate at nearly the speed of light, but the current in the wire actually moves very slowly: pushing electrons into one end of the wire is rather like pushing a marble into one end of a tube filled with marbles—a marble (or electron) gets pushed out the other end almost instantly, even though the marbles (or electrons) inside move only incrementally.

curve (kûrv) **1.** A line or surface that bends in a smooth, continuous way without sharp angles. **2.** The graph of a function on a coordinate plane. In this technical sense, straight lines, circles, and waves are all curves.

curvilinear (kûr′və-lĭn′ē-ər) Formed, bounded, or characterized by curved lines.

Cushing's syndrome (ko͝osh′ĭngz) A syndrome caused by an excess of corticosteroids, especially cortisol, in the blood, characterized by obesity, muscle and skin atrophy, facial fullness (known as *moon facies*), hypertension, and other physical changes. Glucocorticoid excess is usually caused by a tumor of the adrenal cortex, excessive intake of ACTH or glucocorticoids, or increased production of ACTH from a tumor of the anterior lobe of the pituitary gland. Cushing's syndrome is named after its discoverer, American neurosurgeon Harvey Williams Cushing (1869–1939).

cutaneous (kyo͞o-tā′nē-əs) Relating to or involving the skin.

cuticle (kyo͞o′tĭ-kəl) **1.** The noncellular, hardened or membranous protective covering of many invertebrates, such as the transparent membrane that covers annelids. **2.** A layer of wax and cutin that covers the outermost surfaces of a plant. The cuticle is secreted by the epidermis and helps prevent water loss and infection by parasites. **3.** The hard skin around the sides and base of a fingernail or toenail.

cutin (kyo͞ot′n) A waxlike, water-repellent polyester consisting of fatty acids and aromatic compounds that occurs naturally in the walls of many plant cells. Cutin acts together with wax to form the cuticle, a barrier protecting the aboveground surfaces of plants from water loss and microbial attack.

Cuvier (kyo͞o′vē-ā′), Baron **Georges Léopold Chrétien Frédéric Dagobert** 1769–1832. French anatomist who is considered the founder of comparative anatomy. He originated a system of zoological classification that grouped animals according to the structures of their skeletons and organs. Cuvier extended his system to fossils; his reconstructions of the way extinct animals looked, based on their skeletal remains, greatly advanced the science of paleontology.

cwm (ko͝om) See **cirque.**

cyanate (sī′ə-nāt′, -nət) A salt or ester of cyanic acid, containing the group OCN.

cyanic acid (sī-ăn′ĭk) A poisonous, unstable, and highly volatile organic acid used to prepare cyanates. *Chemical formula:* **HOCN.**

cyanide (sī′ə-nīd′) Any of a large group of chemical compounds containing the radical CN, especially the very poisonous salts sodium cyanide and potassium cyanide. Cyanides are used to make plastics and to extract and treat metals.

cyanobacterium (sī′ə-nō-băk-tîr′ē-əm) Plural **cyanobacteria.** Any of a phylum of photosynthetic bacteria that live in water or damp soil and were once thought to be plants. Cyanobacteria have chlorophyll as well as carotenoid and phycobilin pigments, and they conduct photosynthesis in membranes known as thylakoids (which are also found in plant chloroplasts). Cyanobacteria may exist as individual cells or as filaments, and some species live in colonies. Many species secrete a mucilaginous substance that binds the cells or filaments together in colored (often bluish-green) masses. Cyanobacteria exist today in some 7,500 species, many of which are symbiotes, and have lived on Earth for 2.7 billion years. Since all species produce oxygen as a byproduct of metabolism, it is thought that much of Earth's atmospheric oxygen can be attributed to cyanobacteria. Many species can also fix nitrogen and so play an important role in the nitrogen cycle. Also called *blue-green alga.*

cyanocobalamin (sī′ə-nō′kō-băl′ə-mĭn, sī-ăn′ō-) See **vitamin B_{12}.**

cyber– A prefix that means "computer" or "computer network," as in *cyberspace,* the electronic medium in which online communication takes place.

cybernetics (sī′bər-nĕt′ĭks) The scientific study of communication and control processes in biological, mechanical, and electronic systems. Research in cybernetics often involves the comparison of these processes in biological and artificial systems.

cyberspace (sī′bər-spās′) The electronic medium of computer networks, in which online communication takes place.

cycad (sī′kăd′) Any of various evergreen plants that live in tropical and subtropical regions, have large feathery leaves, and resemble palm trees in that most leaves cluster around the top of the stem. Cycads are gymnosperms that bear conelike reproductive structures at the top of the stem, with male and female cones borne on different plants. Cycads were common in many parts of the Earth during the Jurassic Period and survive today in about 250 species. Sago palms are cycads.

cyclamate (sī′klə-māt′, sĭk′lə-) A salt or ester containing the group $C_6H_{12}NO_3S$. Some cyclamates were formerly used as artificial sweeteners.

cycle (sī′kəl) **1.** A single complete execution of a periodically repeated phenomenon. See also **period. 2.** A circular or whorled arrangement of flower parts such as those of petals or stamens.

cyclic (sĭk′lĭk, sī′klĭk) **1.** Occurring or moving in cycles. **2.** Relating to a compound having atoms arranged in a ring or closed-chain structure. Benzene is a cyclic compound. **3.** Having parts arranged in a whorl.

cycloalkane (sī′klō-ăl′kān) Any of various cyclic saturated hydrocarbons having the general formula C_nH_{2n}. Cycloalkanes are alkanes formed into a closed ring. They are one of the three main constituents of crude oil, the other two being alkanes and aromatic compounds. Also called: *naphthene.*

cyclohexane (sī′klō-hĕk′sān′) An extremely flammable, colorless liquid obtained from petroleum and benzene. It is used to make nylon and as a solvent, paint, and varnish remover. *Chemical formula:* **C_6H_{12}.**

cycloid (sī′kloid′) **1.** Resembling a circle. **2.** Thin, rounded, and smooth-edged, like a disk. Used of fish scales. **3.** The curve traced

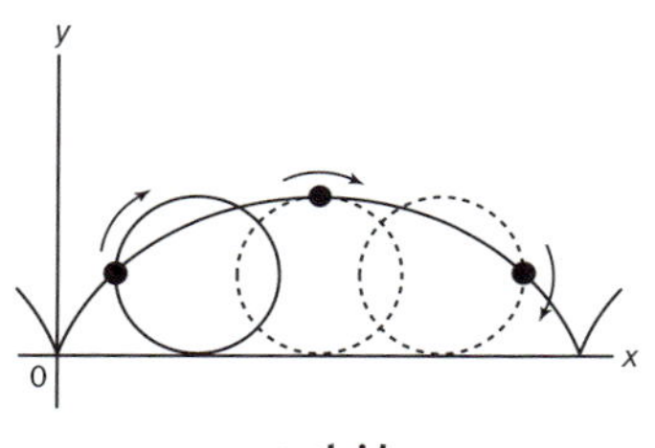

cycloid

by a point on the circumference of a circle that rolls on a straight line.

cyclone (sī′klōn′) **1.** A large-scale system of winds that spiral in toward a region of low atmospheric pressure. A cyclone's rotational direction is opposite to that of an anticyclone. In the Northern hemisphere, a cyclone rotates counterclockwise; in the Southern hemisphere, clockwise. Because low-pressure systems generally produce clouds and precipitation, cyclones are often simply referred to as storms. Compare **anticyclone.** ► An **extratropical cyclone** is one that forms outside the tropics at middle or high latitudes. Extratropical cyclones usually have an organized front and migrate eastward with the prevailing westerly winds of those latitudes. ► A **tropical cyclone** forms over warm tropical waters and is generally smaller than an extratropical cyclone. Such a system is characterized by a warm, well-defined core and can range in intensity from a tropical depression to a hurricane. **2.** A small-scale, violently rotating windstorm, such as a tornado or waterspout. Not in scientific use.

A CLOSER LOOK **cyclone**

Technically, a *cyclone* is nothing more than a region of low pressure around which air flows in an inward spiral. In the Northern Hemisphere the air moves counterclockwise around the low-pressure center, and in the Southern Hemisphere the air travels clockwise. Meteorologists also refer to *tropical cyclones,* which are cyclonic low-pressure systems that develop over warm water. For a tropical cyclone to originate, a large area of ocean must have a surface temperature greater than 27 degrees Celsius (80.6 degrees Fahrenheit). Tropical cyclones are categorized based on the strength of their sustained surface winds. They may begin as a *tropical depression,* with winds less than 39 miles (63 kilometers) per hour. *Tropical storms* are identified and tracked once the winds exceed this speed. Severe tropical cyclones, with winds of 74 miles (119 kilometers) per hour or greater, are better known as *hurricanes* when they occur in the Atlantic Ocean and Gulf of Mexico, or as *typhoons* when they happen in the Pacific Ocean. Because the word *cyclone* broadly defines a kind of air flow, cyclones are not confined to our planet. In 1999 the Hubble Space Telescope photographed a cyclone more than 1,610 kilometers (1,000 miles) across in the northern polar regions of Mars.

cyclopentane (sī′klə-pĕn′tān′, sĭk′lə-) A colorless, flammable, liquid hydrocarbon derived from petroleum and used as a solvent and motor fuel. *Chemical formula:* C_5H_{10}.

cyclopropane (sī′klə-prō′pān′, sĭk′lə-) A highly flammable, explosive, colorless gas that was once in wide use as an anesthetic but has been mostly replaced by less flammable gases. The three carbon atoms of cyclopropane form a triangular ring. *Chemical formula:* C_3H_6.

cyclosporine (sī′klə-spôr′ēn, -ĭn) also **cyclosporin** (sī′klə-spôr′ĭn) A polypeptide obtained from any of various deuteromycete fungi, used as an immunosuppressive drug to prevent the rejection of transplanted organs.

cyclostome (sī′klə-stōm′) Any of various jawless fish of the order Cyclostomata, having a long, eellike body without scales, a cartilaginous skeleton, and a disklike mouth used for sucking juices from prey. Cyclostomes include the hagfish and lampreys, although some scientists classify these two groups as separate orders.

cyclotron (sī′klə-trŏn′) A type of particle accelerator that accelerates charged subatomic particles, such as protons and electrons, in an outwardly spiraling path, greatly increasing

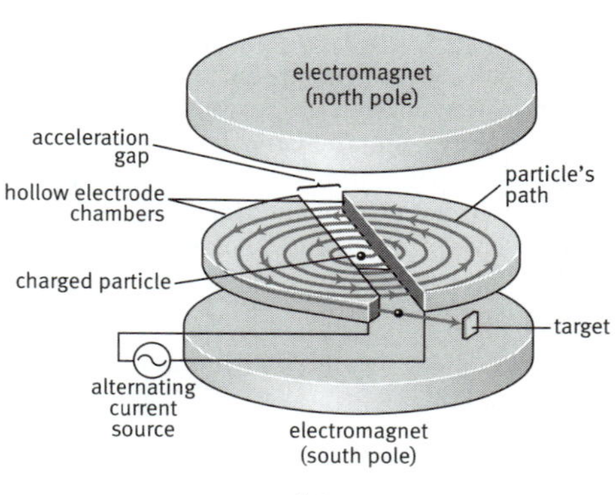

cyclotron

An alternating electric field attracts the particles from one side of the cyclotron to the other. The cyclotron's magnetic field, generated by the two electromagnets, bends each particle's path into a horizontal spiral, forcing it to accelerate in order to keep up with the alternating electric field. When the particle reaches its peak acceleration it is released to collide with the desired target.

their energies. Cyclotrons are used to bring about high-speed particle collisions in order to study subatomic structures. Compare **linear accelerator.** See also **synchrocyclotron.** See Note at **particle accelerator.**

Cygnus (sĭg′nəs) A constellation in the Northern Hemisphere near Cepheus and Lyra. Cygnus (the Swan, or the Northern Cross) contains the bright star Deneb.

cylinder (sĭl′ən-dər) A three-dimensional surface or solid object bounded by a curved surface and two parallel circles of equal size at the ends. The curved surface is formed by all the line segments joining corresponding points of the two parallel circles.

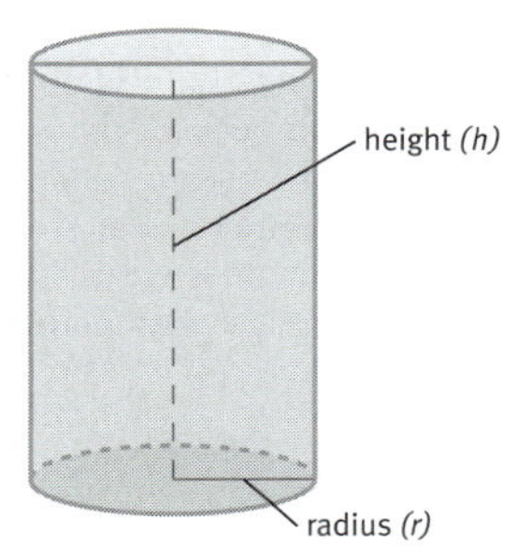

cylinder

The equation for determining the volume (V) of a cylinder is $V = \pi r^2 h$.

cylindrical projection (sə-lĭn′drĭ-kəl) A map projection in which the surface features of a globe are depicted as if projected onto a cylinder typically positioned with the globe centered horizontally inside the cylinder. In flattened form, a cylindrical projection so centered produces a rectangular map with the equator in the middle and the poles at the top and bottom. Parallels and meridians appear as straight lines that intersect each other at right angles in a grid pattern, with the meridians equally spaced and the parallels spaced progressively farther apart moving away form the equator. Distortion of shape and scale in a whole-world cylindrical projection is minimal in equatorial regions and maximal at the poles. Compare **azimuthal projection, conic projection.** See illustration at **Mercator projection.**

cyme (sīm) A usually flat-topped or convex determinate inflorescence in which the central main stem and each side branch end in a flower. The flowers in the cluster begin blooming from the flower on the main stem downwards or outwards. Baby's breath, dogwood, and the tomato have cymes.

cypsela (sĭp′sə-lə) Plural **cypselae** (sĭp′sə-lē′). A type of dry fruit consisting of an achene with a closely adhering calyx, the characteristic fruit of the aster family. Dandelions produce cypselae in which the plume, or pappus, is actually a modified sepal.

cyst (sĭst) **1.** An abnormal membranous sac in the body, containing a gaseous, liquid, or semisolid substance. **2.** A small, capsulelike form of certain organisms that develops in response to adverse or extreme conditions. Under adverse conditions, for instance, dinoflagellates form nonmotile resting cysts that fall to the ocean or lake bottom and can remain there for years before reviving. Certain invertebrates, such as the water bear (phylum Tardigrada), also develop cysts.

cysteine (sĭs′tə-ēn′) A nonessential amino acid. *Chemical formula:* $\mathbf{C_3H_7NO_2S}$. See more at **amino acid.**

cystic fibrosis (sĭs′tĭk fī-brō′sĭs) An inherited disorder of the exocrine glands, usually developing during early childhood and affecting mainly the pancreas, respiratory system, and sweat glands. It is marked by the production of abnormally thick mucus by the affected glands, usually resulting in chronic respiratory infections and impaired pancreatic function.

cyto– A prefix meaning "cell," as in the word *cytoplasm.*

cytochrome (sī′tə-krōm′) Any of a class of usually colored proteins that play important roles in oxidative processes and energy transfer during cell metabolism and cellular respiration. Cytochromes are **electron carriers**. They contain a heme group and are similar in structure to hemoglobin and chlorophyll. The most abundant and stable type is **cytochrome c.** By comparing different kinds of cytochromes, scientists can trace the evolutionary relationships of the organisms in which they occur.

cytogenetics (sī′tō-jə-nĕt′ĭks) The scientific study of the cellular components associated with heredity, especially chromosomes.

cytokine (sī′tə-kīn′) Any of several regulatory proteins, such as the interleukins and lymphokines, that are released by cells of the immune system and act as intercellular mediators in the generation of an immune response. Also called *chemokine.*

cytokinesis (sī′tō-kə-nē′sĭs, -kī-) The division of the cytoplasm of a cell following the division of the nucleus during cell division. See Note at **mitosis.**

cytokinin (sī′tə-kī′nĭn) Any of a class of plant hormones that promote cell division and growth and delay the senescence of leaves. Cytokinins are synthesized mainly in root tips.

cytology (sī-tŏl′ə-jē) The scientific study of the formation, structure, and function of cells.

cytomegalovirus (sī′tə-mĕg′ə-lō-vī′rəs) Any of a group of herpes viruses of the genus *Cytomegalovirus* that cause enlargement and abnormal structures in the cell nucleus (known as *nuclear inclusions*) in infected cells. Although the virus usually causes minor or no symptoms in normal adults, it is a common cause of opportunistic infections in immunosuppressed patients and can cause life-threatening congenital infections.

cytometry (sī-tŏm′ĭ-trē) The counting and measuring of cells, especially the counting and analysis of cell size, morphology, and other characteristics, traditionally performed using a standardized glass slide or small glass chamber of known volume.

cytoplasm (sī′tə-plăz′əm) The jellylike material that makes up much of a cell inside the cell membrane, and, in eukaryotic cells, surrounds the nucleus. The organelles of eukaryotic cells, such as mitochondria, the endoplasmic reticulum, and (in green plants) chloroplasts, are contained in the cytoplasm. The cytoplasm and the nucleus make up the cell's **protoplasm**. See more at **cell.**

cytosine (sī′tə-sēn′) A pyrimidine base that is a component of DNA and RNA, forming a base pair with guanine. *Chemical formula:* $\mathbf{C_4H_5N_3O}$.

cytoskeleton (sī′tə-skĕl′ĭ-tn) The internal framework of a cell, composed of a network of protein filaments and extending throughout the fluid of the cell (the cytosol). The cytoskeleton consists mainly of actin filaments and microtubules and plays an important role in cell movement, shape, growth, division, and differentiation, as well as in the movement of organelles within the cell. All eukaryotic cells have a cytoskeleton.

cytosol (sī′tə-sôl′, -sŏl′) The fluid component of cytoplasm, containing the insoluble, suspended cytoplasmic components. In prokaryotes, all chemical reactions take place in the cytosol. In eukaryotes, the cytosol surrounds the organelles.

cytotoxic T cell (sī′tə-tŏk′sĭk) See **killer T cell.**

D

d Abbreviation of **diameter.**

dacite (dā′sīt′) A fine-grained light gray volcanic rock consisting primarily of quartz, plagioclase feldspar, and potassium feldspar, and also containing biotite, hornblende, or pyroxene. It is the fine-grained equivalent of granodiorite.

D/A converter Short for *digital to analog converter.* An electronic device that takes a digital numerical value as input, and yields a voltage signal at the output proportional in strength to the value of the input. D/A converters are used in digital audio and video technology, in which audio or video information stored in digital form (as on tape or a disk) is converted to electrical voltage signals for amplification or display on a screen. Compare **A/D converter.**

Dale (dāl), Sir **Henry Hallett** 1875–1968. British physiologist who discovered acetylcholine and, with Otto Loewi, investigated the chemical transmission of nerve impulses. For this work they shared the 1936 Nobel Prize for physiology or medicine.

dalton (dôl′tən) See **atomic mass unit.**

Dalton, John 1766–1844. British chemist whose pioneering work on the properties of the atmosphere and gases led him to formulate the atomic theory. Dalton's theory stipulates that all matter is made up of combinations of atoms, the atoms of each element being identical. These atoms can be neither created nor destroyed, but chemical reactions take place through their rearrangement.

damping (dăm′pĭng) The action of a substance or of an element in a mechanical or electrical device that gradually reduces the degree of oscillation, vibration, or signal intensity, or prevents it from increasing. For example, sound-proofing technology dampens the oscillations of sound waves. Built-in damping is a crucial design element in technology that involves the creation of oscillations and vibrations.

damping off Any of various diseases of seedlings that are caused by oomycetes, especially of the genus *Pythium,* or by fungi, and result in wilting and death.

darcy (där′sē′) A unit used to measure the permeability of porous substances such as soil. One darcy is equal to the passage of 1 cubic centimeter of fluid having a viscosity of 1 centipoise for 1 second under the pressure of 1 atmosphere through a medium having a volume of 1 cubic centimeter.

Darcy, Henry Philibert Gaspard 1803–1858. French engineer who formulated the law (now named for him) governing the rate at which a fluid flows through a permeable medium. The darcy unit, used to measure the permeability of porous substances, is also named after him.

Darcy's law (där′sēz′) A law in geology describing the rate at which a fluid flows through a permeable medium. Darcy's law states that this rate is directly proportional to the drop in vertical elevation between two places in the medium and indirectly proportional to the distance between them. The law is used to describe the flow of water from one part of an aquifer to another and the flow of petroleum through sandstone and gravel.

dark cloud (därk) See **absorption nebula.**

dark energy A form of energy hypothesized to reside in the structure of space itself, responsible for the accelerating expansion of the universe. Dark energy theoretically counterbalances the kinetic energy of the universe's expansion, entailing that that the universe has no inherent curvature, as astronomical observations currently suggest. Dark energy appears to account for 73 percent of all the energy and matter in the universe. See also **big bang.**

dark matter Matter that emits little or no detectable radiation. Gravitational forces observed on many astronomical objects suggest the significant presence of such matter in the universe, accounting for approximately 23 percent of the total mass and energy of the universe. Its exact nature is not well understood, but it may be largely composed of varieties of subatomic particles that have not yet been discovered, as well as the mass of black holes and of stars too dim to observe. Also called *missing mass.*

A CLOSER LOOK **dark matter**

What is the universe made of? We know that galaxies consist of planets, stars, and huge gas and dust clouds—all of these objects are observable by the radiation they give off, such as radio, infrared, optical, ultraviolet, x-ray, or gamma-ray radiation, and all can be observed using various kinds of telescopes. But there are reasons to suspect the existence of far more matter than this, matter that is not directly observable. Evidence for such *dark matter* comes from observations of certain gravitational effects. For example, astronomers have found that galaxies rotate much faster than they would be expected to rotate based solely on their observable mass—in fact, they should be flying apart. One explanation for this apparent anomaly is to assume that the galaxies have much more mass than we can see, and this invisible mass holds them together gravitationally. Various theories of the composition of this invisible dark matter have been proposed, from exotic yet-to-be discovered particles to planet-sized objects made of ordinary matter that are too small or far away to be detected by present-day instruments. But none of these theories are entirely satisfactory, and the fundamental question of what makes up most of the universe remains unanswered.

dark nebula See **absorption nebula.**

dark reaction Any of the chemical reactions that take place during the second stage of photosynthesis and do not require light. During the dark reactions, energy released from ATP (created by the light reactions) drives the fixation of carbon from carbon dioxide in organic molecules. The **Calvin cycle** forms part of the dark reactions. As long as ATP is available, the dark reactions can occur in darkness or in light. Compare **light reaction.** See more at **Calvin cycle, photosynthesis.**

darmstadtium (därm′shtät′ē-əm) *Symbol* **Ds** A synthetic radioactive element, first produced by bombarding lead atoms with nickel atoms. It has a mass number of 281, and its most stable isotope has a half-life of over one minute. Atomic number 110. See **Periodic Table.**

Darwin (där′wĭn), **Charles Robert** 1809–1882. British naturalist who proposed the theory of evolution based on natural selection (1858). Darwin's theory, that random variation of traits within an individual species can lead to the development of new species, revolutionized the study of biology.

Charles Darwin

BIOGRAPHY **Charles Darwin**

The flora and fauna of the Galápagos Archipelago, a group of islands 650 miles west of Ecuador in the Pacific Ocean, provided the inspiration for Charles Darwin's *On the Origin of Species by Means of Natural Selection,* which outlined his theory of evolution. Although Darwin spent some time studying medicine and later prepared for the clergy, graduating in 1828 from Christ's College, Cambridge, he couldn't deny his interest in geology and natural history. He spent five years (1831–36) as a naturalist aboard the HMS *Beagle* on an exploration of South America and Australia. In September 1835 the *Beagle* reached the Galápagos Archipelago. "This archipelago," Darwin wrote, "seems to be a little world within itself, the greater number of its inhabitants, both vegetable and animal, being found nowhere else." Darwin observed 26 species of birds, only one of which was known to exist anywhere else, as well as giant tortoises and other unusual reptiles. Each species, he observed, was uniquely adapted to the particular island on which it lived. Upon his return to England, Darwin refined his notes and continued to make scientific observations, this time of his own garden and of the animals kept by his family. In 1859, after 23 years of sustained work, he published *On the*

Origin of Species, in which he argued that traits such as size and color vary from species to species and that individual variations of these traits are passed down from parents to offspring. More progeny are produced than there is available sustenance. Variations that contribute more successfully to attracting a mate and reproducing are passed down to more offspring, eventually influencing the entire species. Through this process of *natural selection,* the highly complex species of today gradually evolved from earlier, simpler organisms.

Darwinism (där′wĭ-nĭz′əm) A theory of biological evolution developed by Charles Darwin and others, stating that all species of organisms arise and develop through the natural selection of small, inherited variations that increase the individual's ability to compete, survive, and reproduce. Darwin's ideas have been refined and modified by subsequent researchers, but his theories still form the foundation of the scientific understanding of the evolution of life. Darwinism is often contrasted with another theory of biological evolution called *Lamarckism,* based on the now-discredited ideas of Jean-Baptiste **Lamarck.** See Note at **evolution.**

database (dā′tə-bās′, dăt′ə-) A collection of data arranged for ease and speed of search and retrieval by a computer.

date line (dāt) The International Date Line.

dative bond (dā′tĭv) See **coordinate bond.**

daughter cell (dô′tər) Either of the two cells formed when a cell undergoes cell division by **mitosis**. Daughter cells are genetically identical to the parent cell because they contain the same number and type of chromosomes.

Davy (dā′vē), Sir **Humphry** 1778–1829. British chemist who was a pioneer of electrochemistry. By means of electrolysis Davy isolated several elements, including sodium and potassium (1807), and barium, boron, calcium, and magnesium (1808). He also proved that diamonds are a form of carbon.

day (dā) See under **sidereal time, solar day.**

daylight-saving time (dā′līt-sā′vĭng) or **daylight-savings time** Time during which clocks are set one hour or more ahead of standard time to provide more daylight at the end of the working day during late spring, summer, and early fall. First proposed by Benjamin Franklin, daylight saving time was instituted in various countries during both world wars in the 20th century and was made permanent in most of the United States beginning in 1973. Arizona, Hawaii, most of eastern Indiana, and certain US territories and possessions do not observe daylight saving time.

day-neutral plant A plant that flowers regardless of the length of the period of light it is exposed to. Rice, corn, and the cucumber are day-neutral plants. Compare **long-day plant, short-day plant.** See more at **photoperiodism.**

dB Abbreviation of **decibel.**

Db The symbol for **dubnium.**

DC Abbreviation of **direct current.**

DDT (dē′dē-tē′) Short for *dichlorodiphenyltrichloroethane.* A powerful insecticide that is also poisonous to humans and animals. It remains active in the environment for many years and has been banned in the United States for most uses since 1972 but is still in use in some countries in which malaria is endemic. *Chemical formula:* $C_{14}H_9Cl_5$.

deafness (dĕf′nĭs) The lack or severe impairment of the ability to hear. Deafness is usually genetic or congenital as a result of prenatal viral infection, birth trauma, or other causes. Acquired deafness is caused mostly by drug toxicity, trauma, and certain diseases. **Cochlear implants** are used to treat some forms of deafness.

death (dĕth) The end of life of an organism or cell. In humans and animals, death is manifested by the permanent cessation of vital organic functions, including the absence of heartbeat, spontaneous breathing, and brain activity. Cells die as a result of external injury or by an orderly, programmed series of self-destructive events known as **apoptosis**. The most common causes of death for humans in well-developed countries are cardiovascular disease, cancer, Alzheimer's disease, certain chronic diseases such as diabetes and emphysema, lung infections, and accidents. See also **brain death.**

DeBakey (də bā′kē), **Michael Ellis** Born 1908. American heart surgeon and physician who developed many of the techniques, procedures, and instruments used in cardiovascular surgery. In 1966 DeBakey implanted the first totally artificial heart in a human. Two years later, he supervised the first successful multi-organ transplant, transferring a heart, both kidneys, and a lung from a single donor to four separate recipients.

Debierne (dəb-yĕrn′), **André** 1874–1949. French chemist who discovered actinium in 1900. In 1910 he collaborated with Marie Curie to isolate pure radium.

deca– A prefix that means "ten," as in *decahedron,* a polygon having ten faces.

decagon (dĕk′ə-gŏn′) A polygon having ten sides.

decahedron (dĕk′ə-hē′drən) Plural **decahedrons** or **decahedra.** A polyhedron having ten faces.

decane (dĕk′ān′) Any of various liquid hydrocarbons containing ten carbon atoms in a chain, attached to hydrogen atoms. Decane is the tenth member of the alkane series. *Chemical formula:* $C_{10}H_{22}$.

decapod (dĕk′ə-pŏd′) **1.** Any of various crustaceans of the order Decapoda, characteristically having ten legs, each joined to a segment of the thorax. Crabs, hermit crabs, lobsters, and shrimp are decapods. **2.** A cephalopod mollusk, such as a squid or cuttlefish, having ten armlike tentacles.

decay (dĭ-kā′) *Noun.* **1.** The breaking down or rotting of organic matter through the action of bacteria, fungi, or other organisms; decomposition. **2.** The spontaneous transformation of a relatively unstable particle into a set of new particles. For example, a pion decays spontaneously into a muon and an antineutrino. The decay of heavy or unstable atomic nuclei (such as uranium or carbon-10) into more stable nuclei and emitted particles is called **radioactive decay**. The study of particle decay is fundamental to subatomic physics. See more at **fundamental force, radioactive decay. 3.** —*Verb.* To undergo decay.

decay chain A sequence of radioactive decay processes, in which the decay of one element creates a new element that may itself be radioactive. The chain ends when stable atoms are formed. For example, uranium-238 decays into thorium-234, which in turn decays into palladium-234, and so on until stable iron is produced at the end of the chain. Also called *decay sequence.*

deci– A prefix that means "one tenth," as in *deciliter,* one tenth of a liter.

decibel (dĕs′ə-bəl) A unit used to measure the power of a signal, such as an electrical signal or sound, relative to some reference level. An increase of ten decibels in the power of a signal is equivalent to increasing its power by a factor of ten. As a measure of sound intensity, a zero-decibel reference is stipulated to be the lowest level audible to the human ear; the speaking voice of most people ranges from 45 to 75 decibels. See Note at **sound**[1].

deciduous (dĭ-sĭj′o͞o-əs) **1.** Shedding leaves at the end of a growing season and regrowing them at the beginning of the next growing season. Most deciduous plants bear flowers and have woody stems and broad rather than needlelike leaves. Maples, oaks, elms, and aspens are deciduous. Compare **evergreen.** See more at **abscission. 2.** Falling off or shed at a particular season or stage of growth, as antlers.

decimal (dĕs′ə-məl) **1.** A representation of a real number using the base ten and decimal notation, such as 201.4, 3.89, or 0.0006. **2.** A decimal fraction.

decimal fraction A decimal having no digits to the left of the decimal point except zero, such as 0.2 or 0.00354.

decimal notation A representation of a fraction or other real number using the base ten and consisting of any of the digits 0, 1, 2, 3, 4, 5, 6, 7, 8, 9, and a decimal point. Each digit to the left of the decimal point indicates a multiple of a positive power of ten, while each digit to the right indicates a multiple of a negative power of ten. For example, the number $26\frac{37}{100}$ can be written in decimal notation as 26.37, where 2 represents 2×10, 6 represents 6×1, 3 represents $3 \times \frac{1}{10}$ (or $\frac{3}{10}$), and 7 represents $7 \times \frac{1}{100}$ (or $\frac{7}{100}$).

decimal place The position of a digit to the right of the decimal point in a number written in decimal notation. In 0.079, for example, 0 is in the first decimal place, 7 is in the second decimal place, and 9 is in the third decimal place.

decimal point A period used in decimal notation to separate whole numbers from fractions, as in the number 1.3, which represents $1 + \frac{3}{10}$.

decimal system 1. A number system based on units of 10 and using decimal notation. **2.** A system of measurement in which all derived units are multiples of 10 of the fundamental units.

declination (dĕk′lə-nā′shən) **1.** On the celestial sphere, the position of a celestial object north or south of the **celestial equator**. Declination is measured in degrees along a great circle drawn through the object being measured and the north and south celestial poles, with positive values north of the celestial equator and negative values south of it, so that the equator itself is 0° and the north and south celestial poles are +90° and –90° declination respectively. See more at **equatorial coordinate system. 2.** See **magnetic declination.**

decoder (dē-kō′dər) An electronic or software device that converts telecommunication signals from their transmitted form into a form interpretable to other devices or to human beings.

decomposer (dē′kəm-pō′zər) See **detritivore.**

decomposition (dē-kŏm′pə-zĭsh′ən) **1.** The separation of a substance into simpler substances or basic elements. Decomposition can be brought about by exposure to heat, light, or chemical or biological activity. **2.** The process of breaking down organic material, such as dead plant or animal tissue, into smaller molecules that are available for use by the organisms of an ecosystem. Decomposition is carried on by bacteria, fungi, protists, worms, and certain other organisms. See more at **detritivore.**

decompression chamber (dē′kəm-prĕsh′ən) A compartment in which atmospheric pressure can be gradually raised or lowered, used especially in readjusting divers or underwater workers to normal atmospheric pressure or in treating decompression sickness.

decompression sickness A common disorder that affects deep-sea divers following a sudden drop in the surrounding pressure, as when ascending rapidly from a dive. When divers are underwater, the amounts of gases such as O_2, CO_2, and N_2 in their blood increase due to the increased pressure. As they ascend to the surface and the pressure decreases, the gases in their blood expand. The extra oxygen is absorbed by the body; the extra CO_2 is excreted efficiently; but nitrogen, which the body does not use, forms bubbles in the blood and tissues. These bubbles cause severe pains in the joints and chest, skin irritation, cramps, and possibly paralysis. Decompression sickness can be avoided by ascending slowly to the surface, or by spending time in a decompression chamber.

decongestant (dē′kən-jĕs′tənt) A medication that reduces congestion of the nose or sinuses, usually by causing vasoconstriction.

deduction (dĭ-dŭk′shən) **1.** The process of reasoning from the general to the specific, in which a conclusion follows necessarily from the premises. **2.** A conclusion reached by this process.

USAGE **deduction/induction**

The logical processes known as deduction and induction work in opposite ways. In *deduction* general principles are applied to specific instances. Thus, using a mathematical formula to figure the volume of air that can be contained in a gymnasium is applying deduction. Similarly, applying a law of physics to predict the outcome of an experiment is reasoning by deduction. By contrast, *induction* makes generalizations based on a number of specific instances. The observation of hundreds of examples in which a certain chemical kills plants might prompt the inductive conclusion that the chemical is toxic to all plants. Inductive generalizations are often revised as more examples are studied and more facts are known. If certain plants that have not been tested turn out to be unaffected by the chemical, the conclusion about the chemical's toxicity must be revised or restricted. In this way, an inductive generalization is much like a hypothesis.

deep scattering layer (dēp) See **scattering layer.**

deet (dēt) A colorless, oily liquid that has a mild odor and is used as an insect repellent. Its chemical name is diethyl toluamide. *Chemical formula:* $C_{12}H_{17}NO$.

defibrillation (dē-fĭb′rə-lā′shən) Termination of fibrillation of the heart muscle and restoration of normal heart rhythm, especially by one or more electric shocks administered by paddles applied to the chest.

definite integral (dĕf′ə-nĭt) The difference between the values of an indefinite integral evaluated at each of two limit points, usually expressed in the form $\int_a^b f(x)dx$. The result of performing the integral is a number that represents the area bounded by the curve of $f(x)$ between the limits and the x-axis if $f(x)$ is greater than or equal to zero between the limits. **2.** The result of an integration performed on a fixed interval.

definitive host (dĭ-fĭn′ĭ-tĭv) See under **host.**

deflation (dĭ-flā′shən) The lifting and removal of fine, dry particles of silt, soil, and sand by the wind. Deflation is common in deserts and in coastal areas that have sand dunes.

DeForest (dĭ-fôr′ĭst), **Lee** 1873–1961. American electrical engineer and inventor who is known as "the father of radio." He patented more than 300 inventions, including the triode electron tube, which made it possible to amplify and detect radio waves.

deforestation (dē-fôr′ĭ-stā′shən) The cutting down and removal of all or most of the trees in a forested area. Deforestation can erode soils, contribute to desertification and the pollution of waterways, and decrease biodiversity through the destruction of habitat.

defragment (dĭ-frăg′mənt) To reorganize the way information is stored on a computer disk so that all of the information belonging to a file is stored in a single, contiguous area the disk.

degauss (dē-gous′) **1.** To neutralize or rebalance the magnetic field of a magnetized object, such as a computer monitor or the read/write head of a disk drive or tape recorder. **2.** To erase information from a magnetic disk, tape, or other magnetic storage device.

degeneracy pressure (dĭ-jĕn′ər-ə-sē) A pressure exerted by dense material consisting of fermions (such as electrons in a white dwarf star). This pressure is explained in terms of the Pauli exclusion principle, which requires that no two fermions be in the same quantum state. The more densely fermions are packed together and must share the same space, the more they must differ from each other in terms of their momentum. In turn, the greater the range in momentum, the greater the fraction of particles with high momentum, and these exert pressure on their surroundings. See also **neutron star, white dwarf.**

degenerative joint disease (dĭ-jĕn′ər-ə-tĭv) See **osteoarthritis.**

deglaciation (dē-glā′shē-ā′shən) The uncovering of land that was previously covered by a glacier. Deglaciation occurs when a glacier melts.

degradable (dĭ-grā′də-bəl) Relating to a compound that breaks down into simpler compounds by stages. During the degradation of a degradable compound, well-defined intermediate products are created.

degree (dĭ-grē′) **1.** A unit division of a temperature scale. **2a.** A unit for measuring an angle or an arc of a circle. One degree is $\frac{1}{360}$ of the circumference of a circle. **b.** This unit used to measure latitude or longitude on the Earth's surface. **3.** The greatest sum of the exponents of the variables in a term of a polynomial or polynomial equation. For example, $x^3 + 2xy + x$ is of the third degree.

degree-day A unit of measurement equal to a difference of one degree between the mean outdoor temperature on a certain day and a reference temperature. The unit is most often used in estimating the energy needs for heating or cooling a building (for example, heating degree-days and cooling degree-days). Originally, degree-days were used to determine the relationship between temperature and plant growth. The term continues to be used in life sciences as a measure of upper- and lower-temperature limits for organisms.

degree of freedom **1.** Any of the independent thermodynamic variables, such as pressure, temperature, or composition, required to specify a system with a given number of phases and components. **2.** Any of the independent terms used to characterize the way a physical system can store energy. For example, a molecule consisting of two atoms can be thought of as having three degrees of freedom: one for its linear motion (as the whole molecule moves through space), one for its angular motion (as it rotates around its center of gravity) and one for its internal vibrational energy (as the atoms pull and push against each other within their chemical bond).

dehiscence (dĭ-hĭs′əns) The spontaneous opening at maturity of a plant structure, such as a fruit, anther, or sporangium, to release its contents. Compare **indehiscence.** —*Verb* **dehisce.**

dehiscence

close-up of a split milkweed pod

dehumidifier (dē′hyo͞o-mĭd**′**ə-fī′ər) A device used to remove water vapor from the air, thereby lowering the humidity of the air.

dehydration (dē′hī-drā**′**shən) **1.** The process of losing or removing water or moisture. **2.** A condition caused by the excessive loss of water from the body, which causes a rise in blood sodium levels. Since dehydration is most often caused by excessive sweating, vomiting, or diarrhea, water loss is usually accompanied by a deficiency of electrolytes. If untreated, severe dehydration can lead to shock.

Deimos (dē**′**mōs, dā**′**-) The satellite of Mars that is second in distance from the planet.

deionization (dē-ī′ə-nĭ-zā**′**shən) The removal of ions from a solution using an ion exchange process. Deionization is used for water purification and for medical purposes.

Delbrück (dĕl**′**bro͝ok′), **Max** 1906–1981. German-born American biologist who was a pioneer in the study of molecular genetics. He discovered in 1946 that viruses can exchange genetic material to create new types of viruses. He shared a 1969 Nobel Prize for his work in viral genetics.

deliquescent (dĕl′ĭ-kwĕs**′**ənt) Relating to a solid substance that absorbs moisture from the air and becomes liquid. Deliquescent substances usually absorb so much moisture from the air that they form a strong solution. Potassium hydroxide is deliquescent.

delirium tremens (dĭ-lîr**′**ē-əm trē**′**mənz) An acute, sometimes fatal episode of delirium that is usually caused by withdrawal or abstinence from alcohol following habitual excessive drinking or an episode of heavy alcohol consumption. It is characterized by trembling, sweating, acute anxiety, confusion, and hallucinations.

delta (dĕl**′**tə) A usually triangular mass of sediment, especially silt and sand, deposited at the mouth of a river. Deltas form when a river flows into a body of standing water, such as a sea or lake, and deposits large quantities of sediment. They are usually crossed by numerous streams and channels and have exposed as well as submerged areas.

delusion (dĭ-lo͞o**′**zhən) A false belief strongly held in spite of invalidating evidence, especially as a symptom of mental illness, as in schizophrenia.

deme (dēm) A small, locally interbreeding group of organisms within a larger population. Demes are isolated reproductively from other members of their species, although the isolation may only be partial and is not necessarily permanent. Because they share a somewhat restricted gene pool, members of a deme generally differ morphologically to some degree from members of other demes. See also **population.**

dementia (dĭ-mĕn**′**shə) Deterioration of intellectual faculties, such as memory, concentration, and judgment, sometimes accompanied by emotional disturbance and personality changes. Dementia is caused by organic damage to the brain (as in Alzheimer's disease), head trauma, metabolic disorders, or the presence of a tumor.

Democritus (dĭ-mŏk**′**rĭ-təs) 460?–370? BCE. Greek philosopher who developed one of the first atomist theories of the universe. Democritus believed that the world consists of an infinite number of extremely small particles whose different characteristics and combinations account for the different qualities of all matter.

denature (dē-nā**′**chər) **1.** To cause the tertiary structure of a protein to unfold, as with heat, alkali, or acid, so that some of its original properties, especially its biological activity, are diminished or eliminated. **2.** To cause the paired strands of DNA to separate into individual strands.

denatured alcohol (dē-nā**′**chərd) Ethyl alcohol to which a poisonous substance, such as acetone or methanol, has been added to make it unfit for consumption.

dendrimer (dĕn′drə-mər) A large, synthetically produced polymer in which the atoms are arranged in many branches and subbranches radiating out from a central core. Dendrimers are being investigated for possible uses in nanotechnology, gene therapy, and other fields.

dendrite (dĕn′drīt′) **1.** Any of several parts branching from the body of a neuron that receive and transmit nerve impulses. **2.** A mineral that has a branching crystal pattern. Dendrites often form within or on the surface of other minerals and often consist of manganese oxides.

dendritic cell (dĕn-drĭt′ĭk) A highly specialized white blood cell found in the skin, mucosa, and lymphoid tissues that initiates a primary immune response by activating lymphocytes and secreting cytokines.

dendrochronology (dĕn′drō-krə-nŏl′ə-jē) The study of growth rings in trees for the purpose of analyzing past climate conditions or determining the dates of past events. Because trees grow more slowly in periods of drought or other environmental stress than they do under more favorable conditions, the size of the rings they produce varies. Analyzing the pattern of a tree's rings provides information about the environmental changes that took place during the period in which it was growing. Matching the pattern in trees whose age is known to the pattern in wood found at an archaeological site can establish the age at which the wood was cut and thus the approximate date of the site. By comparing living trees with old lumber and finding overlapping ring patterns, scientists have established chronological records for some species that go back as far as 9,000 years.

dendrology (dĕn-drŏl′ə-jē) The scientific study of trees and other woody plants.

Deneb (dĕn′ĕb′) A white supergiant star. It is the brightest star in the constellation Cygnus, with an apparent magnitude of 1.26. Deneb, along with Vega and Altair, form the Summer Triangle asterism. *Scientific name:* Alpha Cygnus. See Note at **Rigel.**

dengue (dĕng′gē, -gā) An acute, infectious tropical disease caused by any of several viruses of the genus *Flavivirus.* It is transmitted by mosquitoes, and characterized by high fever, rash, headache, and severe muscle and joint pain.

denominator (dĭ-nŏm′ə-nā′tər) The number below or to the right of the line in a fraction, indicating the number of equal parts into which one whole is divided. For example, in the fraction $\frac{2}{7}$, 7 is the denominator.

density (dĕn′sĭ-tē) A measure of the quantity of some physical property (usually mass) per unit length, area, or volume (usually volume). ▸ **Mass density** is a measure of the mass of a substance per unit volume. Most substances (especially gases such as air) increase in density as their pressure is increases or as their temperature decreases. ▸ **Energy density** is a measure of the amount of energy (often in the form of electromagnetic radiation) per unit volume in a region of space or some material. See also **Boyle's law.**

dental (dĕn′tl) Relating to the teeth.

dentate (dĕn′tāt′) Edged with toothlike projections; toothed. Used of leaves, such as those of birches.

denticle (dĕn′tĭ-kəl) A small tooth or toothlike projection, especially a dermal denticle.

dentin (dĕn′tĭn) The main bony part of a tooth beneath the enamel, surrounding the pulp chamber and root canals.

dentistry (dĕn′tĭ-strē) The branch of medicine that deals with the diagnosis, prevention, and treatment of diseases of the teeth, gums, and other structures of the mouth.

dentition (dĕn-tĭsh′ən) The type, number, and arrangement of teeth in an animal species. In mammals, dentition consists of several different types of teeth, including incisors, canines, and molars. The dentition of toothed fish and reptiles usually consists of only one kind of tooth.

deoxyribonucleic acid (dē-ŏk′sē-rī′bō-no͞o-klē′ĭk) See **DNA.**

deoxyribose (dē-ŏk′sē-rī′bōs′) The sugar found in the side chains of DNA, differing from ribose in having a hydrogen atom instead of an OH group on one of its carbon atoms. *Chemical formula:* $C_5H_{10}O_4$.

dependent variable (dĭ-pĕn′dənt) In mathematics, a variable whose value is determined by the value of an independent variable. For example, in the formula for the area of a circle, $A = \pi r^2$, A is the dependent variable, as its value depends on the value of the radius (r). Compare **independent variable.**

depletion region (dĭ-plē′shən) A region in a semiconductor device, usually at the juncture of P-type and N-type materials, in which there is neither an excess of electrons nor of holes. Large depletion regions inhibit current flow. See also **semiconductor diode.**

deposit (dĭ-pŏz′ĭt) An accumulation or layer of solid material, either consolidated or unconsolidated, left or laid down by a natural process. Deposits include sediments left by water, wind, ice, gravity, volcanic activity, or other agents. A layer of coal formed over many years through the decomposition of plant material is also a deposit.

deposition (dĕp′ə-zĭsh′ən) **1.** The accumulation or laying down of matter by a natural process, as the laying down of sediments in a river or the accumulation of mineral deposits in a bodily organ. **2.** The process of changing from a gas to a solid without passing through an intermediate liquid phase. Carbon dioxide, at a pressure of one atmosphere, undergoes deposition at about –78 degrees Celsius. Compare **sublimation.**

depression (dĭ-prĕsh′ən) **1.** A geographic area, such as a sinkhole or basin, that is lower than its surroundings. **2.** A mood disorder characterized by an inability to experience pleasure, difficulty in concentrating, disturbance of sleep and appetite, and feelings of sadness, guilt, and helplessness. **3.** A reduction in the activity of a physiological process, such as respiration. **4.** A region of low atmospheric pressure. Low pressure systems result in precipitation, ranging from mild to severe in intensity. See also **cyclone.**

depth finder (dĕpth) An instrument used to measure the depth of water, especially by radar, sonar, or ultrasound.

derivative (dĭ-rĭv′ə-tĭv) In calculus, the slope of the tangent line to a curve at a particular point on the curve. Since a curve represents a function, its derivative can also be thought of as the rate of change of the corresponding function at the given point. Derivatives are computed using differentiation.

dermal (dûr′məl) Relating to or involving the skin.

dermal denticle A toothlike or platelike scale of cartilaginous fishes, such as sharks and rays. The teeth of these fish are also dermal denticles of larger size and modified by the presence of serrated edges.

dermapteran (dər-măp′tər-ən) Any of various insects of the order Dermaptera, having an elongated flattened body equipped with a pair of pincerlike appendages at the posterior end. Dermapterans include the earwigs.

dermatogen (dûr-măt′ə-jən) See **protoderm.**

dermatology (dûr′mə-tŏl′ə-jē) The branch of medicine that deals with the diagnosis and treatment of disorders of the skin, hair, and nails.

dermis (dûr′mĭs) The innermost layer of the skin in vertebrate animals. The dermis lies under the epidermis and contains nerve endings and blood and lymph vessels. In mammals, the dermis also contains hair follicles and sweat glands.

DES (dē′ē-ĕs′) Short for *diethylstilbestrol.* A synthetic nonsteroidal substance having estrogenic properties and prescribed between 1938 and 1971 to pregnant women with a history of miscarriage and other problems of pregnancy. It is no longer used due to the incidence of certain vaginal cancers and other disorders in the daughters of women so treated.

desalinization (dē-săl′ə-nĭ-zā′shən) also **desalination** (dē-săl′ə-nā′shən) The removal of salt or other chemicals from something, such as seawater or soil. Desalinization can be achieved by means of evaporation, freezing,

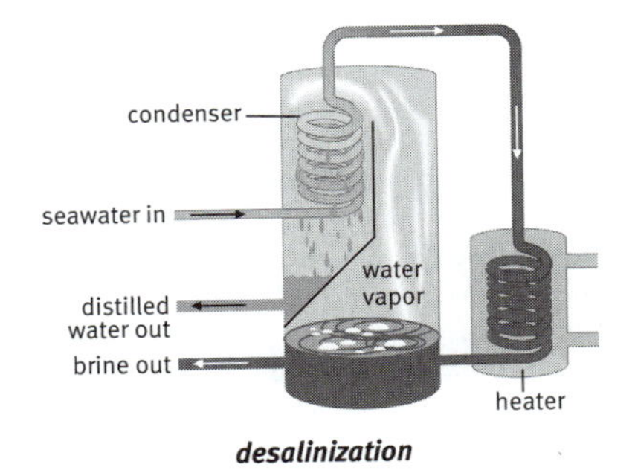

desalinization

diagram of a distillation process showing how seawater is desalinized

reverse osmosis, ion exchange, and electrodialysis. —*Verb* **desalinize.**

Descartes (dā-kärt′), **René** 1596–1650. French mathematician and philosopher who discovered that the position of a point can be determined by coordinates, a discovery that laid the foundation for analytic geometry.

descrambler (dē-skrăm′blər) An electronic device that decodes a scrambled transmission, typically a radio signal, into a signal that is intelligible to the receiving apparatus.

desert (dĕz′ərt) A large, dry, barren region, usually having sandy or rocky soil and little or no vegetation. Water lost to evaporation and transpiration in a desert exceeds the amount of precipitation; most deserts average less than 25 cm (9.75 inches) of precipitation each year, concentrated in short local bursts. Deserts cover about one fifth of the Earth's surface, with the principal warm deserts located mainly along the Tropic of Cancer and the Tropic of Capricorn, where warm, rising equatorial air masses that have already lost most of their moisture descend over the subtropical regions. Cool deserts are located at higher elevations in the temperate regions, often on the lee side of a barrier mountain range where the prevailing winds drop their moisture before crossing the range.

A CLOSER LOOK **desert**

A desert is defined not by temperature but by the sparse amount of water found in a region. An area with an annual rainfall of fewer than 25 centimeters (9.75 inches) generally qualifies as a desert. In spite of the dryness, however, some animals and plants have adapted to desert life and thrive in these harsh environments. While different animals live in different types of deserts, the dominant animals of warm deserts are reptiles, including snakes and lizards, small mammals, such as ground squirrels and mice, and arthropods, such as scorpions and beetles. These animals are usually nocturnal, spending the day resting in the shade of plants or burrowed in the ground, and emerging in the evenings to hunt or eat. Warm-desert plants are mainly ground-hugging shrubs, small wooded trees, and cacti. Plant and animal life is scarcer in the cool desert, where the precipitation falls mainly as snow. Plants are generally scattered mosses and grasses that are able to survive the cold by remaining low to the ground, avoiding the wind, and animal life can include both large and small mammals, such as deer and jackrabbits, as well as a variety of raptors and other birds.

---

desertification (dĭ-zûr′tə-fĭ-kā′shən) The transformation of land once suitable for agriculture into desert. Desertification can result from climate change or from human practices such as deforestation and overgrazing.

desiccate (dĕs′ĭ-kāt′) To remove the moisture from something or dry it thoroughly. ► A **desiccator** is a container that removes moisture from the air within it. ► A desiccator contains a **desiccant,** a substance that traps or absorbs water molecules. Some desiccants include silica gel (silicon dioxide), calcium sulfate (dehydrated gypsum), calcium oxide (calcined lime), synthetic molecular sieves (porous crystalline aluminosilicates), and dried clay.

designer drug (dĭ-zī′nər) Any of various drugs with properties and effects similar to a known hallucinogen or narcotic but having a slightly altered chemical structure, especially such a drug created in order to evade restrictions against illegal substances.

designer gene A gene created or modified by genetic engineering. Designer genes can help produce particular substances needed for research or medical treatment.

destructive distillation (dĭ-strŭk′tĭv) A process by which organic substances such as wood, coal, and oil shale are broken down by heat in the absence of air. This causes them to break down into solids, liquids, and gases, which are then used to make products such as coke, charcoal, oils, and ammonia.

detergent (dĭ-tûr′jənt) A cleaning agent that increases the ability of water to penetrate fabric and break down greases and dirt. Detergents act like soap but, unlike soaps, they are derived from organic acids rather than fatty acids. Their molecules surround particles of grease and dirt, allowing them to be carried away. Compare **soap.**

determinate (dĭ-tûr′mə-nĭt) **1.** Precisely determined, limited, or defined. **2.** Not continuing to grow at an apical meristem. In the cyme, a determinate inflorescence, for example, the first floret develops at the end of the meristem, and no further elongation of the inflorescence can occur.

detritivore (dĭ-trī′tə-vôr′) An organism that feeds on and breaks down dead plant or animal matter, returning essential nutrients to the ecosystem. Detritivores include microorganisms such as bacteria and protists as well as larger organisms such as fungi, insects, worms, and isopod crustaceans. In a food chain, detritivores are primary consumers. Compare **carnivore, herbivore.**

detritus (dĭ-trī′təs) **1.** Loose fragments, such as sand or gravel, that have been worn away from rock. **2.** Matter produced by the decay or disintegration of an organic substance.

deuterium (do͞o-tîr′ē-əm) An isotope of hydrogen whose nucleus has one proton and one neutron and whose atomic mass is 2. Deuterium is used widely as a tracer for analyzing chemical reactions, and it combines with oxygen to form heavy water. Also called *heavy hydrogen*. See Note at **heavy water.**

deuterium oxide See under **heavy water.**

deutero– A prefix meaning "second" or "secondary," as in *deuterostome,* an animal whose mouth is the second opening to develop.

deuteromycete (do͞o′tə-rō-mī′sēt′) Any of various fungi that reproduce only asexually, by means of conidia. The penicillin mold, the fungus that causes athlete's foot, and the fungus that causes root rot and damping off in plants are deuteromycetes. The deuteromycetes do not form a natural phylogenetic grouping. Instead, the classification simply includes the fungal species that have lost the sexual stage in their life cycle. Most deuteromycete species are in fact descended from ascomycetes and can be reclassified as such if their sexual stage is discovered. Also called *imperfect fungus.* Compare **perfect fungus.**

deuteron (do͞o′tə-rŏn′) The nucleus of a deuterium atom, consisting of a proton and a neutron. It is regarded as a subatomic particle with unit positive charge.

deuterostome (do͞o′tə-rō-stōm′) Any of a major group of animals defined by its embryonic development, in which the first opening in the embryo becomes the anus. At this stage in their development, the later specialized function of any given embryonic cell has not yet been determined. Deuterostomes are one of the two groups of animals that have true body cavities (coeloms), and are believed to share a common ancestor. They include the echinoderms, chaetognaths, hemichordates, and chordates. Compare **protostome.**

developmental disability (dĭ-vĕl′əp-mĕn′tl) A mental or physical disability, such as cerebral palsy or mental retardation, that is present during childhood, interferes with normal physical, intellectual, or emotional development, and usually lasts throughout life.

deviation (dē′vē-ā′shən) The difference between one number in a set and the mean of the set.

Devonian (dĭ-vō′nē-ən) The fourth period of the Paleozoic Era, from about 408 to 360 million years ago. During this time there were three major landmasses: most of modern day North America and Europe were located along the equator; a portion of Siberia was located to the north; and a continent consisting of South America, Australia, Africa, India, and Antarctica was located in the Southern Hemisphere. In the early Devonian small plants dominated the landscape, but by the end of the Devonian ferns and seed plants had appeared, as had the first forests. The first tetrapods (terrestrial vertebrates) and terrestrial arthropods appeared, as did many new types of fish. See Chart at **geologic time.**

dew (do͞o) Water droplets condensed from the air, usually at night, onto cool surfaces near the ground. Dew forms when the temperature of the surfaces falls below the dew point of the surrounding air, usually due to radiational cooling. See also **frost.**

dew point The temperature at which the water vapor contained in a volume of air at a given atmospheric pressure reaches saturation and condenses to form dew. The dew point varies depending on how much water vapor the air contains, with humid air having a higher dew point than dry air. When large droplets of condensation form, they are deposited onto surfaces as dew. When smaller droplets form, they remain suspended in the air as mist or fog. If the dew point is below the freezing temperature of water (0°C), the water vapor turns directly into frost by sublimation.

dextrorotation (dĕk′strə-rō-tā′shən) The clockwise rotation of the plane of polarization of light (as observed when looking straight into the incoming light) by certain substances, such as crystals, and by certain solutions. Dextrorotation is caused by a particular arrangement of the atoms in a molecule of the substance. Compare **levorotation.**

dextrorotatory (dĕk′strə-rō**′**tə-tôr′ē) Relating to a substance that causes dextrorotation.

dextrose (dĕk**′**strōs′) A sugar that is the most common form of glucose. It is found in plant and animal tissues and also derived from starch. Dextrose is the dextrorotatory form of glucose.

di– A prefix that means "two," "twice," or "double." It is used commonly in chemistry, as in *dioxide,* a compound having two oxygen atoms.

dia– A prefix meaning "through" or "across," as in *diameter,* the length of a line going through a circle.

diabase (dī**′**ə-bās′) A dark-gray to black, medium-grained igneous rock consisting mainly of labradorite and pyroxene. Diabase is compositionally similar to andesite, but has coarser grains. It is commonly found in sills and dikes. Also called *dolerite.*

diabetes mellitus (dī′ə-bē**′**tĭs mə-lī**′**təs, -tēz) A metabolic disease characterized by abnormally high levels of glucose in the blood, caused by an inherited inability to produce insulin (Type 1) or an acquired resistance to insulin (Type 2). Type 1 diabetes, which typically appears in childhood or adolescence, is marked by excessive thirst, frequent urination, and weight loss and requires treatment with insulin injections. Type 2 diabetes appears during adulthood, usually in overweight or elderly individuals, and is treated with oral medication or insulin. People with either type of diabetes benefit from dietary restriction of sugars and other carbohydrates. Uncontrolled blood glucose levels increase the risk for long-term medical complications including peripheral nerve disease, retinal damage, kidney disease, and progressive atherosclerosis caused by damage to endothelial cells in blood vessels, leading to coronary artery disease and peripheral vascular disease.

diadelphous (dī′ə-dĕl**′**fəs) **1.** Gathered into two groups or bundles of equal or different number. The stamens of certain flowers, such as those of some members of the bean family, are diadelphous. **2.** Having stamens so arranged.

diagenesis (dī′ə-jĕn**′**ĭ-sĭs) The process by which sediment undergoes chemical and physical changes during its lithification (conversion to rock). Compaction, leaching, cementation, and recrystallization are all forms of diagenesis. Erosion and metamorphism are not. Oil, gas, and coal form through the diagenesis of organic sedimentary matter.

diagnosis (dī′əg-nō**′**sĭs) Plural **diagnoses** (dī′əg-nō**′**sēz). The identification by a medical provider of a condition, disease, or injury made by evaluating the symptoms and signs presented by a patient.

diagonal (dī-ăg**′**ə-nəl) *Adjective.* **1.** Connecting two nonadjacent corners in a polygon or two nonadjacent corners in a polyhedron that do not lie in the same face. —*Noun.* **2.** A diagonal line segment.

dialysis (dī-ăl**′**ĭ-sĭs) **1.** The separation of the smaller molecules in a solution from the larger molecules by passing the solution through a membrane that does not allow the large molecules to pass through. **2.** A medical procedure in which this technique of molecular separation is used to remove metabolic waste products or toxic substances from the blood. Dialysis is required for individuals with severe kidney failure.

diamagnetism (dī′ə-măg**′**nĭ-tĭz′əm) The property of being repelled by both poles of a magnet. Most substances commonly considered to be nonmagnetic, such as water, are actually diamagnetic. Though diamagnetism is a very weak effect compared with ferromagnetism and paramagnetism, it can be used to levitate objects. Compare **ferromagnetism, paramagnetism.** See also **Lenz's law.** —*Adjective* **diamagnetic** (dī′ə-măg-nĕt**′**ĭk)

diameter (dī-ăm**′**ĭ-tər) **1.** A straight line segment that passes through the center of a circle or sphere from one side to the other. **2.** The length of such a line segment.

diamond (dī**′**ə-mənd) A form of pure carbon that occurs naturally as a clear, cubic crystal and is the hardest of all known minerals. It often occurs as octahedrons with rounded edges and curved surfaces. Diamond forms under conditions of extreme temperature and pressure and is most commonly found in volcanic breccias and in alluvial deposits. Poorly formed diamonds are used in abrasives and in industrial cutting tools.

diaphragm (dī′ə-frăm′) **1.** The large muscle that separates the chest cavity from the abdominal cavity in mammals and is the principal muscle of respiration. As the diaphragm contracts and moves downward, the lungs expand and air moves into them. As the diaphragm relaxes and moves upward, the lungs contract and air is forced out of them. **2.** A thin, flexible disk, especially in a microphone or telephone receiver, that vibrates in response to sound waves to produce electrical signals, or that vibrates in response to electrical signals to produce sound waves. **3.** A contraceptive device consisting of a thin flexible disk, usually made of rubber, that is designed to cover the cervix of the uterus to prevent the entry of sperm during sexual intercourse. **4.** An optical device in a camera or telescope that regulates the amount of light that enters the lens or optical system. The diaphragm consists of a disk with a circular opening of variable diameter.

diapir (dī′ə-pîr′) A fold or dome, such as an anticline, in which the upper strata of sediment or rock have been ruptured by the upward movement of more plastic rock, such as a body of salt, gypsum, or lava.

diapsid (dī-ăp′sĭd) Any of various reptiles having a skull with two pairs of temporal openings. Diapsids evolved in the Permian Period and grew longer and better-developed hindlimbs than forelimbs (unlike therapsids). Diapsids evolved into the archosaurs (including the dinosaurs and their descendants the birds) and all modern reptiles except turtles. Compare **anapsid, synapsid, therapsid.**

diarrhea (dī′ə-rē′ə) Excessive and frequent evacuation of watery feces, usually a symptom of a gastrointestinal disorder. Severe, prolonged diarrhea can lead to dehydration.

diastole (dī-ăs′tə-lē) The period during the normal beating of the heart in which the chambers of the heart dilate and fill with blood. Diastole of the atria occurs before diastole of the ventricles. Compare **systole.** —*Adjective* **diastolic** (dī′ə-stŏl′ĭk).

diatom (dī′ə-tŏm′) Any of various microscopic protists of the phylum Bacillariophyta that live in both fresh and marine water, have hard bivalve shells (called frustules) composed mostly of silica, and often live in colonies. Most diatoms can perform photosynthesis. They make up a large portion of the marine plankton and are an important food source for many aquatic animals. The skeletal remains of diatoms are the main constituent of diatomite.

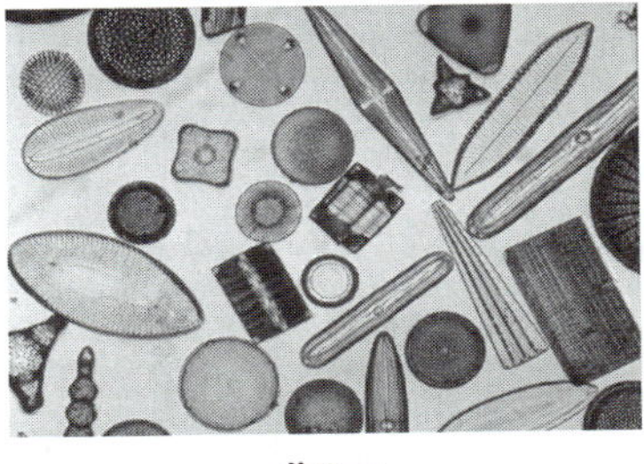

diatom

photomicrograph showing a variety of diatoms

diatomite (dī-ăt′ə-mīt′) A fine, light-colored, friable sedimentary rock consisting mainly of the silica-rich cell walls of diatoms. Diatomite forms both in lacustrine and marine environments. It is used in industry as a filler, filtering agent, absorbent, abrasive, and insulator.

diatreme (dī′ə-trēm′) A volcanic pipe, filled with breccia, formed by a subterranean gaseous explosion. See illustration at **batholith.**

diazepam (dī-ăz′ə-păm′) A drug, $C_{16}H_{13}ClN_2O$, used in the treatment of anxiety and as a sedative, muscle relaxant, and anticonvulsant.

dibasic (dī-bā′sĭk) **1.** Relating to an acid that contains two hydrogen atoms that can be replaced by metal ions. Sulfuric acid (H_2SO_4) is a dibasic acid. The hydrogens can be replaced by sodium and hydrogen to form sodium hydrogen sulfate ($NaHSO_4$) or by two sodium atoms to form sodium sulfate (Na_2SO_4). **2.** Relating to a compound that contains two basic monovalent groups or atoms.

dichogamous (dī-kŏg′ə-məs) Having pistils and stamens that mature at different times, thus promoting cross-pollination rather than self-pollination.

Dick (dĭk), **George Frederick** 1881–1967. American medical researcher who collaborated with his wife, **Gladys Henry Dick** (1881–1963), to isolate the bacterium that causes scarlet fever. They developed a serum for the disease in 1923.

diclinous (dī-klī′nəs) **1.** Bearing imperfect flowers; having carpels and stamens in different flowers. A monoecious plant is diclinous since it bears male and female flowers separately, even though on the same plant. Dioecious plants are also diclinous, since an individual plant will bear flowers of only one sex. Compare **monoclinous. 2.** Having only stamens or only pistils; unisexual. Used of flowers.

dicotyledon (dī′kŏt′l-ēd′n) or **dicot** (dī′kŏt′) An angiosperm that is not a monocotyledon, having two cotyledons in the seed. The term dicotyledon serves as a convenient label for the eudicotyledons, the magnoliids, and a varied group of other angiosperms, but it does not correspond to a single taxonomic group. Compare **monocotyledon.** See more at **eudicotyledon, leaf, magnoliid.**

dictyosome (dĭk′tē-ə-sōm′) The Golgi apparatus of plant cells.

dieldrin (dēl′drĭn) A light tan, toxic, carcinogenic compound used as an insecticide on fruit, soil, and seed, and in controlling tsetse flies and other carriers of tropical diseases. *Chemical formula:* $\mathbf{C_{12}H_8Cl_6O}$.

dielectric (dī′ĭ-lĕk′trĭk) *Adjective.* **1.** Having little or no ability to conduct electricity, generally as a result of having no electrons that are free to move. —*Noun.* **2.** A dielectric substance, such as glass or a ceramic, especially when used in a capacitor to maintain an electric field between the plates.

die-off (dī′ôf′) A sudden, severe decline in a population or community of organisms as a result of natural causes. Local die-offs can be caused by such factors as an unusual or extreme weather pattern, an outbreak of disease, or toxic algal blooms in a body of water. Widespread or global die-offs in which a species or group of species becomes extinct are generally associated with rapid climate change or other large-scale environmental dislocations.

diesel engine (dē′zəl) An internal-combustion engine in which the fuel oil is ignited by the heat of air that has been highly compressed in the cylinder, rather than by a spark. Due to the need for the engine to withstand very high pressures, diesel engines are relatively heavy; however, they are relatively fuel-efficient, especially when running at low power.

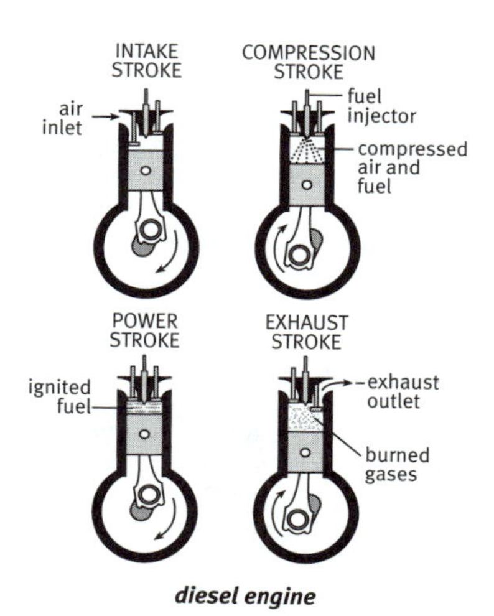

diesel engine

diesel oil A lightweight mixture of liquid hydrocarbons that are derived from petroleum. The hydrocarbons in diesel oil contain between 13 and 25 carbon atoms. Diesel oil is used as a fuel for diesel engines.

diethyl ether (dī-ĕth′əl) See **ether** (sense 2).

diethylstilbestrol (dī-ĕth′əl-stĭl-bĕs′trôl′, -trōl′) See **DES.**

differential (dĭf′ə-rĕn′shəl) **1.** An infinitesimal increment in a variable. **2.** The product of the derivative of a function of one variable and the increment of the independent variable.

differential calculus The mathematics of the variation of a function with respect to changes in independent variables, especially the use of differentiation to calculate rates of change of a function and the maximum and minimum values of a function. Differential calculus has applications such as calculating compound interest, organic growth, and slopes of curves, and studying the acceleration of moving bodies. Compare **calculus of variations, integral calculus.**

differentiation (dĭf′ə-rĕn′shē-ā′shən) **1.** In calculus, the process of computing the derivative of a function. Compare **integration.** **2.** The process by which cells or parts of an organism change during development to serve a specific function. The cells of an animal in its early embryonic phase, for example, are identical at first but develop by differentiation into specific tissues, such as bone, heart muscle, and skin. The factors determining the differentiation of any particular cell are not well understood, but in deuterostomes (vertebrates and other complex animals) they include the location of the cell relative to other cells.

diffraction (dĭ-frăk′shən) The bending and spreading of a wave, such as a light wave, around the edge of an object. See more at **wave.**

diffraction grating A barrier consisting of alternately transparent and opaque stripes, through which radiation such as light is passed and projected onto a screen or other detection device. The interference patterns cast by the diffraction grating on the screen or detector can be analyzed to determine the frequency of the radiation. See also **interferometer.**

diffraction pattern The interference pattern that results when a wave or a series of waves undergoes diffraction, as when passed through a diffraction grating or the lattices of a crystal. The pattern provides information about the frequency of the wave and the structure of the material causing the diffraction. See also **interferometer.**

diffusion (dĭ-fyo͞o′zhən) **1.** The movement of atoms or molecules from an area of higher concentration to an area of lower concentration. Atoms and small molecules can move across a cell membrane by diffusion. Compare **osmosis.** **2.** The reflection or refraction of radiation such as light or sound by an irregular surface, tending to scatter it in many directions.

digestion (dī-jĕs′chən) **1.** The process by which food is broken down into simple chemical compounds that can be absorbed and used as nutrients or eliminated by the body. In most animals, nutrients are obtained from food by the action of digestive enzymes. In humans and other higher vertebrates, digestion takes place mainly in the small intestine. In protists and some invertebrates, digestion occurs by phagocytosis. **2.** The decomposition of organic material, such as sewage, by bacteria.

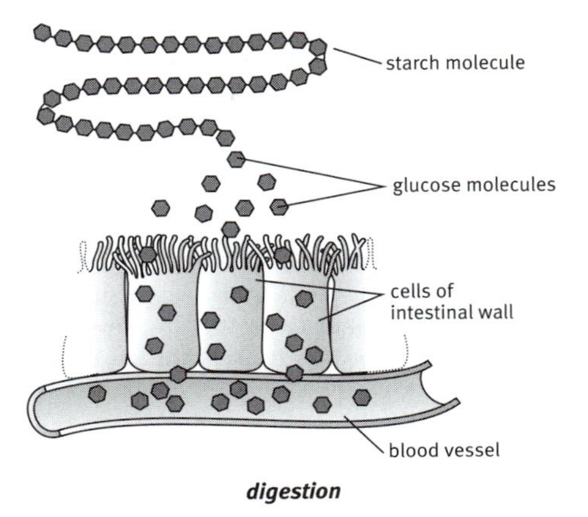

digestion

During digestion, enzymes break down large molecules in foods, such as starch, into simple compounds that are absorbed into the small intestine and the bloodstream.

digestive system (dī-jĕs′tĭv) The alimentary canal together with the salivary glands, liver, pancreas, and other organs of digestion.

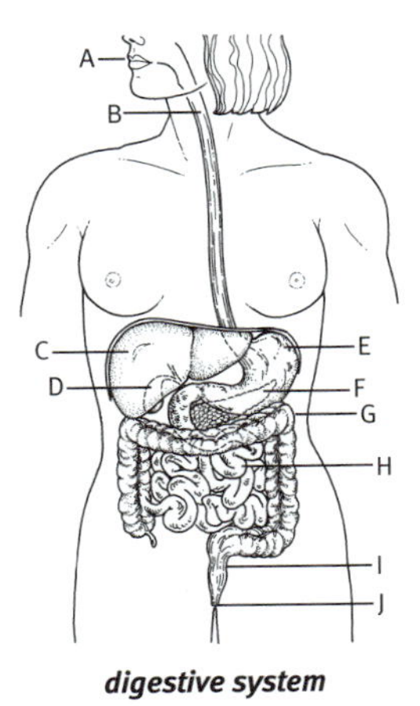

digestive system

A. *mouth*, B. *esophagus*, C. *liver*, D. *gallbladder*, E. *stomach*, F. *pancreas*, G. *large intestine*, H. *small intestine*, I. *rectum*, J. *anus*

digestive tract The series of organs in the digestive system through which food passes, nutrients are absorbed, and waste is eliminated. In higher vertebrates, it consists of the esophagus, stomach, small and large intestines, rectum, and anus.

digit (dĭj′ĭt) **1.** *Anatomy.* A jointed body part at the end of the limbs of many vertebrates. The limbs of primates end in five digits, while the limbs of horses end in a single digit that terminates in a hoof. The fingers and toes are digits in humans. **2.** *Mathematics.* One of the ten Arabic numerals, 0 through 9.

digital (dĭj′ĭ-tl) **1.** *Anatomy.* Relating to or resembling a digit, especially a finger. **2.** *Computer Science.* Representing or operating on data or information in numerical form. A digital clock uses a series of changing digits to represent time at discrete intervals, for example, every second. Modern computers rely on digital processing techniques, in which both data and the instructions for manipulating data are represented as binary numbers. Compare **analog.** See also **logic gate.**

digitalis (dĭj′ĭ-tăl′ĭs) A drug prepared from the seeds and dried leaves of the purple foxglove, *Digitalis purpurea,* and prescribed as a cardiac stimulant in the treatment of congestive heart failure and other disorders of the heart.

digitize (dĭj′ĭ-tīz′) To convert data or signals, such as images, text, or sound, to digital form. See more at **A/D converter.**

digitoxin (dĭj′ĭ-tŏk′sĭn) A highly active glycoside derived from digitalis and prescribed in the treatment of certain cardiac conditions. *Chemical formula:* $C_{41}H_{64}O_{13}$.

digoxin (dĭj-ŏk′sĭn) A cardiac glycoside obtained from the leaves of a foxglove, *Digitalis lanata,* with pharmacological effects that are similar to digitalis. *Chemical formula:* $C_{41}H_{64}O_{14}$.

dihedral (dī-hē′drəl) Formed by a pair of planes or sections of planes that intersect.

dike (dīk) **1.** A body of igneous rock that cuts across the structure of adjoining rock, usually as a result of the intrusion of magma. Dikes are often of a different composition from the rock they cut across. They are usually on the order of centimeters to meters across and up to tens of kilometers long. See illustration at **batholith. 2.** An embankment of earth and rock built to prevent floods or to hold irrigation water in for agricultural purposes.

dilation (dī-lā′shən, dĭ-) The widening or stretching of an opening or a hollow structure in the body.

dilution (dĭ-lo͞o′shən) The process of making a substance less concentrated by adding a solvent, such as water.

diluvial (dĭ-lo͞o′vē-əl) Relating to or produced by a flood.

dimension (dĭ-mĕn′shən) **1a.** Any one of the three physical or spatial properties of length, area, and volume. In geometry, a point is said to have zero dimension; a figure having only length, such as a line, has one dimension; a plane or surface, two dimensions; and a figure having volume, three dimensions. The fourth dimension is often said to be time, as in the theory of General Relativity. Higher dimensions can be dealt with mathematically but cannot be represented visually. **b.** The measurement of a length, width, or thickness. **2.** A unit, such as mass, time, or charge, associated with a physical quantity and used as the basis for other measurements, such as acceleration.

dimensionless number (dĭ-mĕn′shən-lĭs) A number representing a property of a physical system, but not measured on a scale of physical units (as of time, mass, or distance). Drag coefficients and stress, for example, are measured as dimensionless numbers.

dimer (dī′mər) Any of various chemical compounds made of two smaller identical or similar molecules (called monomers) that are linked together. Dimers are linked by hydrogen bonds, coordinate bonds, or covalent bonds. Sucrose is a dimer composed of the monomers glucose and fructose.

dimerous (dĭm′ər-əs) **1.** Consisting of two parts or segments, as the tarsus in certain insects. **2.** Having flower parts, such as petals, sepals, and stamens, in sets of two.

dimetrodon (dī-mĕt′rə-dŏn′) An extinct, carnivorous reptile of the genus *Dimetrodon* of the Permian Period having a body similar to an alligator's but with a tall, curved sail on its back. The sail had a thick network of blood vessels and may have been used to regulate the animal's body temperature. The dimetrodon belonged to the synapsids, an early group of reptiles that was ancestral to mammals.

dimorphism (dī-môr′fĭz′əm) **1.** The existence of two distinct types of individual within a species, usually differing in one or more characteristics such as coloration, size, and shape. The most familiar type of dimorphism is sexual dimorphism, as in many birds (where the male is often more brightly colored than the female), spiders (where the male is often smaller than the female), horned and tusked mammals (where horns and tusks are often present in the male but not the female), and in some species of deep-sea anglerfish (where the male is reduced to a tiny parasitic form attached for life to the much larger female). Fungi also display dimorphism. For example, the same species may exist as a small, budding yeast under some conditions, but as a mass of long hyphae under others. **2.** The occurrence, among plants, of two different forms of the same basic structure, either on the same plant or among individuals of the same species. The common ivy *Hedera helix* produces juvenile leaves with prominent lobes under conditions of low light, but adult leaves of more rounded shape under conditions of greater light. **3.** The characteristic of a chemical compound to crystallize in two different forms. Potassium feldspar, for example, can crystallize as either orthoclase (at higher temperatures) or microcline (at lower temperatures). —*Adjective* **dimorphous.**

dimorphism

female (left) *and male lions*

dinitrobenzene (dī-nī′trō-bĕn′zēn′) Any of three isomeric compounds made from a mixture of nitric acid, sulfuric acid, and heated benzene. Dinitrobenzenes are used in dyes, in synthesizing organic compounds, and in making celluloid. *Chemical formula:* $C_6H_4N_2O_4$.

dinoflagellate (dī′nō-flăj′ə-lĭt) Any of numerous one-celled organisms found mostly in the ocean, usually having two flagella of unequal length and often an armorlike covering of cellulose. Dinoflagellates are one of the main components of plankton. Since dinoflagellates have characteristics of both plants and animals, their classification is controversial. See more at **red tide.**

dinosaur (dī′nə-sôr′) Any of various extinct reptiles of the orders Saurischia and Ornithischia that flourished during the Mesozoic Era. Dinosaurs were carnivorous or herbivorous, dwelled mostly on land, and varied from the size of a small dog to the largest land animals that ever lived. One group of dinosaurs evolved into birds. See more at **ornithischian, saurischian.** See Note at **bird.**

dinotherium (dī′nə-thîr′ē-əm) Any of various extinct elephantlike mammals of the genus *Dinotherium* that existed during the Miocene, Pliocene, and Pleistocene Epochs. Characteristic of the dinotherium were tusks that grew downward from its lower jaw.

diode (dī′ōd′) An electrical device with two active terminals, an anode and a cathode, through which current passes more easily in one direction (from anode to cathode) than in the reverse direction. Diodes have many uses, including conversion of AC power to DC power, and the decoding of audio-frequency signals from radio signals.

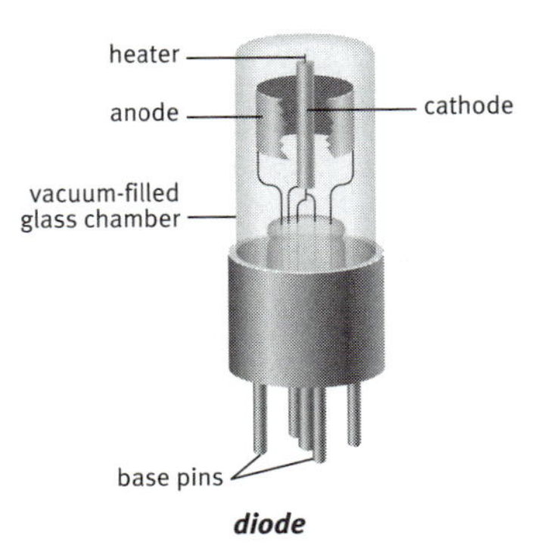

diode

The heated cathode generates a cloud of electrons that are attracted by the anode, causing a current to flow from the cathode to the anode. Because the anode cannot generate electrons of its own, the current cannot flow in the opposite direction.

diode laser See **semiconductor laser.**

dioecious (dī-ē′shəs) Having male flowers on one plant and female flowers on another plant of the same species. The holly and asparagus plants are dioecious. Compare **monoecious.**

diol (dī′ôl, -ŏl) An alcohol containing two hydroxyl groups per molecule.

Diophantine analysis (dī′ə-făn′tīn′) A collection of methods for determining integral solutions of certain algebraic equations, such as those having two unknown variables.

Diophantine equation A type of indeterminate equation in which the coefficients are integers, studied to determine all integral solutions.

diopside (dī-ŏp′sīd′) A light green, monoclinic variety of pyroxene, used as a gemstone and as a refractory material. *Chemical formula:* $CaMgSi_2O_6$.

dioptric (dī-ŏp′trĭk) Relating to the refraction of light, especially by a lens. Dioptric lenses are used in Fresnel lenses and camera viewfinders. Compare **catadioptric, catoptric.**

diorite (dī′ə-rīt′) A gray, coarse-grained plutonic rock. Diorite consists mainly of sodium-rich plagioclase and one or more mafic minerals such as biotite, hornblende, or pyroxene. It is the coarse-grained equivalent of andesite.

dioxane (dī-ŏk′sān′) A flammable, potentially explosive, clear liquid that is used as a solvent for fats, greases, and resins. It is also used in various products including paints, lacquers, glues, cosmetics, and fumigants. *Chemical formula:* $C_4H_8O_2$.

dioxide (dī-ŏk′sīd) A compound containing two oxygen atoms per molecule.

dioxin (dī-ŏk′sĭn) Any of several toxic hydrocarbons that occur as impurities in petroleum-derived herbicides, disinfectants, and other products. Dioxins are composed of two benzene rings connected by two oxygen atoms, and the most familiar kind, called *TCDD,* has two chlorine atoms attached to each benzene ring. TCDD was once thought to cause cancer and birth defects, but subsequent research showed it to have only mild toxic effects except at very high exposure levels.

dip (dĭp) **1.** The downward inclination of a rock stratum or vein in reference to the plane of the horizon. **2.** See **magnetic inclination.**

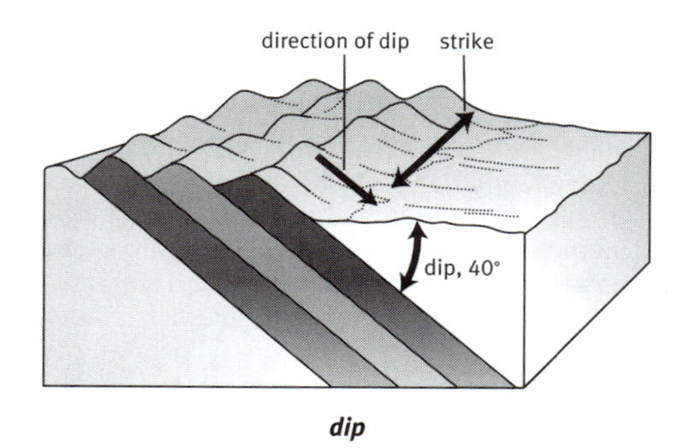

dip

strike and dip of inclined rock strata

diphtheria (dĭf-thîr′ē-ə, dĭp-) An infectious disease caused by the bacterium *Corynebacterium diphtheriae* and characterized by fever, swollen glands, and the formation of a membrane in the throat that prevents breathing. Infants are routinely vaccinated against diphtheria, which was once a common cause of death in children.

diplex (dī′plĕks′) Capable of simultaneous transmission or reception of two electronic or radio signals through a single channel or component, as the simultaneous transmission of two radio signals through a single antenna.

diplodocus (dĭ-plŏd′ə-kəs) A very large herbivorous dinosaur of the genus *Diplodocus* of the late Jurassic Period. Diplodocus had a long, slender neck and tail and a small head with peglike teeth, and could grow to nearly 27 m (90 ft) in length. Fossilized skin impressions show that it probably had dermal spines along its back. Diplodocus is one of the longest known sauropod dinosaurs.

diploid (dĭp′loid′) Having paired sets of chromosomes in a cell or cell nucleus. In diploid organisms that reproduce sexually, one set of chromosomes is inherited from each parent. The somatic cells of most animals are diploid. Compare **haploid.** See Note at **mitosis.**

diploidy (dĭp′loi′dē) The state or condition of being diploid.

diplopia (dĭ-plō′pē-ə) A disorder of vision in which a single object appears double.

dipole (dī′pōl′) **1.** A pair of electric charges or magnetic poles, of equal magnitude but of opposite sign or polarity, separated by a small distance. **2.** A molecule having two such charges or poles. **3.** An antenna consisting of two rods of equal length extending outward in a straight line. Dipole antennas are usually used for frequencies below 30 megahertz.

dipole moment **1.** The product of the strength of either of the charges in an electric dipole and the distance separating the two charges. It is expressed in coulomb meters. Dipole moment is a vector quantity; its direction is defined as toward the positive charge. **2.** The product of the strength of either of the poles in an magnetic dipole and the distance separating the two poles. Dipole moment is a vector quantity; its direction is defined as toward the magnetic north pole. Since magnetic monopoles apparently do not exist, the magnetic moment is usually calculated by analysis of the flow of electric current inducing the magnetic field.

dipteran (dĭp′tər-ən) Any of various insects of the order Diptera, characterized by a single pair of membranous wings, a pair of club-shaped balancing organs, and large compound eyes. Dipterans include the flies, mosquitoes, midges, and gnats. *—Adjective* **dipterous.**

Dirac (dĭ-răk′), **P(aul) A(drien) M(aurice)** 1902–1984. British mathematician and physicist who developed a mathematical interpretation of quantum mechanics with which he was able to provide the first correct description of electron behavior. For this work Dirac shared with Erwin Schrödinger the 1933 Nobel Prize for physics.

Dirac's constant (dĭ-răks′) A physical constant often used in quantum mechanics in place of Planck's constant, equal to $\frac{h}{2}\pi$, where h equals Planck's constant. Its symbol is h-. Also called *h-bar.*

Dirac sea In relativistic quantum mechanics, the completely filled, infinite set of negative energy electron states that makes up a vacuum. If a negative energy electron is promoted to a positive energy state, the hole it leaves behind has the properties of a positron.

direct current (dĭ-rĕkt′) An electric current that moves in one direction with constant strength. Batteries are a source of direct current. Direct current is not used for long-distance power transmission because it is difficult to step up the voltage to a level that is efficient for energy transfer and then to step the voltage back down again for safe domestic use. Compare **alternating current.** See Notes at **current, Tesla.**

directrix (dĭ-rĕk′trĭks) A straight line used in generating a curve such as a parabola.

disaccharide (dī-săk′ə-rīd′) Any of a class of sugars, including lactose and sucrose, that are composed of two monosaccharides.

disc brake also **disk brake** (dĭsk) A brake in which friction is caused by a set of pads, usually made of steel, that press against a rotating disk to slow or stop its rotation.

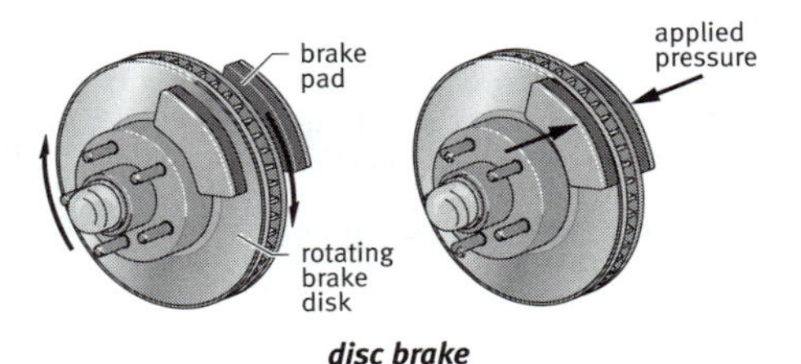

disc brake

discharge (dĭs-chärj′) *Noun.* **1.** The conversion of chemical energy to electric energy within a storage battery. **2.** A flow of electricity in a dielectric, especially in a rarefied gas. **3.** A flowing out or pouring forth, as of a bodily fluid; emission or secretion. **4.** A substance or material that is released, emitted, or excreted, especially from the body. *—Verb.* **5.** To undergo or cause the release of stored energy or electric charge, as from a battery or capacitor. **6.** To release, emit, or excrete a substance, especially from the body.

discharge tube A closed vessel having electrodes at either end and containing a gas at low pressure. When a sufficient voltage is applied to the electrodes, an electric current flows through the gas. Discharge tubes can be used to prevent current flow below a certain voltage; they can also function as lamps by the use of ionizing gas, which glows when current flows through the tube.

disconformity (dĭs′kən-fôr′mĭ-tē) A type of unconformity in which the successive strata are parallel.

discontinuity (dĭs-kŏn′tə-no͞o′ĭ-tē) **1.** A usually uneven surface between two layers of rock or sediment that represents either an interruption in the deposition of the layers, as in an unconformity, or a displacement of one or both layers relative to each other, as in a fault. **2.** A surface within the Earth across which the velocities of seismic waves change. The discontinuities are located at the boundaries between the Earth's various layers and correspond to changes in the elastic properties of the Earth's materials.

discontinuous (dĭs′kən-tĭn′yo͞o-əs) *Mathematics.* Relating to a function that contains one or more points where the function is either discontinuous or undefined.

dish antenna (dĭsh) See **parabolic antenna.**

disk or **disc** (dĭsk) **1a.** See **magnetic disk. b.** See **optical disk. 2.** See **intervertebral disk. 3.** The round, flat center, consisting of many disk flowers, found in the inflorescences of many composite plants such as the daisy.

disk brake Another spelling of **disc brake.**

disk drive *Computer Science.* A device that reads data stored on a magnetic or optical disk and writes data onto the disk for storage.

disk flower Any of the tiny tubular flowers in the central portion of the capitulum or flower head of certain plants of the composite family (Asteraceae or Compositae). In many composite flowers, such as the daisy and sunflower, the disk flowers are surrounded by flattened ray flowers.

disk flower

close-up of a daisy

dislocation (dĭs′lō-kā′shən) **1.** Displacement of a bone from its normal position, especially at a joint. **2.** *Geology.* See **displacement** (sense 5). **3.** An imperfection in the crystal structure of a metal or other solid resulting from an absence of an atom or atoms in one or more layers of a crystal.

dispersion (dĭ-spûr′zhən) The separation by refraction of light or other radiation into individual components of different wavelengths. Dispersion results in most materials because a material's index of refraction depends on the wavelength of the radiation passing through it; thus different wavelengths entering a material along the same path will fan out into different paths within it. Prisms, for example, diffuse white light (which contains an even mixture of visible wavelengths) into its variously colored components; rainbows are an effect of dispersion in water droplets.

displacement (dĭs-plās′mənt) **1.** *Chemistry.* A chemical reaction in which an atom, radical, or molecule replaces another in a compound. **2.** *Physics.* A vector, or the magnitude of a vector, that points from an initial position (of a body or reference frame) to a subsequent position. **3.** The weight or volume of a fluid displaced by a floating body, used especially as a measurement of the weight or bulk of ships. **4.** The volume displaced by a single stroke of a piston in an engine or pump. **5.** *Geology.* **a.** The relative movement between the two sides of a geologic fault. **b.** The distance between the two sides of a fault. Also called *dislocation.*

dissect (dĭ-sĕkt′, dī′sĕkt′) **1.** To cut apart or separate body tissues or organs, especially for anatomical study. **2.** In surgery, to separate different anatomical structures along natural lines by dividing the connective tissue framework.

dissipation (dĭs′ə-pā′shən) The loss of energy from a physical system, most often in the form of heat.

dissipative system (dĭs′ə-pā′tĭv) See under **open system.**

dissociation (dĭ-sō′sē-ā′shən) The separation of a substance into two or more simpler substances, or of a molecule into atoms or ions, by the action of heat or a chemical process. Dissociation is usually reversible.

dissolution (dĭs′ə-lo͞o′shən) The dissolving of a material in a liquid.

dissolve (dĭ-zŏlv′) To pass or cause to pass into solution.

distemper (dĭs-tĕm′pər) **1.** An infectious disease occurring especially in dogs, caused by the canine distemper virus of the genus *Morbil-*

livirus. It is characterized by loss of appetite, a discharge from the eyes and nose, vomiting, fever, lethargy, partial paralysis caused by destruction of myelinated nerve tissue, and sometimes death. **2.** An infectious disease of cats caused by the feline panleukopenia virus of the genus *Parvovirus,* characterized by fever, vomiting, diarrhea leading to dehydration, and sometimes death.

distillation (dĭs′tə-lā′shən) A method of separating a substance that is in solution from its solvent or of separating a liquid from a mixture of liquids having different boiling points. The liquid to be separated is evaporated (as by boiling), and its vapor is then collected after it condenses. Distillation is used to separate fresh water from a salt solution and gasoline from petroleum. ▸ The condensed vapor, which is the purified liquid, is called the **distillate.**

distributary (dĭ-strĭb′yə-tĕr′ē) **1.** A branch of a river that flows away from the main stream. **2.** One of the channels in a braided stream.

distributive (dĭ-strĭb′yə-tĭv) Relating to the property of multiplication over division which states that applying multiplication to a set of quantities that are combined by addition yields the same result as applying multiplication to each quantity individually and then adding those results together. Thus $2 \times (3 + 4)$ is equal to $(2 \times 3) + (2 \times 4)$. See also **associative, commutative.**

diuretic (dī′ə-rĕt′ĭk) A substance or drug that tends to increase the discharge of urine. Diuretics are used in the treatment of high blood pressure, edema, and other medical conditions.

diurnal (dī-ûr′nəl) **1a.** Occurring once in a 24-hour period; daily. **b.** Having a 24-hour cycle. The movement of stars and other celestial objects across the sky are diurnal. **2.** Most active during the daytime. Many animals, including the apes, are diurnal. **3.** Having leaves or flowers that open in daylight and close at night. The morning glory and crocus are diurnal. Compare **nocturnal.**

divergence (dĭ-vûr′jəns) **1.** *Mathematics.* The property or manner of failing to approach a limit, such as a point, line, or value. **2.** *Biology.* The evolution of different forms or structures in related species as they adapt to different environments. An example of divergence is the development of wings in bats from the same bones that form the arm and hand or paw in most other mammals. Also called *divergent evolution.* Compare **convergence.**

divergent plate boundary (dĭ-vûr′jənt) A tectonic boundary where two plates are moving away from each other and new crust is forming from magma that rises to the Earth's surface between the two plates. The middle of the Red Sea and the mid-ocean ridge (running the length of the Atlantic Ocean) are divergent plate boundaries. Also called *passive margin, spreading zone.* See more at **tectonic boundary.** Compare **convergent plate boundary.**

diverticulitis (dī′vûr-tĭk′yə-lī′tĭs) Inflammation of a diverticulum or of diverticula in the intestinal tract, usually causing abdominal pain and fever.

diverticulosis (dī′vûr-tĭk′yə-lō′sĭs) A condition characterized by the presence of numerous diverticula in the colon.

diverticulum (dī′vûr-tĭk′yə-ləm) Plural **diverticula.** A pouch or sac branching out from a portion of the gastrointestinal tract, especially the large intestine. A diverticulum can occur as a normal structure, or it can be caused by a hernia.

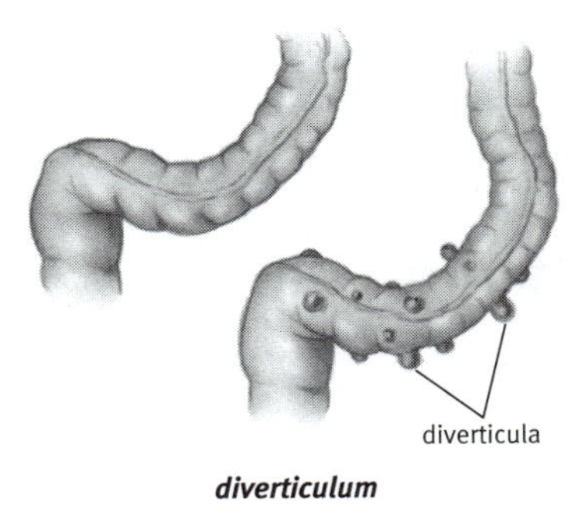

diverticulum

left: *section of human intestine;* right: *section of human intestine with diverticula*

divide (dĭ-vīd′) **1.** To subject (a number) to the process of division. **2.** To be a divisor of. **3.** To use (a number) as a divisor. **4.** To perform the operation of division. **5.** To undergo cell division.

dividend (dĭv′ĭ-dĕnd′) A number divided by another. In the equation $15 \div 3 = 5$, 15 is the dividend.

division (dĭ-vĭzh′ən) **1.** The act, process, or operation of finding out how many times one number or quantity is contained in another. **2.** A taxonomic classification within the plant kingdom that ranks immediately above a class and corresponds to a phylum in other kingdoms. See Table at **taxonomy.**

divisor (dĭ-vī′zər) A number used to divide another. In the equation $15 \div 3 = 5$, 3 is the divisor.

dizygotic (dī′zī-gŏt′ĭk) or **dizygous** (dī-zī′gəs) Derived from two separately fertilized eggs. Used especially of fraternal twins.

Djerassi (djĕ-rä′sē), **Carl** Born 1923. Austrian-born American chemist who pioneered the development of the first contraceptive pill, as well as many commonly used drugs, including antihistamines and anti-inflammatory agents.

D layer See **D region.**

DNA (dē′ĕn-ā′) Short for *deoxyribonucleic acid.* The nucleic acid that is the genetic material determining the makeup of all living cells and many viruses. It consists of two long strands of nucleotides linked together in a structure resembling a ladder twisted into a spiral. In eukaryotic cells, the DNA is contained in the nucleus (where it is bound to proteins known as histones) and in mitochondria and chloroplasts. In the presence of the enzyme DNA polymerase and appropriate nucleotides, DNA can replicate itself. DNA also serves as a template for the synthesis of RNA in the presence of RNA polymerase. Compare **RNA.** See Note at **histone.**

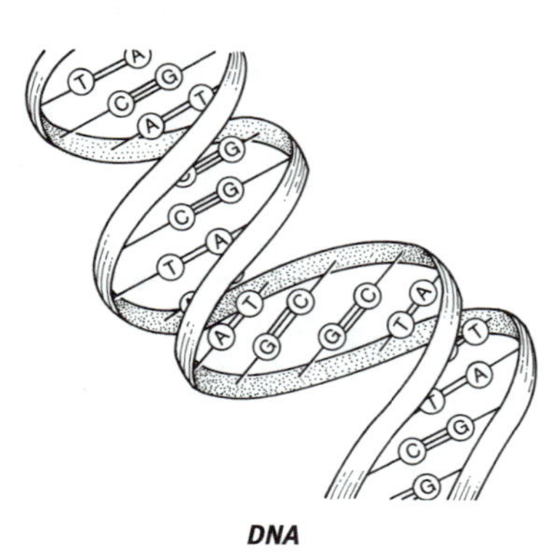

DNA

adenine (a), cytosine (c), guanine (g), and thymine (t)

DNA chip See **DNA microarray.**

DNA fingerprinting The use of a sample of DNA to determine the identity of a person within a certain probability. DNA fingerprinting is done by analyzing repeating patterns of base pairs in DNA sequences that are known to vary greatly among individuals.

DNA methylation (mĕth′ə-lā′shən) The modification of a strand of DNA after it is replicated, in which a methyl (CH_3) group is added to any cytosine molecule that stands directly before a guanine molecule in the same chain. Since methylation of cytosines in particular regions of a gene can cause that gene's suppression, DNA methylation is one of the methods used to regulate the expression of genes.

DNA microarray A small solid support, usually a membrane or glass slide, on which sequences of DNA are fixed in an orderly arrangement. DNA microarrays are used for rapid surveys of the expression of many genes simultaneously, as the sequences contained on a single microarray can number in the thousands. Also called *DNA chip.*

DNA polymerase Any of various enzymes that function in the replication and repair of DNA by catalyzing the linking of nucleotides in a specific order, using single-stranded DNA as a template.

DNA sequencing The determination of the sequence of nucleotides that are in a sample of DNA.

DNA virus A virus whose genome is composed of DNA. Compare **RNA virus.**

dodecahedron (dō′dĕk-ə-hē′drən) Plural **dodecahedrons** or **dodecahedra.** A polyhedron having twelve faces.

doldrums (dōl′drəmz′) A region of the globe found over the oceans near the equator in the **intertropical convergence zone** and having weather characterized variously by calm air, light winds, or squalls and thunderstorms. Hurricanes originate in this region.

dolerite (dŏl′ə-rīt′) See **diabase.**

dolomite (dō′lə-mīt′, dŏl′ə-mīt′) **1.** A gray, pink, or white rhombohedral mineral. Dolomite occurs in curved saddlelike crystals with a pearly to glassy luster, and it is a common rock-forming mineral. *Chemical formula:* $CaMg(CO_3)_2$. **2.** A sedimentary rock containing more than 50 percent of the mineral dolomite by weight.

domain (dō-mān′) **1.** *Mathematics.* The set of all values that an independent variable of a function can have. In the function $y = 2x$, the set of values that x (the independent variable) can have is the domain. Compare **range. 2.** *Computer Science.* A group of networked computers that share a common communications address. **3.** *Biology.* A division of organisms that ranks above a kingdom in systems of classification that are based on shared similarities in DNA sequences rather than shared structural similarities. In these systems, there are three domains: the archaea, the bacteria, and the eukaryotes. **4.** *Physics.* A region in a ferromagnetic substance in which the substance is magnetized with the same polarization throughout.

domain name A series of alphanumeric strings that are separated by periods, such as *www.hmco.com,* that is an address of a computer network connection and that identifies the owner of the address.

dome (dōm) **1.** A circular or elliptical area of uplifted rock in which the rock dips gently away, in all directions, from a central point. **2.** A wedge-shaped mineral crystal that has two nonparallel, similarly inclined faces that intersect along a plane of symmetry.

dominant (dŏm′ə-nənt) **1a.** Relating to the form of a gene that expresses a trait, such as hair color, in an individual organism. The dominant form of a gene overpowers the counterpart, or recessive, form located on the other of a pair of chromosomes. **b.** Relating to the trait expressed by such a gene. See more at **inheritance.** Compare **recessive. 2a.** Being a species that has the greatest effect within its ecological community, especially by determining the presence, abundance, or type of other species. As a plant community progresses through stages of succession, different species may become dominant for a period until the climax community is reached, at which point the dominant species remains stable until a major disruption occurs. Among animals, the dominant species in a community is generally the top predator or the most abundant or widespread species. **b.** Being an animal that occupies the highest position in a social hierarchy and has the greatest access to resources such as food and a mate or mates. Social dominance is gained and maintained through factors such as size and aggressiveness.

donor (dō′nər) **1.** An atom or molecule that releases one or more electrons to another atom or molecule, resulting in a chemical bond or flow of electric current. Compare **acceptor.** See also **electron carrier. 2.** An individual from whom blood, tissue, or an organ is taken for transfusion, implantation, or transplant.

dopamine (dō′pə-mēn′) A monoamine neurotransmitter that is formed during the synthesis of norepinephrine and is essential to the normal functioning of the central nervous system. A reduction of dopamine in the brain is associated with the development of Parkinson's disease. *Chemical formula:* $C_8H_{11}NO_2$.

Doppler (dŏp′lər), **Christian Johann** 1803–1853. Austrian physicist and astronomer who in 1842 explained the effect, now named for him, of variations in the frequency of waves as a result of the relative motion of the wave source with respect to the observer.

Doppler effect The difference between the frequency of a wave (as of sound or light) as measured at its source and as measured by an observer in relative motion. The Doppler effect can be used to determine the relative speed of an object by bouncing a wave (usually a radar wave) off the object and measuring the shift in the frequency of the wave. This technique is the basis of Doppler radar, and its practical applications include its use in traffic control and navigation systems. The Doppler effect is also known as the *Doppler shift.* ▸ If both the source and the observer are getting farther apart, the observed frequency is lower than the source frequency. In the case of light waves, the phenomenon is known as **red shift**. The amount of red shift in the spectra of stars is used in astronomy to determine how quickly the Earth and those stars are moving apart. ▸ If the source and the observer are getting closer together, the observed frequency is higher than the source frequency. In the case of light waves, the phenomenon is known as **blue shift.**

A CLOSER LOOK **Doppler effect**

The whistle of an approaching train has a higher pitch as the train approaches than when it recedes, even though that same whistle, heard by a passenger on the train, maintains a constant pitch. This is an example of the *Doppler effect,* common to all wave phenomena (in this case, a sound wave). Motion toward the source of a wave (or, equivalently, motion of the source toward the observer) entails that the peaks and troughs of the wave are encountered more quickly than if there were no motion, so the frequency of the wave is higher for the moving observer (hence the higher whistle pitch). Similarly, motion away from the source entails following the wave's motion, so the peaks and troughs are encountered less often, and the frequency is lower for the moving observer (hence the lower whistle pitch). The Doppler effect on light waves has enabled scientists to determine that the universe is expanding. The frequencies of light given off by various substances (such as the burning of hydrogen in the fusion reactions of most stars) has been found to be lower in distant galaxies and other celestial objects, a phenomenon called *red shift,* since the visible light is shifted toward the red, low-frequency end of the spectrum. Astronomer Edwin Hubble reasoned that the red shift was due to the Doppler effect. As galaxies speed away from us, the frequency of the light emitted appears lower. Doppler radar and sonar use the Doppler effect on reflected radio and sound waves to distinguish between stationary and moving objects and to determine the velocity of moving ones; the *echolocation* of bats and some whales also exploits the Doppler effect on reflected sound waves for navigating and catching prey.

As a motorcycle speeds forward, the frequency (and pitch) of the sound waves in front of the motorcycle become higher, and the frequency (and pitch) of the sound waves behind it become lower.

Doppler radar Radar that uses the Doppler effect to measure velocity.

Doppler shift See **Doppler effect.**

dormant (dôr′mənt) **1.** Being in an inactive state during which growth and development cease and metabolism is slowed, usually in response to an adverse environment. In winter, some plants survive as dormant seeds or bulbs, and some animals enter the dormant state of hibernation. **2.** Not active but capable of renewed activity. Volcanoes that have erupted within historical times and are expected to erupt again are dormant.

dorsal (dôr′səl) Relating to or on the back or upper surface of an animal.

dosimeter (dō-sĭm′ĭ-tər) An instrument that measures the amount of x-ray or other radiation absorbed in a given period.

dot (dŏt) **1.** A symbol (·) indicating multiplication, as in $2 \cdot 4 = 8$. It is used to indicate the dot product of vectors, for example $\mathbf{A} \cdot \mathbf{B}$. **2.** A period, as used as in URLs and e-mail addresses, to separate strings of words, as in *www.hmco.com.*

dot-matrix printer An impact printer that prints text and graphic images by hammering the ends of pins against an ink ribbon. This produces characters or images made up of a matrix, or pattern, of dots.

double bond (dŭb′əl) A type of covalent bond in which two electron pairs are shared between two atoms. Each atom contributes two electrons to the bond. See more at **covalent bond.**

double fertilization The process in which the two sperm nuclei of a pollen grain unite with nuclei of the embryo sac of an angiosperm plant. One sperm nucleus unites with the egg to form the diploid zygote, from which the embryo develops. The other sperm unites with the two nuclei located in a single cell at the center of the embryo sac. Together these nuclei form the triploid nucleus of the cell from which the endosperm develops. Double fertilization in this form is unique to the angiosperms.

double helix The three-dimensional structure of double-stranded DNA, in which polymeric nucleotide strands whose complementary nitrogen bases are linked by hydrogen bonds form a helical configuration. The two DNA strands are oriented in opposite directions.

double salt A salt that crystallizes from an aqueous solution of a mixture of two different ions. The mineral dolomite ($CaMg(CO_3)_2$), for example, is a double salt that crystallizes from a solution containing both calcium and magnesium ions. Double salts exist only as solids. Compare **complex salt, simple salt.**

double star See **binary star.**

downburst (doun′bûrst′) An extremely powerful downward air current from a cumulonimbus cloud, typically associated with thunderstorm activity. Downbursts can produce effects that resemble those brought about by tornadoes.

downdraft (doun′drăft′) A downward moving current of air in a cumulonimbus cloud. Compare **updraft.**

download (doun′lōd′) To transfer data or programs from a server or host computer to one's own computer or digital device. Compare **upload.**

down quark (doun) A quark with a charge of $-\frac{1}{3}$. Its mass is greater than that of the electron and the up quark, but smaller than that of all the other quarks. See Table at **subatomic particle.**

Down syndrome (doun) or **Down's Syndrome** (dounz) A congenital disorder caused by the presence of an extra 21st chromosome. People with Down syndrome have mild to moderate mental retardation, short stature, and a flattened facial profile. The syndrome is named after its discoverer, British physician John Langdon Haydon Down (1828–1896).

downy mildew (dou′nē) A disease of plants caused by oomycete organisms of the order Peronosporales and characterized by gray, velvety patches of spores on the lower surfaces of leaves. Downy mildew of the grapevine nearly destroyed the French wine industry in the 1870s, spurring the development of the first chemical pesticides used on plants.

D particle Either of two subatomic particles in the meson family, one neutral and one positively charged, having masses 3,649 and 3,658 times that of the electron and average lifetimes of 4.2×10^{-13} and 1.1×10^{-12} seconds, respectively. See Table at **subatomic particle.**

DPT vaccine (dē′pē-tē′) Variant of **DTP vaccine.**

draa (drä) A large dune measuring up to several hundred kilometers in length and hundreds of meters in height. Smaller dunes often form on their leeward and windward faces.

Draco (drā′kō) A constellation (the Dragon) in the polar region of the Northern Hemisphere near Cepheus and Ursa Major.

drag (drăg) A force acting on a moving body, opposite in direction to the movement of the body, caused by the interaction of the body and the medium it moves through. The strength of drag usually depends on the velocity of the body. ► Drag caused by buildup of pressure in front of the moving body and a decrease in pressure behind the body is called **pressure drag**. It is an important factor in the design of aerodynamically efficient shapes for cars and airplanes. ► Drag caused by the viscosity of the medium as the molecules along the body's surface move through it is called **skin drag** or **skin friction**. It is an important factor in the design of efficient surface materials for cars, airplanes, boat hulls, skis, and swimsuits. Compare **lift.** See Note at **aerodynamics.**

drainage basin (drā′nĭj) An area drained by a river system. A drainage basin includes all areas that gather precipitation water and direct it to a particular stream, stream system, lake, or other body of standing water.

Draper (drā′pər), **Henry** 1837–1882. American astronomer who developed methods for photographing celestial objects and phenomena. He became the first to photograph a stellar spectrum (1872) and a nebula (1880).

drawdown (drô′doun′) **1.** A lowering of the water level in a reservoir or other body of water, especially as the result of withdrawal. **2.** The difference in elevation between the level of water in a well and the level of groundwater in the area in which the well is located.

D region The region of the atmosphere that lies in the upper mesosphere, extending from about 70 to 90 km (43 to 56 mi) above the Earth. The D region is responsible for the daytime absorption of high-frequency radio signals; at night, oxygen ions recombine with free electrons and form oxygen molecules that are transparent to such signals, which bounce off of the E region and back to Earth. Also called *D layer*. Compare **E region, F region.**

drone (drōn) A male bee, especially a honeybee whose only function is to fertilize the queen. Drones have no stingers, do no work, and do not produce honey.

drosophila (drō-sŏf′ə-lə) Any of various small fruit flies of the genus *Drosophila,* one species of which (*D. melanogaster*) is used extensively in genetic research to study patterns of inheritance and the functions of genes.

drought (drout) A long period of abnormally low rainfall, lasting up to several years.

drug (drŭg) **1.** A chemical substance, especially one prescribed by a medical provider, that is used in the diagnosis, treatment, or prevention of a condition or disease. Drugs are prescribed for a limited amount of time, as for an acute infection, or on a regular basis for chronic disorders, such as hypertension. **2.** A chemical substance such as a narcotic or a hallucinogen that affects the central nervous system and is used recreationally for perceived desirable effects on personality, perception, or behavior. Many recreational drugs are used illicitly and can be addictive.

drumlin (drŭm′lĭn) An extended, oval hill or ridge of compacted sediment deposited and shaped by a glacier. Drumlins are typically about 30 m (98 ft) high and are longer than they are wide. They have one steep and one gentle slope along their longest axis, which is parallel to the direction of the glacier's movement. The steepest slope faces the direction from which the glacier originated, and the gentler slope faces the direction in which the glacier was advancing.

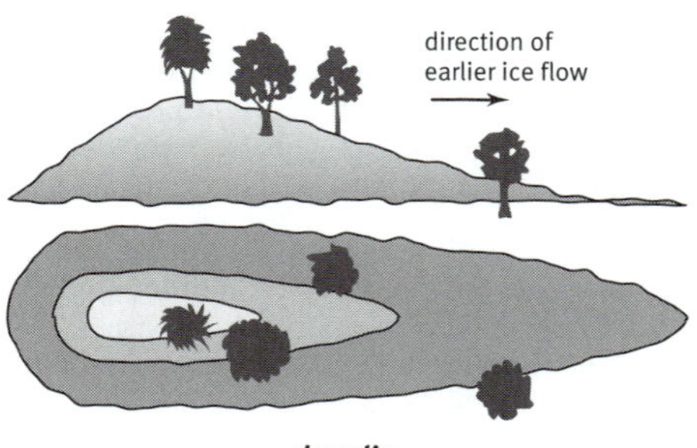

drumlin

drupe (dro͞op) A simple fruit derived from a single carpel. A drupe usually contains a single seed enclosed by a hardened endocarp, which often adheres closely to the seed within. In peaches, plums, cherries, and olives, a fleshy edible mesocarp surrounds the endocarp (the pit or stone). In the coconut, a fibrous mesocarp (the husk) surrounds the endocarp (the shell), while the white edible portion is the endosperm. Compare **berry, pome.** See more at **simple fruit.**

dry adiabatic lapse rate (drī) See under **lapse rate.**

dry cell A galvanic electric cell, such as a flashlight battery, in which the chemicals in the electrolyte are made into a paste so that they cannot easily spill from their container. Compare **wet cell.**

dry deposition See under **acid deposition.**

dry ice Solid carbon dioxide. Dry ice evaporates without first passing through a liquid state by **sublimation** except under moderate pressure (more than 73 atmospheres). It is used for refrigeration and for creating artificial smoke or fog effects.

dry measure A system of units for measuring the volume or capacity of dry commodities, such as grains, fruits, and vegetables. Compare **liquid measure.**

dryopithecine (drī′ō-pĭth′ĭ-sēn′) An extinct ape of the genus *Dryopithecus,* known from fossil remains of the Miocene and Pliocene Epochs in the Eastern Hemisphere. It is believed to be an ancestor of the anthropoid apes and humans.

Ds The symbol for **darmstadtium.**

D-s particle (dē′ĕs′) A positively charged meson having a mass 3,852 times that of the electron and a mean lifetime of approximately 4.7×10^{-13} seconds. See Table at **subatomic particle.**

DST Abbreviation of **daylight-saving time.**

DTP vaccine (dē′tē-pē′) or **DPT vaccine** (dē′pē-tē′) A combination vaccine that is administered intramuscularly to immunize children against diphtheria, tetanus, and pertussis infections. It consists of diphtheria and tetanus toxoids and pertussis vaccine, which contains inactivated parts of the bacterium *Bordetella pertussis.*

Dubhe (dŭb′ē) A pointer star in the constellation Ursa Major and the brightest of the seven stars that form the Big Dipper, with apparent magnitude 1.8. Dubhe and Merak form the outer side of the Dipper's bowl, with Dubhe being the upper of the two stars. A straight line extending northward from these pointer stars leads to the North Star, Polaris.

dubnium (do͞ob′nē-əm) *Symbol* **Db** A synthetic, radioactive element that is produced from californium, americium, or berkelium. Its most long-lived isotopes have mass numbers of 258, 261, 262, and 263 with half-lives of 4.2, 1.8. 34, and 30 seconds, respectively. Atomic number 105. See **Periodic Table.**

Dubois (do͞o-bwä′), **(Marie) Eugène (François Thomas)** 1858–1940. Dutch paleontologist who in 1891 discovered in Java a fossil hominid which he believed to be the so-called missing link between the apes and humans on the evolutionary ladder. He named it *Pithecanthropus erectus,* but it is now known as *Homo erectus.*

Dubos (do͞o-bôs′, -bō′, dü-), **René Jules** 1901–1982. French-born American bacteriologist noted for his research on natural antibiotics, tuberculosis, and environmental factors in disease. In 1939 he discovered tyrothricin, the first commercially produced antibiotic.

duck-billed dinosaur (dŭk′bĭld′) See **hadrosaur.**

duct (dŭkt) A tube or tubelike structure through which something flows, especially a tube in the body for carrying a fluid secreted that is by a gland.

ductile (dŭk′təl) **1.** Easily stretched without breaking or lowering in material strength. Gold is relatively ductile at room temperature, and most metals become more ductile with increasing temperature. Compare **brittle, malleable. 2.** Relating to rock or other materials that are capable of withstanding a certain amount of force by changing form before fracturing or breaking.

ductless gland (dŭkt′lĭs) See **endocrine gland.**

dune (do͞on) A hill or ridge of wind-blown sand. Dunes are capable of moving by the motion of their individual grains but usually keep the same shape. See more at **barchan dune, draa, longitudinal dune, seif dune, transverse dune.**

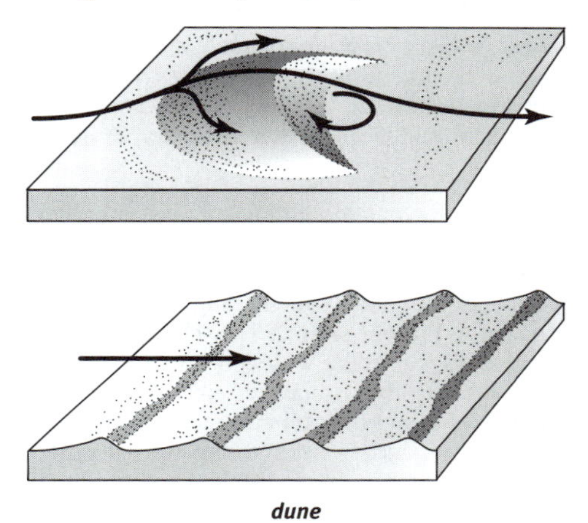

dune

top: *barchan dune;* bottom: *transverse dune*

dunite (do͞o′nīt′, dŭn′īt′) A coarse-grained igneous rock that consists mainly of olivine.

duodecimal (do͞o′ə-dĕs′ə-məl) Relating to or based on the number 12; having 12 as the base. In the duodecimal number system, each digit represents a multiple of a power of 12 instead of 10. Thus the duodecimal number 24 represents $(2 \times 12^1) + (4 \times 12^0)$, or 28.

duodenum (do͞o′ə-dē′nəm, do͞o-ŏd′n-əm) Plural **duodena** or **duodenums.** The beginning part of the small intestine, starting at the lower end of the stomach and extending to the jejunum.

dura mater (do͝or′ə mā′tər, mä′-) The tough fibrous membrane covering the brain and the spinal cord and lining the inner surface of the skull. It is the outermost of the three membranes (meninges) that surround the central nervous system, overlying the arachnoid and pia mater.

Dutch elm disease (dŭch) A disease of elm trees caused by the fungus *Ceratocystis ulmi,* spread by the European elm bark beetle *Scolytus multistriatus* and by the contact of the roots of healthy elms with those of infected trees. It produces brown streaks in the wood and results in the eventual death of the tree. No cure has been discovered, but prevention methods include the injection of insecticide into healthy trees and the destruction of all elms in infected areas.

DVD (dē′vē-dē′) A compact disk designed to store large amounts of data, especially high-resolution audio-visual material.

dwarf (dwôrf) **1.** An abnormally small person, often having limbs and features atypically proportioned or formed. **2.** An atypically small animal or plant. **3.** A dwarf star or dwarf galaxy.

dwarf galaxy A small, dim galaxy, intermediate in size between a regular galaxy and a globular cluster. Like larger galaxies, dwarf galaxies are classified as elliptical, spiral, or irregular based on their shape. The closest known galaxy to the Milky Way is a dwarf galaxy in Canis Major that is believed to be losing stars to ours. Many astronomers think that most of the large galaxies seen today were formed either by collisions of dwarf galaxies, which formed first in the aftermath of the Big Bang, or by larger galaxies attracting material away from dwarf galaxies.

dwarfism (dwôr′fĭz′əm) Abnormally short stature, usually caused by a hereditary disorder.

dwarf planet A celestial body that orbits the sun and is massive enough to assume a nearly spherical shape, but that does not clear other bodies from the neighborhood around its orbit and is not a satellite of a planet. Dwarf planets include Ceres, Pluto, and Eris. This category was created by the International Astronomical Union in 2006.

dwarf star A relatively small, low-mass star that emits an average or below-average amount of light. Most dwarf stars, including the Sun, are main-sequence stars, the principal exception being white dwarfs, which are the remnants of larger collapsed stars. Main-sequence dwarfs burn their hydrogen at a much slower rate than giant and supergiant stars and are consequently less luminous and have longer lifespans than those non-main-sequence stars do.

A CLOSER LOOK **dwarf star**

Despite their diminutive name, most *dwarf stars* are quite normal *main-sequence stars* and come in a wide variety of sizes, formed from protostars with sufficient mass to begin the process of nuclear fusion. But there are other stellar and quasistellar objects called dwarf stars as well. *Brown dwarfs* are formed when insufficient mass accretes for nuclear fusion to take place; brown dwarfs thus never become proper stars. Other kinds of dwarf stars result from the further evolution of main-sequence stars not massive enough to become neutron stars or black holes (which form from the burned-out core of a supernova). The type known as a *white dwarf* is the remnant of a red giant star that has burned nearly all its fuel. The mutual gravitational attraction of its atoms, no longer counterbalanced by the outward pressure of burning fuel within, causes the star to collapse in on itself. After it contracts and blows its outer layers away as a *planetary nebula,* the red giant stabilizes as a white dwarf and slowly fades. Our Sun is of a size and mass that will probably cause it to evolve first into a small red giant and eventually into a white dwarf. *Red dwarfs* have a lower mass and luminosity than white dwarfs, and *black dwarfs,* if any yet exist, are even less luminous, no longer giving off any detectable radiation.

Dy The symbol for **dysprosium.**

dynamic (dī-năm′ĭk) **1a.** Relating to energy or to objects in motion. Compare **static. b.** Relating to the study of dynamics. **2.** Characterized by continuous change or activity.

dynamic RAM A memory chip that stores information as electrical charges in capacitors. Each bit of information is stored by a single capacitor, representing binary 1 with a full charge, and 0 with no charge. Since the charges in the capacitors tends to fade quickly, the value of each bit is dynamically read off at regular intervals, and the charge is refreshed if necessary.

dynamics (dī-năm′ĭks) The branch of physics that deals with the effects of forces on the motions of bodies. Also called *kinetics.* Compare **kinematics.**

dynamic viscosity See under **viscosity.**

dynamite (dī′nə-mīt′) A powerful explosive used in blasting and mining. It typically consists of nitroglycerin and a nitrate (especially sodium nitrate or ammonium nitrate), combined with an absorbent material that makes it safer to handle.

dynamo (dī′nə-mō′) An electric generator, especially one that produces direct current. See more at **generator.**

dynamoelectric (dī′nə-mō′ĭ-lĕk′trĭk) Relating to the conversion of mechanical energy to electrical energy or vice versa. Hydroelectric power plants and electric engines are examples of dynamoelectric technology.

dyne (dīn) The unit of force in the centimeter-gram-second system, equal to the amount of force required to give a mass of one gram an acceleration of one centimeter per second per second.

dynode (dī′nōd′) An electrode used in certain electron tubes, especially photomultipliers, to provide **secondary emission.** Dynodes are arranged by increasing voltage between the cathode and anode of the tube, each attracting the cascade of electrons from the one behind it, and releasing more electrons to the next.

dysentery (dĭs′ən-tĕr′ē) A gastrointestinal disease characterized by severe, often bloody diarrhea, usually caused by infection with bacteria or parasites.

dyslexia (dĭs-lĕk′sē-ə) A learning disability marked by impairment of the ability to recognize and comprehend written words.

dysmenorrhea (dĭs-mĕn′ə-rē′ə) Painful menstruation.

Dysnomia (dĭs-nō′mē-ə) The satellite of the dwarf planet Eris.

dyspepsia (dĭs-pĕp′shə, -sē-ə) Difficulty in digesting food; indigestion.

dysphagia (dĭs-fā′jə) Difficulty in swallowing.

dysplasia (dĭs-plā′zhə) Abnormal development or growth of tissues, organs, or cells. —*Adjective* **dysplastic.**

dyspnea (dĭsp-nē′ə) Difficulty in breathing, often associated with lung or heart disease and resulting in shortness of breath.

dysprosium (dĭs-prō′zē-əm) *Symbol* **Dy** A soft, silvery metallic element of the lanthanide series. Because it has a high melting point and absorbs neutrons well, dysprosium is used to help control nuclear reactions. Atomic number 66; atomic weight 162.50; melting point 1,407°C; boiling point 2,600°C; specific gravity 8.536; valence 3. See **Periodic Table.**

dysthymia (dĭs-thī′mē-ə) A mood disorder characterized by chronic mild depression.

dystrophic (dĭ-strŏf′ĭk, -strō′fĭk) Having brownish acidic waters, a high concentration of humic matter, and a small plant population. Used of a lake, pond, or stream. Compare **eutrophic, oligotrophic.**

E

e (ē) An irrational number, with a numerical value of 2.718281828459.... It is mathematically defined as the limit of $(1 + \frac{1}{n})^n$ as n grows infinitely large. It is the base of natural logarithms and has many applications in mathematics, especially in expressions involving exponential growth and decay.

E 1. The symbol for **energy. 2.** The symbol for **modulus of elasticity.**

ear[1] (îr) **1.** The vertebrate organ of hearing, which in mammals is usually composed of three parts: the outer ear, middle ear, and inner ear. The organs of balance are also located in the ear. **2.** An invertebrate organ analogous to the vertebrate ear.

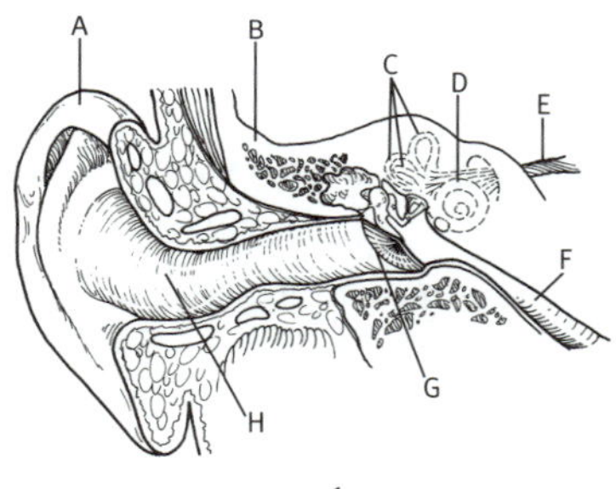

ear[1]

A. *auricle*, B. *bone*, C. *semicircular canals*, D. *cochlea*, E. *acoustic nerve*, F. *eustachian tube*, G. *eardrum*, H. *ear canal*

ear[2] (îr) The seed-bearing spike of a cereal plant, such as corn or wheat.

eardrum (îr′drŭm′) The thin, oval-shaped membrane that separates the middle ear from the outer ear. It vibrates in response to sound waves, which are then transmitted to the ossicles of the middle ear. Also called *tympanic membrane.*

early wood (ûr′lē) The part of the wood in a growth ring of a tree that is produced earlier in the growing season. The cells of early wood are larger and have thinner walls than those produced later in the growing season. Compare **late wood.**

Earth (ûrth) **1.** The third planet from the Sun and the densest planet in the solar system. Earth is a **terrestrial** or **inner planet** consisting of a thin outer crust, an intermediate mantle, and a dense inner core. It has an atmosphere composed primarily of nitrogen and oxygen and is the only planet on which water in liquid form exists, covering more than 70 percent of its surface. It is also the only planet on which life is known to have evolved, occupying the relatively thin region of water, land, and air known as the biosphere. Earth has a single, relatively large natural satellite, the Moon. See more at **atmosphere, core, crust, mantle.** See Table at **solar system. 2. earth** *Electricity.* See **ground** (sense 1).

earthquake (ûrth′kwāk′) A sudden movement of the Earth's lithosphere (its crust and upper mantle). Earthquakes are caused by the release of built-up stress within rocks along geologic faults or by the movement of magma in volcanic areas. They are usually followed by aftershocks. See Note at **fault.**

earth science Any of several sciences, such as geology, oceanography, and meteorology, that study the origin, composition, and physical features of the Earth.

earth station An on-ground radio communication site used for tracking, controlling, or otherwise communicating with spacecraft or satellites.

easterly (ē′stər-lē) A wind, especially a prevailing wind, that blows from the east. The trade winds in tropical regions and the prevailing winds in the polar regions are easterlies. See illustration at **wind.**

Eastern Hemisphere (ē′stərn) The half of the Earth that includes Europe, Africa, Asia, and Australia, as divided roughly by the 0° and 180° meridians. See more at **prime meridian.**

Eastern Standard Time Standard time in the fifth time zone west of Greenwich, England, reckoned at 75° west and used, for example, in the eastern part of North America.

eating disorder (ē′tĭng) Any of several patterns of severely disturbed eating behavior, especially anorexia nervosa and bulimia, seen

A CLOSER LOOK **earthquake**

Fractures in Earth's crust, or *lithosphere,* where sections of rock have slipped past each other are called *faults. Earthquakes* are caused by the sudden release of accumulated strain along these faults, releasing energy in the form of low-frequency sound waves called *seismic waves.* Although thousands of earthquakes occur each year, most are too weak to be detected except by *seismographs,* instruments that detect and record vibrations and movements in the Earth. The point where the earthquake originates is the *seismic focus,* and directly above it on Earth's surface is the earthquake's *epicenter.* Three kinds of waves accompany earthquakes. Primary (P) waves have a push-pull type of vibration. Secondary (S) waves have a side-to-side type of vibration. Both P and S waves travel deep into Earth, reflecting off the surfaces of its various layers. S waves cannot travel through the liquid outer core. Surface (L) waves—named after the nineteenth-century British mathematician A.E.H. Love—travel along Earth's surface, causing most of the damage of an earthquake. The total amount of energy released by an earthquake is measured on the *Richter scale.* Each increase by 1 corresponds to a tenfold increase in strength. Earthquakes above 7 on the Richter scale are considered severe. The famous earthquake that flattened San Francisco in 1906 had a magnitude of 7.8.

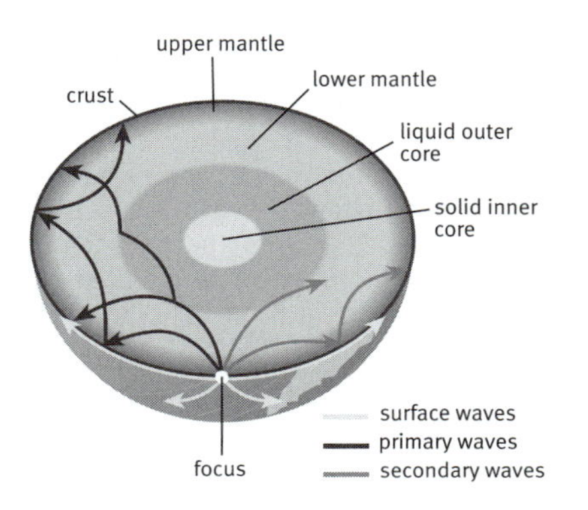

Primary and secondary waves radiate from an earthquake's focus and move through the Earth's interior. As they encounter a boundary, like that between the lower mantle and the liquid outer core, they are reflected and refracted. Secondary waves cannot travel through liquids. Surface waves radiate out from an earthquake's focus and travel only along the Earth's surface.

mainly in female teenagers and young women.

ebb tide (ĕb) The period between high tide and low tide during which water flows away from the shore. Also called *falling tide.* Compare **flood tide.** See more at **tide.**

Ebola virus (ĭ-bō**′**lə) A virus of African origin, belonging to the genus *Filovirus,* that causes a highly contagious infection characterized by fever, respiratory symptoms, bleeding, and sometimes central nervous system involvement with coma. In many patients there is progressive organ failure leading to death.

eccentricity (ĕk′sĕn-trĭs**′**ĭ-tē) **1.** A measure of the deviation of an elliptical path, especially an orbit, from a perfect circle. It is equal to the ratio of the distance between the foci of the ellipse to the length of the major axis of the ellipse (the distance between the two points farthest apart on the ellipse). Eccentricity ranges from zero (for a perfect circle) to values approaching 1 (highly elongated ellipses). **2.** The ratio of the distance of any point on a conic section from a focus to its distance from the corresponding directrix. This ratio is constant for any particular conic section.

echinoderm (ĭ-kī**′**nə-dûrm′) Any of various marine invertebrates of the phylum Echinodermata, having a latticelike internal skeleton composed of calcite and usually a hard, spiny outer covering. The body plans of adult echinoderms show radial symmetry, typically in the pattern of a five-pointed star, while the larvae show bilateral symmetry. Echinoderms probably share a common ancestor with the hemichordates and chordates, and were already quite diversified by the Cambrian Era. They include the starfish, sea urchins, sand dollars, holothurians (sea cucumbers), and crinoids, as well as thousands of extinct forms.

echo (ĕk′ō) **1.** A repeated sound that is caused by the reflection of sound waves from a surface. The sound is heard more than once because of the time difference between the initial production of the sound waves and their return from the reflecting surface. **2.** A wave that carries a signal and is reflected. Echoes of radio signals (carried by electromagnetic waves) are used in **radar** to detect the location or velocity of distant objects.

echocardiogram (ĕk′ō-kär′dē-ə-grăm′) An ultrasound image of the heart that demonstrates the size, motion, and composition of cardiac structures and is used to diagnose various abnormalities of the heart, including valvular dysfunction, abnormal chamber size, congenital heart disease, and cardiomyopathy.

echolocation (ĕk′ō-lō-kā′shən) Sonar, especially of animals, such as bats and toothed whales. See more at **sonar.**

Eckert (ĕk′ərt), **John Presper** 1919–1995. American engineer who contributed to the development of ENIAC (Electronic Numeral Integrator and Calculator), the first electronic computer (1946). He later helped develop one of the first computers to be sold commercially.

eclipse (ĭ-klĭps′) The partial or total blocking of light of one celestial object by another. An eclipse of the Sun or Moon occurs when the Earth, Moon, and Sun are aligned. ► In a **solar eclipse** the Moon comes between the Sun and Earth. During a total solar eclipse the disk of the Moon fully covers that of the Sun, and only the Sun's corona is visible. ► An **annular eclipse** occurs when the Moon is farthest in its orbit from the Earth so that its disk does not fully cover that of the Sun, and part of the Sun's photosphere is visible as a ring around the Moon. ► In a **lunar eclipse** all or a part of the Moon's disk enters the umbra of the Earth's shadow and is no longer illuminated by the Sun. Lunar eclipses occur only during a full moon, when the Moon is directly opposite the Sun.

eclipsing binary (ĭ-klĭp′sĭng) See under **binary star.**

eclipsing variable star A variable star whose change in luminosity is caused by two or

A CLOSER LOOK **eclipse**

The Sun is about 400 times wider than the Moon and 400 times farther from Earth, causing the two to appear to be almost exactly the same size in our sky. This relationship is also responsible for the phenomenon of the *total solar eclipse,* an eclipse of the Sun in which the disk of the Moon fully covers that of the Sun, blocking the Sun's light and causing the Moon's shadow to fall across the Earth. A total solar eclipse can be viewed only from a very narrow area on Earth, or zone of *totality,* where the dark central shadow of the Moon, or *umbra,* falls. From this perspective one can view the Sun's delicate *corona*—tendrils of charged gases that surround the Sun but are invisible to the unaided eye in normal daylight. This is also the only time when stars are visible in the day sky. Those viewing the eclipse from where the edges of the Moon's shadow, or *penumbra,* fall to Earth will see only a partial solar eclipse. The orbits of the Earth around the Sun and of the Moon around the Earth are not perfect circles, causing slight variations in how large the Sun and Moon appear to us and in the length of solar eclipses. The maximum duration of a total solar eclipse when the Earth is farthest from the Sun and the Moon is closest to the Earth is seven and a half minutes.

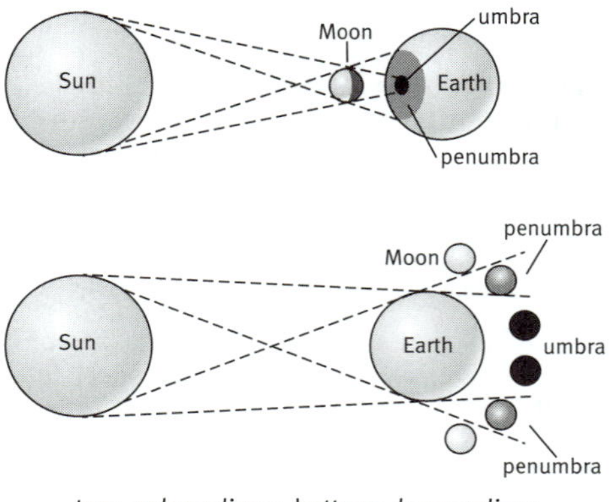

top: *solar eclipse;* bottom: *lunar eclipse*

more stars in a binary or multiple system eclipsing each other rather than by any intrinsic property of the star itself. The period of variation coincides with the orbital period of the system and can range from a few minutes to several years. See more under **binary star, multiple star.**

ecliptic (ĭ-klĭp′tĭk) The great circle on the celestial sphere that represents the Sun's apparent path among the background stars in one year. The northernmost point this path reaches on the celestial sphere is the Tropic of Cancer, its southernmost point is the Tropic of Capricorn, and it crosses the celestial equator at the points of vernal and autumnal equinox. ► The **plane of the ecliptic** is the imaginary plane that intersects the celestial sphere along the ecliptic, and the north and south **ecliptic poles** are the points where a perpendicular line through the middle of this plane intersect the sphere. The plane of the ecliptic corresponds to the plane in which the Earth orbits the Sun. If the Earth's axis were not tilted, the ecliptic would be identical to the celestial equator and the ecliptic poles identical to the celestial poles. In this case, the Sun's path would not move northward or southward from the equator during the year. As it is, the plane of the celestial equator is tilted 23.45° to the plane of the ecliptic, corresponding to the tilt of the Earth's axis with respect to its orbital plane, giving the Sun its apparent northward and southward movement among the background stars. See illustration at **celestial sphere.**

ecliptic coordinate system The coordinate system in which a celestial object's position on the celestial sphere is described in terms of its **celestial latitude** and **celestial longitude**, measured with respect to the **ecliptic**. Celestial latitude is measured in degrees north or south of the ecliptic, and celestial longitude is measured in degrees eastward from the **vernal equinox**. Because the ecliptic is fixed among the stars (that is, it is unaffected by the precession of the Earth's axis), an object's celestial latitude does not alter over time. However, due to the gradual precession of the equinoxes westward along the ecliptic, an object's celestial longitude increases by about 1.4° each century. The ecliptic system is used by astronomers especially in describing the position of objects within the solar system. Compare **altazimuth coordinate system, equatorial coordinate system.**

eclogite (ĕk′lə-jīt′) A greenish, coarse-grained metamorphic rock consisting of pyroxene, quartz, and feldspar with large red garnet inclusions. Eclogites form under conditions of high pressure and moderate to high temperatures.

E. coli (ē kō′lī) A bacillus *(Escherichia coli)* normally found in the human gastrointestinal tract and occurring in numerous strains, some of which are responsible for diarrheal diseases. Other strains have important experimental uses in molecular biology.

ecological efficiency (ē′kə-lŏj′ĭ-kəl) See under **trophic level.**

ecology (ĭ-kŏl′ə-jē) **1.** The scientific study of the relationships between living things and their environments. Also called *bionomics.* **2.** A system of such relationships within a particular environment.

ecospecies (ē′kō-spē′shēz, -sēz) A species considered in terms of its ecological characteristics and usually including several ecotypes capable of interbreeding.

ecosphere (ē′kō-sfîr′) The regions of the Earth that are capable of supporting life, together with the ecosystems they contain; the biosphere.

ecosystem (ē′kō-sĭs′təm) A community of organisms together with their physical environment, viewed as a system of interacting and interdependent relationships and including such processes as the flow of energy through trophic levels and the cycling of chemical elements and compounds through living and nonliving components of the system.

ecotone (ē′kə-tōn′) A transitional zone between two ecological communities, as between a forest and grassland or a river and its estuary. An ecotone has its own characteristics in addition to sharing certain characteristics of the two communities. See also **edge effect.**

ecotype (ē′kə-tīp′) A subdivision of an ecospecies, comparable to a subspecies or geographic race and consisting of an isolated population selectively adapted to a particular set of environmental conditions.

ectoderm (ĕk′tə-dûrm′) The outermost of the primary germ layers of an animal embryo. In vertebrates, the ectoderm gives rise to the epidermis and associated tissues (such as hair and sweat glands), enamel of the teeth, sense organs, nervous system, and lining of the nose, mouth, and anus. Compare **endoderm, mesoderm.**

ectogenous (ĕk-tŏj′ə-nəs) also **ectogenic** (ĕk′tə-jĕn′ĭk) **1.** Able to live and develop outside a host, as certain pathogenic bacteria. **2.** Originating or produced from outside an organism, tissue, or cell; exogenous.

ectoparasite (ĕk′tə-păr′ə-sīt′) See under **parasite.**

ectopic (ĕk-tŏp′ĭk) **1.** Out of place, as of an organ not in its proper position, or of a pregnancy occurring elsewhere than in the cavity of the uterus. **2.** Of or relating to a heartbeat that has its origin elsewhere than in the sinoatrial node.

ectotherm (ĕk′tə-thûrm′) A cold-blooded organism. Also called *poikilotherm.*

eczema (ĕk′sə-mə) An acute or chronic noncontagious inflammation of the skin, often caused by allergy and characterized by itching, scaling, and blistering.

edaphic (ĭ-dăf′ĭk) **1.** Relating to soil, especially as it affects living organisms. Edaphic characteristics include such factors as water content, acidity, aeration, and the availability of nutrients. **2.** Influenced by factors inherent in the soil rather than by climatic factors.

Eddington (ĕd′ĭng-tən), Sir **Arthur Stanley** 1882–1944. British mathematician, astronomer, and physicist who founded modern astrophysics. He conducted research on the evolution, structure, and motion of stars and was one of the first scientists to promote the theory of relativity. He also wrote a series of scientific books for the layperson.

eddy (ĕd′ē) A current, as of water or air, moving in a direction that is different from that of the main current. Eddies generally involve circular motion; unstable patterns of eddies are often called **turbulence**. See also **vortex.**

Edelman (ĕd′l-mən), **Gerald Maurice** Born 1929. American biochemist who shared with Rodney Porter the 1972 Nobel Prize for physiology or medicine for their research on the chemical structure and nature of antibodies.

edema (ĭ-dē′mə) An accumulation of an excessive amount of watery fluid in cells, tissues, or body cavities. Edema can be mild and benign as in pregnancy or prolonged standing in the elderly, or a serious sign of heart, liver, or kidney failure, or of other diseases.

edentate (ē-dĕn′tāt′) *Adjective.* **1.** Lacking teeth. —*Noun.* **2.** Any of various mammals belonging to the order Xenarthra (or Edentata), having no front teeth and few or no back teeth. The lumbar vertebrae have extra joints, which add support during digging. Sloths, armadillos, and anteaters are edentates.

edge effect (ĕj) The influence that two ecological communities have on each other along the boundary (called the ecotone) that separates them. Because such an area contains habitats common to both communities as well as others unique to the transition zone itself, the edge effect is typically characterized by greater species diversity and population density than occur in either of the individual communities.

Ediacaran (ē′dē-ä′kə-rən) Relating to a group of fossilized organisms that are the earliest known remains of multicellular life. They are soft-bodied marine life forms that date from between 560 and 545 million years ago, during the late Precambrian Eon.

Edison (ĕd′ĭ-sən), **Thomas Alva** 1847–1931. American inventor and physicist who took out more than 1,000 patents in his lifetime. His inventions include the telegraph (1869), microphone (1877), and light bulb (1879). He also designed the first power plant (1881–82), making possible the widespread distribution of electricity. During World War I, Edison worked on a number of military devices, including flamethrowers, periscopes, and torpedoes.

EEG Abbreviation of **electroencephalogram.**

efferent (ĕf′ər-ənt) Carrying motor impulses away from a central organ or part, as a nerve that conducts impulses from the central nervous system to the periphery of the body. Compare **afferent.**

effervescence (ĕf′ər-vĕs′əns) The bubbling of a solution due to the escape of gas. The gas may form by a chemical reaction, as in a fermenting liquid, or by coming out of solution after having been under pressure, as in a carbonated drink.

efficiency (ĭ-fĭsh′ən-sē) **1.** The ratio of the energy delivered (or work done) by a machine to the energy needed (or work required) in operating the machine. The efficiency of any machine is always less than one due to forces such as friction that use up energy unproductively. See also **mechanical advantage. 2.** The ratio of the effective or useful output to the total input in any system.

efflorescence (ĕf′lə-rĕs′əns) **1.** A whitish, powdery deposit on the surface of rocks or soil in dry regions. It is formed as mineral-rich water rises to the surface through capillary action and then evaporates and usually consists of gypsum, salt, or calcite. **2.** See **anthesis.**

effluent (ĕf′lo͞o-ənt) *Adjective.* **1.** Flowing out or forth. *Noun.* **2.** A stream flowing out of a body of water. **3.** An outflow or discharge of liquid waste, as from a sewage system, factory, or nuclear plant.

effort (ĕf′ərt) **1.** Force applied against inertia. **2.** The force needed by a **machine** in order to accomplish work on a load. Compare **load.**

egg (ĕg) **1.** The larger, usually nonmotile female reproductive cell of most organisms that reproduce sexually. Eggs are haploid (they have half the number of chromosomes as the other cells in the organism's body). During fertilization, the nucleus of an egg cell fuses with the nucleus of a sperm cell (the male reproductive cell) to form a new diploid organism. In animals, eggs are spherical, covered by a membrane, and usually produced by the ovaries. In some simple aquatic animals, eggs are fertilized and develop outside the body. In some terrestrial animals, such as insects, reptiles, and birds, eggs are fertilized inside the body but are incubated outside the body, protected by durable, waterproof membranes (shells) until the young hatch. In mammals, eggs produced in the ovaries are fertilized inside the body and (except in the cases of monotremes) develop in the reproductive tract until birth. The human female fetus possesses all of the eggs that she will ever have; every month after the onset of puberty, one of these eggs matures and is released from the ovary into the fallopian tube. In many plants (such as the bryophytes, ferns, and gymnosperms) eggs are produced by flasked-shaped structures known as archegonia. In gymnosperms and angiosperms, eggs are enclosed within ovules. In angiosperms, the ovules are enclosed within ovaries. See also **oogenesis. 2.** In many animals, a structure consisting of this reproductive cell together with nutrients and often a protective covering. The embryo develops within this structure if the reproductive cell is fertilized. The egg is often laid outside the body, but the female of ovoviviparous species may keep it inside the body until after hatching.

egg apparatus A group of three cells in the seven-celled embryo sac of an angiosperm (flowering plant) consisting of the egg cell and two associated cells called **synergids.** The egg apparatus is located at the end of the embryo sac closer to the micropyle (the opening through which pollen nuclei enter the ovule.) The synergids are thought to help direct the pollen nucleus to the egg cell as part of the process of **double fertilization** characteristic of angiosperms. The synergids degenerate after the egg cell has been fertilized. See more at **embryo sac.**

egg case 1. A protective capsule of certain animals, such as insects and mollusks, that contains eggs. **2.** See **egg sac.**

egg sac The silken pouch in which many spiders deposit their eggs. Also called *egg case.*

egg tooth A hard, toothlike projection from the beak of embryonic birds, or from the upper jaw of embryonic reptiles, that is used to cut the egg membrane and shell upon hatching and that later falls off.

Ehrlich (âr′lĭk), **Paul** 1854–1915. German bacteriologist who was a pioneer in the study of

Paul Ehrlich

the blood and the immune system, and in the development of drugs to fight specific disease-causing agents. He discovered a compound that was effective in combating sleeping sickness as well as a drug, called salvarsan, that cured syphilis.

Einstein (īn′stīn′), **Albert** 1879–1955. German-born American theoretical physicist whose theories of Special Relativity (1905) and General Relativity (1916) revolutionized modern thought on the nature of space and time and formed a theoretical base for the exploitation of atomic energy. He won the 1921 Nobel Prize for physics for his explanation of the photoelectric effect.

Albert Einstein

BIOGRAPHY **Albert Einstein**

By around 1900, the increased precision of new measuring instruments had shown that the laws of motion and gravity established by Galileo and Newton were unable to explain certain phenomena. The observed orbit of Mercury, for example, differed slightly from that predicted by Newton, and laws describing the motion of electromagnetic waves left many electrical effects unexplained. In 1905, an unknown 26-year-old patent office clerk named Albert Einstein published four papers that not only solved these problems, but revolutionized physics. The first two presented his Special Theory of Relativity, which departed from the classical Newtonian concepts of space and time in its assertion that all reference frames (all coordinate systems) do not measure space and time equivalently. That is, space and time are not the same throughout the universe, but depend on the motion of the observer. But for Einstein, not everything was relative. Following the electromagnetic theory of Maxwell, Einstein argued that the speed of light is the same for all observers, and introduced a new concept of *space-time* to reconcile this with concepts of relative motion. He also introduced the famous equation expressing a direct relation between mass and energy, $E = mc^2$, known as *mass-energy equivalence.* A third paper analyzed electromagnetic radiation such as light in terms of particles called photons, and explained how some substances, when exposed to such radiation, eject electrons in a quantum process called the *photoelectric effect.* A fourth paper explained the random movement of particles suspended in a fluid, now known as *Brownian motion.* In 1916, in his General Theory of Relativity, Einstein described gravity as a warping of space-time (as opposed to Newton's force) caused by the mere presence of objects possessing mass. Einstein's new conception of gravity correctly predicted Mercury's observed orbit, and his work on photons led to a more accurate description of electromagnetic radiation. In his later years, Einstein devoted himself to a search for a theory that would unify gravity with the other three fundamental forces in nature: the strong force, the electromagnetic force, and the weak force. This search is still ongoing.

einsteinium (īn-stī′nē-əm) *Symbol* **Es** A synthetic, radioactive metallic element of the actinide series that is usually produced by bombarding plutonium or another element with neutrons. It was first isolated in a region near the explosion site of a hydrogen bomb. Its longest-lived isotope is Es 254 with a half-life of 276 days. Atomic number 99; melting point 860°C. See **Periodic Table.**

ejaculation (ĭ-jăk′yə-lā′shən) The act of discharging semen from the urethra during orgasm.

ejecta (ĭ-jĕk′tə) Ejected matter, especially from an erupting volcano. Volcanic ejecta usually includes volcanic glass, ash, and rock fragments.

EKG Abbreviation of **electrocardiogram.**

elasmobranch (ĭ-lăz′mə-brăngk′) Any of numerous cartilaginous fishes of the subclass Elasmobranchii, having five to seven gill slits on each side, dermal denticles for scales, and a small respiratory opening (spiracle) behind each eye. The pectoral fins of elasmobranchs are often greatly enlarged. Elasmobranchs include the sharks, rays, and skates.

elastic collision (ĭ-lăs′tĭk) *Physics.* A collision between bodies in which the total kinetic energy of the bodies is conserved. In a perfectly elastic collision, no energy is dissipated as heat energy internal to the bodies, and none is spent on permanently deforming the bodies or radiated away in some other fashion. Elastic collisions, such as the collision of a rubber ball on a hard surface, result in the reflection or "bouncing" of bodies away from each other. Comapre **inelastic collision.**

elasticity (ĭ-lă-stĭs′ĭ-tē) The ability of a solid to return to its original shape or form after being subject to strain. Most solid materials display elasticity, up to a load point called the **elastic limit**; loads higher than this limit cause permanent deformation of the material. See also **Hooke's law.**

elastic limit The stress point at which a material, if subjected to higher stress, will no longer return to its original shape. Brittle materials tend to break at or shortly past their elastic limit, while ductile materials deform at stress levels beyond their elastic limit.

elastic strain A form of strain in which the distorted body returns to its original shape and size when the deforming force is removed. See more at **strain.**

elater (ĕl′ə-tər) A tiny elongated structure that helps disperse plant spores by coiling and uncoiling in response to changes in humidity. The elaters of horsetails are bands attached to the spore wall, while those of liverworts are sterile cells occurring among the spores.

E layer See **E region.**

Eldredge (ĕl′drĕdj), **Niles** Born 1943. American paleontologist who with Stephen Jay Gould developed the theory of punctuated equilibrium in 1972.

electric (ĭ-lĕk′trĭk) also **electrical** (ĭ-lĕk′trĭ-kəl) Relating to or operated by electricity. Compare **electronic.**

electrical engineering The branch of engineering that specializes in the design, construction, and practical uses of electrical systems.

electrical storm A thunderstorm.

electric arc An electric current, often strong, brief, and luminous, in which electrons jump across a gap. Electric arcs across specially designed electrodes can produce very high heats and bright light, and are used for such purposes as welding and illumination in spotlights. Unwanted arcs in electrical circuits can cause fires. Lightning is a case of an electric arc between one cloud and the earth or another cloud, as are sparks caused by discharges of static electricity.

electric cell A device, such as a battery, that is capable of changing some form of energy, such as chemical energy or radiant energy, into electricity. Also called *voltaic cell.* ▸ An electric cell that converts light energy into electrical energy using the photoelectric effect is called a **photoelectric** or **photovoltaic cell**; such cells are used in the generation of solar power and are called **solar cells**. See also **galvanic.**

electric charge A form of charge, designated positive, negative, or zero, found on the elementary particles that make up all known matter. Particles with electric charge interact with each other through the **electromagnetic force**, creating electric fields, and when they are in motion, magnetic fields. The electric fields tend to result in a repulsive force between particles with charges of the same sign, and an attractive force between charges of opposite sign. The electron is defined to have an electric charge of −1; the protons in an atomic nucleus have charge of +1, and the neutrons have charge of 0.

A CLOSER LOOK **electric charge**

Electric charge is a basic property of elementary particles of matter. The protons in an atom, for example, have a positive charge, the electrons have a negative charge, and the neutrons have zero charge. In an ordinary atom, the number of protons equals the number of electrons, so the atom normally has no net electric charge. An atom becomes negatively charged if it gains extra electrons, and it becomes positively charged if it loses electrons; atoms with net charge are called ions. Every charged particle is surrounded by an *electric field*, the area in which the charge exerts a force. Particles with nonzero electric charge interact with each other by exchanging *photons,* the carriers of the *electromagnetic force.* The strength and direction of the force charged particles exert on each other depends on the product of their charges: they attract each other if the product of their charges is negative and repel each other if the product is positive. Thus two electrons, each with charge -1, will repel each other, since $-1 \times -1 = +1$, a positive number. Static electricity consists of charged particles at rest, while electric *current* consists of moving charged particles, especially electrons or ions.

---

electric displacement See **electric flux density.**

electric field The distribution in space of the strength and direction of forces that would be exerted on an electric charge at any point in that space. Electric fields themselves result directly from other electric charges or from changing magnetic fields. The strength of an electric field at a given point in space near an electrically charged object is proportional to the amount of charge on the object, and inversely proportional to the distance between the point and the object. See also **electromagnetism, electrostatic force.**

electric field strength A measure of the strength of an electric field at a given point in space, equal to the force the field would induce on a unit electric charge at that point. Also called *electric field intensity, electric intensity.*

electric flux The electric flux density across a given cross-sectional area in an electric field.

electric flux density A measure of the strength of an electric field generated by a free electric charge, corresponding to the number of electric lines of force passing through a given area. It is equal to the **electric field strength** multiplied by the **permittivity** of the material through which the electric field extends. It is measured in coulombs per square meter. Also called *electric displacement.*

electricity (ĭ-lĕk-trĭs′ĭ-tē) **1.** The collection of physical effects related to the force and motion of electrically charged particles, typically electrons, through or across matter and space. See also **circuit, conductor, electric potential. 2.** Electric current, or a source of electric current. **3.** A buildup of electric charge. See also **static electricity.**

electric moment See **dipole moment** (sense 1).

electric potential A measure of the work required by an electric field to move electric charges. Its unit is the volt. Also called *voltage.* See more at **Ohm's law.**

electrocardiogram (ĭ-lĕk′trō-kär′dē-ə-grăm′) A graphic recording of the electrical activity of the heart, used to evaluate cardiac function and to diagnose arrhythmias and other disorders. ▸ An **electrocardiograph** is the apparatus used to generate electrocardiograms. The machine functions as a portable set of galvanometers that measure electric potentials at different anatomic sites on the chest and extremities, and contains internal circuitry for computing calculations based on these measurements. Twelve electrodes act as transducers to pick up the electrical signals. Various combinations of signals from the electrodes can be selected for output, each of which provides information about electrical activity in the heart from a different anatomical perspective. For example, electrodes placed on the right arm, left leg and left arm record variations in potential in the frontal plane of the heart. The signals are converted to waveform tracings that are recorded and printed for diagnostic interpretation.

electrochemistry (ĭ-lĕk′trō-kĕm′ĭ-strē) The scientific study of the electrical aspects of chemical reactions, especially the changes they bring about in the arrangement and energy of electrons. Electrochemistry is vital to the study of electrolysis, power generation by electric cells, and the transmission of electrical signals by neurons.

electroconvulsive therapy (ĭ-lĕk′trō-kən-vŭl′sĭv) Administration of electric current to the brain through electrodes placed on the head, usually near the temples, in order to induce unconsciousness and brief seizures. It is used in the treatment of certain psychiatric disorders, especially severe depression.

electrode (ĭ-lĕk′trōd′) A conductor through which an electric current enters or leaves a substance (or a vacuum) whose electrical characteristics are being measured, used, or manipulated. Electrodes can be used to detect electrical activity such as brain waves. Terminal points in electrical components such as transistors, diodes, and batteries are electrodes.

electrodialysis (ĭ-lĕk′trō-dī-ăl′ĭ-sĭs) A process by which ionized materials dissolved in a liquid, such as the anions and cations of dissolved salts, are moved across a membrane by the application of an electric field, separating them from liquids or ions of opposite charge. Electrodialysis can be use for the desalinization of brackish water.

electrodynamic (ĭ-lĕk′trō-dī-năm′ĭk) Related to or employing the effects of changing electric and magnetic fields, along with the forces and motions those fields induce on objects with electric charge. Compare **electrostatic.**

electrodynamics (ĭ-lĕk′trō-dī-năm′ĭks) The scientific study of electric charge and electric and magnetic fields, along with the forces and motions those fields induce. See also **electromagnetism.**

electroencephalogram (ĭ-lĕk′trō-ĕn-sĕf′ə-lə-grăm′) A graphic record of brain waves representing electrical activity in the brain, used especially in the diagnosis of seizures and other neurological disorders. ► The instrument used to record an electroencephalogram is called an **electroencephalograph.**It generates a record of the electrical activity of the brain by measuring electric signals using a set of electrodes attached to the scalp that act as transducers. Differences of electric potential between different parts of the brain are measured by a portable set of galvanometers and printed as a wide paper strip with multiple simultaneous waveform tracings that have standard configurations in the normal brain.

electrolysis (ĭ-lĕk-trŏl′ĭ-sĭs) A process in which a chemical change, especially decomposition, is brought about by passing an electric current through a solution of electrolytes so that the electrolyte's ions move toward the negative and positive electrodes and react with them. If negative ions move toward the anode, they lose electrons and become neutral, resulting in an oxidation reaction. This also happens if atoms of the anode lose electrons and go into the electrolyte solution as positive ions. If positive ions move toward the cathode and gain electrons, becoming neutral, a reduction reaction takes place. Electrolysis is used for many purposes, including the extraction of metals from ores, the cleaning of archaeological artifacts, and the coating of materials with thin layers of metal (electroplating).

electrolyte (ĭ-lĕk′trə-līt′) **1.** A melted or dissolved compound that has broken apart into ions (anions and cations). Applying an electric field across an electrolyte causes the anions and cations to move in opposite directions, thereby conducting electrical current while gradually separating the ions. See also **electrodialysis, electrolysis. 2.** Any of these ions found in body fluids. Electrolytes are needed by cells to regulate the flow of water molecules across cell membranes.

electrolytic capacitor (ĭ-lĕk′trə-lĭt′ĭk) A type of capacitor in which one plate is coated through electrolysis with an oxide to serve as the dielectric, while the other plate is replaced by an electrolyte. Electrolytic capacitors can achieve very high capacitance with very small sizes, but only act as capacitors as long as the current flows in one direction.

electrolytic cell A device that contains two electrodes in contact with an electrolyte and that brings about a chemical reaction when connected to an outside source of electricity. The electrodes are made of metal or carbon, and when connected to direct current, one electrode becomes positively charged, and the other becomes negatively charged. This initiates the movement of ions in the electrolyte toward the electrodes: positive ions move toward the negative electrode and negative ions move toward the positive electrode. A chemical reaction then takes place at each electrode, with ions changing from positive to negative (or vice versa), or becoming neutralized. Electrolytic cells have many practical uses, including the recovery of pure metal from alloys, the plating of one metal with another, and the manufacture of chlorine and sodium hydroxide. Compare **voltaic cell.**

electromagnet (ĭ-lĕk′trō-măg′nĭt) A device consisting of a coil of insulated wire wrapped around an iron core that becomes magnetized when an electric current flows through the wire. Electromagnets are used to convert electrical control signals into mechanical movements. See Note at **magnetism.**

electromagnetic force (ĭ-lĕk′trō-măg-nĕt′ĭk) The fundamental force associated with electric and magnetic fields. The electromagnetic force is carried by the photon and is responsible for atomic structure, chemical reactions, the attractive and repulsive forces associated with electrical charge and magnetism, and all other electromagnetic phenomena. Like gravity, the electromagnetic force has an infinite range and obeys the inverse-square law. The electromagnetic force is weaker than the strong nuclear force but stronger than the weak force and gravity. Some scientists believe that the electromagnetic force and the weak nuclear force are both aspects of a single force called the electroweak force.

electromagnetic pulse A pulse of intense electromagnetic radiation generated by certain physical events, especially by a nuclear explosion. The pulse includes a continuous spectrum of low frequencies and can interfere with the operation of electrical equipment.

electromagnetic radiation Energy in the form of transverse magnetic and electric waves. In a vacuum, these waves travel at the speed of light (which is itself a form of electromagnetic radiation). The acceleration of electric charges (such as alternating current in a radio transmitter) gives rise to electromagnetic radiation. Other common examples of electromagnetic radiation are x-rays, microwaves, and radio waves. A single unit, or quantum, of electromagnetic radiation is called a **photon**. See also **electromagnetism, polarization.**

A CLOSER LOOK **electromagnetic radiation**

In the nineteenth century, physicists discovered that a changing electric field creates a magnetic field and vice versa. Thus a variation in an electric field (for example, the changing field created when a charged particle such as an electron moves up and down) will generate a magnetic field, which in turn induces an electric field. Equations formulated by James Clerk Maxwell predicted that these fields could potentially reinforce each other, creating an electromagnetic ripple that propagates through space. In fact, visible light and all other forms of *electromagnetic radiation* consist exactly of such waves of mutually reinforcing electric and magnetic fields, traveling at the speed of light. The frequency of the radiation determines how it interacts with charged particles, especially with the electrons of atoms, which absorb and reemit the radiation. The energy of the electromagnetic radiation is proportional to its frequency: the greater the frequency of the waves, the greater their energy. Electromagnetic radiation can also be conceived of as streams of particles known as *photons*. The photon is the quantum (the smallest possible unit) of electromagnetic radiation. In quantum mechanics, all phenomena in which charged particles interact with one another, as in the binding of protons and electrons in an atom or the formation of chemical bonds between atoms in a molecule, can be understood as an exchange of photons by the charged particles.

electromagnetic spectrum The entire range of electromagnetic radiation. At one end of the spectrum are gamma rays, which have the shortest wavelengths and high frequencies. At the other end are radio waves, which have the longest wavelengths and low frequencies. Visible light is near the center of the spectrum.

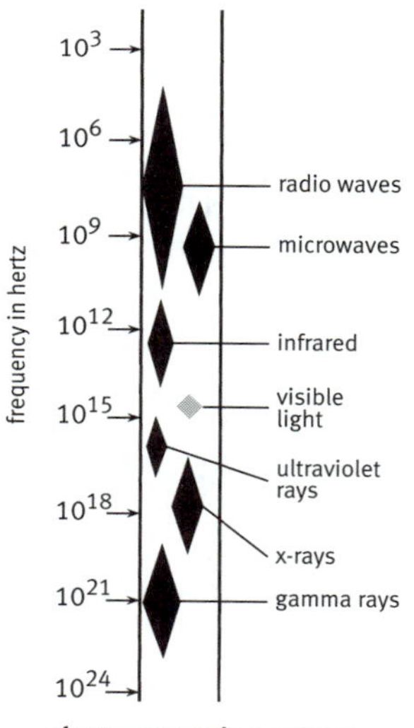

electromagnetic spectrum

electromagnetic unit Any of various units, such as the **abampere** and the **stilb**, used in the centimeter-gram-second system of units to describe electric and magnetic field strengths, electric current strengths, and other quantities associated with electromagnetism.

electromagnetic wave A wave of energy consisting of electric and magnetic fields, oscillating at right angles to each other. See more at **electromagnetic radiation.**

electromagnetism (ĭ-lĕk′trō-măg′nĭ-tĭz′əm) Any of the wide range of phenomena associated with the behavior and interaction of electric charges and electric and magnetic fields, such as electricity, magnetism, chemical bonds, and all forms of electromagnetic radiation, including light.

electromotive (ĭ-lĕk′trō-mō′tĭv) Capable of accelerating electric charges and creating electric current; having electric potential.

electromotive force Electric potential or voltage. Electromotive force is not really a force, but a measure of how much work would be done by moving an electric charge.

electron (ĭ-lĕk′trŏn′) **1.** A stable elementary particle in the lepton family having a mass at rest of 9.107×10^{-28} grams and an electric charge of approximately -1.602×10^{-19} coulombs. Electrons orbit about the positively charged nuclei of atoms in distinct **orbitals** of different energy levels, called **shells**. Electrons are the primary charge carriers in electric current. Compare **positron.** See also **electromagnetism, elementary particle, ion.** See Table at **subatomic particle.** **2.** A positron or a negatron. See more at **negatron.**

electron amplifier See **electron multiplier.**

electron carrier Any of various molecules that are capable of accepting one or two electrons from one molecule and donating them to another in the process of **electron transport**. As the electrons are transferred from one electron carrier to another, their energy level decreases, and energy is released. Cytochromes and quinones (such as coenzyme Q) are some examples of electron carriers.

electronegative (ĭ-lĕk′trō-nĕg′ə-tĭv) **1.** Tending to attract electrons and to form a negative ion. Nonmetals are generally electronegative. **2.** Having a negative electric charge. **3.** Capable of acting as a negative electrode.

electron gun The part of a cathode-ray tube that emits a narrow beam of electrons, consisting of a cathode, control grids, and usually a heater. Electrons are emitted from the cathode, which is typically heated by electric current to give the electrons escape energy. The electrons are then focused into a beam by the electric fields of the control grids.

electron hole See **hole.**

electronic (ĭ-lĕk′trŏn′ĭk) Relating to electrical devices that amplify and process electrical signals. Audio amplifiers, radios, and digital circuitry are electronic devices. ▸ The scientific study of the behavior and design of electronic devices and circuits is called **electronics.** Compare **electric.**

electron microscope A microscope that produces images of extremely small objects by using beams of electrons rather than visible light. Since electrons have a much shorter wavelength than light, the use of electron beams rather than light beams can resolve much finer structural details in the sample. Electrons are beamed at the sample and focused by magnets; a detector then converts the refracted or reflected beams into a black and white image. Powerful electron microscopes can create accurate images of objects as small as individual molecules. See also **scanning electron microscope.**

electron multiplier An electrical component in which a single electron can create a large current flow. Electron multipliers are used in photomultipliers, particle detectors, and electron microscopes. ▸ In **tube electron multipliers**, electrons released from a cathode collide with a dynode or anode, causing release of more electrons through **secondary emission**; this process is generally repeated through a number of stages to achieve great amplification of current. Electron multipliers are also called *electron amplifiers.*

electron neutrino A type of neutrino associated with the electron, often created in particle interactions involving electrons (such as beta decay). The electron neutrino has a mass of zero or nearly zero, and has no electric charge. See more at **neutrino.** See Table at **subatomic particle.**

electron pair **1.** Two electrons functioning or regarded as functioning together, especially two electrons that form a nonpolar covalent bond between atoms. **2.** An electron and a positron produced by a high-energy photon.

electron transport The movement of electrons from one electron carrier to another in a series of oxidation-reduction reactions, with each electron transferred to a slightly lower level of energy, so that energy is released. Electron transport is used in the light reactions of photosynthesis and in the final stage of cellular respiration to produce ATP from ADP and phosphate.

electron tube A sealed glass tube containing either a vacuum or a small amount of gas, in which electrons move from a negatively charged electrode, the **cathode**, to a positively charged one, the **anode**. The cathode is usually heated by an electric current to free the electrons. Other electrodes in the tube can vary the electric or magnetic fields in the tube to control the strength and direction of the moving electrons. Electron tubes amplify signals, rectify AC currents, and produce x-rays. They have been mostly replaced by transistors but are still used in television screens, computer monitors, and microwave technology. Also called *valve.*

electron volt A unit used to measure the energy of subatomic particles. One electron volt is defined as the energy needed to move an electron (which has an electric charge equal to –1) across an electric potential of one volt.

electrophoresis (ĭ-lĕk′trō-fə-rē′sĭs) The migration of electrically charged molecules through a fluid or gel under the influence of an electric field. Electrophoresis is used especially to separate combinations of compounds, such as fragments of DNA, for the purpose of studying their components.

electrophorus (ĭ-lĕk′trŏf′ər-əs, ē′lĕk-) Plural **electrophori.** An electrostatic generator, constructed like one half of a large capacitor, the other half being any grounded surface, such as a table. A small charge is given to the electrophorus when it is near the grounded surface, effectively charging it like capacitor. As the electrophorous is lifted away from the surface, its voltage relative to the surface increases (as the capacitance decreases).

electroplating (ĭ-lĕk′trō-plā′tĭng) The process of coating the surface of a conducting material with a metal. During the process, the surface to be covered acts as a cathode in an electrolytic cell, and the metal that is to cover it acts as an anode. Electroplating is usually used to cover a cheaper metal with a more expensive one, or to cover a corrosive metal with a less corrosive or noncorrosive one.

electropositive (ĭ-lĕk′trō-pŏz′ĭ-tĭv) **1.** Tending to donate electrons and to form a positive ion. **2.** Having a positive electric charge. **3.** Capable of acting as a positive electrode.

electrostatic (ĭ-lĕk′trō-stăt′ĭk) Relating to or caused by electric charges that are not in motion. Compare **electrodynamic.**

electrostatic force See **Coulomb force.**

electrostatic generator Any of various devices, including the electrophorus, the Wimshurst machine, and especially the Van de Graaff generator, that generate or store high voltages by accumulating static electric charge.

electroweak force (ĭ-lĕk′trō-wēk′) A hypothetical force postulated to explain both the electromagnetic force and the weak nuclear force as two aspects of a single force.

element (ĕl′ə-mənt) **1.** A substance that cannot be broken down into simpler substances by chemical means. An element is composed of atoms that have the same atomic number, that is, each atom has the same number of protons in its nucleus as all other atoms of that element. Today 117 elements are known, of which 92 are known to occur in nature, while the remainder have only been made with particle accelerators. Eighty-one of the elements have isotopes that are stable. The others, including technetium, promethium, and those with atomic numbers higher than 83, such as plutonium, are radioactive. See table at **Periodic Table.** **2.** *Mathematics.* A member of a set.

WORD HISTORY **element**

When Russian chemist Dmitri Mendeleev devised the Periodic Table in 1869, 63 *elements* were known, which he classified by atomic weight, and arranged in a table by vertical rows corresponding to shared chemical characteristics. Gaps in the table suggested the possibility of elements that would later be discovered, or in some cases, artificially created, that would fill the gaps and had the expected chemical properties. The striking correlation between the atomic weight of an element and its chemical properties was later explained by quantum mechanical theories of the atom. The weight of an atom of any given element depends on the number of protons (and neutrons) in its nucleus, but the number of protons also determines the number and arrangement of electrons that can orbit the nucleus, and it is these outer shells of electrons that largely determine the element's chemical properties.

element 112 *Symbol* **Uub** An artificially produced radioactive element whose most stable isotope has a mass number of 285 and a half-life of about 34 seconds. Also called *ununbium.* See **Periodic Table.**

element 113 *Symbol* **Uut** An artificially produced radioactive element whose most stable isotope has a mass number of 284 and a half-life of slightly less than half a second. Also called *ununtrium.* See **Periodic Table.**

element 114 *Symbol* **Uuq** An artificially produced radioactive element whose most stable isotope has a mass number of 289 and a half-life of approximately 2.7 seconds. Radioactive decay of all isotopes is chiefly by alpha-particle transmission. Also called *ununquadium.* See **Periodic Table.**

element 115 *Symbol* **Uup** An artificially produced radioactive element whose most stable confirmed isotopes have mass numbers of 287 and 288 and half-lives of less than one second. Also called *ununpentium.* See **Periodic Table.**

element 116 *Symbol* **Uuh** An artificially produced radioactive element whose most stable isotope has a mass number of 293 and a half-life of less than one-tenth of a second. Also called *ununhexium.* See **Periodic Table.**

element 118 *Symbol* **Uuo** An artificially produced radioactive element, detected indirectly by decay, whose most stable isotope has a mass number of 294 and a half-life of less than two milliseconds. Also called *ununoctium.* See **Periodic Table.**

elementary particle (ĕl′ə-mĕn′tə-rē) Any of the smallest, discrete entities of which the universe is composed, including the quarks, leptons, and gauge bosons, which are not themselves made up of other particles. Most types of elementary particles have mass, though at least one, the photon, does not. Also called *fundamental particle.* See also **composite particle, subatomic particle.**

A CLOSER LOOK **elementary particles**

The smallest known units of matter, or elementary particles, are classified under three distinct groups: the quarks, the leptons, and the bosons. The six types or "flavors" of quarks are the up quark, the down quark, the charm quark, the strange quark, the top quark, and the bottom quark. All quarks have mass, electric charge, and a special kind of charge called color, and each is associated with a distinct antiparticle, making twelve quarks in all. The leptons include the electron, the muon, the tau particle, the electron neutrino, the muon neutrino, and the tau neutrino. These particles also have distinct antiparticles; the neutrinos are electrically neutral and, if they do have mass, are extremely light. Each of these elementary particles interacts with other elementary particles through one or more forces: the electromagnetic force (between particles with electric charge), the strong force (between particles with color charge, such as the quarks), the weak force (between all leptons and quarks), and the gravitational force (between all particles). These forces are mediated by yet another set of elementary particles, the gauge bosons: when two particles interact, they exchange one or more gauge bosons. The gauge bosons include the W and Z bosons, which mediate the weak nuclear force, the gluon, which mediates the strong nuclear force, and the photon, which mediates the electromagnetic force. The hypothetical graviton, which would mediate the gravitational force, has not been isolated. A sixth boson, the Higgs boson, is believed to interact with the other elementary particles in such a way as to impart mass to them; it too has not been experimentally isolated. Though these particles are believed to be elementary, they can under certain circumstances change into other elementary particles. In beta decay, for example, an up quark turns into a down quark, emitting an electron and an electron antineutrino in the process. All known forms of matter and energy are made of combinations of and interactions between elementary particles; atoms, for example, are made of electrons orbiting a nucleus composed of quarks bound together into larger particles, the protons and neutrons. Whether these particles might themselves be composed of more fundamental building blocks is an open question, and the construction of a "theory of everything" that would explain the properties of all of the known particles and forces is the ultimate goal for modern physics.

elevation (ĕl′ə-vā′shən) The vertical distance between a standard reference point, such as sea level, and the top of an object or point on the Earth, such as a mountain. At 8,850 m

(29,028 ft), the summit of Mount Everest is the highest elevation on Earth.

Elion (ĕl′ē-ən, -ŏn′), **Gertrude Belle** 1918–1999. American pharmacologist who, with George Hitchings, developed drugs to treat leukemia and malaria, gout, herpes, and urinary and respiratory tract infections. She and Hitchings shared with Sir James Black the 1988 Nobel Prize for physiology or medicine.

ellipse (ĭ-lĭps′) A closed, symmetric curve shaped like an oval, which can be formed by intersecting a cone with a plane that is not parallel or perpendicular to the cone's base. The sum of the distances of any point on an ellipse from two fixed points (called the foci) remains constant no matter where the point is on the curve.

ellipsoid (ĭ-lĭp′soid′) A three-dimensional geometric figure resembling a flattened sphere. Any cross section of an ellipsoid is an ellipse or circle. An ellipsoid is generated by rotating an ellipse around one of its axes.

elliptical galaxy (ĭ-lĭp′tĭ-kəl) The most common type of galaxy, ranging in shape from nearly spherical (classified as E0) to greatly elongated (classified as E7). Elliptical galaxies vary greatly in size and include some of the largest and smallest known galaxies. They do not have spiral arms and have considerably less interstellar gas and dust than spiral galaxies, with little or no star formation taking place within them. Their stars follow individual elliptical orbits around the center of the galaxy. Long thought to be older galaxies which had used up all the material for star formation, it has also been suggested that elliptical galaxies actually form from collisions between spiral galaxies. Compare **irregular galaxy, lenticular galaxy, spiral galaxy.** See more at **Hubble classification system.**

El Niño (ĕl nēn′yō) A warming of the surface water of the eastern and central Pacific Ocean, occurring every 4 to 12 years and causing unusual global weather patterns. An El Niño is said to occur when the trade winds that usually push warm surface water westward weaken, allowing the warm water to pool as far eastward as the western coast of South America. When this happens, the typical pattern of coastal upwelling that carries nutrients from the cold depths to the ocean surface is disrupted, and fish and plankton die off in large numbers. El Niño warming is associated with the atmospheric phenomenon known as the **southern oscillation**, and their combined effect brings heavy rain to western South American and drought to eastern Australia and Indonesia. El Niño also affects the weather in the United States, but not as predictably. Compare **La Niña.**

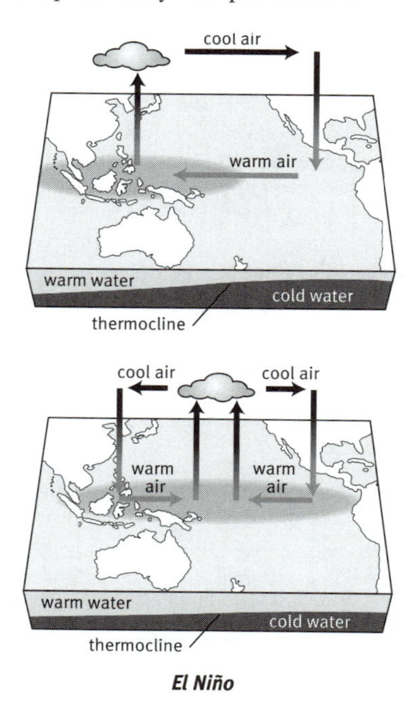

El Niño

top: *normal water temperatures, with warm water concentrated in the western tropical Pacific;* bottom: *El Niño conditions, with warm water extending from the western tropical Pacific to the eastern Pacific*

elongation (ĭ-lông′gā′shən) The angular distance between two celestial bodies as seen from a third. Elongation is normally conceived as a measure of the angle formed between the Sun and a celestial body, such as a planet or the Moon, with Earth at the vertex. In terms of the celestial sphere, elongation is the distance between the Sun and the body as measured in degrees of celestial longitude. When the body lies on a direct line drawn from Earth to or through the Sun, its elongation is 0° and it is said to be in **conjunction**. It is said to be in **quadrature** when it lies at a right angle to a line between the Earth and Sun with an elongation of 90°, and it is in **opposition** when it lies on the opposite side of Earth from the Sun with an elongation of 180°. Superior planets (those that are farther from the Sun than Earth) have a full range of elongations between 0° and 180°. Inferior planets (those closer to the Sun than Earth) have limited elongations due to their smaller orbits; Venus has a greatest elongation of about 48°, while Mercury's greatest elongation is about 28°. See more at **conjunction, opposition.**

elution (ĭ-lo͞o′shən) The process of extracting a substance that is adsorbed to another by washing it with a solvent. ▸ The substance used as a solvent in elution is called an **eluent.** ▸ The solution of the solvent and the substance that was adsorbed to another is called the **eluate.**

eluviation (ĭ-lo͞o′vē-ā′shən) The lateral or downward movement of the suspended material in soil through the percolation of water. Eluviation differs from **leaching** in that it affects suspended, not dissolved, material and usually results only in the movement of the material from one soil horizon to another.

eluvium (ĭ-lo͞o′vē-əm) **1.** Residual deposits of soil, dust, and sand produced by the action of the wind. **2.** Residual deposits of soil, dust, and rock particles produced by the in-situ decomposition and disintegration of rock.

elve (ĕlv) An extremely dim, short-lived, expanding disk of reddish light above thunderstorms, believed to be caused by electromagnetic pulses from intense lightning in the lower ionosphere. Elves last less than a second and can be as wide as 500 km (310 mi) in diameter.

elytron (ĕl′ĭ-trŏn′) Plural **elytra.** Either of the modified forewings of a beetle or related insect that encase the thin hind wings used in flight.

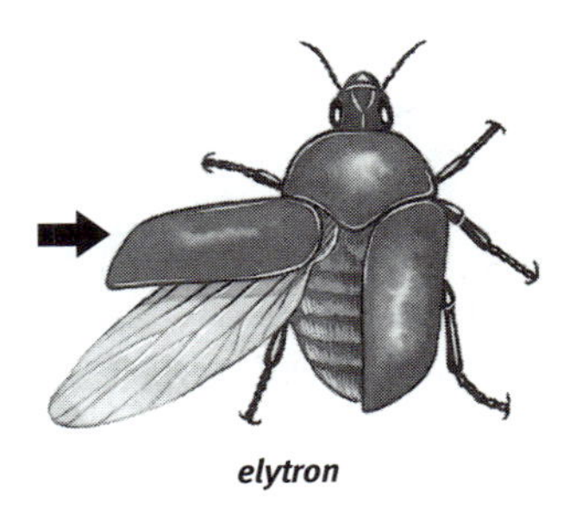

elytron

e-mail or **email** (ē′māl′) **1.** A system for sending and receiving messages electronically over a computer network. E-mail is asynchronous and does not require the receiver of the message to be online at the time the message is sent or received. E-mail also allows a user to distribute messages to large numbers of recipients instantaneously. **2.** A message or messages sent or received by such a system.

emarginate (ĭ-mär′jə-nĭt, -nāt′) Having a shallow notch at the tip, as in some petals and leaves or the tails of some birds.

embolism (ĕm′bə-lĭz′əm) **1.** A mass, such as an air bubble, detached blood clot, or foreign body, that travels in the bloodstream, lodges in a blood vessel, and obstructs or occludes it. Also called *embolus.* **2.** The obstruction or occlusion of a blood vessel by such a mass.

embolus (ĕm′bə-ləs) Plural **emboli** (ĕm′bə-lī). See **embolism** (sense 1).

embryo (ĕm′brē-ō′) **1a.** An animal in its earliest stage of development, before all the major body structures are represented. In humans, the embryonic stage lasts through the first eight weeks of pregnancy. In humans, other placental mammals, and other viviparous animals, young born as embryos cannot thrive. In marsupials, the young are born during the embryonic stage and complete their development outside the uterus, attached to a teat within the mother's pouch. **b.** The developing young of an egg-laying animal before hatching. **2.** The sporophyte of a plant in its earliest stages of development, such as the miniature, partially developed plant that is contained within a seed before germination. —*Adjective* **embryonic** (ĕm′brē-ŏn′ĭk)

embryology (ĕm′brē-ŏl′ə-jē) The scientific study of embryos and their development.

embryo sac An oval structure within an ovule of an angiosperm that contains the egg. Together with the fertilized egg, it develops into a seed. The embryo sac is the female gametophyte of angiosperms, consisting of eight nuclei: the egg and two adjacent and short-lived synergids that are near the micropyle (the opening where the pollen nuclei will enter), two central nuclei (which will combine with one of the pollen nuclei to form the endosperm), and three antipodal nuclei at the end of the embryo sac opposite the micropyle. Like the synergids, these nuclei degenerate at or shortly after fertilization. See more at **gametophyte, pollination.**

embryo transfer The transfer into a recipient's uterus of an egg that has been fertilized in vitro and is at the blastula stage of development.

emerald (ĕm′ər-əld) A transparent, green form of the mineral beryl. It is valued as a gem.

emergent (ĭ-mûr′jənt) Rooted below a body of water or in an area that is periodically submerged but extending above the water level. Used of aquatic plants such as cattails, rushes, or cord grass.

EMF 1. Abbreviation of **electromotive force. 2.** Abbreviation of **electromagnetic field.**

emission nebula (ĭ-mĭsh′ən) A nebula that absorbs ionizing ultraviolet radiation from nearby hot stars or star remnants and reemits it as visible light. **HII regions**, **planetary nebulae**, and nebulae created by the explosion of **supernovae** are different kinds of emission nebulae. See more at **nebula.**

emission spectrum The distribution of electromagnetic radiation released by a substance whose atoms have been excited by heat or radiation. A spectroscope can be used to determine which frequencies have been emitted by a substance. The emission spectrum is a combination of the atomic spectra of the various kinds of atoms making up the substance and can be analyzed to determine the substance's chemical or atomic composition. See more at **atomic spectrum.**

emotion (ĭ-mō′shən) A psychological state that arises spontaneously rather than through conscious effort and is sometimes accompanied by physiological changes; a feeling.

emphysema (ĕm′fĭ-sē′mə) A chronic lung disease characterized by progressive, irreversible expansion of the alveoli with eventual destruction of alveolar tissue, causing obstruction to airflow. Patients with emphysema often have labored breathing, wheezing, chronic fatigue, and increased susceptibility to infection, and may require oxygen therapy. Long-term smoking is a common cause of emphysema.

empirical (ĕm-pîr′ĭ-kəl) Relying on or derived from observation or experiment.

empirical formula A chemical formula that indicates the relative proportions of the elements in a molecule rather than the actual number of atoms of the elements. The empirical formula of a compound may be simpler than its molecular formula, which is a multiple of the empirical formula. For example, glucose has the molecular formula $C_6H_{12}O_6$ but the empirical formula CH_2O. Compare **molecular formula, structural formula.**

empty set (ĕmp′tē) *Mathematics.* The set that has no members or elements.

EMU Abbreviation of **electromagnetic unit.**

emulsion (ĭ-mŭl′shən) A suspension of tiny droplets of one liquid in a second liquid. By making an emulsion, one can mix two liquids that ordinarily do not mix well, such as oil and water. Compare **aerosol, foam.** —*Verb* **emulsify.**

enamel (ĭ-năm′əl) The hard, translucent substance covering the exposed portion of a tooth in mammals. Enamel is the hardest substance in the body, and consists mostly of calcium salts.

enantiomer (ĭ-năn′tē-ə-mər) Either of two stereoisomers that are mirror images of one another but cannot be superimposed on one another and that rotate the plane of polarized light in opposite directions. Enantiomers usually behave the same chemically but differ in optical behavior and sometimes in how quickly they react with other enantiomers. Also called *optical isomer, enantiomorph.* Compare **geometric isomer.**

encephalitis (ĕn-sĕf′ə-lī′tĭs) Inflammation of the brain, usually caused by infection with a virus.

encephalopathy (ĕn-sĕf′ə-lŏp′ə-thē) Degeneration of brain function, caused by any of various acquired disorders, including metabolic disease, organ failure, inflammation, and chronic infection.

encode (ĕn-kōd′) To specify the genetic code for the synthesis of a protein molecule or a part of a protein molecule.

encrypt (ĕn-krĭpt) To alter information using a code or mathematical algorithm so as to be unintelligible to unauthorized readers.

endangered species (ĕn-dān′jərd) A plant or animal species existing in such small numbers that it is in danger of becoming extinct, especially such a species placed in jeopardy as a result of human activity. One of the principal factors in the endangerment or extinction of a species is the destruction or pollution of its native habitat. Other factors include overhunting, intentional extermination, and the accidental or intentional introduction of alien species that outcompete the native species for environmental resources.

endemic (ĕn-dĕm′ĭk) **1.** Relating to a disease or pathogen that is found in or confined to a particular location, region, or people. Malaria, for example, is endemic to tropical regions. See also **epidemic, pandemic. 2.** Native to a specific region or environment and not occurring naturally anywhere else. The giant sequoia is endemic to the western slopes of the Sierra Nevada. Compare **alien, indigenous.**

USAGE **endemic/epidemic**

A disease that occurs regularly in a particular area, as malaria does in many tropical countries, is said to be *endemic.* The word *endemic,* built from the prefix *en–*, "in or within," and the Greek word *demos,* "people," means "within the people (of a region)." A disease that affects many more people than usual in a particular area or that spreads into regions in which it does not usually occur is said to be *epidemic.* This word, built from the prefix *epi–*, meaning "upon," and *demos,* means "upon the people." In order for a disease to become epidemic it must be highly contagious, that is, easily spread through a population. Influenza has been the cause of many epidemics throughout history. Epidemics of waterborne diseases such as cholera often occur after natural disasters such as earthquakes and severe storms that disrupt or destroy sanitation systems and supplies of fresh water.

endergonic (ĕn′dər-gŏn′ĭk) Relating to a chemical reaction that requires the absorption of energy. The production of sugars by plants during photosynthesis is an endergonic reaction, where sunlight provides the necessary energy. Compare **exergonic.**

endo– A prefix meaning "inside," as in *endoskeleton.*

endocarp (ĕn′də-kärp′) The hard inner layer of the pericarp of many fruits, such as the layer that forms the pit or stone of a cherry, peach, or olive. Compare **exocarp, mesocarp.**

endocrine gland (ĕn′də-krĭn, -krēn′) A gland of the body that produces hormones and secretes them directly into the bloodstream. In mammals, the thyroid gland, adrenal glands, and pituitary gland, as well as the ovaries, testes, and the islets of Langerhans of the pancreas are endocrine glands.

endocrine system The bodily system that consists of the endocrine glands and the hormones that they secrete. See Note at **hormone.**

endocrinology (ĕn′də-krə-nŏl′ə-jē) The branch of medicine that deals with the diagnosis and treatment of diseases of the endocrine system.

endoderm (ĕn′də-dûrm′) The innermost of the primary germ layers of an animal embryo. In vertebrates, the endoderm gives rise to the respiratory tract, gastrointestinal tract (except mouth and anus), glands associated with the gastrointestinal tract, bladder, and urethra. Compare **ectoderm, mesoderm.**

endodermis (ĕn′də-dûr′mĭs) The innermost layer of the cortex that forms a sheath around the vascular tissue of roots and some stems. In the roots the endodermis helps regulate the intake of water and minerals into the vascular tissues from the cortex.

endoergic (ĕn′dō-ûr′jĭk) Relating to a process, such as a chemical or nuclear reaction, that absorbs energy. Endothermic reactions are endoergic. Compare **exoergic.**

endogenous (ĕn-dŏj′ə-nəs) Originating or produced within an organism, tissue, or cell. Compare **exogenous.**

endometriosis (ĕn′dō-mē′trē-ō′sĭs) A gynecologic disorder characterized by the abnormal presence of functional endometrial tissue outside the uterus, often resulting in pelvic pain and dysmenorrhea.

endometrium (ĕn′dō-mē′trē-əm) Plural **endometria.** The mucous membrane that lines the uterus. A fertilized egg must embed itself in the endometrium in order to develop into an embryo. See more at **menstrual cycle.**

endoparasite (ĕn′dō-păr′ə-sīt′) See under **parasite.**

endoplasmic reticulum (ĕn′də-plăz′mĭk) An organelle consisting of a network of membranes within the cytoplasm of eukaryotic cells that is important in protein synthesis and folding and is involved in the transport of cellular materials. The endoplasmic reticulum can be continuous in places with the membrane of the cell nucleus. The function of the endoplasmic reticulum can vary greatly with cell type, and even within the same cell it can have different functions depending on whether it is rough or smooth. ▸ The **rough endoplasmic reticulum** is a series of connected flattened sacs that have many ribosomes on their outer surface. Rough endoplasmic reticulum synthesizes and secretes serum proteins (such as albumin) in the liver, and hormones (such as insulin) and other substances (such as milk) in the glands. ▸ The **smooth endoplasmic reticulum** is tubular in form and is involved in the synthesis of phospholipids, the main lipids in cell membranes. Smooth endoplasmic reticulum is the site of the breakdown of toxins and carcinogens in the liver, the conversion of cholesterol into steroids in the gonads and adrenal glands, and the release of calcium ions in the muscles, causing muscle contraction. The smooth endoplasmic reticulum also transports the products of the rough endoplasmic reticulum to other cell parts, notably the Golgi apparatus. See more at **cell.**

endorphin (ĕn-dôr′fĭn) Any of a group of peptide substances secreted by the anterior portion of the pituitary gland that inhibit the perception of painful stimuli. Endorphins act as neurotransmitters in the pain pathways of the brain and spinal cord. Narcotic drugs may stimulate the secretion of endorphins.

A CLOSER LOOK **endorphins**

Endorphins are long chains of amino acids, or *polypeptides,* that are able to bind to the neuroreceptors in the brain and are capable of relieving pain in a manner similar to that of morphine. There are three major types of endorphins: *beta-endorphins* are found almost entirely in the pituitary gland, while *enkephalins* and *dynorphins* are both distributed throughout the nervous system. Scientists had suspected that analgesic opiates, such as morphine and heroin, worked effectively against pain because the body had receptors that were activated by such drugs. They reasoned that these receptors probably existed because the body itself had natural painkilling compounds that also bonded to those receptors. When scientists in the 1970s isolated a biochemical from a pituitary gland hormone that showed analgesic properties, Choh Li, a chemist from Berkeley, California, named it *endorphin,* meaning "the morphine within." Besides behaving as a pain reducer, endorphins are also thought to be connected to euphoric feelings, appetite modulation, and the release of sex hormones. Prolonged, continuous exercise contributes to an increased production of endorphins and, in some people, the subsequent "runner's high."

---

endoscope (ĕn′də-skōp′) A medical instrument used for visual examination of the interior of a body cavity or a hollow organ such as the colon, bladder, or stomach. It is a rigid or flexible tube fitted with lenses, a fiber-optic light source, and often a probe, forceps, suction device, or other apparatus for examination or retrieval of tissue. —*Noun* **endoscopy** (ĕn-dŏs′kə-pē)

endoskeleton (ĕn′dō-skĕl′ĭ-tn) The internal supporting framework of humans and other vertebrates, usually made of bone. Certain invertebrates, such as sponges and echinoderms, also have endoskeletons. Compare **exoskeleton.**

endosperm (ĕn′də-spûrm′) The tissue that surrounds and provides nourishment to the embryo in the seeds of many angiosperms. The cells of the endosperm arise from a process similar to that of fertilization. The pollen of angiosperms contains two sperm, one of which fertilizes the egg cell in the female gametophyte. The second unites with two other nuclei in the female gametophyte, producing cells that are triploid (having three sets of chromosomes) and that develop into the endosperm. In some species of angiosperms, the endosperm is absorbed by the embryo before germination, while in others it is consumed during germination. Embyros that lack an endosperm (such as peas and beans) have absorbed most of their food storage tissues before becoming dormant and develop large, fleshy cotyledons.

endospore (ĕn′də-spôr′) A rounded, inactive form that certain bacteria assume under conditions of extreme temperature, dryness, or lack of food. The bacterium develops a waterproof cell wall that protects it from being dried out or damaged. Bacteria have been known to remain dormant but alive in the form of endospores for long periods of time, even thousands of years. Also called *endosporium.*

endosporium (ĕn′də-spôr′ē-əm) Plural **endosporia.** See **endospore.**

endothelium (ĕn′dō-thē′lē-əm) Plural **endothelia.** A thin layer of flat epithelial cells that lines the lymph vessels, blood vessels, and the inner cavities of the heart. Compare **mesothelium.**

endotherm (ĕn′də-thûrm′) A warm-blooded organism. Also called *homeotherm.*

endothermic (ĕn′dō-thûr′mĭk) **1.** Relating to a chemical reaction that absorbs heat. Compare **exothermic. 2.** Warm-blooded.

-ene A suffix used to form the names of hydrocarbons having one or more double bonds, such as *benzene.*

energy (ĕn′ər-jē) The capacity or power to do work, such as the capacity to move an object (of a given mass) by the application of force. Energy can exist in a variety of forms, such as electrical, mechanical, chemical, thermal, or nuclear, and can be transformed from one form to another. It is measured by the amount of work done, usually in joules or watts. See also **conservation of energy, kinetic energy, potential energy.** Compare **power, work.**

energy density See under **density.**

energy level One of a set of states of a physical system associated with a range of energies. Electrons in an atom, for example, can shift between the different energy levels corresponding to orbitals in different shells. Also called *energy state.*

engine (ĕn′jĭn) A machine that turns energy into mechanical force or motion, especially one that gets its energy from a source of heat, such as the burning of a fuel. The efficiency of an engine is the ratio between the kinetic energy produced by the machine and the energy needed to produce it. See more at **internal-combustion engine, steam engine.** See also **motor.**

engineering (ĕn′jə-nîr′ĭng) The application of science to practical uses such as the design of structures, machines, and systems. Engineering has many specialities such as civil engineering, chemical engineering, and mechanical engineering.

englacial (ĕn-glā′shəl) Located or occurring within a glacier, as certain meltwater streams, till deposits, and moraines.

enstatite (ĕn′stə-tīt′) A glassy, usually yellowish gray orthorhombic variety of pyroxene. It is usually found in igneous rocks and meteorites. *Chemical formula:* $\mathbf{Mg_2Si_2O_6}$.

enterovirus (ĕn′tə-rō-vī′rəs) Any of various viruses of the genus *Enterovirus* in the family Picornaviridae, including polioviruses, coxsackieviruses, and echoviruses. Enteroviruses affect the intestinal tract and also cause respiratory, neurologic and other infections.

enthalpy (ĕn′thăl′pē) A partial measure of the internal energy of a system. Enthalpy cannot be directly measured, but changes in it can be. If an outside pressure on a system is held constant, a change in enthalpy entails a change in the system's internal energy, plus a change in the system's volume (meaning the system exchanges energy with the outside world). For example, in **endothermic** chemical reactions, the change in enthalpy is the amount of energy absorbed by the reaction; in **exothermic** reactions, it is the amount given off. See also **thermodynamics.**

entomology (ĕn′tə-mŏl′ə-jē) The scientific study of insects.

WORD HISTORY **entomology**

Scientists who study insects (there are close to a million that can be studied!) are called entomologists. Why are they not called "insectologists"? Well, in a way they are. The word *insect* comes from the Latin word *insectum,* meaning "cut up or divided into segments." (The plural of *insectum,* namely *insecta,* is used by scientists as the name of the taxonomic class that insects belong to.) This Latin word was created in order to translate the Greek word for "insect," which is *entomon.* This Greek word also literally means "cut up or divided into segments," and it is the source of the word *entomology.* The Greeks had coined this term for insects because of the clear division of insect bodies into three segments, now called the head, thorax, and abdomen.

entomophilous (ĕn′tə-mŏf′ə-ləs) Pollinated by insects.

entropy (ĕn′trə-pē) A measure of the amount of energy in a physical system not available to do work. As a physical system becomes more disordered, and its energy becomes more evenly distributed, that energy becomes less able to do work. For example, a car rolling along a road has kinetic energy that could do work (by carrying or colliding with something, for example); as friction slows it down and its energy is distributed to its surroundings as heat, it loses this ability. The amount of entropy is often thought of as the amount of disorder in a system. See also **heat death.**

environment (ĕn-vī′rən-mənt) All of the biotic and abiotic factors that act on an organism, population, or ecological community and influence its survival and development. Biotic factors include the organisms themselves, their food, and their interactions. Abiotic factors include such items as sunlight, soil, air, water, climate, and pollution. Organisms respond to changes in their environment by evolutionary adaptations in form and behavior.

enzyme (ĕn′zīm) Any of numerous proteins produced in living cells that accelerate or catalyze the metabolic processes of an organism. Enzymes are usually very selective in the molecules that they act upon, called **substrates,** often reacting with only a single substrate. The substrate binds to the enzyme at a location called the active site just before the reaction catalyzed by the enzyme takes place. Enzymes can speed up chemical reactions by up to a millionfold, but only function within a narrow temperature and pH range, outside of which they can lose their structure and become **denatured.** Enzymes are involved in such processes as the breaking down of the large protein, starch, and fat molecules in food into smaller molecules during digestion, the joining together of nucleotides into strands of DNA, and the addition of a phosphate group to ADP to form ATP. The names of enzymes usually end in the suffix *-ase*. Common enzymes include protease, amylase, and DNA polymerase.

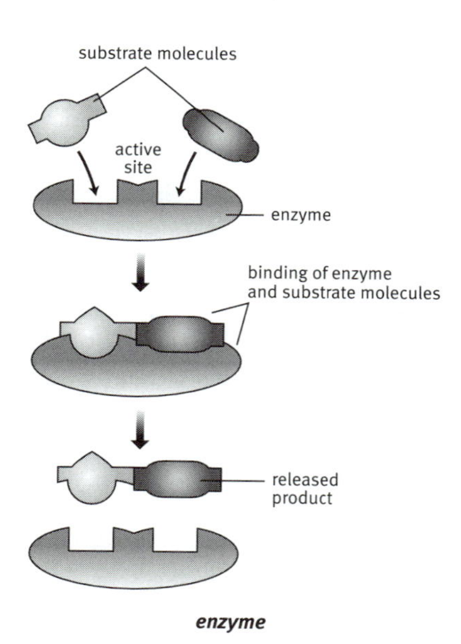

enzyme

Eocene (ē′ə-sēn′) The second epoch of the Tertiary Period, from about 58 to 37 million years ago. During the earliest part of this epoch, land connections existed between Antarctica and Australia, between Europe and North America, and between North America and Asia, and the climate was warm. The land connection between Antarctica and Australia disappeared in the mid-Eocene and early Oligocene, resulting in a change in the predominant oceanic currents and a cooler climate. With this change, the average size of mammals changed from less than 10 kg (22 lbs) to more than 10 kg. The Himalayas also formed during the Eocene, and most modern orders of mammals appeared. See Chart at **geologic time.**

eohippus (ē′ō-hĭp′əs) See **hyracotherium.**

eolian also **aeolian** (ē-ō′lē-ən) Relating to, caused by, or carried by the wind. Loess is an eolian deposit.

eon (ē′ŏn′) The longest division of **geologic time,** containing two or more eras.

epeirogeny (ĕp′ī-rŏj′ə-nē) Uplift or depression of the Earth's crust, affecting large areas of land or ocean bottom. Basins and plateaus are often formed as the result of epeirogeny. Epeirogeny differs from **orogeny** in that it affects larger regions of the Earth's crust and is not as frequently associated with folding and faulting of rocks.

ephedrine (ĭ-fĕd′rĭn, ĕf′ĭ-drēn′) A white, odorless, powdered or crystalline alkaloid isolated from shrubs of the genus *Ephedra* or made synthetically. It is used in the treatment of allergies and asthma. *Chemical formula:* $C_{10}H_{15}NO$.

ephemeris (ĭ-fĕm′ər-ĭs) Plural **ephemerides** (ĕf′ə-mĕr′ə-dēz′). A table giving the coordinates of a celestial body at specific times during a given period. Ephemerides can be used by navigators to determine their longitude while at sea and by astronomers in following objects such as comets. The use of computers has allowed modern ephemerides to determine celestial positions with far greater accuracy than in earlier publications.

ephemeris time A highly accurate astronomical system for the measurement of time based in theory on the period of Earth's orbit, but in practice relying on lunar observations and an accurate lunar ephemeris so as to not be subject to the irregularities of Earth's rotation. The year 1900 CE is considered a standard year, containing 31,556,925.9747 seconds of ephemeris time.

epicanthic fold (ĕp′ĭ-kăn′thĭk) or **epicanthal fold** (ĕp′ĭ-kăn′thəl) A fold of skin of the upper eyelid that partially covers the inner corner of the eye.

epicarp (ĕp′ĭ-kärp′) See **exocarp.**

epicenter (ĕp′ĭ-sĕn′tər) The point on the Earth's surface that is directly above the focus (the point of origin) of an earthquake. The epicenter is usually the location where the greatest damage associated with an earthquake occurs. See Note at **earthquake.**

epicotyl (ĕp′ĭ-kŏt′l) The stem of a seedling or embryo located between the cotyledons and the first true leaves.

epicycle (ĕp′ĭ-sī′kəl) **1.** In Ptolemaic cosmology, a small circle representing a temporary adjustment to the position of a planet as it orbits the Earth. The five known planets, along with the Sun and Moon, were conceived as moving through the sky in large circular paths with the Earth at their center. As a planet moved along its path, it occasionally departed from its regular motion to follow a much smaller circle centered on the orbital path itself. These smaller circles, or epicycles, were necessary to reconcile the observed motions of the planets with a geocentric model of the universe. The epicycles of the inferior planets Mercury and Venus were fixed to the orbit of the Sun and explained why those planets were never observed far from it in the sky. The epicycles of the superior planets Mars, Jupiter, and Saturn explained why those bodies were sometimes observed to move backward in their orbits, a phenomenon known as retrograde motion and explained in a heliocentric model by the differing orbital velocities of the Earth and the planet being observed. See illustration at **Ptolemaic system. 2.** A circle whose circumference rolls along the circumference of a fixed circle, thereby generating an epicycloid or a hypocycloid.

epicycloid (ĕp′ĭ-sī′kloid′) The curve described by a point on the circumference of a circle as the circle rolls on the outside of the circumference of a second, fixed circle.

epidemic (ĕp′ĭ-dĕm′ĭk) An outbreak of a disease or illness that spreads rapidly among individuals in an area or population at the same time. See also **endemic, pandemic.**

epidemiology (ĕp′ĭ-dē′mē-ŏl′ə-jē) The scientific study of the causes, distribution, and control of disease in populations.

epidermis (ĕp′ĭ-dûr′mĭs) **1.** The protective outer layer of the skin. In invertebrate animals, the epidermis is made up of a single layer of cells. In vertebrates, it is made up of many layers of cells and overlies the dermis. Hair and feathers grow from the epidermis. **2.** The outer layer of cells of the stems, roots, and leaves of plants. In most plants, the epidermis is a single layer of cells set close together to protect the plant from water loss, invasion by fungi, and physical damage. The epidermis that is exposed to air is covered with a protective substance called **cuticle.** See more at **photosynthesis.**

epidote (ĕp′ĭ-dōt′) A yellowish-green or blackish-green monoclinic mineral. Epidote occurs as formless grains or as prism-shaped crystals, and is found in limestones that have undergone slight metamorphism or in igneous rocks. *Chemical formula:* $\mathbf{Ca_2(Al,Fe)_3(SiO_4)_3OH}$.

epidural (ĕp′ĭ-do͝or′əl) *Adjective.* **1.** Located on or over the dura mater. —*Noun.* **2.** An injection into the epidural space of the spine, as an epidural anesthetic.

epifauna (ĕp′ə-fô′nə) Benthic animals that live on the surface of a substrate, such as rocks, pilings, marine vegetation, or the sea or lake floor itself. Epifauna may attach themselves to such surfaces or range freely over them, as by crawling or swimming. Mussels, crabs, starfish, and flounder are epifaunal animals. Compare **infauna.**

epigeous (ĕp′ə-jē′əs) Relating to the germination of a seed in which the cotyledons emerge above the surface of the ground. Compare **hypogeous.**

epiglottis (ĕp′ĭ-glŏt′ĭs) A thin, triangular plate of cartilage at the base of the tongue that covers the glottis during swallowing to keep food from entering the trachea.

epigynous (ĭ-pĭj′ə-nəs) Having floral parts (such as the petals and stamens) attached to or near the upper part of the ovary, as in the flower of the apple, cucumber, or daffodil. Compare **hypogynous, perigynous.**

epilepsy (ĕp′ə-lĕp′sē) Any of various neurological disorders characterized by recurrent seizures. Epilepsy is caused by abnormal electrical activity in the brain.

epinephrine (ĕp′ə-nĕf′rĭn) A hormone that is secreted by the adrenal gland in response to physical or mental stress, as from fear, and is regulated by the autonomic nervous system. The release of epinephrine causes an increase in heart rate, blood pressure, and respiratory rate. Epinephrine also raises glucose levels in the blood for use as fuel when more alertness or greater physical effort is needed. Also called *adrenaline. Chemical formula:* $C_9H_{13}NO_3$.

epipelagic zone (ĕp′ə-pə-lăj′ĭk) The uppermost part of the oceanic zone, lying above the **mesopelagic zone**, that receives enough sunlight to allow photosynthesis. The epipelagic zone can reach depths of about 200 m (656 ft) in tropical and subtropical latitudes and about 100 m (328 ft) in higher latitudes or where upwellings or other conditions cause turbidity. In general, the clearer the water, the deeper the epipelagic layer at that location. The epipelagic zone is more abundant in marine life than the lower zones; in particular, phytoplankton cannot live at any lower depths. ► The epipelagic zone is divided into the **neritic epipelagic zone**, which lies over the continental shelf, and the **oceanic epipelagic zone**, which is the upper layer of water in the part of the open ocean that is not over the continental shelf.

epiphyte (ĕp′ə-fīt′) A plant that grows on another plant and depends on it for support but not food. Epiphytes get moisture and nutrients from the air or from small pools of water that collect on the host plant. Spanish moss and many orchids are epiphytes. Also called *aerophyte, air plant. —Adjective* **epiphytic** (ĕp′ə-fĭt′ĭk)

epiphyte

close-up of an ant plant (Myrmecodia beccarii)

epiphytotic (ĕp′ə-fī-tŏt′ĭk) Relating to or characterized by a sudden or abnormally destructive outbreak of a plant disease, usually over an extended geographic area.

episiotomy (ĭ-pĭz′ē-ŏt′ə-mē, ĭ-pē′zē-) Surgical incision of the perineum during vaginal childbirth to facilitate delivery.

epithelium (ĕp′ə-thē′lē-əm) Plural **epithelia.** The thin, membranous tissue that lines most of the internal and external surfaces of an animal's body. Epithelium is composed of one or more layers of densely packed cells. In vertebrates, it lines the outer layer of the skin (epidermis), the surface of most body cavities, and the lumen of fluid-filled organs, such as the gut or intestine. *—Adjective* **epithelial.**

epizoic (ĕp′ĭ-zō′ĭk) Living or growing on the external surface of an animal.

epizootic (ĕp′ĭ-zō-ŏt′ĭk) *Adjective.* **1.** Relating to a rapidly spreading disease that affects a large number of animals at the same time within a particular area. *—Noun.* **2.** An epizootic disease.

epoch (ĕp′ək, ē′pŏk′) The shortest division of **geologic time**. An epoch is a subdivision of a period.

epoetin alfa (ĭ-pō′ĭ-tĭn ăl′fə) A synthetic preparation of human erythropoietin, manufactured using **recombinant DNA** technology, and used to treat some forms of anemia.

epoxide (ĕ-pŏk′sīd) A ring-shaped compound consisting of an oxygen atom bonded to two other atoms, usually of carbon, that are already bonded to each other. Epoxides are used to make epoxies.

epoxy (ĭ-pŏk′sē) Any of various artificial resins made of chains of epoxide rings. Epoxies are tough, very adhesive, and resistant to chemicals. They are used to make protective coatings and glues. Also called *epoxy resin.*

Epsom salts (ĕp′səm) A bitter, colorless, crystalline salt, used in making textiles, in fertilizers, for medical purposes, and as an additive to bath water to soothe the skin. *Chemical formula:* **$MgSO_4 \cdot 7H_2O$.**

Epstein-Barr virus (ĕp′stīn-bär′) A virus of the family Herpesviridae and the genus *Lymphocryptovirus* that causes infectious mononucleosis and is associated with several types of human cancers. It is named after two of its discoverers, British pathologist Michael Anthony Epstein (born 1921) and British virologist Yvonne M. Barr (born 1932).

equalizer (ē′kwə-lī′zər) An electronic device made of filters and amplifiers, used to alter the relative strengths of different frequencies in an electronic signal. Equalizers are used primarily in audio equipment, allowing fine-tuning of the signal to compensate for distortions such as weak response or oversensitivity at various frequencies. ► A **graphic equalizer** uses a set of controls that determine the level of boost or suppression of individual frequencies. The controls are usually sliding faders, set up in a row from lowest frequency to highest frequency, so that the final settings resemble a graph of the frequency response of the equalizer. ► A **parametric equalizer** consists of one or more filters whose characteristics can be controlled, such as the frequency to be manipulated, whether to boost or suppress the frequency, the amount of boost or suppression, and how much nearby frequencies are also affected.

equation (ĭ-kwā′zhən) **1.** *Mathematics.* A written statement indicating the equality of two expressions. It consists of a sequence of symbols that is split into left and right sides joined by an equal sign. For example, $2 + 3 + 5 = 10$ is an equation. **2.** *Chemistry.* A written representation of a chemical reaction, in which the symbols and amounts of the reactants are separated from those of the products by an equal sign, arrow, or a set of opposing arrows. For example, $Ca(OH)_2 + H_2SO_4 = CaSO_4 + 2H_2O$, is an equation.

equator (ĭ-kwā′tər) **1.** An imaginary line forming a great circle around the Earth's surface, equidistant from the poles and in a plane perpendicular to the Earth's axis of rotation. It divides the Earth into the Northern and Southern hemispheres and is the basis from which latitude is measured. **2.** A similar circle on the surface of any celestial body. **3.** The celestial equator.

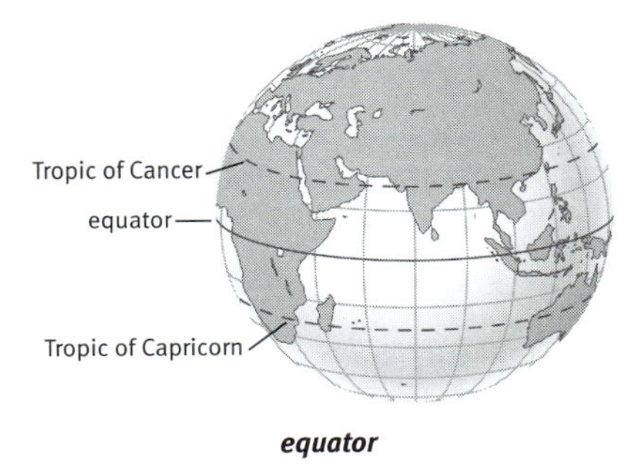

equator

equatorial coordinate system (ē′kwə-tôr′ē-əl, ĕk′wə-) The coordinate system in which a celestial object's position on the celestial sphere is described in terms of its **declination** and **right ascension**, measured with respect to the **celestial equator**. Declination and right ascension correspond directly to geographic latitude and longitude as projected outward onto the celestial sphere. Declination is measured in degrees north or south of the celestial equator, the same as geographic latitude, but right ascension is measured in hours, minutes, and seconds eastward along the celestial equator from the point of the **vernal equinox**. Because the celestial equator moves among the fixed stars with the precession of the Earth's poles, an object's declination and right ascension change gradually over time, and coordinates in the equatorial system must be specified for particular years. The equatorial system is the system most used in astronomy for describing the position of objects outside the solar system. Compare **altazimuth coordinate system, ecliptic coordinate system.**

equatorial current Either of two oceanic currents (the **North** and **South Equatorial Currents**) flowing westward on either side of the equator. The equatorial currents are driven primarily by the easterly trade winds and are separated by a narrower countercurrent flowing eastward along the equator itself. Equatorial currents are relatively shallow, involving the epipelagic zone and part of the mesopelagic zone to depths of less than 500 m (1,640 ft). They are deflected by the continental landmasses in their path, usually splitting into northward and southward flows that feed other ocean currents and form part of the oceanic gyres (large-scale spiral current systems). See more at **gyre.**

equatorial mount A mounting for astronomical telescopes having two axes, one of which revolves about an axis that is parallel to the axis of the Earth's rotation. This axis can be driven by a motor at a rate equal to the diurnal motion of the celestial object being viewed, allowing the telescope to keep the object in view for long periods without adjustment. The other axis, perpendicular to the first, has free movement that allows the telescope to be pointed anywhere in the sky. Compare **altazimuth mount.**

equi– A prefix that means "equal" or "equally," as in *equidistant.*

equiangular (ē′kwē-ăng**′**gyə-lər, ĕk′wē-) Having all angles equal.

equid (ĕk**′**wĭd, ē**′**kwĭd) Any of various hoofed mammals of the family Equidae, which includes horses, donkeys, and zebras. Equids have muscular bodies with long, slender legs adapted for running and a single hoofed digit at the end of each limb.

equidistant (ē′kwĭ-dĭs**′**tənt) Equally distant.

equilateral (ē′kwə-lăt**′**ər-əl) Having all sides of equal length, as a triangle that is neither scalene nor isosceles.

equilibrium (ē′kwə-lĭb**′**rē-əm) Plural **equilibriums** or **equilibria. 1.** *Physics.* The state of a body or physical system that is at rest or in constant and unchanging motion. A system that is in equilibrium shows no tendency to alter over time. ▸ If a system is in **static equilibrium**, there are no net forces and no net torque in the system. ▸ If a system is in **stable equilibrium**, small disturbances to the system cause only a temporary change before it returns to its original state. **2.** *Chemistry.* The state of a reversible chemical reaction in which its forward and reverse reactions occur at equal rates so that the concentration of the reactants and products remains the same.

equine (ē**′**kwīn′, ĕk**′**wīn′) Characteristic of or resembling horses or related animals, such as donkeys.

equinoctial (ē′kwə-nŏk**′**shəl, ĕk′wə-) *Adjective.* **1.** Relating to an equinox. **2.** Relating to the celestial equator. —*Noun.* **3.** See **celestial equator.**

equinoctial circle See **celestial equator.**

equinox (ē**′**kwə-nŏks′) **1.** Either of the two points on the celestial sphere where the **ecliptic** (the apparent path of the Sun) crosses the **celestial equator.** ▸ The point at which the Sun's path crosses the celestial equator moving from south to north is called the **vernal equinox**. The vernal equinox marks the zero point in both the equatorial and ecliptic coordinate systems; horizontal angular distances (right ascension in the equatorial system and celestial longitude in the ecliptic system) are measured eastward from this point. The vernal equinox is also known as *the first point of Aries* because when first devised some 2,000 years ago this point occurred at the beginning of Aries in the zodiac. Because of the westward precession of the equinoxes, the vernal equinox is now located at the beginning of Pisces. ▸ The point at which the Sun's path crosses the celestial equator moving from north to south is called the **autumnal equinox**. **2.** Either of the two corresponding moments of the year when the Sun is directly above the Earth's equator. The vernal equinox occurs on March 20 or 21 and the autumnal equinox on September 22 or 23, marking the beginning of spring and autumn, respectively, in the Northern Hemisphere (and the reverse in the Southern Hemisphere). The days on which an equinox falls have about equal periods of sunlight and darkness. Compare **solstice.**

equipotential (ē′kwə-pə-tĕn**′**shəl, ĕk′wə-) A surface within a region containing a potential (typically electric potential), such that all the points on the surface have equal potential.

equivalence principle (ĭ-kwĭv**′**ə-ləns) A principle central to General Relativity stating that a gravitational field is locally indistinguishable from the effects of inertial forces. For example, according to the equivalence principle, it is impossible for someone in a box who experiences a force pushing him to the bottom of the box to know, from the force alone, whether that force is the result of a gravitational field (the box is standing on the surface of a planet) or an acceleration (the box is being pushed by a rocket).

equivalent (ĭ-kwĭv′ə-lənt) **1.** Equal, as in value, meaning, or force. **2a.** Of or relating to a relation between two elements that is reflexive, symmetric, and transitive. **b.** Having a one-to-one correspondence, as between parts. Two triangles having the same area are equivalent, as are two congruent geometric figures.

Er The symbol for **erbium.**

era (îr′ə) A division of **geologic time**, longer than a period and shorter than an eon.

Eratosthenes (ĕr′ə-tŏs′thə-nēz′) 276?–194 BCE. Greek mathematician and astronomer who is best known for making an accurate estimate of the circumference of the Earth by measuring the angle of the Sun's rays at two different locations at the same time. He also invented a method for listing the prime numbers that are less than any given number.

BIOGRAPHY **Eratosthenes**

Had he been born and raised farther north or south, the ancient astronomer Eratosthenes might never have come to think about the circumference of the Earth. It so happened that in his hometown of Syene (now Aswan), Egypt—which lies just north of the Tropic of Cancer—the Sun's rays were almost exactly perpendicular to the ground at noon on the summer solstice. One year in Alexandria, about 500 miles away, he noticed that the Sun's rays hit the ground at a deviation of about 7 degrees from the vertical on the same date and time. He believed, correctly, that the Sun was very distant and that its rays were essentially parallel when they hit the Earth. Therefore, he reasoned that the difference between the angles of incidence in Syene and Alexandria could only be due to curvature of the Earth's surface. Fairly basic geometry allowed him to use these figures to calculate the circumference of the Earth. Although one usually reads that his calculation was very close to that of modern scientists, we do not know this for certain; the units of length that he used (called stadia) were not fixed throughout the ancient world, and there is no record of precisely which length he had in mind. His measurement could have been anywhere from 0.5 to 17 percent off from modern measurements. Whatever the case, Eratosthenes's calculations were remarkably good.

erbium (ûr′bē-əm) *Symbol* **Er** A soft, silvery, metallic element of the lanthanide series. It is used as a neutron absorber in nuclear technology and in light amplification for fiber-optic telecommunications. Atomic number 68; atomic weight 167.26; melting point 1,497°C; boiling point 2,900°C; specific gravity 9.051; valence 3. See **Periodic Table.**

E region The region of the ionosphere that lies approximately 90 to 150 km (56 to 93 mi) above the Earth's surface. This region may affect long-distance communications because it influences radio waves in the 1-3 megahertz range. Also called the *E layer, Heaviside layer.* Compare **D region, F region.**

erg[1] (ûrg) The unit of energy or work in the centimeter-gram-second system, equal to the force of one dyne over a distance of one centimeter. This unit has been mostly replaced by the joule.

erg[2] (ûrg) An extensive area of desert covered with shifting sand dunes.

ergocalciferol (ûr′gō-kăl-sĭf′ə-rôl′, -rōl′) See **vitamin D_2.**

ergonomics (ûr′gə-nŏm′ĭks) The scientific study of equipment design, as in office furniture or transportation seating, to improve efficiency, comfort, or safety.

ergot (ûr′gət) A fungus *(Claviceps purpurea)* that infects rye as well as other cereal grasses fed to livestock. Ergot forms sclerotia (masses of hyphae) that replace individual seeds in the spike of the infected plant and contain a complex mixture of alkaloids, several of which are medicinally important. Ergot is the basic source of ergotamine and lysergic acid. Ingestion of infected rye produces convulsions, hallucinations, and severe vasoconstriction that can lead to gangrene. Ergot poisoning may have been responsible for outbreaks of mass hysteria and reports of demonic visions in medieval Europe.

ergotamine (ûr-gŏt′ə-mēn′, -mĭn) A crystalline alkaloid derived from ergot that induces vasoconstriction and is used especially in the treatment of migraine headaches. *Chemical formula:* $C_{33}H_{35}N_5O_5$.

Eris (îr′ĭs, ĕr′-) A dwarf planet with a diameter of about 2,400 kilometers (1,500 miles). When Eris was discovered in 2005, it and its satellite Dysnomia were the most distant known celestial bodies in the Solar System.

erosion (ĭ-rō′zhən) The gradual wearing away of land surface materials, especially rocks, sediments, and soils, by the action of water, wind, or a glacier. Usually erosion also involves the transport of eroded material from one place to another, as from the top of a mountain to an adjacent valley, or from the upstream portion of a river to the downstream portion.

erosion

eruption (ĭ-rŭp′shən) The release of gas, ash, molten materials, or hot water into the atmosphere or onto the Earth's surface from a volcano or other opening in the Earth's surface.

erythrocyte (ĭ-rĭth′rə-sīt′) See **red blood cell.**

erythromycin (ĭ-rĭth′rə-mī′sĭn) An antibiotic obtained from the bacteria *Streptomyces erythreus,* effective against many gram-positive bacteria and some gram-negative bacteria.

erythropoietin (ĭ-rĭth′rō-poi-ē′tĭn) A glycoprotein hormone, secreted mostly by the kidneys in adults and the liver in children, that stimulates stem cells in the bone marrow to produce red blood cells.

Es The symbol for **einsteinium.**

Esaki (ĭ-sä′kē), **Leo** Born 1925. Japanese physicist who developed a very fast and very small type of diode that is now widely used in many electronic devices, especially computers and microwave devices.

escape velocity (ĭ-skāp′) The velocity needed for a celestial body to overcome the gravitational pull of another, larger body and not fall back to that body's surface. Escape velocity is determined by the mass of the larger body and by the distance of the smaller body from the larger one's center. Depending on its initial trajectory, a smaller body traveling at the escape velocity will either enter a periodic **orbit** around the larger body or recede from the surface of the larger body indefinitely. The escape velocity at the Earth's surface is about 11.2 kilometers per second (25,000 miles per hour); the escape velocity on the Moon's surface is 2.4 kilometers per second (5,300 miles per hour). The escape velocity within the event horizon of a black hole is higher than the speed of light; since nothing can exceed the speed of light, nothing—even light—can escape from within the event horizon of a black hole.

escarpment (ĭ-skärp′mənt) A steep slope or long cliff formed by erosion or by vertical movement of the Earth's crust along a fault. Escarpments separate two relatively level areas of land. The term is often used interchangeably with *scarp* but is more accurately associated with cliffs produced by erosional processes rather than those produced by faulting.

esker (ĕs′kər) A long, narrow, steep-sided ridge of coarse sand and gravel deposited by a stream flowing in or under a melting sheet of glacial ice. Eskers range in height from 3 m (9.8 ft) to more than 200 m (656 ft) and in length from less than 100 m (328 ft) to more than 500 km (310 mi).

esophagus (ĭ-sŏf′ə-gəs) Plural **esophagi** (ĭ-sŏf′ə-jī′, -gī′). The muscular tube in vertebrates through which food passes from the pharynx to the stomach.

Espy (ĕs′pē), **James Pollard** 1785–1860. American meteorologist who is credited with the first correct explanation of the role heat plays in cloud formation and growth. His use of the telegraph in relaying meteorological observations and tracking storms laid the foundation for modern weather forecasting.

essential oil (ĭ-sĕn′shəl) Any of various volatile liquids, such as rose oil or lavender oil, that have a characteristic odor and are produced by plants. Essential oils are composed primarily of terpenes and of lesser quantities of alcohols, aldehydes, esters, phenols, and other compounds that impart particular odors or flavors. They are used to make perfumes, soaps, flavorings, and other products.

ester (ĕs′tər) An organic compound formed when an acid and an alcohol combine and release water. Esters formed from carboxylic acids are the most common, and have the general formula RCOOR′, where R and R′ are organic radicals. Esters formed from simple hydrocarbon groups are colorless, volatile liquids with pleasant aromas and create the fragrances and flavors of many flowers and fruits. They are also used as food flavorings. Larger esters, formed from long-chain carboxylic acids, commonly occur as animal and vegetable fats, oils, and waxes. Esters have a wide range of uses in industry.

estivation (ĕs′tə-vā′shən) An inactive state resembling deep sleep, in which some animals living in hot climates, such as certain snails, pass the summer. Estivation protects these animals against heat and dryness. Compare **hibernation.**

estrogen (ĕs′trə-jən) Any of a group of steroid hormones that primarily regulate the growth, development, and function of the female reproductive system. The main sources of estrogen in the body are the ovaries and the placenta. Estrogen-like compounds are also formed by certain plants.

estrogen replacement therapy The therapeutic administration of estrogen to postmenopausal women in order to reduce symptoms and signs of estrogen deficiency, such as hot flashes and osteoporosis.

estrous cycle (ĕs′trəs) The series of changes that occur in the female of most mammals from one period of estrus to another. The estrous cycle usually takes place during the period known as the breeding season, which ensures that young are born at a time when the chance of survival is greatest. The length of the cycle varies from species to species.

estrus (ĕs′trəs) A regularly recurring period in female mammals other than humans during which the animal is sexually receptive. Estrus occurs around the time of ovulation. Also called *heat.*

estuary (ĕs′cho͞o-ĕr′ē) **1.** The wide lower course of a river where it flows into the sea. Estuaries experience tidal flows and their water is a changing mixture of fresh and salt. **2.** An arm of the sea that extends inland to meet the mouth of a river.

eta-c particle (ā′tə-sē′, ē′tə-) An electrically neutral meson having a mass 5,832 times that of the electron and a mean lifetime of approximately 3.1×10^{-22} seconds. See Table at **subatomic particle.**

eta particle (ā′tə, ē′tə) An electrically neutral meson having a mass 1,071 times that of the electron and a mean lifetime of approximately 3.5×10^{-8} seconds. See Table at **subatomic particle.**

ethane (ĕth′ān′) A colorless, odorless, flammable gas occurring in natural gas. It is used as a fuel and in refrigeration. Ethane is the second member of the alkane series. *Chemical formula:* C_2H_6.

ethanethiol (ĕth′ə-nĕth′ē-ôl′, -ōl′) See **ethyl mercaptan.**

ethanol (ĕth′ə-nôl′) An alcohol obtained from the fermentation of sugars and starches or by chemical synthesis. It is the intoxicating ingredient of alcoholic beverages, and is also used as a solvent, in explosives, and as an additive to or replacement for petroleum-based fuels. Also called *ethyl alcohol, grain alcohol. Chemical formula:* C_2H_6O.

ethanolamine (ĕth′ə-nŏl′ə-mēn′, -nō′lə-) A colorless liquid used in the purification of petroleum, as a solvent in dry cleaning, and as an ingredient in paints and pharmaceuticals. *Chemical formula:* C_2H_7NO.

ethene (ĕth′ēn′) See **ethylene.**

ether (ē′thər) **1.** An organic compound in which two hydrocarbon groups are linked by an oxygen atom, having the general structure ROR′, where R and R′ are the two hydrocarbon groups. At room temperature, ethers are pleasant-smelling liquids resembling alcohols but less dense and less soluble in water. Ethers are part of many naturally occurring organic compounds, such as starches and sugars, and are widely used in industry and in making pharmaceuticals. **2.** A colorless, flammable liquid used as a solvent and formerly used as an anesthetic. Ether consists of two ethyl groups joined by an oxygen atom. Also called *diethyl ether, ethyl ether. Chemical formula:* $\mathbf{C_4H_{10}O}$. **3.** A hypothetical medium formerly believed to permeate all space, and through which light and other electromagnetic radiation were thought to move. The existence of ether was disproved by the American physicists Albert Michelson and Edward Morley in 1887.

ethnography (ĕth-nŏg′rə-fē) The branch of anthropology that deals with the scientific description of specific human cultures.

ethology (ĭ-thŏl′ə-jē, ē-thŏl′-) The scientific study of animal behavior, especially as it occurs in a natural environment.

ethyl (ĕth′əl) The radical C_2H_5, derived from ethane.

ethyl acetate A colorless, volatile, flammable liquid used as a solvent and in perfumes, lacquers, pharmaceuticals, and rayon. *Chemical formula:* $\mathbf{C_4H_8O_2}$.

ethyl alcohol See **ethanol.**

ethyl carbamate (kär′bə-māt′) See **urethane.**

ethylene (ĕth′ə-lēn′) A colorless, flammable gas that occurs naturally in certain plants and can be obtained from petroleum and natural gas. As a plant hormone, it ripens and colors fruit, and it is manufactured for use in agriculture to speed these processes. It is also used as a fuel and in making plastics. Ethylene is the simplest alkene, consisting of two carbon atoms joined by a double bond and each attached to two hydrogen atoms. Also called *ethene. Chemical formula:* $\mathbf{C_2H_4}$.

ethylene glycol (glī′kôl′) A poisonous, syrupy, colorless alcohol used as an antifreeze in heating and cooling systems that use water. Ethylene glycol is chemically like ethanol but has two hydroxyl (OH) groups instead of one. Also called *glycol. Chemical formula:* $\mathbf{C_2H_6O_2}$.

ethylene series The alkene series. See under **alkene.**

ethyl ether See **ether** (sense 2).

ethyl mercaptan A colorless organic liquid that has a very strong odor. It is added to odorless fuel, such as natural gas, and fuel systems as a warning agent in the event of leakage or spills. Also called *ethanethiol. Chemical formula:* $\mathbf{C_2H_6S}$.

ethyne (ĕth′īn′) See **acetylene.**

etiolation (ē′tē-ə-lā′shən) A pathological condition of plants that grow in places that provide insufficient light, as under stones. It is characterized by elongated stems and pale color due to lack of chlorophyll. —*Verb* **etiolate.**

etiology (ē′tē-ŏl′ə-jē) The cause or origin of a disease, condition, or constellation of symptoms or signs, as determined by medical diagnosis or research.

Eu The symbol for **europium.**

Euclid (yo͞o′klĭd) fl. 300 BCE. Greek mathematician whose book, *Elements,* was used continuously until the 19th century. In it he organized and systematized all that was known about geometry. Euclid's systematic use of deductions and axioms was widely regarded as a model working method and influenced mathematicians and scientists for over two thousand years.

Euclidean (yo͞o-klĭd′ē-ən) Relating to geometry of plane figures based on the five postulates (axioms) of Euclid, involving the derivation of theorems from those postulates. The five postulates are: **1.** Any two points can be joined by a straight line. **2.** Any straight line segment can be extended indefinitely in a straight line. **3.** Given any straight line segment, a circle can be drawn having the line segment as radius and an endpoint as center. **4.** All right angles are congruent. **5.** (Also called the *parallel postulate.*) If two lines are drawn that intersect a third in such a way that the sum of inner angles on one side is less than the sum of two right triangles, then the two lines will intersect each other on that side if the lines are extended far enough. Compare **non-Euclidean.**

eudicotyledon (yo͞o′dī-kŏt′l-ēd′n) or **eudicot** (yo͞o-dī′kŏt′) An angiosperm having two cotyledons in the seed, leaves with a network of veins radiating from a central main vein, flower parts in multiples of four or five, and a ring of vascular cambium in the stem. In contrast to most monocotyledons, eudicotyledons undergo secondary growth. Eudicotyledons are the largest angiosperm group; they include most cultivated plants and many trees.

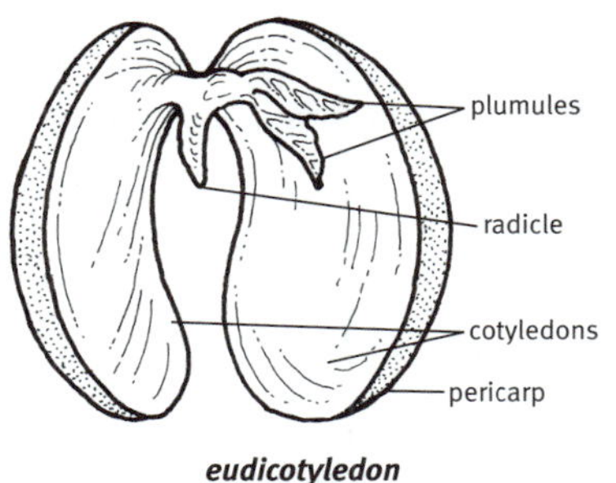

eudicotyledon

a bean seed

euglena (yo͞o-glē′nə) Any of various unicellular protist organisms of the genus *Euglena* that live in fresh water, have a cylindrical or sausage-like shape, and move by means of a flagellum. Euglenas contain chloroplasts and can produce their own food by photosynthesis. They can also absorb nutrients directly into the cell from the environment. Euglenas have no rigid covering or cell wall, such as the cellulose cell walls of green algae or plants, over the membrane enclosing the plasma of their cells. They also have a reddish, light-sensitive eyespot which helps them navigate in relation to light sources. In warm weather, euglenas multiply rapidly and form scum on the surfaces of bodies of water.

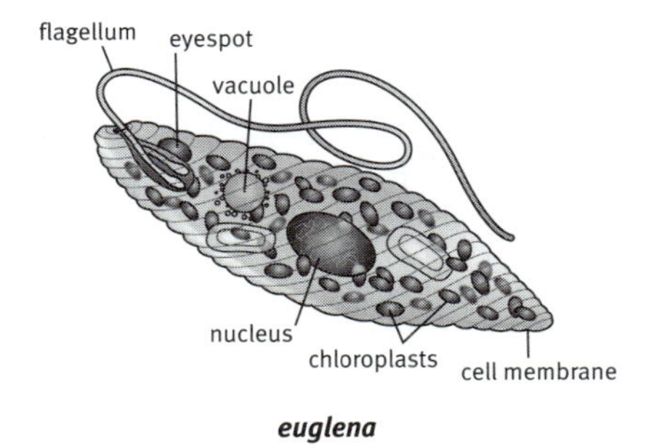

euglena

eukaryote (yo͞o-kăr′ē-ōt) An organism whose cells contain a nucleus surrounded by a membrane and whose DNA is bound together by proteins (histones) into chromosomes. The cells of eukaryotes also contain an endoplasmic reticulum and numerous specialized organelles not present in prokaryotes, especially mitochondria, Golgi bodies, and lysosomes. The organelles are enclosed in a three-part membrane (called a unit membrane) consisting of a lipid layer sandwiched between two protein layers. All organisms except for bacteria and archaea are eukaryotes. Compare **prokaryote.** —*Adjective* **eukaryotic**

Euler (oi′lər), **Leonhard** 1707–1783. Swiss mathematician who made many contributions to numerous areas of pure and applied mathematics, physics, and astronomy. He was one of the first to develop the methods used in differential and integral calculus, and he introduced much of the basic mathematical notation still used today.

euphotic (yo͞o-fŏt′ĭk) See **photic.**

Europa (yo͝o-rō′pə) One of the four brightest satellites of Jupiter and the sixth in distance from the planet. It was originally sighted by Galileo. See Note at **moon.**

europium (yo͝o-rō′pē-əm) *Symbol* **Eu** A very rare, silvery-white metallic element that is the softest member of the lanthanide series. It is used in making color television tubes and lasers and as a neutron absorber in nuclear research. Atomic number 63; atomic weight 151.96; melting point 826°C; boiling point 1,439°C; specific gravity 5.259; valence 2, 3. See **Periodic Table.**

eustachian tube (yo͞o-stā′shən) A slender tube that connects the middle ear with the upper part of the pharynx, serving to equalize air pressure on either side of the eardrum.

eustasy (yo͞o′stə-sē) A uniform worldwide change in sea level caused especially by fluc-

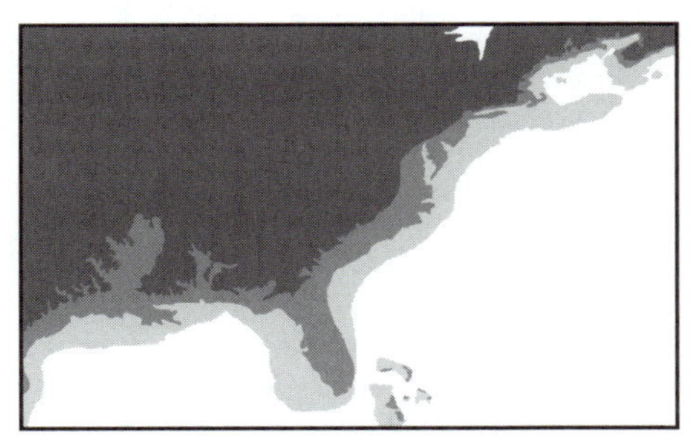

eustasy

The medium gray shading shows the modern-day coastlines of eastern North America and the Gulf of Mexico. The dark gray shading shows the same coastlines as they may have appeared during the warm climate interval of the Pliocene Epoch. The light gray shading shows the same coastlines as they may have appeared during the most recent ice age when much of the world's ocean water was locked up in continental ice sheets.

tuations in the amount of water taken up by continental and polar icecaps, or by a change in the capacity of ocean basins. — **eustatic** (yo͞o-stăt′ĭk)

eutectic (yo͞o-tĕk′tĭk) **1.** The proportion of constituents in an alloy or other mixture that yields the lowest possible complete melting point. In all other proportions, the mixture will not have a uniform melting point; some of the mixture will remain solid and some liquid. At the eutectic, the **solidus** and **liquidus** temperatures are the same. **2.** An alloy or other mixture with constituents in the proportions of the eutectic. **3.** The melting point of the eutectic.

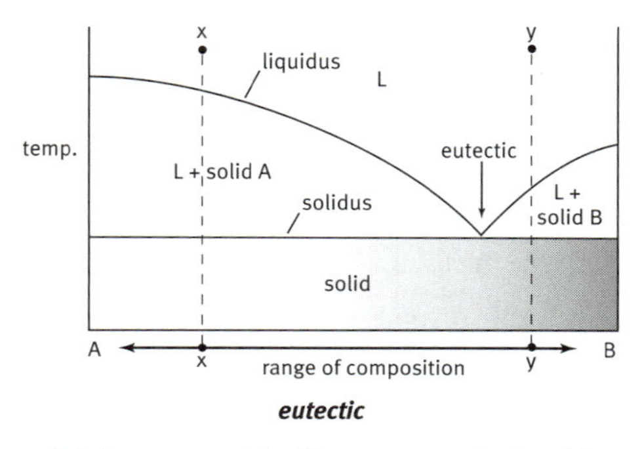

eutectic

Substance x *consists of two components,* A *and* B *(approximately 80%* A *and 20%* B*). Above the liquidus (the temperature at which the first solid begins to form) both components are liquid. As the temperature drops to the liquidus, component* A *starts to solidify, and the remaining liquid becomes less rich in component* A *and more rich in component* B*. When the temperature has dropped to the solidus, which is the same as the eutectic temperature, solid* B *starts to form as well. Below the solidus, the entire mixture is solid. A liquid of composition* y *(consisting of approximately 80%* B *and 20%* A*) would cool in a similar manner, but with solid* B *forming first. A mixture of eutectic proportions is always either entirely solid or entirely liquid.*

eutectoid (yo͞o-tĕk′toid′) *Adjective.* Relating to a eutectic mixture or alloy. —*Noun.* A eutectic mixture or alloy.

euthanasia (yo͞o′thə-nā′zhə) The act or practice of painlessly ending the life of an animal or a willing individual who has a terminal illness or incurable condition, as by giving a lethal drug.

eutrophic (yo͞o-trŏf′ĭk, -trō′fĭk) Having waters rich in phosphates, nitrates, and organic nutrients that promote a proliferation of plant life, especially algae. Used of a lake, pond, or stream. Compare **dystrophic, oligotrophic.**

eutrophication (yo͞o-trŏf′ĭ-kā′shən) The process by which a lake, pond, or stream becomes eutrophic, typically as a result of mineral and organic runoff from the surrounding land. The increased growth of plants and algae that accompanies eutrophication depletes the dissolved oxygen content of the water and often causes a die-off of other organisms.

eV Abbreviation of **electron volt.**

Evans (ĕv′ənz), **Herbert McLean** 1882–1971. American biologist who discovered vitamin E in 1922 and conducted research that led to the discovery of the growth hormone in the pituitary gland.

evaporation (ĭ-văp′ə-rā′shən) The change of a liquid into a vapor at a temperature below the boiling point. Evaporation takes place at the surface of a liquid, where molecules with the highest kinetic energy are able to escape. When this happens, the average kinetic energy of the liquid is lowered, and its temperature decreases.

evaporation pressure See **vapor pressure.**

evaporimeter (ĭ-văp′ə-rĭm′ĭ-tər) See **atmometer.**

evaporite (ĭ-văp′ə-rīt′) A sedimentary rock that consists of one or more minerals formed as precipitates from the evaporation of a saline solution, such as saltwater. Usually, the first evaporite to form from saltwater is calcium, followed by gypsum and then halite.

even (ē′vən) Divisible by 2 with a remainder of 0, such as 12 or 876.

event horizon (ĭ-vĕnt′) A spatial boundary around a black hole inside which gravity is strong enough to prevent all matter and radiation from escaping. The inability of even light to escape this region is what gives black holes their name.

even-toed ungulate See **artiodactyl.**

evergreen (ĕv′ər-grēn′) *Adjective.* **1.** Having green leaves or needles all year. Evergreen trees lose their leaves individually on an ongoing basis, rather than losing all of them in a short period at the end of a growing season in the manner of deciduous trees. Compare **deciduous.** —*Noun.* **2.** An evergreen tree, shrub, or plant, such as the pine, holly, or rhododendron.

evolution (ĕv′ə-lo͞o′shən) **1.** The process by which species of organisms arise from earlier life forms and undergo change over time through **natural selection**. The modern understanding of the origins of species is based on the theories of Charles Darwin combined with a modern knowledge of genetics based on the work of Gregor Mendel. Darwin observed there is a certain amount of variation of traits or characteristics among the different individuals belonging to a population. Some of these traits confer fitness—they allow the individual organism that possesses them to survive in their environment better than other individuals who do not possess them and to leave more offspring. The offspring then inherit the beneficial traits, and over time the adaptive trait spreads through the population. In twentieth century, the development of the the science of genetics helped explain the origin of the variation of the traits between individual organisms and the way in which they are passed from generation to generation. This basic model of evolution has since been further refined, and the role of **genetic drift** and **sexual selection** in the evolution of populations has been recognized. See also **natural selection, sexual selection.** See Notes at **adaptation, Darwin. 2.** A process of development and change from one state to another, as of the universe in its development through time.

A CLOSER LOOK **evolution**

Darwin's theory of *evolution* by natural selection assumed that tiny adaptations occur in organisms constantly over millions of years. Gradually, a new species develops that is distinct from its ancestors. In the 1970s, however, biologists Niles Eldredge and Stephen Jay Gould proposed that evolution by natural selection may not have been such a smooth and consistent process. Based on fossils from around the world that showed the abrupt appearance of new species, Eldredge and Gould suggested that evolution is better described through *punctuated equilibrium.* That is, for long periods of time species remain virtually unchanged, not even gradually adapting. They are in equilibrium, in balance with the environment. But when confronted with environmental challenges—sudden climate change, for example—organisms adapt quite quickly, perhaps in only a few thousand years. These active periods are *punctuations,* after which a new equilibrium exists and species remain stable until the next punctuation.

evolve (ĭ-vŏlv′) **1.** To undergo biological evolution, as in the development of new species or new traits within a species. **2.** To develop a characteristic through the process of evolution. **3.** To undergo change and development, as the structures of the universe.

exchange force (ĭks-chānj′) A force that results from the continuous interchange of particles between two or more bodies. The exchange of electrons between two atoms or molecules, for example, is an exchange force, as is the exchange of gluons between quarks in the strong nuclear force.

excitation (ĕk′sī-tā′shən) The activity produced in an organ, tissue, or cell of the body that is caused by stimulation, especially by a nerve or neuron. Compare **inhibition.**

excited state (ĭk-sī′tĭd) A state of a physical system in which the system has more than the minimum possible potential energy. Excited states tend to be unstable and easily or spontaneously revert to lower energy states, giving off energy. Compare **ground state.**

exclusion principle (ĭk-sklo͞o′zhən) See **Pauli exclusion principle.**

excretion (ĭk-skrē′shən) The elimination by an organism of waste products that result from metabolic processes. In plants, waste is minimal and is eliminated primarily by diffusion to the outside environment. Animals have specific organs of excretion. In vertebrates, the kidney filters blood, conserving water and producing urea and other waste products in the form of urine. The urine is then passed through the ureters to the bladder and discharged through the urethra. The skin and lungs, which eliminate carbon dioxide, are also excretory organs.

exercise physiology (ĕk′sər-sīz′) The scientific study of the acute and chronic metabolic responses of the human body to exercise, including biochemical and physiologic changes in the heart and skeletal muscles.

exergonic (ĕk′sər-gŏn**′**ĭk) Relating to a chemical reaction that releases energy. The breakdown of sugars by cellular metabolism is an exergonic reaction that releases energy used to drive endergonic reactions. Compare **endergonic.**

exfoliation (ĕks-fō′lē-ā**′**shən) The process in which layers of tissue peel or are peeled off an organism, such as the distinctive ways in which bark peels off a tree in strips or flakes.

exhalation (ĕks′hə-lā**′**shən) The act of breathing out air. During exhalation, the diaphragm relaxes and moves upward, causing compression of the lungs and an outward flow of air. Also called *expiration.* Compare **inhalation.**

exine (ĕk**′**sēn′, -sīn′) The outer layer of the wall of a pollen grain. The exine is composed of the most durable organic polymer known, sporopollenin. Also called *extine.*

exo– A prefix that means "outside" or "external," as in *exoskeleton.*

exobiology (ĕk′sō-bī-ŏl**′**ə-jē) The branch of biology that deals with the search for extraterrestrial life and the effects of extraterrestrial surroundings on living organisms. Also called *astrobiology.*

exocarp (ĕk**′**sō-kärp′) The outermost layer of the pericarp. In a fleshy drupe such as a peach, the skin constitutes the exocarp. Also called *epicarp.* Compare **endocarp, mesocarp.**

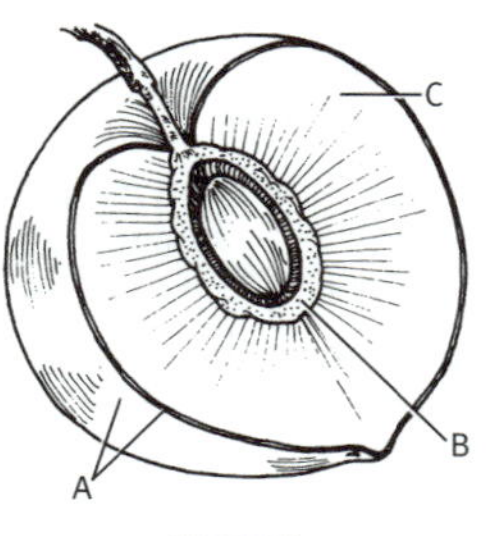

exocarp

A longitudinal section of a peach showing:
A. *exocarp*, B. *endocarp*, C. *mesocarp*

exocrine gland (ĕk**′**sə-krĭn, -krēn) Any gland of the body that produces secretions and discharges them into a cavity or through a duct to the surface of the body. In mammals, the sweat glands and mammary glands are exocrine glands.

exoergic (ĕk′sō-ûr**′**jĭk) Relating to a process, such as a chemical or nuclear reaction, that releases energy. Exothermic reactions are exoergic. Compare **endoergic.**

exogenous (ĕk-sŏj**′**ə-nəs) Originating or produced from outside an organism, tissue, or cell. Compare **endogenous.**

exon (ĕk**′**sŏn) A segment of a gene that contains information used in coding for protein synthesis. Genetic information within genes is discontinuous, split among the exons that encode for messenger RNA and absent from the DNA sequences in between, which are called **introns**. Genetic splicing, catalyzed by enzymes, results in the final version of messenger RNA, which contains only genetic information from the exons. Compare **intron.**

exoplanet (ĕk′sō-plăn**′**ĭt) See **extrasolar planet.**

exoskeleton (ĕk′sō-skĕl**′**ĭ-tn) A hard, protective outer body covering of an animal, such as an insect, crustacean, or mollusk. The exoskeletons of insects and crustaceans are largely made of chitin. Compare **endoskeleton.**

exosphere (ĕk**′**sō-sfîr′) The outermost region of the Earth's atmosphere, beginning at an altitude of approximately 550 km to 700 km (341 to 434 mi) and merging with the **interplanetary medium** at around 10,000 km (6,200 mi). The exosphere consists chiefly of ionized hydrogen, which creates the **geocorona** by reflecting far-ultraviolet light from the Sun. On the remote edges of the exosphere, hydrogen atoms are so sparse that each cubic centimeter might contain only one atom; furthermore, the pressure and gravity are weak enough that atoms in the exosphere can escape entirely and drift into space. Artificial satellites generally orbit in this region. See also **mesosphere, stratosphere, thermosphere, troposphere.** See illustration at **atmosphere.**

exothermic (ĕk′sō-thûr**′**mĭk) Relating to a chemical reaction that releases heat. Compare **endothermic.**

expanding universe theory (ĭk-spăn**′**dĭng) **1.** The cosmological theory based on the work of Edwin Hubble that holds that the universe is expanding. Central to the theory is the interpretation of the color shift in the spectra of all observed galaxies as being the result of the Doppler effect, indicating that the galaxies are moving away from one another. According to the theory, galaxies are not moving through space but rather with space as it expands. The expansion of the universe implies that all of the matter of the universe was once concentrated in one place, which lends support to the **big bang theory.** **2.** The cosmogonical theory holding that a violent eruption from a singularity led to the formation of elementary particles, the subsequent formation of hydrogen and helium, and the dispersion of the galaxies from these elements.

expansion (ĭk-spăn′shən) **1.** An increase in the volume of a substance while its mass remains the same. Expansion is usually due to heating. When substances are heated, the molecular bonds between their particles are weakened, and the particles move faster, causing the substance to expand. **2.** A number or other mathematical expression written in an extended form. For example, $a^2 + 2ab + b^2$ is the expansion of $(a + b)^2$.

expansion board A printed circuit board that can be installed to enhance a computer's capabilities, as by increasing memory or improving graphics.

expectorant (ĭk-spĕk′tər-ənt) A drug that promotes the discharge of phlegm or mucus from the respiratory tract.

experiment (ĭk-spĕr′ə-mənt) A test or procedure carried out under controlled conditions to determine the validity of a hypothesis or make a discovery. See Note at **hypothesis.**

expiration (ĕk′spə-rā′shən) See **exhalation.**

explosion (ĭk-splō′zhən) A violent blowing apart or bursting caused by energy released from a very fast chemical reaction, a nuclear reaction, or the escape of gases under pressure.

exponent (ĕk′spō′nənt, ĭk-spō′nənt) A number or symbol, placed above and to the right of the expression to which it applies, that indicates the number of times the expression is used as a factor. For example, the exponent 3 in 5^3 indicates $5 \times 5 \times 5$; the exponent x in $(a + b)^x$ indicates $(a + b)$ multiplied by itself x times.

exponential (ĕk′spə-nĕn′shəl) Relating to a mathematical expression containing one or more exponents. ▸ Something is said to increase or decrease **exponentially** if its rate of change must be expressed using exponents. A graph of such a rate would appear not as a straight line, but as a curve that continually becomes steeper or shallower.

exponentiation (ĕk′spə-nĕn′shē-ā′shən) The act of raising a quantity to a power.

extension (ĭk-stĕn′shən) **1.** *Mathematics.* A set that includes a given and similar set as a subset. **2.** *Computer Science.* A set of characters that follow a filename and are separated from it by a period, used to identify the kind of file.

extensor (ĭk-stĕn′sər) A muscle that extends or straightens a limb or joint. Compare **flexor.**

exterior angle (ĭk-stîr′ē-ər) The angle formed between a side of a polygon and an extended adjacent side. Compare **interior angle.**

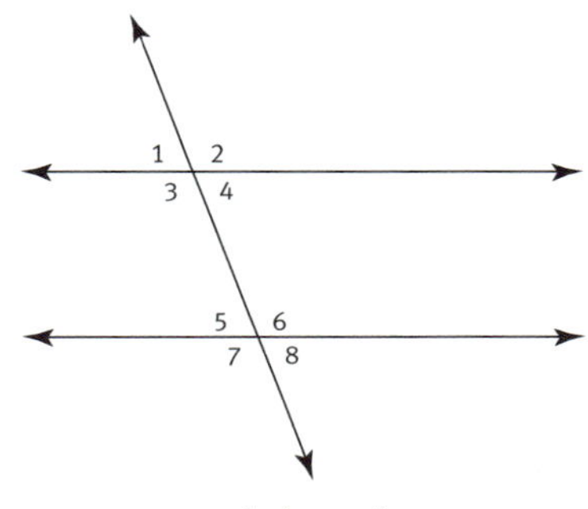

exterior angle

Angles 1, 2, 7, and 8 are exterior angles. Angles 1 and 8 and angles 2 and 7 are alternate exterior angles.

extinct (ĭk-stĭngkt′) **1.** Having no living members. Species become extinct for many reasons, including climate change, disease, destruction of habitat, local or worldwide natural disasters, and development into new species (speciation). The great majority of species that have ever lived—probably more than 99 percent—are now extinct. **2.** No longer active or burning, as an extinct volcano.

extinction (ĭk-stĭngk′shən) **1.** The fact of being extinct or the process of becoming extinct. See more at **background extinction, mass extinction.** **2.** A progressive decrease in the strength of a conditioned response, often resulting in its elimination, because of withdrawal of a specific stimulus.

extine (ĕk′stēn′, -stīn′) See **exine.**

extracellular (ĕk′strə-sĕl′yə-lər) Located or occurring outside a cell or cells.

extracellular matrix See **matrix** (sense 2).

extranet (ĕk′strə-nĕt′) An extension of an institution's intranet, enabling communication between the institution and people it deals with, often by providing limited access to its intranet.

extrapolate (ĭk-străp′ə-lāt′) To estimate the value of a quantity that falls outside the range in which its values are known.

extrasolar planet (ĕk′strə-sō′lər) A planet that orbits a star other than the Sun. The first such planet to be discovered, in 1991, was found orbiting a pulsar, although most of the more than 100 extrasolar planets that have since been identified orbit normal stars. Many of them, known as **hot Jupiters**, are very large and revolve around their star in extremely close orbits, at less than the distance of Mercury's orbit around the Sun. Other Jupiter-sized and larger planets have been found in highly eccentric orbits. Evidence suggests that extrasolar planets may be relatively common throughout the universe. In 2004, astronomers located the first extrasolar planet with an atmosphere containing oxygen and carbon. The planet, HD 209458b (also called Osiris), orbits a star 150 light-years from Earth. The apparent lack of terrestrial, Earth-sized planets among those that have so far been discovered may simply be the result of the much greater difficulty in identifying smaller, less massive bodies at such great distances. Also called *exoplanet.*

extraterrestrial (ĕk′strə-tə-rĕs′trē-əl) Originating, located, or occurring outside the Earth or its atmosphere.

extratropical cyclone (ĕk′strə-trŏp′ĭ-kəl) See under **cyclone.**

extreme (ĭk-strēm′) **1.** Either the first or fourth term of a proportion of four terms. In the proportion $\frac{2}{3} = \frac{4}{6}$, the extremes are 2 and 6. Compare **mean** (sense 2). **2.** A maximum or minimum value of a function.

extremophile (ĭk-strē′mə-fīl′) An organism adapted to living in conditions of extreme temperature, pressure, or chemical concentration, as in highly acidic or salty environments. Many extremophiles are unicellular organisms known as **archaea**.

extrorse (ĕk′strôrs′) Facing outward, away from the central axis around which a flower is arranged. Used of anthers and the direction in which they open to release pollen.

extrusion (ĭk-strōō′zhən) **1.** The emission of lava onto the surface of the Earth. ► Rocks that form from the cooling of lava are generally fine-grained (because they cool quickly, before large crystals can grow) and are called **extrusive rocks.** Compare **intrusion. 2.** The process of making a shaped object, such as a rod or tube, by forcing a material into a mold.

eye (ī) **1.** *Anatomy.* The vertebrate organ of sight, composed of a pair of fluid-filled spherical structures that occupy the orbits of the skull. Incoming light is refracted by the cornea of the eye and transmitted through the pupil to the lens, which focuses the image onto the retina. **2.** *Zoology.* An organ in invertebrates that is sensitive to light. See more at **compound eye, eyespot. 3.** *Botany.* A bud on a tuber, such as a potato. **4.** *Meteorology.* The relatively calm area at the center of a hurricane or similar storm. See more at **hurricane.**

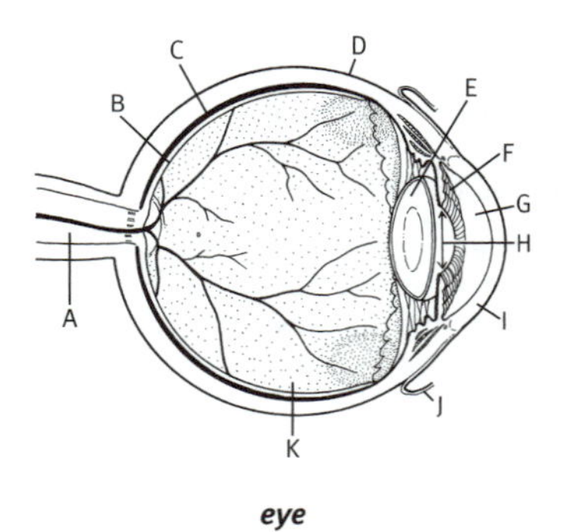

eye

A. *optic nerve,* B. *retina,* C. *choroid,* D. *sclera,* E. *lens,* F. *iris,* G. *aqueous humor,* H. *pupil,* I. *cornea,* J. *conjunctiva,* K. *vitreous humor*

eyepiece (ī′pēs′) The lens or group of lenses closest to the eye in an optical instrument such as a telescope or microscope.

eyespot (ī′spŏt′) **1.** An area that is sensitive to light and functions somewhat like an eye, found in certain single-celled organisms as well as many invertebrate animals. **2.** A round marking resembling an eye, as on the tail feather of a peacock

eyestalk (ī′stôk′) A movable stalk having a compound eye on its tip, found on crabs, lobsters, and other crustaceans.

F **1.** Abbreviation of **Fahrenheit. 2.** The symbol for **farad. 3.** The symbol for **fluorine.**

face (fās) **1.** A plane surface of a geometric solid. A cube has 6 faces; a dodecahedron, 12. **2.** Any of the surfaces of a rock or crystal.

facies (fā′shē-ēz′, -shēz) Plural **facies. 1a.** A body of sedimentary rock distinguished from others by its lithology, geometry, sedimentary structures, proximity to other types of sedimentary rock, and fossil content, and recognized as characteristic of a particular depositional environment. **b.** For a metamorphic rock, the particular combination of pressure and temperature under which metamorphism occurred. **2.** The general aspect or makeup of an ecological community, especially a local modification of a community characterized by a conspicuous or abundant species that is absent or less concentrated in other locations. **3.** The appearance or expression of the face, especially when typical of a certain disorder or disease.

factor (făk′tər) *Noun.* **1.** One of two or more numbers or expressions that are multiplied to obtain a given product. For example, 2 and 3 are factors of 6, and $a + b$ and $a - b$ are factors of $a^2 - b^2$. **2.** A substance found in the body, such as a protein, that is essential to a biological process. For example, growth factors are needed for proper cell growth and development. —*Verb.* **3.** To find the factors of a number or expression. For example, the number 12 can be factored into 2 and 6, or 3 and 4, or 1 and 12.

factorial (făk-tôr′ē-əl) The product of all of the positive integers from 1 to a given positive integer. It is written as the given integer followed by an exclamation point. For example, the factorial of 4 (written 4!) is $1 \times 2 \times 3 \times 4$, or 24.

facula (făk′yə-lə) Plural **faculae** (făk′yə-lē′). A bright, cloudlike structure on the Sun's surface, ascending several hundred kilometers above the photosphere and often associated with sunspots. Faculae are formed when a strong magnetic field heats a region of the photosphere to higher temperatures than the surrounding area. They occur all over the Sun but are usually only visible near the limb (the outer edge of the Sun's apparent disk), where the photosphere appears dimmer than in the center.

facultative (făk′əl-tā′tĭv) Capable of existing under varying environmental conditions or by assuming various behaviors. Bacteria that are facultative aerobes can live in both aerobic and anaerobic environments. A facultative parasite can live independently of its usual host. Compare **obligate.**

Fahrenheit (făr′ən-hīt′) Relating to or based on a temperature scale that indicates the freezing point of water as 32° and the boiling point of water as 212° under standard atmospheric pressure.

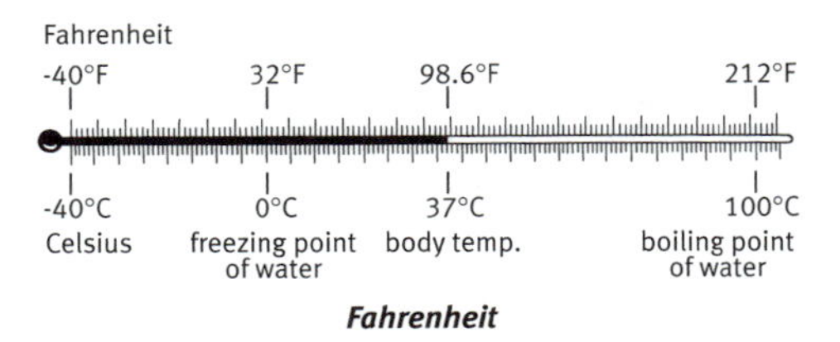

Fahrenheit

Fahrenheit and Celsius readings on a mercury thermometer.

Fahrenheit, Gabriel Daniel 1686–1736. German physicist who invented the mercury thermometer in 1714 and devised the Fahrenheit temperature scale.

falling tide (fô′lĭng) See **ebb tide.**

fall line (fôl) A line connecting the waterfalls of nearly parallel rivers that marks a drop in land level.

fallopian tube (fə-lō′pē-ən) Either of a pair of long, slender tubes found in female mammals that carry egg cells from the ovaries to the uterus.

false fruit (fôls) **1.** See **accessory fruit. 2.** See **pome.**

family (făm′ə-lē) A group of organisms ranking above a genus and below an order. The names of families end in *–ae,* a plural ending in Latin. In the animal kingdom, family names end in *–idae,* as in Canidae (dogs and their kin), while those in the plant kingdom usually end in *–aceae,* as in Rosaceae (roses and their kin). See Table at **taxonomy.**

fang (făng) A long, pointed tooth in vertebrate animals or a similar structure in spiders, used to seize prey and sometimes to inject venom. The fangs of a poisonous snake, for example, have a hollow groove through which venom flows.

farad (făr′əd) The SI derived unit used to measure electric capacitance. A capacitor in which a stored charge of one coulomb provides an electric potential difference of one volt across its plates has a capacitance of one farad.

faraday (făr′ə-dā′) A measure of electric charge equal to the charge carried by one **mole** of electrons, about 96,494 coulombs per mole. The faraday is used in measurements of the electricity required to break down a compound by electrolysis.

Faraday (făr′ə-dā′, -dē), **Michael** 1791–1867. British physicist and chemist whose experiments into the connections between electricity, magnetism, and light laid the foundation for modern physics. In addition to discovering electromagnetic induction, he invented the electric motor, generator, and transformer, and he discovered the carbon compound benzene.

BIOGRAPHY **Michael Faraday**

The nineteenth century saw rapid growth in the understanding of electricity and magnetism, and much of this progress was due to Michael Faraday. There was no hint from his humble beginnings that he was to become a great scientist. Born in 1791, the son of an English blacksmith, Faraday received little formal schooling. At 14 he was apprenticed to a bookbinder, and it was during this time that he developed an interest in science. In 1812 he attended a series of lectures by Humphry Davy, the well-known chemist. Later in the year, Faraday sent Davy his notes on the talks, asking to become his assistant. When an opening became available, Davy took him on. Faraday, a truly gifted experimenter, started amassing an impressive body of work, converting electrical into mechanical energy (1821), liquefying chlorine (1823), and isolating benzene (1825). He made perhaps his greatest discovery—electromagnetic induction—in 1831, when he produced electricity from magnetism by moving a magnet inside a wire coil. Faraday also came up with the concept of electric and magnetic fields. When James Clerk Maxwell put Faraday's ideas into mathematical form (Faraday knew little mathematics), they became a cornerstone of physics. It was Faraday's research that helped transform electricity from a scientific curiosity into a workable technology. But he also transformed the language, helping to coin the words *anode, cathode, ion,* and *electrode,* among others. It is only fitting that there are now two words named after him: *farad,* the unit of capacitance, and *faraday,* a unit used to measure the amount of electrical charge.

Faraday cage A container made of a conductor, such as wire mesh or metal plates, shielding what it encloses from external electric fields. Since the conductor is an equipotential, there are no potential differences inside the container. The metal hull of an aircraft acts as a faraday cage, protecting its occupants from lightning. Faraday cages are used to protect electronic equipment from such electrical interference as electromagnetic interference. Also called *Faraday shield.*

Faraday effect The rotation of the plane of polarization of polarized light when subject to a magnetic field parallel to the direction in which the light is propagating. Also called *Faraday rotation.*

Faraday shield See **Faraday cage.**

far-infrared radiation (fär′ĭn′frə-rĕd′) Infrared electromagnetic radiation of very long wavelengths, in the range of 2–50 microns. It is used in medicine to provide a penetrating heat to the body. Also called *far-red radiation.*

farsightedness (fär′sī′tĭd-nĭs) See **hyperopia.**

fascia (făsh′ē-ə) Plural **fasciae** (făsh′ē-ē′, fā′shē-ē). A sheet or band of fibrous connective tissue. Fascia envelops, separates, or binds together muscles, organs, and other soft structures of the body.

fascicle (făs′ĭ-kəl) A bundle or cluster of stems, flowers, or leaves, such as the bundles in which pine needles grow.

fascicular cambium (fə-sĭk′yə-lər) Cambium that develops within the vascular bundles in the stem of a plant. Among the eudicotyledons, layers of fascicular cambium within the individual vascular bundles are connected by cambium tissue between the bundles (the interfascicular cambium) to form a ring of tissue within the stem. This ring of cambium matures to become the vascular cambium, which permits secondary growth of the eudicotyledon stem. See also **interfascicular cambium, vascular bundle.**

fast-breeder reactor (făst′brē′dər) A breeder reactor that uses high-speed neutrons to produce fissionable material. See also **nuclear reactor.**

fast neutron A neutron that is not in thermal equilibrium with the surrounding medium, especially one produced by fission. Compare **slow neutron.** See also **fast-breeder reactor.**

fat (făt) Any of a large number of oily compounds that are widely found in plant and animal tissues and serve mainly as a reserve source of energy. In mammals, fat or adipose tissue is deposited beneath the skin and around the internal organs, where it also protects and insulates against heat loss. Fat is a necessary, efficient source of energy. An ounce of fat contains more than twice as much stored energy as does an ounce of protein or carbohydrates and is digested more slowly, resulting in the sensation of satiety after eating. It also enhances the taste, aroma, and texture of food. Fats are made chiefly of triglycerides, each molecule of which contains three fatty acids. Dietary fat supplies humans with essential fatty acids, such as *linoleic acid* and *linolenic acid.* Fat also regulates cholesterol metabolism, and is a precursor of *prostaglandins.* See more at **saturated fat, unsaturated fat.**

fatty acid (făt′ē) Any of a large group of organic acids, especially those found in animal and vegetable fats and oils. Fatty acids are mainly composed of long chains of hydrocarbons ending in a carboxyl group. A fatty acid is saturated when the bonds between carbon atoms are all single bonds. It is unsaturated when any of these bonds is a double bond.

fault (fôlt) A fracture in a rock formation along which there has been movement of the blocks of rock on either side of the plane of fracture. Faults are caused by plate-tectonic forces. See

A CLOSER LOOK **faults**

Bedrock, the solid rock just below the soil, is often cracked along surfaces known as planes. Cracks can extend up to hundreds of kilometers in length. When tensional and compressional stresses cause rocks separated by a crack to move past each other, the crack is known as a *fault.* Faults can be horizontal, vertical, or oblique. The movement can occur in the sudden jerks known as earthquakes. *Normal faults,* or *tensional faults,* occur when the rocks above the fault plane move down relative to the rocks below it, pulling the rocks apart. Where there is compression and folding, such as in mountainous regions, the rocks above the plane move upward relative to the rocks below the plane; these are called *reverse faults. Strike-slip faults* occur when shearing stress causes rocks on either side of the crack to slide parallel to the fault plane between them. *Transform faults* are strike-slip faults in which the crack is part of a boundary between two tectonic plates. A well-known example is the San Andreas Fault in California. Geologists use sightings of displaced outcroppings to infer the presence of faults, and they study faults to learn the history of the forces that have acted on rocks.

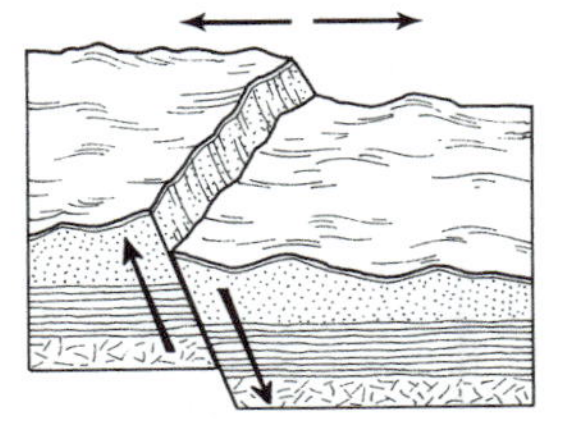
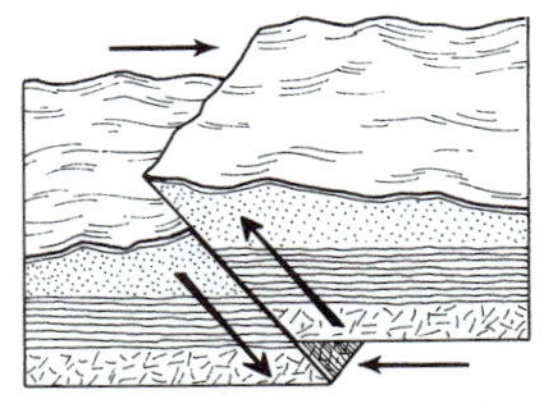
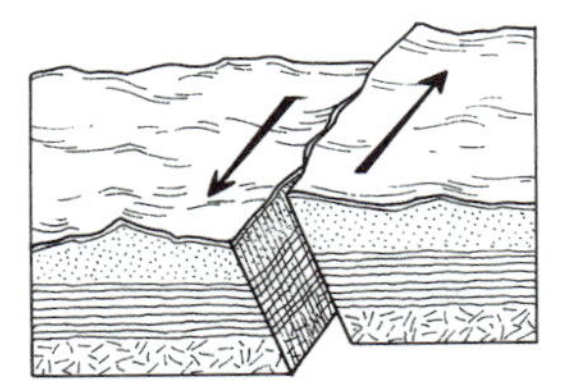

left to right: *normal, reverse, and strike-slip faults*

more at **normal fault, reverse fault, strike-slip fault, thrust fault, transform fault.** See Note at **earthquake.**

fauna (fô′nə) Plural **faunas** or **faunae** (fô′nē′). The animals of a particular region or time period.

fax machine (făks) A device that sends and receives printed pages or images over telephone lines by digitizing the material with an internal optical scanner and transmitting the information as electronic signals.

fayalite (fā′ə-līt′) See under **olivine.**

Faye correction (fā) See **free-air correction.**

Fe The symbol for **iron.**

feather (fĕ*th*′ər) One of the light, flat structures that cover the skin of birds. A feather is made of a horny substance and has a narrow, hollow shaft bearing flat vanes formed of many parallel barbs. The barbs of outer feathers are formed of even smaller structures (called barbules) that interlock. The barbs of down feathers do not interlock. Evolutionarily, feathers are modified scales, first seen in certain dinosaurs.

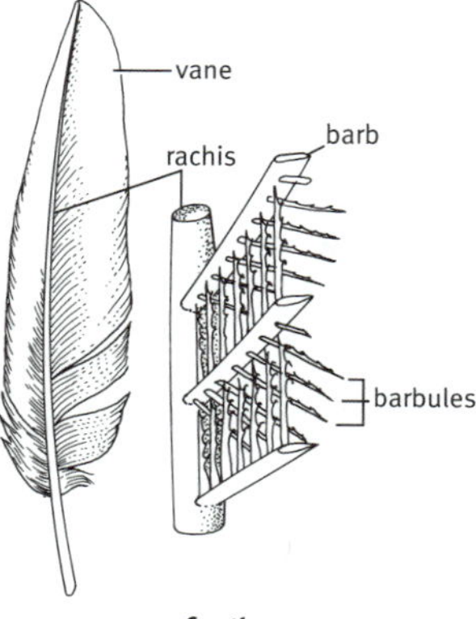

feather

feces (fē′sēz) Waste matter eliminated from the intestinal tract.

feedback (fēd′băk′) The supply of an input to some process or system as a function of its output. See more at **negative feedback, positive feedback.**

feeler (fē′lər) A slender body part used for touching or sensing. The antennae of insects and the barbels of catfish are feelers.

feldspar (fĕld′spär′, fĕl′-) Any of a group of abundant monoclinic or triclinic minerals having the general formula $MAl(Al,Si)_3O_8$, where M is either potassium (K), sodium (Na), or calcium (Ca) or less commonly barium (Ba), rubidium (Rb), strontium (Sr), or iron (Fe). Feldspars range from white, pink, or brown to grayish blue in color. They occur in igneous, sedimentary, and metamorphic rocks and make up more than 60 percent of the Earth's crust. When they decompose, feldspars form clay or the mineral kaolinite.

feldspathic (fĕld-spăth′ĭk, fĕl-) Relating to or containing feldspar.

feldspathoid (fĕld-spăth′oid′, fĕl-) Any of a group of relatively rare minerals that are chemically very similar to feldspars, but have too little silica to qualify as feldspars. Feldspathoids occur in igneous rocks with little or no silica.

felid (fē′lĭd) Any of various carnivorous mammals of the family Felidae, which includes the domesticated cat and big cats such as lions, tigers, panthers, lynxes, leopards, pumas, and cheetahs.

feline (fē′līn′) Characteristic of or resembling felids.

felsic (fĕl′sĭk) Relating to an igneous rock that contains a group of light-colored silicate minerals, including feldspar, feldspathoid, quartz, and muscovite. Compare **mafic.**

felsite (fĕl′sīt′) A fine-grained, light-colored igneous rock, consisting mainly of feldspar and quartz.

female (fē′māl′) *Adjective.* **1.** In organisms that reproduce sexually, being the gamete that is larger and less motile than the other corresponding gamete (the male gamete) of the same species. The egg cells of higher animals and plants are female gametes. **2.** Possessing or being a structure that produces only female gametes. The ovaries of humans are female reproductive organs. Female flowers possess only carpels and no stamens. **3.** Having the genitalia or other structures typical of a female organism. Worker ants are female but sterile. —*Noun.* **4.** A female organism.

femur (fē′mər) The long bone of the thigh or of the upper portion of the hind leg. See more at **skeleton.**

feral (fîr′əl, fĕr′-) Existing in a wild or untamed state, either naturally or having returned to such a state from domestication.

Fermat (fĕr-mä′), **Pierre de** 1601–1665. French mathematician who is best known for his work on probability and on the properties of numbers. He formulated Fermat's last theorem, which remained unsolved for over three hundred years.

Fermat's last theorem (fĕr-mäz′) A theorem stating that the equation $a^n + b^n = c^n$ has no solution if *a, b,* and *c* are positive integers and if *n* is an integer greater than 2. The theorem was first stated by the French mathematician Pierre de Fermat around 1630, but not proved until 1994.

fermentation (fûr′mĕn-tā′shən) The process by which complex organic compounds, such as glucose, are broken down by the action of enzymes into simpler compounds without the use of oxygen. Fermentation results in the production of energy in the form of two ATP molecules, and produces less energy than the aerobic process of **cellular respiration**. The other end products of fermentation differ depending on the organism. In many bacteria, fungi, protists, and animals cells (notably muscle cells in the body), fermentation produces lactic acid and lactate, carbon dioxide, and water. In yeast and most plant cells, fermentation produces ethyl alcohol, carbon dioxide, and water.

Fermi (fĕr′mē), **Enrico** 1901–1954. Italian-born American physicist who won a 1938 Nobel Prize for his research on neutrons. In 1942, with Leo Szilard, Fermi built the world's first nuclear reactor. He also discovered over 40 new isotopes, including the element fermium, which is named for him.

Enrico Fermi

Fermi-Dirac statistics See under **statistical mechanics.**

fermion (fûr′mē-ŏn′, fĕr′-) An elementary or composite particle, such as an electron, quark, or proton, whose spin is an integer multiple of $\frac{1}{2}$. Fermions act on each other by exchanging bosons and are subject to the Pauli exclusion principle, which requires that no two fermions be in the same quantum state. Fermions are named after the physicist Enrico Fermi, who along with Paul Dirac developed quantum statistical models of their behavior. Compare **boson.**

fermium (fûr′mē-əm) *Symbol* **Fm** A synthetic, radioactive metallic element of the actinide series that is produced from plutonium or uranium. Its most stable isotope is Fm 257 with a half-life of approximately 100 days. Atomic number 100. See **Periodic Table.**

fern (fûrn) Any of numerous seedless vascular plants belonging to the phylum Pterophyta that reproduce by means of spores and usually have feathery fronds divided into many leaflets. Most species of ferns are homosporous (producing only one kind of spore). The haploid spore grows into a small, usually flat gametophyte known as a **prothallus**, which is undifferentiated into roots, stems, and leaves. The green prothallus anchors itself with hairlike extensions known as **rhizoids** and bears both archegonia (organs producing female gametes) and antheridia (organs producing male gametes). The male gametes require the presence of water to swim to the female gametes and fertilize the eggs. Normally only one embryo is produced, and it then grows out of the gametophyte plant as a diploid sporophyte plant that has roots, stems, and leaves and conducts photosynthesis, while the smaller gametophyte withers away. The leaves of these sporophytes eventually produce sporangia (in some species occurring in clusters known as **sori**). Under dry conditions, the sori burst releasing hundreds of thousands or millions of spores. Ferns were abundant in the Carboniferous period and exist today in about 11,000 species, about three-quarters of which live in tropical climates.

ferrate (fĕr′āt′) A compound containing ferric oxide and another oxide. Ferrates are stable only in strongly alkaline conditions and impart a purple color to aqueous solutions.

ferric (fĕr′ĭk) Containing iron, especially iron with a valence of 3. Compare **ferrous.**

ferric oxide A reddish-brown to silver or black compound which occurs naturally as the mineral hematite and as rust. It is often used as a pigment and a metal polish. *Chemical formula:* Fe_2O_3.

ferroelectric (fĕr′ō-ĭ-lĕk′trĭk) **1.** Relating to a typically crystalline **dielectric** that can be given a permanent electric polarization by application of an electric field. **2.** A ferroelectric substance.

ferromagnesian (fĕr′ō-măg-nē′zhən) Containing iron and magnesium. Magnetite and hornblende are ferromagnesian minerals.

ferromagnetism (fĕr′ō-măg′nĭ-tĭz′əm) The property of being strongly attracted to either pole of a magnet. Ferromagnetic materials, such as iron, contain unpaired electrons, each with a small magnetic field of its own, that align readily with each other in response to an external magnetic field. This alignment tends to persists even after the magnetic field is removed, a phenomenon called **hysteresis**. Ferromagnetism is important in the design of electromagnets, transformers, and many other electrical and mechanical devices, and in analyzing the history of the earth's magnetic reversals. Compare **diamagnetism, paramagnetism.** —*Adjective* **ferromagnetic** (fĕr′ō-măg-nĕt′ĭk).

ferrous (fĕr′əs) Containing iron, especially iron with a valence of 2. Compare **ferric.**

ferrous oxide A black powder used to make steel, green heat-absorbing glass, and enamels. *Chemical formula:* **FeO.**

ferrous sulfate A bluish-green crystalline compound that is used in sewage and water treatment, and as a pigment and fertilizer. It is also used in medicine to treat iron deficiency. Also called *green vitriol. Chemical formula:* $FeSO_4 \cdot 7H_2O$.

fertile (fûr′tl) **1.** Capable of producing offspring, seeds, or fruit. **2.** Capable of developing into a complete organism; fertilized. **3.** Capable of supporting plant life; favorable to the growth of crops and plants. —*Noun* **fertility.**

fertilization (fûr′tl-ĭ-zā′shən) **1.** The process by which two gametes (reproductive cells having a single, haploid set of chromosomes) fuse to become a zygote, which develops into a new organism. The resultant zygote is diploid (it has two sets of chromosomes). In cross-fertilization, the two gametes come from two different individual organisms. In self-fertilization, the gametes come from the same individual. Fertilization includes the union of the cytoplasm of the gametes (called plasmogamy) followed by the union of the nuclei of the two gametes (called karyogamy). Among many animals, such as mammals, fertilization occurs inside the body of the female. Among fish, eggs are fertilized in the water. Among plants, fertilization of eggs occurs within the reproductive structures of the parent plant, such as the ovules of gymnosperms and angiosperms. See Note at **pollination. 2.** The process of making soil more productive of plant growth, as by the addition of organic material or fertilizer. —*Verb* **fertilize.**

fertilizer (fûr′tl-ī′zər) Any of a large number of natural and synthetic materials, including manure and compounds containing nitrogen, phosphorus, and potassium, spread on or worked into soil to increase its capacity to support plant growth. Synthetic fertilizers can greatly increase the productivity of soil but have high energy costs, since fossil fuels are required as a source of hydrogen, which is necessary to fix nitrogen in ammonia.

fetal alcohol syndrome (fē′təl) A syndrome of congenital defects caused by maternal alcohol abuse during pregnancy and characterized by any of various abnormalities, including growth and mental retardation, cranial and facial malformations, and cardiac defects.

fetus (fē′təs) The unborn offspring of a mammal at the later stages of its development, especially a human from eight weeks after fertilization to its birth. In a fetus, all major body organs are present. —*Adjective* **fetal.**

fever (fē′vər) A body temperature that is higher than normal. Fever is the body's natural response to the release of substances called pyrogens by infectious agents such as bacteria and viruses. The pyrogens stimulate the hypothalamus in the brain to conserve heat and increase the basal metabolic rate.

Feynman (fīn′mən), **Richard Phillips** 1918–1988. American physicist who developed the theory of quantum electrodynamics, laying the foundation for all other quantum field theories. His approach combined quantum mechanics and relativity theory, and exploited a method using diagrams of particle interactions to greatly simplify calculations. For this work he shared with American physicist Julian Schwinger and Japanese physicist Sin-Itiro Tomonaga the 1965 Nobel Prize for physics.

Feynman diagram A diagram used to help describe and visualize the possible interactions between particles in quantum electrodynamics and quantum chromodynamics. Fermions, such as electrons, are represented with straight lines and bosons, such as photons, with wavy lines. Points of intersection indicate an interaction, such as an electromagnetic interaction, between the particles.

fiber (fī′bər) **1.** The parts of grains, fruits, and vegetables that contain cellulose and are not digested by the body. Fiber helps the intestines absorb water, which increases the bulk of the stool and causes it to move more quickly through the colon. **2.** One of the elongated, thick-walled cells, often occurring in bundles, that give strength and support to tissue in vascular plants. Fibers are one type of sclerenchyma cell. **3a.** Any of the elongated cells of skeletal or cardiac muscle, made up of slender threadlike structures called *myofibrils.* **b.** The axon of a neuron. —*Adjective* **fibrous.**

fiberglass (fī′bər-glăs′) A material made up of very fine fibers of glass. Fiberglass is resistant to heat and fire and is used to make various products, such as building insulation and boat hulls. Because the fibers in fiberglass are capable of transmitting light around curves, fiberglass is an important component of fiber optics.

fiber optics Technology based on the use of hair-thin, transparent fibers to transmit light or infrared signals. The fibers are flexible and consist of a core of optically transparent glass or plastic surrounded by a glass or plastic cladding that reflects the light signals back into the core. Light signals can be modulated to carry almost any other sort of signal, including sounds, electrical signals, and computer data, and a single fiber can carry hundreds of such signals simultaneously, literally at the speed of light. Signals that have weakened after travelling very long distances in the fibers can be optically pumped with lasers, amplifying them without the need to convert them into electrical signals. Optical fibers are relatively inexpensive to manufacture and install when compared with wire cables, and they require very little power and are easily laid out underground. Optical fibers are also used to transmit images focused on one end to the other end through circuitous paths, as in bronchoscopes and colonoscopes used in medical examinations.

Fibonacci (fē′bə-nä′chē), **Leonardo** 1170?–1250? Italian mathematician who popularized the modern Arabic system of numerals in the western world and discovered the Fibonacci sequence of integers.

Fibonacci sequence A sequence of numbers, such as 1, 1, 2, 3, 5, 8, 13 ..., in which each successive number is equal to the sum of the two preceding numbers. Many shapes occurring in nature, such as certain spirals, have proportions that can be described in terms of the Fibonacci sequence. See also **golden section.**

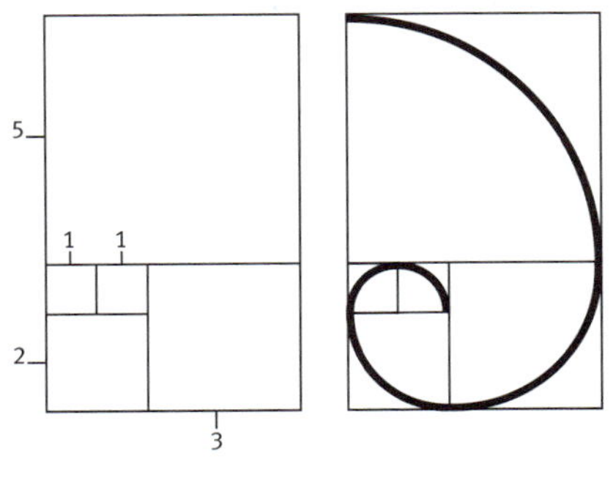

Fibonacci sequence

fibril (fī′brəl, fĭb′rəl) Any of various threadlike fibers or filaments that are constituent parts of a cell or larger structure. Cellulose fibrils are the main component of cell walls in plants. Fibrils make up the contractile part of striated muscle fiber in the body.

fibrillation (fĭb′rə-lā′shən) A rapid twitching of muscle fibers, as of the heart, that is caused by the abnormal discharge of electrical nerve impulses. Ventricular fibrillation is life-threatening.

fibrin (fī′brĭn) A fibrous protein produced by the action of thrombin on fibrinogen and essential to the coagulation of blood. Fibrin works by forming a fibrous network in which blood cells become trapped, thereby producing a clot.

fibrinogen (fī-brĭn′ə-jən) A protein in the blood plasma that is essential for the coagulation of blood. It is converted to fibrin by the action of thrombin in the presence of calcium ions.

fibroid (fī′broid′) A benign tumor composed of fibrous or muscle tissue, especially one that develops in the uterus.

fibromyalgia (fī′brō-mī-ăl′jə) A syndrome characterized by chronic pain in any of various muscles and surrounding soft tissues (such as tendons and ligaments), point tenderness at specific sites in the body, and fatigue. Inflammation is absent, and the cause is unknown.

fibronectin (fī′brə-nĕk′tn) Any of several glycoproteins that occur especially in plasma and in soft connective tissue. Fibronectins are important for the adhesion of fibrous extracellular tissue matrices and also play roles in cellular adhesion, embryonic cellular differentiation, phagocytosis, and the aggregation of platelets in blood clotting.

fibrous root (fī′brəs) Any of the roots in a system that is made up of many threadlike members of more or less equal length and is characteristic of monocotelydons. Fibrous roots develop from adventitious roots arising from the plant's stem and usually do not penetrate the soil very deeply. Because their roots attach themselves firmly to soil particles, plants with fibrous root systems are especially useful in preventing soil erosion. Compare **taproot.**

fibrovascular (fī′brō-văs′kyə-lər) Having fibrous tissue and vascular tissue, as in the woody tissue of plants. The veins of leaves are made of fibrovascular tissue.

fibrovascular bundle See **vascular bundle.**

fibula (fĭb′yə-lə) Plural **fibulae** (fĭb′yə-lē′) or **fibulas.** The smaller of the two bones of the lower leg or lower portion of the hind leg. See more at **skeleton.**

field (fēld) **1.** A distribution in a region of space of the strength and direction of a force, such as the electrostatic force near an electrically charged object, that would act on a body at any given point in that region. See also **electric field, magnetic field. 2.** The region whose image is visible to the eye or accessible to an optical instrument. **3.** A set of elements having two operations, designated addition and multiplication, satisfying the conditions that multiplication is distributive over addition, that the set is a group under addition, and that the elements with the exception of the additive identity (0) form a group under multiplication. The set of all rational numbers is a field. **4a.** In a database, a space for a single item of information contained in a record. **b.** An interface element in a graphical user interface that accepts the input of text.

field capacity The maximum amount of water that a soil or rock can hold, as by capillary action, before the water is drawn away by gravity.

field effect transistor A type of transistor, usually made of semiconductors, in which the flow of current from a source on one side to a drain on the other is regulated by the strength of an electric field. This field is produced by a voltage at a third point called the gate, which effectively squeezes or opens the channel to the flow of current. Field effect transistors are especially useful for amplifying or switching very low power signals, as found in portable wireless technology, microprocessors, and digital memory circuits. Compare **bipolar transistor.**

field emission The emission of electrons from the surface of a conductor, caused by a strong electric field. Field emission is used to create electron beams in certain electron microscopes, as well as in flat-panel computer and television displays, in which the electron beams produce light by striking a phosphor-coated screen.

field ion microscope A microscope that produces an image of a sample of molecules, or even individual atoms, on the surface of a metal tip. Gas atoms absorbed in the tip are positively ionized by an electric field, and the tip is given a strong positive electric charge, causing the ions to be repelled and thus fly away from the tip. The pattern that the ions form on a collecting surface provides an image of the sample on the tip.

field magnet A permanent magnet used to produce a base magnetic field against which other magnetized materials react, as in the operation of electrical devices such as motors, generators, and solenoids.

field theory **1.** An explicit mathematical description of physical phenomena that models physical forces using fields. **2.** The study of fields and field extensions in algebra.

fifth force (fĭfth) Any of various hypothetical, very weak forces thought to cause bodies to repel each other. Such forces are occasionally considered as a solution to unexpected deviations in the measured value of the gravitational constant or to explain the apparent acceleration of very distant galaxies away from earth.

fight-or-flight response (fīt′ôr-flīt′) A physiological reaction in response to stress, characterized by an increase in heart rate and blood pressure, elevation of glucose levels in the blood, and redistribution of blood from the digestive tract to the muscles. These changes are caused by activation of the sympathetic nervous system by epinephrine (adrenaline), which prepares the body to challenge or flee from a perceived threat.

filament (fĭl′ə-mənt) **1.** A fine or slender thread, wire, or fiber. **2.** The part of a stamen that supports the anther of a flower; the stalk of a stamen. See more at **flower. 3a.** A fine wire that gives off radiation when an electric current is passed through it, usually to provide light, as in an incandescent bulb, or to provide heat, as in a vacuum tube. **b.** A wire that acts as the cathode in some electron tubes when it is heated with an electric current. **4.** Any of the dark, sinuous lines visible through certain filters on the disk of the Sun. Filaments are solar prominences that are viewed against the solar surface rather than being silhouetted along the outer edges of the disk. See more at **prominence.**

filaria (fə-lâr′ē-ə) Plural **filariae** (fə-lâr′ē-ē′). Any of various slender, threadlike nematode worms of the superfamily Filarioidea that are parasitic in vertebrates and are often transmitted as larvae by mosquitoes and other biting insects. The adult form lives in the blood and lymphatic tissues and can cause inflammation and obstruction of lymphatic vessels.

filariasis (fĭl′ə-rī′ə-sĭs) Any of various infections, often of the skin, eyes, and lymph nodes, caused by infestation of tissue with filariae.

file (fīl) A collection of related data or program records stored as a unit with a single name. Files are the basic units that a computer works with in storing and retrieving data.

filename (fīl′nām′) A name given to a computer file to distinguish it from other computer files. Filenames often contain an extension that classifies the file by type.

File Transfer Protocol See **FTP.**

filter (fĭl′tər) **1.** A material that has very tiny holes and is used to separate out solid particles contained in a liquid or gas that is passed through it. **2.** A device that allows signals with certain properties, such as signals lying in a certain frequency range, to pass while blocking the passage of others. For example, filters on photographic lenses allow only certain frequencies of light to enter the camera, while polarizing filters allow only light polarized along a given plane to pass. Radio tuners are filters that allow frequencies of only a narrow range to pass into an amplification circuit.

filter feeder An aquatic animal, such as a clam or sponge, that feeds by filtering tiny organisms or fine particles of organic material from currents of water that pass through it.

filtration (fĭl-trā′shən) The act or process of filtering, especially the process of passing a liquid or gas, such as air, through a filter in order to remove solid particles.

fin (fĭn) One of the winglike or paddlelike parts of a fish, dolphin, or whale that are used for propelling, steering, and balancing in water.

finite (fī′nīt′) **1.** Relating to a set that cannot be put into a one-to-one correspondence with any proper subset of its own members. **2.** Relating to or being a numerical quantity describing the size of such a set. **3.** Being a member of the set of real or complex numbers. **4.** Being a quantity that is non-zero and not infinite.

finite state machine A model of a computational system, consisting of a set of states (including a start state), an alphabet of symbols that serves as a set of possible inputs to the machine, and a transition function that maps each state to another state (or to itself) for any given input symbol. The machine operates by being fed a string of symbols, and moves through a series of states. The study of the computational power of finite state machines and other related machines is important in computer science and linguistics. The computational core of a **Turing machine** is a finite state machine. Also called *finite state automaton.*

firewall (fîr′wôl′) A software program or hardware device that restricts communication between a private network or computer system and outside networks.

firn (fîrn) Granular, partially consolidated snow that has passed through one summer melt season but is not yet glacial ice. Firn becomes glacial ice once it has become impermeable to liquid water.

firth (fûrth) A long, narrow inlet of the sea. Firths are usually the lower part of an estuary, but are sometimes fjords.

first quantum number (fîrst) See under **quantum number.**

fish (fĭsh) Plural **fish** or **fishes.** Any of numerous cold-blooded vertebrate animals that live in water. Fish have gills for obtaining oxygen, a lateral line for sensing pressure changes in the water, and a vertical tail. Most fish are covered with scales and have limbs in the form of fins. Fish were once classified together as a single group, but are now known to compose numerous evolutionarily distinct classes, including the **bony fish, cartilaginous fish, jawless fish, lobe-finned fish,** and **placoderms**.

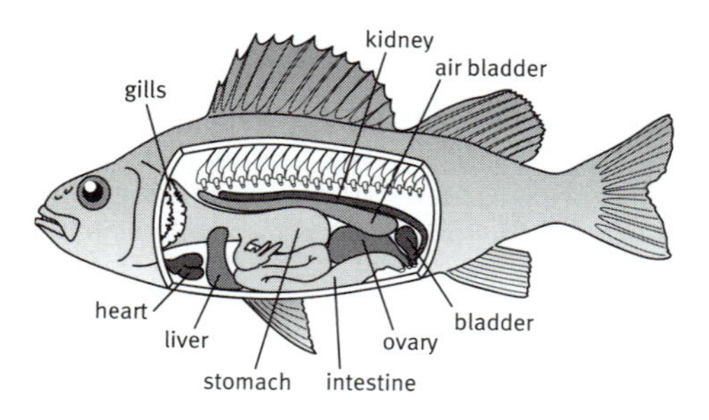

fish *(anatomy of a female bony fish)*

fission (fĭsh′ən) **1.** The splitting of the nucleus of an atom into two or more nuclei. Fission occurs spontaneously, generally when a nucleus has an excess of neutrons, resulting in the inability of the **strong force** to bind the protons and neutrons together. The fission reaction used in many nuclear reactors and bombs involves the absorption of neutrons by uranium-235 nuclei, which immediately undergo fission, releasing energy and **fast neutrons.** Compare **fusion. 2.** A process of asexual reproduction in which a single cell splits to form two identical, independent cells. In fission, the chromosomal DNA replicates before the cell divides. Most bacteria and other prokaryotes reproduce by means of fission. Also called *binary fission.*

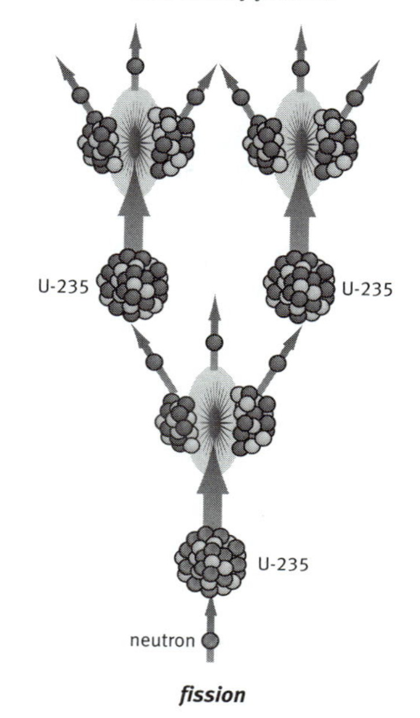

fission

three nuclei of uranium-235 undergoing fission in a chain reaction

fission track A track left by the decay of uranium atoms in a mineral. By analyzing the number and nature of these tracks in a sample mineral and inducing further fission of the remaining uranium by irradiation with neutrons to create new tracks for comparison, one can estimate the approximate age of the sample.

fissure (fĭsh′ər) A long, narrow crack or opening in the face of a rock. Fissures are often filled with minerals of a different type from those in the surrounding rock.

fix (fĭks) **1.** To convert inorganic carbon or nitrogen into stable, organic compounds that can be assimilated into organisms. Photosynthetic organisms such as green plants fix carbon in carbohydrates as food; certain bacteria fix nitrogen as ammonia that can be absorbed directly or through nitrification by plant roots. See more at **carbon fixation, nitrogen fixation.** **2.** To convert a substance, especially a gas, into solid or liquid form by chemical reactions. **3.** To kill and preserve a tissue specimen rapidly to retain as nearly as possible the characteristics it had in the living body.

fixed star (fĭkst) A star or other celestial object so distant from Earth that its position in relation to other stars appears not to change over time. The fixed stars, which include virtually all visible objects beyond the solar system, form the background against which the motions of the Sun, planets, and other bodies of the solar system are measured, and they provide the reference for determining sidereal time. In actuality, no celestial object has a fixed position in relation to any other, and the movement of so-called fixed stars can be measured by precise observation over long periods of time. See more at **proper motion.**

fjeld (fyĕld) A high, barren plateau. The most well-known fjeld is on the Scandinavian Peninsula.

fjord (fyôrd) A long, narrow, deep inlet from the sea between steep slopes of a mountainous coast. Fjords usually occur where ocean water flows into valleys formed near the coast by glaciers.

flagellate (flăj′ə-lāt′) Any of various protozoans of the subphylum Mastigophora that move by means of one or more flagella. Some flagellates can make food by photosynthesis (such as euglenas and volvox), and are often classified as green algae by botanists. Others are symbiotic or parasitic (such as trypanosomes). Flagellates are related to amoebas. Also called *mastigophoran.*

flagellum (flə-jĕl′əm) Plural **flagella.** A slender whiplike part extending from some single-celled organisms, such as the dinoflagellates, that moves rapidly back and forth to impart movement to the organism.

flake (flāk) **1.** A relatively thin, sharp-edged stone fragment removed from a core or from another flake by striking or prying, serving as a tool or blade itself or as a blank for making other tools. See more at **flake tool.** **2.** A small, symmetrical, six-sided crystal of snow. Flakes can be large or small and wet or dry, depending on weather conditions. They are white in color because of their large number of reflecting surfaces.

flake tool A stone tool consisting of a flake that is often modified by further chipping or flaking. Although sharp-edged flakes removed in the production of **core tools** were presumably used as cutting tools by early humans, true flake tools are those that were purposefully removed from the core and then modified for a particular purpose. Flake tools include scrapers, awls, projectile points, and blades. Compare **core tool.**

flame (flām) The hot, glowing mixture of burning gases and tiny particles that arises from combustion. Flames get their light either from the fluorescence of molecules or ions that have become excited, or from the incandescence of solid particles involved in the combustion process, such as the carbon particles from a candle.

flare star (flâr) A variable star that sporadically displays sudden increases in brightness, sometimes becoming six times as bright in a matter of only a few minutes, after which it gradually fades back to its normal brightness. The cause of this phenomenon is thought to be an event similar to a solar flare, but on a more intense scale with a much higher emission of x-rays.

flash flood (flăsh) A sudden, localized flood of great volume and short duration, typically caused by unusually heavy rain in a semiarid area. Flash floods can reach their peak volume in a matter of a few minutes and often carry large loads of mud and rock fragments.

flash memory A kind of ROM that retains data when power is turned off and that can be electronically erased and reprogrammed without being removed from the circuit board.

flash point The lowest temperature at which the vapor of a flammable liquid will ignite in air. The flash point is generally lower than the temperature needed for the liquid itself to ignite.

flask (flăsk) A rounded container with a long neck, used in laboratories.

flat-panel display (flăt′păn′əl) A thin, often lightweight video display used in computer monitors and televisions as an alternative to the cathode-ray tube. Flat-panel displays often employ liquid crystals or electroluminescent materials such as light-emitting diodes. Also called *flat screen.*

flatworm (flăt′wûrm′) Any of various parasitic and nonparasitic worms of the phylum Platyhelminthes, characteristically having a soft, flat, bilaterally symmetrical body. Flatworms lack a coelom (body cavity), respiratory system, and circulatory system, but are the most primitive invertebrates to have a brain. The evolutionary history of flatworms is uncertain, but they share some basic characteristics with rotifers, nematodes, and a few other invertebrate phyla. Cestodes (tapeworms), planarians, and trematodes (flukes) are flatworms.

flavonoid (flā′və-noid′) Any of a large group of water-soluble plant pigments that are beneficial to health. Flavonoids are polyphenols and have antioxidant, anti-inflammatory, and antiviral properties. They also help to maintain the health of small blood vessels and connective tissue, and some are under study as possible treatments of cancer. Also called *bioflavonoid.*

flavor (flā′vər) Any of six classifications of quark varieties, distinguished by mass and electric charge. The flavors have the names up, down, strange, charm, top, and bottom. Protons in atomic nuclei are composed of two up quarks and one down quark, while neutrons consist of one up quark and two down quarks. The flavor of a quark may be changed in interactions involving the **weak force**.

F layer See **F region.**

fledgling (flĕj′lĭng) A young bird that has just grown the feathers needed to fly and is capable of surviving outside the nest.

Fleming (flĕm′ĭng), Sir **Alexander** 1881–1955. Scottish bacteriologist who discovered penicillin in 1928. The drug was developed and purified 11 years later by Howard Florey and Ernst Chain, with whom Fleming shared a 1945 Nobel Prize for physiology or medicine. Fleming was also the first to administer typhoid vaccines to humans.

BIOGRAPHY **Alexander Fleming**

Many famous scientific discoveries come about by accident, and such was the case with penicillin. The first and still best-known antibiotic, penicillin is a natural substance excreted by a type of mold of the genus *Penicillium.* It so happened that a Scottish bacteriologist, Alexander Fleming, was doing research on staphylococcal bacteria in the late 1920s and noticed that one culture had become contaminated with some mold. What was curious was that there was a circular area around the mold that was free of bacterial growth. After some investigation, Fleming discerned that the mold was excreting a substance deadly to the bacteria, and he named it penicillin in the mold's honor. Fleming had already discovered another natural antibacterial substance a few years earlier in 1921—lysozyme, an enzyme contained in tears and saliva. But the discovery of penicillin was of far greater importance, although its impact was not fully felt right away because Fleming lacked the equipment necessary to isolate the active compound and to synthesize it in quantities that could be used medicinally. This happened a dozen years later during World War II and stimulated the development of new drugs that could fight infections transmitted on the battlefield. Two other scientists, Ernst Chain and Howard Florey, were responsible for this further work, and together with Fleming the three shared the 1945 Nobel Prize for physiology or medicine.

Fleming, Sir **John Ambrose** 1849–1945. British physicist and electrical engineer who devised the first electron tube in 1904. His invention was essential to the development of radio, television, and early computer circuitry. Fleming also helped develop electric devices designed for large-scale use, such as the electric lamp.

flexor (flĕk′sər) A muscle that bends or flexes a joint. Compare **extensor.**

flint (flĭnt) **1.** A very hard, gray to black variety of chalcedony that makes sparks when struck with steel. It breaks with a conchoidal fracture. **2.** The dark gray to black variety of chert.

flipper (flĭp′ər) A wide, flat limb adapted for swimming, found on aquatic animals such as whales, seals, and sea turtles. Flippers evolved from legs.

float (flōt) An air-filled sac in certain aquatic organisms, such as kelp, that helps maintain buoyancy. Also called *air bladder, air vesicle.*

floating-point (flō′tĭng-point′) Relating to a method of representing numerical quantities that uses two sets of integers, a mantissa and a characteristic, in which the value of the number is understood to be equal to the mantissa multiplied by a base (often 10) raised to the power of the characteristic. Scientific notation is one means of displaying floating-point numbers.

flocculation (flŏk′yə-lā′shən) The process by which individual particles of clay aggregate into clotlike masses or precipitate into small lumps. Flocculation occurs as a result of a chemical reaction between the clay particles and another substance, usually salt water.

floe (flō) A mass or sheet of floating ice.

flood (flŭd) A temporary rise of the water level, as in a river or lake or along a seacoast, resulting in its spilling over and out of its natural or artificial confines onto land that is normally dry. Floods are usually caused by excessive runoff from precipitation or snowmelt, or by coastal storm surges or other tidal phenomena. ► Floods are sometimes described according to their statistical occurrence. A **fifty-year flood** is a flood having a magnitude that is reached in a particular location on average once every fifty years. In any given year there is a two percent statistical chance of the occurrence of a fifty-year flood and a one percent chance of a **hundred-year flood**.

floodplain (flŭd′plān′) Flat land bordering a river and made up of alluvium (sand, silt, and clay) deposited during floods. When a river overflows, the floodplain is covered with water.

flood tide The period between low tide and high tide, during which water flows toward the shore. Compare **ebb tide.** See more at **tide.**

floodway (flŭd′wā′) A channel for an overflow of water caused by flooding.

floppy disk (flŏp′pē) A flexible plastic disk coated with magnetic material and covered by a protective jacket, used for storing data. Floppy disks were once the principal storage medium for personal computers, but inexpensive hard disks and writable compact disks have greatly diminished their role.

flora (flôr′ə) Plural **floras** or **florae** (flôr′ē′). **1.** The plants of a particular region or time period. **2.** The bacteria and other microorganisms that normally inhabit a bodily organ or part, such as the intestine.

floral envelope (flôr′əl) See **perianth.**

floral tube A tube formed by the fusion of parts of the perianth and often the stamens at their bases, as in flowers of the daffodil or morning glory.

floret (flôr′ĭt) A small or reduced flower, especially one that is part of a larger inflorescence, such as those of the grasses and plants of the composite family.

Florey (flôr′ē), Sir **Howard Walter.** Baron Florey of Adelaide. 1898–1968. Australian-born British pathologist who developed and purified penicillin with Ernst Chain in 1939. For this work, Florey and Chain shared a 1945 Nobel Prize with Alexander Fleming, who first discovered the antibiotic in 1928. Florey also supervised the clinical testing and mass production of the drug in the United States.

flower (flou′ər) The reproductive structure of the seed-bearing plants known as angiosperms. A flower may contain up to four whorls or arrangements of parts: carpels, stamens, petals, and sepals. The female reproductive organs consist of one or more **carpels.** Each carpel includes an ovary, style, and stigma. A single carpel or a group of fused carpels is sometimes called a pistil. The male reproductive parts are the **stamens,** made up of a filament and anther. The reproductive organs may be enclosed in an inner whorl of **petals** and an outer whorl of **sepals.** Flowers first appeared over 120 million years ago and have evolved a great diversity of forms and coloration in response to the agents that pollinate them. Some flowers produce nectar to attract animal pollinators, and these flowers are often highly adapted to specific groups of pollinators. Flowers pollinated by moths, such as species of jasmine and nicotiana, are

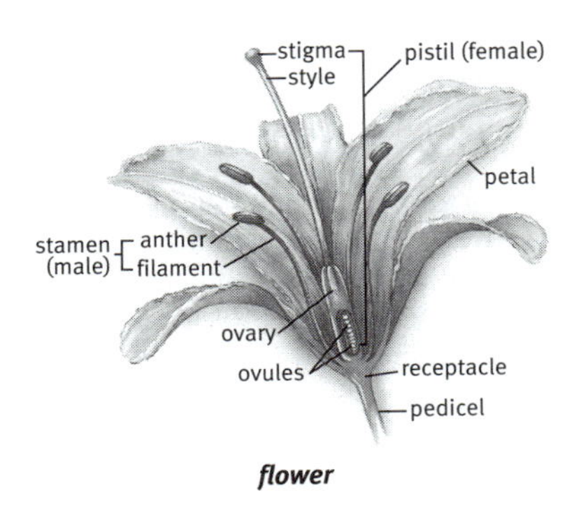

flower

often pale and fragrant in order to be found in the evening, while those pollinated by birds, such as fuschias, are frequently red and odorless, since birds have good vision but a less developed sense of smell. Wind-pollinated flowers, such as those of oak trees or grass, are usually drab and inconspicuous. See Note at **pollination.**

flower cluster See **inflorescence.**

flower head **1.** A short, dense, indeterminate inflorescence of sessile flowers, as of composite plants or clover. See more at **composite family.** **2.** A very dense grouping of flower buds, as in broccoli and cauliflower.

flowering plant (flou′ər-ĭng) A plant that produces flowers and fruit; an angiosperm. See more at **angiosperm.**

fl oz Abbreviation of **fluid ounce.**

flu (flo͞o) See **influenza.**

fluid (flo͞o′ĭd) A state of matter, such as liquid or gas, in which the component particles (generally molecules) can move past one another. Fluids flow easily and conform to the shape of their containers. See also **state of matter, viscosity.**

fluid dram A unit of liquid volume or capacity in the US Customary System equal to $\frac{1}{8}$ of a fluid ounce (3.70 milliliters).

fluid mechanics The scientific study of the mechanical properties of fluids, especially their behavior when subject to internal and external forces. Compare **aerodynamics, hydrodynamics.**

fluid ounce A unit of liquid volume or capacity in the US Customary System equal to $\frac{1}{16}$ of a pint (29.57 milliliters). See Table at **measurement.**

fluke (flo͝ok) **1.** Either of the two flattened fins of a whale's tail. **2.** See **trematode.**

fluorescein (flo͝o-rĕs′ē-ĭn, flô-) An orange-red crystalline compound that exhibits intense fluorescence in alkaline solution. It is used in medicine for diagnostic purposes, in oceanography as a tracer, and as a textile dye. *Chemical formula:* $C_{20}H_{12}O_5$.

fluorescence (flo͝o-rĕs′əns) **1.** The giving off of light by a substance when it is exposed to electromagnetic radiation, such as visible light or x-rays. As long as electromagnetic radiation continues to bombard the substance, electrons in the fluorescent material become excited but return very quickly to lower energy, giving off light, always of the same frequency. Fluorescent dyes are often used in microscopic imaging, where different dyes can penetrate and illuminate different parts of the sample being examined, helping to distinguish its structures. Compare **phosphorescence** (sense 1). **2.** The light produced in this way.

fluorescent lamp (flo͝o-rĕs′ənt) An electric lamp that produces light through fluorescence. In most fluorescent lamps, a mixture of argon and mercury gas contained in a glass bulb is stimulated by an electric current, producing ultraviolet rays. These rays strike a fluorescent phosphor coating on the interior surface of the bulb, causing it to emit visible light. Fluorescent lamps are much more efficient than incandescent lamps because very little energy is lost as heat. Compare **incandescent lamp.**

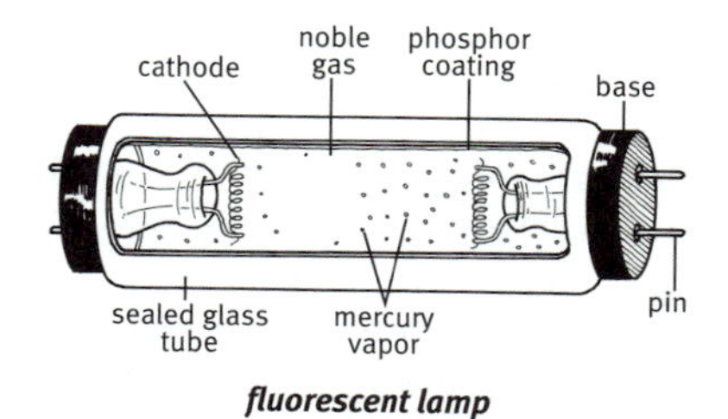

fluorescent lamp

fluoridate (flo͝or′ĭ-dāt′) To add fluorine or a fluoride to something, especially to drinking water in order to prevent tooth decay.

fluoride (flo͝or′īd′) A compound containing fluorine and another element or radical. Fluorine combines readily with nearly all the other elements, except the noble gases, to form fluorides. In some countries, fluoride is added to the drinking water as a preventive measure against tooth decay.

fluorine (flo͝or′ēn′) *Symbol* **F** A pale-yellow, poisonous, gaseous element of the halogen group. It is highly corrosive and is used to separate certain isotopes of uranium and to make refrigerants and high-temperature plastics. Fluorine is also added in fluoride form to the water supply to prevent tooth decay. Atomic number 9; atomic weight 18.9984; melting point –223°C; boiling point

−188.14°C; specific gravity of liquid 1.108 (at boiling point); valence 1. See **Periodic Table.**

fluorite (flo͝or′īt′) A transparent to translucent mineral occurring in many colors, especially yellow and purple, and usually in cube-shaped crystals with octahedral cleavage. It is found in sedimentary rocks and in ore deposits within igneous rocks. It is often fluorescent in ultraviolet light. *Chemical formula:* **CaF_2.**

fluorocarbon (flo͝or′ō-kär′bən) An inert, liquid or gaseous organic compound similar to a hydrocarbon but having fluorine atoms in the place of hydrogen atoms. Fluorocarbons are used in aerosol propellants and refrigerants.

fluoroscope (flo͝o-rŏs′kə-pē) A radiologic instrument that is equipped with a fluorescent screen on which opaque internal structures can be viewed as moving shadow images formed by the differential transmission of x-rays through the body. —*Noun* **fluoroscopy** (flo͝o-rŏs′kə-pē).

fluvial (flo͞o′vē-əl) **1.** Relating to or inhabiting a river or stream. **2.** Produced by the action of a river or stream.

fluviomarine (flo͞o′vē-ō-mə-rēn′) Relating to deposits, especially near the mouth of a river, formed by the combined action of river and sea.

flux (flŭks) **1.** The rate of flow of fluids, particles, or energy across a given surface or area. **2.** The presence of a field of force in a region of space, represented as a set of lines indicating the direction of the force. The density of the lines indicates the strength of the force. Lines used to represent magnetic fields in depictions of magnets, for example, follow the lines of flux of the field. See also **field, magnetic flux. 3.** A measure of the strength of such a field. **4.** A readily fusible glass or enamel used as a base in ceramic work. **5.** An additive that improves the flow of plastics during fabrication. **6.** A substance applied to a surface to be joined by welding, soldering, or brazing to facilitate the flowing of solder and prevent formation of oxides. **7.** A substance used in a smelting furnace to make metals melt more easily.

flux density Field flux per unit area.

fly (flī) Any of numerous insects of the order Diptera, having one pair of wings and large compound eyes. Flies include the houseflies, horseflies, and mosquitoes. See more at **dipteran.**

Fm The symbol for **fermium.**

FM Abbreviation of **frequency modulation.**

foam (fōm) **1.** Small, frothy bubbles formed in or on the surface of a liquid, as from fermentation or shaking. **2.** A colloid in which particles of a gas are dispersed throughout a liquid. Compare **aerosol, emulsion.**

focal length (fō′kəl) The distance between the optical center of a lens or mirror to its **focal point**.

focal point The point at which all radiation coming from a single direction and passing through a lens or striking a mirror converges. Also called *focus.*

focus (fō′kəs) Plural **focuses** or **foci** (fō′sī′, fō′kī′). **1.** The degree of clarity with which an eye or optical instrument produces an image. **2.** See **focal point. 3.** A central point or region, such as the point at which an earthquake starts. **4.** *Mathematics.* A fixed point or one of a pair of fixed points used in generating a curve such as an ellipse, parabola, or hyperbola. **5.** The region of a localized bodily infection or disease.

foehn also **föhn** (fœn, fān) A warm, dry, and often strong wind coming off the lee slopes of a mountain range, especially off the northern slopes of the Alps. A foehn is a katabatic wind that warms as it descends because it has dropped its moisture before crossing the mountain range and is put under greater atmospheric pressure as it moves downward. Various local names are also used for foehns (such as *chinook* in the Rocky Mountain regions). A foehn can cause sudden and dramatic increases in the temperature—from 10° to 20°C (50° to 68°F) in a few minutes—which can cause snow to melt rapidly and even trigger flooding. See also **chinook.**

fog (fôg) **1.** A dense layer of cloud lying close to the surface of the ground or water and reducing visibility to less than 1 km (0.62 mi). Fog occurs when the air temperature becomes identical, or nearly identical, to the dew point. **2.** An opaque or semiopaque condensation of a substance floating in a region or forming on a surface.

fogbow (fôg′bō′) A faint white or yellowish arc-shaped light that sometimes appears in fog opposite the Sun. A fogbow is similar to a rainbow but forms in a cloud or in fog. Also called **cloudbow.**

fold (fōld) A bend in a layer of rock or in another planar feature such as foliation or the cleavage of a mineral. Folds occur as the result of deformation, usually associated with plate-tectonic forces.

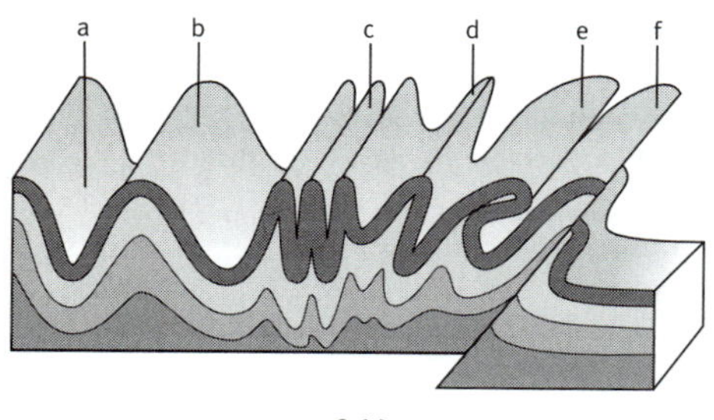

fold

types of folds: a. *anticline,* b. *syncline,* c. *isocline,* d. *overturned,* e. *recumbent,* f. *overthrust*

foliation (fō′lē-ā′shən) The set of layers visible in many metamorphic rocks as a result of the flattening and stretching of mineral grains during metamorphism. —*Adjective* **foliated.**

folic acid (fō′lĭk, fŏl′ĭk) A water-soluble vitamin belonging to the vitamin B complex that is necessary for the formation of red blood cells and important in embryonic development. It is also the parent compound of coenzymes in various metabolic reactions. Folic acid is found especially in green leafy vegetables, liver, and fresh fruit. Deficiency of folic acid in the diet results in anemia. *Chemical formula:* $\mathbf{C_{19}H_{19}N_7O_6}$.

folium (fō′lē-əm) Plural **folia. 1.** A thin, leaflike layer or stratum occurring especially in metamorphic rock. **2.** A plane cubic curve having a single loop, a node, and two ends asymptotic to the same line. Also called *folium of Descartes.*

follicle (fŏl′ĭ-kəl) **1.** A small, protective sac, gland, or cluster of cells in the body. In mammals, unfertilized eggs develop in follicles located in the ovaries. Hair grows from follicles in the skin. **2.** A dry, dehiscent fruit that develops from a single carpel, has a single chamber, and splits open along only one seam to release its seeds. The pod of the milkweed and the fruit of the magnolia are follicles.

follicle-stimulating hormone A glycoprotein hormone secreted by the anterior portion of the pituitary gland. It stimulates the growth of follicles in the ovary and induces the formation of sperm in the testis.

food chain (fo͞od) The sequence of the transfer of food energy from one organism to another in an ecological community. A food chain begins with a **producer**, usually a green plant or alga that creates its own food through photosynthesis. In the typical predatory food chain, producers are eaten by **primary consumers** (herbivores) which are eaten by **secondary consumers** (carnivores), some of which may in turn be eaten by **tertiary consumers** (the top carnivore in the chain). ► Many species of animals in an ecological community feed on both plants and animals and thus play multiple roles in the chain. Parasites feed on living tissues, generally without killing their hosts, and may themselves be hosts to smaller parasites. In addition, organisms that die without being eaten are consumed by detritivores, some of which serve as prey for other consumers. The complex system of interrelated food chains in an environment is known as a **food web.** See more at **trophic level.**

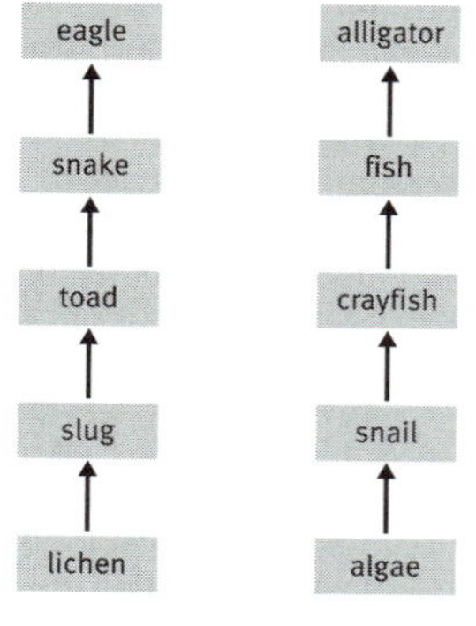

food chain

examples of terrestrial (left) *and aquatic* (right) *food chains*

food poisoning An acute gastrointestinal disorder characterized by vomiting and diarrhea, caused by eating food contaminated with bacteria, usually of the genus *Salmonella* or the genus *Staphylococcus*, which produces a toxin.

food pyramid 1. A graphic representation of the structure of a food chain, depicted as a pyramid having a broad base formed by producers and tapering to a point formed by end consumers. Between successive trophic levels, total biomass decreases as energy is lost from the system. See more at **trophic level. 2.** A pyramid-shaped diagram representing a set of dietary guidelines for humans, typically based on a recommended number of servings from each of several food groups. Foods along the broadest row, at the bottom, are considered basic to human nutrition and have the highest recommended number of servings. Foods in the narrowest part, at the top, are considered to be nonessential and have the fewest number of recommended servings. In the middle row or rows are foods whose recommended servings fall between those two groups.

food web See under **food chain.**

fool's gold (fo͞olz) Any of several minerals, especially pyrite and chalcopyrite, sometimes mistaken for gold.

foot (fo͝ot) Plural **feet** (fēt). A unit of length in the US Customary System equal to $\frac{1}{3}$ of a yard or 12 inches (30.48 centimeters). See Table at **measurement.**

foot-and-mouth disease A highly contagious disease of cattle and other hoofed animals caused by any of various viruses of the family Picornaviridae and the genus *Aphthovirus*, characterized by fever and the presence of blisters around the mouth and hooves.

foot-pound 1. A unit of work equal to the work or energy needed to lift a one-pound weight a distance of one foot against the force of the Earth's gravity. One foot pound is equivalent to 1.3558 joules. **2.** A unit of torque equal to a pound of force acting perpendicularly to an axis of rotation at a distance of one foot. Also called *pound-foot.*

footwall (fo͝ot**′**wôl′) The block of rock lying under an inclined geologic fault plane. See more at **fault.** Compare **hanging wall.**

foramen (fə-rā**′**mən) Plural **foramina** (fə-răm**′**ə-nə) or **foramens.** An opening or short passage, especially in the body. ► The large opening in the base of the skull through which the spinal cord passes is called the **foramen magnum** (măg**′**nəm). ► The opening in the septum between the right and left atria of the heart, present in the fetus but usually closed soon after birth, is the **foramen ovale** (ō-văl**′**ē, -vā**′**lē, -vä**′**-).

foraminiferan (fə-răm′ə-nĭf**′**ər-ən) Any of various chiefly marine protozoans of the order Foraminiferida or Foraminifera, having a body enclosed by a shell called a *test* and making up an important constituent of plankton. Perforations in a foraminiferan's test allow the protrusion of numerous long extensions (pseudopods), which form a net used to trap food. The tests of foraminiferans grow throughout the organism's life, and can exceed 5 cm (2 inches) in diameter. The tests of dead organisms form ooze found on the ocean floor. Extinct foraminiferans are important index fossils.

forb (fôrb) A broad-leaved herb (as opposed to a grass), especially one growing in a field, prairie, or meadow.

force (fôrs) **1.** Any of various factors that cause a body to change its speed, direction, or shape. Force is a vector quantity, having both magnitude and direction. Contributions of force from different sources can be summed to give the net force at any given point. **2.** Any of the four natural phenomena involving the interaction between particles of matter. From the strongest to the weakest, the four forces are the **strong nuclear force**, the **electromagnetic force**, the **weak nuclear force**, and **gravity**.

force carrier Any of four elementary particles that mediate one of the four fundamental forces. Force carriers are **bosons**. See more at **fundamental force.**

forebrain (fôr**′**brān′) The forwardmost part of the vertebrate brain. In humans, it consists of the thalamus, the hypothalamus, and the cerebrum. Compare **hindbrain, midbrain.**

foreland basin (fôr**′**lənd) A low-lying region that is adjacent and parallel to a mountain belt formed as the result of the collision of tectonic plates. Foreland basins form when the lithosphere flexes downward in front of a mountain belt in response to the added load of thickened crust that results from the collision of the two plates. Sediments eroded from the mountain belt accumulate in the foreland basin, causing it to further subside and make room for additional sediments.

forensic medicine (fə-rĕn′sĭk) The branch of medicine that interprets or establishes the medical facts in civil or criminal law cases.

foreshore (fôr′shôr′) The seaward-sloping area of a shore that lies between the average high tide mark and the average low tide mark. Compare **backshore.**

forest (fôr′ĭst) A dense growth of trees and underbrush covering a large area. Forests exist in all regions of the Earth except for regions of extreme cold or dryness.

forestry (fôr′ĭ-strē) The scientific study of the cultivation, maintenance, and management of forests.

formaldehyde (fôr-măl′də-hīd′) A colorless gas having a sharp, suffocating odor. It is used in making plastics and, when dissolved in a solution of water and methanol, to preserve biological specimens. *Chemical formula:* CH_2O.

format (fôr′măt′) *Noun.* **1.** The arrangement of data for storage or display. —*Verb.* **2.** To divide a disk into marked sectors so that it may store data. **3.** To determine the arrangement of data for storage or display.

formate (fôr′māt′) A salt or ester of formic acid, containing the group HCOO.

formation (fôr-mā′shən) A long, mappable body of rock that is recognizable by its physical characteristics and by its location within the rock record.

formic acid (fôr′mĭk) A colorless, caustic, fuming liquid that occurs naturally as the poison of ants and stinging nettles. It is used in making textiles and paper and in insecticides. Formic acid is the simplest organic acid, containing a carboxyl (COOH) group attached to a hydrogen. *Chemical formula:* CH_2O_2.

formula (fôr′myə-lə) Plural **formulas** or **formulae** (fôr′myə-lē′). **1.** A set of symbols showing the composition of a chemical compound. A formula lists the elements contained within it and indicates the number of atoms of each element with a subscript numeral if the number is more than 1. For example, H_2O is the formula for water, where H_2 indicates two atoms of hydrogen and O indicates one atom of oxygen. **2.** A set of symbols expressing a mathematical rule or principle. For example, the formula for the area of a rectangle is $a = lw$, where a is the area, l the length, and w the width.

formyl (fôr′mĭl′) The radical HCO, derived from formic acid.

Forrester (fôr′ĭs-tər), **Jay Wright** Born 1918. American computer engineer who pioneered the development of computer storage devices. In 1949 he devised the first magnetic core memory for an electronic digital computer.

forsterite (fôr′stə-rīt′) See under **olivine.**

Fossey (fŏs′ē), **Dian** 1932–1985. American zoologist who conducted extensive studies of mountain gorillas in Rwanda, Africa. Her research brought about a new understanding of the gorilla's behavior and habitat and supported conservation efforts in Africa.

fossil (fŏs′əl) The remains or imprint of an organism from a previous geologic time. A fossil can consist of the preserved tissues of an organism, as when encased in amber, ice, or pitch, or more commonly of the hardened relic of such tissues, as when organic matter is replaced by dissolved minerals. Hardened fossils are often found in layers of sedimentary rock and along the beds of rivers that flow through them. See also **index fossil, microfossil, trace fossil.** —*Verb* **fossilize.**

fossil

left: *fossilized fern of the Devonian Period, from Quebec, Canada;* right: *fossilized bird of the Eocene Epoch, from Wyoming*

fossil fuel A hydrocarbon deposit, such as petroleum, coal, or natural gas, derived from the accumulated remains of ancient plants and animals and used as fuel. Carbon dioxide and other greenhouse gases generated by burning fossil fuels are considered to be one of the principal causes of global warming.

Foucault (fo͞o-kō′), **Jean Bernard Léon** 1819–1868. French physicist who determined that light travels more slowly in water than in air, confirming predictions made by the wave theory of light. In 1850 Foucault also measured the absolute velocity of light. In 1851, by using a type of pendulum that is now named after him, Foucault demonstrated the rotation of the Earth, and in 1852 perfected a gyroscope for the same purpose.

Foucault pendulum A pendulum suspended from a long wire, set into motion, and sustained in motion over long periods. Due to the axial rotation of the earth, the plane of motion of the pendulum shifts at a rate and direction dependent on its latitude, clockwise in the Northern Hemisphere and counterclockwise in the Southern Hemisphere. At the poles the plane rotates once per day, while at the equator it does not rotate at all.

Fourier (fo͝or′ē-ā′, fo͞o-ryā′), Baron **Jean Baptiste Joseph** 1768–1830. French mathematician and physicist who introduced the expansion of periodic functions in the trigonometric series that is now named for him. He also studied the conduction of heat in solid bodies.

Fourier analysis The branch of mathematics concerned with the approximation of periodic functions by the Fourier series and with generalizations of such approximations to a wider class of functions.

Fourier series An infinite series whose terms are constants multiplied by sine and cosine functions and that can, if uniformly convergent, approximate a wide variety of functions.

Fourier's theorem Any of a set of theorems stating that a function may be represented by a Fourier series if it meets certain general continuity and periodicity conditions.

fourth dimension (fôrth) Time regarded as a coordinate dimension. A fourth dimension is required by relativity theory, along with three spatial dimensions, to specify completely the location of any event.

fourth quantum number See under **quantum number.**

Fr The symbol for **francium.**

fractal (frăk′təl) A complex geometric pattern

A CLOSER LOOK **fractal**

Fractals are often associated with recursive operations on shapes or sets of numbers, in which the result of the operation is used as the input to the same operation, repeating the process indefinitely. The operations themselves are usually very simple, but the resulting shapes or sets are often dramatic and complex, with interesting properties. For example, a fractal set called a Cantor dust can be constructed beginning with a line segment by removing its middle third and repeating the process on the remaining line segments. If this process is repeated indefinitely, only a "dust" of points remains. This set of points has zero length, even though there is an infinite number of points in the set. The Sierpinski triangle (or Sierpinski gasket) is another example of such a recursive construction procedure involving triangles (see the illustration). Both of these sets have subparts that are exactly the same shape as the entire set, a property known as *self-similarity.* Under certain definitions of dimension, fractals are considered to have non-integer dimension: for example, the dimension of the Sierpinski triangle is generally taken to be around 1.585, higher than a one-dimensional line, but lower than a two-dimensional surface. Perhaps the most famous fractal is the Mandelbrot set, which is the set of complex numbers C for which a certain very simple function, $Z^2 + C$, iterated on its own output (starting with zero), eventually converges on one or more constant values. Fractals arise in connection with nonlinear and chaotic systems, and are widely used in computer modeling of regular and irregular patterns and structures in nature, such as the growth of plants and the statistical patterns of seasonal weather.

Each figure above is constructed from the figure to its left by removing an inverted triangle from the center of each solid triangle. Repeating this process indefinitely results in a fractal comprised of a set of points known as a Sierpinski triangle.

exhibiting **self-similarity** in that small details of its structure viewed at any scale repeat elements of the overall pattern. See also **chaos.**

fraction (frăk′shən) **1.** A number that compares part of an object or a set with the whole, especially the quotient of two whole numbers written in the form $\frac{a}{b}$. The fraction $\frac{1}{2}$, which means 1 divided by 2, can represent such things as 10 pencils out of a box of 20, or 50 cents out of a dollar. See also **decimal fraction, improper fraction, proper fraction. 2.** A chemical component separated by fractionation.

fractional crystallization (frăk′shə-nəl) A process by which a chemical compound is separated into components by crystallization. In fractional crystallization the compound is mixed with a solvent, heated, and then gradually cooled so that, as each of its constituent components crystallizes, it can be removed in its pure form from the solution.

fractional distillation A process by which a chemical compound is separated into components by distillation. In fractional distillation the compound is heated and, as each of its constituent components comes to a boil, its vapors are separated and cooled, so it can be removed in its pure form. Fractional distillation is used to refine petroleum. See also **distillation.**

fractionation (frăk′shə-nā′shən) The separation of a chemical compound into components by fractional crystallization or distillation.

fracture (frăk′chər) A break or rupture in bone tissue. ► A **comminuted fracture** results in more than two fragments. ► Although most fractures are caused by a direct blow or sudden, twisting force, **stress fractures** result from repetitive physical activity. ► In an **incomplete fracture**, the fracture line does not completely traverse the bone.

fragmentation (frăg′mən-tā′shən) The scattering of parts of a computer file across different regions of a disk. Fragmentation occurs when the operating system breaks up the file and stores it in locations left vacant by previously deleted files. The more fragmented the file, the slower it is to retrieve, since each piece of the file must be identified and located on the disk.

frame of reference (frām) See **reference frame.**

frameshift (frām′shĭft′) Relating to a mutation that occurs when one or two nucleotides are added or deleted, with the result that every codon beyond the point of insertion or deletion is read incorrectly during translation. See more at **point mutation.**

francium (frăn′sē-əm) *Symbol* **Fr** An extremely unstable, radioactive element of the alkali group. It is the heaviest metal of the group. Francium occurs in nature, but less than 28.35 g (1 oz) is present in the Earth's crust at any time. It has approximately 19 isotopes, the most stable of which is Fr 223 with a half-life of 21 minutes. Atomic number 87; valence 1. See **Periodic Table.**

Franklin (frăngk′lĭn), **Benjamin** 1706–1790. American public official, scientist, inventor, and writer who fully established the distinction between negative and positive electricity, proved that lightning and electricity are identical, and suggested that buildings could be protected by lightning conductors. He also invented bifocal glasses, established the direction of the prevailing storm track in North America and determined the existence of the Gulf Stream.

Franklin, Rosalind Elsie 1920–1958. British x-ray crystallographer whose diffraction images, made by directing x-rays at DNA, provided crucial information that led to the discovery of its structure as a double helix by Francis Crick and James D. Watson.

Rosalind Franklin

BIOGRAPHY **Rosalind Elsie Franklin**

James D. Watson and Francis Crick's famous double helix model of the structure of DNA is rightly considered one of the greatest scientific discoveries ever made. While Watson and Crick became famous the world over, later sharing the Nobel Prize for physiology or medicine, the contributions of Rosalind Franklin are less well-known, even though her work was crucial to their discovery. Franklin's x-ray photograph depicting the double-helix shape of DNA gave Watson and Crick the essential experimental evidence they needed to determine DNA's structure. Born in London in 1920 to a wealthy Anglo-Jewish family, Franklin attended the University of Cambridge, where she earned a doctorate in physical chemistry. It was there that she learned x-ray crystallography, a process used to determine the structure of molecules by bombarding them with x-rays and analyzing the resultant diffraction patterns. Franklin later accepted a post at King's College London in 1951 to study DNA, thus entering the race to discover the molecule's structure. Without her knowledge, a close colleague at King's, Maurice Wilkins, showed her unpublished research to Watson and Crick, who were then able to establish DNA's configuration and soon after published their findings in the journal *Nature*. When Franklin saw the model produced by Watson and Crick, she accepted it immediately, as it fit with her experimental data. Franklin left King's in 1953 and continued a distinguished career, studying the structure of viruses. She died of ovarian cancer at 37, never knowing how her own work had contributed to their important discovery.

---

free-air correction (frē′âr′) A compensation factor used in gravitational surveys that takes into account the decrease in the force of gravity with increasing altitude, assuming only air intervenes between the observer and sea level. It is equal to –0.3086 mGal per meter above sea level. Also called *Faye Correction*. Compare **Bouguer correction.**

free energy A thermodynamic quantity that is the difference between the internal energy of a system and the product of its absolute temperature and entropy. Free energy is a measure of the capacity of the system to do work. If its value is negative, the system will have a tendency to do work spontaneously, as in an exothermic chemical reaction. Free energy is measured in kilojoules per mole. Also called *Gibbs free energy*.

free radical An atom or group of atoms that has at least one unpaired electron and is therefore unstable and highly reactive. In animal tissues, free radicals can damage cells and are believed to accelerate the progression of cancer, cardiovascular disease, and age-related diseases.

freeware (frē′wâr′) Software that is available to users for free. Freeware is often made available for downloading over the Internet.

freeze (frēz) To change from a liquid to a solid state by cooling or being cooled to the freezing point.

freeze-dry To preserve a substance, such as food, by freezing it rapidly and placing it in a vacuum chamber, where the water frozen in the substance evaporates through sublimation.

freezing point (frē′zĭng) The temperature at which a liquid, releasing sufficient heat, becomes a solid. For a given substance, the freezing point of its liquid form is the same as the melting point of its solid form, and depends on such factors as the purity of the substance and the surrounding pressure. The freezing point of water at a pressure of one atmosphere is 0°C (32°F); that of liquid nitrogen is –209.89°C (–345.8°F). See also **state of matter.**

F region The highest region of the ionosphere, consisting of free electrons generated by the ionizing effect of solar radiation. This region is subdivided into two layers, designated the F_1 layer, and the F_2 layer. The F_1 layer extends from about 150 km (93 mi) to 240 km (149 mi) above the surface of the Earth and exists only during sunlight hours. At night it merges into the F_2 layer. The F_2 layer extends from about 240 km (149 mi) to 400 km (248 mi). The density of free electrons in this layer is much higher than it is in the F_1 layer. It is the main reflecting layer for high frequency communications, both during sunlight hours and during the night. Also called *F layer*. See also **ionosphere.**

frequency (frē′kwən-sē) **1.** *Physics.* The rate at which a repeating event occurs, such as the full cycle of a wave. Frequencies are usually measured in hertz. Compare **amplitude.** See also **period. 2.** *Mathematics.* The ratio of the number of occurrences of some event to the number of opportunities for its occurrence.

frequency distribution A range of conditions, often represented in graph form, in which each item is paired with the number of observed events or measurements meeting those conditions. For example, a list of price ranges paired with the number of cars available in each price range is a frequency distribution. Frequency distributions are commonly used in statistical analysis.

frequency modulation A method of transmitting signals, especially in radio broadcasting, in which the value of the signal is given by the frequency of a high frequency carrier wave. In FM radio transmission, for example, the signal to be carried is a sound wave, and its increasing and decreasing value is reflected in the increasing and decreasing frequency of a radio frequency carrier wave. Compare **amplitude modulation.**

freshwater (frĕsh′wô′tər) Consisting of or living in water that is not salty.

Fresnel (frā-nĕl′), **Augustin Jean** 1788–1827. French physicist whose investigations of the interference, diffraction, and polarization of light helped establish the theory that light moves in a wavelike motion. Fresnel also made great contributions to the field of optics, including the development of a compound lens for use in lighthouses.

Fresnel lens (frə-nĕl′) A thin optical lens consisting of concentric rings of segmental lenses and having a short focal length. Placing a light source at the focal point of the lens gives rise to a strong beam of nearly parallel rays. Fresnel lenses are used primarily in spotlights, lighthouses, and the headlights of motor vehicles.

friction (frĭk′shən) A force on objects or substances in contact with each other that resists motion of the objects or substances relative to each other. ▸ **Static friction** arises between two objects that are not in motion with respect to each other, as for example between a cement block and a wooden floor. It increases to counterbalance forces that would move the objects, up to a certain maximum level of force, at which point the objects will begin moving. It is measured as the maximum force the bodies will sustain before motion occurs. ▸ **Kinetic friction** arises between bodies that are in motion with respect to each other, as for example the force that works against sliding a cement block along a wooden floor. Between two hard surfaces, the kinetic friction is usually somewhat lower than the static friction, meaning that more force is required to set the objects in motion than to keep them in motion. See also **drag.**

Frigid Zone (frĭj′ĭd) Either of two regions of the Earth of extreme latitude, the **North Frigid Zone,** extending north of the Arctic Circle, or the **South Frigid Zone,** extending south of the Antarctic Circle.

fringing reef (frĭn′jĭng) A coral reef formed close to the shoreline of an island or continent. Fringing reefs usually have a rough, tablelike surface that is exposed during low tide and a steep edge sloping toward the open water. Compare **atoll, barrier reef.**

frond (frŏnd) **1.** A leaf of a fern or cycad, usually consisting of multiple leaflets. **2.** A large, fanlike leaf of a palm tree. **3.** A leaflike structure such as the thallus of a lichen or a seaweed.

front (frŭnt) The boundary between two air masses that have different temperatures or humidity. In the mid-latitude areas of the Earth, where warm tropical air meets cooler polar air, the systems of fronts define the weather and often cause precipitation to form. Warm air, being lighter than cold air,

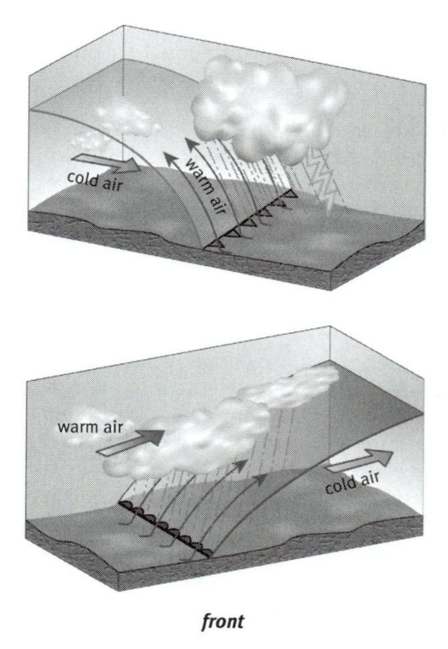

front

top: *cold front;* bottom: *warm front*

tends to rise, cool, and condense along such boundaries, forming rain or snow. See also **cold front, occluded front, polar front, stationary front, warm front.**

frontal lobe (frŭn′tl) The largest and forwardmost lobe of each cerebral hemisphere, responsible for the control of skilled motor activity, including speech. Mood and the ability to think are also controlled by the frontal lobe.

frontogenesis (frŭn′tō-jĕn′ĭ-sĭs) The formation or intensification of a meteorological front caused by an increase in the horizontal thermal gradient (the difference in temperature of adjacent air masses) over time. Compare **frontolysis.**

frontolysis (frŭn-tŏl′ĭ-sĭs) The weakening or complete dissipation of a meteorological front caused by a decrease in the horizontal thermal gradient (the difference in temperature of adjacent air masses) over time. Compare **frontogenesis.**

frost (frôst) A deposit of tiny, white ice crystals on a surface. Frost forms through sublimation, when water vapor in the air condenses at a temperature below freezing. It gets its white color from tiny air bubbles trapped in the ice crystals. See more at **dew point.**

frostbite (frôst′bīt′) Damage to a part of the body as a result of exposure to freezing temperatures. It is caused by a loss of blood supply and the formation of ice crystals in the affected body part.

frost line 1. In regions where there is no permafrost, the maximum depth to which frost penetrates the ground in the winter. **2.** The lower limit of permafrost. **3.** In tropical regions, the elevation below which frost never occurs.

fructification (frŭk′tə-fĭ-kā′shən) **1.** The producing of fruit by an angiosperm. **2.** A seed-bearing or spore-bearing structure.

fructose (frŭk′tōs′) A simple sugar (monosaccharide) found in honey, many fruits, and some vegetables. Fructose linked to glucose is the structure of table sugar, or *sucrose.* Fructose is an important source of energy for cellular processes. *Chemical formula:* $C_6H_{12}O_6$.

fruit (fro͞ot) The ripened ovary of a flowering plant that contains the seeds, sometimes fused with other parts of the plant. Fruits can be dry or fleshy. Berries, nuts, grains, pods, and drupes are fruits. ► Fruits that consist of ripened ovaries alone, such as the tomato and pea pod, are called **true fruits.** ► Fruits that consist of ripened ovaries and other parts such as the receptacle or bracts, as in the apple, are called **accessory fruits** or **false fruits.** See also **aggregate fruit, multiple fruit, simple fruit.** See Note at **berry.**

USAGE **fruit/vegetable**

To most of us, a *fruit* is a plant part that is eaten as a dessert or snack because it is sweet, but to a botanist a fruit is a mature ovary of a plant, and as such it may or may not taste sweet. All species of flowering plants produce fruits that contain seeds. A peach, for example, contains a pit that can grow into a new peach tree, while the seeds known as peas can grow into another pea vine. To a botanist, apples, peaches, peppers, tomatoes, pea pods, cucumbers, and winged maple seeds are all fruits. A *vegetable* is simply part of a plant that is grown primarily for food. Thus, the leaf of spinach, the root of a carrot, the flower of broccoli, and the stalk of celery are all vegetables. In everyday, nonscientific speech we make the distinction between sweet plant parts (fruits) and nonsweet plant parts (vegetables). This is why we speak of peppers and cucumbers and squash—all fruits in the eyes of a botanist—as vegetables.

fruitcover (fro͞ot′kŭv′ər) See **indusium** (sense 1).

fruitdot (fro͞ot′dŏt′) See **sorus** (sense 1).

fruiting body (fro͞o′tĭng) A specialized spore-producing structure, especially of a fungus. A mushroom is the fruiting body of a fungus.

frustule (frŭs′cho͞ol) The silica-rich cell wall of a diatom. Frustules are divided into two halves, and the intricate patterns of depressions and projections on each half help to identify individual diatom species.

FSH Abbreviation of **follicle stimulating hormone.**

ft Abbreviation of **foot.**

FTP (ĕf′tē-pē′) Short for *File Transfer Protocol.* A communications protocol governing the transfer of files between computers over a network. Compare **HTTP, SMTP.**

fucoxanthin (fyo͞o'kō-zăn**'**thĭn) A brown or olive-green carotenoid pigment, $C_{40}H_{60}O_6$, found in brown algae.

fuel (fyo͞o**'**əl) A substance that produces useful energy when it undergoes a chemical or nuclear reaction. Fuel such as coal, wood, oil, or gas provides energy when burned. Compounds in the body such as glucose are broken down into simpler compounds to provide energy for metabolic processes. Some radioactive substances, such as plutonium and tritium, provide energy by undergoing nuclear fission or fusion.

fuel cell A device that produces electricity by combining a fuel, usually hydrogen, with oxygen. In this reaction, electrons are freed from the hydrogen in the fuel cell by a catalyst, and gain energy from the chemical reaction binding hydrogen and oxygen; this provides a source for electric current. The exhaust of hydrogen fuel cells consists simply of water. Fuel cells are currently used in spacecraft, and increasingly in ground transportation, with potential use everywhere electricity is required.

fulcrum (fo͝ol**'**krəm) The point or support on which a lever turns. The position of the fulcrum, relative to the positions of the load and effort, determines the type of lever.

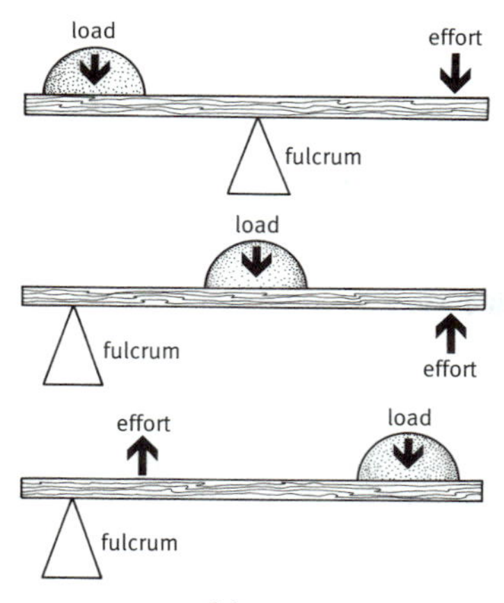

fulcrum

top: *first-class lever, with fulcrum between load and effort, as in a crowbar;* center: *second-class lever, with load between fulcrum and effort, as in a wheelbarrow;* bottom: *third-class lever, with effort between fulcrum and load, as in a person's forearm, where the fulcrum is the elbow and the load is something held in the hand.*

fulgurite (fo͝ol**'**gyə-rīt', -gə-) A slender, usually tubular body of glassy rock produced by lightning striking and then fusing dry sandy soil.

fullerene (fo͝ol**'**ə-rēn') Any of various carbon molecules that are nearly spherical in shape, are composed of hexagonal, pentagonal, or heptagonal groups of atoms, and constitute the third form of pure carbon after diamond and graphite. See more at **buckminsterfullerene.**

full moon (fo͝ol) The phase of the Moon in which it is visible as a fully illuminated disk. This phase occurs when the Moon is on the opposite side of Earth as the Sun and is not in Earth's shadow. See more at **moon.** Compare **new moon.**

Fulton (fo͝ol**'**tən), **Robert** 1765–1815. American engineer and inventor who developed the first useful submarine and torpedo (1800) and produced the first practical steamboat (1807).

fumarate (fyo͞o**'**mə-rāt') A salt or ester of fumaric acid, in which one or both of the hydrogen atoms in the carboxyl groups of the fumaric acid have been replaced with another element or group.

fumaric acid (fyo͞o-măr**'**ĭk) A colorless crystalline compound found in various plants and produced synthetically. It is used mainly in resins, paints, varnishes, and inks. Fumaric acid is a geometric isomer of maleic acid, having two carboxyl (COOH) groups attached on opposite sides of an ethylene chain. *Chemical formula:* $\mathbf{C_4H_4O_2}$.

fumarole (fyo͞o**'**mə-rōl') A vent in the surface of the Earth from which hot smoke and gases escape. Fumaroles are found on or near volcanoes, especially in areas where volcanic activity is in its later stages.

fume (fyo͞om) Smoke, vapor, or gas, especially if irritating, harmful, or smelly.

function (fŭngk**'**shən) **1.** A relationship between two sets that matches each member of the first set with a unique member of the second set. Functions are often expressed as an equation, such as $y = x + 5$, meaning that y is a function of x such that for any value of x, the value of y will be 5 greater than x. **2.** A quantity whose value depends on the value given to one or more related quantities. For example, the area of a square is a function of the length of its sides.

functional (fŭngk′shə-nəl) Affecting bodily functions but not organic structure, as a disorder such as irritable bowel syndrome. Compare **organic** (sense 4).

functional group An atom or group of atoms that replaces hydrogen in an organic compound. Functional groups define the structure of a family of compounds and determine its properties. The carboxyl and hydroxyl groups are functional groups that define the compounds they are part of as organic acids and alcohols, respectively.

fundamental force (fŭn′də-mĕn′tl) One of four forces that act between bodies of matter and that are mediated by one or more particles. In order of decreasing strength, the four fundamental forces are the strong force, the electromagnetic force, the weak force, and gravity. The particles associated with these forces, known as **force carriers**, are the gluon, the photon, the intermediate vector bosons (the Z boson and the W boson), and the graviton, respectively. Some scientists believe that the weak force and the electromagnetic force are both aspects of a single force called the **electroweak force**. Decay processes in which a subatomic particle is converted into other particles are mediated by the fundamental forces, which relate the decaying particle to the resulting particles; for example, beta decay is mediated by the weak force.

fundamental particle See **elementary particle.**

fungicide (fŭn′jĭ-sīd′, fŭng′gĭ-sīd′) A pesticide used to kill fungi, especially those that cause disease. Streptomycin is a fungicide. Compare **herbicide, insecticide, rodenticide.**

fungus (fŭng′gəs) Plural **fungi** (fŭn′jī, fŭng′gī). Any of a wide variety of organisms that reproduce by spores, including the mushrooms, molds, yeasts, and mildews. The spores of most fungi grow a network of slender tubes called hyphae that spread into and feed off of dead organic matter or living organisms. Fungi absorb food by excreting enzymes that break down complex substances into molecules that can be absorbed into the hyphae. The hyphae also produce reproductive structures, such as mushrooms and other growths. Some fungi (called **perfect fungi**) can reproduce by both sexually produced spores and asexual spores; other fungi (called **imperfect fungi** or **deuteromycetes**) are thought to have lost their sexual stage and can only reproduce by asexual spores. Fungi can live in a wide variety of environments, and fungal spores can survive extreme temperatures. Fungi exist in over 100,000 species, nearly all of which live on land. They can be extremely destructive, feeding on almost any kind of material and causing food spoilage and many plant diseases. Although fungi were once grouped with plants, they are now considered a separate kingdom in taxonomy. See Table at **taxonomy.** —*Adjective* **fungal.**

funiculus (fyo͝o-nĭk′yə-ləs) Plural **funiculi** (fyo͝o-nĭk′yə-lī). **1.** A stalk connecting an ovule or a seed with the placenta (the ovary wall). In some plants, the funiculus develops into a fleshy seed covering called an **aril. 2.** A slender, cordlike strand or band, especially a bundle of nerve fibers in a nerve trunk. **3.** Any of three major divisions of white matter in the spinal cord. **4.** The umbilical cord.

Funk (fŭngk, fo͝ongk), **Casimir** 1884–1967. Polish-born American biochemist who is credited with the discovery of vitamins. In 1912 he postulated the existence of four organic bases he called *vitamines* which were necessary for normal health and the prevention of deficiency diseases. He also contributed to the knowledge of the hormones of the pituitary gland and the sex glands.

furan (fyo͝or′ăn′, fyo͝o-răn′) **1.** Any of a group of colorless, volatile, organic compounds containing a ring of four carbon atoms and one oxygen atom. Furans are obtained from wood oils and used in the synthesis of many organic compounds. **2.** The simplest such compound, consisting of a furan ring with two double bonds and attached to four hydrogen atoms. *Chemical formula:* C_4H_4O.

furanose (fyo͝or′ə-nōs′) Any of a class of simple sugars (monosaccharides) that has a ring containing four carbon atoms and one oxygen atom (a furan ring). Fructose and ribose are furanoses.

furfural (fûr′fə-răl′) A colorless, sweet-smelling, liquid made from corncobs and used as a solvent in petroleum refining and as a fungicide and weed killer. It turns reddish brown when exposed to air and light. Furfural is an aldehyde of furan. *Chemical formula:* $\mathbf{C_5H_4O_2}$.

fuse (fyo͞oz) *Noun.* **1.** A safety device that protects an electric circuit from becoming overloaded. Fuses contain a length of thin wire (usually of a metal alloy) that melts and breaks the circuit if too much current flows through it. They were traditionally used to protect electronic equipment and prevent fires, but have largely been replaced by circuit breakers. **2.** A cord of readily combustible material that is lighted at one end to carry a flame along its length to detonate an explosive at the other end. —*Verb.* **3.** To melt something, such as metal or glass, by heating. **4.** To blend two or more substances by melting.

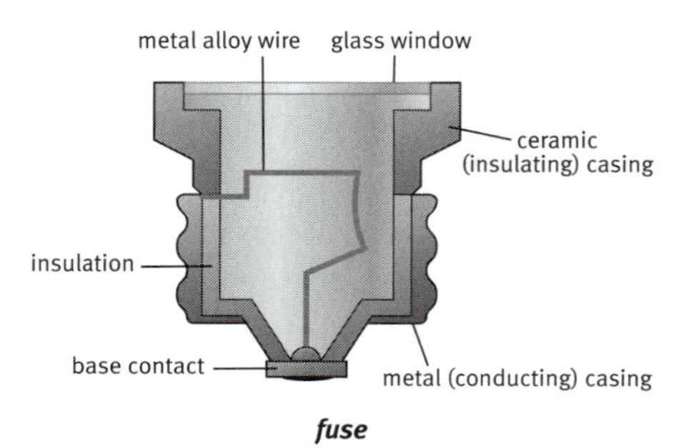

fuse

fusel oil (fyo͞o′zəl) An acrid, oily, poisonous liquid occurring in the distillation products of fermented alcoholic liquids. Fusel oil is a mixture of amyl alcohols, fatty acids, and esters. It is used in paints, plastics, and varnishes, and in the manufacture of explosives.

fusion (fyo͞o′zhən) **1.** The joining together of atomic nuclei, especially hydrogen or other light nuclei, to form a heavier nucleus, especially a helium nucleus. Fusion occurs when plasmas are heated to extremely high temperatures, forcing the nuclei to collide at great speed. The resulting unstable nucleus emits one or more neutrons at very high speeds, releasing more energy than was required to fuse the nuclei, thereby making chain-reactions possible, since the reaction is exothermic. Fusion reactions are the source of the energy in the Sun and in other stars, and in hydrogen bombs. See also **fission. 2.** A mixture or blend formed by fusing two or more things.

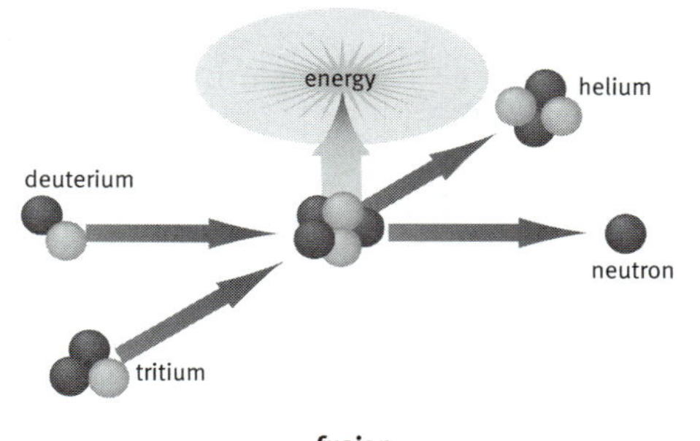

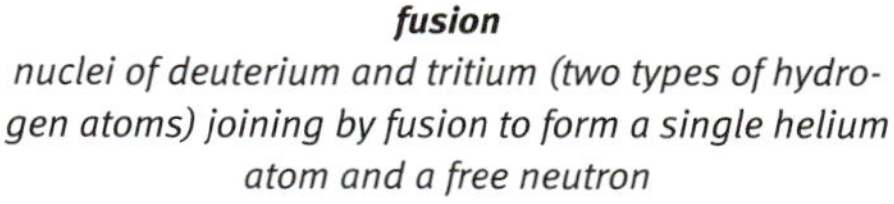

fusion

nuclei of deuterium and tritium (two types of hydrogen atoms) joining by fusion to form a single helium atom and a free neutron

fusor (fyo͞o′zôr) Any of a proposed category of celestial objects that undergo nuclear fusion in their cores at some point during their lifetimes, the least massive of which are about 13 times Jupiter's mass and sustain the fusion of deuterium atoms into heavier elements.

fuzzy logic (fŭz′ē) A form of algebra employing a range of values from "true" to "false" that is used in making decisions with imprecise data. The outcome of an operation is assigned a value between 0 and 1 corresponding to its degree of truth. Fuzzy logic is used, for example, in artificial intelligence systems.

g **1.** The symbol for **acceleration of gravity.** **2.** A symbol for **g-force.** **3.** Abbreviation of **gram.**

G **1.** The symbol for **gauss.** **2.** A symbol for **g-force.** **3.** The symbol for **gravitational constant.** **4.** Abbreviation of **guanine.**

Ga The symbol for **gallium.**

GABA Abbreviation of **gamma-aminobutyric acid.**

gabbro (găb′rō) A usually dark, coarse-grained igneous rock composed mostly of plagioclase feldspar and clinopyroxene, and sometimes olivine. Gabbro is the coarse-grained equivalent of basalt.

Gabor (gä′bôr, gə-bôr′), **Dennis** 1900–1979. Hungarian-born British physicist who invented the technique of holography in 1947.

gadolinium (găd′l-ĭn′ē-əm) *Symbol* **Gd** A silvery-white, malleable, ductile metallic element of the lanthanide series that has seven natural isotopes and 11 artificial isotopes. Two of the natural isotopes, Gd 155 and Gd 157, are the best known neutron absorbers. Gadolinium is used to improve the heat and corrosion resistance of iron, chromium, and various alloys and in medicine as a contrast medium for magnetic resonance imaging and as a radioisotope in bone mineral analysis. Atomic number 64; atomic weight 157.25; melting point 1,312°C; boiling point approximately 3,000°C; specific gravity from 7.8 to 7.896; valence 3. See **Periodic Table.**

gal. Abbreviation of **gallon.**

galactic cluster (gə-lăk′tĭk) See **open cluster.**

galactic coordinate system The coordinate system in which a celestial object's position on the celestial sphere is described in relation to the structure of the Milky Way galaxy. ► An object's **galactic longitude** is measured along the **galactic equator**, a great circle on the celestial sphere that follows the band of the Milky Way. The galactic equator, also called the *galactic circle,* is inclined at an angle of approximately 62° to the celestial equator; distances are measured along it beginning at a point in the constellation Sagittarius lying in the direction of the Milky Way's nucleus. The **galactic poles** are the two points where a perpendicular line through the middle of the plane of the galactic equator intersect the celestial sphere. ► An object's **galactic latitude** is measured in degrees north or south of the galactic equator toward the galactic poles.

galactic halo A large, spherical region of relatively dust-free space surrounding a spiral galaxy such as the Milky Way. The inner regions of the galactic halo contain globular clusters of very old stars, while the outer regions are apparently occupied by large amounts of dark matter. See also **MACHO.**

galactic nucleus The center of a galaxy, usually small, very bright, and containing a high density of stars and other objects. ► An **active galactic nucleus** is a nucleus that emits more radiation than can be explained in terms of stars alone. **Blazars** and **quasars** are thought to be active galactic nuclei; the nuclei of Seyfert galaxies are also active, and are thought to contain an extremely massive **black hole**.

galactose (gə-lăk′tōs′) A monosaccharide commonly occurring in lactose and in certain pectins, gums, and mucilages. *Chemical formula:* $C_6H_{12}O_6$.

galaxy (găl′ək-sē) **1.** Any of numerous large-scale collections of stars, gas, and dust that make up the visible universe. Galaxies are

galaxy

left: *polar-ring galaxy NGC 4650A;* right: *Whirlpool spiral galaxy*

held together by the gravitational attraction of the material contained within them, and most are organized around a galactic nucleus into elliptical or spiral shapes, with a small percentage of galaxies classed as irregular in shape. A galaxy may range in diameter from some hundreds of light-years for the smallest dwarfs to hundreds of thousands of light-years for the largest ellipticals, and may contain from a few million to several trillion stars. Many galaxies are grouped into clusters, with the clusters themselves often grouped into larger superclusters. See more at **active galaxy.** See also **elliptical galaxy, irregular galaxy, lenticular galaxy, spiral galaxy. 2. the Galaxy.** The Milky Way.

Galen (gā′lən) 130?–200? CE. Greek anatomist, physician, and writer who developed numerous theories about the structures and functions of the human body, many of which were based on information he gained from dissecting animals. Galen's theories formed the basis of European medicine until the Renaissance.

galena (gə-lē′nə) A gray, metallic mineral occurring in cube-shaped crystals within veins of igneous rock or in sedimentary rocks. It is the main ore of lead. *Chemical formula:* **PbS.**

Galileo Galilei (găl′ə-lā′ō găl′ə-lā′) 1564–1642. Italian astronomer, mathematician, and physicist. He was the first to use a telescope to study the stars and planets, and he discovered various astronomical phenomena and physical principles.

Galileo Galilei

BIOGRAPHY **Galileo Galilei**

Galileo Galilei is considered to be the father of modern experimental science. His most significant experiments concerned gravitation. Galileo conducted a series of experiments to measure the effects of gravity on motion, such as measuring the speed of balls of different weights rolling down inclined planes. He found that all objects accelerate at the same, constant rate. He is also famous for the probably apocryphal experiment in which he dropped balls of different masses from the Tower of Pisa. Had the experiment actually taken place, air resistance might have caused the balls to fall at different rates, defying the principle of acceleration that Galileo was trying to demonstrate. In 1609, having heard of the invention of the spyglass, a tube with a piece of glass at each end that made objects appear closer and larger, Galileo set about making his own. Using his telescope, he observed mountains on the Moon's surface (which was thought to be flat), Jupiter's four largest moons, and sunspots. Because he openly supported Copernicus's theory that Earth and all the planets orbit the Sun, Galileo was called before authorities of the Catholic Church and forced to declare the theory false. He was put under house arrest on his own farm, where he continued his scientific work until the end of his life.

gall (gôl) An abnormal swelling of plant tissue, caused by injury or by parasitic organisms such as insects, mites, nematodes, and bacteria. Parasites stimulate the production of galls by secreting chemical irritants on or in the plant tissue. Galls stimulated by egg-laying parasites typically provide a protective environment in which the eggs can hatch and the pupae develop, and they usually do only minor damage to the host plant. Gall-stimulating fungi and microorganisms, such as the bacterium that causes crown gall, are generally considered to be plant diseases.

gallbladder (gôl′blăd′ər) A small, pear-shaped muscular sac in most vertebrates in which bile is stored. The gallbladder is located beneath the liver and secretes bile into the duodenum of the small intestine.

gallium (găl′ē-əm) *Symbol* **Ga** A rare, silvery metallic element that is found as a trace element in coal, in bauxite, and in several minerals. It is liquid near room temperature and expands when it solidifies. It is used in thermometers and semiconductors. Atomic number 31; atomic weight 69.72; melting point 29.78°C; boiling point 2,403°C; specific gravity 5.907; valence 2, 3. See **Periodic Table.**

gallium arsenide A dark-gray crystalline compound used in transistors, solar cells, and semiconductor lasers. *Chemical formula:* **GaAs.**

gallon (găl′ən) A unit of liquid volume or capacity in the US Customary System equal to 4 quarts (3.79 liters). See Table at **measurement.**

gallstone (gôl′stōn′) A small, hard, abnormal mass composed chiefly of cholesterol, calcium salts, and bile pigments, formed in the gallbladder or in a bile duct. The presence of gallstones can lead to painful obstruction or infection and is sometimes treated with **cholecystectomy**.

galvanic (găl-văn′ĭk) Relating to electricity generated by a chemical reaction. A **galvanic cell** is an **electric cell**, such as found in household and car batteries, that makes use of galvanic reactions to act as a power source. See Note at **battery.**

galvanometer (găl′və-nŏm′ĭ-tər) An instrument that detects small electric currents and indicates their direction and relative strength. Current flowing through the galvanometer passes through a coil near a magnetized needle on a pivot; the strength of the current in the coil regulates the strength of a magnetic field that displaces the needle. Galvanometers can be used directly as ammeters, and are the core element of many ohmmeters and voltmeters.

gametangium (găm′ĭ-tăn′jē-əm) Plural **gametangia.** An organ or a cell in which gametes are produced, such as those of fungi, algae, mosses, and ferns.

gamete (găm′ēt′) A cell whose nucleus unites with that of another cell to form a new organism. A gamete contains only a single (haploid) set of chromosomes. Animal egg and sperm cells, the nuclei carried in grains of pollen, and egg cells in plant ovules are all gametes. Also called *germ cell, reproductive cell, sex cell.* See Note at **mitosis.**

gamete intrafallopian transfer A technique of assisted reproduction in which eggs and sperm are inserted directly into a woman's fallopian tubes, where fertilization may occur.

game theory (gām) A mathematical method of making decisions in which a competitive situation is analyzed to determine the optimal course of action for an interested party. Game theory is often used in political, economic, and military planning.

gametocyte (gə-mē′tə-sīt′) A cell from which gametes develop by meiosis. Oocytes and spermatocytes are gametocytes.

gametogenesis (gə-mē′tə-jĕn′ĭ-sĭs) The formation or production of gametes. In most multicellular organisms, gametogenesis takes place by **meiosis**.

gametophore (gə-mē′tə-fôr′) A structure, as in liverworts and mosses, on which gametangia are borne.

gametophyte (gə-mē′tə-fīt′) Among organisms which display an **alternation of generations** as part of their life cycle (such as plants and certain algae), the haploid organism that produces gametes. Each of its cells has only one, unpaired set of chromosomes, as opposed to the corresponding diploid form of the organism, called the **sporophyte.** A gametophyte develops from spores produced by the sporophyte. The gametophytes of **homosporous** plants are bisexual (produce both eggs and sperm), while the gametophytes of **heterosporous** plants, such as all seeds plants, are unisexual (produce only eggs or only sperm). See more at **alternation of generations, sporophyte.**

A CLOSER LOOK **gametophyte**

The evolution of plants has seen a steady reduction in the independence and size of the *gametophyte* generation when compared to the *sporophyte* generation. In the more primitive nonvascular plants, such as the mosses and liverworts, the gametophyte is the main plant form. It is relatively large and not dependent on the sporophyte for nutrition. In some species, the gametophytes are bisexual; in others, the sexes are separate. Female gametes remain in the gametophyte, where they are fertilized by sperm, usually from another gametophyte. The resultant embryos develop into sporophytes. These grow as smaller plants in or upon the gametophytes. The diploid sporophytes then produce haploid spores by meiosis, and these are dispersed and grow into new gametophytes. Among the vascular plants, however, the sporophyte is the larger plant form. For example, the familiar pine tree represents the diploid sporophyte generation. By meiosis, the pine eventually produces male and female gametophytes. The male gametophytes of the pine are small—they are the pollen grains released by male cones to be dispersed to female cones. The larger female gametophytes remain within the ovules of the female cones of the pine and are dependent on the surrounding sporophyte tissues for their nutrition. Each female gametophyte develops two egg cells, both of which are fertilized by a male gamete from a pollen grain. However, usually only one embryo will complete its development. This embryo is a new diploid sporophyte, housed within the developing pine seed. A pine seed thus contains three generations of the pine plant: the seed coat that arises from diploid tissue belonging to the ovule, the haploid gametophyte tissue that will serve as a food reserve for the embryo after germination, and the diploid embryo itself. In angiosperms, the female gametophyte has evolved into an even more reduced form. It consists of only the seven cells of the embryo sac.

---

gamma-aminobutyric acid (găm′ə-ə-mē′nō-byo͞o-tîr′ĭk, -ăm′ə-) An amino acid occurring in the brain as a neurotransmitter that acts to inhibit the transmission of nerve impulses. Certain antianxiety drugs, called *benzodiazepines,* mimic the actions of gamma-aminobutyric acid. *Chemical formula:* $C_4H_9NO_2$.

gamma decay A radioactive process in which an atomic nucleus loses energy by emitting a gamma ray (a stream of high-energy photons). When an element undergoes gamma decay its atomic number and mass number do not change.

gamma globulin 1. A class of globulins in the blood plasma of humans and other mammals that function as part of the body's immune system and include most antibodies. **2.** A solution of this substance prepared from human blood and administered for immunization against measles, German measles, hepatitis A, and other infections.

gamma ray A stream of high-energy electromagnetic radiation given off by an atomic nucleus undergoing radioactive decay. Because the wavelengths of gamma rays are shorter than those of x-rays, gamma rays have greater energy and penetrating power than x-rays. Gamma rays are emitted by pulsars, quasars, and radio galaxies but cannot penetrate the Earth's atmosphere. See more at **radioactive decay.**

gamma ray astronomy The study of astronomical objects by analyzing the gamma rays they emit. Because gamma rays are subject to atmospheric interference that makes them difficult to observe using ground-based telescopes, high-altitude balloons and orbiting observatories are often used to detect them. Most gamma rays in the universe are produced by events such as **supernovas** and **black holes.**

gamma ray burst Intense high-energy electromagnetic radiation, lasting between a fraction of a second and several minutes, emanating from distant regions of the universe. Recent theory suggests that they result from **supernova** explosions.

ganglion (găng′glē-ən) Plural **ganglia.** A compact group of neurons enclosed by connective tissue and having a specific function. In invertebrate animals, pairs of ganglia occur at intervals along the axis of the body, with the forwardmost pair functioning like a brain. In vertebrates, ganglia are usually located outside the brain or spinal cord, where they regulate the functioning of the body's organs and glands as part of the autonomic nervous system.

gangrene (găng′grēn′) Death of tissue in a living body, especially in a limb, caused by a bacterial infection resulting from a blockage of the blood supply to the affected tissue.

Ganymede (găn′ə-mēd′) One of the four brightest satellites of Jupiter and the seventh in distance from the planet. Originally sighted by Galileo, it is the largest satellite in the solar system.

garnet (gär′nĭt) Any of several common red, brown, black, green, or yellow minerals having the general chemical formula **$A_3B_2SiO_8$**, where A is either calcium (Ca), magnesium (Mg), iron (Fe), or manganese (Mn) and B is either aluminum (Al), manganese, iron, chromium (Cr), or vanadium (V). Garnet crystals are dodecahedral in shape, transparent to semitransparent, and have a vitreous luster. They usually occur in metamorphic rocks but also occur in igneous and sedimentary rocks.

gas (găs) One of four main **states of matter**, composed of molecules in constant random motion. Unlike a solid, a gas has no fixed shape and will take on the shape of the space available. Unlike a liquid, the intermolecular forces are very small; it has no fixed volume and will expand to fill the space available. —*Adjective* **gaseous** (găs′ē-əs, găsh′əs).

gas constant A constant, equal to 8.314 joules per mole degree Kelvin or 1.985 calories per mole degree Celsius, that is the constant of proportionality (R) in the **universal gas law**.

gas exchange The diffusion of gases from an area of higher concentration to an area of lower concentration, especially the exchange of oxygen and carbon dioxide between an organism and its environment. In plants, gas exchange takes place during photosynthesis. In animals, gases are exchanged during respiration.

gas giant A large, massive, low-density planet composed primarily of hydrogen, helium, methane, and ammonia in either gaseous or liquid state. Gas giants have swirling atmospheres primarily of hydrogen and helium, with no well-defined planetary surface; they are assumed to have rocky cores. They are also characterized by ring systems, although only Saturn's is readily visible from Earth. Our solar system contains four gas giants: Jupiter, Saturn, Uranus, and Neptune. The majority of extrasolar planets discovered so far are the size of the solar system's gas giants, although they orbit their stars much more closely and may differ in composition from ours. Also called *Jovian planet*. Compare **terrestrial planet.**

gas laws A series of laws in physics that predict the behavior of an ideal gas by describing the relations between the temperature, volume, and pressure. The laws include Boyle's law, Charles' law, and the pressure law, and are combined in the ideal gas law.

gasohol (găs′ə-hôl′) A fuel consisting of a blend of ethanol and unleaded gasoline, especially a blend of 10 percent ethanol and 90 percent gasoline. The ethanol is obtained by the fermentation and subsequent distillation of sugar cane, maize, or potatoes. Gasohol has a high octane rating and produces lower levels of pollutants than ordinary gasoline.

gasoline (găs′ə-lēn′) A highly flammable mixture of liquid hydrocarbons that are derived from petroleum. The hydrocarbons in gasoline contain between five and eight carbon atoms. Gasoline is used as a fuel for internal-combustion engines in automobiles, motorcycles, and small trucks.

gastric (găs′trĭk) Relating to or involving the stomach.

gastric juice A fluid secreted by glands lining the inside of the stomach. It contains hydrochloric acid and enzymes, such as pepsin, that aid in digestion.

gastroenterology (găs′trō-ĕn′tə-rŏl′ə-jē) The branch of medicine dealing with the diagnosis and treatment of disorders of the gastrointestinal tract.

gastroesophageal reflux disease (găs′trō-ĭ-sŏf′ə-jē′əl) A condition in which relaxation of the lower esophageal sphincter allows gastric acids to move up into the esophagus, causing heartburn, dyspepsia, and possible injury to the mucous membranes lining the esophagus.

gastrointestinal tract (găs′trō-ĭn-tĕs′tə-nəl) The digestive tract, especially of a human.

gastropod (găs′trə-pŏd′) Any of various carnivorous or herbivorous mollusks of the class Gastropoda, having a head with eyes and feelers and a muscular foot on the underside of its body with which it moves. Most gastropods are aquatic, but some have adapted to life on land. Gastropods include snails, which have a coiled shell, and slugs, which have a greatly reduced shell or none at all.

WORD HISTORY **gastropod**

Snails, conchs, whelks, and many other similar animals with shells are all called gastropods by scientists. The word *gastropod* comes from Greek and means "stomach foot," a name that owes its existence to the unusual anatomy of snails. Snails have a broad flat muscular "foot" used for support and for forward movement. This foot runs along the underside of the animal—essentially along its belly. The Greek elements *gastro–*, "stomach," and *–pod*, "foot," are found in many other scientific names, such as *gastritis* (an inflammation of the stomach) and *sauropod* ("lizard foot," a type of dinosaur).

---

gastrula (găs′trə-lə) Plural **gastrulas** or **gastrulae** (găs′trə-lē′). An animal embryo at the stage following the blastula. The gastrula develops from the blastula by invagination (inpocketing), forming an inner cavity with an opening and causing the cells to be distributed into an outer layer (ectoderm) and an inner layer (endoderm). In complex animals such as vertebrates, a third layer (mesoderm) also forms. These layers later develop into the organs and tissues of the body. In vertebrates and other deuterostomes, the opening of the gastrula becomes the anus, while in protostomes (such as arthropods), it becomes the mouth. Compare **blastula.** ▸ The development of an embryo from blastula to gastrula is called **gastrulation.**

gas turbine An internal-combustion engine consisting of an air compressor, combustion chamber, and turbine wheel that is turned by the expanding products of combustion. The four major types of gas turbine engines are the **turboprop**, **turbojet**, **turbofan,** and **turboshaft**. See more at **turbojet.**

gauge boson (gāj) A boson that acts as a mediator of one of the fundamental forces of nature. The gauge bosons are the photon, which mediates the electromagnetic force, the gluon, which mediates the strong nuclear force, the intermediate vector bosons (the Z boson and the W boson), which mediate the weak nuclear force, and the hypothetical graviton, which mediates gravity. See Table at **subatomic particle.**

Gause's law (gô′zĭz) See **competitive exclusion principle.**

gauss (gous) The unit of magnetic flux density in the centimeter-gram-second system, equal to one maxwell per square centimeter, or 10^{-4} tesla.

Gauss, Carl Friedrich 1777–1855. German mathematician, astronomer and physicist who introduced significant and rapid advances to mathematics with his contributions to algebra, geometry, statistics and theoretical mathematics. He also correctly calculated the orbit of the asteroid **Ceres** in 1801 and studied electricity and magnetism, developing the magnetometer in 1832. The gauss unit of magnetic flux density is named for him.

Gay-Lussac (gā′lə-săk′), **Joseph Louis** 1778–1850. French chemist and physicist who in 1808 developed a law governing the ratio of volumes of gases participating in chemical reactions. In that same year, with Louis Jacques Thénard, he discovered the element boron.

GB Abbreviation of **gigabyte.**

Gd The symbol for **gadolinium.**

Ge The symbol for **germanium.**

gear (gîr) A wheel with teeth around its rim that mesh with the teeth of another wheel to transmit motion. Gears are used to transmit power (as in a car transmission) or change the direction of motion in a mechanism (as in a differential axle). Fixed ratios of speed in various parts of a machine is often established by the arrangement of gears.

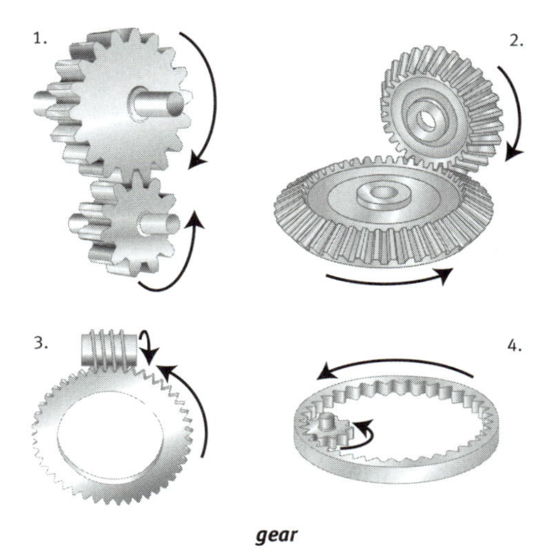

gear

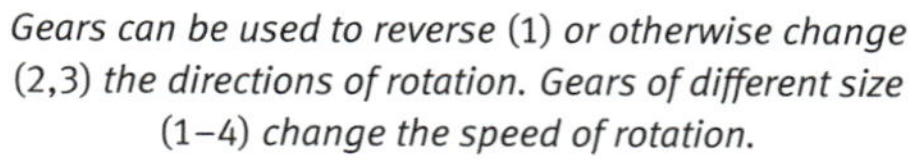

Gears can be used to reverse (1) *or otherwise change* (2,3) *the directions of rotation. Gears of different size* (1–4) *change the speed of rotation.*

Geiger (gī′gər), **Hans Wilhelm** 1882–1945. German physicist who was a pioneer in nuclear physics. He invented numerous instruments and techniques used to detect charged particles, including the Geiger counter (1908).

Geiger counter An electronic instrument that detects and measures nuclear radiation, such as x-rays or gamma rays. The Geiger counter consists of a gas-filled tube with a charged electrode connected to a counter. As radiation passes through the gas it ionizes atoms along its path. The ions are attracted to the charged electrode, creating pulses of electric current that are registered by the counter.

gelatin (jĕl′ə-tn) An odorless, colorless protein substance obtained by boiling a mixture of water and the skin, bones, and tendons of animals. The preparation forms a gel when allowed to cool. It is used in foods, drugs, glue, and film.

Gell-Mann (gĕl′măn′), **Murray** Born 1929. American physicist who helped introduce the concept of quarks. He received the Nobel Prize for physics in 1969 for his contributions to the description and classification of subatomic particles.

BIOGRAPHY **Murray Gell-Mann**

Physicists have long sought the fundamental building blocks of matter. The atoms that make up the elements appeared to be good candidates, but further investigation into the structure of the atom led to the identification of even smaller subunits, such as the proton and the neutron. With the advent of particle accelerators, dozens more hitherto unknown subatomic particles were discovered, flying out of high-energy particle collisions. Physicist *Murray Gell-Mann* helped bring order into the chaos of what was called the particle zoo, eventually winning the Nobel Prize for his work in 1969. Gell-Mann, born in 1929, was a child prodigy who entered Yale at age 15 and is today considered to be one of the greatest physicists of the 20th century. Much of his career was spent at the California Institute of Technology, where his sometime collaborator and rival Richard Feynman also taught. Gell-Mann saw that the hundreds of new subatomic particles could be arranged in patterns, similar to the patterns of the periodic table. Always ready with an evocative name for a new discovery, he called his classification scheme "The Eightfold Way," after Buddha's Eightfold Path to Enlightenment. Gell-Mann found that the patterns he saw could be explained by assuming that these unexpected particles were simply combinations of a few fundamental building blocks. Gell-Mann called these particles *quarks,* after a line in James Joyce's notoriously difficult novel *Finnegans Wake,* and showed how properties of quarks corresponded with certain abstract algebraic structures.

Gemini (jĕm′ə-nī′) A constellation in the Northern Hemisphere near Cancer and Auriga. Gemini (the Twins) contains the bright stars Pollux and Castor and is the third sign of the zodiac.

gemma (jĕm′ə) Plural **gemmae** (jĕm′ē′). A budlike mass of undifferentiated tissue which serves as a means of vegetative reproduction among mosses and liverworts. The gemmae, often formed in structures called **gemma cups,** are usually dispersed from the parent plant by the splashing of raindrops, after which they develop into new individuals.

gemmule (jĕm′yo͞ol) A small gemma or similar structure, especially a reproductive structure in some sponges that remains dormant through the winter and later develops into a new individual.

gene (jēn) A segment of DNA, occupying a specific place on a chromosome, that is the basic unit of heredity. Genes act by directing the production of RNA, which determines the synthesis of proteins that make up living matter and are the catalysts of all cellular processes. The proteins that are determined by genetic DNA result in specific physical traits, such as the shape of a plant leaf, the coloration of an animal's coat, or the texture of a person's hair. Different forms of genes, called **alleles**, determine how these traits are expressed in a given individual. Humans are thought to have 20,000 to 25,000 genes; bacteria have between 500 and 6,000. See also **dominant, recessive.** See Note at **Mendel.**

gene amplification An increase in the number of copies of a gene in a cell, resulting in an elevation in the level of the RNA or protein encoded for by the gene and a corresponding amplification of the phenotype that the gene confers on the cell. Drug resistance in cancer cells is linked to amplification of the gene that prevents absorption of the chemotherapeutic agent by the cell.

gene map Variant of **genetic map.**

gene pool The collective genetic information contained within a population of sexually reproducing organisms.

General Relativity (jĕn′ər-əl) A geometrical theory of gravity developed by Albert Einstein in which gravity's effects are a consequence of the curvature of four-dimensional space-time. According to this theory, the energy and momentum of all matter and radiation cause curvature in space-time, in a way similar to the creation of electric and magnetic fields by electric charges and currents. This curvature also opens the possibility that the universe is closed, having finite volume but without any boundary. Among the many experimentally confirmed consequences of General Relativity are the perihelion precession of the planet Mercury, the bending of light in a gravitational field, and the slowing of time in a gravitational field. See also **closed universe, equivalence principle, Special Relativity.**

generation (jĕn′ə-rā′shən) **1a.** All of the offspring that are at the same stage of descent from a common ancestor. **b.** The average interval of time between the birth of parents and the birth of their offspring. **2.** A form or stage in the life cycle of an organism. See more at **alternation of generations. 3.** The formation of a line or geometric figure by the movement of a point or line.

generative cell (jĕn′ər-ə-tĭv) A cell of the male gametophyte (pollen grain) of seed plants that divides to give rise directly or indirectly to sperm. See more at **pollination.**

generator (jĕn′ə-rā′tər) **1.** A machine that converts mechanical energy into electricity to serve as a power source for other machines. Electrical generators found in power plants use water turbines, combustion engines, windmills, or other sources of mechanical energy to spin wire coils in strong magnetic fields, inducing an electric potential in the coils. A generator that provides alternating current power is called an **alternator**. See also **induction** (sense 2a). **2.** See **generatrix.**

generatrix (jĕn′ə-rā′trĭks) Plural **generatrices** (jĕn′ə-ə-rā′trĭ-sēz′, -ər-ə-trī′sēz). A geometric element that generates a geometric figure, especially a straight line that generates a surface by moving in a specified fashion. Also called *generator.*

gene-splicing The process in which fragments of DNA from one or more different organisms are combined to form recombinant DNA.

gene therapy The treatment of a disorder or disease, especially one caused by the inheritance of a defective gene, by replacing defective genes with healthy ones through genetic engineering.

genetically modified organism (jə-nĕt′ĭ-kə-lē) An organism whose genetic characteristics have been altered using the techniques of genetic engineering.

A CLOSER LOOK **genetically modified organism**

Scientists today have the ability to modify the genetic makeup of plants and animals, and even to transfer genes from one species to another. Not since nuclear power has a technology been so controversial, with opponents concerned about the creation of so-called Frankenfoods and proponents promising a better tomorrow through science. The term *genetically modified organism* (GMO) is used to refer to any microorganism, plant, or animal in which genetic engineering techniques have been used to introduce, remove, or modify specific parts of its genome. Examples include plants being modified for pest resistance; lab animals being manipulated to exhibit human diseases, such as sickle cell anemia; and even glowing jellyfish genes inserted in a rabbit for an art piece. GMOs show great promise in improving agriculture. Plants could be engineered to better tolerate temperature or weather extremes, to contain various vitamins, or to dispense medicines and vaccines. Many think genetically modified foods have the potential to end world hunger. On the other hand, there are fears that the disease- or pest-resistant genes inserted into crop plants might escape into other plants, creating hard-to-control superweeds. There is also the possibility of unexpected effects on other flora and fauna, the risk of agriculture being controlled by biotech companies, and, as with any new technology, problems that are yet unknown.

genetic code (jə-nĕt′ĭk) The sequence of nucleotides in DNA and RNA that serve as instructions for synthesizing proteins. The genetic code is based on an "alphabet" consisting of sixty-four triplets of nucleotides called **codons.** The order in which codons are strung together determines the order in which the amino acids for which they code are arranged in a protein.

genetic counseling The counseling of individuals, especially prospective parents with respect to their offspring, about the statistical probabilities of inheriting genetic diseases and the nature, diagnosis, and treatment of such diseases.

genetic disorder A disease or condition caused by an absent or defective gene or by a chromosomal aberration, as in **Down Syndrome**.

genetic drift Variation in the frequency of a gene in a small isolated population, thought to be due to random chance rather than natural selection.

genetic engineering The science of altering and cloning genes to produce a new trait in an organism or to make a biological substance, such as a protein or hormone. Genetic engineering mainly involves the creation of recombinant DNA, which is then inserted into the genetic material of a cell or virus.

genetic load The aggregate of deleterious genes that are carried, mostly hidden, in the genome of a population and may be transmitted to descendants. Inbreeding usually causes an increase in genetic load.

genetic map or **gene map** A graphic representation of the arrangement of a gene or a DNA sequence on a chromosome. A genetic map is used to locate and identify the gene or group of genes that determines a particular inherited trait. ► Locating and identifying genes in a genetic map is called **genetic mapping.**

genetic marker A gene or DNA sequence having a known location on a chromosome. Genetic markers associated with certain diseases can be used to determine whether an individual is at risk for developing an inherited disease.

genetics (jə-nĕt′ĭks) The scientific study of the principles of heredity and the variation of inherited traits among related organisms.

genitals (jĕn′ĭ-tlz) The organs of reproduction in animals, especially the external sex organs.

genome (jē′nōm) The total amount of genetic information in the chromosomes of an organism, including its genes and DNA sequences. The genome of eukaryotes is made up of a single, haploid set of chromosomes that is contained in the nucleus of every cell and exists in two copies in all cells except reproductive and red blood cells. The human genome is made up of about 20,000 to 25,000 genes. Compare **proteome.**

genomics (jə-nō′mĭks) The scientific study of genomes.

genomic sequencing (jə-nō′mĭk) The sequencing of the entire genome of an organism.

A CLOSER LOOK **genomic sequencing**

The technique that allows researchers to read and decipher the genetic information found in the DNA of anything from bacteria to plants to animals is called *genomic sequencing.* Once a tedious, painstaking process, today, thanks to new techniques and the advent of powerful computers, it has been sped up a hundredfold. Sequencing involves determining the order of bases, the nucleotide subunits (adenine, guanine, cytosine and thymine, referred to by the letters A, G, C and T) found in DNA. (Humans have about 3 billion base pairs.) Three nucleotides in a row specify one kind of amino acid, and a gene is defined by a long string of these triplets. There are various techniques to read the sequence of these "letters" in genes and genomes. So-called shotgun sequencing was used to help decode the human genome. Using this method, the DNA to be sequenced is first broken apart with enzymes, since gene-sequencing machines can only handle small stretches of DNA at a time. Once these random fragments are sequenced, powerful supercomputers compare overlapping sections and recreate the original, long strand. By decoding genomes and sequencing genes, researchers are beginning to understand how those strings of letters work together to create a living organism.

genotype (jĕn′ə-tīp′, jē′nə-tīp′) The genetic makeup of an organism as distinguished from its physical characteristics. Compare **phenotype.**

genus (jē′nəs) Plural **genera** (jĕn′ər-ə). A group of organisms ranking above a species and below a family. The names of genera, like those of species, are written in italics. For example, *Periplaneta* is the genus of the American cockroach, and comes from the Greek for "wandering about." See Table at **taxonomy.**

geo– or **ge–** A prefix that means "earth," as in *geochemistry,* the study of the Earth's chemistry.

geobotany (jē′ō-bŏt′n-ē) See **phytogeography.**

geocentric (jē′ō-sĕn′trĭk) **1.** Relating to or measured from the Earth's center. **2.** Relating to a model of the solar system or universe having the Earth as the center. Compare **heliocentric.**

geocentric latitude The angular distance between a point on the Earth's surface and the equator, using the center of the Earth as the vertex. It differs from geodetic latitude in that it accounts for the ellipsoidal shape of the Earth rather than considering it as a perfect sphere. Compare **geodetic latitude.**

geocentric longitude The angular distance between the prime meridian and another meridian passing through a point on the Earth's surface, having the center of the Earth as its vertex. Compare **geodetic longitude.**

geochemistry (jē′ō-kĕm′ĭ-strē) The scientific study of the chemical composition of the Earth or other celestial body and of the reactions that control the distribution of chemical elements in its minerals, rocks, soil, waters, and atmosphere.

geochronometry (jē′ō-krə-nŏm′ĭ-trē) The measurement of geologic time using absolute measures, such as radiometric dating, or relative measures, such as determining the sequence of deposition of rock strata.

geocline (jē′ō-klīn′) See **geosyncline.**

geocorona (jē′ō-kə-rō′nə) The halo of far-ultraviolet solar light that reflects off the Earth's **exosphere**.

geode (jē′ōd′) A small, hollow, usually rounded rock lined on the inside with inward-pointing crystals. Geodes form when mineral-rich water entering a cavity in a rock undergoes a sudden change in pressure or temperature, causing crystals to form from the solution and line the cavity's walls.

geodesic (jē′ə-dĕs′ĭk, -dē′sĭk) *Noun.* **1.** A curve that locally minimizes the distance between two points on any mathematically defined space, such as a curved manifold. Equivalently, it is a path of minimal curvature. In noncurved three-dimensional space, the geodesic is a straight line. In **General Relativity**, the trajectory of a body with negligible mass on which only gravitational forces are acting (i.e. a free falling body) is a geodesic in (curved) 4-dimensional spacetime. *—Adjective.* **2.** Of or relating to the branch of geometry that deals with geodesics.

geodesic dome A domed or vaulted structure of straight elements that form interlocking polygons.

geodesy (jē-ŏd′ĭ-sē) The scientific study of the size and shape of the Earth, its field of gravity, and such varying phenomena as the motion of the magnetic poles and the tides. *—Adjective* **geodetic.**

geodetic latitude (jē′ə-dĕt′ĭk) The angular distance between a point on the Earth's surface and the equator, using as the vertex the intersection with the equatorial plane of a perpendicular line drawn from the surface point. Compare **geocentric latitude.**

geodetic longitude The angular distance between the prime meridian and another meridian passing through a point on the Earth's surface. Because the vertex of the angle is the center of the Earth, the same as for geocentric longitude, the value is the same for both types of longitude. Compare **geocentric longitude.**

geodetic survey A survey of a large area of land in which corrections are made to account for the curvature of the Earth.

Geographic Information System (jē′ə-grăf′ĭk) A computer application used to store, view, and analyze geographical information, especially maps.

geographic mile See **nautical mile.**

geographic north The direction from any point on Earth toward the North Pole. Also called *true north.* Compare **magnetic north.**

geographic south The direction from any point on Earth toward the South Pole. Compare **magnetic south.**

geography (jē-ŏg′rə-fē) **1.** The scientific study of the Earth's surface and its various climates, countries, peoples, and natural resources. **2.** The physical characteristics, especially the surface features, of an area.

geoid (jē′oid′) The hypothetical surface of the Earth that coincides everywhere with mean sea level and is perpendicular, at every point, to the direction of gravity. The geoid is used as a reference surface for astronomical measurements and for the accurate measurement of elevations on the Earth's surface.

geologic time (jē′ə-lŏj′ĭk) The period of time covering the formation and development of the Earth, from about 4.6 billion years ago to today. *See chart,* pages 262–263.

geology (jē-ŏl′ə-jē) **1.** The scientific study of the origin of the Earth along with its rocks, minerals, land forms, and life forms, and of the processes that have affected them over the course of the Earth's history. **2.** The structure of a specific region of the Earth, including its rocks, soils, mountains, fossils, and other features.

geomagnetic pole (jē′ō-măg-nĕt′ĭk) One of two regions of the Earth with very high magnetic field strength, taken to be the points at which a line, drawn between the poles of an idealized magnetic dipole generating the Earth's magnetic field and extending out in both directions, would cross the earth's surface. The north pole of a magnet, such as a compass needle, is attracted to the geomagnetic north pole because the Earth's north pole is actually a magnetic south pole (and its geomagnetic south pole is a magnetic north pole). **Magnetic reversals** over the course of the Earth's history have switched the polarities of the geomagnetic poles many times. The geomagnetic poles are close to the geographical North and South poles, and to the Earth's magnetic poles. See also **magnetic pole.**

geomagnetic reversal See **magnetic reversal.**

geomagnetism (jē′ō-măg′nĭ-tĭz′əm) **1.** The magnetic properties of the Earth and its atmosphere. **2.** The study of these properties.

geometric isomer (jē′ə-mĕt′rĭk) Any of two or more stereoisomers that differ in the arrangement of atoms or groups of atoms around a structurally rigid bond, such as a double bond or a ring. Geometric isomers differ from one another in physical properties like melting and boiling points. Compare **enantiomer.**

geometric mean The positive *n*th root of the product of a set of *n* numbers. For example, the geometric mean of $2 \times 3 \times 6 \times 8$ is the 4th root of 288, or approximately 4.12.

geometric progression A sequence of numbers in which each number is multiplied by the same factor to obtain the next number in the sequence. In a geometric progression, the ratio of any two adjacent numbers is the same. An example is 5, 25, 125, 625, ..., where each number is multiplied by 5 to obtain the following number, and the ratio of any number to the next number is always 1 to 5. Compare **arithmetic progression.**

geometry (jē-ŏm′ĭ-trē) The mathematical study of the properties, measurement, and relationships of points, lines, planes, surfaces, angles, and solids.

geomorphology (jē′ō-môr-fŏl′ə-jē) The scientific study of the formation, alteration, and configuration of landforms and their relationship with underlying structures.

geophone (jē′ə-fōn′) An electronic receiver designed to pick up seismic vibrations on or below the Earth's surface and to convert them into electric impulses that are proportional to the displacement, velocity, and acceleration of ground movement. Geophones detect motion in only one direction and are usually used in groups of at least three, oriented at different angles, so that a three-dimensional record of ground movement can be obtained.

geophysics (jē′ō-fĭz′ĭks) The scientific study of the physical characteristics of the Earth, including its hydrosphere and atmosphere, and of the Earth's relationship to the rest of the universe.

geophyte (jē′ə-fīt′) A perennial plant with an underground food storage organ, such as a bulb, tuber, corm, or rhizome. The parts of the plant that grow above ground die away during adverse conditions, as in winter or during the dry season, and grow again from buds that are on or within the underground portion when conditions improve. Crocuses and tulips are geophytes.

geostationary orbit (jē′ō-stā′shə-nĕr′ē) A circular orbit positioned approximately 35,900 km

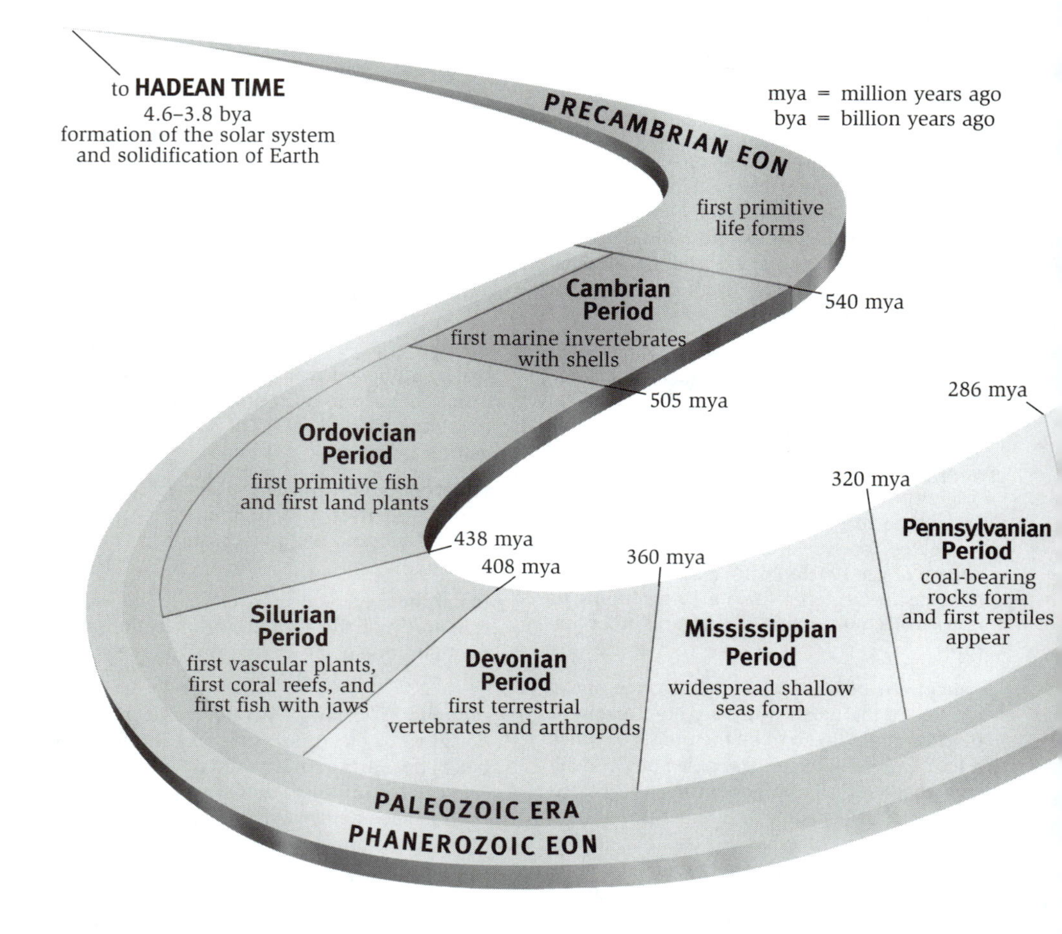

Geologic Time

The Earth formed approximately 4.6 billion years ago. The first 800 million years of Earth's history are referred to as Hadean Time. During this time, the solar system was forming and Earth was solidifying. The next 3.8 billion years are subdivided into eons, eras, periods, and epochs, with eons being the longest divisions of time, and epochs the shortest.

The oldest currently known rocks date to the beginning of the Precambrian Eon, 3.8 billion years ago. Rocks that formed earlier than this, during Hadean Time, have likely been eroded or drawn deep into Earth and melted through the processes of plate tectonics. Despite the absence of these rocks, the age of Earth has been determined through the study of meteorites and lunar rocks that formed contemporaneously with it.

Because much of our understanding of Earth's history is based on what we learn by studying its rocks and fossils, we know a lot more about its recent history than we do about its earlier history. This is why most of the geologic time divisions correspond to the last 540 million years (the Phanerozoic Eon), even though most of the Earth's history occurred before that time.

Most of the boundaries between the divisions of geologic time correspond to visible changes in the types of life forms preserved in the corresponding rocks. The boundary between the Precambrian and Phanerozoic eons, for example, was designated based on the appearance of an abundance of fossils at the beginning of the Phanerozoic. The boundary between the Cretaceous Period and the Tertiary Period was designated based on the disappearance of dinosaurs from the fossil record. Humans have been part of Earth's history for only the past million years or so.

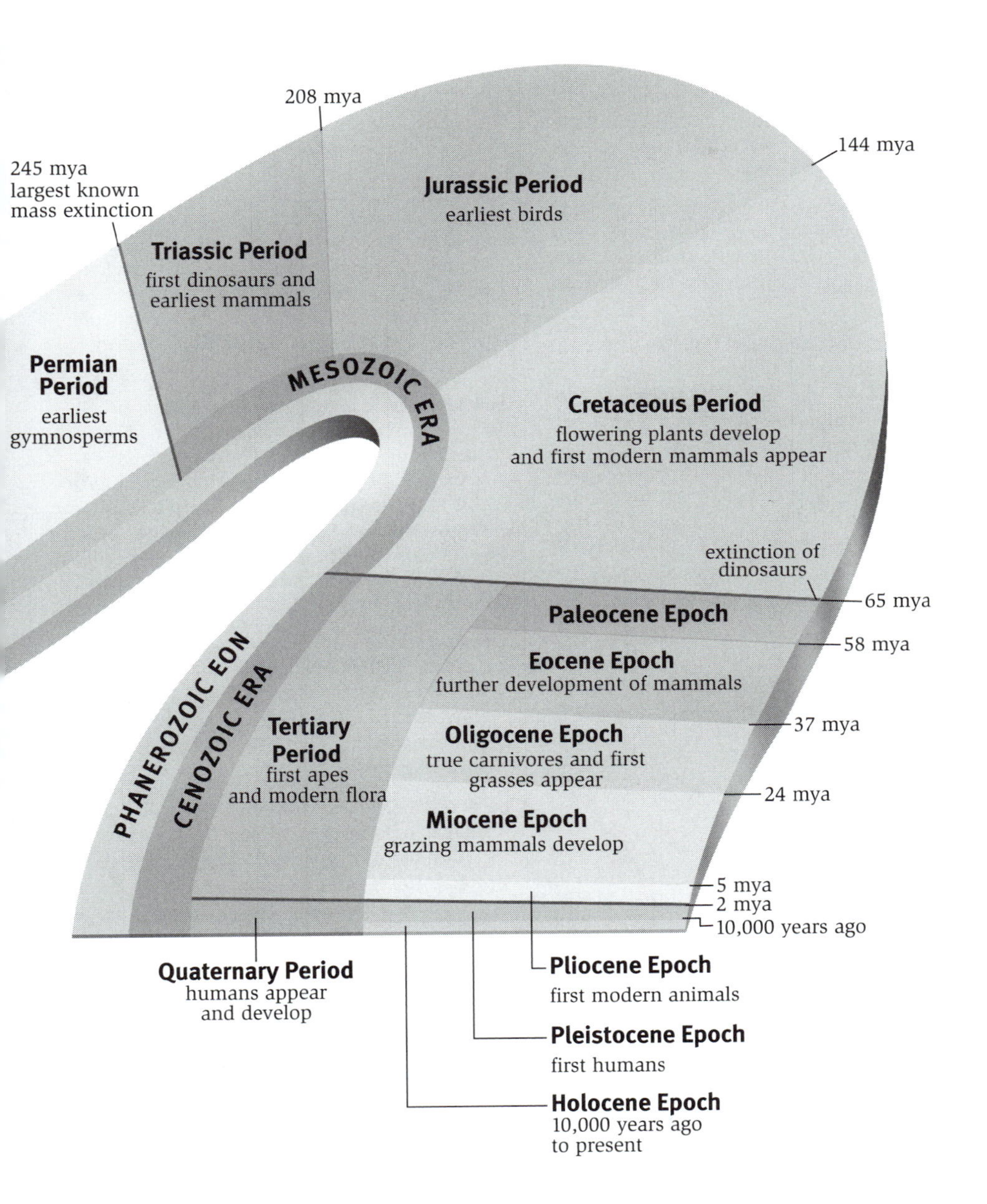
208 mya
144 mya
245 mya
largest known
mass extinction
Jurassic Period
earliest birds
Triassic Period
first dinosaurs and
earliest mammals
Permian
Period
earliest
gymnosperms
MESOZOIC ERA
Cretaceous Period
flowering plants develop
and first modern mammals appear
extinction of
dinosaurs
65 mya
Paleocene Epoch
58 mya
Eocene Epoch
further development of mammals
PHANEROZOIC EON
CENOZOIC ERA
Tertiary
Period
first apes
and modern flora
Oligocene Epoch
true carnivores and first
grasses appear
37 mya
24 mya
Miocene Epoch
grazing mammals develop
5 mya
2 mya
10,000 years ago
Quaternary Period
humans appear
and develop
Pliocene Epoch
first modern animals
Pleistocene Epoch
first humans
Holocene Epoch
10,000 years ago
to present

(22,258 mi) above Earth's equator and having a period of the same duration and direction as the rotation of the Earth. An object in this orbit will appear stationary relative to the rotating Earth. Communications and weather satellites are usually placed in a geostationary orbit. See also **synchronous orbit.**

geosyncline (jē′ō-sĭn′klīn′) A usually elongate, basinlike depression along the edge of a continent, in which a thick sequence of sediments and volcanic deposits has accumulated.

geothermal (jē′ō-thûr′məl) Relating to the internal heat of the Earth. The water of hot springs and geysers is heated by geothermal sources. ► **Geothermal energy** is power generated from natural steam, hot water, hot rocks, or lava in the Earth's crust. In general, geothermal power is produced by pumping water into cracks in the Earth's crust and then conveying the heated water or steam back to the surface so that its heat can be extracted through a heat exchanger, or its pressure can be used to drive turbines.

geotropism (jē-ŏt′rə-pĭz′əm) The directional growth of an organism in response to gravity. Roots display positive geotropism when they grow downwards, while shoots display negative geotropism when they grow upwards. Also called *gravitropism.* —*Adjective* **geotropic** (jē′ə-trō′pĭk, jē′ə-trŏp′ĭk).

geriatrics (jĕr′ē-ăt′rĭks) The branch of medicine that deals with the diagnosis and treatment of diseases and conditions of the elderly.

germ (jûrm) A microscopic organism or agent, especially one that is pathogenic, such as a bacterium or virus.

USAGE **germ/microbe/pathogen**

The terms *germ* and *microbe* have been used to refer to invisible agents of disease since the nineteenth century, when scientists introduced the *germ theory of disease,* the idea that infections and contagious diseases are caused by *microorganisms. Microbe,* a shortening and alteration of *microorganism,* comes from the Greek prefix *mikro–,* "small," and the word *bios,* "life." Scientists no longer use the terms *germ* and *microbe* very much. Today they can usually identify the specific agents of disease, such as individual species of bacteria or viruses. To refer generally to agents of disease, they use the term *pathogen,* from the Greek *pathos,* "suffering," and the suffix *–gen,* "producer." They use *microorganism* to refer to any unicellular organism, whether disease-causing or not.

Germain (zhĕr-măn′), **Sophie** 1776–1831. French mathematician who made significant advances in theoretical mathematics. Her researches into number theory in particular provided the first partial solution to Fermat's last theorem (1820).

germanium (jər-mā′nē-əm) *Symbol* **Ge** A brittle, crystalline, grayish-white metalloid element that is found in coal, in zinc ores, and in several minerals. It is used as a semiconductor and in wide-angle lenses. Atomic number 32; atomic weight 72.59; melting point 937.4°C; boiling point 2,830°C; specific gravity 5.323 (at 25°C); valence 2, 4. See **Periodic Table.**

German measles (jûr′mən) An infectious disease caused by the rubella virus of the genus *Rubivirus,* characterized by mild fever and skin rash. German measles can cause congenital defects if a woman is exposed during early pregnancy. Also called *rubella.*

germ cell 1. One of the diploid cells in the reproductive organs of a multicellular organism that undergo division and are the precursors of haploid gametes, such as a spermatogonium or oogonium. **2.** See **gamete.**

germination (jûr′mə-nā′shən) The beginning of growth, as of a seed, spore, or bud. The germination of most seeds and spores occurs in response to warmth and water.

germ layer Any of three cellular layers, the ectoderm, endoderm, or mesoderm, into which most animal embryos differentiate in the gastrula stage and from which the organs and tissues of the body develop through further differentiation.

gestation (jĕ-stā′shən) The period of time spent in the uterus between conception and birth. Gestation in humans is about nine months.

geyser (gī′zər) A natural hot spring that regularly ejects a spray of steam and boiling water into the air.

A CLOSER LOOK **germination**

Dormant seeds are very dry and require the absorption of water to initiate the metabolic processes of respiration and begin to digest their stored food. Respiration requires the presence of oxygen, which must be sufficiently available in the soil for *germination* to proceed, so the soil must be wet but not so waterlogged as to make oxygen inaccessible. Temperatures must be above freezing (zero degrees Celsius) but not excessively hot (not more than about 45 degrees Celsius). If conditions are right, a radicle (an embryonic root) emerges from the seed coat, anchoring the seed; it then grows and puts out lateral roots. In most eudicots, a part of the developing stem, either the epicotyl (the stem above the cotyledons) or the hypocotyl (the stem below the cotyledons) elongates, forming a hook and gradually pulling the seed coat and the delicate shoot tip above the soil surface. Germination of eudicot seeds is normally divided into two types, designated epigeous and hypogeous. In *epigeous* germination, the cotyledons emerge above the soil surface, and wither and drop off after their food stores have been used up; in *hypogeous* germination, the cotyledons remain below the surface and decompose after their food stores have been used up. In most monocots, food is stored in the seed's endosperm (rather than the cotyledon), and it is the single tubular cotyledon that elongates and draws the seed coat out of the soil. The cotyledon conducts photosynthesis, making more food, while the shoot grows up inside the tube.

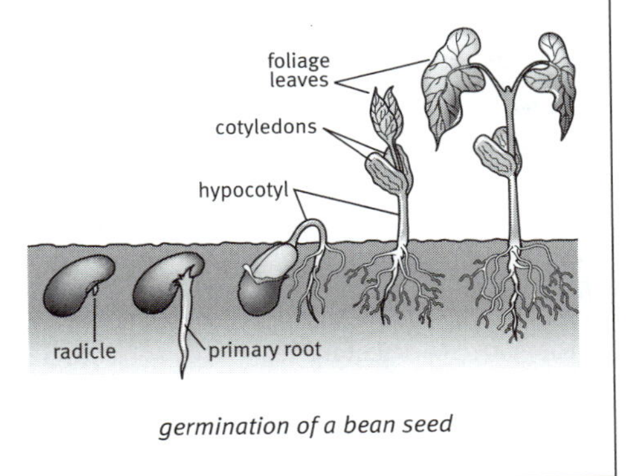

germination of a bean seed

g-force A force acting on a body as a result of acceleration or gravity, informally described in units of acceleration equal to one g. For example, a 12 pound object undergoing a g-force of 2g experiences 24 pounds of force. See more at **acceleration of gravity.**

giant impact theory (jī′ənt) A theory that explains the origins of Earth's moon, postulating that an asteroid roughly the size of Mars impacted the Earth during its formation. This impact resulted in rocky surface material being blown outward from the Earth, some of which accreted to form the Moon. This theory explains the similar oxygen isotope ratios between the Earth and the Moon as well as the Moon's lack of an iron core. It also accounts for the angular momentum that was necessary for the Moon to reach its current position.

giant molecular cloud A cloud of molecular hydrogen typically located in the arm of a spiral galaxy, believed to be an area of active star formation. Protostars are located in the densest regions of such clouds. See more at **protostar.**

giant star A very large, bright non-main-sequence star that burns hydrogen at a much faster rate than **a dwarf star.** Giant stars are much more luminous and have shorter lifespans than the slower-burning dwarfs. The larger the giant, the shorter its lifespan; the largest stars, with solar mass of around 100, blaze at several hundred thousand times the energy of the Sun and will last only a few million years, a very brief time when compared with the Sun's 10-billion-year lifespan. Giant stars usually end their lives as **supernovae**, but even before that event the immense ultraviolet radiation they produce has a dramatic impact on their stellar surroundings; the presence of a giant star in a star system prevents the formation of new protostars because the radiation from the giant star breaks apart any nearby nebulae.

giardiasis (jē′är-dī′ə-sĭs) An intestinal infection caused by the protozoan *Giardia lamblia.* It is usually asymptomatic in humans but can cause diarrhea, nausea, and abdominal cramping. Giardiasis is most commonly transmitted by contaminated water and by direct contact among individuals in group settings.

gibberellic acid (jĭb′ə-rĕl**′**ĭk) A plant hormone that promotes growth and that is the most important of the gibberellins. It is used commercially to break dormancy of plants, promote germination, and increase or retard the development of fruit. Gibberellic acid is also a metabolic byproduct of certain fungi. *Chemical formula:* $C_{19}H_{22}O_6$.

gibberellin (jĭb′ə-rĕl**′**ĭn) Any of numerous plant hormones, especially gibberellic acid, that promote stem elongation. The seeds, young shoots, and roots of plants contain gibberellins, and they are also found in fungi.

gibbous (gĭb**′**əs) More than half but less than fully illuminated. Used to describe the Moon or a planet. Compare **crescent.**

Gibbs (gĭbz), **Josiah Willard** 1839–1903. American physicist known especially for his investigations of thermodynamics. He developed methods for analyzing the thermodynamic properties of substances, and his findings established the basic theory for physical chemistry.

Gibbs free energy See **free energy.**

Gibbs phase rule See **phase rule.**

giga– A prefix that means: **1.** One billion (10^9), as in *gigahertz,* one billion hertz. **2.** 2^{30} (that is, 1,073,741,824), which is the power of 2 closest to a billion, as in *gigabyte.*

gigabit (gĭg**′**ə-bĭt′) **1.** One billion bits. **2.** 1,073,741,824 (that is, 2^{30}) bits. See Note at **megabyte.**

gigabyte (gĭg**′**ə-bīt′) **1.** A unit of computer memory or data storage capacity equal to 1,024 megabytes (2^{30} bytes). **2.** One billion bytes. See Note at **megabyte.**

gigaflop (gĭg**′**ə-flŏp′) A measure of computing speed equal to one billion floating-point operations per second.

gigahertz (gĭg**′**ə-hûrtz′) A unit of frequency equal to one billion (10^9) hertz.

Gilbert (gĭl**′**bərt), **Walter** Born 1932. American biologist who, building upon the work of Frederick Sanger, formulated a method for determining the sequence of bases in DNA that made it possible to manufacture genetic materials in the laboratory. For this work he shared with Sanger and American biologist Paul Berg the 1980 Nobel Prize for chemistry.

Gilbert, William 1544–1603. English court physician and physicist whose book *De Magnete* (1600) was the first comprehensive scientific work published in England. Gilbert demonstrated that the Earth itself is a magnet, with lines of force running between the North and South Poles. He theorized that magnetism and electricity were two types of a single force and was the first to use the words *electricity* and *magnetic pole.*

gill (gĭl) **1.** The organ that enables most aquatic animals to take dissolved oxygen from the water. It consists of a series of membranes that have many small blood vessels. Oxygen passes into the bloodstream and carbon dioxide passes out of it as water flows across the membranes. **2.** One of the thin strips of tissue on the underside of the cap of many species of basidiomycete fungi. Gills produce the spore-bearing structures known as basidia.

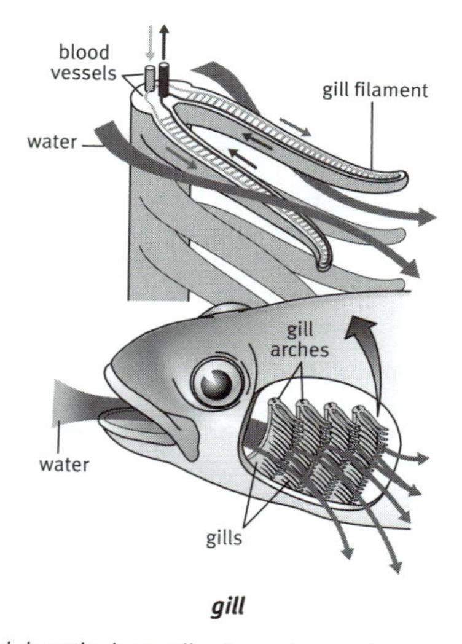

gill

Fish breathe by swallowing water, and passing it through gill slits on each side of their head. Blood-filled filaments on the gills extract oxygen from the water as it flows through.

gill slit In primitive chordates and cartilaginous fish, such as sharks, one of several narrow external openings connecting with the pharynx through which water passes to the exterior, thereby bathing the gills. Bony fish and all other vertebrates, including humans, have rudimentary gill slits during their embryonic stage.

gingiva (jĭn′jə-və) Plural **gingivae** (jĭn′jə-vē′). The gums of the mouth. The gingiva are made up of epithelial tissue that is attached to the bones of the jaw and surrounds and supports the bases of the teeth. Also called *gum.*

gingivitis (jĭn′jə-vī′tĭs) Inflammation of the gums, characterized by redness and swelling.

ginkgo also **gingko** (gĭng′kō) A deciduous, dioecious tree *(Ginkgo biloba)* which is the sole surviving member of the Ginkgoales, an order of gymnosperms that was extremely widespread in the Mesozoic era. It belongs to a genus which has changed very little since the end of the Jurassic period. The tree, a native of China, has fan-shaped leaves and fleshy yellowish seeds containing an edible kernel. Ginkgoes are often grown as ornamental street trees.

girdle (gûr′dl) To kill a tree or woody shrub by removing or destroying a band of bark and cambium from its circumference. The plants die because the distribution of food down from the leaves (through the phloem) and sometimes the flow of water and nutrients up from the roots (through the xylem) is disrupted, and the cambium can no longer regenerate these vascular tissues to repair the damage. Unwanted trees, such as invasive or nonnative species, are often eliminated by girdling. Some plant diseases kill trees by destroying a ring of cambium and so girdling them. Gnawing animals, especially rodents, can also girdle trees.

GIS Abbreviation of **Geographic Information System.**

gizzard (gĭz′ərd) A muscular pouch behind the stomach in birds. It has a thick lining and often contains swallowed sand or grit, which helps in the mechanical breakdown of food.

glacial (glā′shəl) **1.** Relating to or derived from a glacier. **2.** Characterized or dominated by the existence of glaciers, as the Pleistocene Epoch.

glacial striation One of several, long, straight, parallel lines or grooves in a bedrock surface, formed by boulders, gravel, and pebbles embedded in a glacier that has passed over the surface.

glacial valley A steep-sided, U-shaped valley formed by the erosional forces of a moving glacier.

glacier (glā′shər) A large mass of ice moving very slowly through a valley or spreading outward from a center. Glaciers form over many years from packed snow in areas where snow accumulates faster than it melts. A glacier is always moving, but when its forward edge melts faster than the ice behind it advances, the glacier as a whole shrinks backward.

gland (glănd) An organ or group of specialized cells in the body that produces and secretes a specific substance, such as a hormone. See also **endocrine gland, exocrine gland.**

Glashow (glăsh′ō), **Sheldon Lee** Born 1932. American physicist who developed one of the first theories of an electroweak force, unifying two of the four fundamental forces of nature—the electromagnetic force and the weak force—as two aspects of a single underlying force. He developed the theories of Abdus Salam and Steven Weinberg by introducing a new particle property known as **charm**, and for this work he shared with Salam and Weinberg the 1979 Nobel Prize for physics.

glass (glăs) A usually transparent or translucent material that has no crystalline structure yet behaves like a solid. Common glass is generally composed of a silicate (such as silicon oxide, or quartz) combined with an alkali and sometimes other substances. The glass used in windows and windshields, called soda glass, is made by melting a silicate with sodium carbonate (soda) and calcium oxide (lime). Other types of glass are made by adding other chemical compounds. Adding boron oxide causes some silicon atoms to be replaced by boron atoms, resulting in a tougher glass that remains solid at high temperatures and is used for cooking utensils and scientific apparatuses. Glass used for decorative purposes often has iron in it to alter its optical properties.

A CLOSER LOOK **glass**

Common sand and *glass* are both made primarily of silicon and oxygen, yet sand is opaque and glass is transparent. Glass owes its transparency partly to the fact that it is not a typical solid. On the molecular level, solids usually have a highly regular, three-dimensional crystalline structure; the regularities distributed throughout the solid act as mirrors that scatter incoming light. Glass, however, consists of molecules which, though relatively motionless like a typical solid, are not arranged in regular patterns and thus exhibit little scattering; light passes directly through. At a specific temperature, called the *melting point,* the intermolecular forces holding together the components of a typical solid can no longer maintain the regular structure, which then breaks down, and the material undergoes a phase transition from solid to liquid. The phase transition in glass, however, depends on how quickly the glass is heated (or how quickly it cools), due to its irregular solid structure.

---

glaucoma (glou-kō′mə, glô-) A disease of the eye in which the pressure of fluid inside the eyeball is abnormally high, caused by obstructed outflow of the fluid. The increased pressure can damage the optic nerve and lead to partial or complete loss of vision.

glaucophane (glou′kə-fān′) A blue to grayish-blue or bluish-black monoclinic mineral of the amphibole group. Glaucophane occurs as fibrous prisms in schists (especially blueschists) formed by the regional metamorphism of sodium-rich rocks. *Chemical formula:* $Na_2(Mg,Fe)_3Al_2Si_8O_{22}(OH)_2$.

gleba (glē′bə) Plural **glebae** (glē′bē′). The fleshy, spore-bearing inner mass of the basidiomycete fungus known as the puffball.

glia (glē′ə, glī′ə) The delicate network of branched cells and fibers that supports the tissue of the central nervous system. —*Adjective* **glial.**

glial cell (glē′əl, glī′əl) Any of the cells making up the glia, such as the star-shaped cells called astrocytes.

Global Positioning System (glō′bəl) A system of satellites combined with receivers on the Earth that determines the latitude and longitude of any particular receiver through triangulation. The distance of the receiver to three of the satellites is ascertained by measuring the time-delay of a predetermined radio signal (called a **pseudo-random code**). Errors in timing can be corrected by checking the signals against the signal from a fourth satellite. Current systems can pinpoint the location of the receiver with an accuracy of around 5 m (16 ft). The system is used for navigation, surveying, and many other applications. Compare **loran.**

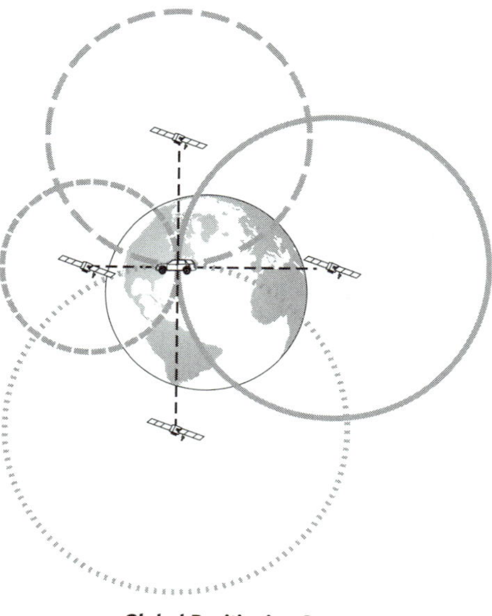

Global Positioning System

Signals received from four satellites are used to determine the three-dimensional position of the receiver (vehicle). The receiver interprets the arrival times of the signals in terms of latitude, longitude, and altitude.

global warming An increase in the average temperature of the Earth's atmosphere, especially a sustained increase great enough to cause changes in the global climate. The Earth has experienced numerous episodes of global warming through its history, and currently appears to be undergoing such warming. The present warming is generally attributed to an increase in the **greenhouse effect**, brought about by increased levels of **greenhouse gases**, largely due to the effects of human industry and agriculture. Expected long-term effects of current global warming are rising sea levels, flooding, melting of polar ice caps and glaciers, fluctuations in temperature and precipitation, more frequent and stronger El Niños and La Niñas, drought, heat waves, and forest fires. See more at **greenhouse effect.**

globular cluster (glŏb′yə-lər) A spherical mass made up of thousands to hundreds of thousands of densely packed stars of nearly the same age (typically very old). Globular clusters occupy the inner regions of a galactic halo and revolve around the nucleus of galaxies in highly elliptical orbits inclined to the disk of the spiral arms. There are approximately 150 of these clusters in the Milky Way. Most known galaxies are orbited by numerous globular clusters. It is believed that globular clusters can provide information about the evolution and lifespan of stars. Compare **open cluster.**

globulin (glŏb′yə-lĭn) A major class of proteins found in the seeds of plants and in various tissues and substances of vertebrate and invertebrate animals, including blood, muscle, and milk. The globulins in blood comprise all the plasma proteins besides albumin. Two kinds, *alpha* and *beta globulin,* are primarily transport proteins or serve as substrates for forming other substances, and include lipoproteins and enzymes. A third kind, the *gamma globulins,* consists almost entirely of the **immunoglobulins.** Most globulins are insoluble in water but soluble in saline solution.

glomerulus (glō-mĕr′yə-ləs) Plural **glomeruli** (glō-mĕr′yə-lī′). A knot of highly permeable capillaries located within the Bowman's capsule of a nephron. Waste products are filtered from the blood in the glomerulus, initiating the process of urine formation.

glottis (glŏt′ĭs) Plural **glottises** or **glottides** (-ĭ-dēz′). The part of the larynx that contains the vocal cords and the space between them.

glucagon (glo͞o′kə-gŏn′) A polypeptide hormone produced by the pancreas that stimulates an increase in blood glucose levels, thus opposing the action of insulin.

glucosamine (glo͞o-kō′sə-mēn′, glo͞o′kō-) An amino derivative of glucose in which an amino group replaces a hydroxyl group. It is a component of many polysaccharides and is the basic structural unit of chitin. Glucosamine is used as an over-the-counter dietary supplement by some people with symptoms of arthritis. *Chemical formula:* $C_6H_{13}NO_5$.

glucose (glo͞o′kōs′) A monosaccharide sugar found in plant and animal tissues. Glucose is a product of photosynthesis, mostly incorporated into the disaccharide sugar sucrose rather than circulating free in the plant. Glucose is essential for energy production in animal cells. It is transported by blood and lymph to all the cells of the body, where it is metabolized to form carbon dioxide and water along with ATP, the main source of chemical energy for cellular processes. Glucose molecules can also be linked into chains to form the polysaccharides cellulose, glycogen, and starch. *Chemical formula:* $C_6H_{12}O_6$. See more at **cellular respiration, Krebs cycle, photosynthesis.**

glucoside (glo͞o′kə-sīd′) A glycoside in which the sugar component is glucose.

glume (glo͞om) One of the two chaffy bracts at the base of a grass spikelet.

gluon (glo͞o′ŏn) The subatomic particle that mediates the **strong force**. The exchange of gluons between two quarks changes the color of the quarks and results in the attractive force holding them together in hadrons. Gluons are **bosons**. See Table at **subatomic particle.**

glutamic acid (glo͞o-tăm′ĭk) A nonessential amino acid. *Chemical formula:* $C_5H_9NO_4$. See more at **amino acid.**

glutamine (glo͞o′tə-mēn′) A nonessential amino acid. *Chemical formula:* $C_5H_{10}N_2O_3$. See more at **amino acid.**

glutathione (glo͞o′tə-thī′ōn′) A polypeptide consisting of glycine, cysteine, and glutamic acid that occurs widely in plant and animal tissues. It is important in cellular respiration in both plants and animals, and serves as a cofactor for many enzymes. It is a major protective mechanism against oxidative stress. For example, it protects red blood cells from hydrogen peroxide, a toxic byproduct of certain metabolic reactions.

gluten (glo͞ot′n) **1.** The mixture of proteins, found in wheat grains, which are not soluble in water and which give wheat dough its elastic texture. **2.** Any of the prolamins found in cereal grains, especially the prolamins in wheat, rye, barley, and possibly oats, that cause digestive disorders such as celiac disease.

glycemic index (glī-sē′mĭk) A numerical index that is given to a carbohydrate-rich food that is based on the average increase in blood glucose levels occurring after the food is eaten.

glyceraldehyde (glĭs′ə-răl′də-hīd′) A sweet colorless syrupy liquid that is an intermediate compound in carbohydrate metabolism. *Chemical formula:* $C_3H_6O_3$.

glyceride (glĭs′ə-rīd′) Any of various esters formed when glycerol reacts with a fatty acid. The fatty acids can react with one, two, or all three of the hydroxyl groups of the glycerol, resulting in mono-, di-, and triglycerides, respectively. Triglycerides are the main components of plant and animal oils and fats.

glycerin also **glycerine** (glĭs′ər-ĭn) See **glycerol.**

glycerol (glĭs′ə-rôl′) A sweet, syrupy liquid obtained from animal fats and oils or by the fermentation of glucose. It is used as a solvent, sweetener, and antifreeze and in making explosives and soaps. Glycerol consists of a propane molecule attached to three hydroxyl (OH) groups. Also called *glycerin, glycerine. Chemical formula:* $C_3H_8O_3$.

glyceryl (glĭs′ər-əl) **1.** The radical CH_2CHCH_2, obtained from glycerol by removing all three hydroxyl (OH) groups. This radical is a component of many natural oils and fats. **2.** The radical $CH_2CH(OH)CH_2OH$, obtained from glycerol by removing one hydroxyl (OH) group.

glycine (glī′sēn′, -sĭn) A nonessential amino acid. Glycine is the simplest amino acid. *Chemical formula:* $C_2H_5NO_2$. See more at **amino acid.**

glycogen (glī′kə-jən) A polysaccharide stored in animal liver and muscle cells that is easily converted to glucose to meet metabolic energy requirements. Most of the carbohydrate energy stored in animal cells is in the form of glycogen.

glycol (glī′kôl′, -kōl′) **1.** See **ethylene glycol. 2.** Any of various alcohols containing two hydroxyl groups (OH).

glycolic acid (glī-kŏl′ĭk) A colorless crystalline compound that occurs naturally in sugar beets and sugarcane. It is used in leather dyeing and tanning, and in making pharmaceuticals, pesticides, adhesives, and plasticizers. *Chemical formula:* $C_2H_4O_3$.

glycolysis (glī-kŏl′ə-sĭs) The process in cell metabolism by which carbohydrates and sugars, especially glucose, are broken down, producing ATP and pyruvic acid. See more at **cellular respiration.**

glycoprotein (glī′kō-prō′tēn′) Any of a group of cellular macromolecules that are made up of proteins bonded to one or more carbohydrate chains.

glycosaminoglycan (glī′kōs-ə-mē′nō-glī′kăn) Any of a group of polysaccharides with high molecular weight that contain amino sugars and often form complexes with proteins. Heparin is a glycosaminoglycan. Also called *mucopolysaccharide.*

glycoside (glī′kə-sīd′) Any of various organic compounds formed from a simple sugar (monosaccharide) by replacing the hydrogen atom of one of its hydroxyl groups (OH) with the bond to another biologically active molecule. Glycosides occur abundantly in plants, especially as pigments, and are used in medicines, dyes, and cleansing agents.

GMO Abbreviation of **genetically modified organism.**

gneiss (nīs) A highly foliated, coarse-grained metamorphic rock consisting of light-colored layers, usually of quartz and feldspar, alternating with dark-colored layers of other minerals, usually hornblende and biotite. Individual grains are often visible between layers. Gneiss forms as the result of the regional metamorphism of igneous, sedimentary, or other metamorphic rocks.

gnetophyte (nē′tə-fīt′) Any of a small but diverse phylum (Gnetophyta) of gymnosperm plants with some features similar to those of angiosperms, such as xylem with vessels, strobili resembling inflorescences, and the absence of archegonia. The genus *Ephedra,* well-known as a source of stimulant compounds, is classified among the gnetophytes.

goblet cell (gŏb′lĭt) Any of the specialized epithelial cells, such as those found in the mucous membrane of the stomach, intestines, and respiratory passages, that secrete mucus. The goblet cells distend with mucin before secretion and collapse to a goblet shape after secretion.

Goddard (gŏd′ərd), **Robert Hutchings** 1882–1945. American physicist who developed numerous rockets and rocket devices, including the first successful liquid-fueled rocket (1926), the first instrument-carrying rocket that could make observations in flight (1929), and the first rockets to exceed the speed of sound.

Gödel (gŭd′l), **Kurt** 1906–1978. Austrian-born American mathematician who in 1931 published the most important axiom in modern mathematics, known as Gödel's proof. It states that in any finite mathematical system, there will always be statements that cannot be proved or disproved. Gödel's proof ended efforts by mathematicians to find a mathematical system that was entirely consistent in itself.

Goeppert-Mayer (gŭp′ûrt-mā′ər), **Maria** 1906–1972. German-born American physicist who, with Hans Jensen, developed a model of the atomic nucleus that explained why certain nuclei were stable and had an unusual number of stable isotopes. For this work, Goeppert-Mayer and Jensen shared the 1963 Nobel Prize for physics with American physicist Eugene Wigner.

goiter (goi′tər) An enlarged thyroid gland, visible as a swelling at the front of the neck. It is often associated with thyroid disease, especially in areas of the world outside of North America where iodine deficiency is endemic.

gold (gōld) *Symbol* **Au** A soft, shiny, yellow element that is the most malleable of all the metals. It occurs in veins and in alluvial deposits. Because it is very durable, resistant to corrosion, and a good conductor of heat and electricity, gold is used as a plated coating on electrical and mechanical components. It is also an international monetary standard and is used in jewelry and for decoration. Atomic number 79; atomic weight 196.967; melting point 1,063.0°C; boiling point 2,966.0°C; specific gravity 19.32; valence 1, 3. See **Periodic Table.** See Note at **element.**

Goldberger (gōld′bər-gər), **Joseph** 1874–1929. Hungarian-born American physician who investigated the cause and treatment of pellagra. He demonstrated that contrary to common belief, pellagra was not an infectious disease but a nutritional disorder caused by an unbalanced diet and cured by the addition of fresh milk, meat, or yeast.

golden section (gōl′dən) The ratio between two numbers *a* and *b* chosen such that the ratio of *a* to *b* is equal to the ratio of *a*+*b* to *a*. Its value is approximately 1.618. Shapes with proportions equal to the golden section are observed especially in the fine arts and in architecture, as between the two dimensions of a plane figure such as a rectangle. The ratio between consecutive numbers in a Fibonacci sequence approximates the golden section with increasing precision as the series progresses. Also called *golden mean, golden ratio.*

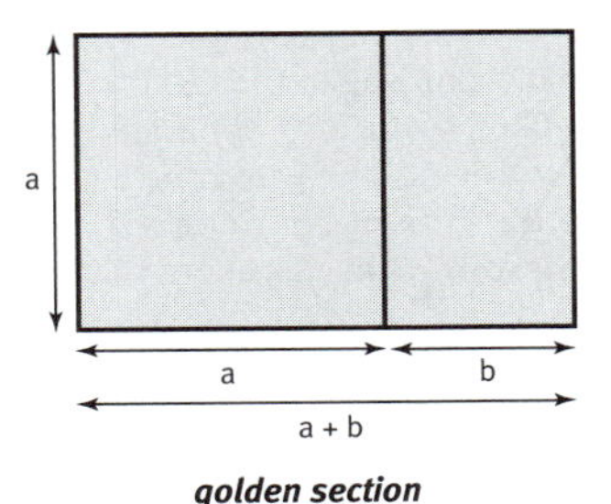

golden section

Two rectangles with the proportions of the golden section: the large rectangle, and the small one on the right inside it. The smaller rectangle has proportions $\frac{a}{b}$, equal to the proportions of the larger one, $\frac{(a+b)}{a}$.

Golgi apparatus (gōl′jē) An organelle in eukaryotic cells that stores and modifies proteins for specific functions and prepares them for transport to other parts of the cell. The Golgi apparatus is usually near the cell nucleus and consists of a stack of flattened sacs. Proteins secreted by the **endoplasmic reticulum** are transported into and across the Golgi apparatus by **vesicles** and may be combined with sugars to form glycoproteins. The modified products are stored in vesicles (such as lysosomes) for later use or transported by vesicles to the plasma membrane, where they are excreted from the cell. The Golgi apparatus is named for its identifier, Italian cytologist Camillo Golgi (1843–1926). It is also called the *Golgi body* or, in plant cells, the *dictyosome.* ► Collectively in the cell, these organelles are known as the **Golgi complex**. See more at **cell.**

gonad (gō′năd′) An organ in animals that produces reproductive cells (gametes). The ovary and testis are gonads.

gonadotropin (gō-năd′ə-trō′pĭn, -trŏp′ĭn) also **gonadotrophin** (gō-năd′ə-trō′fĭn, -trō′pĭn) Any of several hormones that stimulate the growth and activity of the gonads, especially follicle-stimulating hormone and luteinizing hormone.

gonadotropin-releasing hormone A peptide hormone produced by the hypothalamus that signals the anterior pituitary gland to begin secreting luteinizing hormone and follicle-stimulating hormone.

Gondwanaland (gŏnd-wä′nə-lănd′) A supercontinent of the Southern Hemisphere made up of the landmasses that currently correspond to India, Australia, Antarctica, and South America. According to the theory of plate tectonics, Gondwanaland separated from **Pangaea** at the end of the Paleozoic Era and broke up into the current continents in the middle of the Mesozoic Era. Compare **Laurasia.**

gonidium (gō-nĭd′ē-əm) Plural **gonidia. 1.** An asexual reproductive cell found in certain algae that form colonies. Gonidia undergo repeated mitoses to form new colonies, which then hatch out of the parent colonies. **2.** A chlorophyll-bearing, photosynthetic algal cell housed in the thallus of a lichen.

gonophore (gŏn′ə-fôr′) A structure bearing or consisting of a reproductive organ or part, such as the one of the buds that produce sperm or eggs on a cnidarian polyp.

gonopore (gŏn′ə-pôr′) A reproductive aperture or pore, especially of certain insects and worms.

gonorrhea (gŏn′ə-rē′ə) A sexually transmitted disease caused by the bacterium *Neisseria gonorrhoeae,* characterized by inflammation of the mucous membranes of the genital and urinary tracts, an acute discharge containing pus, and painful urination, especially in men. Women often have few or no symptoms.

Goodall (go͝od′ôl), **Jane** Born 1934. British zoologist whose study of the life and habitat of the chimpanzee has greatly increased understanding of primate behavior. Goodall's research demonstrated that chimpanzees are capable of complex emotional relationships, and have the skill and intelligence to make tools. She has been a leader in international conservation efforts.

Jane Goodall

googol (go͞o′gôl′, go͞o′gəl) The number 10 raised to the 100th power (10^{100}), written out as 1 followed by 100 zeros.

googolplex (go͞o′gôl-plĕks′) The number 10 raised to the power googol, written out as the numeral 1 followed by a googol (10^{100}) of zeros.

Gorgas (gôr′gəs), **William Crawford** 1854–1920. American army surgeon who directed programs to eradicate the *Aedes aegypti* mosquito in Havana, Cuba (1901), and in the Panama Canal Zone (1904–1906). The mosquito had been shown by Dr. Walter Reed and others to be responsible for the transmission of yellow fever.

gorge (gôrj) A deep, narrow valley with steep rocky sides, often with a stream flowing through it. Gorges are smaller and narrower than canyons and are often a part of a canyon.

Gould (go͞old), **Stephen Jay** 1941–2002. American paleontologist and evolutionary biologist who with Niles Eldredge developed the theory of punctuated equilibrium in 1972. He published numerous books which popularized his sometimes controversial ideas on evolutionary theory among the general public.

gout (gout) An inherited disorder of uric acid metabolism occurring predominantly in men, characterized by painful inflammation of the joints. Elevated levels of uric acid in the blood result in deposition of crystals of uric acid salts (known as *urates*) around the joints, causing arthritis. The condition can become chronic and result in deformity.

G-protein (jē′prō′tēn′) Any of a class of cell membrane proteins that function as intermediaries between hormone receptors and enzymes that enable the cell to regulate its metabolism in response to hormonal changes.

GPS Abbreviation of **Global Positioning System.**

graben (grä′bən) A usually elongated block of rock that is bounded by parallel geologic faults along its two longest sides, and has a lower elevation than the rock at its sides.

Grabens form where rock is being pulled apart by tectonic forces. The East African Rift Valley is a graben.

gradation (grā-dā′shən) **1.** The process by which land is leveled off through erosion or the transportation or deposition of sediments, especially the process by which a riverbed is brought to a level where it is just able to transport the amount of sediment delivered to it. **2.** The proportion of particles (such as sand grains) of a given size within a sample of particulate material, such as soil or sandstone.

grade (grād) **1.** The degree of inclination of a slope, road, or other surface. **2.** A grouping of organisms done purely on the basis of shared features and without regard to evolutionary relationships. Grades may include organisms that do not share a common ancestor, or may exclude some organisms having the same common ancestor as the other organisms in the grade. For this reason, many taxonomists do not accept grades as formal classifications. The class Reptilia (reptiles) is a grade since it includes dinosaurs but not birds, even though birds are descended from dinosaurs. Compare **clade.**

gradient (grā′dē-ənt) **1.** The degree to which something inclines; a slope. A mountain road with a gradient of ten percent rises one foot for every ten feet of horizontal length. **2.** The rate at which a physical quantity, such as temperature or pressure changes over a distance. **3.** A operator on scalar fields yielding a vector function, where the value of the vector evaluated at any point indicates the direction and degree of change of the field at that point.

gradient

The gradient on the left is gentle enough for trees to take root. The gradient on the right is too steep.

gradualism (grăj′o͞o-ə-lĭz′əm) The theory that new species evolve from existing species through gradual, often imperceptible changes rather than through abrupt, major changes. The small changes are believed to result in perceptible changes over long periods of time. Compare **punctuated equilibrium.**

graduated (grăj′o͞o-ā′tĭd) Divided into or marked with intervals indicating measures, as of length, volume, or temperature.

graft (grăft) *Noun.* **1.** A shoot or bud of one plant that is inserted into or joined to the stem, branch, or root of another plant so that the two grow together as a single plant. Grafts are used to strengthen or repair plants, create dwarf trees, produce seedless fruit, and increase fruit yields without requiring plants to mature from seeds. **2.** A piece of body tissue that is surgically removed and then transplanted or implanted to replace a damaged part or compensate for a defect. See also **allograft, autograft,** and **xenograft.** —*Verb.* **3.** To join a graft to another plant. **4.** To transplant or implant a graft.

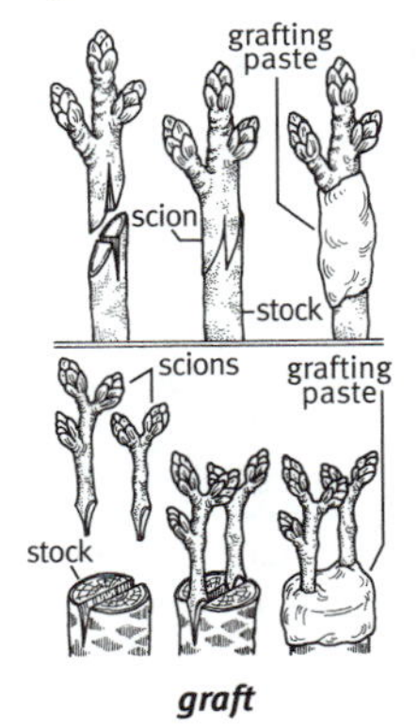

graft

top: *whip grafting;* bottom: *cleft grafting*

grain (grān) **1.** See **caryopsis. 2.** A small particle of something, such as salt, pollen, or sand. **3.** A unit of weight in the US Customary System, equal to $\frac{2}{1000}$ of an ounce (0.07 gram).

gram (grăm) A unit of mass in the metric system, equal to 0.001 kilogram or 0.035 ounce. See Table at **measurement.**

Gram (gräm, grăm), **Hans Christian Joachim** 1853–1938. Danish bacteriologist who in 1884 developed a method of staining bacteria, called Gram's stain or Gram's dye, that is used to identify and classify bacteria, often from samples of infected body fluids. The classification, called gram-negative or gram-positive, can be useful in the initial selection of antibiotics to treat the infection.

gram-negative Relating to a group of bacteria that do not change color when subjected to the laboratory staining method known as Gram's method or Gram's stain. Gram-negative bacteria have relatively thin cell walls and are generally resistant to the effects of antibiotics or the actions of the body's immune cells. Gram-negative bacteria include *E. coli* and the bacteria that cause gonorrhea, typhoid fever, rickettsial fever, cholera, syphilis, plague, and Lyme disease.

gram-positive Relating to a group of bacteria that turn a dark-blue color when subjected to a laboratory staining method known as Gram's method. Gram-positive bacteria have relatively thick cell walls and are generally sensitive to the destructive effects of antibiotics or the actions of the body's immune cells. Gram-positive bacteria include beneficial nitrogen-fixing bacteria in soil, as well as the bacteria that cause anthrax, botulism, leprosy, tuberculosis, scarlet fever, and strep throat.

grand unified theory (grănd) A theory of elementary forces that unites the weak, strong, electromagnetic, and gravitational interactions into one field theory and views the known interactions as low-energy manifestations of a single unified interaction.

granite (grăn′ĭt) A usually light-colored, coarse-grained igneous rock consisting mostly of quartz, orthoclase feldspar, sodium-rich plagioclase feldspar, and micas. Quartz usually makes up 10 to 50 percent of the light-colored minerals in granite, with the remaining minerals consisting of the feldspars and muscovite. The darker minerals in granite are usually biotite and hornblende. Granite is one of the most common rocks in the crust of continents, and is formed by the slow, underground cooling of magma.

granodiorite (grăn′ə-dī′ə-rīt′) A coarse-grained igneous rock consisting primarily of quartz, plagioclase, and potassium feldspar, and also containing biotite, hornblende, or pyroxene.

granophyre (grăn′ə-fīr′) A fine-grained igneous rock that has large intergrown crystals of quartz and feldspar in the matrix.

granule (grăn′yo͞ol) **1.** A rock or mineral fragment larger than a sand grain and smaller than a pebble. Granules have a diameter between 2 and 4 mm (0.08 and 0.16 in) and are often rounded. **2.** Any of the small, transient convective cells within the Sun's photosphere where hot gases rise and quickly dissipate. Granules are generally between a few hundred and 1,500 km in width. They completely cover the Sun's surface, giving it its characteristic grainy or stippled look, and form and break up within a matter of minutes. **3.** An aggregate of enclosed grainy matter found in a cell. Granulocytes, mast cells and other cells contain granules in their cytoplasm, which differ in size and can often be identified by a characteristic laboratory stain based on their composition. Granules produce and store biologically active substances, the release of which is called *degranulation*. The granules of granulocytes contain mostly multiple enzymes and other proteins; those of mast cells contain histamine and other chemical mediators.

granulite (grăn′yə-līt′) A fine-grained metamorphic rock consisting of similarly sized, interlocking minerals. Unlike most metamorphic rocks, granulites do not exhibit foliation or textural or mineralogical layering.

granulocyte (grăn′yə-lō-sīt′) Any of various white blood cells that contain granular material in the cytoplasm and are immunologically active, especially in phagocytosis. Granulocytes are the most numerous of the white blood cells in humans.

granum (grā′nəm) Plural **grana** (grā′nə). A stacked membranous structure within the chloroplasts of plants and green algae that contains the chlorophyll and is the site of the light reactions of photosynthesis. The saclike membranes that make up grana are known as **thylakoids.** See more at **chloroplast.**

graph (grăf) **1.** A diagram showing the relationship of quantities, especially such a diagram in which lines, bars, or proportional areas represent how one quantity depends on or changes with another. **2.** A curve or line showing a mathematical function or equation, typically drawn in a Cartesian coordinate system. The graph of the function $y = x^2$ is a parabola.

graphical user interface (grăf′ĭ-kəl) See **GUI.**

graphic equalizer (grăf′ĭk) See under **equalizer.**

graphics (grăf′ĭks) The representation of data in a way that includes images in addition to or instead of text. Computer-aided design, typesetting, and video games, for example, involve the use of graphics.

graphite (grăf′īt′) A naturally occurring, steel-gray to black, crystalline form of carbon. The carbon atoms in graphite are strongly bonded together in sheets. Because the bonds between the sheets are weak, other atoms can easily fit between them, causing graphite to be soft and slippery to the touch. Graphite is used in pencils and paints and as a lubricant and electrode. It is also used to control chain reactions in nuclear reactors because of its ability to absorb neutrons.

graptolite (grăp′tə-līt′) Any of numerous hemichordates of the class Graptolithina. Graptolites form colonies consisting of interlocked cuplike chambers arranged in one or more branches and covered by an exoskeleton. They flourished from the late Cambrian to the early Mississippian Period, and were thought to be extinct until 1992 when scientists discovered what is believed to be a living species. Graptolites are important index fossils used to date the rocks of the Silurian and Ordovician Periods.

grass (grăs) Any of a large family (*Gramineae* or *Poaceae*) of monocotyledonous plants having narrow leaves, hollow stems, and clusters of very small, usually wind-pollinated flowers. Grasses include many varieties of plants grown for food, fodder, and ground cover. Wheat, maize, sugar cane, and bamboo are grasses. See more at **leaf.**

grassland (grăs′lănd′) An area that is dominated by grass or grasslike vegetation. Moderately dry climatic conditions and seasonal disturbances, such as floods or fires, are generally conducive to the growth of grasses and prohibitive of that of trees and shrubs. Grasslands are found in tropical, subtropical, and temperate regions and typically occupy regions between forests and deserts.

graupel (grou′pəl) A small, white ice particle that falls as precipitation and breaks apart easily when it lands on a surface. Also called *snow pellet, soft hail.*

Graves' disease (grāvz) An autoimmune disease of the thyroid gland characterized by excessive production of thyroid hormone, goiter, protrusion of the eyeballs (*exophthalmos*), and symptoms of hyperthyroidism, such as rapid heartbeat and weight loss. The disease is named after its discoverer, Irish physician Robert James Graves (1796–1853).

gravimeter (gră-vĭm′ĭ-tər) **1.** An instrument used to measure variations in a gravitational field, typically by measuring the rate of acceleration of a falling body. Gravimeters are used to survey geological features with different densities beneath the Earth's surface, such as ore-laden rock or oil fields, that affect the local strength of gravity above them. **2.** An instrument, such as a hydrometer, used to measure the specific gravity of a liquid or solid.

gravimeter

gravitational field gravimeter

gravitation (grăv′ĭ-tā′shən) See **gravity.**

gravitational collapse (grăv′ĭ-tā′shə-nəl) **1.** The implosion of a star or other celestial body as a result of its own gravity, resulting in a body that is many times smaller and denser than the original body. **2.** The process by which stars, star clusters, and galaxies form from interstellar gas under the influence of gravity. Clusters of matter are drawn together by gravitational pull, with additional matter continuing to accumulate until the growing nebula develops into even denser gaseous bodies such as stars or groups of stars.

gravitational constant A constant relating the force of the gravitational attraction between two bodies to their masses and their distance from each other in Newton's law of gravitation. The gravitational constant equals approximately 6.67259×10^{-11} newton square meters per square kilogram. Its symbol is G. See more at **Newton's law of gravitation.**

gravitational lens A massive celestial object, such as a galaxy, whose gravity can act as a lens that functions to bend and focus the light of a more distant object. This results in a magnified, distorted, or multiple image of the original light source for a distant observer.

gravitino (grăv′ĭ-tē′nō) A hypothetical particle postulated in supergravity theory to correspond to the graviton, just as neutrinos correspond to other fermions. See also **neutrino, supergravity.**

graviton (grăv′ĭ-tŏn′) A hypothetical particle postulated in supergravity theory to be the quantum of gravitational interaction, mediating the gravitational force. Like all **force carriers**, the graviton is a boson. It is presumed to have an indefinitely long lifetime, zero electric charge, a spin of 2, and zero rest mass (thus travelling at the speed of light). The graviton has never been detected. See also **supersymmetry.** See Table at **subatomic particle.**

gravitropism (gră-vĭt′rə-pĭz′əm) See **geotropism.**

gravity (grăv′ĭ-tē) The fundamental force of attraction that all objects with mass have for each other. Like the electromagnetic force, gravity has effectively infinite range and obeys the inverse-square law. At the atomic level, where masses are very small, the force of gravity is negligible, but for objects that have very large masses such as planets, stars, and galaxies, gravity is a predominant force, and it plays an important role in theories of the structure of the universe. Gravity is believed to be mediated by the graviton, although the graviton has yet to be isolated by experiment. Gravity is weaker than the strong force, the electromagnetic force, and the weak force. Also called *gravitation.* See more at **acceleration, relativity.**

A CLOSER LOOK **gravity**

With his law of universal gravitation, Sir Isaac Newton described *gravity* as the mutual attraction between any two bodies in the universe. He developed an equation describing an instantaneous gravitational effect that any two objects, no matter how far apart or how small, exert on each other. These effects diminish as the distance between the objects gets larger and as the masses of the objects get smaller. His theory explained both the trajectory of a falling apple and the motion of the planets—hitherto completely unconnected phenomena—using the same equations. Albert Einstein developed the first revision of these ideas. Einstein needed to extend his theory of *Special Relativity* to be able to understand cases in which bodies were subject to forces and acceleration, as in the case of gravity. According to Special Relativity, however, the instantaneous gravitational effects in Newton's theory would not be possible, for to act instantaneously, gravity would have to travel at infinite velocities, faster than the speed of light, the upper limit of velocity in Special Relativity. To overcome these inconsistencies, Einstein developed the theory of *General Relativity,* which connected gravity, mass, and acceleration in a new manner. Imagine, he said, an astronaut standing in a stationary rocket on the Earth. Because of the Earth's gravity, his feet are pressed against the rocket's floor with a force equal to his weight. Now imagine him in the same rocket, this time accelerating in outer space, far from any significant gravity. The accelerating rocket pushing against his feet creates a force indistinguishable from that of a gravitational field. Developing this *principle of equivalence,* Einstein showed that mass itself forms curves in space and time and that the effects of gravity are related to the trajectories taken by objects—even objects without mass, such as light. Whether gravity can be united with the other fundamental forces understood in quantum mechanics remains unclear.

---

gray (grā) The SI derived unit used to measure the energy absorbed by a substance per unit weight of the substance when exposed to radiation. One gray is equal to one joule per kilogram, or 100 rads. The gray is named after British physicist Louis Harold Gray (1905–1965).

gray matter The brownish-gray tissue of the vertebrate brain and spinal cord, made up chiefly of the cell bodies and dendrites of neurons. Compare **white matter.**

graywacke (grā′wăk′, -wăk′ə) Any of various dark gray, coarse-grained sandstones that contain abundant feldspar and rock fragments and often have a clay-rich matrix. Graywackes are thought to originate in environments where erosion, transportation, and deposition happen so quickly that minerals and rock fragments do not have sufficient time to break down into finer constituents.

graywater (grā′wô′tər) Wastewater from baths, bathroom sinks, and washing machines that does not contain body or food wastes. Graywater can be recycled especially for use in gardening or for flushing toilets. Compare **blackwater.**

great ape (grāt) See **anthropoid ape.**

great circle A circle on the surface of a sphere whose plane passes through the center of the sphere. The Earth's equator is a great circle on the sphere of the globe.

Great Red Spot A very large, high-pressure atmospheric feature on the planet Jupiter, characterized by anticyclonic winds circulating at a speed of approximately 400 km (248 mi) per hour. The storm has persisted on Jupiter's surface for more than 300 years since first observed through telescopes; the cause of its reddish color is unknown.

green alga Any of various photosynthetic protists belonging to the phylum Chlorophyta. The green algae share many characteristics with plants, notably in their having chlorophylls a and b, in their storage of food as starch, and in the composition of their cell walls from cellulose or other polysaccharides. The green algae show a great variety of body types, ranging from unicellular forms to filaments to leaflike thalli, and many species live in colonies. Green algae also show a variety of reproductive processes, both sexual, by the formation of conjugating gametes or the exchange of nuclei through conjugation tubes, and asexual, by means of spores. Green algae are mostly aquatic, in both freshwater and marine environments. However, many species live on land or in the soil, and even in extreme environments, such as the surface of snow. Green algae are not always green, since they produce carotenoid pigments that can give them orange or red colors. Some lichens consist of a symbiotic relationship between a fungus and a green alga. Sea moss and the common pond scum *Spirogyra* are green algae. Also called *chlorophyte.* See more at **alga.**

greenfield (grēn′fēld) A piece of usually semirural property that is undeveloped except for agricultural use, especially one considered as a site for expanding urban development. Compare **brownfield.**

greenhouse effect (grēn′hous′) The retention of part of the Sun's energy in the Earth's atmosphere in the form of heat as a result of the presence of greenhouse gases. Solar energy, mostly in the form of short-wavelength visible radiation, penetrates the atmosphere and is absorbed by the Earth's surface. The heated surface then radiates some of that energy into the atmosphere in the form of longer-wavelength infrared radiation. Although some of this radiation escapes into space, much of it is absorbed by greenhouse gases in the lower atmosphere, which in turn re-radiate a portion back to the Earth's surface. The atmosphere thus acts in a manner roughly analagous to the glass in a greenhouse, which allows sunlight to penetrate and warm the plants and soil but which traps most of the resulting heat energy inside. The greenhouse effect is essential to life on Earth; however, the intensification of its effect due to increased levels of greenhouse gases in the atmosphere is considered to be the main contributing factor to **global warming**.

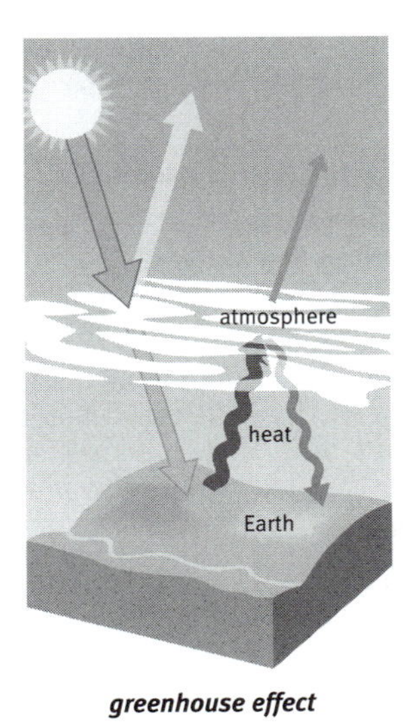

greenhouse effect

greenhouse gas Any of the atmospheric gases that contribute to the greenhouse effect by absorbing infrared radiation produced by solar warming of the Earth's surface. They include carbon dioxide (CO_2), methane (CH_4), nitrous oxide (N_2O), and water vapor. Although greenhouse gases occur naturally in the atmosphere, the elevated levels especially of carbon dioxide and methane that have been observed in recent decades are directly related, at least in part, to human activities such as the burning of fossil fuels and the deforestation of tropical forests.

green manure A growing crop, such as clover or grass, that is plowed under the soil to improve fertility. Green manure can also reduce erosion and, if the crop is leguminous, add nitrogen to the soil.

green revolution The application of science to increasing agricultural productivity, including the breeding of high-yield varieties of grains, the effective use of pesticides, and improved fertilization, irrigation, mechanization, and soil conservation techniques.

greenschist (grēn′shĭst′) A green, schistose metamorphic rock that gets its color from the presence of chlorite, epidote, or actinolite. Greenschist forms under conditions of low temperature and low pressure.

greenstone (grēn′stōn′) Any of various green metamorphic rocks formed from igneous rocks that have a relatively low silica content and owe their color to the presence of a green mineral such as chlorite, hornblende, or epidote.

green vitriol See **ferrous sulfate.**

Greenwich Mean Time (grĕn′ĭch) See **universal time.**

ground (ground) **1.** A connection between an electrical conductor and the Earth. Grounds are used to establish a common zero-voltage reference for electric devices in order to prevent potentially dangerous voltages from arising between them and other objects. Also called *earth.* **2.** The set of shared points in an electrical circuit at which the measured voltage is taken to be zero. The ground is usually connected directly to the power supply and acts as a common "sink" for current flowing through the components in the circuit.

ground meristem The primary meristem in vascular plants that gives rise to the nonvascular tissues, such as cortex, pericycle, and pith. Within the seeds of angiosperms, it surrounds the procambium.

ground state The state of a physical system having the lowest possible potential energy. For example, an electron in the lowest energy orbital in a hydrogen atom is in a ground state. The ground state of a physical system tends to be stable unless energy is applied to it from the outside; states that are not the ground state have a tendency to revert to the ground state, giving off energy in the process. Compare **excited state.**

ground substance **1.** The intercellular material in which the cells and fibers of connective tissue are embedded, composed largely of glycosaminoglycans, metabolites, water, and ions. **2.** The clear, fluid portion of cytoplasm as distinguished from the organelles and other cell components.

ground tissue The tissue of a plant other than the epidermis, periderm, and vascular tissues, consisting primarily of **parenchyma**, and (in lesser amounts) of **collenchyma** and **sclerenchyma**. Cortex and pith are subtypes of ground tissue.

groundwater (ground′wô′tər) Water that collects or flows beneath the Earth's surface, filling the porous spaces in soil, sediment, and rocks. Groundwater originates from rain and from melting snow and ice and is the source of water for aquifers, springs, and wells. The upper surface of groundwater is the **water table**.

group (gro͞op) **1.** *Chemistry.* **a.** Two or more atoms that are bound together and act as a unit in a number of chemical compounds, such as a hydroxyl (OH) group. **b.** In the Periodic Table, a vertical column that contains elements having the same number of electrons in the outermost shell of their atoms. Elements in the same group have similar chemical properties. See **Periodic Table. 2.** *Mathematics.* A set with an operation whose domain is all ordered pairs of members of the set, such that the operation is binary (operates on two elements) and associative, the set contains the identity element of the operation, and each element of the set has an inverse element for the operation. The positive and negative integers and zero form a set that is a group under the operation of ordinary addition, since zero is the identity element of addition and the negative of each integer is its inverse. Groups are used extensively in quantum physics and chemistry to model phenomena involving symmetry and invariance.

group theory The branch of mathematics concerned with groups and the description of their properties.

growth (grōth) An increase in the size of an organism or part of an organism, usually as a result of an increase in the number of cells. Growth of an organism may stop at maturity, as in the case of humans and other mammals, or it may continue throughout life, as in many plants. In humans, certain body parts, like hair and nails, continue to grow throughout life.

growth hormone 1. A polypeptide hormone secreted by the anterior portion of the pituitary gland that promotes growth by stimulating protein synthesis. Growth hormone also acts on the liver to produce peptides called **somatomedins**, which stimulate growth of bone, cartilage, and muscle. Also called *somatotropin.* **2.** Any of various natural or synthetic substances that regulate the growth of plants. Auxins in plants are growth hormones.

growth ring 1. A layer of wood formed in a plant during a single period of growth. Growth rings are visible as concentric circles of varying width when a tree is cut crosswise. They represent layers of cells produced by vascular cambium. ▸ Most growth rings reflect a full year's growth and are called **annual rings.** But abrupt changes in the environment, especially in the availability of water, can cause a plant to produce more than one growth ring in a year. See more at **dendrochronology. 2.** A similar layer in a part of an animal marking a period of growth, such as an annulus in a fish scale.

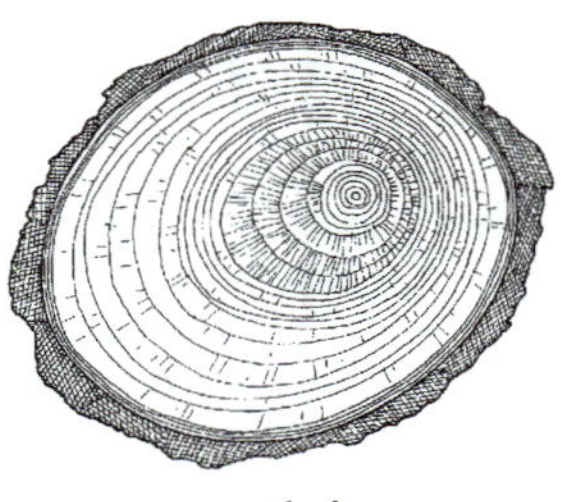

growth ring

guanine (gwä′nēn′) A purine base that is a component of DNA and RNA, forming a base pair with cytosine. It also occurs in guano, fish scales, sugar beets, and other natural materials. *Chemical formula:* $C_5H_5ON_5$.

guano (gwä′nō) **1.** A substance composed chiefly of the dung of sea birds or bats, accumulated along certain coastal areas or in caves and used as fertilizer. **2.** Any of various similar substances, such as a fertilizer prepared from ground fish parts.

guard cell (gärd) One of the paired cells in the epidermis of a plant that control the opening and closing of a stoma of a leaf. When swollen with water, guard cells pull apart from each other, opening the stoma to allow the escape of water vapor and the exchange of gases. When drier, guard cells become more flaccid and move closer together, allowing the plant to conserve water. Unlike the other cells in the epidermis, guard cells have chloroplasts and conduct photosynthesis. See more at **stoma.**

GUI (go͞o′ē) Short for *graphical user interface.* An interface that is used to issue commands to a computer by means of a device such as a mouse that manipulates and activates onscreen images.

gulf (gŭlf) A large body of ocean or sea water that is partly surrounded by land.

Gulf Stream A warm ocean current of the northern Atlantic Ocean off eastern North America. It flows northward and eastward from the Gulf of Mexico, eventually dividing into several branches. A major branch continues eastward to warm the coast and moderate the climate of northwest Europe.

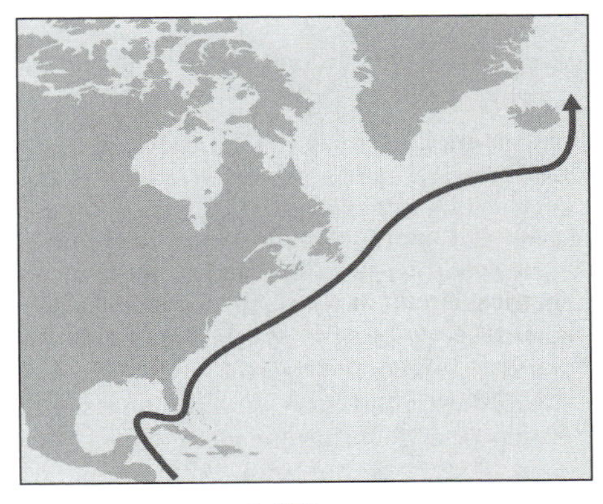

Gulf Stream

gully (gŭl′ē) A narrow, steep-sided channel formed in loose earth by running water. A gully is usually dry except after periods of heavy rainfall or after the melting of snow or ice.

gum[1] (gŭm) Any of various sticky substances that are produced by certain plants and trees and dry into brittle solids soluble in water. Gums typically are colloidal mixtures of polysaccharides and mineral salts.

gum[2] (gŭm) See **gingiva.**

gum arabic A gum exuded by various African trees of the genus *Acacia,* especially *A. senegal.* Gum arabic is used in the preparation of pills and emulsions, in the manufacture of adhesives and candies, and as a thickener and stabilizer of colloids. Gum arabic consists mostly of a mixture of oligosaccharides and heavy glycoproteins.

gut (gŭt) **1.** The intestine of a vertebrate animal. **2.** The alimentary canal of an invertebrate animal. **3.** The tube in a vertebrate embryo that later develops into the alimentary canal, lungs, and liver.

GUT Abbreviation of **grand unified theory** (another name for **unified field theory**).

guttation (gə-tā′shən) The exudation of water from leaves as a result of root pressure. Compare **transpiration.**

guyot (gē′ō) A flat-topped, extinct submarine volcano having an elevation of over 1,000 m (3,280 ft) above the ocean floor. Guyots are thought to form as volcanos in sea-floor spreading zones and to become extinct as they move away from the spreading zones through plate tectonic forces. Their flat tops are believed to form by the erosional action of waves when they initially project above sea level.

gymnosperm (jĭm′nə-spûrm′) Any of a group of seed-bearing plants whose ovules are not enclosed in an ovary, but are exposed on the surface of sporophylls or similar structures. Each ovule may contain several eggs, all of which may be fertilized and start to develop in a process known as **polyembryony.** In most seeds, however, only a single embryo survives. The reproductive structures of many gymnosperms are arranged in cones. The gymnosperms do not form a distinct monophyletic grouping, but simply include all the seed-bearing plants that are not angiosperms. In addition to several extinct groups, there are four very diverse living gymnosperm phyla: the conifers, the cycads, the ginkgo (surviving in a single species), and the gnetophytes. Compare **angiosperm.** See more at **seed-bearing plant.**

gynecology (gī′nĭ-kŏl′ə-jē) The scientific study of the female reproductive system, its diseases, and their treatment.

gynoecium (gī-nē′sē-əm, jĭ-) Plural **gynoecia.** The female reproductive organs of a flower considered as a group; the pistil or pistils. Compare **androecium.**

gynophore (gī′nə-fôr′, jĭn′ə-) The stalk of a pistil.

gypsum (jĭp′səm) A colorless, white, or pinkish mineral. Gypsum occurs as individual blade-shaped crystals or as massive beds in sedimentary rocks, especially those formed through the evaporation of saline-rich water. It is used in manufacturing plasterboard, cement, and fertilizers. *Chemical formula:* $\mathbf{CaSO_4{\cdot}2H_2O}$**.**

gyre (jīr) A spiral oceanic surface current driven primarily by the global wind system and constrained by the continents surrounding the three ocean basins (Atlantic, Pacific, and Indian). Each ocean basin has a large gyre in the subtropical region, centered around 30° north and south latitude. Smaller gyres occur at 50° north latitude in the North Atlantic and Pacific Oceans. The direction of a gyre's rotation is determined by the prevailing winds in the region, with the large subtropical gyres rotating clockwise in the Northern Hemisphere and counterclockwise in the Southern Hemisphere.

gyro pilot (jī′rō) An automatic pilot that incorporates a gyroscope and initiates corrections to control surfaces on aircraft, thus maintaining a preset course and altitude.

gyroscope (jī′rə-skōp′) An instrument consisting of a heavy disk or wheel spun rapidly about an axis like a top. The angular momentum of the disk causes it to resist changes in the direction of its axis of rotation, due to the principle of conservation of angular momentum. Because of the gyroscope's tendency to remain oriented in one direction, it is used as a stabilizing device in missiles, as well as in the navigation and piloting systems of airplanes, ships, rockets, and other vehicles.

gyrus (jī′rəs) Plural **gyri.** A rounded ridge, as on the surfaces of the cerebral hemispheres.

H

h **1.** Abbreviation of **height. 2.** The symbol for **Planck's constant.**

H **1.** The symbol for **henry. 2.** The symbol for **hydrogen.**

Haber (hä′bər), **Fritz** 1868–1934. German chemist who received a 1918 Nobel Prize for developing a method of producing ammonia synthetically from nitrogen and hydrogen gases.

habit (hăb′ĭt) **1.** The characteristic shape of a crystal, such as the cubic habit that is characteristic of pyrite. **2.** The characteristic manner of growth of a plant. For example, grape plants and ivy display a vining habit.

habitat (hăb′ĭ-tăt′) The area or natural environment in which an organism or population normally lives. A habitat is made up of physical factors such as soil, moisture, range of temperature, and availability of light as well as biotic factors such as the availability of food and the presence of predators. A habitat is not necessarily a geographic area—for a parasitic organism it is the body of its host or even a cell within the host's body.

habituation (hə-bĭch′o͞o-ā′shən) **1.** The gradual decline of a response to a stimulus resulting from repeated exposure to the stimulus. **2.** Physiological tolerance for a drug resulting from repeated use. **3.** Psychological dependence on a drug resulting from repeated use.

hadal (hād′l) Relating to the deepest regions of the ocean, below about 6,000 m (19,680 ft). The hadal zone mostly comprises trenches that can reach depths greater than 10,000 m (32,800 ft).

Hadean Time (hā-dē′ən) The period of geologic time between 4.6 and 3.8 billion years ago, when the solar system was forming and the Earth was solidifying. No rocks are known from this time, as they were presumably eroded or drawn deep into the Earth and melted through the processes of plate tectonics. See Chart at **geologic time.**

hadopelagic zone (hā′də-pə-lăj′ĭk) The bottommost layer of the oceanic zone, lying below the **abyssopelagic zone** at depths greater than about 6,000 m (19,680 ft).

hadron (hăd′rŏn′) Any of a class of subatomic particles composed of a combination of two or more quarks or antiquarks. Quarks (and antiquarks) of different colors are held together in hadrons by the strong nuclear force. Hadrons include both baryons (composed of three quarks or three antiquarks) and mesons (composed of a quark and an antiquark). The combination of quark colors in a hadron must be neutral, for example, red and antired (as in a pion) or red, blue, and green (as in a proton). Compare **baryon, lepton.**

hadrosaur (hăd′rə-sôr′) Any of various medium-sized to large dinosaurs of the group Hadrosauroidea of the Cretaceous Period. Hadrosaurs had a duck-like bill and a mouth containing many series of rough grinding teeth for chewing tough plants. They walked on two legs or on all fours and had hoofed feet. Many hadrosaurs bore hollow crests on their skulls. They were the last and largest ornithopod dinosaurs. Also called *duck-billed dinosaur.*

Haeckel (hĕk′əl), **Ernst Heinrich** 1834–1919. German naturalist who was the first to attempt a genealogical tree of all animals. He was also one of the first scientists to publicly support Darwin's theory of evolution. His own ideas about evolution attracted popular attention, and though they were later disproved, they helped to stimulate biological research.

hafnium (hăf′nē-əm) *Symbol* **Hf** A bright, silvery metallic element that occurs in zirconium ores. Because hafnium absorbs neutrons better than any other metal and is resistant to corrosion, it is used to control nuclear reactions. Atomic number 72; atomic weight 178.49; melting point 2,220°C; boiling point 5,400°C; specific gravity 13.3; valence 4. See **Periodic Table.**

Hahn (hän), **Otto** 1879–1968. German chemist who investigated radioactive elements and helped discover several new ones. His research on the irradiation on uranium and thorium with neutrons led to the 1938 discovery of nuclear fission.

hail (hāl) Precipitation in the form of rounded pellets of ice and hard snow that usually falls during thunderstorms. Hail forms when raindrops are blown up and down within a cloud, passing repeatedly through layers of warm and freezing air and collecting layers of ice until they are too heavy for the winds to keep them from falling.

hair (hâr) **1.** One of the fine strands that grow from the skin of mammals, usually providing insulation against the cold. Modified hairs sometimes serve as protective defenses, as in the quills of a porcupine or hedgehog, or as tactile organs, as in the whiskers (called *vibrissae*) of many nocturnal mammals. Hair filaments are a modification of the epidermis of the skin and are composed primarily of keratin. Hair also contains melanin, which determines hair color. **2.** A slender growth resembling a mammalian hair, found on insects and other animals. **3.** A fine, threadlike growth from the epidermis of plants. See more at **trichome.**

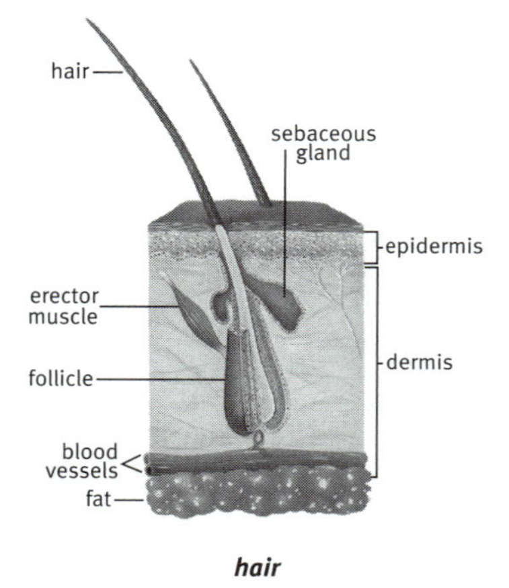

hair

hair cell A cell found in the organs of balance in the inner ear of mammals that senses the body's position with respect to gravity. Hair cells are found in both the semicircular canals and the vestibule.

half-life (hăf′līf′) The average time needed for half the nuclei in a sample of a radioactive substance to undergo radioactive decay. The half-life of a substance does not equal half of its full duration of radioactivity. For example, if one starts with 100 grams of radium 229, whose half-life is 4 minutes, then after 4 minutes only 50 grams of radium will be left in the sample, after 8 minutes 25 grams will be left, after 12 minutes 12.5 grams will be left, and so on.

halide (hăl′īd′, hā′līd′) A chemical compound consisting of a halogen and another element, especially a strongly electropositive metal such as sodium or potassium. Salt is a halide.

halite (hăl′īt′, hā′līt′) A colorless or white mineral occurring as cubic crystals. Halite is found in dried lakebeds in arid climates and is used as table salt. *Chemical formula:* **NaCl.** See more at **salt.**

Hall effect (hôl) A phenomenon that occurs when an electric current moving through a conductor is exposed to an external magnetic field applied at a right angle, in which an electric potential develops in the conductor at a right angle to both the direction of current and the magnetic field. The Hall effect is a direct result of Lorentz forces acting on the charges in the current, and is named after physicist Edwin Herbert Hall (1855–1938).

Halley (hăl′ē), **Edmond** 1656–1742. English astronomer and mathematician best known for his study of comets. Using Newton's laws of motion, he determined mathematically that comets move in elliptical orbits around the Sun. Based on his findings, he accurately predicted that a comet observed in 1583 would return in 1758, 1835, and 1910. This comet is now named after him. In 1679 Halley became the first person to catalog the stars in the Southern Hemisphere.

Halley's comet (hăl′ēz, hā′lēz) A short-period comet that orbits the Sun once every 76 years. It is visible to the unaided eye and last appeared in 1986, when close observation by spacecraft showed that its nucleus measures about 16 km (10 mi) by 8 km (5 mi) and is composed of water ice, stony minerals, and organic compounds. Its next appearance will be in the year 2061.

hallucinogen (hə-lo͞o′sə-nə-jən) A drug or chemical that causes a person to have hallucinations. Mescaline, LSD, and psilocybin are hallucinogens.

halo (hā′lō) A hazy ring of colored light in the sky around the Sun, Moon, or a similar bright object. A halo is caused by the reflection and refraction of light through atmospheric ice crystals.

halocarbon (hăl′ə-kär′bən) A compound, such as a fluorocarbon, that consists of carbon combined with one or more halogens. Halocarbons are typically nonflammable and nonreactive, though some halocarbons are broken down by ultraviolet radiation in the upper atmosphere, and this process releases free halogen atoms that damage the ozone layer. Some halocarbons have also been implicated as greenhouse gases.

halocline (hăl′ə-klīn′) A relatively sharp discontinuity in ocean salinity at a particular depth. In general, water with a higher concentration of salinity sinks below water that is less saline; therefore, saltier haloclines lie below less salty ones. An exception is the surface halocline of the Arctic Ocean, which is both colder and more saline than the warmer Atlantic water beneath it and which protects the polar ice from melting from below.

halogen (hăl′ə-jən) Any of a group of five nonmetallic elements with similar properties. The halogens are fluorine, chlorine, bromine, iodine, and astatine. Because they are missing an electron from their outermost shell, they react readily with most metals to form salts. See **Periodic Table.**

halon (hā′lŏn) Any of several compounds consisting of one or two carbon atoms combined with bromine and one or more other halogens. Halons are gases and are used as fire-extinguishing agents. They are between three and ten times more destructive to the ozone layer than CFCs are.

halophyte (hăl′ə-fīt′) A plant adapted to living in salty soil, as along the seashore or in salt flats. Mangroves, salt-marsh grasses, and saltbushes are halophytes.

halothane (hăl′ə-thān′) A colorless nonflammable liquid, $C_2HBrClF_3$, used as an inhalant anesthetic.

Halsted (hôl′stĕd′), **William Stewart** 1852–1922. American surgeon who discovered the technique of local anesthesia by injecting cocaine into specific nerves in 1885. He administered what is believed to be the first blood transfusion in the United States in 1881. Halsted also developed new surgical techniques for treating cancers and other abnormalities and introduced the use of rubber gloves during surgery.

Hamiltonian (hăm′əl-tō′nē-ən) A mathematical function or operator that can be used to describe the state of a physical system. In classical mechanics, the Hamiltonian is a **function** of coordinates and momenta of bodies in the system, treated as independent variables. It is equal to the sum of the kinetic and potential energies of the system, and can be used to derive the equations of motion for the system. In quantum mechanics, the Hamiltonian is an **operator** corresponding to the total energy of the system. The Hamiltonian is named after Irish mathematician William Rowan Hamilton (1805–1865).

hammerstone (hăm′ər-stōn′) A hand-held stone or cobble used by hominids perhaps as early as 2.5 million years ago as a crude pounding or pecking tool. Hammerstones were also used by early humans in striking flakes from stone cores to produce **core tools.**

hamstring (hăm′strĭng′) A powerful group of muscles at the back of the thigh that arise in the hip and pelvis and insert as strong tendons behind the knee. The hamstring bends the knee and helps to straighten the hip.

hand ax also **handax** (hănd′ăks′) A cutting or chopping tool, especially of the Lower Paleolithic Period, typically consisting of a piece of flint or other coarse stone that has been flaked on both sides to produce a sharp edge running all around the perimeter. Hand axes are core tools (produced from a found stone rather than from a processed flake) and have been found in several basic, often pointed shapes, including oval, triangular, and cordate (heart-shaped). The most common Paleolithic tool, they are especially associated with the Acheulian and some Mousterian tool cultures.

handedness (hăn′dĭd-nĭs) **1.** A preference for using one hand rather than the other to perform most manual tasks and activities. Most people are right-handed. Historically, it has been theorized that handedness is associated with a dominance of the opposite cerebral hemisphere of the brain, but this has not been conclusively proven. Although the scientific basis for handedness is unknown, the fact that left-handed parents more frequently have left-handed offspring suggests at least a partial genetic component. Some experts believe that children are trained to favor one hand over the other (usually the right hand). Handedness is usually established in the first few years of life. **2.** See **chirality.**

hanging valley (hăng′ĭng) A side valley that enters a main valley at an elevation high above the main valley floor. Hanging valleys are typically formed when the main valley has been widened and deepened by glacial erosion, leaving the side valley cut off abruptly from the main valley below. The steep drop from the hanging valley to the main valley floor usually creates cascading waterfalls.

hanging wall The block of rock lying above an inclined geologic fault plane. See more at **fault.** Compare **footwall.**

hantavirus (hăn′tə-vī′rəs) Any of a group of viruses of the genus *Hantavirus,* carried by rodents, that cause severe respiratory infections in humans and, in some cases, hemorrhaging, kidney disease, and death.

haploid (hăp′loid′) Having a single set of each chromosome in a cell or cell nucleus. In most animals, only the gametes (reproductive cells) are haploid. Compare **aneuploid, diploid.** See Note at **mitosis.**

haploidy (hăp′loi′dē) The state or condition of being haploid.

hard disk (härd) A rigid magnetic disk fixed within a disk drive and used for storing computer data. Hard disks hold more data than floppy disks, and data on a hard disk can be accessed faster than data on a floppy disk.

hard drive A disk drive that reads data stored on hard disks. Also called *hard disk drive.*

hardness (härd′nĭs) A measure of how easily a mineral can be scratched. Hardness is measured on the Mohs scale.

hard palate See under **palate.**

hardpan (härd′păn′) A hard, usually clay-rich layer of soil lying at or just below the ground surface, in which soil particles are cemented together by silica, iron oxide, calcium carbonate, or organic matter that has precipitated from water percolating through the soil. Hardpans do not soften when exposed to water. Also called *caliche.*

hardware (härd′wâr′) A computer, its components, and its related equipment. Hardware includes disk drives, integrated circuits, display screens, cables, modems, speakers, and printers. Compare **software.**

hardwood (härd′wo͝od′) **1.** An angiosperm tree, especially as distinguished from a coniferous, or softwood, tree. **2.** The wood of an angiosperm tree. Hardwoods are in general harder than softwood. However, some hardwoods, such as basswood, are comparatively soft, while some softwoods, such as yew, are comparatively hard.

Hardy-Weinberg law (här′dē-wīn′bûrg) A fundamental principle in population genetics stating that the genotype frequencies and gene frequencies of a large, randomly mating population remain constant provided immigration, mutation, and selection do not take place. In the simple case of a chromosome locus with two alleles, *A* and *a,* with frequencies *p* and *q* respectively, the frequency of the homozygotic genotype *AA* under random mating will be p^2, of heterozygotic *Aa* will be $2pq$, and of homozygotic *aa* will be q^2. The law is named for its formulators, British mathematician Godfrey Harold Hardy (1877–1947) and German physician Wilhelm Weinberg (1862–1937).

harmonic (här-mŏn′ĭk) *Noun.* Periodic motion whose frequency is a whole-number multiple of some fundamental frequency. The motion of objects or substances that vibrate or oscillate in a regular fashion, such as the strings of musical instruments, can be analyzed as a combination of a fundamental frequency and higher harmonics. ► Harmonics above the first harmonic (the fundamental frequency) in sound waves are called **overtones.** The first

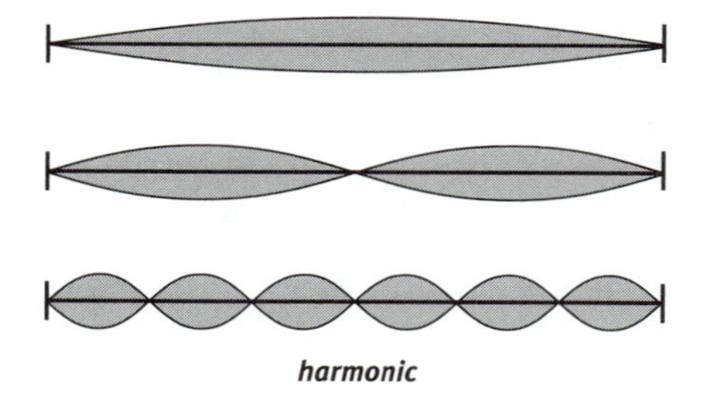

harmonic

Possible harmonics in the periodic motion of a vibrating guitar string. Top: *first harmonic (or fundamental),* center: *second harmonic,* bottom: *sixth harmonic.*

overtone is the second harmonic, the second overtone is the third harmonic, and so on. —*Adjective.* Related to or having the properties of such periodic motion.

harmonic analysis The study of functions given by a Fourier series or analogous representations, such as periodic functions and functions on topological groups.

harmonic mean The reciprocal of the arithmetic mean of the reciprocals of a specified set of numbers.

harmonic motion A periodic vibration, as of a violin string, in which the motions are symmetrical about a region of equilibrium. Such a vibration may have only one frequency and amplitude or may be a combination of two or more components called harmonics. Also called *periodic motion.*

harmonic oscillator A physical system in which some value oscillates above and below a mean value at one or more characteristic frequencies. Such systems often arise when a contrary force results from displacement from a force-neutral position, and gets stronger in proportion to the amount of displacement. For example, pulling or pushing the end of a spring from its rest position results in a force pushing back toward the rest position. Letting the spring go from a position of tension results in harmonic motion of the spring; the spring is now a harmonic oscillator. Other examples include a swinging pendulum, a vibrating violin string, or an electronic circuit that produces radio waves.

harmonic progression A sequence of quantities whose reciprocals form an arithmetic progression, such as $1, \frac{1}{3}, \frac{1}{5}, \frac{1}{7}, \ldots$ and so on.

harmonic series A series whose terms are in harmonic progression, especially the series $1 + \frac{1}{2} + \frac{1}{3} + \frac{1}{4} + \ldots$ and so on.

Harvey (här′vē), **William** 1578–1657. English physician and physiologist who in 1628 demonstrated the function of the heart and the circulation of blood throughout the human body.

BIOGRAPHY **William Harvey**

In the second century CE, the Greek physician Galen theorized that blood is created in the liver, passes once through the heart, and is then absorbed by bodily tissues. Galen's ideas were widely accepted in European medicine until 1628, when William Harvey published a book describing the circulation of blood throughout the body. Through his observations of human and animal dissections, Harvey saw that blood flows from one side of the heart to the other and that it flows through the lungs and returns to the heart to be pumped elsewhere. There was one missing part of the cycle: How did the blood pumped to distant body tissues get into the veins to be carried back to the heart? As an answer, Harvey offered his own, unproven theory, one that has since been shown to be true: blood passes from small, outlying arteries through tiny vessels called capillaries into the outlying veins. Harvey's views were so controversial at the time that many of his patients left his care, but his work became the basis for all modern research on the heart and blood vessels.

hassium (hä′sē-əm) *Symbol* **Hs** A synthetic, radioactive element that is produced by bombarding lead with iron ions. Its most long-lived isotopes have mass numbers of 264 and 265 with half-lives of 0.08 millisecond and 2 milliseconds, respectively. Atomic number 108. See **Periodic Table.**

haustorium (hô-stôr′ē-əm) Plural **haustoria.** A specialized absorbing structure of a parasitic plant, such as the rootlike outgrowth of the dodder, that obtains food from a host plant. In parasitic fungi, haustoria are specialized hyphae that penetrate the cells of other organisms and absorb nutrients directly from them.

Hawking (hô′kĭng), **Stephen William** Born 1942. British physicist noted for his study of black holes and the origin of the universe, especially the big bang theory. His work has provided much of the mathematical basis for scientific explanations of the physical properties of black holes.

BIOGRAPHY **Stephen Hawking**

The world-renowned theoretical physicist and cosmologist Stephen Hawking needs little introduction to those familiar with the bespectacled man who uses a wheelchair and lectures around the world with the aid of a computerized speech synthesizer. The condition that has left him all but totally paralyzed, amyotrophic lateral sclerosis, is usually fatal within a few years, but Hawking has beaten the odds by living with the disease for all his adult life, since its onset when he was a 20-year-old college student. Hawking's story is a testament to a determined person's ability to overcome unexpected adversity—his career in fact did not take off until after the disease had been diagnosed. Hawking partly credits the disease for giving him a sense of purpose and the ability to enjoy life. His academic position at Oxford is a chaired professorship in mathematics that was also held by Isaac Newton, in 1669. He originally set out to study mathematics, but it is for his discoveries in physics that he is best known. With his collaborator Roger Penrose, he theorized that Einstein's Theory of General Relativity predicts that space and time have a definite origin and conclusion, providing mathematical support for the Big Bang theory. This led to further attempts to unify General Relativity with quantum theory, one consequence of which is the intriguing view that black holes are not entirely "black," as originally thought, but emit radiation and should eventually evaporate and disappear.

Hawking radiation A form of radiation believed to emanate from black holes, emerging from the region just beyond the black hole's event horizon (from which no radiation can emerge). Pairs of virtual particles and antiparticles, created naturally in the **vacuum fluctuation** near the black hole, are split apart, one particle falling into the black hole and the other radiating away. The energy lost to such radiated particles is believed to come from the mass of the black hole.

hay fever (hā) An seasonal allergic condition characterized by a sensitivity to airborne pollen, resulting in nasal discharge, sneezing, and itchy, watery eyes. It occurs especially during late spring, late summer, and early fall and can be caused by the pollens of various plants, especially ragweed and certain trees and grasses.

hazardous waste (hăz′ər-dəs) A used or discarded material that can damage the environment and be harmful to health. Hazardous wastes include heavy metals and toxic chemicals used in industrial products and processes as well as infectious medical wastes and radioactive materials such as spent nuclear fuel rods.

h-bar (āch′bär′) See **Dirac's constant.**

HDL Abbreviation of **high-density lipoprotein.**

He The symbol for **helium.**

headache (hĕd′āk′) Pain in the head, caused by stimulation of or pressure to any of various structures of the head, such as tissue covering the cranium, cranial nerves, or blood vessels. Headache can be a primary disorder, as in **migraine** or **cluster headaches**, or a common symptom associated with head injury or many illnesses such as acute infection, brain tumor or abscess, eye disorders such as glaucoma, dental disease, and hypertension. See also **cluster headache, migraine.**

headwall (hĕd′wôl′) A steep slope or precipice rising at the head of a valley or glacial cirque.

headwind or **head wind** (hĕd′wĭnd′) A wind blowing directly against the course of a moving object, especially an aircraft or ship.

heart (härt) **1.** The hollow, muscular organ that pumps blood through the body of a vertebrate animal by contracting and relaxing. In humans and other mammals, it has four chambers, consisting of two atria and two ventricles. The right side of the heart collects blood with low oxygen levels from the veins and pumps it to the lungs. The left side receives blood with high oxygen levels from the lungs and pumps it into the aorta, which carries it to the arteries of the body. The heart in other vertebrates functions similarly but often has fewer chambers. **2.** A similar but simpler organ in invertebrate animals.

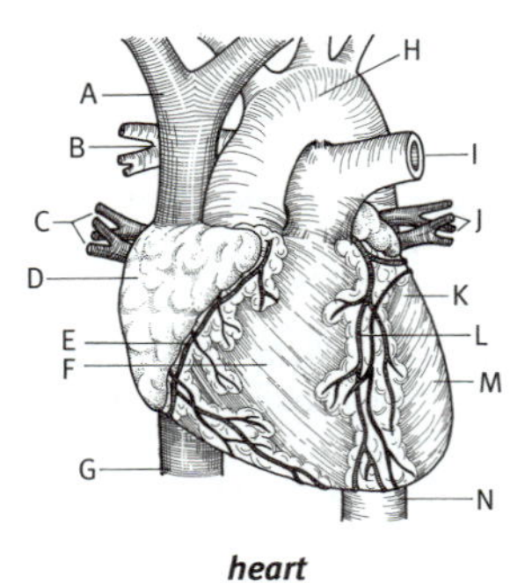

heart

anatomy of a human heart: A. *superior vena cava,* B. *right pulmonary artery,* C. *right pulmonary veins,* D. *right atrium,* E. *right coronary artery,* F. *right ventricle,* G. *inferior vena cava,* H. *aorta,* I. *left pulmonary artery,* J. *left pulmonary veins,* K. *left atrium,* L. *left coronary artery,* M. *left ventricle,* N. *aorta*

heart attack Necrosis of a region of the heart muscle caused by an interruption in the supply of blood to the heart, usually as a result of occlusion of a coronary artery resulting from **coronary artery disease**. Symptoms typically include sudden, crushing chest pain, nausea, and sweating. Characteristic changes in the **electrocardiogram** are used to diagnose heart attacks. Also called *myocardial infarction.*

heartburn (härt′bûrn′) A burning sensation, usually centered in the middle of the chest near the sternum, caused by the reflux of acidic stomach fluids that enter the lower end of the esophagus. Also called *acid reflux.*

heart failure An acute or chronic inability of the heart to maintain adequate blood circulation to the peripheral tissues and the lungs, usually characterized by fatigue, edema, and shortness of breath. Heart failure has many causes, including coronary artery disease, hypertension, and cardiomyopathy. Also called *congestive heart failure.*

heart-lung machine An apparatus through which blood is temporarily diverted, especially during heart surgery, in order to oxygenate and pump it, thus maintaining circulation.

heart rate The number of heartbeats per unit of time, usually expressed as beats per minute.

heartwood (härt′wo͝od′) The older, nonliving central wood of a tree or woody plant, usually darker and harder than the younger sapwood. Unlike the sapwood, it no longer conducts water, and its main function is the support of the tree.

heat (hēt) **1.** Internal energy that is transferred to a physical system from outside the system because of a difference in temperature and does not result in work done by the system on its surroundings. Absorption of energy by a system as heat takes the form of increased kinetic energy of its molecules, thus resulting in an increase in temperature of the system. Heat is transferred from one system to another in the direction of higher to lower temperature. See also **thermodynamics.** See Note at **temperature. 2.** See **estrus.**

heat capacity The ratio of the heat energy absorbed by a substance to its increase in temperature. Heat capacity is also called *thermal capacity.* ▸ The **specific heat** or **specific heat capacity** of a substance is the heat capacity per unit mass, usually measured in joules per kilogram per degree Kelvin. See also **latent heat, thermodynamics.**

heat death The eventual dispersion of all of the energy within a physical system to a completely uniform distribution of heat energy, that is, to maximum entropy. Heat death for all macroscopic physical systems, including the universe, is predicted by the Second Law of Thermodynamics. See more at **entropy, thermodynamics.**

heat exchanger A device used to transfer heat from one fluid to another without direct contact of the fluids. Heat exchangers usually maximize the transfer of heat by maximizing the contact surface area between fluids, as when the warmer fluid is passed through a series of coils or thin plates. A car radiator is a heat exchanger, transferring the heat in a liquid that has circulated through the engine to the air.

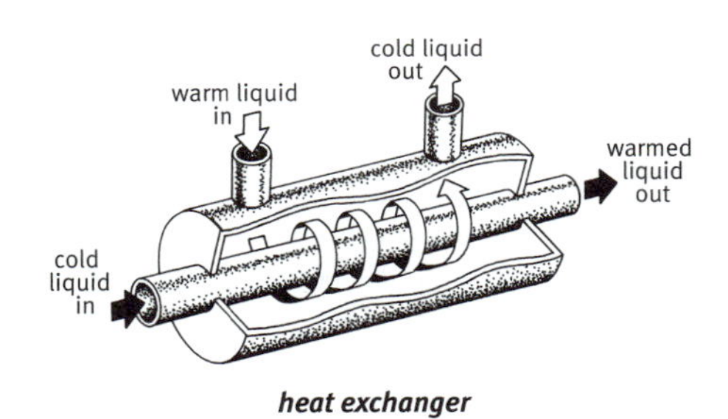

heat exchanger

parallel-flow heat exchanger

heat index A measurement of the air temperature in relation to the relative humidity, used as an indicator of discomfort. The heat index is higher when high air temperatures occur with high humidity, and lower when they occur with low humidity. The heat index is based on studies of skin cooling caused by the evaporation of sweat. Also called *apparent temperature.*

heat lightning Intermittent flashes of reddish or orange light near the horizon, usually seen on a hot summer evening and unaccompanied by thunder. Heat lightning is thought to be the reflection of distant lightning on clouds. Its color is believed to be due to the refraction of light in the atmosphere. See also **sheet lightning.** See more at **lightning.**

heat of combustion The amount of heat released when one mole of a substance is completely oxidized.

heat of fusion The amount of heat required to convert a solid at its melting point into a liquid without an increase in temperature. Liquids release the same amount of heat when they solidify. See also **heat of vaporization.**

heat of vaporization The amount of heat required to convert a liquid into a gas at constant temperature and pressure. A gas releases the same amount of heat when it becomes liquid. Compare **heat of fusion.**

heat shield A barrier that prevents a substance from absorbing heat energy from an outside source by absorbing and dissipating, or simply reflecting, that heat. Heat shields are commonly used to protect parts of a device from heat generated by its energy source, as in isolating the cabin of a car from its motor. Many spacecraft dissipate heat generated by friction with the atmosphere upon reentry using heat shields that melt and vaporize, dissipating the energy back into the atmosphere.

heat sink 1. A protective device that absorbs and dissipates the excess heat generated by a system. **2.** An environment capable of absorbing heat from substances within it (and with which it is in thermal contact) without an appreciable change in its own temperature and without a change in its own phase.

Heaviside layer (hĕv′ē-sīd′) See **E region.**

heavy chain (hĕv′ē) The larger of the two types of polypeptide chains in an immunoglobulin molecule. A heavy chain consists of an antigen-binding portion having a variable amino acid sequence, and a constant region that is different for each class of imunoglobulin. Each heavy chain is one of a Y-shaped identical pair and is attached along its variable region to a smaller light chain. Compare **light chain.**

heavy hydrogen See **deuterium.**

heavy ion 1. The nucleus of a heavy element, such as gold. Collisions of heavy ions are induced in particle accelerators to explore the internal structure quark structure of the nuclei; recent experiments suggest that the extremely high temperatures generated by such collisions can create a short-lived plasma of quarks and gluons. **2.** A charged microscopic particle that forms when ions attach to a dust mote or similar object.

heavy particle A subatomic particle with relatively high mass, especially a baryon.

heavy water Water in which deuterium, a heavy isotope of hydrogen, takes the place of hydrogen. Heavy water has physical and chemical properties that are like those of ordinary water, but heavy water is 10 percent heavier and has higher freezing and boiling points. Also called *deuterium oxide. Chemical formula:* **D_2O.** ▸ **Semiheavy water** is similar to heavy water, but only one of the two hydrogen atoms in each molecule is replaced with deuterium. *Chemical formula:* **DHO.**

A CLOSER LOOK **heavy water**

The nucleus of most hydrogen atoms consists of a single proton, but in one isotope of hydrogen, called *deuterium* or *heavy hydrogen,* the nucleus also contains a neutron and thus weighs nearly twice as much as standard hydrogen. The substance called *heavy water* is chemically identical to ordinary water (H_2O), except that the hydrogen atoms in the molecule are the deuterium isotopes (D_2O). Heavy water makes up a small percentage (0.02%) of water naturally occurring on Earth. It is an excellent *moderator* for nuclear reactions, slowing down the fast neutrons produced in a nuclear fission reaction, increasing the likelihood that the neutrons will successfully collide with heavy nuclei to cause further fission. Although heavy water is chemically nearly identical with ordinary water, it is about ten percent heavier and interferes with cell mitosis if consumed in place of normal water. Ice cubes made of heavy water are denser than ordinary liquid water and will sink to the bottom of a glass of cola.

Heisenberg (hī′zən-bûrg′), **Werner Karl** 1901–1976. German physicist who founded the field of quantum mechanics in 1925 and elaborated the uncertainty principle in 1927. He was awarded the Nobel Prize for physics in 1932. See more at **uncertainty principle.**

BIOGRAPHY **Werner Karl Heisenberg**

Philosophical problems concerning what it means to know something about the world have always been of interest to many scientists, but philosophy underwent an unexpected twist with the advent of what we now call the uncertainty principle or the Heisenberg uncertainty principle, after its discoverer. A brilliant physicist, Werner Heisenberg had made discoveries by the age of 24 that would garner him a Nobel Prize a few years later (in 1932), namely, a way of formulating quantum mechanics using the then-new branch of mathematics called matrix algebra. In 1927, he formulated a quantum mechanical indeterminacy or uncertainty principle, which concerns how accurately certain properties of subatomic particles can be measured. Earlier physical theories had held that the accuracy of such measurements was limited only by the accuracy of available instruments. Heisenberg overturned this notion by demonstrating that no matter how accurate the instruments, the quantum mechanical nature of the universe itself prevents us from having complete knowledge of every measurable property of a physical system simultaneously. For example, the more precise our knowledge of a subatomic particle's position, the less precise our knowledge of its momentum; more profoundly, the particle does not merely have a momentum that we simply cannot accurately measure, but literally does not have a determinate momentum. This principle had profound implications not just for physics, but also for twentieth-century philosophy, as it threw into question certain basic principles such as causality and determinacy, and suggested that the very act of observing the universe profoundly shapes it. Nonetheless, Heisenberg's quantum mechanical equations have led to physical theories with vast practical applications, bringing us everything from the transistor to new drugs.

---

helicity (hə-lĭs′ĭ-tē, hē′) The projection of a particle's spin vector in the direction of its momentum vector, being positive if it points in the same direction, and negative if it points in the opposite direction. Helicity is in effect the handedness, or **chirality**, of the spin of a particle.

helicoid (hĕl′ĭ-koid′, hē′lĭ-) *Adjective.* **1.** Arranged in or having the approximate shape of a flattened coil or spiral curve. *—Noun.* **2.** A surface in the form of a coil or screw. A helicoid is generated mathematically by rotating a plane or twisted curve about an axis at a fixed rate and simultaneously translating it in the direction of the line of axis, also at a fixed rate.

heliocentric (hē′lē-ō-sĕn′trĭk) **1.** Relating to or measured from the center of the Sun. **2.** Relating to a model of the solar system or universe having the Sun as the center. Compare **geocentric.** See Note at **Copernicus.**

heliopause (hē′lē-ə-pôz′) The region surrounding the solar system at which pressure from the outgoing solar wind equals the pressure from the interstellar medium (made up mostly of hydrogen and helium), and the solar wind can penetrate no further. It is considered to be the outer boundary of our solar system. See more at **solar wind.**

heliosphere (hē′lē-ə-sfîr′) The large, roughly elliptical region of space around the Sun through which the solar wind extends and through which the Sun exerts a magnetic influence. The heliosphere extends well beyond the orbits of the planets, and its shape and extent fluctuate with changes to the solar wind and other influences. ▸ The boundary between the heliosphere and interstellar space is known as the **heliopause**.

heliotropism (hē′lē-ŏt′rə-pĭz′əm) The growth or movement of a fixed organism, especially a plant, toward or away from sunlight. Heliotropism can be easily seen in sunflowers, which slowly turn their large flowers so that they continually face the sun. *—Adjective* **heliotropic** (hēl′lē-ə-trō′pĭk, -trŏp′ĭk).

helium (hē′lē-əm) *Symbol* **He** A very lightweight, colorless, odorless element in the noble gas group. Helium occurs in natural gas, in radioactive ores, and in small amounts in the atmosphere. It has the lowest boiling point of any substance and is the second most abundant element in the universe. Helium is used to provide lift for balloons and blimps and to create artificial air that will not react chemically. Atomic number 2; atomic weight 4.0026; boiling point –268.9°C; density at 0°C 0.1785 gram per liter. See **Periodic Table.**

WORD HISTORY **helium**

The second most abundant element in the universe after hydrogen, Helium (symbol He) is a colorless, odorless, nonmetallic gas that is produced abundantly by the nuclear fusion in all stars and is found in smaller amounts on

Earth. It was discovered by the British scientist—and founding editor of the journal *Nature*—Joseph Norman Lockyer in 1868, while he was studying a solar eclipse with a spectroscope, an instrument that breaks light up into a spectrum. If an element is heated up enough to glow, the emitted light produces a unique spectrum when refracted through a prism. Lockyer noticed that the spectrum of the Sun's corona, which is visible only during a solar eclipse, contained lines produced by an unknown element. He named the element helium from *helios,* the Greek word for "sun." *Helios* gives us many other words pertaining to the Sun, such as *heliocentric* and *perihelion.*

---

helix (hē′lĭks) **1.** A three-dimensional spiral curve. In mathematical terms, a helix can be described as a curve turning about an axis on the surface of a cylinder or cone while rising at a constant upward angle from a base. **2.** Something, such as a strand of DNA, having a spiral shape.

Helmholtz (hĕlm′hōlts′), **Hermann Ludwig Ferdinand von** 1821–1894. German physicist and physiologist who was a founder of the law of conservation of energy. Helmholtz did pioneering research on vision and invented an instrument for examining the interior of the eye in 1851.

helminth (hĕl′mĭnth′) A worm, especially a parasitic roundworm or tapeworm.

helper T cell also **helper cell** (hĕl′pər) Any of various T cells that, when stimulated by a specific antigen, release lymphokines that promote the activation and function of B cells and killer T cells.

hematite (hē′mə-tīt′) A reddish-brown to silver-gray metallic mineral. Hematite occurs as rhombohedral crystals, as reniform (kidney-shaped) crystals, or as fibrous aggregates in igneous, metamorphic, and sedimentary rocks. It is the most abundant ore of iron, and it is usually slightly magnetic. *Chemical formula:* Fe_2O_3.

hemato– Variant of **hemo–.**

hematocrit (hĭ-măt′ə-krĭt′) **1.** The percentage by volume of red blood cells in a given sample of blood after it has been spun in a centrifuge. **2.** A centrifuge used to determine the relative volumes of blood cells and plasma in a given sample of blood.

hematology (hē′mə-tŏl′ə-jē) The branch of medicine that deals with the blood and blood-producing organs.

hematoma (hē′mə-tō′mə) Plural **hematomas** or **hematomata** (hē′mə-tō′mə-tə). The abnormal buildup of blood in an organ or other tissue of the body, caused by a break in a blood vessel.

heme (hēm) The deep red, nonprotein, iron-containing component of hemoglobin that carries oxygen. Heme is a porphyrin with an iron atom at its center. One of the free valence electrons of the iron atom of heme is bound to the hemoglobin molecule, while the other is available for binding to an oxygen atom. A hemoglobin molecule contains four hemes. *Chemical formula:* $C_{34}H_{32}FeN_4O_4$.

hemi– A prefix meaning "half," as in *hemisphere,* half a sphere.

hemichordate (hĕm′ĭ-kôr′dāt′, -dĭt) Any of various mostly small, wormlike marine invertebrates once thought to be chordates but now considered more closely related to echinoderms. They may constitute their own phylum, the Hemichordata. The bodies of hemichordates are divided into a feeding organ called a proboscis, a ringlike section called a collar, and a trunk. Hemichordates have a gut, circulatory system, and nervous system and are filter feeders. Acorn worms and graptolites are hemichordates.

hemiparasite (hĕm′ĭ-păr′ə-sīt′) A plant, such as mistletoe, that obtains some nourishment from its host but also carries on photosynthesis.

hemiplegia (hĕm′ĭ-plē′jə) Paralysis of one side of the body, usually resulting from a stroke or other brain injury.

hemipteran (hĭ-mĭp′tər-ən) Any of various insects of the order Hemiptera, having biting or sucking mouthparts and two pairs of wings. Hemipterans include the leafhoppers, treehoppers, cicadas, aphids, scales, and true bugs.

hemisphere (hĕm′ĭ-sfîr′) **1.** One half of a sphere, formed by a plane that passes through the center of the sphere. **2.** Either the northern or southern half of the Earth as divided by the equator, or the eastern or western half as divided by a meridian, especially the prime meridian. **3.** One half of the celestial sphere as divided by any of various great circles, especially the celestial equator and the ecliptic. See more at **celestial sphere. 4.** See **cerebral hemisphere.**

hemo– or **hemato–** A prefix meaning "blood," as in *hemophilia,* a disorder in which blood fails to clot, or *hematology,* the scientific study of blood.

hemoglobin (hē′mə-glō′bĭn) An iron-containing protein present in the blood of many animals that, in vertebrates, carries oxygen from the lungs to the tissues of the body and carries carbon dioxide from the tissues to the lungs. Hemoglobin is contained in the red blood cells of vertebrates and gives these cells their characteristic color. Hemoglobin is also found in many invertebrates, where it circulates freely in the blood. It consists of four peptide units, each attached to a nonprotein compound called **heme** that binds to oxygen. See Note at **red blood cell.**

A CLOSER LOOK **hemoglobin**

Ninety percent of the protein in red blood cells is made up of *hemoglobin,* the main oxygen transport molecule in mammals. A protein with four iron-containing subunits called *hemes,* hemoglobin is a complex molecule with a complex function. It must bind to oxygen in the lungs, then release that oxygen in the tissues, then bind to carbon dioxide in the tissues and release it in the lungs. Hemoglobin accomplishes oxygen transport by changing its structure, and even its substructures, around the oxygen-binding heme groups, making them more or less accessible to the environment. When oxygen binds to at least one of the heme groups (as happens in the oxygen-rich lungs), all of the heme groups become exposed to the environment and bind oxygen easily. The bond between oxygen and heme is a loose one, however, so that the oxygen can break free in the tissues, where the concentration of oxygen is relatively low, and thereby become available for use in the cells. When the last of the four heme subunits loses its oxygen, the structure of hemoglobin changes again, so that the size of the opening from the environment to the heme groups decreases, making it difficult for an oxygen molecule to rebind to the hemoglobin. In this way, hemoglobin stops itself from competing with the tissues for needed oxygen. When the red blood cell carrying hemoglobin returns to the lungs, where oxygen concentration is high, the cycle of oxygen binding, transport, and release starts again. Normally, iron binds with oxygen to form rust (iron oxide), but the structure of hemoglobin prevents this from happening, since it would inactivate the heme subunits. Carbon dioxide does not bind the heme in hemoglobin, but rather the amino groups at the ends of the hemoglobin's protein subunits. Hemoglobin transport is only one of a number of bodily mechanisms by which carbon dioxide travels from the tissues to the lungs for release to the air.

hemolymph (hē′mə-lĭmf′) The circulatory fluid of invertebrates, including all arthropods and most mollusks, that have an open circulatory system. Hemolymph is analogous to blood and lymph in vertebrate animals and is not confined in a system of vessels. Hemolymph consists of water, amino acids, inorganic salts, lipids, and sugars. See more at **circulatory system.**

hemolysis (hĭ-mŏl′ĭ-sĭs, hē′mə-lī′sĭs) The destruction of red blood cells, caused by disruption of the cell membrane and resulting in the release of hemoglobin. Hemolysis is seen in some types of anemia, which can be either inherited or acquired, as by exposure to toxins or by the presence of antibodies that attack red blood cells. —*Adjective* **hemolytic.**

hemolytic anemia (hē′mə-lĭt′ĭk) Anemia resulting from the lysis of red blood cells, as in response to certain toxic or infectious agents and in certain inherited blood disorders.

hemophilia (hē′mə-fĭl′ē-ə) Any of several hereditary coagulation disorders, seen almost exclusively in males, in which the blood fails to clot normally because of a deficiency or an abnormality of one of the **clotting factors**.

hemorrhage (hĕm′ər-ĭj) Excessive or uncontrollable bleeding, often caused by trauma, surgical or obstetrical complications, or the advanced stages of certain illnesses, such as cirrhosis and peptic ulcer disease.

hemorrhagic fever (hĕm′ə-răj′ĭk) Any of a group of viral infections, including **dengue**, **Ebola virus** infection and **yellow fever**, that occur primarily in tropical climates, are usually transmitted to humans by arthropods or rodents, and are characterized by high fever, internal bleeding, hypotension, and eventual shock.

henry (hĕn′rē) A SI derived unit of electrical inductance, especially of transformers and inductance coils. A current changing at the rate of one ampere per second in a circuit with an inductance of one henry induces an electromotive force of one volt.

Henry, Joseph 1797–1878. American physicist who studied electromagnetic phenomena. He discovered electrical induction independently of Michael Faraday, and constructed a small electromagnetic motor in 1829. He also developed a system of weather forecasting based on meteorological observations. The henry unit of inductance is named for him.

heparin (hĕp′ər-ĭn) An acidic glycosaminoglycan found especially in lung and liver tissue that prevents the clotting of blood and is used intravenously in the treatment of thrombosis and embolism.

hepatic (hĭ-păt′ĭk) Relating to or involving the liver.

hepatitis (hĕp′ə-tī′tĭs) Inflammation of the liver, usually caused by any of various infectious agents or toxins, including alcohol and numerous chemical compounds. Symptoms usually include jaundice, fatigue, fever, liver enlargement, and abdominal pain. There are five types of viral hepatitis: A, B, C, D, and E. *Hepatitis A,* an acute infection caused by a virus of the genus *Hepatovirus* is transmitted by contaminated food and water. *Hepatitis B,* caused by a virus of the genus *Orthohepadnavirus* and Hepatitis C, caused by a virus of the genus *Hepacivirus,* are more serious infections that are transmitted through infected bodily fluids such as blood and semen.

heptachlor (hĕp′tə-klôr′) A white or tan powder used as a pesticide. Because it is highly toxic to humans and is a suspected carcinogen, its use has been largely discontinued. *Chemical formula:* $C_{10}H_5Cl_7$.

heptagon (hĕp′tə-gŏn′) A polygon having seven sides.

heptane (hĕp′tān′) A volatile, colorless, highly flammable liquid hydrocarbon obtained in the fractional distillation of petroleum. It is used as a standard in determining octane ratings (combustion characteristics), as an anesthetic, and as a solvent. Heptane is the seventh member of the alkane series. *Chemical formula:* C_7H_{16}.

herb (ûrb) A flowering plant whose stem does not produce woody tissue and generally dies back at the end of each growing season. Both grasses and forbs are herbs. —*Adjective* **herbaceous** (hûr-bā′shəs, ûr-).

herbicide (hûr′bĭ-sīd′, ûr′-) A pesticide used to kill weeds. Paraquat is a herbicide. Compare **fungicide, insecticide, rodenticide.**

herbivore (hûr′bə-vôr′, ûr′-) An animal that feeds mainly or only on plants. In a food chain, herbivores are primary consumers. Compare **carnivore, detritivore.**

Hercules (hûr′kyə-lēz′) A constellation in the Northern Hemisphere near Lyra and Corona Borealis.

hereditary (hə-rĕd′ĭ-tĕr′ē) Passed or capable of being passed from parent to offspring by means of genes.

heredity (hə-rĕd′ĭ-tē) The passage of biological traits or characteristics from parents to offspring through the inheritance of genes.

heritable (hĕr′ĭ-tə-bəl) Capable of being passed from one generation to the next through the genes.

hermaphrodite (hər-măf′rə-dīt′) An organism, such as an earthworm or flowering plant, having both male and female reproductive organs in a single individual.

hernia (hûr′nē-ə) A condition in which an organ or body part, such as the intestine, protrudes through an opening in the body structure that normally contains it.

Hero (hē′rō) First century CE. Greek mathematician who wrote on mechanics and invented many water-driven and steam-driven machines. He also developed a formula for determining the area of a triangle.

heroin (hĕr′ō-ĭn) A white, odorless, bitter crystalline compound, $C_{17}H_{17}NO(C_2H_3O_2)_2$, that is derived from morphine and is a highly addictive narcotic.

herpes (hûr′pēz) Any of several infections caused by the herpes simplex virus of the genus *Simplexvirus* or by the varicella-zoster virus, a herpes virus of the genus *Varicellavirus.* Herpes infections are characterized by painful blisters on the skin or a mucous membrane and are highly contagious. Genital herpes is a sexually transmitted disease. Varicella-zoster infection is also called *shingles.*

herpetology (hûr′pĭ-tŏl′ə-jē) The scientific study of reptiles and amphibians.

Herschel (hûr′shəl) Family of British astronomers led by Sir **William Herschel** (1738–1822), who discovered Uranus (1781) and cataloged more than 800 binary stars and 2,500 nebulae. His sister **Caroline Herschel** (1750–1848) discovered eight comets and several nebulae and star clusters, and published at least two astronomical catalogs which are still currently used. His son Sir **John Frederick William Herschel** (1792–1871) discovered 525 nebulae and pioneered celestial photography. See Note at **infrared.**

William and Caroline Herschel

BIOGRAPHY **William and Caroline Herschel**

Brother and sister William Herschel and Caroline Herschel began their professional careers as musicians. Born in Germany, they moved to England, where Caroline became a soprano soloist in performances conducted by her brother. William's background in music spurred him to study mathematics and astronomy, which he then taught his sister, and they each went on to produce a string of important scientific discoveries. William was the first astronomer to study binary stars and, while searching for comets in 1781, he discovered Uranus, the first new planet to be discovered since ancient times. He also discovered two satellites of Uranus (Titania and Oberon, 1787), and two of Saturn (Mimas and Enceladus, 1789–90). Caroline observed her first comet in 1786 and eventually discovered seven others, as well as nebulae and star clusters. King George III appointed William his Astronomer Royal in 1787, and Caroline was made assistant astronomer. After William's death, Caroline returned to Germany and published a catalog of 2,500 nebulae, for which the (British) Royal Astronomical Society awarded her its gold medal in 1828.

hertz (hûrts) The SI derived unit used to measure the frequency of vibrations and waves, such as sound waves and electromagnetic waves. One hertz is equal to one cycle per second. The hertz is named after German physicist Heinrich Hertz (1857–1894).

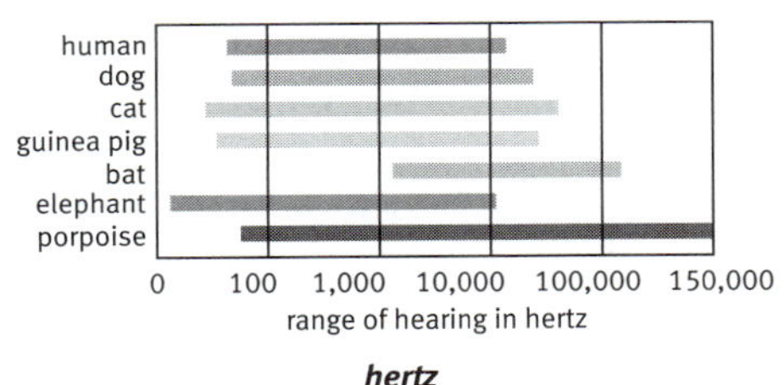

hertz

the ranges of sound frequencies, measured in hertz, that humans and a variety of animals are able to hear

Hertzsprung (hĕrts′sprŏŏng), **Ejnar** 1877–1967. Danish astronomer who specialized in photographing the stars and introduced the concept of **absolute magnitude**. Hertzsprung also demonstrated the relationship between the surface temperature of stars and their absolute magnitude, but his work was ignored until Henry Russell independently developed a similar correlation, which is now named after both of them.

Hertzsprung-Russell diagram A graph in which the absolute magnitude (intrinsic luminosity) of stars is plotted vertically against their surface temperatures (or corresponding spectral types). The diagram shows a strong correlation between luminosity and surface temperature among the average-size stars known as dwarfs, with hot, blue stars having the highest luminosities and relatively cool, red stars having the lowest. The roughly diagonal line (running from the upper left of the diagram to the lower right) that shows this correlation is called the **main sequence.** Giant and supergiant stars have relatively high luminosities for their surface temperatures and are positioned on the diagram above the main sequence. The faint white dwarfs have relatively low luminosities for their surface temperatures and cluster below the main sequence. See more at **main sequence.**

Hess (hĕs), **Harry Hammond** 1906–1969. American geologist who studied the sea floor and developed the theory of sea-floor spreading in 1960. Hess theorized that sea floors were constantly renewed by the flow of magma from the Earth's mantle through the oceanic rifts. This hypothesis became an important component of the theory of plate tectonics.

hetero– A prefix that means "different" or "other," as in *heterophyllous,* having different kinds of leaves.

heterochrony (hĕt′ə-rō-krŏn′ē) A change or set of changes in the timing or duration of an organism's ontogenetic development compared with an ancestral species, resulting in morphological differences between ancestor and descendant.

heterocyclic (hĕt′ə-rō-sī′klĭk, -sĭk′lĭk) Of or relating to a compound containing a closed ring structure made of more than one kind of atom. Pyridine (C_5H_5N), for example, is a heterocyclic compound, as its ring contains five carbon atoms and one nitrogen atom. Compare **homocyclic.**

heteroecious (hĕt′ə-rē′shəs) Relating to a parasite that spends different stages of its life cycle on different, usually unrelated hosts. The term is used especially of certain kinds of rust fungi, but may also be applied to other parasites like tapeworms. Compare **autoecious.**

heterogamy (hĕt′ə-rŏg′ə-mē) **1.** Alternation of sexual and parthenogenetic generations, as in the life cycles of some aphids. **2.** A reproductive system that is characterized by the fusion of gametes (called **heterogametes**) that are dissimilar in size and structure, such as the egg and sperm of animals. *—Adjective* **heterogamous.**

heterologous (hĕt′ə-rŏl′ə-gəs) Derived or transplanted from a different species or source.

heterophyte (hĕt′ər-ə-fīt′) A plant that obtains its nourishment from other living or dead organisms. For example, Indian pipe is a pale white plant without chlorophyll that obtains its nutrition from fungal symbionts.

heterosporous (hĕt′ər-ə-spôr′əs, hĕt′ə-rŏs′pər-əs) Producing two types of spores differing in size and sex, the male microspore and the female megaspore, which develop into separate male and female gametophytes. All seed-bearing plants, as well as some ferns and other seedless plants, are heterosporous. Compare **homosporous.**

heterotroph (hĕt′ər-ə-trŏf′) An organism that cannot manufacture its own food and instead obtains its food and energy by taking in organic substances, usually plant or animal matter. All animals, protozoans, fungi, and most bacteria are heterotrophs. Compare **autotroph.** *—Adjective* **heterotrophic** (hĕt′ər-ə-trŏf′ĭk).

heterozygous (hĕt′ər-ə-zī′gəs) Relating to a cell that has two different alleles at corresponding positions on homologous chromosomes. Compare **homozygous.**

Hewish (hyo͞o′ĭsh), **Antony** Born 1924. British astronomer. In 1967, working with the astronomer Susan Bell Burnell, he discovered the first pulsar.

hexachlorobenzene (hĕk′sə-klôr′ō-bĕn′zēn′) See **benzene hexachloride.**

hexadecimal (hĕk′sə-dĕs′ə-məl) Of, relating to, or based on the number 16. ► The **hexadecimal number system** is a way of representing numbers where each successive digit or number represents a multiple of a power of 16. It uses the digits 0–9 plus the letters A, B, C, D, E, and F to represent the decimal values 10–15. For example, 4B7E represents $(4 \times 16^3) + (11 \times 16^2) + (7 \times 16^1) + (15 \times 16^0)$, or 19,327 in the decimal system.

hexagon (hĕk′sə-gŏn′) A polygon having six sides.

hexagonal (hĕk-săg′ə-nəl) **1.** Having six sides. **2.** Relating to a crystal having three axes of equal length intersecting at angles of 60° in one plane, and a fourth axis of a different length that is perpendicular to this plane. The mineral calcite has hexagonal crystals. See illustration at **crystal.**

hexahedron (hĕk′sə-hē′drən) Plural **hexahedrons** or **hexahedra.** A polyhedron, such as a cube, that has six faces.

hexane (hĕk′sān′) A colorless flammable liquid derived from the fractional distillation of petroleum. It is used as a solvent and in low-temperature thermometers. Hexane is the sixth member of the alkane series. *Chemical formula:* C_6H_{14}.

hexose (hĕk′sōs′) Any of various simple sugars (monosaccharides), such as glucose and fructose, that have six carbon atoms per molecule.

hexyl (hĕk′səl) The radical C_6H_{13}, derived from hexane.

Hf The symbol for **hafnium.**

Hg The symbol for **mercury.**

hibernaculum (hī′-bər-năk′yə-ləm) **1.** A protective case, covering, or structure, such as a plant bud, in which an organism remains dormant for the winter. **2.** The shelter of a hibernating animal.

hibernation (hī′bər-nā′shən) An inactive state resembling deep sleep in which certain animals living in cold climates pass the winter. In hibernation, the body temperature is lowered and breathing and heart rates slow down. Hibernation protects the animal from cold and reduces the need for food during the season when food is scarce. Compare **estivation.**

Hib vaccine (hĭb) A vaccine administered to immunize young children against the bacterium *Haemophilus influenzae* type b, which causes meningitis, pneumonia, and other infections.

Higgs boson (hĭgz) A hypothetical, massive subatomic particle with zero electric charge. The Higgs boson is postulated to interact with other particles in such a way as to impart mass to them. It is predicted by the **standard model**, but has yet to be isolated experimentally. The Higgs boson is named after its discoverer, British theoretical physicist Peter Ware Higgs (born 1929).

high blood pressure (hī) See **hypertension.**

high-density lipoprotein A complex of lipids and proteins in approximately equal amounts that functions as a transporter of cholesterol in the blood. High levels are associated with a decreased risk of atherosclerosis and coronary heart disease.

high-tension Having a high voltage, or designed to work at or sustain high voltages. High-tension wires used to carry electrical power over long distances sustain voltages over 200,000 volts. Compare **low-tension.**

high-test Relating to highly volatile high-octane gasoline.

high tide 1. The tide when it is at its highest level at a particular time and place. The highest tides reached under normal meteorological conditions (the **spring tides**) take place when the Moon and Sun are directly aligned with respect to Earth. High tides are less extreme (the **neap tides**) when the Moon and Sun are at right angles. Storms and other meteorological conditions can greatly affect the height of the tides as well. See more at **tide. 2.** The time at which a high tide occurs.

HII region (āch′to͞o′) A cloudlike region in space where the interstellar medium consists of ionized rather than neutral hydrogen. HII regions are centered around stars with very high surface temperatures and brilliant luminosities, whose radiant energy is sufficient to heat the interstellar medium to the point that hydrogen is ionized. The most conspicuous and best-known HII region is the Orion Nebula.

hilum (hī′ləm) Plural **hila. 1.** A mark or scar on a seed, such as a bean, showing where it was formerly attached to the plant. The hilum indicates the point of attachment of the **funiculus. 2.** A depression or opening through which nerves, ducts, or blood vessels pass in an organ or a gland, as in the medial aspect of the lungs or the kidneys.

hindbrain (hīnd′brān′) The rearmost part of the vertebrate brain. In humans, it consists of the pons and the medulla oblongata. Compare **forebrain, midbrain.**

hinge joint (hĭnj) A joint, such as the elbow, in which a convex part of one bone fits into a concave part of another, allowing motion in only one plane.

hipbone (hĭp′bōn′) Either of two large, flat bones, each forming one of the outer borders of the pelvis in mammals and consisting of the fused ilium, ischium, and pubis.

Hipparchus also **Hipparchos** (hĭ-pär′kəs) 190?–120? BCE. Greek astronomer who mapped the positions of about 850 stars in the earliest known star chart (129 BCE). His observations of the heavens formed the basis of Ptolemy's Earth-centered model of the universe. He was also a pioneer of trigonometry.

hippocampus (hĭp′ə-kăm′pəs) Plural **hippocampi** (hĭp′ə-kăm′pī′). A convoluted, seahorse-shaped structure in the cerebral cortex of the temporal lobe of the brain, composed of two gyri with white matter above gray matter. It forms part of the limbic system and is involved in the processing of emotions and memory.

Hippocrates (hĭ-pŏk′rə-tēz′) 460?–377? BCE. Greek physician who is credited with establishing the foundations of scientific medicine. He and his followers worked to distinguish medicine from superstition and magic beliefs by basing their treatment of illness on close observation and rational deduction.

histamine (hĭs′tə-mēn′) An organic compound found widely in animals and plants that in humans and other mammals is released as part of the body's immune response, causing physiological changes including dilation of the blood vessels, contraction of smooth muscle (as in the airways), and increased gastric acid secretion. The itching and sneezing typical of respiratory allergies are caused by the release of histamine. *Chemical formula:* $C_5H_9N_3$

histidine (hĭs′tĭ-dēn′) An amino acid that is essential for children but not for adults. *Chemical formula:* $C_6H_9N_3O_2$.

histocompatibility (hĭs′tō-kəm-păt′ə-bĭl′ĭ-tē) A state or condition in which the absence of immunological interference permits the grafting of tissue or the transfusion of blood without rejection.

histology (hĭ-stŏl′ə-jē) The scientific study of the microscopic structure of plant and animal tissues.

histone (hĭs′tōn′) Any of several proteins that, together with DNA, make up most of the chromatin in a cell nucleus.

A CLOSER LOOK **histone**

DNA is normally conceived of as a spiral ladder, but in eukaryotic cells (cells with nuclei) the DNA in the nucleus is strung around a series of spool-shaped proteins known as *histones.* Their chief functions are to compact and control the long threads of DNA. They compact the DNA by interacting with each other to form a structure like a compact spool. Two turns of DNA are wrapped around this spool, forming the subunits known as **nucleosomes** and decreasing the effective length of DNA eightfold. At high magnification these DNA-histone complexes look like a series of beads on a string. The complexes are further compacted by a factor of four by a linker histone that binds the DNA between the nucleosomes, organizing them into a coil. In this way a chromosome containing 20 million base pairs of DNA is organized into approximately 100,000 nucleosome core particles. Histones are also involved in controlling which sequences of DNA are turned on for transcription of RNA. When histones are chemically modified in certain ways, they may loosen their hold on the DNA and allow it to become accessible to proteins that activate transcription, or they may tighten their hold on the DNA and make it inaccessible. DNA itself may be chemically modified in the process known as *DNA methylation,* which is another mechanism for regulating gene expression. It is thought that the histones stay with the same sequences of DNA after cell replication, so the modifications of the histones and DNA allow the same sets of genes to be turned on and off in the daughter cells as in the parent cell. This is one way that multicellular organisms can make multiple types of cells (such as muscle, liver, and skin), even though the different types of cells all contain the same DNA in their nuclei. The histones are among the most well-conserved proteins known. There are only two minor changes in the amino acid sequences of the histone designated H4 in the pea and cow, for example. This near uniformity across species suggests that the entire surface of each histone is important to its function and that all plants and animals use histones for the same functions.

Hitchings (hĭch′ĭngz), **George Herbert** 1905–1998. American pharmacologist who with his colleague Gertrude Elion developed drugs to treat leukemia and malaria, gout, herpes, and urinary and respiratory tract infections. He and Elion shared with Sir James Black the 1988 Nobel Prize for physiology or medicine.

HIV (āch′ī-vē′) Short for *human immunodeficiency virus.* Any of various strains of a retrovirus of the genus *Lentivirus* that cause AIDS by infecting the body's immune system.

hives (hīvz) A skin condition characterized by transient, itching welts, usually resulting from an allergic reaction.

Ho The symbol for **holmium.**

hoarfrost (hôr′frôst′) Frozen dew that forms a white coating on a surface.

Hodgkin (hŏj′kĭn), **Dorothy Mary Crowfoot** 1910–1994. British chemist who used x-ray techniques to determine the structure of several complex molecules, including penicillin (1942–45) and vitamin B_{12} (1948–56). For this work she received the 1964 Nobel Prize for chemistry. She later used more advanced computing methods to analyze the structure of insulin.

Dorothy Crowfoot Hodgkin

Hodgkin's disease (hŏj′kĭnz) A progressive neoplastic disease, marked by proliferation of cells arising from the lymph nodes and bone marrow; enlargement of the lymph nodes, spleen, and liver; fever; and anemia. The disease is most common in teenagers and young adults. Hodgkin's disease is named after its identifier, English pathologist Thomas Hodgkin (1798–1866).

Hofstadter (hŏf′stăt′ər), **Robert** 1915–1990. American physicist who determined the inner structure of protons and neutrons (1948) and in 1961 shared with German physicist Rudolf Ludwig Mössbauer the 1961 Nobel Prize for physics.

hole (hōl) A gap, usually the valence band of an insulator or semiconductor, that would normally be filled with one electron. If an electron accelerated by a voltage moves into a gap, it leaves a gap behind it, and in this way the hole itself appears to move through the substance. Even though holes are in fact the absence of a negatively charged particle (an electron), they can be treated theoretically as positively charged particles, whose motion gives rise to electric current.

Hollerith (hŏl′ə-rĭth′), **Herman** 1860–1929. American inventor who in 1880 created a system of recording and retrieving information on punched cards, an important step in the development of modern computer science. In 1896 he founded his own company, and later merged with two others to form the International Business Machines Corporation (IBM) in 1924.

Holmes (hōmz, hōlmz), **Arthur** 1890–1965. British geologist who pioneered a method of determining the age of rocks by measuring their radioactive components. He was also an early supporter of Alfred Wegener's theory of continental drift.

holmium (hōl′mē-əm) *Symbol* **Ho** A soft, silvery, malleable metallic element of the lanthanide series. Its compounds are highly magnetic. It is mainly used in scientific research but has also been used to make electronic devices. Atomic number 67; atomic weight 164.930; melting point 1,461°C; boiling point 2,600°C; specific gravity 8.803; valence 3. See **Periodic Table.**

Holocene (hŏl′ə-sēn′, hō′lə-sēn′) The more recent of the two epochs of the Quaternary Period, beginning at the end of the last major Ice Age, about 10,000 years ago. It is characterized by the development of human civilizations. Also called *Recent.* See Chart at **geologic time.**

hologram (hŏl′ə-grăm′, hō′lə-) A three-dimensional image of an object made by holography.

A CLOSER LOOK hologram

To produce a simple *hologram,* a beam of coherent, monochromatic light, such as that produced by a laser, is split into two beams. One part, the object or illumination beam, is directed onto the object and reflected onto a high-resolution photographic plate. The other part, the reference beam, is beamed directly onto the photographic plate. The interference pattern of the two light beams is recorded on the plate. When the developed hologram is illuminated from behind (in the same direction as the original reference beam) by a beam of coherent light, it projects a three-dimensional image of the original object in space, shifting in perspective when viewed from different angles. Appropriately enough, the word hologram comes from the Greek words *holos,* "whole," and *gramma,* "message." If a hologram is cut into pieces, each piece projects the entire image, but as if viewed from a smaller subset of angles. The large amount of information contained in holograms makes them harder to forge than two-dimensional images. Many credit cards, CDs, sports memorabilia, and other items include holographic stickers as indicators of authenticity. Holography is used in many fields, including medicine, data storage, architecture, engineering, and the arts.

---

holography (hə-lŏg′rə-fē) A method of creating a three-dimensional image of an object on film by encoding not just the intensity but also the phase information of the light striking the film. See Note at **hologram.**

holophyte (hŏl′ə-fīt′) An organism that produces its own food through photosynthesis.

holoplankton (hŏl′ə-plăngk′tən, hō′lə-) Plankton that remains free-swimming through all stages of its life cycle.

holothurian (hŏl′ə-thŏŏr′ē-ən, hō′lə-) Any of various echinoderms of the class Holothuroidea, characterized by elongated bodies with tentacles around the mouth and five rows of tube feet. Holothurians are softer than other echinoderms because they lack spines and the skeleton is greatly reduced. Holothurians can extrude their internal organs to distract predators and grow a new set within a few weeks. Also called *sea cucumber.*

holotype (hŏl′ə-tīp′, hō′lə-) The single specimen or illustration designated as the type for naming a species or subspecies or used as the basis for naming a species or subspecies when no type has been selected. Also called *type species.*

homeobox (hō′mē-ə-bŏks′) Any of various DNA sequences containing about 180 nucleotides that encode for corresponding sequences of usually 60 amino acids, called *homeodomains,* found in proteins that bind DNA and regulate gene transcription. Genes containing homeoboxes are found in all eukaryotic genomes and are associated with cell differentiation and bodily segmentation during embryologic development.

homeomorphism (hō′mē-ə-môr′fĭz′əm) **1.** A close similarity in the crystal forms of unlike compounds. **2.** A one-to-one correspondence between the points of two geometric figures such that open sets in the first geometric figure correspond to open sets in the second figure and conversely. If one figure can be transformed into another without tearing or folding, there exists a homeomorphism between them. Topological properties are defined on the basis of homeomorphisms.

homeopathy (hō′mē-ŏp′ə-thē) A nontraditional system for treating and preventing disease, in which minute amounts of a substance that in large amounts causes disease symptoms are given to healthy individuals. This is thought to enhance the body's natural defenses.

homeostasis (hō′mē-ō-stā′sĭs) The tendency of an organism or cell to regulate its internal conditions, such as the chemical composition of its body fluids, so as to maintain health and functioning, regardless of outside conditions. The organism or cell maintains homeostasis by monitoring its internal conditions and responding appropriately when these conditions deviate from their optimal state. The maintenance of a steady body temperature in warm-blooded animals is an example of homeostasis. In human beings, the homeostatic regulation of body temperature involves such mechanisms as sweating when the internal temperature becomes excessive and shivering to produce heat, as well as the generation of heat through metabolic processes when the internal temperature falls too low.

homeotherm (hō′mē-ə-thûrm′) See **endotherm.**

hominid (hŏm′ə-nĭd) Any of various primates of the family Hominidae, whose only living members are modern humans. Hominids are characterized by an upright gait, increased brain size and intelligence compared with other primates, a flattened face, and reduction in the size of the teeth and jaw. Besides the modern species *Homo sapiens,* hominids also include extinct species of *Homo* (such as *H. erectus*) and the extinct genus *Australopithecus.* In some classifications, the family Hominidae also includes the anthropoid apes.

hominoid (hŏm′ə-noid′) A primate belonging to the superfamily Hominoidea, which includes apes and humans.

homo– A prefix meaning "same," as in *homogamous,* having the same kind of flower.

homocyclic (hō′mə-sĭk′lĭk) Containing a closed ring structure made of only one kind of atom. Benzene, for example, is a homocyclic compound, having a ring consisting of six carbon atoms. Compare **heterocyclic.**

homocysteine (hō′mə-sĭs′tə-ēn′, -ĭn, -tē-) An amino acid used normally by the body in cellular metabolism and the manufacture of proteins. Elevated concentrations in the blood are thought to increase the risk for heart disease by damaging the lining of blood vessels and increasing the risk of blood clot formation. High homocysteine levels are associated with certain vitamin deficiencies and metabolic disorders. *Chemical formula:* $\mathbf{C_4H_9NO_2S}$.

homoecious (hō-mē′shəs, hŏ-) Autoecious.

Homo erectus (hō′mō ĭ-rĕk′təs) An extinct species of humans that lived during the Pleistocene Epoch from about 1.6 million years ago to 250,000 years ago. *Homo erectus* is associated mainly with stone tools of the Acheulian culture and was the first species of humans to master fire, although this skill may not have been widely practiced until late in its existence. Its remains have been found in Africa, Europe, and Asia, and it is widely thought to be the direct ancestor of modern humans. See also **pithecanthropus, sinanthropus.** ▸ The *H. erectus* remains from Africa are thought by some to evince significant differences in comparison to other *H. erectus* populations and thus to constitute a separate species called **Homo ergaster.** *H. ergaster* is sometimes further claimed to be the true ancestor of modern humans. The fossil evidence is not complete enough to definitively support these or many other claims concerning early *Homo* populations.

Homo habilis (hăb′ə-ləs) An extinct species of early humans, known from fossils found in eastern Africa and often considered to be the first member of the genus *Homo.* It is associated with stone tools of the Oldowan culture. *Homo habilis* existed between about 2.5 and 1.6 million years ago and overlapped with late australopithecines and other hominids whose relationship to each other and to the later *Homo erectus* are uncertain.

homologous (hə-mŏl′ə-gəs) **1.** Similar in structure and evolutionary origin but having different functions, as a human's arm and a seal's flipper. **2.** Being one of a pair of chromosomes, one from the female parent and one from the male parent, that have genes for the same traits in the same positions. Genes on homologous chromosomes may not have the same form, however. For example, one chromosome in a pair of homologous chromosomes may contain a gene for brown eyes, and the other a gene for blue eyes. Human females have 23 pairs of homologous chromosomes (including the two X chromosomes), while human males have 22 because the Y chromosome is not paired. **3.** Belonging to or being a series of organic compounds, each successive member of which differs from the preceding member by a constant increment, especially by an added CH_2 group. The alkanes (methane, ethane, propane, and others) are a homologous series of compounds. **4.** Involving organisms of the same species, as in grafted body tissues.

homologue or **homolog** (hŏm′ə-lôg′, hō′mə-) **1.** A homologous organ or part. **2.** A homologous chromosome. **3.** A homologous chemical compound. Compare **analog.**

homology (hə-mŏl′ə-jē) **1.** A homologous relationship or correspondence. **2.** The relation of the chemical elements of a periodic family or group. **3.** The relation of the organic compounds forming a homologous series. **4.** A topological classification of configurations into distinct types that imposes an algebraic structure or hierarchy on families of geometric figures.

homolosine projection (hə-mŏl′ə-sīn′) A map projection with interruptions in the oceans, designed so that the continents appear with their proper size with respect to each other. A homolosine projection map presents the entire world in one view, with the landmasses uninterrupted except for Antarctica and Greenland. Distance and direction are not accurate for all areas of the map. Compare **conic projection, Mercator projection, sinusoidal projection.**

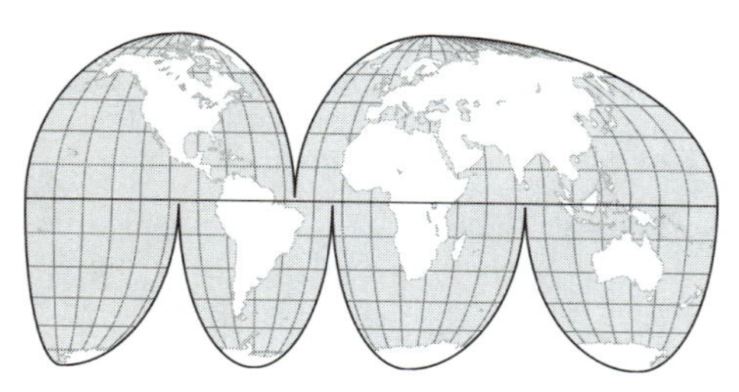

homolosine projection

homomorphism (hō′mə-môr′fĭz′əm, hŏm′ə-) A transformation of one set into another that preserves in the second set the operations between the members of the first set.

homophyly (hō′mə-fī′lē, hŏm′ə-, hō-mŏf′ə-lē) Resemblance arising from common ancestry.

homopteran (hə-mŏp′tər-ən) Any of various insects belonging to the group Homoptera. Homopterans suck sap from plants and can be very destructive. They include the cicadas, treehoppers, leafhoppers, aphids, scale insects, whiteflies, and mealybugs. Some scientists consider the homopterans to be a suborder of the order Hemiptera, while others consider them to be a separate insect order.

Homo sapiens (sā′pē-ənz) The modern species of humans. Archaic forms of *Homo sapiens* probably evolved around 300,000 years ago or earlier in Africa, and anatomically modern fossils are known from about 100,000 years ago. All humans now living belong to the subspecies *Homo sapiens sapiens*. The closest living relative of *Homo sapiens* is the chimpanzee. See more at **archaic Homo sapiens, Cro-Magnon, Neanderthal.**

homosporous (hō′mə-spôr′əs, hŏm′ə-, hō-mŏs′pər-əs) Producing spores of one kind only that are not differentiated by sex. The spores of homosporous plants, such as horsetails and most ferns, grow into bisexual gametophytes (producing both male and female gametes). Compare **heterosporous.**

homozygous (hō′mō-zī′gəs) Relating to a cell that has two identical alleles for a particular trait at corresponding positions on homologous chromosomes. Compare **heterozygous.**

Hooke (ho͝ok), **Robert** 1635–1703. English physicist, inventor, and mathematician who contributed to many aspects of science. With Robert Boyle he demonstrated that both combustion and respiration require air and that sound does not travel in a vacuum. Hooke studied plants and other objects under microscopes and was the first to use the word *cell* to describe the patterns he observed. He also identified fossils as a record of changes among organisms on the planet throughout history.

Hooke's law (ho͝oks) A law stating that the stress applied to a material is proportional to the strain on that material. For example, if a stress on a metal bar of ten newtons per square centimeter causes it to be compressed by four millimeters, then a stress of 20 newtons per square centimeter will cause the bar to be compressed by eight millimeters. Hooke's law generally holds only up to the **elastic limit** of stress for that material. See also **modulus of elasticity.**

hookworm (ho͝ok′wûrm′) Any of numerous small, parasitic nematode worms of the family Ancylostomatidae, having hooked mouthparts with which they fasten themselves to the intestinal walls of various animals, including humans.

Hopper (hŏp′ər), **Grace Murray** 1906–1992. American mathematician and computer programmer who in 1951 conceived the idea for an internal computer program, called a **compiler**, that scanned a set of alphanumeric instructions (such as words and symbols) and compiled a set of binary instructions executed by the machine. Her ideas were widely influential in the development of programming languages, in particular COBOL.

horizon (hə-rī′zən) **1a.** The apparent intersection of the Earth and sky as seen by an observer. Also called *apparent horizon.* **b.** See **celestial horizon. c.** See **sensible horizon. 2.** *Geology.* **a.** A specific position in a stratigraphic column, such as the location of one or more fossils, that serves to identify the stratum with a particular period. **b.** A specific layer of soil or subsoil in a vertical cross-section of land. **3.** *Archaeology.* A period during which the influence of a particular culture spread rapidly over a defined area.

hormone (hôr′mōn′) **1.** A chemical substance secreted by an endocrine gland or group of endocrine cells that acts to control or regulate specific physiological processes, including growth, metabolism, and reproduction. Most hormones are secreted by endocrine cells in one part of the body and then transported by the blood to their target site of action in another part, though some hormones act only in the region in which they are secreted. Many of the principal hormones of vertebrates, such as growth hormone and thyrotropin, are secreted by the pituitary gland, which is in turn regulated by neurohormone secretions of the hypothalamus. Hormones also include the endorphins, androgens, and estrogens. See more at **endocrine gland. 2.** A substance that is synthesized by a plant part and acts to control or regulate the growth and development of the plant. The action and effectiveness of a hormone can depend on the hormone's chemical structure, its amount in relation to other hormones that have competing or opposing effects, and the ways in which it interacts with chemical receptors in various plant parts. Auxins, cytokinins, gibberellins, abscisic acid, and ethylene are plant hormones.

A CLOSER LOOK **hormones**

Among the most abundant and influential chemicals in the human body are the *hormones,* found also throughout the entire animal and plant kingdoms. The endocrine glands alone, including the thyroid, pancreas, adrenals, ovaries, and testes, release more than 20 hormones that travel through the bloodstream before arriving at their targeted sites. The pea-sized pituitary gland, located at the base of the brain below the hypothalamus, is considered the most crucial part of the endocrine system, producing growth hormone and hormones that control other endocrine glands. Specialized cells of the nervous system also produce hormones. The brain itself releases endorphins, hormones that act as natural painkillers. Hormones impact almost every cell and organ of the human body, regulating mood, growth, tissue function, metabolism, and sexual and reproductive function. Compared to the nervous system, the endocrine system regulates slower processes such as metabolism and cell growth, while the nervous system controls more immediate functions, such as breathing and movement. The action of hormones is a delicate balancing act, which can be affected by stress, infection, or changes in fluids and minerals in the blood. The pituitary hormones are influenced by a variety of factors, including emotions and fluctuations in light and temperature. When hormone levels become abnormal, disease can result, such as diabetes from insufficient insulin or osteoporosis in women from decreased estrogen. On the other hand, excessive levels of growth hormone may cause uncontrolled development. Treatment for hormonal disorders usually involves glandular surgery or substitution by synthetic hormones.

hormone replacement therapy The therapeutic administration of estrogen and often progesterone to postmenopausal women in order to reduce symptoms and signs of estrogen deficiency, such as hot flashes and osteoporosis.

horn (hôrn) **1.** Either of the bony growths projecting from the upper part of the head of certain hoofed mammals, such as cattle, sheep, and goats. The horns of these animals are never shed, and they consist of bone covered by keratin. **2.** A hard growth that looks like a horn, such as an antler or a growth on the head of a giraffe or rhinoceros. Unlike true horns, antlers are shed yearly and have a velvety covering, and the horns of a rhinoceros are made not of bone but of hairy skin fused with keratin. **3.** The hard durable substance that forms the outer covering of true horns. It consists of keratin.

hornblende (hôrn′blĕnd′) A common, green to black mineral of the amphibole group. It has a variable composition and occurs in monoclinic crystals with a hexagonal cross-section, in fibrous forms, or in granular forms. Hornblende is found in many kinds of metamorphic and igneous rocks. *Chemical formula:* $\mathbf{(Ca,Na)_{2\text{-}3}(Mg,Fe,Al)_5(Al,Si)_8O_{22}(OH)_{22}}$.

hornfels (hôrn′fĕlz′) A fine-grained metamorphic rock having a uniform grain size and formed by contact metamorphism.

hornito (hôr-nē′tō) A low mound of matter ejected from a volcano, sometimes emitting smoke or vapor.

hornwort (hôrn′wûrt′, -wôrt′) Any of about 100 species of small bryophyte plants belonging to the phylum Anthocerophyta. Unlike liverworts but like mosses, hornwort sporophytes have stomata. The hornwort gametophyte consists of a low thallus, out of which numerous slender, upright sporophytes tipped with sporangia grow. The sporophyte has a meristem that elongates the sporophyte with new growth, a feature that distinguishes the plant from the other bryophytes. The name of the hornworts was suggested by the hornlike appearance of the sporophytes. See more at **bryophyte.**

horology (hô-rŏl′ə-jē) The science of measuring time.

horse latitudes (hôrs) Either of two regions of the globe, found over the oceans about 30 degrees north and south of the equator, where winds are light and the weather is hot and dry. They are associated with high atmospheric pressure and with the large-scale descent of cool dry air that spreads either toward the equator, as the trade winds, or toward the poles, as the westerlies.

horsepower (hôrs′pou′ər) A unit that is used to measure the power of engines and motors. One unit of horsepower is equal to the power needed to lift 550 pounds one foot in one second. This unit has been widely replaced by the watt in scientific usage; one horsepower is equal to 745.7 watts.

horsetail (hôrs′tāl′) A member of a genus, *Equisetum,* of seedless vascular plants having a jointed hollow stem and narrow, sometimes much reduced leaves. Plants extremely similar to modern horsetails are known from fossils 300 million years old. The horsetails are the last surviving members of the phylum Sphenophyta, which dominated the forests of the Devonian and Carboniferous periods.

horst (hôrst) A usually elongated block of rock that is bounded by parallel geologic faults along its two longest sides and has a higher elevation than the rock at its sides. Horsts form where rock is being compressed by tectonic forces.

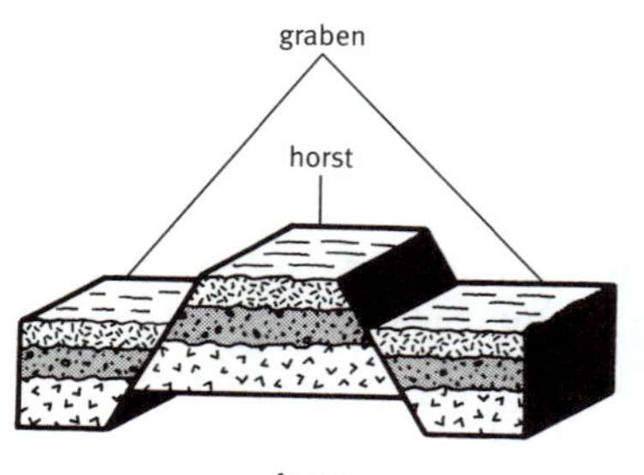

horst

host (hōst) **1a.** The larger of two organisms in a symbiotic relationship. **b.** An organism or cell on or in which a parasite lives or feeds. ► A **definitive host** is an organism in which a parasite reaches sexual maturity. The anopheles mosquito is the definitive host for the malaria plasmodium because, while the mosquito is not adversely affected by the plasmodium's presence, it is the organism in which the plasmodium matures and reproduces. ► An **intermediate host** is an organism in which a parasite develops but does not attain sexual maturity. Humans and certain other vertebrates are the intermediate host of the malaria plasmodium. ► A **paratenic host** is an organism which may be required for the completion of a parasite's life cycle but in which no development of the parasite occurs. The unhatched eggs of nematodes are sometimes carried in a paratenic host such as a bird or rodent. When a predator eats the paratenic host, the eggs are ingested as well. **2.** The recipient of a transplanted tissue or organ. **3.** A computer containing data or programs that another computer can access by means of a network or modem.

hot Jupiter (hŏt) See under **extrasolar planet.**

hot spot A volcanic area that forms as a tectonic plate moves over a point heated from deep within the Earth's mantle. The source of the heat is thought to be the decay of radioactive elements. The Hawaiian Islands formed as a series of hot spots. See more at **tectonic boundary.**

hot spring A spring of warm water, usually having a temperature greater than that of the human body.

hot-wire anemometer See under **anemometer.**

hour (our) **1.** A unit of time equal to one of the 24 equal parts of a day; 60 minutes. ► A **sidereal hour** is $\frac{1}{24}$ of a sidereal day, and a **mean solar hour** is $\frac{1}{24}$ of a mean solar day. See more at **sidereal time, solar time. 2.** A unit of measure of longitude or right ascension, equal to 15° or $\frac{1}{24}$ of a great circle.

hour angle The angular distance, measured westward along the **celestial equator**, between the celestial meridian of the observer and the hour circle passing through a celestial body. A body's hour angle is measured in hours, minutes, and seconds, and corresponds to its right ascension as measured with respect to the observer's meridian (which changes with time) rather than the vernal equinox (which is fixed on the celestial equator). A celestial object that crossed the observer's meridian 3 hours and 20 minutes ago has an hour angle of +3 hours 20 minutes. An object that will not cross the meridian for another 3 hours and 20 minutes has an hour angle of –3 hours 20 minutes.

hour circle A great circle passing through the poles of the celestial sphere and intersecting the **celestial equator** at right angles. An hour circle is equivalent to a meridian on Earth and is used in describing the position of a celestial body with respect to an observer's celestial meridian. See more at **hour angle.**

Hs The symbol for **hassium.**

HTML (āch′tē-ĕm-ĕl′) A markup language used to structure text and multimedia documents and to set up hypertext links between documents, used extensively on the World Wide Web.

HTTP (āch′tē-tē-pē′) Short for *Hypertext Transfer Protocol.* A protocol used to request and transmit files, especially webpages and webpage components, over the Internet or other computer network.

Hubble (hŭb′əl), **Edwin Powell** 1889–1953. American astronomer who demonstrated that there are galaxies beyond our own and that they are receding from ours, providing strong evidence that the universe is expanding. Hubble also established the first measurements for the age and radius of the known universe, and his methods for determining them remain in use today. See Note at **big bang, Doppler effect.**

Hubble classification system A system developed by Edwin Hubble in 1936 that classifies galaxies as **elliptical, spiral, barred spiral, lenticular,** and **irregular.** Hubble speculated that galaxies might fit into an evolutionary sequence from earlier elliptical galaxies to later spiral forms, but most astronomers no longer believe this is the case. Our galaxy, the **Milky Way,** is a spiral galaxy. Also called *Hubble sequence.*

Hubble's law (hŭb′əlz) A law of cosmology stating that the rate at which astronomical objects in the universe move apart from each other is proportional to their distance from each other. Current estimates of the value of this proportion, known as **Hubble's constant,** put its value at approximately 71 kilometers per second per megaparsec.

hue (hyo͞o) The property of colors by which they are seen as ranging from red through orange, yellow, green, blue, indigo, and violet, as determined by the dominant wavelength of the light. Compare **saturation, value.**

hull (hŭl) **1.** The dry outer covering of a fruit, seed, or nut; a husk. **2.** The enlarged calyx of a fruit, such as a strawberry, that is usually green and easily detached.

human (hyo͞o′mən) **1.** A member of the species *Homo sapiens;* a human being. **2.** A member of any of the extinct species of the genus *Homo,* such as *Homo erectus* or *Homo habilis,* that are considered ancestral or closely related to modern humans.

human chorionic gonadotropin A glycoprotein hormone that is produced by the placenta and maintains the corpus luteum during the first few weeks of pregnancy.

Human Genome Project An international scientific research project designed to study and identify all of the genes in the human genome, to determine the base-pair sequences in human DNA, and to store this information in computer databases. The Human Genome Project began in the United States in 1990 and was completed in 2003.

human immunodeficiency virus See **HIV.**

human papillomavirus (păp′ə-lō′mə-vī′rəs) A virus of the genus *Papillomavirus,* certain strains of which cause skin and genital warts in humans or are associated with various cancers, including cancers of the cervix, vagina, and larynx.

Humboldt (hŭm′bōlt′, ho͝om′bôlt′), Baron **(Friedrich Heinrich) Alexander von** 1769–1859. German naturalist and writer who explored South America, Cuba, and Mexico (1799–1804) and recorded a wide range of species, particularly plants, and attempted to explain their geographic distribution with respect to their environment. His work laid the foundation the science of ecology.

Humboldt Current (hŭm′bōlt′) A cold ocean current of the South Pacific, flowing north along the western coast of South America from Chile to Peru. Extending up to 1,000 km (620 mi) offshore, the Humboldt Current results in significant cooling of the marine environment and influences the weather pattern that makes this section of coast one of the driest regions in the world. The current is also the world's largest upwelling current, bringing cold, nutrient-rich waters to the surface and creating an ecosystem abundant in plankton, fish, and other marine life. It is named after Baron Alexander von Humboldt, who explored this coast in 1802. Also called *Peru current.*

humerus (hyo͞o′mər-əs) Plural **humeri** (hyo͞o′mər-ī′). The bone of the upper arm or the upper portion of the foreleg. See more at **skeleton.**

humidistat (hyo͞o-mĭd′ĭ-stăt′) See **hygrostat.**

humidity (hyo͞o-mĭd′ĭ-tē) The amount of water vapor in the atmosphere, usually expressed as either **absolute humidity** or **relative humidity.**

humiture (hyo͞o′mĭ-chər) See **heat index.**

humor (hyo͞o′mər) **1.** See **aqueous humor. 2.** See **vitreous humor. 3.** One of the four fluids of the body—blood, phlegm, black bile, and yellow bile—whose relative proportions were thought in ancient and medieval medicine to determine general health and character.

WORD HISTORY **humor**

Doctors in ancient times and in the Middle Ages thought the human body contained a mixture of four substances, called humors, that determined a person's health and character. The humors were fluids (*humor* means "fluid" in Latin), and they differed from each other in being either warm or cold and moist or dry. Each humor was also associated with one of the four elements, the basic substances that made up the universe in ancient schemes of thought. Blood was the warm, moist humor associated with the element fire, and phlegm was the cold, moist humor associated with water. Black bile was the cold, dry humor associated with the earth, and yellow bile was the warm, dry humor associated with the air. Illnesses were thought to be caused by an imbalance in the humors within the body, as were defects in personality, and some medical terminology in English still reflects these outmoded concepts. For example, too much black bile was thought to make a person gloomy, and nowadays symptoms of depression such as insomnia and lack of pleasure in enjoyable activities are described as *melancholic* symptoms, ultimately from the Greek word *melancholia,* "excess of black bile," formed from *melan-,* "black," and *khole,* "bile." The old term for the cold, clammy humor, *phlegm,* lives on today as the word for abnormally large accumulations of mucus in the upper respiratory tract. Another early name of yellow bile in English, *choler,* is related to the name of the disease *cholera,* which in earlier times denoted stomach disorders thought to be due to an imbalance of yellow bile. Both words are ultimately from the Greek word *chole,* "bile."

humoral immune response (hyo͞o′mər-əl) The immune response involving the transformation of **B cells** into **plasma cells** that produce and secrete antibodies to a specific antigen. See Note at **antibody.**

humoral immunity Immunity resulting from a **humoral immune response.**

humus (hyo͞o′məs) A dark-brown or black organic substance made up of decayed plant or animal matter. Humus provides nutrients for plants and increases the ability of soil to retain water.

hurricane (hûr′ĭ-kān′) A severe, rotating tropical storm with heavy rains and cyclonic winds exceeding 74 mi (119 km) per hour, especially such a storm occurring in the Northern Hemisphere. Hurricanes originate in the tropical parts of the Atlantic Ocean or the Caribbean Sea and move generally northward. They lose force when they move over land or colder ocean waters. See Note at **cyclone.**

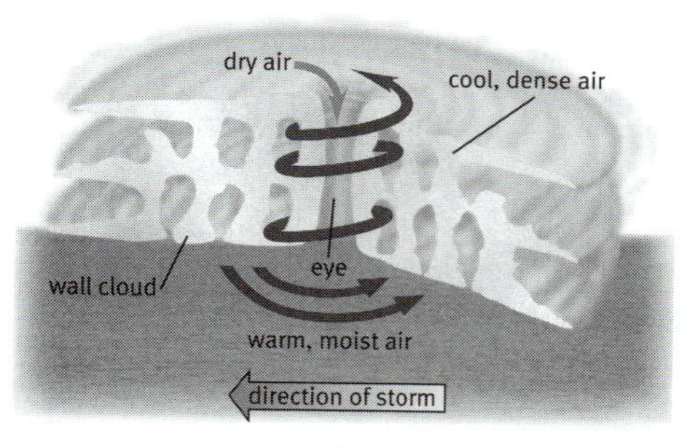

hurricane

husbandry (hŭz′bən-drē) The application of scientific principles to agriculture, especially to animal breeding.

Hutton (hŭt′n), **James** 1726–1797. Scottish geologist whose theories of rock and land formation laid the foundation for modern geology. He showed that, over long periods of time, the erosion of rocks produces sediments, which are transported by water, ice, and air to locations at or near sea level. These sediments eventually become solidified into other rocks.

James Hutton

BIOGRAPHY James Hutton

The father of modern geology did not start out as a geologist. He first apprenticed with a lawyer and then earned a degree in medicine. But after he inherited some land, he decided to devote himself to agriculture, and this led him to think about the origin of soil and its relation to the rest of the landscape, especially the rocks. Based upon his observations, he concluded that old rocks are pushed upwards to become mountains, that new rocks form from the emergence and solidification of lava, and that the driving energy for both of these processes must be the Earth's internal heat. He also concluded that soil forms from rocks through the long process of weathering. In this way Hutton developed the idea that the soil, the rocks, and the landscape were all connected in a single process, which he called *Plutonism,* in honor of Pluto, the Greek god of the underworld. Hutton realized that the cycle of uplift and erosion required a long time and that the Earth must therefore be much older than a few thousand years, as was widely believed at the time. But it was not until the twentieth century that Hutton's theory was proven correct when geologists, using a technique called *radiometric dating,* demonstrated that the Earth is in fact more than four billion years old.

Huygens (hī′gənz, hoi′gĕns), **Christiaan** 1629–1695. Dutch physicist and astronomer who in 1655 discovered Saturn's rings and its fourth satellite, using a telescope he constructed with his brother. In 1657 he built the first pendulum clock. Huygens also proposed that light consists of transverse waves that vibrate up and down perpendicular to the direction in which the light travels. This theory, which explained some properties of light better than Newton's theory, was made public in 1690.

Huygens' principle The principle that any point on a wave front may be regarded as a source of spherical or circular wavelets. The sum of all such wavelets emanating from the front is same as the wavefront itself. This principle can be used to derive the laws of reflection and refraction. See also **superposition.**

hybrid (hī′brĭd) An organism that is the offspring of two parents that differ in one or more inheritable characteristics, especially the offspring of two different varieties of the same species or the offspring of two parents belonging to different species. In agriculture and animal husbandry, hybrids of different varieties and species are bred in order to combine the favorable characteristics of the parents. Hybrids often display **hybrid vigor**. The mule, which is the offspring of a male donkey and a female horse, is an example of a hybrid. It is strong for its size and has better endurance and a longer useful lifespan than its parents. However, mules are sterile, as are many animals that are hybrids between two species.

hybrid vigor The increased vigor or general health, resistance to disease, and other superior qualities that are often manifested in hybrid organisms, especially plants and animals. Compare **inbreeding depression.**

hydra (hī′drə) Plural **hydras** or **hydrae** (hī′drē). See under **hydroid.**

hydrate (hī′drāt′) *Noun.* **1.** A compound produced by combining a substance chemically with water. Many minerals and crystalline substances are hydrates. —*Verb.* **2.** To combine a compound with water, especially to form a hydrate. **3.** To supply water to a person in order to restore or maintain a balance of fluids.

hydraulic (hī-drô′lĭk) **1.** Operated by the pressure of water or other liquids. Hydraulic systems, such as hydraulic brakes, allow mechanical force to be transferred along curved paths (through pipes or tubes) that would be difficult for solid mechanisms, such as levers or cables, to negotiate efficiently. **2.** Relating to hydraulics. **3.** Capable of hardening under water, as cement.

hydraulics (hī-drô′lĭks) **1.** The scientific study of water and other liquids, in particular their behavior under the influence of mechanical forces and their related uses in engineering. **2.** A mechanical device or system using hydraulic components.

hydrazide (hī′drə-zīd′) A compound formed by combining hydrazine with an acyl compound. Hydrazides are important in the manufacture of certain medicines.

hydrazine (hī′drə-zēn′, -zĭn) A colorless, fuming, corrosive liquid with an odor like ammonia that is a powerful reducing agent. It can be combined with organic compounds to form jet and rocket fuels and is also used to make explosives, fungicides, medicines, and photographic chemicals. *Chemical formula:* N_2H_4.

hydric (hī′drĭk) Relating or adapted to a wet but not flooded habitat. Cottonwoods, willows, and hemlocks are hydric plants. Compare **mesic, xeric.**

hydride (hī′drīd′) A compound of hydrogen with another element or radical.

hydro– A prefix that means: "water" (as in *hydroelectric*) or "hydrogen," (as in *hydrochloride*).

hydrocarbon (hī′drə-kär′bən) Any of numerous organic compounds, such as benzene, that contain only carbon and hydrogen.

hydrocast (hī′drə-kăst′) The process of using a device consisting of several water-collection bottles, such as Nansen bottles, that are wired and clamped together and used to collect data on water characteristics at various depths.

hydrocephalus (hī′drō-sĕf′ə-ləs) also **hydrocephaly** (hī′drō-sĕf′ə-lē) A usually congenital condition in which an abnormal accumulation of cerebrospinal fluid in the cerebral ventricles causes enlargement of the skull and compression of and injury to brain tissue. If hydrocephalus becomes progressive, a shunt is surgically placed to reduce pressure by conducting fluid away from the brain, usually to the peritoneum. —*Adjective* **hydrocephalic** (hī′drō-sə-făl′ĭk).

hydrochloric acid (hī′drə-klôr′ĭk) A solution of hydrogen chloride in water, forming a very strong, poisonous, corrosive acid with a sharp odor. It is used in food processing, metal cleaning, and dyeing. Small amounts of hydrochloric acid are also secreted by the stomachs of animals for digestion. Also called *muriatic acid.*

hydrochloride (hī′drə-klôr′īd′) A salt containing the group HCl. Many important drugs are hydrochlorides.

hydrocortisone (hī′drə-kôr′tĭ-sōn′, -zōn′) A preparation of the hormone cortisol that is obtained naturally or that is produced synthetically. Hydrocortisone is used widely in the treatment of inflammatory conditions and allergies.

hydrocracking (hī′drə-krăk′ĭng) A process by which the hydrocarbon molecules of petroleum are broken into simpler molecules, as of gasoline or kerosene, by the addition of hydrogen under high pressure and in the presence of a catalyst. See also **cracking.**

hydrocyanic acid (hī′drō-sī-ăn′ĭk) An aqueous solution of hydrogen cyanide, having a characteristic smell of bitter almonds. Also called *prussic acid.*

hydrodynamics (hī′drō-dī-năm′ĭks) The scientific study of the motion of fluids, especially noncompressible liquids, under the influence of internal and external forces. Hydrodynamics is a branch of **fluid mechanics** and has many applications in engineering. Compare **aerodynamics, hydrostatics.**

hydroelectric (hī′drō-ĭ-lĕk′trĭk) Using the power of water currents to generate electric power. Generally, hydroelectric power is created by directing water flow through a turbine, where the water causes fans to turn, creating the torque needed to drive an electric generator.

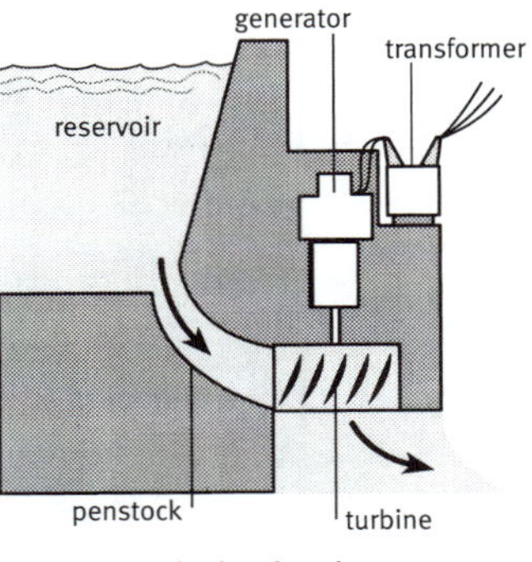

hydroelectric

low head hydroelectric power plant

hydrofluoric acid (hī′drō-flo͝or**′**ĭk, -flôr**′**-) A colorless, fuming, aqueous solution of hydrogen fluoride. It is corrosive and is used to etch or polish glass, to clean certain metals before plating, and to clean masonry. It is very poisonous.

hydrogen (hī**′**drə-jən) *Symbol* **H** The lightest and most abundant element in the universe, normally consisting of one proton and one electron. It occurs in water in combination with oxygen, in most organic compounds, and in small amounts in the atmosphere as a gaseous mixture of its three isotopes (protium, deuterium, and tritium) in the colorless, odorless compound H_2. Hydrogen atoms are relatively electropositive and form **hydrogen bonds** with electronegative atoms. In the Sun and other stars, the conversion of hydrogen into helium by nuclear **fusion** produces heat and light. Hydrogen is used to make rocket fuel, synthetic ammonia, and methanol, to hydrogenate fats and oils, and to refine petroleum. The development of physical theories of electron **orbitals** in hydrogen was important in the development of quantum mechanics. Atomic number 1; atomic weight 1.00794; melting point –259.14°C; boiling point –252.8°C; density at 0°C 0.08987 gram per liter; valence 1. See **Periodic Table.** See Note at **oxygen.**

hydrogenate (hī**′**drə-jə-nāt′, hī-drŏj**′**ə-nāt′) **1.** To treat or combine chemically an unsaturated compound with hydrogen. Liquid vegetable oils are often hydrogenated to turn them into solids. **2.** To turn coal into oil by combining its carbon with hydrogen to form hydrocarbons.

hydrogen bomb An extremely destructive bomb whose explosive power is derived from the energy released when hydrogen atoms are fused to form helium. This atomic **fusion** reaction is the same reaction that takes place in stars like the Sun, where the pressure of gravity forces hydrogen atoms to fuse; a hydrogen bomb uses the force of an atomic explosion (the **fission** reaction exploited in **atomic bombs**) to compress the hydrogen to the point where fusion takes place. Hydrogen bombs are many times more powerful than atomic bombs.

hydrogen bond A chemical bond formed between an electropositive atom (typically hydrogen) and a strongly electronegative atom, such as oxygen or nitrogen. Hydrogen bonds are responsible for the bonding of water molecules in liquid and solid states, and are weaker than covalent and ionic bonds.

hydrogen chloride A colorless, corrosive, suffocating gas used in making plastics and in many industrial processes. When mixed with water, it forms **hydrochloric acid.** *Chemical formula:* **HCl.**

hydrogen cyanide A colorless, flammable, extremely poisonous liquid. Salts derived from it have many industrial uses, such as hardening iron and steel, extracting metals from ores, electroplating metallic surfaces, and making acrylonitrile, from which acrylic fibers and plastics are produced. It is also used to make dyes and poisons. A solution of hydrogen cyanide in water forms a colorless acid called **hydrocyanic acid.** *Chemical formula:* **HCN.**

hydrogen fluoride A corrosive compound that exists as a colorless, fuming liquid or a highly soluble gas. Hydrogen fluoride is used as a reagent, catalyst, and fluorinating agent, in the refining of uranium, and in making many fluorine compounds. An aqueous solution of hydrogen fluoride is called **hydrofluoric acid.** *Chemical formula:* **HF.**

hydrogen peroxide A colorless, dense liquid, that is often used as a bleach or is diluted with water for use as an antiseptic. *Chemical formula:* H_2O_2.

hydrogen sulfide A colorless, poisonous gas that smells like rotten eggs. It is formed naturally by decaying organic matter and is the smelly component of intestinal gas. It is also emitted by volcanoes and fumaroles. Hydrogen sulfide is used in the petroleum, rubber, and mining industries, and in making sulfur. *Chemical formula:* H_2S.

hydrogeology (hī′drō-jē-ŏl′ə-jē) The scientific study of the occurrence, distribution, and effects of groundwater.

hydrography (hī-drŏg′rə-fē) **1.** The scientific description and analysis of the physical characteristics of Earth's surface waters, including temperature, salinity, oxygen saturation, and the chemical content of water. Oceanography (the study of saltwater bodies) and limnology (the study of freshwater bodies) are subsets of hydrography. **2.** The mapping of bodies of water.

hydroid (hī′droid′) Any of numerous, usually colonial marine coelenterates of the order Hydroida, having a polyp rather than a medusoid form as the dominant stage of the life cycle. Hydroids have a simple cylindrical body with a mouthlike opening surrounded by tentacles. Most species form colonies with individual hydroids branching off from a common hollow tube that is probably used to share ingested food. The young develop from eggs or from buds. The most well-known hydroids are the **hydras** (genus *Hydra*), which are atypical in being both freshwater and solitary.

hydroid

scanning electron micrograph

hydrologic cycle (hī′drə-lŏj′ĭk) The continuous process by which water is circulated throughout the Earth and its atmosphere. The Earth's water enters the atmosphere through evaporation from bodies of water and from ground surfaces. Plants and animals also add water vapor to the air by transpiration. As it rises into the atmosphere, the water vapor condenses to form clouds. Rain and other forms of precipitation return it to the Earth, where it flows into bodies of water and into the ground, beginning the cycle again. Also called *water cycle.*

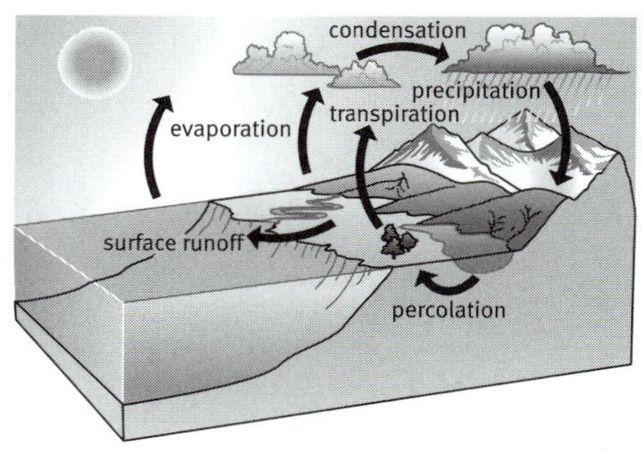

hydrologic cycle

Water that evaporates from the Earth's surface forms clouds and returns to the Earth as rain or snow.

hydrology (hī-drŏl′ə-jē) The scientific study of the properties, distribution, and effects of water as a liquid, solid, or gas on the Earth's surface, in the soil and underlying rocks, and in the atmosphere.

hydrolysis (hī-drŏl′ĭ-sĭs) The breaking down of a chemical compound into two or more simpler compounds by reacting with water. The proteins, fats, and complex carbohydrates in food are broken down in the body by hydrolysis that is catalyzed by enzymes in the digestive tract.

hydrometeor (hī′drō-mē′tē-ər) Any of various forms of water or ice that occur in the atmosphere or on the Earth's surface as a result of condensation. Rain, snow, fog, clouds, and dew are all hydrometeors.

hydrometeorology (hī′drō-mē′tē-ə-rŏl′ə-jē) The scientific study of the interaction between meteorological and hydrologic phenomena, including the occurrence, motion, and changes of state of atmospheric water, and the land surface and subsurface phases of the hydrologic cycle. Hydrometeorologic studies address questions regarding land use, the long-term effects of climate change on water resources, and regional precipitation.

hydrometer (hī-drŏm′ĭ-tər) An instrument used to measure the density of a liquid as compared to that of water. Hydrometers consist of a calibrated glass tube ending in a weighted glass sphere that makes the tube stand upright when placed in a liquid. The lower the density of the liquid, the deeper the tube sinks.

hydrophobia (hī′drə-fō′bē-ə) **1.** Abnormal fear of water. **2.** Rabies.

WORD HISTORY **hydrophobia**

Hydrophobia is an older term for the disease rabies, and it means "fear of water." Because of this name, many people think that rabies makes one afraid of water. In fact, this is not the case (although rabies does cause mental confusion of other kinds). The name hydrophobia comes from the fact that animals and people with rabies get spasms in their throat muscles that are so painful that they cannot eat or drink, and so will refuse water in spite of being very thirsty.

hydrophone (hī′drə-fōn′) A device used to detect or monitor sound under water. Hydrophones are often installed or towed in arrays that can be used to pinpoint a sound source or provide sea-floor imaging as part of a sonar system.

hydrophyte (hī′drə-fīt′) A plant that grows wholly or partly submerged in water. Because they have less need to conserve water, hydrophytes often have a reduced cuticle and fewer stomata than other plants. Floating leaves have stomata only on their upper surfaces, and underwater leaves generally have no stomata at all. Because water is readily available, hydrophytes also have a reduced root system and less vascular tissue than other plants (which also makes plant parts less dense and helps them float). Hydrophytes tend to have less supportive tissue as well, since they are buoyed by water. Many species of hydrophytes (such as the Eurasian milfoil) have divided leaves that have less resistance to flowing water. The lotus, water lily, and cattail are hydrophytes. Compare **mesophyte, xerophyte.**

hydroponics (hī′drə-pŏn′ĭks) The cultivation of plants in a nutrient-rich solution, rather than in soil, and under controlled conditions of light, temperature, and humidity. Also called *aquaculture.*

hydroponics

hydrospace (hī′drə-spās′) The regions beneath the ocean's surface, especially when considered as an area to be studied, as in marine mammal research or sonar mapping of the ocean floor.

hydrosphere (hī′drə-sfîr′) All of the Earth's water, including surface water (water in oceans, lakes, and rivers), groundwater (water in soil and beneath the Earth's surface), snowcover, ice, and water in the atmosphere, including water vapor. Compare **asthenosphere, atmosphere, lithosphere.**

hydrostatic pressure (hī′drə-stăt′ĭk) The pressure exerted by a fluid at equilibrium at a given point within the fluid, due to the force of gravity. Hydrostatic pressure increases in proportion to depth measured from the surface because of the increasing weight of fluid exerting downward force from above.

hydrostatics (hī′drə-stăt′ĭks) The scientific study of fluids, especially noncompressible liquids, in equilibrium with their surroundings and hence at rest. Hydrostatics has many applications in biology and engineering, as in the design of dams. Compare **hydrodynamics.**

hydrothermal (hī′drə-thûr′məl) Relating to or produced by hot water, especially water heated underground by the Earth's internal heat. ► **Hydrothermal energy** is power that is generated using the Earth's hot water.

hydrothermal vent A fissure on the floor of a sea out of which flows water that has been heated by underlying magma. The water can be as hot as 400°C (752°F) and usually contains dissolved minerals that precipitate out of it upon contact with the colder seawater, building a stack of minerals, or **chimney**. Hydrothermal vents form an ecosystem for microbes and animals, such as tubeworms, giant clams, and blind shrimp, that can withstand the hostile environment. ▸ The hottest hydrothermal vents are called **black smokers** because they spew iron and sulfide which combine to form iron monosulfide, a black compound.

hydrotropism (hī-drŏt′rə-pĭz′əm) The growth or movement of a fixed organism, especially a plant, or a part of an organism toward or away from water. Roots often display hydrotropism in growing towards a water source. *—Adjective* **hydrotropic** (hī′drə-trō′pĭk, hī′drə-trŏp′ĭk).

hydrous (hī′drəs) Containing water.

hydroxide (hī-drŏk′sīd′) A chemical compound containing one or more hydroxyl radicals (OH). Inorganic hydroxides include hydroxides of metals, some of which, like sodium hydroxide (caustic soda) and calcium hydroxide, are strong bases that are important industrial alkalis. Some metal hydroxides, such as those of zinc and lead, are amphoteric (they act like both acids and bases). Organic hydroxides include the alcohols.

hydroxy (hī-drŏk′sē) Containing the hydroxyl group (OH).

hydroxyl (hī-drŏk′sĭl) The group OH. Hydroxyl is present in bases, certain acids, hydroxides, and alcohols.

hyetal (hī′ĭ-tl) Relating to rain or rainy regions, used especially in regard to isohyetal rainfall maps.

Hygiea (hī-jē′ə) The fourth largest asteroid, having a diameter of about 430 km (267 mi). See more at **asteroid.**

hygrogram (hī′grə-grăm′) The permanent record made by a hygrograph.

hygrograph (hī′grə-grăf′) A hygrometer that records variations in atmospheric humidity.

hygrometer (hī-grŏm′ĭ-tər) Any of several instruments that measure humidity. The most common type of hygrometer consists of two, side-by-side mercury or electronic thermometers, one of which has a dry bulb, and one of which has a bulb wrapped with a wet cotton or linen wick. As water evaporates from the wet bulb, it absorbs heat from the thermometer, driving down its temperature reading. The difference in temperature between the two thermometers is then used to calculate the relative humidity. This type of hygrometer is also called a **psychrometer.** Other hygrometers make use of the temperatures at which dew forms and disappears to calculate the relative humidity. Older hygrometers used the length of a strand of hair, which stretches when it absorbs moisture, to measure relative humidity.

hygroscope (hī′grə-skōp′) An instrument that records changes in atmospheric humidity. Unlike a hygrometer, a hygroscope only indicates a change in relative humidity, without measuring the magnitude of the change.

hygroscopic (hī′grə-skŏp′ĭk) Relating to a compound that easily absorbs moisture from the atmosphere.

hygrostat (hī′grə-stăt′) A device used for regulating the relative humidity of an enclosed space. A hygrostat contains a sensor that detects a preset level of humidity and is connected to a heater that is turned off and on as needed in order to keep the humidity at the preset level. Hygrostats are used in humidifiers and dehumidifiers, and are placed in the ductwork of laboratories and other facilities where climate control is necessary. Also called *humidistat.*

hymen (hī′mən) A mucous membrane that partly closes the opening of the vagina.

hymenium (hī-mē′nē-əm) Plural **hymenia.** The spore-bearing layer of the fruiting body of certain fungi, containing asci or basidia.

hymenopteran (hī′mə-nŏp′tər-ən) Any of various insects of the order Hymenoptera, having two pairs of wings and a characteristic thin constriction that separates the abdomen from the thorax. Some hymenopterans live in complex social groups. Hymenopterans include the ants, bees, wasps, and sawflies.

hypanthium (hī-păn′thē-əm) Plural **hypanthia.** The ringlike, cup-shaped, or tubular structure of a flower on which the sepals, petals, and stamens are borne, as in the flowers of the rose or cherry. It is formed by the enlargement of the receptacle.

Hypatia (hī-pā′shə) 370?–415 CE. Greek philosopher who was the first notable woman mathematician and astronomer. She invented instruments used to view the stars and wrote commentaries on mathematics and astronomy, though none of them survives.

hyper– A prefix that means "excessive" or "excessively," especially in medical terms like *hypertension* and *hyperthyroidism.*

hyperactivity (hī′pər-ăk-tĭv**′**ĭ-tē) **1.** An abnormally high level of activity or excitement shown by a person, especially a child, that interferes with the ability to concentrate or interact with others. **2.** Abnormally high activity in a body part, especially a gland.

hyperbola (hī-pûr**′**bə-lə) Plural **hyperbolas** or **hyperbolae** (hī-pûr**′**bə-lē). A plane curve having two separate parts or branches, formed when two cones that point toward one another are intersected by a plane that is parallel to the axes of the cones.

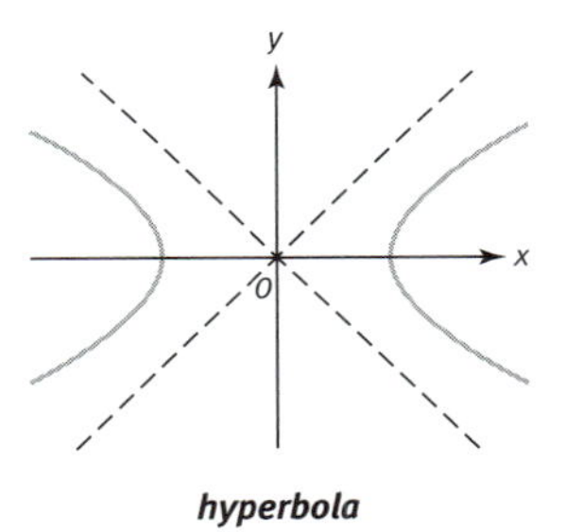

hyperbola

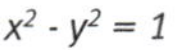
$x^2 - y^2 = 1$

hyperbolic function (hī′pər-bŏl**′**ĭk) Any of a set of six functions related, for a real or complex variable *x,* to the hyperbola in a manner analogous to the relationship of the trigonometric functions to a circle, including: **1.** The **hyperbolic sine,** defined by the equation $\sinh x = \frac{1}{2}(e^x - e^{-x})$. **2.** The **hyperbolic cosine,** defined by the equation $\cosh x = \frac{1}{2}(e^x + e^{-x})$. **3.** The **hyperbolic tangent,** defined by the equation $\tanh x = \frac{\sinh x}{\cosh x}$. **4.** The **hyperbolic cotangent,** defined by the equation $\coth x = \frac{\cosh x}{\sinh x}$. **5.** The **hyperbolic secant,** defined by the equation $\operatorname{sech} x = \frac{1}{\cosh x}$. **6.** The **hyperbolic cosecant,** defined by the equation $\operatorname{csch} x = \frac{1}{\sinh x}$.

hyperboloid (hī-pûr**′**bə-loid′) Either of two surfaces generated by rotating a hyperbola about either of its main axes and having a finite center, with certain plane sections that are hyperbolas and others that are ellipses or circles.

hypercube (hī**′**pər-kyo͞ob′) An object resembling a three dimensional cube but having an arbitrary number of dimensions (typically more than three, although cubes and squares can be considered hypercubes in three and two dimensions). Each corner or node of a hypercube is equidistant from every other. The number of corners in a hypercube is equal to 2^n, where n is the number of dimensions. Diagrams and models of hypercubes of four or more dimensions are not real hypercubes any more than a diagram of a cube is an actual cube, but they do depict the manner in which the corner points are connected. See also **tesseract.**

hypergolic (hī′pər-gŏl**′**ĭk) Relating to or using a rocket propellant consisting of liquid fuel and an oxidizer that ignite spontaneously on contact.

hyperlink (hī**′**pər-lĭngk′) See **link.**

hyperon (hī**′**pə-rŏn′) Any of various baryons, other than the proton and neutron, that do not decay via the strong force. The lambda particle is such an example. Hyperons are unstable or semistable, heavier than protons and neutrons, and have nonzero **strangeness.**

hyperopia (hī′pə-rō**′**pē-ə) A defect of the eye that causes light to focus behind the retina instead of directly on it, resulting in an inability to see near objects clearly. Hyperopia is often caused by a shortened eyeball or a misshapen lens. Also called *farsightedness.* Compare **myopia.**

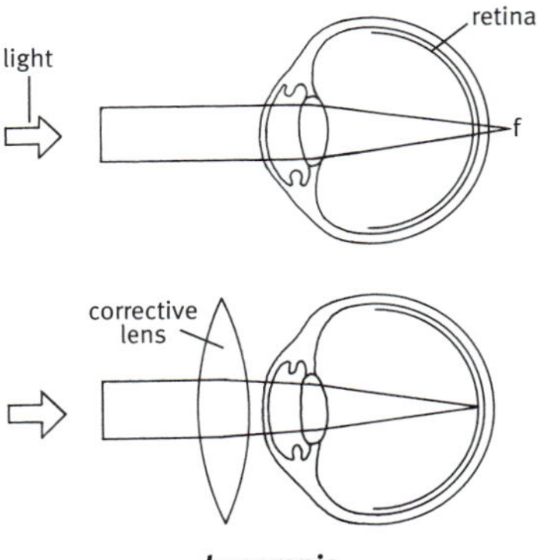

hyperopia

top: *When the focal point* (f) *of light extends beyond the retina, vision is blurred.* bottom: *A convex-shaped lens corrects the defect by focusing light on the retina.*

hypersonic (hī′pər-sŏn**′**ĭk) Relating to or capable of speeds equal to or exceeding five times the speed of sound (Mach 5 and above). Compare **subsonic, supersonic, transonic.**

hypertension (hī′pər-tĕn′shən) Abnormally high blood pressure, especially in the arteries. High blood pressure increases the risk for heart attack and stroke. Also called *high blood pressure.*

hypertext (hī′pər-tĕkst′) A computer-based text retrieval system that enables a user to access particular locations or files in webpages or other electronic documents by clicking on links within specific webpages or documents.

hyperthermia (hī′pər-thûr′mē-ə) An abnormally high body temperature, usually resulting from infection, certain drugs and medications, or head injury. Hyperthermia is sometimes created intentionally to treat diseases, especially some cancers. Compare **hypothermia.**

hyperthyroidism (hī′pər-thī′roi-dĭz′əm) An abnormality of the thyroid gland characterized by excessive production of thyroid hormone, which can result in an increased basal metabolic rate, causing weight loss, heart palpitations, and tremors. Compare **hypothyroidism.**

hypha (hī′fə) Plural **hyphae** (hī′fē). One of the long slender tubes that develop from germinated spores and form the structural parts of the body of a fungus. In many species of fungi, hyphae are divided into sections by cross walls called **septa**. Each section contains at least one haploid nucleus, and the septa usually have perforations that allow cytoplasm to flow through the hypha. A large mass of hyphae is known as a **mycelium**, which is the growing form of most fungi. From time to time, hyphae develop reproductive structures that are partitioned from the hypha by holeless septa. In many species, these structures are microscopic; in others, they are visible and large. Mushrooms and shelf fungi are visible reproductive structures of fungi.

hypnosis (hĭp-nō′sĭs) A trancelike state resembling sleep, usually induced by a therapist by focusing a subject's attention, that heightens the subject's receptivity to suggestion. The uses of hypnosis in medicine and psychology include recovering repressed memories, modifying or eliminating undesirable behavior (such as smoking), and treating certain chronic disorders, such as anxiety.

hypo– or **hyp–** A prefix that means "beneath" or "below," as in *hypodermic,* below the skin. It also means "less than normal," especially in medical terms like *hypoglycemia.* In the names of chemical compounds, it means "at the lowest state of oxidation," as in *sodium hypochlorite.*

hypobaric (hī′pə-băr′ĭk) Relating to conditions of low air pressure and low oxygen content, such as atmospheric conditions at high altitudes, or in special chambers used to establish low-pressure conditions.

hypochlorite (hī′pə-klôr′īt′) A salt or ester of hypochlorous acid, containing the group OCl.

hypochlorous acid (hī′pə-klôr′əs) A weak, unstable acid occurring only in solution and used as a bleach, oxidizer, deodorant, and disinfectant. *Chemical formula:* **HOCl.**

hypochondria (hī′pə-kŏn′drē-ə) A psychiatric disorder characterized by the conviction that one is ill or soon to become ill, often accompanied by physical symptoms, when illness is neither present nor likely. ► A person with hypochondria is called a **hypochondriac.**

hypocotyl (hī′pə-kŏt′l) The part of a plant embryo or seedling that lies between the radicle and the cotyledons. Upon germination, the hypocotyl pushes the cotyledons above the ground to develop. It eventually becomes part of the plant stem. Most seed-bearing plants have hypocotyls, but the grasses have different, specialized structures.

hypocycloid (hī′pō-sī′kloid′) The curve described by a point on the circumference of a circle as the circle rolls on the inside of the circumference of a second, fixed circle.

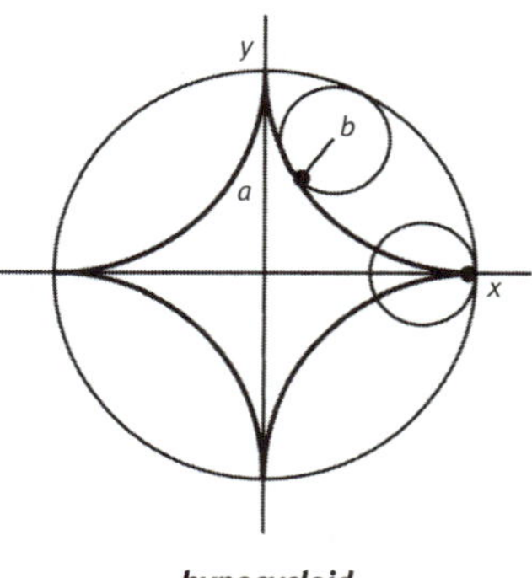

hypocycloid

a = radius of fixed circle; b = radius of rotating circle

hypodermic needle (hī′pə-dûr′mĭk) A hollow needle used in medical syringes to inject fluids into the body or draw fluids from it.

hypogeous (hī′pə-jē′əs) Relating to the germination of a seed in which the cotyledons remain below the surface of the ground. Compare **epigeous.**

hypoglycemia (hī′pō-glī-sē′mē-ə) An abnormally low level of sugar in the blood, most commonly caused by excessive doses of insulin in people with diabetes or by excessive ingestion of alcohol or certain other drugs. It can cause weakness, dizziness, disorientation, and, if prolonged, permanent brain damage.

hypogynous (hī-pŏj′ə-nəs) Having the floral parts, such as sepals, petals, and stamens, borne on the receptacle beneath the ovary. Compare **epigynous, perigynous.**

hypotension (hī′pə-tĕn′shən) Abnormally low blood pressure, especially in the arteries. Also called *low blood pressure.*

hypotenuse (hī-pŏt′n-o͞os′) The side of a right triangle opposite the right angle. It is the longest side, and the square of its length is equal to the sum of the squares of the lengths of the other two sides.

hypothalamus (hī′pō-thăl′ə-məs) The part of the brain in vertebrate animals that lies below the thalamus and cerebrum. The hypothalamus controls the autonomic nervous system and the secretion of hormones by the pituitary gland. Through these nerve and hormone channels, the hypothalamus regulates many vital biological processes, including body temperature, blood pressure, thirst, hunger, and the sleep-wake cycle.

hypothermia (hī′pə-thûr′mē-ə) An abnormally low body temperature, often caused by prolonged exposure to cold. Compare **hyperthermia.**

hypothesis (hī-pŏth′ĭ-sĭs) Plural **hypotheses** (hī-pŏth′ĭ-sēz′). A statement that explains or makes generalizations about a set of facts or principles, usually forming a basis for possible experiments to confirm its viability.

USAGE **hypothesis/law/theory**

The words *hypothesis, law,* and *theory* refer to different kinds of statements, or sets of statements, that scientists make about natural phenomena. A *hypothesis* is a proposition that attempts to explain a set of facts in a unified way. It generally forms the basis of experiments designed to establish its plausibility. Simplicity, elegance, and consistency with previously established hypotheses or laws are also major factors in determining the acceptance of a hypothesis. Though a hypothesis can never be proven true (in fact, hypotheses generally leave some facts unexplained), it can sometimes be verified beyond reasonable doubt in the context of a particular theoretical approach. A scientific *law* is a hypothesis that is assumed to be universally true. A law has good predictive power, allowing a scientist (or engineer) to model a physical system and predict what will happen under various conditions. New hypotheses inconsistent with well-established laws are generally rejected, barring major changes to the approach. An example is the law of conservation of energy, which was firmly established but had to be qualified with the revolutionary advent of quantum mechanics and the uncertainty principle. A *theory* is a set of statements, including laws and hypotheses, that explains a group of observations or phenomena in terms of those laws and hypotheses. A theory thus accounts for a wider variety of events than a law does. Broad acceptance of a theory comes when it has been tested repeatedly on new data and been used to make accurate predictions. Although a theory generally contains hypotheses that are still open to revision, sometimes it is hard to know where the hypothesis ends and the law or theory begins. Albert Einstein's theory of relativity, for example, consists of statements that were originally considered to be hypotheses (and daring at that). But all the hypotheses of relativity have now achieved the authority of scientific laws, and Einstein's theory has supplanted Newton's laws of motion. In some cases, such as the germ theory of infectious disease, a theory becomes so completely accepted it stops being referred to as a theory.

hypothesize (hī-pŏth′ĭ-sīz′) To form a hypothesis.

hypothyroidism (hī′pō-thī′roi-dĭz′əm) An abnormality of the thyroid gland characterized by insufficient production of thyroid hormone, which can result in a decreased basal metabolic rate, causing weight gain and fatigue. Compare **hyperthyroidism.**

hypsography (hĭp-sŏg′rə-fē) **1a.** The scientific study of the Earth's topologic features above sea level, especially the measurement and mapping of land elevations. **b.** A representation or description of these features, as on a map. **2.** See **hypsometry.**

hypsometer (hĭp-sŏm′ĭ-tər) An instrument used to determine land elevation by observing the atmospheric pressure as measured by the change in the boiling point of a liquid, usually water. Liquids boil at progressively lower temperatures as the atmospheric pressure decreases, and since atmospheric pressure decreases with altitude, the temperature at which the liquid boils is an indicator of the atmospheric pressure at that location and hence of the location's altitude.

hypsometry (hĭp-sŏm′ĭ-trē) The measurement of elevation relative to sea level. Also called *hypsography.*

hyracotherium (hī′rə-kō-thîr′ē-əm) Plural **hyracotheria.** A small primitive horse that lived about 50 million years ago during the early Eocene Epoch. It had three or four hoofed toes on each foot and is considered by some to be the ancestor of modern horses. It is sometimes called the "dawn horse," a translation of its earlier scientific name, *Eohippus.*

hysterectomy (hĭs′tə-rĕk′tə-mē) Surgical removal of part or all of the uterus.

hysteresis (hĭs′tə-rē′sĭs) The dependence of the state of a system on the history of its state. For example, the magnetization of a material such as iron depends not only on the magnetic field it is exposed to but on previous exposures to magnetic fields. This "memory" of previous exposure to magnetism is the working principle in audio tape and hard disk devices. Deformations in the shape of substances that last after the deforming force has been removed, as well as phenomena such as **supercooling**, are examples of hysteresis.

Hz Abbreviation of **hertz.**

i (ī) The number whose square is equal to –1. Numbers expressed in terms of *i* are called imaginary or complex numbers.

I 1. The symbol for electric **current. 2.** The symbol for **iodine.**

Ibn al-Haytham (ĭb′ən ĕl-hī′thəm), **Abu 'Ali al-Hasan** Also known as **Alhazen** (ăl-hăz′ən) 965?–1040? Arab mathematician who wrote almost 100 works on mathematics, astronomy, philosophy, and medicine, but who is best known for his book on optics, which became very influential in Europe after it was translated in the 13th century. It contained a detailed description of the eye and disproved the older Greek idea that vision is the result of the eye sending out rays to the object being looked at.

Ibn Sina (ĭb′ən sē′nä), **Hakim Abu Ali al-Husain ibn Abdallah** Also known as **Avicenna.** (ăv′ĭ-sĕn′ə) 980–1037. Persian physician and philosopher whose medical textbook, *The Canon of Medicine,* is a comprehensive medical encyclopedia that remained a standard work in European medical studies until the 17th century.

ibuprofen (ī′byo͞o-prō′fən) An anti-inflammatory drug used to reduce fever or pain.

Icarus (ĭk′ər-əs) A small asteroid with a highly eccentric, Earth-crossing orbit that takes it to within 30 million km (19 million mi) of the Sun, or closer than the planet Mercury. In 1968 Icarus approached within 6 million km (4 million mi) of the Earth. See more at **asteroid.**

ice (īs) **1.** A solid consisting of frozen water. Ice forms at or below a temperature of 0°C (32°F). Ice expands during the process of **freezing**, with the result that its density is lower than that of water. **2.** A solid form of a substance, especially of a substance that is a liquid or a gas at room temperature at sea level on Earth. The nuclei of many comets contain methane ice.

ice age 1. Any of several cold periods during which glaciers covered much of the Earth. **2. Ice Age.** The most recent glacial period, which occurred during the Pleistocene Epoch and ended about 10,000 years ago. During the Pleistocene Ice Age, great sheets of ice up to two miles thick covered most of Greenland, Canada, and the northern United States as well as northern Europe and Russia.

iceberg (īs′bûrg′) A massive body of floating ice that has broken away from a glacier or ice field. Most of an iceberg lies underwater, but because ice is not as dense as water, about one ninth of it remains above the surface.

iceberg

icecap (īs′kăp′) **1.** A dome-shaped body of ice and snow that covers a mountain peak or a large area and spreads out under its own weight. Ice caps have an area of less than 50,000 square km (19,500 square mi). Compare **ice sheet. 2.** A polar cap.

ice field 1. A large expanse of ice covering a mountainous region and consisting of several interconnected glaciers. **2.** An extensive area of ice on the surface of the ocean, consisting of multiple ice floes and covering an area that is greater than 10 km (6.2 mi) across.

ice floe A large, flat expanse of floating ice smaller than a marine ice field.

Iceland spar (īs′lənd) A form of calcite that is transparent and causes light passing through it to refract in two directions (forming a double image of an object seen through it). Iceland spar occurs in perfect rhombohedrons and is used in optical instruments.

ice point The temperature, equal to 0°C (32°F), at which pure water and ice coexist in equilibrium at one atmosphere of pressure.

ice sheet A large sheet of ice and snow that covers an entire region and spreads out under its own weight. Ice sheets have an area of more than 50,000 square km (19,500 square mi). Compare **icecap** (sense 1).

ichthyology (ĭk′thē-ŏl′ə-jē) The scientific study of fish.

ichthyosaur (ĭk′thē-ə-sôr′) Any of various extinct sea reptiles of the genus *Ichthyosaurus* and related genera, that had a medium-sized to large dolphin-like body with a dorsal fin, four flippers, and a large, crescent-shaped tail. The head had a long beak with sharp teeth, large eyes and earbones, and nostrils near the eyes on top of the skull. Ichthyosaurs were most common and diverse in the Triassic and Jurassic Periods and died out well before the end of the Cretaceous.

icon (ī′kŏn′) In a graphical user interface, a picture on the screen that represents a specific file, directory, window, or program. Clicking on an icon will start the associated program or open the associated file, directory, or window.

icosahedron (ī-kō′sə-hē′drən) Plural **icosahedrons** or **icosahedra.** A polyhedron having twenty faces. —*Adjective* **icosahedral.**

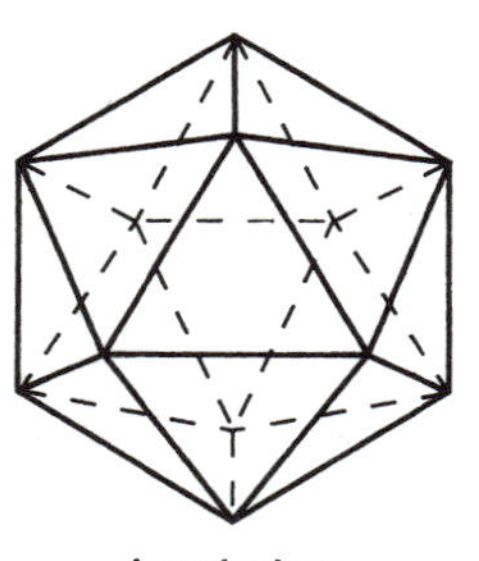

icosahedron

-ide A suffix used to form the names of various chemical compounds, especially the second part of the name of a compound that has two members (such as sodium *chloride*) or the name of a general type of compound (such as *polysaccharide*).

ideal gas (ī-dē′əl) A hypothetical gas whose molecules bounce off each other (and the boundaries of their container) with perfect elasticity and have negligible size, and in which the intermolecular forces acting between molecules not in contact with each other are also negligible. Such a gas would obey the gas laws (such as Charles's law and Boyle's law) exactly at all temperatures and pressures. Most actual gases behave approximately as ideal gases, except at very low temperatures (when the potential energy of their intermolecular forces is high relative to the kinetic energy of the molecules and becomes significant), and under very high pressures (when the molecules are packed so close together that close-range intermolecular forces become significant).

ideal gas law A law that describes the relationships between measurable properties of an ideal gas. The law states that $P \times V = n \times (R) \times T$, where P is pressure, V is volume, n is the number of moles of molecules, T is the absolute temperature, and R is the gas constant (8.314 joules per degree Kelvin or 1.985 calories per degree Celsius). A consequence of this law is that, under constant pressure and temperature conditions, the volume of a gas depends solely on the number of moles of its molecules, not on the type of gas. Also called *universal gas law.* See also **Boyle's law, Charles's law, van der Waal's equation.**

identity element (ī-dĕn′tĭ-tē) The element of a set of numbers that when combined with another number under a particular binary operation leaves the second number unchanged. For example, 0 is the identity element under addition for the real numbers, since for any real number a, $a + 0 = a$, and 1 is the identity element under multiplication for the real numbers, since $a \times 1 = a$.

idiopathic (ĭd′ē-ə-păth′ĭk) Relating to or being a disease having no known cause.

igneous (ĭg′nē-əs) **1.** Relating to rocks or minerals formed by the cooling and hardening of magma or molten lava. Basalt and granite are examples of igneous rocks. **2.** Relating to the processes, such as volcanism, by which such rocks and minerals form.

ignimbrite (ĭg′nĭm-brīt′) A volcanic rock formed by the consolidation of volcanic ash and other material ejected by an explosive volcanic eruption.

ignition point (ĭg-nĭsh′ən) The minimum temperature at which a substance will continue to burn on its own without the application of additional external heat.

ileum (ĭl′ē-əm) Plural **ilea.** The lower part of the small intestine, connecting the jejunum to the cecum of the large intestine. *—Adjective* **ileal.**

ilium (ĭl′ē-əm) Plural **ilia.** The uppermost and widest of the three bones that fuse together to form each of the hipbones. See more at **skeleton.** *—Adjective* **iliac.**

illuminance (ĭ-lo͞o′mə-nəns) The **luminous flux** per unit area at any point on a surface exposed to incident light. It is measured in luxes. Also called *illumination.*

illuviation (ĭ-lo͞o′vē-ā′shən) The deposition of colloids, soluble salts, and suspended mineral particles in a lower soil horizon through the process of eluviation (downward movement) from an upper soil horizon.

ilmenite (ĭl′mə-nīt′) A lustrous black to brownish rhombohedral mineral that is an ore of titanium. Ilmenite occurs in igneous rocks and is one of the principal dark minerals observed in beach sands. *Chemical formula:* $FeTiO_3$.

imaginary number (ĭ-măj′ə-nĕr′ē) A type of complex number in which the multiple of i (the square root of –1) is not equal to zero. Examples of imaginary numbers include $4i$ and $2 - 3i$, but not $3 + 0i$ (which is just 3). See more at **complex number.**

imaging (ĭm′ĭ-jĭng) The creation of visual representations of objects, such as a body parts or celestial bodies, for the purpose of medical diagnosis or data collection, using any of a variety of usually computerized techniques. Within the field of medicine, important imaging technologies include **compuertized axial tomography, magnetic resonance imaging,** and **ultrasonography.**

imago (ĭ-mā′gō) Plural **imagoes** or **imagines** (ĭ-mā′gə-nēz′). An insect in its sexually mature adult stage after metamorphosis. Compare **larva, nymph, pupa.**

imide (ĭm′īd′) A compound derived from ammonia and containing the bivalent NH group combined with a bivalent acid group or two monovalent acid groups. Peptides and proteins are chains of imides formed when two amino acids are joined by a peptide bond.

imido (ĭm′ĭ-dō′) Of or relating to imides or an imide.

imine (ĭm′ēn′, -ĭn, ĭ-mēn′) **1.** A compound derived from ammonia and containing an NH group attached by a double bond to a carbon atom in another group. **2.** The radical CNH, having a double bond between the carbon and nitrogen atoms.

imino (ĭm′ə-nō′) Of or relating to imines or an imine.

immiscible (ĭ-mĭs′ə-bəl) Incapable of being mixed or blended together. Immiscible liquids that are shaken together eventually separate into layers. Oil and water are immiscible. Compare **miscible.**

immittance (ĭ-mĭt′ns) **1.** Electrical impedance or admittance. **2.** In acoustic testing, the ease with which sound travels from one medium to another, as from air to bone.

immune response (ĭ-myo͞on′) A protective response of the body's immune system to an antigen, especially a microorganism or virus that causes disease. The immune response involves the action of lymphocytes that deactivate antigens either by stimulating the production of antibodies (humoral immune response) or by a direct attack on foreign cells (cell-mediated immune response.) An inability to produce a normal immune response results in immunodeficiency diseases such as AIDS. See also **cell-mediated immune response, humoral immune response.**

immune system The body system in humans and other animals that protects the organism by distinguishing foreign tissue and neutralizing potentially pathogenic organisms or substances. The immune system includes organs such as the skin and mucous membranes, which provide an external barrier to infection, cells involved in the immune response, such as lymphocytes, and cell products such as lymphokines. See also **autoimmune disease, immune response.**

immunity (ĭ-myo͞o′nĭ-tē) The protection of the body from a disease caused by an infectious agent, such as a bacterium or virus. Immunity may be natural (that is, inherited) or acquired. See also **acquired immunity.**

immunization (ĭm′yə-nĭ-zā′shən) **1.** The process of inducing **immunity** to an infectious organism or agent in an individual or animal through **vaccination**. **2.** A vaccination that induces immunity. A recommended schedule of immunizations for infants and young children includes vaccines against diphtheria, polio, tetanus, measles, mumps, and rubella. —*Verb* **immunize.**

immunoassay (ĭm′yə-nō-ăs′ā, ĭ-myo͞o′-) A laboratory technique that identifies and quantifies (usually in minute amounts) a protein such as a hormone or an enzyme, based on its ability to act as an antigen or antibody in a chemical reaction.

immunodeficiency (ĭm′yə-nō-dĭ-fĭsh′ən-sē, ĭ-myo͞o′-) The inability to produce a normal immune response, caused by an acquired or inherited disease.

immunoglobulin (ĭm′yə-nō-glŏb′yə-lĭn, ĭ-myo͞o′-) See **antibody.**

immunology (ĭm′yə-nŏl′ə-jē) The scientific study of the structure and function of the immune system.

immunosuppression (ĭm′yə-nō-sə-prĕsh′ən, ĭ-myo͞o′-) Suppression of the body's immune response, as by drugs or radiation, in order to prevent the rejection of grafts or transplants or to treat autoimmune diseases, such as systemic lupus erythematosus.

impedance (ĭm-pēd′ns) A measure of the opposition to the flow of alternating current through a circuit. Impedance is measured in ohms. The **resistance** of a circuit to direct current (also measured in ohms) is generally not the same as its impedance, due to the effects of **capacitance** and **induction** in and among the components of the circuit. See also **impedance matching.**

impedance matching A technique of electric circuit design in which one component provides power to another, and the output circuit of the first component has the same impedance as the input circuit of the second component. Maximum power transfer is achieved when the impedances in both circuits are exactly the same. Impedance matching is important wherever power needs to be transmitted efficiently, as in the design of power lines, transformers, and signal-processing devices such as audio and computer circuits.

imperfect flower (ĭm-pûr′fĭkt) A flower that lacks either stamens or carpels. Compare **perfect flower.** See also **incomplete flower.**

imperfect fungus See **deuteromycete.**

impermeable (ĭm-pûr′mē-ə-bəl) Relating to a material through which substances, such as liquids or gases, cannot pass. Some substances, such as some types of contact lenses, are permeable to gas but impermeable to liquid.

impetigo (ĭm′pĭ-tī′gō) A contagious skin infection caused by staphylococcal or streptococcal bacteria and seen most commonly in children. Impetigo is characterized by superficial pustules that rupture and form thick yellow crusts, usually on the face.

implant *Noun.* (ĭm′plănt′) **1.** Something that is placed, usually surgically, within a living body, as grafted tissue or a medical device, such as a **pacemaker.** —*Verb.* (ĭm-plănt′) **2.** To become attached to and embedded in the maternal uterine lining. Used of a fertilized egg.

imprinting (ĭm′prĭn′tĭng) A rapid learning process by which a newborn or very young animal establishes a behavior pattern of recognition and attraction towards other animals of its own kind, as well as to specific individuals of its species, such as its parents, or to a substitute for these. Ducklings, for example, will imprint upon and follow the first large moving object they observe. In nature, this is usually their mother, but they can be made to imprint upon other moving objects, such as a soccer ball.

improper fraction (ĭm-prŏp′ər) A fraction in which the numerator is greater than or equal to the denominator, such as $\frac{3}{2}$. Compare **proper fraction.**

impulse (ĭm′pŭls′) **1.** A sudden flow of electrical current in one direction. **2.** An electrical signal traveling along the axon of a neuron. Nerve impulses excite or inhibit activity in other neurons or in the tissues of the body, such as muscles and glands. **3.** The change of momentum of a body or physical system over a time interval in classical mechanics, equal to the force applied times the length of the time interval over which it is applied.

in. Abbreviation of **inch.**

In The symbol for **indium.**

inbreeding (ĭn′brē′dĭng) The breeding or mating of related individuals within an isolated or closed group of organisms or people. Inbreeding can result in **inbreeding depres-**

sion. However, in agriculture and animal husbandry, the continued breeding of closely related individuals can help to preserve desirable traits in a stock.

inbreeding depression The loss of vigor and general health that sometimes characterizes organisms that are the product of inbreeding. Compare **hybrid vigor.**

incandescence (ĭn′kən-dĕs**′**əns) The emission of visible light from a substance or object as a result of heating it to a high temperature. The color of the light emitted from solids and liquids is a function of their chemical structure and their temperature; the higher the temperature, the more intense and even the distribution of frequencies is (that is, higher temperatures create brighter and whiter light than lower temperatures). Compare **fluorescence.** See also **blackbody radiation**

incandescent lamp (ĭn′kən-dĕs**′**ənt) A lamp that produces light by heating up a filament of wire inside a bulb with an electric current, causing incandescence. The glass bulb containing the filament is filled with a nonreactive gas, such as argon, to prevent the wire from burning. Compare **fluorescent lamp.**

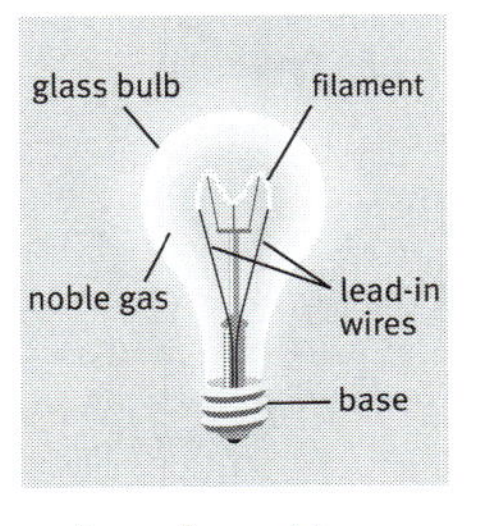

incandescent lamp

inch (ĭnch) A unit of length in the US Customary System equal to $\frac{1}{12}$ of a foot (2.54 centimeters). See Table at **measurement.**

incisor (ĭn-sī**′**zər) A sharp-edged tooth in mammals that is adapted for cutting or gnawing. The incisors are located in the front of the mouth between the canine teeth.

inclination (ĭn′klə-nā**′**shən) A deviation or the degree of deviation from the horizontal or vertical.

inclined plane (ĭn**′**klīnd′) A plane surface, such as a ramp or a blade, set at an acute angle to a horizontal surface, a direction of motion, or a direction of force. Inclined planes are used to increase the distance over which work is done, reducing the amount of force needed to impart energy to a system. Rolling a car up a hill, for example, requires less force than lifting it straight up off the ground. Many tools, such as the ax, wedge, chisel, and highway ramp, exploit the mechanical properties of the inclined plane.

incomplete dominance (ĭn′kəm-plēt**′**) *Genetics.* The fact or condition of being codominant.

incomplete flower A flower lacking one or more of the four parts found in a complete flower: sepals, petals, stamens, and pistils. Compare **complete flower.** See also **imperfect flower.**

incomplete fracture See under **fracture.**

incubation (ĭn′kyə-bā**′**shən) **1.** The act of warming eggs in order to hatch them, as by a bird sitting upon a clutch of eggs in a nest. **2.** The act of keeping an organism, a cell, or cell culture in conditions favorable for growth and development. **3.** The maintenance of an infant, especially one that is ill or born before the usual gestation period, in an environment of controlled temperature, humidity, and oxygen concentration in order to provide optimal conditions for growth and development. **4.** The development of an infection from the time the pathogen enters the body until signs or symptoms first appear. —*Verb* **incubate.**

incubator (ĭn**′**kyə-bā′tər) **1.** An apparatus in which environmental conditions, such as temperature and humidity, can be controlled, often used for growing bacterial cultures, hatching eggs artificially, or providing suitable conditions for a chemical or biological reaction. **2.** An apparatus for maintaining an infant, especially one that is ill or born before the usual gestation period, in an environment of controlled temperature, humidity, and oxygen concentration.

incus (ĭng-kyo͞o**′**dēz) Plural **incudes** (ĭng-kyo͞o**′**dēz). **1.** The anvil-shaped bone (ossicle) that lies between the malleus and the stapes in the middle ear. **2.** The elongated, often anvil-shaped upper portion of a fully developed cumulonimbus cloud; a thunderhead.

indefinite integral (ĭn-dĕf**′**ə-nĭt) A function whose derivative is a given function.

indehiscence (ĭn′dĭ-hĭs**′**əns) The condition of not splitting open spontaneously at maturity to disperse contents. Used of plant structures such as seed pods. Compare **dehiscence.**

indene (ĭn**′**dēn′) A colorless organic liquid obtained from coal tar and used in preparing synthetic resins. *Chemical formula:* $\mathbf{C_9H_8}$.

independent variable (ĭn′dĭ-pĕn**′**dənt) In mathematics, a variable whose value determines the value of other variables. For example, in the formula for the area of a circle, $A = \pi r^2$, r is the independent variable, as its value determines the value of the area (A). Compare **dependent variable.**

indeterminate (ĭn′dĭ-tûr**′**mə-nĭt) Continuing to grow at the apical meristem or the terminal bud indefinitely, allowing for the development of an ever-increasing number of plant organs such as leaves, stems, or flowers to the side.

indeterminate equation An equation having more than one variable and an infinite number of solutions, such as $5x^2 + 3y = 10$.

index fossil (ĭn**′**dĕks′) The fossil remains of an organism that lived in a particular geologic age, used to identify or date the rock or rock layer in which it is found. The best type of index fossils are usually those of swimming or floating organisms that evolved quickly (and therefore did not cover a long span of geologic history) and were able to spread over large areas. Ammonites and graptolites are good index fossils.

index mineral A mineral that forms only under specific pressure and temperature conditions, and that therefore provides information about the metamorphic history of the rock in which it is found. The minerals kyanite, andalusite, and sillimanite are examples of index minerals.

index of refraction A measure of the extent to which a substance slows down light waves passing through it. The index of refraction of a substance is equal to the ratio of the velocity of light in a vacuum to its speed in that substance. Its value determines the extent to which light is refracted when entering or leaving the substance.

indicator (ĭn**′**dĭ-kā′tər) A chemical compound that changes color and structure when exposed to certain conditions and is therefore useful for chemical tests. Litmus, for example, is an indicator that becomes red in the presence of acids and blue in the presence of bases.

indicator species A species whose presence, absence, or relative well-being in a given environment is a sign of the overall health of its ecosystem. By monitoring the condition and behavior of an indicator species, scientists can determine how changes in the environment are likely to affect other species that are more difficult to study. Compare **keystone species.**

indigenous (ĭn-dĭj**′**ə-nəs) Native to a particular region or environment but occurring naturally in other places as well. The American black bear is indigenous to many different parts of North America. Compare **alien, endemic.**

indigestion (ĭn′dĭ-jĕs**′**chən) See **dyspepsia.**

indium (ĭn**′**dē-əm) *Symbol* **In** A soft, malleable, silvery-white metallic element that occurs mainly in ores of zinc and lead. It is used in the manufacture of semiconductors, in bearings for aircraft engines, and as a plating over silver in mirrors. Atomic number 49; atomic weight 114.82; melting point 156.61°C; boiling point 2,080°C; specific gravity 7.31; valence 1, 2, 3. See **Periodic Table.**

indole (ĭn**′**dōl′) **1.** A white crystalline compound obtained from coal tar or various plants and produced by the bacterial decomposition of tryptophan in the intestine. It is used in the perfume industry and as a reagent. *Chemical formula:* $\mathbf{C_8H_7N}$. **2.** Any of various derivatives of this compound.

indricotherium (ĭn′drə-kō-thîr**′**ē-əm) Plural **indricotheria.** A very large, extinct land mammal of the genus *Indricotherium* (formerly *Baluchitherium*) of the Oligocene and Miocene Epochs. It stood 5.5 m (18 ft) high at the shoulder and weighed four times as much as an elephant. It was related to the rhinoceros but had a long neck, long legs, and no horns. The indricotherium is the largest land mammal known to have existed.

induced emission (ĭn-do͞ost) See **stimulated emission.**

induced reaction A chemical reaction caused or accelerated by another simultaneous chemical reaction.

inductance (ĭn-dŭk′təns) A measure of the reaction of electrical components (especially coils) to changes in current flow by creating a magnetic field and inducing a voltage. Its unit is the henry.

induction (ĭn-dŭk′shən) **1a.** The process of deriving general principles from particular facts or instances. **b.** A conclusion reached by this process. See Note at **deduction. 2a.** The creation of a voltage difference across a conductive material (such as a coil of wire) by exposing it to a changing magnetic field. Induction is fundamental to hydroelectric power, in which water-powered turbines spin wire coils through strong magnetic fields. It is also the working principle underlying transformers and induction coils. **b.** The generation of an electric current in a conductor, such as a copper wire, by exposing it to the electric field of an electrically charged conductor. **c.** The building up of a net electric charge on a conductive material by separating its charge to create two oppositely charged regions, then bleeding off the charge from one region.

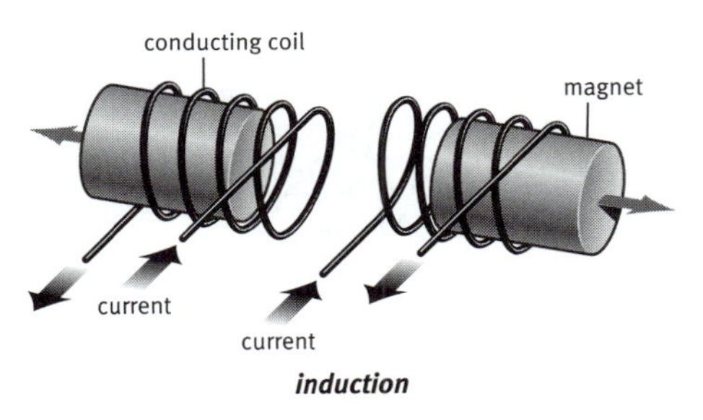

induction

When a magnet moves through a conducting coil, it induces a voltage across the coil that can cause electric current to flow. The direction of the current depends on the direction in which the magnet moves. In the diagram on the left, the current runs from right to left. In the diagram on the right, the current moves from left to right.

induction coil 1. An electrical device consisting of a single coil of conductive material, often surrounding a metallic core, designed to establish a strong magnetic field around the coil. Changes in the current flow through the coil cause fluctuations in the magnetic field that induce a voltage across the coil. Induction coils have many applications, especially in circuits that tune to signals of specific frequencies, as in radios. The ability of an induction coil to induce a voltage is called **inductance**, and is measured in henrys. Compare **capacitor. 2.** A type of transformer that changes a low-voltage direct current to a high-voltage alternating current. Induction coils are used for many purposes, especially as **spark coils** for firing spark plugs in automobile engines and starting oil burners.

inductor (ĭn-dŭk′tər) **1.** An electrical component or circuit, especially an induction coil, that introduces inductance into a circuit. **2.** A substance that causes an induced reaction. Unlike a catalyst, an inductor is irreversibly transformed in the reaction.

indusium (ĭn-do͞o′zē-əm, -zhē-) Plural **indusia. 1.** A thin membrane covering the sorus of a fern. The indusium often shrivels away when spores are ready to be dispersed. Also called *fruitcover.* **2.** A cuplike structure fringed with hairs and located at the top of the style in flowers of the family Goodeniaceae (which includes the garden flowers lobelia and scaevola). Pollen is deposited into the indusium by the anthers of the same flower and, as the style grows, carried up for dispersal by pollinating insects.

inelastic collision (ĭn′ĭ-lăs′tĭk) *Physics.* A collision between bodies in which the total kinetic energy of the bodies is not conserved. In an inelastic collision, the total momentum of the two bodies remains the same, but some of the initial kinetic energy is transformed into heat energy internal to the bodies, used up in deforming the bodies, or radiated away in some other fashion. Inelastic collisions, such as the collision of two balls of clay, tend to result in the slowing and sometimes the joining together of the colliding bodies. Comapre **elastic collision.**

inert (ĭn-ûrt′) Not chemically reactive.

inert gas See **noble gas.**

inertia (ĭ-nûr′shə) The resistance of a body to changes in its momentum. Because of inertia, a body at rest remains at rest, and a body in motion continues moving in a straight line and at a constant speed, unless a force is applied to it. Mass can be considered a measure of a body's inertia. See more at **Newton's laws of motion.** See also **mass.**

inertial force (ĭ-nûr′shəl) An apparent force that appears to affect bodies within a non-inertial frame, but is absent from the point of view of an inertial frame. Centrifugal forces and Coriolis forces, both observed in rotating systems, are inertial forces. Inertial forces are proportional to the body's mass. See also **General Relativity.**

inertial frame A reference frame in which the observers are not subject to any accelerating force. In Special Relativity, time measurements in inertial frames that are not at rest with respect to each other are not equivalent; each inertial frame must have its own time coordinate, the value of which is the time as read off a standard clock at rest in that frame. Also called *inertial frame of reference, inertial reference frame, inertial system.* Compare **non-inertial frame.** See also **Special Relativity.**

infarct (ĭn′färkt′, ĭn-färkt′) An area of living tissue that undergoes necrosis as a result of obstruction of local blood supply, as by a thrombus. See also **heart attack, stroke.**

infauna (ĭn′fô′nə) Benthic animals that live in the substrate of a body of water, especially in a soft sea bottom. Infauna usually construct tubes or burrows and are commonly found in deeper and subtidal waters. Clams, tubeworms, and burrowing crabs are infaunal animals. Compare **epifauna.**

infection (ĭn-fĕk′shən) The invasion of the body of a human or an animal by a pathogen such as a bacterium, fungus, or virus. Infections can be localized, as in **pharyngitis**, or widespread as in **sepsis**, and are often accompanied by fever and an increased number of white blood cells. Individuals with **immunodeficiency** syndromes are predisposed to certain infections. See also **infectious disease, opportunistic infection.**

infectious (ĭn-fĕk′shəs) Capable of causing infection. See Note at **contagious.**

infectious disease A disease caused by a microorganism or other agent, such as a bacterium, fungus, or virus, that enters the body of an organism.

inferior conjunction (ĭn-fîr′ē-ər) See under **conjunction.**

inferior planet Either of the planets Mercury or Venus, whose orbits lie between Earth and the Sun. Because these planets lie in the general direction of the Sun, they can only be seen a few hours before sunrise or after sunset and are always positioned relatively near the horizon, never overhead. Inferior planets go through a complete cycle of phases as viewed from Earth, although their full phase, which occurs on the far side of the Sun, is lost in its glare. Compare **superior planet.** See also **inner planet.**

infertile (ĭn-fûr′tl) **1.** Not capable of reproducing. **2.** Not capable of developing into a complete organism, as infertile eggs. **3.** Relating to soil or land that is not capable of supporting or is unfavorable to the growth of plants.

infertility (ĭn′fər-tĭl′ĭ-tē) The inability to achieve conception after persistent attempts over a given period of time, usually one year in humans.

infinite (ĭn′fə-nĭt) **1.** Relating to a set that can be put into a one-to-one correspondence with some proper subset of its own members. **2.** Relating to or being a numerical quantity describing the size of such a set. **3.** Being without an upper or lower numerical bound.

infinitesimal (ĭn′fĭn-ĭ-tĕs′ə-məl) *Adjective.* **1.** Capable of having values approaching zero as a limit. —*Noun.* **2.** A function or variable continuously approaching zero as a limit.

infinity (ĭn-fĭn′ĭ-tē) A space, extent of time, or quantity that has no limit.

inflammation (ĭn′flə-mā′shən) The reaction of a part of the body to injury or infection, characterized by swelling, heat, redness, and pain. The process includes increased blood flow with an influx of white blood cells and other chemical substances that facilitate healing.

inflammatory bowel disease (ĭn-flăm′ə-tôr′ē) Any of several chronic disorders of the gastrointestinal tract, especially Crohn's disease or an ulcerative form of colitis, characterized by inflammation of the intestine and resulting in abdominal cramping and persistent diarrhea.

inflation theory (ĭn-flā′shən) A theory according to which the universe underwent extremely rapid expansion after an original event called the **big bang**, and has been expanding ever since. The basic homogeneity in the distribution of matter in the universe was established as a consequence of the first phase of inflation. Compare **steady state theory.** See more at **big bang.**

inflorescence (ĭn′flə-rĕs′əns) A group of flowers growing from a common stem, often in a characteristic arrangement. Also called *flower cluster.*

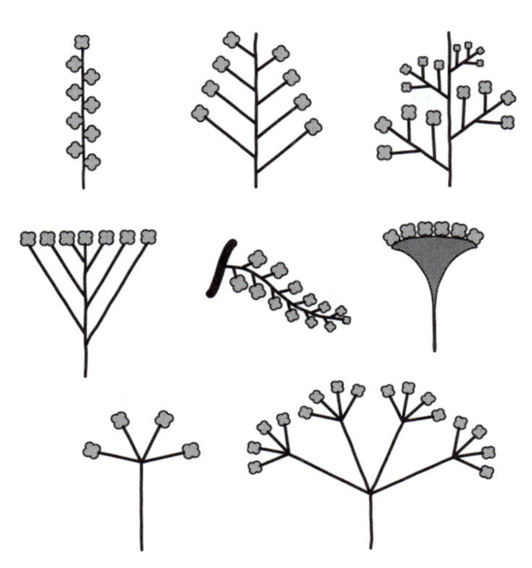

inflorescence

top left to right: *spike, raceme, panicle;* center left to right: *corymb, catkin, head;* bottom left to right: *umbel and compound umbel*

influenza (ĭn′flo͞o-ĕn′zə) A highly contagious infectious disease that is caused by any of various viruses of the family Orthomyxoviridae and is characterized by fever, respiratory symptoms, fatigue, and muscle pain. It commonly occurs in epidemics, one of which killed 20 million people between 1917 and 1919.

WORD HISTORY **influenza**

Since ancient times, influenza has periodically swept the world. Until recently, people could not tell how this illness, which we call the flu, could spread so widely. Before people knew that organisms cause disease, they thought the stars influenced the spread of influenza. *Influenza* comes ultimately from the Latin word *influentia,* meaning "influence of the stars." Today, however, the stars are no longer blamed for the flu. Inhaling influenza viruses causes the spread of the illness.

information science (ĭn′fər-mā′shən) The scientific study of the gathering, manipulation, classification, storage, and retrieval of recorded knowledge. Information science involves the development and analysis of methods of classifying information, as in a library's card catalog, as well as the use of computer systems for archiving information and identifying and retrieving information relevant to specific purposes. Information science also deals with the history of recorded information in various writing systems and media such as papyrus, paper, and microfiche.

information technology The technology involved with the transmission and storage of information, especially the development, installation, implementation, and management of computer systems within companies, universities, and other organizations.

information theory A branch of mathematics that mathematically defines and analyzes the concept of information. Information theory involves statistics and probability theory, and applications include the design of systems that have to do with data transmission, encryption, compression, and other information processing.

infraclass (ĭn′frə-klăs′) A taxonomic category of related organisms ranking below a subclass and containing one or more orders.

infraorder (ĭn′frə-ôr′dər) A taxonomic category of related organisms ranking below a suborder and containing one or more families.

infrared (ĭn′frə-rĕd′) Relating to the invisible part of the electromagnetic spectrum with wavelengths longer than those of visible red light but shorter than those of microwaves. See more at **electromagnetic spectrum.**

A CLOSER LOOK **infrared**

In 1800 the astronomer Sir William Herschel discovered *infrared light* while exploring the relationship between heat and light. Herschel used a prism to split a beam of sunlight into a spectrum and then placed a thermometer in each of the bands of light. When he placed the thermometer just outside the red band, where there was no visible color, the temperature rose, as if light were shining on the thermometer. Further experiment showed that this invisible radiation behaved like visible light in many ways; for example, it could be reflected by a mirror. Infrared radiation is simply electromagnetic radiation with a lower frequency than visible light, having longer wavelengths of 0.7 micrometer to 1 millimeter. Ultraviolet radiation, like infrared radiation, lies just outside the visible part of the spectrum, but with higher frequencies; some animals, such as bees, are capable of seeing such radiation. Both infrared and ultraviolet radiation are often referred to as forms of light, though they cannot be seen by human beings. Heat energy is often transferred in the form of infrared radiation, which is given off from an object as a result of molecular collisions within it. Molecules typically have a characteristic infrared *absorption spectrum,* and infrared spectroscopy is a common technique for identifying the molecular structure of substances. Astronomers similarly analyze the infrared radiation emitted by celestial bodies to determine their temperature and composition.

---

infrared astronomy The study of celestial objects by means of the analysis of the infrared radiation they emit. Because infrared radiation can pass through clouds of interstellar dust or gas that absorb visible light, infrared astronomy enables the study of celestial objects or features that are undetectable by optical telescopes.

infrared telescope A telescope, similar in operation to an optical telescope, that is designed to detect infrared radiation. Because infrared radiation is emitted by warm objects, infrared telescopes need to be shielded from local heat sources, as by chilling them with liquid nitrogen or locating them in polar regions. Many are placed on high mountains or are mounted on balloons or satellites in order to place them above the lower atmosphere, where water vapor absorbs much of the incoming infrared radiation.

infrasound (ĭn**′**frə-sound′) Sound whose wave frequency is too low (under 15–20 hertz) to be heard by humans. Some animals, such as elephants and whales, emit calls at infrasound frequencies. See Note at **ultrasound.** —*Adjective* **infrasonic.**

infraspecific (ĭn′frə-spĭ-sĭf**′**ĭk) Occurring within a species; intraspecific.

inhalation (ĭn′hə-lā**′**shən) The act of taking in breath. Inhalation results from the negative pressure in the lungs caused by contraction of the diaphragm, which causes it to move downwards and to expand the chest cavity. The resulting flow of air into the lungs restores a pressure equal to that of the atmosphere. Also called *inspiration.* Compare **exhalation.**

inheritance (ĭn-hĕr**′**ĭ-təns) The process by which traits or characteristics pass from parents to offspring through the genes.

inhibition (ĭn′hə-bĭsh**′**ən) The blocking or limiting of the activity of an organ, tissue, or cell of the body, caused by the action of a nerve or neuron or by the release of a substance such as a hormone or neurotransmitter. Compare **excitation.**

injection (ĭn-jĕk**′**shən) **1.** A substance that is introduced into a organism, especially by means of a hypodermic syringe, as a liquid into the veins or muscles of the body. **2.** A function that maps each member of one set (the domain) to exactly one member of another set (the range). Compare **bijection, surjection.**

injection well A deep well into which water or pressurized gas is pumped in order to push petroleum resources out of underground reservoirs toward production wells so as to increase their yield.

ink (ĭngk) A dark liquid ejected for protection by most cephalopods, including the octopus and squid. Ink consists of highly concentrated melanin.

inner ear (ĭn**′**ər) The innermost part of the ear in many vertebrate animals, consisting of the cochlea, the semicircular canals, and the vestibule. Sound vibrations are transmitted from the cochlea of the inner ear to the brain by the auditory nerve. The semicircular canals and the vestibule are the body's organs of balance. See more at **ear.**

inner planet Any of the four planets Mercury, Venus, Earth, and Mars, whose orbits lie nearest the Sun. The inner planets are of similar size and have high densities compared to the larger gas giants among the outer planets. They are composed mostly of rock and metal and are relatively slow to rotate, with solid surfaces, no rings, and few moons. Also called *terrestrial planet.* Compare **outer planet.** See also **inferior planet.**

inoculation (ĭ-nŏk′yə-lā′shən) **1.** The introduction of a serum, a vaccine, or an antigenic substance into the body of a person or an animal, especially as a means to produce or boost immunity to a specific disease. **2.** The introduction of a microorganism or an agent of disease into an host organism or a growth medium. —*Verb* **inoculate.**

inorganic (ĭn′ôr-găn′ĭk) **1.** Not involving organisms or the products of their life processes. **2.** Relating to chemical compounds that occur mainly outside of living or once living organisms, such as those in rocks, minerals, and ceramics. Most inorganic compounds lack carbon, such as salt (NaCl) and ammonia (NH_3); a few, such as carbon dioxide (CO_2), do contain it, but never attached to hydrogen atoms as in hydrocarbons. Inorganic molecules tend to have a relatively small number of atoms as compared with organic molecules.

inorganic chemistry The branch of chemistry that deals with inorganic compounds.

inositol (ĭ-nō′sĭ-tôl′, -tōl′, ī-nō′-) Any of nine isomeric alcohols, especially one found in plant and animal tissue and classified as a member of the vitamin B complex. Inositol is necessary for the growth of yeasts and other fungi, and in humans is especially abundant as part of a phospholipid found in the brain. *Chemical formula:* $C_6H_{12}O_6$.

input device (ĭn′po͝ot′) A device, such as a keyboard or a mouse, that is used to enter information into a computer.

insect (ĭn′sĕkt′) Any of very numerous, mostly small arthropods of the class Insecta, having six segmented legs in the adult stage and a body divided into three parts (the head, thorax, and abdomen). The head has a pair of antennae and the thorax usually has one or two pairs of wings. Most insects undergo substantial change in form during development from the young to the adult stage. More than 800,000 species are known, most of them beetles. Other insects include flies, bees, ants, grasshoppers, butterflies, cockroaches, aphids, and silverfish. See Notes at **biomass, bug, entomology.**

insecticide (ĭn-sĕk′tĭ-sīd′) A pesticide used to kill insects. Chlordane and DDT are insecticides. Compare **fungicide, herbicide, rodenticide.**

insectivore (ĭn-sĕk′tə-vôr′) **1.** An animal or plant that feeds mainly on insects. **2.** Any of various small, usually nocturnal mammals of the order Insectivora that feed on insects and other invertebrates. Insectivores have long snouts and resemble rodents, but lack gnawing incisors. Moles, shrews, hedgehogs, and tenrecs are insectivores. —*Adjective* **insectivorous.**

insemination (ĭn-sĕm′ə-nā′shən) The introduction of semen into the reproductive tract of a female either through sexual intercourse or through use of an instrument such as a syringe in the process known as *artificial insemination.*

insolation (ĭn′sō-lā′shən) **1.** The solar radiation striking Earth or another planet. **2.** The rate of delivery of solar radiation per unit of horizontal surface.

insoluble (ĭn-sŏl′yə-bəl) Not capable of being fully dissolved. Fats and oils are insoluble in water.

insomnia (ĭn-sŏm′nē-ə) Chronic inability to fall asleep or remain asleep for an adequate length of time.

inspiration (ĭn′spə-rā′shən) See **inhalation.**

instinct (ĭn′stĭngkt′) An inherited tendency of an organism to behave in a certain way, usually in reaction to its environment and for the purpose of fulfilling a specific need. The development and performance of instinctive behavior does not depend upon the specific details of an individual's **learning** experiences. Instead, instinctive behavior develops in the same way for all individuals of the same species or of the same sex of a species. For example, birds will build the form of nest typical of their species although they may never have seen such a nest being built before. Some butterfly species undertake long migrations to wintering grounds that they have never seen. Behavior in animals often reflects the influence of a combination of instinct and learning. The basic song pattern of many bird species is inherited, but it is often refined by learning from other members of the species. Dogs that naturally seek to gather animals such as sheep or cattle into a group are said to have a herding instinct, but the effective use of this instinct by the dog also requires learning on the dog's part. Instinct, as opposed to **reflex**, is usually used of inherited behavior patterns that are more complex or sometimes involve a degree of interaction with learning processes.

instruction (ĭn-strŭk′shən) A sequence of bits that tells a computer's central processing unit to perform a particular operation. An instruction can also contain data to be used in the operation.

insulator (ĭn′sə-lā′tər) A material or an object that does not easily allow heat, electricity, light, or sound to pass through it. Air, cloth and rubber are good electrical insulators; feathers and wool make good thermal insulators. Compare **conductor.**

insulin (ĭn′sə-lĭn) **1.** A hormone produced in the pancreas that regulates the amount of sugar in the blood by stimulating cells, especially liver and muscle cells, to absorb and metabolize glucose. Insulin also stimulates the conversion of blood glucose into glycogen and fat, which are the body's chief sources of stored carbohydrates. **2.** A drug containing this hormone, obtained from the pancreas of animals or produced synthetically and used to treat diabetes.

insulinlike growth factor (ĭn′sə-lĭn-līk′) See **somatomedin.**

integer (ĭn′tĭ-jər) A positive or negative whole number or zero. The numbers 4, –876, and 5,280 are all integers.

integral (ĭn′tĭ-grəl) *Adjective.* **1.** Relating to, involving, or expressed as an integer or integers. —*Noun.* **2.** See **definite integral, indefinite integral.**

integral calculus The study of integration and its uses, such as in calculating areas that are bounded by curves, volumes that are bounded by surfaces, and solutions to differential equations. Compare **calculus of variations, differential calculus.**

integrand (ĭn′tĭ-grănd′) A function to be integrated.

integrated circuit (ĭn′tĭ-grā′tĭd) A device made of interconnected electronic components, such as transistors and resistors, that are etched or imprinted onto a tiny slice of a semiconducting material, such as silicon or germanium. An integrated circuit smaller than a fingernail can hold millions of circuits. Also called *chip, microchip.*

integration (ĭn′tĭ-grā′shən) In calculus, the process of calculating an integral. Integration is the inverse of differentiation, since integrating a given function results in a function whose derivative is the given function. Integration is used in the calculation of such things as the areas and volumes of irregular shapes and solids. Compare **differentiation.**

integument (ĭn-tĕg′yo͞o-mənt) A natural outer covering of an animal or plant or of one of its parts, such as skin, a shell, or the part of a plant ovule that develops into a seed coat.

inter– A prefix meaning "between" or "among," as in *interplanetary,* located between planets.

intercellular (ĭn′tər-sĕl′yə-lər) Located between or among cells.

intercept (ĭn′tər-sĕpt′) In a Cartesian coordinate system, the coordinate of a point at which a line, curve, or surface intersects a coordinate axis. If a curve intersects the x-axis at (4,0), then 4 is the curve's x-intercept; if the curve intersects the y-axis at (0,2), then 2 is its y-intercept.

interface (ĭn′tər-fās′) **1.** The point of interaction or communication between a computer and any other entity, such as a printer or human operator. **2.** The layout of an application's graphic or textual controls in conjunction with the way the application responds to user activity. See more at **GUI.**

interfascicular cambium (ĭn′tər-fə-sĭk′yə-lər) The cambium arising between the vascular bundles in the stem of a plant. See also **fascicular cambium, vascular bundle.**

interference (ĭn′tər-fîr′əns) **1.** The superposition of two or more waves propagating through a given region. Depending on how the peaks and troughs of the interacting waves coincide with each other, the resulting wave amplitude can be higher or smaller than the amplitudes of the individual waves. ▸ When two waves interact so that they rise and fall together more than half the time, the amplitude of the resulting wave is greater than that of the larger wave. This is called **constructive interference.** ▸ When two waves interact such that they rise and fall together less than half the time, the resulting amplitude is smaller than the amplitude of the stronger wave. This interference is called **destructive interference.** It is possible for two waves of the same magnitude to completely cancel out in destructive interference where their sum is always zero, that is, where their peaks and troughs are perfectly opposed. See more at **wave. 2.** In electronics, the distortion or interruption of one broadcast signal by others.

interference pattern An overall pattern that results when two or more waves interfere with each other, generally showing regions of constructive and of destructive interference. Optical interference patterns are analyzed in devices such as interferometers; the acoustic effect of **beats** is an example of an interference pattern.

interferometer (ĭn′tər-fə-rŏm**′**ĭ-tər) Any of several optical, acoustic, or radio frequency instruments that use interference phenomena between a reference wave and an experimental wave or between two parts of an experimental wave to determine wavelengths and wave velocities, measure very small distances and thicknesses, and calculate indices of refraction.

interferon (ĭn′tər-fîr**′**ŏn′) Any of a group of glycoproteins that are involved in blocking viral replication in newly infected cells and are cytokines that modulate the body's immune response. Alpha interferon is used a treatment for viral hepatitis and certain cancers, such as leukemia. Beta interferon is used as a treatment for some types of multiple sclerosis.

interfluve (ĭn′tər-flo͞ov**′**) The region of higher land between two rivers that are in the same drainage system.

intergalactic space (ĭn′tər-gə-lăk**′**tĭk) See under **space.**

interglacial (ĭn′tər-glā**′**shəl) *Adjective.* **1.** Occurring between glacial epochs. *—Noun.* **2.** A comparatively short period of warmth during an overall period of glaciation. Interglacials are characterized both by the melting of ice and by a change in vegetation.

interior angle (ĭn-tîr**′**ē-ər) **1.** Any of the four angles that are formed inside two straight lines when these lines are intersected by a third straight line. **2.** An angle that is formed by two adjacent sides of a polygon and that is included within the polygon. Compare **exterior angle.**

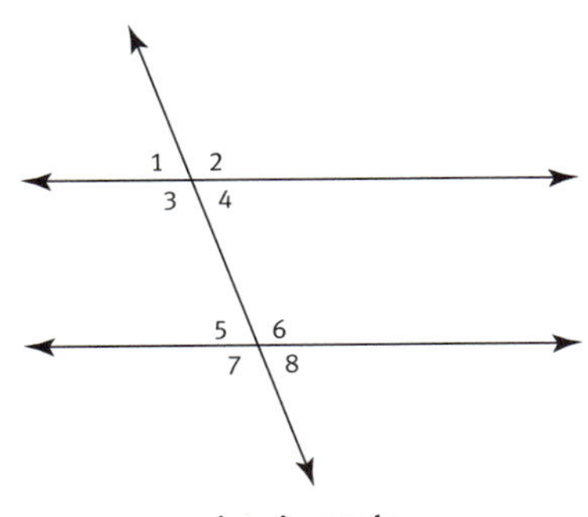

interior angle

Angles 3, 4, 5, and 6 are interior angles. Angles 3 and 6 and angles 4 and 5 are alternate interior angles.

interleukin (ĭn**′**tər-lo͞o′kĭn) Any of a class of cytokines that act to stimulate, regulate, or modulate lymphocytes such as T cells. *Interleukin-1,* which has two subtypes, is released by macrophages and certain other cells, and regulates cell-mediated and humoral immunity. It induces the production of interleukin-2 by helper T cells and also acts as a pyrogen. *Interleukin-2* stimulates the proliferation of helper T cells, stimulates B cell growth and differentiation, and has been used experimentally to treat cancer. *Interleukin-3* is released by mast cells and helper T cells in response to an antigen and stimulates the growth of blood stem cells and lymphoid cells such as macrophages and mast cells. There are many other interleukins that are part of the immune system.

intermediate host (ĭn′tər-mē**′**dē-ĭt) See under **host.**

intermediate vector boson Either of two subatomic particles, the W boson or the Z boson, that mediate the weak nuclear force.

internal-combustion engine (ĭn-tûr**′**nəl-kəm-bŭs**′**chən) An engine whose fuel is burned inside the engine itself rather than in an outside furnace or burner. Gasoline and diesel engines are internal-combustion engines, as are gas turbine engines such as turbojets. Compare **steam engine.**

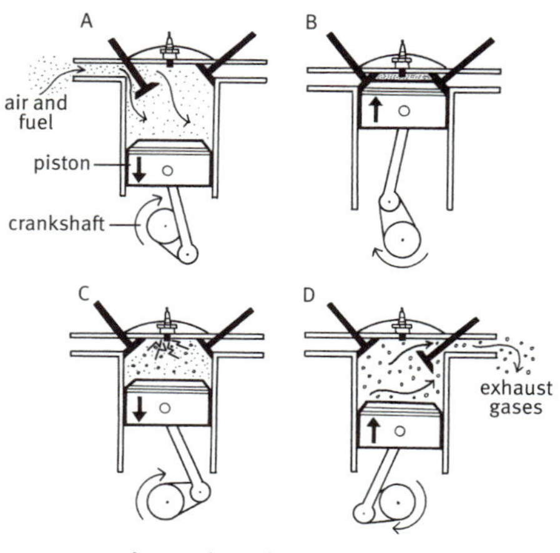

internal-combustion engine

cycles of a four-stroke gasoline-fueled engine: A. intake stroke: *the piston moves down, drawing air and fuel into the cylinder;* B. compression stroke: *the piston moves up, compressing and heating the air and fuel mixture;* C. power stroke: *the hot air and fuel mixture ignites, forcing the piston down;* D. exhaust stroke: *the piston moves up, forcing the exhaust gases out of the cylinder*

internal medicine The branch of medicine that deals with the diagnosis and nonsurgical treatment of diseases in adults.

International Date Line (ĭn′tər-năsh′ə-nəl) An imaginary line on the Earth's surface where each new calendar day begins. The line extends from the North to the South Pole through the Pacific Ocean, roughly along the 180th meridian. The calendar day to the east of the line is one day earlier than it is to the west of the line. The International Date Line was established at the International Meridian Conference in 1884 in order to standardize time, especially for the purpose of travel.

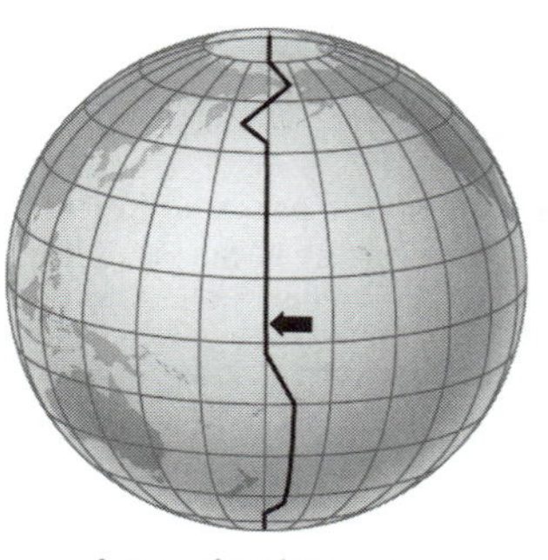

International Date Line

International System The English name for *Système International [d'Unites]*, a decimal system of units used mainly in scientific work, in which the basic quantities are length, mass, time, electric current, temperature, amount of matter, and luminous intensity. In addition, the International System uses two supplementary units to measure plane angles and solid angles. See Table at **measurement.**

international unit A unit for measuring a biologically active substance, such as a hormone or vitamin.

Internet (ĭn′tər-nĕt′) A system connecting computers around the world using **TCP/IP,** which stands for Transmission Control Protocol/Internet Protocol, a set of standards for transmitting and receiving digital data. The Internet consists primarily of the collection of billions of interconnected webpages that are transferred using HTTP (Hypertext Transfer Protocol), and are collectively known as the World Wide Web. The Internet also uses FTP (File Transfer Protocol) to transfer files, and SMTP (Simple Mail Transfer Protocol) to transfer e-mail.

interphase (ĭn′tər-fāz′) The stage of a cell following mitosis or meiosis, during which the nucleus is not dividing. In cells that will undergo further division, the DNA in the nucleus is duplicated in preparation for the next division.

interplanetary space (ĭn′tər-plăn′ĭ-tĕr′ē) See under **space.**

interpreted language (ĭn-tûr′prĭ-tĭd) See under **programming language.**

intersection (ĭn′tər-sĕk′shən) **1.** The point or set of points where one line, surface, or solid crosses another. **2.** The set that contains only those elements shared by two or more sets. The intersection of the sets {3,4,5,6} and {4,6,8,10} is the set {4,6}. The symbol for intersection is ∩. Compare **union.**

interspecific (ĭn′tər-spĭ-sĭf′ĭk) also **interspecies** (ĭn′tər-spē′shēz, -sēz) Arising or occurring between species.

interstellar medium (ĭn′tər-stĕl′ər) Material, mostly hydrogen gas, other gases, and dust, occupying the space between the stars and providing the raw material for the formation of new stars. Nebulae are the most distinct areas of the interstellar medium; they appear when the clouds of gas and dust cluster due to interaction with nearby stars or star remnants.

interstellar space See under **space.**

interstice (ĭn-tûr′stĭs) An opening or space, especially a small or narrow one between mineral grains in a rock or within sediments or soil.

intertidal (ĭn′tər-tīd′l) Relating to the region between the high and low tide marks.

intertropical convergence zone (ĭn′tər-trŏp′ĭ-kəl) A broad area of low atmospheric pressure located in the equatorial region where the northeasterly and southeasterly trade winds converge, extending approximately 10° north and south of the equator. As warm, humid air converges on this zone, it rises and cools, forming clouds and frequent, heavy showers. The **doldrums** occur within the intertropical convergence zone.

intervertebral disk (ĭn′tər-vûr′tə-brəl) A broad disk of cartilage that separates adjacent vertebrae of the spine and acts as a shock absorber during movement.

intestine (ĭn-tĕs′tĭn) The muscular tube that forms the part of the digestive tract extending from the stomach to the anus and consisting of the small and large intestines. In the intestine, nutrients and water from digested food are absorbed and waste products are solidified into feces. See also **large intestine, small intestine.**

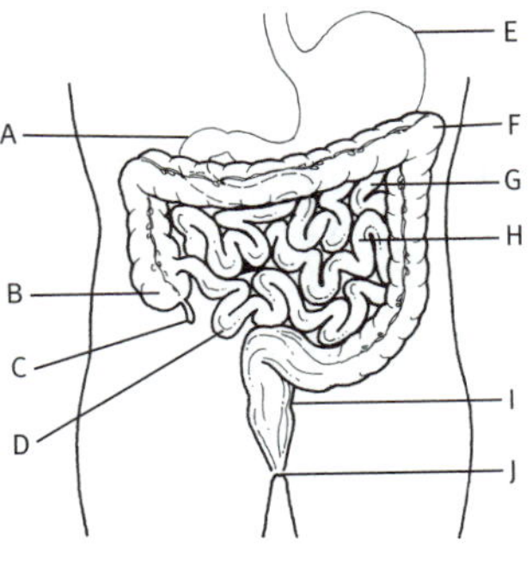

intestine

A. *duodenum*, B. *cecum*, C. *appendix*, D. *ileum*, E. *stomach*, F. *large intestine*, G. *small intestine*, H. *jejunum*, I. *rectum*, J. *anus*

intra– A prefix meaning "inside" or "within," as in *intravenous,* within a vein.

intracellular (ĭn′trə-sĕl′yə-lər) Occurring or situated within a cell or cells.

intranet (ĭn′trə-nĕt′) A privately maintained computer network that only authorized persons can access. Many corporations and institutions, for example, communicate with employees or members through the use of a private intranet.

intraspecific (ĭn′trə-spĭ-sĭf′ĭk) also **intraspecies** (ĭn′trə-spē′shēz, -sēz) Arising or occurring within a species or between members of the same species.

intrauterine device (ĭn′trə-yo͞o′tər-ĭn, -tə-rīn′) A birth control device that is inserted into the uterus to prevent implantation.

intravenous (ĭn′trə-vē′nəs) **1.** Existing or occurring within a vein. **2.** Administered into a vein, as an injection.

intron (ĭn′trŏn) A segment of a gene situated between exons that does not function in coding for protein synthesis. After transcription of a gene to messenger RNA, the transcriptions of introns are removed, and the exons are spliced together by enzymes before translation and assembly of amino acids into proteins. Compare **exon.**

introrse (ĭn′trôrs′) Facing inwards toward the axis around which a flower is arranged. Used of anthers and the direction in which they open to release pollen.

intrusion (ĭn-tro͞o′zhən) The movement of magma through cracks in underground rocks within the Earth, usually in an upward direction. ► Rocks that form from the underground cooling of magma are generally coarse-grained (because they cool slowly so that large crystals have time to grow) and are called **intrusive rocks.** Compare **extrusion.**

invariance (ĭn-vâr′ē-əns) The property of remaining unchanged regardless of changes in the conditions of measurement. For example, the area of a surface remains unchanged if the surface is rotated in space; thus the area exhibits rotational invariance. In physics, invariance is related to **conservation laws**. For example, conservation of angular momentum is directly related to rotational invariance (the laws of physics don't depend on the angle of the reference point), conservation of energy is related to invariance over time (the laws of physics remain the same over time), and conservation of momentum is related to invariance over translations through space (the laws of physics don't depend on the position of the reference point). A form of invariance called **Lorenz invariance** is fundamental to the theory of Special Relativity. Invariance is also called *symmetry.* See also **Noether's theorem.**

invariant mass (ĭn-vâr′ē-ənt) See **rest mass.**

invasive (ĭn-vā′sĭv) **1a.** Relating to a disease or condition that has a tendency to spread, especially a malignant cancer that spreads into healthy tissue. **1b.** Relating to a medical procedure in which a part of the body is entered, as by puncture or incision. **2.** Not native to and tending to spread widely in a habitat or environment. Invasive species often have few natural predators or other biological controls in their new environment. Although not always considered harmful to an environment, invasive species can become agricultural or ecological pests and can displace native species from their habitats. Invasive species are often introduced to an environment unintentionally, as the zebra mussel was to the Great Lakes, but are sometimes introduced for a purpose, as kudzu was to the southern US, where it was originally planted to control erosion.

inverse *Adjective.* (ĭn-vûrs′) **1.** Relating to a mathematical operation whose nature or effect is the opposite of another operation. For example, addition and subtraction are inverse operations, as are multiplication and division. *—Noun.* (ĭn′vûrs′) **2.** An inverse operation. Subtraction is the inverse of addition. **3.** Either of a pair of elements in a set whose result under the mathematical operation of the set is the identity element. For example, the inverse of 5 under multiplication is $\frac{1}{5}$, since $5 \times \frac{1}{5} = 1$, the identity element under multiplication. The inverse of 5 under addition is –5, since 5 + –5 = 0.

inverse-square law The principle in physics that the effect of certain forces on an object varies by the inverse square of the distance between the object and the source of the force. The magnitude of light, sound, and gravity obey this law, as do other quantities. For example, an object placed three feet away from a light source will receive only one ninth ($\frac{1}{32}$, the inverse of 3 squared) as much illumination as an object placed one foot from the light.

inversion (ĭn-vûr′zhən) A departure from the normal effect of altitude on a meteorological property, especially an atmospheric condition in which the air temperature rises with increasing altitude. ► A layer of air that is warmer than the air below it is called an **inversion layer.** Such a layer traps the surface air in place and prevents dispersion of any pollutants it contains.

invertebrate (ĭn-vûr′tə-brĭt, -brāt′) *Adjective.* **1.** Having no backbone or spinal column. *—Noun.* **2.** An animal that has no backbone or spinal column and therefore does not belong to the subphylum Vertebrata of the phylum Chordata. Most animals are invertebrates. Corals, insects, worms, jellyfish, starfish, and snails are invertebrates.

inverter (ĭn-vûr′tər) **1.** An electronic device that reverses the sign of the current or voltage of a signal or power source. Also called *phase inverter.* **2.** An electrical device used to convert direct current into alternating current.

in vitro (ĭn vē′trō) In an artificial environment, such as a test tube. Compare **in vivo.**

in vitro fertilization Fertilization of an egg outside the body of a female by the addition of sperm, as a means of producing a zygote.

in vivo (ĭn vē′vō) Inside a living organism. Compare **in vitro.**

involucre (ĭn′-və-lo͞o′kər) A series of bracts beneath or around a flower or flower cluster. The cupule, the cuplike structure holding an oak acorn, is a modified, woody involucre.

involucre

close-up of the underside of a Gerbera daisy

involuntary (ĭn-vŏl′ən-tĕr′ē) Not under conscious control. Most of the biological processes in animals that are vital to life, such as contraction of the heart, blood flow, breathing, and digestion, are involuntary and controlled by the autonomic nervous system.

involution (ĭn′və-lo͞o′shən) **1.** A mathematical operation, such as negation, which, when applied to itself, returns the original number. **2.** The ingrowth and curling inward of a group of cells, as in the formation of a gastrula from a blastula. **3.** A decrease in size of an organ, as of the uterus following childbirth.

Io (ī′ō, ē′ō) One of the four brightest satellites of Jupiter and the fifth in distance from the planet. It was first sighted by Galileo. See Note at **moon.**

iodide (ī′ə-dīd′) A chemical compound consisting of iodine together with another element or radical.

iodine (ī′ə-dīn′) *Symbol* **I** A shiny, grayish-black element of the halogen group. It is corrosive and poisonous and occurs in very small amounts in nature except for seaweed, in which it is abundant. Iodine compounds are used in medicine, antiseptics, and dyes. Atomic number 53; atomic weight 126.9045; melting point 113.5°C; boiling point 184.35°C; specific gravity (solid, at 20°C) 4.93; valence 1, 3, 5, 7. See **Periodic Table.**

iodize (ī′ə-dīz′) To treat or combine with iodine or an iodide. Salt is iodized to prevent goiter.

ion (ī′ən, ī′ŏn′) An atom or a group of atoms that has an electric charge. Positive ions, or cations, are formed by the loss of electrons; negative ions, or anions, are formed by the gain of electrons.

ion engine A rocket engine that develops thrust by expelling ions fired by an electron gun, used in some spacecraft. Though ion engines create less thrust than chemical combustion engines, they are considerably more efficient and require very little propellant, resulting in much lighter weight and less need for high thrust. See also **solar-electric propulsion.**

ion exchange A reversible chemical reaction between two substances (usually a relatively insoluble solid and a solution) during which ions of equal charge may be interchanged. Ion exchange is used in water softening and purification, chemical analysis, the separation of radioactive isotopes, and kidney dialysis. ► **Cation exchange** is the exchange of positively charged ions. ► **Anion exchange** is the exchange of negatively charged ions.

ionic bond (ī-ŏn′ĭk) A chemical bond formed between two ions with opposite charges. Ionic bonds form when one atom gives up one or more electrons to another atom. These bonds can form between a pair of atoms or between molecules and are the type of bond found in salts.

ionic propulsion Propulsion by the **reactive thrust** of a high-speed beam of similarly charged ions ejected by an **ion engine**. See also **solar-electric propulsion.**

ionize (ī′ə-nīz′) **1.** To give an atom or group of atoms a net electric charge by adding or removing one or more electrons. **2.** To form ions in a substance. Lightning ionizes air, for example.

ionosphere (ī-ŏn′ə-sfîr′) A region of the Earth's upper atmosphere, extending from a height of 70 km (43 mi) to 400 km (248 mi) and containing atoms that have been ionized by radiation from the Sun. The ionosphere lies mostly in the lower thermosphere and is subdivided into three regions, the D region (70 km to 90 km; 43 to 56 mi), the E region (90 km to 150 km; 56 to 93 mi), and the F region (150 km to 400 km; 93 to 248 mi). The concentration of ionized atoms is lowest in the D region, intermediate in the E region, and highest in the F region. The ionosphere is useful for radio transmission because radio waves, which normally propagate in straight lines, are reflected off the ionized gas particles, thereby being transmitted long distances across the Earth's curved surface. See more at **D region, E region, F region.**

ion trap A device, such as a magnet, used to prevent ions in an electron beam from striking another apparatus, such as a cathode-ray tube screen, which could cause it to lose luminescence.

IP Short for *Internet Protocol.* A protocol that specifies the way data is broken into packets and the way those packets are addressed for transmission. See more at **TCP/IP.**

IP address The numerical sequence that serves as an identifier for an Internet server. An IP address appears as a series of four groups of numbers separated by dots. The first group is a number between 1 and 255 and the other groups are a number between 0 and 255, such as 192.135.174.1. Every server has its own unique address.

Ir The symbol for **iridium.**

iridium (ĭ-rĭd′ē-əm) *Symbol* **Ir** A rare, whitish-yellow element that is the most corrosion-resistant metal known. It is very dense, hard, and brittle, and is is used to make hard alloys of platinum for jewelry, pen points, and electrical contacts. Atomic number 77; atomic weight 192.2; melting point 2,410°C; boiling point 4,130°C; specific gravity 22.42 (at 17°C); valence 3, 4. See **Periodic Table.**

A CLOSER LOOK **iridium**

In 1978 geologist Walter Alvarez discovered a high concentration of *iridium* in a layer of clay that had formed between the Mesozoic and Cenozoic eras, a period about 65 million years ago during which dinosaurs and many other organisms became extinct. This finding was significant as iridium is rare at Earth's surface (an unusually high concentration is called an *iridium anomaly*). Most surface iridium is thought to come from dust created when meteors disintegrate in the atmosphere or collide with Earth. Alvarez's father, the physicist Luis Alvarez, suggested that the iridium might have come from the impact of a meteor about 10 km (6.2 mi) across. Such an impact would have caused an enormous explosion, sending huge clouds of dust into the atmosphere. The dust, blocking out the Sun and causing extensive acid rain, would have triggered a worldwide ecological disaster. Many scientists think that such a disaster caused the extinction of the dinosaurs and at least 70 percent of all other species alive at the time, including most of Earth's land plants. Geologists have since found iridium deposits in rocks of a similar age in more than 100 places worldwide. Scientists in the early 1990s identified a large impact crater in the Yucatán peninsula of central Mexico that is the same age as the iridium deposit found by Alvarez. It is 200 km (125 mi) wide and may have been caused by the same impact.

iris (ī′rĭs) Plural **irises** or **irides** (ī′rĭ-dēz′, ĭr′ĭ-). The colored, muscular ring around the pupil of the eye in vertebrate animals, located between the cornea and lens. Contraction and expansion of the iris controls the size of the pupil, thereby regulating the amount of light reaching the retina.

iron (ī′ərn) *Symbol* **Fe** A silvery-white, hard metallic element that occurs abundantly in minerals such as hematite, magnetite, pyrite, and ilmenite. It is malleable and ductile, can be magnetized, and rusts readily in moist air. It is used to make steel and other alloys important in construction and manufacturing. Iron is a component of hemoglobin, which allows red blood cells to carry oxygen and carbon dioxide through the body. Atomic number 26; atomic weight 55.845; melting point 1,535°C; boiling point 2,750°C; specific gravity 7.874 (at 20°C); valence 2, 3, 4, 6. See **Periodic Table.** See Note at **element.**

Iron Age The period in cultural development succeeding the Bronze Age in Asia, Europe, and Africa, characterized by the introduction of iron metallurgy. In southeastern Europe and the Middle East the beginning of the Iron Age is generally dated to around 1200 BCE, with later dates for other parts of Europe and the other continents. Although not as hard or durable as bronze, iron is a more abundant resource, and the Iron Age saw a rapid expansion of metalworking wherever the technology was introduced. See Note at **Three Age system.**

iron meteorite See under **meteorite.**

iron oxide Any of various oxides of iron, such as ferric oxide or ferrous oxide.

irradiate (ĭ-rā′dē-āt′) To expose to or treat with radiation. For example, meat sold as food is often irradiated with x-rays or other radiation to kill bacteria; uranium 238 can be irradiated with neutrons to create fissionable plutonium 239.

irrational number (ĭ-răsh′ə-nəl) A real number that cannot be expressed as a ratio between two integers. If written in decimal notation, an irrational number would have an infinite number of digits to the right of the decimal point, without repetition. Pi and the square root of 2 ($\sqrt{2}$) are irrational numbers.

irregular flower (ĭ-rĕg′yə-lər) A flower in which one or more members of a whorl, or of several floral whorls, differ in form from other members. Irregular flowers, such as those of the violet or the pea, are often bilaterally symmetric. The pea has one large upper petal above, two free petals on the each side, and two petals fused together in a keel shape below. Compare **regular flower.**

irregular galaxy A galaxy that does not have the clearly defined shape and structure of typical elliptical, lenticular, or spiral galaxies. Irregular galaxies typically contain large amounts of gas and dust, and their stars are often young. They account for only a small percentage of known galaxies. Some irregular galaxies are the result of gravitational interactions or collisions between formerly regular galaxies. Many irregular galaxies orbit larger regular ones; the Magellanic Cloud galaxies orbiting the Milky Way are examples. Compare **elliptical galaxy, lenticular galaxy, spiral galaxy.** See more at **Hubble classification system.**

irritable bowel syndrome (ĭr′ĭ-tə-bəl) A gastrointestinal disorder without demonstrable organic pathology, characterized by abdominal cramping, constipation, diarrhea, and mucus in the stool.

isallobar (ī-săl′ə-bär′) A line on a weather map connecting places having equal changes in atmospheric pressure within a given period of time. Isallobars measure pressure tendency, a value that describes both the difference in barometric pressure at the beginning and end of a given time period as well as the pattern—falling, rising, or steady—of pressure change.

ischium (ĭs′kē-əm) Plural **ischia.** The lowest of the three major bones that constitute each half of the pelvis, distinct at birth but later becoming fused with the ilium and pubis.

isinglass (ī′zən-glăs′, ī′zĭng-) **1.** A transparent, almost pure gelatin prepared from the inner membrane of the swim bladder of the sturgeon and certain other fishes. It is used as an adhesive and a clarifying agent. **2.** Mica, especially in the form of the mineral muscovite.

island (ī′lənd) A land mass, especially one smaller than a continent, entirely surrounded by water.

island arc A usually curved chain of volcanic islands bounded on the convex side by a deep oceanic trench. Island arcs form in the overriding tectonic plates of subduction zones as the result of rising melt from the downgoing plate. The arcs are curved because of the curvature of the Earth. The Aleutian Islands, in Alaska, are an island arc. An island arc is a kind of *volcanic arc.*

islets of Langerhans (ī′lĭts əv läng′ər-häns′) Irregular clusters of endocrine cells that are scattered throughout the tissue of the pancreas. The islets of Langerhans contain cells that produce and secrete the hormones insulin and glucagon.

ISO (ī′ĕs-ō′) An organization, the International Organization for Standardization, that sets standards in many businesses and technologies, including computing and communications. The term ISO is not an abbreviation, but instead derives from the Greek word *īsos,* meaning *equal.*

iso– **1.** A prefix that means "equal," as in *isometric,* "having equal measurements." **2.** A prefix used to indicate an isomer of an organic compound, especially a branched isomer of a compound that normally consists of a straight chain. The isomer is characterized by a Y-shaped branch at the end of the chain that consists of two "prongs". Each prong consists of one carbon atom. Thus *isopentane* contains five carbon atoms like normal pentane, but arranged as a chain of three carbons plus a Y-shaped branch of two carbons at the end.

isobar (ī′sə-bär′) A line drawn on a weather map connecting places having the same atmospheric pressure. The distance between isobars indicates the barometric gradient (the degree of change in atmospheric pressure) across the region shown on the map. When the lines are close together, a strong pressure gradient is indicated, creating conditions for strong winds. When the lines are far apart, a weak pressure gradient is indicated and calm weather is forecast.

isocline (ī′sə-klīn′) A geologic fold that has two parallel limbs. See illustration at **fold.**

isoclinic line (ī′sə-klĭn′ĭk) A line on a map connecting points of equal magnetic inclination.

isodynamic line (ī′sō-dī-năm′ĭk) A line on a map connecting points on the Earth where the strength of the Earth's magnetic field is the same.

isogamy (ī-sŏg′ə-mē) A reproductive system characterized by the fusion of gametes (called **isogametes**) that are not distinguished in size or structure. Isogamy is found in some algae and other protist groups. Compare **heterogamy.** —*Adjective* **isogamous.**

isogonic line (ī′sə-gŏn′ĭk) A line on a map connecting points of equal magnetic declination.

isograd (ī′sə-grăd′) A line on a map connecting points on the Earth where metamorphism of rocks occurred under the same pressure and temperature conditions. These lines are established on the basis of the distribution of index minerals and are useful in reconstructing the tectonic history of a given region.

isogram (ī′sə-grăm′) See **isoline.**

isohel (ī′sō-hĕl′) A line drawn on a weather map connecting points that receive equal amounts of sunlight.

isohyet (ī′sō-hī′ĭt) A line drawn on a weather map connecting points that receive equal amounts of precipitation during a given period of time.

isolated system (ī′sə-lā′tĭd) See **closed system.**

isoleucine (ī′sə-lo͞o′sēn′) An essential amino acid. *Chemical formula:* $C_6H_{13}NO_2$. See more at **amino acid.**

isoline (ī′sə-līn′) A line on a map, chart, or graph connecting points of equal value.

isomer (ī′sə-mər) Any of two or more compounds, such as lactose and sucrose, composed of the same elements in the same proportions but differing in structure and other properties. There are two types of isomers, structural isomers and stereoisomers.

isometric (ī′sə-mĕt′rĭk) *Adjective.* **1.** See **cubic** (sense 3). **2.** Of or involving muscular contraction against resistance in which the length of the muscle remains the same. **3.** A graph showing the relationship between two quantities, such as pressure and temperature, when a third quantity, such as volume, is held constant.

isometry (ī-sŏm′ĭ-trē) **1.** Equality of measure. **2.** Equality of elevation above sea level. **3.** A function between two metric spaces (such as two coordinate systems) which preserves distances. A rotation or translation in a plane is an isometry, since the distances between two points on the plane remain the same after the rotation or translation.

isomorphism (ī′sə-môr′fĭz′əm) **1.** Similarity in form, as in organisms of different ancestry. **2.** A one-to-one correspondence between the elements of two sets such that the result of an operation on elements of one set corresponds to the result of the analogous operation on their images in the other set. **3.** A close similarity in the crystalline structure of two or more substances of different chemical composition. Isomorphism is seen, for example, in the group of minerals known as garnets, which can vary in chemical composition but always have the same crystal structure.

isopod (ī′sə-pŏd′) Any of numerous mostly small crustaceans of the order Isopoda, characterized by a flattened body and a series of wide, armorlike plates covering the back. Isopods include the sow bugs, pill bugs, and gribbles.

isoprene (ī′sə-prēn′) A colorless, volatile liquid obtained from petroleum or coal tar and occurring naturally in many plants. It is used chiefly to make synthetic rubber. The isoprene in plants occurs in the chloroplasts and is used to build terpenes and other biologically important chemicals. *Chemical formula:* C_5H_8.

isopropyl alcohol (ī′sə-prō′pəl) A clear, colorless, flammable, mobile liquid that is one of the two isomers of propyl alcohol. It is used in antifreeze compounds, rubbing alcohol, lotions, and cosmetics, and also as a solvent for gums, shellac, and essential oils. *Chemical formula:* C_3H_8O.

isopteran (ī-sŏp′tə-rən) See **termite.**

isosceles (ī-sŏs′ə-lēz′) Of or relating to a geometric figure having at least two sides of equal length.

isospin (ī′sə-spĭn′) A vector quantity of quantum systems (especially subatomic particles) that is conserved in strong interactions and is used to characterize particles that are affected equally by the strong force but have other dissimilar properties, such as different electric charges. For example, protons and neutrons, which are indistinguishable with respect to strong force interactions, have the same isospin. Also called *isotopic spin.* See also **baryon number, strangeness.**

isostasy (ī-sŏs′tə-sē) Equilibrium in the Earth's crust, in which an elevated part in one area is counterbalanced by a depressed part in another. Isostasy exists because the Earth's crust is relatively light compared to the denser mantle over which it lies, and therefore behaves as if it is floating. Areas of the Earth's crust rise or subside to accommodate added load (as from a glacier) or diminished load (as from erosion), so that the forces that elevate landmasses balance the forces that depress them. —*Adjective* **isostatic** (ī′sō-stăt′ĭk).

isotherm (ī′sə-thûrm′) A line drawn on a weather map connecting points that have the same temperature. Each point can mark one temperature reading or an average of several readings.

isothermal (ī′sə-thûr′məl) **1.** Relating to or indicating equal or constant temperatures. **2.** Relating to a process, usually changes of pressure and volume, occurring at a constant temperature and following Boyle's Law. **3.** Relating to an isotherm.

isotope (ī′sə-tōp′) One of two or more atoms that have the same atomic number (the same number of protons) but a different number of neutrons. Carbon 12, the most common form of carbon, has six protons and six neutrons, whereas carbon 14 has six protons and eight neutrons. Isotopes of a given element typically behave alike chemically. With the exception of hydrogen, elements found on Earth generally have the same number of protons and neutrons; heavier and lighter isotopes (with more or fewer neutrons) are often unstable and undergo **radioactive decay**.

isotopic spin (ī′sə-tŏp′ĭk) See **isospin.**

isotropic (ī′sə-trō′pĭk, -trŏp′ĭk) Identical in all directions; invariant with respect to direction. For example, isotropic scattering of light by a substance entails that the intensity of light radiated is the same in all directions. Compare **anisotropic.**

isthmus (ĭs′məs) Plural **isthmuses** or **isthmi** (ĭs′mī′). A narrow strip of land connecting two larger masses of land.

–ite **1.** A suffix used to form the names of minerals, such as *hematite* and *malachite.* **2.** A suffix used to form the name of a salt or ester of a specified acid whose name ends in *–ous.* Such salts or esters have one oxygen atom fewer than corresponding salts or esters with names ending in *–ate.* For example, a *nitrite* is a salt of nitrous acid and contains the group NO_2, while a nitrate contains NO_3. Compare **–ate.**

–itis A suffix meaning "inflammation," as in *bronchitis,* inflammation of the bronchial tubes.

ivory (ī′və-rē) The hard, smooth, yellowish-white substance forming the teeth and tusks of certain animals, such as the tusks of elephants and walruses and the teeth of certain whales. Ivory is composed of dentin.

J

J Abbreviation of **joule.**

Jacob (zhä-kôb′), **François** Born 1920. French geneticist who studied how genes control cellular activity by directing the synthesis of proteins. With Jacques Monod, he theorized that there are genes that regulate the activity of other, neighboring genes. They also proposed the existence of messenger RNA.

jade (jād) A hard gemstone that is pale green or white and consists either of the mineral jadeite (a pyroxene) or the mineral nephrite (an amphibole). It usually forms within metamorphic rocks.

jadeite (jād′īt′) A rare, usually green mineral of the pyroxene group. Jadeite can also occur in white, auburn, buff, or violet varieties. The most highly valued form of jade consists of jadeite. *Chemical formula:* $\mathbf{NaAlSi_2O_6}$.

Japan Current (jə-păn′) A branch of the North Equatorial Current that flows northeast from the Philippine Sea, north along the eastern shore of Taiwan, past southeastern Japan, and onward to the North Pacific. Its warm air moderates the climate of Taiwan and Japan. Also called *Kuroshio Current.* See more at **equatorial current, gyre.**

jasper (jăs′pər) A reddish, brown, or yellow variety of chert. Jasper usually occurs in association with iron ores and contains iron impurities that give it its color.

jaundice (jôn′dĭs) Yellowish discoloration of the whites of the eyes, skin, or mucous membranes caused by the deposition of bile salts in these tissues, occurring as a sign of disorders that interfere with normal metabolism or transport of bile. Liver diseases such as **hepatitis** commonly cause jaundice.

Java man (jä′və) See **pithecanthropus.**

jaw (jô) **1.** Either of two bony or cartilaginous structures that in most vertebrate animals form the framework of the mouth, hold the teeth, and are used for biting and chewing food. The lower, movable part of the jaw is the mandible. The upper, fixed part is the maxilla. **2.** Any of various structures of invertebrate animals, such as the pincers of spiders or mites, that function similarly to the jaws of vertebrates.

jawless fish (jô′lĭs) Any of various primitive fish of the class Agnatha that lack jaws. Living jawless fish (lampreys and hagfish) have a long, cylindrical body and a cartilaginous skeleton. The numerous extinct species were often heavily armored and are among the earliest vertebrate fossils known. Jawless fish occupy an intermediate evolutionary position between primitive chordates and jawed fish. Also called *agnathan.* See more at **cyclostome.** Compare **bony fish, cartilaginous fish.**

jejunum (jə-jo͞o′nəm) Plural **jejuna.** The middle part of the small intestine, connecting the duodenum and the ileum.

Jenner (jĕn′ər), **Edward** 1749–1823. British physician who pioneered the practice of vaccination. His experiments proved that individuals who had been inoculated with the virus that caused cowpox, a mild skin disease of cattle, became immune to smallpox. Jenner's discovery laid the foundations for the science of immunology.

Edward Jenner

BIOGRAPHY **Edward Jenner**

In 1980 the World Health Organization declared that the deadly disease smallpox had been eradicated, an accomplishment attributed to the success of the smallpox vaccine. The vaccine had been developed almost 200 years earlier by the British physician Edward Jenner, who had based his work on a piece of folk wisdom from the countryside that few doctors had taken seriously: people who caught cowpox, a mild viral infection of cattle, never got smallpox. In 1796 Jenner proved the truth of this scientifically in a famous experiment he conducted on an eight-year-old boy named James Phipps. Jenner exposed Phipps to a person with cowpox, then two months later exposed him to smallpox (this would be considered unethical by today's standards). As Jenner expected, the boy warded off the smallpox without any complications. Prior to this, there existed a form of vaccination against smallpox that consisted of exposing people to a mild form of the disease. Although this method often worked, it was risky, and the exposed person sometimes died. Jenner, who devised the word *vaccination* from the Latin *vacca,* for "cow," is considered to be the father of immunology. He also did significant research on heart disease.

Jensen (yĕn′zən), **(Johannes) Hans Daniel** 1907–1973. German physicist who, with Maria Goeppert-Mayer, developed a model of the atomic nucleus that explained why certain nuclei were stable and had an unusual number of stable isotopes. For this work, Jensen and Goeppert-Mayer shared the 1963 Nobel Prize for physics with American physicist Eugene Wigner.

jet (jĕt) **1.** A rapid stream of liquid or gas forced through a small opening or nozzle under pressure. **2.** An aircraft or other vehicle propelled by one or more jet engines. **3.** A jet engine.

jet engine An engine, generally a **gas turbine**, that develops thrust from a jet of hot gases produced by burning fuel in a combustion chamber.

jet lag A temporary disruption of normal circadian rhythm caused by high-speed travel across several time zones typically in a jet aircraft, resulting in fatigue, disorientation, and disturbed sleep patterns.

jet propulsion 1. The driving of an aircraft by the powerful thrust developed when a jet of gas is forced out of a jet engine. **2.** Propulsion by means of any fluid that is forced out in a stream in the opposite direction. Squids, octopuses, and cuttlefish, for example, jet their way through the ocean by taking in and then quickly expelling water.

jet stream A narrow current of strong wind circling the Earth from west to east at altitudes of about 11 to 13 km (7 to 8 mi) above sea level. There are usually four distinct jet streams, two each in the Northern and Southern hemispheres. Jet stream wind speeds average 56 km (34 mi) per hour in the summer and 120 km (74 mi) in the winter. They are caused by significant differences in the temperatures of adjacent air masses. These differences occur where cold, polar air meets warmer, equatorial air, especially in the latitudes of the westerlies.

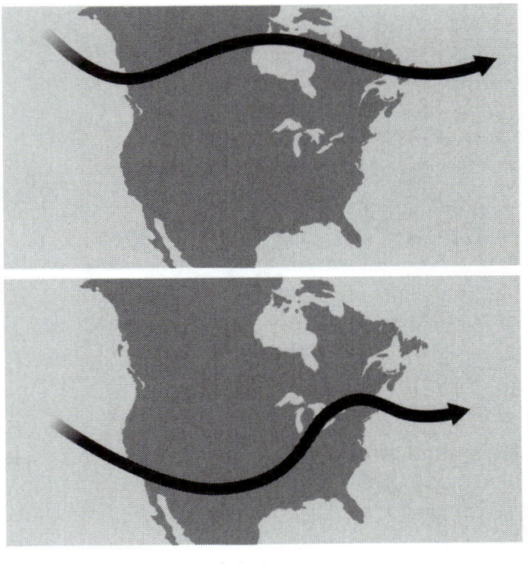

jet stream

summer (top) *and winter* (bottom) *jet streams across North America*

joint (joint) **1.** *Anatomy.* A usually movable body part in which adjacent bones are joined

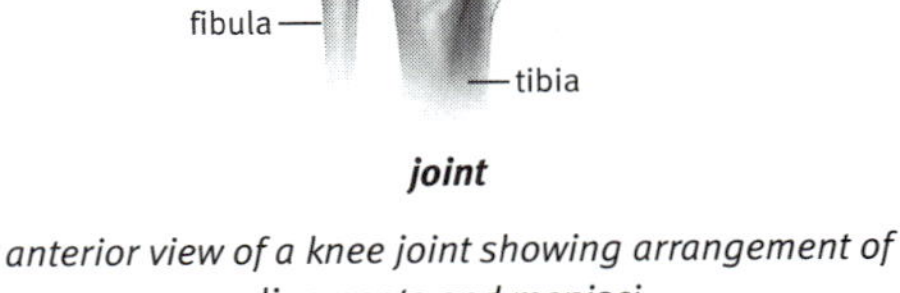

joint

anterior view of a knee joint showing arrangement of ligaments and menisci

by ligaments and other fibrous tissues. See also **ball-and-socket joint, hinge joint. 2.** *Zoology.* A point in the exoskeleton of an invertebrate at which movable parts join, as along the leg of an arthropod. **3.** *Botany.* A point on a plant stem from which a leaf or branch grows.

Joliot-Curie (zhô-lyō′kyŏŏr**′**ē), **Irène** 1897–1956. French physicist who with her husband, **Frédéric Joliot-Curie** (1900–1958), made the first artificial radioactive isotope. They also contributed to the development of nuclear reactors.

Joliot-Curie

Irène (left) *and Frédéric*

Josephson effect (jō**′**zəf-sən) An effect in which electron pairs undergo **quantum tunneling** with zero resistance across a barrier separating two superconductors. The effect can be manipulated by varying a magnetic field at the junction where the tunneling occurs and is being investigated as a possible part of the design of high-speed switches for computer microprocessors. The Josephson effect is named after its discoverer, Welsh physicist Brian David Josephson (b. 1940).

joule (jo͞ol, joul) The SI derived unit used to measure energy or work. One joule is equal to the energy used to accelerate a body with a mass of one kilogram using one newton of force over a distance of one meter. One joule is also equivalent to one watt-second.

Joule, James Prescott 1818–1889. British physicist who demonstrated that heat is a form of energy. His work established the law of conservation of energy, stating that energy is never destroyed but may be converted from one form into another. The joule unit of energy is named for him.

Jovian planet (jō**′**vē-ən) See **gas giant.**

J particle See **J/psi particle.**

JPEG (jā′pĕg′) Short for *Joint Photographic Experts Group.* **1.** A standard algorithm for the compression of digital images, making it easier to store and transmit them. **2.** A digital image that has been compressed using this algorithm.

J/psi particle (jā**′**sī**′**, -psī**′**) An electrically neutral meson having a mass 6,060 times that of the electron and a mean lifetime of approximately 1×10^{-20} seconds. The observation of this unusually long lifetime, in 1974, is what led to the hypothesis that other quarks existed in addition to the up, down, and strange quarks. It was the first firm experimental evidence for the charm quark and is considered to be a charm-anticharm quark pair. Also called *J particle, psi particle.* See Table at **subatomic particle.**

jugular vein (jŭg**′**yə-lər) Either of the two large veins on either side of the neck in mammals that drain blood from the head and return it to the heart.

Julian (jo͞ol**′**yən), **Percy Lavon** 1899–1975. American physician noted for developing cortisone and also physostigmine, a drug used to treat glaucoma and memory loss.

jumping gene (jŭm**′**pĭng) See **transposon.**

junction box (jŭngk**′**shən) An enclosure within which electric circuits, such as the electrical wiring for different sections of a building, are connected to other circuits, such as outside power sources. Junction boxes are very common in telecommunications circuitry and are used to protect the connections and provide easy access to them.

junk DNA (jŭngk) DNA that serves no known biological purpose, such as coding for proteins or their regulation. Junk DNA makes up the vast majority of the DNA in the cells of most plants and animals, composing, for example, about 95 percent of the human genome.

Juno (jo͞o**′**nō) An asteroid having a diameter of about 240 km (149 mi). It was the third to be discovered, in 1804. See more at **asteroid.**

Jupiter (jo͞o**′**pĭ-tər) The fifth planet from the Sun and the largest, with a diameter about 11 times that of Earth. Jupiter is a **gas giant** made up mostly of hydrogen and helium. It turns on its axis faster than any other planet in the solar system, taking less than ten hours to complete one rotation; this rapid rotation draws its atmospheric clouds into distinct belts parallel to its equator. Jupiter has more known moons by far than any other planet in the solar system—as many as 63, with new ones being discovered regularly in recent years—and it has a faint ring system that was unknown until 1979, when the Voyager space probe investigated the planet. A persistent anticyclonic storm known as the **Great Red Spot** is Jupiter's most prominent feature. See Table at **solar system.**

Jurassic (jo͝o-răs′ĭk) The second and middle period of the Mesozoic Era, from about 208 to 144 million years ago. During this time the supercontinent Pangaea continued to split up and numerous shallow seas inundated the new continents. Dinosaurs were the dominant form of terrestrial animal life, and the earliest birds appeared. Marine life was dominated by ammonites and belemnites, and sponges, corals, bryozoa, and gastropods all flourished. Gymnosperms and cycads were the dominant land plants. See Chart at **geologic time.**

K

K 1. Abbreviation of **kelvin. 2.** The symbol for **potassium.**

kame (kām) A small hill or ridge consisting of layers of sand and gravel deposited by a meltwater stream at the margin of a melting glacier.

kaolinite (kā′ə-lĭ-nīt′) A soft, white triclinic mineral occurring in friable masses. Kaolinite forms as the result of the hydrothermal alteration or weathering of feldspar, and it is used in the ceramic industry. *Chemical formula:* $Al_2Si_2O_5(OH)_4$.

kaon (kā′ŏn′) Any of three unstable mesons, one having charge +1 and a mass of 966 electron masses, and two being electrically neutral, with a mass 974 electron masses. Their half-life is approximately 10^{-8} seconds, and they decay through the weak force. Their decay patterns suggest that CP invariance may be violated. Also called K-meson, K particle. See Table at **subatomic particle.**

Kapitsa (kä′pyĭ-tsə), **Pyotr Leonidovich** 1894–1984. Russian physicist who developed equipment capable of generating powerful magnetic fields, which he used to make several discoveries in the area of low-temperature physics. For this work he shared with American physicists Arno Penzias and Robert Wilson the 1978 Nobel Prize for physics.

karoo also **karroo** (kə-ro͞o′) A semiarid plateau. The most well-known karoos cover nearly 400,000 sq km (156,000 sq mi) of southern Africa. Karoos are terrace shaped and characterized by low scrub vegetation.

Karrer (kär′ər), **Paul** 1889–1971. Russian-born Swiss chemist who researched the structure of carotene, which led to important discoveries concerning vitamin A. He also described the structures of riboflavin and vitamins E and K. For these achievements, he shared with British chemist Sir Norman Haworth the 1937 Nobel Prize for chemistry.

karst topography (kärst) A landscape that is characterized by numerous caves, sinkholes, fissures, and underground streams. Karst topography usually forms in regions of plentiful rainfall where bedrock consists of carbonate-rich rock, such as limestone, gypsum, or dolomite, that is easily dissolved. Surface streams are usually absent from karst topography.

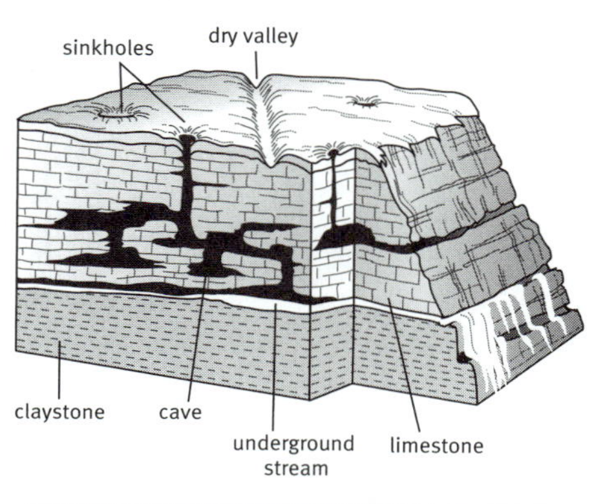

karst topography

top: *cutaway diagram;* bottom: *aerial photograph of karst terrain north of Lewisburg, West Virginia*

karyotype (kăr′ē-ə-tīp′) *Noun.* **1.** An organized visual profile of the chromosomes in the nucleus of a body cell of an organism. Karyotypes are prepared using cells in the metaphase stage of cell division, when chromosomal strands have coiled together and duplicated, rendering them easily visible under a microscope after staining. Photomicrographs of the stained chromosomes are then arranged in a standard format according to size, the relative position of the centromere, and other criteria. The normal human karyotype consists of 46 chromosomes. —*Verb.* **2.** To prepare the karyotype of an organism.

katabatic (kăt′ə-băt′ĭk) Relating to wind currents that blow down a gradient, especially down the slopes of a mountain or glacier. When air comes in contact with the cool surface of a glacier or the upper regions of a mountain or slope, the air cools, becomes dense, and blows downward. Katabatic winds are usually cool and are especially common at night in polar regions. Compare **anabatic.**

Kb Abbreviation of **kilobit.**

KB Abbreviation of **kilobyte.**

Kekulé von Stradonitz (kā′ko͞o-lā′ fôn shträ′dō-nĭts), **Friedrich August** 1829–1896. German chemist who was a founder of organic chemistry. His discovery of the structure of benzene, a basic unit of organic chemistry, was fundamental to understanding many other organic compounds.

kelp (kĕlp) Any of various brown, often very large seaweeds that grow in colder ocean regions. Kelps are varieties of brown algae of the order Laminariales, with some species growing over 61 m (200 ft) long. Kelps are harvested as food (primarily in eastern Asia), as fertilizer, and for their sodium and potassium salts, used in industrial processes. Kelps are also a source of thickening agents and colloid stabilizers used in many commercial products. See more at **brown alga.**

kelvin (kĕl′vĭn) The SI unit used to measure temperature, the basic unit of the Kelvin scale. A difference of one degree Kelvin corresponds to the same temperature difference as a difference of one degree Celsius. See Table at **measurement.** See also **absolute zero.**

Kelvin, First Baron. Title of **William Thomson.** 1824–1907. British mathematician and physicist known especially for his work on heat and electricity. In 1848 he proposed a scale of temperature independent of any physical substance, which became known as the Kelvin scale.

Kelvin scale A scale of temperature beginning at absolute zero (–273.15°C or –459.67°F). Each degree, or kelvin, represents the same temperature increment as one degree on the Celsius scale. On the Kelvin scale water freezes at 273.15 K and boils at 373.15 K.

Kendrew (kĕn′dro͞o′), Sir **John Cowdery** 1917–1997. British molecular biologist who studied the chemistry of the blood and determined by x-ray crystallography the structure of the muscle protein myoglobin. For this work he shared with Max Perutz the 1962 Nobel Prize for chemistry.

Kennelly-Heaviside layer (kĕn′ə-lē-hĕv′ē-sīd′) See **E layer.**

Kepler (kĕp′lər), **Johannes** 1571–1630. German astronomer and mathematician who is considered the founder of celestial mechanics. He was first to accurately describe the elliptical orbits of Earth and the planets around the Sun and demonstrated that planets move fastest when they are closest to the Sun. He also established that a planet's distance from the Sun can be calculated if its period of revolution is known.

Kepler's laws of planetary motion (kĕp′lərz) Three laws devised by Johannes Kepler to define the mechanics of planetary motion. The *first law* states that planets move in an elliptical orbit, with the Sun being one focus of the ellipse. This law identifies that the distance between the Sun and Earth is constantly changing as the Earth goes around its orbit. The *second law* states that the radius of the vector joining the planet to the Sun sweeps out equal areas in equal times as the planet travels around the ellipse. As such, the planet moves quickest when the vector radius is shortest (closest to the Sun), and moves more slowly when the radius vector is long (furthest from the Sun). The *third law* states that the ratio of the squares of the orbital period for two planets is equal to the ratio of the cubes of their mean orbit radius. This indicates that the length of time for a planet to orbit the Sun increases rapidly with the increase of the radius of the planet's orbit.

keratin (kĕr′ə-tĭn) Any of a class of tough, fibrous proteins that are the main structural component of hair, nails, horns, feathers, and hooves. Keratins are rich in sulfur-containing amino acids, especially cysteine. Individual keratin molecules are entwined helically around each other in long filaments, which are cross-linked by bonds between sulfur atoms on different chains. The twining and cross-linking produce strength and toughness.

kernel (kûr′nəl) **1.** A grain or seed, as of a cereal grass, enclosed in a husk. **2.** The inner, usually edible seed of a nut or fruit stone.

kerogen (kĕr′ə-jən) A fossilized mixture of insoluble organic material that, when heated, breaks down into petroleum and natural gas. Kerogen consists of carbon, hydrogen, oxygen, nitrogen, and sulfur and forms from compacted organic material, including algae, pollen, spores and spore coats, and insects. It is usually found in sedimentary rocks, such as shale.

kerosene (kĕr′ə-sēn′) A thin, light-colored oil that is a mixture of hydrocarbons derived from petroleum. The hydrocarbons in kerosene contain between 11 and 12 carbon atoms. Kerosene is used as a fuel in lamps, home heaters and furnaces, and jet engines.

ketamine (kē′tə-mēn′) A general anesthetic given intravenously or intramuscularly in the form of its hydrochloride salt, used especially for minor surgical procedures in which skeletal muscle relaxation is not required.

ketone (kē′tōn′) Any of a class of organic compounds having the general formula RCOR′, where R and R′ are hydrocarbon radicals that are both attached to the carbon atom of the carbonyl (CO) group. Acetone is a ketone.

ketone body A ketone-containing substance that is an intermediate product of the metabolism of fatty acids. Ketone bodies tend to accumulate in the blood and urine of individuals affected by starvation or uncontrolled diabetes mellitus.

ketose (kē′tōs′) Any of a class of simple sugars (monosaccharides) containing a ketone group. Fructose is a ketose. Compare **aldose.**

kettle (kĕt′l) A steep, bowl-shaped hollow in ground once covered by a glacier. Kettles are believed to form when a block of ice left by a glacier becomes covered by sediments and later melts, leaving a hollow. They are usually tens of meters deep and up to tens of kilometers in diameter and often contain surface water.

key (kē) See **cay.**

keystone species (kē′stōn′) A species whose presence and role within an ecosystem has a disproportionate effect on other organisms within the system. A keystone species is often a dominant predator whose removal allows a prey population to explode and often decreases overall diversity. Other kinds of keystone species are those, such as coral or beavers, that significantly alter the habitat around them and thus affect large numbers of other organisms. Compare **indicator species.**

kg Abbreviation of **kilogram.**

Khorana (kō-rä′nə), **Har Gobind** Born 1922. Indian-born American biochemist. He developed one of the first artificial genes and determined the sequence of nucleic acids for each of the 20 amino acids in the human body. He shared a 1968 Nobel prize for his studies of the genetic code.

Khwarizmi (kwär′ĭz-mē), **al-** Full name **Muhammad ibn-Musa al-Khwarizmi.** 780?–850? Arab mathematician and astronomer who compiled an early work on arithmetic and the oldest astronomical tables. His work was widely translated into Latin, introducing Arabic numerals and algebraic concepts to Western mathematics. The word *algorithm* is derived from his name.

kidney (kĭd′nē) Either of a pair of organs that are located in the rear of the abdominal cavity in vertebrates. The kidneys regulate fluid balance in the body and filter out wastes from the blood in the form of urine. The functional unit of the kidney is the nephron. Wastes filtered from the blood by the nephrons drain into the ureters, muscular tubes that connect each kidney to the bladder. See also **nephron.**

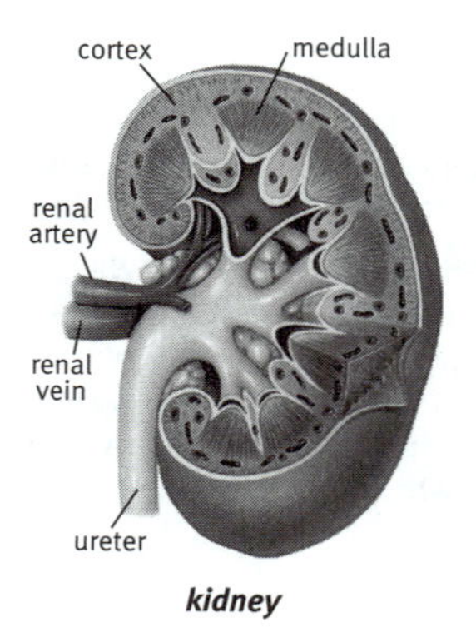

kidney

kidney stone A calculus that originates in the kidney and is usually composed of calcium salts, uric acid, cystine, and other compounds. Kidney stones cause extreme pain and bleeding if they obstruct the passage of urine in the kidney or in the ureter. They can often be treated with **lithotripsy.**

killer T cell (kĭl′ər) A large differentiated T cell that functions in cell-mediated immunity by attacking and lysing target cells that have specific surface antigens. Also called *cytotoxic T cell, killer cell.*

kilo– A prefix that means: **1.** One thousand, as in *kilowatt,* one thousand watts. **2.** 2^{10} (that is, 1,024), which is the power of 2 closest to 1,000, as in *kilobyte.*

kilobit (kĭl′ə-bĭt′) **1.** One thousand bits. **2.** 1,024 (that is, 2^{10}) bits. See Note at **megabyte.**

kilobyte (kĭl′ə-bīt′) **1.** A unit of computer memory or data storage capacity equal to 1,024 (that is, 2^{10}) bytes. **2.** One thousand bytes. See Note at **megabyte.**

kilocalorie (kĭl′ə-kăl′ə-rē) See **calorie** (sense 2).

kilogram (kĭl′ə-grăm′) The basic unit of mass in the metric system, equal to 1,000 grams (2.2 pounds). See Table at **measurement.**

kilohertz (kĭl′ə-hûrts′) A unit of frequency equal to 1,000 cycles per second (1,000 hertz). It is used in the measurement of radio, sound, and other waves.

kilometer (kĭ-lŏm′ĭ-tər, kĭl′ə-mē′tər) A unit of length in the metric system equal to 1,000 meters (0.62 mile). See Table at **measurement.**

kilowatt (kĭl′ə-wŏt′) A unit of power equal to 1,000 watts.

kilowatt-hour A unit used to measure energy, especially electrical energy in commercial applications. One kilowatt-hour is equal to one kilowatt of power produced or consumed over a period of one hour, or 3.6×10^6 joules.

kimberlite (kĭm′bər-līt′) A type of peridotite consisting of a fine-grained matrix of calcite and olivine and containing phenocrysts of olivine, garnet, and sometimes diamonds. Kimberlites are found in long, vertical volcanic pipes, especially in South Africa.

kinase (kī′nās′, -nāz′, kĭn′ās′, -āz′) Any of various enzymes that catalyze the transfer of a phosphate group from a donor, such as ADP or ATP, to an acceptor.

kinematics (kĭn′ə-măt′ĭks) The branch of physics that deals with the characteristics of motion without regard for the effects of forces or mass. Compare **dynamics.**

kinematic viscosity (kĭn′ə-măt′ĭk) See under **viscosity.**

kinetic energy (kə-nĕt′ĭk) The energy possessed by a system or object as a result of its motion. The kinetic energy of objects with mass is dependent upon the velocity and mass of the object, while the energy of waves depends on their velocity, frequency, and amplitude, as well as the density of the medium if there is one (as with ocean waves). Compare **potential energy.**

kinetic friction See under **friction.**

kinetics (kə-nĕt′ĭks) See **dynamics.**

kinetic theory A fundamental theory of matter that explains physical properties in terms of the motion of atoms and molecules. In kinetic theory, properties such as pressure and temperature are viewed as statistical properties of the overall behavior of large numbers of particles. For example, the pressure exerted by a gas on an object is the net result of the numerous collisions of the gas molecules against the object. See also **pressure, statistical mechanics, temperature, thermodynamics.**

kingdom (kĭng′dəm) The highest classification into which living organisms are grouped in Linnean taxonomy, ranking above a phylum. One widely accepted system of classification divides life into five kingdoms: prokaryotes, protists, fungi, plants, and animals. See Table at **taxonomy.**

Kinsey (kĭn′zē), **Alfred Charles** 1894–1956. American biologist and zoologist noted for his studies of human sexuality. He published *Sexual Behavior in the Human Male* in 1948 and *Sexual Behavior in the Human Female* in 1953 (commonly known as the Kinsey reports), based on interviews with 18,500 Americans about their sexual practices. The reports revealed a greater variety of sexual behavior than had previously been suspected and received widespread attention in the scientific community and among the general public. The reports have been criticized for their statistical limitations and especially for the restricted nature of the sample, consisting almost exclusively of white, middle-class men and women, primarily under age 35.

Kirchhoff (kîr′kôf′), **Gustav Robert** 1824–1887. German chemist who with Robert Bunsen discovered the elements cesium and rubidium. He also investigated the solar spectrum and researched electrical circuits and the flow of currents. His electromagnetic theory of diffraction is still the most commonly used in optics.

Kirkwood gaps (kûrk′wo͝od) Specific regions in the asteroid belt where few asteroids are found. The Kirkwood gaps are believed to have formed as a consequence of gravitational interactions of asteroids with Jupiter, which resulted in the movement of the asteroids from within this area into another orbit, leaving the area sparsely populated. The Kirkwood gaps are named after the American astronomer Daniel Kirkwood (1814–1895), who first observed them in 1866.

Klebs (klāps), **Edwin** 1834–1913. German bacteriologist who described the diphtheria bacillus in 1883 although he did not demonstrate it to be the cause of the disease. It wasn't until a year later that Friedrich Löffler made the causal link between the disease and the bacillus, which is now named after both of them. Klebs also demonstrated the presence of bacteria in infected wounds and showed that tuberculosis can be transmitted through infected milk.

Klein bottle (klīn) A smooth surface that has no inside or outside. It is often pictured in ordinary space as a tube that bends back upon itself, entering through the side and joining with the open end. A true Klein bottle, which cannot be constructed in ordinary three-dimensional space, would not actually intersect itself. The Klein bottle is named after the German mathematician Felix Klein (1849–1925). Compare **Möbius strip.**

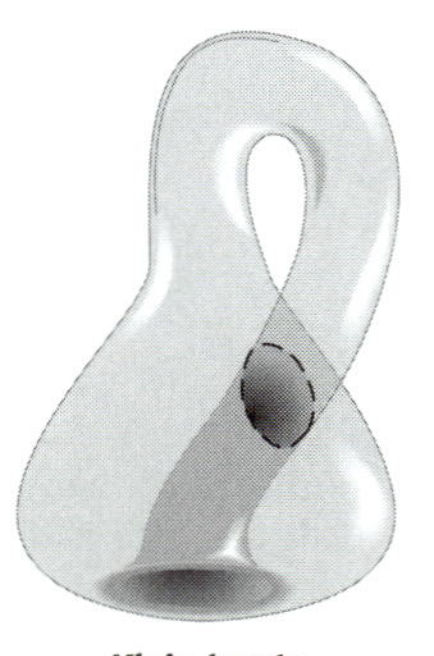

Klein bottle

km Abbreviation of **kilometer.**

K-meson See **kaon.**

kneecap (nē′kăp′) See **patella.**

knob (nŏb) A prominent, rounded hill or mountain.

Koch (kôk), **Robert** 1843–1910. German bacteriologist who demonstrated that specific diseases are caused by specific microorganisms. He identified the bacilli that cause anthrax, tuberculosis, and cholera, and he showed that fleas and rats are responsible for transmission of the bubonic plague and that the tsetse fly is responsible for transmitting sleeping sickness. Koch won the Nobel Prize for physiology or medicine in 1905.

Robert Koch

BIOGRAPHY Robert Koch

Robert Koch is deservedly famous for his discovery of the bacteria that cause tuberculosis and many other diseases, and his illumination of the life cycle of the anthrax bacillus in 1876 showed how a particular microorganism caused a particular disease, definitively establishing the modern germ theory of disease. What Koch is less well-known for is his equally important and pioneering work in laboratory methods, especially in culture techniques and microscopy. Some attempts before Koch had already been made to grow microorganisms outside the body, but it was he who, through ingenious experiments, devised cheap, reliable, and duplicable techniques for growing pure cultures of single species of bacteria in the lab. Except for the lid, he invented the petri dish and a jellylike culture medium for it (the lid was later added by one of his assistants, Julius Petri). For years a passionate amateur photographer, Koch soon applied that interest to his lab work: he devised methods for preparing and culturing bacteria in thin layers on glass slides so that they could be photographed under a microscope. He invented ways of staining bacterial cultures to make poorly visible bacteria stand out under magnification. All of these innovations allowed the life cycles of bacteria to be studied and documented for the first time—an advance that bore its first and perhaps most dramatic fruit in Koch's demonstration of the life cycle of the anthrax bacillus, which was accompanied by dramatic photographs that took the scientific world by storm.

Köhler (kŭ′lər), **Georges Jean Franz** 1946–1995. German immunologist who with Cesar Milstein developed a method of fusing together different cells to maintain antibody production. For the discovery of this technique, which is widely used in the development of drugs and in diagnostic tests for cancer and other diseases, Köhler and Milstein shared with British immunologist Niels K. Jerne the 1984 Nobel Prize for physiology or medicine.

Kolbe (kôl′bə), **(Adolphe Wilhelm) Hermann** 1818–1884. German chemist who was the first to synthesize an organic compound from inorganic materials (1843–45). He also developed an electrolytic procedure for the preparation of alkanes, and in 1859 synthesized salicylic acid from coal tar, which laid the foundation for the development of aspirin.

Kornberg (kôrn′bûrg′), **Arthur** Born 1918. American biochemist who discovered DNA polymerase, the enzyme that synthesizes new DNA. For this work, he shared with Severo Ochoa the 1959 Nobel Prize for physiology or medicine. In 1967 Kornberg became the first person to synthesize viral DNA.

Kovalevsky (kŏv′ə-lĕv′skē), **Sonya** 1850–1891. Russian mathematician who made important contributions to calculus. Her mathematical determination of the shape of Saturn's rings became a model for other scientists.

K particle See **kaon.**

Kr The symbol for **krypton.**

Krebs (krĕbz), Sir **Hans Adolf** 1900–1981. German-born British biochemist who in 1936 discovered the process that came to be known as the Krebs cycle. For this work he shared with American biochemist Fritz Lipmann the 1953 Nobel Prize for physiology or medicine.

Krebs cycle A series of chemical reactions that occur in most aerobic organisms and are part of the process of aerobic cell metabolism, by which glucose and other molecules are broken down in the presence of oxygen into carbon dioxide and water to release chemical energy in the form of ATP. The Krebs cycle is the intermediate stage, occurring between glycolysis and phosphorylation, and results in the enzymatic breaking down, rearranging, and recombination of byproducts of glycolysis. The combination of glycolysis and the Krebs cycle ultimately allows 36 ATP molecules to be produced from the energy contained in one molecule of glucose and six molecules of oxygen. Also called *citric acid cycle.* See more at **cellular respiration.**

krill (krĭl) Small crustaceans that float in the ocean in huge numbers and are one of the most important parts of zooplankton. Krill are the main food of baleen whales.

krypton (krĭp′tŏn′) *Symbol* **Kr** A colorless, odorless element in the noble gas group. It is used in certain fluorescent lamps and photographic flash lamps. Atomic number 36; atomic weight 83.80; melting point −156.6°C; boiling point −152.30°C; density 3.73 grams per liter (0°C). See **Periodic Table.**

Kuiper belt (kī′pər) A disk-shaped region in the outer solar system lying beyond the orbit of Neptune and extending to a distance of about 50 astronomical units, containing thousands of small, icy celestial bodies. It is believed to be a reservoir for short-period comets (comets that make one complete orbit of the Sun in less than 200 years). The Kuiper belt is named after American astronomer Gerard Kuiper (1905–1973), who first predicted its existence. ► The bodies populating this region are known as **Kuiper belt objects**, and unlike the bodies in the Oort cloud, these objects are believed to have originated in situ. There are an estimated 70,000 such objects having diameters of more than 100 km (62 mi). The dwarf planet Pluto and its moons are also found in the Kuiper belt. Compare **Oort cloud.**

Kuroshio Current (ko͝o-rō′shē-ō′) See **Japan Current.**

kurtosis (kər-tō′sĭs) The general form or a quantity indicative of the general form of a statistical frequency curve near the mean of the distribution.

kuru (ko͝or′o͞o) A rare, progressive, degenerative neurological disease found in certain peoples of New Guinea and associated with cannibalism. It is thought to be caused by a prion and results in a fatal encephalopathy. See Note at **prion.**

kwashiorkor (kwä′shē-ôr′kôr′) A severe malnutrition, seen primarily in children of tropical and subtropical regions, caused by deficiency in the quality and quantity of dietary protein. Kwahiorkor is characterized by failure to grow, anemia, liver damage, edema, discoloration of the skin or hair, and bulky stools containing undigested food.

Kwolek (kwŏl′ĕk′), **Stephanie** Born 1923. American chemist who pioneered the use of polymers to make synthetic fibers. She developed the first liquid crystal polymer fiber, now used to make many products, including bulletproof vests.

kyanite (kī′ə-nīt′) A bluish-green to colorless triclinic mineral. Kyanite occurs as long, thin, blade-shaped crystals in metamorphic rocks. It is unique among minerals in having two grades of hardness, one along its length, and one along its width. It is a polymorph of andalusite and sillimanite, but can form at lower temperatures than either of these. *Chemical formula:* Al_2SiO_5.

l Abbreviation of **length, liter.**

La The symbol for **lanthanum.**

label (lā′bəl) See **tracer.**

labium (lā′bē-əm) Plural **labia.** Either of two pairs of folds of tissue that make up part of the external genitalia of female mammals.

labor (lā′bər) The process by which the birth of a mammal occurs, beginning with contractions of the uterus and ending with the expulsion of the fetus and the placenta.

Labrador Current (lăb′rə-dôr′) A cold ocean current flowing southward from Baffin Bay along the coast of Labrador and turning east after intersecting with the Gulf Stream. The fog characteristic of Labrador is created when warm Gulf Stream air meets the cold waters of the Labrador Current. The Labrador Current also brings down icebergs from the Arctic into transatlantic shipping lanes. Also called the *Arctic Stream.*

labradorite (lăb′rə-dôr′īt′, -dô-rīt′) A blue, gray, green, or brown triclinic mineral that is a variety of plagioclase feldspar. It occurs in igneous rocks.

labyrinth (lăb′ə-rĭnth′) The system of interconnecting canals and spaces that make up the inner ear of many vertebrates. The labyrinth has both a bony component, made up of the cochlea, the semicircular canals, and the vestibule, and a membranous one.

labyrinthodont (lăb′ə-rĭn′thə-dŏnt′) Any of various extinct amphibians of the group Labyrinthodontia, which were the dominant animals of the late Paleozoic Era. Labyrinthodonts had stocky, lizardlike bodies with short limbs, and fishlike teeth with labyrinthine structure (with complex infolding of the enamel). They varied from the size of a salamander to that of a crocodile. One early genus, *Ichthyostega,* was probably the first terrestrial vertebrate.

laccolith (lăk′ə-lĭth′) A body of igneous rock intruded between layers of sedimentary rock, resulting in uplift. Laccoliths are usually plano-convex in cross-section, having a flat bottom and a convex top, and are roughly circular in plan. They are usually connected to a dike and are typically up to 8 km (5 mi) in diameter and tens to hundreds of meters thick. See illustration at **batholith.**

lacertilian (lăs′ər-tĭl′ē-ən) Relating to or characteristic of lizards and closely related reptiles.

lacrimal gland (lăk′rə-məl) Either of two glands, one in each eye, that produce and secrete tears.

lactase (lăk′tās′) An enzyme that is found in the small intestine, liver, and kidneys of mammals and catalyzes the breakdown of lactose into galactose and glucose.

lactate (lăk′tāt′) A salt or ester of lactic acid. Lactate is a product of fermentation and is produced during cellular respiration as glucose is broken down.

lactation (lăk-tā′shən) The secretion or production of milk by the mammary glands in female mammals after giving birth.

lactic acid (lăk′tĭk) A syrupy, water-soluble organic acid produced when milk sours or certain fruits ferment. It is also produced in the body during the anaerobic metabolism of glucose, as in muscle tissue during exercise, where its buildup can cause cramping pains. A synthetic form of lactic acid is used as a flavoring and preservative, in dyeing and textile printing, and in pharmaceuticals. *Chemical formula:* $C_3H_6O_3$.

lactone (lăk′tōn′) Any of various organic esters derived from organic acids by removal of water. Lactones are formed when the carboxyl (COOH) group of the acid reacts with a hydroxyl (OH) group in the same acid, releasing water and causing the carbon atom to join to the hydroxyl's remaining oxygen atom, forming a ring. Vitamin C, the antibiotic erythromycin, and many commercially important substances are lactones.

lactose (lăk′tōs′) A white crystalline disaccharide consisting of a glucose and a galactose molecule, found in milk and used in the manufacture of various other foods. *Chemical for-*

mula: $C_{12}H_{22}O_{11}$. The inability to digest lactose properly is called *lactose intolerance.* It is caused by a deficiency of the enzyme lactase and marked by abdominal cramping and other symptoms after ingesting lactose.

lacustrine (lə-kŭs′trĭn) **1.** Relating to lakes. **2.** Relating to a system of inland wetlands and deep-water habitats associated with freshwater lakes and reservoirs, characterized by the absence of trees, shrubs, or emergent vegetation. Compare **marine, palustrine, riverine.**

lagoon (lə-go͞on′) **1.** A shallow body of salt water close to the sea but separated from it by a narrow strip of land, such as a barrier island, or by a coral reef. **2.** A shallow pond or lake close to a larger lake or river but separated from it by a barrier such as a levee.

lagoon

aerial photograph of Kayangel Atoll, Palau

Lagrange (lə-grānj′, lə-gränj′), Comte **Joseph Louis** 1736–1813. Italian-born French mathematician and astronomer who made important contributions to algebra and calculus. His work on celestial mechanics extended scientific understanding of planetary and lunar motion. In 1772 he discovered the points in space that are now named for him.

Lagrangian point (lə-grăn′jē-ən) A point in space where a small body with negligible mass under the gravitational influence of two large bodies will remain at rest relative to the larger ones. In a system consisting of two large bodies (such as the Sun-Earth system or the Moon-Earth system), there are five Lagrangian points (L1 through L5). Knowledge of these points is useful in deciding where to position orbiting bodies.

lahar (lä′här′) **1.** A wet mass of volcanic fragments flowing rapidly downhill. Lahars usually contain ash, breccia, and boulders mixed with rainwater or with river or lake water displaced by the lava flow associated with the volcano. **2.** The deposit produced by such a flowing mass.

lake (lāk) A large inland body of standing fresh or salt water. Lakes generally form in depressions, such as those created by glacial or volcanic action; they may also form when a section of a river becomes dammed or when a channel is isolated by a change in a river's course.

Lamarck (lə-märk′, lä-), Chevalier de **Jean-Baptiste Pierre Antoine de Monet** 1744–1829. French naturalist who introduced the taxonomic distinction between vertebrates and invertebrates. His theory that the acquired characteristics of a species could be inherited by later generations was a forerunner to Charles Darwin's theory of evolution, although it was eventually discredited.

lambda-b particle (lăm′də-bē′) An electrically neutral baryon having a mass 11,000 times that of the electron and a mean lifetime of approximately 1.1×10^{-12} seconds. See Table at **subatomic particle.**

lambda-c particle A positively charged baryon having a mass 4,471 times that of the electron and a mean lifetime of approximately 2.1×10^{-13} seconds. See Table at **subatomic particle.**

lambda particle An electrically neutral baryon having a mass 2,183 times that of the electron and a mean lifetime of approximately 2.6×10^{-10} seconds. See Table at **subatomic particle.**

lambert (lăm′bərt) A unit of luminance in the centimeter-gram-second system, equivalent to the luminance of a perfectly diffusing surface that emits or reflects one lumen per square centimeter. The lambert is named after the Swiss mathematician and physicist Johann Heinrich Lambert (1728–1777).

lamellibranch (lə-mĕl′ə-brănk′) **1.** Any of the bivalve mollusks of the subclass Lamellibranchia, the largest group of bivalves. Lamellibranchs have gills that consist of long, folded filaments with cilia at the end that draw in water for respiration and also filter food particles. Lamellibranchs include clams, scallops, and oysters. **2.** In former classifications, any bivalve mollusk.

lamina (lăm′ə-nə) Plural **laminae** (lăm′ə-nē′) or **laminas. 1.** The expanded area of a leaf or petal; a blade. See more at **leaf.** **2.** A thin layer of bone, membrane, or other tissue. **3.** The thinnest recognizable layer of sediment, differing from other layers in color, composition, or particle size. Laminae are usually less than 1 cm (0.39 inches) thick.

laminar flow (lăm′ə-nər) Smooth, orderly movement of a fluid, in which there is no turbulence, and any given subcurrent moves more or less in parallel with any other nearby subcurrent. Laminar flow is common in viscous fluids, especially those moving at low velocities. Compare **turbulent flow.**

laminarin (lăm′ə-nâr′ĭn) A polymer of glucose that is stored as food in brown algae.

lamprophyre (lăm′prə-fīr′) A dark igneous rock, having a porphyritic texture in which both the phenocrysts (larger crystals) and the matrix consist primarily of pyroxene, hornblende, and biotite.

LAN (lăn) Short for *local area network.* A network that links together computers and peripheral equipment within a limited area, such as a building or a group of buildings. The computers in an LAN have independent central processing units, but they are able to exchange data with each other and to share resources such as printers. See also **client/server network, peer-to-peer network.**

lancelet (lăns′lĭt) Any of various small, transparent, fishlike marine organisms of the subphylum Cephalochordata that are related to vertebrates but have a notochord instead of a true backbone. Unlike other primitive chordates, lancelets have a body divided into serially repeated muscular segments. Also called *amphioxus.*

lanceolate (lăn′sē-ə-lāt′) Tapering from a rounded base toward an apex; lance-shaped. Many willows have lanceolate leaves.

Landé factor (län-dā′) The Bohr magneton. See more at **magneton.**

landfill (lănd′fĭl′) A disposal site where solid waste, such as paper, glass, and metal, is buried between layers of dirt and other materials in such a way as to reduce contamination of the surrounding land. Modern landfills are often lined with layers of absorbent material and sheets of plastic to keep pollutants from leaking into the soil and water. Also called *sanitary landfill.*

landform (lănd′fôrm′) A recognizable, naturally formed feature on the Earth's surface. Landforms have a characteristic shape and can include such large features as plains, plateaus, mountains, and valleys, and smaller features such as hills, eskers, and canyons.

landmass (lănd′măs′) A large, continuous area of land, such as a continent or large island.

Landsat (lănd′săt′) Any of various satellites used to gather data for images of the Earth's land surface and coastal regions. These satellites are equipped with sensors that respond to Earth-reflected sunlight and infrared radiation. The first Landsat satellite was launched in 1972. Currently, the seventh satellite (Landsat 7) is orbiting Earth.

landslide (lănd′slīd′) **1.** The rapid downward sliding of a mass of earth and rock. Landslides usually move over a confined area. Many kinds of events can trigger a landslide, such as the oversteepening of slopes by erosion associated with rivers, glaciers, or ocean waves, heavy snowmelt that saturates soil and rock, or earthquakes that lead to the failure of weak slopes. **2.** The mass of soil and rock that moves in this way.

Landsteiner (lănd′stī′nər), **Karl** 1868–1943. Austrian-born American pathologist who discovered the human blood groups A, B, and O in 1901. In 1902, his colleagues discovered a fourth group, AB, and in 1927 Landsteiner discovered two more groups, M and N. For this work Landsteiner received the 1930 Nobel Prize for physiology or medicine. In 1940 he discovered the Rh factor. See Note at **blood type.**

Langevin (länzh-văn′), **Paul** 1872–1946. French physicist who during World War I pioneered the use of sonar techniques to detect submarines. He also advanced the theory of magnetism and the study of the molecular structure of gases.

langley (lăng′lē) A unit equal to one gram calorie per square centimeter of irradiated surface, used to measure solar radiation. The langley is named after American astronomer and aeronautical pioneer Samuel Pierpont Langley (1834–1906).

language (lăng′gwĭj) **1.** A system of objects or symbols, such as sounds or character sequences, that can be combined in various ways following a set of rules, especially to communicate thoughts, feelings, or instructions. See also **machine language, programming language. 2.** The set of patterns or structures produced by such a system.

La Niña (lä nēn′yä) A cooling of the surface water of the eastern and central Pacific Ocean, occurring somewhat less frequently than El Niño events but causing similar, generally opposite disruptions to global weather patterns. La Niña conditions occur when the Pacific trade winds blow more strongly than usual, pushing the sun-warmed surface water farther west and increasing the upwelling of cold water in the eastern regions. Together with the atmospheric effects of the related **southern oscillation**, the cooler water brings drought to western South America and heavy rains to eastern Australia and Indonesia. Compare **El Niño.**

lanolin (lăn′ə-lĭn) A yellowish-white wax secreted by the sebaceous glands of sheep to coat wool. Lanolin is composed of esters and polyesters of almost seventy alcohols and fatty acids. Since it is easily absorbed by the skin, it is used in soaps, cosmetics, and ointments.

lanthanide (lăn′thə-nīd′) Any of a series of 15 naturally occurring metallic elements. The lanthanides include elements having atomic numbers 57 (lanthanum) through 71 (lutetium). They are grouped apart from the rest of the elements in the Periodic Table because they all behave in a similar way in chemical reactions. Also called *rare-earth element.* See **Periodic Table.**

lanthanum (lăn′thə-nəm) *Symbol* **La** A soft, malleable, silvery-white metallic element of the lanthanide series. It is used to make glass for lenses and lights for movie and television studios. Atomic number 57; atomic weight 138.91; melting point 920°C; boiling point 3,469°C; specific gravity 5.98 to 6.186; valence 3. See **Periodic Table.**

lanugo (lə-no͞o′gō) A covering of fine, soft hair or hairlike structures, as on a leaf, insect, or human fetus.

laparoscope (lăp′ər-ə-skōp′) A slender, tubular endoscope that is inserted through an incision in the abdominal wall to examine or perform minor surgery within the abdomen or pelvis. —*Noun* **laparoscopy** (lăp′ə-rŏs′kə-pē).

lapillus (lə-pĭl′əs) Plural **lapilli** (lə-pĭl′ī′). A small fragment of lava, between 2 and 64 mm in size, blown out from a volcano.

Laplace (lə-pläs′, lä-), Marquis **Pierre Simon de** 1749–1827. French mathematician and astronomer noted for his theory of a nebular origin of the solar system and his investigations into gravity and the stability of planetary motion. He also made important contributions to the theory of probability. See more at **nebular hypothesis.**

lapse rate (lăps) The rate of change of any meteorological phenomenon, especially atmospheric temperature with altitude. The lapse rate varies depending on the ground temperature, time of year (for example, in the Northern hemisphere it is lower in the winter), whether the air is over land or ocean water, and what the degree of moisture is. ▸ The **dry adiabatic lapse rate** is the lapse rate of a dry mass of air which expands and cools as it rises. This rate is typically –9.8°C (–14.36°F) per 1,000 m (3,280 ft). ▸ The **saturated adiabatic lapse rate** is the lapse rate of a wet mass of air, which slows down once the dew point has been reached and condensation has started to form. This rate ranges from 4°C (39.2°F) per 1,000 m (3,280 ft) to 9°C (48.2°F) per 1,000 m (3,280 ft).

large calorie (lärj) See **calorie** (sense 2).

large intestine The wide lower section of the intestine that extends from the end of the small intestine to the anus. The large intestine acts mainly to absorb water from digested materials and solidify feces. In most vertebrate animals, it includes the cecum, colon, and rectum.

larva (lär′və) Plural **larvae** (lär′vē) or **larvas. 1.** An animal in an early stage of development that differs greatly in appearance from its adult stage. Larvae are adapted to a different environment and way of life from those of adults and go through a process of metamorphosis in changing to adults. Tadpoles are the larvae of frogs and toads. **2.** The immature, wingless, and usually wormlike feeding form of those insects that undergo three stages of metamorphosis, such as butterflies, moths, and beetles. Insect larvae hatch from eggs, later turn into pupae, and finally turn into adults. Compare **imago, nymph, pupa.**

laryngitis (lăr′ĭn-jī′tĭs) Inflammation of the larynx, usually caused by a virus and characterized by hoarseness.

larynx (lăr′ĭngks) Plural **larynges** (lə-rĭn′jēz) or **larynxes.** The upper part of the trachea in most vertebrate animals, containing the vocal cords. The walls of the larynx are made of cartilage. Sound is produced by air passing through the larynx on the way to the lungs, causing the walls of the larynx to vibrate. The pitch of the sound that is produced can be altered by the pull of muscles, which changes the tension of the vocal cords. Also called *voice box.* —*Adjective* **laryngeal.**

laser (lā′zər) Short for *light amplification by stimulated emission of radiation.* A device that creates and amplifies electromagnetic radiation of a specific frequency through the process of **stimulated emission**. The radiation emitted by a laser consists of a coherent beam of photons, all in phase and having the same polarization. Lasers have many uses, such as cutting hard or delicate substances, reading data from compact disks and other storage devices, and establishing straight lines in geographical surveying.

laser diode See **semiconductor laser.**

laser pump A device used to excite atoms for stimulated emission in the production or amplification of laser signals. Optical resonators of lasers use laser pumps to induce stimulated emission; relay stations in fiber optic communications use laser pumps to boost signals. See more at **laser.**

laser trap A device that uses magnetic coils and lasers with tunable frequencies to suspend atoms or particles in a small region of space and to slow them down, reducing their temperature. Laser traps have been used to form Bose-Einstein condensates.

LASIK (lā′zĭk) Eye surgery in which the surface of the cornea is reshaped using a laser, per-

A CLOSER LOOK **laser**

A *laser* emits a thin, intense beam of nearly monochromatic visible or infrared light that can travel long distances without diffusing. Most light beams consist of many waves traveling in roughly the same direction, but the phases and polarizations of each individual wave (or photon) are randomly distributed. In laser light, the waves are all precisely in step, or in phase, with each other, and have the same polarization. Such light is called *coherent.* All of the photons that make up a laser beam are in the same quantum state. Lasers produce coherent light through a process called *stimulated emission.* The laser contains a chamber in which atoms of a medium such as a synthetic ruby rod or a gas are excited, bringing their electrons into higher orbits with higher energy states. When one of these electrons jumps down to a lower energy state (which can happen spontaneously), it gives off its extra energy as a photon with a specific frequency. But this photon, upon encountering another atom with an excited electron, will stimulate that electron to jump down as well, emitting another photon with the same frequency as the first and in phase with it. This effect cascades through the chamber, constantly stimulating other atoms to emit yet more coherent photons. Mirrors at both ends of the chamber cause the light to bounce back and forth in the chamber, sweeping across the entire medium. If a sufficient number of atoms in the medium are maintained by some external energy source in the higher energy state—a condition called *population inversion*—then emission is continuously stimulated, and a stream of coherent photons develops. One of the mirrors is partially transparent, allowing the laser beam to exit from that end of the chamber. Lasers have many industrial, military, and scientific uses, including welding, target detection, microscopic photography, fiber optics, surgery, and optical instrumentation for surveying.

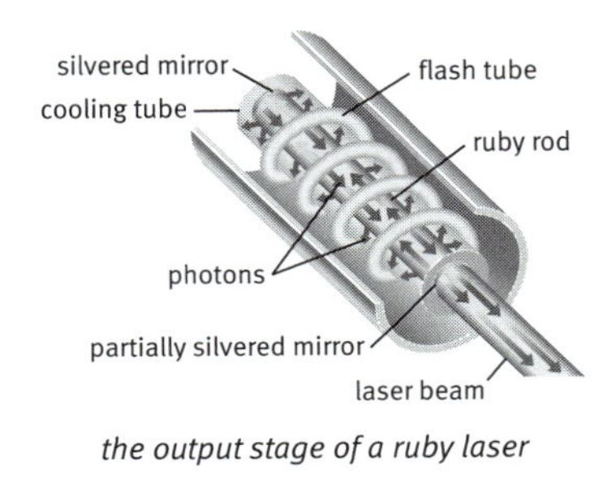

the output stage of a ruby laser

formed to correct certain refractive disorders such as myopia.

late blight (lāt) A disease of potato plants caused by the oomycete organism *Phytophthora infestans* and characterized by decay of the foliage and tubers. An outbreak of late blight led to widespread famine in Ireland in 1845-1850.

latent heat (lāt′nt) The quantity of heat absorbed or released by a substance undergoing a change of state, such as ice changing to liquid water or liquid water changing to ice, at constant temperature and pressure. The latent heat absorbed by air when water vapor condenses is ultimately the source of the power of thunderstorms and hurricanes. See also **heat capacity.**

latent period 1. The interval between exposure to an infectious organism or a carcinogen and the appearance of symptoms or signs of disease. **2.** The period elapsing between the application of a stimulus and the expected response, such as the contraction of a muscle.

lateral bud (lăt′ər-əl) See under **bud.**

lateral line A series of tubelike canals along the head and sides of fish and some amphibians by which vibrations, as from water currents, and changes in pressure are detected.

lateral meristem One of the two meristems in vascular plants (the cork cambium and the vascular cambium) in which secondary growth occurs, resulting in increase in stem girth. Also called *secondary meristem.* Compare **primary meristem.** See more at **cambium, secondary growth.**

lateral moraine See under **moraine.**

laterite (lăt′ə-rīt′) A red, porous, claylike soil formed by the leaching of silica-rich components and enrichment of aluminum and iron hydroxides. They are especially common in humid climates. Laterites that are poor in iron oxides and rich in aluminum oxides are called bauxites. Also called *latosol.* See more at **bauxite.**

laterization (lăt′ə-rĭ-zā′shən) The weathering process by which soils and rocks are depleted of soluble substances, such as silica-rich and alkaline components and enriched with insoluble substances, such as hydrated aluminum and iron oxides. Laterization is especially common in tropical regions that have a pronounced dry season and a water table that is close to the surface.

late wood The part of the wood in a growth ring of a tree that is produced later in the growing season. The cells of late wood are smaller and have thicker cell walls than those produced earlier in the season. Within a growth ring, the change of early wood to late wood is gradual, but each layer of early wood from the next growing season makes an abrupt contrast with the late wood before it, thus leading to the perception of rings. Compare **early wood.**

latex (lā′tĕks′) **1.** The colorless or milky sap of certain trees and plants, such as the milkweed and the rubber tree, that hardens when exposed to the air. Latex usually contains gum resins, waxes, and oils, and sometimes toxic substances. **2.** A manufactured emulsion of synthetic rubber or plastic droplets in water that resembles the latex of plants. It is used in paints, adhesives, and synthetic rubber products.

latitude (lăt′ĭ-to͞od′) **1.** A measure of relative position north or south on the Earth's surface, measured in degrees from the equator, which has a latitude of 0°, with the poles having a latitude of 90° north and south. The distance of a degree of latitude is about 69 statute miles or 60 nautical miles (111 km). Latitude and longitude are the coordinates that together identify all positions on the Earth's surface. Compare **longitude. 2.** Celestial latitude.

latosol (lăt′ə-sôl′) See **laterite.**

latrorse (lăt′rôrs′) Relating to anthers that open or split toward the side, toward other anthers, and not toward or away from the central axis around which a flower is arranged.

lattice (lăt′ĭs) A set of points that, when joined together, form the geometric shape of a mineral crystal. The lattice of the mineral halite, for example, is in the shape of a cube. See more at **crystal** (sense 1).

laughing gas (lăf′ĭng) Nitrous oxide used as a mild anesthetic.

Laurasia (lô-rā′zhə) A supercontinent of the Northern Hemisphere made up of the landmasses that currently correspond to North America, Greenland, Europe, and Asia (except

India). According to the theory of plate tectonics, Laurasia separated from **Pangaea** at the end of the Paleozoic Era and broke up into the current continents in the middle of the Mesozoic Era. Compare **Gondwanaland.**

lauric acid (lôr′ĭk) A saturated fatty acid obtained chiefly from coconut and laurel oils and used in making soaps, cosmetics, esters, and lauryl alcohol. It is combustible and forms colorless needles that have a waxy odor and taste. *Chemical formula:* $\mathbf{C_{12}H_{24}O_2}$.

lauryl alcohol (lôr′əl) A colorless solid alcohol used in synthetic detergents and pharmaceuticals. *Chemical formula:* $\mathbf{C_{12}H_{26}O}$.

lava (lä′və) **1.** Molten rock that flows from a volcano or from a crack in the Earth. Most lava flows at a rate of a few kilometers per hour, but rates as high as 60 km (37 mi) per hour have been observed. Lava that contains abundant iron- and magnesium-rich components usually erupts with temperatures between 1,000°C and 1,200°C (1,832°F and 2,192°F). Lava that contains abundant silica- and feldspar-rich components usually erupts with temperatures between 800°C and 1,000°C (1,472°F and 1,832°F). Compare **magma. 2.** The igneous rock formed when this substance cools and hardens. Depending on its composition and the rate at which it cools, lava can be glassy, very finely grained, ropelike, or coarsely grained. When it cools underwater, it cools in pillow-shaped masses. See also **aa, pahoehoe, pillow lava.**

Antoine Lavoisier

Lavoisier (lä-vwä-zyā′), **Antoine Laurent** 1743–1794. French chemist who is regarded as one of the founders of modern chemistry. In 1778 he discovered that air consists of a mixture of two gases, which he called oxygen and nitrogen. Lavoisier also discovered the law of conservation of mass and devised the modern method of naming chemical compounds. His wife **Marie** (1758–1836) assisted him with his laboratory work and translated a number of important chemistry texts. See Notes at **oxygen, Priestley.**

BIOGRAPHY **Antoine Lavoisier**

Antoine Lavoisier's superior organizational skills made it possible for him to interpret and extend the research of other scientists, leading to the important experiments and discoveries that designate him as one of the founders of modern chemistry. He introduced a rigorous experimental approach to the field based on the determination of the weights of reagents and products in chemical reactions. In his *Elementary Treatise of Chemistry,* published in 1789, he presented a systematic and unified view of new theories and established a system of nomenclature for chemical compounds. His classification of substances laid the foundation for the modern distinction between chemicals and compounds. Lavoisier also disproved the longstanding phlogiston theory of combustion, which for centuries held that a substance called phlogiston, a volatile part of all combustible substances, was released during the process of combustion. By repeating the experiments of Joseph Priestley, Lavoisier demonstrated that during combustion the burning substance combines with a constituent of the air, the gas he named oxygen. He also described the role of oxygen in the respiration of both animals and plants, and he proved that water is made up of oxygen and hydrogen.

law (lô) A statement that describes invariable relationships among phenomena under a specified set of conditions. Boyle's law, for instance, describes what will happen to the volume of an **ideal gas** if its pressure changes and its temperature remains the same. The conditions under which some physical laws hold are idealized (for example, there are no ideal gases in the real world), thus some physical laws apply universally but only approximately. See Note at **hypothesis.**

law of conservation For the laws of conservation, see under **conservation.**

law of dominance See under **Mendel's law.**

law of gravitation See **Newton's law of gravitation.**

law of independent assortment See under **Mendel's law.**

law of inertia See under **Newton's laws of motion.**

law of large numbers The rule or theorem that the average of a large number of independent measurements of a random quantity tends toward the theoretical average of that quantity. Also called *Bernoulli's law.*

law of segregation See under **Mendel's law.**

law of superposition 1. *Physics.* See **superposition. 2.** *Geology.* A general law stating that in any sequence of sediments or rocks that has not been overturned, the youngest sediments or rocks are at the top of the sequence and the oldest are at the bottom.

law of thermodynamics See under **thermodynamics.**

law of universal gravitation See **Newton's law of gravitation.**

Lawrence (lôr′əns), **Ernest Orlando** 1901–1958. American physicist who in 1929 built the first cyclotron, which he used to study the structure of the atom, transmute elements, and produce artificial radiation. His work laid the foundation for the development of the atomic bomb.

lawrencium (lô-rĕn′sē-əm) *Symbol* **Lr** A synthetic, radioactive metallic element of the actinide series that is produced by bombarding californium with boron ions. Its most stable isotope is Lr 262 with a half-life of 3.6 hours. Atomic number 103. See **Periodic Table.**

laws of motion (lôz) See **Newton's laws of motion.**

lb Abbreviation of **pound.**

LCD (ĕl′sē-dē′) Short for *liquid-crystal display.* A low-power, flat-panel display used in many digital devices to display numbers or images. It is made of a liquid containing crystals that are affected by electric current, sandwiched between filtering layers of glass or plastic. LCDs do not produce light of their own; instead, when electric current is passed through the material, the molecules of the "liquid crystal" twist so that they either reflect or transmit light from an external source.

LDL Abbreviation of **low-density lipoprotein.**

L-dopa (ĕl-dō′pə) An amino acid that is the metabolic precursor of dopamine, is converted in the brain to dopamine, and is used in synthetic form to treat Parkinson's disease. *Chemical formula:* $C_9H_{11}NO_4$.

leaching (lē′chĭng) The removal of soluble material from a substance, such as soil or rock, through the percolation of water. Organic matter is typically removed from a soil horizon and soluble metals or salts from a rock by leaching. Leaching differs from **eluviation** in that it affects soluble, not suspended, material and often results in the complete removal of the material from the soil or rock.

lead (lĕd) *Symbol* **Pb** A soft, ductile, heavy, bluish-gray metallic element that is extracted chiefly from galena. It is very durable and resistant to corrosion and is a poor conductor of electricity. Lead is used to make radiation shielding and containers for corrosive substances. It was once commonly used in pipes, solder, roofing, paint, and antiknock compounds in gasoline, but its use in these products has been curtailed because of its toxicity. Atomic number 82; atomic weight 207.2; melting point 327.5°C; boiling point 1,744°C; specific gravity 11.35; valence 2, 4. See **Periodic Table.** See Note at **element.**

lead acetate A poisonous, white crystalline compound used in hair dyes, waterproofing compounds, and varnishes. *Chemical formula:* $C_4H_6O_4Pb$.

leaf (lēf) An appendage growing from the stem of a plant. Leaves are extremely variable in form and function according to species. For example, the needles of pine trees, the spines of cacti, and the bright red parts of the poinsettia plant are all leaves modified for different purposes. However, most leaves are flat and green and adapted to capturing sunlight and carbon dioxide for photosynthesis. They consist of an outer tissue layer (the epidermis) through which water and gases are exchanged, a spongy inner layer of cells that contain chloroplasts, and veins that supply water and minerals and carry out food. Some leaves are simple, while others are compound, consisting of multiple leaflets. The flat part of the leaf, the blade, is often attached to the stem by a leafstalk.

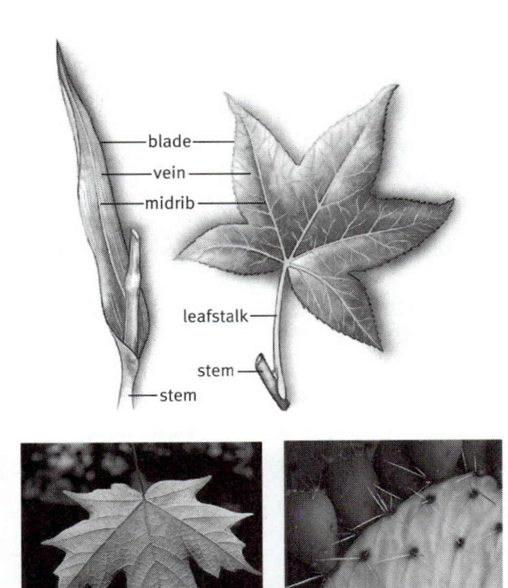

leaf

top to bottom: *renderings of monocotyledon* (left) *and eudicotyledon leaves; maple leaf* (left) *and cactus spines; fir needles, and field grass*

leaflet (lē′flĭt) A small leaf or leaflike part, especially one of the blades or divisions of a compound leaf.

leaf scar The mark left on a stem after a leaf falls. Leaf scars can be used to identify tree species in winter or from specimens of their twigs.

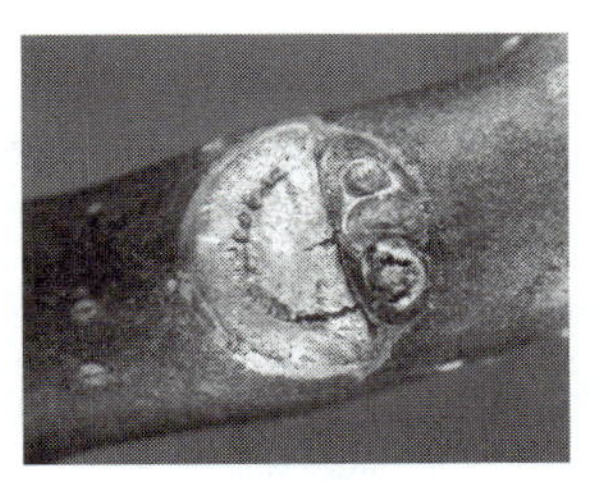

leaf scar

on a white ash (Fraxinus americana)

leaf spot Any of various plant diseases resulting in well-defined necrotic areas on the leaves.

leafstalk (lēf′stôk′) The slender, elongated structure by which the leaves of most plants are attached to the stem. Also called *petiole.*

Leakey (lē′kē) Family of British scientists. **Louis S(eymour) B(azett)** (1903–1972) is known for fossil discoveries of early humans that he made in close collaboration with his wife **Mary** (1913–1996). In 1959, while working in Tanzania, Africa, Mary Leakey uncovered skull and teeth fragments of a species the Leakeys named *Zinjanthropus,* since renamed *Australopithecus boisei.* The next year the Leakeys discovered remains of a larger-brained species, *Homo habilis.* Their discoveries provided powerful evidence that human ancestors were of greater age than was previously thought, and that they had evolved in Africa rather than in Asia. Their son **Richard** (born 1944) and his wife **Meave** (born 1942) have continued the family's research and discoveries. In 2001 Meave Leakey discovered a skull belonging to an entirely new genus, called *Kenyanthropus platyops* and believed to be 3.5 million years old.

Leakey

Louis (left) *and Mary*

BIOGRAPHY **Louis Leakey**

The discoveries made by the famous Leakey family of anthropologists and paleontologists are nowadays so familiar (at least in their general import) that we can easily forget how much they changed our views of hominid evolution. Before Louis and his wife Mary made their first major discoveries, it was widely thought that humans had originated in Asia. This was because our immediate ancestor, *Homo erectus,* had been discovered in East and Southeast Asia, and for a long time its best and oldest fossil remains came from there (such as the famous "Java Man" and "Peking Man"). The Leakeys' discoveries of an earlier hominid species, *Homo habilis,* showed not only that hominid evolution was a good deal earlier than previously thought, but also that it had been centered in East Africa. Interestingly, Louis Leakey had (at least indirectly) almost as much influence on the study of modern primates as he did on the study of ancient ones: he persuaded Jane Goodall and Dian Fossey to live among gorillas and chimpanzees to study their behavior over long periods of time.

leap second (lēp) A second of time, as measured by an atomic clock, added to or omitted from official timekeeping systems annually to compensate for changes in the rotation of the Earth. See more at **coordinated universal time.**

learning disability (lûr′nĭng) Any of various disabilities of the basic cognitive and psychological processes involved in using language or performing mathematical calculations. Learning disabilities are not caused by low intelligence, emotional disturbance, or physical impairment (as of hearing). Dyslexia is a common learning disability.

Leavitt (lē′vĭt), **Henrietta Swan** 1868–1921. American astronomer who discovered four novae and over 2,400 variable stars. She also developed a mathematical formula used to measure intergalactic distances.

lecithin (lĕs′ə-thĭn) A fatty substance present in most plant and animal tissues that is an important structural part of cell membranes, particularly in nervous tissue. It consists of a mixture of diglycerides of fatty acids (especially linoleic, palmitic, stearic, and oleic acid) linked to a phosphoric acid ester. Lecithin is used commercially in foods, cosmetics, paints, and plastics for its ability to form emulsions.

LED (ĕl′ē-dē′, lĕd) Short for *light-emitting diode.* An electronic semiconductor device that emits light when an electric current passes through it. They are considerably more efficient than incandescent bulbs, and rarely burn out. LEDs are used in many applications such as flat-screen video displays, and increasingly as general sources of light. See also **semiconductor laser.**

Lederberg (lĕd′ər-bûrg′, lā′dər-), **Joshua** 1925–2008. American geneticist who made important discoveries concerning the organization of the genetic material of bacteria and developed techniques for the manipulation and combination of genes. For this work he shared with American biochemists George Beadle and Edward Tatum the 1958 Nobel Prize for physiology or medicine.

Leeuwenhoek (lā′vən-ho͝ok′), **Anton van** 1632–1723. Dutch naturalist and pioneer of microscopic research. He was the first to describe protozoa, bacteria, and spermatozoa. He also made observations of yeasts, red blood cells, and blood capillaries, and traced the life histories of various animals, including the flea, ant, and weevil.

BIOGRAPHY **Anton van Leeuwenhoek**

As a young man Anton van Leeuwenhoek worked in a drapery store, where he used magnifying glasses to count thread densities. Perhaps inspired by Robert Hooke's *Micrographia* (an account of Hooke's microscopic investigations in botany, chemistry, and other branches of science, published in 1665), he began building microscopes. He examined hair, blood, insects, and other things around him, keeping detailed records and drawings of his observations. Although compound microscopes with more than one lens had been invented at the end of the fourteenth century, they were able to magnify objects only 20 to 30 times. Van Leeuwenhoek's single-lens microscopes were basically powerful magnifying glasses, but his superior lens-grinding skills and acute eyesight enabled him to magnify objects up to 200 times. Van Leeuwenhoek made each microscope for a specific investigation, and he had his specimens permanently mounted so he could study them as long as he wanted. His discoveries include protozoans (1674), blood cells (1674), bacteria (1676), spermatozoa (1677), and the structure of nerves (1717). By the time of his death at the age of ninety, van Leeuwenhoek had constructed more than 400 microscopes.

---

left brain (lĕft) The cerebral hemisphere located on the left side of the corpus callosum. The left brain controls activities on the right side of the body, and in humans, usually controls speech and language functions. The thought processes of logic and calculation are generally associated with the left brain.

legume (lĕg′yo͞om′, lə-gyo͞om′) **1.** Any of a large number of eudicot plants belonging to the family Leguminosae (or Fabaceae). Their characteristic fruit is a seed pod. Legumes live in a symbiotic relationship with bacteria in structures called nodules on their roots. These bacteria are able to take nitrogen from the air, which is in a form that plants cannot use, and convert it into compounds that the plants can use. Many legumes are widely cultivated for food, as fodder for livestock, and as a means of improving the nitrogen content of soils. Beans, peas, clover, alfalfa, locust trees, and acacia trees are all legumes. **2.** The seed pod of such a plant. —*Adjective* **leguminous.**

Leibniz (līb′nĭts), Baron **Gottfried Wilhelm von** 1646–1716. German philosopher and mathematician who invented the mathematical processes of differentiation and integration, which greatly expanded the field of calculus. Leibniz also established the foundations of probability theory and conceived the idea for a practical calculating machine.

leishmaniasis (lēsh′mə-nī′ə-sĭs) An infection or disease caused by any of the flagellate protozoans of the genus *Leishmania,* transmitted to humans and animals by bloodsucking sand flies and characterized by skin ulcerations or an acute illness marked by fever, anemia, and enlargement of the liver and spleen.

lemma (lĕm′ə) Plural **lemmas** or **lemmata** (lĕm′ə-tə). The outer or lower of the two bracts enclosing one of the flowers within a grass spikelet.

Lenoir (lĕ-nwär′), **Jean-Joseph-Étienne** 1822–1900. French inventor who in 1859 constructed the first practical internal-combustion engine, which was fueled by coal gas and air. He later built a car and a boat that were powered by this engine.

lens (lĕnz) **1.** A transparent structure behind the iris of the eye that focuses light entering the eye on the retina. **2a.** A piece of glass or plastic shaped so as to focus or spread light rays that pass through it, often for the purpose of forming an image. **b.** A combination of two or more such lenses, as in a camera or telescope. Also called *compound lens.* **3.** A device that causes radiation to converge or diverge by an action analogous to that of an optical lens. The system of electric fields used to focus electron beams in electron microscopes is an example of a lens.

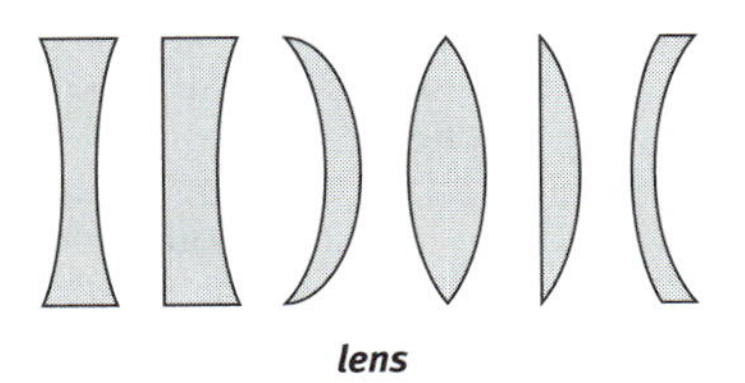

lens

left to right: *biconcave, plano-concave, concavo-convex, biconvex, plano-convex, and convexo-concave lenses*

lenticel (lĕn′tĭ-sĕl′) One of the small areas on the surface of the stems and roots of woody plants that allow the interchange of gases between the metabolically active interior tissue and the surrounding air or pockets of air in the soil. Lenticels are portions of the periderm that have numerous pores or intercellular spaces. They appear as raised circular or elongated areas. The dark lines in birch bark and the tiny dots sometimes seen on skin of apples and pears are lenticels.

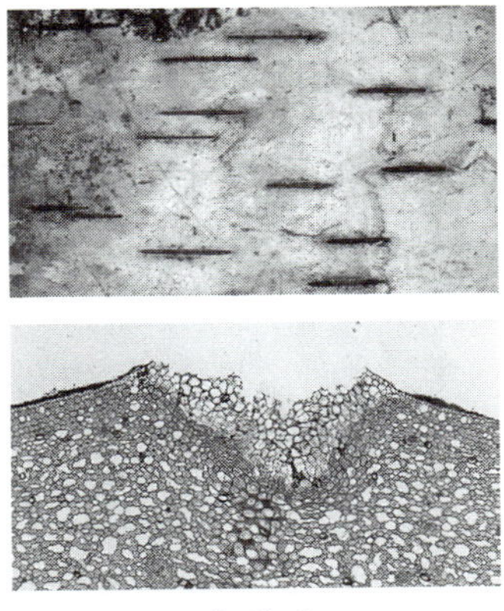

lenticel

top: *in the bark of a white birch;* bottom: *photomicrograph of a lenticel in a rhizome of a sweet flag*

lenticular galaxy (lĕn-tĭk′yə-lər) A galaxy having a central bulge surrounded by a flattened disk with no pattern of spiral arms. Lenticular galaxies are intermediate in the Hubble classification system between elliptical and spiral forms and are classified using the designation S0, or SB0 if they have a bar of stars, gas, and dust through the nucleus. Like spiral galaxies, they are disk-shaped with a central bulge, but like elliptical galaxies they contain relatively little gas and dust. Compare **elliptical galaxy, irregular galaxy, spiral galaxy.** See more at **Hubble classification system.**

Lenz's law (lĕnt′sĭz) A principle stating that an electric current, induced by a source such as a changing magnetic field, always creates a counterforce opposing the force inducing it. This law explains such phenomena as diamagnetism and the electrical properties of inductors. The law is named after its discoverer, German physicist Heinrich Lenz (1804–1865).

Leo (lē′ō) A constellation in the Northern Hemisphere near Cancer and Virgo. Leo (the Lion) contains the bright star Regulus and is the fifth sign of the zodiac.

Leonardo da Vinci (lē′ə-när′dō də vĭn′chē) 1452–1519. Italian artist, scientist, and inventor whose scientific insights were far ahead of their time. He investigated anatomy, geology, botany, hydraulics, optics, mathematics, meteorology, and mechanics. He also drew up designs for the first workable helicopter, parachute, and bicycle, all of which were eventually constructed centuries after his death using modern materials and technology.

Leonardo da Vinci

lepidopteran (lĕp′ĭ-dŏp′tər-ən) Any of various insects of the order Lepidoptera, characterized by four large, flat, membranous wings covered with small scales. The larvae of lepidopterans are caterpillars. Lepidopterans include butterflies, moths, and skippers.

leprosy (lĕp′rə-sē) A slowly progressive, chronic infectious disease caused by the bacterium *Mycobacterium leprae,* that damages nerves, skin, and mucous membranes, and can lead to loss of sensation, paralysis, gangrene, and deformity if untreated.

leptin (lĕp′tĭn′) A peptide hormone and neurotransmitter produced by fat cells and involved in the regulation of appetite.

lepton (lĕp′tŏn′) Any of a family of elementary particles that interact through the weak force and do not participate in the strong force. Leptons include electrons, muons, tau particles, and their respective neutrinos, the electron neutrino, the muon neutrino, and the tau neutrino. The antiparticles of these six particles are also leptons. Compare **hadron.** See Note at **elementary particle.** See Table at **subatomic particle.**

leucine (lo͞o′sēn′) An essential amino acid. *Chemical formula:* $C_6H_{13}NO_2$. See more at **amino acid.**

leukemia (lo͞o-kē′mē-ə) Any of various acute or chronic neoplastic diseases of the bone marrow in which unrestrained proliferation of white blood cells occurs, usually accompanied by anemia, impaired blood clotting, and enlargement of the lymph nodes, liver, and spleen. Certain viruses, genetic defects, chemicals, and ionizing radiation, are associated with an increased risk of leukemia, which is classified according to the cellular maturity of the involved white blood cells.

leukocyte also **leucocyte** (lo͞o′kə-sīt′) See **white blood cell.**

leukoplast (lo͞o′kə-plăst′) A colorless plastid in the cytoplasm of plant cells that makes and stores starch.

Levalloisian (lĕv′ə-loi′zē-ən) Relating to a technique for producing flaked stone tools that appeared in the late stages of the Acheulian (Lower Paleolithic) tool culture, characterized by a distinctive method of striking off flake tools from a prepared stone core. In the Levalloisian technique, large, sharp flakes were struck from the core with a single blow and could be used, possibly for skinning and butchering, without further flaking or finishing. Later refinements to the Levalloisian technique formed the basis for the Mousterian (Middle Paleolithic) technology.

levee (lĕv′ē) **1.** A long ridge of sand, silt, and clay built up by a river along its banks, especially during floods. **2.** An artificial embankment along a rivercourse or an arm of the sea, built to protect adjoining land from inundation.

lever (lĕv′ər) A simple machine consisting of a bar that pivots on a fixed support, or **fulcrum**, and is used to transmit **torque**. A force applied by pushing down on one end of the lever results in a force pushing up at the other end. If the fulcrum is not positioned in the middle of the lever, then the force applied to one end will not yield the same force on the other, since the torque must be the same on either side of the fulcrum. Levers, like gears, can thus be used to increase the force available from a mechanical power source. See more at **fulcrum.** See also **mechanical advantage.**

Levi-Montalcini (lē′vē-mŏn′tl-chē′nē, lĕ′vē-mōn′täl-), **Rita** Born 1909. Italian-born American developmental biologist who discovered the *nerve growth factor* (NGF), a bodily substance that stimulates the growth of nerve cells. For this work she shared with American biochemist Stanley Cohen the 1986 Nobel Prize for physiology or medicine.

levorotation (lē′və-rō-tā′shən) The counterclockwise rotation of the plane of polarization of light (as observed when looking straight into the incoming light) by certain substances, such as crystals, and by certain solutions. Levorotation is caused by a particular arrangement of the atoms in a molecule of the substance. Compare **dextrorotation.**

levorotatory (lē′və-rō′tə-tôr′ē) Relating to a substance that causes levorotation.

Leyden jar (līd′n) An early device for storing electric charge that uses the same principle as a modern **capacitor**. It consists of a glass jar with conductive metal foil covering its inner and outer surfaces, with the glass insulating these surfaces from each other. The inner surface is charged (by an external source) through an electrode penetrating the top of the jar; the inner and outer foil layers can then hold an equal and opposite charge.

Li The symbol for **lithium.**

Libra (lē′brə) A constellation in the Southern Hemisphere near Scorpius and Virgo. Libra (the Scales or Balance) is the seventh sign of the zodiac.

lichen (lī′kən) The mutualistic symbiotic association of a fungus with an alga or a cyanobacterium, or both. The fungal component of a lichen absorbs water and nutrients from the surroundings and provides a suitable environment for the alga or cyanobacterium. These live protected among the dense fungal hyphae and produce carbohydrates for the fungus by photosynthesis. Owing to this partnership, lichens can thrive in harsh environments such as mountaintops and polar regions. The more familiar lichens grow slowly as crusty patches, but lichens are found in a variety of forms, such as the tall, plantlike reindeer moss. The association between the different organisms in a lichen is so close that lichens are routinely referred to as a single organism, and scientists classify lichens using the name of the fungal component.

lidar (lī′där) **1.** A method of detecting distant objects and determining their position, velocity, or other characteristics by analysis of pulsed laser light reflected from their surfaces. Lidar operates on the same principles as **radar** and **sonar**. **2.** The equipment used in such detection. See also **Doppler effect, radar, sonar.**

lidocaine (lī′də-kān′) A synthetic amide, $C_{14}H_{22}N_2O$, used chiefly in the form of its hydrochloride as a local anesthetic.

Liebig (lē′bĭg), Baron **Justus von** 1803–1873. German chemist who was one of the first to investigate organic compounds and to develop techniques for their analysis. Liebig also first described the process now known as photosynthesis, and he made observations about the use of fertilizers that led to many improvements in agricultural practices.

life (līf) **1.** The properties or qualities that distinguish living plants and organisms from dead or inanimate matter, including the capacity to grow, metabolize nutrients, respond to stimuli, reproduce, and adapt to the environment. The definitive beginning and end of human life are complex concepts informed by medical, legal, sociological, and religious considerations. **2.** Living organisms considered as a group, such as the plants or animals of a given region.

life cycle The series of changes in the growth and development of an organism from its beginning as an independent life form to its mature state in which offspring are produced. In simple organisms, such as bacteria, the life cycle begins when an organism is produced by fission and ends when that organism in turn divides into two new ones. In organisms that reproduce sexually, the life cycle may be thought of as beginning with the fusion of reproductive cells to form a new organism. The cycle ends when that organism produces its own reproductive cells, which then begin the cycle again by undergoing fusion with other reproductive cells. The life cycles of plants, algae, and many protists often involve an alternation between a generation of organisms that reproduces sexually and another that reproduces asexually. See more at **alternation of generations.**

life science Any of several branches of science, such as biology, medicine, and ecology, that study the structural and functional organization of living organisms and their relationships to each other and the environment. Compare **physical science.**

lift (lĭft) An upward force acting on an object. Lift can be produced in many ways; for example, by creating a low-pressure area above an object, such an airplane wing or other airfoil that is moving through the air, or by lowering the overall density of an object relative to the air around it, as with a hot air balloon. Compare **drag.** See also **airfoil, buoyancy.** See Note at **aerodynamics.**

ligament (lĭg′ə-mənt) A sheet or band of tough fibrous tissue that connects two bones or holds an organ of the body in place.

light (līt) **1.** Electromagnetic radiation that can be perceived by the human eye. It is made up of electromagnetic waves with wavelengths between 4×10^{-7} and 7×10^{-7} meters. Light, and all other electromagnetic radiation, travels at a speed of about 299,728 km (185,831 mi) per second in a vacuum. See also **photon.** **2.** Electromagnetic energy of a wavelength just outside the range the human eye can detect, such as infrared light and ultraviolet light. See Note at **electromagnetic radiation.**

light chain The smaller of the two types of polypeptide chains in an antibody molecule. A light chain consists of an antigen-binding portion with a variable amino acid sequence, and a constant region with an amino acid sequence that is relatively unchanging. The light chains are attached to the heavy chains so that the variable regions of both lie alongside each other. Compare **heavy chain.**

light-emitting diode See **LED.**

light mill See **radiometer.**

lightning (līt′nĭng) A flash of light in the sky caused by an electrical discharge between clouds or between a cloud and the Earth's surface. The flash heats the air and usually causes thunder. Lightning may appear as a jagged streak, as a bright sheet, or in rare cases, as a glowing red ball.

lightning rod A grounded metal rod placed high on a structure to conduct electrical current from a lightning strike directly to the ground, preventing the currents from injuring people or animals or from damaging objects. Lightning rods usually have a sharp, pointed tip, since electric lines of force are more highly concentrated around pointed objects, in this case increasing the attractiveness of the rod compared with other nearby objects. See also **Saint Elmo's fire.**

light reaction Any of the chemical reactions that take place during the first stage of photosynthesis and require the presence of light. During the light reactions, energy captured from light by chlorophyll and its accessory pigments drives the production of ATP, the source of energy that is later used to drive the production of carbohydrates. Compare **dark reaction.** See more at **photosynthesis.**

light water Ordinary water, H_2O. Compare **heavy water.**

light-year The distance that light travels in a vacuum in one year, equal to about 9.46 trillion km (5.88 trillion mi). Light-years are used in measuring interstellar and intergalactic distances. Compare **astronomical unit, parsec.**

lignify (lĭg′nə-fī′) To make or become stiffer and stronger by the deposition of lignin.

lignin (lĭg′nĭn) A complex organic compound that binds to cellulose fibers and hardens and

A CLOSER LOOK lightning

As storm clouds develop, the temperature at the top of the cloud becomes much cooler than that at the bottom. For reasons that scientists still do not understand, this temperature difference results in the accumulation of negatively charged particles near the base and positively charged particles near the top of the storm cloud. The negatively charged particles repel the electrons of atoms in nearby objects, such as the bases of other storm clouds or tall objects on the ground. Consequently, these nearby objects take on a positive *charge.* The difference in charge, or *voltage,* builds until an electric current starts to flow between the objects along a pathway of charged atoms in the air. The current flow heats up the air to such a degree that it glows, generating *lightning.* Initially, a bolt of lightning carrying a negative charge darts from one storm cloud to another or from a storm cloud to the ground, leaving the bottom of the cloud with a positive charge. In response, a second bolt (*reverse lightning*) shoots in the opposite direction (from the other storm cloud or the ground) as the mass of negative charges on it moves back to neutralize the positive charge on the bottom of the first cloud. The heat generated by the lightning causes the air to expand, in turn creating very large sound waves, or *thunder.*

strengthens the cell walls of plants. Lignin is a polymer consisting of various aromatic alcohols, and is the chief noncarbohydrate constituent of wood.

lignite (lĭg′nīt′) A soft, brownish-black form of coal having more carbon than peat but less carbon than bituminous coal. Lignite is easy to mine but does not burn as well as other forms of coal. It is a greater polluter than bituminous coal because it has a higher sulphur content. Compare **anthracite, bituminous coal.**

ligule (lĭg′yo͞ol) A straplike structure, such as the corolla of fused petals in a ray flower or a membranous or hairy appendage between the sheath and blade of a grass leaf.

limb (lĭm) **1.** One of the appendages of an animal, such as an arm of a starfish, the flipper of dolphins, or the arm and leg of a human, used for locomotion or grasping. **2.** The expanded tip of a plant organ, such as a petal or corolla lobe. **3.** The circumferential edge of the apparent disk of a celestial body.

limbic system (lĭm′bĭk) A group of interconnected structures of the brain including the hypothalamus, amydala, and hippocampus that are located beneath the cortex, are common to all mammals, and are associated with emotions such as fear and pleasure, memory, motivation, and various autonomic functions.

lime (līm) A white, lumpy, caustic powder made of calcium oxide sometimes mixed with other chemicals. It is made industrially by heating limestone, bones, or shells. Lime is used as an industrial alkali, in waste treatment, and in making glass, paper, steel, insecticides, and building plaster. It is also added to soil to lower its acidity.

limestone (līm′stōn′) A sedimentary rock consisting primarily of calcium carbonate, often in the form of the minerals calcite or aragonite, and sometimes with magnesium carbonate in the form of dolomite. Minor amounts of silica, feldspar, pyrite, and clay may also be present. Limestone can occur in many colors but is usually white, gray, or black. It forms either through the accumulation and compaction of fossil shells or other calcium-carbonate based marine organisms, such as coral, or through the chemical precipitation of calcium carbonate out of sea water.

limit (lĭm′ĭt) A number or point for which, from a given set of numbers or points, one can choose an arbitrarily close number or point. For example, for the set of all real numbers greater than zero and less than one, the numbers one and zero are limit points, since one can pick a number from the set arbitrarily close to one or zero (even though one and zero are not themselves in the set). Limits form the basis for **calculus**, where a number L is defined to be the limit approached by a function $f(x)$ as x approaches a if, for every positive number ε, there exists a number δ such that $|f(x)-L| < \varepsilon$ if $0 < |x-a| < \delta$.

limit-cycle A periodic attractor. See more at **attractor.**

limnology (lĭm-nŏl′ə-jē) The scientific study of the organisms living in and the phenomena of fresh water, especially lakes and ponds.

linac (lĭn′ăk′) Short for **linear accelerator.**

lindane (lĭn′dān) A white crystalline powder that is an isomer of benzene hexachloride, banned as an agricultural pesticide because of its toxicity but still used topically to treat scabies and pediculosis. *Chemical formula:* $\mathbf{C_6H_6Cl_6}$.

line (līn) A geometric figure formed by a point moving in a fixed direction and in the reverse direction. The intersection of two planes is a line. ► The part of a line that lies between two points on the line is called a **line segment.**

linear (lĭn′ē-ər) Being or resembling a line.

linear accelerator A type of **particle accelerator** that accelerates charged subatomic particles, such as protons and electrons, in a straight line by means of alternating negative and positive impulses from electric fields. Linear accelerators were largely supplanted by cyclotrons and other architectures that require less path length to achieve the same or higher particle velocities.

linear algebra The branch of mathematics that deals with the theory of systems of linear equations, matrices, vector spaces, and linear transformations.

linear equation An algebraic equation, such as $y = 4x + 3$, in which the variables are of the first degree (that is, raised only to the first power). The graph of such an equation is a straight line.

linear momentum See **momentum.**

line of force A line used to indicate the direction of a field, especially an electric or magnetic field, at various points in space. The tangent of a line of force at each point indicates the orientation of the field at that point. Arrows are usually used to indicate the direction of the force. See Note at **magnetism.**

line spectrum An image of colored lines or bands of light formed in optical **spectroscopy**, each line representing one of the frequencies in the spectrum of a light source. The light source is usually broken into individual bands by a prism or a diffraction grating.

link (lĭngk) A segment of text or a graphical item that serves as a cross-reference between parts of a webpage or other hypertext documents or between webpages or other hypertext documents.

linkage group (lĭng′kĭj) A pair or set of genes that are close to each other on a chromosome and tend to be transmitted together.

Linnaeus (lĭ-nē′əs, lĭ-nā′əs), **Carollus.** Originally **Carl von Linné.** 1707–1778. Swedish naturalist who in 1735 introduced a method for classifying plants and animals using generic and specific designations. This laid the foundation for the modern system of binomial nomenclature. See more at **binomial nomenclature.**

Linnaeus

Linnean also **Linnaean** (lĭ-nē′ən) Relating to the system of taxonomic classification and binomial nomenclature originated by Carolus Linnaeus. In the Linnean system, organisms are grouped according to shared characteristics into a hierarchical series of fixed categories ranging from subspecies at the bottom to kingdom at the top. Compare **cladistics.**

linoleic acid (lĭn′ə-lē**′**ĭk) An unsaturated fatty acid that has two double bonds and is a nutrient essential for prostaglandin production in the human body. It is an important component of many vegetable oils, such as linseed, soybean, peanut, corn, and safflower oil, and is also found in meat and dairy products. *Chemical formula:* $C_{18}H_{32}O_2$.

linolenic acid (lĭn′ə-lĕn**′**ĭk) An unsaturated fatty acid that has three double bonds and is a nutrient essential to the formation of prostaglandins in the human body. Linolenic acid belongs to the class of omega-3 fatty acids. It is an important component of natural drying oils (such as linseed oil) and is also found in some fish oils. *Chemical formula:* $C_{18}H_{30}O_2$.

lipase (lĭp**′**ās′, lī**′**pās′) Any of various enzymes that catalyze the hydrolysis of fats, especially triglycerides and phospholipids, into glycerol and fatty acids.

lipid (lĭp**′**ĭd) Any of a large group of organic compounds that are oily to the touch and insoluble in water. Lipids include fatty acids, oils, waxes, sterols, and triglycerides. They are a source of stored energy and are a component of cell membranes.

lipoprotein (lĭp′ō-prō**′**tēn′, lī′pō-) Any of a group of conjugated proteins in which at least one of the components is a lipid. Lipoproteins, classified according to their densities and chemical qualities, are the principal means by which lipids are transported in the blood. See also **high-density lipoprotein, low-density lipoprotein.**

liquefaction (lĭk′wə-făk**′**shən) **1.** *Chemistry.* The act or process of turning a gas into a liquid. Liquefaction is usually achieved by compression of vapors (provided the temperature of the gas is below the critical temperature), by refrigeration, or by adiabatic expansion. **2.** *Geology.* The process by which sediment that is very wet starts to behave like a liquid. Liquefaction occurs because of the increased pore pressure and reduced effective stress between solid particles generated by the presence of liquid. It is often caused by severe shaking, especially that associated with earthquakes.

liquid (lĭk**′**wĭd) One of four main **states of matter**, composed of molecules that can move about in a substance but are bound loosely together by intramolecular forces. Unlike a solid, a liquid has no fixed shape, but instead has a characteristic readiness to flow and therefore takes on the shape of any container. Because pressure transmitted at one point is passed on to other points, a liquid usually has a volume that remains constant or changes only slightly under pressure, unlike a gas.

liquid crystal Any of various liquids in which molecules are regularly arrayed like a solid crystal along one or two dimensions, but are free in the other dimensions as with typical liquids. Liquid crystals often display unusual and often manipulable optical properties such as anisotropic scattering. See more at **LCD.**

liquid-crystal display See **LCD.**

liquid measure A system of units for measuring liquid volume or capacity. Compare **dry measure.**

liquidus (lĭk**′**wĭ-dəs) The minimum temperature at which all components of a mixture (such as an alloy) can be in a liquid state. Below the liquidus the mixture will be partly or entirely solid. See illustration at **eutectic.** Compare **solidus.**

Lister (lĭs**′**tər), First Baron. Title of **Joseph Lister.** 1827–1912. British surgeon who, influenced by Pasteur's germ theory of disease, established in 1865 a system of antiseptic measures in hospitals to combat infections. His practices dramatically decreased the number by deaths caused by infection and were gradually adopted in hospitals throughout Europe.

Joseph Lister

liter (lē′tər) **1.** The basic unit of liquid volume or capacity in the metric system, equal to 1.06 quart or 2.12 pints. See Table at **measurement. 2.** The basic unit of dry volume or capacity in the metric system, equal to 0.90 quart or 1.82 pint. See Table at **measurement.**

lithium (lĭth′ē-əm) *Symbol* **Li** A soft, silvery metallic element of the alkali group that occurs in small amounts in some minerals. It is the lightest of all metals and is highly reactive. Lithium is used to make alloys, batteries, glass for large telescopes, and ceramics. Atomic number 3; atomic weight 6.941; melting point 179°C; boiling point 1,317°C; specific gravity 0.534; valence 1. See **Periodic Table.**

lithology (lĭ-thŏl′ə-jē) **1.** The scientific study and description of rocks, especially at the macroscopic level, in terms of their color, texture, and composition. **2.** The gross physical character of a rock or rock formation.

lithophyte (lĭth′ə-fīt′) A plant that grows on rock and derives its nourishment chiefly from the atmosphere.

lithosphere (lĭth′ə-sfîr′) The outer part of the Earth, consisting of the crust and upper mantle. It is about 55 km (34 mi) thick beneath the oceans and up to about 200 km (124 mi) thick beneath the continents. The high velocity with which seismic waves propagate through the lithosphere suggests that it is completely solid. Compare **asthenosphere, atmosphere, hydrosphere.**

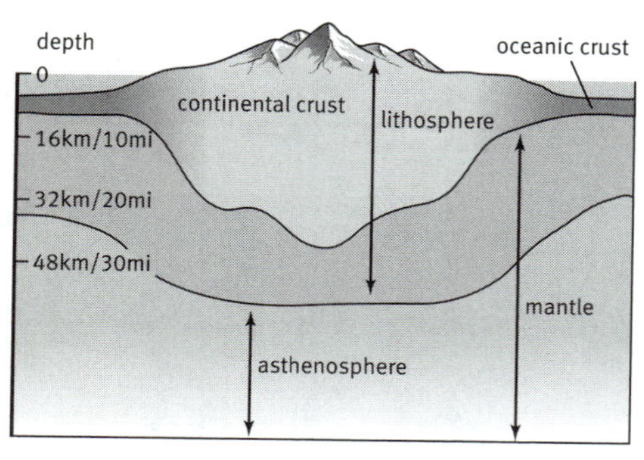

lithosphere

lithostratigraphy (lĭth′ō-strə-tĭg′rə-fē) The scientific study and categorization of rock strata based on their lithology (color, texture, and composition). Compare **biostratigraphy.**

lithotripsy (lĭth′ə-trĭp′sē) The procedure of crushing a stone in the urinary bladder or urethra by means of a **lithotriptor**, a device that passes shock waves through a water-filled tub in which the patient sits. The resulting stone fragments are small enough to be expelled in the urine.

litmus (lĭt′məs) A colored powder, obtained from certain lichens, that changes to red in an acid solution and to blue in an alkaline solution. Litmus is a mixture of various closely related heterocyclic organic compounds. ▸ Litmus is typically added to paper to make **litmus paper**, which can be used to determine whether a solution is basic or acidic by dipping a strip of the paper into the solution and seeing how the paper changes color.

Little Dipper (lĭt′l) An asterism composed of seven stars in the constellation Ursa Minor that form the outline of a dipper.

Little Ice Age The period from about 1400 to 1900, characterized by expansion of mountain glaciers and cooling of global temperatures, especially in the Alps, Scandinavia, Iceland, and Alaska. The Little Ice Age followed the Medieval Warm Period. See also **Maunder minimum.**

littoral (lĭt′ər-əl) Relating to the coastal zone between the limits of high and low tides. The littoral zone is subject to a wide range of environmental conditions, including high-energy wave action and intermittent periods of flooding and drying along with the associated fluctuations in exposure to solar radiation and extremes of temperature. Compare **sublittoral.**

liver (lĭv′ər) **1.** A large glandular organ in the abdomen of vertebrate animals that is essen-

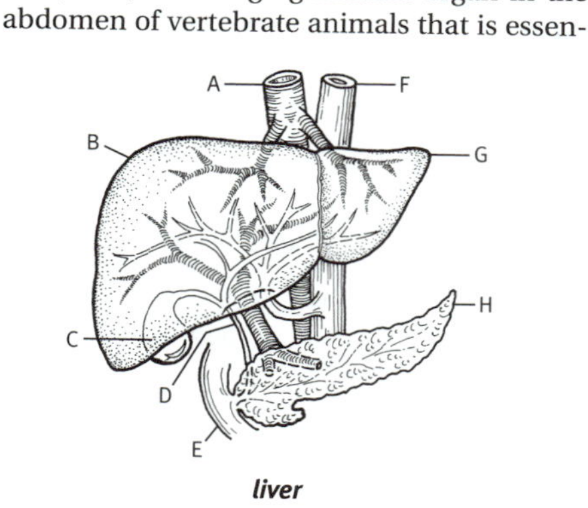

liver

A. *inferior vena cava*, B. *right lobe*, C. *gallbladder*, D. *bile duct*, E. *duodenum*, F. *abdominal aorta*, G. *left lobe*, H. *pancreas*

tial to many metabolic processes. The liver secretes bile, stores fat and sugar as reserve energy sources, converts harmful substances to less toxic forms, and regulates the amount of blood in the body. **2.** A similar organ of invertebrate animals.

liverwort (lĭv′ər-wûrt′, -wôrt′) Any of about 6,000 species of bryophyte plants belonging to the phylum Hepatophyta. Many liverworts reproduce asexually by means of **gemmae.** They also reproduce sexually, and their free-swimming sperm, produced in structures called antheridia, require liquid water, such as splashing raindrops, to reach the egg-producing archegonia. After fertilization, the small sporophyte grows directly on or in the gametophyte and is nourished by it. Liverworts are common in the tropics and often grow in moist soil, on damp rocks, and on tree trunks. Some liverworts have leafy bodies, while others have only a simple thallus. The name *liverwort* comes from the liverlike shape of the thalli of some species. See more at **bryophyte.**

liverwort

Gametophytes of a liverwort, including a flat thallus (foreground). *The disc-shaped structures on stalks* (right) *contain antheridia, while the umbrella-shaped structures* (left) *contain archegonia.*

ln The symbol for **natural logarithm.**

load (lōd) **1.** The resistance, weight, or power drain sustained by a machine or electrical circuit. Compare **effort. 2.** The power output of a generator or power plant. **3.** The amount of a pathogen or toxic substance present in an organism.

loam (lōm) Soil composed of approximately equal quantities of sand, silt, and clay, often with variable amounts of decayed plant matter.

Lobachevski (lō′bə-chĕf′skē), **Nikolai Ivanovich** 1792–1856. Russian mathematician who was a pioneer of non-Euclidean geometry.

lobe (lōb) **1.** A rounded projection, as on a leaf or petal. The leaves of many oak species have prominent lobes. **2.** An anatomical division of an organ of the body. The liver, lungs, and brain are all characterized by lobes that are held in place by connective tissue.

lobe-finned fish Any of various fishes of the class Sarcopterygii, having fins that are rounded and fleshy, suggesting limbs. One group of lobe-finned fish are thought to be ancestors of amphibians and other land-dwelling vertebrate animals. They first appeared in the Ordovician Period and are extinct except for the coelacanth and lungfish. Also called *sarcopterygian.* Compare **ray-finned fish.** See also **crossopterygian.**

lobotomy (lə-bŏt′ə-mē) Surgical incision into the frontal lobe of the brain to sever one or more nerve tracts, a technique formerly used to treat certain psychiatric disorders but now rarely performed.

local area network (lō′kəl) See **LAN.**

Local Group A group of more than 30 galaxies that includes 2 large spiral galaxies—the Milky Way and Andromeda—as well as numerous smaller galaxies, many of which are dwarfs. The Local Group has a diameter of approximately 10 million light-years, with its gravitational center located between our galaxy and the Andromeda galaxy. It is part of a larger grouping known as the Virgo supercluster.

local mean time Solar time as measured by the position of the mean sun with respect to an observer's local meridian. Like apparent time, local mean time differs continuously with the observer's longitude and is not standardized over a time zone. However, a day as measured by local mean time does not vary in length throughout the year—it is always 24 hours. Compare **apparent time.** See more at **mean time, solar time.**

local meridian See **celestial meridian** (sense 1).

locomotion (lō′kə-mō′shən) The movement of an organism from one place to another, often by the action of appendages such as flagella, limbs, or wings. In some animals, such as fish, locomotion results from a wavelike series of muscle contractions.

locule (lŏk′yo͞ol) A small cavity or compartment within an organ or part of an animal or plant, as any of the cavities within a plant ovary in which the ovules develop.

locus (lō′kəs) Plural **loci** (lō′sī′, -kē, -kī′). **1.** The set or configuration of all points whose coordinates satisfy a single equation or one or more algebraic conditions. **2.** The position that a given gene occupies on a chromosome.

lode (lōd) A vein of mineral ore that is deposited between clearly demarcated layers of rock or that fills a fissure in a rock formation.

lodestar also **loadstar** (lōd′stär′) A star, especially Polaris, that is used as a point of reference.

lodestone also **loadstone** (lōd′stōn′) A piece of the mineral magnetite that acts like a magnet.

lodicule (lŏd′ĭ-kyo͞ol′) One of two or three small rounded bodies at the base of the carpel of a grass flower. The swelling of the lodicules forces apart the flower's bracts, exposing the flower's reproductive organs.

loess (lō′əs, lĕs, lŭs) A very fine grained silt or clay, thought to have formed as the result of grinding by glaciers and to have been deposited by the wind. Most loess is believed to have originated during the Pleistocene Epoch from areas of land covered by glaciers and from desert surfaces.

Loewi (lĕv′ē), **Otto** 1873–1961. German pharmacologist who, with Sir Henry Dale, investigated the chemical transmissions of nerve impulses. For this work they shared the 1936 Nobel Prize for physiology or medicine.

Löffler (lŭf′lər), **Friedrich** 1852–1915. German bacteriologist who in 1884 demonstrated that diphtheria was caused by a bacillus described by Edwin Klebs a year earlier. This bacillus is now named after both scientists. Löffler also isolated an organism that causes food poisoning and developed a vaccine against foot-and-mouth disease (1899).

log (lôg) A logarithm.

logarithm (lô′gə-rĭ*th*′əm) The power to which a base must be raised to produce a given number. For example, if the base is 10, then the logarithm of 1,000 (written log 1,000 or $\log_{10}$ 1,000) is 3 because $10^3 = 1{,}000$. See more at **common logarithm, natural logarithm.**

logic (lŏj′ĭk) The study of the principles of reasoning, especially of the structure of propositions as distinguished from their content and of method and validity in deductive reasoning.

logical operation (lŏj′ĭ-kəl) also **logic operation** A function on binary variables whose output is also a binary variable. Logical operations are the function of logic gates in digital circuits. Logical operations include AND, OR, NOT, and combinations of those operations. See more at **Boolean algebra.**

logical operator also **logic operator** A symbol, as in a programming language, or a function that denotes a logical operation.

logic circuit A computer switching circuit, consisting of one or more logic gates, used to process digital information such as computer instructions and data.

logic gate A device, usually an electrical circuit, that performs one or more logical operations on one or more input signals. Logic gates are the building blocks of digital technology.

loment (lō′mĕnt′) An indehiscent legume (a seed pod that does not split open) that is divided into separate seed-bearing segments, giving it a jointed appearance. The segments break off one by one at maturity. It is the characteristic fruit of many leguminous plants, such as the tick trefoil and the crown vetch.

long-day plant (lông′dā′) A plant that flowers only after being exposed to light periods longer than a certain critical length, as in summer. Spinach, lettuce, and some varieties of wheat are long-day plants. Compare **day-neutral plant, short-day plant.** See more at **photoperiodism.**

longitude (lŏn′jĭ-to͞od′) **1.** A measure of relative position east or west on the Earth's surface, given in degrees from a certain meridian, usually the prime meridian at Greenwich, England, which has a longitude of 0°. The distance of a degree of longitude is about 69 statute miles or 60 nautical miles (111 km) at the equator, decreasing to zero at the poles. Longitude and latitude are the coordinates used to identify any point on the Earth's surface. Compare **latitude. 2.** Celestial longitude.

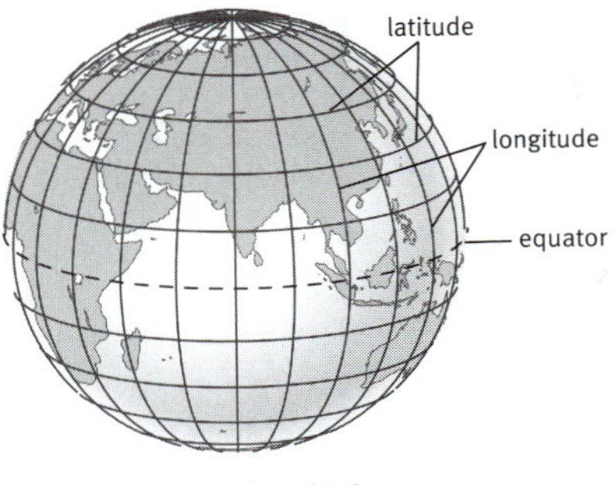

longitude

longitudinal dune (lŏn′jĭ-to͞od′n-əl) A large, elongated dune lying parallel to the prevailing wind direction. Longitudinal dunes usually have symmetrical cross sections. They generally form in areas that are located behind an obstacle where sand is abundant and the wind is constant and strong. They are usually tens of meters high and up to 100 km (62 mi) long. Compare **transverse dune.**

longitudinal wave A wave that oscillates back and forth on an axis that is the same as the axis along which the wave propagates. Sound waves are longitudinal waves, since the air molecules are displaced forward and backward on the same axis along which the sound travels. Compare **transverse wave.** See more at **wave.**

long ton See **ton** (sense 2).

Lonsdale (lŏnz′dāl′), Dame **Kathleen Yardley** 1903–1971. Irish physicist noted for using x-ray analysis to show that the carbon atoms in the benzene ring are arranged hexagonally and are in the same plane (1929).

loop of Henle (lo͞op əv hĕn′lē) The loop-shaped segment of the nephron of a vertebrate kidney. It plays a role in the transport of ions and water and the concentrating of urine.

lophophore (lŏf′ə-fôr′) A horseshoe-shaped ciliated organ located near the mouth of brachiopods, bryozoans, and phoronids that is used to gather food.

lopolith (lŏp′ə-lĭth) A large, bowl-shaped body of igneous rock intruded between layers of sedimentary rock. Lopoliths are usually connected to a dike and are typically tens of kilometers thick and hundreds of kilometers wide.

loran (lôr′ăn′) A long-range navigational system, in which a receiver's position is determined by an analysis involving the time intervals between pulsed radio signals from two or more pairs of ground stations of known position. The difference in the timing of the received signals corresponds to differences in distance from the transmitters, and the position of the receiver can be calculated by **triangulation.** Compare **Global Positioning System.**

Lorentz (lôr′ənts), **Hendrik Antoon** 1853–1928. Dutch physicist who was one of the first to develop theories of the electron, for which he shared the 1902 Nobel Prize for physics with Pieter Zeeman. His ideas on the invariance of physical laws with respect to time and space paved the way for Albert Einstein's theory of Special Relativity.

Lorentz-FitzGerald contraction (lôr′ənts-fĭts-jĕr′ld) The shortening of an object along its direction of motion as its speed approaches the speed of light, as measured by an observer at rest with respect to the body. Lorentz-FitzGerald contraction is an effect predicted by Einstein's theory of Special Relativity. It is named for Dutch physicist Hendrik Lorentz and Irish physicist George Francis FitzGerald (1851–1901), who independently proposed such a contraction. See more at **relativity.**

Lorentz force The total force exerted on a charged particle by electric and magnetic fields. All charged particles encounter a force from an electric field, oriented in the direction of the field (or the opposite direction, depending on the sign of the charge), while moving charged particles also encounter a force oriented at a right angle to both the direction of motion and the magnetic field. The Lorentz force is the driving force in electromagnets and is responsible for the Hall effect. The Lorentz force is named for Henrik Lorentz. See also **electromagnetism.**

Lorentz transformation A linear map that expresses the time and space coordinates of one reference frame in terms of those of another one. Much like simple rotations, which leave the lengths of objects unchanged while transforming their coordinates, Lorentz transformations leave unchanged the expression $c^2t^2 - x^2 - y^2 - z^2$, where c is the speed of light, and the other variables are space-time coordinates. ▸ A **pure Lorentz transformation**, a kind of **boost**, relates the reference frames of two inertial systems that are moving with a constant relative velocity. ► **General Lorentz transformations** include pure Lorentz transformations as well as rotations of the spatial coordinate system. See also **invariance, Special Relativity.**

Lou Gehrig's disease (lo͞o′ gĕr′ĭgz) See **amyotrophic lateral sclerosis.**

Love wave (lŭv) A type of seismic surface wave in which particles move with a side-to-side motion perpendicular to the main propagation of the earthquake. The amplitude of this motion decreases with depth. Love waves cause the rocks they pass through to change in shape. They travel faster than **Rayleigh waves.** Love waves are named after their discoverer, British mathematician Augustus Love (1863–1940).

low blood pressure (lō) See **hypotension.**

low-density lipoprotein A lipoprotein that contains relatively high amounts of cholesterol and is associated with an increased risk of atherosclerosis and coronary artery disease.

lower (lō′ər) Being an earlier division of the geological or archaeological period named. Compare **upper.**

lower bound A number that is less than or equal to every number in a given set.

lowest common denominator (lō′ĭst) The least common multiple of the denominators of a set of fractions. For example, the lowest common denominator of $\frac{1}{3}$ and $\frac{3}{4}$ is 12.

low-tension Having a low voltage, or designed to work at low voltages. Compare **high-tension.**

low-test Relating to gasoline with low volatility and a high boiling point.

low tide 1. The tide at its lowest level at a particular time and place. The lowest tides reached under normal meteorological conditions (the **spring tides**) take place when the Moon and Sun are directly aligned with respect to Earth. Low tides are less extreme when the Moon and Sun are at right angles (the **neap tides**). Storms and other meteorological conditions can greatly affect the height of the tides as well. See more at **tide. 2.** The time at which a low tide occurs.

Lr The symbol for **lawrencium.**

LSD (ĕl′ĕs-dē′) Short for *lysergic acid diethylamide.* A crystalline compound that is a synthetic derivative of lysergic acid. It is used as a powerful hallucinogenic drug. *Chemical formula:* $C_{20}H_{25}N_3O$.

Lu The symbol for **lutetium.**

lumbar (lŭm′bər) Located at or near the part of the back lying between the lowest ribs and the hips.

lumbar puncture The insertion of a hollow needle beneath the arachnoid membrane of the spinal cord in the lumbar region to withdraw cerebrospinal fluid for diagnostic purposes or to administer medication.

lumen (lo͞o′mən) Plural **lumens** or **lumina. 1.** The central space within a tube-shaped body part or organ, such as a blood vessel or the intestine. **2.** The SI derived unit used to measure the amount of light passing through a given area per second. One lumen is equal to the luminous flux passing per unit solid angle from a light source with a strength of one candela.

luminance (lo͞o′mə-nəns) The luminous intensity of a light source per unit area. Occasionally the **lambert** unit is used to measure luminance. Also called *photometric brightness.*

luminescence (lo͞o′mə-nĕs′əns) **1.** The emission of light as a result of the excitation of atoms by energy other than heat. Bioluminescence, fluorescence, and phosphorescence are examples of luminescence that can be produced by biological or chemical processes. **2.** The light produced in this way.

luminous flux (lo͞o′mə-nəs) A measure of the radiant power of light emitted from a source without regard for the direction in which it is emitted. It is measured in lumens. See also **luminous intensity.**

luminous intensity The luminous flux per unit solid angle (per steradian), as measured in a given direction relative to a light source. Its unit is the candela. See also **luminance.**

lunar (lo͞o′nər) **1.** Relating to the Moon. **2.** Measured by the revolution of the Moon around the Earth.

lunar eclipse See under **eclipse.**

lunar month The average time between successive occurrences of the same phase of the Moon, equal to approximately 29 days, 12 hours, 44 minutes. Depending on the lunar calendar being used, a lunar month can begin with the new moon or the full moon. Also called *lunation, synodic month.* ▸ A period of 12 lunar months is a **lunar year**.

lung (lŭng) **1.** Either of two spongy organs in the chest of air-breathing vertebrate animals that serve as the organs of gas exchange. Blood flowing through the lungs picks up oxygen from inhaled air and releases carbon dioxide, which is exhaled. Air enters and leaves the lungs through the bronchial tubes. **2.** A similar organ found in some invertebrates.

lungfish (lŭng′fĭsh′) Any of several tropical freshwater fish of the order or subclass Dipnoi that, in addition to having gills, have lunglike organs for breathing air. Lungfish have a long, narrow body, and certain species can survive periods of drought inside a mucus-lined cocoon in the mud. The lungfish and the coelacanths are the only living lobe-finned fishes.

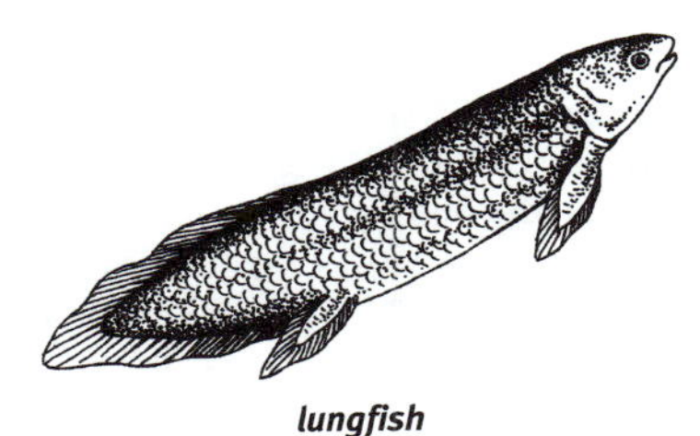

lungfish

Australian lungfish

lunisolar (lo͞o′nĭ-sō′lər) Relating to or caused by both the Sun and the Moon. Lunisolar gravitational attraction is the principle force influencing the Earth's tides and the cause of the Earth's orbital precession.

lunitidal (lo͞o′nĭ-tīd′l) Relating to tidal phenomena caused by the Moon. ▸ The length of time between the transit of the Moon over the meridian at a particular location and the following high or low tide at that location is known as the **lunitidal interval**.

lupine (lo͞o′pīn′) Characteristic of or resembling wolves.

lupus (lo͞o′pəs) See **systemic lupus erythematosus.**

Luria (lo͝or′ē-ə), **Salvador Edward** 1912–1991. Italian-born American biologist whose research on gene mutation and bacteria increased scientific understanding of the role of DNA in bacterial viruses.

luster (lŭs′tər) The shine from the surface of a mineral. Luster is important in describing different kinds of minerals. It is usually characterized as metallic, glassy, pearly, or dull.

luteinizing hormone (lo͞o′tē-ə-nī′zĭng) A glycoprotein hormone produced by the anterior portion of the pituitary gland. Luteinizing hormone stimulates ovulation and the development of the corpus luteum in female mammals and the production of testosterone in males.

lutetium (lo͞o-tē′shē-əm) *Symbol* **Lu** A silvery-white metallic element of the lanthanide series that is used in nuclear technology. Its radioactive isotope is used in determining the age of meteorites. Atomic number 71; atomic weight 174.97; melting point 1,663°C; boiling point 3,395°C; specific gravity 9.840 (at 25°C); valence 3. See **Periodic Table.**

lux (lŭks) Plural **luxes** or **luces** (lo͞o′sēz). A SI derived unit of **illuminance** in photometry, equal to one lumen per square meter.

L wave See **Love wave.**

Lwoff (lwôf), **André Michel** 1902–1994. French microbiologist who studied the genetics of bacterial viruses and explained how they reproduce. His findings have been important in cancer research and in understanding how viruses resist drugs.

lycopene (lī′kə-pēn′) A red carotenoid found chiefly in blood, the reproductive organs, tomatoes, and palm oils. It is an antioxidant and is the parent substance from which all natural carotenoids are derived. *Chemical formula:* $C_{40}H_{56}$.

lycophyte (lī′kə-fīt′) Any of various seedless vascular plants belonging to the phylum Lycophyta and characterized by microphylls (primitive leaves found in ancient plants). Among lycophytes, the sporophyte is the dominant generation (the large plant). Lycophytes first appeared in the Devonian period, and lycophyte trees were abundant in the ancient forests of the Carboniferous period. Modern lycophytes include such plants as the club mosses and the resurrection plant (*Selaginella lepidophylla*).

lye (lī) A strong alkaline solution or solid of potassium hydroxide or sodium hydroxide, made by allowing water to wash through wood ashes. It is used to make soap and drain and oven cleaners. *Chemical formula:* **KOH** or **NaOH.**

Lyell (lī′əl), Sir **Charles** 1797–1875. Scottish geologist who is considered one of the founders of modern geology. He is most famous for his principle of **uniformitarianism,** as first set forth in his three-volume *Principles of Geology* (1830–1833).

BIOGRAPHY **Sir Charles Lyell**

As a boy, Charles Lyell collected butterflies. This hobby might seem a far cry from his later professional work in geology, but in fact the two were closely linked. At that young age, he was already a keen observer of nature with excellent instincts for comparison. Throughout his life, Lyell traveled and collected observations on natural phenomena as he had once collected butterflies. These observations convinced him even as a young man that nature was to be understood through genuinely *natural* processes rather than as the result of supernatural forces or catastrophic events. Thus was born the theory of *uniformitarianism,* which maintained that the same geological processes had been at work in the same way throughout Earth's history, and that major features such as mountains showed that the Earth was very old, since geological processes worked very slowly. To convince people of his notions, Lyell needed to back them up with facts, and the three volumes of his pathbreaking *Principles of Geology* (1830–1833) are notable for being chock-full of geological facts. This principled method of thinking and marshaling evidence, no less than his radical ideas about the Earth's history, was itself revolutionary for the time (something perhaps not appreciated today). By claiming that the Earth was many millions rather than a few thousands of years old, Lyell opened up vast new possibilities for other thinkers, most notably Charles Darwin, whose theory of evolution by natural selection also required time for slow, incremental changes in the history of life.

Lyme disease (līm) A disease caused by the bacterial spirochete *Borrelia burgdorferi,* transmitted by deer ticks and characterized initially by a bull's-eye-shaped rash followed by flu-like symptoms such as fever, joint pain, and headache. If untreated, it can result in chronic arthritis and neurologic or cardiac dysfunction. It is named after Lyme, Connecticut, where it was first reported.

lymph (lĭmf) The clear fluid flowing through the lymphatic system that serves to bathe and nourish the tissues of the body. It is composed of blood plasma that has leaked out through the capillaries into the tissues.

lymphatic system (lĭm-făt′ĭk) A network of vessels, tissues, and organs in vertebrate animals that helps the body regulate fluid balance and fight infection. The vessels of the lymphatic system drain excess fluid, called lymph, from the tissues and return it to the circulating blood. Lymphocytes circulate throughout the lymphatic system.

lymph node A bean-shaped mass of tissue found at intervals along the vessels of the lymphatic system. Lymph nodes filter foreign substances from the blood.

lymphocyte (lĭm′fə-sīt′) Any of various white blood cells, including B cells and T cells, that function in the body's immune system by recognizing and deactivating specific foreign substances called antigens. B cells act by stimulating the production of antibodies. T cells contain receptors on their cell surfaces that are capable of recognizing and binding to specific antigens. Lymphocytes are found in the lymph nodes and spleen and circulate continuously in the blood and lymph.

lymphokine (lĭm′fə-kīn′) Any of various cytokines released by T cells that have been activated by specific antigens. Lymphokines act as mediators in the immune response by activating macrophages and stimulating lymphocyte production. Interferon is a lymphokine.

lymphoma (lĭm-fō′mə) Plural **lymphomas** or **lymphomata** (lĭm-fō′mə-tə). Any of various usually malignant tumors that arise in the lymph nodes or other lymphatic tissues, often manifested by painless enlargement of one or more lymph nodes. See also **Hodgkin's disease.**

Lyra (lī′rə) A constellation in the Northern Hemisphere near Cygnus and Hercules. Lyra (the Lyre) contains the bright star Vega.

lysergic acid (lĭ-sûr′jĭk, lī-) A crystalline alkaloid that is a major constituent of ergot. It is used in medical research to induce hallucinations, delusions, and other symptoms of psychosis. The drug LSD is a derivative of lysergic acid. *Chemical formula:* $\mathbf{C_{16}H_{16}N_2O_2}$.

lysine (lī′sēn′) An essential amino acid. *Chemical formula:* $\mathbf{C_6H_{14}N_2O_2}$. See more at **amino acid.**

lysis (lī′sĭs) The disintegration of a cell resulting from destruction of its membrane by a chemical substance, especially an antibody or enzyme.

lysosome (lī′sə-sōm′) A cell organelle that is surrounded by a membrane, has an acidic interior, and contains hydrolytic enzymes that break down food molecules, especially proteins and other complex molecules. Lysosomes fuse with vacuoles to digest their contents. The digested material is then transported across the organelle's membrane for use in or transport out of the cell. See more at **cell.**

M

m Abbreviation of **mass, meter.**

maar (mär) A broad, shallow, generally flat-floored volcanic crater that is often filled with water. Maars usually form from explosive eruptions caused by the heating and boiling of groundwater that is invaded by rising magma.

Mach (mäk, mäKH), **Ernst** 1838–1916. Austrian physicist and philosopher who experimented with supersonic projectiles and the flow of gases, obtaining early photographs of shock waves and gas jets. His work laid an important foundation for later developments in the science of projectiles and aeronautical design, and the Mach number and Mach bands were named for him.

Mach bands (mäk) Illusory bands of intense lightness and darkness perceived adjacent to borders of light and dark in a visual image, caused by early image-processing in the retina and optic nerve.

machine (mə-shēn′) A device that applies force, changes the direction of a force, or changes the strength of a force, in order to perform a task, generally involving work done on a **load**. Machines are often designed to yield a high **mechanical advantage** to reduce the **effort** needed to do that work. ► A **simple machine** is a wheel, a lever, or an inclined plane. All other machines can be built using combinations of these simple machines; for example, a drill uses a combination of gears (wheels) to drive helical inclined planes (the drill-bit) to split a material and carve a hole in it.

machine language The set of instructions, encoded as strings of binary bits, interpreted directly by a computer's central processing unit. Each different type of central processing unit has its own machine language. For a given machine language, each unique combination of 1's and 0's in an instruction has a unique interpretation, including such operations as arithmetical operations, incrementing a counter, saving data to memory, testing if data has a certain value, and so on. Computer programs are rarely written directly in machine language; instead, higher-level programming languages are used. See more at **programming language.**

Mach number The ratio of the speed of a body to the speed of sound in a particular medium, usually the Earth's atmosphere. For example, an aircraft flying through air at twice the speed of sound has a Mach number of 2. The Mach number of an aircraft travelling at a given velocity depends on the altitude of the aircraft and other atmospheric conditions that affect the speed of sound near the aircraft. See also **subsonic, supersonic, transonic.**

MACHO (mä′chō) Short for *massive astrophysical compact halo object.*Any of various massive dark objects, such as a brown dwarf star, black hole, or large planet, found in a **galactic halo**. MACHOs are thought to make up at least a part of the **dark matter** that apparently pervades much of the universe.

macro– A prefix meaning "large," as in *macromolecule,* a large molecule.

macroclimate (măk′rō-klī′mĭt) The climate of a large geographic area. Compare **microclimate.**

macroevolution (măk′rō-ĕv′ə-lo͞o′shən) Evolution that results in the formation of a new taxonomic group above the level of a species.

macrogamete (măk′rō-găm′ēt, -gə-mēt′) The larger of a pair of conjugating gametes, usually the female, in an organism that reproduces by heterogamy.

macrolide (măk′rə-līd′) **1.** Any of a class of organic compounds containing a large lactone ring made up of twelve or more members. **2.** Any of a class of antibiotics having a macrolide ring structure linked to one or more sugars. Macrolides are produced by actinomycete bacteria of the genus *Streptomyces* and act by inhibiting protein synthesis. Erythromycin is a macrolide.

macromolecule (măk′rō-mŏl′ĭ-kyo͞ol′) A large molecule, such as a protein, consisting of many smaller molecules linked together.

macrophage (măk′rə-fāj′) Any of various large white blood cells that play an essential immunologic role in vertebrates and some lower organisms by eliminating cellular debris and particulate antigens, including bacteria, through phagocytosis. Macrophages develop from circulating monocytes that migrate from the blood into tissues throughout the body, especially the spleen, liver, lymph nodes, lungs, brain, and connective tissue. Macrophages also participate in the immune response by producing and responding to inflammatory cytokines.

macrophyll (măk′rə-fĭl′) See **megaphyll.**

macula (măk′yə-lə) Plural **maculae** (măk′yə-lē′) or **maculas. 1.** A minute yellowish area located near the center of the retina of the eye, at which visual perception is most acute. **2.** A discolored spot on the skin that is not elevated above the surface.

macular degeneration (măk′yə-lər) A progressive condition, usually seen in the elderly, characterized by a gradual loss of vision in the central area of the visual field and eventual blindness.

mad cow disease (măd) A degenerative neurologic disease of cattle, thought to be caused by infection-causing agents called **prions**, in which brain tissues deteriorate and take on a spongy appearance, resulting in abnormal behaviors and loss of muscle control. A variant form of **Creutzfeldt-Jakob disease** is transmitted to humans through the eating of infected cattle tissue. Also called *bovine spongiform encephalopathy.*

mafic (măf′ĭk) Relating to an igneous rock that contains a group of dark-colored minerals, composed chiefly of magnesium and iron. Compare **felsic.**

Magdalenian (măg′də-lē′nē-ən) Relating to the final Upper Paleolithic culture of Europe, succeeding the Solutrean and dating from about 17,000 to 11,000 years ago. Magdalenian tools and weapons are highly specialized and demonstrate skilled craftsmanship in bone and antler as well as flaked stone. The Magdalenians are best known for their sophisticated artwork, including engravings, sculpture, and polychrome wall paintings such as those found in the Altamira caverns in northern Spain.

Magellanic Clouds (măj′ə-lăn′ĭk) Two small, irregular dwarf galaxies that orbit the **Milky Way.** They are among the galaxies closest to the Milky Way and are faintly visible near the south celestial pole. See also **irregular galaxy.**

magic number (măj′ĭk) Any of the numbers, 2, 8, 20, 28, 50, 82, or 126, that represent the number of neutrons or protons in strongly bound and exceptionally stable atomic nuclei. The existence of such stable nuclei is explained by assuming a shell structure for nucleons, much like the shell structure of electron orbitals around the nucleus.

magic square A square that contains numbers arranged in equal rows and columns such that the sum of each row, column, and sometimes diagonal is the same.

16	3	2	13
5	10	11	8
9	6	7	12
4	15	14	1

magic square

the number square from Albrecht Dürer's engraving Melencolia I

magma (măg′mə) Plural **magmata** (măg-mä′tə) or **magmas.** The molten rock material that originates under the Earth's crust and forms igneous rock when it has cooled. When magma cools and solidifies beneath the Earth's surface, it forms what are known as intrusive rocks. When it reaches the Earth's surface, it flows out as lava and forms extrusive (or volcanic) rocks.

magmatic differentiation (măg-măt′ĭk) The process by which chemically different igneous rocks, such as basalt and granite, can form from the same initial magma. Magmatic differentiation can occur by the chemical reaction between the magma and the first crystals to solidify out of it, or by the physical separation of the first crystals that form from the remaining magma, either through settling to the bottom of a magma chamber or through crustal deformations that cause the remaining magma to be squeezed out to cool in veins and dikes.

magnesia (măg-nē′zhə) A white powder with a very high melting point. It is used to make heat-resistant materials, electrical insulators, cements, fertilizer, and plastics. It is also used in medicine as an antacid and laxative. *Chemical formula:* **MgO.**

magnesium (măg-nē′zē-əm) *Symbol* **Mg** A lightweight, moderately hard, silvery-white metallic element of the alkaline-earth group that burns with an intense white flame. It is an essential component of chlorophyll and is used in lightweight alloys, flash photography, and fireworks. Atomic number 12; atomic weight 24.305; melting point 649°C; boiling point 1,090°C; specific gravity 1.74 (at 20°C); valence 2. See **Periodic Table.**

magnet (măg′nĭt) A material or object that produces a magnetic field. Lodestones are natural magnets, though many materials, especially metals, can be made into magnets by exposing them to a magnetic field. See also **electromagnet, ferromagnetism, magnetic pole.** See Note at **magnetism.**

magnetar (măg′nə-tär′) A neutron star with a very strong magnetic field. Magnetars are the proposed sources of gamma ray bursts that have been observed.

magnetic (măg-nĕt′ĭk) Producing, caused by, or making use of magnetic fields.

magnetic charge A theoretical property of matter manifesting magnetic property, analogous to electric charge, arising from magnetic monopoles.

magnetic declination The horizontal angle between the true geographic North Pole and the magnetic north pole, as figured from a specific point on the Earth. Compare **magnetic inclination.**

magnetic dip See **magnetic inclination.**

magnetic dipole A model of an object that generates a magnetic field in which the field is considered to emanate from two opposite poles, as in the north and south poles of a magnet, much as an electric field emanates from a positive and a negative charge (each of which is a **monopole**) in an electric dipole. Even though the existence of magnetic monopoles as isolable particles has not been established, the magnetic dipole remains a useful simplification of the electrodynamics involved in magnetism. Magnetic dipoles experience **torque** in the presence of magnetic fields.

magnetic dipole moment A vector quantity associated with the magnetic properties of electric current loops or, more generally, magnets. It is equal to the amount of current flowing through the loop multiplied by the area encompassed by the loop, and its direction is established by the **right hand rule** for rotations. It can be thought of as a vector pointing from the south to the north of a magnetic dipole, and is then equal to the length of the dipole times the strength of either of its poles. Also called *magnetic moment.*

magnetic disk A memory device, such as a floppy disk or a hard disk, that is covered with a magnetic coating. Digital information is stored on magnetic disks in the form of microscopically small, magnetized needles, each of which encodes a single bit of information by being polarized in one direction (representing 1) or the other (representing 0).

magnetic equator A line connecting all points on the earth's surface at which the magnetic field is parallel to the Earth's surface. A balanced magnetic needle on the magnetic equator stabilizes in a perfectly horizontal position.

magnetic field 1. A field of force associated with changing **electric fields**, as when electric charges are in motion. Magnetic fields exert deflective forces on moving electric charges. Most **magnets** have magnetic fields as a result of the spinning motion of the electrons orbiting the atoms of which they are composed; **electromagnets** create such fields from electric current moving through coils. Large objects, such as the earth, other planets, and stars, also produce magnetic fields. See Note at **magnetism. 2.** See **magnetic field strength.**

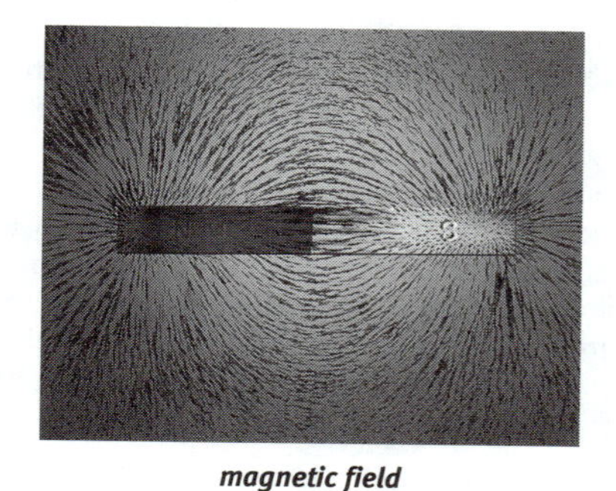

magnetic field

iron filings show the magnetic field surrounding a bar magnet

magnetic field strength A vector quantity indicating the ability of a magnetic field to exert a force on moving electric charges. It is equal to the **magnetic flux density** divided by the **magnetic permeability** of the space where the field exists. It is measured in amperes per meter. Also called *magnetic intensity.* See more at **magnetic flux density.**

magnetic flux 1. The lines of force associated with a magnetic field. The strength of magnetic flux is equivalent to its **magnetic flux density** per unit area. The SI unit of magnetic flux is the weber. **2.** See **magnetic flux density.**

magnetic flux density A vector quantity measuring the strength and direction of the magnetic field around a magnet or an electric current. Magnetic flux density is equal to **magnetic field strength** times the **magnetic permeability** in the region in which the field exists. Electric charges moving through a magnetic field are subject to a force described by the equation $\mathbf{F} = q\mathbf{v} \times \mathbf{B}$, where q is the amount of electric charge, v is the velocity of the charge, B is the magnetic flux density at the position of the charge, and $\times$ is the **vector product**. Magnetic flux density also can be understood as the density of magnetic lines of force, or magnetic flux lines, passing through a specific area. It is measured in units of tesla. Also called *magnetic flux, magnetic induction.*

magnetic inclination The angle that a magnetic needle makes with the horizontal plane at any specific location. The magnetic inclination is 0° at the magnetic equator and 90° at each of the magnetic poles. Also called *dip, magnetic dip.* Compare **magnetic declination.**

magnetic induction 1. The process by which a substance, such as iron, becomes magnetized by a magnetic field. **2.** See **magnetic flux density.**

magnetic intensity See **magnetic field strength.**

magnetic meridian An imaginary reference line passing through both geomagnetic poles of the Earth, used in models and maps of the Earth's magnetic field.

magnetic moment See **magnetic dipole moment.**

magnetic north The direction toward which the north-seeking arrow of a compass points. ► The **magnetic north pole** is the northern pole of the Earth's magnetic field and changes slightly in response to variations in the Earth's magnetism. The current magnetic north pole is located in the Arctic Islands of Canada. Compare **geographic north.**

magnetic permeability A measure of the ability of a substance to sustain a magnetic field, equal to the ratio between magnetic flux density and magnetic field strength. For a vacuum, its value is 1.257×10^{-6} henries per meter. Highly magnetizable materials, such as ferromagnetic materials, have higher magnetic permeability. See also **ferromagnetism.**

magnetic pole 1. Either of two regions of a magnet, designated north and south, where the magnetic field is strongest. Electromagnetic interactions cause the north poles of magnets to be attracted to the south poles of other magnets, and conversely. The **north pole** of a magnet is the pole out of which magnetic lines of force point, while the **south pole** is the pole into which they point. The Earth's geomagnetic "north" and "south" poles are, in fact, magnetically the opposite of what their names suggest; this is why the north end of a compass needle is attracted to the geomagnetic "north" pole. See Note at **magnetism.** See also **magnetic. 2.** Either of two regions of the Earth's surface at which magnetic lines of force are perpendicular to the Earth's surface. The Earth's magnetic poles are close to, but not identical with, both its geographic poles (the North and South Poles) and its **geomagnetic poles**. See Note at **magnetic reversal.**

magnetic quantum number The quantum number corresponding to the shape of electron subshells that are in an atom. Also called *third quantum number.* See Note at **quantum number.**

magnetic recording The recording of a signal, such as sound or computer instructions, in the form of a magnetic pattern on a magnetizable surface, such as tape or a disk coated with metal oxides. In the recording process, small electromagnets are used to convert an electrical signal into a magnetic one that magnetizes that surface. The value of the signal at any given point corresponds to the degree and polarity of magnetization of the surface at a corresponding point. During playback, the opposite process occurs, and the magnetic fields along the surface are converted into electrical signals for further processing.

magnetic resonance See **nuclear magnetic resonance.**

magnetic resonance imaging See **MRI.**

magnetic reversal A change in the Earth's magnetic field resulting in the magnetic north being aligned with the geographic south, and the magnetic south being aligned with the geographic north. Also called *geomagnetic reversal.*

magnetic south The direction toward which the south-seeking arrow of a compass points. ▸ The **magnetic south pole** is the southern pole of the Earth's magnetic field and changes slightly in response to variations in the Earth's magnetism. The current magnetic south pole is located in the Antarctic Ocean. Compare **geographic south.**

magnetic storm A disturbance or fluctuation in the Earth's outer magnetosphere, usually caused by streams of charged particles (plasma) given off by solar flares. The entry of large amounts of plasma into the upper

A CLOSER LOOK **magnetic reversal**

When magma rises to the Earth's surface at a mid-ocean ridge, it flows out onto both sides of the ridge, gradually cooled by the seawater. Like tiny compass needles, the magnetic minerals in the hot magma are at first free to align themselves with the Earth's magnetic field when the magma settles into the tectonic plate, but once the lava cools below the *Curie point,* their orientation becomes fixed. When readings of the strength of the magnetic field are taken along sections of the ocean floor near such ridges, segments where it is anomalously high alternate with segments where it is anomalously low. Anomalously high readings occur because the magnetometer is picking up both the reading from today's magnetic field and that from the minerals in the rock that are aligned with it, adding to the total strength of the field, while anomalously low readings occur when the magnetic minerals are aligned against the Earth's magnetic field, diminishing the total strength. The rocks that yield these anomalously low readings therefore must have formed at a time when the Earth's magnetic field was reversed—oriented in such a way that the north magnetic pole was roughly where today's south magnetic pole is, and vice versa. These *magnetic reversals,* in which the direction of the field is flipped, are believed to occur when small, complex fluctuations of magnetic fields in the Earth's outer liquid core interfere with the Earth's main dipolar magnetic field to the point where they overwhelm it, causing it to reverse. The length of time between magnetic reversals is not always the same, but is on the order of 200,000 to 1,000,000 years; the last magnetic reversal was about 750,000 years ago. Because the pattern of positive and negative readings is more or less symmetrical about the axis of the mid-ocean ridge and remains the same throughout the length of the ridge, geophysicists have been able to construct a calendar of the Earth's magnetic record dating back to as far as 150–200 million years ago. It is not known when the next magnetic reversal will be, or how long the process will take, though it will certainly have a significant impact on the artificial and biological navigational systems of humans and animals.

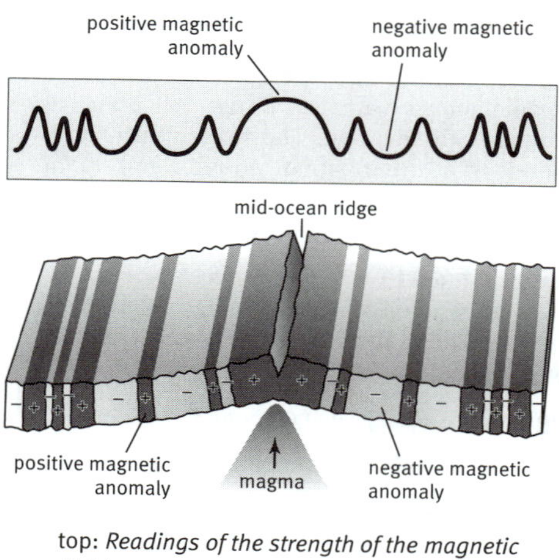

top: *Readings of the strength of the magnetic field across a section of the mid-ocean ridge show anomalously high* (positive) *and anomalously low* (negative) *results.* bottom: *The magnetic record patterns are symmetrical about the axis of the mid-ocean ridge.*

atmosphere results in intense auroral displays and other magnetic phenomena in the polar regions of the Earth. See also **aurora.**

magnetic tape A plastic tape coated with iron oxide for use in magnetic recording.

magnetism (măg′nĭ-tĭz′əm) **1.** The properties or effects of magnetic fields. **2.** The force produced by a magnetic field. See more at **magnetic field.**

A CLOSER LOOK **magnetism**

Magnetism is intimately linked with electricity, in that a magnetic field is established whenever electric charges are in motion, as in the flow of electrons in a wire, or the movement of electrons around an atomic nucleus. In atoms, this invisible field consists of closed loops called *lines of force* that surround and run through the atom. Magnetic regions where lines of force come together densely are called north and south poles. In substances in which the magnetic fields of each atom are aligned, the magnetic field causes the entire substance to act like single magnet—with north and south poles and a surrounding magnetic field. *Permanent magnets* are made of substances that retain this alignment. If a magnet is cut in two, each piece becomes a separate magnet with two poles. A coil of wire wrapped around an iron core can be made magnetic by running electric current through it; the looping electrons then create a magnetic field in just the same way as the spinning electrons in individual atoms. As long as current flows, the coil remains magnetized. Such magnets, called *electromagnets,* are used in many devices such as doorbells and switches. The connection between electric and magnetic fields is not one of cause and effect, however. Einstein showed that both the magnetic and electric fields are part of a single electromagnetic field, described by a single mathematical object called a *tensor.* Observers in different reference frames will not observe the same separate values for electric and magnetic fields, but will observe identical electromagnetic tensors. Whether or not magnetic monopoles (elementary particles carrying an isolated north or south magnetic "charge," analogous to positive or negative electric charge) actually exist remains unknown; though they are predicted by some theories, none have been detected.

magnetite (măg′nĭ-tīt′) A brown to black mineral that is strongly magnetic. It crystallizes in the cubic system and commonly occurs as small octahedrons. Magnetite occurs in many different types of rock and is an important source of iron. *Chemical formula:* Fe_3O_4.

magnetize (măg′nĭ-tīz′) To cause an object to become temporarily or permanently magnetic. For example, an unmagnetized object made of **ferromagnetic** material consists of molecules that are magnetic but randomly aligned, producing no net magnetic field; exposure to a magnetic field causes the molecules to align themselves with the field, producing their own net field, so that the object as a whole becomes magnetized.

magnetograph (măg-nē′tō-grăf′) A recording of the strength and direction of a magnetic field made by a magnetometer. Also called *magnetogram.* See **magnetometer.**

magnetometer (măg′nĭ-tŏm′ĭ-tər) An instrument for measuring the magnitude and direction of a magnetic field. Magnetometers are often used in archaeological and geological investigations to determine the intensity and direction of the Earth's magnetic field at various times in the past by examining the strength and direction of magnetization of ferromagnetic materials in different geological strata. See also **magnetic reversal.**

magneton (măg′nĭ-tŏn′) A unit of the magnetic dipole moment of a molecular, atomic, or subatomic particle. ► The **Bohr magneton** is defined as the magnetic dipole moment of the electron due to its inherent spin angular momentum, and is equal to 9.2741×10^{-24} joules per tesla. Also called *Landé factor.* ► The **nuclear magneton** is defined using the electric charge and rest mass of the proton, and is equal to 5.0508×10^{-27} joules per tesla.

magnetosphere (măg-nē′tō-sfîr′) A highly asymmetrical region surrounding the Earth, beginning about 100 km (62 mi) above the surface on the side of the Earth facing the Sun and extending hundreds of thousands of kilometers into space on the opposite side. In this region the Earth's magnetic field exerts a significant influence on any charged particles that encounter it. The magnetosphere deflects most of the charged particles in the **solar wind**, but also traps and deflects some of these particles toward the Earth's magnetic poles, causing **magnetic storms** and **auroras**.

magnetostriction (măg-nē′tō-strĭk′shən) The change in shape and density of a substance, especially a ferromagnetic substance, when exposed to a magnetic field. The change depends on the direction and strength of the magnetic field. Rapid, alternating magnetostriction causes the iron cores of household transformers, which are subject to a changing magnetic field, to hum or buzz.

magnetotail (măg-nē′tō-tāl′) The elongated extension of the Earth's magnetosphere on the side facing away from the Sun. The magnetotail is shaped by the pressure of the solar wind as it streams around the magnetosphere, compressing it on the side facing the Sun and stretching it on the opposite side into a long, taillike shape trailing far into interplanetary space. See more at **magnetosphere.**

magnetron (măg′nĭ-trŏn′) An electron tube that produces coherent microwave radiation. Magnetrons are **diodes** in which the electrons traveling to the anode are set in spiraling paths by a magnetic field created by permanent magnets. The circular component of the electrons' motion causes microwave-frequency oscillations in the voltage induced in resonating cavities built into the anode, which is connected to an antenna that emits the microwaves. Magnetrons are used in radar and in microwave ovens.

magnitude (măg′nĭ-to͞od′) **1.** The degree of brightness of a star or other celestial body, measured on a logarithmic scale in which lower numbers mean greater brightness, such that a decrease of one unit represents an increase in brightness by a factor of 2.512. An object that is 5 units less than another object on the magnitude scale is 100 times more luminous. Because of refinements in measurement after the zero point was assigned, very bright objects have negative magnitudes. ▸ The brightness of a celestial body as seen from Earth is called its **apparent magnitude**. (When unspecified, an object's magnitude is normally assumed to be its apparent magnitude.) The dimmest stars visible to the unaided eye have apparent magnitude 6, while the brightest star in the night sky, Sirius, has apparent magnitude –1.4. The full Moon and the Sun have apparent magnitudes of –12.7 and –26.8 respectively. ▸ The brightness of a celestial body computed as if viewed from a distance of 10 parsecs (32.6 light-years) is called its **absolute magnitude**. Absolute magnitude measures the intrinsic brightness of a celestial object rather than how bright it appears on Earth, using the same logarithmic scale as for apparent magnitude. Sirius has an absolute magnitude of 1.5, considerably dimmer than Rigel which, though its apparent magnitude is 0.12, has an absolute magnitude of –8.1. Stars that appear dim in the night sky but have bright absolute magnitudes are much farther from Earth than stars that shine brightly at night but have relatively dim absolute magnitudes. The Sun, a star of only medium brightness, has an absolute magnitude of 4.8. ▸ The degree of total radiation emitted by a celestial body, including all infrared and ultraviolet radiation in addition to visible light, is called its **bolometric magnitude**. Bolometric magnitude is generally measured by applying a standard correction to an object's absolute magnitude. **2.** A measure of the total amount of energy released by an earthquake, as indicated on the Richter scale. See more at **Richter scale.**

magnoliid (măg-nō′lē-ĭd) Any of a heterogenous group of angiosperms that are neither eudicotyledons nor monocotyledons, and are considered to retain the characteristics of more primitive angiosperms, such as flowers with fewer or less differentiated parts. Magnoliids have embryos with two cotyledons (like the eudicotyledons), and they usually have long, broad leaves and large flowers. Magnolias, laurel, and the cinnamon tree are magnoliids.

maiasaura (mī′ə-sôr′ə) or **maiasaur** (mī′ə-sôr′) A duck-billed dinosaur of the genus *Maiasaura* of the late Cretaceous Period of North America. Its remains suggest that the adults lived in herds and cared for their young in large nesting sites.

Maiman (mā′mən), **Theodore Harold** 1927–2007. American physicist who constructed the first working laser in 1960.

mainframe (mān′frām′) A large, often powerful computer, usually dedicated to lengthy, complex calculations or set up for use by many people simultaneously. Compare **personal computer.**

main sequence (mān) The continuous, generally diagonal line or band in the **Hertzsprung-**

Russell diagram ranging from the upper left to the lower right and representing stars of average size whose luminosities correspond predictably to their surface temperatures. Stars in this grouping maintain a stable nuclear reaction and experience only small fluctuations in luminosity and temperature. Main-sequence stars are believed to be in the stable, middle phase of their development; they are expected to move off the main sequence once the hydrogen in their core is exhausted. At that point, depending on its size, a main-sequence star will become a giant star, a supergiant star, or a white dwarf. The more massive the star, the faster it burns its nuclear fuel and the shorter it remains in the main sequence. See more at **Hertzsprung-Russell diagram, star.**

major histocompatibility complex (mā′jər) A group of genes that code for cell-surface histocompatibility antigens and are the principal determinants of tissue type and transplant compatibility. They are the most diverse genes in humans and are used to determine if a sample of DNA comes from a specific person.

malachite (măl′ə-kīt′) A bright-green monoclinic mineral occurring as a mass of crystals (an aggregate) with smooth or botryoidal (grape-shaped) surfaces. It is often concentrically banded in different shades of green. Malachite often occurs together with the mineral azurite in copper deposits. *Chemical formula:* $\mathbf{Cu_2CO_3(OH)_2}$.

malacology (măl′ə-kŏl′ə-jē) The scientific study of mollusks.

malaria (mə-lâr′ē-ə) An infectious disease of tropical areas caused by the parasitic infection of red blood cells by a protozoan of the genus *Plasmodium,* which is transmitted by the bite of an infected female mosquito. Malaria is characterized by recurrent episodes of chills, fever, sweating, and anemia and is endemic in Africa, Central America, and much of Southern Asia and northern South America.

malate (măl′āt′, mā′lāt′) **1.** A salt or ester of malic acid, containing the group $C_4H_4O_5$. **2.** See **malic acid.**

male (māl) *Adjective.* **1.** In organisms that reproduce sexually, being the gamete that is smaller and more motile than the other corresponding gamete of the same species (the female gamete). The sperm cells of higher animals and plants are male gametes. **2.** Possessing or being a structure that produces only male gametes. The testicles of humans are male reproductive organs. Male flowers possess only stamens and do not possess carpels. —*Noun.* **3.** A male organism.

maleate (mā′lē-āt′, mə-lē′ət) A salt or ester of maleic acid.

maleic acid (mə-lē′ĭk) A colorless crystalline acid used in textile processing and as an oil and fat preservative. Maleic acid is a geometric isomer of fumaric acid, having two carboxyl groups (COOH) on the same side of an ethylene chain. *Chemical formula:* $\mathbf{C_4H_4O_4}$.

malic acid (măl′ĭk, mā′lĭk) A colorless, crystalline compound that occurs naturally in a wide variety of unripe fruits, including apples, cherries, and tomatoes, and is an intermediate product of the Krebs cycle. It is used as a flavoring and in the aging of wine. Also called *malate. Chemical formula:* $\mathbf{C_4H_6O_5}$.

malignant (mə-lĭg′nənt) **1.** Tending to have a destructive clinical course, as a malignant illness. **2.** Relating to cancer cells that are invasive and tend to metastasize. Malignant tumor cells are histologically more primitive than normal tissue. Compare **benign.**

malleable (măl′ē-ə-bəl) Capable of great deformation without breaking, when subject to compressive **stress**. Gold is the most malleable metal. Compare **ductile.**

malleus (măl′ē-əs) Plural **mallei** (măl′ē-ī′). The hammer-shaped bone that is the largest and outermost of the three small bones (ossicles) in the middle ear.

malnutrition (măl′no͞o-trĭsh′ən) Poor nutrition caused by an insufficient, oversufficient, or poorly balanced diet or by a medical condition, such as chronic diarrhea, resulting in inadequate digestion or utilization of foods.

malocclusion (măl′ə-klo͞o′zhən) Misalignment between the upper and lower teeth when the jaw is closed, resulting in a faulty bite.

malonate (măl′ə-nāt′, -nĭt) A salt or ester of malonic acid.

malonic acid (mə-lō′nĭk, -lŏn′ĭk) A white crystalline acid derived from malic acid and used in making barbiturates. *Chemical formula:* $\mathbf{C_3H_4O_4}$.

Malpighi (măl-pē′gē), **Marcello** 1628–1694. Italian anatomist who was the first to use a microscope in the study of anatomy. He discovered the capillary system, extending the work of William Harvey. He is also noted for his studies of the structure of the lungs, spleen, liver, kidneys, skin, brain, and spinal cord.

maltose (môl′tōs′) A sugar made by the action of various enzymes on starch. It is formed in the body during digestion. Maltose is a disaccharide consisting of two linked glucose molecules. *Chemical formula:* $C_{12}H_{22}O_{11}$.

malware (măl′wâr′) Software that is written and distributed for malicious purposes, such as impairing or destroying computer systems. Computer viruses are malware.

mammal (măm′əl) Any of various warm-blooded vertebrate animals of the class Mammalia, whose young feed on milk that is produced by the mother's mammary glands. Unlike other vertebrates, mammals have a diaphragm that separates the heart and lungs from the other internal organs, red blood cells that lack a nucleus, and usually hair or fur. All mammals but the monotremes bear live young. Mammals include rodents, cats, dogs, ungulates, cetaceans, and apes.

mammalogy (mă-măl′ə-jē, -mŏl′-) The scientific study of mammals.

mammary gland (măm′ə-rē) One of the glands in female mammals that produces milk. It is present but undeveloped in the male. In most animals, the gland opens onto the surface by means of a nipple or teat. Mammary glands number from 2 to 20, depending on the species.

mammogram (măm′ə-grăm′) An x-ray image of the human breast, used to detect tumors or other abnormalities. *—Noun* **mammography** (mă-mŏg′rə-fē).

mammoth (măm′əth) Any of various extinct elephants of the genus *Mammuthus,* having long, upwardly curving tusks and thick hair. Mammoths grew to great size and lived throughout the Northern Hemisphere during the Ice Age.

Mandelbrot set (män′dəl-brŏt′) The set of complex numbers C for which the iteration $z_{n+1} = z_n^2 + C$ produces finite z_n for all n when started at $z_0 = 0$. The boundary of the Mandelbrot set is a fractal.

mandible (măn′də-bəl) **1.** The lower part of the jaw in vertebrate animals. See more at **skeleton. 2.** One of the pincerlike mouthparts of insects and other arthropods.

manganese (măng′gə-nēz′) *Symbol* **Mn** A grayish-white, hard, brittle metallic element that occurs in several different minerals and in nodules on the ocean floor. It is used to increase the hardness and strength of steel and other important alloys. Atomic number 25; atomic weight 54.9380; melting point 1,244°C; boiling point 1,962°C; specific gravity 7.21 to 7.44; valence 1, 2, 3, 4, 6, 7. See **Periodic Table.**

manic-depressive illness (măn′ĭk-dĭ-prĕs′ĭv) See **bipolar disorder.**

manifold (măn′ə-fōld′) A topological space or surface.

mannitol (măn′ĭ-tôl′, -tōl′) A white, crystalline, water-soluble, slightly sweet alcohol that is used as a dietary supplement and dietetic sweetener and in medical tests of kidney function. Mannitol occurs naturally as an important food storage and transportation molecule in brown algae like kelp. *Chemical formula:* $C_6H_{14}O_6$.

mannose (măn′ōs′) A naturally occurring simple sugar that is a stereoisomer of glucose. *Chemical formula:* $C_6H_{12}O_6$.

manometer (mə-nŏm′ĭ-tər) An instrument used to measure the pressure exerted by liquids and gases. Pressure is exerted on one end of a U-shaped tube partially filled with liquid; the liquid is displaced upwards on the other side

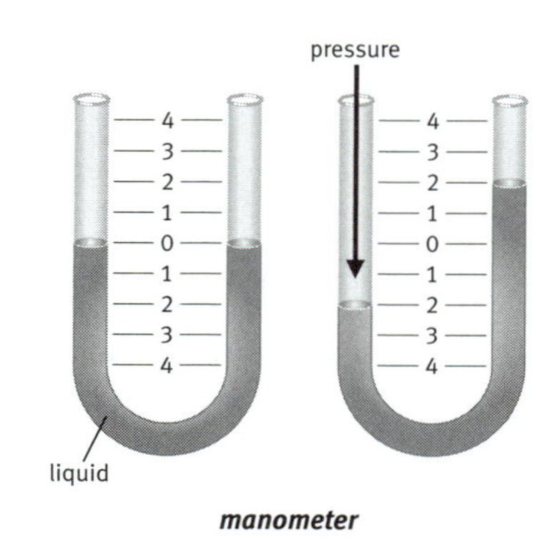

manometer

To calculate pressure in a U-tube manometer, the readings above and below zero are added. The manometer on the left is at equilibrium. The manometer on the right shows readings of 2 above zero and 2 below zero, indicating a pressure of 4.

of the tube by a distance proportional to the pressure difference on each side of the tube.

mantissa (măn-tĭs′ə) The part of a logarithm to the base ten that is to the right of the decimal point. For example, if 2.749 is a logarithm, .749 is the mantissa. Compare **characteristic.**

mantle (măn′tl) **1.** The layer of the Earth between the crust and the core. It is about 2,900 km (1,798 mi) thick and consists mainly of magnesium-iron silicate minerals, such as olivine and pyroxene. It has an upper, partially molten part, which is about 660 km (409 mi) thick, and a lower, solid part. The upper mantle is the source of magma and volcanic lava. **2.** The layer of soft tissue that covers the body of a clam, oyster, or other mollusk and secretes the material that forms the shell.

MAOI Abbreviation of **monoamine oxidase inhibitor.**

map (măp) **1.** A representation of a region of three-dimensional space, such as of the Earth or a part of the universe, usually on a two-dimensional plane surface. See also **projection. 2.** See **genetic map.**

marble (mär′bəl) A metamorphic rock consisting primarily of calcite and dolomite. Marble is formed by the metamorphism of limestone. Although it is usually white to gray in color, it often has irregularly colored marks due to the presence of impurities such as silica and clay. Marble is used especially in sculpture and as a building material.

marcasite (mär′kə-sīt′, -zīt′) A light yellow to gray, metallic, orthorhombic mineral. Marcasite is a polymorph of pyrite and looks similar to it but has a lower specific gravity, is paler in color, and often has a radiating fibrous structure. *Chemical formula:* FeS_2.

marcescent (mär-sĕs′ənt) Withering but not falling off, as a blossom that persists on a twig after flowering. Many oaks have marcescent foliage that stays on the tree through winter.

Marconi (mär-kō′nē), **Guglielmo** 1874–1937. Italian physicist and inventor who was the first to use radio waves to transmit signals in Morse code across the Atlantic Ocean (1901). Soon after his experiment, he developed shortwave radio equipment and helped establish radio as a widely used medium for communications.

Guglielmo Marconi

mare (mä′rā) Plural **maria** (mä′rē-ə). Any of the large, low-lying dark areas on the Moon or on Mars or other inner planets. The lunar maria are believed to consist of volcanic basalts, and many are believed to be basins formed initially by large impacts with meteoroids and later filled with lava flows. Compare **terra.**

marine (mə-rēn′) **1.** Relating to the sea. **2.** Relating to a system of open-ocean and unprotected coastal habitats, characterized by exposure to wave action, tidal fluctuation, and ocean currents and by the absence of trees, shrubs, or emergent vegetation. Water in the marine system is at or near the full salinity of seawater. Compare **lacustrine, palustrine, riverine.**

marine biology The scientific study of organisms living in or dependent on the oceans.

markup language (märk′ŭp′) A coding system, such as HTML and SGML, used to structure, index, and link computer files.

marl (märl) A crumbly mixture of clays, calcium and magnesium carbonates, and remnants of shells that forms in both freshwater and marine environments.

marrow (măr′ō) See **bone marrow.**

Mars (märz) The fourth planet from the Sun and the second smallest in the solar system, with a diameter about half that of Earth. Mars is the last of the **terrestrial** or **inner planets** and has notable similarities to Earth, including polar ice caps and a tilted axis that gives it seasons. However, it is significantly less dense than Earth and has no magnetic field, suggesting that it lacks a metallic core, and its atmosphere, made up mostly of carbon dioxide, is much thinner than Earth's. Mars has no surface water apart from a layer of permanent ice that underlies the seasonally changing caps of frozen carbon dioxide at its poles; there is, however, clear evidence of earlier water flows in the form of channels, outwashes, and canyons. Other surface features include numerous craters, especially in the southern hemisphere, along with very large volcanoes and extensive windblown dunes. Mar's reddish color is due to the abundance of hematite in its surface rocks. Its two small, irregular moons, Phobos and Deimos, may be asteroids captured earlier by gravitational attraction. See Table at **solar system.**

marsh (märsh) An area of low-lying wetland in which the level of water is generally shallow and often fluctuating. The water may be either standing or slow-moving. The water in a marsh is also more or less neutral or alkaline, in contrast to the water in a bog, which is acidic. The environment of a marsh is in general well-oxygenated and nutrient-rich and allows a great variety of organisms to flourish. In contrast to a **swamp,** in which there is an abundance of woody plants, the plants in a marsh are mostly herbaceous. Reeds and rushes dominate the vegetation of marshes. See also **salt marsh.**

marsupial (mär-so͞o′pē-əl) Any of various mammals of the order Marsupialia, whose young are very undeveloped when born and continue developing outside their mother's body attached to one of her nipples. Most marsupials have longer hindlegs than forelimbs, and the females usually have pouches in which they carry their young. Kangaroos, opossums, and koalas are marsupials.

maser (mā′zər) Short for *microwave amplification by stimulated emission of radiation.* A device that generates coherent microwaves using the same principles as a **laser**. Masers are used in a variety of applications, including in **atomic clocks**. Natural masers are found in outer space when water or other substances are excited by radiation from a star or by the energy of a collision.

mass (măs) A measure of the amount of matter contained in or constituting a physical body. In classical mechanics, the mass of an object is related to the force required to accelerate it and hence is related to its **inertia**, and is essential to **Newton's laws of motion**. Objects that have mass interact with each other through the force of **gravity**. In Special Relativity, the observed mass of an object is dependent on its velocity with respect to the observer, with higher velocity entailing higher observed mass. Mass is measured in many different units; in most scientific applications, the SI unit of kilogram is used. See Note at **weight.** See also **rest energy, General Relativity.**

mass-energy equivalence An equation derived from Einstein's theory of Special Relativity expressing the relationship between the mass and energy of objects with mass. The equation is $E = mc^2$, where E is the energy of the object in joules, m is its **relativistic mass** in kilograms, and c is the speed of light (approximately 3×10^8 meters per second). Mass-energy equivalence entails that the total mass of a system may change, although the total energy and momentum remain constant; for example, the collision of an electron and a proton annihilates the mass of both particles, but creates energy in the form of photons. The discovery of mass-energy equivalence was essential to the development of theories of atomic **fission** and **fusion** reactions.

mass extinction The extinction of a large number of species within a relatively short period of geological time, thought to be due to factors such as a catastrophic global event or widespread environmental change that occurs too rapidly for most species to adapt. At least five mass extinctions have been identified in the fossil record, coming at or toward the end of the Ordovician, Devonian, Permian, Triassic, and Cretaceous Periods. The Permian extinction, which took place 245 million years ago, is the largest known mass extinction in the Earth's history, resulting in the extinction of an estimated 90 percent of marine species. In the Cretaceous extinction, 65 million years ago, an estimated 75 percent of species, including the dinosaurs, became extinct, possibly as the result of an asteroid colliding with the Earth. Compare **background extinction.**

massif (mă-sēf′) A large mountain mass or compact group of connected mountains forming an independent portion of a range. A massif often consists of rocks that are more rigid than the surrounding rocks.

mass number The total number of protons and neutrons in the nucleus of an atom. For example, nitrogen has 7 protons and 7 neutrons in its nucleus, giving it a mass number of 14. **Isotopes** of elements are distinguished by their mass number; for example, carbon-12 and carbon-14 have mass numbers of 12 and 14 respectively. Also called *nucleon number.* Compare **atomic mass, atomic weight.**

mass spectroscope See under **spectroscope.**

mass spectroscopy See under **spectroscopy.**

mast cell (măst) A granular cell found in body tissue, especially connective tissue, that activates inflammation by releasing a variety of chemical substances including histamine, tumor necrosis factor, and interleukins. Mast cells have membrane receptors that bind to bacteria, triggering the release of inflammatory mediators from the mast cell's cytoplasmic granules. Mast cells also play an important role in allergic reactions. Other receptors on their membranes bind to specific antibodies that, combined with certain antigens, initiate granular release of chemical mediators that cause allergic signs and symptoms.

mastectomy (mă-stĕk′tə-mē) Surgical removal of all or part of a breast, performed as a treatment for cancer. ► A **radical mastectomy** includes excision of the underlying pectoral muscles and regional lymph nodes.

mastigophoran (măs′tĭ-gŏf′ər-ən) See **flagellate.**

mastodon (măs′tə-dŏn′) Any of several extinct mammals of the genus *Mastodon* (or *Mammut*). Mastodons resembled elephants and mammoths except that their molar teeth had conelike cusps rather than parallel ridges for grinding. Like elephants, mastodons had a pair of long, curved tusks growing from their upper jaw, but males also sometimes had a second pair from the lower jaw. Like mammoths, mastodons were covered with hair. They lived from the Oligocene Epoch to the end of the Ice Age.

mastoiditis (măs′toid-ī′tĭs) Inflammation of the mastoid process, usually resulting from an acute infection of the middle ear.

mastoid process (măs′toid′) A protruding bony area in the lower part of the skull that is located behind the ear in humans and many other vertebrates and serves as a site of muscle attachment. The mastoid process contains small air-filled cavities called *mastoid cells* that communicate with the middle ear.

mathematics (măth′ə-măt′ĭks) The study of the measurement, relationships, and properties of quantities and sets, using numbers and symbols. Arithmetic, algebra, geometry, and calculus are branches of mathematics.

matrix (mā′trĭks) Plural **matrices** (mā′trĭ-sēz′, măt′rĭ-) or **matrixes. 1.** *Geology.* The mineral grains of a rock in which fossils are embedded. **2.** *Biology.* The component of an animal or plant tissue that is outside the cells. Bone cells are embedded in a matrix of collagen fibers and mineral salts. Connective tissue consists of cells and extracellular fibers in a liquid called **ground substance**. Also called *extracellular matrix.* **3.** *Mathematics.* A rectangular array of numeric or algebraic quantities subject to mathematical operations. **4.** *Anatomy.* The formative cells or tissue of a fingernail, toenail, or tooth.

matter (măt′ər) Something that has mass. Most of the matter in the universe is composed of **atoms** which are themselves composed of **subatomic particles**. See also **energy, state of matter.**

Maunder minimum (môn′dər, män′-) A period of unusually low sunspot activity lasting from approximately 1645 to 1715, as noted in records kept by contemporary observers. The Maunder minimum corresponds roughly to the middle and coldest portion of the climatic period known as the **Little Ice Age**, and although no definitive link has yet been proved, many scientists believe that the two phenomena are likely related. The Maunder minimum is named after its discoverer, British astronomer Edward Walter Maunder (1851–1924).

Maury (môr′ē), **Matthew Fontaine** 1806–1873. American naval officer and oceanographer who charted the currents and winds of the Atlantic, Pacific, and Indian Oceans and wrote the pioneering book *Physical Geography of the Sea* (1855).

maxilla (măk-sĭl′ə) Plural **maxillae** (măk-sĭl′ē) or **maxillas.** The upper part of the jaw in vertebrate animals.

maximum (măk′sə-məm) Plural **maximums** or **maxima. 1.** The greatest known or greatest possible number, measure, quantity, or degree. **2.** The greatest value of a mathematical function, if it has such a value.

maxwell (măks′wĕl′, -wəl) The unit of magnetic flux in the centimeter-gram-second system, equal to the flux perpendicularly intersecting an area of one square centimeter in a region where the magnetic intensity is one gauss.

Maxwell (măks′wĕl′), **James Clerk** 1831–1879. Scottish physicist who developed four laws of electromagnetism showing that light is composed of electromagnetic waves. He also investigated heat and the kinetic theory of gases, and he experimented with color vision, producing the first color photograph in 1861.

BIOGRAPHY **James Clerk Maxwell**

James Clerk Maxwell was only fourteen years old when he published his first paper—an accomplishment for anyone, but especially for one who was thought by his first tutor to be slow-witted. His precocious talents, especially in mathematics, did not go unrecognized by others, however, and he started making lasting contributions to science while still very young. In his 20s, he wrote a prize-winning essay in which he showed, based on laws of classical physics, that Saturn's rings were not a single object, but a collection of small objects—a finding not confirmed until over 120 years later, when the Voyager space probe reached the planet. His most famous work was his demonstration, done while he was in his 30s, of the existence of electromagnetic waves and his conclusion that light was also part of the electromagnetic spectrum. This set of discoveries was of fundamental importance for 20th-century physics, as it paved the road for Einstein's theories of relativity and for quantum theory. Other novel ideas of Maxwell's led to the establishment of such diverse fields as information theory and cybernetics. Little wonder, then, that Einstein said, on the centenary of Maxwell's birth in 1931, that his work had been "the most profound and the most fruitful that physics has experienced since the time of Newton."

Maxwell's demon (măks′wĕlz′, -wəlz) An imaginary creature who is able to sort fast-moving molecules from slow-moving molecules without adding any energy to the system. Such a creature could separate a gas into two containers, one containing hot gas and the other cold gas, bringing about a general decrease in entropy and violating the second law of **thermodynamics**; however, such a creature is impossible, since the act of observing the molecules to detect their speed must add energy to the system.

Mayr (mī′ər), **Ernst Walter** 1904–2005. German-born American zoologist who synthesized Gregor Mendel's theories of inheritance with Charles Darwin's natural selection theory to develop the concept of speciation, which expanded scientific understanding of the formation of species and how they adapt to their their environment.

mb Abbreviation of **millibar.**

Mb Abbreviation of **megabit.**

MB Abbreviation of **megabyte.**

Mbps Abbreviation of **megabits per second.**

McClintock (mə-klĭn′tək), **Barbara** 1902–1992. American geneticist who researched the chromosome theory of heredity and demonstrated how genes can control other genes and jump from chromosome to chromosome. These "jumping genes" were later called **transposons**. She was awarded the Nobel Prize for physiology or medicine in 1983.

Barbara McClintock

Md The symbol for **mendelevium.**

mean (mēn) **1.** A number or quantity having a value that is intermediate between other numbers or quantities, especially an arithmetic mean or average. See more at **arithmetic mean. 2.** Either the second or third term of a proportion of four terms. In the proportion $\frac{2}{3}=\frac{4}{6}$, the means are 3 and 4. Compare **extreme.**

meander (mē-ăn′dər) A sinuous curve, bend, or loop along the course of a stream or river.

meandering stream (mē-ăn′dər-ĭng) A stream consisting of successive meanders. Meandering streams develop in areas that are relatively flat, such as a floodplain, and where sediment consists primarily of fine sands, silts, and muds.

mean deviation In a statistical distribution, the average of the absolute values of the differences between individual numbers and their mean.

mean sea level See under **sea level.**

mean solar day See under **solar time.**

mean sun A hypothetical Sun defined as moving at a uniform rate along the celestial equator at the mean speed with which the real Sun apparently moves along the ecliptic, used in computing the mean solar day.

mean time Solar time as measured by the mean sun, resulting in equal 24-hour days throughout the year. If days were measured by the actual movement of the Sun, they would vary slightly in length at different times of the year due to differences in Earth's orbital speed and other factors. Mean time is used as the basis for standard clock time throughout most of the world. See more at **solar time, universal time.**

measles (mē′zəlz) An infectious disease caused by the rubeola virus of the genus *Morbillivirus*, characterized by fever, cough, and a rash that begins on the face and spreads to other parts of the body. Vaccinations, usually given in early childhood, confer immunity to measles. Also called *rubeola.*

measurement (mĕzh′ər-mənt) A method of determining quantity, capacity, or dimension. Several systems of measurement exist, each one comprising units whose amounts have been arbitrarily set and agreed upon by specific groups. While the **United States Customary System** remains the most commonly used system of measurement in the United States, the **International System** is accepted all over the world as the standard system for use in science. *See table,* pages 386–387.

mechanical advantage (mĭ-kăn′ĭ-kəl) The ratio of the output force (acting on a **load**) produced by a machine to the applied **effort** (the input force). See also **efficiency.**

mechanical engineering The branch of engineering that specializes in the design, production, and uses of **machines**. The physics of mechanics is widely used in mechanical engineering.

mechanics (mĭ-kăn′ĭks) **1.** The branch of physics concerned with the relationships between matter, force, and energy, especially as they affect the motion of objects. See also **classical physics, quantum mechanics. 2.** The functional aspect of a system, such as the mechanics of blood circulation.

mecopteran (mĭ-kŏp′tər-ən) Any of various carnivorous insects of the order Mecoptera, characterized by long, membranous wings and an elongated, beaklike head having chewing mouthparts at the tip. Scorpion flies are mecopterans.

medial moraine (mē′dē-əl) See under **moraine.**

median (mē′dē-ən) **1.** In a sequence of numbers arranged from smallest to largest: **a.** The middle number, when such a sequence has an odd number of values. For example, in the sequence 3, 4, 14, 35, 280, the median is 14. **b.** The average of the two middle numbers, when such a sequence has an even number of values. For example, in the sequence 4, 8, 10, 56, the median is 9 (the average of 8 and 10). Compare **arithmetic mean, average, mode. 2.** A line joining a vertex of a triangle to the midpoint of the opposite side.

mediate (mē′dē-āt′) To effect or convey a force between subatomic particles. The gauge bosons, for example, mediate the four fundamental forces of nature.

medicine (mĕd′ĭ-sĭn) **1.** The scientific study or practice of diagnosing, treating, and preventing diseases or disorders of the body or mind of a person or animal. **2.** An agent, such as a drug, used to treat disease or injury.

Medieval Warm Period (mē′dē-ē′vəl, mĕd′ē-, mĭ-dē′vəl) The period from about 1000 to 1400 in which global temperatures are thought to have been a few degrees warmer than those of the preceding and following periods. The climatic effects of this period were confined primarily to Europe and North America. The period following the Medieval Warm Period is known as the Little Ice Age.

medium (mē′dē-əm) Plural **media. 1.** A substance, such as agar, in which bacteria or other microorganisms are grown for scientific purposes. **2.** A substance that makes possible the transfer of energy from one location to another, especially through waves. For example, matter of sufficient density can be a medium for sound waves, which transfer mechanical energy. See more at **wave.**

Measurement Table

UNITS OF THE INTERNATIONAL SYSTEM

The **International System** (abbreviated **SI,** for Système International, the French name for the system) was adopted in 1960 by the 11th General Conference on Weights and Measures. An expanded and modified version of the metric system, the International System addresses the needs of modern science for additional and more accurate units of measurement. The key features of the International System are decimalization, a system of prefixes, and a standard defined in terms of an invariable physical measure.

BASE UNITS

The International System has base units from which all others in the system are derived. The standards for the base units, except for the kilogram, are defined by unchanging and reproducible physical occurrences. For example, the meter is defined as the distance traveled by light in a vacuum in 1/299,792,458 of a second. The standard for the kilogram is a platinum-iridium cylinder kept at the International Bureau of Weights and Standards in Sèvres, France.

Unit	Quantity	Symbol
meter	length	m
kilogram	mass	kg
second	time	s
ampere	electric current	A
kelvin	temperature	K
mole	amount of matter	mol
candela	luminous intensity	cd

SUPPLEMENTARY UNITS

The International System uses two supplementary units that are based on abstract geometrical concepts rather than physical standards.

Unit	Quantity	Symbol
radian	plane angles	rad
steradian	solid angles	sr

PREFIXES

A multiple of a unit in the International System is formed by adding a prefix to the name of that unit. The prefixes change the magnitude of the unit by orders of ten.

Prefix	Symbol	Multiplying Factor
yotta-	Y	10^{24} = 1,000,000,000,000,000,000,000,000
zetta-	Z	10^{21} = 1,000,000,000,000,000,000,000
exa-	E	10^{18} = 1,000,000,000,000,000,000
peta-	P	10^{15} = 1,000,000,000,000,000
tera-	T	10^{12} = 1,000,000,000,000
giga-	G	10^{9} = 1,000,000,000
mega-	M	10^{6} = 1,000,000
kilo-	k	10^{3} = 1,000
hecto-	h	10^{2} = 100
deca-	da	10 = 10
deci-	d	10^{-1} = 0.1
centi-	c	10^{-2} = 0.01
milli-	m	10^{-3} = 0.001
micro-	μ	10^{-6} = 0.000,001
nano-	n	10^{-9} = 0.000,000,001
pico-	p	10^{-12} = 0.000,000,000,001
femto-	f	10^{-15} = 0.000,000,000,000,001
atto-	a	10^{-18} = 0.000,000,000,000,000,001
zepto-	z	10^{-21} = 0.000,000,000,000,000,000,001
yocto-	y	10^{-24} = 0.000,000,000,000,000,000,000,001

ADDITIONAL UNITS

Listed below are a few of the non-SI units that are commonly used with the International System.

Unit	Quantity	Symbol
angstrom (= 10^{-10} m)	length	Å
electron-volt (= 0.160 aJ)	energy	eV
hectare (= 10,000 m^2)	land area	ha
liter (= 1.0 dm^3)	volume or capacity	l
standard atmosphere (= 101.3 kPa)	pressure	atm

DERIVED UNITS

Most of the units in the International System are derived units, that is, units defined in terms of base units and supplementary units. Derived units can be divided into two groups—those that have a special name and symbol and those that do not.

Without Names and Symbols

Measure of	*Derivation*
acceleration	m/s^2
angular acceleration	rad/s^2
angular velocity	rad/s
density	kg/m^3
electric field strength	V/m
luminance	cd/m^2
magnetic field strength	A/m
velocity	m/s

With Names and Symbols

Unit	*Measure of*	*Symbol*	*Derivation*
coulomb	electric charge	C	A • s
farad	electric capacitance	F	A • s/V
henry	inductance	H	V • s/A
hertz	frequency	Hz	cycles/s
joule	quantity of energy	J	N • m
lumen	flux of light	lm	cd • sr
lux	illumination	lx	lm/m^2
newton	force	N	$kg \cdot m/s^2$
ohm	electric resistance	Ω	V/A
pascal	pressure	Pa	N/m^2
tesla	magnetic flux density	T	Wb/m^2
volt	voltage	V	W/A
watt	power	W	J/s
weber	magnetic flux	Wb	V • s

UNITS OF THE US CUSTOMARY SYSTEM

LENGTH

Unit	Relation to Other US Customary Units
inch	$\frac{1}{12}$ foot
foot	12 inches or $\frac{1}{3}$ yard
yard	36 inches or 3 feet
rod	$16\frac{1}{2}$ feet or $5\frac{1}{2}$ yards
furlong	220 yards or $\frac{1}{8}$ mile
mile (statute)	5,280 feet or 1,760 yards
mile (nautical)	6,076 feet or 2,025 yards

VOLUME OR CAPACITY (LIQUID MEASURE)

Unit	Relation to Other US Customary Units
fluid ounce	$\frac{1}{16}$ pint
pint	16 ounces
quart	2 pints or $\frac{1}{4}$ gallon
gallon	128 ounces or 8 pints

VOLUME OR CAPACITY (DRY MEASURE)

Unit	Relation to Other US Customary Units
pint	$\frac{1}{2}$ quart
quart	2 pints
peck	8 quarts
bushel	4 pecks

WEIGHT

Unit	Relation to Other US Customary Units
grain	$\frac{1}{7000}$ pound
dram	$\frac{1}{16}$ ounce
ounce	16 drams
pound	16 ounces
ton (short)	2,000 pounds
ton (long)	2,240 pounds

CONVERSION BETWEEN METRIC AND US CUSTOMARY SYSTEMS

FROM US CUSTOMARY TO METRIC

When you know	multiply by	to find
inches	25.4	millimeters
	2.54	centimeters
feet	30.48	centimeters
yards	0.91	meters
miles	1.61	kilometers
fluid ounces	29.57	milliliters
pints (liquid)	0.47	liters (liquid)
quarts (liquid)	0.95	liters (liquid)
gallons	3.79	liters (liquid)
pints (dry)	0.55	liters (dry)
quarts (dry)	1.10	liters (dry)
ounces	28.35	grams
pounds	0.45	kilograms
short tons (2,000 lbs)	0.91	metric tons
square inches	6.45	square centimeters
square feet	0.09	square meters
square yards	0.84	square meters
square miles	2.59	square kilometers
acres	0.40	hectares

FROM METRIC TO US CUSTOMARY

When you know	multiply by	to find
millimeters	0.04	inches
centimeters	0.39	inches
meters	3.28	feet
	1.09	yards
kilometers	0.62	miles
milliliters (liquid)	0.03	fluid ounces
liters (liquid)	1.06	quarts (liquid)
	0.26	gallons
	2.12	pints (liquid)
liters (dry)	1.82	pints (dry)
	0.90	quarts (dry)
grams	0.035	ounces
kilograms	2.20	pounds
metric tons (1,000 kg)	1.10	short tons
square centimeters	0.155	square inches
square meters	1.20	square yards
square kilometers	0.39	square miles
hectares	2.47	acres

BRITISH IMPERIAL SYSTEM

VOLUME OR CAPACITY (LIQUID MEASURE)

Unit	Relation to Other British Imperial Units	Conversion to US Customary Units	Conversion to Metric Units
pint	$\frac{1}{2}$ quart	1.201 pints	0.5683 liter
quart	2 pints $\frac{1}{4}$ gallon	1.201 quarts	1.137 liters
gallon	8 pints 4 quarts	1.201 gallons	4.546 liters

VOLUME OR CAPACITY (DRY MEASURE)

Unit	Relation to Other British Imperial Units	Conversion to US Customary Units	Conversion to Metric Units
peck	$\frac{1}{4}$ bushel	1.0314 pecks	9.087 liters
bushel	4 pecks	1.0320 bushels	36.369 liters

TEMPERATURE CONVERSION BETWEEN CELSIUS AND FAHRENHEIT

°C = (°F −32) ÷ 1.8

Condition	Fahrenheit	Celsius
Boiling point of water	212°	100°
Normal body temperature	98.6°	37°
A warm day	86°	30°
A cool day	45°	7°

°F = (°C × 1.8) + 32

Condition	Fahrenheit	Celsius
Freezing point of water	32°	0°
Lowest temperature that Gabriel Fahrenheit could obtain mixing salt and ice	0°	−17.8°

medulla (mĭ-dŭl′ə, -do͞o′lə) Plural **medullas** or **medullae** (mĭ-dŭl′ē). **1.** See **medulla oblongata. 2.** The central portion of an anatomical structure, such as the adrenal gland or the kidney.

medulla oblongata (ŏb′lông-gä′tə) Plural **medulla oblongatas** or **medullae oblongatae** (ŏb′lông-gä′tē). The lowermost portion of the brainstem in humans and other mammals. It is important in the reflex control of involuntary processes, including respiration, heartbeat, and blood pressure.

medusa (mĭ-do͞o′sə) Plural **medusas** or **medusae** (mĭ-do͞o′sē). A cnidarian in its free-swimming stage. Medusas are bell-shaped, with tentacles hanging down around a central mouth. Jellyfish are medusas, while corals and sea anemones lack a medusa stage and exist only as polyps. Compare **polyp.**

mega– A prefix that means: **1.** Large, as in *megadose,* a large dose. **2.** One million, as in *megahertz,* one million hertz. **3.** 2^{20} (that is, 1,048,576), which is the power of 2 closest to a million, as in *megabyte.*

megabit (mĕg′ə-bĭt′) **1.** One million bits. **2.** 1,048,576 (or 2^{20}) bits. See Note at **megabyte.**

megabyte (mĕg′ə-bīt′) **1.** A unit of computer memory or data storage capacity equal to 1,048,576 bytes (1,024 kilobytes or 2^{20} bytes). **2.** One million bytes.

USAGE **megabyte**

In computer science and industry usage, the prefix *mega–* often does not have its standard scientific meaning of 1,000,000, but refers instead to the power of two closest to 1,000,000, which is 2^{20}, or 1,048,576. The calculation of data storage capacity (measured in bytes) is based on powers of two because of the binary nature of bits (1 byte is 8, or 2^3, bits). Thus, a megabyte is 1,048,576 bytes, although it is also used less technically to refer to a million bytes. Other numerical prefixes are interpreted similarly. With data transmission rates (measured in bits per second), a bit is considered as a signal pulse, and calculations are generally based on powers of ten. Thus, a rate of one megabit per second is equal to one million bits per second. However, in certain technical contexts, megabit can also refer to 1,048,576 bits. Similarly, the prefix *kilo–* refers to 1,000 or 2^{10} (1,024); *giga–* to 1,000,000,000 (one billion) or 2^{30} (1,073,741,824); and *tera–* to 1,000,000,000,000 (one trillion) or 2^{40} (1,099,511,627,776).

megafauna (mĕg′ə-fô′nə) Large or relatively large animals of a particular place or time period. Saber-toothed tigers and mastodons belong to the extinct megafauna of the Oligocene and Pleistocene Epochs.

megagametophyte (mĕg′ə-gə-mē′tə-fīt′) The female gametophyte that develops from the megaspores of heterosporous plants. Among heterosporous species of the lycophyte plants, for example, the sporophyte plant produces megaspores stocked with food. These spores grow into megagametophytes that produce eggs. After fertilization of the eggs, the embryos grow into new sporophytes inside the megagametophyte and are nourished by the food within it. In the seed plants, the megagametophyte generation is very small and lives its entire existence within the sporophyte plant's ovules, where some of its cells become eggs, while others become food reserves for the resultant embryos. See more at **alternation of generations, gametophyte.**

megahertz (mĕg′ə-hûrts′) A unit of frequency equal to one million hertz.

megalopteran (mĕg′ə-lŏp′tər-ən) Any of various relatively large carnivorous insects of the order Megaloptera, having biting mouthparts and two pairs of large wings. Megalopterans include the alderflies and dobsonflies, and used to be classified as neuropterans.

megaparsec (mĕg′ə-pär′sĕk) One million parsecs.

megaphyll (mĕg′ə-fĭl′) A leaf with several or many large veins branching apart or running parallel and connected by a network of smaller veins. The fronds of ferns and the leaves of gymnosperms and angiosperms are megaphylls. Megaphylls are thought to have evolved from groups of branched stems that have become fused together. Also called *macrophyll.* Compare **microphyll.**

megasporangium (mĕg′ə-spə-răn′jē-əm) Plural **megasporangia.** A plant structure in which megaspores are formed, such as those of the female cones of pines.

megaspore (mĕg′ə-spôr) One of the two types of haploid spores produced by a heterosporous plant. Megaspores develop into female gametophytes and are usually larger than microspores.

megasporocyte (mĕg′ə-spôr′ə-sīt′) A diploid cell that undergoes meiosis to produce megaspores as part of **megasporogenesis.** Also called *megaspore mother cell.*

megasporogenesis (mĕg′ə-spôr′ə-jĕn′ĭ-sĭs) The formation of megaspores inside the ovules of seed plants. A diploid cell in the ovule, called a megasporocyte or a megaspore mother cell, undergoes meiosis and gives rise to four haploid megaspores. In most plants, only one of the megaspores then goes on to develop into a megagametophyte within the ovule, while the other three disintegrate. In the ovules of angiosperms, megasporogenesis takes place within a structure called a nucellus, and it is the megaspore farthest from the micropyle of the ovary that survives.

megasporophyll (mĕg′ə-spôr′ə-fĭl′) A leaflike structure that bears megasporangia.

megatherium (mĕg′ə-thîr′ē-əm) Plural **megatheriums** or **megatheria.** A large, extinct ground sloth of the genus *Megatherium* that lived from the Miocene through the Pleistocene Epochs, primarily in South America. It was as large as an elephant, had long curved claws and teeth only in the sides of its jaws, and ate plants.

meiosis (mī-ō′sĭs) The process in cell division in sexually reproducing organisms that reduces the number of chromosomes from

A CLOSER LOOK **meiosis**

Meiosis is the process by which the nucleus divides in all sexually reproducing organisms during the production of *spores* or *gametes.* These cells have single chromosomes and are called *haploid,* as opposed to *diploid* cells with two sets. In humans, for example, gametes have one set of 23 chromosomes and are formed through meiosis from special diploid cells found in the testes and ovaries. When meiosis begins, each of the 46 chromosomes in these cells consists of two identical chromatids, just as in body cells about to divide by mitosis. However, in meiosis, there are two cell divisions instead of one, so that four daughter cells are produced, instead of two. At the start of the first meiotic division, *homologous* chromosomes (which have genes for the same traits in the same position) form pairs and exchange genetic material in the process known as *crossing over.* This process does not occur in mitosis. Then during the first meiotic division, one member of each pair of homologous chromosomes moves to each end of the cell, and the cell itself divides. Each of the two cells produced by the first division has just one set of 23 chromosomes. However, every chromosome still consists of two chromatids at this stage. The two daughter cells then undergo the second meiotic division, which is similar to mitosis. One chromatid from each of the 23 chromosomes moves to each end of the cell, and the daughter cell itself divides. The chromatids form the chromosomes of the new cells produced by the second meiotic division, which become gametes, and each cell has a single set of 23 chromosomes, normally with slight genetic variation from the original parent cell. In the human female, just one of the four daughter cells will become a functional gamete (the ovum), but in the human male, all four cells develop into gametes (the sperm). At fertilization, the union of the male and female gametes restores the two full sets of chromosomes in the human zygote.

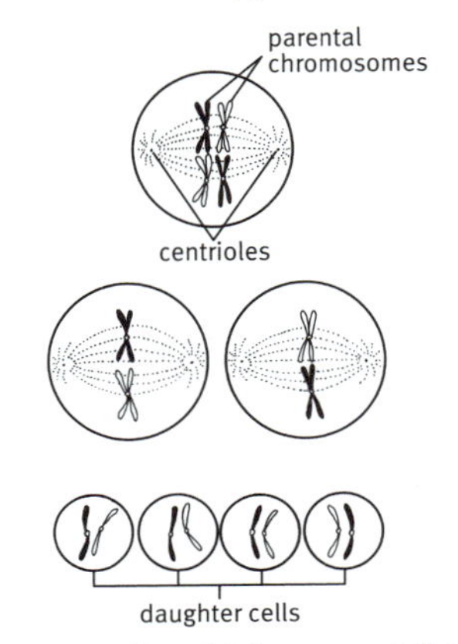

top to bottom: *First division, second division, and reproductive cells. Light and dark chromosomes distinguish chromosomes from the two parents.*

diploid to haploid (half the original number). Meiosis involves two consecutive divisions of the nucleus and leads to the production of reproductive cells (gametes) in animals and the formation of spores in plants, fungi, and most algae (the haploid spores grow into organisms that produce gametes by mitosis). Meiosis begins when the chromosomes, which have already duplicated, condense along the center of the nucleus, and pairs of homologous chromosomes undergo **crossing over**, whereby some of their genetic material is exchanged. The pairs of chromosomes then separate and move to opposite ends of the cell, and the cell itself divides into two cells. In the second stage, each of these two cells also divides into two cells. Meiosis thus produces four cells, each of which contain half the number of chromosomes as the original cell. Some or all of the four cells may become functional gametes or spores. Compare **mitosis.**

Meitner (mīt′nər), **Lise** 1878–1968. Austrian-born Swedish physicist who contributed to the first theories of nuclear fission. Her contributions to the field of nuclear physics led to the development of the atomic bomb and nuclear energy.

meitnerium (mīt-nûr′ē-əm) *Symbol* **Mt** A synthetic, radioactive element that is produced by bombarding bismuth with iron ions. Its most long-lived isotopes have mass numbers of 266 and 268 with half-lives of 3.4 milliseconds and 70 milliseconds, respectively. Atomic number 109. See **Periodic Table.**

mélange (mā-länzh′) A metamorphic rock formation created from materials scraped off the top of a downward moving tectonic plate in a subduction zone. Mélanges occur where plates of oceanic crust subduct beneath plates of continental crust, as along the western coast of South America. They consist of intensely deformed marine sediments and ocean-floor basalts and are characterized by the lack of regular strata, the inclusion of fragments and blocks of various rock types, and the presence of minerals that form only under high pressure, low temperature conditions.

melanin (mĕl′ə-nĭn) Any of various pigments that are responsible for the dark color of the skin, hair, scales, feathers, and eyes of animals and are also found in plants, fungi, and bacteria. Melanins are polymers, often bound to proteins, and in the animal kingdom are built from compounds produced by the oxidation of the amino acid tyrosine.

melanism (mĕl′ə-nĭz′əm) Dark coloration of the skin, hair, fur, or feathers because of a high concentration of melanin.

melanocyte (mĕl′ə-nō-sīt′, mə-lăn′ə-) An epidermal cell that synthesizes melanin.

melanocyte-stimulating hormone A hormone that is secreted by the pituitary gland and regulates skin color by stimulating melanin synthesis in humans and other vertebrates. Melanocyte-stimulating hormone stimulates the melanin-producing cells (called **melanocytes**) of mammals and the melanin-containing cells (called **melanophores**) of reptiles and amphibians.

melanoma (mĕl′ə-nō′mə) Plural **melanomas** or **melanomata** (mĕl′ə-nō′mə-tə). A dark-pigmented benign or malignant tumor that arises from a melanocyte and occurs most commonly in the skin. Malignant melanoma metastasizes quickly and is associated with sun exposure.

melanophore (mĕl′ə-nə-fôr′, mə-lăn′ə-) A pigmented cell that contains melanin, found especially in the skin of fishes, amphibians, and reptiles.

melatonin (mĕl′ə-tō′nĭn) A hormone produced in the pineal gland that plays a role in regulating biological rhythms, including sleep and reproductive cycles. In many animals, melatonin also regulates the physiological effects that occur in response to seasonal changes, such as the growth of a winter coat of fur. *Chemical formula:* $C_{13}H_{16}N_2O_2$.

A CLOSER LOOK melatonin

Melatonin, a natural hormone manufactured by the pineal gland in the brain, communicates information about light to different parts of the body. It helps regulate biological rhythms and plays an important role in the reproductive cycles of many animals. In humans it is best known for helping to regulate the body's circadian sleep-wake cycle. Melatonin production is affected by light exposure to the eyes; melatonin levels rise during the night and fall during the day, becoming almost undetectable. Though it does not actually induce sleep, melatonin can have sleep-promoting effects. Experiments have shown that at high doses melatonin lowers body temperature, decreases motor activity, and increases fatigue. Melatonin production starts falling after puberty, and it can virtually disappear in the elderly, a phenomenon which could help to explain why sleep disturbances are more prevalent among older adults. Marketed as a dietary supplement and touted as a cure-all for insomnia, jet lag, and even cancer and aging, the overall effects of melatonin on human health are still largely unknown.

---

melt (mĕlt) To change from a solid to a liquid state by heating or being heated with sufficient energy at the melting point. See also **heat of fusion.**

meltdown (mĕlt′doun′) Severe overheating of a nuclear reactor core, resulting in melting of the core and escape of radiation.

melting point (mĕl′tĭng) The temperature at which a solid, given sufficient heat, becomes a liquid. For a given substance, the melting point of its solid form is the same as the freezing point of its liquid form, and depends on such factors as the purity of the substance and the surrounding pressure. The melting point of ice at a pressure of one atmosphere is 0°C (32°F); that of iron is 1,535°C (2,795°F). See also **state of matter.**

member (mĕm′bər) *Mathematics.* **1.** A quantity that belongs to a set. **2.** The expression on either side of an equal sign.

membrane (mĕm′brān′) **1.** A thin, flexible layer of tissue that covers, lines, separates, or connects cells or parts of an organism. Membranes are usually made of layers of phospholipids containing suspended protein molecules and are permeable to water and fat-soluble substances. **2.** See **cell membrane. 3.** *Chemistry.* A thin sheet of natural or synthetic material that is permeable to substances in solution.

membranous (mĕm′brə-nəs) **1.** Relating to, made of, or similar to a membrane. **2.** Characterized by the formation of a membrane or a layer like a membrane.

memory (mĕm′ə-rē) **1a.** The ability to remember past experiences or learned information, involving advanced mental processes such as learning, retention, recall, and recognition and resulting from chemical changes between neurons in several different areas of the brain, including the hippocampus. Immediate memory lasts for just a few seconds. Short-term memory stores information that has been minimally processed and is available only for a few minutes, as in remembering a phone number just long enough to use it. Short-term memory is transferred into long-term memory, which can last for many years, only when repeated use of the information facilitates neurochemical changes that allow it to be retained. The loss of memory because of disease or injury is called **amnesia. b.** The collection of information gained from past learning or experience that is stored in a person's mind. **c.** A piece of information, such as the mental image of an experience, that is stored in the memory. **2a.** A part of a computer in which data is stored for later use. **b.** The capacity of a computer, chips, and storage devices to preserve data and programs for retrieval. Memory is measured in bytes. See more at **hard disk, RAM, ROM. 3.** The capacity of a material, such as plastic or metal, to return to a previous shape or condition. **4.** The capacity of the immune system to produce a specific immune response to an antigen it has previously encountered.

menaquinone (mĕn′ə-kwĭn′ōn′, -kwī′nōn′) See **vitamin K_2.**

menarche (mə-när′kē) The first menstrual period.

Mendel (mĕn′dl), **Gregor Johann** 1822–1884. Austrian botanist and founder of the science of genetics. He formulated the important principles, known as Mendel's laws, that form the basis of modern genetics.

Gregor Mendel

BIOGRAPHY Gregor Mendel

In 1851 Austrian monk Gregor Mendel was sent by his monastery to the University of Vienna to study mathematics and science.

Upon his return in 1854, be began teaching science, and several years later he began his now-famous series of hybridization experiments with garden peas in the monastery's small garden. There Mendel cross-pollinated plants of different sizes and shapes as well as plants that produced different colored flowers or peas. Analyzing each generation of new plants, he observed that some characteristics remained constant (or *dominant*) in every generation, while others remained hidden (or *recessive*), appearing only in later generations. Mendel called the units of inheritance *factors* and proved that two such factors, one from each parent, exist for each trait. A dominant trait requires only one factor for it to appear, but a recessive trait requires both. We now call the units of inheritance *genes*, and we know that each gene, which is composed of a DNA sequence, occupies a specific place on the chromosome. Although Mendel's rules of inheritance were published in 1865 in his article *"Versuche über Pflanzenhybriden" ("Experiments with Plant Hybrids")*, his work was ignored until 1900, when it was rediscovered by the Dutch botanist Hugo de Vries (1878–1918).

Mendeleev (mĕn′də-lā′əf), **Dmitri Ivanovich** 1834–1907. Russian chemist who devised the Periodic Table, which shows the relationships between the chemical elements. He first published the Table in 1869 and continued to refine it over the next 20 years.

Dmitri Mendeleev

mendelevium (mĕn′də-lē′vē-əm) *Symbol* **Md** A synthetic, radioactive metallic element of the actinide series that is produced by bombarding einsteinium with helium ions. Its most stable isotope is Md 258 with a half-life of approximately 51.5 days. Atomic number 101. See **Periodic Table.**

Mendel's law (mĕn′dlz) Any of the principles first proposed by Gregor Mendel to describe the inheritance of traits passed from one generation to the next. ▸ Mendel's first law (also called **the law of segregation**) states that during the formation of reproductive cells (gametes), pairs of hereditary factors (genes) for a specific trait separate so that offspring receive one factor from each parent. ▸ Mendel's second law (also called the **law of independent assortment**) states that chance determines which factor for a particular trait is inherited. ▸ Mendel's third law (also called the **law of dominance**) states that one of the factors for a pair of inherited traits will be dominant and the other recessive, unless both factors are recessive. See more at **inheritance.**

meninges (mə-nĭn′jēz) The three membranes that enclose the vertebrate brain and spinal cord: the pia mater, arachnoid, and dura mater.

meningitis (mĕn′ĭn-jī′tĭs) Inflammation of the meninges of the brain and the spinal cord, usually resulting from a bacterial or viral infection and often characterized by fever, vomiting, an intense headache, and a stiff neck.

meniscus (mə-nĭs′kəs) Plural **menisci** (mə-nĭs′ī, -kī, -kē) or **meniscuses. 1.** A lens that is concave on one side and convex on the other. **2.**

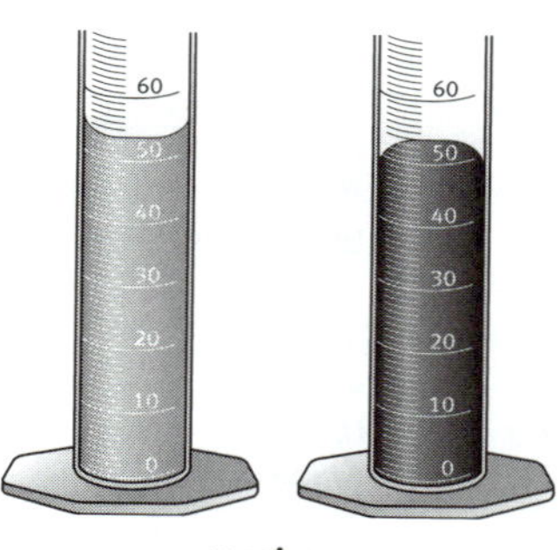

meniscus

left: *Water and glass molecules are attracted to each other, causing the surface of the water to attach to the glass and curve upward.* right: *Mercury and glass molecules are repulsed by each other, causing the surface of the mercury to curve downward and away from the glass.*

The curved upper surface of a column of liquid in a container. The surface is concave if the molecules of the liquid are attracted to the container walls and convex if they are not. See also **surface tension.** **3.** A piece of cartilage shaped like a crescent and located at the junction of two bones in a joint. The meniscus acts to absorb shock.

menopause (mĕn′ə-pôz′) The time at which menstruation ceases, occurring usually between 45 and 55 years of age in humans.

menses (mĕn′sēz) See **menstruation.**

menstrual cycle (mĕn′stro͞o-əl) The recurring monthly series of physiological changes in women and other female primates in which an egg is produced in the process known as ovulation, and the uterine lining thickens to allow for implantation if fertilization occurs. If the egg is not fertilized, the lining of the uterus breaks down and is discharged during menstruation. See also **ovulation, menstruation.**

menstruation (mĕn′stro͞o-ā′shən) The monthly flow of blood from the uterus beginning at puberty in girls and other female primates. Also called *menses.* See also **menstrual cycle.**

mental illness (mĕn′tl) Any of various psychiatric disorders or diseases, usually characterized by impairment of thought, mood, or behavior.

mental retardation Below-average intellectual ability resulting from a genetic defect, brain injury, or disease, and usually present from birth or early infancy.

menthol (mĕn′thôl′) A white, crystalline compound obtained from peppermint oil. It is used as a flavoring and as a mild anesthetic. *Chemical formula:* $\mathbf{C_{10}H_{20}O}$.

Merak (mē′răk′) A pointer star in the constellation Ursa Major, with an apparent magnitude of 2.4. See more at **Dubhe.**

Mercalli scale (mər-kä′lē, mĕr-) A scale of earthquake intensity based on observed effects and ranging from I (detectable only with instruments) to XII (causing almost total destruction). It is named after the Italian seismologist Giuseppe Mercalli.

mercaptan (mər-kăp′tăn′) See **thiol.**

Mercator (mər-kā′tər), **Gerhardus** Originally **Gerhard Kremer.** 1512–1594. Flemish cartographer who in 1568 developed the Mercator projection. In 1585 he began work on a book of maps of Europe, a project that was later completed by his son and published in 1595. As a result of the drawing of the Greek titan Atlas carrying the globe on his shoulders on the book's cover, the term "atlas" was subsequently applied to any book of maps.

Mercator projection A cylindrical projection of the Earth's surface developed by Gerhardus Mercator. As in other such projections, the areas farther from the equator appear larger, making the polar regions greatly distorted. However, the faithful representation of direction in a Mercator projection makes it ideal for navigation. See more at **cylindrical projection.**

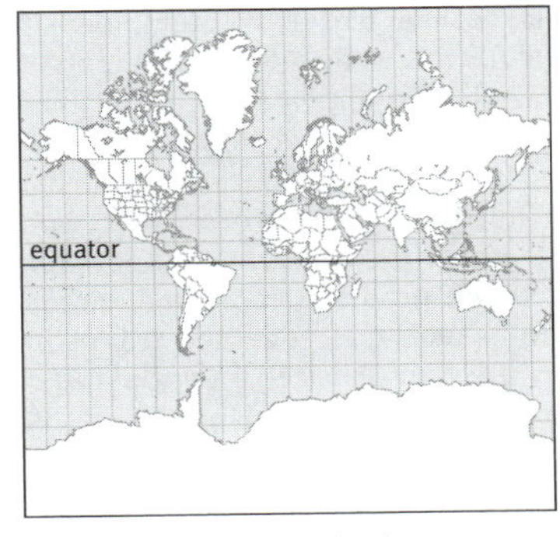

Mercator projection

mercuric (mər-kyo͝or′ĭk) Containing mercury, especially mercury with a valence of 2. Compare **mercurous.**

mercurous (mər-kyo͝or′əs, mûr′kyər-əs) Containing mercury, especially mercury with a valence of 1. Compare **mercuric.**

mercury (mûr′kyə-rē) *Symbol* **Hg** A silvery-white, dense, poisonous metallic element that is a liquid at room temperature and is used in thermometers, barometers, batteries, and pesticides. Atomic number 80; atomic weight 200.59; melting point –38.87°C; boiling point 356.58°C; specific gravity 13.546 (at 20°C); valence 1, 2. See **Periodic Table.**

WORD HISTORY **mercury**

Like a few other elements, mercury has a chemical symbol, Hg, that bears no resemblance to its name. This is because Hg is an abbreviation of the Latin name of the element, which was *hydrargium.* This word in turn was taken over from Greek, where it literally meant "water-silver." With this name the Greeks were referring to the fact that mercury is a silvery liquid at room temperature, rather than a solid like other metals. Similarly, an older English name for this element is *quicksilver,* which means "living silver," referring to its ability to move like a living thing. (The word *quick* used to mean "alive," as in the Biblical phrase "the quick and the dead.") The name *mercury* refers to the fact that the element flows about quickly: the name comes from the Roman god Mercury, who was the swift-footed messenger of the gods.

Mercury The planet closest to the Sun and the smallest in the solar system. Mercury is a **terrestrial** or **inner planet**, second in density only to Earth, with a rugged, heavily-cratered surface similar in appearance to Earth's Moon. Its rotational period of 58.6 days is two-thirds of its 88-day orbital period, thus, it makes three full axial rotations every two years. Mercury's atmosphere is almost nonexistent; this fact, which produces rapid radiational cooling on its dark side, together with its proximity to the Sun, gives it a temperature range greater than any other planet in the solar system, from 466° to –184°C (870° to –300°F). Because it is so close to the Sun, Mercury is only visible shortly before sunrise or after sunset, and observation is further hindered by the fact that its light must pass obliquely through the lower atmosphere where it is distorted or filtered by dust and pollution. See Table at **solar system.**

–mere or **–mer** A suffix meaning "part" or "segment," as in *blastomere,* one of the cells that form a blastula.

meridian (mə-rĭd**′**ē-ən) **1a.** An imaginary line forming a great circle that passes through the Earth's North and South geographic poles. **b.** Either half of such a circle from pole to pole. All the places on the same meridian have the same longitude. See illustration at **longitude. 2.** See **celestial meridian** (sense 1).

meridional (mə-rĭd**′**ē-ə-nəl) Of or relating to meridians or a meridian.

meristem (mĕr**′**ĭ-stĕm′) Plant tissue whose cells actively divide to form new tissues that cause the plant to grow. The originally undifferentiated cells of the meristem can produce specialized cells to form the tissues of roots, leaves, and other plant parts. The meristem includes the growing tips of roots and stems (the apical meristems) and the tissue layer known as **cambium.** —*Adjective* **meristematic** (mĕr′ĭ-stə-măt**′**ĭk).

mesa (mā**′**sə) An area of high land with a flat top and two or more steep, clifflike sides. Mesas are larger than buttes and smaller than plateaus, and are common in the southwest United States.

mesa

Black Mesa, near Los Alamos, New Mexico

mesic (mĕz**′**ĭk, mĕs**′**-, mē**′**zĭk, -sĭk) Relating or adapted to a moderately moist habitat. The sugar maple, white ash, and basswood are mesic plants. Compare **hydric, xeric.**

mesocarp (mĕz**′**ə-kärp′) The middle, often fleshy layer of the pericarp, such as the yellow flesh of the peach. Compare **endocarp, exocarp.**

mesoderm (mĕz**′**ə-dûrm′) The middle of the three primary germ layers of the embryos of vertebrates and other complex animals. In vertebrates, the mesoderm gives rise to the muscles, bones, cartilage, connective tissue, blood, blood and lymph vessels, dermis, kidneys, and gonads. The mesoderm develops during gastrulation from either the ectoderm or the endoderm. The embryos of simpler animals lack a mesoderm. Compare **ectoderm, endoderm.**

Mesolithic (mĕz′ə-lĭth**′**ĭk) The cultural period of the Stone Age that developed primarily in Europe between the Paleolithic and Neolithic periods, beginning around 10,000 years ago and lasting in various places as late as 3000 BCE. The Mesolithic is marked by the appearance of small-bladed, often hafted stone tools and weapons and by the beginnings of settled communities. European Mesolithic cultures existed contemporaneously with the early

Neolithic cultures of the Middle East and disappeared in any particular region with the introduction of agriculture. Also called *Middle Stone Age*. Compare **Neolithic, Paleolithic.**

meson (mĕz′ŏn′, mĕs′-, mē′zŏn′, -sŏn′) Any of a family of subatomic particles that are composed of a quark and an antiquark. Their masses are generally intermediate between leptons and baryons, and they can have positive, negative, or neutral charge. Mesons form a subclass of hadrons and include the kaon, pion and J/psi particles. Mesons were originally believed to be the particles that mediated the strong nuclear force, but it has since been shown that the gluon mediates this force. See Table at **subatomic particle.**

mesopause (mĕz′ə-pôz′) The boundary between the upper **mesosphere** and the lower **thermosphere**, approximately 80 km (50 mi) above the Earth's surface. It is the site of the coldest temperatures in the Earth's atmosphere, with temperatures of –100°C (–148°F).

mesopelagic zone (mĕz′ə-pə-lăj′ĭk) A layer of the oceanic zone lying beneath the **epipelagic zone** and above the **bathypelagic zone**, at depths generally between 200 and 1,000 m (656 and 3,280 ft). The mesopelagic zone receives very little sunlight and is home to many bioluminescent organisms. Because food is scarce in this region, most mesopelagic organisms migrate to the surface to feed at night or live off the falling detritus from the epipelagic ecosystem.

mesophyll (mĕz′ə-fĭl′) The tissues of a leaf that are located in between the layers of epidermis and carry on photosynthesis, consisting of the **palisade layer** and the **spongy parenchyma**. Most mesophyll cells contain chloroplasts.

mesophyte (mĕz′ə-fīt′) A plant that grows in an environment having a moderate supply of water. Mesophytes tend to have root systems and vascular tissues that are well-developed, and many species can grow to great heights as a result. Most crop plants, grasses, and broadleaved bushes and trees growing in temperate climates are mesophytes. Compare **hydrophyte, xerophyte.**

mesosphere (mĕz′ə-sfîr′) The region of the Earth's atmosphere lying above the stratosphere and below the thermosphere, from a height of about 50 km (31 mi) to about 80 km (50 mi) above the Earth's surface. In the mesosphere temperatures decrease with increasing altitude due to the decreasing absorption of ultraviolet radiation from the Sun. At the top of this region temperatures are around –95°C (–135.4°F). Most of the meteors that enter Earth's atmosphere burn up while passing through the mesosphere. See also **exosphere, stratosphere, thermosphere, troposphere.** See illustration at **atmosphere.**

mesothelium (mĕz′ə-thē′lē-əm) Plural **mesothelia.** A layer of flattened epithelial cells that lines the membranes of closed body cavities, including the pericardium, pleurae, and peritoneum. Compare **endothelium.**

mesozoan (mĕz′ə-zō′ən) Any of a variety of very small parasitic marine animals of the group Mesozoa. Mesozoans resemble worms and have simple bodies divided into two layers. The outer layer is covered with cilia, and there are no internal organs. The taxonomic status of the Mesozoa is unclear, as some scientists believe the mesozoans to be an ancient and primitive group and others regard them as degenerate forms of more complex animals.

Mesozoic (mĕz′ə-zō′ĭk) The era of geologic time from about 245 to 65 million years ago. The Mesozoic Era was characterized by a drastic change in plants and animals. In the early part of the Mesozoic, ferns, cycads, and ginkgos were dominant; later, gymnosperms and angiosperms developed. Dinosaurs also first appeared in the Mesozoic and, with the exception of birds, became extinct at the end of the era. See Chart at **geologic time.**

messenger RNA (mĕs′ən-jər) See under **RNA.**

Messier catalog (mĕs′ē-ā′, mĕ-syā′) A group of fixed nonstellar celestial objects originally cataloged by the French astronomer Charles Messier (1730–1817) and since expanded from 103 to 110, although several are now considered questionable. Items in the Messier catalog are numbered from M1 (the Crab Nebula) through M110 and include what are now known to be galaxies, nebulae, and globular and open clusters. Messier's purpose was to further the search for comets by listing the indistinct celestial objects that might be mistaken for them; he had no understanding of what the items in his catalog actually were.

metabolism (mĭ-tăb′ə-lĭz′əm) The chemical processes by which cells produce the substances and energy needed to sustain life. As part of metabolism, organic compounds are broken down to provide heat and energy in the process called **catabolism**. Simpler molecules are also used to build more complex compounds like proteins for growth and repair of tissues as part of **anabolism**. Many metabolic processes are brought about by the action of enzymes. The overall speed at which an organism carries out its metabolic processes is termed its metabolic rate (or, when the organism is at rest, its basal metabolic rate). Birds, for example, have a high metabolic rate, since they are warm-blooded, and their usual method of locomotion, flight, requires large amounts of energy. Accordingly, birds usually need large amounts of high-quality, energy-rich foods such as seeds or meat, which they must eat frequently. See more at **cellular respiration.** —*Adjective* **metabolic** (mĕt′ə-bŏl′ĭk).

metabolite (mĭ-tăb′ə-līt′) **1.** A substance produced by metabolism. **2.** A substance necessary for or taking part in a particular metabolic process. Examples of metabolites are glucose in the metabolism of sugars and starches, amino acids in the biosynthesis of proteins, and squalene in the biosynthesis of cholesterol.

metabolize (mĭ-tăb′ə-līz′) To subject a substance to metabolism or produce a substance by metabolism.

metacarpal (mĕt′ə-kär′pəl) Any of the bones of the hands in humans or the forelimbs in animals that are located between the carpal bones and the phalanges.

metal (mĕt′l) **1.** Any of a large group of chemical elements, including iron, gold, copper, lead, and magnesium, that readily become **cations** and form **ionic bonds**, having relatively free **valence electrons** (electrons in the outer shells). Metals are generally good conductors of electricity because of the freedom of their valence electrons. Metals generally conduct heat well, and in solid form are relatively malleable and ductile compared to other solids. They are usually shiny and opaque. All metals except mercury are solid at room temperature. **2.** An alloy, such as steel or bronze, made of two or more metals. **3.** In astronomy, any atom except hydrogen and helium. **4.** Small stones or gravel, mixed with tar to form tarmac for the surfacing of roads.

USAGE **metal**

Most metallic elements are lustrous or colorful solids that are good conductors of heat and electricity, and readily form ionic bonds with other elements. Many of their properties are due to the fact that their outermost electrons, called *valence electrons,* are not tightly bound to the nucleus. For instance, most metals form ionic bonds easily because they readily give up valence electrons to other atoms, thereby becoming positive ions (cations). The electrical *conductivity* of metals also stems from the relative freedom of valence electrons. In a substance composed of metals, the atoms are in a virtual "sea" of valence electrons that readily jump from atom to atom in the presence of an electric potential, creating electric current. With the exception of hydrogen, which behaves like a metal only at very high pressures, the elements that appear in the left-hand column of the Periodic Table are called *alkali metals.* Alkali metals, such as sodium and potassium, have only one electron in their outermost shell, and are chemically very reactive. (Hydrogen is exceptional in that, although it is highly reactive, its other metallic properties are manifest only at very high pressures.) Metals farther toward the right side of the Periodic Table, such as tin and lead, have more electrons in their outermost shell, and are not as reactive. The somewhat reactive elements that fall between the two extremes are the *transition elements,* such as iron, copper, tungsten, and silver. In most atoms, inner electron shells must be maximally occupied by electrons before an outer shell will accept electrons, but many transition elements have electron gaps in the shell just inside the valence shell. This configuration leads to a wide variety of available energy levels for electrons to move about in, so in the presence of electromagnetic radiation such as light, a variety of frequencies are readily emitted or absorbed. Thus transition metals tend to be very colorful, and each contributes different colors to different compounds.

metallic (mĭ-tăl′ĭk) Relating to or having the characteristics of a metal.

metallic bond The chemical bonding that holds the atoms of a metal together. Metallic bonds are formed from the attraction between mobile electrons and fixed, positively charged metallic atoms. Whereas most chemical bonds are localized between specific neighboring atoms, metallic bonds extend over the entire molecular structure.

metalloid (mĕt′l-oid′) **1.** An element that is not a metal but that has some properties of metals. Arsenic, for example, is a metalloid that has the visual appearance of a metal, but is a poor conductor of electricity; metalloids are generally **semiconductors**. The elements classified as metalloids are boron, silicon, germanium, arsenic, antimony, tellurium, and polonium. Metalloids can be viewed as a diagonal section on the Period Table, separating metals from nonmetals. **2.** A nonmetallic element, such as carbon, that can form alloys with metals.

metallurgy (mĕt′l-ûr′jē) The scientific study and technology of extracting metals from ores, refining them for use, and creating alloys and useful objects from them.

metamorphic (mĕt′ə-môr′fĭk) **1.** *Zoology.* Relating to metamorphosis. **2.** *Geology.* Relating to rocks that have undergone metamorphism. Metamorphic rocks are formed when igneous, sedimentary, or other metamorphic rocks undergo a physical change due to extreme heat and pressure. These changes often produce folded layers or banding in the rocks, and they can also cause pockets of precious minerals to form. The folds and banding can be produced by incomplete segregation of minerals during recrystallization, or they can be inherited from preexisting beds in sedimentary rocks or preexisting layers in igneous rocks. The precious minerals can form as the result of recrystallization when the rocks undergoing metamorphism are subjected to changes in pressure and temperature.

metamorphic

gneiss formation in Godthåbsfjord, Greenland

metamorphism (mĕt′ə-môr′fĭz′əm) The process by which rocks are changed in composition, texture, or structure by extreme heat and pressure. ▸ In **prograde metamorphism** metamorphic rocks that were formed under low pressure and temperature conditions undergo a second metamorphic event in which they are exposed to higher pressures and temperatures. ▸ In **retrograde metamorphism** metamorphic rocks that were formed under high pressure and temperature conditions undergo a second metamorphic event in which they are exposed to lower pressures and temperatures. See more at **contact metamorphism, regional metamorphism.**

metamorphosis (mĕt′ə-môr′fə-sĭs) Dramatic change in the form and often the habits of an animal during its development after birth or hatching. The transformation of a maggot into an adult fly and of a tadpole into an adult frog are examples of metamorphosis. The young of such animals are called larvae.

metamorphosis

top to bottom: *bullfrog tadpole, bullfrog tadpole with developing legs and webbed feet, and mature bullfrog*

metaphase (mĕt′ə-fāz′) The stage of cell division in which the duplicated chromosomes become aligned along the center of the cell, called the *equatorial plate* or *metaphase plate.* Metaphase lasts up to an hour, and ends in mitosis and the second division of meiosis when separation of the paired chromosomal strands (called chromatids) begins. In the first division of meiosis, the paired chromosomes separate from one another. Metaphase is preceded by prophase and followed by anaphase. See more at **meiosis, mitosis.**

metasomatism (mĕt′ə-sō**′**mə-tĭz′əm) also **metasomatosis** (mĕt′ə-sō′mə-tō**′**sĭs) The process by which the chemical composition of a rock is changed through the introduction or extraction of chemicals dissolved in fluids that migrate through the rock's pores. Metasomatism often results in the formation of new minerals, especially metal ore deposits.

metastasis (mə-tăs**′**tə-sĭs) A cancerous tumor formed by transmission of malignant cells from a primary cancer located elsewhere in the body. —*Verb* **metastasize.**

metatag (mĕt**′**ə-tăg′) An HTML tag that contains descriptive information about a webpage and does not appear when the webpage is displayed in a browser. A word that is in a metatag of a webpage will cause that webpage to turn up as a result of a search engine's search on that word, even if that word is not found in the webpage as viewed in a browser.

metatarsal (mĕt′ə-tär**′**səl) Any of the bones of the feet in humans or the back feet in animals that are located between the tarsal bones and the phalanges.

metazoan (mĕt′ə-zō**′**ən) **1.** Any of the animals belonging to the subkingdom Metazoa, having a body made up of differentiated cells arranged in tissues and organs. All multicellular animals besides sponges are metazoans. **2.** A multicellular animal. No longer in scientific use.

meteor (mē**′**tē-ər) **1.** A bright trail or streak of light that appears in the night sky when a meteoroid enters the Earth's atmosphere. The friction with the air causes the rock to glow with heat. Also called *shooting star*. **2.** A rocky body that produces such light. Most meteors burn up before reaching the Earth's surface. See Note at **solar system.**

USAGE **meteor/meteorite/meteoroid**

The streaks of light we sometimes see in the night sky and call *meteors* were not identified as interplanetary rocks until the 19th century. Before then, the streaks of light were considered only one of a variety of atmospheric phenomena, all of which bore the name *meteor.* Rain was an *aqueous meteor,* winds and storms were *airy meteors,* and streaks of light in the sky were *fiery meteors.* This general use of *meteor* survives in our word *meteorology,* the study of the weather and atmospheric phenomena. Nowadays, astronomers use any of three words for rocks from interplanetary space, depending on their stage of descent to the Earth. A *meteoroid* is a rock in space that has the potential to collide with the Earth's atmosphere. Meteoroids range in size from a speck of dust to a chunk about 100 meters in diameter, though most are smaller than a pebble. When a meteoroid enters the atmosphere, it becomes a *meteor.* The light that it gives off when heated by friction with the atmosphere is also called a *meteor.* If the rock is not obliterated by the friction and lands on the ground, it is called a *meteorite.* For this term, scientists borrowed the *–ite* suffix used in the names of minerals like malachite and pyrite.

meteorite (mē**′**tē-ə-rīt′) A meteor that reaches the Earth's surface because it has not been burned up by friction with the atmosphere. Meteorites are believed to be fragments of comets and asteroids. ► Meteorites that consist mostly of silicates are called **stony meteorites** and are classified as either **chondrites** or **achondrites.** ► Meteorites that consist mostly of iron are called **iron meteorites.** ► Meteorites that consist of a mixture of silicates and iron are called **stony-iron meteorites.**

meteorograph (mē′tē-ôr**′**ə-grăf′) An instrument that records simultaneously several meteorological conditions, such as temperature, barometric pressure, rainfall, humidity, and wind direction.

meteoroid (mē**′**tē-ə-roid′) A small, rocky or metallic body revolving in interplanetary space around the Sun. A meteoroid is significantly smaller than an asteroid, ranging from small grains or particles to the size of large boulders. The clustered meteoroids associated with regular annual meteor showers are believed to be very small particles of cometary debris. Meteoroids that survive their passage through the Earth's atmosphere and land as meteorites are somewhat larger, solitary bodies and are encountered in no predictable pattern. See Note at **meteor.**

meteorology (mē′tē-ə-rŏl′ə-jē) The scientific study of the atmosphere and of atmospheric conditions, especially as they relate to weather and weather forecasting.

meteor shower A brief period of heightened meteor activity, often occurring regularly in a particular part of the sky at a particular time of year. Meteor showers are generally named after the constellation in which they appear to originate; thus the Perseids appear to originate in the constellation Perseus and the Leonids in the constellation Leo. The showers occur when the Earth passes through a region having a greater than usual concentration of interplanetary debris, such as particles left by a disintegrating comet, at certain points in its orbit. Although the meteors enter the Earth's atmosphere on parallel trajectories, perspective makes it appear as if they originate from the same point in the sky, known as the radiant.

meter (mē′tər) The basic unit of length in the metric system, equal to 39.37 inches. See Table at **measurement.**

meter-kilogram-second system A system of measurement in which the basic units of length, mass, and time are the meter, the kilogram, and the second. It is used in the United States chiefly in mechanics. Also called *mks system.*

methacrylate (mĕth-ăk′rə-lāt′) **1.** An ester of methacrylic acid, having the general formula $C_4H_3O_2R$, where R is an organic radical. Methacrylates are used in the manufacture of plastics. **2.** A resin derived from methacrylic acid.

methacrylic acid (mĕth′ə-krĭl′ĭk) A colorless liquid used in the manufacture of methacrylate esters, resins and plastics. *Chemical formula:* $\mathbf{C_4H_6O_2}$.

methamphetamine (mĕth′ăm-fĕt′ə-mēn′, -mĭn) An amine derivative of amphetamine, $C_{10}H_{15}N$, used in the form of its crystalline hydrochloride both as a central nervous system stimulant in the medical treatment of obesity and attention deficit hyperactivity disorder and illicitly as a recreational drug.

methane (mĕth′ān′) A colorless, odorless, flammable gas that is the simplest hydrocarbon. It is the major constituent of natural gas and is released during the decomposition of plant or other organic compounds, as in marshes and coal mines. Methane is the first member of the alkane series. *Chemical formula:* $\mathbf{CH_4}$.

methane series The alkane series. See under **alkane.**

methanol (mĕth′ə-nôl′) A colorless, toxic, flammable liquid used as a general solvent, antifreeze, and fuel. Also called *methyl alcohol, wood alcohol. Chemical formula:* $\mathbf{CH_4O}$.

methionine (mə-thī′ə-nēn′) An essential amino acid. *Chemical formula:* $\mathbf{C_5H_{11}NO_2S}$. See more at **amino acid.**

methotrexate (mĕth′ə-trĕk′sāt) A toxic drug that acts as a folic acid antagonist to interfere with cellular reproduction and is used to treat psoriasis, certain cancers, and certain inflammatory diseases, such as rheumatoid arthritis. *Chemical formula:* $\mathbf{C_{20}H_{22}N_8O_5}$.

methyl (mĕth′əl) The radical CH_3, derived from methane.

methyl alcohol See **methanol.**

methylamine (mĕth′ə-lə-mēn′, -lăm′ēn, mə-thĭl′ə-mēn′) A toxic, flammable gas produced naturally by the decomposition of organic matter and also made synthetically. It is used as a solvent and in the manufacture of many products, such as dyes and insecticides. *Chemical formula:* $\mathbf{CH_5N}$.

methylate (mĕth′ə-lāt′) *Noun.* **1.** An organic compound having the general formula CH_3OR, in which R is a metal. Methylates are formed by replacing the hydrogen of the hydroxyl group (OH) of methyl alcohol with a metal. —*Verb.* **2.** To combine with the methyl radical.

methylbenzene (mĕth′əl-bĕn′zēn′) See **toluene.**

methylene (mĕth′ə-lēn′) A bivalent hydrocarbon radical, CH_2. Because it has two unshared electrons, it is extremely reactive and occurs only as an intermediate byproduct in chemical reactions. Methylene is a component of unsaturated hydrocarbons.

methylene blue A basic dye that forms a deep blue solution when dissolved in water. It is used as an antidote for cyanide poisoning and as a bacteriological stain. *Chemical formula:* $\mathbf{C_{16}H_{18}N_3SCl}$.

methyl ethyl ketone See **butanone.**

metric (mĕt′rĭk) Relating to the meter or the metric system.

metric system A decimal system of weights and measures based on the meter as a unit of length, the kilogram as a unit of mass, and the liter as a unit of volume. Compare **US Customary System.** See Table at **measurement.**

metric ton A unit of mass or weight in the metric system equal to 1,000 kilograms (2,205 pounds). See Table at **measurement.**

mg Abbreviation of **milligram.**

Mg The symbol for **magnesium.**

mho (mō) The SI derived unit of electrical conductance, equal to one ampere per volt. It is equivalent to the reciprocal of the **ohm** unit. Also called *siemens.*

MHz Abbreviation of **megahertz.**

mi Abbreviation of **mile.**

mica (mī′kə) Any of a group of hydrous aluminosilicate minerals with the general formula $(K,Na,Ca)(Mg,Fe,Li,Al)_{2-3}(Al,Si)_4O_{10}(OH,F)_2$ that can be split easily into thin, partly transparent sheets. Mica is common in igneous and metamorphic rocks and often occurs as flakes or sheets. It is highly resistant to heat and is used in electric fuses and other electrical equipment. Muscovite and biotite are types of mica

Michelson (mī′kəl-sən), **Albert Abraham** 1852–1931. German-born American physicist who (with Edward Morley) disproved the existence of ether, the hypothetical medium of electromagnetic waves. Their work served as the starting point for Albert Einstein's development of the theory of relativity.

micro– A prefix that means: "small" (as in *microorganism*) or "one millionth" (as in *microsecond*).

microbar (mī′krō-bär′) A unit of pressure equal to one millionth (10^{-6}) of a bar.

microbarograph (mī′krō-băr′ə-grăf′) A high-precision **barograph**, capable of resolving pressure differences with an accuracy of microbars.

microbe (mī′krōb′) A microorganism, especially a bacterium that causes disease. See Note at **germ.**

microbiology (mī′krō-bī-ŏl′ə-jē) The scientific study of microorganisms.

microburst (mī′krō-bûrst′) A sudden, violent downdraft of air over a small area (less than 16 sq km or 6.24 sq mi) that lasts at least 25 minutes. Microbursts can cause winds with speeds as high as 270 km (167 mi) per hour. They are difficult to detect and predict with standard weather instruments. They are especially hazardous to airplanes during landing or taking off.

microchip (mī′krə-chĭp′) See **integrated circuit.**

microclimate (mī′krō-klī′mĭt) The climate of a small, specific place within a larger area. An area as small as a yard or park can have several different microclimates depending on how much sunlight, shade, or exposure to the wind there is at a particular spot. Compare **macroclimate.**

microcline (mī′krō-klīn′) A white, pink, red-brown, or green type of potassium feldspar. It is dimorphous with orthoclase feldspar, differing from it in shape and in the fact that it forms at lower temperatures.*Chemical formula:* $KAlSi_3O_8$.

microcrystalline (mī′krō-krĭs′tə-lĭn) Having a crystalline structure visible only under a microscope.

microelectronics (mī′krō-ĭ-lĕk-trŏn′ĭks) The branch of electronics that deals with miniature components generally too small to be seen by the naked eye.

microenvironment (mī′krō-ĕn-vī′rən-mənt) The environment of a very small, specific area, distinguished from its immediate surroundings by such factors as the amount of incident light, the degree of moisture, and the range of temperatures. The side of a tree that is shaded from sunlight is a microenvironment that typically supports a somewhat different community of organisms than is found on the side that receives regular light. Also called *microhabitat.*

microevolution (mī′krō-ĕv′ə-lo͞o′shən) Evolutionary change below the level of the species, resulting from relatively small genetic variations. Microevolution produces new strains of microorganisms, for example, or the rise of a new subspecies. The accumulation of many microevolutionary changes results in macroevolution.

microfarad (mī′krō-făr′əd, -ăd) A unit of capacitance equal to one millionth (10^{-6}) of a farad. The microfarad is used for most electronic applications. Its symbol is μF.

microfossil (mī′krō-fŏs′əl) A microscopic fossil, as of a pollen grain or unicellular organism.

microgamete (mī′krō-găm′ēt′, -gə-mēt′) The smaller of a pair of conjugating gametes, usually the male, in an organism that reproduces by heterogamy.

microgametophyte (mī′krō-gə-mē′tə-fīt′) The male gametophyte that develops from the microspores of heterosporous plants. The pollen grains of gymnosperms and angiosperms are microgametophytes. See more at **gametophyte, pollination.**

micrograph (mī′krə-grăf′) A drawing or photograph taken from an image formed by a microscope.

microgravity (mī′krō-grăv′ĭ-tē) A condition in which an object in the gravitational field of some other body (such as the Earth) is accelerated freely as a result of the gravitational force. Free-falling objects, such as a skydiver or a satellite orbiting the Earth, are in a condition of microgravity, while objects held up by forces resisting gravity (as in the case of objects resting on the Earth's surface) or held up by aerodynamic forces (as in the case of birds or aircraft) are not. Since the normal experience of weight on Earth is the result of forces that resist gravity, objects in microgravity appear weightless. Not all effects of gravity are eliminated in such conditions; **tidal forces,** for example, still affect bodies in microgravity, especially large bodies such as the Earth and the Moon.

microhabitat (mī′krō-hăb′ĭ-tăt′) See **microenvironment.**

microlith (mī′krō-lĭth′) A very small blade made of flaked stone and used as a tool, especially in the European Mesolithic Period.

micrometeoroid (mī′krō-mē′tē-ə-roid′) An extremely small meteoroid, typically the size of a grain of dust. Particles measuring less than 0.05 mm (0.002 inch) in diameter are classed as micrometeoroids. Despite their small size, their high velocities mean that micrometeoroids pose a significant threat to spacecraft and artificial satellites. ▸ A micrometeoroid that reaches the surface of the Earth, the Moon, or another celestial body is called a **micrometeorite**. The tiny size of micrometeoroids allows them to radiate away the heat generated by friction with the atmosphere, with the result that the proportion of micrometeoroids becoming micrometeorites is greater than the proportion of meteoroids becoming meteorites.

micrometeorology (mī′krō-mē′tē-ə-rŏl′ə-jē) The branch of meteorology that deals with weather conditions on a small scale, both in terms of space and time. For example, weather conditions lasting less than a day in the area immediately surrounding a smokestack, a building, or a mountain are studied in micrometeorology.

micrometer[1] (mī-krŏm′ĭ-tər) A device for measuring very small distances, angles, or objects, especially one based on the rotation of a finely threaded screw, as in relation to a microscope.

micrometer[2] (mī′krō-mē′tər) A unit of length in the metric system equal to one millionth (10^{-6}) of a meter. Also called *micron.*

micron (mī′krŏn′) See **micrometer[2].**

microorganism (mī′krō-ôr′gə-nĭz′əm) An organism that can be seen only with the aid of a microscope and that typically consists of only a single cell. Microorganisms include bacteria, protozoans, and certain algae and fungi. See Note at **germ.**

micropaleontology (mī′krō-pā′lē-ŏn-tŏl′ə-jē) The branch of paleontology that deals with microfossils.

microphyll (mī′krə-fĭl′) A leaf with only one vascular bundle and no complex network of veins. Horsetails and lycophytes (such as club mosses) have microphylls. Microphylls on modern plants are generally small but in extinct phyla the same structures could grow quite large. In contrast to **megaphylls,** microphylls are thought to have evolved from modifications of a single stem.

microprocessor (mī′krō-prŏs′ĕs-ər) An integrated circuit that contains a processor, such as a central processing unit.

micropyle (mī′krə-pīl′) **1.** A minute opening in the ovule of a seed plant through which the pollen tube usually enters. **2.** A pore in the membrane covering the ovum of some animals through which a spermatozoon can enter.

microscope (mī′krə-skōp′) Any of various instruments used to magnify small objects that are difficult or impossible to observe the naked eye. ▸ **Optical microscopes** use light reflected from or passed through the sample being observed to form a magnified image of the object, refracting the light with an arrangement of lenses and mirrors similar to those found in telescopes. See also **atomic force microscope, electron microscope, field ion microscope.**

microsecond (mī′krō-sĕk′ənd) A unit of time equal to one millionth (10^{-6}) of a second.

microseism (mī′krə-sī′zəm) A faint Earth tremor, unrelated to earthquakes, caused by natural phenomena, such as winds and strong ocean waves.

microsporangium (mī′krō-spə-răn′jē-əm) Plural **microsporangia.** A plant structure in which microspores are formed. The pollen-producing male cones of a pine consist of many microsporangia, where the microspores of the pine develop into pollen grains to be dispersed.

microspore (mīk′rə-spôr′) One of the two types of haploid spores produced by a heterosporous plant. Microspores develop into male gametophytes and are usually smaller than megaspores. In angiosperms, the microspore develops into the pollen grain.

microsporocyte (mī′krə-spôr′ə-sīt′) A diploid cell that undergoes meiosis to produce microspores as part of **microsporogenesis.** Also called *microspore mother cell, pollen mother cell.*

microsporogenesis (mī′krə-spôr′ə-jĕn′ĭ-sĭs) The formation of microspores inside the microsporangia (or pollen sacs) of seed plants. A diploid cell in the microsporangium, called a microsporocyte or a pollen mother cell, undergoes meiosis and gives rise to four haploid microspores. Each microspore then develops into a pollen grain (the microgametophyte).

microsporophyll (mī′krə-spôr′ə-fĭl′) A leaflike structure that bears microsporangia, such as those of in the strobili of lycophytes or in the male cones of conifers. The stamens of flowering plants are highly modified microsporophylls.

microsurgery (mī′krō-sûr′jə-rē) Surgery on tiny body structures or cells that is performed with the aid of a microscope and other specialized instruments, such as a laser.

microtubule (mī′krō-to͞o′byo͞ol) Any of the tube-shaped protein structures that help eukaryotic cells maintain their shape and assist in forming the cell spindle during cell division. Microtubules and actin filaments are the main components of the cell's supporting matrix or **cytoskeleton**.

microwave (mī′krō-wāv′) An electromagnetic wave with a frequency in the range of 100 megahertz to 30 gigahertz (lower than infrared but higher than other radio waves). Microwaves are used in radar, radio transmission, cooking, and other applications. Microwaves are generated naturally by many astronomical phenomena and are found in **cosmic background radiation**. See more at **electromagnetic spectrum.**

microwave background See **cosmic background radiation.**

midbrain (mĭd′brān′) The middle part of the vertebrate brain. In most animals except mammals, the midbrain processes sensory information. In mammals, it serves primarily to connect the forebrain with the hindbrain. Compare **forebrain, hindbrain.**

middle ear (mĭd′l) The part of the ear in most mammals that contains the eardrum (tympanic membrane) and the three ossicles (malleus, incus, and stapes) which transmit sound vibrations from the eardrum to the inner ear. See more at **ear.**

middle lamella The pectin-rich intercellular material cementing together the primary walls of adjacent plant cells.

Middle Stone Age See **Mesolithic.**

mid-ocean ridge (mĭd′ō′shən) A long mountain range on the ocean floor, extending almost continuously through the North and South Atlantic Oceans, the Indian Ocean, and the South Pacific Ocean. A deep rift valley is located at its center, from which magma flows and forms new oceanic crust. As the magma cools and hardens it becomes part of the mountain range. The mid-ocean ridge is approximately 1,500 km (930 mi) wide, 1 to 3 km (0.62 to 1.86 mi) high, and over 84,000 km

(52, 080 mi) long. See more at **sea-floor spreading.**

midrib (mĭd′rĭb′) The central or main vein of a leaf, as in eudicots, magnoliids, and ferns. Midribs generally protrude from the underside of leaves with pinnate venation. See more at **venation.**

migmatite (mĭg′mə-tīt′) A rock of both metamorphic and igneous origin, that exhibits characteristics of both rock types. Migmatites probably form through the heating (but not melting) of rocks in the presence of abundant fluids.

migraine (mī′grān′) A severe recurring headache, usually affecting only one side of the head, that is characterized by sharp, throbbing pain and is often accompanied by nausea, vomiting, sensitivity to light, and visual disturbances. Vasodilation in the brain causes inflammation that results in pain, but the exact cause of migraine is unknown.

migration (mī-grā′shən) **1.** The seasonal movement of a complete population of animals from one area to another. Migration is usually a response to changes in temperature, food supply, or the amount of daylight, and is often undertaken for the purpose of breeding. Mammals, insects, fish, and birds all migrate. The precise mechanism of navigation during migration is not fully understood, although for birds it is believed that sharp eyesight, sensibility to the Earth's magnetic field, and the positions of the Sun and other stars may play a role. **2.** The movement of one atom or more, or of a double bond, from one position to another within a molecule. **3.** The movement of ions between electrodes during electrolysis.

migratory (mī′grə-tôr′ē) Traveling from one place to another at regular times of year, often over long distances. Salmon, whales, and swallows are all migratory animals.

mil (mĭl) A unit of length in the US Customary System equal to $\frac{1}{1000}$ of an inch (0.03 millimeter), used chiefly to measure the diameter of wires.

mildew (mĭl′do͞o′) Any of various fungi or oomycete organisms that form a white or grayish coating on surfaces, such as plant leaves, cloth, or leather, especially under damp, warm conditions. *Powdery mildews* are important plant diseases usually caused by ascomycete fungi, while *downy mildews*, including a serious disease of grapevines, are caused by oomycetes.

mile (mīl) **1.** A unit of length in the US Customary System, equal to 5,280 feet or 1,760 yards (about 1.61 kilometers). Also called *statute mile.* **2.** See **nautical mile.** See Table at **measurement.**

milk (mĭlk) A white liquid produced by the mammary glands of female mammals for feeding their young beginning immediately after birth. Milk is an emulsion of proteins, fats, vitamins, minerals, and sugars, especially lactose, in water. The proteins in milk contain all the essential amino acids.

Milky Way (mĭl′kē) The spiral galaxy that contains our solar system. Made up of an estimated two hundred billion stars or more, it is seen from Earth as an irregular band of hazy light across the night sky. The solar system is located in one of the revolving spiral arms, about 50 light-years north of the galactic plane and some 27,700 light-years from the galaxy's center, which lies in the direction of the constellation Sagittarius. It takes approximately 250 million years for the solar system to orbit the galactic center, which is believed to contain a massive black hole. The Milky Way measures about 100,000 light-years in diameter and is the second largest galaxy, after the Andromeda Galaxy, in the cluster known as the Local Group. See also **spiral galaxy.**

milli– A prefix that means "one thousandth," as in *millimeter,* one thousandth of a meter.

millibar (mĭl′ə-bär′) A unit of pressure equal to 0.001 bars. It is equivalent to 100 newtons per square meter, or 0.0145 pounds per square inch. Standard atmospheric pressure at sea level is 1,013.2 millibars.

milligram (mĭl′ĭ-grăm′) A unit of mass or weight in the metric system equal to 0.001 gram. See Table at **measurement.**

Millikan (mĭl′ĭ-kən), **Robert Andrews** 1868–1953. American physicist who measured the electron charge and experimentally verified Einstein's equation describing the photoelectric effect. For this work he received the 1923 Nobel Prize for physics. Millikan also proved the existence of (and coined the term for) cosmic rays.

BIOGRAPHY Robert Andrews Millikan

Robert Millikan calculated the charge of an electron with his famous *oil-drop experiment* in 1910, which took advantage of the fact that droplets of oil can carry an electric charge on their surfaces. He took a closed transparent chamber with two parallel horizontal metal plates, one passing through the middle of the chamber and one on the bottom. The upper plate had a small hole in it, and the plates were connected by an electric current, giving them a charge. Millikan sprayed tiny droplets of oil into the chamber's upper half; these floated downward, with some falling through the hole in the upper plate. Their mass could be calculated by measuring how fast they fell. Millikan then ionized the air in the lower half by beaming x-rays at it, which stripped electrons from the air molecules; the electrons attached themselves to the droplets, giving them a negative charge. By changing the voltage between the two plates, which changed the electric differential between them, he could modulate the rate of the droplets' fall. If the voltage equaled the known gravitational force acting on a droplet, the droplet remained stationary. This voltage, together with the droplet's mass, he then used to calculate the droplet's charge. Millikan found through many experiments that the charge was always a whole-number multiple of a particular quantity, which he deduced was the charge of a single electron (1.602×10^{-19} coulombs). For this discovery, he was awarded the Nobel Prize for physics in 1923.

milliliter (mĭl′ə-lē′tər) A unit of liquid volume or capacity in the metric system equal to 0.001 liter. See Table at **measurement.**

millimeter (mĭl′ə-mē′tər) A unit of length in the metric system equal to 0.001 meter. See Table at **measurement.**

millipede (mĭl′ə-pēd′) Any of various wormlike arthropods of the class Diplopoda, having a long body composed of many narrow segments, most of which have two pairs of legs. Millipedes feed on plants and, unlike centipedes, do not have venomous pincers. Compare **centipede.**

Milstein (mĭl′stēn′), **Cesar** 1927–2002. Argentinean-born British immunologist who conducted important research into antibodies. With Georges Köhler he developed a method of fusing together different cells to maintain antibody production. For the discovery of this technique, which is widely used in the development of drugs and in diagnostic tests for cancer and other diseases, Milstein and Köhler shared with British immunologist Niels K. Jerne the 1984 Nobel Prize for physiology or medicine.

milt (mĭlt) Fish sperm, together with the milky liquid that contains them.

mimicry (mĭm′ĭ-krē) The resemblance of one organism to another or to an object in its surroundings for concealment or protection from predators. See also **aggressive mimicry, Batesian mimicry, Müllerian mimicry.**

mine (mīn) An underground excavation in the Earth from which ore, rock, or minerals can be extracted.

mineral (mĭn′ər-əl) **1.** A naturally occurring, solid, inorganic element or compound having a uniform composition and a regularly repeating internal structure. Minerals typically have a characteristic hardness and color, or range of colors, by which they can be recognized. Rocks are made up of minerals. **2.** A natural substance of commercial value, such as iron ore, coal, or petroleum, that is obtained by mining, quarrying, or drilling.

mineralogy (mĭn′ə-rŏl′ə-jē) The scientific study of minerals, their composition and properties, and the places where they are likely to occur.

mineral oil A colorless, odorless, tasteless oil distilled from petroleum. It is used as a lubricant and, in medicine, as a laxative.

minimum (mĭn′ə-məm) Plural **minimums** or **minima. 1.** The lowest known or lowest possible number, measure, quantity, or degree. **2.** The lowest value of a mathematical function, if it has such a value.

Minkowski space-time (mĭn-kôf′skē) A four-dimensional space-time with no curvature, assumed in the theory of Special Relativity. The curved space-time of General Relativity appears to be a Minkowski space-time at any given point near an observer, much as the Earth appears basically flat in the vicinity of someone who is standing on a plain. Minkowski space-time is named for Russian-born German mathematician Hermann Minkowski (1864–1909).

minor planet (mī′nər) See **asteroid.**

minuend (mĭn′yo͞o-ĕnd′) A number from which another is subtracted. For example, in the numerical expression 100 – 23 = 77, the minuend is 100.

minute (mĭn′ĭt) **1.** A unit of time equal to $\frac{1}{60}$ of an hour or 60 seconds. ► A **sidereal minute** is $\frac{1}{60}$ of a sidereal hour, and a **mean solar minute** is $\frac{1}{60}$ of a mean solar hour. See more at **sidereal time, solar time. 2.** A unit of angular measurement, such as longitude or right ascension, that is equal to $\frac{1}{60}$ of a degree or 60 seconds.

Miocene (mī′ə-sēn′) The fourth epoch of the Tertiary Period, from about 24 to 5 million years ago. During this time the climate was warmer than it had been in the Oligocene, and kelp forests and grasslands first developed. With the isolation of Antarctica, a circumpolar ocean current was established in the southern Hemisphere, reducing the amount of mixing of cold polar water and warm equatorial water and causing a buildup of ice sheets in Antarctica. The African-Arabian plate became connected to Asia, closing the seaway which had previously separated Africa from Asia. Mammalian diversity was at its peak. See Chart at **geologic time.**

mirage (mĭ-räzh′) An image formed under certain atmospheric conditions, in which objects appear to be reflected or displaced or in which nonexistent objects seem to appear. For example, the difference in the **index of refraction** between a low layer of very hot air and a higher level of cold air can cause light rays, travelling down from an object (such as the sky or a cloud) and passing through ever warmer air, to be refracted back up again. An observer viewing these light rays perceives them coming up off the ground, and thus sees the inverted image of the object, which appears lower than the object really is. In this way the sky itself can be reflected, resulting in the mirage of a distant lake.

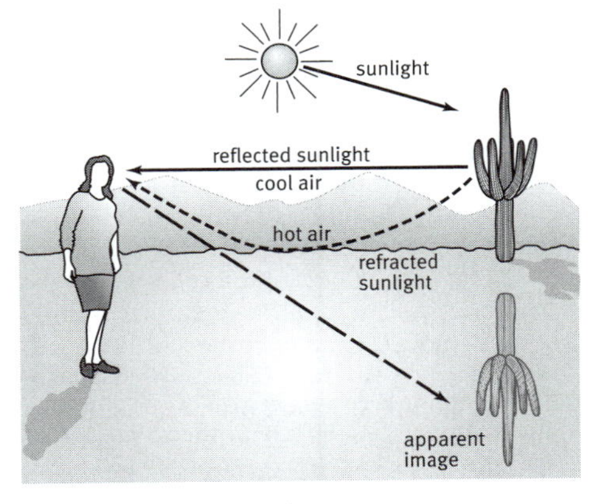

mirage

The hot air near the ground surface warms the air above it causing a light ray passing through it to be refracted upwards. Because the light ray reaching the viewer's eye appears to be coming from the ground surface, and because the reflected ray follows the same path, an apparent image is seen.

mirror (mĭr′ər) An object that causes light or other radiation to be reflected from its surface, with little or no diffusion. Common mirrors consist of a thin sheet or film of metal, such as silver, behind or covering a glass pane. Mirrors are used extensively in telescopes, microscopes, lasers, fiber optics, measuring instruments, and many other devices. See more at **reflection.**

miscarriage (mĭs′kăr′ĭj) The premature, spontaneous expulsion of the products of pregnancy from the uterus, usually in the first trimester. Also called *spontaneous abortion.*

miscible (mĭs′ə-bəl) Relating to two or more substances, such as water and alcohol, that can be mixed together or can dissolve into one another in any proportion without separating. Compare **immiscible.**

missense (mĭs′sĕns′) Relating to a mutation that changes a codon for one amino acid into a codon for a different amino acid. See more at **point mutation.**

missing mass (mĭs′ĭng) See **dark matter.**

Mississippian (mĭs′ĭ-sĭp′ē-ən) The fifth period of the Paleozoic Era, from about 360 to 320 million years ago. During this time shallow seas spread over former land areas and the first primitive conifers appeared. See Chart at **geologic time.**

mist (mĭst) A mass of fine droplets of water in the atmosphere near or in contact with the Earth. Mist reduces visibility to not less than 1 km (0.62 mi). Compare **fog.**

Mitchell (mĭch′əl), **Maria** 1818–1889. American astronomer and educator noted for her study of sunspots and nebulae and for her 1847 discovery of a comet.

BIOGRAPHY Maria Mitchell

Maria Mitchell, the first acknowledged woman astronomer in the United States, was born in 1818, in an era when women were discouraged from pursuing scientific careers. It was her good fortune to have a father who himself was an astronomer and who delighted in fostering his daughter's abilities in mathematics and astronomy. Already assisting her father's research by age twelve and becoming an apprentice schoolteacher at sixteen, Mitchell went on to gain immediate worldwide fame in 1847 when she became the first person to discover a comet using a telescope and established its orbit. For this she was awarded a medal by the King of Denmark, and in 1848 she became the first woman admitted to the American Academy of Arts and Sciences. In the ensuing decades, she made many discoveries about nebulae, double stars, the paths taken by meteors, the surface features of the bodies of the solar system, and many other celestial phenomena. She was a pathbreaker in telescope photography, and made pioneering daily photographs of sunspots, which she demonstrated were cavities in the sun's surface rather than clouds as had previously been thought. In 1865 she became a professor of astronomy at Vassar College, director of its observatory, and its most distinguished faculty member. Her accomplishments and brilliance as a teacher were inspirational to many other women. An outspoken supporter of women's education, Mitchell was able to break numerous barriers to women in the sciences, cofounding the Association for the Advancement of Women in 1873.

Maria Mitchell

mite (mīt) Any of various very small arachnids of the subclass Acari that often live as parasites on other animals or plants. Like ticks and unlike spiders, mites have no division between the cephalothorax and abdomen.

mitochondrion (mī′tə-kŏn′drē-ən) Plural **mitochondria.** A structure in the cytoplasm of all cells except bacteria in which food molecules (sugars, fatty acids, and amino acids) are broken down in the presence of oxygen and converted to energy in the form of ATP. Mitochondria have an inner and outer membrane. The inner membrane has many twists and folds (called cristae), which increase the surface area available to proteins and their associative reactions. The inner membrane encloses a liquid containing DNA, RNA, small ribosomes, and solutes. The DNA in mitochondria is genetically distinct from that in the cell nucleus, and mitochondria can manufacture some of their own proteins independent of the rest of the cell. Each cell can contain thousands of mitochondria, which move about producing ATP in response to the cell's need for chemical energy. It is thought that mitochondria originated as separate, single-celled organisms that became so symbiotic with their hosts as to be indispensible. Mitochondrial DNA is thus considered a remnant of a past existence as a separate organism. See more at **cell, cellular respiration.**

mitosis (mī-tō′sĭs) The process in cell division in eukaryotes in which the nucleus divides to produce two new nuclei, each having the same number and type of chromosomes as the original. Prior to mitosis, each chromosome is replicated to form two identical strands (called chromatids). As mitosis begins, the chromosomes line up along the center of the cell by attaching to the fibers of the cell **spindle**. The pairs of chromatids then separate, each strand of a pair moving to an opposite end of the cell. When a new membrane forms around each of the two groups of chromosomes, division of the nucleus is complete. The four main phases of mitosis are prophase, metaphase, anaphase, and telophase. Compare **meiosis.** *—Adjective* **mitotic** (mī-tŏt′ĭk).

mitral valve (mī′trəl) A valve of the heart composed of two triangular flaps, that is located between the left atrium and left ventricle. The mitral valve regulates blood flow between the two chambers.

A CLOSER LOOK **mitosis**

Mitosis is the process by which the nucleus divides in eukaryotic organisms, producing two new nuclei that are genetically identical to the nucleus of the parent cell. It occurs in cell division carried on by human somatic cells—the cells used for the maintenance and growth of the body. These cells have two paired sets of 23 chromosomes, or 46 chromosomes in total. (Cells with two sets of chromosomes are called *diploid.)* Before cell division occurs, the genetic material in each chromosome is duplicated as part of the normal functioning of the cell. Each chromosome then consists of two chromatids, identical strands of DNA. When a cell undergoes mitosis, the chromosomes condense into 46 compact bodies. The chromatids then separate, and one chromatid from each of the 46 chromosomes moves to each side of the cell as it prepares to divide. The chromatids form the chromosomes of the daughter cells, so that each new cell has 46 chromosomes (two complete sets of 23), just like the parent cell.
▸ While both mitosis and meiosis refer properly to types of nuclear division, they are often used as shorthand to refer to the entire processes of cell division themselves. When mitosis and meiosis are used to refer specifically to nuclear division, they are often contrasted with *cytokinesis,* the division of the cytoplasm.

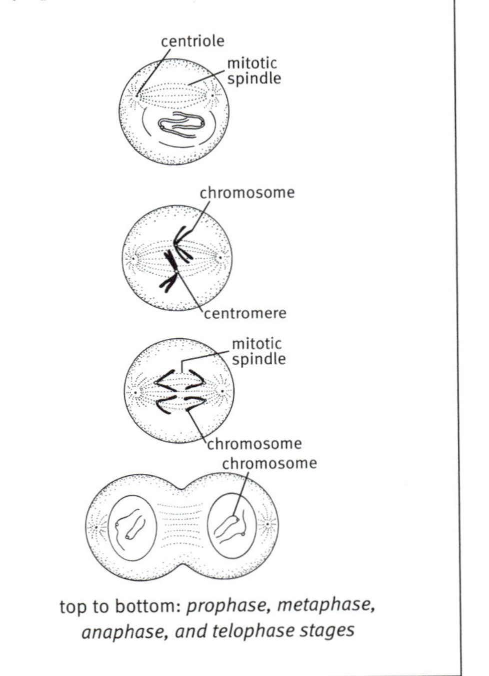

top to bottom: *prophase, metaphase, anaphase, and telophase stages*

mixed decimal (mĭkst) A number, such as 7.125, consisting of an integer (in this case, 7) plus a decimal (0.0125).

mixed number 1. A number, such as $7\frac{3}{8}$, consisting of a whole number and a fraction. **2.** An expression, such as $7x + 6 - \frac{5}{(x+4)}$, consisting of a polynomial and a rational algebraic fraction.

mixture (mĭks′chər) A composition of two or more substances that are not chemically combined with each other and are capable of being separated.

mks system (ĕm′kā′ĕs′) See **meter-kilogram-second system.**

ml or **mL** Abbreviation of **milliliter.**

mm Abbreviation of **millimeter.**

Mn The symbol for **manganese.**

Mo The symbol for **molybdenum.**

Möbius strip (mō′bē-əs) A continuous one-sided surface formed by rotating one end of a rectangular strip 180° and attaching it to the other end. Compare **Klein bottle.**

Möbius strip

mode (mōd) The value that occurs most frequently in a data set. For example, in the set 125, 140, 172, 164, 140, 110, the mode is 140. Compare **arithmetic mean, average, median.**

model (mŏd′l) A systematic description of an object or phenomenon that shares important characteristics with the object or phenomenon. Scientific models can be material, visual, mathematical, or computational and are often used in the construction of scientific theories. See also **hypothesis, theory.**

modem (mō′dəm) A device for transmitting and receiving digital data over telephone wires. Modems send data by converting it into audio signals and receive it by converting audio signals back into digital form. The speed at which modems transmit data is measured in bps (bits per second).

moderator (mŏd′ə-rā′tər) A substance, such as graphite, water, or heavy water, placed in a nuclear reactor to slow neutrons down to speeds at which they are more likely to be captured by fissionable components of a fuel (such as uranium-235) and less likely to be absorbed by nonfissionable components of a fuel (such as uranium-238). Also called *neutron moderator.* See also **slow neutron.**

modification (mŏd′ə-fĭ-kā′shən) A change in an organism that results from external influences and cannot be inherited.

modulate (mŏj′ə-lāt′) To vary the amplitude, frequency, or some other characteristic of a signal or power source. See also **amplitude modulation, frequency modulation.**

modulo (mŏj′ə-lō) With respect to a specified modulus. Eighteen is congruent to 42 modulo 12 because both 18 and 42 leave 6 as remainder when divided by 12.

modulus (mŏj′ə-ləs) Plural **moduli** (mŏj′ə-lī′). **1.** A number by which two given numbers can be divided and produce the same remainder. **2.** The numerical length of the vector that represents a complex number. For a complex number $a + bi$, the modulus is the square root of $(a^2 + b^2)$. **3.** The number by which a logarithm to one base must be multiplied to obtain the corresponding logarithm to another base.

modulus of elasticity The ratio of the stress applied to a body to the strain that results in the body in response to it. The modulus of elasticity of a material is a measure of its stiffness and for most materials remains constant over a range of stress. ► The ratio of the longitudinal strain to the longitudinal stress is called **Young's modulus.** ► The ratio of the stress on the body to the body's fractional decrease in volume is the **bulk modulus.** ► The ratio of the tangential force per unit area to the angular deformation in radians is the **shear modulus.** See also **Hooke's law.**

Moho (mō′hō′) The Mohorovičić discontinuity.

Mohorovičić discontinuity (mō′hə-rō′və-chĭch) The boundary between the Earth's crust and mantle, located at an average depth of 8 km (5 mi) under the oceans and 32 km (20 mi) under the continents. The velocity of seismic primary waves across this boundary changes abruptly from 6.7 to 7.2 km (4.1 to 4.5 mi) per second in the lower crust to 7.6 to 8.6 km (4.7 to 5.3 mi) per second in the upper mantle. The boundary is estimated to be between 0.2 and 3 km (0.1 and 1.9 mi) thick and is believed to coincide with a change in rock type from basalts (above) to peridotites and dunites (below). It is named after its discoverer, Croatian seismologist Andrija Mohorovičić (1857– 1936).

Mohs scale (mōz) A scale used to measure the relative hardness of a mineral by its resistance to scratching. From softest to hardest, the ten minerals of the Mohs scale are talc (measuring 1 on the scale), gypsum, calcite, fluorite, apatite, orthoclase, quartz, topaz, corundum, and diamond (measuring 10 on the scale).

molar[1] (mō′lər) *Chemistry.* **1.** Relating to a mole. **2.** Containing one mole of solute per liter of solution.

molar[2] (mō′lər) Any of the teeth located toward the back of the jaws, having broad crowns for grinding food. Adult humans have 12 molars.

mold (mōld) Any of various fungi that often form a fuzzy growth (called a **mycelium**) on the surface of organic matter. Some molds cause food to spoil, but others are beneficial, such as those used to make certain cheeses and those from which antibiotics like penicillin are developed. The molds do not form a distinct phylogenetic grouping but belong to various phyla including the ascomycetes and the zygomycetes. See also **slime mold.**

mole[1] (mōl) A small, usually pigmented, benign growth on the skin.

mole[2] (mōl) The amount of an element, compound, or other substance that has the same number of basic particles as 12 grams of Carbon-12. The number of particles making up a mole is Avogadro's number. For elements and compounds, the mass of one mole, in grams, is roughly equal to the atomic or molecular weight of the substance. For example, carbon dioxide, CO_2, has a molecular weight of 44; therefore, one mole of it weighs 44 grams.

molecular biology (mə-lĕk′yə-lər) The branch of biology that deals with the formation, structure, and function of macromolecules essential to life, such as nucleic acids and proteins, including their roles in cell replication and the transmission of genetic information.

molecular formula A chemical formula that shows the total number and kinds of atoms in a molecule, but not their structural arrangement. For example, the molecular formula of aspirin is $C_9H_8O_4$. Compare **empirical formula, structural formula.**

molecular weight The sum of the atomic weights of the atoms contained in a molecule. Also called *relative molecular mass.*

molecule (mŏl′ĭ-kyo͞ol′) A group of two or more atoms linked together by sharing electrons in a chemical bond. Molecules are the fundamental components of chemical compounds and are the smallest part of a compound that can participate in a chemical reaction.

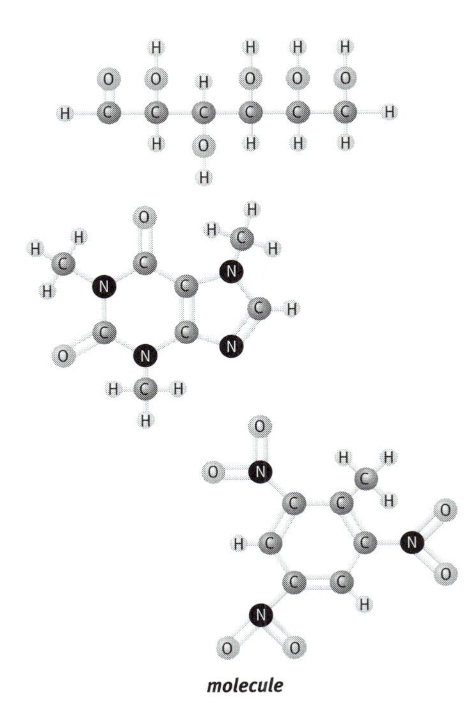

molecule

top to bottom: *glucose, caffeine, and TNT molecule models*

mollusk or **mollusc** (mŏl′əsk) Any of numerous invertebrate animals of the phylum Mollusca, usually living in water and often having a hard outer shell. They have a muscular foot, a well-developed circulatory and nervous system, and often complex eyes. Mollusks include gastropods (snails and shellfish), slugs, octopuses, squids, and the extinct ammonites. Mollusks appear in the fossil record in the early Cambrian Period, but it is not known from what group they evolved.

molt (mōlt) To shed an outer covering, such as skin or feathers, for replacement by a new growth. Many snakes, birds, and arthropods molt.

molybdenite (mə-lĭb′də-nīt′) A soft, lead-gray hexagonal mineral that is the principal ore of molybdenum. It occurs as sheetlike masses in pegmatites and in areas where contact metamorphism has taken place. *Chemical formula:* $\mathbf{MoS_2}$.

molybdenum (mə-lĭb′də-nəm) *Symbol* **Mo** A hard, silvery-white metallic element that resists corrosion and retains its strength at high temperatures. It is used to harden and toughen steel and to make high-temperature wiring. Molybdenum is an essential trace element in plant metabolism. Atomic number 42; atomic weight 95.94; melting point 2,617°C; boiling point 4,612°C; specific gravity 10.22 (at 20°C); valence 2, 3, 4, 5, 6. See **Periodic Table.**

moment of inertia (mō′mənt) A measure of a body's resistance to angular acceleration, equal to the product of the mass of the body and the square of its distance from the axis of rotation. See also **angular momentum, torque.**

momentum (mō-mĕn′təm) Plural **momenta** or **momentums.** A vector quantity that expresses the relation of the velocity of a body, wave, field, or other physical system, to its energy. The direction of the momentum of a single object indicates the direction of its motion. Momentum is a conserved quantity (it remains constant unless acted upon by an outside force) and is related by **Noether's theorem** to translational **invariance**. In classical mechanics, momentum is defined as mass times velocity. The theory of Special Relativity uses the concept of **relativistic mass**. The momentum of photons, which are massless, is equal to their energy divided by the speed of light. In quantum mechanics, **momentum** more generally refers to a mathematical operator applied to the wave equation describing a physical system and corresponding to an **observable**; solutions to the equation using this operator provide the vector quantity traditionally called momentum. In all of these uses, momentum is sometimes called *linear momentum*. See also **angular momentum, impulse.**

monadelphous (mŏn′ə-dĕl′fəs, mō′nə-) Related to stamens whose filaments are united into a single tubelike group. The stamens of flowers of leguminous plants are often monadelphous.

monadnock (mə-năd′nŏk′) A mountain or rocky mass that has resisted erosion and stands isolated in an essentially level area.

monazite (mŏn′ə-zīt′) A yellow or reddish-brown monoclinic mineral that is a principal ore of several lanthanide (rare-earth) elements. It occurs as tabular crystals in pegmatites, granites, and metamorphic rocks, as well as in sand. *Chemical formula:* **(Ce, La, Nd, Th)PO_4.**

moneran (mə-nîr′ən) See **prokaryote.**

monitor (mŏn′ĭ-tər) A device that accepts video signals from a computer and displays information on a screen. Monitors generally employ cathode-ray tubes or flat-panel displays to project the image. See Note at **pixel.**

mono– A prefix that means "one, only, single," as in *monochromatic,* having only one color. It is often found in chemical names where it means "containing just one" of the specified atom or group, as in *carbon monoxide,* which is carbon attached to a single oxygen atom.

monoamine (mŏn′ō-ăm′ēn, -ə-mēn′) An amine compound containing one amino group (NH_2), especially such a compound that functions as a neurotransmitter. The catecholamines and serotonin are monoamines. Abnormal levels of monoamines in the brain have been implicated in mood disorders.

monoamine oxidase inhibitor Any of a class of antidepressant drugs that block the action of monoamine oxidase in the brain, thereby allowing the accumulation of monoamines such as norepinephrine, which can alter mood.

monobasic (mŏn′ə-bā′sĭk) **1.** Relating to an acid that contains only one hydrogen atom that can be replaced in an acid-base reaction. Hydrochloric acid (HCl) and nitric acid (HNO_3) are monobasic acids. **2.** Of or relating to a compound that contains one metal ion or positive radical.

monocarp (mŏn′ə-kärp′) A plant that produces fruit only once in its lifetime. *—Adjective* **monocarpic.**

monochromatic (mŏn′ə-krō-măt′ĭk) **1.** Consisting of a single wavelength of light or other radiation. Lasers, for example, usually produce monochromatic light. **2.** Having or appearing to have only one color. Compare **polychromatic.**

monocline (mŏn′ə-klīn′) A set of rock layers that all slope downward from the horizontal in the same direction.

monoclinic (mŏn′ə-klĭn′ĭk) Relating to a crystal having three axes of different lengths. Two of the axes are at oblique angles to each other, and the third axis is perpendicular to the plane that is made by the other two. The mineral gypsum has monoclinic crystals. See illustration at **crystal.**

monoclinous (mŏn′ə-klī′nəs) Having pistils and stamens in the same flower; bearing perfect (hermaphroditic) flowers. Compare **diclinous.**

monocotyledon (mŏn′ə-kŏt′l-ēd′n) or **monocot** (mŏn′ə-kŏt′) Any of a class of angiosperm plants having a single cotyledon in the seed. Monocotyledons have leaves with parallel veins, flower parts in multiples of three, and fibrous root systems. Their primary vascular bundles are scattered throughout the stem, not arranged in a ring as in eudicotyledons. Grasses, palms, lilies, irises, and orchids are

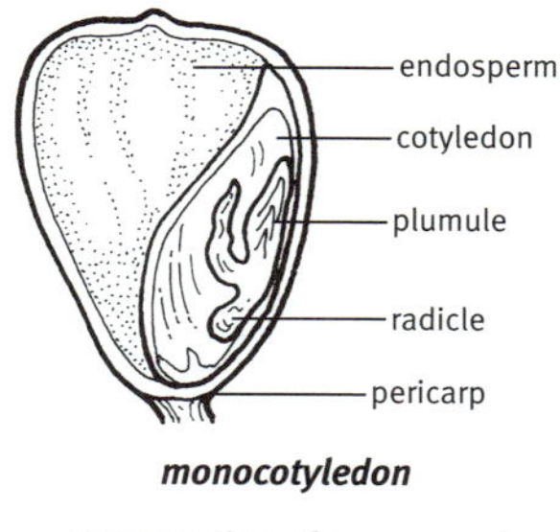

monocotyledon

cross section of a corn seed

monocotyledons. See more at **leaf.** Compare **eudicotyledon.** —*Adjective* **monocotyledonous.**

monocyclic (mŏn′ə-sī′klĭk, -sĭk′lĭk) **1.** Having a single cycle, as of activity or development. **2.** Having a single whorl, as certain flowers and the shells of certain invertebrates. **3.** Having a molecular structure with only one ring. Benzene and toluene are monocyclic molecules.

monocyte (mŏn′ə-sīt′) Any of various large white blood cells that are formed in the bone marrow, circulate in the blood, and destroy pathogenic bacteria by phagocytosis. Monocytes develop into macrophages in various body tissues.

Monod (mô-nō′), **Jacques Lucien** 1910–1976. French biochemist who, with François Jacob, proposed the existence of messenger RNA. Monod and Jacob also studied how genes control cellular activity by directing the synthesis of proteins.

monoecious (mə-nē′shəs) Having separate male flowers and female flowers on the same plant. Maize and oaks are monoecious plants. Compare **dioecious.**

monomer (mŏn′ə-mər) A molecule that can combine with others of the same kind to form a polymer. Glucose molecules, for example, are monomers that can combine to form the polymer cellulose. Polymers can also be composed of different kinds of monomers.

monomial (mŏ-nō′mē-əl) An algebraic expression consisting of a single term, especially one of the form ax^n, where a is a number and n is an integer greater than or equal to 0.

mononucleosis (mŏn′ō-no͞o′klē-ō′sĭs) A common infectious disease usually affecting young people, caused by the Epstein-Barr virus and characterized by fever, sore throat, swollen lymph nodes, and fatigue. The symptoms may last for several weeks.

monophyletic (mŏn′ō-fī-lĕt′ĭk) Relating to a taxonomic group that contains all the descendants of a single common ancestor. All **clades**, such as birds and placental mammals, are monophyletic. Compare **paraphyletic, polyphyletic.**

monopodium (mŏn′ə-pō′dē-əm) Plural **monopodia.** A main axis of a plant, such as the trunk of a spruce, that maintains a single line of growth, giving off lateral branches.

monopole (mŏn′ə-pōl′) The minimal region for which lines of force, as from an electric or magnetic field, either all enter or all leave the region. Particles with electric charge, such as electrons, are monopoles; though magnetic fields can behave as if generated by sets of monopoles (as in the case of **magnetic dipoles**), it is not known whether isolable magnetic monopoles exist.

monosaccharide (mŏn′ə-săk′ə-rīd′) Any of a class of carbohydrates that cannot be broken down to simpler sugars by hydrolysis and that constitute the building blocks of oligosaccharides and polysaccharides. Monosaccharides consist of at least three carbon atoms, one of which is attached to an oxygen atom to form an aldehyde group (CHO) or a ketone, and the others of which are each attached to a hydroxyl group (OH). Monosaccharides can occur as chains or rings. Fructose, glucose, and ribose are monosaccharides. Also called *simple sugar.* Compare **oligosaccharide, polysaccharide.** See more at **aldose, ketose.**

monosodium glutamate (mŏn′ə-sō′dē-əm glo͞o′tə-māt′) A white, crystalline salt used to flavor food, especially in China and Japan. It occurs naturally in tomatoes, Parmesan cheese, and seaweed. *Chemical formula:* $C_5H_8NNaO_4$.

monotreme (mŏn′ə-trēm′) Any of various mammals of the order Monotremata. Monotremes are the most primitive type of living mammal. They lay eggs and have a single opening (cloaca) for reproduction and elimination of wastes. The females have no teats but provide milk directly through the skin to their young. The only living monotremes are the duck-billed platypus, found in Australia and New Guinea, and the echidnas, found in New Guinea. Monotremes may have evolved already in the Jurassic Period, but the precise nature of their relationship to marsupials and placental mammals is disputed.

monotypic (mŏn′ə-tĭp′ĭk) Having a single form or member, especially containing no more than one taxonomic category of the next lower rank. A monotypic genus contains a single species, while a monotypic species consists of a single population that is not divided into subspecies. Compare **polytypic.**

monounsaturated (mŏn′ō-ŭn-săch′ə-rā′tĭd) Relating to an organic compound, usually a fatty acid, having only one double bond per molecule. See more at **unsaturated.**

monovalent (mŏn′ə-vā′lənt) **1.** Having a valence of 1; univalent. **2.** Containing antibodies with specific activity against a single strain of a microorganism or against antigens from such a strain.

monoxide (mə-nŏk′sīd′) A compound consisting of two elements, one of which is a single oxygen atom. Carbon monoxide, for example, contains a carbon atom bound to a single oxygen atom.

monsoon (mŏn-so͞on′) **1.** A system of winds that influences the climate of a large area and that reverses direction with the seasons. Monsoons are caused primarily by the much greater annual variation in temperature over large areas of land than over large areas of adjacent ocean water. This variation causes an excess of atmospheric pressure over the continents in the winter, and a deficit in the summer. The disparity causes strong winds to blow between the ocean and the land, bringing heavy seasonal rainfall. **2.** In southern Asia, a wind that is part of such a system and that blows from the southwest in the summer and usually brings heavy rains.

monticule (mŏn′tĭ-kyo͞ol′) A minor cone of a volcano.

monzonite (mŏn-zō′nīt′, mŏn′zə-nīt′) An igneous rock composed chiefly of plagioclase and orthoclase, with small amounts of amphibole, pyroxene, and biotite. Monzonite contains little or no quartz.

mood disorder (mo͞od) Any of a group of psychiatric disorders, including depression and bipolar disorder, characterized by a pervasive disturbance of mood that is not caused by an organic abnormality. Also called *affective disorder.*

moon (mo͞on) **1.** Often **Moon.** The natural satellite of Earth, visible by reflection of sunlight and traveling around Earth in a slightly elliptical orbit at an average distance of about 381,600 km (237,000 mi). The Moon's average diameter is 3,480 km (2,160 mi), and its mass is about $\frac{1}{80}$ that of Earth. See more at **giant impact theory. 2.** A natural satellite revolving around a planet.

Moore's law (môrz) The observation that steady technological improvements in miniaturization leads to a doubling of the density of transistors on new integrated circuits every 18 months. In the mid-1960s, Gordon Moore (born 1929), one of the founders of Intel Corporation, observed that the density of transistors had been doubling every year, although the pace slowed slightly in the following years. The 18-month pattern held true into the 21st century, though as technology approaches the point where circuits are only a few atoms wide, new technologies, possibly not involving transistors at all, may be required for further miniaturization.

moraine (mə-rān′) A mass of till (boulders, pebbles, sand, and mud) deposited by a glacier, often in the form of a long ridge. Moraines typically form because of the plowing effect of a moving glacier, which causes it to pick up rock fragments and sediments as it moves, and because of the periodic melting of the ice, which causes the glacier to deposit these materials during warmer intervals. ► A moraine deposited in front of a glacier is a **terminal moraine.** ► A moraine deposited along the side of a glacier is a **lateral moraine.** ► A moraine deposited down the middle of a glacier is a **medial moraine.** Medial moraines are actually the combined lateral moraines of two glaciers that have merged.

moraine

medial moraine in the McBride Glacier, Glacier Bay National Park, Alaska

Morgan (môr′gən), **Thomas Hunt** 1866–1945. American zoologist whose experiments with fruit flies demonstrated that hereditary traits are carried by genes on chromosomes and that traits can cross over from one chromosome to another. He was awarded the Nobel Prize for physiology or medicine in 1933.

A CLOSER LOOK **moons**

The Earth's *Moon* is a desolate and quiet place. The only natural satellite of Earth, it consists almost entirely of rock, shows no signs of ongoing geologic activity, has no water, and has a very thin atmosphere consisting primarily of sodium. But our Moon does not present a typical case for planetary satellites. Over the last 50 years, over a hundred more moons have been discovered in the solar system, so that they now total almost 170, nearly all of them orbiting the larger planets Jupiter, Saturn, and Uranus. (Mercury and Venus have no moons; an additional four moons orbit dwarf planets.) Because they are so far from the Sun, these moons are for the most part extremely cold. Io, one of Jupiter's 63 known moons, is an exception. It is the most geologically active body in the solar system, with almost constant volcanic activity and a surface covered by cooling lava. Some scientists think that another moon of Jupiter, Europa, may have liquid water capable of supporting life underneath a thick layer of surface ice. Titan, one of Saturn's moons, may also be capable of supporting primitive life in the ocean of liquid methane on its frigid surface.

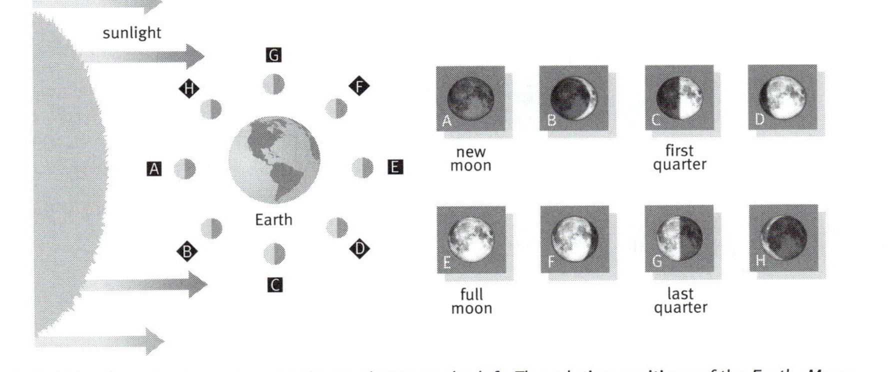

Half of the Moon is always in sunlight, as shown on the left. The relative positions of the Earth, Moon, and Sun determine the Moon's phase as seen from the Earth, as shown on the right.

Morley (môr′lē), **Edward Williams** 1838–1923. American chemist and physicist who with Albert Michelson disproved the existence of ether, the hypothetical medium of electromagnetic waves. Their work served as the starting point for Albert Einstein's development of the theory of relativity.

morph (môrf) A phenotypically distinct form of an organism or species.

morphine (môr′fēn′) A highly addictive drug derived from opium and used to treat intractable pain, as in severe injury or metastatic cancer.

morphogenesis (môr′fō-jĕn′ĭ-sĭs) Formation of the structure of an organism or part involving differentiation and growth of tissues and organs during development.

morphology (môr-fŏl′ə-jē) The size, shape, and structure of an organism or one of its parts. Biologists usually describe the morphology of an organism separately from its physiology. In traditional systems of taxonomy, classifications were based on the morphological characteristics of organisms. However, a method of classification based purely on morphology runs the risk of grouping together organisms that are actually relatively unrelated but have evolved similar features. In more modern systems of taxonomy, the genetic similarity of organisms, studied through the methods of molecular biology, is considered in addition to morphology when establishing taxa.

Morse (môrs), **Samuel Finley Breese** 1791–1872. American inventor who was a pioneer in the field of telegraphy and in 1844 introduced a telegraphic code for transmitting messages, which became known as Morse code.

Morse code A code used for transmitting messages in which letters of the alphabet and numbers are represented by various sequences of written dots and dashes, or short and long signals such as electric tones or voltages. Morse code was used extensively in telegraphy. In a format that has been standardized for international use, it is still sometimes used for long distance radio communication.

morula (môr′yə-lə) Plural **morulae** (môr′yə-lē′). The spherical embryonic mass of blastomeres that results from cleavage of the fertilized ovum and develops into the **blastula**.

mosaic (mō-zā′ĭk) Any of various viral diseases of plants, resulting in light and dark areas in the leaves, which often become shriveled and dwarfed.

mosasaur (mō′sə-sôr′) Any of various medium-sized to large extinct aquatic lizards of the genus *Mosasaurus* of the Cretaceous Period, having modified limbs that served as paddles for swimming. Mosasaurs were related to the modern monitor lizard, with which they share a similar skull structure.

MOSFET (mŏs′fĕt′) Short for *metal-oxide semiconductor field effect transistor.* A type of field effect transistor, used predominantly in microprocessor and related technology, that consumes very little power and can be highly miniaturized.

moss (môs) **1.** Any of numerous small bryophyte plants belonging to the phylum Bryophyta. Mosses, unlike liverworts, have some tissues specialized for conducting water and nutrients. As in the other bryophytes, the diploid **sporophyte** grows on the haploid **gametophyte** generation, which supplies it with nutrients. Mosses often live in moist, shady areas and grow in clusters or mats. Sphagnum mosses play a crucial role in the ecology of peat bogs. See more at **bryophyte.** **2.** Any of a number of plants that look like mosses but are not related to them. For instance, reindeer moss is a lichen, Irish moss is an alga, and Spanish moss is a bromeliad, a flowering plant.

moss

capsule-bearing sporophytes growing out of the gametophyte generation of a moss

motherboard (mŭ*th*′ər-bôrd′) The main circuit board of a computer, usually containing the central processing unit and main system memory as well as circuitry that controls the disk drives, keyboard, monitor, and other peripheral devices.

mother cell (mŭ*th*′ər) See **parent cell.**

mother-of-pearl The hard, smooth, pearly layer on the inside of certain seashells, such as abalones and certain oysters. It is used to make buttons and jewelry. Also called *nacre.*

motile (mōt′l, mō′tīl′) Moving or able to move by itself. Sperm and certain spores are motile. —*Noun* **motility** (mō-tĭl′ĭ-tē).

motor (mō′tər) *Noun.* **1.** A machine that uses energy, such as electric or chemical energy (as from burning a fuel), to produce mechanical motion. See also **engine.** —*Adjective.* **2.** Involving the muscles or the nerves that are connected to them. Compare **sensory.**

mountain (moun′tən) A generally massive and usually steep-sided, raised portion of the Earth's surface. Mountains can occur as single peaks or as part of a long chain. They can form through volcanic activity, by erosion, or by uplift of the continental crust when two tectonic plates collide. The Himalayas, which are the highest mountains in the world, were formed when the plate carrying the landmass of India collided with the plate carrying the landmass of China.

mouse (mous) Plural **mice** (mīs) or **mouses.** A hand-held input device that is moved about on a flat surface to direct the cursor on a computer screen. It also has buttons for activating computer functions. The underside of a mechanical mouse contains a rubber-coated ball that rotates as the mouse is moved; optical sensors detect the motion and move the screen pointer correspondingly. An optical mouse is cordless and uses reflections from an LED to track the mouse's movement over a special reflective mat which is marked with a grid that acts as a frame of reference.

Mousterian (mo͞o-stîr′ē-ən) Relating to a Middle Paleolithic tool culture that succeeded the Acheulian, ending around 35,000 years ago and traditionally associated with Neanderthals. While Mousterian tools show improvements in stone-flaking techniques over Acheulian tools, they remain largely unchanged over long periods of time and show little of the rapid innovation and specialization that characterize tool cultures of the Upper Paleolithic associated with modern humans.

mouthpart (mouth′pärt′) Either of the appendages occurring in pairs in insects and other arthropods that extend from the head and are used for feeding.

MP3 (ĕm′pē-thrē′) An MPEG standard used especially for transmitting music digitally over the Internet. Many programs are available to facilitate the transfer of these files, but there are numerous legal issues regarding the swapping of files and violation of copyright laws.

MPEG (ĕm′pĕg′) Short for *Moving Pictures Expert Group.* Any of a set of standards established for the compression of digital audio and video data.

MRI (ĕm′är′ī′) Short for *magnetic resonance imaging.* The use of **nuclear magnetic resonance** to produce images of the molecules that make up a substance, especially the soft tissues of the human body. Magnetic resonance imaging is used in medicine to diagnose disorders of body structures that do not show up well on x-rays. See more at **nuclear magnetic resonance.**

mRNA Abbreviation of **messenger RNA.**

A CLOSER LOOK **MRI**

A picture is worth a thousand words, and nowhere is this more apparent than in the powerful diagnostic technique known as magnetic resonance imaging (MRI), which has revolutionized many areas of medicine. Compared to imaging techniques that use x-rays, such as computerized axial tomography (CAT), MRI generates far more detailed three-dimensional images of the soft tissues of the body, especially of the nervous system from the brain to the spine. These images greatly improve the ability of doctors to distinguish abnormal from healthy tissues. MRI can also be used to observe and measure dynamic physiological changes inside a patient without cutting into or penetrating the body. To produce an image, an MRI machine uses a powerful magnet to generate a magnetic field. When a patient lies within this field, the nuclei of atoms within the body align themselves with the magnetic field (much as iron filings line up around a magnet). Radio waves are then pulsed through the body, causing the nuclei to change their alignment with respect to the axis of the magnetic lines of force. As they return to their previous state after each pulse, they produce faint, distinctive radio signals; the rate at which they emit signals and the frequency of the signals depend on the type of atom, the temperature, the chemical environment, position, and other factors. These signals are detected by coils around the body and processed by a computer to produce images of internal structures. MRI holds yet another significant advantage over CAT in that exposure to potentially harmful x-ray radiation is avoided.

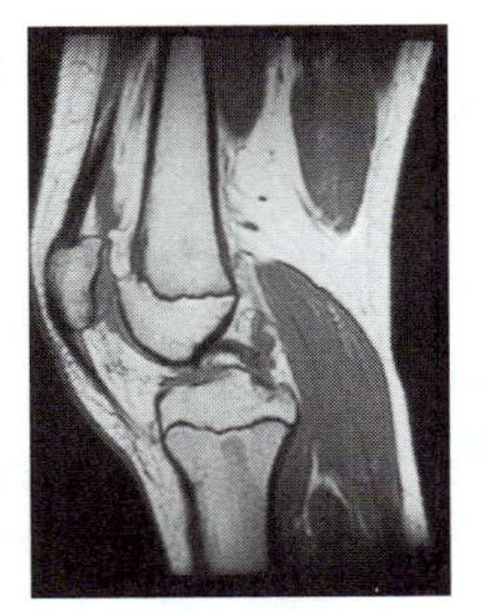

profile view of a child's knee

MSG Abbreviation of **monosodium glutamate.**

Mt The symbol for **meitnerium.**

mucigel (myo͞o′sĭ-jĕl′) The slimy, viscous substance secreted by the roots of plants, consisting of a hydrated polysaccharide. Mucigel lubricates the tips of roots as they push their way through the soil during growth. Soil particles adhere to the mucigel layer around the root, improving the uptake of moisture and nutrients. The mucigel also encourages the growth of beneficial fungi and bacteria that fix nitrogen in the soil.

mucopolysaccharide (myo͞o′kō-pŏl′ē-săk′ə-rīd′) See **glycosaminoglycan.**

mucous membrane (myo͞o′kəs) Any of the membranes lining the passages of the body, such as the respiratory and digestive tracts, that open to the outside. Cells in the mucous membranes secrete mucus, which lubricates the membranes and protects against infection.

mucus (myo͞o′kəs) The slimy, viscous substance secreted as a protective lubricant by mucous membranes. Mucus is composed chiefly of large glycoproteins called *mucins* and inorganic salts suspended in water.

mud flat (mŭd) Low-lying land consisting of silt or sand that is covered at high tide and exposed at low tide.

mudflow (mŭd′flō′) A downhill movement of soft, wet, unconsolidated earth and debris, made fluid by rain or melted snow and often building up great speed.

mudslide (mŭd′slīd′) A slow-moving mudflow.

mudstone (mŭd′stōn′) A fine-grained, dark gray sedimentary rock consisting primarily of compacted and hardened silt and clay, similar to shale but without laminations. The proportions of silt and clay in mudstone are approximately equal.

Müllerian mimicry (myo͞o-lîr′ē-ən, mə-) A form of protective mimicry in which two or more poisonous or unpalatable species closely resemble each other and are therefore avoided equally by all their natural predators. The similarity in coloration between the monarch and viceroy butterflies, once considered an example of Batesian mimicry, is now generally considered as Müllerian mimicry because the viceroy is thought to be as bad-tasting to birds as the monarch. Müllerian mimicry is named after the German-born Brazilian zoologist Fritz Müller (1821–97). Compare **aggressive mimicry, Batesian mimicry.**

multi– A prefix that means "many" or "much," as in *multicellular,* having many cells.

multicellular (mŭl′tē-sĕl′yə-lər) Having or consisting of many cells. Compare **unicellular.**

multimeter (mŭl-tĭm′ĭ-tər) An electrical measuring device that combines an ammeter, an ohmmeter, a voltmeter, and occasionally other measurement or testing devices into one unit.

multinomial (mŭl′tĭ-nō′mē-əl) See **polynomial.**

multinomial theorem The theorem that establishes the rule for forming the terms of the *n*th power of a sum of numbers in terms of products of powers of those numbers.

multiple (mŭl′tə-pəl) A number that may be divided by another number with no remainder. For example, 4, 10, and 32 are multiples of 2.

multiple fruit A fruit, such as a fig, mulberry, or pineapple, that consists of the ripened ovaries of more than one flower that are combined into a single structure. Compare **accessory fruit, aggregate fruit, simple fruit.**

multiple sclerosis (sklə-rō′sĭs) A chronic degenerative disease of the central nervous system in which gradual destruction of myelin occurs in the brain or spinal cord or both, interfering with the nerve pathways and causing muscular weakness, loss of coordination, and speech and visual disturbances. It occurs chiefly in young adults and is thought to be caused by a defect in the immune system that may be of genetic or viral origin.

multiple star A system of three or more stars that are bound together by gravity and orbit a common center of mass. The group generally appears as a single star to the naked eye. Astronomers believe that most stars in the universe are part of multiple or binary systems. **Alpha Centauri,** the closest star to our Sun, is a multiple star system containing three bodies. See also **binary star.**

multiplicand (mŭl′tə-plĭ-kănd′) A number that is multiplied by another number.

multiplication (mŭl′tə-plĭ-kā′shən) **1.** A mathematical operation performed on a pair of numbers in order to derive a third number called a product. For positive integers, multiplication consists of adding a number (the multiplicand) to itself a specified number of times. Thus multiplying 6 by 3 means adding 6 to itself three times. The operation of multiplication is extended to other real numbers according to the rules governing the multiplicational properties of positive integers. **2.** Any of certain analogous operations involving expressions other than real numbers.

multiplier (mŭl′tə-plī′ər) The number by which another number is multiplied.

multiply (mŭl′tə-plī′) To perform multiplication on a pair of quantities.

multitasking (mŭl′tē-tăs′kĭng) The concurrent operation by one central processing unit of two or more processes.

multivalent (mŭl′tĭ-vā′lənt, mŭl-tĭv′ə-lənt) **1.** Polyvalent. **2.** Relating to the association of three or more homologous chromosomes during the first division of meiosis.

multivariate (mŭl′tē-vâr′ē-ĭt, -āt′) Having or involving more than one variable.

mumps (mŭmps) An infectious disease caused by a virus of the family Paramyxoviridae and the genus *Rubulavirus*, characterized by swelling of the salivary glands, especially the parotid glands, and sometimes of the pancreas, testes, or ovaries. Vaccinations, usually given in early childhood, confer immunity to mumps.

muon (myo͞o′ŏn′) An elementary particle in the lepton family having a mass 209 times that of the electron, a negative electric charge, and a mean lifetime of 2.2×10^{-6} seconds. The muon was originally called the mu-meson and was once thought to be a meson. See Table at **subatomic particle.**

muon neutrino A type of neutrino associated with the muon, often created in particle interactions involving muons (such as muon or pion decay). The muon neutrino has a mass no greater than 0.49 times that of the electron and has no charge. See more at **neutrino.** See Table at **subatomic particle.**

muriatic acid (myo͝or′ē-ăt′ĭk) See **hydrochloric acid.**

murine (myo͝or′īn′) **1.** Of or relating to a rodent of the family Muridae or subfamily Murinae, including rats and mice. **2.** Caused, transmitted, or affected by such a rodent.

muscarine (mŭs′kə-rēn′) A highly toxic, hallucinogenic alkaloid related to the cholines, derived from the red form of the mushroom *Amanita muscaria* and other mushrooms and found in decaying animal tissue. *Chemical formula:* $C_9H_{20}NO_2$.

muscle (mŭs′əl) A body tissue composed of sheets or bundles of cells that contract to produce movement or increase tension. Muscle cells contain filaments made of the proteins actin and myosin, which lie parallel to each other. When a muscle is signaled to contract, the actin and myosin filaments slide past each other in an overlapping pattern. ▸ **Skeletal muscle** effects voluntary movement and is made up of bundles of elongated cells (muscle fibers), each of which contains many nuclei. ▸ **Smooth muscle** provides the contractile force for the internal organs and is controlled by the autonomic nervous system. Smooth muscle cells are spindle-shaped and each contains a single nucleus. ▸ **Cardiac muscle** makes up the muscle of the heart and consists of a meshwork of striated cells. —*Adjective* **muscular.**

muscovite (mŭs′kə-vīt′) A usually colorless to yellow or pale-gray mica. Muscovite is a monoclinic mineral and is found in igneous rocks, such as granites and pegmatites, metamorphic rocks, such as schists and gneisses, and in many sedimentary rocks. *Chemical formula:* $KAl_2(AlSi_3)O_{10}(OH)_2$.

muscular dystrophy (mŭs′kyə-lər dĭs′trə-fē) Any of a group of inherited progressive muscle disorders caused by a defect in one or more genes that control muscle function and characterized by gradual irreversible wasting of skeletal muscle.

mushroom (mŭsh′ro͞om′) Any of various basidiomycete fungi whose mycelium produces a spore-dispersing body (called a **basidioma**) that usually consists of a stalk topped by a fleshy, often umbrella-shaped cap. Some species of mushrooms are edible, though many are poisonous. The term *mushroom* is often applied to the stalk and cap alone. See more at **basidiomycete.**

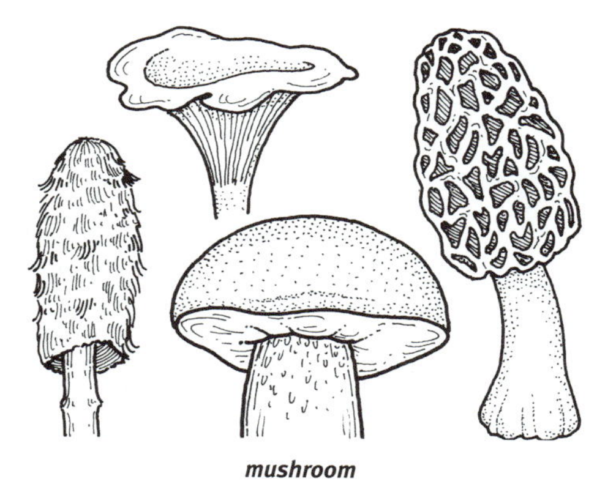

mushroom

clockwise from top: *chanterelle, yellow morel, King boletus, and shaggy mane mushrooms*

mustard gas (mŭs′tərd) An oily, volatile liquid that is corrosive to the skin and mucous membranes and causes severe, sometimes fatal respiratory damage. It was introduced in World War I as a chemical warfare agent. *Chemical formula:* $C_4H_8Cl_2S$.

mustelid (mŭs′tə-lĭd′) Any of various small to midsize carnivorous mammals of the family Mustelidae, usually having long, slender bodies, short legs, and well-developed anal scent glands. The pelts of many mustelids have been important for use in clothing. Weasels, skunks, badgers, wolverines, ferrets, mink, martens, and otters are mustelids.

musth also **must** (mŭst) An annual period of heightened aggressiveness and sexual activity in male elephants.

mutation (myo͞o-tā′shən) A change in the structure of the genes or chromosomes of an organism. Mutations occurring in the reproductive cells, such as an egg or sperm, can be passed from one generation to the next. Most mutations occur in junk DNA and have no discernible effects on the survivability of an organism. Of the remaining mutations, the majority have harmful effects, while a minority can increase an organism's ability to survive. A mutation that benefits a species may evolve by means of natural selection into a trait shared by some or all members of the species. See Note at **sickle cell anemia.**

mutual inductance (myo͞o′cho͞o-əl) A measure of the relation between the change of current flow in one circuit to the electric potential generated in another by mutual induction. Like **inductance**, mutual inductance is measured in henries.

mutual induction The production of an electric potential in a circuit resulting from a change of current in a neighboring circuit. Mutual induction lies behind the operation of **transformers.**

mutualism (myo͞o′cho͞o-ə-lĭz′əm) A symbiotic relationship in which each of the organisms benefits. ► In **obligate mutualism** the interacting species are interdependent and cannot survive without each other. The fungi and algae that combine to form lichen are obligate mutualists. ► In the more common **facultative mutualism** the interacting species derive benefit without being fully dependent. Many plants produce fruits that are eaten by birds, and the birds later excrete the seeds of these fruits far from the parent plant. While both species benefit, the birds have other food available to them, and the plants can disperse their seeds when the uneaten fruit drops. Compare **amensalism, commensalism, parasitism.**

mutualism

In this mutualistic relationship, the impala (Buphagus erythrorhynchus) *gets rid of ticks and other parasites and the oxpecker* (Aepyceros melampus) *gets a meal.*

mycelium (mī-sē′lē-əm) Plural **mycelia.** The mass of fine branching tubes (known as **hyphae**) that forms the main growing structure of a fungus. Visible structures like mushrooms are reproductive structures produced by the mycelium.

mycology (mī-kŏl′ə-jē) The scientific study of fungi.

mycoplasma (mī'kō-plăz**'**mə) Any of a phylum of extremely small, parasitic bacteria that have a flexible cell membrane instead of a rigid cell wall, can assume a variety of shapes, and are capable of forming colonies. Too small to be seen with a light microscope, mycoplasmas are thought to be the smallest organisms capable of independent growth. They cause a number of important plant diseases, notably among citrus fruits. Mycoplasmas of the genus *Mycoplasma* are dependent upon sterols such as cholesterol for growth and cause several types of pneumonia in humans and animals. See also **phytoplasma.**

mycorrhiza (mī'kə-rī**'**zə) The symbiotic association of the mycelium of a fungus with the roots of plants. The majority of vascular plants have mycorrhizae. The fungus assists in the absorption of minerals and water from the soil and defends the roots from other fungi and nematodes, while the plant provides carbohydrates to the fungus. There are two kinds of mycorrhizae: *endomycorrhizae,* in which the fungal hyphae enter the cells of the root cortex, and *ectomycorrhizae,* in which they surround the cells.

myelin (mī**'**ə-lĭn) A whitish, fatty substance that forms a sheath around many vertebrate nerve fibers. Myelin insulates the nerves and permits the rapid transmission of nerve impulses. The white matter of the brain is composed of nerve fibers covered in myelin.

mylonite (mī**'**lə-nīt') A fine-grained laminated metamorphic rock in which preexisting minerals have been partially pulverized and drawn out into bands. Mylonite forms along geologic faults where shearing and grinding of rocks takes place.

myocardial infarction (mī'ō-kär**'**dē-əl) See **heart attack.**

myoglobin (mī**'**ə-glō'bĭn) An iron-containing protein found in muscle fibers, consisting of heme connected to a single peptide chain that resembles one of the subunits of hemoglobin. Myoglobin combines with oxygen released by red blood cells and transfers it to the mitochondria of muscle cells, where it is used to produce energy.

myopia (mī-ō**'**pē-ə) A defect of the eye that causes light to focus in front of the retina instead of directly on it, resulting in an inability to see distant objects clearly. Myopia is often caused by an elongated eyeball or a misshapen lens. Also called *nearsightedness.* Compare **hyperopia.**

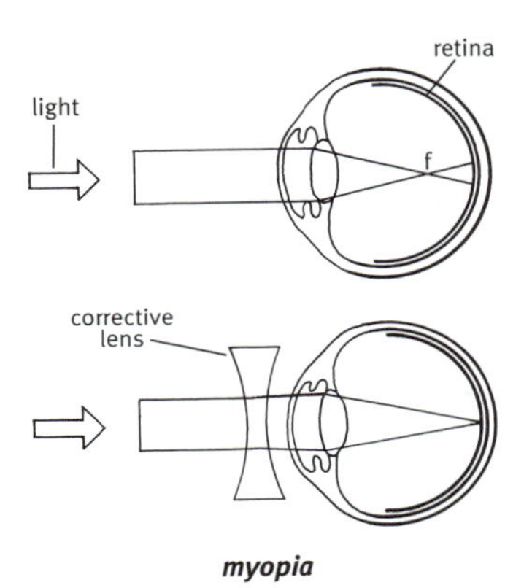

myopia

top: *When the focal point* (f) *of light falls in front of the retina, vision is blurred.* bottom: *A concave-shaped lens corrects the defect by focusing light on the retina.*

myosin (mī**'**ə-sĭn) A protein found in muscle tissue as a thick filament made up of an aggregate of similar proteins. Myosin and the protein actin form the contractile units (sarcomeres) of skeletal muscle. In the sarcomere, actin and myosin filaments slide past each other to cause the shortening of a muscle fiber.

myriapod (mĭr**'**ē-ə-pŏd') Any of various arthropods belonging to several closely related groups, having long segmented bodies, one pair of antennae, and at least nine pairs of legs. Centipedes and millipedes are myriapods.

myrmecology (mûr'mĭ-kŏl**'**ə-jē) The scientific study of ants.

myxomycete (mĭk'sō-mī**'**sēt) A plasmodial slime mold. See under **slime mold.**

N

N 1. Abbreviation of **newton. 2.** The symbol for **nitrogen.**

Na The symbol for **sodium.**

nacre (nā′kər) See **mother-of-pearl.**

nacreous cloud (nā′krē-əs) A cloud resembling a cirrus, showing iridescent coloration when the Sun is several degrees below the horizon. These clouds appear mostly during winter in regions of high latitude.

NAD (ĕn′ā-dē′) Short for *nicotinamide adenine dinucleotide.* A coenzyme that occurs in many living cells and that functions as an electron acceptor. NAD is used alternately with NADH as either an oxidizing agent or a reducing agent in metabolic reactions. *Chemical formula:* $C_{21}H_{27}N_7O_{14}P_2$.

NADH (ĕn′ā-dē-āch′) The reduced form of NAD. NADH has one more electron than NAD.

nadir (nā′dər) The point on the **celestial sphere** that is directly below the observer (90 degrees below the **celestial horizon**). Compare **zenith.**

NADP (ĕn-ā′dē-pē′) Short for *nicotinamide adenine dinucleotide phosphate.* A coenzyme that occurs in many living cells and functions as an electron acceptor like NAD but reacts with different metabolites. NADP is similar in structure to NAD but has an extra phosphate group. *Chemical formula:* $C_{21}H_{28}N_7O_{14}P_3$.

NADPH (ĕn′ā-dē′pē-āch′) The reduced form of NADP. NADPH has one more electron than NADP.

naked (nā′kĭd) **1.** *Zoology.* Lacking outer covering such as scales, fur, feathers, or a shell. **2.** *Botany.* **a.** Lacking a pericarp, as the seeds of the pine. **b.** Lacking a perianth, as the flowers of spurge. **c.** Unprotected by scales, as a bud. **d.** Having no leaves, as a branch or stem. **e.** Having no covering of fine, hairlike structures, as a stalk or leaf; glabrous.

nannoplankton also **nanoplankton** (năn′ə-plăngk′tən) Plankton of minute size, especially plankton composed of organisms measuring from 2 to 20 micrometers.

nano– A prefix that means: **1.** Very small or at a microscopic level, as in *nanotube.* In this sense, this prefix is sometimes spelled **nanno–**, as in *nannoplankton.* **2.** One billionth, as in *nanosecond,* one billionth of a second.

nanometer (năn′ə-mē′tər) One billionth (10^{-9}) of a meter.

nanoscale (năn′ə-skāl′) Relating to or occurring on a scale of nanometers.

nanosecond (năn′ə-sĕk′ənd) One billionth (10^{-9}) of a second.

nanotechnology (năn′ə-tĕk-nŏl′ə-jē) The science and technology of devices and materials, such as electronic circuits or drug delivery systems, constructed on extremely small scales, as small as individual atoms and molecules.

A CLOSER LOOK **nanotechnology**

Nanotechnology is the science and technology of precisely manipulating the structure of matter at the molecular level. The term nanotechnology embraces many different fields and specialties, including engineering, chemistry, electronics, and medicine, among others, but all are concerned with bringing existing technologies down to a very small scale, measured in nanometers. A nanometer—a billionth of a meter—is about the size of six carbon atoms in a row. (The prefix *nano-* comes from the Greek word *nanos,* which meant "little old man" or "dwarf.") Today, as in the past, most industrial products are created by pushing piles of millions of atoms together—by mixing, grinding, heating—a very imprecise process. However, scientists can now pick up individual atoms to assemble them into simple structures or cause specific chemical reactions. Propellers have been attached to molecular motors, and electricity has been conducted through nanowires. Nanotubes made of carbon are being investigated for a variety of industrial and research purposes. In the future, nanotechnology may be able to harness the forces that operate at the scale of the nanometer, such as the van der Waals force, as well as changes in the quantum states of particles, for new engineering purposes. The development of nanotechnology holds out great promise of improvements in the quality of life, including new treatments for disease and greater efficiency in computer data storage and processing. For example, tiny autonomous robots, or nanobots, may one day be sent into human bodies to repair cells and cure cancers, perhaps even extending the human life span by many years. The simple devices created by nanotechnology so far have not yet approached the complexity of the envisioned nanomachines and nanobots. Some scientists even see a dark side to the technology, emphasizing the need for caution in its development, particularly in attempts to create nanobots that can replicate themselves like living organisms.

---

nanotube (năn′ə-to͝ob′) A hollow cylindrical or toroidal molecule made of one element, usually carbon. Nanotubes are being investigated as semiconductors and for uses in nanotechnology. See also **fullerene.**

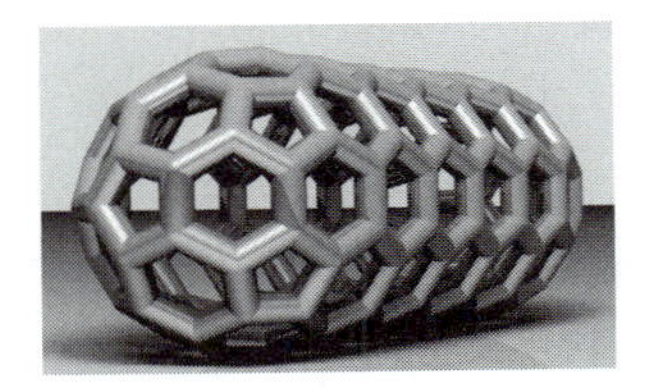

nanotube

computer graphic of a nanotube

Nansen bottle (năn′sən, nän′-) An ocean-water sampling bottle with spring-loaded valves at both ends that are closed at an appropriate depth by a messenger device sent down the wire connecting the bottle to the surface. The Nansen bottle has been replaced by the **Niskin bottle**, which is made of plastic and thus does not corrode like Nansen's metal bottle. These plastic bottles, however, are frequently referred to as Nansen bottles because their basic design is the same. The Nansen bottle was named for its inventor, Norwegian explorer Fridtjof Nansen (1861–1930).

napalm (nā′päm′) A firm jelly made by mixing gasoline with aluminum salts (made of fatty acids). It is used in some bombs and in flamethrowers. Napalm was developed during World War II.

naphtha (năf′thə) Any of several liquid mixtures of hydrocarbons made by refining petroleum or by breaking down coal tar. Naphtha is usually flammable, and is used as a solvent and as an ingredient in gasoline. It is also used to make plastics.

naphthalene (năf′thə-lēn′) A white crystalline compound made from coal tar or petroleum and used to make dyes, mothballs, explosives, and solvents. Naphthalene consists of two benzene rings fused together. *Chemical formula:* $C_{10}H_8$.

naphthene (năf′thēn′, năp′-) See **cycloalkane.**

naphthol (năf′thôl′, -thōl′, năp′-) A poisonous organic compound occurring in two isomeric forms. Both isomers are important in the manufacture of dyes, and also in making antiseptics, insecticides, and tanning agents. Naphthol consists of a hydroxyl group (OH) attached to naphthalene. *Chemical formula:* $C_{10}H_8O$.

Napierian logarithm (nə-pîr′ē-ən, nā-) See **natural logarithm.**

narcotic (När-kŏt′ĭk) Any of a group of highly addictive analgesic drugs derived from opium or opiumlike compounds. Narcotics can cause drowsiness and significant alterations of mood and behavior.

nasal (nā′zəl) Relating to or involving the nose.

Nathans (nā′thənz), **Daniel** 1928–1999. American microbiologist who pioneered the use of *restriction enzymes*—enzymes that break DNA molecules down into manageable fragments—to create the first genetic map on which the location of specific genes on the DNA could be identified. For this work, which revolutionized genetic engineering, Nathans shared the 1978 Nobel Prize for physiology or medicine with Werner Arber and Hamilton Smith.

native (nā′tĭv) **1.** Living or growing naturally in a particular place or region; indigenous. **2.** Occurring in nature on its own, uncombined with other substances. Copper and gold are often found in native form. **3.** Of or relating to the naturally occurring conformation of a macromolecule, such as a protein.

natural gas (năch′ər-əl) A mixture of hydrocarbon gases that occurs naturally beneath the

Earth's surface, often with or near petroleum deposits. Natural gas contains mostly of methane but also has varying amounts of ethane, propane, butane, and nitrogen. It is used as a fuel and in making organic compounds.

natural history The study and description of living things and natural objects, especially their origins, evolution, and relationships to one another. Natural history includes the sciences of zoology, mineralogy, geology, and paleontology.

naturalist (năch′ər-ə-lĭst) A person who specializes in natural history, especially in the study of plants and animals in their natural surroundings.

naturalize (năch′ər-ə-līz′) To establish a nonnative species in a region where it is able to reproduce successfully and live alongside native species in the wild. Naturalized species may be introduced intentionally or unintentionally. Eucalyptus trees are native to Australia but have become naturalized in many other parts of the world.

natural logarithm A **logarithm** using base *e*. Natural logarithms are frequently used in calculus. They are also called *Napierian logarithms*, after their inventor, English mathematician John Napier (1550–1617). See more at **e.** Compare **common logarithm.**

natural number A positive integer.

natural resource Something, such as a forest, a mineral deposit, or fresh water, that is found in nature and is necessary or useful to humans. See more at **nonrenewable, renewable.**

natural science A science, such as biology, chemistry, or physics, that deals with the objects, phenomena, or laws of nature and the physical world.

natural selection The process by which organisms that are better suited to their environment than others produce more offspring. As a result of natural selection, the proportion of organisms in a species with characteristics that are adaptive to a given environment increases with each generation. Therefore, natural selection modifies the originally random variation of genetic traits in a species so that alleles that are beneficial for survival predominate, while alleles that are not beneficial decrease. Originally proposed by Charles Darwin, natural selection forms the basis of the process of evolution. See Notes at **adaptation, evolution.** Compare **artificial selection.**

nature (nā′chər) **1.** The world and its naturally occurring phenomena, together with all of the physical laws that govern them. **2.** Living organisms and their environments.

nausea (nô′zē-ə, -zhə) A symptom characterized by gastrointestinal distress and an urge to vomit.

nautical mile (nô′tĭ-kəl) A unit of length in the US Customary System, used in air and sea navigation and equal to 6,076 feet or 2,025 yards (1,852 meters). Also called *geographic mile*. See Table at **measurement.**

nautiloid (nôt′l-oid′) Any of various cephalopod mollusks of the subclass Nautiloidea, having a straight or coiled shell divided internally into a series of chambers of increasing size connected by a central tube. The nautiloids include the modern nautiluses as well as numerous extinct species dating back as far as the Cambrian Period.

Nb The symbol for **niobium.**

Nd The symbol for **neodymium.**

Ne The symbol for **neon.**

Neanderthal (nē-ăn′dər-thôl′, -tôl′) or **Neandertal** (nē-ăn′dər-tôl′) An extinct variety of human that lived throughout Europe and in parts of western Asia and northern Africa during the late Pleistocene Epoch, until about 30,000 years ago. Neanderthals had a stocky build and large skulls with thick eyebrow ridges and big teeth. They usually lived in caves, made flaked stone tools, and were the earliest humans known to bury their dead. Neanderthals were either a subspecies of modern humans (*Homo sapiens neanderthalensis*) or a separate, closely related species (*Homo neanderthalensis*). They coexisted with early modern humans (Cro-Magnons) for several thousand years before becoming extinct, but are not generally believed to have interbred with them. See also **Mousterian.**

neanthropic (nē′ən-thrŏp′ĭk) Relating to members of the extant species *Homo sapiens* as compared with other, now extinct species of *Homo*.

neap tide (nēp) A tide in which the difference between high and low tide is the least. Neap tides occur twice a month when the Sun and Moon are at right angles to the Earth. When this is the case, their total gravitational pull on the Earth's water is weakened because it comes from two different directions. Compare **spring tide.** See more at **tide.**

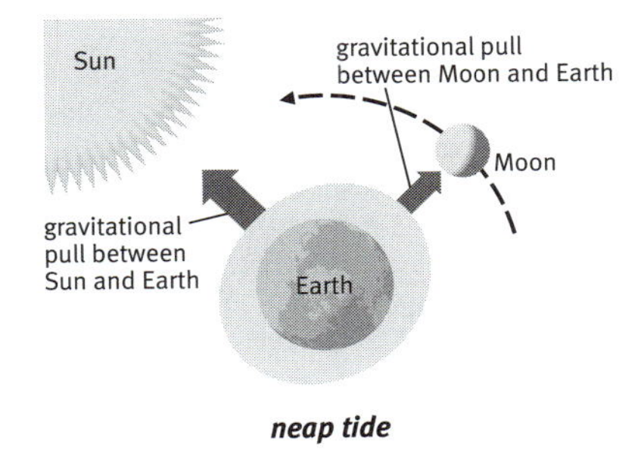

neap tide

Neap tides occur when the Sun and Moon are at right angles to the Earth and their gravitational pulls counteract each other.

near-Earth object (nîr′ûrth′) A comet or asteroid with an orbit or trajectory that comes near Earth's orbit, often drawn into such a path by the gravitational effect of the Earth and other planets. While there is currently no known near-Earth object on a collision course with Earth, collisions with such objects have taken place in the past and may be associated with mass extinctions such as those that took place at the end of the Permian and Cretaceous Periods.

nearshore (nîr′shôr′) The region of land extending between the backshore, or shoreline, and the beginning of the offshore zone. Water depth in this area is usually less than 10 m (33 ft).

nearsightedness (nîr′sī′tĭd-nĭs) See **myopia.**

nebula (nĕb′yə-lə) Plural **nebulae** (nĕb′yə-lē′) or **nebulas.** A visible, thinly spread cloud of interstellar gas and dust. Some nebulae are the remnants of a supernova explosion, others are gravity-induced condensations of the gases in the interstellar medium which in certain cases may become a site for the formation of new stars. The term was formerly used of any hazy, seemingly cloudlike object, including what are now recognized as other galaxies beyond the Milky Way; it is restricted now to actual clouds of gas and dust within our own galaxy. See more at **star.** ▸ Nebulae are generally classified as bright or dark. Among the bright nebulae are cold clouds that reflect light from nearby stars (**reflection nebulae**) and hot, ionized clouds that glow with their own light (**emission nebulae**). Dark nebulae—cold clouds that absorb the passing light from background stars—are called **absorption nebulae**.

nebular hypothesis (nĕb′yə-lər) A model of star and planet formation in which a nebula contracts under the force of gravity, eventually flattening into a spinning disk with a central bulge. A protostar forms at the nebula's center. As matter condenses around the protostar in the bulge, planets are formed from the spinning matter in the disk. This theory is widely accepted to account for the formation of stars and planetary systems such as ours. The first version of the nebular hypothesis was proposed in 1755 by the German philosopher Immanuel Kant and modified in 1796 by Pierre Laplace. ▸ The nebula that according to this hypothesis condensed to form the solar system is called the **solar nebula**.

nebulizer (nĕb′yə-lī′zər) A device that reduces liquid to an extremely fine cloud, especially used for delivering medication to the deep part of the respiratory tract.

necrosis (nə-krō′sĭs) The death of cells or tissues from severe injury or disease, especially in a localized area of the body. Causes of necrosis include inadequate blood supply (as in infarcted tissue), bacterial infection, traumatic injury, and hyperthermia. *—Adjective* **necrotic** (nə-krŏt′ĭk).

nectar (nĕk′tər) A sweet liquid secreted by plants as food to attract animals that will benefit them. Many flowers produce nectar to attract pollinating insects, birds, and bats. Bees collect nectar to make into honey. Nectar is produced in structures called *nectaries.* Some plants have nectaries located elsewhere, outside the flower. These provide a food source for animals such as ants which in turn defend the plant from harmful insects. Nectar consists primarily of water and varying concentrations of many different sugars, including fructose, glucose, and sucrose.

needle (nēd′l) **1.** A narrow, stiff leaf, as of firs, pines, and other conifers. The reduced surface area of needles minimizes water loss and allows needle-bearing plants to live in dry climates. See more at **leaf. 2.** See **hypodermic needle.**

negative (nĕg′ə-tĭv) **1.** Less than zero. **2.** Having the electric charge or voltage less than zero. **3.** Devoid of evidence of a suspected condition or disease, as a diagnostic test.

negative feedback Feedback in which the output quantity or signal lowers the input quantity or signal. Negative feedback is used in natural and artificial regulatory mechanisms, as well as in the design of oscillators.

negative pressure See under **pressure.**

negatron (nĕg′ə-trŏn′) An electron with a negative charge; the antiparticle of the **positron.** Most branches of particle physics construe each particle along with its antiparticle to be two different forms of one underlying phenomenon, and the term *electron* is sometimes used as a precisely such a general term, with *positron* and *negatron* referring to the forms of the electron as they are manifested in nature. See more at **electron.**

nekton (nĕk′tən, -tŏn′) The collection of marine and freshwater organisms that can swim freely and are generally independent of currents, ranging in size from microscopic organisms to whales. Compare **benthos, plankton.**

nematocyst (nĕm′ə-tə-sĭst′, nĭ-măt′ə-sĭst′) One of the minute capsules in the tentacles of cnidarians, such as jellyfish, hydras, or sea anemones, used for stinging. The capsule is produced by a special cell (called a cnidoblast) and contains a tightly coiled barbed thread that quickly shoots forth if the capsule's lid is disturbed. The thread often contains poison.

nematode (nĕm′ə-tōd′) Any of several slender, cylindrical worms of the group Nematoda, which some scientists consider to be a class of the aschelminths and others to be a separate phylum. Most nematodes are tiny and live in enormous numbers in water, soil, plants, and animals. They have a simple structure, with a long hollow gut separated from the body wall by a fluid-filled space. Several nematodes, such as pinworm, roundworm, filaria, and hookworm, are parasites on animals and humans and cause disease. One species, *Caenorhabditis elegans* (usually called *C. elegans*), was one of the first animals to have its entire genome sequenced and is important in biological research as a model organism.

Neo-Darwinism (nē′ō-där′wə-nĭz′əm) Darwinism as modified by the findings of modern genetics, stating that mutations due to random copying errors in DNA cause variation within a population of individual organisms and that natural selection acts upon these variations.

neodymium (nē′ō-dĭm′ē-əm) *Symbol* **Nd** A shiny, silvery metallic element of the lanthanide series. It is used to make glass for welders' goggles and purple glass for lasers. Atomic number 60; atomic weight 144.24; melting point 1,024°C; boiling point 3,027°C; specific gravity 6.80 or 7.004 (depending on allotropic form); valence 3. See **Periodic Table.**

Neogene (nē′ō-jēn′) The youngest of two subdivisions of the Tertiary Period, including the Miocene and Pliocene Epochs.

Neolithic (nē′ə-lĭth′ĭk) The period of human culture that began around 10,000 years ago in the Middle East and later in other parts of the world. It is characterized by the beginning of farming, the domestication of animals, the development of crafts such as pottery and weaving, and the making of polished stone tools. The Neolithic Period is generally considered to end for any particular region with the introduction of metalworking, writing, or other developments of urban civilization. Also called *New Stone Age*. Compare **Mesolithic, Paleolithic.**

neon (nē′ŏn′) *Symbol* **Ne** A rare colorless element in the noble gas group that occurs naturally in extremely small amounts in the atmosphere. It glows reddish orange when electricity passes through it, as in a tube in an electric neon light. Neon is also used for refrigeration. Atomic number 10; atomic weight 20.180; melting point –248.67°C; boiling point –245.95°C. See **Periodic Table.**

neonatology (nē′ō-nā-tŏl′ə-jē) The branch of medicine that deals with the diseases and care of newborn infants.

neoplasm (nē′ə-plăz′əm) An abnormal growth of tissue in animals or plants. Neoplasms can be benign or malignant. Also called *tumor.* —*Adjective* **neoplastic** (nē′ə-plăs′tĭk).

neoprene (nē′ə-prēn′) A tough, synthetic rubber that is resistant to the effects of oils, solvents, heat, and weather. Neoprene is a polymer whose basic constituent is chlorinated butadiene. Neoprene was one of the first synthetic rubbers to be developed.

neoteny (nē-ŏt′n-ē) **1.** The retention of juvenile characteristics in the adults of a species. Humans, for example, are sometimes said to demonstrate neoteny by retaining through adulthood the relatively large head and hairlessness characteristic of very young primates. The body proportions of flightless birds, which resemble those of fetal flying birds, are also considered to be evidence of neoteny. **2.** The attainment of sexual maturity by an organism still in its larval stage, seen in certain amphibians and insects. Certain species of salamanders, for instance, demonstrate neoteny as they become sexually mature but remain aquatic and do not develop legs. Neoteny sometimes occurs in response to environmental stresses such as low temperature or lack of iodine (which is essential for the thyroid gland). If environmental conditions improve, however, the organism can often develop into a fully mature adult form.

nephrite (nĕf′rīt′) A green to blue variety of jade. Nephrite is an amphibole and is the least precious form of jade. *Chemical formula:* $\mathbf{Ca_2(Mg,Fe)_5Si_8O_{22}(OH)_2}$.

nephritis (nə-frī′tĭs) Inflammation of the kidneys.

nephrology (nə-frŏl′ə-jē) The branch of medicine that deals with functions and diseases of the kidneys.

nephron (nĕf′rŏn) The functional unit of the kidney, in which waste products are filtered from the blood and urine is produced. The nephron consists of a system of tubules in close association with a network of blood vessels. As fluid that is filtered through the glomerulus of the nephron enters the tubules, its composition is gradually changed by the absorption and secretion of solutes, and it eventually leaves the nephron as urine. See more at **Bowman's capsule, glomerulus, loop of Henle.**

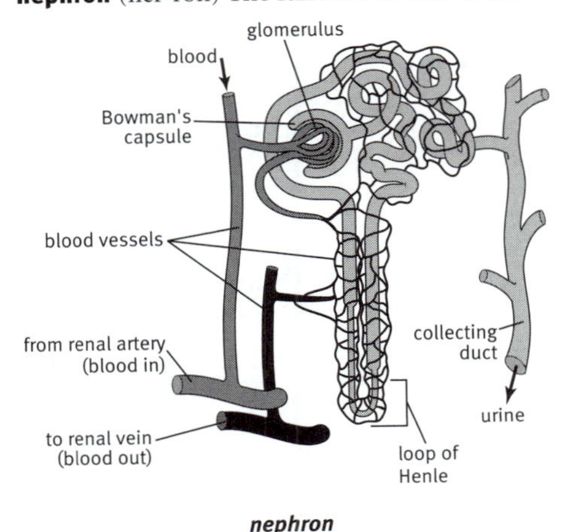

nephron

Neptune (nĕp′to͞on′) The eighth planet from the Sun and the fourth largest, with a diameter almost four times that of Earth. Neptune is a **gas giant** with a very active weather system, exhibiting extremely long and powerful storms with the fastest winds observed in the solar system. Neptune's axis is tilted 28.8° from the plane of its orbit, and its summer and winter seasons each last 40 years. For a period of 20 years out of every 248, Pluto's highly elliptical orbit crosses within that of Neptune. Neptune has four faint rings and 13 known moons and appears blue due to the absorption of red light by the methane within its atmosphere. See Table at **solar system.**

neptunium (nĕp-to͞o′nē-əm) *Symbol* **Np** A silvery, radioactive metallic element of the actinide series. It occurs naturally in minute amounts in uranium ores and is produced artificially as a byproduct of plutonium production. Its longest-lived isotope is Np 237 with a half-life of 2.1 million years. Atomic number 93. See **Periodic Table.**

neritic (nə-rĭt′ĭk) Relating to the ocean waters over the sublittoral region of the ocean floor, ranging in depth between the low tide mark to about 200 m (656 ft). See more at **epipelagic zone.**

nerve (nûrv) Any of the bundles of fibers made up of neurons that carry sensory and motor information throughout the body in the form of electrical impulses. Afferent nerves carry information to the central nervous system, and efferent nerves carry information from the central nervous system to the muscles, organs, and glands. Efferent nerves include the nerves of the peripheral nervous system, which control voluntary motor activity and of the autonomic nervous system, which controls involuntary motor activity.

nerve cell See **neuron.**

nerve fiber See **axon.**

nerve gas Any of various poisonous gases that interfere with the functioning of nerves by inhibiting the breakdown of the neurotransmitter acetylcholine. Increased levels of acetylcholine stimulate the parasympathetic nervous system, leading to cardiac and respiratory arrest.

nervous system (nûr′vəs) The system of neurons and tissues that regulates the actions and responses of vertebrates and many invertebrates. The nervous system of vertebrates is a complex information-processing system that consists mainly of the brain, spinal cord, and peripheral and autonomic nerves. It receives chemical information from hormones in the circulating blood and can also regulate secretions of the endocrine system by the action of neurohormones. The nervous systems of invertebrates vary from a simple network of nerves to a complex nerve network under the control of a primitive brain. See also **autonomic nervous system, central nervous system, peripheral nervous system.**

network (nĕt′wûrk′) A system of computers and peripherals, such as printers, that are linked together. A network can consist of as few as two computers connected with cables or millions of computers that are spread over a large geographical area and are connected by telephone lines, fiberoptic cables, or radio waves. The Internet is an example of very large network. See more at **LAN, WAN.**

Neumann (noi′män′), **John von** 1903–1957. Hungarian-born American mathematician who contributed to mathematical theories about numbers and games. He was a leader in the design and development of high-speed electronic computers and also contributed to the field of cybernetics, a term he coined.

neural (no͝or′əl) Relating to the nerves or nervous system.

neural tube A tubular structure that results from the folding of tissue along the back of vertebrate embryos and develops into the brain and spinal cord. Improper folding of the neural tube is the cause of spina bifida and other birth defects.

neuristor (no͞o-rĭs′tər) An electronic component that processes electrical signals in a manner modeling the behavior of **neurons.**

neuroanatomy (no͝or′ō-ə-năt′ə-mē) **1.** The scientific study of the anatomy of the nervous system. Neuroanatomy is a branch of neurology. **2.** The neural structure of an organism or part of an organism.

neurofibromatosis (no͝or′ō-fī′brō-mə-tō′sĭs) A genetic disease characterized by multiple benign tumors of peripheral nerves, called *neurofibromas*, and pigmented spots on the skin, sometimes accompanied by bone deformity and a predisposition to cancers, especially of the brain.

neurohormone (no͝or′ō-hôr′mōn) A hormone that is produced and secreted by neurons and that effects its action on the nervous system. The hormones secreted by the hypothalamus that in turn control the secretions of the pituitary gland are neurohormones.

neurology (no͞o-rŏl′ə-jē) The scientific study of the diagnosis and treatment of disorders of nerves and the nervous system.

neuron (no͝or′ŏn′) A cell of the nervous system. Neurons typically consist of a cell body, which contains a nucleus and receives incoming nerve impulses, and an axon, which carries impulses away from the cell body. Also called *nerve cell.*

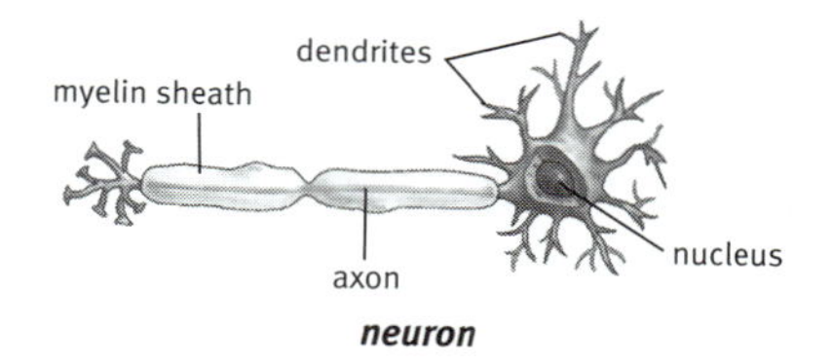

neuron

multipolar neuron

neuropteran (no͞o-rŏp′tər-ən) Any of various carnivorous insects of the order Neuroptera, having four net-veined wings and mouthparts adapted for chewing. Neuropterans include the lacewings and antlions. Formerly, the dobsonflies and alderflies, now classified as a separate order Megaloptera, and the snakeflies, now classified as the order Raphidiodea, were considered neuropterans.

neurosis (no͞o-rō′sĭs) A psychological state characterized by excessive anxiety or insecurity without evidence of neurologic or other organic disease, sometimes accompanied by defensive or immature behaviors. This term is no longer used in psychiatric diagnosis.

neurosurgery (no͝or′ō-sûr′jə-rē) Surgery on any part of the nervous system, such as the brain or spinal cord.

neurotransmitter (no͝or′ō-trănz′mĭt-ər) A chemical substance that is produced and secreted by a neuron and then diffuses across a synapse to cause excitation or inhibition of another neuron. Acetylcholine, norepinephrine, dopamine, and serotonin are examples of neurotransmitters.

neutral (no͞o′trəl) **1.** Neither acid nor alkaline. **2.** Having no inherent or net charge, especially electric charge.

neutralize (no͞o′trə-līz′) To cause an acidic solution to become neutral by adding a base to it or to cause a basic solution to become neutral by adding an acid to it. Salt and water are usually formed in the process.

neutrino (no͞o-trē′nō) Any of three electrically neutral subatomic particles with extremely low mass. These include the electron-neutrino, the muon-neutrino, and the tau-neutrino. ▸ The study of neutrinos that come to the earth as cosmic rays suggests that neutrinos can transform into each other in a process called **neutrino oscillation.** For this phenomenon to be theoretically possible, the three neutrinos must have distinct masses; for this reason, many scientists believe that they have mass. See Table at **subatomic particle.**

A CLOSER LOOK **neutrino**

Neutrinos were not observed until 1955, roughly a quarter of a century after the physicist Wolfgang Pauli first hypothesized their existence on theoretical grounds. Pauli was studying certain radioactive decay processes called *beta decay,* processes now known to involve the decay of a neutron into a proton and an electron. A certain amount of energy that was lost in these processes could not be accounted for. Pauli suggested that the energy was carried away by a very small, electrically neutral particle that was not being detected. (He originally wanted to name the particle a *neutron* but didn't publish the suggestion, and a few years later the particle we now know as the neutron was discovered and named in print. The Italian physicist Enrico Fermi then coined the term *neutrino,* which means "little neutron" in Italian.) Neutrinos are hard to detect because their mass, if they indeed have any, is extremely low, and they possess no electric charge; a chunk of iron a few light-years thick would absorb only about half of the neutrinos that struck it. Nevertheless, neutrinos can be detected, and three different types have been distinguished, each of which is associated with a particular lepton (the electron, the muon, and the taon) with which it is often paired in interactions involving the *weak force.* Recent analysis of neutrinos emanated by the Sun has suggested that each type of neutrino can spontaneously turn into one of the others in a process of *neutrino oscillation,* and for theoretical reasons this in turn would require that neutrinos have mass. If so, then despite their light weight, their abundance may in fact mean that neutrinos contribute significantly to the overall mass of the universe.

neutron (no͞o′trŏn′) An electrically neutral subatomic particle in the baryon family, having a mass of 1.674×10^{-24} grams (1,838 times that of the electron and slightly greater than that of the proton). Neutrons are part of the nucleus of all atoms, except hydrogen, and have a mean lifetime of approximately 1.0×10^3 seconds as free particles. They consist of a triplet of quarks, including two down quarks and one up quark, bound together by gluons. In radioactive atoms, excess neutrons are converted to protons by beta decay. Beams of neutrons from nuclear reactors are used to bombard the atoms of various elements to produce fission and other nuclear reactions and to determine the atomic arrangements in molecules. See Table at **subatomic particle.**

neutron moderator See **moderator.**

neutron star A celestial object consisting of an extremely dense mass of neutrons, formed at the core of a supernova, where electrons and nuclei are compressed together so intensely by the force of gravity that protons and electrons merge together into neutrons. Though their mass is close to that of the Sun, the density of neutron stars is much higher—about 3×10^{11} kg per cubic cm (by comparison, the density of steel is 7.7 gm per cubic cm). Neutron stars are typically about 10 km across, and rotate very rapidly. Due to the spinning of electrically charged protons and electrons at their surfaces, their rotation gives rise to strong magnetic fields. The existence of neutron stars was predicted in the 1930s but was not confirmed until the discovery of the first **pulsar** in 1967. See more at **pulsar.**

névé (nā-vā′) **1.** The upper part of a glacier, consisting of hardened snow. **2.** The granular snow typically found in such a field.

Newcomen (no͞o′kə-mən), **Thomas** 1663–1729. English inventor who developed an early steam engine (1711) that was was widely used to pump water in coal mines.

new moon (no͞o) The phase of the Moon that occurs when it passes between Earth and the Sun, making it either invisible or visible only as a thin crescent at sunset. See more at **moon.** Compare **full moon.**

New Stone Age See **Neolithic.**

newton (no͞ot′n) The SI derived unit used to measure force. One newton is equal to the force needed to accelerate a mass of one kilogram one meter per second per second. See also **joule.**

Newton, Sir **Isaac** 1642–1727. English mathematician and scientist. He invented a form of calculus and formulated principles of physics that remained basically unchallenged until the work of Albert Einstein, including the law of universal gravitation, a theory of the nature of light, and three laws of motion. His treatise on gravitation, presented in *Principia Mathematica* (1687), was in his own account inspired by the sight of a falling apple. See **Newton's law of gravitation, Newton's laws of motion.**

Isaac Newton

BIOGRAPHY **Isaac Newton**

The British mathematician and physicist Sir Isaac Newton stands as one of the greatest scientists of all time. Newton spent most of his working life at Cambridge University. In 1665, the year he received his bachelor's degree, an outbreak of the bubonic plague caused Cambridge to close for two years. Newton returned to his family home in Lincolnshire and, working alone, did some of his most important scientific work. Perhaps his greatest achievement was to demonstrate that scientific principles have universal applications. His universal law of gravitation states that there is an attractive force acting between all bodies in the universe. According to the famous—and possibly true—story, he observed an apple falling from a tree and, remarkably, connected the force drawing the apple to the ground with that keeping the Moon in its orbit. Along with his law of gravitation, Newton's three laws of motion, which laid the basis for the science of mechanics, bridged the gap between scientific thinking about terrestrial and celestial dynamics. The laws are: (1) A body at rest or moving in a straight line will continue to do so unless acted upon by an external force; (2) The acceleration of a moving object is proportional to and in the same direction as the force acting on it and inversely proportional to the object's mass; and (3) For every action there is an equal and opposite reaction. For nearly 400 years these laws have remained unchallenged; even Einstein's Theory of Relativity is consistent with them. Newton stated his laws of motion in his 1687 masterpiece, the *Principia Mathematica,* in which he also introduced his formulation of the calculus (what we now call simply "calculus," a different version of which was simultaneously developed by Leibnitz). In optics, Newton demonstrated that white light contains all the colors of the spectrum and provided strong evidence that light was composed of particles.

Newton's law of gravitation (no͞ot′nz) The principle that expresses the force of gravitational attraction between two bodies as a function of their mass and their distance. Expressed mathematically, $F = \frac{Gm_1m_2}{d^2}$ where F is the force in Newtons, m_1 and m_2 are the masses of the bodies in kilograms, G is the gravitational constant, and d is the distance between the bodies in meters. Newton's Principle of Gravitation is an example of an **inverse square law.** Also called *law of gravitation, law of universal gravitation.* See Note at **gravity.**

Newton's laws of motion The three laws proposed by Sir Isaac Newton concerning relations between force, motion, acceleration, mass, and inertia. These laws form the basis of classical mechanics and were elemental in solidifying the concepts of force, mass, and inertia. ► **Newton's first law** states that a body at rest will remain at rest, and a body in motion will remain in motion with a constant velocity, unless acted upon by a force. This law is also called the *law of inertia.* ► **Newton's second law** states that a force acting on a body is equal to the acceleration of that body times its mass. Expressed mathematically, F = ma, where F is the force in Newtons, m is the mass of the body in kilograms, and a is the acceleration in meters per second per second. ► **Newton's third law** states that for every action there is an equal and opposite reaction. Thus, if one body exerts a force F on a second body, the first body also undergoes a force of the same strength but in the opposite direction. This law lies behind the design of rocket propulsion, in which matter forced out of a burner at high speeds creates an equal force driving the rocket forward.

Ni The symbol for **nickel.**

niacin (nī′ə-sĭn) A water-soluble organic acid belonging to the vitamin B complex that is important in carbohydrate metabolism. It is a pyridine derivative and is a precursor of the coenzyme NAD. Niacin is found in liver, fish, and whole-grain foods. Deficiency of niacin in the diet causes **pellagra.** Also called *nicotinic acid. Chemical formula:* $C_6H_5NO_2$.

niche (nĭch, nēsh) The function or position of a species within an ecological community. A species's niche includes the physical environment to which it has become adapted as well as its role as producer and consumer of food resources. See also **competitive exclusion principle.**

nickel (nĭk′əl) *Symbol* **Ni** A silvery, hard, ductile metallic element that occurs in ores along with iron or magnesium. It resists oxidation and corrosion and is used to make alloys such as stainless steel. It is also used as a coating for other metals. Atomic number 28; atomic weight 58.69; melting point 1,453°C; boiling point 2,732°C; specific gravity 8.902; valence 0, 1, 2, 3. See **Periodic Table.**

nicotine (nĭk′ə-tēn′) A colorless, poisonous compound occurring naturally in the tobacco plant. It is used in medicine and as an insecticide, and it is the substance in tobacco products to which smokers can become addicted. Nicotine is an alkaloid. *Chemical formula:* $C_{10}H_{14}N_2$.

nicotinic acid (nĭk′ə-tĭn′ĭk, -tē′nĭk) See **niacin.**

nictitating membrane (nĭk′tĭ-tā′tĭng) A transparent inner eyelid in birds, reptiles, amphibians, and some mammals that protects and moistens the eye without blocking vision.

nimbostratus (nĭm′bō-străt′əs) Plural **nimbostrati** (nĭm′bō-străt′ī′). A dark, gray, mid-altitude cloud that often covers the entire sky and precipitates rain, snow, or sleet. Nimbostratus clouds generally form around 2,000 m (6,560 ft) but often extend to much higher and lower altitudes. See illustration at **cloud.**

nimbus (nĭm′bəs) Plural **nimbi** (nĭm′bī′) or **nimbuses.** A rain cloud.

Niño (nēn′yō) See **El Niño.**

niobium (nī-ō′bē-əm) *Symbol* **Nb** A soft, silvery, ductile metallic element that usually occurs in nature together with the element tantalum. It is used to build nuclear reactors, to make steel alloys, and to allow magnets to conduct electricity with almost no resistance. Atomic number 41; atomic weight 92.906; melting point 2,468°C; boiling point 4,927°C; specific gravity 8.57; valence 2, 3, 5. See **Periodic Table.**

nipple (nĭp′əl) A small projection near the center of the mammary gland. In females, the nipple contains the outlets of the milk ducts.

Niskin bottle (nĭs′kĭn) See under **Nansen bottle.**

niter (nī′tər) A naturally occurring mineral form of potassium nitrate. It is used to make gunpowder.

nitrate (nī′trāt′) A salt or ester of nitric acid, containing the group NO_3. Nitrates dissolve extremely easily in water and are an important component of the nitrogen cycle. Compare **nitrite.**

nitric (nī′trĭk) Containing nitrogen, especially nitrogen with a valence of 5. Compare **nitrous.**

nitric acid A clear, colorless to yellow liquid that is very corrosive and can dissolve most metals. It is used to make fertilizers, explosives, dyes, and rocket fuels. *Chemical formula:* **HNO_3.**

nitric oxide A colorless, poisonous gas produced as an intermediate compound during the manufacture of nitric acid from ammonia or from atmospheric nitrogen. It is also produced through cellular metabolism. In the body, nitric oxide is involved in oxygen transport to the tissues, the transmission of nerve impulses, and other physiological activities. *Chemical formula:* **NO.**

A CLOSER LOOK **nitric oxide**

While *nitric oxide* (NO) was once regarded solely as a poisonous air pollutant, responsible for the formation of photochemical smog and acid rain leading to the destruction of the ozone layer, today it is also appreciated as a molecule essential to human health. Nitric oxide is the first gas discovered to act as a signaling molecule, a transmitter of important signals to cells in various systems of the human body. Even though NO continues to be detrimental to the environment, it was heralded as *Science Magazine*'s Molecule of the Year in 1992, and the Nobel Prize for physiology or medicine was awarded in 1998 to the three scientists who discovered that NO works as a signaling molecule in the cardiovascular system. It is now known that the cells of a blood vessel's inner walls use NO to signal the vessel to relax and dilate, increasing blood flow. Nitroglycerin, whose effectiveness in treating heart problems was once a mystery, is now known to work by releasing NO. NO has a variety of other important biological functions, including destroying bacteria within the immune system and acting as a neurotransmitter.

nitrification (nī′trə-fĭ-kā′shən) The process by which bacteria in soil and water oxidize ammonia and ammonium ions and form nitrites and nitrates. Because the nitrates can be absorbed by more complex organisms, as by the roots of green plants, nitrification is an important step in the **nitrogen cycle.**

nitrifying bacteria (nī′trə-fī′ĭng) Any of various soil bacteria that change ammonia or ammonium into nitrite or change nitrite into nitrate as part of the **nitrogen cycle.** Bacteria of the genus *Nitrosomonas* are the primary converters of ammonium into nitrite (which is actually toxic to plants), and bacteria of the genus *Nitrobacter* oxidize the nitrite to form nitrate ions (which are readily absorbed and usable by plants). Most of the nitrogen contained in fertilizer is made available to plants by these bacteria.

nitrile also **nitril** (nī′trəl) An organic compound, such as acrylonitrile, containing the cyanide group CN. Nitriles are typically colorless solids or liquids and have a distinctive smell.

nitrite (nī′trīt′) A salt or ester of nitrous acid, containing the group NO_2. Nitrites are an important component of the nitrogen cycle and are used as food preservatives. Compare **nitrate.**

nitrocellulose (nī′trō-sĕl′yə-lōs′) A pulpy or cottonlike polymer derived from cellulose treated with sulfuric and nitric acids. It is used in the manufacture of explosives, plastics, and solid propellants.

nitrofuran (nī′trō-fyo͝or′ăn′) Any of several drugs derived from furan that are used to inhibit bacterial growth.

nitrogen (nī′trə-jən) *Symbol* **N** A nonmetallic element that makes up about 78 percent of the atmosphere by volume, occurring as a colorless, odorless gas. It is a component of all proteins, making it essential for life, and it is also found in various minerals. Nitrogen is used to make ammonia, nitric acid, TNT, and fertilizers. Atomic number 7; atomic weight 14.0067; melting point −209.86°C; boiling point −195.8°C; valence 3,5. See **Periodic Table.** See Note at **oxygen.**

nitrogenase (nī-trŏj′ə-nās′, nī′trə-jə-) An enzyme of nitrogen-fixing bacteria that catalyzes the conversion of nitrogen to ammonia.

nitrogen base One of the nitrogen-containing purines (adenine or guanine) or pyrimidines (cytosine, thymine, or uracil) found in the nucleic acids DNA and RNA. The bases may be attached to a sugar (deoxyribose in DNA, ribose in RNA) to form **nucleosides.**The addition of a phosphate to a nucleoside results in the formation of a **nucleotide,** which is the basic constituent of nucleic acids.

nitrogen cycle The continuous process by which nitrogen is exchanged between organisms and the environment. Nitrogen is an essential nutrient, needed to make amino acids and other important organic compounds, but most organisms cannot use free nitrogen, which is abundant as a gas in the atmosphere. Gaseous nitrogen is broken apart and fixed (converted to stable, biologically assimilable inorganic compounds) in the process of **nitrogen fixation.** Some atmospheric nitrogen is fixed naturally during lightning strikes and some by industrial processes. Cyanobacteria and certain other species of bacteria, especially those living as symbionts in the roots of legumes, fix atmospheric nitrogen biologically in ammonium ions. Ammonia and ammonium ions are also produced by the ongoing decay of organic materials. Ammonia can be absorbed directly by plant cells, and certain bacteria living in soil and water convert ammonia and ammonium ions into nitrites and nitrates in the process known as **nitrification.** The nitrates are easily absorbed by plant roots. In this way, nitrogen is passed into the food chain and ultimately returned to the soil, water, and atmosphere by the metabolism and decay of plants and animals.

nitrogen fixation The process by which free nitrogen from the air is combined with other elements to form inorganic compounds, such as ammonium ions, which can then be converted by **nitrification** into nutrients that can be readily absorbed by plants and other organisms for incorporation into more complex organic compounds. During lighting strikes, the atmosphere's free nitrogen molecules (N_2) are broken apart and combine with oxygen to form nitrogen oxides that dissolve in rain to form nitric acid; atmospheric nitrogen is also fixed industrially under high pressure and heat to form ammonia, as in the production of fertilizers. Many species of cyanobacteria and certain other forms of bacteria, especially those that live in the roots of legumes, conduct nitrogen fixation as part of their metabolism, using the enzyme nitrogenase to combine nitrogen with hydrogen as ammonia. All living organisms are dependent on nitrogen fixation and would ultimately die without it. See more at **nitrogen cycle.**

nitrogenous (nī-trŏj′ə-nəs) Containing nitrogen or a compound of nitrogen.

nitroglycerin (nī′trō-glĭs′ər-ĭn) A thick, pale-yellow, explosive liquid formed by treating glycerin with nitric and sulfuric acids. It is used to make dynamite and in medicine to dilate blood vessels. *Chemical formula:* $C_3H_5N_3O_9$.

nitrosamine (nī-trō′sə-mēn′, nī′trōs-ăm′ēn) Any of a class of organic compounds with the general formula R_2NNO or $RNHNO$, where R is an organic radical. Nitrosamines are present in various foods and other products, and certain ones are very carcinogenic.

nitrous (nī′trəs) Containing nitrogen, especially nitrogen with a valence of 3. Compare **nitric.**

nitrous acid A weak inorganic acid existing only in solution or in the form of its salts. It can act either as an oxidizing agent or as a reducing agent. *Chemical formula:* HNO_2.

nitrous oxide A colorless, sweet-smelling gas. It is used as a mild anesthetic, often called *laughing gas.* Nitrous oxide occurs naturally in the atmosphere and is a greenhouse gas. *Chemical formula:* N_2O.

NMR Abbreviation of **nuclear magnetic resonance.**

No The symbol for **nobelium.**

nobelium (nō-bĕl′ē-əm) *Symbol* **No** A synthetic, radioactive metallic element in the actinide series that is produced by bombarding curium with carbon ions. Its longest-lived isotope is No 255 with a half-life of 3.1 minutes. Atomic number 102. See **Periodic Table.**

noble gas (nō′bəl) Any of the six gases helium, neon, argon, krypton, xenon, and radon. Because the outermost electron shell of atoms of these gases is full, they do not react chemically with other substances except under certain special conditions. Also called *inert gas.* See **Periodic Table.**

noctilucent (nŏk′tə-lo͞o′sənt) Luminous at night. The term is used especially to describe certain high atmospheric cloud formations visible during summer nights at high latitudes.

nocturnal (nŏk-tûr′nəl) **1.** Occurring at night. **2.** Most active at night. Many animals, such as owls and bats, are nocturnal. **3.** Having flowers that open during the night and close at daylight. Nocturnal plants are often pollinated by moths. Compare **diurnal.**

Noddack (nŏd′ăk′), **Ida Eva Tacke** 1896–1979. German chemist who with her husband,

Walter Karl Friedrich Noddack (1893–1960), discovered rhenium and an element they called masurium (later named technetium) in 1925.

node (nōd) **1.** *Anatomy.* A small mass or lump of body tissue that either occurs naturally, as in the case of lymph nodes, or is a result of disease. **2.** *Botany.* **a.** A point on a stem where a leaf is or has been attached. **b.** A swelling or lump on a tree; a knob or knot. **3.** *Physics.* A point or region of a vibrating or oscillating system, such as the standing wave of a vibrating guitar string, at which the amplitude of the vibration or oscillation is zero. Harmonic frequencies in oscillating systems always have nodes. Compare **antinode. 4.** *Astronomy.* **a.** Either of the two points on the celestial sphere at which the path of a revolving body, such as the Moon, a planet, or a comet, intersects the ecliptic. ▸ The point at which the body traverses from south of the ecliptic to north is the **ascending node.** The opposite point, when the body traverses the ecliptic from north to south, is the **descending node. b.** Either of the two points at which the orbit of an artificial satellite intersects the equatorial plane of the planet it is orbiting. **5.** *Computer Science.* A computer or a peripheral that is connected to a network.

nodule (nŏj′o͞ol) **1.** A small, usually hard mass of tissue in the body. **2.** A small, knoblike outgrowth found on the roots of many legumes, such as alfalfa, beans, and peas. Nodules grow after the roots have been infected with nitrogen-fixing bacteria of the genus *Rhizobium.* See more at **legume. 3.** A small, rounded lump of a mineral or mixture of minerals that is distinct from and usually harder than the surrounding rock or sediment. Nodules often form by replacement of a small part of the rocks in which they form.

Noether (nŭ′tər), **Amalie** Known as **Emmy.** 1882–1935. German mathematician who was a major contributor to the development of modern algebra and geometry. She is best noted for introducing Noether's theorem (1915).

Noether's theorem (nŭ′tərz) A theorem of physics, stating that for every case of invariance in a physics, there exists a unique conservation law, and for every conservation law, there exists a unique invariance.

noise pollution (noiz) Environmental noise, as from vehicles or machinery, that is annoying, distracting, or physically harmful. The physical effects can include hearing loss, tinnitus, stress, and sleeplessness. Noise pollution is usually considered in terms of its effects on human populations, though it is known to affect wildlife as well.

non-Euclidean (nŏn′yo͞o-klĭd′ē-ən) Relating to any of several modern geometries that are based on a set of postulates other than the set proposed by Euclid, especially one in which all of the postulates of Euclidean geometry hold except the **parallel postulate**. Compare **Euclidean.**

non-inertial frame A reference frame in which the observers are undergoing some accelerating force, such as gravity or a mechanical acceleration. Also called *non-inertial frame of reference, non-inertial reference frame, non-inertial system.* Compare **inertial frame.** See also **inertial force, Special Relativity.**

non-REM sleep (nŏn′rĕm′) A period of sleep characterized by decreased metabolic activity, slowed breathing and heart rate, and the absence of dreaming. In humans and certain other animals, the sleep cycle occurs in five stages, the first four consisting of non-REM sleep and the last stage consisting of REM sleep. This cycle repeats itself about five times during a normal episode of sleep. In non-REM sleep, Stage I is characterized by drowsiness, Stage II by light sleep, and Stages III and IV by deep sleep. In adult humans, non-REM sleep accounts for about 75–80 percent of total sleep. Also called *NREM sleep.* Compare **REM sleep.** See more at **sleep.**

nonrenewable (nŏn′rĭ-no͞o′ə-bəl) Relating to a natural resource, such as petroleum or a mineral ore, that cannot be replaced once it has been extracted or procured. Nonrenewable resources that are not significantly altered by their use, including most metals, can often be recovered and their usefulness extended by recycling. Compare **renewable.**

nonsense (nŏn′sĕns′) Relating to a mutation in a structural gene that changes a nucleotide triplet into a stop codon, thus prematurely terminating the polypeptide chain during protein synthesis. See more at **point mutation.**

nonvascular plant (nŏn-văs′kyə-lər) Any of various plants that lack vascular tissue; a bryophyte. Compare **vascular plant.**

nonzero (nŏn-zîr′ō, -zē′rō) Not equal to zero.

noradrenaline (nôr′ə-drĕn′ə-lĭn) See **norepinephrine.**

norepinephrine (nôr′ĕp-ə-nĕf′rĭn) A substance that acts both as a neurotransmitter and hormone, secreted in the central nervous system, at the nerve endings of the sympathetic nervous system, and by the adrenal gland. Norepinephrine is similar to epinephrine in its physiological effects but acts to regulate regular physiologic activity rather than being released in response to stress. Also called *noradrenaline. Chemical formula:* $C_8H_{11}NO_3$.

norite (nôr′īt) **1.** See **gabbro. 2.** A coarse-grained igneous rock, very similar to gabbro but containing orthopyroxene instead of clinopyroxene.

normal distribution (nôr′məl) A theoretical frequency distribution for a set of variable data, usually represented by a bell-shaped curve symmetrical about the mean.

normal fault A geologic fault in which the hanging wall has moved downward relative to the footwall. Normal faults occur where two blocks of rock are pulled apart, as by tension. Compare **reverse fault.** See Note and illustration at **fault.**

Northern Hemisphere (nôr′*th*ərn) **1.** The half of the Earth north of the equator. **2.** The half of the celestial sphere north of the celestial equator.

northern lights See under **aurora.**

North Frigid Zone (nôrth) See **Frigid Zone.**

North Pole The northern end of the Earth's axis of rotation, located at 90° north latitude at a point in the Arctic Ocean. See more at **axis.**

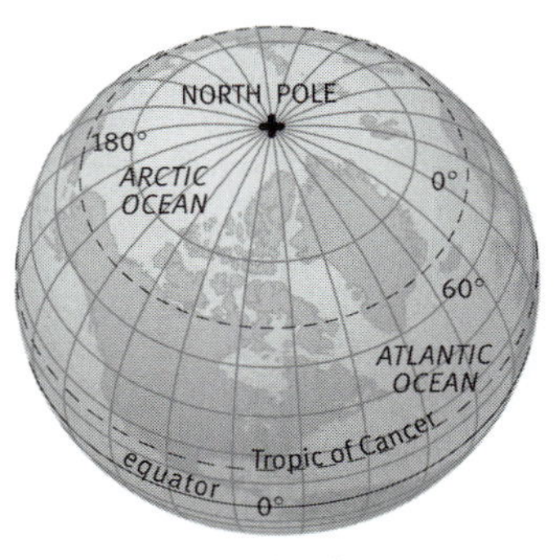

North Pole

North Star See **Polaris.**

North Temperate Zone See **Temperate Zone.**

notochord (nō′tə-kôrd′) A flexible rodlike structure that forms the main support of the body in all chordates during some stage of their development. In vertebrates, the notochord develops into a true backbone in the embryonic phase. Primitive chordates, such as lancelets and tunicates, retain a notochord throughout their lives.

nova (nō′və) Plural **novae** (nō′vē) or **novas.** A white dwarf star that suddenly and temporarily becomes extremely bright as a result of the explosion at its surface of material accreted from an expanding companion star. The material, mostly hydrogen and helium, is attracted by the white dwarf's gravity and accumulates under growing pressure and heat until nuclear fusion is ignited. Unlike a supernova, a nova is not blown apart by the explosion and gradually returns to its original brightness over a period of weeks to years. Because of their sudden appearance where no star had been previously visible, novae were long thought to be new stars. Since 1925, novae have been classified as **variable stars.** Compare **supernova.**

Np The symbol for **neptunium.**

NREM sleep (ĕn′rĕm′) See **non-REM sleep.**

NSAID (ĕn′săd′, -sĕd′) A nonsteroidal anti-inflammatory drug, such as aspirin or ibuprofen.

N-type Made of material, usually a semiconductor such as silicon, that has been doped with impurities so that it has an excess of conductive electrons. Compare **P-type.**

nucellus (no͞o-sĕl′əs) Plural **nucelli** (no͞o-sĕl′ī). The megasporangium of a seed-bearing plant, located in the ovule. In angiosperms, it is the central portion of the ovule in which the embryo sac develops. In some cases of apomixis (production of seeds without fertilization), cells of the nucellus develop into an embryo.

nuclear (no͞o′klē-ər) **1.** Relating to or forming a cell nucleus. **2.** Relating to atomic nuclei. **3.** Using energy derived from the nuclei of atoms through fission or fusion reactions.

nuclear energy **1.** The energy released by the nucleus of an atom as the result of nuclear fission, nuclear fusion, or radioactive decay. The amount of energy released by the nuclear fission of a given mass of uranium is about 2,500,000 times greater than that released by the combustion of an equal mass of carbon. And the amount of energy released by the nuclear fusion of a given mass of deuterium is about 400 times greater that that released by the nuclear fission of an equal mass of uranium. Also called *atomic energy.* **2.** Electricity generated by a nuclear reactor.

nuclear envelope The double-layered membrane enclosing the nucleus of a eukaryotic cell. The nuclear envelope has pores that allow the passage of materials into and out of the nucleus. Also called *nuclear membrane.*

nuclear magnetic resonance The absorption of electromagnetic energy (typically radio waves) by the nuclei of atoms placed in a strong magnetic field. The nuclei of different atoms absorb unique frequencies of radiation depending on their environment, thus by observing which frequencies are absorbed by a sample placed in a strong magnetic field (and later emitted again, when the magnetic field is removed), it is possible to learn much about the sample's makeup and structure. Nuclear magnetic resonance has no known side effects on the human body, and is therefore used to analyze soft body tissues in magnetic resonance imaging (MRI).

nuclear magneton See under **magneton.**

nuclear membrane See **nuclear envelope.**

nuclear physics The scientific study of the structure and behavior of atomic nuclei. See also **neutron, proton, strong force.**

nuclear reaction A process, such as fission, fusion, or radioactive decay, in which the structure of an atomic nucleus is altered through release of energy or mass or by being broken apart. See more at **fission, fusion.**

nuclear reactor A device used to generate power, in which nuclear **fission** takes place as a controlled chain reaction, producing heat energy that is generally used to drive turbines and provide electric power. Nuclear reactors are used as a source of power in large power grids and in submarines.

A CLOSER LOOK **nuclear reactor**

A *nuclear reactor* uses a *nuclear fission chain reaction* to produce energy. The cylindrical core of a reactor consists of *fuel rods* containing pellets of fissionable material, usually uranium 235 or plutonium 239. These unstable isotopes readily split apart into smaller nuclei (in the fission reaction) when they absorb a neutron; they release large quantities of energy upon splitting, along with more neutrons that may be absorbed by the nuclei of other isotopes, causing a chain reaction. The neutrons are expelled from the fission reaction at very high speeds, and are not likely to be absorbed at such speeds. *Moderators* such as heavy water are therefore needed to slow the neutrons to a speed at which they are readily absorbed. The fuel rods contain enough fissionable material arranged in close enough proximity to start a self-sustaining chain reaction. To regulate the speed of the reaction, the fuel rods are interspersed with *control rods* made of a material (usually boron or cadmium) that absorbs some of the neutrons given off by the fuel. The deeper the control rods are inserted into the reactor core, the more the reaction is slowed down. If the control rods are fully inserted, the reaction stops. The chain reaction releases enormous amounts of heat, which is transferred through a closed loop of radioactive water to a separate, nonradioactive water system, creating pressurized steam. The steam drives turbines to turn electrical generators.

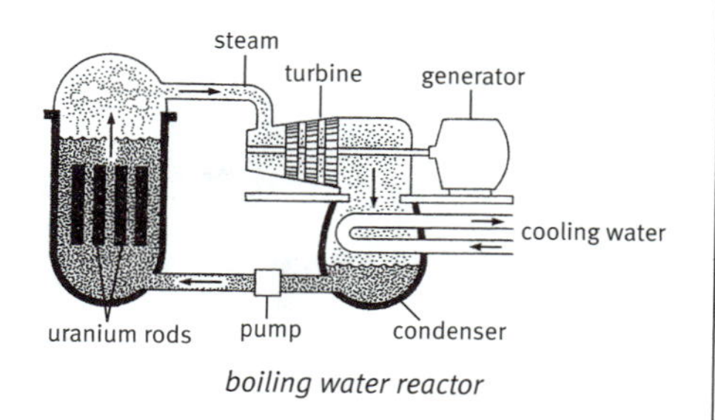

boiling water reactor

nuclear weapon A weapon whose destructive power comes from nuclear energy; an atomic bomb or a hydrogen bomb.

nucleic acid (no͞o-klē′ĭk) Any of a group of very large polymeric nucleotides that constitute the genetic material of living cells and viruses and that code for the amino acid sequences of proteins. Nucleic acids consist of either one or two long chains of repeating units called **nucleotides,** which consist of a nitrogen base (a purine or pyrimidine) attached to a sugar phosphate. The two main nucleic acids are DNA and RNA. In DNA, the nitrogen bases along the length of one chain are linked to complementary bases in the other chain by hydrogen bonds, and both chains coil around each other in a **double helix.** Particular sequences of nucleotides constitute **genes** and encode instructions for sequences of amino acids when proteins are synthesized. In RNA, which is usually single-stranded, complementary bases within the single strand may pair with each other, forming structures other than a double helix. See more at **DNA, RNA.**

nucleolus (no͞o-klē′ə-ləs) Plural **nucleoli** (no͞o-klē′ə-lī′). A small, typically spherical granular body located in the nucleus of a eukaryotic cell, composed largely of protein and RNA. When the cell is not undergoing division, loops of DNA from one or more chromosomes extend into the nucleolus and direct the synthesis of ribosomal RNA and the formation of ribosomes. The ribosomes are eventually transferred out of the nucleus via pores in the nuclear envelope into the cytoplasm.

nucleon (no͞o′klē-ŏn′) A proton or a neutron, especially as part of an atomic nucleus.

nucleonics (no͞o′klē-ŏn′ĭks) **1.** The study of the quantum behavior of atomic nuclei, in particular of the transitions they make between discrete energy levels as they emit and give off radiation. **2.** Development of instruments for use in nuclear research.

nucleon number See **mass number.**

nucleoplasm (no͞o′klē-ə-plăz′əm) The jellylike material within a cell nucleus, containing the nucleolus and chromatin. See more at **nucleus** (sense 2).

nucleoside (no͞o′klē-ə-sīd′) Any of various compounds consisting of a sugar, usually ribose or deoxyribose, and a nitrogen base (a purine or pyrimidine). Nucleosides are constituents of the nucleotides of nucleic acids. Adenosine and thymidine are nucleosides.

nucleoside analogue Any of a group of antiviral drugs that interfere with the activity of the viral enzyme reverse transcriptase and are used in the treatment of retroviral infections, especially HIV.

nucleosome (no͞o′klē-ə-sōm′) Any of the repeating subunits of chromatin in eukaryotic cells, consisting of a DNA chain coiled around a core of histones. See Note at **histone.**

nucleosynthesis (no͞o′klē-ō-sĭn′thĭ-sĭs) The process by which heavier chemical elements are synthesized in the interiors of stars from hydrogen nuclei and other previously synthesized elements. Precisely which elements are involved in nucleosynthesis depends on the age and mass of the star. The most prevalent reaction in smaller stars like our Sun is the fusion of hydrogen into helium by the **proton-proton chain;** in more massive stars this fusion occurs via the **carbon cycle.** When a star is burning hydrogen in its core, it is a main-sequence star. In older stars such as the red giants, nucleosynthesis involves the burning of heavier elements created by earlier fusion; for example, helium may burn via the **triple alpha process.** More massive stars—over eight solar masses—also fuse carbon into neon and magnesium, oxygen into silicon and sulfur, and silicon into iron. The nucleosynthesis of iron is the precursor to the transition into the **supernova** phase.

nucleotide (no͞o′klē-ə-tīd′) Any of a group of organic compounds composed of a nucleoside linked to a phosphate group. Nucleotides are the basic building blocks of nucleic acids.

nucleus (no͞o′klē-əs) Plural **nuclei** (no͞o′klē-ī′). **1.** The positively charged central region of an atom, composed of one or more protons and (for all atoms except hydrogen) one or more neutrons, containing most of the mass of the atom. The **strong force** binds the protons and neutrons, also known as **nucleons,** to each other, overcoming the mutual repulsion of the positively charged protons. In nuclei with many nucleons, however, the forces of repulsion may overcome the strong force, and the nucleus breaks apart in the process of

radioactive decay. The protons and neutrons are arranged in the nucleus in energy levels known as shells analogous to those of the electrons orbiting the nucleus. The number of protons in the nucleus determines the atom's **atomic number** and its position in the Periodic Table. See more at **atom. 2.** An organelle in the cytoplasm of eukaryotic cells (all cells except prokaryotes) that contains nearly all the cell's DNA and controls its metabolism, growth, and reproduction. The nucleus is surrounded by a pair of membranes called the **nuclear envelope,** which can be continuous in places with the membranes of the endoplasmic reticulum. The membranes of the nuclear envelope have interconnected pores that allow the exchange of substances with the cell's cytoplasm. The nuclear DNA is wrapped around proteins (called histones) in strands of **chromatin,** which exists in a matrix known as **nucleoplasm** (analogous to the cytoplasm outside the nucleus). Just prior to cell division, the chromatin condenses into individual chromosomes, which contain the cell's hereditary information. The nucleus also contains at least one spherical **nucleolus**, which mainly contains RNA and proteins, and directs the construction of the cell's ribosomes. See more at **cell. 3a.** The solid central part of a comet, typically several kilometers in diameter and composed of ice, frozen gases, and embedded chunks of rock and dust. It is the permanent part of a comet from which the coma and tail are generated as the comet approaches the Sun. See more at **comet. b.** See **galactic nucleus.**

nuclide (no͞o′klīd′) An atomic nucleus identified by its atomic element and its mass number. For example, a carbon-14 nuclide is the nucleus of a carbon atom, which has six protons, with mass number 14 (thus having eight neutrons). See also **isotope.**

nudibranch (no͞o′də-brăngk′) See **sea slug.**

nuée ardente (no͞o-ā′ ăr-dänt′) Plural **nuées ardentes** (no͞o-āz′ ăr-dänt′). A fast moving gaseous cloud of hot ashes and other material thrown out from an erupting volcano. Nuées ardentes are often incandescent.

null (nŭl) Of or relating to a set having no members or to zero magnitude.

number (nŭm′bər) **1.** A member of the set of positive integers. Each number is one of a series of unique symbols, each of which has exactly one predecessor except the first symbol in the series (1), and none of which are the predecessor of more than one number. **2.** A member of any of the further sets of mathematical objects defined in terms of such numbers, such as negative integers, real numbers, and complex numbers.

numeral (no͞o′mər-əl) A symbol or mark used to represent a number.

numerator (no͞o′mə-rā′tər) A number written above or to the left of the line in a common fraction to indicate the number of parts of the whole. For example, 2 is the numerator in the fraction $\frac{2}{7}$.

numerical taxonomy (no͞o-mĕr′ĭ-kəl) A method of taxonomy in which classification is made on the basis of a multivariate analysis of observable differences and similarities between taxonomic groups. Classifications based on numerical taxonomy reflect degrees of similarity rather than evolutionary relationships.

nut (nŭt) A dry, indehiscent **simple fruit** consisting of one seed surrounded by a hard and thick pericarp (fruit wall). The seed does not adhere to the pericarp but is connected to it by the funiculus. A nut is similar to an **achene** but larger. Acorns, beechnuts, chestnuts, and hazelnuts are true nuts. Informally, other edible seeds or dry fruits enclosed in a hard or leathery shell are also called nuts, though they are not true nuts. For instance, an almond kernel is actually the seed of a **drupe.** Its familiar whitish shell is an endocarp found within the greenish fruit of the almond tree. Peanuts are actually individual seeds from a seed pod called a **legume.**

nutation (no͞o-tā′shən) **1.** A small, cyclic variation of the Earth's axis of rotation with a period of 18.6 years, caused by tidal forces (mostly due to the gravity of the Moon). Nutation is a small and relatively rapid oscillation of the axis superimposed on the larger and much slower oscillation known as **precession.** Although discovered in 1728 by the British astronomer James Bradley (1693–1762), nutation was not explained until two decades later. **2.** A slight curving or circular movement in a stem, as of a twining plant, caused by irregular growth rates of different parts.

nutrient (no͞o′trē-ənt) A substance that provides nourishment for growth or metabolism. Plants absorb nutrients mainly from the soil in the form of minerals and other inorganic compounds, and animals obtain nutrients from ingested foods.

nutrigenomics (no͞o′trə-jə-nō**′**mĭks) *Used with a singular verb.* The study of how the genome and the diet interact to influence human and animal health and disease.

nutrition (no͞o-trĭsh**′**ən) **1.** The process by which living organisms obtain food and use it for growth, metabolism, and repair. The stages of nutrition include ingestion, digestion, absorption, transport, assimilation, and excretion. **2.** The scientific study of food and nourishment, including food composition, dietary guidelines, and the roles that various nutrients have in maintaining health.

nylon (nī**′**lŏn′) Any of various materials made of synthetic polyamides (a type of nitrogen-containing polymer). Nylon is very strong and elastic, and can be formed into fibers, sheets, or bristles. It is used to make fabrics, plastics, and molded products.

nymph (nĭmf) The immature form of those insects that do not pass through a pupal stage. Nymphs usually resemble the adults, but are smaller, lack fully developed wings, and are sexually immature. Compare **imago, larva, pupa.**

O

O The symbol for **oxygen.**

oasis (ō-ā′sĭs) Plural **oases** (ō-ā′sēz). A small area in a desert that has a supply of water and is able to support vegetation. An oasis forms when groundwater lies close enough to the surface to form a spring or to be reached by wells.

oasis

obduction (ŏb-dŭk′shən) A geologic process in which the edge of a tectonic plate consisting of oceanic crust is thrust over the edge of an adjacent plate consisting of continental crust. Compare **subduction.**

obesity (ō-bē′sĭ-tē) The condition of being obese; increased body weight caused by excessive accumulation of fat.

object code (ŏb′jĭkt) The code produced by a compiler from the source code, usually in the form of machine language that a computer can execute directly. It may, however, be in assembly language, an intermediate code that is then translated into machine language. Compare **source code.**

objective (əb-jĕk′tĭv) The lens or mirror in a microscope or other optical instrument that first receives light rays from the object and forms the image.

object-oriented programming A schematic paradigm for computer programming in which the linear concepts of procedures and tasks are replaced by the concepts of objects and messages. An object includes a package of data and a description of the operations that can be performed on that data. A message specifies one of the operations, but unlike a procedure, does not describe how the operation should be carried out. C++ is an example of an object-oriented programming language.

obligate (ŏb′lĭ-gĭt, -gāt′) Capable of existing only in a particular environment or by assuming a particular role. An obligate aerobe, such as certain bacteria, can live only in the presence of oxygen. An obligate parasite cannot survive independently of its host. Compare **facultative.**

observable (əb-zûr′və-bəl) A measurable property of a physical system, such as mass or momentum. In quantum mechanics, observables correspond to mathematical operators used in the calculation of measurable quantities. Operators that do not **commute,** having a nonzero **commutator,** correspond to observables that cannot be precisely measured at the same time, such as momentum and position. See also **uncertainty principle.**

obsessive-compulsive disorder (əb-sĕs′ĭv-kəm-pŭl′sĭv) A psychiatric disorder characterized by the persistent intrusion of repetitive, unwanted thoughts which may be accompanied by compulsive actions, such as handwashing. The individual cannot voluntarily prevent these thoughts or actions, which interfere with normal functioning.

obsidian (ŏb-sĭd′ē-ən) A shiny, usually black, volcanic glass. Obsidian forms above ground from lava that is similar in composition to the magma from which granite forms underground, but cools so quickly that minerals do not have a chance to form within it.

obstetrics (ŏb-stĕt′rĭks) The branch of medicine that deals with the care of women during pregnancy and childbirth.

obtuse angle (ŏb-to͞os′) An angle whose measure is between 90° and 180°. Compare **acute angle.**

Occam's razor or **Ockham's razor** (ŏk′əmz) A rule in science and philosophy stating that entities should not be multiplied needlessly. This rule is interpreted to mean that the simplest of two or more competing theories is preferable and that an explanation for unknown phenomena should first be attempted in terms of what is already known. Occam's razor is named after the deviser of the rule, English philosopher and theologian William of Ockham (1285?–1349?).

occipital lobe (ŏk-sĭp′ĭ-tl) The rearmost lobe of each cerebral hemisphere, containing the main visual centers of the brain.

occlude (ə-klo͞od′) To force air upward from the Earth's surface, as when a cold front overtakes and undercuts a warm front.

occluded front (ə-klo͞o′dĭd) The front formed when a cold front occludes a warm front. Compare **cold front, warm front.**

occlusion (ə-klo͞o′zhən) **1.** An obstruction in a passageway, especially of the body. **2.** The alignment of the upper and lower sets of teeth with each other.

occultation (ŏk′ŭl-tā′shən) The passage of one celestial object in front of another, temporarily blocking the more distant object from view. Occultations can provide information about the existence and measurements of the obscuring object. For example, when an asteroid passes in front of a star, the star is temporarily obscured to an observer on Earth, thus revealing the presence and approximate size of the asteroid. In 1977, astronomers were able to identify the rings around the planet Uranus when the otherwise invisible rings were observed to occult a background star. Occultations have also led to the discovery of more distant objects in space, such as binary stars and extrasolar planets. Compare **transit** (sense 1).

occupational medicine (ŏk′yə-pā′shə-nəl) The branch of medicine that deals with the prevention and treatment of occupational injuries and diseases. ▸ An **occupational disease** is one that is associated with a particular occupation and occurs in the workplace. Some occupations confer specific risks, such as the prevalence of black lung in coal miners.

ocean (ō′shən) **1.** The continuous body of salt water that covers 72 percent of the Earth's surface. The average salinity of ocean water is approximately three percent. The deepest known area of the ocean, at 11,034 m (36,192 ft) is the Mariana Trench, located in the western Pacific Ocean. **2.** Any of the principal divisions of this body of water, including the Atlantic, Pacific, Indian, and Arctic Oceans.

USAGE **ocean/sea**

The word *ocean* refers to one of the Earth's four distinct, large areas of salt water, the Pacific, Atlantic, Indian, and Arctic Oceans. The word can also mean the entire network of water that covers almost three quarters of our planet. It comes from the Greek *Okeanos,* a river believed to circle the globe. The word *sea* can also mean the vast ocean covering most of the world. But it more commonly refers to large landlocked or almost landlocked salty waters smaller than the great oceans, such as the Mediterranean Sea or the Bering Sea. Sailors have long referred to all the world's waters as *the seven seas.* Although the origin of this phrase is not known for certain, many people believe it referred to the Red Sea, the Mediterranean Sea, the Persian Gulf, the Black Sea, the Adriatic Sea, the Caspian Sea, and the Indian Ocean, which were the waters of primary interest to Europeans before Columbus.

oceanic (ō′shē-ăn′ĭk) **1.** Relating to the ocean. **2.** Relating to the ocean waters that lie beyond the continental shelf and exceed 200 m (656 ft) in depth. Compare **neritic.** See more at **epipelagic zone.**

oceanic crust See under **crust.**

oceanography (ō′shə-nŏg′rə-fē) The scientific study of oceans, the life that inhabits them, and their physical characteristics, including the depth and extent of ocean waters, their movement and chemical makeup, and the topography and composition of the ocean floors. Oceanography also includes ocean exploration. Also called *oceanology.*

ocellus (ō-sĕl′əs) Plural **ocelli** (ō-sĕl′ī′). **1.** A small, simple eye or eyespot, found in many

ocellus

on the wing of an owl butterfly (Caligo memnon)

invertebrates. **2.** A marking that resembles an eye, as on the wings of some butterflies.

Ochoa (ō-chō′ə), **Severo** 1905–1993. Spanish-born American geneticist who in 1955 discovered an enzyme that was used in the first synthesis of artificial RNA. For this work he shared with Arthur Kornberg the 1959 Nobel Prize for physiology or medicine.

octagon (ŏk′tə-gŏn′) A polygon having eight sides. *—Adjective* **octagonal** (ŏk-tăg′ə-nəl).

octahedron (ŏk′tə-hē′drən) Plural **octahedrons** or **octahedra.** A polyhedron that has eight faces.

octal (ŏk′təl) Relating to a number system having a base of 8. Each place in an octal number represents a power of 8. Octal notation is used in computer programming because three-digit binary numbers are readily converted into one-digit octal numbers from 0 to 7.

octane (ŏk′tān′) Any of several hydrocarbons having eight carbon atoms connected by single bonds. It is commonly added to gasoline to prevent knocking from uneven burning of fuel in internal-combustion engines. Octane is the eighth member of the alkane series. *Chemical formula:* C_8H_{18}.

octane number A numerical representation of the ability of a fuel to resist knocking when ignited in the cylinder of an internal-combustion engine. The octane number of a given fuel is determined by comparing the amount of knocking it causes, when combusted, with that caused by two standard reference fuels, isooctane (which resists knocking and has an octane number of 100) and heptane (which causes knocking and has an octane number of 0). The octane number is then assigned as the percentage of isooctane required in a blend with normal heptane to match the knocking behavior of the fuel being tested.

ocular (ŏk′yə-lər) *Adjective.* **1.** Of or relating to the eye or the sense of vision. — *Noun.* **2.** The eyepiece of a microscope, telescope, or other optical instrument.

odd (ŏd) Divisible by 2 with a remainder of 1, such as 17 or –103.

odd-toed ungulate See **perissodactyl.**

odometer (ō-dŏm′ĭ-tər) An instrument for indicating the distance traveled by a vehicle, typically by measuring the number of rotations of a wheel or fan whose rate of rotation depends on the speed of the vehicle. Compare **speedometer.**

oersted (ûr′stĕd′) The unit of magnetic field strength in the centimeter-gram-second system. A unit magnetic monopole in a magnetic field with a strength of one oersted would be subjected to a force of one dyne. It is equal to 79.577 amperes per meter.

Oersted, Hans Christian 1777–1851. Danish physicist who is credited as the founder of the science of electromagnetism. Oersted established the connection between electric current and magnetic force when he accidentally discovered that a compass's magnetic needle is deflected at right angles when placed next to a conductor carrying an electric current. The oersted unit of magnetic field strength is named after him.

offset (ôf′sĕt′) A shoot that develops laterally at the base of a plant, often rooting to form a new plant. Many succulents and cacti are propagated by removing offsets and planting them elsewhere. See more at **vegetative reproduction.**

offshore (ôf′shôr′) The relatively flat, irregularly shaped zone that extends outward from the breaker zone to the edge of the continental shelf. The water depth in this area is usually at least 10 m (33 ft). The offshore is continually submerged.

ohm (ōm) The SI derived unit used to measure the electrical resistance of a material or an electrical device. One ohm is equal to the resistance of a conductor through which a current of one ampere flows when a potential difference of one volt is applied to it.

Ohm, Georg Simon 1789–1854. German physicist who discovered the relationship between voltage, current, and resistance in an electrical circuit, now known as Ohm's law. The ohm unit of electrical resistance is named for him.

ohmmeter (ōm′mē′tər) An instrument used for direct measurement of the electrical resistance of a material or electronic component, usually in ohms. Ohmmeters typically use an ammeter to measure current through the material after it has been given some set voltage by the ohmmeter; the direct-current resistance of the sample can then be directly determined through Ohm's law. Compare **voltmeter.**

Ohm's law (ōmz) A law relating the voltage difference between two points, the electric current flowing between them, and the resistance of the path of the current. Mathematically, the law states that V = IR, where V is the voltage difference, I is the current in amperes, and R is the resistance in ohms. For a given voltage, higher resistance entails lower current flow.

-oid A suffix meaning "like" or "resembling," as in *ellipsoid,* a geometric solid that resembles an ellipse.

oil (oil) Any of a large class of viscous liquids that are typically very slippery and greasy. Oils are composed mostly of glycerides. They are flammable, do not mix with water, and include animal and vegetable fats as well as substances of mineral or synthetic origin. They are used in food, soap, and candles, and make good lubricants and fuels. See **essential oil, mineral oil, petroleum.**

oil field An area with reserves of recoverable petroleum, especially one with several oil-producing wells.

oil sand A stratum of sand or sandstone containing petroleum that can be extracted by way of wells. The term is also applied to limestone and dolomite strata that contain petroleum.

oil shale A fine-grained black or dark brown shale containing hydrocarbons in the form of kerogen. When heated, oil shale yields petroleum or natural gas.

-ol A suffix used to form the names of chemical compounds having a hydroxyl (OH) group, such as *ethanol.*

Oldowan (ōl′də-wən, ôl′-) Relating to the earliest recognized stage of Paleolithic tool culture, dating from around 2.5 to 1.5 million years ago and characterized by crude cores of quartz or basalt from which flakes were removed with blows from a hammerstone. Both the flaked cores and the flakes themselves were probably used as tools for such tasks as chopping, cutting, and scraping. Oldowan tools are associated with early *Homo habilis* sites at Olduvai Gorge, in Tanzania, and other East African locations; they may also have been made by late australopithecines. Oldowan tools show little change during the million years they were in use, and were gradually replaced by the Acheulian tools associated with *Homo erectus.*

Old Stone Age (ōld) See **Paleolithic.**

olefin (ō′lə-fĭn) See **alkene.**

oleic acid (ō-lē′ĭk) An oily liquid occurring in animal and vegetable oils and used in making soap. *Chemical formula:* $C_{18}H_{34}O_2$.

olfactory (ŏl-făk′tə-rē, ōl-) Relating to or involving the organs or sense of smell.

olfactory nerve Either of the first pair of cranial nerves that carries sensory information relating to smell from the nose to the brain.

Oligocene (ŏl′ĭ-gō-sēn′) The third epoch of the Tertiary Period, from about 37 to 24 million years ago. During this time there was an increase in volcanic activity, and Australia and South America separated from Antarctica. The climate started to cool and a glacier started to form in Antarctica. Modern mammalian groups continued to develop, and the first cats, dogs, horses, and related mammals appeared. Artiodactyls (even-toed ungulates) took over from the perissodactyls (uneven-toed ungulates) as the dominant medium-sized herbivores. Many types of grass also first appeared at this time. See Chart at **geologic time.**

oligochaete or **oligochete** (ŏl′ĭ-gō-kēt′, ō′lĭ-) Any of various annelid worms of the class Oligochaeta. Oligochaetes, unlike polychaetes, have relatively few bristles (called setae) along the body, and often have a thickened, ringlike region (called a clitellum) that secretes a substance used for enclosing eggs in a cocoon. Oligochaetes include the earthworms and a few small freshwater forms. Compare **polychaete.**

oligoclase (ŏl′ĭ-gō-klās′, ō′lĭ-) A white to gray triclinic mineral that is of the plagioclase feldspar group. Oligoclase occurs in igneous rocks that have a relatively high silica content, for example, granite. *Chemical formula:* $(Na,Ca)(Al,Si)AlSi_2O_8$.

oligomer (ə-lĭg′ə-mər) A molecule that consists of a relatively small and specifiable number of monomers (usually less than five). Unlike a polymer, if one of the monomers is removed from an oligomer, its chemical properties are altered.

oligosaccharide (ŏl′ĭ-gō-săk′ə-rīd′, ō′lĭ-) A carbohydrate consisting of a relatively small and specifiable number of monosaccharides joined together. Lactose, maltose, and sucrose are oligosaccharides consisting of two simple sugars. Raffinose is an oligosaccharide consisting of three simple sugars. Compare **monosaccharide, polysaccharide.**

oligotrophic (ŏl′ĭ-gō-trō′fĭk, -trŏf′ĭk, ō′lĭ-) Lacking in plant nutrients such as phosphates, nitrates, and organic matter, and consequently having few plants and a large amount of dissolved oxygen throughout. Used of a lake, pond, or stream. Compare **dystrophic, eutrophic.**

olivine (ŏl′ə-vēn′) An olive-green to brownish-green orthorhombic mineral. Olivine is a common mineral in the igneous rocks, such as basalt and gabbro, that make up most of the Earth's crust beneath the oceans. *Chemical formula:* $(Mg,Fe)_2SiO_4$. ► Olivine in which the mafic component consists entirely of magnesium is called **forsterite**. *Chemical formula:* Mg_2SiO_4. ► Olivine in which the mafic component consists entirely of iron is called **fayalite**. *Chemical formula:* Fe_2SiO_4.

-oma A suffix meaning "tumor" or "cancer," as in *carcinoma.* Often, the suffix is added to the name of the affected body part, as in *lymphoma,* cancer of the lymph tissue.

omasum (ō-mā′səm) Plural **omasa.** The third division of the stomach in ruminant animals. It removes excess water from food and further reduces the size of food particles before passing them to the abomasum for digestion by enzymes. See more at **ruminant.**

omega (ō-mĕg′ə, ō-mē′gə, ō-mā′-) **1.** An omega baryon. **2.** An omega meson.

omega-3 fatty acid Any of several polyunsaturated fatty acids found in leafy green vegetables, vegetable oils, and cold-water fish such as salmon and mackerel. These acids are capable of reducing serum cholesterol levels and have anticoagulant properties. Omega-3 fatty acids are chemically characterized by having a double bond three carbon atoms away from one end of their carbon chain.

omega baryon A subatomic particle in the baryon family, consisting of three strange quarks. It has a mass 3,272 times that of the electron, a negative electric charge, and an average lifetime of 8×10^{-11} seconds. The 1964 experimental observation of this particle was a great triumph in the history of particle physics, because its existence, mass, and method of decay had been predicted based on the model of baryons, but it had not yet been isolated. See Table at **subatomic particle.**

omega-c baryon An electrically neutral baryon having a mass 5,292 times that of the electron and an average lifetime of approximately 6.4×10^{-14} seconds. See Table at **subatomic particle.**

omega meson A neutral meson having a mass 1,532 times that of the electron and an average lifetime of 6.6×10^{-23} seconds. See Table at **subatomic particle.**

ommatidium (ŏm′ə-tĭd′ē-əm) Plural **ommatidia.** One of the tiny light-sensitive parts of the compound eye of insects and other arthropods. An ommatidium resembles a single simplified eye. See more at **compound eye.**

omnivore (ŏm′nə-vôr′) An organism that eats both plants and animals. —*Adjective* **omnivorous.**

oncogene (ŏn′kə-jēn) A gene that causes normal cells to become cancerous either because the gene is mutated or because the gene is expressed at the wrong time in development. See Note at **cancer.**

oncolite (ŏn′kə-līt′) A stromatolite that is spherical rather than dome-shaped. Oncolites usually form around a nucleus consisting of a detrital fragment (such as a shell fragment or a grain of sand) and are less than 10 cm (3.9 inches) in diameter.

oncology (ŏn-kŏl′ə-jē) The branch of medicine that deals with the diagnosis and treatment of cancer.

-one A suffix used to form the names of chemical compounds containing an oxygen atom attached to a carbon atom, such as *acetone.*

online (ŏn′līn′) Connected to or accessible by means of a computer or computer network.

ontogeny (ŏn-tŏj′ə-nē) The origin and development of an individual organism from embryo to adult.

onyx (ŏn′ĭks) A type of chalcedony that occurs in straight and parallel bands of different colors, often black and white.

oocyte (ō′ə-sīt′) A diploid cell that undergoes meiosis to form eggs.

oogamy (ō-ŏg′ə-mē) A system of sexual reproduction in which one gamete (called the egg) is large and nonmotile, while the other (called the sperm) is small and motile. Oogamy is a type of **heterogamy.** Compare **isogamy.** —*Adjective* **oogamous.**

oogenesis (ō′ə-jĕn′ĭ-sĭs) The formation, development, and maturation of an ovum or egg cell. —*Adjective* **oogenetic** (ō′ə-jə-nĕt′ĭk).

A CLOSER LOOK **oogenesis**

The details of the exact nature of *oogenesis* vary by species, since the females of some species produce thousands of eggs at a time, while in others, females produce relatively few mature eggs. The human female, for example, ovulates only about 400 times during her lifetime. Oogenesis in humans begins in embryonic and fetal development, when diploid germ cells called oogonia divide by mitosis to produce cells called primary oocytes. The primary oocytes of the female fetus enlarge and begin to undergo meiosis. But they are suspended in an early phase of meiosis called *prophase* until the female reaches puberty. The human female has about 700,000 such primary oocytes at birth. After puberty, one of the oocytes resumes development each month in response to changes initiated by follicle-stimulating hormone. The primary oocyte undergoes the first meiotic division, producing a cell called a secondary oocyte and another called the first polar body. During cytokinesis, most of the cytoplasm of the primary oocyte moves to the secondary oocyte. The first polar body undergoes the second meiotic division and its daughter cells degenerate. The secondary oocyte is released from the ovary during ovulation. If it encounters a spermatozoon and fertilization is initiated, the secondary oocyte undergoes the second meiotic division, producing the ovum and a second polar body that degenerates. The spermatozoon then fertilizes the ovum. If the secondary oocyte does not encounter a spermatozoon, it does not undergo the second meiotic division and simply degenerates.

oogonium (ō′ə-gō′nē-əm) Plural **oogonia** or **oogoniums. 1.** A female reproductive structure in certain algae and fungi. It is usually a rounded cell or sac containing one or more oospheres. **2.** A cell that arises from a primordial germ cell and differentiates into an oocyte in the ovary of female animals.

oolite (ō′ə-līt′) A sedimentary rock made of ooliths that are cemented together by calcium carbonate.

oolith (ō′ə-lĭt′) A small, round grain consisting of calcium carbonate, silica, or dolomite. Ooliths have concentric layers that form around a nucleus, such as a shell fragment, a sand grain, or a pellet of alga. They typically have diameters of 0.25 to 2 mm. Ooliths usually form by inorganic precipitation.

oomycete (ō′ə-mī′sēt′) Any of various nonphotosynthetic protists belonging to the phylum Oomycota and living in marine, freshwater, and soil environments. Oomycetes have cell walls made of cellulose or similar substances, and store their food as glycogen. They reproduce asexually by the formation of diploid spores (called zoospores) with two flagellae. Sexual reproduction involves the formation of a number of eggs within a structure called an oogonium. The eggs are fertilized by antheridia that penetrate the oogonium and deliver nuclei to the eggs. The oomycetes were formerly classified among the fungi because their filamentous bodies resembled fungal hyphae. The phylum includes several species that cause important plant diseases, such as **late blight.**

Oort cloud (ôrt) A sphere-shaped mass of more than 100 billion comets that makes up the outer edge of the solar system, surrounding the Kuiper belt and the planets. Some comets from this area are drawn into the inner solar system by passing stars and other forces and take more than 200 years to make one complete orbit of the Sun. Compare **Kuiper belt.**

oosphere (ō′ə-sfîr′) A large nonmotile female gamete formed in an oogonium, such as the eggs formed in the oogonium of an oomycete.

oospore (ō′ə-spôr′) A fertilized female cell or zygote, especially one with thick chitinous walls, developed from a fertilized oosphere.

opal (ō′pəl) A usually transparent mineral consisting of hydrous silica. Opal can occur in almost any color, but it is often pinkish white with a milky or pearly appearance. It typically forms within cracks in igneous rocks, in limestones, and in mineral veins. It also occurs in the silica-rich shells of certain marine organisms. *Chemical formula:* **$SiO_2 \cdot nH_2O$.**

opalescent (ō′pə-lĕs**′**ənt) Exhibiting a milky iridescence like that of opal.

opaque (ō-pāk**′**) Resistant to the transmission of certain kinds of radiation, usually light. Metals and many minerals are opaque to light, while being transparent to radio waves and neutrinos. Compare **translucent, transparent.**

open circuit (ō**′**pən) An electric circuit in which the normal path of current has been interrupted, either by the disconnection of one part of its conducting pathway from another, or by the intervention of an electric component, such as a transistor. Compare **closed circuit.**

open chain An arrangement of atoms of the same type that does not form a ring. Aliphatic compounds, such as ethane and isopropyl alcohol, are open-chain compounds.

open cluster A loose, irregular grouping of stars that originated from a single nebula in the arms of a spiral galaxy. Compared to globular clusters, open clusters generally contain younger and fewer (from a hundred to several thousand) stars and are confined to the disk of the galaxy. Because they are young, open clusters are sometimes still surrounded by the leftover gas and dust from which they formed. Visible from Earth with just a pair of binoculars and containing over 3,000 stars, the Pleiades is the best known open cluster. Also called *galactic cluster.* Compare **globular cluster.**

open-heart surgery Surgery in which the thoracic cavity is opened to expose the heart and the blood is recirculated and oxygenated by a heart-lung machine.

open interval A set of numbers consisting of all the numbers between a pair of given numbers but not including the endpoints.

open-source Relating to source code that is available to the public without charge. Open-source code is often enhanced, improved, and adapted for specific purposes by interested programmers, with the revised versions of the code are made available to the public. For example, most of the code in the Linux operating system is open-source.

open system A physical system that interacts with other systems. The physical description of an open system can appear to violate conservation laws; for example, in a good description of the mechanism of energy transfer in a car engine (gears, driveshaft, and so on), energy will appear to be lost from the system over time, despite the law of conservation of energy. This is because the system is open, losing energy (in the form of heat) to surrounding systems (through friction). A system that loses energy in this way also called a **dissipative system.** Compare **closed system.**

open universe Model of the universe in which the curvature of space is flat or curved away from itself, entailing that the size of the universe is infinite. According to this model, gravity between objects is not able to stop or reverse the expansion of the universe, thus objects continue to move farther and farther apart as space moves outward. An object moving in a straight line in an open universe would never return to its starting point. According to current cosmological theories, the universe is open if it is insufficiently dense. Such a universe will never end, but will eventually become very cold and dark because stars gradually lose all of their energy. Compare **closed universe.** See Note at **big bang.**

operating system (ŏp**′**ə-rā′tĭng) Software designed to handle basic elements of computer operation, such as sending instructions to hardware devices like disk drives and computer screens, and allocating system resources such as memory to different software applications being run. Given uniformly designed operating systems that run on many different computers, developers of software do not need to concern themselves with these problems, and are provided with a standard platform for new programs.

operation (ŏp′ə-rā**′**shən) **1.** *Medicine.* A surgical procedure for remedying an injury, ailment, defect, or dysfunction. **2.** *Mathematics.* A process or action, such as addition, substitution, transposition, or differentiation, performed in a specified sequence and in accordance with specific rules. **3.** A logical operation. **4.** *Computer Science.* An action resulting from a single instruction.

operator (ŏp**′**ə-rā′tər) **1.** *Mathematics.* A function, especially one from a set to itself, such as differentiation of a differentiable function or rotation of a vector. In quantum mechanics, measurable quantities of a physical system, such as position and momentum, are related to unique operators applied to the wave equation describing the system. **2.** A logical operator. **3.** *Genetics.* A segment of chromosomal DNA that regulates the activity of the structural genes of an operon by interacting with a specific repressor.

operculum (ō-pûr′kyə-ləm) Plural **opercula** or **operculums.** A lid or flap covering an opening, such as the gill cover in some fish or the horny flap covering the opening of a snail.

operon (ŏp′ə-rŏn′) A sequence of genetic material that functions in a coordinated manner, consisting of an operator, a promoter, and one or more structural genes that are transcribed together. Operons were first found in prokaryotes.

ophidian (ō-fĭd′ē-ən) A member of the suborder Ophidia or Serpentes; a snake.

ophiolite suite (ŏf′ē-ə-līt′, ō′fē-) A sequence of rocks consisting of deep-sea marine sediments overlying (from top to bottom) pillow basalts, sheeted dikes, gabbro, dunite, and peridotite. Ophiolite sequences are indicators of sea-floor spreading where the oceanic crust is thinned by the forces of plate tectonics, and magma rises to the sea's floor and generates new crust. As lava spills out into the cold ocean water it forms pillow basalts; magma that cools before reaching the surface forms sheeted dikes (dikes that appear to be vertically stratified because of the way the magma cools). Magma cooling deeper in the crust forms gabbro, dunite, and peridotite.

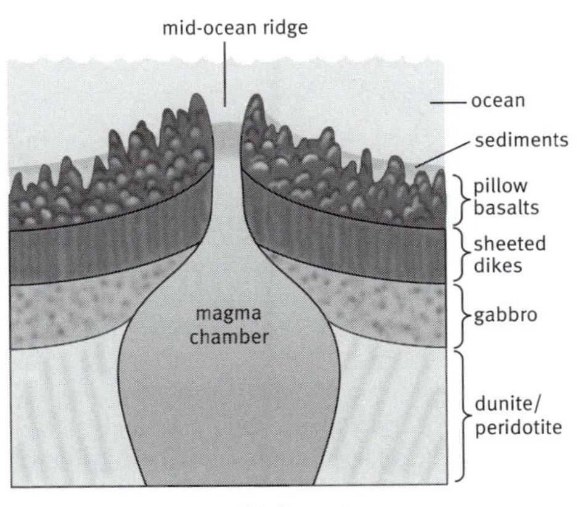

ophiolite suite

ophthalmology (ŏf′thəl-mŏl′ə-jē, ŏp′-) The scientific study of the eye, its diseases, and their treatment.

opium (ō′pē-əm) A highly addictive, yellowish-brown drug obtained from the pods of a variety of poppy, from which other drugs, such as morphine, are prepared.

Oppenheimer (ŏp′ən-hī′mər), **J(ulius) Robert** 1904–1967. American physicist who directed the Los Alamos, New Mexico, laboratory during the development of the first atomic bomb (1942–1945). After World War II, he became an advocate for the peaceful use of atomic energy and opposed the development of the hydrogen bomb.

J. Robert Oppenheimer

opportunistic infection (ŏp′ər-to͞o-nĭs′tĭk) An infection by a microorganism that normally does not cause disease but does so when lowered resistance to infection is caused by the impairment of the body's immune system.

opposable thumb (ə-pō′zə-bəl) A thumb that can be placed opposite the fingers of the same hand. Opposable thumbs allow the digits to grasp and handle objects and are characteristic of primates.

opposite (ŏp′ə-zĭt) Arranged as one of a pair on either side of a stem or twig. Maple and ash

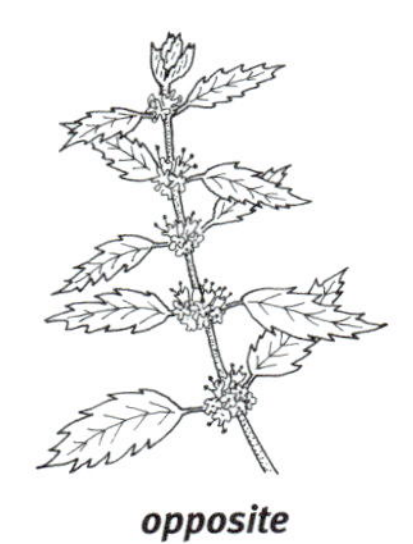

opposite

detail of a field mint (Menta arvensis) *plant*

trees have opposite leaves. Compare **alternate.**

opposition (ŏp′ə-zĭsh′ən) **1.** A characteristic movement of the primate thumb, in which the pad of the thumb can be placed in contact with the pads of the fingers of the same hand. **2.** The position of two celestial bodies when their celestial longitude differs by 180°, especially a configuration in which Earth lies on a straight line between the Sun and a superior planet or the Moon. Planets in this position rise as the Sun sets and are visible all night long, reaching their highest point in the sky at midnight; the Moon in this position is full. Compare **conjunction.** See more at **elongation.**

optic (ŏp′tĭk) Relating to or involving the eye or vision.

optical (ŏp′tĭ-kəl) **1.** Relating to vision or the eyes. **2.** Relating to optics. **3.** Relating to or using visible light.

optical binary See under **binary star.**

optical character recognition The electronic identification and digital encoding of printed or handwritten characters by means of an optical scanner and specialized software.

optical disk or **optical disc** A plastic-coated disk that stores digital data, such as music or text, as tiny pits on its surface. The disc is rotated rapidly, and a laser is reflected off the disc to an optical sensor, which is sampled periodically. If the laser strikes the flat area between pits, the sensor picks up the reflection and registers the digital value 1. If the laser strikes one of the pits, it is deflected from the sensor, and the sample value is a digital 0.

optical fiber A flexible transparent fiber of extremely pure glass or plastic, generally between 10 and 200 micrometers in diameter, used especially to carry light signals for telecommunication purposes. See more at **fiber optics.**

optical isomer See **enantiomer.**

optical maser A laser that produces visible radiation.

optical microscope See under **microscope.**

optical resonator A part of a laser, consisting of two mirrors, one highly reflective and one partly reflective, placed on either side of a laser pump. Amplified light bounces back and forth between the mirrors, enhancing stimulated emission within the pump, eventually being emitted through the partly reflective mirror. See more at **laser.**

optical scanner A device that converts printed images and text into digital information that can be edited, transmitted, and stored. Optical scanners work by electronically measuring the intensity of color at a large number of individual locations across the page (often using **phototransistors**), and converting these measurements into digital numerical values usable by computers and other digital devices. See also **A/D converter.**

optic nerve Either of the second pair of cranial nerves, which carry sensory information relating to vision from the retina of the eye to the brain. Disease or injury of the optic nerve can result in partial or total blindness.

optics (ŏp′tĭks) The scientific study of light and vision. The study of optics led to the development of more general theories of electromagnetic radiation and theories of color.

oral (ôr′əl) Relating to or involving the mouth.

orbit (ôr′bĭt) *Noun.* **1a.** The path followed by a celestial body or artificial satellite as it revolves around another body due to the force of gravity. Orbits are nearly elliptical or circular in shape and are very closely approximated by **Kepler's laws of planetary motion. b.** One complete revolution of such a body. See Note at **solar system. 2.** A stable quantum state of an electron (or other particle) in motion around an atomic nucleus. See more at **orbital. 3.** Either of two bony hollows in the skull containing the eye and its associated structures. *—Verb.* **4.** To move in an orbit around another body. **5.** To put into an orbit, as a satellite is put into orbit around Earth.

orbital (ôr′bĭ-tl) A partial description of the quantum state of an electron (or other particle) orbiting the nucleus of an atom. Different orbitals have different shapes and orientations, depending on the energy of the electron, its angular momentum, and its magnetic number. Orbitals have no clear boundaries; the shape of an orbital, as depicted graphically, shows only the regions around the nucleus in which an electron has a relatively high probability of being found. No more than two electrons (each with opposite spin) can coexist in a single orbital because of the Pauli exclusion principle. See also **quantum number, shell** (sense 2).

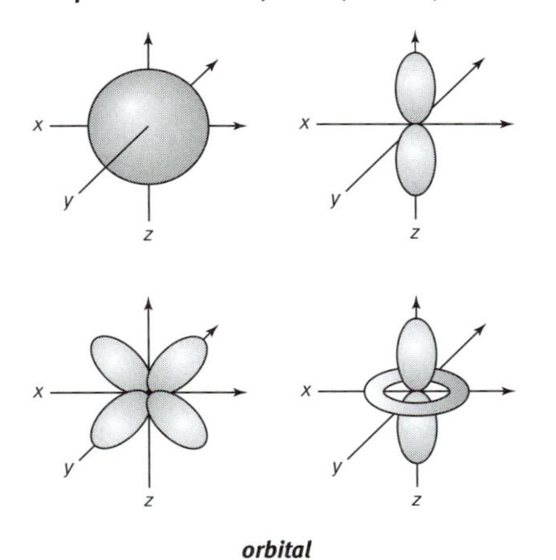

orbital

The shape of atomic orbitals is governed by several factors, the most important of which are the quantum numbers which individually characterize an electron and collectively specify its state in an atom. In a hydrogen atom, for example, the electron is most likely to be found within a sphere around the nucleus. In atoms with higher energy states, the space in which the electrons are most likely to be found are represented by more complex shapes.

orbital quantum number See under **quantum number.**

order (ôr**′**dər) A group of organisms ranking above a family and below a class. See Table at **taxonomy.**

ordinal number (ôr**′**dn-əl) A number, such as 3rd, 11th, or 412th, used in counting to indicate position in a series but not quantity. Compare **cardinal number.**

ordinate (ôr**′**dn-ĭt) The distance of a point from the x-axis on a graph in the Cartesian coordinate system. It is measured parallel to the y-axis. For example, a point having coordinates (2,3) has 3 as its ordinate. Compare **abscissa.**

Ordovician (ôr′də-vĭsh**′**ən) The second period of the Paleozoic Era, from about 505 to 438 million years ago. During this time most of the Earth's landmasses were gathered in the supercontinent **Gondwanaland,** located in the Southern Hemisphere. Much of this continent was submerged under shallow seas, and marine invertebrates, including trilobites, brachiopods, graptolites, and conodonts were widespread. The first primitive fishes appeared; some evidence suggests the first land plants may also have appeared at this time. By the end of the Ordovician massive glaciers formed on Gondwanaland, causing sea levels to drop and approximately 60 percent of all known marine invertebrates to become extinct. See Chart at **geologic time.**

ore (ôr) A naturally occurring mineral or rock from which a valuable or useful substance, especially a metal, can be extracted at a reasonable cost.

organ (ôr**′**gən) A distinct part of an organism that performs one or more specialized functions. Examples of organs are the eyes, ears, lungs, and heart of an animal, and the roots, stems, and leaves of a plant.

organelle (ôr′gə-nĕl**′**) A structure or part that is enclosed within its own membrane inside a cell and has a particular function. Organelles are found only in eukaryotic cells and are absent from the cells of prokaryotes such as bacteria. The nucleus, the mitochondrion, the chloroplast, the Golgi apparatus, the lysosome, and the endoplasmic reticulum are all examples of organelles. Some organelles, such as mitochondria and chloroplasts, have their own genome (genetic material) separate from that found in the nucleus of the cell. Such organelles are thought to have their evolutionary origin in symbiotic bacteria or other organisms that have become a permanent part of the cell.

organic (ôr-găn**′**ĭk) **1.** Involving organisms or the products of their life processes. **2.** Relating to chemical compounds containing carbon, especially hydrocarbons. See Table on page 448. **3.** Using or produced with fertilizers or pesticides that are strictly of animal or vegetable origin. **4.** Relating to or affecting organs or an organ of the body. An organic disease is one in which there is a demonstrable abnormality on physical examination, laboratory testing, or other diagnostic studies. Compare **functional.**

organic chemistry The branch of chemistry that deals with carbon and organic compounds, especially hydrocarbons.

organism (ôr**′**gə-nĭz′əm) An individual form of life that is capable of growing, metabolizing nutrients, and usually reproducing. Organ-

Organic Compounds

Chemical compounds containing one or more carbon atoms are called **organic compounds.** Hundreds of thousands of organic compounds are found in nature or have been artificially made. They range from the very simple, like methane with its five atoms, to the very complex, like DNA, which has millions of atoms.

A very common class of organic compounds, called the **alkanes,** all have one or more carbon atoms arranged in a row, or **chain.** In an alkane, each carbon atom (C) is bonded to a neighboring carbon by sharing a single electron with it; this is called a **single bond.** A carbon atom must share a total of four electrons with other atoms; the electrons that are not shared with other carbon atoms are shared with hydrogen atoms (H) (hence the generic name **hydrocarbon** for alkanes and many other organic compounds). The diagrams below illustrate the first four compounds in the alkane series—***methane, ethane, propane,*** and ***butane.*** The single lines between the element symbols represent single bonds.

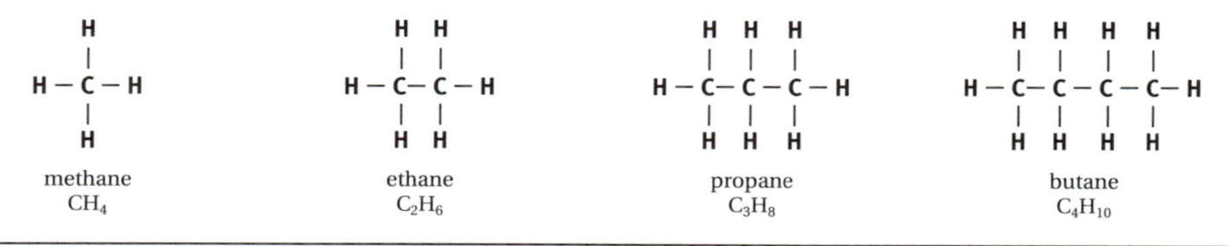

methane CH_4 ethane C_2H_6 propane C_3H_8 butane C_4H_{10}

A large number of compounds can be created by replacing one or more of the hydrogen atoms of an alkane with another atom or group of atoms. For example, alcohols are formed by replacing a hydrogen atom with a group consisting of oxygen and hydrogen (OH), called a hydroxy group. Replacing a hydrogen atom in ethane with a hydroxy group produces ***ethanol,*** the most common form of alcohol. A further modification, in which an additional two hydrogen atoms are replaced with an oxygen atom, produces ***acetic acid,*** the main acid in vinegar (see the diagrams below). The double line in the second diagram indicates a bond where two electrons are shared, called a **double bond.**

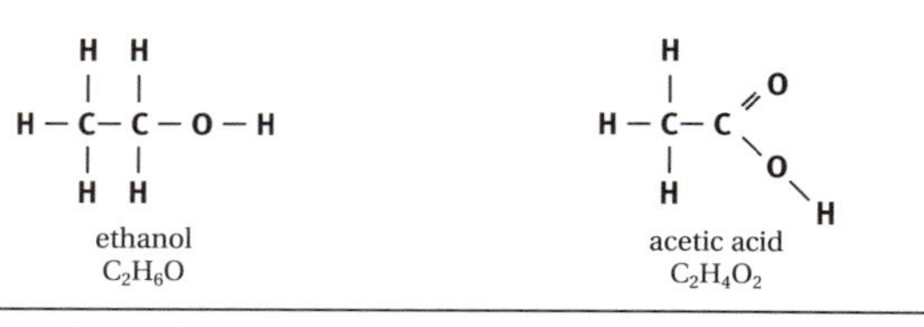

ethanol C_2H_6O acetic acid $C_2H_4O_2$

Other families of organic compounds have double or, more rarely, triple bonds between the carbons in a chain. The simplest such molecules are ***ethylene*** and ***acetylene,*** illustrated below. The **triple bond** in acetylene, where three electrons are shared between the atoms, is indicated by a triple line. Longer chains may have alternating single and double bonds, as in ***butadiene*** below, a compound used to make synthetic rubber.

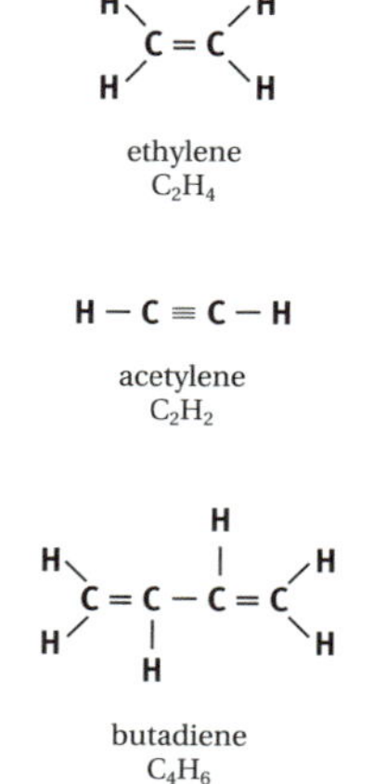

ethylene C_2H_4

acetylene C_2H_2

butadiene C_4H_6

Very often, chains of carbon atoms loop to form ***rings.*** The basic ring compound in organic chemistry is ***benzene,*** which has six carbon atoms joined to each other by alternating single and double bonds, with each carbon further joined to a hydrogen atom.

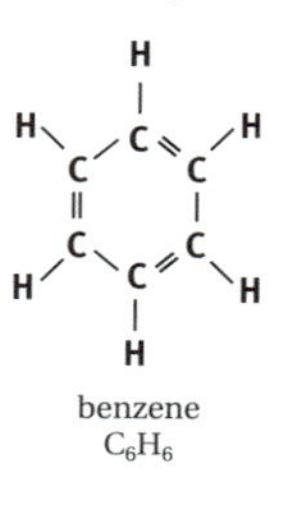

benzene C_6H_6

isms can be unicellular or multicellular. They are scientifically divided into five different groups (called kingdoms) that include prokaryotes, protists, fungi, plants, and animals, and that are further subdivided based on common ancestry and homology of anatomic and molecular structures.

organ of Corti (kôr′tē) A spiral-shaped organ on the inner surface of the cochlea containing sensory receptors called **hair cells** that convert sound vibrations into nerve impulses. The organ of Corti is named after its discoverer, Italian anatomist Alfonso Corti (1822–1888).

orgasm (ôr′găz′əm) The peak of sexual excitement, characterized by strong feelings of pleasure and by a series of involuntary contractions of the muscles of the genitals, usually accompanied by the ejaculation of semen by the male.

origin (ôr′ə-jĭn) The point at which the axes of a Cartesian coordinate system intersect. The coordinates of the origin are (0,0) in two dimensions and (0,0,0) in three dimensions.

Orion (ō-rī′ən) A constellation in the equatorial region of the celestial sphere, near Taurus and Gemini. Orion (the Hunter) contains the bright stars Betelgeuse and Rigel.

ornithine (ôr′nə-thēn′) An amino acid not found in proteins, formed by hydrolyzing arginine and important in the formation of urea. *Chemical formula:* $C_5H_{12}N_2O_2$.

ornithischian (ôr′nə-thĭs′kē-ən) Any of various dinosaurs belonging to the group Ornithischia, one of the two main divisions of dinosaurs. Ornithischians had a pelvis similar to that of modern birds (although birds are not descended from them), in which part of the pubis pointed backwards and parallel to the ischium. They also had a special bone in front of the lower jaw (called the *predentary bone*), and frequently had armor plating or bony outgrowths of the skull. Ornithischians comprise all the herbivorous dinosaurs except the sauropods, and include ankylosaurs, hadrosaurs, stegosaurs, and ceratopsians. Compare **saurischian.**

ornithology (ôr′nə-thŏl′ə-jē) The scientific study of birds.

ornithophilous (ôr′nə-thŏf′ə-ləs) Pollinated by birds.

ornithopod (ôr-nĭth′ə-pŏd′) One of the main types of ornithischian dinosaurs, including the hadrosaurs. Ornithopods walked on their hind legs and had three blunt toes on each foot. They lived from the late Triassic Period until the end of the Cretaceous Period.

orogeny (ô-rŏj′ə-nē) also **orogenesis** (ôr′ə-jĕn′ĭ-sĭs) The process of mountain formation, especially by folding and faulting of the Earth's crust and by plastic folding, metamorphism, and the intrusion of magmas in the lower parts of the lithosphere. Unlike **epeirogeny,** orogeny usually affects smaller regions and is associated with evidence of folding and faulting. The long chains of mountains often seen on the edges of continents form through orogeny.

orthoclase (ôr′thə-klās′) A white to yellowish-red monoclinic mineral of the potassium feldspar group that forms from medium- to low-temperature magmas. *Chemical formula:* $KAlSi_3O_8$.

orthogonal (ôr-thŏg′ə-nəl) **1.** Relating to or composed of right angles. **2.** Relating to a matrix whose transpose equals its inverse. **3.** Relating to a linear transformation that preserves the length of vectors.

orthopedics (ôr′thə-pē′dĭks) The branch of medicine that deals with the treatment of disorders or injuries of the bones, joints, and associated muscles.

orthopyroxene (ôr′thə-pī-rŏk′sēn′) Any variety of the mineral pyroxene that crystallizes in the orthorhombic system and contains no calcium and little or no aluminum. Enstatite is an orthopyroxene.

orthorhombic (ôr′thō-rŏm′bĭk) Relating to a crystal having three axes of different lengths intersecting at right angles. The mineral topaz has orthorhombic crystals. See illustration at **crystal.**

Os The symbol for **osmium.**

oscillating universe (ŏs′ə-lā′tĭng) A model of a closed universe in which the expansion of the universe slows and reverses, thereby causing a collapse into a **singularity,** which then expands into a new universe, repeating the cycle. Such a model implies that our universe is just one of a long line of universes. Scientists have not been able to develop a theory of how a collapsed universe could expand into a new one.

oscillation (ŏs′ə-lā′shən) **1.** A repeating fluctuation in a physical object or quantity. See also **attractor, harmonic motion. 2.** A single cycle of such fluctuation.

oscilloscope (ə-sĭl′ə-skōp′) An electronic instrument used to observe and measure changing electrical signals. The amplitude of the signal as it varies with time is displayed graphically on a screen as a line stretching from left to right, with displacements up and down indicating the amplitude of the signal. Oscilloscopes are used to diagnose problems in electronic signal-processing devises, such as computers or stereos, and to monitor electrical activity in the body, such as that of heartbeats.

–ose A suffix used to form the chemical names of carbohydrates, such as *glucose.*

–osis A suffix that means: **1.** Diseased condition, as in *tuberculosis.* **2.** Condition or process, as in *osmosis.*

osmium (ŏz**′**mē-əm) *Symbol* **Os** A hard, brittle, bluish-white metallic element that is the densest naturally occurring element. It is used to make very hard alloys for fountain pen points, electrical contacts, and instrument pivots. Atomic number 76; atomic weight 190.2; melting point 3,000°C; boiling point 5,000°C; specific gravity 22.57; valence 2, 3, 4, 8. See **Periodic Table.**

osmosis (ŏz-mō**′**sĭs) The movement of a solvent through a membrane separating two solutions of different concentrations. The solvent from the side of weaker concentration usually moves to the side of the stronger concentration, diluting it, until the concentrations of the solutions are equal on both sides of the membrane. ▸ The pressure exerted by the molecules of the solvent on the membrane they pass through is called **osmotic pressure.** Osmotic pressure is the energy driving osmosis and is important for living organisms because it allows water and nutrients dissolved in water to pass through cell membranes.

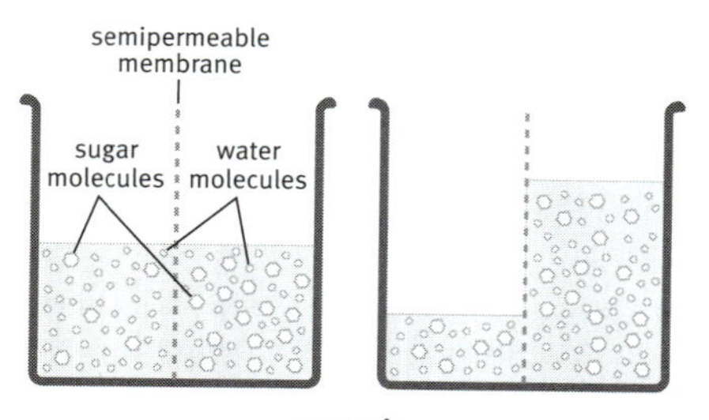

osmosis

left: *The concentration of sugar molecules is greater on the right side of the membrane than on the left. The water molecules are small enough to move across the membrane, but the larger sugar molecules cannot pass through.* right: *The water molecules move across the membrane until the water and sugar molecules are of equal concentration on both sides. This lowers the water level on the left side and raises it on the right side.*

ossicle (ŏs**′**ĭ-kəl) A small bone, especially one of the three located in the middle ear (the incus, malleus, and stapes) that transmit sound vibrations from the eardrum to the inner ear.

ossification (ŏs′ə-fĭ-kā**′**shən) The process of bone formation, brought about by the action of specialized bone cells called **osteoclasts,** which absorb old bone tissue, and **osteoblasts,** which form from osteoclasts and produce new bone tissue. This remodeling of bone is a constant process that maintains bone strength. See more at **osteoblast, osteoclast.**

osteoarthritis (ŏs′tē-ō-är-thrī**′**tĭs) A form of arthritis, occurring mainly in older people, that is characterized by chronic degeneration of the cartilage of the joints.

osteoblast (ŏs**′**tē-ə-blăst′) A specialized bone cell that produces and deposits the matrix that is needed for the development of new bone and consists primarily of collagen fibers. Osteoblasts are formed from osteoclasts on the outer surfaces of bone and in bone cavities, and bone deposition takes place constantly in living bone. As new bone grows and hardens with the addition of calcium and phosphate, osteoblasts become embedded in the bone matrix and develop into osteocytes.

osteoclast (ŏs**′**tē-ə-klăst′) A specialized bone cell that absorbs bone, allowing for the deposition of new bone and maintenance of bone strength. Osteoclasts secrete enyzmes that dissolve the matrix of old bone tissue and acids that dissolve bone salts, which contain calcium and phosphorus. Except in growing bone, the rate of bone deposition and bone absorption equal each other so that bone mass remains constant. A mass of osteoclasts absorbs bone from the outer surfaces inward for about three weeks. The osteoclasts are then converted into osteoblasts that form new bone to fill in the cavities. See also **osteoblast.**

osteocyte (ŏs**′**tē-ə-sīt′) A cell characteristic of mature bone tissue. It is derived from osteoblasts and embedded in the calcified matrix of bone. Osteocytes are found in small, round cavities called *lacunae* and have thin, cytoplasmic branches.

osteogenesis imperfecta (ŏs′tē-ə-jĕn**′**ĭ-sĭs ĭm′pər-fĕk**′**tə) A hereditary disease characterized by abnormally brittle, easily fractured bones.

osteopathy (ŏs′tē-ŏp′ə-thē) A system of medicine based on the theory that disturbances in the musculoskeletal system can cause disorders in other bodily parts and can be corrected by various manipulative techniques. These are used in conjunction with conventional medical and surgical treatments. —*Adjective* **osteopathic** (ŏs′tē-ə-păth′ĭk).

osteoporosis (ŏs′tē-ō-pə-rō′sĭs) A bone disease characterized by decrease in bone mass and density, resulting in a predisposition to fractures and bone deformities such as the collapse of one or more vertebrae. It occurs most commonly in women after menopause as a result of estrogen deficiency. Calcium supplementation and weight-bearing exercise are used to treat and prevent osteoporosis.

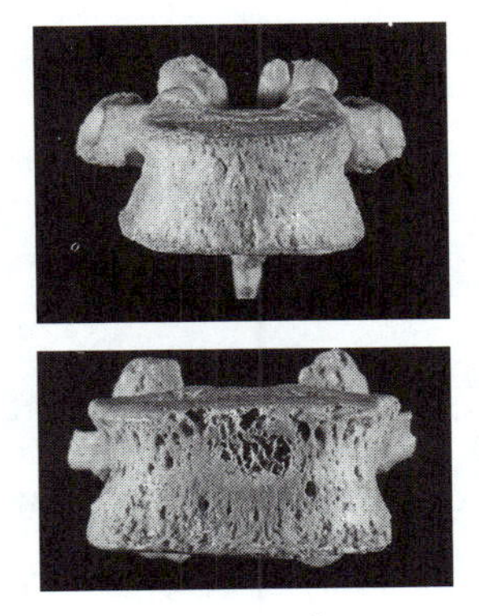

osteoporosis

top: *healthy human vertebra;* bottom: *osteoporosis of a human vertebra*

ostracoderm (ŏs′trə-kō-dûrm′) Any of three orders of small, extinct primitive jawless fishes of the Ordovician through the Devonian Period. Most ostracoderms had their heads or entire bodies encased in fused bony plates. Some or all ostracoderms are sometimes classified together with the modern cyclostomes (lampreys and hagfish).

otitis media (ō-tī′tĭs mē′dē-ə) Inflammation of the middle ear, occurring commonly in children as a result of infection and often causing pain and temporary hearing loss.

ounce (ouns) **1.** A unit of weight in the US Customary System equal to $\frac{1}{16}$ of a pound or 437.5 grains (28.35 kilograms). See Table at **measurement. 2.** See **fluid ounce.**

outbreeding (out′brē′dĭng) The mating or breeding of distantly related or unrelated individuals. Outbreeding often produces offspring of superior quality because it increases homozygosity (the occurrence of two alleles for the same trait at corresponding positions on homologous chromosomes), thereby sharply reducing the risk of deleterious recessive genes being expressed. Crossbreeding is the most common form of outbreeding.

outcrop (out′krŏp′) An area of visible bedrock that is not covered with soil.

outer ear (ou′tər) The part of the ear in many vertebrates that is external to the eardrum (tympanic membrane) and contains the canal leading to the eardrum as well as the ear lobe and other visible structures. In mammals, the outer ear, which is made mostly of cartilage, gathers and focuses incoming sound waves and transmits them to the eardrum. See more at **ear.**

outer planet Any of the four planets Jupiter, Saturn, Uranus, and Neptune, whose orbits lie outside that of Mars. The outer planets are large **gas giants.** Compare **inner planet.** See also **superior planet.**

outfall (out′fôl′) The place where a sewer, drain, or stream discharges.

output (out′po͝ot′) **1.** The energy, power, or work that is produced by a system or device. **2.** The information that a computer produces by processing a specific input. Compare **input.**

ovary (ō′və-rē) **1.** The reproductive organ in female animals that produces eggs and the sex hormones estrogen and progesterone. In most vertebrate animals, the ovaries occur in pairs. In mammals, the ovaries contain numerous follicles, which house the developing eggs (oocytes). See more at **menstrual cycle, ovulation. 2.** The part of a carpel or of a gynoecium made of fused carpels that contains the ovules in a flower. The ovary is located at the base of the carpel and ripens into a fruit after fertilization of one or more of the ovules. See more at **flower.** —*Adjective* **ovarian** (ō-vâr′ē-ən).

overpopulation (ō′vər-pŏp′yə-lā′shən) The population of an environment by a particular species in excess of the environment's **carrying capacity.** The effects of overpopulation can include the depletion of resources, environmental deterioration, and the prevalence of famine and disease.

overtone (ō′vər-tōn′) See under **harmonic.**

oviduct (ō′vĭ-dŭkt′) A tube through which eggs or egg cells (oocytes) are carried to the uterus in mammals or to the outside of the body in other animals. The fallopian tubes are oviducts.

ovine (ō′vīn′) Relating to or characteristic of sheep.

oviparous (ō-vĭp′ər-əs) Producing eggs that hatch outside the body. Amphibians, birds, and most insects, fish, and reptiles are oviparous. Compare **ovoviviparous, viviparous.**

ovipositor (ō′və-pŏz′ĭ-tər) A tube in many female insects that extends from the end of the abdomen and is used to lay eggs.

ovoviviparous (ō′vō-vī-vĭp′ər-əs) Producing eggs that hatch within the female's body. Some fish and reptiles are ovoviviparous. Compare **oviparous, viviparous.**

ovulation (ō′vyə-lā′shən, ŏv′yə-lā′shən) The release of an egg cell (ovum) from the ovary in female animals, regulated in mammals by hormones produced by the pituitary gland during the menstrual cycle.

ovule (ō′vyo͞ol, ŏv′yo͞ol) The female reproductive structure that develops into a seed in a seed-bearing plant. An ovule consists of a megasporangium surrounded by one or two layers of tissue called integuments. The megasporangium produces spores that develop into megagametophytes. These megagametophytes remain within the tissues of the ovule and produce one or more egg cells. Sperm from pollen grains enter the ovule through an opening called a micropyle and fertilize the egg cells. The resulting embryo then begins to develop within the ovule, which becomes a seed. Among the conifers and cycads, the ovules are typically found in pairs on scales in the female cones. The ovules of angiosperms are contained in a structure called the **ovary** within in the flower. See more at **flower, gametophyte, megasporogenesis, pollination.**

ovum (ō′vəm) Plural **ova.** The mature reproductive cell of female animals, produced in the ovaries. See more at **egg** (sense 1).

oxalate (ŏk′sə-lāt′) A salt or ester of oxalic acid.

oxalic acid (ŏk-săl′ĭk) A poisonous, crystalline acid found in a number of plants such as sorrel and the leaf blades of rhubarb. It is used for many industrial purposes, including rust removal and bleaching. *Chemical formula:* $C_2H_2O_4$.

oxaloacetic acid (ŏk′sə-lō-ə-sē′tĭk, ŏk-săl′ō-) or **oxalacetic acid** (ŏk-săl′ə-sē′tĭk, ŏk′sə-lə-) A colorless crystalline organic acid that is formed by oxidation of malic acid in the Krebs cycle and by transamination from aspartic acid. It is important as an intermediate in the metabolism of carbohydrates and a precursor in the synthesis of amino acids. *Chemical formula:* $C_4H_4O_5$.

oxbow (ŏks′bō′) A sharp, U-shaped bend in a river. The bend is so sharp that only a narrow neck of land is left between the two parts of the river.

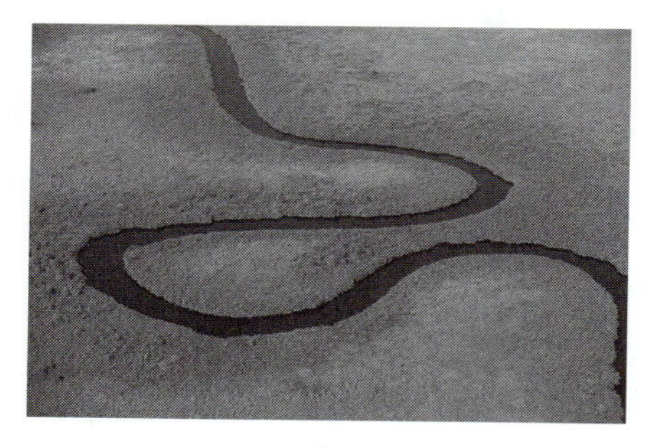

oxbow

aerial photograph of the Okavango River, Botswana

oxbow lake A crescent-shaped lake formed when a river changes its course and cuts through the strip of land in the middle of an oxbow, abandoning its previous course and isolating the water in the oxbow.

oxidation (ŏk′sĭ-dā′shən) **1.** The chemical combination of a substance with oxygen. **2.** A chemical reaction in which an atom or ion loses electrons, thus undergoing an increase in valence. Removing an electron from an iron atom having a valence of +2 changes the valence to +3. Compare **reduction.** —*Adjective* **oxidative.**

oxidation-reduction A chemical reaction in which an atom or ion loses electrons to another atom or ion.

oxidation state See **valence.**

oxidative stress (ŏk′sĭ-dā′tĭv) A condition of increased oxidant production in animal cells characterized by the release of free radicals and resulting in cellular degeneration.

oxide (ŏk′sīd′) A compound of oxygen and another element or radical. Water (H_2O) is an oxide.

oxidize (ŏk′sĭ-dīz′) To undergo or cause to undergo oxidation.

oxyacetylene torch (ŏk′sē-ə-sĕt′l-ĭn, -ēn′) A gas torch that burns a mixture of acetylene and oxygen to produce a high-temperature flame (3,000°C or 5,400°F) that can weld or cut metal.

oxygen (ŏk′sĭ-jən) *Symbol* **O** A nonmetallic element that exists in its free form as a colorless, odorless gas and makes up about 21 percent of the Earth's atmosphere. It is the most abundant element in the Earth's crust and occurs in many compounds, including water, carbon dioxide, and iron ore. Oxygen combines with most elements, is required for combustion, and is essential for life in most organisms. Atomic number 8; atomic weight 15.9994; melting point −218.4°C; boiling point −183.0°C; gas density at 0°C 1.429 grams per liter; valence 2. See **Periodic Table.**

WORD HISTORY **oxygen, hydrogen, and nitrogen**

In 1786, the French chemist Antoine Lavoisier coined a term for the element *oxygen* (*oxygène* in French). He used Greek words for the coinage: *oxy–* means "sharp," and *–gen* means "producing." Oxygen was called the "sharp-producing" element because it was thought to be essential for making acids. Lavoisier also coined the name of the element *hydrogen,* the "water-producing" element, in 1788. Soon after, in 1791, another French chemist, J. A. Chaptal, introduced the word *nitrogen,* the "niter-producing" element, referring to its discovery from an analysis of nitric acid.

oxygenate (ŏk′sĭ-jə-nāt′) To combine or mix with oxygen, as in a physical, chemical, or biological system. Blood is oxygenated in the lungs.

oxyhemoglobin (ŏk′sē-hē′mə-glō′bĭn) The compound formed when a molecule of hemoglobin binds with a molecule of oxygen. In vertebrate animals, oxyhemoglobin forms in the red blood cells as they take up oxygen in the lungs. See Note at **hemoglobin.**

oxytocin (ŏk′sĭ-tō′sĭn) A polypeptide hormone secreted by the posterior portion of the pituitary gland. Oxytocin stimulates the contraction of smooth muscle of the uterus during childbirth and facilitates ejection of milk from the mammary glands.

oz Abbreviation of **ounce.**

ozone (ō′zōn′) An unstable, poisonous allotrope of oxygen having the chemical formula O_3. Ozone forms in the atmosphere through the process of photolysis, when ultraviolet radiation from the Sun strikes oxygen molecules (O_2), causing them to split apart. When freed oxygen atoms bump into and join other O_2 molecules, they form ozone. Although ozone is broken down naturally in the atmosphere through chemical reactions with other atmospheric gases (such as nitrogen, hydrogen, and chlorine), in an unpolluted atmosphere the formation and breakdown of ozone is generally balanced, and the total concentration of ozone is relatively constant. The formation and destruction rates of ozone vary with altitude in the atmosphere, and with latitude. Most ozone forms in the 15 to 30 km (10 to 19 mi) altitude range and in latitudes closest to the equator where sunshine strikes the Earth the most. The ozone is then transported northward and southward by wind and is generally most concentrated in areas above the Canadian Arctic and Siberia and above Antarctica. Ozone is used commercially in water purification, in air conditioning, and as a bleach.

A CLOSER LOOK **ozone**

Ozone is both beneficial for and threatening to all of Earth's organisms, including human beings, depending on how high in the atmosphere it is found. Ozone is naturally produced in the stratospheric portion of Earth's atmosphere (in the ozone layer) by the action of high-energy ultraviolet radiation on molecular oxygen (O_2). By absorbing much of the Sun's ultraviolet radiation, the ozone layer serves as a sunscreen for organisms on Earth. In recent years the ozone has thinned or disappeared in parts of the ozone layer, creating an *ozone hole* that lets in dangerous amounts of ultraviolet radiation. Ozone holes are caused in part by the release into the atmosphere of industrial and commercial chemicals, in particular the chlorofluorocarbons (such as freon) used in aerosols, refrigerants, and certain cleaning solvents. Closer to Earth's surface, ozone is one of the so-called greenhouse gases that are produced by the burning of fossil fuels and cause the *greenhouse effect.* Ozone at ground level is also an air pollutant, contributing to respiratory diseases such as asthma.

---

ozone hole A severe depletion of ozone in a region of the ozone layer, particularly over Antarctica and over the Arctic. The depletion is caused by the destruction of ozone by CFCs and by other compounds, such as carbon tetrachloride (CCl_4) and carbon tetrafluoride (CF_4). The amount of ozone in ozone holes is about 55 to 60 percent of the normal concentration in the ozone layer. Although the full effect of increased ozone depletion is not yet known, the amount of ultraviolet radiation the Earth receives is greatly increased by ozone depletion, creating a heightened risk of skin cancers and likely contributing to global warming. See Note at **ozone.**

ozone layer A region of the upper atmosphere containing relatively high levels of ozone, located mostly within the stratosphere, with the greatest concentrations occurring from about 15 to 30 km (10 to 19 mi) above the Earth's surface. The ozone absorbs large amounts of solar ultraviolet radiation, preventing it from reaching the Earth's surface. The concentration of ozone in the ozone layer is usually under 10 parts per million. Also called *ozonosphere.* See Note at **ozone.**

ozonosphere (ō-zō′nə-sfîr′) See **ozone layer.**

P

P **1.** The symbol for **parity. 2.** The symbol for **phosphorus. 3.** The symbol for **power. 4.** The symbol for **pressure.**

Pa **1.** Abbreviation of **pascal. 2.** The symbol for **protactinium.**

PABA (pä′bə) Short for *para-aminobenzoic acid.* A crystalline form of aminobenzoic acid that is part of the vitamin B complex. It is a growth factor required by some microorganisms, and it is often used in sunscreens.

pacemaker (pās′mā′kər) **1.** Any of several usually miniaturized and surgically implanted electronic devices used to stimulate or regulate contractions of the heart muscle. Electrodes attached to the heart muscle conduct electrical signals generated from the pacemaker, which either provides constant electrical stimuli to regularize the heartbeat (fixed-rate pacemakers) or provides a stimulus only when electrical activity within the heart is abnormal (demand pacemakers). Pacemakers are used in the treatment of various arrhythmias. **2.** A part of the body, such as the specialized mass of cardiac muscle fibers of the sinoatrial node, that sets the pace or rhythm of physiological activity.

pachycephalosaurus (păk′ĭ-sĕf′ə-lə-sôr′əs) A large herbivorous dinosaur of the genus *Pachycephalosaurus* of the late Cretaceous Period. It grew to about 7.6 m (25 ft) long and had a domed skull up to 25.4 cm (10 inches) thick that was lined with small bumps and spikes. The thick skull may have been used for head-butting during mating displays.

pachyderm (păk′ĭ-dûrm′) Any of various large, thick-skinned mammals, such as the elephant, rhinoceros, or hippopotamus.

packet switching (păk′ĭt) A method of network data transmission, in which small blocks of data, or **packets**, are transmitted over a channel which, for the duration of the packet's transmission, is dedicated to that packet alone and is not interrupted to transmit other packets. This strategy is used in transmitting data over the Internet and often over a LAN, and it capitalizes on the increase in efficiency that is obtained when there are many paths available and there is a large volume of traffic over these paths.

pack ice (păk) The floating sea-ice cover of the polar regions. Driven by winds and ocean currents, pack ice is a mixture of ice fragments of varying size and age that are squeezed together and cover the sea surface with little or no open water. At maximum expansion during the winter, pack ice covers about five percent of Arctic waters and about eight percent of Antarctic waters.

packing fraction (păk′ĭng) **1.** The ratio of the total volume of a set of objects packed into a space to the volume of that space. **2.** The difference between the isotopic mass of a nuclide and its mass number, divided by its mass number. The packing fraction is often interpreted as a measure of the stability of the nucleus.

Paget (păj′ĭt), Sir **James** 1814–1899. British surgeon and researcher who was a pioneer of modern pathology. He discovered the cause of trichinosis in 1834, reported on diseases of the bones and joints, and described Paget's disease.

Paget's disease (păj′ĭts) **1.** A disease, occurring chiefly in the elderly, in which the bones become enlarged and weakened, often resulting in fracture or deformity. **2.** A breast cancer manifested by inflammatory changes of the nipple and surrounding skin.

pahoehoe (pə-hoi′hoi′) A type of lava having a smooth, swirled surface. It is highly fluid and spreads out in shiny sheets. Compare **aa.**

pahoehoe

pahoehoe (foreground) *and aa types of lava*

WORD HISTORY **pahoehoe/aa**

The islands that make up Hawaii were born and bred from volcanoes that rose up over thousands of years from the sea floor. Volcanoes are such an important part of the Hawaiian landscape and environment that the people who originally settled Hawaii, the Polynesians, worshiped a special volcano goddess, Pele. Not surprisingly, two words have entered English from Hawaiian that are used by scientists in naming different kinds of lava flows. One, *pahoehoe,* refers to lava with a smooth, shiny, or swirled surface and comes from the Hawaiian verb *hoe,* "to paddle" (since paddles make swirls in the water). The other, *aa,* refers to lava having a rough surface and comes from the Hawaiian word meaning "to burn."

pair production (pâr) The simultaneous creation of a subatomic particle and its antiparticle from some other particle or a field fluctuation. A high-energy gamma ray photon, for example, can spontaneously create an electron-positron pair in a strong electric field. Pair production is essential to theories of particle physics; pair production from gamma rays is used to detect such rays in **pair production telescopes,** since electrons and positrons are easily detected.

palate (păl′ĭt) The roof of the mouth in vertebrate animals, separating the mouth from the passages of the nose. ► The bony part of the palate is called the **hard palate.** ► A soft, flexible, rear portion of the palate, called the **soft palate,** is present in mammals only and serves to close off the mouth from the nose during swallowing.

palea (pā′lē-ə) Plural **paleae** (pā′lē-ē′). **1.** The inner or upper of the two bracts enclosing one of the small flowers within a grass spikelet. **2.** The chaffy scales on the receptacle of a flower head in a plant of the composite family.

paleo– A prefix that means "prehistoric" (as in *paleontology*) or "early or primitive" (as in *Paleolithic*).

paleoanthropology (pā′lē-ō-ăn′thrə-pŏl′ə-jē) The scientific study of extinct members of the genus *Homo sapiens* by means of their fossil remains.

paleobotany (pā′lē-ō-bŏt′n-ē) The branch of paleontology that deals with plant fossils and ancient vegetation.

Paleocene (pā′lē-ə-sēn′) The earliest epoch of the Tertiary Period, from about 65 to 58 million years ago. During this time, the Rocky Mountains formed and sea levels dropped, exposing dry land in North America, Australia, and Africa. Many new types of small mammals evolved and filled the niches left empty after the extinctions that ended the Cretaceous Period. Soft-bodied squid replaced the ammonites as the dominant form of mollusk. See Chart at **geologic time.**

paleoclimatology (pā′lē-ō-klī′mə-tŏl′ə-jē) The scientific study of climatic conditions, along with their causes and effects, in the geologic past. These conditions are reconstructed on the basis of evidence found in the geologic record, especially in the form of glacial deposits, fossils, sediments, and rock and ice cores. Because much of the geologic record studied in paleoclimatology predates humans, this research is valuable for weighing the relative influence of human and natural causes of global climate change. It also provides test situations for computerized climate modeling systems used to predict present-day climate changes.

paleoecology (pā′lē-ō-ĭ-kŏl′ə-jē) The branch of ecology that deals with the interaction between ancient organisms and their environment.

Paleogene (pā′lē-ə-jēn′) The oldest of two subdivisions of the Tertiary Period, including the Paleocene, Eocene, and Oligocene Epochs.

paleolith (pā′lē-ə-lĭth′) A stone implement of the Paleolithic Period.

Paleolithic (pā′lē-ə-lĭth′ĭk) The cultural period of the Stone Age that began about 2.5 to 2 million years ago, marked by the earliest use of tools made of chipped stone. The Paleolithic Period ended at different times in different parts of the world, generally around 10,000 years ago in Europe and the Middle East. Also called *Old Stone Age.* ► The **Lower Paleolithic** is by far the longest division of this period, lasting until about 200,000 years ago and characterized by hammerstones and simple core tools such as hand axes and cleavers. The earliest tools belong to the Oldowan tool culture and may have been made by australopithecines as well as by *Homo habilis*. Later Lower Paleolithic cultures include the Abbevilian, Clactonian, Acheulian, and Levalloisian, associated with early *Homo erectus*. ▸ The **Middle Paleolithic** is generally dated to about 40,000 years ago and is associated with archaic *Homo sapiens*, primarily the Neanderthals and their Mousterian tool culture. The tools produced during this period represent improvements on those of the Lower Paleolithic, especially in flaking techniques, but remain little changed throughout the duration of the period. ▸ The **Upper Paleolithic** dates to about 10,000 years ago in Europe and the Middle East and is associated with modern *Homo sapiens*. Various distinctive local tool cultures such as the Aurignacian, Solutrean, and Magdalenian flourished during this relatively brief period, producing a great variety of skillfully flaked tools as well as tools made of bone, antler, wood, and other materials. Compare **Mesolithic, Neolithic.**

paleomagnetism (pā′lē-ō-măg**′**nĭ-tĭz′əm) **1.** The fixed orientation of a rock's magnetic minerals as originally aligned at the time of the rock's formation. Paleomagnetism is usually the result of **thermoremanent magnetization** (magnetization that occurs in igneous rocks as they cool). Examination of the paleomagnetism of the Earth's ocean floors revolutionized the field of geology by providing evidence for the existence and movement of tectonic plates. See Note at **magnetic reversal. 2.** The scientific study of such magnetic remanence.

paleontology (pā′lē-ŏn-tŏl**′**ə-jē) The scientific study of life in the geologic past, especially through the study of animal and plant fossils.

paleosol (pā**′**lē-ə-sôl′) A soil horizon from the geologic past, usually buried beneath other rocks or recent soil horizons.

Paleozoic (pā′lē-ə-zō**′**ĭk) The era of geologic time from about 540 to 245 million years ago. The beginning of the Paleozoic Era is characterized by a great diversity of marine invertebrate animals. Primitive fish and reptiles, land plants, and insects also first appeared during this time. The end of the Paleozoic is marked by the largest recorded mass extinction in the Earth's history, which wiped out nearly 90% of known marine life forms. See Chart at **geologic time.**

paleozoology (pā′lē-ō-zō-ŏl**′**ə-jē) The scientific study of animal fossils and the evolution of animals.

palisade cell (păl′ĭ-sād**′**) One of the columnar cells of palisade layer.

palisade layer A layer of cells just below the upper surface of most leaves, consisting of cylindrical cells that contain many chloroplasts and stand at right angles to the leaf surface. It is the principal region of the leaf in which photosynthesis is carried out and lies above or to the outside of the **spongy parenchyma**. Also called *palisade parenchyma.* See more at **photosynthesis.**

palisades (păl′ĭ-sādz**′**) A line of steep, high cliffs, especially of basalt, usually along a river.

palladium (pə-lā**′**dē-əm) *Symbol* **Pd** A malleable, ductile, grayish-white metallic element that occurs naturally with platinum. It is used as a catalyst in hydrogenation and in alloys for making electrical contacts and jewelry. Atomic number 46; atomic weight 106.4; melting point 1,552°C; boiling point 3,140°C; specific gravity 12.02 (20°C); valence 2, 3, 4. See **Periodic Table.**

Pallas (păl**′**əs) The second largest asteroid, measuring about 570 km (353 mi) at its greatest diameter, and the second to be discovered, in 1802. See more at **asteroid.**

palmate (păl**′**māt′, päl**′**-) **1.** Having a shape similar to that of a hand with the fingers extended. Some kinds of coral and the antlers of moose and certain deer are palmate. **2.** Having three or more veins, leaflets, or lobes radiating from one point. Maples have palmately lobed leaves. **3.** Having webbed toes. The feet of many swimming and diving birds are palmate.

palmitate (păl**′**mĭ-tāt′, päl**′**-, pä**′**mĭ-) A salt or ester of palmitic acid, containing the group $C_{16}H_{31}O_2$.

palmitic acid (păl-mĭt**′**ĭk, päl-, pä-mĭt**′**-) A saturated fatty acid occurring as combustible white crystals in many natural oils (such as spermaceti and palm oil) and fats. It is used in making soaps. *Chemical formula:* $\mathbf{C_{16}H_{32}O_2}$.

palp (pălp) A segmented organ extending from the mouthparts of arthropods, used for touch or taste. Also called *palpus.*

palustrine (pə-lŭs′trēn) Relating to a system of inland, nontidal wetlands characterized by the presence of trees, shrubs, and emergent vegetation (vegetation that is rooted below water but grows above the surface). Palustrine wetlands range from permanently saturated or flooded land (as in marshes, swamps, and lake shores) to land that is wet only seasonally (as in vernal pools). Compare **lacustrine, marine, riverine.**

palynology (păl′ə-nŏl′ə-jē) The scientific study of spores and pollen, both living and fossilized. Palynology helps improve knowledge of ecosystems in both the recent and distant past, since pollen and spores are extremely durable, unlike many other plant parts.

palytoxin (păl′ə-tŏk′sĭn) An extremely powerful toxin occurring in corals of the genus *Palythoa* of the South Pacific. Palytoxin disrupts the flow of ions across cell membranes and is the most potent known naturally occurring poison, being rapidly fatal to humans in doses of 4 micrograms. One of the most complex known nonprotein substances, it has more contiguous carbon atoms (115) than any other naturally occurring molecule. *Chemical formula:* $C_{129}H_{223}N_3O_{54}$.

pampa (păm′pə) An extensive, treeless grassland of southern South America.

pancreas (păng′krē-əs) A long, irregularly shaped gland in vertebrate animals that is located behind the stomach and is part of the digestive system. It secretes hormones (insulin, glucagon, and somatostatin) into the bloodstream and digestive enzymes into the small intestine or gut. The pancreas also secretes sodium bicarbonate, which protects the lining of the intestine by neutralizing acids from the stomach.

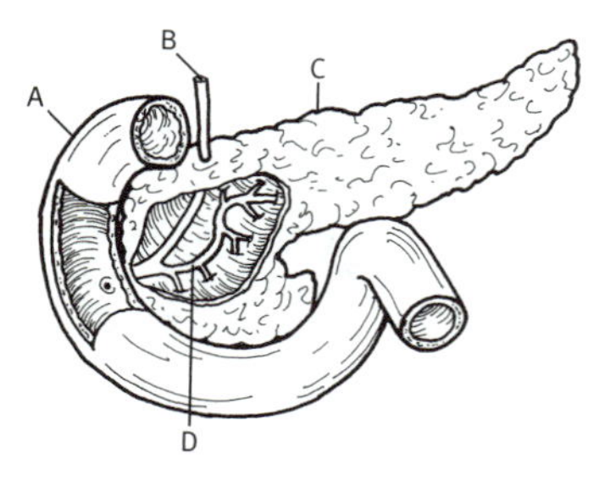

pancreas
A. *duodenum*, B. *bile duct*, C. *pancreas*, D. *pancreatic duct*

pancreatic duct (păng′krē-ăt′ĭk) The excretory duct of the pancreas, extending through the gland from tail to head, where it empties into the duodenum.

pandemic (păn-dĕm′ĭk) An epidemic that spreads over a very wide area, such as an entire country or continent.

Pangaea (păn-jē′ə) A supercontinent made up of all the world's present landmasses joined together in the configuration they are thought to have had during the Permian and Triassic Periods. According to the theory of plate tectonics, Pangaea later broke up into **Laurasia** and **Gondwanaland**, which eventually broke up into the continents we know today.

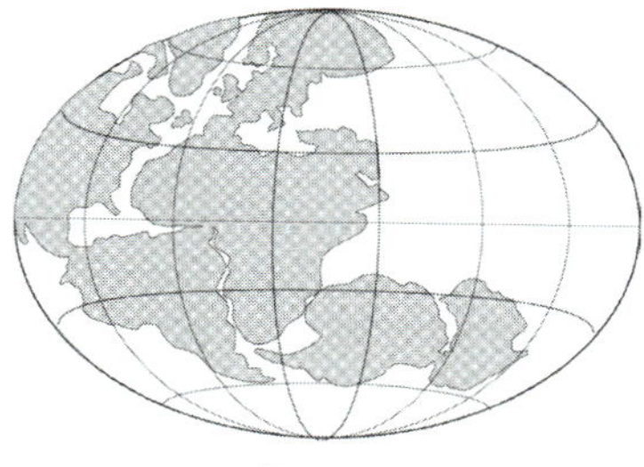

Pangaea

panicle (păn′ĭ-kəl) A branched indeterminate inflorescence in which the branches are racemes, so that each flower has its own stalk (called a pedicel) attached to the branch. Oats and sorghum have panicles. See illustration at **inflorescence.**

Panthalassa (păn′thə-lăs′ə) The ocean that surrounded the supercontinent Pangaea.

pantothenic acid (păn′tə-thĕn′ĭk) A water-soluble organic acid belonging to the vitamin B complex that is an essential component of coenzyme A. It is a derivative of the amino acid alanine, and it is important in the metabolism of fats and carbohydrates. Pantothenic acid is found in all animal and plant cells, but it is particularly abundant in liver, rice bran, molasses, and many vegetables. *Chemical formula:* $C_9H_{17}NO_5$.

papilla (pə-pĭl′ə) Plural **papillae** (pə-pĭl′ē). A small part projecting from the surface of an organism. In mammals, the nipples of the mammary glands and the taste buds of the tongue are papillae. Papillae are often seen on the undersurfaces of mosses and ferns.

pappus (păp′əs) Plural **pappi** (păp′ī). A structure made of scales, bristles, or featherlike hairs that is attached to the seeds (called cypselae) of plants of the composite family and that aids in dispersal by the wind. The downy part of a dandelion or thistle seed is a pappus. The pappus is derived from a modified calyx.

Pap smear (păp) A screening test, especially for cervical cancer, in which a smear of cells scraped from the cervix or vagina is treated with a chemical stain and examined under a microscope for pathological changes. Pap smears are performed routinely during gynecological exams.

para-aminobenzoic acid (păr′ə-ə-mē′nō-bĕn-zō′ĭk, -ăm′ə-) See **PABA.**

parabola (pə-răb′ə-lə) The curve formed by the set of points in a plane that are all equally distant from both a given line (called the directrix) and a given point (called the focus) that is not on the line.

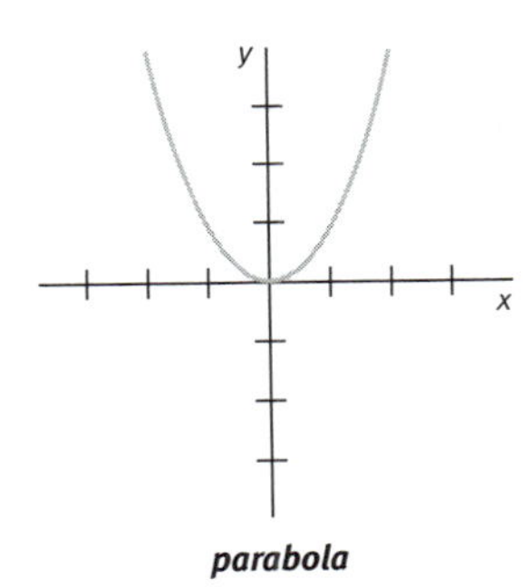

parabola

The parabola of the function $y = x^2$.

parabolic antenna (păr′ə-bŏl′ĭk) An antenna used for sending or receiving radio signals that uses the principle of a parabolic mirror to focus incoming signals onto one reception point or direct the emissions of signals from a focal source point into a directed beam. Parabolic antennae are used extensively in radio communications, especially with satellites, and in radar. Also called *dish antenna.*

parabolic mirror A cone-shaped concave mirror with a rounded-off tip, whose cross-section is shaped like the tip of a parabola. Most of the light, radio waves, sound, and other radiation that enter the mirror straight on is reflected by the surface and converges on the focus of the parabola, where being concentrated, it can be easily detected. Conversely, radiation emanating from the focal point reflects from the inner surface of the mirror into a fairly direct beam of nearly parallel radiation that can be aimed at a target. Parabolic mirrors are the basis of parabolic antennae, as well as some megaphones and telescopic mirrors.

paraboloid (pə-răb′ə-loid′) A surface having parabolic sections parallel to a single coordinate axis and elliptic sections perpendicular to that axis.

paradichlorobenzene (păr′ə-dī-klôr′ə-bĕn′zēn′) A white crystalline compound used as a germicide and an insecticide. *Chemical formula:* $\mathbf{C_6H_4Cl_2}$.

paraffin (păr′ə-fĭn) **1.** A waxy, white or colorless solid mixture of hydrocarbons made from petroleum and used to make candles, wax paper, lubricants, and waterproof coatings. Also called *paraffin wax.* **2.** See **alkane.**

paraffin series See under **alkane.**

paraffin wax See **paraffin.**

parallax (păr′ə-lăks′) An apparent shift in the position of an object, such as a star, caused by a change in the observer's position that provides a new line of sight. The parallax of nearby stars caused by observing them from opposite points in Earth's orbit around the Sun is used in estimating the stars' distance from Earth through triangulation.

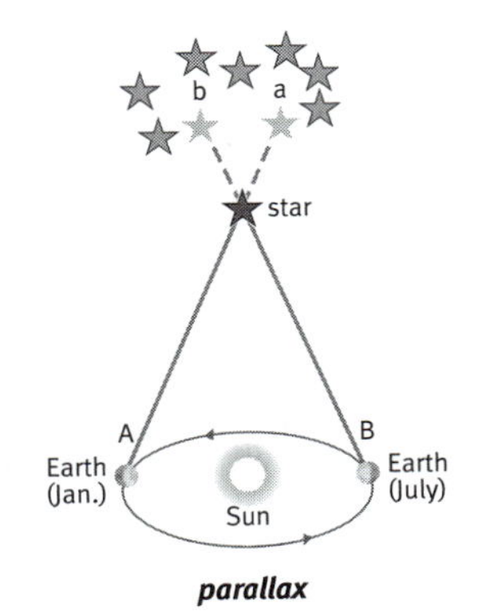

parallax

Viewed from point A, *a nearby star appears to occupy position* a *against a background of more distant stars. Six months later, from point* B, *the star appears to occupy position* b.

parallel (păr′ə-lĕl′) *Adjective.* **1.** Of or relating to lines or surfaces that are separated everywhere from each other by the same distance. —*Noun.* **2.** Any of the imaginary lines encircling the surface of the Earth that are parallel to the plane of the equator, used to represent degrees of latitude. See illustration at **longitude.**

parallel circuit See under **circuit.**

parallelepiped (păr′ə-lĕl′ə-pī′pĭd, -pĭp′ĭd) A polyhedron with six faces, each a parallelogram and each being parallel to the opposite face.

parallelogram (păr′ə-lĕl′ə-grăm′) A four-sided plane figure with opposite sides parallel. Rhombuses and rectangles are parallelograms.

parallel postulate See under **Euclidean.**

paralysis (pə-răl′ĭ-sĭs) Loss or impairment of voluntary movement or sensation in a part of the body, usually as a result of neurologic injury or disease.

paramagnetism (păr′ə-măg′nĭ-tĭz′əm) The property of being weakly attracted to either pole of a magnet. Paramagnetic materials, such as aluminum and platinum, become slightly magnetized when exposed to a magnetic field, but they lose their magnetism when the field is removed. When these materials are magnetized by a nearby magnet, the orientation of their north and sole poles results in their being attracted to the closest pole of the magnet. Compare **diamagnetism, ferromagnetism.**

paramecium (păr′ə-mē′sē-əm) Plural **paramecia** or **parameciums.** Any of various freshwater protozoans of the genus *Paramecium* that are usually oval in shape and that move by means of cilia. Although they consist of a single cell, paramecia are large enough to be visible to the naked eye. Like other ciliates, paramecia contain two nuclei, a macronucleus and a micronucleus. On the cellular surface is a groove that opens into a gullet, into which food particles are absorbed.

parametric equalizer (păr′ə-mĕt′rĭk) See under **equalizer.**

paraphyletic (păr′ə-fī-lĕt′ĭk) Relating to a taxonomic group that includes some but not all of the descendants of a common ancestor. In the traditional taxonomy of vertebrates, where fish are a separate class from the classes of terrestrial vertebrates, the class of fish is paraphyletic, since the terrestrial vertebrates are descended from a type of fish. Compare **monophyletic, polyphyletic.**

paraphysis (pə-răf′ĭ-sĭs) Plural **paraphyses** (pə-răf′ĭ-sēz′). One of the erect sterile filaments often occurring among the reproductive organs of certain fungi, algae, and mosses.

paraplegia (păr′ə-plē′jē-ə) Paralysis of the lower part of the body, caused by injury to the spinal cord.

paraquat (păr′ə-kwŏt′) A toxic compound used as a herbicide, especially in its colorless, dichloride form ($C_{12}H_{14}Cl_2N_2$) or in its yellow, bismethyl sulfate form ($C_{14}H_{20}N_2O_8S_2$). Paraquat is used primarily to control the growth of grass and weeds. Its use in the United States is restricted because of its high level of toxicity.

parasite (păr′ə-sīt′) An organism that lives on or in a different kind of organism (the host) from which it gets some or all of its nourishment. Parasites are generally harmful to their hosts, although the damage they do ranges widely from minor inconvenience to debilitating or fatal disease. ► A parasite that lives or feeds on the outer surface of the host's body, such as a louse, tick, or leech, is called an **ectoparasite.** Ectoparasites do not usually cause disease themselves although they are frequently a vector of disease, as in the case of ticks, which can transmit the organisms that cause such diseases as Rocky Mountain spotted fever and Lyme disease. ► A parasite that lives inside the body of its host is called an **endoparasite.** Endoparasites include organisms such as tapeworms, hookworms, and trypanosomes that live within the host's organs or tissues, as well as organisms such as sporozoans that invade the host's cells. See more at **host.**

parasitism (păr′ə-sĭ-tĭz′əm) A symbiotic relationship in which one organism (the parasite) benefits and the other (the host) is generally harmed. Parasites derive nutrition from their host and may also gain other benefits such as shelter and a habitat in which they are able to grow and reproduce. See more at **parasite.** Compare **amensalism, commensalism, mutualism.**

parasitoid (păr′ə-sĭ-toid′, -sī′toid) Any of various insects, such as the ichneumon fly, whose larvae are parasites that eventually kill their hosts. The adult parasitoid deposits an egg on or inside the body of its host, typically the larva of another arthropod. When the egg hatches, the parasitoid larva feeds on the host's tissues, gradually killing it.

parasympathetic nervous system (păr′ə-sĭm′pə-thĕt′ĭk) The part of the autonomic nervous system that tends to act in opposition to the sympathetic nervous system, as by slowing down the heartbeat and dilating the blood vessels. It regulates the function of many glands, such as those that produce tears and saliva, and stimulates motility and secretions of the digestive system. Compare **sympathetic nervous system.**

paratenic host (păr′ə-tĕn′ĭk) See under **host.**

parathyroid gland (păr′ə-thī′roid) Any of four kidney-shaped glands located behind or within the thyroid gland of many vertebrate animals. ► The parathyroid glands secrete **parathyroid hormone,** which regulates the amount of calcium in the blood.

parenchyma (pə-rĕng′kə-mə) The basic tissue of plants, consisting of cells with thin cellulose walls. The cortex and pith of the stem, the internal layers of leaves, and the soft parts of fruits are made of parenchyma. In contrast to sclerenchyma cells, parenchyma cells remain alive at maturity. They perform various functions, such as water storage, replacement of damaged tissue, and physical support of plant structures. Chloroplasts, the organelles in which photosynthesis takes place, are found in parenchyma cells. Compare **collenchyma, sclerenchyma.**

parent cell (pâr′ənt) A cell that is the source of other cells, as a cell that divides to produce two or more daughter cells, or a stem cell that is a progenitor of other cells or is the first in a line of developing cells. Also called *mother cell.*

parhelic circle (pär-hē′lĭk) A luminous halo visible at the height of the Sun and parallel to the horizon, caused by the Sun's rays reflecting off atmospheric ice crystals.

parhelion (pär-hē′lē-ən) Plural **parhelia.** A white spot appearing at times in the **parhelic circle.** White parhelia are believed to form from light that is reflected off of atmospheric ice crystals; colored parhelia are believed to form from light that is refracted by atmospheric ice crystals. Multiple parhelia can often be seen simultaneously. Compare **anthelion.**

parietal lobe (pə-rī′ĭ-təl) The upper middle lobe of each cerebral hemisphere, located above the temporal lobe. Complex sensory information from the body is processed in the parietal lobe, which also controls the ability to understand language.

parity (păr′ĭ-tē) **1.** The property of a physical system that entails how the system would behave if the coordinate system were reversed, each dimension changing sign from x, y, z to $-x, -y, -z$. If a system behaves in the same way when the coordinate system is reversed, then it is said to have ***even parity***; if it does not, it is said to have ***odd parity.*** For bosons, the antiparticle of any given particle has the same parity, odd or even, as that particle. For fermions, the antiparticle has the opposite parity. See also **conservation law, parity conjugation. 2.** A quantum number, either +1 or −1, that mathematically describes this property. **3.** The number of 1's in a piece of binary code, generally taken as the quality of odd or even rather than as a specific number. The parity of packets of binary data is often transmitted along with the data to help detect whether the value of any bits has been altered.

parity conjugation An operation, denoted P in the mathematics of quantum mechanics, in which the sign of each dimension of the coordinate system is reversed. See more at **parity.**

Parkinson's disease (pär′kĭn-sənz) A progressive neurologic disease occurring most often after the age of 50, associated with the destruction of brain cells that produce dopamine. Individuals with Parkinson's disease exhibit tremors while at rest, slowing of movement, stiffening of gait and posture, and weakness. The disease is named after its discoverer, British physician and paleontologist James Parkinson (1755–1824).

parotid gland (pə-rŏt′ĭd) Either of the pair of salivary glands situated below and in front of each ear.

parsec (pär′sĕk′) A unit of astronomical length equal to 3.26 light-years. It is based on the distance from Earth at which a star would have a parallax of one second of arc. Its metric equivalent is about 30.8 trillion km (19.1 trillion mi). It is used in measuring distances in interstellar and intergalactic space. The closest star to Earth, Alpha Centauri, is about 1.3 parsecs away. Compare **astronomical unit, light-year.**

parthenocarpy (pär′thə-nō-kär′pē) The production of fruit without fertilization. Many varieties of the common fig produce fruit through parthenocarpy.

parthenogenesis (pär′thə-nō-jĕn′ĭ-sĭs) Reproduction in which an egg develops into a new individual without being fertilized. Aphids and certain other insects can reproduce by parthenogenesis. Parthenogenesis does not necessarily produce clones of the parent. Among hymenopterans such as honeybees and ants, the haploid males develop from unfertilized eggs laid by the queen, who is diploid. Parthenogenesis is a form of **apomixis.** —*Adjective* **parthenogenetic** (pär′thə-nō-jə-nĕt′ĭk).

partial derivative (pär′shəl) The derivative with respect to a single variable of a function of two or more variables, regarding other variables as constants.

partial eclipse An eclipse in which only part of the surface of a celestial object is obscured. See more at **eclipse.**

partially dominant (pär′shə-lē) Relating to two alleles of a gene pair in a heterozygote that are both fully expressed. When alleles for both white and red are present in a carnation, for example, the result is a pink carnation, since both alleles are partially dominant.

partial product A product formed by multiplying the multiplicand by one digit of the multiplier when the multiplier has more than one digit. Partial products are used as intermediate steps in calculating larger products. For example, the product of 67 and 12 can be calculated as the sum of two partial products, 134 (67×2) + 670 (67×10), or 804.

particle (pär′tĭ-kəl) **1.** A very small piece of solid matter. **2.** An elementary particle, subatomic particle, or atomic nucleus. Also called *corpuscle.*

particle accelerator Any of several machines, such as the cyclotron and linear accelerator, that increase the speed and energy of protons, electrons, or other atomic particles, and direct them at atomic nuclei or other particles to cause high-energy collisions. Such collisions produce other particles, whose paths are tracked and analyzed. Particle accelerators are used to study the nature of the atomic nucleus, subatomic particles, and the forces relating them, and to create radioactive isotopes.

A CLOSER LOOK **particle accelerator**

The *particle accelerators* used by physicists are not as remote from our everyday experience as one might imagine. The cathode ray tubes of televisions and computer monitors, commonly known as picture tubes, are in fact small, low-energy particle accelerators, creating beams of electrons guided and focused by magnets that hit a phosphorescent screen to produce light. The electrons, having an electric charge, are accelerated by an electric field produced by a voltage difference of about a thousand volts. Accelerating electrons to higher velocities, using voltages in the tens of thousands, allows higher-energy radiation to be released; the x-ray tubes used in diagnostic imaging operate on this principle. Today's high-energy particle accelerators, such as *synchrocyclotrons* and *synchrotrons,* accelerate charged particles such as electrons and protons using the same basic principles as ordinary picture tubes, but to much higher velocities. These machines are ring-shaped, often extremely large (some more than ten miles in length), and they accelerate particles to velocities so close to the speed of light that the effects of relativity, such as time dilation and increased particle mass, become important factors. For theoretical physicists, these high speeds are generated to smash the particles against other particles as hard as possible—just like smashing a rock against a wall—just to see what happens. For example, particles once thought to be elementary, like protons, have been shown to consist of yet smaller constituents (quarks, in this case) by observing the scattering patterns that follow certain collisions. A large variety of exotic particles have been created as well in the shower of particles that result from some collisions, and explaining their existence and behavior has deepened theories of fundamental physics. From the explosive aftermath of these artificial high-energy particle collisions, robust theories of the most fundamental constituents of the natural world are being developed.

---

particle physics The branch of physics that deals with **subatomic particles**. See also **quantum field theory, quantum mechanics.**

particulate (pər-tĭk′yə-lĭt) *Adjective.* **1.** Formed of very small, separate particles. Dust and soot are forms of particulate matter. —*Noun.* **2.** A very small particle, as of dust or soot. Particulates that are given off by the burning of oil, gasoline, and other fuels can remain suspended in the atmosphere for long periods, where they are a major component of air pollution and smog. **3.** A substance or suspension composed of such particles, such as sand or smoke.

parvovirus (pär′vō-vī′rəs) Any of a group of small DNA viruses of the family Parvoviridae that cause disease in many vertebrates, especially a febrile infection in dogs resulting in vomiting, diarrhea, and sometimes death. In humans it commonly causes an acute contagious infection called *fifth disease,* marked by a facial rash.

pascal (pă-skăl′, pä-skäl′) The SI derived unit used to measure pressure. One pascal is equal to one newton per square meter.

Pascal, Blaise 1623–1662. French mathematician, physicist, and philosopher who, with Pierre de Fermat, developed the mathematical theory of probability. He also contributed to the development of differential calculus, and he invented the mechanical calculator and the syringe. The pascal unit of pressure is named after him.

Blaise Pascal

Pascal's law (pă-skălz′, pä-skälz′) The principle that external static pressure exerted on a fluid is distributed evenly throughout the fluid. Differences in static pressure within a fluid thus arise only from sources within the fluid (such as the fluid's own weight, as in the case of atmospheric pressure).

passerine (păs′ə-rīn′) Belonging to the avian order Passeriformes, which includes the perching birds. Passerine birds make up more than half of all living birds. They are of small to medium size, have three toes pointing forward and one pointing back, and are often brightly colored. Larks, swallows, jays, crows, wrens, thrushes, cardinals, finches, sparrows, and blackbirds are all passerine birds.

passive immunity (păs′ĭv) See under **acquired immunity.**

passive margin See **divergent plate boundary.**

Pasteur (păs-tûr′), **Louis** 1822–1895. French chemist who founded modern microbiology. His early work with fermentation led him to invent the process of pasteurization. Pasteur established that microorganisms cause communicable diseases and infections.

Louis Pasteur

BIOGRAPHY **Louis Pasteur**

Through his experiments with bacteria in the 1860s, French chemist Louis Pasteur disproved the centuries-old belief that disease was caused by *spontaneous generation,* the idea that disease-causing parasites arise spontaneously in an organism. Pasteur demonstrated that the fermentation of wine to vinegar was caused by living agents that entered the wine from the air surrounding it, proving instead that microorganisms were able to reproduce. Drawing the conclusion that airborne agents could enter the bodies of humans and animals and cause disease, he then devoted his research to isolating the organisms that cause specific diseases and finding treatments to prevent them. He developed vaccines for anthrax, chicken cholera, and rabies. Pasteur's germ theory of disease was not immediately accepted, but thanks to the work of other pioneering scientists, such as Robert Koch, it eventually provided the foundation for modern branches of medicine such as microbiology, bacteriology, virology, and immunology. Pasteur is also known for developing pasteurization (originally for wine), a process of heating and rapidly cooling liquids that is used to kill disease-causing bacteria, particularly in dairy products.

---

pasteurization (păs′chər-ĭ-zā′shən) **1.** A process in which an unfermented liquid, such as milk, or a partially fermented one, such as beer, is heated to a specific temperature for a certain amount of time in order to kill pathogens that could cause disease, spoilage, or undesired fermentation. During pasteurization, the liquid is not allowed to reach its boiling point so as to avoid changing its molecular structure. **2.** The process of destroying most pathogens in certain foods, such as fish or clams, by irradiating them with gamma rays or other radiation to prevent spoilage. See Note at **Pasteur.**

patch (păch) **1.** A temporary, removable electronic connection, as one between two components in a communications system. **2.** A piece of code added to software in order to fix a bug, especially as a temporary correction between two versions of the same software.

patella (pə-tĕl′ə) Plural **patellae** (pə-tĕl′ē) The small, flat, movable bone at the front of the knee in most mammals. Also called *kneecap.* See more at **skeleton.**

pathname (păth′nām′) The fully specified name of a computer file, including the position of the file in the file system's directory structure.

pathogen (păth′ə-jən) An agent that causes infection or disease, especially a microorganism, such as a bacterium or protozoan, or a virus. See Note at **germ.**

pathology (pə-thŏl′ə-jē) **1.** The scientific study of disease and its causes, processes, and effects. **2.** The physical and mental abnormalities that result from disease or trauma.

pathway (păth′wā′) **1.** A chain of nerve fibers along which impulses normally travel. **2.** A sequence of enzymatic or other reactions by which one biological material is converted to another.

Pauli (pou′lē), **Wolfgang** 1900–1958. Austrian-born American physicist who in 1924 formulated a principle stating that no two fermions, such as two electrons in an atom, can have identical energy, mass, and angular momentum at the same time. This principle is known as the Pauli Exclusion Principle. He also hypothesized the existence of the neutrino in 1931, which was confirmed in 1956.

Pauli exclusion principle (pô′lē, pou′-) The principle that two fermions of a given type, such as electrons, protons, or neutrons, cannot occupy the same quantum state. It does not apply to bosons. This principle plays a key role in the electron orbital structure of atoms, since it prevents more than two electrons from occupying any given orbital (two are allowed, since they may have opposite spin, and thus be in different quantum states). See also **orbital, degeneracy pressure.**

Pauling (pô′lĭng), **Linus Carl** 1901–1994. American chemist noted for his work on the structure and nature of chemical bonding. By applying quantum physics to chemistry, he discovered the structure of many molecules found in living tissue, especially proteins and amino acids. Pauling also discovered the genetic defect that causes sickle cell anemia. He was awarded the Nobel Prize for chemistry in 1954.

Linus Pauling

BIOGRAPHY Linus Pauling

After American chemist Linus Pauling completed his major theoretical work on chemical bonding in the 1930s, culminating with the publication of his influential *The Nature of the Chemical Bond, and the Structure of Molecules and Crystals* (1939), he then turned his attention to examining protein structure. In 1948, while toying with folded pieces of paper, he had the sudden insight that the polypeptide chain of amino acids was coiled into a helical shape, which he later named the alpha helix. Around this same time Pauling realized that many diseases, in particular sickle cell anemia, might be molecular in origin, and his work laid the foundation for later human genome research. Late in his life Pauling devoted much of his time to the field he called *orthomolecular medicine,* which entailed studies of the health benefits of megadoses of vitamins and minerals, especially of vitamin C. Pauling is as well known for his efforts to make the world a better place as he is for advancing the frontiers of scientific knowledge. He campaigned tirelessly on behalf of world peace, and in the 1950s when his studies of the harmful effects of nuclear fallout from atomic weapons made him draw the conclusion that they should be banned, he was accused of being a Communist and prevented from traveling abroad, almost missing the award ceremony for the Nobel Prize for chemistry that he was awarded in 1954. He continued his work, nonetheless, circulating a petition against atmospheric nuclear testing that eventually was signed by more than 11,000 scientists. On October 10, 1963, the day the Nuclear Test Ban Treaty went into effect, Pauling was awarded the 1962 Nobel Peace Prize. He is the only person to receive two unshared Nobel Prizes.

Pavlov (păv′lôv′, -lôf′), **Ivan Petrovich** 1849–1936. Russian physiologist who studied the digestive system of dogs, investigating the nervous control of salivation and the role of enzymes. His experiments showed that if a bell is rung whenever food is presented to a dog, the dog will eventually salivate when it hears the bell, even if no food is presented. This demonstration of what is known as a *conditioned response* prompted later scientific studies of human and animal behavior.

Pb The symbol for **lead.**

PCB (pē′sē-bē′) Short for *polychlorinated biphenyl.* Any of a family of very stable industrial compounds used as lubricants, heat-transfer fluids, and plasticizers. The manufacture and use of PCBs has been restricted since the 1970s because they are very harmful to the environment, being especially deadly to fish and invertebrates, and stay in the food chain for many years.

Pd The symbol for **palladium.**

PDA (pē′dē-ā′) Short for *personal digital assistant.* A lightweight, handheld computer, typically employing a touch-sensitive screen rather than a keyboard, generally used for storing information such as addresses or schedules. Many PDAs include handwriting recognition software, some support voice recognition, and some have an internal cell phone and modem to link with other computers or networks.

pearl (pûrl) A smooth, slightly iridescent, white or grayish rounded growth inside the shells of some mollusks. Pearls form as a reaction to the presence of a foreign particle, and consist of thin layers of mother-of-pearl that are deposited around the particle. The pearls of oysters are often valued as gems.

peat (pēt) Partially decayed vegetable matter, especially peat moss, found in bogs. The low levels of oxygen and the acidic environment in bogs prevent the degradation of peat. Peat is burned as fuel and also used as fertilizer. See more at **bog.**

peat bog See under **bog.**

peat moss Any of various mosses of the genus *Sphagnum,* growing in very wet places, especially bogs, around the world. The leaves of peat moss have large dead cells surrounded by smaller living ones that contain chloroplasts. The walls of the dead cells are perforated and readily absorb water, up to 20 times their dry weight. The walls also contain phenol compounds that resist decay and have antiseptic properties. Peat moss releases hydrogen ions that increase the acidity of the water in bogs. Because of its ability to absorb liquids, peat moss is sometimes used as diaper material by traditional peoples and was once used in making bandages. Peat moss is now used primarily to increase the water-holding capacity of soil. Also called *sphagnum.* See more at **bog.**

pebble (pĕb′əl) A rock fragment larger than a granule and smaller than a cobble. Pebbles have a diameter between 4 and 64 mm (0.16 and 2.56 inches) and are often rounded.

pectin (pĕk′tĭn) Any of a group of carbohydrate substances found in the cell walls of plants and in the tissue between certain plant cells. Pectin is produced by the ripening of fruit and helps the ripe fruit remain firm. As the fruit overripens, the pectin breaks down into simple sugars (monosaccharides) and the fruit loses its shape and becomes soft. Pectins can be made to form gels, and are used in certain medicines and cosmetics and in making jellies.

pectoral (pĕk′tər-əl) Located in or attached to the chest, as a pectoral fin or a pectoral muscle.

pedalfer (pĭ-dăl′fər) Soil that is characterized by an abundance of aluminum and iron oxides. Pedalfers are common in humid regions and are deposited in the B horizon of ABC soils, through leaching.

pediatrics (pē′dē-ăt′rĭks) The branch of medicine that deals with the care of infants and children.

pedicel (pĕd′ĭ-səl) A small stalk supporting a single flower in an inflorescence.

pediment (pĕd′ə-mənt) A broad, gently sloping rock surface at the base of a steeper slope such as a mountain, often covered with alluvium. Pediments are formed through the exposure of bedrock by erosional processes, such as the flow of water. Pediments are usually found in arid regions where there is little vegetation to hold the overlying soil.

pedocal (pĕd′ə-kăl′) Soil that is characterized by an abundance of calcium carbonate and calcium oxide. Pedocals are common in arid or semiarid regions where the rate of evaporation is greater than the rate of leaching. They are deposited in the A horizon of ABC soils.

pedogenesis (pĕd′ə-jĕn′ĭ-sĭs) The process whereby soil forms from unconsolidated rocks.

pedology (pĭ-dŏl′ə-jē) The scientific study of soils, including their origins, characteristics, and classification.

pedon (pĕd′ən) The smallest unit or volume of soil that contains all the soil horizons of a particular soil type. It usually has a surface area of approximately 1 sq m (10.76 sq ft) and extends from the ground surface down to bedrock.

peduncle (pĭ-dŭng′kəl, pē′dŭng′kəl) **1.** The stalk that attaches a single flower, flower cluster, or fruit to the stem. See more at **flower.** **2.** A stalk supporting an animal organ, such as the eyestalk of a lobster. **3.** A slender stalk by which the base of a nonsessile tumor is attached to normal tissue. **4.** Any of several stalklike connecting structures in the brain, composed either of white matter or of white and gray matter.

peer-to-peer network (pîr′tə-pîr′) A network of personal computers, each of which acts as both client and sever, so that each can exchange files and email directly with every other computer on the network. Each computer can access any of the others, although access can be restricted to those files that a computer's user chooses to make available. Peer-to-peer networks are less expensive than client/server networks but less efficient when large amounts of data need to be exchanged. Compare **client/server network.**

Pegasus (pĕg′ə-səs) A constellation in the Northern Hemisphere near Aquarius and Andromeda.

pegmatite (pĕg′mə-tīt′) Any of various coarse-grained igneous rocks that often occur as wide veins cutting across other types of rock. Pegmatites form from water-rich magmas or lavas that cool slowly, allowing the crystals to grow to large sizes. Although pegmatites can be compositionally similar to a number of rocks, they most often have the composition of granite.

Peking man (pē′kĭng′) See **sinanthropus.**

pelagic (pə-lăj′ĭk) Relating to or living in or on oceanic waters. The pelagic zone of the ocean begins at the low tide mark and includes the entire oceanic water column. The pelagic ecosystem is largely dependent on the **phytoplankton** inhabiting the upper, sunlit regions, where most ocean organisms live. Biodiversity decreases sharply in the unlit zones where water pressure is high, temperatures are cold, and food sources scarce. Pelagic waters are divided, in descending order, into the **epipelagic,**

mesopelagic, bathypelagic, abyssopelagic, and **hadopelagic zones.**

pelite (pē′līt′) A sediment or sedimentary rock composed of fine fragments, as of clay or mud.

pellagra (pə-lăg′rə, -lā′grə) A disease caused by a lack of niacin in the diet, characterized by skin and digestive disorders and mental deterioration.

pelvic inflammatory disease (pĕl′vĭk) An infection of the female genital tract, especially of the fallopian tubes, caused by any of several sexually transmitted microorganisms, especially the bacteria *Chlamydia trachomatis* and *Neisseria gonorrhoeae,* and characterized by severe abdominal pain, fever, vaginal discharge, and, in some cases, scarring that can cause infertility.

pelvis (pĕl′vĭs) Plural **pelvises** or **pelves** (pĕl′vēz). The basin-shaped structure in vertebrate animals that joins the spine and lower or hind limbs. In primates, the pelvis is composed of the two hipbones joined to the sacrum. It contains, protects, and supports the intestines, bladder, and internal reproductive organs.

pendulum (pĕn′jə-ləm) A mass hung from a fixed support so that it is able to swing freely under the influence of gravity. Since the motion of pendulums is regular and periodic, they are often used to regulate the action of various devices, especially clocks.

penetrometer (pĕn′ĭ-trŏm′ĭ-tər) also **penetrameter** (pĕn′ĭ-trăm′ĭ-tər) **1.** A device for measuring the penetrating power of electromagnetic radiation, especially x-rays. **2.** A device for measuring the denseness, compaction, or penetrability of a substance, such as soil, agricultural produce, or semisolid petroleum products. A penetrometer typically measures the resistance of the substance to penetration to a given depth by a weight-driven cone or needle of a given shape.

Penfield (pĕn′fēld′), **Wilder Graves** 1891–1976. American-born Canadian neurosurgeon noted for his experimental work on the exposed brains of conscious humans. His findings increased scientific understanding of the functions of the brain, brain diseases such as epilepsy, and the mechanisms involved in speech.

penicillin (pĕn′ĭ-sĭl′ĭn) An antibiotic drug obtained from molds of the genus *Penicillium* and used to treat or prevent various infections caused by **gram-positive** bacteria such as **streptococcus.** Penicillin was the first of a class of antibiotics (whose names end in *–icillin*) that are derived from it and are active against a broader spectrum of bacteria. See Note at **Fleming** (Alexander).

penicillium (pĕn′ĭ-sĭl′ē-əm) Plural **penicilliums** or **penicillia.** Any of various bluish-green fungi of the genus *Penicillium,* that grow as molds on decaying fruits, ripening cheeses, and bread, and are used to produce penicillin and certain other antibiotics.

peninsula (pə-nĭn′syə-lə) A piece of land that projects into a body of water and is connected with a larger landmass.

penis (pē′nĭs) Plural **penises** or **penes** (pē′nēz). **1.** The male reproductive organ of mammals and some reptiles and birds. In mammals, the penis contains the urethra, which carries urine from the bladder and releases sperm during reproduction. **2.** A similar organ found in the males of some invertebrate animals.

Pennsylvanian (pĕn′səl-vān′yən) The sixth period of the Paleozoic Era, from about 320 to 286 million years ago. Rock deposits from this period of time alternate between being marine and terrestrial in origin, and are rich in coal. Reptiles first appeared in the Pennsylvanian Period. See Chart at **geologic time.**

Penrose (pĕn′rōz′), **Roger** Born 1931. British mathematical astronomer and physicist who worked with Stephen Hawking to explain the physics of black hole formation. He is also known for his theories on artificial intelligence.

penstock (pĕn′stŏk′) **1.** A sluice or gate used to control a flow of water. **2.** A pipe or conduit used to carry water to a water wheel or turbine.

pentagon (pĕn′tə-gŏn′) A polygon having five sides. —*Adjective* **pentagonal** (pĕn-tăg′ə-nəl).

pentane (pĕn′tān′) A colorless, flammable hydrocarbon derived from petroleum and used as a solvent. Pentane occurs in three isomers and is the fifth member of the alkane series. *Chemical formula:* $\mathbf{C_5H_{12}}$.

pentastome (pĕn′tə-stōm′) also **pentastomid** (pĕn′tə-stō′mĭd) See **tongue worm.**

pentode (pĕn′tōd′) Any electron tube with the basic structure and functionality of a triode, but including two extra electrodes, a **screen** and a **suppressor grid.** The screen helps the tube respond well at high frequencies (as in a **tetrode**), while a negatively charged suppressor grid adjacent to the **plate** prevents **secondary emission** of electrons from the plate, increasing the efficiency of the tube. See more at **suppressor.**

pentose (pĕn′tōs′) Any of a class of simple sugars (monosaccharides) having five carbon atoms per molecule. Ribose and deoxyribose are pentoses.

pentyl (pĕn′təl) See **amyl.**

penumbra (pĭ-nŭm′brə) Plural **penumbras** or **penumbrae** (pĭ-nŭm′brē). **1.** A partial shadow between regions of full shadow (the umbra) and full illumination, especially as cast by Earth, the Moon, or another body during an eclipse. During a partial lunar eclipse, a portion of the Moon's disk remains within the penumbra of Earth's shadow while the rest is darkened by the umbra. See Note at **eclipse. 2.** The grayish outer part of a sunspot. Compare **umbra.**

pepsin (pĕp′sĭn) Any of various digestive enzymes found in vertebrate animals that catalyze the hydrolysis of proteins to peptides.

peptic (pĕp′tĭk) **1.** Relating to the process of digestion or the secretions associated with it. **2.** Relating to or involving pepsin.

peptide (pĕp′tīd′) A chemical compound that is composed of a chain of two or more amino acids and is usually smaller than a protein. The amino acids can be alike or different. Many hormones and antibiotics are peptides.

peptide bond The chemical bond formed between amino acids, constituting the primary linkage in all protein structures. In a peptide bond, the carboxyl group (COOH) of one amino acid bonds with the amino group (NH_2) of another, forming the sequence CONH and releasing water (H_2O).

percent also **per cent** (pər-sĕnt′) One part in a hundred. For example, 62 percent (also written 62%) means 62 parts out of 100.

percentile (pər-sĕn′tīl′) Any of the 100 equal parts into which the range of the values of a set of data can be divided in order to show the distribution of those values. The percentile of a given value is determined by the percentage of the values that are smaller than that value. For example, a test score that is higher than 95 percent of the other scores is in the 95th percentile.

perchlorate (pər-klôr′āt′) A salt of perchloric acid, containing the group ClO_4.

perchloric acid (pər-klôr′ĭk) A clear, colorless liquid that is very corrosive and, under some conditions, extremely explosive. It is a powerful oxidant and is used as a catalyst and in explosives. *Chemical formula:* **$HClO_4$.**

perchloroethylene (pər-klôr′ō-ĕth′ə-lēn′) See **tetrachloroethylene.**

perennial (pə-rĕn′ē-əl) *Adjective.* **1.** Living for three or more years. —*Noun.* **2.** A perennial plant. Herbaceous perennials survive winter and drought as underground roots, rhizomes, bulbs, corms, or tubers. Woody perennials, including vines, shrubs, and trees, usually stop growing during winter and drought. Asters, irises, tulips, and peonies are familiar garden perennials. Compare **annual, biennial.**

Perey (pĕ-rā′), **Marguerite Catherine** 1909–1975. French physicist who discovered the element francium in 1939.

Marguerite Perey

perfect flower (pûr′fĭkt) A flower having both stamens and carpels. Most angiosperms have perfect flowers. Compare **imperfect flower.** See also **complete flower.**

perfect fungus Any of various fungi that have both a sexual phase and an asexual phase and reproduce by both sexually and asexually produced spores. Ascomycetes, basidiomycetes, and zygomycetes are perfect fungi. Compare **deuteromycete.**

perfect number A positive integer that equals the sum of all of its divisors other than itself. An example is 28, whose divisors (not counting itself) are 1, 2, 4, 7, and 14, which added together give 28.

peri– A prefix that means: "around" (as in *pericardium*), or "near" (as in *perihelion*).

perianth (pĕr′ē-ănth′) The sepals and petals of a flower considered together.

periapsis (pĕr′ē-ăp′sĭs) Plural **periapsides** (pĕr′ē-ăp′sĭ-dēz′) The point at which an orbiting object is closest to the center of mass of the body it is orbiting. This point is sometimes given a name that is specific to the body being orbited. For example, the periapsis of an object that is orbiting Earth is its **perigee** (from *gaia*, the Greek word for Earth), and the periapsis of an object orbiting the Sun is its **perihelion** (from *hēlios*, the Greek word for Sun). According to **Kepler's laws of planetary motion,** an object moves at its greatest velocity when it is at the periapsis. Compare **apoapsis.**

periastron (pĕr′ē-ăs′trən, -trŏn) Plural **periastra.** The point at which an object, such as a planet or comet, is closest to the center of mass of the star it is orbiting. Compare **apastron.**

pericardium (pĕr′ĭ-kär′dē-əm) Plural **pericardia.** The membrane sac that encloses the heart in vertebrate animals.

pericarp (pĕr′ĭ-kärp′) The tissue that arises from the ripened ovary wall of a fruit; the fruit wall. In fleshy fruits, the pericarp can often be divided into the exocarp, the mesocarp, and the endocarp. For example, in a peach, the skin is the exocarp, the yellow flesh is the mesocarp, and the stone or pit surrounding the seed is the endocarp.

pericycle (pĕr′ĭ-sī′kəl) A layer of nonvascular tissue that surrounds the vascular tissues in the roots of plants and is in turn surrounded by the endodermis. New lateral roots begin growth from the pericycle.

periderm (pĕr′ĭ-dûrm′) The outer, protective layers of tissue of woody roots and stems, consisting of the cork cambium and the tissues produced by it. See more at **cork cambium.**

peridotite (pĕr′ĭ-dō-tīt′, pə-rĭd′ə-) A coarse-grained igneous rock that consists mainly of olivine and pyroxene. It is believed to be one of the main constituent rocks of the Earth's mantle.

perigee (pĕr′ə-jē) **1.** The point nearest Earth's center in the orbit of the Moon or an artificial satellite. **2.** The point in an orbit that is nearest to the body being orbited. Compare **apogee, perihelion.**

perigynous (pə-rĭj′ə-nəs) Having sepals, petals, and stamens around the edge of a cuplike receptacle (the **hypanthium**) containing the pistil, as in flowers of the rose or cherry. Compare **epigynous, hypogynous.**

perihelion (pĕr′ə-hē′lē-ən) The point at which an orbiting object, such as a planet or a comet, is closest to the Sun. Compare **aphelion, perigee.**

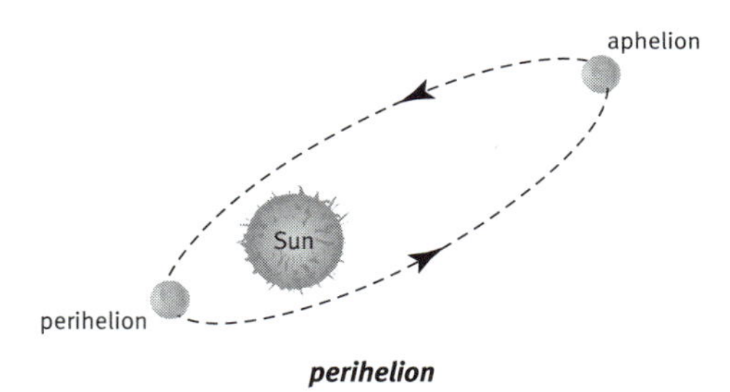

perihelion

perilune (pĕr′ĭ-lo͞on′) The point at which an object orbiting the Moon, such as an artificial satellite, is closest to the Moon's center.

perimeter (pə-rĭm′ĭ-tər) **1.** The sum of the lengths of the segments that form the sides of a polygon. **2.** The total length of any closed curve, such as the circumference of a circle.

perineum (pĕr′ə-nē′əm) Plural **perinea.** The region between the scrotum and the anus in males, and between the posterior vulva and the anus in females.

period (pĭr′ē-əd) **1.** A division of **geologic time** that is longer than an epoch and shorter than an era. **2.** The duration of one cycle of a regularly recurring action or event. See also **cycle, frequency.** **3.** An occurrence of menstruation. **4.** In the Periodic Table, any of the seven horizontal rows that contain elements arranged in order of increasing atomic number. All the elements in a particular period have the same number of electron shells in their atoms, equal to the number of the period. Thus, atoms of nickel, copper, and zinc, in period four, each have four electron shells. See **Periodic Table.**

periodic attractor (pîr′ē-ŏd′ĭk) See under **attractor.**

periodic motion See **harmonic motion.**

Periodic Table A table in which the chemical elements are arranged in order of increasing atomic number. Elements with similar properties are arranged in the same column (called a group), and elements with the same number of electron shells are arranged in the same row (called a period). See Table on pages 472–473.

periodontics (pĕr′ē-ə-dŏn′tĭks) The branch of dentistry that deals with the study and treatment of the tissues surrounding and supporting the teeth, especially the gums.

peripheral (pə-rĭf′ər-əl) *Adjective. Anatomy.* **1a.** Relating to or being the surface or outer part of a body or organ. **b.** Relating to or being part of the peripheral nervous system. *—Noun. Computer Science.* **2.** An auxiliary device, such as a printer or modem, distinct from a computer's central processing unit and working memory, and often connected externally.

peripheral nervous system The part of the nervous system in vertebrate animals that lies outside of the brain and spinal cord (central nervous system). It includes the nerves that extend to the limbs and many sense organs. Compare **central nervous system.**

periscope (pĕr′ĭ-skōp′) An instrument that has angled mirrors or prisms and allows objects not in the direct line of sight to be seen, often used on submarines and in military reconnaissance.

perissodactyl (pə-rĭs′ō-dăk′təl) Any of various hoofed mammals of the order Perissodactyla, having one or three hoofed toes on each hindfoot. During the Tertiary Period, perissodactyls were the dominant herbivorous fauna. Horses, tapirs, and rhinoceroses are perissodactyls. Also called *odd-toed ungulate.*

peristalsis (pĕr′ĭ-stôl′sĭs) The wavelike muscular contractions in tubular structures, especially organs of the digestive system such as the esophagus and the intestines. Peristalsis is characterized by alternate contraction and relaxation, which pushes ingested food through the digestive tract towards its release at the anus. Worms propel themselves through peristaltic movement.

peristome (pĕr′ĭ-stōm′) **1.** A fringe of toothlike appendages surrounding the mouth of the spore capsule of some mosses. The teeth unfold under damp conditions and curl up under dry conditions to disperse spores gradually. **2.** The area or parts around the mouth in certain invertebrates, such as the echinoderms.

perithecium (pĕr′ə-thē′shē-əm, -sē-əm) Plural **perithecia.** A small flask-shaped fruiting body in some ascomycete fungi that encloses the asci (spore sacs).

peritoneum (pĕr′ĭ-tn-ē′əm) Plural **peritonea.** The membrane that lines the walls of the abdomen and the pelvis (called the *parietal* peritoneum) and encloses the abdominal and pelvic organs (called the *visceral* peritoneum.) The space between the two, the peritoneal cavity, fills with inflammatory cells and pus when the peritoneum becomes infected.

peritonitis (pĕr′ĭ-tn-ī′tĭs) Infection or inflammation of the peritoneal cavity, usually caused by a ruptured organ, such as the appendix, in the gastrointestinal tract.

permafrost (pûr′mə-frôst′) A layer of soil or bedrock that has been continuously frozen for at least two years and as long as tens of thousands of years. Permafrost can reach depths of up to 1,524 m (4,999 ft). It is found throughout most of the polar regions and underlies about one fifth of the Earth's land surface.

permanent magnet (pûr′mə-nənt) A piece of ferromagnetic material that retains its magnetism after it has been magnetized.

permanganate (pər-măng′gə-nāt′) A salt of permanganic acid, containing the group MnO_4. Permanganates are strong oxidizing agents.

permanganic acid (pûr′măn-găn′ĭk) An unstable inorganic acid existing only in dilute solution. Its purple aqueous solution is used as an oxidizing agent. *Chemical formula:* $\mathbf{HMnO_4}$.

permeability (pûr′mē-ə-bĭl′ĭ-tē) **1.** The ability of

a substance to allow another substance to pass through it, especially the ability of a porous rock, sediment, or soil to transmit fluid through pores and cracks. Geologic permeability is usually measured in millidarcies. See more at **darcy. 2.** Magnetic permeability.

permeable (pûr′mē-ə-bəl) Capable of being passed through or permeated, especially by liquids or gases.

Permian (pûr′mē-ən) The seventh and last period of the Paleozoic Era, from about 286 to 245 million years ago. During the Permian Period the supercontinent Pangaea, comprising almost all of today's landmasses, formed. Gymnosperms evolved, the first modern conifers appeared, and reptiles diversified. The Permian Period ended with the largest known mass extinction in the history of life. It wiped out nearly 90 percent of known marine life forms. See Chart at **geologic time.**

permittivity (pûr′mĭ-tĭv′ĭ-tē) A measure of the ability of a material to resist the formation of an electric field within it, equal to the ratio between the electric flux density and the electric field strength generated by an electric charge in the material.

peroxide (pə-rŏk′sīd′) **1.** A compound containing the group O_2. Peroxides are strong oxidizers and are used as industrial bleaches. When any peroxide is combined with an acid, one of the products is hydrogen peroxide. **2.** Hydrogen peroxide.

peroxisome (pə-rŏk′sĭ-sōm′) A cell organelle containing enzymes that catalyze the production and breakdown of hydrogen peroxide.

perpendicular (pûr′pən-dĭk′yə-lər) *Adjective.* **1.** Intersecting at or forming a right angle or right angles. —*Noun.* **2.** A line or plane that is perpendicular to a given line or plane.

Perseus (pûr′sē-əs) A constellation in the Northern Hemisphere near Andromeda and Auriga.

personal computer (pûr′sə-nəl) A computer built around a microprocessor for use by an individual. Personal computers have their own operating systems, software, and peripherals, and can generally be linked to networks. Compare **mainframe.**

personal digital assistant See **PDA.**

personality disorder (pûr′sə-năl′ĭ-tē) Any of a group of psychiatric disorders in which a person's abnormal self-perception or ability to relate to others results in undesirable behaviors and interferes with normal social and emotional functioning.

perturbation (pûr′tər-bā′shən) **1.** A small change in a physical system, most often in a physical system at equilibrium that is disturbed from the outside. **2.** Variation in a designated orbit, as of a planet, that results from the influence of one or more external bodies. Gravitational attraction between planets can cause perturbations and cause a planet to deviate from its expected orbit. Perturbations in Neptune's orbit led to the discovery of the object—Pluto—that was causing the perturbations. Perturbations in the orbits of stars have led to the discovery of planetary systems outside of our Solar system.

perturbation theory A set of mathematical methods for obtaining approximate solutions to complex equations for which no exact solution is possible or known, generally involving an iterative algorithm in which each new term contributing to the solution has less significance than the last. In quantum physics, Feynman diagrams are used to calculate the terms for the perturbation theory solution of interactions between particles.

pertussis (pər-tŭs′ĭs) See **whooping cough.**

Peru Current (pə-ro͞o′) See **Humboldt Current.**

Perutz (pə-ro͞ots′, pĕr′əts), **Max Ferdinand** 1914–2002. Austrian-born British biochemist who determined the structure of hemoglobin, demonstrating that it is composed of four chains of molecules. For this work he shared with John Kendrew the 1962 Nobel Prize for chemistry.

pervasive developmental disorder (pər-vā′sĭv) Any of several developmental disorders, such as autism or Asperger's syndrome, characterized by severe deficits in social interaction and communication or by the presence of repetitive, stereotyped behaviors. Most of these disorders are evident in the first years of life.

pesticide (pĕs′tĭ-sīd′) A chemical used to kill harmful animals or plants. Pesticides are used especially in agriculture and around areas where humans live. Some are harmful to humans, either from direct contact or as

Periodic Table

The Periodic Table arranges the chemical elements in two ways. The first is by **atomic number,** starting with hydrogen (atomic number = 1) in the upper left-hand corner and continuing in ascending order from left to right. The second is by the number of electrons in the outermost shell. Elements having the same number of electrons in the outermost shell are placed in the same column. Since the number of electrons in the outermost shell in large part determines the chemical nature of an element, elements in the same column have similar chemical properties.

This arrangement of the elements was devised by **Dmitri Mendeleev** in 1869,

Key	
1	atomic number
H	symbol
Hydrogen	
1.00794	atomic weight (or mass number of most stable isotope if in parentheses)

	Group 1a								
Period 1	1 H Hydrogen 1.00794	Group 2a							
Period 2	3 Li Lithium 6.941	4 Be Beryllium 9.0122							
Period 3	11 Na Sodium 22.9898	12 Mg Magnesium 24.305	Group 3b	Group 4b	Group 5b	Group 6b	Group 7b	Group 8	Group 8
Period 4	19 K Potassium 39.098	20 Ca Calcium 40.08	21 Sc Scandium 44.956	22 Ti Titanium 47.87	23 V Vanadium 50.942	24 Cr Chromium 51.996	25 Mn Manganese 54.9380	26 Fe Iron 55.845	27 Co Cobalt 58.9332
Period 5	37 Rb Rubidium 85.47	38 Sr Strontium 87.62	39 Y Yttrium 88.906	40 Zr Zirconium 91.22	41 Nb Niobium 92.906	42 Mo Molybdenum 95.94	43 Tc Technetium (98)	44 Ru Ruthenium 101.07	45 Rh Rhodium 102.905
Period 6	55 Cs Cesium 132.905	56 Ba Barium 137.33	57–71 * Lanthanides	72 Hf Hafnium 178.49	73 Ta Tantalum 180.948	74 W Tungsten 183.84	75 Re Rhenium 186.2	76 Os Osmium 190.2	77 Ir Iridium 192.2
Period 7	87 Fr Francium (223)	88 Ra Radium (226)	89–103 ** Actinides	104 Rf Rutherfordium (261)	105 Db Dubnium (262)	106 Sg Seaborgium (266)	107 Bh Bohrium (264)	108 Hs Hassium (265)	109 Mt Meitnerium (268)

*LANTHANIDES	57 La Lanthanum 138.91	58 Ce Cerium 140.12	59 Pr Praseodymium 140.908	60 Nd Neodymium 144.24	61 Pm Promethium (145)	62 Sm Samarium 150.36	63 Eu Europium 151.96
**ACTINIDES	89 Ac Actinium (227)	90 Th Thorium 232.038	91 Pa Protactinium 231.036	92 U Uranium 238.03	93 Np Neptunium (237)	94 Pu Plutonium (244)	95 Am Americium (243)

ALPHABETICAL TABLE OF THE ELEMENTS

Element	Symbol	Atomic Number	Element	Symbol	Atomic Number	Element	Symbol	Atomic Number	Element	Symbol	Atomic Number
Actinium	Ac	89	Calcium	Ca	20	Element 113	Uut	113	Hassium	Hs	108
Aluminum	Al	13	Californium	Cf	98	Element 114	Uuq	114	Helium	He	2
Americium	Am	95	Carbon	C	6	Element 115	Uup	115	Holmium	Ho	67
Antimony	Sb	51	Cerium	Ce	58	Element 116	Uuh	116	Hydrogen	H	1
Argon	Ar	18	Cesium	Cs	55	Element 118	Uuo	118	Indium	In	49
Arsenic	As	33	Chlorine	Cl	17	Erbium	Er	68	Iodine	I	53
Astatine	At	85	Chromium	Cr	24	Europium	Eu	63	Iridium	Ir	77
Barium	Ba	56	Cobalt	Co	27	Fermium	Fm	100	Iron	Fe	26
Berkelium	Bk	97	Copper	Cu	29	Fluorine	F	9	Krypton	Kr	36
Beryllium	Be	4	Curium	Cm	96	Francium	Fr	87	Lanthanum	La	57
Bismuth	Bi	83	Darmstadtium	Ds	110	Gadolinium	Gd	64	Lawrencium	Lr	103
Bohrium	Bh	107	Dubnium	Db	105	Gallium	Ga	31	Lead	Pb	82
Boron	B	5	Dysprosium	Dy	66	Germanium	Ge	32	Lithium	Li	3
Bromine	Br	35	Einsteinium	Es	99	Gold	Au	79	Lutetium	Lu	71
Cadmium	Cd	48	Element 112	Uub	112	Hafnium	Hf	72	Magnesium	Mg	12

of the Elements

before many of the elements now known were discovered. To maintain the overall logic of the table, Mendeleev allowed space for undiscovered elements whose existence he predicted. This space has since been mostly filled in. Elements 112–116 and 118 have been isolated experimentally but not yet officially named.

The **lanthanide** series (elements 57–71) and the **actinide** series (elements 89–103) are composed of elements with Group 3b chemical properties. They are placed below the main body of the table to make it easier to read.

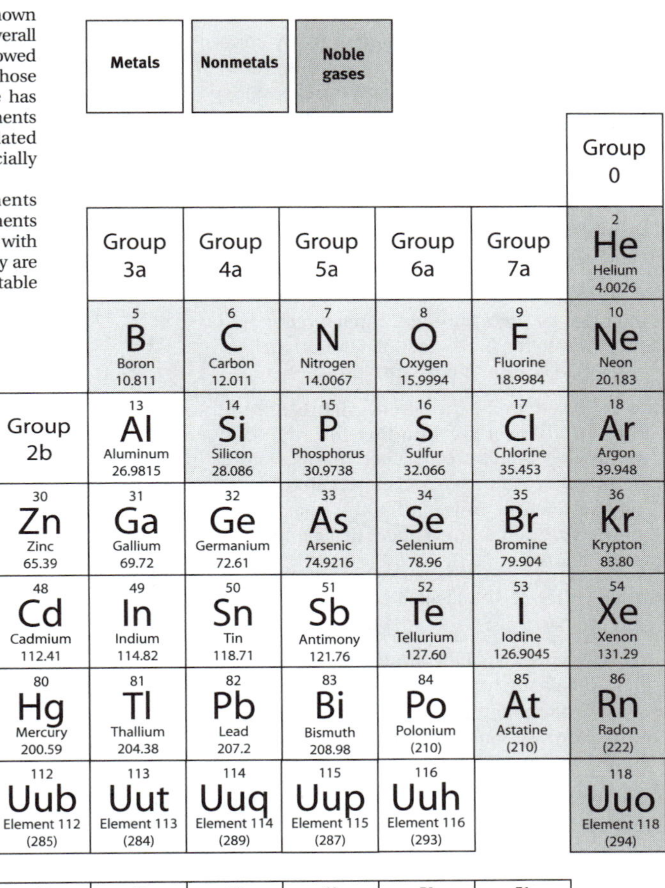

64 Gd Gadolinium 157.25	65 Tb Terbium 158.925	66 Dy Dysprosium 162.50	67 Ho Holmium 164.930	68 Er Erbium 167.26	69 Tm Thulium 168.934	70 Yb Ytterbium 173.04	71 Lu Lutetium 174.97
96 Cm Curium (247)	97 Bk Berkelium (247)	98 Cf Californium (251)	99 Es Einsteinium (252)	100 Fm Fermium (257)	101 Md Mendelevium (258)	102 No Nobelium (259)	103 Lr Lawrencium (262)

Element	Symbol	Atomic Number	Element	Symbol	Atomic Number	Element	Symbol	Atomic Number	Element	Symbol	Atomic Number
Manganese	Mn	25	Phosphorus	P	15	Rutherfordium	Rf	104	Thorium	Th	90
Meitnerium	Mt	109	Platinum	Pt	78	Samarium	Sm	62	Thulium	Tm	69
Mendelevium	Md	101	Plutonium	Pu	94	Scandium	Sc	21	Tin	Sn	50
Mercury	Hg	80	Polonium	Po	84	Seaborgium	Sg	106	Titanium	Ti	22
Molybdenum	Mo	42	Potassium	K	19	Selenium	Se	34	Tungsten	W	74
Neodymium	Nd	60	Praseodymium	Pr	59	Silicon	Si	14	Uranium	U	92
Neon	Ne	10	Promethium	Pm	61	Silver	Ag	47	Vanadium	V	23
Neptunium	Np	93	Protactinium	Pa	91	Sodium	Na	11	Xenon	Xe	54
Nickel	Ni	28	Radium	Ra	88	Strontium	Sr	38	Ytterbium	Yb	70
Niobium	Nb	41	Radon	Rn	86	Sulfur	S	16	Yttrium	Y	39
Nitrogen	N	7	Roentgenium	Rg	111	Tantalum	Ta	73	Zinc	Zn	30
Nobelium	No	102	Rhenium	Re	75	Technetium	Tc	43	Zirconium	Zr	40
Osmium	Os	76	Rhodium	Rh	45	Tellurium	Te	52			
Oxygen	O	8	Rubidium	Rb	37	Terbium	Tb	65			
Palladium	Pd	46	Ruthenium	Ru	44	Thallium	Tl	81			

residue on food, or are harmful to the environment because of their high toxicity, such as DDT (which is now banned in many countries). Pesticides include fungicides, herbicides, insecticides, and rodenticides. See more at **fungicide, herbicide, insecticide.**

peta– A prefix that means: **1.** One quadrillion (10^{15}), as in *petahertz,* one quadrillion hertz. **2.** 2^{50} (that is, 1,125,899,906,842,624), which is the power of two closest to a quadrillion, as in *petabyte.*

petabyte (pĕtʹə-bīt) **1.** A unit of computer memory or data storage capacity equal to 1,024 terabytes (2^{50} bytes). **2.** One quadrillion bytes. See Note at **megabyte.**

petal (pĕtʹl) One of the often brightly colored parts of a flower surrounding the reproductive organs. Petals are attached to the receptacle underneath the carpels and stamens and may be separate or joined at their bases. As a group, the petals are called the **corolla**. See more at **flower.**

petiole (pĕtʹē-ōl′) See **leafstalk.**

petri dish (pēʹtrē) A shallow, circular dish with a loose cover, usually made of transparent glass or plastic and used to grow cultures of microorganisms. The petri dish is named after German bacteriologist Julius Richard Petri (1852–1921).

petrifaction (pĕt′rə-făkʹshən) also **petrification** (pĕt′rə-fĭ-kāʹshən) The process by which organic materials are turned into rock. Petrifaction occurs when water that is rich with inorganic minerals, such as calcium carbonate or silica, passes slowly through organic matter, such as wood or bone, replacing its cellular structure with minerals.

petrochemical (pĕt′rō-kĕmʹĭ-kəl) Any of a large number of chemicals made from petroleum or natural gas. Important petrochemicals include benzene, ammonia, acetylene, and polystyrene. Petrochemicals are used to produce a wide variety of materials, such as plastics, explosives, fertilizers, and synthetic fibers.

petrogenesis (pĕt′rō-jĕnʹĭ-sĭs) The branch of petrology that deals with the origin and formation of rocks, especially igneous rocks.

petrography (pə-trŏgʹrə-fē) The description and classification of rocks, especially by means of microscopic analysis.

petroleum (pə-trōʹlē-əm) A thick, flammable, yellow-to-black mixture of gaseous, liquid, and solid hydrocarbons that occurs naturally beneath the Earth's surface. It can be separated into fractions including natural gas, gasoline, naphtha, kerosene, paraffin wax, asphalt, and fuel and lubricating oils, and is used as raw material for a wide variety of derivative products. It is believed to originate from the accumulated remains of fossil plants and animals, especially in shallow marine environments.

petrology (pə-trŏlʹə-jē) The scientific study of the origin, composition, and structure of rocks.

PET scan (pĕt) Short for *positron emission tomography scan.* A cross-sectional image of a metabolic process in a human or animal body produced by positron emission tomography.

pH (pēʹāchʹ) A numerical measure of the acidity or alkalinity of a solution, usually measured on a scale of 0 to 14. Neutral solutions (such as pure water) have a pH of 7, acidic solutions have a pH lower than 7, and alkaline solutions have a pH higher than 7. The pH of lemon juice is 2.4; that of household ammonia is 11.5. The normal pH for human blood is 7.4. ► The letters *pH* stand for *potential of hydrogen,* since pH is effectively a measure of the concentration of hydrogen ions (that is, protons) in a substance. The pH scale was devised in 1923 by Danish biochemist Søren Peter Lauritz Sørensen (1868–1969).

phagocyte (făgʹə-sīt′) Any of various organisms or specialized cells that engulf and ingest other cells or particles. In vertebrate animals, phagocytes are white blood cells that break down bacteria and other microorganisms, foreign particles, and cellular debris. These include monocytes, macrophages, and most granulocytes. ► The process by which phagocytes engulf and break down bacteria or particles is called **phagocytosis** (făg′ə-sī-tōʹsĭs). During phagocytosis the cell encloses foreign material and the extracellular fluid surrounding it by an infolding of a part of the cell membrane, which then pinches off to form a vesicle, called a *phagosome.* The phagosomes fuse with lysosomes, resulting in digestion of the ingested matter. Unicellular protists such as amoebas ingest food by the process of phagocytosis.

phalanx (fā′lăngks′) Plural **phalanges** (fə-lăn′jēz). Any of the small bones of the fingers or toes in humans or the digits of many other vertebrates.

phaneritic (făn′ə-rĭt′ĭk) Of or relating to an igneous rock in which the crystals are so coarse that individual minerals can be distinguished with the naked eye. Phaneritic rocks are intrusive rocks that cooled slowly enough to allow significant crystal growth. Compare **aphanitic.**

Phanerozoic (făn′ər-ə-zō′ĭk) The period of geologic time from about 540 million years ago to the present, including the Paleozoic, Mesozoic, and Cenozoic Eras. When this period of time was first defined, it was thought to coincide with the first appearance of life in the fossil record. It is now known that bacterial and other forms of life were present in the Precambrian Eon, and the Phanerozoic is understood to coincide with the appearance of life forms that evolved external skeletons. See Chart at **geologic time.**

pharmacogenetics (fär′mə-kō-jə-nĕt′ĭks) The study of the genetic factors that influence an organism's reaction to a drug.

pharmacology (fär′mə-kŏl′ə-jē) The scientific study of drugs and their effects, especially in the treatment of disease.

pharyngitis (făr′ĭn-jī′tĭs) Inflammation of the pharynx, often a result of viral or bacterial infection, especially streptococcal bacteria. See also **strep throat.**

pharynx (făr′ĭngks) Plural **pharynges** (fə-rĭn′jēz) or **pharynxes.** The passage that leads from the cavities of the nose and mouth to the larynx (voice box) and esophagus. Air passes through the pharynx on the way to the lungs, and food enters the esophagus from the pharynx.

phase (fāz) **1.** Any of the forms, recurring in cycles, in which the Moon or a planet appears in the sky. **2.** One of a set of possible homogenous, discrete states of a physical system. States of matter such as solid and liquid are examples of phases, as are different crystal lattice structures in metals such as iron. See also **phase transition, state of matter.** **3.** A measure of how far some cyclic behavior, such as wave motion, has proceeded through its cycle, measured in degrees or radians. At the beginning of the phase, its value is zero; at one quarter of its cycle, its phase is 90 degrees ($\pi/2$ radians); halfway through the cycle its value is 180 degrees (π radians), and so on. ► The **phase angle** between two waves is a measure of their difference in phase. Two waves of the same frequency that are perfectly in phase have phase angle zero; if one wave is ahead of the other by a quarter cycle, its phase angle 90 degrees ($\pi/2$ radians); waves that are perfectly out of phase have phase angle 180 degrees (π radians), and so on. See more at **wave.**

phase inverter See **inverter** (sense 1).

phase modulation A method of transmitting signals in which the value of the signal is proportional to the phase angle of a carrier wave. It is often combined with **amplitude modulation** to transmit digital information via modems.

phase rule A rule used in thermodynamics stating that the number of **degrees of freedom** in a physical system at equilibrium is equal to the number of chemical components in the system minus the number of phases plus the constant 2. Also called *Gibbs phase rule.* See also **phase transition, state of matter.**

phase transition A change in a feature of a physical system that results in a discrete transition of that system to another state. For example, the melting of ice is a phase transition of water from a solid phase to a liquid phase. Phase transitions often involve the absorption or emission of energy from the system; ice, at 0 ° Celsius, must absorb a considerable amount of heat energy to become water. See also **state of matter, thermodynamics.**

phellem (fĕl′əm) See **cork** (sense 1).

phelloderm (fĕl′ə-dûrm′) The layer of tissue, often very thin, produced on the inside of the cork cambium in woody plants. It forms a secondary cortex. See more at **cork cambium.**

phellogen (fĕl′ə-jən) See **cork cambium.**

phenanthrene (fə-năn′thrēn′) A colorless, crystalline hydrocarbon obtained by fractional distillation of coal-tar oils and used in dyes, drugs, and explosives. Phenanthrene is an isomer of anthracene and has three benzene rings fused together but not arranged in a straight line. *Chemical formula:* $C_{14}H_{10}$.

phenazine (fĕn′ə-zēn′) **1.** Any of a class of organic compounds containing two benzene rings joined to each other by an inner benzene ring in which two of the carbon atoms have been replaced by nitrogen atoms. Phenazines are a type of pyrazine, and are used especially in making dyes and in pharmaceuticals. **2.** The simplest of this class of compounds. It forms yellow crystals and is used to make dyes. *Chemical formula:* $C_{12}H_8N_2$.

phenobarbital (fē′nō-bär′bĭ-tôl′, -tăl′) A crystalline barbiturate that is used as a sedative and an anticonvulsant. *Chemical formula:* $C_{12}H_{12}N_2O_3$.

phenocryst (fē′nə-krĭst′) A large crystal that is surrounded by a finer-grained matrix in an igneous rock. Phenocrysts are usually the first crystals to form from a cooling magma, and therefore have sufficient room to grow to a large size. They are analogous to *porphyroblasts* in metamorphic rock.

phenol (fē′nôl′, -nōl′) **1.** Any of a class of organic compounds that contain a hydroxyl group (OH) attached to a carbon atom that is part of an aromatic ring. Phenols are similar to alcohols but are more soluble in water, and occur as colorless solids or liquids at room temperature. Some phenols occur naturally in the essential oils of plants. Phenols are used in industry to make plastics and detergents. **2.** The simplest phenol, consisting of a benzene ring attached to a hydroxyl group (OH). It is a poisonous, white, crystalline compound and is used to make plastics and drugs. Also called *carbolic acid. Chemical formula:* C_6H_6O.

phenology (fĭ-nŏl′ə-jē) The scientific study of cyclical biological events, such as flowering, breeding, and migration, in relation to climatic conditions. Phenological records of the dates on which seasonal phenomena occur provide important information on how climate change affects ecosystems over time.

phenolphthalein (fē′nōl-thăl′ēn′) A white or pale-yellow, crystalline powder used as an indicator for acid and basic solutions. In solutions that are either neutral or basic, it is colorless, while it is pink or red in solutions that are alkali. It is also used as a laxative and in making dyes. *Chemical formula:* $C_{20}H_{14}O_4$.

phenothiazine (fē′nō-thī′ə-zēn′) **1.** A yellow or green, toxic organic compound used in insecticides and dyes and to treat infections with worms and other parasites in livestock. *Chemical formula:* $C_{12}H_9NS$. **2.** Any of a group of drugs derived from this compound. Phenothiazines are dopamine antagonists and are used in the treatment of psychiatric disorders, such as schizophrenia.

phenotype (fē′nə-tīp′) The physical appearance of an organism as distinguished from its genetic makeup. The phenotype of an organism depends on which genes are dominant and on the interaction between genes and environment. Compare **genotype.**

phenyl (fĕn′əl, fē′nəl) The radical C_6H_5, derived from benzene by the removal of one hydrogen atom.

phenylalanine (fĕn′əl-ăl′ə-nēn′) An essential amino acid. *Chemical formula:* $C_9H_{11}NO_2$. See more at **amino acid.**

phenylene (fĕn′ə-lēn′, fē′nə-) The radical C_6H_4, derived from benzene by removal of two hydrogen atoms.

phenylketonuria (fĕn′əl-kēt′n-o͝or′ē-ə, fē′nəl-) A genetic disorder in which the body lacks an enzyme necessary to metabolize phenylalanine to tyrosine. If untreated, the disorder can cause brain damage and progressive mental retardation as a result of the accumulation of phenylalanine and its breakdown products.

phenylthiocarbamide (fĕn′əl-thī′ō-kär′bə-mīd′, -kär-băm′īd, fē′nəl-) A crystalline compound that tastes somewhat or intensely bitter to people with a specific dominant gene and is used to test for the presence of the gene. Also called *phenylthiourea. Chemical formula:* $C_6H_5NHCSNH_2$.

phenylthiourea (fĕn′əl-thī′ō-yo͝o-rē′ə, fē′nəl-) See **phenylthiocarbamide.**

pheromone (fĕr′ə-mōn′) A chemical secreted by an animal that influences the behavior or development of other members of the same species. Queen bees, for example, give off a pheromone that prevents other females in the hive from becoming sexually mature, with the result that only the queen bee mates and lays eggs. In many animal species, pheromones are used to establish territory and attract mates.

A CLOSER LOOK **pheromone**

The release of *pheromones* is one of various forms of nonverbal communication many animals use to transmit messages to other members of the same species. The complex molecular structure of pheromones allows these chemical messages to contain a great deal of often very specific information. The pheromone released by sexually receptive silkworm moths, first isolated in the 1950s, is one of the best-studied examples. The pheromone *bombykol,* released by the female from a gland in her belly, is detectable by male silkworm moths up to several kilometers away. The male identifies the chemical in the environment with tiny receptors at the tip of his antennae and is then able to hone in on the female. Hornets, when disturbed, release an alarm pheromone that calls other hornets to their aid. Female mice pheromones may excite a male mouse to mate immediately. In addition to producing instinctive behavioral responses, pheromones can also produce changes in an animal's physiology, spurring the onset of puberty or bringing on estrus. Pheromones used by animals, such as cats and dogs, to mark territory can convey information about an animal's species, gender, age, social and reproductive status, size, and even when it was last in the area. But can humans communicate via chemicals, too? In the 1970s Martha McClintock showed that the menstrual cycles of women living closely together in dormitories tended to become synchronized, an effect thought by some to be mediated by pheromones. Despite such evidence, no pheromone receptors have yet been found in humans.

phlebotomy (flĭ-bŏt′ə-mē) The act or practice of opening a vein by incision or puncture to remove blood.

phlegm (flĕm) Thick mucus produced by the mucous membranes of the respiratory tract, as during a cold or other respiratory infection.

phloem (flō′ĕm′) A tissue in vascular plants that conducts food from the leaves and other photosynthetic tissues to other plant parts. Phloem consists of several different kinds of cells: sieve elements, parenchyma cells, sclereids, and fibers. In mature woody plants it forms a sheathlike layer of tissue in the stem, just inside the bark. See more at **cambium, photosynthesis.** Compare **xylem.**

phlogiston (flō-jĭs′tən) A hypothetical colorless, odorless, weightless substance once believed to be the combustible part of all flammable substances and to be given off as flame during burning. In the 18th century, Antoine Lavoisier proved that phlogiston does not exist. See Note at **Lavoisier.**

phlogopite (flŏg′ə-pīt′) A yellow to dark-brown mica. Phlogopite is monoclinic and is usually found in limestone. It is used in insulation. *Chemical formula:* $K(Mg,Fe)_3AlSi_3O_{10}(OH)_2$.

Phobos (fō′bəs) The larger and inner of the two planetary satellites of Mars.

phon (fŏn) A unit of apparent loudness. The loudness of a signal in phons is equal to the intensity in decibels of a 1,000-hertz tone judged to be as loud as the signal being measured.

phonon (fō′nŏn′) The quantum of acoustic or vibrational energy. Phonons, like all quanta in quantum mechanics, have wavelike and particlelike properties. Phonons propagate through the vibrating material at the speed of sound in that material. Phonons are especially useful in mathematical models for calculating thermal and vibrational properties of solids.

phoronid (fə-rō′nĭd) Any of various small, mostly solitary, wormlike marine invertebrates of the phylum Phoronida. As adults, phoronids live in the ocean floor in a tube that they secrete made of chitin. They have a U-shaped digestive tract and feed by filtering food particles with a ciliated structure called a lophophore. The larvae of phoronids are very different in appearance and are free-swimming. Phoronids are thought to be related to the brachiopods and bryozoans.

phosgene (fŏs′jēn′) A colorless, volatile gas that has the odor of freshly mowed hay. When it reacts with water (as in the lungs during respiration), phosgene produces hydrochloric acid and carbon monoxide. It is used in making glass, dyes, resins, and plastics, and was used as a poisonous gas during World War I. Also called *carbonyl chloride. Chemical formula:* $COCl_2$.

phosphate (fŏs′fāt′) A salt or ester of phosphoric acid, containing the group PO_4. Phosphates are important in metabolism and are frequently used in fertilizers.

phospholipid (fŏs′fō-lĭp′ĭd) Any of various phosphorus-containing lipids, such as lecithin, that are composed mainly of fatty acids, a phosphate group, and a simple organic molecule such as glycerol. Phospholipids are the main lipids in cell membranes.

phosphoprotein (fŏs′fō-prō′tēn′) Any of a group of proteins, such as casein, containing chemically bound phosphoric acid.

phosphor (fŏs′fər) Any of various substances that can emit light after absorbing some form of radiation. Television screens and fluorescent lamp tubes are coated on the inside with phosphors. See Note at **cathode-ray tube.**

phosphorescence (fŏs′fə-rĕs′əns) **1.** The emission of light by a substance as a result of having absorbed energy from a form of electromagnetic radiation, such as visible light or x-rays. Unlike fluorescence, phosphorescence continues for a short while after the source of radiation is removed. Glow-in-the-dark products are phosphorescent. Compare **fluorescence** (sense 1). **2.** The light produced in this way.

phosphoric acid (fŏs-fôr′ĭk) A clear liquid, or a solid that forms colorless, rhombus-shaped crystals, that is used in fertilizers, detergents, food flavoring, and pharmaceuticals. *Chemical formula:* H_3PO_4.

phosphorous acid (fŏs′fər-əs, fŏs-fôr′əs) A white or yellowish crystalline solid used as a reducing agent. It is hygroscopic (easily absorbs moisture from the atmosphere). *Chemical formula:* H_3PO_3.

phosphorus (fŏs′fər-əs) *Symbol* **P** A highly reactive, poisonous nonmetallic element occurring naturally in phosphates, especially in the mineral apatite. It exists in white (or sometimes yellow), red, and black forms, and is an essential component of protoplasm. Phosphorus is used to make matches, fireworks, and fertilizers and to protect metal surfaces from corrosion. Atomic number 15; atomic weight 30.9738; melting point (white) 44.1°C; boiling point 280°C; specific gravity (white) 1.82; valence 3, 5. See **Periodic Table.**

phosphorylation (fŏs′fər-ə-lā′shən) The addition of a phosphate group to an organic molecule. Phosphorylation is important for many processes in living cells. ATP is formed during cell respiration from ADP by phosphorylation, as in the mitochondria of eukaryotic cells (oxidative phosphorylation) and the chloroplasts of plant cells (photosynthetic phosphorylation). Phosphorylation also regulates the activity of proteins, such as enzymes, which are often activated by the addition of a phosphate group and deactivated by its removal (called *dephosphorylation*).

photic (fō′tĭk) **1.** Of or relating to light. **2.** Penetrated by or receiving light. **3.** Relating to the layer of a body of water that is penetrated by sufficient sunlight for photosynthesis. The depth of the photic zone is dependent on the clarity of the water and the amount and intensity of direct sunlight, although it does not usually exceed 200 m. Also called *euphotic.* Compare **aphotic.**

photo– A prefix that means "light," as in *photoreceptor.*

photoautotroph (fō′tō-ô′tə-trŏf′, -trōf′) See **phototroph.**

photocell (fō′tō-sĕl′) See **photoresistor.**

photochemical smog (fō′tō-kĕm′ĭ-kəl) See **smog** (sense 1).

photochemistry (fō′tō-kĕm′ĭ-strē) The scientific study of the effects of light and ultraviolet radiation on chemical reactions.

photodegradable (fō′tō-dĭ-grā′də-bəl) Capable of decomposing when exposed to light. Photodegradable plastic, for example, becomes brittle and breaks into smaller pieces when exposed to sunlight, helping reduce litter and environmental damage.

photodiode (fō′tō-dī′ōd′) A diode that exhibits sensitivity to light, either by varying its electrical resistance like a photoresistor, or generating a electric potential in the manner of a photoelectric cell. See more at **photoelectric.**

photodynamics (fō′tō-dī-năm′ĭks) The scientific study of the effects of light on organic compounds and on the metabolisms of living things.

photoelectric (fō′tō-ĭ-lĕk′trĭk) Relating to or exhibiting to electrical effects upon exposure to light. For example, some photoelectric materials emit electrons called **photoelectrons** upon exposure to certain frequencies of light;

others, such as photoresistors and phototransistors, change their electrical properties. See also **photoelectric effect.**

photoelectric cell An electronic device having an electrical output that varies in response to the strength of incident electromagnetic radiation, especially visible light. Photoelectric cells make use of the **photoelectric effect,** in which electrons are displaced by photons in substances such as silicon or selenium to generate a voltage in response to radiation. Photoelectric cells are used to detect light electronically in cameras and night vision apparatus and to generate electrical power in solar cells. Also called *photovoltaic cell.* See also **solar cell.**

photoelectric effect The emission of electrons from a material, such as a metal, as a result of being struck by photons. Some substances, such as selenium, are particularly susceptible to this effect. The photoelectric effect is used in photoelectric and solar cells to create an electric potential. Also called *photoemission.*

photoelectron (fō′tō-ĭ-lĕk′trŏn′) An electron released or ejected from a photoelectric substance, having absorbed energy from incoming light.

photoemission (fō′tō-ĭ-mĭsh′ən) See **photoelectric effect.**

photolysis (fō-tŏl′ĭ-sĭs) Chemical decomposition induced by light or other radiant energy. Photolysis plays an important role in photosynthesis, during which it produces energy by splitting water molecules into gaseous oxygen and hydrogen ions.

photometric brightness (fō′tə-mĕt′rĭk) See **luminance.**

photometry (fō-tŏm′ĭ-trē) The measurement of the intensity, brightness, or other properties of light. Also called *photometrics.* See also **luminous intensity.**

photomicrograph (fō′tō-mī′krə-grăf′) A photograph made through a microscope.

photomultiplier (fō′tō-mŭl′tə-plī′ər) An electrical device designed for the detection of weak electromagnetic radiation, usually light, by amplifying the energy of the photons that strike it into stronger electrical signals. Photomultipliers are used in night-vision technology and in telescopes to detect light not strong enough to be visible by the unaided eye. ▸ The most common photomultiplier is the **tube photomultiplier**; it exploits secondary emission of electrons in a vacuum tube in the manner of an electron multiplier. When radiation strikes the cathode of a tube photomultiplier, electrons called **photoelectrons** are emitted and attracted to positively charged electrodes called dynodes. When they collide with the dynode, more electrons are released; these are in turn attracted to another dynode at a higher voltage to release yet more electrons, and so on. At the end of this process, there is a current flow at the anode that is strong enough to be easily detected.

photon (fō′tŏn′) The subatomic particle that carries the electromagnetic force and is the quantum of electromagnetic radiation. The photon has a rest mass of zero, but has measurable momentum, exhibits deflection by a gravitational field, and can exert a force. It has no electric charge, has an indefinitely long lifetime, and is its own antiparticle. See Note at **electromagnetic radiation.** See Table at **subatomic particle.**

photonics (fō-tŏn′ĭks) The scientific study or application of electromagnetic energy whose basic unit is the photon, incorporating optics, laser technology, electrical engineering, materials science, and information storage and processing.

photoperiod (fō′tō-pîr′ē-əd) The duration of an organism's daily exposure to light, considered especially with regard to the phenomena of photoperiodism.

photoperiodism (fō′tō-pîr′ē-ə-dĭz′əm) also **photoperiodicity** (fō′tō-pîr′ē-ə-dĭs′ĭ-tē) The response of an organism to changes in its photoperiod, especially as indicated by vital processes. For example, many plants exhibit photoperiodism by flowering only after being exposed to a set amount of daylight, as by requiring either a long or short day to flower. Plant growth, seed germination, and fruiting are also affected by day length. Photoperiodic responses in plants are regulated by special pigments known as **phytochromes**. In animals, migration, mating, amount of sleep, and other behaviors are also photoperiodic. In many animals, photoperiodism is regulated by the hormone **melatonin**.

photoreceptor (fō′tō-rĭ-sĕp′tər) **1.** A specialized structure or cell that is sensitive to light. In vertebrate animals, the photoreceptors are the rods and cones of the eye's retina. See Note at **circadian rhythm.** **2.** An electronic device that converts light energy into electrical signals. Photoreceptors are used in photocopy and facsimile machines, cameras, and solar cells.

photoresistor (fō′tō-rĭ-zĭs′tər) A resistor whose resistance varies as a function of the intensity of light it is exposed to. Also called *photocell.* See also **photoelectric.**

photorespiration (fō′tō-rĕs′pə-rā′shən) The chemical combination of carbohydrates with oxygen in plants with the release of carbon dioxide. Photorespiration requires the presence of light, is catalyzed in the chloroplasts by the same enzymes that catalyze the combination of carbohydrates with carbon dioxide during photosynthesis, and occurs when oxygen concentrations in the cell are high. Photorespiration typically takes place during conditions of high light intensity, dryness, and heat (often resulting in the closure of stomata), when the amount of carbon dioxide entering the plant is reduced, and the amount of oxygen produced by photosynthesis accumulates. Photorespiration thus acts to produce carbon dioxide when it is unavailable and acts as a check on photosynthesis and on the productivity of the plant. Unlike cellular respiration, photorespiration does not produce any ATP or NADH, and so consumes chemical energy rather than produces it. Many angiosperms have a supplemental method of carbon-dioxide uptake that minimizes losses from photorespiration.

photosensitivity (fō′tō-sĕn′sĭ-tĭv′ĭ-tē) **1.** Sensitivity or responsiveness to light. **2.** An abnormally heightened response, especially of the skin or eyes, to sunlight or ultraviolet radiation, caused by certain disorders, medications, or chemicals.

photosphere (fō′tə-sfîr′) The lowest visible layer of a star, lying beneath the chromosphere and the corona. Stars are made entirely of gas and thus have no surface per se, but the gas beneath the photosphere is opaque, so the photosphere acts as their effective visible surface; it is also the boundary from which the Sun's diameter is measured. The Sun's photosphere is a very thin layer made up of numerous granules (transient convective cells) where hot gases rise and give off light and heat. The photosphere of the Sun has a temperature of around 6,000°K and is the region in which sunspot activity is located.

photosynthesis (fō′tō-sĭn′thĭ-sĭs) The process by which green plants, algae, diatoms, and certain forms of bacteria make carbohydrates from carbon dioxide and water in the presence of chlorophyll, using energy captured from sunlight by chlorophyll, and releasing excess oxygen as a byproduct. In plants and algae, photosynthesis takes place in organelles called **chloroplasts.** Photosynthesis is usually viewed as a two-step process. First, in the **light reactions,** the energy-providing molecule ATP is synthesized using light energy absorbed by chlorophyll and accessory pigments such as carotenoids and phycobilins, and water is broken apart into oxygen and a hydrogen ion, with the electron of the hydrogen transferred to another energy molecule, NADPH. The ATP and NADPH molecules power the second part of photosynthesis by the transfer of electrons. In these light-independent or **dark reactions,** carbon is broken away from carbon dioxide and combined with hydrogen via the **Calvin cycle** to create carbohydrates. Some of the carbohydrates, the sugars, can then be transported around the organism for immediate use; others, the starches, can be stored for later use. Compare **chemosynthesis.** See Note at **transpiration.**

phototransistor (fō′tō-trăn-zĭs′tər) A transistor that regulates current or switches it on and off based on the intensity of the light it is exposed to rather than an external electric signal. Phototransistors are used in many electric-eye applications, including digital cameras, in which millions of tiny phototransistors map an image into an array of electrical signals. See also **photoelectric effect.**

phototroph (fō′tə-trŏf′) An organism that manufactures its own food from inorganic substances using light for energy. Green plants, certain algae, and photosynthetic bacteria are phototrophs. Also called *photoautotroph.*

phototropism (fō-tŏt′rə-pĭz′əm) The growth or movement of a fixed organism toward or away from light. In plants, phototropism is a

A CLOSER LOOK **photosynthesis**

Almost all life on Earth depends on food made by organisms that can perform *photosynthesis,* such as green plants, algae, and cyanobacteria. These organisms make carbohydrates from carbon dioxide and water using light energy from the Sun. They capture this energy with various pigments which absorb different wavelengths of light. The most important pigment, *chlorophyll a,* captures mainly blue and red light frequencies, but reflects green light. In plants, the other pigments are *chlorophyll b* and *carotenoids.* The carotenoids are usually masked by the green color of chlorophyll, but in temperate environments they can be seen as the bright reds and yellows of autumn after the chlorophyll in the leaves has broken down. The energy gathered by these pigments is passed to chlorophyll a. During the light reactions, the plant uses this energy to break water molecules into oxygen (O_2), hydrogen ions, and electrons. The light reactions produce more oxygen than is needed for cellular respiration, so it is released as waste. All of the oxygen in the Earth's atmosphere today was produced as waste by photosynthetic organisms, especially cyanobacteria, which have been producing oxygen for some three billion years, since their first appearance in the Precambrian Eon. During the dark reactions, the plant uses hydrogen ions and the electrons to make carbon dioxide into carbohydrates. Within the leaf of a green plant, photosynthesis takes place in chlorophyll-containing chloroplasts in the columnlike cells of the palisade layer and in the cells of the spongy parenchyma. The cells obtain carbon dioxide from air that enters the leaf through holes called *stomata,* which also allow excess oxygen to escape. Water from the roots is brought to the leaf by the vascular tissues called *xylem,* while the carbohydrates made by the leaf are distributed to the rest of the plant by the vascular tissue called *phloem.*

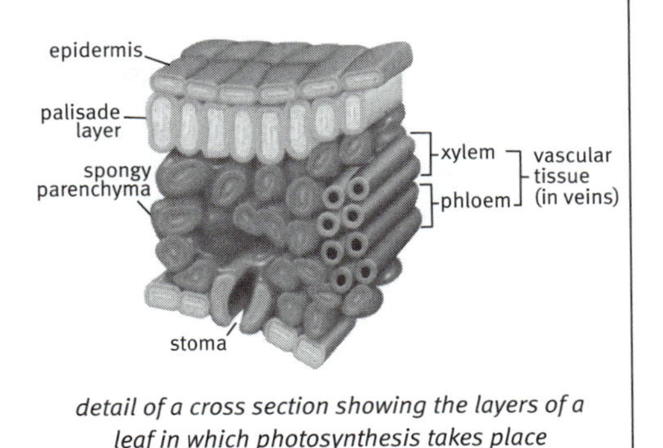

detail of a cross section showing the layers of a leaf in which photosynthesis takes place

response to blue wavelengths of light and is caused by a redistribution of auxin from the illuminated side to the darker side of the shoot, resulting in quicker growth on the darker side and bending of the shoot toward the source of light. Certain sessile invertebrates also exhibit phototropism.

photovoltaic (fō′tō-vŏl-tā′ĭk, -vōl-) Capable of producing a voltage, usually through photoemission, when exposed to radiant energy, especially light. See more at **photoelectric, solar cell.**

photovoltaic cell See **photoelectric cell.**

phreatic zone (frē-ăt′ĭk) A subsurface zone of soil or rock in which all pores and interstices are filled with fluid. Because of the weight of the overlying groundwater, the fluid pressure in the phreatic zone is greater than the atmospheric pressure. Compare **vadose zone.**

phreatophyte (frē-ăt′ə-fīt′) A deep-rooted plant that obtains water from a permanent ground supply or from the water table, such as many tamarisk species. Phreatophytes are often found in arid environments.

phrenology (frĭ-nŏl′ə-jē) The study of the shape of the skull as a means of determining character and intelligence. Phrenology has been disproven as a science.

phthalein (thăl′ēn′, thā′lēn′, fthăl′-) Any of a group of chemical compounds formed by a reaction of phthalic anhydride with a phenol. They are used to derive certain synthetic dyes.

phthalic acid (thăl′ĭk, fthăl′-) A colorless, crystalline organic acid prepared from naphthalene and used in the synthesis of dyes, perfumes, and other organic compounds. *Chemical formula:* $C_8H_6O_4$.

phthalic anhydride A white crystalline compound prepared by oxidizing naphthalene and used in the manufacture of phthaleins and other dyes, resins, plasticizers, and insecticides. *Chemical formula:* $C_8H_4O_3$.

phycobilin (fī′kō-bī′lĭn) Any of a class of water-soluble, mostly red, orange, and blue pigments found in cyanobacteria and red algae. Phycobilins absorb the blue and blue-green frequencies of light which penetrate deep water, and allow red algae to carry on photosynthesis at greater depths than other organisms. Phycobilins are chemically similar to chlorophyll and, like the heme in hemoglobin, are often found bound to proteins.

phycology (fī-kŏl′ə-jē) The scientific study of algae.

phyllite (fĭl′īt′) A green, gray, or red metamorphic rock, similar to slate but often having a wavy surface and a distinctive luster imparted by the presence of mica.

phylloclade (fĭl′ə-klād′) A flattened, photosynthetic branch or stem that resembles or performs the function of a leaf, as in certain cacti such as the prickly pear.

phyllode (fĭl′ōd) A flattened leafstalk that functions as a leaf, as in an acacia.

phyllome (fĭl′ōm′) A leaf or a plant part that evolved from a leaf.

phylloquinone (fĭl′ə-kwĭ-nōn′, -kwĭ′nōn) See **vitamin** K_1.

phyllotaxy (fĭl′ə-tăk′sē) The pattern of leaf distribution and arrangement on a stem. —*Adjective* **phyllotactic.**

phylogeny (fī-lŏj′ə-nē) The evolutionary development and history of a species or higher taxonomic grouping of organisms.

phylum (fī′ləm) Plural **phyla.** A group of organisms ranking above a class and below a kingdom. See Table at **taxonomy.**

physical anthropology (fĭz′ĭ-kəl) The branch of anthropology that deals with human evolutionary biology, physical variation, and classification. Compare **cultural anthropology.**

physical chemistry The branch of chemistry that is concerned with the physical structure of chemical compounds, the amount of energy they have, the way they react with other compounds, and the bonds that hold their atoms together.

physical geography The scientific study of the natural features of the Earth's surface, especially in its current aspects, including land formation, climate, currents, and distribution of flora and fauna. Also called *physiography.*

physical medicine The branch of medicine that deals with the diagnosis, treatment, and prevention of disease and disability by physical means such as manipulation, massage, and exercise, often with mechanical devices, and the application of heat, cold, electricity, radiation, and water.

physical science Any of several branches of science, such as physics, chemistry, and astronomy, that study the nature and properties of energy and nonliving matter. Compare **life science.**

physics (fĭz′ĭks) **1.** The scientific study of matter, energy, space, and time, and of the relations between them. **2.** The behavior of a given physical system, especially as understood by a physical theory.

physiography (fĭz′ē-ŏg′rə-fē) See **physical geography.**

physiology (fĭz′ē-ŏl′ə-jē) The scientific study of an organism's vital functions, including growth and development, the absorption and processing of nutrients, the synthesis and distribution of proteins and other organic molecules, and the functioning of different tissues, organs, and other anatomic structures. Physiology studies the normal mechanical, physical, and biochemical processes of animals and plants.

phytochemical (fī′tō-kĕm′ĭ-kəl) A plant-derived chemical that is not considered an essential nutrient in the human diet but is believed to have beneficial health effects.

phytochrome (fī′tə-krōm′) Any of a group of cytoplasmic pigments found in green plants and some green algae that absorb red light and regulate dormancy, seed germination, and flowering. Phytochromes consist of a bile pigment attached to a protein, and occur in an active and inactive form, each of which can be converted into the other depending on the wavelength of red light that is absorbed.

phytogeography (fī′tō-jē-ŏg′rə-fē) The scientific study of the geographic distribution of plants. Also called *geobotany.* See more at **biogeography.**

phytohormone (fī′tō-hôr′mōn′) See **plant hormone.**

phytology (fī-tŏl′ə-jē) The study of plants; botany.

phytopathology (fī′tō-pə-thŏl′ə-jē) The scientific study of plant diseases and their causes, processes, and effects.

phytoplankton (fī′tō-plăngk′tən) Plankton consisting of free-floating algae, protists, and cyanobacteria. Phytoplankton form the beginning of the food chain for aquatic animals and fix large amounts of carbon, which would otherwise be released as carbon dioxide.

phytoplasma (fī′tə-plăz′mə) Any of a group of extremely small bacteria that are similar to **mycoplasmas** in that they have a cell membrane instead of cell walls and can assume a variety of shapes, but are parasitic solely in plants. In many plants, phytoplasmas invade cells of the food-carrying tissue known as phloem and are usually spread by plant-sucking insects, such as the leafhopper, which draws its food from phloem. Phytoplasmas cause some 200 plant diseases affecting several hundred genera of plants. See also **mycoplasma.**

phytoremediation (fī′tō-rĭ-mē′dē-ā′shən) See under **bioremediation.**

phytotoxic (fī′tō-tŏk′sĭk) Poisonous to plants.

pi (pī) An irrational number that has a numerical value of 3.14159265358979… and is represented by the symbol π. It expresses the ratio of the circumference to the diameter of a circle and appears in many mathematical expressions.

pia mater (pī′ə mā′tər, pē′ə mä′tər) The fine vascular membrane that closely envelops the brain and spinal cord under the arachnoid and the dura mater.

Pickering (pĭk′ər-ĭng), **Edward Charles** 1846–1919. American astronomer who made many innovations in the equipment used to observe and measure the distance of stars. In 1884 he published the first catalog of stellar magnitudes. His brother **William Henry Pickering** (1858–1938) discovered Phoebe, the ninth moon of Saturn (1899), and predicted the existence of Pluto (1919).

picric acid (pĭk′rĭk) A poisonous, yellow crystalline solid used in explosives, dyes, and antiseptics. *Chemical formula:* $C_6H_3N_3O_7$.

piedmont (pēd′mŏnt′) An area of land, glacier, or other feature formed or lying at the foot of a mountain or mountain range.

piezoelectric effect (pī-ē′zō-ĭ-lĕk′trĭk) The generation of an electric charge in certain nonconducting materials, such as quartz crystals and ceramics, when they are subjected to mechanical stress (such as pressure or vibration), or the generation of vibrations in such materials when they are subjected to an electric field. Piezoelectric materials exposed to a fairly constant electric field tend to vibrate at a precise frequency with very little variation, making them useful as time-keeping devices in electronic clocks, as used in wristwatches and computers.

piezometer (pī′ĭ-zŏm′ĭ-tər) **1.** An instrument for measuring fluid pressure, such as the pressure of water or gas in a pipe. **2.** An instrument for measuring the compressibility of a material, especially in terms of the change in volume it undergoes when subjected to hydrostatic pressure.

pigment (pĭg′mənt) **1.** An organic compound that gives a characteristic color to plant or animal tissues and is involved in vital processes. Chlorophyll, which gives a green color to plants, and hemoglobin, which gives blood its red color, are examples of pigments. **2.** A substance or material used as coloring.

pileus (pī′lē-əs) Plural **pilei** (pī′lē-ī′). The umbrellalike fruiting structure forming the top of a fleshy fungus. It is supported by the **stipe.** The cap of a mushroom is a pileus.

pillow lava (pĭl′ō) Lava that forms from an underwater eruption and is characterized by pillow-shaped masses. Pillow lava forms when hot lava is suddenly exposed to cold water, forming a sacklike membrane that is filled with additional cooling and solidifying lava. The outer membrane is usually finer grained than the inner lava.

Piltdown man (pĭlt′doun′) A presumed early species of human, *Eoanthropus dawsoni,* postulated from a skull supposedly found in a gravel bed in about 1912 but determined in 1953 to be a fake constructed from a human cranium and the jawbone of an ape.

pi meson See **pion.**

pincers (pĭn′sərz) A jointed grasping claw of certain animals, such as lobsters and scorpions.

pineal eye (pĭn′ē-əl, pī′nē-əl) A sensory structure capable of light reception, appearing as a spot on the top of the head in lampreys, hagfish, amphibians, and some reptiles, especially the tuatara. Pineal eyes can contain a rudimentary cornea, lens, cone cells, and retina, and are thought to be sensitive to light and dark but not to be able to form images.

pineal gland A small gland located near the brain and primarily involved in the production and secretion of the hormone melatonin, which regulates circadian rhythms in many animals, including humans.

pingo (pĭng′gō) Plural **pingos** or **pingoes.** A large mound or dome of ice covered with soil. Pingos are about 30 to 50 m (98 to 164 ft) high and up to 400 m (1,312 ft) in diameter and are found in Arctic regions. They are believed to form in basins (such as drained lake beds) as a result of the freezing and upward expansion of water held in subsurface soil, which initiates the doming, as well as by the rising and freezing of water trapped beneath or within the permafrost, as a result of hydrostatic pressure.

pinna (pĭn′ə) Plural **pinnae** (pĭn′ē) or **pinnas.** A leaflet or primary division of a pinnately compound leaf, especially of a fern frond.

pinnate (pĭn′āt′) Having parts or divisions arranged on each side of a common axis in the manner of a feather. Ash, hickory, and walnut trees have pinnate leaves.

pinnate

detail of a white ash (Fraxinus americana) *tree*

pinnigrade (pĭn′ĭ-grād′) Walking by means of finlike organs or flippers, as seals and walruses.

pinniped (pĭn′ə-pĕd′) Any of various carnivorous, aquatic mammals of the group Pinnipedia, which some believe is a suborder of the Carnivora but others consider a separate mammalian order. Pinnipeds have long, smooth bodies and finlike flippers for swimming. Seals and walruses are pinnipeds.

pinnule (pĭn′yo͞ol) Any of the smaller leaflets into which each leaflet of a bipinnately compound leaf is subdivided. The leaves of many ferns are divided into pinnules.

pint (pīnt) **1.** A unit of liquid volume or capacity in the US Customary System, equal to 16 fluid ounces or 28.88 cubic inches (about 0.47 liter). **2.** A unit of dry volume or capacity used in the US Customary System, equal to $\frac{1}{2}$ of a quart or 34.6 cubic inches (about 0.55 liter). See Table at **measurement.**

pion (pī′ŏn′) A meson occurring either in a neutral form with a mass 264 times that of an electron and a mean lifetime of 8.4×10^{-17} seconds or in a positively or negatively charged form with a mass 273 times that of an electron and a mean lifetime of 2.6×10^{-8} seconds. The pion was once believed to be the particle that mediates the **strong force,** which holds nucleons together in the nucleus; it is now believed that the gluon is the mediator particle. Pions do interact with nucleons, however, and are able to transform neutrons into protons and vice versa. Also called *pi-meson.* See Table at **subatomic particle.**

pipe (pīp) **1.** A vertical cylindrical vein of ore. **2.** See **volcanic pipe.**

piperazine (pī-pĕr′ə-zēn′, pĭ-) A colorless crystalline compound used as a hardener for epoxy resins, as an antihistamine, and as an agent for expelling or destroying parasitic intestinal worms. Piperazine belongs to the class of chemicals called pyrazines. *Chemical formula:* $C_4H_{10}N_2$.

pipette (pī-pĕt′) A graduated narrow glass tube, often with an enlarged bulb, used for transferring measured volumes of liquids.

Pisces (pī′sēz) A constellation in the Northern Hemisphere near Aries and Pegasus. Pisces (the Fishes) is the twelfth sign of the zodiac.

piscine (pī′sēn′, pĭs′īn′) Relating to or characteristic of fishes.

piscivorous (pĭ-sĭv′ər-əs, pī-) Habitually feeding on fish. Terns and cormorants are piscivorous birds.

pistil (pĭs′təl) One of the female reproductive organs of a flower, consisting of a single carpel or of several carpels fused together. A flower may have one pistil or more than one, though some flowers lack pistils and bear only the male reproductive organs known as **stamens**. See more at **carpel, flower.**

pistillate (pĭs′tə-lāt′) Having pistils but no stamens. Female flowers are pistillate.

piston (pĭs′tən) A solid cylinder or disk that fits snugly into a hollow cylinder and moves back and forth under the pressure of a fluid (typically a hot gas formed by combustion, as in many engines), or moves or compresses a fluid, as in a pump or compressor.

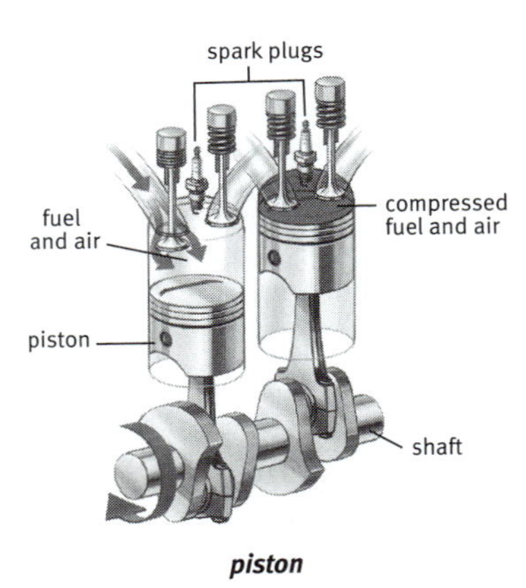

piston

pit (pĭt) The hard, inner layer (the endocarp) of certain drupes that are valued for their flesh, such as peaches, cherries, or olives. Not in scientific use.

pitch (pĭch) **1.** A thick, tarlike substance obtained by distilling coal tar, used for roofing, waterproofing, and paving. **2.** Any of various natural bitumens, such as asphalt, having similar uses. **3.** A resin derived from the sap of a cone-bearing tree, such as a pine.

pitchblende (pĭch′blĕnd′) A brown to black, often crusty, cubic mineral that is a principal ore of uranium. It is highly radioactive. *Chemical formula:* UO_2.

pith (pĭth) *Noun.* **1.** The soft, spongy tissue in the center of the stems of most flowering plants, gymnosperms, and ferns. Pith is composed of parenchyma cells. In plants that undergo secondary growth, such as angiosperms, the pith is surrounded by the vascular tissues and is gradually compressed by the inward growth of the vascular tissue known as xylem. In plants with woody stems, the pith dries out and often disintegrates as the plant grows older, leaving the stem hollow. See illustration at **xylem.** —*Verb.* **2.** To remove the pith from a plant stem. **3.** To sever or destroy the spinal cord of an animal for the purpose of dissecting it, usually by inserting a needle into the spinal canal.

pithecanthropus (pĭth′ĭ-kăn′thrə-pəs, -kăn-thrō′pəs) An extinct hominid postulated from bones found in Java in 1891 and originally designated *Pithecanthropus erectus* because it was thought to represent a species evolutionarily between apes and humans. Pithecanthropus is now classified as *Homo erectus.* Also called *Java man.* See more at **Homo erectus.**

pithecoid (pĭth′ĭ-koid′, pĭ-thē′koid) Resembling or relating to the apes, especially the anthropoid apes.

pituitary gland (pĭ-to͞o′ĭ-tĕr′ē) A gland at the base of the brain in vertebrate animals that is divided into two regions, anterior and posterior, each of which secretes important hormones. The anterior portion, whose secretions are directly controlled by the hypothalamus, produces hormones that regulate the function of most of the body's hormone-producing glands and organs, including the thyroid and adrenal glands. Growth hormone is also produced by the anterior pituitary. The posterior pituitary releases antidiuretic hormone (ADH) and oxytocin.

pit viper Any of various very venomous snakes of the family Crotalidae. Pit vipers are characterized by a triangular head with a small sensory pit below each eye. The pits detect infrared radiation and help the snake locate prey. Some scientists classify the pit vipers as a subfamily Crotalinae within the Viperidae, the family of vipers, rather than as a separate family. Copperheads, rattlesnakes, and fer-de-lances are pit vipers.

pixel (pĭk′səl) The most basic unit of an image

A CLOSER LOOK pixels

The images on a computer screen are composed of tiny dots called *pixels* (short for *picture element*). The computer controls each pixel individually. Most monitors have hundreds of thousands, or often millions, of pixels that are lit or dimmed to create an image. Each pixel of a color screen is made out of one red, one blue, and one green *subpixel*, generally arranged in a triangle, adjusted individually to create the combined effect of a single color but treated as a unit pixel for determining resolution. Pixels vary in size according to the size and resolution of the monitor. Smaller pixels provide higher resolution, and therefore sharper images, but require more memory to store the color and intensity data of each pixel and more processing time to refresh the screen. Resolution is frequently referred to in terms of *dpi*, or *dots per inch*.

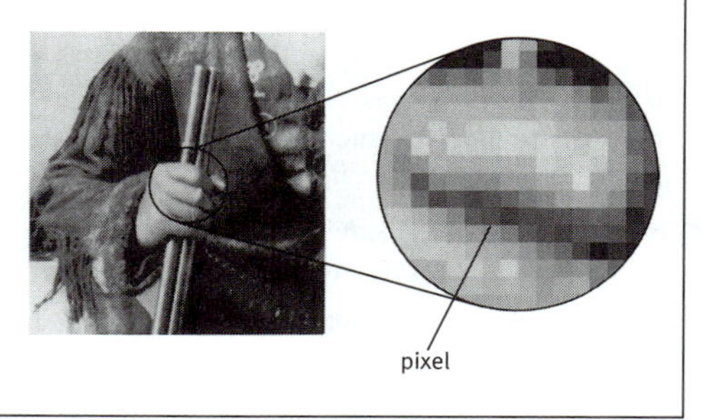

displayed on a computer or television screen or on a printer. Pixels are generally arranged in rows and columns; a given combination among the pixels of various brightness and color values forms an image. ▸ A **subpixel** is one of three components of a pixel used in the representation of a color image. Each subpixel represents the contribution of a single color—red, green, or blue—to the overall color and brightness of the pixel.

placebo (plə-sē′bō) A substance containing no medication and prescribed to reinforce a patient's expectation of getting well or used as a control in a clinical research trial to determine the effectiveness of a potential new drug.

placenta (plə-sĕn′tə) **1.** The sac-shaped organ that attaches the embryo or fetus to the uterus during pregnancy in most mammals. Blood flows between mother and fetus through the placenta, supplying oxygen and nutrients to the fetus and carrying away fetal waste products. The placenta is expelled after birth. **2.** The part of the ovary of a flowering plant to which the ovules are attached. In a green pepper, for example, the whitish tissue to which the seeds are attached is the placenta.

placer (plăs′ər) A surface deposit of minerals, such as gold or magnetite, laid down by a river. The minerals are usually concentrated in one area because they are relatively heavy and therefore settle out of the river's currents more quickly than lighter sediments such as silt and sand. ▸ The extraction of minerals from placers, as by panning, washing, or dredging, is called **placer mining**.

placoderm (plăk′ə-dûrm′) Any of various extinct fishes of the class Placodermi of the Silurian and Devonian Periods, characterized by bony plates of armor covering the head and flanks. The bodies of placoderms were spindle-shaped or flattened, and their skeletons were usually partially bony and included a cranium. Placoderms were the first group of fish to evolve jaws, but are not closely related to the jawed fish of today.

plage (pläzh) A bright and intensely hot area in the Sun's chromosphere, usually associated with a sunspot. It is typically brighter than its surroundings but may be indistinguishable due to lack of contrast. Plages are sources of strong ultraviolet radiation.

plagioclase (plā′jē-ə-klās′) Any of a series of common feldspar minerals, consisting of differing mixtures of sodium and calcium aluminum silicates. Plagioclase is typically white, yellow, or reddish-gray, but it can also be blue to black. It occurs in igneous rocks. The minerals albite, oligoclase, andesine, labradorite, bytownite, and anorthite are all examples of plagioclases. *Chemical formula:* $\mathbf{(Na,Ca)Al(Si,Al)Si_2O_8}$.

plague (plāg) **1.** Any of various highly infectious, usually fatal epidemic diseases. **2.** An often fatal disease caused by the bacterium *Yersinia pestis,* transmitted to humans usually by fleas that have bitten infected rats or other rodents. ▸ **Bubonic plague,** the most common type, is characterized by the tender, swollen lymph nodes called *buboes,* fever, clotting abnormalities of the blood, and tissue necrosis. An epidemic of bubonic plague in fourteenth-century Europe and Asia was known as the Black Death.

plain (plān) **1.** An extensive, relatively level area of land. Plains are present on all continents except Antarctica and are most often located in the interior regions. Because they can occur at almost any altitude or latitude, plains can be humid and forested, semiarid and grass-covered, or arid. **2.** A broad, level expanse, such as an area of the sea floor or a lunar mare.

planarian (plə-nâr′ē-ən) Any of various small, chiefly freshwater flatworms of the class Turbellaria, having soft, broad, ciliated bodies shaped like a leaf. Planarians have a mouth on their lower side that is often closer to the tail than the head, and a three-branched digestive cavity. If a planarian is cut into several pieces, each piece can grow into a whole new organism.

Planck (plängk), **Max Karl Ernst Ludwig** 1858–1947. German physicist who in 1900 formulated quantum theory, which explained and predicted certain phenomena that could not be accounted for in classical physics. Planck's theory was essential to the work of Albert Einstein, Niels Bohr, and many other modern physicists. In 1918 he won the Nobel Prize for physics.

Max Planck

Planck's constant (plängks) A physical constant that is used extensively in quantum mechanics and fixes the scale of quantization of many phenomena, such as the relation between the energy of a photon (a quantum of light) and its wavelength. Its value is approximately 6.626×10^{-34} joule-seconds (equivalent to units of angular momentum). Planck's constant is fundamental to phenomena as the quantization of angular momentum and is used in Heisenberg's Uncertainty Principle. See also **Dirac's constant, quantize.**

plane (plān) *Noun.* **1.** A two-dimensional surface, any two of whose points can be joined by a straight line that lies entirely in the surface. —*Adjective.* **2.** Lying in a plane: *a plane curve.*

plane geometry The mathematical study of geometric figures whose parts lie in the same plane, such as polygons, circles, and lines.

planet (plăn′ĭt) **1.** In the traditional model of solar systems, a celestial body larger than an asteroid or comet, illuminated by light from a star, such as the sun, around which it revolves. **2.** A celestial body that orbits the sun, has sufficient mass to assume nearly a round shape, clears out dust and debris from the neighborhood around its orbit, and is not a satellite of another planet. The eight planets are Mercury, Venus, Earth, Mars, Jupiter, Saturn, Uranus, and Neptune. Pluto was considered to be a planet until its reclassification in 2006 as a **dwarf planet.** A planetlike body with more than about ten times the mass of Jupiter would be considered a **brown dwarf** rather than a planet. See also **extrasolar planet, inner planet, outer planet.** —*Adjective* **planetary.**

planetary nebula (plăn′ĭ-tĕr′ē) A nebula consisting of a rapidly expanding shell of glowing gas, mostly hydrogen, ejected from a red giant upon its collapse into a white dwarf. Ultraviolet radiation from the hot, luminous white dwarf ionizes the expanding gas and causes it to glow. The nebula disappears once the cooling dwarf can no longer ionize it, and its material eventually returns to the interstellar medium. See more at **white dwarf.**

planetesimal (plăn′ĭ-tĕs′ə-məl) Any of innumerable small bodies of accreted gas and dust thought to have orbited the Sun during the formation of the planets. ▸ The theory that explains the formation of the solar system in terms of the aggregation of such bodies is known as the **planetesimal hypothesis.**

According to this theory, first proposed in 1900, the planetesimals formed within a spiral disk of dust and gas surrounding a central nucleus. Their gravitational attraction eventually caused the planetesimals to coalesce into **protoplanetary disks** from which larger objects such as planets, asteroids, and satellites were formed, while the nucleus coalesced into the Sun.

planetoid (plăn**′**ĭ-toid′) See **asteroid.**

plankton (plăngk**′**tən) Small organisms that float or drift in great numbers in bodies of salt or fresh water. Plankton is a primary food source for many animals, and consists of bacteria, protozoans, certain algae, cnidarians, tiny crustaceans such as copepods, and many other organisms. Compare **benthos, nekton.**

plant (plănt) Any of a wide variety of multicellular eukaryotic organisms, belonging to the kingdom Plantae and including the bryophytes and vascular plants. Plant cells have cell walls made of cellulose. Except for a few specialized symbionts, plants have chlorophyll and manufacture their own food through photosynthesis. Most plants grow in a fixed location and reproduce sexually, showing an alternation of generations between a diploid stage (with each cell having two sets of chromosomes) and haploid stage (with each cell having one set of chromosomes) in their life cycle. The first fossil plants date from the Silurian period. Formerly the algae, slime molds, dinoflagellates, and fungi, among other groups, were classified as plants, but now these are considered to belong to other kingdoms. See table at **taxonomy.**

plant hormone Any of various hormones produced by plants that control or regulate germination, growth, metabolism, or other physiological activities. Auxins, cytokinins, gibberellins, and abscisic acid are examples of plant hormones. Also called *phytohormone.*

plantigrade (plăn**′**tĭ-grād′) Walking with the entire sole of the foot on the ground, as humans, bears, raccoons, and rabbits.

planula (plăn**′**yə-lə) Plural **planulae** (plăn**′**yə-lē′). The flat, free-swimming, ciliated larva of a cnidarian.

plaque (plăk) **1.** A small disk-shaped formation or growth; a patch. **2.** A film of mucus and bacteria on the surface of the teeth. **3.** A deposit of material in a bodily tissue or organ, especially one of the fatty deposits that collect on the inner lining of an artery wall in atherosclerosis or one of the amyloid deposits that accumulate in the brain in Alzheimer's disease.

plasma (plăz**′**mə) **1.** See **blood plasma. 2.** Protoplasm or cytoplasm. **3.** One of four main **states of matter,** similar to a gas, but consisting of positively charged ions with most or all of their detached electrons moving freely about. Plasmas are produced by very high temperatures, as in the Sun and other stars, and also by the ionization resulting from exposure to an electric current, as in a fluorescent light bulb or a neon sign. See more at **state of matter.**

plasma cell A lymphocyte that originates from a **B cell** and produces antibodies as part of a humoral immune response. See Note at **antibody.**

plasma membrane See **cell membrane.**

plasma tail See under **tail** (sense 2).

plasmid (plăz**′**mĭd) A small, circular unit of DNA that replicates within a cell independently of the chromosomal DNA and is most often found in bacteria. Certain plasmids can insert themselves into chromosomes in places where there is a common sequence of nucleotides. Plasmids contain a few genes, which usually code for proteins, especially enzymes, some of which confer resistance to antibiotics. Plasmids are used in recombinant DNA research, especially to transform bacterial cells. See more at **transformation.**

plasmodial slime mold (plăz-mō**′**dē-əl) See under **slime mold.**

plasmodium (plăz-mō**′**dē-əm) Plural **plasmodia. 1.** A mass of protoplasm having many cell nuclei but not divided into separate cells. It is formed by the combination of many amoeba-like cells and is characteristic of the active, feeding phase of certain slime molds. **2.** Any of various single-celled organisms (called protozoans) that exist as parasites in vertebrate animals, one of which causes malaria.

plasmolysis (plăz-mŏl**′**ĭ-sĭs) Plural **plasmolyses** (plăz-mŏl**′**ĭ-sēz′). Shrinkage or contraction of the protoplasm away from the wall of a living plant or bacterial cell, caused by loss of water through osmosis. It results in loss of turgor and, in plants, wilting of the stems and leaves. If too severe, it can be fatal to the cell.

plaster of Paris (plăs′tər) A form of calcium phosphate derived from gypsum. It is mixed with water to make casts and molds.

plastic (plăs′tĭk) *Noun.* **1.** Any of numerous substances that can be shaped and molded when subjected to heat or pressure. Plastics are easily shaped because they consist of long-chain molecules known as polymers, which do not break apart when flexed. Plastics are usually artificial resins but can also be natural substances, as in certain cellular derivatives and shellac. Plastics can be pressed into thin layers, formed into objects, or drawn into fibers for use in textiles. Most do not conduct electricity well, are low in density, and are often very tough. Polyvinyl chloride, methyl methacrylate, and polystyrene are plastics. See more at **thermoplastic, thermosetting.** —*Adjective.* **2.** Capable of being molded or formed into a shape.

plastic strain Strain in which the distorted body does not return to its original size and shape after the deforming force has been removed. See more at **strain.**

plastic surgery The branch of surgery that deals with the remodeling, repair, or restoration of body parts, especially by the transfer of tissue.

plastid (plăs′tĭd) An organelle found in the cells of plants, green algae, red algae, and certain other protists. Like mitochondria, plastids have an inner and outer membrane, and contain their own DNA and ribosomes. Some plastids, such as the chloroplasts in plant leaves, contain pigments.

plate (plāt) *Noun.* **1.** A thin, flat sheet of metal or other material, especially one used as an electrode in a storage battery or capacitor, or as the anode of an electron tube. **2.** In plate tectonics, one of the sections of the Earth's lithosphere (crust and upper mantle) that is in constant motion along with other sections. It is the interaction of the plates that causes mountains, volcanos, and other land features to form and that causes earthquakes to occur. Six major plates and numerous smaller ones are recognized. See more at **tectonic boundary.** —*Verb.* **3.** To coat or cover with a thin layer of metal.

plateau (plă-tō′) An elevated, comparatively level expanse of land. Plateaus make up about 45 percent of the Earth's land surface.

platelet (plāt′lĭt) Any of the numerous small, round cell fragments found in the blood of mammals that function in the clotting of blood. Platelets contain no nuclei and are formed in the bone marrow from precursor cells called *megakaryocytes.* Platelets contribute to the coagulation process by adhering to damaged blood vessels, fibrinogen, and other platelets. An inadequate number of platelets leads to uncontrolled bleeding.

plate tectonics In geology, a theory that the Earth's lithosphere (the crust and upper mantle) is divided into a number of large, platelike sections that move as distinct masses. The movement of the plates is believed to result from the presence of large convection cells in the Earth's mantle which allow the rigid plates to move over the relatively plastic asthenosphere. The theory of plate tectonics was developed in the 1960s in an effort to explain the jigsawlike pattern of the Earth's continents. See Note at **fault.** See more at **tectonic boundary.**

A CLOSER LOOK **plate tectonics**

Although German physicist, meteorologist, and explorer Alfred Wegener proposed the theory of *continental drift* in 1912, suggesting that the continents were once joined as one large landmass, the explanation for the movement of such large landmasses into their current positions was not developed for several more decades. According to the theory of *plate tectonics,* which was proposed in the 1960s, the continents (and ocean floors) ride atop about a dozen semirigid plates—huge slabs of Earth's lithosphere—that are much larger than the continents themselves. The plates' constant movement is powered by huge convection currents of molten rock in Earth's mantle, thought by many geologists to be heated by the decay of radioactive elements deep within Earth. Although the plates move only a few inches per year, over the hundreds of millions of years of geological time, the continents are carried thousands of miles. Along their margins, the independently moving plates interact in three main ways. Where plates pull apart, new crust is formed. Where they collide, one plate is submerged beneath the other, and material from the bottom one returns to Earth's mantle. If the converging plates have land masses on them, the boundaries crumple, forming mountains. Plates also slide past each other, creating the faults that produce earthquakes. The six major plates are the Eurasian, American, African, Pacific, Indian, and Antarctic. See more at note at **Wegener.**

---

platform (plăt′fôrm′) **1.** The basic technology of a computer system's hardware and software, defining how a computer is operated and determining what other kinds of software can be used. Additional software or hardware must be compatible with the platform. **2.** The part of a continent's craton (the ancient, relatively undisturbed portion of a continental plate) that is covered by flat or nearly flat strata of sediment.

platinum (plăt′n-əm) *Symbol* **Pt** A soft, ductile, malleable, silver-white metallic element that usually occurs with osmium, iridium, palladium, or nickel. It has a high melting point and does not corrode in air. Platinum is used as a catalyst and in making jewelry, electrical contacts, and dental crowns. Atomic number 78; atomic weight 195.08; melting point 1,772°C; boiling point 3,827°C; specific gravity 21.45; valence 2, 3, 4. See **Periodic Table.**

platyhelminth (plăt′ĭ-hĕl′mĭnth) See **flatworm.**

platyrrhine (plăt′ĭ-rīn′) Of or relating to the New World monkeys, distinguished from the Old World monkeys by a broad nasal septum and widely separated nostrils that generally open to the side.

playa (plī′ə) A dry lake bed at the bottom of a desert basin, sometimes temporarily covered with water. Playas have no vegetation and are among the flattest geographical features in the world. Also called *sink.*

plecopteran (plĭ-kŏp′tər-ən) Any of various insects of the order Plecoptera, having two pairs of large wings, long antennae, and chewing mouthparts. Plecopterans cannot fly well, and generally have dull coloring that allows them to blend in with their surroundings. Plecopteran nymphs can live up to four years before turning into adults. The stoneflies are plecopterans.

Pleiades (plē′ə-dēz′) A loose collection of several hundred stars in the constellation Taurus, at least six of which are visible to the unaided eye.

Pleistocene (plī′stə-sēn′) The earlier of the two epochs of the Quaternary Period, from about 2 million to 10,000 years ago. The Pleistocene Epoch was characterized by the formation of widespread glaciers in the Northern Hemisphere and by the appearance of humans. Mammals included both small forms, such as saber-toothed tigers and horses and giant ones, such as mammoths and mastodons. Almost all the giant mammals, including woolly mammoths, giant wolves, giant ground sloths, and massive wombats disappeared at the end of the Pleistocene and the start of the Holocene. See Chart at **geologic time.**

pleopod (plē′ə-pŏd′) See **swimmeret.**

plesiosaur (plē′sē-ə-sôr′) Any of various large, extinct marine reptiles of the genus *Plesiosaurus* and related genera of the Mesozoic Era. Most plesiosaurs had a small head on a long neck and a broad body with paddlelike limbs; one group had a large head on a short neck. The exact relationship between plesiosaurs and other reptiles is not known.

pleura (plo͝or′ə) Plural **pleurae** (plo͝or′ē). A membrane that encloses each lung and lines the chest cavity.

pleurisy (plo͝or′ĭ-sē) An inflammation of the pleura, usually occurring because of complications of a respiratory disease or condition such as pneumonia, tuberculosis, pleural injury, or asbestos exposure. Pleurisy is usually accompanied by the accumulation of fluid between the pleurae, chills, fever, and painful breathing and coughing.

Pliocene (plī′ə-sēn′) The fifth and last epoch of the Tertiary Period, from about 5 to 2 million years ago. During this time the global climate became cooler and the number and expanse of grasslands and savannas increased greatly. This change in vegetation was accompanied by an increase in long-legged grazers. The land bridge between North America and South America also formed at this time, and massive ice sheets accumulated at the poles. In the later part of the epoch many of the species living in polar regions became extinct. See Chart at **geologic time.**

plug-in (plŭg′ĭn′) An accessory software or hardware package that is used in conjunction with an existing application or device to extend its capabilities or provide additional functions.

plumage (plo͞o′mĭj) The covering of feathers on a bird.

plumb line (plŭm) A line that is regarded as directed exactly toward the Earth's center of gravity.

plume (plo͞om) **1.** A feather, especially a large one. **2.** A body of magma that rises from the Earth's mantle into the crust. ► If a plume rises to the Earth's surface, it erupts as **lava.** ► If it remains below the Earth's surface, it eventually solidifies into a body of rock known as a **pluton. 3.** An area in air, water, soil, or rock containing pollutants released from a single source. A plume often spreads in the environment due to the action of wind, currents, or gravity.

plumule (plo͞om′yo͞ol) The developing bud of a plant embryo, situated above the cotyledons and consisting of the epicotyl and immature leaves. See more at **germination.**

plutino (plo͞o-tē′nō) A trans-Neptunian **Kuiper belt object** that orbits the Sun in the same time period as Pluto, exactly two orbits for every three orbits that Neptune makes. Too small to be considered planets, plutinos make up approximately one fourth of the Kuiper Belt objects.

Pluto (plo͞o′tō) A **dwarf planet** that until 2006 was classified as the ninth planet in the solar system. Pluto was not discovered until 1930, when Clyde Tombaugh noticed it while searching for an unknown planet thought to influence Uranus's orbit. Pluto's surface is covered with frozen methane and other ices, and its extremely thin atmosphere consists primarily of methane and nitrogen. Between 1979 and 1999 the orbit of Pluto crossed inside Neptune's orbit. Pluto has three moons: Charon (discovered in 1978) and Hydra and Nix (both discovered in 2005). See Table at **solar system.**

pluton (plo͞o′tŏn′) A large body of igneous rock formed when a plume of magma cools and solidifies underground. Although most plutons are deep within the Earth's crust, some become exposed at the surface due to plate-tectonic processes.

plutonium (plo͞o-tō′nē-əm) *Symbol* **Pu** A silvery, radioactive metallic element of the actinide series that has the highest atomic number of all naturally occurring elements. It is found in minute amounts in uranium ores and is produced artificially by bombarding uranium with neutrons. It is absorbed by bone marrow and is highly poisonous. Plutonium is used in nuclear weapons and as a fuel in nuclear reactors. Its longest-lived isotope is Pu 244 with a half-life of 76 million years. Atomic number 94; melting point 640°C; boiling point 3,232°C; specific gravity 19.84; valence 3, 4, 5, 6. See **Periodic Table.**

pluvial (plo͞o′vē-əl) Of, relating to, or caused by rain.

pluviometer (plo͞o′vē-ŏm′ĭ-tər) An instrument for measuring the amount of precipitation at a given location over a specified period of time. Also called *udometer.*

Pm The symbol for **promethium.**

pneumatic (no͞o-măt′ĭk) **1.** Of or relating to gases, especially air. **2.** Filled with or operated by compressed air. Pneumatic machines often involve the transmission of force through air pressure in pipes or tubes. See also **hydraulic.**

pneumatophore (no͞o-măt′ə-fôr′, no͞o′mə-tə-) A specialized root that grows upwards out of the water or mud to reach the air and obtain oxygen for the root systems of trees that live in swampy or tidal habitats. The "knees" of mangroves and the bald cypress are pneumatophores. Also called *air root.*

pneumoconiosis (no͞o′mō-kō′nē-ō′sĭs) A disease of the lungs, such as asbestosis or black lung, caused by chronic inhalation of especially mineral or metallic dust.

pneumonia (no͞o-mōn′yə) An acute or chronic disease that is marked by inflammation of the lungs, especially an infectious disease that is caused by viruses, bacteria, or other pathogens, such as mycoplasmas. Individuals with pneumonia often have abnormal chest x-rays that show areas with fluid in the infected part of the lungs.

pneumothorax (no͞o′mō-thôr′ăks′) Accumulation of a gas, such as air, in the space between the pleurae of the lungs and the pleurae lining the chest wall (called the pleural cavity), occurring as a result of disease or injury or induced to collapse the lung in the treatment of tuberculosis and other lung diseases. A large pneumothorax is treated by inserting a syringe or a tube into the pleural cavity to aspirate air, which helps the collapsed lung to expand.

Po The symbol for **polonium.**

pod (pŏd) A fruit or seed case that usually splits along two seams to release its seeds when mature. Legumes, such as peas and beans, produce pods.

-pod A suffix meaning "foot." It is used in the scientific names of the members of many groups of organisms, such as *arthropod,* an organism having "jointed feet," and *sauropod,* a dinosaur having "lizard feet." It is also used in the names of different kinds of limbs or limblike body parts, such as *pseudopod,* the "false foot" of an amoeba.

podzol (pŏd′zôl′) Soil that is characterized by an upper dark organic zone overlying a white to gray zone formed by leaching, overlying a reddish-orange zone formed by the deposition of iron oxide, alumina, and organic matter. Podzols form in coniferous areas or under heath in cool, humid climates.

pogonophoran (pō′gə-nŏf′ər-ən) also **pogonophore** (pō-gŏn′ə-fôr′) Any of various wormlike marine invertebrates of the phylum Pogonophora that grow in upright chitin tubes, usually at great depths. Pogonophorans have tentacles that are often featherlike and are attached to the head region. An intestine is present in the embryonic stage but disappears as the animal matures. Pogonophorans are preserved in the fossil record as early as the Cambrian Period, and are thought to be related to chaetognaths, hemichordates, and echinoderms.

poikilotherm (poi-kĭl′ə-thûrm′) See **ectotherm.**

point (point) A geometric object having no dimensions and no property other than its location. The intersection of two lines is a point.

point attractor See under **attractor.**

point bar A low, curved ridge of sand and gravel along the inner bank of a meandering stream. Point bars form through the slow accumulation of sediment deposited by the stream when its velocity drops along the inner bank.

point mutation A mutation in which one nucleotide is added, deleted, or replaced by another. Point mutations include missense, nonsense, frameshift, and silent mutations.

poise (poiz, pwäz) The unit of dynamic viscosity in the centimeter-gram-second system, equal to one dyne-second per square centimeter, or 0.1 pascal-seconds.

Poisson distribution (pwä-sôɴ′) A probability distribution which arises when counting the number of occurrences of a rare event in a long series of trials. It is named after its discoverer, French mathematician and physicist Siméon Denis Poisson (1781–1840).

polar (pō′lər) **1.** Relating to a pole, such as the pole of a magnet or one of the electrodes of an electrolytic cell. **2.** Relating to the North Pole or the South Pole of Earth, or analogous regions of another planet. **3.** Relating to a molecule or substance that has polar bonds.

polar body One of the small cells that are produced during the development of an oocyte and ultimately degenerate. A polar body contains one of the nuclei derived from the first or second meiotic division but little or no cytoplasm. See Note at **oogenesis.**

polar bond A type of covalent bond between two atoms in which electrons are shared unequally. Because of this, one end of the molecule has a slightly negative charge and the other a slightly positive charge. See more at **covalent bond.**

polar cap 1. The mass of ice that covers either of the Earth's polar regions year-round. **2.** The mass of frozen carbon dioxide and water that covers either of Mars's polar regions.

polar circle The Arctic Circle or the Antarctic Circle.

polar coordinate system A system of coordinates in which the location of a point is determined by its distance from a fixed point at the center of the coordinate space (called the pole) and by the measurement of the angle formed by a fixed line (the polar axis, corresponding to the x-axis in Cartesian coordinates) and a line from the pole through the given point. The polar coordinates of a point are given as (r, θ), where r is the distance of the point from the pole, and θ is the measure of the angle. Compare **Cartesian coordinate system.**

polar front The region or boundary separating air masses of polar origin from those of tropical or subtropical origin. The position of the polar front during winter in a particular hemisphere is at approximately 30° latitude, while in the summer it is at approximately 60° latitude. The convergence of warm and cool air masses along the polar front often produces cyclonic wind systems.

Polaris (pə-lăr′ĭs) A bright star at the end of the handle of the Little Dipper in the constellation Ursa Minor. Polaris is 1° from the north celestial pole, and it remains in the same location in the sky all year, making it a useful navigation tool. Polaris is actually a double star with a faint companion star and has an apparent magnitude of 2.04. Also called *North Star. Scientific name:* Alpha Ursae Minoris.

polarity (pō-lăr′ĭ-tē) The condition of having poles or being aligned with or directed toward poles, especially magnetic or electric poles.

polarization (pō′lər-ĭ-zā′shən) **1.** A condition in which transverse waves vibrate consistently in a single plane, or along a circle or ellipse. Electromagnetic radiation such as light is composed of transverse waves and can be polarized. Certain kinds of light filters, including sunglasses that reduce glare, work by filtering out light that is polarized in one direction. **2.** The displacement of positive and negative electric charge to opposite ends of a nuclear, atomic, molecular, or chemical system, especially by subjection to an electric field. Atoms and molecules have some inherent polarization. **3.** An increased resistance to the flow of current in a voltaic cell, caused by chemical reactions at the electrodes. Polarization results in a reduction of the electric potential across the voltaic cell.

polarize (pō′lə-rīz′) **1.** To separate or accumulate positive and negative electric charges in two distinct regions. Polarized objects have an electric dipole moment and will undergo torque when placed in an external electric field. **2.** To magnetize a substance so that it has the properties of a magnetic dipole, such as having a north and south pole. **3.** To cause the electrical and magnetic fields associated with electromagnetic waves, especially light, to vibrate in a particular direction or path. The transverse electric and magnetic waves always vibrate at right angles to each other, but in ordinary unpolarized light sources, the direction of polarization of each wave is randomly distributed. Light can be polarized by reflection, and by passing through certain materials. See more at **polarization.**

polarizing microscope (pō′lə-rī′zĭng) A microscope in which the object viewed is illuminated by polarized light. Since some materials act like polarizing filters, viewing objects with polarized light can reveal such materials when ordinary light would fail to bring them into relief.

pole (pōl) **1.** *Mathematics.* **a.** Either of the points at which an axis that passes through the center of a sphere intersects the surface of the sphere. **b.** The fixed point used as a reference in a system of polar coordinates. It corresponds to the origin in the Cartesian coordinate system. **2a.** *Geography.* Either of the points at which the Earth's axis of rotation intersects the Earth's surface; the North Pole or South Pole. **b.** Either of the two similar points on another planet. **3.** *Physics.* A magnetic pole. **4.** *Electricity.* Either of two oppositely charged terminals, such as the two electrodes of an electrolytic cell or the electric terminals of a battery. **5.** *Biology.* **a.** Either of the two points at the extremities of the axis of an organ or body. **b.** Either end of the spindle formed in a cell during mitosis.

poliomyelitis (pō′lē-ō-mī′ə-lī′tĭs) A highly communicable infectious disease caused by the poliovirus of the genus *Enterovirus* that causes inflammation of motor neurons of the spinal cord and brainstem, leading to paralysis, muscular atrophy, and often disability and deformity. Childhood vaccinations are given to prevent infection. Also called *polio.*

pollen (pŏl′ən) Powdery grains that contain the male reproductive cells of most plants. In gymnosperms, pollen is produced by male cones or conelike structures. In angiosperms, pollen is produced by the anthers at the end of stamens in flowers. Each pollen grain contains a generative cell, which divides into two nuclei (one of which fertilizes the egg), and a tube cell, which grows into a pollen tube to conduct the generative cell or the nuclei into the ovule. The pollen grain is the male gametophyte generation of seed-bearing plants. In gymnosperms, each pollen grain also contains two sterile cells (called prothallial cells), thought to be remnants of the vegetative tissue of the male gametophyte.

pollen

top: *mimosa* (Acacia dealbata); center: *loosestrife* (Lythrum sp.); bottom: *hollyhock* (Alcea rosea)

pollen mother cell See **microsporocyte.**

pollen sac The microsporangium of a seed plant in which pollen is produced. The pollen sacs of angiosperms are located in the anthers, while those of conifers are located in the male cones.

pollen tube The slender tube that is formed after pollination by division of the tube cell in a pollen grain. The pollen tube penetrates the ovule and releases the male gametes.

pollination (pŏl′ə-nā′shən) The process by which plant pollen is transferred from the male reproductive organs to the female reproductive organs to form seeds. In flowering plants, pollen is transferred from the anther to the stigma, often by the wind or by insects. In cone-bearing plants, male cones release pollen that is usually borne by the wind to the ovules of female cones. —*Verb* **pollinate.**

A CLOSER LOOK **pollination and fertilization**

When a pollen grain lands on or is carried to the receptive tissue of a pistil known as the stigma, the flower has been *pollinated.* But this is only the first step in a complicated process that, if successful, leads to *fertilization.* The pollen grain contains two cells—a generative cell and a tube cell. The generative nucleus divides to form two sperm nuclei. The tube cell grows down into the pistil until it reaches one of the ovules contained in the ovary. The two sperm travel down the tube and enter the ovule. There, one sperm nucleus unites with the egg. The other sperm nucleus combines with the polar nuclei that exist in the ovule, completing the process known as double fertilization. These fertilized nuclei then develop into the endocarp, the tissue that feeds the embryo. The ovule itself develops into a seed that is contained in the flower's ovary (which ripens into a fruit). In gymnosperms, the ovule is exposed (that is, not contained in an ovary), and the pollen produced by the male reproductive structures lands directly on the ovule in the female reproductive structures. Fertilization in conifers can be slow in comparison to flowering plants—the pollen nuclei of pines, for example, take as long as 15 months to reach the ovule after landing on the female cone. And there are variations: In the ginkgo, the ovules fall off the tree and pollination occurs on the ground.

pollinium (pō-lĭn′ē-əm) Plural **pollinia.** A mass or packet of pollen grains specialized for transfer to other flowers as a unit by pollinating insects. Orchids and milkweeds produce pollinia.

pollutant (pə-lo͞ot′nt) A substance or condition that contaminates air, water, or soil. Pollutants can be artificial substances, such as pesticides and PCBs, or naturally occurring substances, such as oil or carbon dioxide, that occur in harmful concentrations in a given environment. Heat transmitted to natural waterways through warm-water discharge from power plants and uncontained radioactivity from nuclear wastes are also considered pollutants.

pollution (pə-lo͞o′shən) The contamination of air, water, or soil by substances that are harmful to living organisms. Pollution can occur naturally, for example through volcanic eruptions, or as the result of human activities, such as the spilling of oil or disposal of industrial waste. ► Light from cities and towns at night that interferes with astronomical observations is known as **light pollution.** It can also disturb natural rhythms of growth in plants and other organisms. ► Continuous noise that is loud enough to be annoying or physically harmful is known as **noise pollution.** ► Heat from hot water that is discharged from a factory into a river or lake, where it can kill or endanger aquatic life, is known as **thermal pollution.**

Pollux (pŏl′əks) A bright giant star in the constellation Gemini, with an apparent magnitude of 1.15. *Scientific name:* Beta Geminorum.

polonium (pə-lō′nē-əm) *Symbol* **Po** A very rare, naturally radioactive, silvery-gray or black metalloid element. It is produced in extremely small amounts by the radioactive decay of radium or the bombardment of bismuth or lead with neutrons. Atomic number 84; melting point 254°C; boiling point 962°C; specific gravity 9.32; valence 2, 4. See **Periodic Table.**

poly– A prefix meaning "many," as in *polygon,* a figure having many sides. In chemistry, it is used to form the names of polymers by being attached to the name of the base unit of which the polymer is made, as in *polysaccharide,* a polymer made of repeating simple sugars (monosaccharides).

polyacrylamide (pŏl′ē-ə-krĭl′ə-mīd′) A white, water-soluble polymer containing repeating units of acrylamide (C_3H_5NO)and related to acrylic acid. Polyacrylamide is used in food packaging, adhesives, coatings, and paper manufacturing. It is also used to reduce soil erosion and as a gel for electrophoresis in the laboratory analysis of protein and DNA structures.

polyamide (pŏl′ē-ăm′īd′) A polymer produced by the reaction of the amino group (NH_2) from one molecule with the carboxylic acid group (CO_2H) from another molecule. The resulting structure is similar to that of a protein. Silk is a naturally occurring polyamide, and nylon is a synthetic polyamide.

polyandrous (pŏl′ē-ăn′drəs) **1.** Relating to a species of animals in which the females mate with more than one male in a single breeding season. **2.** Relating to an angiosperm plant that has an indefinite number of stamens in its flowers.

polybasic (pŏl′ē-bā′sĭk) Of or relating to an acid that has two or more hydrogen atoms per molecule that can be replaced by basic atoms or radicals.

polycarpic (pŏl′ē-kär′pĭk) **1.** Relating to a plant that produces fruit more than once during its lifetime. **2.** Variant of **polycarpous.**

polycarpous (pŏl′ē-kär′pəs) or **polycarpic** (pŏl′ē-kär′pĭk) Having fruit or pistils with two or more carpels.

polychaete or **polychete** (pŏl′ĭ-kēt′) Any of various often brightly colored annelid worms of the class Polychaeta. Each segment of a polychaete has a pair of fleshy appendages that are tipped with bristles (setae), used for swimming or burrowing. Most species of polychaetes live in saltwater, feed on tiny aquatic animals and plants, and range in size from a few millimeters to 3 m (10 ft) in length. Compare **oligochaete.**

polychlorinated biphenyl (pŏl′ē-klôr′ə-nā′tĭd) See **PCB.**

polychromatic (pŏl′ē-krō-măt′ĭk) **1.** Consisting of or related to radiation of more than one wavelength. **2.** Of or having many colors. Compare **monochromatic.**

polycyclic (pŏl′ĭ-sī′klĭk, -sĭk′lĭk) Having two or more atomic rings in a molecule. Steroids are polycyclic compounds.

polyembryony (pŏl′ē-ĕm′brē-ə-nē, -ĕm-brī′-) Development from a single fertilized egg cell or, in plants, from a single ovule. In human beings, identical twins are the result of polyembryony. In gymnosperm plants, polyembryony involves the fertilization of more than one egg, though usually only one embryo survives in the ovule.

polyene (pŏl′ē-ēn′) An organic compound containing at least four carbon atoms and at least two double bonds. Numerous naturally occurring plant pigments, such as carotenes, are polyenes.

polyester (pŏl′ē-ĕs′tər) Any of various mostly

synthetic polymers that are light, strong resins resistant to weather and corrosion. Polyesters are long chains of esters and are used to make fibers and plastics. They are thermosetting. Some polyesters, such as suberin, occur naturally.

polyethylene (pŏl′ē-ĕth′ə-lēn′) Any of various artificial resins consisting of many ethyl groups (CH_2CH_2) joined end to end or in branched chains. Polyethylenes are easily molded and are resistant to other chemicals. They can be repeatedly softened and hardened by heating and cooling, and are used for many purposes, such as making containers, tubes, and packaging.

polyethylene glycol Any of a family of polymers that are either colorless liquids or waxy solids and are soluble in water. They are present in many organic solvents. Polyethylene glycols are used in detergents, cosmetics, and as emulsifiers and plasticizers.

polygon (pŏl′ē-gŏn′) A closed plane figure having three or more sides. Triangles, rectangles, and octagons are all examples of polygons. ► A **regular polygon** is a polygon all of whose sides are the same length and all of whose interior angles are the same measure.

polyhedron (pŏl′ē-hē′drən) Plural **polyhedrons** or **polyhedra.** A three-dimensional geometric figure whose sides are polygons. A tetrahedron, for example, is a polyhedron having four triangular sides. ► A **regular polyhedron** is a polyhedron whose faces are all congruent regular polygons. The regular tetrahedron (pyramid), hexahedron (cube), octahedron, dodecahedron, and icosahedron are the five regular polyhedrons. Regular polyhedrons are a type of Archimedean solid. —*Adjective* **polyhedral.**

polymer (pŏl′ə-mər) Any of various chemical compounds made of smaller, identical molecules (called monomers) linked together. Some polymers, like cellulose, occur naturally, while others, like nylon, are artificial. Polymers have extremely high molecular weights, make up many of the tissues of organisms, and have extremely varied and versatile uses in industry, such as in making plastics, concrete, glass, and rubber. ► The process by which molecules are linked together to form polymers is called **polymerization** (pŏl′ə-lĭm′ər-ĭ-zā′shən).

polymerase (pŏl′ə-mə-rās′) Any of various enzymes, such as DNA polymerase, RNA polymerase, or reverse transcriptase, that catalyze the formation of sequences of DNA or RNA using an existing strand of DNA or RNA as a template.

polymorphism (pŏl′ē-môr′fĭz′əm) **1.** The existence of two or more different forms in an adult organism of the same species, as of an insect. In bees, the presence of queen, worker, and drone is an example of polymorphism. Differences between the sexes and between breeds of domesticated animals are not considered examples of polymorphism. **2.** The crystallization of a compound in at least two distinct forms. Diamond and graphite, for example, are polymorphs of the element carbon. They both consist entirely of carbon but have different crystal structures and different physical properties.

polymorphism

two forms of carbon: diamonds and graphite (in powdered form and in its common use as the "lead" in pencils)

polynomial (pŏl′ē-nō′mē-əl) An algebraic expression that is the sum of two or more monomials. The expressions $x^2 - 4$ and $5x^4 + 2x^3 - x + 7$ are both polynomials.

polynya (pŏl′ən-yä′, pə-lĭn′yə) An area of open water surrounded by sea ice. A polynya can be formed by the presence of a heat source that keeps the area from freezing. A more complex process involves wind or ocean currents that carry ice away from the polynya, constantly exposing more ocean water to ice formation and resulting in the release of very salty, dense water that sinks to become part of a **halocline.**

polyp (pŏl′ĭp) **1.** A cnidarian in its sedentary stage. Polyps have hollow, tube-shaped bodies with a central mouth on top surrounded by tentacles. Some cnidarians, such as corals and sea anemones, only exist as polyps after their larval stage, while others turn into medusas as adults or lack a polyp stage completely. Compare **medusa. 2.** An abnormal growth extending from a mucous membrane, as of the intestine.

polypeptide (pŏl′ē-pĕp′tīd′) A peptide, such as a small protein, containing many molecules of amino acids, typically between 10 and 100.

polyphenol (pŏl′ē-fē′nôl′, -nōl′) Any of various alcohols containing two or more benzene rings that each have at least one hydroxyl group (OH) attached. Many polyphenols occur naturally in plants and some kinds, such as the flavonoids and tannins, are believed to be beneficial to health.

polyphyletic (pŏl′ē-fī-lĕt′ĭk) Relating to a taxonomic group that does not include the common ancestor of the members of the group and whose members have two or more separate origins. A group that consists of flying vertebrates would be polyphyletic, as bats and birds independently evolved flight and do not share a common ancestor. Compare **monophyletic, paraphyletic.**

polyploid (pŏl′ē-ploid′) Having more than two complete sets of chromosomes. Many plants that are polyploid, such as dandelions, are sterile but can reproduce by apomixis or other asexual means. Other polyploid plants are fertile. For example, durum wheat (*Triticum turgidum durum*), which is used to make pasta, is tetraploid (it has four sets of chromosomes), while bread wheat (*Triticum aestivum*) is hexaploid (six sets of chromosomes). Polyploid plants, if viable, are often larger or more productive than diploid plants, and plant breeders often deliberately produce such plants by crossing species or other means. In the animal kingdom, polyploidy is abnormal and often fatal.

polyploidy (pŏl′ē-ploi′dē) The state or condition of being polyploid.

polypropylene (pŏl′ē-prō′pə-lēn′) Any of various thermoplastic resins that are polymers consisting of repeated branched units derived from propane and having the formula $CH_2CH(CH_3)$, usually with the methane groups (CH_3) all on one side of the chain. Polypropylenes are similar to polyethylenes but are harder and tougher and are used to make molded articles and fibers. **2.** A fabric of fibers made from any of these resins.

polysaccharide (pŏl′ē-săk′ə-rīd′) Any of a class of carbohydrates that are made of long chains of simple carbohydrates (called monosaccharides). Starch and cellulose are polysaccharides. Compare **monosaccharide, oligosaccharide.**

polysorbate (pŏl′ē-sôr′bāt′) Any of a class of emulsifiers used in food preparation and in some pharmaceuticals. Polysorbates are fatty acid esters with short polyethylene chains branching out from a central ring.

polystyrene (pŏl′ē-stī′rēn) A brittle synthetic polymer composed of repeated styrene units. Polystyrene is transparent and rigid because the benzene rings in each styrene unit prevent the polystyrene chains from arranging themselves into a tight crystalline structure. Polystyrene has a wide variety of uses, especially as a solid foam for insulation and packaging.

polytetrafluoroethylene (pŏl′ē-tĕt′rə-flo͝or′ō-ĕth′ə-lēn′, -flôr′-) A synthetic polymer consisting of a chain of fluorinated ethane units (C_2F_4). It is a thermoplastic resin that is resistant to heat and chemicals and has an extremely low coefficient of friction (resistance to objects sliding over its surface). It is used as a coating on cookware, gaskets, seals, and hoses.

polytypic (pŏl′ē-tĭp′ĭk) Having several variant forms, especially containing more than one taxonomic category of the next lower rank. A polytypic genus contains two or more different species, while a polytypic species consists of two or more subspecies. Compare **monotypic.**

polyunsaturated (pŏl′ē-ŭn-săch′ə-rā′tĭd) Relating to an organic compound, especially a fat, in which more than one pair of carbon atoms are joined by double or triple bonds. See more at **unsaturated.**

polyurethane (pŏl′ē-yo͝or′ə-thăn′) Any of various synthetic resins used to make tough resistant coatings, adhesives, foams, and electrical insulation.

polyvalent (pŏl′ē-vā′lənt) **1.** Acting against or

interacting with more than one kind of antigen, antibody, toxin, or microorganism. **2.** Having more than one valence. Iron and manganese are polyvalent elements.

polyvinyl chloride (pŏl′ē-vī′nəl) A synthetic resin, composed of repeating units of vinyl chloride (C_2H_3Cl). Polyvinyl chloride is very versatile and is used in a wide variety of products, including rainwear, garden hoses, audio discs, and floor tiles.

pome (pōm) A fleshy simple fruit that has several seed chambers developed from a compound ovary and an outer fleshy part developed from the enlarged base of the flower. The pome is an *accessory fruit* and is characteristic of certain plants in the rose family, such as the apple and pear. Also called *false fruit.* Compare **berry, drupe.** See more at **accessory fruit, simple fruit.**

pond (pŏnd) An inland body of standing water that is smaller than a lake. Natural ponds form in small depressions and are usually shallow enough to support rooted vegetation across most or all of their areas.

pongid (pŏn′jĭd) See **anthropoid ape.**

pons (pŏnz) Plural **pontes** (pŏn′tēz). A thick band of nerve fibers in the brainstem of humans and other mammals that links the brainstem to the cerebellum and upper portions of the brain. It is important in the reflex control of involuntary processes, including respiration and circulation. All neural information transmitted between the spinal cord and the brain passes through the pons.

population (pŏp′yə-lā′shən) A group of individuals of the same species occupying a particular geographic area. Populations may be relatively small and closed, as on an island or in a valley, or they may be more diffuse and without a clear boundary between them and a neighboring population of the same species. For species that reproduce sexually, the members of a population interbreed either exclusively with members of their own population or, where populations intergrade, to a greater degree than with members of other populations. See also **deme.**

population genetics The scientific study of the inheritance and prevalence of genes in populations, usually using statistical analysis.

population inversion The condition of having enough excited or high-energy states distributed throughout a substance to sustain a chain reaction of stimulated emission. Lasers, for example, need a constant power source that maintains population inversion in order to generate radiation continuously, since each stimulated emission reduces the population of high-energy states. See also **stimulated emission.**

porcine (pôr′sīn′) Relating to or resembling pigs.

pore (pôr) **1.** A tiny opening, as one in an animal's skin or on the surface of a plant leaf or stem, through which liquids or gases may pass. **2.** A space in soil, rock, or loose sediment that is not occupied by mineral matter and allows the passage or absorption of fluids, such as water, petroleum, or air.

poriferan (pə-rĭf′ər-ən) See **sponge** (sense 1).

porosity (pə-rŏs′ĭ-tē, pô-) **1.** The condition of being porous. **2.** The ratio of the volume of all the pores in a material to the volume of the whole.

porous (pôr′əs) Having many pores or other small spaces that can hold a gas or liquid or allow it to pass through.

porphyrin (pôr′fə-rĭn) Any of various organic pigments containing four pyrrole rings bonded to one another. The rings form the corners of a large flat square, in the middle of which is a cavity that often contains a metal atom. Porphyrins occur universally in protoplasm and function with bound metals such as iron in hemoglobin and magnesium in chlorophyll.

porphyroblast (pôr-fîr′ə-blăst′) A large crystal that is surrounded by a finer-grained matrix in a metamorphic rock. Porphyroblasts form by the recrystallization of existing mineral crystals during metamorphism. They are analogous to *phenocrysts* in igneous rock.

porphyry (pôr′fə-rē) An igneous rock containing the large crystals known as phenocrysts embedded in a fine-grained matrix. *—Adjective* **porphyritic** (pôr′fə-rĭt′ĭk).

port (pôrt) **1.** An opening, as in a cylinder or valve face, for the passage of steam or fluid. **2.** A place where data can pass into or out of a central processing unit, computer, or peripheral. With central processing units, a port is a fixed set of connections for incoming and outgoing data or instructions. With computers and peripherals, a port is generally a socket into which a connector can be plugged.

Porter (pôr′tər), **Rodney Robert** 1917–1985. British biochemist who shared with George Edelman the 1972 Nobel Prize for physiology or medicine for their study of the chemical structure of antibodies.

positive (pŏz′ĭ-tĭv) **1.** Greater than zero. **2.** Having an electric charge or voltage greater than zero. **3.** Indicating the presence of a disease, condition, or organism, as a diagnostic test.

positive feedback Feedback in which the output quantity or signal adds to the input quantity or signal. Positive feedback is responsible for the squealing of microphones when placed too close to the speaker through which their input signals are amplified. Compare **negative feedback.**

positron (pŏz′ĭ-trŏn′) The antiparticle that corresponds to an electron. Also called *antielectron.*

positron emission tomography Tomography in which a computer-generated image of metabolic or physiologic activity within the body is produced through the detection of gamma rays that are emitted when introduced radionuclides decay and release positrons. The images are used in the evaluation of coronary artery disease, epilepsy, and other medical disorders.

post– A prefix that means "after," as in *postoperative,* after an operation, or "behind," as in *postnasal,* behind the nose or nasal passages.

postglacial (pōst-glā′shəl) Relating to or occurring during the time following a glacial period.

posttraumatic stress disorder (pōst′trô-măt′ĭk, -trou-) A psychological disorder affecting individuals who have experienced or witnessed profoundly traumatic events, such as torture, murder, rape, or wartime combat, characterized by recurrent flashbacks of the traumatic event, nightmares, irritability, anxiety, fatigue, forgetfulness, and social withdrawal.

postulate (pŏs′chə-lĭt) See **axiom.**

potash (pŏt′ăsh′) Any of several chemical compounds that contain potassium, especially potassium carbonate (K_2CO_3), which is a strongly alkaline material obtained from wood ashes and used in fertilizers.

potash feldspar Potassium feldspar.

potassium (pə-tăs′ē-əm) *Symbol* **K** A soft, highly reactive, silvery-white metallic element of the alkali group occurring in nature only in compounds. It is essential for the growth of plants and is used especially in fertilizers and soaps. Atomic number 19; atomic weight 39.098; melting point 63.65°C; boiling point 774°C; specific gravity 0.862; valence 1. See **Periodic Table.**

potassium-argon dating A method of **radiometric dating**, involving analysis of the ratio of potassium 40 (a radioactive isotope of potassium) to argon (the product of radioactive decay of potassium 40) in a given sample.

potassium chloride A white crystalline solid or powder used widely in fertilizers and in the preparation of most potassium compounds. It occurs naturally as the mineral *sylvite.* A unique property of potassium chloride is that it is more soluble in hot water, but less soluble in cold water, than sodium chloride is. *Chemical formula:* **KCl.**

potassium feldspar A type of alkali feldspar that contains a high proportion of potassium relative to sodium. Microcline and orthoclase are types of potassium feldspar.

potassium hydroxide A white, corrosive, solid compound used in bleaches and to make soaps and detergents. It is deliquescent, soluble in water and very soluble in alcohol. In solution, it forms lye. *Chemical formula:* **KOH.**

potassium iodide A white crystalline compound used in photography and medicine and as a reagent in chemical analysis. It is also added to table salt to prevent goiter and other iodide-deficiency disorders. *Chemical formula:* **KI.**

potassium nitrate A transparent, white, crystalline compound and strong oxidizing agent. It is used in gunpowder and fireworks, in making glass, and in fertilizer. Also called *saltpeter. Chemical formula:* **KNO_3.** See also **niter.**

potassium permanganate A dark purple crystalline compound used as an oxidizing agent and disinfectant and in deodorizers and dyes. *Chemical formula:* **$KMnO_4$.**

potential difference (pə-tĕn′shəl) The difference, measured in volts, in electric potential between two points, especially two points in an electric circuit.

potential energy The energy possessed by a body as a result of its position or condition rather than its motion. A raised weight, coiled spring, or charged battery has potential energy. Compare **kinetic energy.**

potentiometer (pə-tĕn′shē-ŏm′ĭ-tər) **1.** A mechanical variable resistor. See more at **resistor. 2.** An instrument for measuring an unknown voltage by comparison with a known voltage, such as that of a generator.

potentiometric surface (pə-tĕn′shē-ə-mĕt′rĭk) A hypothetical surface representing the level to which groundwater would rise if not trapped in a confined aquifer (an aquifer in which the water is under pressure because of an impermeable layer above it that keeps it from seeking its level). The potentiometric surface is equivalent to the water table in an unconfined aquifer. See illustration at **artesian well.**

pound (pound) A unit of weight in the US Customary System equal to 16 ounces (0.45 kilograms). See Table at **measurement.** See Note at **weight.**

pound-foot See **foot-pound** (sense 2).

powdery mildew (pou′də-rē) Any of various important plant diseases caused by fungi, especially those of the family Erysiphaceae, which produce powdery conidia on the surface of the hosts leaves and stems.

power (pou′ər) **1.** The source of energy used to operate a machine or other system. **2.** The rate at which work is done, or energy expended, per unit time. Power is usually measured in watts (especially for electrical power) or horsepower (especially for mechanical power). For a path conducting electrical current, such as a component in an electric circuit, $P = VI$, where P is the power dissipated along the path, V is the voltage across the path, and I is the current through the path. Compare **energy, work. 3.** *Mathematics.* The number of times a number or expression is multiplied by itself, as shown by an exponent. Thus ten to the sixth power, or 10^6, equals one million. **4.** A number that represents the magnification of an optical instrument, such as a microscope or telescope. A 500-power microscope can magnify an image to 500 times its original size.

power series A sum of successively higher integral powers of a variable or combination of variables, each multiplied by a constant coefficient.

Pr The symbol for **praseodymium.**

prairie (prâr′ē) An extensive area of flat or rolling grassland, especially the large plain of central North America.

praseodymium (prā′zē-ō-dĭm′ē-əm) *Symbol* **Pr** A soft, malleable, silvery metallic element of the lanthanide series that develops a green tarnish in air. It is used to add a yellow tint to glass and ceramics and to make the glass used in welding goggles. Atomic number 59; atomic weight 140.908; melting point 935°C; boiling point 3,127°C; specific gravity 6.8; valence 3, 4. See **Periodic Table.**

Precambrian (prē-kăm′brē-ən, -kām′-) The period of geologic time between Hadean Time and the Phanerozoic Eon, from about 3.8 billion to 540 million years ago. During the Precambrian Eon, which is divided into the Archean and Proterozoic, primitive forms of life first appeared on Earth. See Chart at **geologic time.**

precession (prē-sĕsh′ən) **1.** The rotational motion of the axis of a spinning body, such as the wobbling of a spinning top, caused by torque applied to the body along its axis of rotation. **2.** The motion of this kind made by the Earth's axis, caused mainly by the gravitational pull of the Sun, Moon, and other planets. The precession of Earth's axis has a period of nearly 25,800 years, during which time the reference points on the **equatorial coordinate system** (the celestial poles and celestial equator) will gradually shift their positions on the celestial sphere. ▸ The **precession of the equinoxes** is the slow westward shift of the autumnal and vernal equinoxes along the ecliptic, resulting from precession of the Earth's axis. See also **nutation.**

precipitate *Verb.* (prĭ-sĭp′ĭ-tāt′) **1.** To fall from the atmosphere as rain, snow, or another form of precipitation. **2.** To separate as a solid from a solution in chemical precipitation. —*Noun.* (prĭ-sĭp′ĭ-tāt′, -tĭt) **3.** A solid material precipitated from a solution.

precipitation (prĭ-sĭp′ĭ-tā′shən) **1.** A form of water, such as rain, snow, or sleet, that condenses from the atmosphere, becomes too heavy to remain suspended, and falls to the Earth's surface. Different atmospheric conditions are responsible for the different forms of precipitation. **2.** The process by which a substance is separated out of a solution as a solid. Precipitation occurs either by the action of gravity or through a chemical reaction that forms an insoluble compound out of two or more soluble compounds.

precocial (prĭ-kō′shəl) Born or hatched in a condition requiring relatively little parental care, as by having hair or feathers, open eyes, and the ability to move about. Water birds, reptiles, and herd animals usually have precocial young. Compare **altricial.**

precocious (prĭ-kō′shəs) Relating to or having flowers that blossom before the leaves emerge. Some species of magnolias are precocious.

predator (prĕd′ə-tər) An animal that lives by capturing and eating other animals.

pregnancy (prĕg′nən-sē) **1.** The condition of carrying developing offspring within the body. **2.** The time period during which this condition exists; gestation.

prehensile (prē-hĕn′səl) Adapted for seizing, grasping, or holding, especially by wrapping around an object. The feet of many birds, the tails of monkeys, and the trunks of elephants are prehensile.

premolar (prē-mō′lər) Any of eight bicuspid teeth in mammals, arranged in pairs on both sides of the upper and lower jaws between the canines and molars. Premolars are used to tear and grind food.

pressure (prĕsh′ər) The force per unit area that one region of a gas, liquid, or solid exerts on another. Pressure is usually measured in Pascal units, atmospheres, or pounds per square inch. ► A substance is said to have **negative pressure** if some other substance exerts more force per unit area on it than vice versa. Its value is simply the negative of the pressure exerted by the other substance.

pressure drag See under **drag.**

pressure-tube anemometer See under **anemometer.**

prevailing wind (prĭ-vā′lĭng) A wind that blows predominantly from a single general direction. The trade winds of the tropics, which blow from the east throughout the year, are prevailing winds. See illustration at **wind.**

Priestley (prēst′lē), **Joseph** 1733–1804. British chemist who discovered oxygen (1774) and 10 other gases, including hydrogen chloride, sulphur dioxide, and ammonia.

BIOGRAPHY **Joseph Priestley**

Raised a strict Calvinist, Joseph Priestley originally hoped to become a minister, but his exposure to and interest in more liberal theological and philosophical issues ultimately led him to the calling of science. When Priestley met Benjamin Franklin in 1766, Franklin's enthusiasm for experimentation with electricity inspired Priestley to conduct his own experiments. One of Priestley's first discoveries was that graphite conducts electricity. Intrigued by the quality of the air emitted by fermentation at a nearby brewery, he later developed an improved technique for isolating and storing gases—at the time understood as varieties of air—in sealed glass vessels. Priestley also noted that the "damage" done to air by the respiration of animals, which slowly rendered it less and less life-sustaining for animals, was reversed by the respiration of plants. Using a magnifying glass to focus the Sun's rays on a piece of mercuric oxide and capturing the emitted gas, he discovered that this gas made a candle burn more brightly and could keep a mouse alive while all the other gases he tested extinguished the candle's flame and killed the mice. Priestley did not appreciate the full implications of his discovery, however. After he discussed his results with the French chemist Antoine Lavoisier, Lavoisier repeated Priestley's experiments, showing that combustion required the presence of Priestley's gas and implied that air was not an element but was made up of various parts. Lavoisier named the gas *oxygen,* and the modern theory of combustion was born.

primary (prī′mĕr′ē) **1.** Relating to a primary color. **2.** Relating to plant tissues or growth derived from the **apical meristem** in the tips of roots and shoots, whose cells divide and elongate to cause the plant to grow lengthwise. **3a.**

Relating to or having a carbon atom that is attached to only one other carbon atom in a molecule. **3b.** Relating to an organic molecule, such as an alcohol, in which the functional group is attached to a primary carbon. A primary alcohol, for example, has the hydroxyl (OH) group attached to the last carbon in a chain. **4a.** Arising first and spontaneously, as a disease, disorder, or tumor, and not as a result of a known medical condition or injury. **4b.** Relating to the first set of teeth that develops in humans. **5.** Relating to the initial medical care given by a healthcare provider to a patient, especially in a setting of ambulatory, continuous care, and sometimes followed by referral to other medical providers. Compare **secondary, tertiary.**

primary color Any of a group of colors from which all other colors can be made by mixing. See more at **additive, subtractive.** See Note at **color.**

primary consumer See under **consumer.**

primary growth Growth in vascular plants resulting from the production of primary tissues by an **apical meristem.** The plant body grows lengthwise chiefly by the enlargement of cells produced by the apical meristem (rather than by cell division). Because they lack secondary tissues, most monocots and herbaceous plants grow solely by primary growth until they reach maturity, when growth stops. Compare **secondary growth.**

primary meristem One of the three meristems in vascular plants (the procambium, protoderm, or ground meristem) derived from an **apical meristem.** Compare **lateral meristem.** See more at **apical meristem, primary growth.**

primary root The first root of a plant, originating in the embryo. In gymnosperms, eudicotyledons, and magnoliids, the primary root develops into the **taproot**. In monocotyledons, the primary root disintegrates as the lateral roots develop into a system of **fibrous roots**.

primary structure The linear sequence of amino acids in a protein. See also **quaternary structure, secondary structure, tertiary structure.**

primary succession See under **succession.**

primary wave A type of seismic body wave in which rock particles vibrate parallel to the direction of wave travel. Primary waves are alternatingly compressional and extensional, and cause the rocks they pass through to change in volume. These waves are the fastest traveling seismic waves and can travel through solids, liquids, and gases. Also called *P wave.* See Note at **earthquake.**

primate (prī′māt′) Any of various mammals of the order Primates, having a highly developed brain, eyes facing forward, a shortened nose and muzzle, and opposable thumbs. Primates usually live in groups with complex social systems, and their high intelligence allows them to adapt their behavior successfully to different environments. Lemurs, monkeys, apes, and humans are primates.

primatology (prī′mə-tŏl′ə-jē) The scientific study of primates.

prime meridian (prīm) The meridian with a longitude of 0°, adopted officially in 1884 as a reference line from which longitude east and west are measured and as a basis for standardized time zones. It passes through Greenwich, England, the site of the Royal Greenwich Observatory, which was founded in 1675 and which closed except as a museum in 1998. The prime meridian, together with its opposite meridian having a longitude of 180°, divide the Earth roughly into the Eastern and Western Hemispheres, with those portions of the British Isles, Europe, and Africa that lie west of the prime meridian, considered for practical purposes as belonging to the Eastern Hemisphere. See illustration at **time zone.**

prime number A positive integer greater than 1 that can only be divided by itself and 1 without leaving a remainder. Examples of prime numbers are 7, 23, and 67. Compare **composite number.**

primitive (prĭm′ĭ-tĭv) **1.** Relating to an early or original stage. **2.** Having evolved very little from an early type. Lampreys and sturgeon are primitive fishes.

primordial soup (prī-môr′dē-əl) A liquid rich in organic compounds and providing favorable conditions for the emergence and growth of life forms. Oceans of primordial soup are thought to have covered the Earth during the Precambrian Eon billions of years ago. The organic compounds in the primordial soup, such as amino acids, may have been produced by reactions in the Earth's early atmosphere, which was probably rich in methane and ammonia. The complex self-replicating organic molecules that were precursors to life on Earth may have developed in this liquid.

principal quantum number (prĭn′sə-pəl) See under **quantum number.**

printed circuit board (prĭn′tĭd) A flat plastic or fiberglass board on which interconnected circuits and components are laminated or etched. Chips and other electronic components are mounted on the circuits. Computers consist of one or more printed circuit boards, usually called **cards** or **adapters**.

prion (prē′ŏn, prī′-) A particle that is thought to be able to self-replicate and to be the agent of infection in a variety of diseases of the nervous system, such as mad cow disease. Prion replication (in which strings of amino acids are reproduced) stands as an exception to a central tenet of biology stating that only nucleic acids, such as DNA, can self-replicate. The mechanism of prion replication is not clearly understood.

A CLOSER LOOK **prion**

In 1997 Stanley Prusiner was awarded the Nobel Prize for physiology or medicine for his theory that a deviant form of a harmless protein could be an infectious agent, a transmitter of disease. Named *prions* (short for *proteinaceous infectious particle),* these misshapen proteins cause healthy proteins to misfold, fatally clumping together in the brain. Unlike other disease-causing agents, prions lack genetic material (DNA and RNA). Neurodegenerative prion diseases are often called *spongiform encephalopathies* because they leave the brain riddled with holes like a sponge. In animals, prion diseases include scrapie in sheep and bovine spongiform encephalopathy, commonly known as mad cow disease in cattle. In humans, diseases such as kuru and Creutzfeldt-Jakob disease (CJD) are also thought to be caused by prions. All the diseases are characterized by loss of motor control, dementia, paralysis, and eventual death due to massive destruction of brain tissue. Humans are thought to contract prion disease most commonly by eating prion-contaminated flesh. Kuru, a rare and fatal brain disorder, brought prion disease to the forefront. First described in the 1950s, kuru was most common among the Fore people of Papua New Guinea, who had a custom of eating the brains of their dead during funeral feasts. It is speculated that a tribe member developed CJD, his or her contaminated brain tissue was ingested, and the disease spread. Kuru reached epidemic levels in the 1960s, but the disease declined after the government discouraged the practice of cannibalism and now it has almost completely disappeared.

---

prism (prĭz′əm) **1.** A geometric solid whose bases are congruent polygons lying in parallel planes and whose sides are parallelograms. **2.** A solid of this type, often made of glass with triangular ends, used to disperse light and break it up into a spectrum. **3.** A crystal form having 3, 4, 6, 8, or 12 faces parallel to the vertical axis and intersecting the horizontal axis.

prismatic (prĭz-măt′ĭk) **1.** Relating to or resembling a prism. **2.** Formed by refraction of light through a prism, used especially of a spectrum of light.

probability (prŏb′ə-bĭl′ĭ-tē) A number expressing the likelihood of the occurrence of a given event, especially a fraction expressing how many times the event will happen in a given number of tests or experiments. For example, when rolling a six-sided die, the probability of rolling a particular side is 1 in 6, or $\frac{1}{6}$.

probability distribution A function of a discrete random variable (that is, a variable whose values are obtained from a finite or countable set) yielding the probability that the variable will have a given value.

probability wave A quantum state of a particle or system, as characterized by a wave propagating through space, in which the square of the magnitude of the wave at any given point corresponds to the probability of finding the particle at that point. Mathematically, a probability wave is described by the wave function, which is a solution to the wave equation describing the system.

probable error (prŏb′ə-bəl) The amount by which the arithmetic mean of a sample is expected to vary because of chance alone.

proboscidean also **proboscidian** (prō′bə-sĭd′ē-ən, prō-bŏs′ĭ-dē′ən) **1.** Any of various mammals of the order Proboscidea, having a long trunk, large tusks, and a massive body. The elephants and its extinct relatives, such as the mastodons, are proboscidians. **2.** Of or belonging to the order Proboscidea.

proboscis (prō-bŏs′ĭs) Plural **proboscises** or **proboscides** (prō-bŏs′ĭ-dēz′). **1.** A long, flexible snout or trunk, as of an elephant. **2.** The slender, tubular feeding and sucking organ of certain invertebrates, such as butterflies and mosquitoes.

procambium (prō-kăm′bē-əm) The primary meristem in vascular plants that gives rise to primary vascular tissues (phloem and xylem).

processor (prŏs′ĕs′ər, prō′sĕs′-) **1.** A part of a computer, such as the central processing unit, that performs calculations or other manipulations of data. **2.** A program that translates another program into a form acceptable by the computer being used.

proctology (prŏk-tŏl′ə-jē) The branch of medicine that deals with the diagnosis and treatment of disorders affecting the colon, rectum, and anus.

Procyon (prō′sē-ŏn′) A very bright binary star in the constellation Canis Minor, with an apparent magnitude of 0.34. *Scientific name:* Alpha Canis Minoris.

producer (prə-dōō′sər) An autotrophic organism that serves as a source of food for other organisms in a food chain. Producers include green plants, which produce food through photosynthesis, and certain bacteria that are capable of converting inorganic substances into food through chemosynthesis. Compare **consumer.**

product (prŏd′əkt) A number or quantity obtained by multiplication. For example, the product of 3 and 7 is 21.

production well (prə-dŭk′shən) A well used to retrieve petroleum or gas from an underground reservoir.

progenitor cell (prō-jĕn′ĭ-tər) See **stem cell.**

progesterone (prō-jĕs′tə-rōn′) A steroid hormone that prepares the uterus for pregnancy, maintains pregnancy, and promotes development of the mammary glands. The main sources of progesterone are the ovary and the placenta. *Chemical formula:* $\mathbf{C_{21}H_{30}O_2}$.

prograde (prō′grād′) Having a rotational or orbital movement that is the same as most bodies within a celestial system. In our solar system, prograde movement for both rotating and orbiting bodies is in a counterclockwise direction when viewed from a vantage point above the Earth's north pole. Compare **retrograde.**

program (prō′grăm′) A organized system of instructions and data interpreted by a computer. Programming instructions are often referred to as code. See more at **source code.** See also **programming language.**

programmed cell death (prō′grămd′) See **apoptosis.**

programming language (prō′grăm′ĭng) An artificial language used to write instructions that can be translated into machine language and then executed by a computer. English and other natural languages are not used as programming languages because they cannot be easily translated into machine language. ► A **compiled language** is a language in which the set of instructions (or **code**) written by the programmer is converted into **machine language** by special software called a **compiler** prior to being executed. C++ and SmallTalk are examples of compiled languages. ► An **interpreted language** is a language in which the set of instructions (or **code**) written by the programmer is converted into **machine language** by special software called a **compiler-** prior to being executed. Most scripting and macro languages are interpreted languages. See also **program.**

progression (prə-grĕsh′ən) See **sequence** (sense 1).

projection (prə-jĕk′shən) **1.** The image of a geometric figure reproduced on a line, plane, or surface. **2.** A system of intersecting lines, such as the grid of a map, on which part or all of the globe or another spherical surface is represented as a plane surface. See more at **azimuthal projection, conic projection, cylindrical projection.**

prokaryote (prō-kăr′ē-ōt′) Any of a wide variety of one-celled organisms of the kingdom Monera (or Prokaryota) that are the most primitive and ancient known forms of life. Prokaryotes lack a distinct cell nucleus and their DNA is not organized into chromosomes. They also lack the internal structures bound by membranes called organelles, such as mitochondria. At the molecular level, prokaryotes differ from eukaryotes in the structure of their lipids and of certain metabolic enzymes, and in how genes are expressed for protein synthesis. Prokaryotes reproduce asexually and include the bacteria and blue-green algae. Also called *moneran.* Compare **eukaryote.** See Table at **Taxonomy.**

prolactin (prō-lăk′tĭn) A protein hormone secreted by the anterior portion of the pituitary gland that stimulates and maintains the secretion of milk in mammals.

prolamin (prō′lə-mĭn) Any of a class of simple proteins soluble in alcohol and usually having a high proline and glutamine content, found in the grains of cereal crops such as wheat, rye, barley, corn, and rice.

proline (prō′lēn′) A nonessential amino acid. *Chemical formula:* $C_5H_9NO_2$.

promethium (prə-mē′thē-əm) *Symbol* **Pm** A radioactive metallic element of the lanthanide series. Promethium does not occur in nature but is prepared through the fission of uranium. It has 17 isotopes; the longest-lived isotope, Pm 147, has a half-life of 2.5 years and is used as a source of beta rays. Atomic number 61; melting point 1,168°C; boiling point 2,460°C; valence 3. See **Periodic Table.**

prominence (prŏm′ə-nəns) An eruption of tonguelike clouds of glowing ionized gas extending from the Sun's chromosphere and sometimes reaching hundreds of thousands of kilometers into space. When viewed against the solar surface instead of along the edges of its disk, prominences appear as dark, sinuous lines known as **filaments**. Solar prominences can influence Earth's atmosphere by interfering with electromagnetic activity. ► **Active prominences** erupt suddenly and usually disappear within minutes or hours. **Quiescent prominences** form more smoothly and can last for several months. See also **solar flare.**

promontory (prŏm′ən-tôr′ē) A high ridge of land or a rock cliff jutting over a body of water.

promoter (prə-mō′tər) The region of an operon that acts as the initial binding site for RNA polymerase.

proof (pro͞of) A demonstration of the truth of a mathematical or logical statement, based on axioms and theorems derived from those axioms.

propagule (prŏp′ə-gyo͞ol′) **1.** Any of various structures that can give rise to a new individual organism, especially parts of a plant that serve as means of **vegetative reproduction,** such as corms, tubers, offsets, or runners. Seeds and spores are also propagules. **2.** An elongated, dart-shaped seedling of various mangrove species growing in swampy habitats. A propagule develops from a seed that germinates while still attached to the parent tree. The parent supplies the seedling with nutrients and water until it becomes heavy and drops off. Its pointed end sticks in the mud or it floats away to colonize another area.

propane (prō′pān′) A colorless, gaseous hydrocarbon found in petroleum and natural gas. It is widely used as a fuel. Propane is the third member of the alkane series. *Chemical formula:* C_3H_8.

propanoic acid (prō′pə-nō′ĭk) See **propionic acid.**

propanol (prō′pə-nôl′, -nōl′) A clear colorless liquid used as a solvent and antiseptic. Also called *propyl alcohol. Chemical formula:* C_3H_8O.

propeller (prə-pĕl′ər) A device consisting of a set of two or more twisted, airfoil-shaped blades mounted around a shaft and spun to provide propulsion of a vehicle through water or air, or to cause fluid flow, as in a pump. The **lift** generated by the spinning blades provides the force that propels the vehicle or the fluid—the lift does not have to result in an actual upward force; its direction is simply parallel to the rotating shaft.

propene (prō′pēn′) See **propylene.**

proper fraction (prŏp′ər) A fraction in which the numerator is less than the denominator, such as $\frac{1}{2}$. Compare **improper fraction.**

proper motion Movement of a celestial object in the sky that is the result of the object's own motion in space rather than of how it is observed from Earth. All celestial objects are in motion with regard to each other, but because objects outside the solar system are so distant from Earth most of them seem fixed in the sky. Over long periods of time, however, their proper motions result in gradual changes in their relative positions as viewed from Earth. Measurements of these motions by modern instruments can be extrapolated forward or backward in time to produce a celestial sphere on which the stars have somewhat different positions than they have today. In general, objects nearest the Earth have the greatest proper motions and will move the farthest on the celestial sphere in such extrapolations. Extremely distant objects, although they may be moving through space at equal or higher speeds than nearby objects, will appear to move little in the sky even over thousands of years.

prophase (prō′fāz′) **1.** The first stage in the process of mitosis. Before prophase begins, the chromosomes duplicate to form two long, thin strands called chromatids. During prophase itself, the chromatids condense and thicken to form distinct bodies. Chromatids making up a single chromosome are joined at the middle in an area called the centromere. The membrane surrounding the nucleus disappears, and the **spindle** begins to form. In prophase and the later stages of mitosis until separation of the individual chromatids during anaphase, each chromosome consists of two chromatids, and each chromatid contains a complete copy of the genetic information belonging to the chromosome. For example, human beings have 23 pairs of chromosomes in all somatic cells, or 46 chromosomes in total. At the end of prophase, each of these 46 chromosomes contains two identical chromatids. **2.** One of the two stages in meiosis that resemble prophase in mitosis. However, there are important distinctions between prophase of mitosis and prophase of meiosis. The prophase of meiosis occurring during the first meiotic division of the cell is usually called *prophase I.* In prophase I of meiosis, pairs of homologous chromosomes intertwine and the process called **crossing over** occurs as chromatids from homologous pairs of chromosomes swap genetic information. This process creates genetic diversity among the gametes formed through meiosis. In mitosis, by contrast, pairs of homologous chromosomes remain separate and there is no crossing over, since the purpose of mitosis is to produce cells with identical genetic material rather than gametes. At the beginning of *prophase II* of meiosis, which occurs after telophase during the first meiotic division, the chromosomes of each daughter cell are grouped together in a mass. During prophase II, the individual chromosomes of the daughter cells become distinct again and begin to prepare for the second meiotic division. If a membrane has formed around the chromosomes at the end of the first division, it disappears during prophase II. See more at **meiosis, mitosis.**

propionate (prō′pē-ə-nāt′) A salt or an ester of propionic acid, containing the group CH_3CH_2COO.

propionic acid (prō′pē-ŏn′ĭk) A liquid fatty acid found naturally in sweat and milk products and as a product of bacterial fermentation. It is also prepared synthetically from ethanol and carbon monoxide, and is used chiefly in the form of its propionates as a mold inhibitor in bread and as an ingredient in perfume. *Chemical formula:* $C_3H_6O_2$.

proportion (prə-pôr′shən) A statement of equality between two ratios. Four quantities, *a, b, c,* and *d,* are said to be in proportion if $\frac{a}{b}=\frac{c}{d}$.

proprioception (prō′prē-ō-sĕp′shən) The unconscious perception of movement and spatial orientation arising from stimuli within the body itself. In humans, these stimuli are detected by nerves within the body itself, as well as by the semicircular canals of the inner ear.

prop root (prŏp) An adventitious root that arises from the stem, penetrates the soil, and helps support the stem, as in corn.

prop root

of a corn plant (Zea mays)

propyl (prō′pĭl) The radical C_3H_7, derived from propane.

propyl alcohol See **propanol.**

propylene (prō′pə-lēn′) A flammable gas produced by cracking (breaking down) petroleum and used to make plastics and isopropyl alcohol. Propylene is the second member of the alkene series. Also called *propene. Chemical formula:* C_3H_6.

propylene glycol A colorless, viscous liquid used in antifreeze solutions, in hydraulic fluids, and as a solvent. Unlike ethylene glycol, it is not toxic and is also used in foods, cosmetics, and oral hygiene products. *Chemical formula:* $C_3H_8O_2$.

prosimian (prō-sĭm′ē-ən) Any of various primates of the suborder Strepsirrhini (formerly Prosimii), considered the most primitive primates. Prosimians have a moist, bare muzzle and a retina that lacks a fovea but is backed by a reflective layer that increases night vision. Unlike other primates, female prosimians do not menstruate because the lining of their uteri is not built up each month to prepare for possible pregnancy. Prosimians are mostly small in size, and include the lemurs, aye-ayes, indris, and lorises. The tarsiers were once classified as prosimians but are now considered more closely related to the monkeys and apes. Compare **simian.**

prostaglandin (prŏs′tə-glăn′dĭn) Any of a group of substances that are derived from fatty acids and have a wide range of effects in the body. Prostaglandins influence the contraction of the muscles lining many internal organs and can lower or raise blood pressure.

prostate gland (prŏs′tāt′) A gland in male mammals located at the base of the bladder. The prostate gland opens into the urethra and secretes a milky fluid that is a major component of semen. *—Adjective* **prostatic** (prŏ-stăt′ĭk).

prostate-specific antigen A protease secreted by the epithelial cells of the prostate gland. Blood levels are elevated in patients with prostate enlargement and prostate cancer and are used as a screening test for prostate cancer.

prosthesis (prŏs-thē′sĭs) Plural **prostheses** (prŏs-thē′sēz). An artificial device used to replace a missing or defective body part, such as a limb or a heart valve. *—Adjective* **prosthetic** (prŏs-thĕt′ĭk).

prosthetic group The nonprotein component of a conjugated protein, as the heme group in hemoglobin.

prostrate (prŏs′trāt′) Growing flat along the ground. Creeping jenny, pennyroyal, and many species of ivy have a prostrate growth habit.

protactinium (prō′tăk-tĭn′ē-əm) *Symbol* **Pa** A rare, extremely toxic, radioactive metallic element of the actinide series that occurs in uranium ores. It has 13 known isotopes, the most stable of which is protactinium 231 with a half-life of 32,760 years. Atomic number 91; approximate melting point 1,550°C; specific gravity 15.37; valence 4, 5. See **Periodic Table.**

protease (prō′tē-ās′) Any of various enzymes that bring about the breakdown of proteins into peptides or amino acids by hydrolysis. Pepsin is an example of a protease.

protein (prō′tēn′) Any of a large class of complex organic chemical compounds that are essential for life. Proteins play a central role in biological processes and form the basis of living tissues. They consist of long chains of amino acids connected by peptide bonds and have distinct and varied three-dimensional structures, usually containing **alpha helices** and **beta sheets** as well as looping and folded chains. Enzymes, antibodies, and hemoglobin are examples of proteins.

A CLOSER LOOK **protein folding**

Proteins are the true workhorses of the body, carrying out most of the chemical processes and making up the majority of cellular structures. Proteins are made up of long chains of amino acids, but they don't resemble linear pieces of spaghetti. The atoms in these long chains have their own attractive and repulsive properties. Some of the amino acids can form bonds with other molecules in the chain, kinking and twisting and folding into complicated, three-dimensional shapes, such as helices or densely furrowed globular structures. These folded shapes are immensely important because they define the protein's function in the cell. Some protein shapes fit perfectly in cell receptors, turning chemical processes on and off, like a key in a lock, whereas others work to transport molecules throughout the body (hemoglobin's shape is ideal for carrying oxygen). When proteins fail to take on their preordained shapes, there can be serious consequences: misfolded proteins have been implicated in diseases such as Alzheimer's, mad cow, and Parkinson's, among others. Exactly how proteins are able to fold into their required shapes is poorly understood and remains a fundamental question in biochemistry. See more at **prion.**

---

proteinase (prōt′n-ās′) A protease that begins the hydrolytic breakdown of proteins, usually by splitting them into polypeptide chains.

proteome (prō′tē-ōm′) The complete set of proteins that can be expressed by the genetic material of an organism. Compare **genome.**

proteomics (prō′tē-ō′mĭks) The analysis of the expression, localizations, functions, and interactions of the proteins expressed by the genetic material of an organism.

Proterozoic (prŏt′ər-ə-zō′ĭk) The later of the two divisions of the Precambrian Eon, from about 2.5 billion to 540 million years ago. The Proterozoic was characterized by the formation of stable continents, the appearance of abundant bacteria and archaea, and the buildup of oxygen in the atmosphere. By about 1.8 billion years ago the oxygen buildup was significant enough to cause many types of bacteria to die out. At this time eukaryotes, including multicellular algae and the first animals, first appear in the fossil record. See Chart at **geologic time.**

prothallus (prō-thăl′əs) Plural **prothalli** (prō-thăl′ī). The gametophyte of homosporous ferns and some other plants. Prothalli have chlorophyll for photosynthesis, but they are not differentiated into roots, stems, or leaves. They are usually small, flat, and delicate. Prothalli develop from germinated spores, and they bear both archegonia for producing eggs and antheridia for producing sperm. See more at **alternation of generations.**

prothrombin (prō-thrŏm′bĭn) A glycoprotein that is converted to thrombin during blood clotting. Prothrombin is formed by and stored in the liver.

protist (prō′tĭst) Any of a large variety of usually one-celled organisms belonging to the kingdom Protista (or Protoctista). Protists are eukaryotes and live in water or in watery tissues of organisms. Some protists resemble plants in that they produce their own food by photosynthesis, while others resemble animals in consuming organic matter for food. Protist cells are often structurally much more elaborate than the cells of multicellular plants and animals. Protists include the protozoans, most algae, diatoms, oomycetes, and the slime molds. Also called *protoctist.* See Table at **taxonomy.**

protium (prō′tē-əm, -shē-əm) The most abundant isotope of hydrogen, having an atomic mass of 1. Its nucleus consists of a single proton. See more at **hydrogen.**

protocol (prō′tə-kôl′, -kōl′) **1.** The plan for a course of medical treatment or for a scientific experiment. **2.** A set of standardized procedures for transmitting or storing data, especially those used in regulating data transmission between computers or peripherals.

protocontinent (prō′tō-kŏn′tə-nənt) A landmass that existed at some point in the past. Protocontinents that are thought to have comprised some or all of the present-day continents in a single landmass are also known as **supercontinents**.

protoctist (prə-tŏk′tĭst) See **protist.**

protoderm (prō′tə-dûrm′) The primary meristem in vascular plants that gives rise to epidermis. Also called *dermatogen.*

protohuman (prō′tō-hyo͞o′mən) Any of various extinct hominids or other extinct primates that were primitive predecessors or ancestors of humans.

proton (prō′tŏn′) A stable subatomic particle in the baryon family having a mass of 1.672×10^{-24} grams (1,836 times that of the electron) and a positive electric charge of approximately 1.602×10^{-19} coulombs. Protons make up part of the nucleus of all atoms except hydrogen, whose nucleus consists of a single proton. In neutral atoms, the number of protons is the same as the number of electrons. In positively charged atoms, the number of protons is greater than the number of electrons, and in negatively charged atoms electrons outnumber protons. Protons are believed to be composed of two up quarks and one down quark. See Table at **subatomic particle.**

protonema (prō′tə-nē′mə). Plural **protonemata** (prō′tə-nē′mə-tə, -nĕm′ə-). The green filamentous or flat, thallus-like structure that grows from the germinated spores of liverworts and mosses and eventually gives rise to a mature gametophyte.

proton-proton chain A set of thermonuclear reactions in which hydrogen nuclei (protons) fuse to form deuterium; these then fuse to form light helium isotopes, releasing more hydrogen nuclei that undergo further fusion with each other and with other nuclei. The main product of the proton-proton chain is He 4, the most common isotope of helium. The proton-proton chain releases large amounts of energy and involves the conversion of protons into neutrons, positrons, and electron neutrinos. The proton-proton chain is the main reaction in most main-sequence stars, including the Sun. See also **carbon cycle, triple alpha process.**

protophloem (prō′tə-flō′ĕm′) The first formed phloem that differentiates from the procambium.

protoplanetary disk (prō′tə-plăn′ĭ-tĕr′ē) A rotating disk of dust and gas that surrounds the core of a developing solar system. It may eventually develop into orbiting celestial bodies such as planets and asteroids. See more at **planetesimal.**

protoplasm (prō′tə-plăz′əm) The semifluid, translucent substance that forms the living matter in all plant and animal cells. Composed of proteins, fats, and other substances suspended in water, it includes the cytoplasm and (in eukaryotes) the nucleus.

protostar (prō′tə-stär′) A celestial object made of a contracting cloud of interstellar medium (mostly hydrogen gas) that eventually becomes a main-sequence star. Disturbances in some region of interstellar medium can cause fluctuations of density through that region, and the denser areas, having more mass, begin to attract more and more of the medium through the force of gravity (a process known as *accretion*). Ever increasing densities of such protostar regions lead to ever higher temperatures within the accreting body, until the point is reached when thermal energy is sufficient to promote the fusion reactions typical of main-sequence stars. Less massive protostars may take hundreds of millions of years to evolve into stars; massive ones contract more quickly and may take only a few hundred thousand years.

protostele (prō′tə-stēl′, prō′tə-stē′lē) The most primitive form of stele, consisting of a solid core of xylem encased by phloem or of xylem interspersed with phloem. The roots of all vascular plants, as well as the stems of lycophytes, have protosteles.

protostome (prōt′tə-stōm′) Any of a major group of animals defined by its embryonic development, in which the first opening in the embryo becomes the mouth. At this stage of development, the later specialization of any given embryonic cell has already been determined. Protostomes are one of the two groups of animals having a true body cavity (coelom) and are believed to share a common ancestor. They include the mollusks, annelids, and arthropods. Compare **deuterostome.**

protoxylem (prō′tə-zī′ləm) The first formed xylem that differentiates from the procambium.

protozoan (prō′tə-zō′ən) Plural **protozoans** or **protozoa.** Any of a large group of one-celled organisms (called protists) that live in water or as parasites. Many protozoans move about by means of appendages known as cilia or flagella. Protozoans include the amoebas, flagellates, foraminiferans, and ciliates. Their traditional classification as the subkingdom Protozoa is still used for convenience, but it is now known that protozoans represent several evolutionarily distinct groups. See more at **protist.**

protozoology (prō′tə-zō-ŏl′ə-jē) The scientific study of protozoans.

Proxima Centauri (prŏk′sə-mə sĕn-tôr′ē) See under **Alpha Centauri.**

Prusiner (pro͞o′sĭ-nər), **Stanley Ben** Born 1942. American biochemist who received the 1997 Nobel Prize for physiology or medicine for his discovery of prions. See Note at **prion.**

prussic acid (prŭs′ĭk) See **hydrocyanic acid.**

PSA Abbreviation of **prostatic-specific antigen.**

pseudobulb (so͞o′dō-bŭlb′) A thickened, bulblike, fleshy stem located above the ground, as in many orchids. See also **bulb, corm.**

pseudocarp (so͞o′də-kärp′) See **accessory fruit.**

pseudo force (so͞o′dō) A physically apparent but nonexistent force felt by an observer in a noninertial frame (that is, a frame undergoing acceleration). Newton's laws of motion hold true within such a reference frame only if the existence of such a force is presumed. The centrifugal force is an example of a pseudo force.

pseudopod (so͞o′də-pŏd′) also **pseudopodium** (so͞o′də-pō′dē-əm) Plural **pseudopods** or **pseudopodia.** A temporary footlike extension of a one-celled organism, such as an amoeba, used for moving about and for surrounding and taking in food.

psilocybin (sĭl′ə-sī′bĭn, sī′lə-) A hallucinogenic compound obtained from certain mushrooms, especially of the genus *Psilocybe. Chemical formula:* $C_{12}H_{17}N_2O_4P$.

psi particle (sī, psī) See **J/psi particle.**

psi′ particle (sī′prīm′, psī′-) An electrically neutral meson having a mass 7,213 times that of the electron and a mean lifetime of approximately 1×10^{-20} seconds, composed, like the J/psi particle, of a charm-anticharm quark pair. See Table at **subatomic particles.**

psoriasis (sə-rī′ə-sĭs) A chronic, inflammatory skin disease in which recurring reddish patches, often covered with silvery scales, appear especially on the knees, elbow, scalp and trunk. Psoriasis is noncontagious and is inheritable.

psychiatry (sĭ-kī′ə-trē) The branch of medicine that deals with the diagnosis, treatment, and prevention of mental and emotional disorders.

psychology (sī-kŏl′ə-jē) **1.** The scientific study of mental processes and behavior. **2.** The behavioral and cognitive characteristics of a specific individual, group, activity, or circumstance. ► **Clinical psychology** is the application of psychological knowledge to the diagnosis and treatment of patients.

psychopharmacology (sī′kō-fär′mə-kŏl′ə-jē) The study and clinical use of drugs that affect the mind, especially those that are used to treat psychiatric disorders.

psychosis (sī-kō′sĭs) Plural **psychoses** (sī-kō′sēz). A mental state caused by psychiatric or organic illness, characterized by a loss of contact with reality and an inability to think rationally. A psychotic person often behaves inappropriately and is incapable of normal social functioning. *—Adjective* **psychotic** (sī-kŏt′ĭk).

psychrometer (sī-krŏm′ĭ-tər) See under **hygrometer.**

Pt The symbol for **platinum.**

pt. Abbreviation of **pint.**

pteranodon (tə-răn′ə-dŏn′) Any of several very large, extinct flying reptiles (pterosaurs) of the genus *Pteranodon* of the Cretaceous Period. Pteranodon had a long pointed crest on its head, no teeth, a very short tail, and a wingspan upwards of 9 m (30 ft).

pteridine (tĕr′ĭ-dēn′) Any of a group of organic compounds having two fused six-member rings each containing two nitrogen atoms and four carbon atoms. One of the rings is a pyrimidine, the other a pyrazine. Pteridines include folic acid and the pigments of butterfly wings.

pteridophyte (tə-rĭd′ə-fīt′, tĕr′ĭ-dō-) Any of various vascular plants that reproduce by spores rather than seeds; a seedless vascular plant. The pteridophytes include the lycophytes, the horsetails, and the ferns, as well as a small group of related plants called the psilophytes and several extinct phyla.

pterodactyl (tĕr′ə-dăk′təl) Any of various small, extinct flying reptiles (pterosaurs) of the genus *Pterodactylus* of the late Jurassic and Cretaceous Periods. Pterodactyls had long, narrow jaws with sharp teeth, and a wingspan of 1 m (3.3 ft) or less.

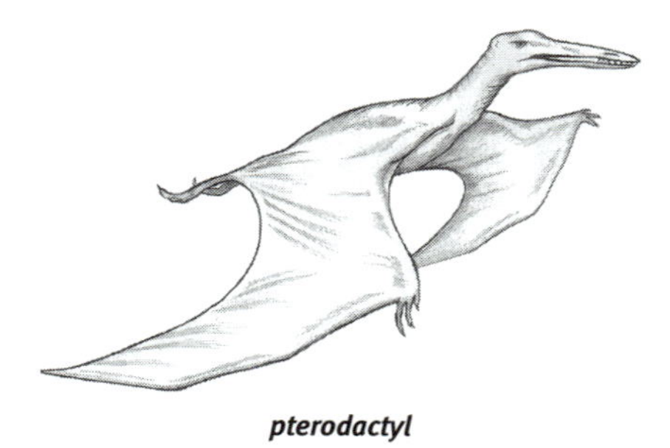

pterodactyl

pterosaur (tĕr′ə-sôr′) Any of various extinct flying reptiles of the Jurassic and Cretaceous

Periods with wings consisting of a flap of skin supported by an elongated fourth digit on each forelimb (rather than an elongated second digit as in birds). Some pterosaurs were unique among reptiles in being covered with hair. Pterosaurs had wingspans ranging from less than 0.3 m (1 ft) to close to 15.2 m (50 ft).

Ptolemaic system (tŏl′ə-mā′ĭk) The astronomical system of Ptolemy, in which Earth is at the center of the universe and all celestial bodies revolve around it. The Sun, Moon, and planets revolve at different levels in circular orbits, and the stars lie in fixed locations on a sphere that revolves beyond these orbits. See more at **epicycle.**

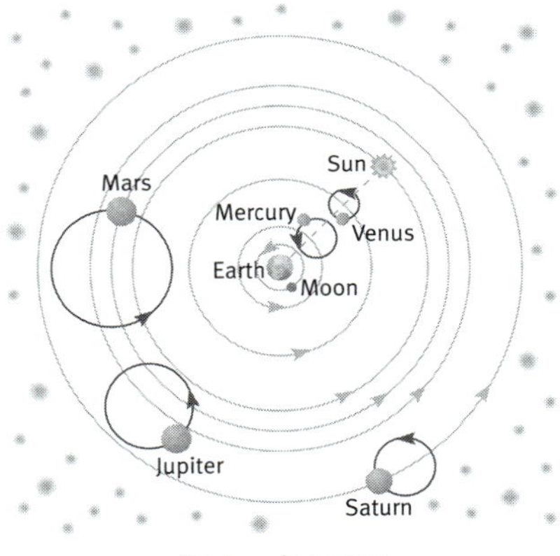

Ptolemaic system

Ptolemy believed that Earth was at the center of the universe and that the Sun and planets orbited Earth (with the planets also moving in smaller circles called epicycles). In this system, the centers of Mercury's and Venus's epicycles always lie on the line shown in the diagram between the Earth and the Sun.

Ptolemy (tŏl′ə-mē), 90?–168 CE. Greek astronomer and mathematician who based his astronomy on the belief that all heavenly bodies revolved around Earth. Ptolemy's model of the solar system endured until the 16th century when Nicolaus Copernicus proposed that the heavenly bodies in the solar system orbited the Sun. See Note at **Copernicus.**

ptomaine (tō′mān′) Any of various toxic nitrogenous organic compounds produced by bacterial decomposition of protein, especially in dead animal tissue. Ptomaines are bases and are formed by removing the carboxyl group (COOH) from amino acids. They do not cause food poisoning, as was previously thought, but the term *ptomaine poisoning* is still used to describe food poisoning caused by bacteria.

ptyalin (tī′ə-lĭn) An enzyme found in the saliva of humans and herbivorous animals that helps in the predigestion of starches. Ptyalin is a type of amylase.

P-type Made of material, usually a semiconductor such as silicon, that has been doped with impurities so that it has an excess of electron holes. Compare **N-type.**

Pu The symbol for **plutonium.**

puberty (pyo͞o′bər-tē) The stage in the development of humans and other primates marked by the development of **secondary sex characteristics,** including menarche in females. In humans, puberty occurs at the onset of adolescence, between the ages of about 11 and 14 in girls and 13 and 16 in boys.

pubis (pyo͞o′bĭs) Plural **pubes** (pyo͞o′bēz). The forwardmost of the three bones that fuse together to form each of the hipbones. See more at **skeleton.** —*Adjective* **pubic.**

pulley (po͝ol′ē) A machine consisting of a wheel over which a pulled rope or chain runs to change the direction of the pull used for lifting a load. Combinations of two or more pulleys working together reduce the force needed to lift a load. See also **block and tackle.**

pulmonary (po͝ol′mə-nĕr′ē) Relating to or involving the lungs.

pulmonary artery The artery that carries blood with low levels of oxygen from the right ventricle of the heart to the lungs.

pulmonary vein Any of the veins that carry blood with high levels of oxygen from the lungs to the left atrium of the heart.

pulmonology (po͝ol′mə-nŏl′ə-jē) The branch of medicine that deals with diseases of the respiratory system.

pulp (pŭlp) **1.** The soft tissue forming the inner structure of a tooth and containing nerves and blood vessels. **2.** The soft moist part of a fruit, especially a drupe or pome. **3.** The soft pith forming the contents of the stem of a plant.

pulsar (pŭl′sär′) A rapidly spinning neutron

star that emits radiation, usually radio waves, in narrow beams focused by the star's powerful magnetic field and streaming outward from its magnetic poles. Because the pulsar's magnetic poles do not align with the poles of its rotational axis, the beams of radiation sweep around like the beacon of a lighthouse and are thus observed on Earth as short, regular pulses, with periods anywhere between 1 millisecond and 4 seconds.

pulse (pŭls) **1.** The rhythmic expansion and contraction of the arteries as blood is pumped through them by the heart. The pulse can be felt at several parts of the body, as over the carotid and radial arteries. **2.** A dose of a medication or other substance given over a short period of time, usually repetitively. **3a.** A brief sudden change in a normally constant quantity, such as an electric current or field. **b.** Any of a series of intermittent occurrences characterized by a brief sudden change in a quantity.

pumice (pŭm′ĭs) A usually light-colored, porous, lightweight rock of volcanic origin. The pores form when water vapor and gases escape from the lava during its quick solidification into rock.

pump (pŭmp) **1.** A device used to raise or transfer fluids. Most pumps function either by compression or suction. **2.** A molecular mechanism for the active transport of ions or molecules across a cell membrane.

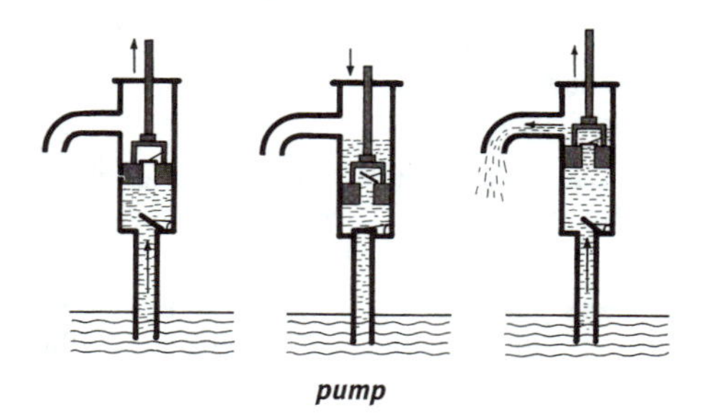

pump

punctuated equilibrium (pŭngk′cho͞o-ā′tĭd) The theory that new species evolve suddenly over relatively short periods of time (a few hundred to a thousand years), followed by longer periods in which little genetic change occurs. Punctuated equilibrium is a revision of Darwin's theory that evolution takes place at a slow, constant rate over millions of years. Compare **gradualism.** See Note at **evolution.**

pupa (pyo͞o′pə) Plural **pupae** (pyo͞o′pē). An insect in the nonfeeding stage of development between the larva and adult, during which it typically undergoes a complete transformation within a protective cocoon or hardened case. Only certain kinds of insects, such as moths, butterflies, ants, and beetles, develop as larvae and pupae. Compare **imago, larva, nymph.**

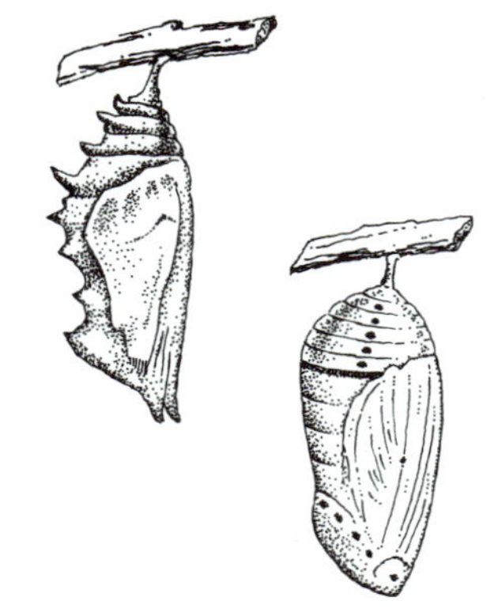

pupa

top: *a mourning cloak butterfly;* bottom: *a monarch butterfly*

pupil (pyo͞o′pəl) The opening in the center of the iris through which light enters the eye.

purine (pyo͝or′ēn′) Any of a group of organic compounds containing two fused rings of carbon and nitrogen atoms. One ring has six members, the other has five, and each has two nitrogens. Purines include a number of biologically important compounds, such as adenosine, caffeine, uric acid, and the two bases adenine and guanine, which are components of DNA and RNA.

pus (pŭs) A thick, yellowish-white liquid that forms in infected body tissues, consisting of white blood cells, dead tissue, and cellular debris.

pustule (pŭs′cho͞ol) A small inflamed swelling of the skin that is filled with pus.

PVC See **polyvinyl chloride.**

P wave See **primary wave.**

pylorus (pī-lôr′əs, pĭ-) Plural **pylori** (pī-lôr′ī′, pĭ-). The passage at the lower end of the stomach that opens into the small intestine.

pyran (pī′răn′) **1.** Any of a class of compounds having a ring of five carbon atoms and one oxygen atom and two double bonds. The pyran ring structure forms part of many organic compounds, especially sugars. **2.** An unstable compound that is the simplest member of this class. *Chemical formula:* C_5H_6O.

pyrazine (pĭr′ə-zēn′) Any of a group of organic compounds having a six-member ring in which the first and fourth atoms are nitrogen and the rest are carbon. Pyrazines are components of many important compounds, including pteridines, some vitamins and antibiotics, and numerous dyes called phenazines.

pyrheliometer (pīr′hē-lē-ŏm′ĭ-tər) An instrument used to measure the total amount of radiation from the Sun that hits a surface. It works by converting the Sun's heat into a voltage that is then recorded by a voltmeter.

pyridine (pĭr′ĭ-dēn′) **1.** Any of a class of organic compounds containing a six-member ring in which one of the carbon atoms has been replaced by a nitrogen atom. Pyridines include compounds used as water repellents, herbicides, and various drugs. The pyridine ring structure is also part of many larger compounds, including niacin and nicotine. **2.** The simplest of these compounds, a flammable, colorless or yellowish liquid base having a penetrating odor. It is used as a solvent and waterproofing agent and in the manufacture of various drugs and vitamins. *Chemical formula:* C_5H_5N.

pyridoxine (pĭr′ĭ-dŏk′sēn) A pyridine derivative that is the main form of vitamin B_6. *Chemical formula:* $C_8H_{11}NO_3$.

pyrimidine (pī-rĭm′ĭ-dēn′) Any of a group of organic compounds having a single six-member ring in which the first and third atoms are nitrogen and the rest are carbon. Pyrimidines include the bases cytosine, thymine, and uracil, which are components of DNA and RNA. Pyrimidine rings are also components of several larger compounds, such as thiamine and some synthetic barbiturates.

pyrite (pī′rīt′) A silver to yellow, metallic, cubic mineral. Pyrite often crystallizes in cubes or octahedrons but also occurs as shapeless masses of grains. It occurs in most types of rocks, and is used as a source of iron and in making sulfur dioxide. It is a polymorph of marcasite. Because of its shiny look and often yellow color, it is sometimes mistaken for gold and for this reason is also called *fool's gold. Chemical formula:* FeS_2.

pyrochemical (pī′rō-kĕm′ĭ-kəl) Relating to or designating chemical activity at elevated temperatures. Pyrochemical reactions are used, for example, to bring about the transmutation of certain transuranic elements.

pyroclastic (pī′rō-klăs′tĭk) Composed chiefly of rock fragments of explosive origin, especially those associated with explosive volcanic eruptions. Volcanic ash, obsidian, and pumice are examples of pyroclastic materials.

pyrogen (pī′rə-jən) A polypeptide that produces fever by causing metabolic changes in the hypothalamus. Pyrogens are either exogenous (produced by infectious agents) or endogenous (produced by cells in the body such as macrophages).

pyrophosphate (pī′rə-fŏs′fāt′) A salt or ester of pyrophosphoric acid, containing the group P_2O_7.

pyrophosphoric acid (pī′rō-fŏs-fôr′ĭk) A viscous liquid used as a catalyst and in the manufacture of organic chemicals. *Chemical formula:* $H_4P_2O_7$.

pyrosulfate (pī′rō-sŭl′fāt′) A salt or ester of pyrosulfuric acid, containing the group S_2O_7.

pyrosulfuric acid (pī′rō-sŭl-fyo͝or′ĭk) A heavy, oily, colorless to dark brown liquid produced by adding sulfur trioxide to concentrated sulfuric acid. It is used in petroleum refining and the manufacture of explosives. *Chemical formula:* $H_2S_2O_7$.

pyroxene (pī-rŏk′sēn′) Any of a series of dark silicate minerals having the general chemical formula $ABSi_2O_6$, where A is either calcium (Ca), sodium (Na), magnesium (Mg), or iron (Fe), and B is either magnesium, iron, chromium (Cr), manganese (Mn), or aluminum (Al). Pyroxenes vary in color from white to dark green or black and are characterized by a rectangular-shaped cross section. They can be either monoclinic or orthorhombic and occur in igneous and metamorphic rocks. The minerals enstatite, diopside, and augite are pyroxenes.

pyrrole (pîr′ōl′) **1.** Any of a class of organic

compounds having a five-member ring composed of four carbon atoms and one nitrogen atom. The pyrrole ring structure is a component of many biologically important compounds, including porphyrins, alkaloids, and certain amino acids. **2.** The simplest of this class of compounds, in which a single hydrogen atom is attached to each atom in the ring. It has a pleasant odor similar to that of chloroform. *Chemical formula:* C_4H_5N.

pyruvate (pī-ro͞o′vāt, pĭ-) A salt, ester, or ionized form of pyruvic acid, containing the group CH_3COCOO.

pyruvic acid (pī-ro͞o′vĭk) A colorless organic liquid formed by the breakdown of carbohydrates and sugars during cell metabolism. It is the final product of glycolysis and is converted into acetyl coenzyme A, which is required for the Krebs cycle. It is also used in the body to synthesize the amino acid alanine. *Chemical formula:* $C_3H_4O_3$.

Pythagoras (pĭ-thăg′ər-əs), 580?–500? BCE. Greek philosopher and mathematician who theorized that numbers constitute the essence of all natural things. He developed the Pythagorean theorem and was one of the first to apply mathematical order to observations of the stars.

Pythagorean theorem (pĭ-thăg′ə-rē′ən) A theorem stating that the square of the length of the hypotenuse of a right triangle is equal to the sum of the squares of the lengths of the other sides. It is mathematically stated as $c^2 = a^2 + b^2$, where c is the length of the hypotenuse and a and b the lengths of the other two sides.

Q

qt. Abbreviation of **quart.**

quadrant (kwŏd′rənt) **1.** An arc equal to one quarter of the circumference of a circle; an arc of 90°. **2.** Any of the four regions into which a plane is divided by the axes of a Cartesian coordinate system. The quadrants are numbered counterclockwise one through four, beginning with the quadrant in which both the x- and y-coordinates are positive (usually the upper right quadrant). **3.** A navigational instrument similar to a sextant but with an arc of 90° rather than 60°. See more at **sextant.**

quadratic (kwŏ-drăt′ĭk) Relating to a mathematical expression containing a term of the second degree, such as $x^2 + 2$. ► A **quadratic equation** is an equation having the general form $ax^2 + bx + c = 0$, where a, b, and c are constants. ► The **quadratic formula** is

$$x = \frac{-b \pm \sqrt{(b^2 - 4ac)}}{2a}$$

It is used in algebra to calculate the roots of quadratic equations.

quadrature (kwŏd′rə-cho͝or′) **1.** The process of constructing a square equal in area to a given surface. **2.** A configuration in which the position of one celestial body is 90° from another celestial body, as measured from a third. For example, the half moon lies in quadrature from the Sun when Earth is the reference point. See more at **elongation.**

quadriceps (kwŏd′rĭ-sĕps′) The large, four-part muscle at the front of the thigh that arises in the hip and pelvis and inserts as a strong tendon below the kneecap (patella). The quadriceps straightens and helps stabilize the knee.

quadrilateral (kwŏd′rə-lăt′ər-əl) A polygon that has four sides, such as a rectangle or rhombus.

quadriplegia (kwŏd′rə-plē′jē-ə) Paralysis of the body from the neck down, caused by injury to the spinal cord.

quadrumanous (kwŏ-dro͞o′mə-nəs) Having four feet and using all four feet as hands, as primates other than humans do. The big toes as well as the thumbs of quadrumanous species are opposable.

quadruped (kwŏd′rə-pĕd′) An animal having four feet, such as most reptiles and mammals. —*Adjective* **quadrupedal** (kwŏ-dro͞o′pə-dəl, kwŏd′rə-pĕd′l).

qualitative analysis (kwŏl′ĭ-tā′tĭv) See under **analysis.**

quantasome (kwŏn′tə-sōm′) One of numerous granules located on the inner lamellar surface of a chloroplast and thought to be involved in the light reactions of photosynthesis.

quantitative analysis (kwŏn′tĭ-tā′tĭv) See under **analysis.**

quantitative genetics The scientific study of the statistical analysis of the effects that heredity and environment have on phenotypic variation.

quantity (kwŏn′tĭ-tē) Something, such as a number or symbol that represents a number, on which a mathematical operation is performed.

quantize (kwŏn′tīz′) To limit a variable or variables describing a physical system to discrete, distinct values. For example, the energy of electromagnetic radiation such as light at a given frequency must be an integer multiple of hv, where v is the frequency and h is a Planck's constant; electromagnetic energy is thus inherently quantized (in this case, photons are the **quanta** of energy). The distinct orbitals of electrons in an atom are also a case of quantized energy. Many apparently continuous phenomena turn out to be quantized at a very fine level or very small scale; quantum mechanics was developed in large part to explain many unexpected cases of quantization in the natural world.

quantum (kwŏn′təm) Plural **quanta.** A discrete, indivisible manifestation of a physical property, such as a force or angular momentum. Some quanta take the form of elementary particles; for example, the quantum of electromagnetic radiation is the photon, while the quanta of the weak force are the W and Z particles. See also **quantum state.**

quantum bit The smallest unit of information in a quantum computer. Unlike bits in classical systems, which are in one of two possible states labelled 1 and 0, a quantum bit exists in a **superposition** of these two states, settling on one or the other only when a measurement of the state is made. Also called *qubit.*

quantum chromodynamics A quantum field theory of the **strong force** that explains the interaction between particles with **color charge**, such as quarks and gluons. In quantum chromodynamics, particles interact through the strong force by exchanging gluons, which are the carriers of the strong force (much as photons are the carriers of the electromagnetic force in quantum electrodynamics). The theory is particularly important in theories of the atomic nucleus, whose nucleons are composed of quarks.

quantum computer A computer that exploits the quantum mechanical properties of superposition in order to allow a single operation to act on a large number of pieces of data. In a quantum computer, the data to be manipulated, represented in quantum bits, exists in all possible states simultaneously, in superposition. This allows a single operation to operate over all of these states at once, in contrast with a classical computer, which must carry out an operation for each state separately. Because of the difficulty of creating environments small enough for quantum effects to emerge but sufficiently isolated to prevent interaction with outside influences such as heat, only extremely rudimentary quantum computers currently exist, though algorithms for possible future devices are being developed.

quantum electrodynamics A quantum field theory of electromagnetism that explains the interactions between electrically charged particles and photons. According to quantum electrodynamics, charged particles interact with one other by emitting or absorbing photons, which are the carriers of the electromagnetic force. Quantum electrodynamics correctly predicts such phenomena as the structure of atoms, and the creation and annihilation of particles (for example, when matter and antimatter collide). Quantum electrodynamics is one of the most well-tested, accurate, and successful theories in physics.

quantum field theory The application of quantum mechanics to physical systems described by fields, such as electromagnetic fields. Quantum field theory was developed to overcome certain deficiencies in **Schrödinger's equation**, in particular the fact that it was not consistent with special relativity and was difficult to apply to systems involving many particles or to the creation and destruction of particles. See also **quantum chromodynamics, quantum electrodynamics.**

quantum jump A change from one quantum state to another, as when an electron orbiting a nucleus moves from one shell to another with the loss or gain of a quantum of energy.

quantum mechanics A fundamental theory of matter and energy that explains facts that previous physical theories were unable to account for, in particular the fact that energy is absorbed and released in small, discrete quantities (quanta), and that all matter displays both wavelike and particlelike properties, especially when viewed at atomic and subatomic scales. Quantum mechanics suggests that the behavior of matter and energy is inherently probabilistic and that the effect of the observer on the physical system being observed must be understood as a part of that system. Also called *quantum physics, quantum theory.* Compare **classical physics.** See also **probability wave, quantum, uncertainty principle, wave-particle duality.**

quantum number Any of a set of numbers that together fully determine the state of a quantum mechanical system by quantifying its individual properties. For example, four quantum numbers are used to specify the quantum state of an electron orbiting the nucleus of an atom: one characterizes its basic orbital energy level (principal or first quantum number), one the shape of its orbit (its azimuthal, orbital or second quantum number), one the orientation of its orbit relative to other orbits (magnetic or third quantum number), and one its spin (spin or fourth magnetic number).

A CLOSER LOOK **quantum number**

Quantum numbers are used in quantum mechanics to describe the possible states of a physical system. Because many physical properties are *quantized*, taking on only discrete, distinct values, quantum numbers are generally integers or simple fractions, rather than continuous ranges. One of the great successes of quantum mechanics is its account of the structure of electron orbits around atomic nuclei, and the state of an electron in this particular system can be described using four quantum numbers. These are the principal or first quantum number, the orbital, azimuthal, or second quantum number, the magnetic quantum number, and the spin or spin magnetic quantum number. The *principal quantum number,* designated *n,* characterizes the basic energy level for the electron, and indicates in which shell the electron is located. It has integer values starting at 1; the higher the number, the farther the electron is from the atom's nucleus. The principal quantum numbers correspond to the traditional orbital shell designations K, L, M, and so on, used in chemistry. The *orbital quantum number,* designated *l,* characterizes the electron's angular momentum and determines the shape of its orbit. Its possible values for a given electron depend on the value of that electron's principal quantum numbers, ranging from 0 to n–1. Because of these different possibilities, shells (other than the first shell) include *subshells.* These are traditionally designated *s* (where l=0), *p* (where l=1), *d* (where l=2), and *f* (where l=3). The *magnetic quantum number,* designated *m* or *ml,* takes on integer values between –l and +l, and indicates the orientation of the electron's orbit within the subshell. For example, there are three orbitals in the *p* subshell, designated as p_x, p_y, and p_z. Finally, the *spin quantum number,* designated *ms,* characterizes the spin direction of the electron. It can have values of $+\frac{1}{2}$ or $-\frac{1}{2}$. Electrons are fermions, meaning that no two electrons can be in the same quantum state (due to the *Pauli exclusion principle*); therefore, each electron in an atom is uniquely characterized by a set of these four quantum numbers. In fact, the chemical properties of atoms depend almost entirely on the quantum numbers associated with their electrons. Other quantum numbers are used to describe other physical systems, such as the shell structure of the atomic nucleus.

---

quantum physics See **quantum mechanics.**

quantum state A description in quantum mechanics of a physical system or part of a physical system. Different quantum states for a physical system show discrete differences in the value of the variables used to define the state. For example, the spin of an isolated electron can take on one of only two values; there are no other quantum states available for the electron and no intermediate values, since spin is **quantized.** The quantum state is sometimes described by a set of **quantum numbers** that pick out the appropriate values for describing the state.

quantum statistics Any of various kinds of **statistical mechanics** that assume quantum mechanical behavior of particles.

quantum theory 1. See **quantum mechanics. 2.** Any of various theories that makes use of the assumptions, principles, and laws of quantum mechanics.

quantum tunneling A quantum mechanical effect in which particles have a finite probability of crossing an energy barrier, such as the energy needed to break a bond with another particle, even though the particle's energy is less than the energy barrier. Quantum tunneling has no counterpart in classical mechanics, in which a particle can never cross an energy barrier with a higher energy level than the particle has. The emission of alpha rays in radioactive decay is a case of quantum tunneling; though the alpha particles are strongly bound to the nucleus and don't have as much energy as the bond does, they still have a finite probability of escaping the nucleus. The design of transistors and many diodes makes use of this effect. See also **radioactivity.**

Quaoar (kwä′ə-wär′) A **Kuiper belt object** that, with a diameter of about 1,288 km (800 mi), is the largest such object so far discovered. Quaoar is approximately the size of Charon, Pluto's moon, and has a nearly circular orbit at a distance of some 1.6 billion km (1 billion mi) beyond that of Pluto.

quark (kwôrk, kwärk) Any of a group of elementary particles supposed to be the fundamental units that combine to make up the subatomic particles known as hadrons (baryons, such as neutrons and protons, and mesons). There are six different flavors (or types) of quark: up quark, down quark, top quark, bottom quark, charm quark, and strange quark. Quarks have fractional electric charges, such as $\frac{1}{3}$ the charge of an electron. See Note at **elementary particle.** See Table at **subatomic particle.**

quark star A superdense celestial object that is formed when the remnants of an old star collapses on itself, denser than a neutron star but not dense enough to become a black hole. Quark stars were first hypothesized in the 1980s, but the first was not discovered until early 2002. Like neutron stars, quark stars are composed of neutrons that have undergone enough pressure by the collapse of the star to have lost their differentiation and dissolved into a mass of quarks and gluons. The up and down quarks of which neutrons are composed then change into strange quarks, with the resulting **strange matter** compacting into an even denser mass than a neutron star. Also called *strange star.*

quart (kwôrt) **1.** A unit of volume or capacity in the US Customary System, used in liquid measure and equal to $\frac{1}{4}$ of a gallon or 32 ounces (0.95 liter). See Table at **measurement. 2.** A unit of volume or capacity in the US Customary System, used in dry measure and equal to $\frac{1}{8}$ of a peck or 2 pints (1.10 liter). See Table at **measurement.**

quartz (kwôrts) A hard, transparent trigonal mineral that, after feldspar, is the most common mineral on the surface of the Earth. It occurs as a component of igneous, metamorphic, and sedimentary rocks as well as in a variety of other forms such as rock crystal, flint, and agate. Some crystalline forms, such as amethyst, are considered gemstones. *Chemical formula:* $\mathbf{SiO_2}$.

quartzite (kwôrt′sīt′) A metamorphic rock consisting entirely of quartz. Quartzite forms when sandstone or chert are heated during metamorphism.

quasar (kwā′zär′) Short for *quasi-stellar radio source.* A compact, starlike celestial body with a power output greater than our entire galaxy. Believed to be the oldest and most distant objects ever detected, quasars are billions of light-years from Earth and moving away from us at nearly 80 percent of the speed of light. For this reason, quasars are highly important to astronomers' understanding of the early universe. Little is currently understood about the nature of quasars; one theory suggests that they are produced by giant black holes destroying enormous amounts of matter, causing the subsequent ejection of radiation along their north and south poles. Many astronomers believe that quasars represent an early stage in the evolution of galaxies such as our own. See also **blazar, Seyfert galaxy.**

quaternary (kwŏt′ər-nĕr′ē) *Noun.* **1. Quaternary.** The second and last period of the Cenozoic Era, from about 2 million years ago to the present. During this time the continents were situated approximately as they are today, although the geography was different due to fluctuations in sea levels caused by the advance and retreat of ice sheets. Humans first appeared during this period. See Chart at **geologic time.** *—Adjective.* **2.** Relating to or having a carbon atom that is attached to four other carbon atoms in a molecule. Quaternary pentane, for example, contains five carbon atoms of which one is in the center and the other four are each attached to it.

quaternary structure The structure that is formed by the joining together of two or more proteins or nucleic acids. The functions of the proteins and nucleic acids are only expressed correctly when they are joined together. See also **primary structure, secondary structure, tertiary structure.**

quaternion (kwə-tûr′nē-ən) Any number of the form $a + bi + cj + dk$ where a, b, c, and d are real numbers, $ij = k$, $i^2 = j^2 = -1$, and $ij = -ji$. Under addition and multiplication, quaternions have all the properties of a field, except that multiplication is not commutative.

qubit (kyo͞o′bĭt′) See **quantum bit.**

qubyte (kyo͞o′bīt′) A sequence of eight quantum bits operated on as a unit by a quantum computer.

quicksand (kwĭk′sănd′) A deep bed of loose, smoothly rounded sand grains, saturated with water and forming a soft, shifting mass that yields easily to pressure and tends to engulf objects resting on its surface. Although it is possible for a person to drown while mired in quicksand, the human body is less dense than any quicksand and is thus not drawn or sucked beneath the surface as is sometimes popularly believed.

quill (kwĭl) **1.** The hollow shaft of a feather, the bottom of which attaches to the bird's skin. **2.** One of the sharp hollow spines of a porcupine or hedgehog.

quinine (kwī′nīn′) A bitter-tasting, colorless drug derived from the bark of certain cinchona trees and used medicinally to treat malaria. For hundreds of years quinine was the only drug known to effectively combat malarial infection. It has since been largely replaced by synthetic compounds that not only relieve the symptoms of malaria but also rid the body of the malarial parasite, which quinine does not do. See Note at **aspirin.**

quinoline (kwĭn′ə-lēn′, -lĭn) An aromatic organic liquid having a pungent, tarlike odor. Quinoline is a base and is obtained from coal tar or is synthesized. It is used as a food preservative and in making antiseptics and dyes. *Chemical formula:* C_9H_7N.

quinolone (kwĭn′ə-lōn′) Any of a class of synthetic antibiotics that inhibit the replication of bacterial DNA.

quinone (kwĭ-nōn′, kwĭn′ōn′) **1.** Any of a class of organic compounds that occur naturally as pigments in bacteria, plants, and certain fungi. Quinones have two carbonyl groups (CO) in an unsaturated six-member carbon ring. **2.** A yellow crystalline compound belonging to this class, used in photography, to make dyes and to tan hides. *Chemical formula:* $C_6H_4O_2$.

quotient (kwō′shənt) The number that results when one number is divided by another. If 6 is divided by 3, the quotient can be represented as 2, or as $6 \div 3$, or as the fraction $\frac{6}{3}$.

R

r Abbreviation of **radius.**

R The symbol for **resistance.**

Ra The symbol for **radium.**

rabies (rā′bēz) A usually fatal infectious disease of warm-blooded animals caused by a virus of the genus *Lyssavirus* that causes inflammation of the brain and spinal cord. It is transmitted by the bite of an infected animal, such as a dog or bat and can be prevented in humans by a vaccine. See Note at **hydrophobia.**

race (rās) **1a.** An interbreeding, usually geographically isolated population of organisms differing from other populations of the same species in the frequency of hereditary traits. A race that has been given formal taxonomic recognition is known as a **subspecies. b.** A breed or strain, as of domestic animals. **2.** Any of several extensive human populations associated with broadly defined regions of the world and distinguished from one another on the basis of inheritable physical characteristics, traditionally conceived as including such traits as pigmentation, hair texture, and facial features. Because the number of genes responsible for such physical variations is tiny in comparison to the size of the human genome and because genetic variation among members of a traditionally recognized racial group is generally as great as between two such groups, most scientists now consider race to be primarily a social rather than a scientific concept.

raceme (rə-sēm′) An indeterminate inflorescence in which each flower grows on its own stalk from a common stem. The lily of the valley and snapdragon have racemes. See illustration at **inflorescence.**

racemic (rə-sē′mĭk) **1.** Relating to a chemical compound that contains equal quantities of the dextrorotatory and levorotatory forms of the compound and therefore does not rotate the plane of incident polarized light. **2.** Relating to or consisting of racemes.

racemic acid An optically inactive form of tartaric acid that can be separated into dextrorotatory and levorotatory components and is sometimes found in grape juice. *Chemical formula:* $C_4H_6O_6$.

rachilla (rə-kĭl′ə) Plural **rachillae** (rə-kĭl′ē). The stalk that bears the florets in the spikelets of grasses and similar plants, such as rushes and sedges. The rachilla often has a zigzag shape, with florets at each point at which the orientation of the rachilla turns.

rachis (rā′kĭs) Plural **rachises** or **rachides** (răk′ĭ-dēz′, rā′kĭ-). A main axis or shaft, such as the main stem of an inflorescence, the stalk of a pinnately compound leaf, the shaft of a feather, or the spinal column.

rad (răd) A unit used to measure energy absorbed by a material from radiation. One rad is equal to 100 ergs per gram of material. Many scientists now measure this energy in grays rather than in rads.

radar (rā′där) **1.** A method of detecting distant objects and determining their position, speed, material composition, or other characteristics by causing radio waves to be reflected from them and analyzing the reflected waves. The waves can be converted into images, as for use on weather maps. **2.** The equipment used in such detecting. See also **Doppler effect, lidar, sonar.**

radar

A radio antenna sends out a stream of radio waves (light gray). When they reach the airplane, they bounce off of it and send reflected waves (dark gray) back to a receiver in the antenna. The reflected waves are then processed electronically and analyzed to determine the airplane's distance, speed, and position. An image of the airplane is sometimes also generated.

radar astronomy The scientific study of nearby astronomical objects by reflecting microwaves off the objects and analyzing the echoes. The techniques of radio astronomy can be tailored to answer specific questions, such as the rotation period of the inner planets and the Moon. They have has also been used to test General Relativity, measure distances and features of planets in the solar system, and determine the composition, position, and spin of asteroids.

radar gun A usually hand-held device that measures the velocity of a moving object by sending out a continuous radio wave and measuring the frequency of reflected waves. See more at **Doppler effect.**

radial artery (rā′dē-əl) A major artery that follows the course of the radius on the ventral aspect of the forearm to the wrist, where it is felt as a pulse, and enters the hand. It supplies blood mostly to the thumb, index finger, and muscles of the forearm.

radial keratotomy (kĕr′ə-tŏt′ə-mē) A surgical procedure consisting of a radial pattern of corneal incisions, performed to reduce or correct myopia.

radial symmetry Symmetrical arrangement of parts of an organism around a single main axis, so that the organism can be divided into similar halves by any plane that contains the main axis. The body plans of echinoderms, ctenophores, cnidarians, and many sponges and sea anemones show radial symmetry. Compare **bilateral symmetry.**

radian (rā′dē-ən) A supplementary unit of the International System used in angular measure. One radian is equal to the angle subtended at the center of a circle by an arc equal in length to the radius of the circle, approximately 57°17'44.6".

radiant (rā′dē-ənt) *Adjective.* **1.** Transmitting light, heat, or other radiation. Stars, for example, are radiant bodies. **2.** Consisting of or transmitted as radiation. *—Noun.* **3.** The apparent celestial origin of a **meteor shower.** For example, a point in the constellation Gemini is the radiant of the Geminid meteor shower.

radiant energy Energy in the form of waves, especially electromagnetic waves. Radio waves, x-rays, and visible light are all forms of radiant energy.

radiation (rā′dē-ā′shən) **1a.** Streams of photons, electrons, small nuclei, or other particles. Radiation is given off by a wide variety of processes, such as thermal activity, nuclear reactions (as in fission), and by radioactive decay. **b.** The emission or movement of such particles through space or a medium, such as air. See Notes at **conduction, electromagnetic radiation. 2.** The use of such energy, especially x-rays, in medical diagnosis and treatment.

radiational cooling (rā′dē-ā′shə-nəl) The cooling of the Earth's surface and the air near the surface, occurring chiefly at night. It is caused by the emission of infrared radiation from the Earth's surface and from the tops of clouds and the atmosphere. Because infrared radiation is absorbed by water vapor, cloudless nights usually allow for greater radiational cooling than overcast nights. Radiational cooling occurs in all regions of the Earth and is important in maintaining the Earth's energy balance.

radiation pressure Force per unit area exerted by waves or particles of radiation, especially photons. Though photons have no mass, they do have momentum, and can transfer that momentum to other particles upon impact. The amount of pressure exerted by a given amount of radiation depends on whether the radiation is absorbed or reflected. Radiation pressure is responsible for the Casimir effect; solar radiation pressure is exploited in the design of **solar sails.**

radiative zone (rā′dē-ə-tĭv) The layer of a star that lies just outside the core, to which radiant energy is transferred from the core in the form of photons. In this layer, photons bounce off other particles, following fairly random paths until they enter the convection zone. Despite the high speed of photons, it can take hundreds of thousands of years for radiant energy in the Sun's radiative zone to escape and enter the convection zone.

radiator (rā′dē-ā′tər) A body that emits radiation. Radiators are commonly designed to transfer heat energy from one place to another, as in an automobile, in which the radiator cools the engine by transferring heat energy from the engine to the air, or in buildings, where radiators transfer heat energy from a furnace to the air and objects in the surrounding room.

radical (răd′ĭ-kəl) **1.** A root, such as $\sqrt{2}$, especially as indicated by a radical sign ($\sqrt{\ }$). **2.** A group of atoms that behaves as a unit in chemical reactions and is often not stable except as part of a molecule. The hydroxyl, ethyl, and phenyl radicals are examples. Radicals are unchanged by chemical reactions.

radical mastectomy See under **mastectomy.**

radicand (răd′ĭ-kănd′) The number or expression that is written under a radical sign, such as the 3 in $\sqrt{3}$.

radicle (răd′ĭ-kəl) **1.** The part of a plant embryo that develops into a root. In most seeds, the radicle is the first structure to emerge on germination. **2.** A small anatomical structure, such as a fibril of a nerve, that resembles a root.

radio (rā′dē-ō) *Noun.* **1.** The equipment used to generate, alter, transmit, and receive radio waves so that they carry information. —*Adjective.* **2.** Relating to or involving the emission of radio waves.

radioactive decay (rā′dē-ō-ăk′tĭv) The spontaneous transformation of an unstable atomic nucleus into a lighter one, in which radiation is released in the form of alpha particles, beta particles, gamma rays, and other particles. The rate of decay of radioactive substances such as carbon 14 or uranium is measured in terms of their **half-life**. See also **decay, radioisotope.**

radioactivity (rā′dē-ō-ăk-tĭv′ĭ-tē) The emission of **radiation** by unstable atomic nuclei undergoing **radioactive decay.**

A CLOSER LOOK **radioactivity**

In the nuclei of stable atoms, such as those of lead, the force binding the protons and neutrons to each other individually is great enough to hold together each nucleus as a whole. In other atoms, especially heavy ones such as those of uranium, this energy is insufficient, and the nuclei are unstable. An unstable nucleus spontaneously emits particles and energy in a process known as *radioactive decay.* The term *radioactivity* refers to the particles emitted. When enough particles and energy have been emitted to create a new, stable nucleus (often the nucleus of an entirely different element), radioactivity ceases. Uranium 238, a very unstable element, goes through 18 stages of decay before becoming a stable isotope of lead, lead 206. Some of the intermediate stages include the heavier elements thorium, radium, radon, and polonium. All known elements with atomic numbers greater than 83 (bismuth) are radioactive, and many isotopes of elements with lower atomic numbers are also radioactive. When the nuclei of isotopes that are not naturally radioactive are bombarded with high-energy particles, the result is *artificial radioisotopes* that decay in the same manner as natural isotopes. Each element remains radioactive for a characteristic length of time, ranging from mere microseconds to billions of years. An element's rate of decay is called its *half-life.* This refers to the average length of time it takes for half of its nuclei to decay.

radio astronomy The study of celestial objects by measurement of the radio waves they emit. Radio astronomy has enabled the detection and study of objects such as pulsars, quasars, radio galaxies, and other objects, some of which emit considerably less radiation at other wavelengths. Radio astronomy has contributed to the discovery of cosmic background radiation and has enhanced the understanding of solar activity and the structure of galaxies. See also **radio telescope.**

radiocarbon (rā′dē-ō-kär′bən) A radioactive isotope of carbon, especially carbon 14. Other radiocarbons include carbon 10, carbon 11, carbon 15, and carbon 16.

radiocarbon dating A technique for measuring the age of organic remains based on the rate of decay of carbon 14. Because the ratio of carbon 12 to carbon 14 present in all living organisms is the same, and because the decay rate of carbon 14 is constant, the length of time that has passed since an organism has died can be calculated by comparing the ratio of carbon 12 to carbon 14 in its remains to the known ratio in living organisms. Also called *carbon-14 dating.*

A CLOSER LOOK **radiocarbon dating**

In the late 1940s, American chemist Willard Libby developed a method for determining when the death of an organism had occurred. He first noted that the cells of all living things contain atoms taken in from the organism's environment, including carbon; all organic compounds contain carbon. Most carbon consists of the isotopes carbon 12 and carbon 13, which are very stable. A very small percentage of carbon, however, consists of the isotope carbon 14, or *radiocarbon*, which is unstable. Carbon 14 has a half-life of 5,780 years, and is continuously created in Earth's atmosphere through the interaction of nitrogen and gamma rays from outer space. Because atmospheric carbon 14 arises at about the same rate that the atom decays, Earth's levels of carbon 14 have remained fairly constant. Once an organism is dead, however, no new carbon is actively absorbed by its tissues, and its carbon 14 gradually decays. Libby thus reasoned that by measuring carbon 14 levels in the remains of an organism that died long ago, one could estimate the time of its death. This procedure of *radiocarbon dating* has been widely adopted and is considered accurate enough to study remains up to 50,000 years old.

---

radiochemistry (rā′dē-ō-kĕm′ĭ-strē) The scientific study of the chemical behavior of radioactive materials.

radio frequency A frequency of electromagnetic radiation in the range at which radio signals are transmitted, ranging from approximately 3 kilohertz to 300 gigahertz. Many astronomical bodies, such as pulsars, quasars, and possibly black holes, emit radio frequency radiation.

radio galaxy A galaxy that emits large amounts of radio energy. Radio galaxies are typically elliptical galaxies with large symmetrical lobes. The centers of radio galaxies are **active galactic nuclei**, and many expel one or more jets of matter directly from the nucleus that can be observed in the visible spectrum and extend for millions of light-years. See also **Seyfert galaxy.**

radiogenic (rā′dē-ō-jĕn′ĭk) **1.** Being a stable element that is product of radioactive decay. For example, carbon 12 is often radiogenic, having been produced by radioactive decay of carbon 14. **2.** Relating to the relation between radiogenic and radioactive elements, especially as a means of determining the age of objects, as in **radiometric dating.**

radioimmunoassay (rā′dē-ō-ĭm′yə-nō-ăs′ā, -ĭm-yo͞o′-) An immunoassay in which the substance to be identified or quantified is labelled with a radioactive substance (called a **tracer**), such as an ion. See also **immunoassay, tracer.**

radioisotope (rā′dē-ō-ī′sə-tōp′) A radioactive isotope of a chemical element. Carbon 14 and radon 222 are examples of naturally occurring radioactive isotopes.

radiolarian (rā′dē-ō-lâr′ē-ən) Any of various marine protozoans of the group Radiolaria, having rigid skeletons usually made of silica. The skeletons are usually spherically symmetrical and structurally complex, containing elaborate patterns of perforations (through which pseudopods extend) and often spicules. Skeletal remains of radiolarians sink to form ooze on the ocean floor, and prehistoric radiolarian ooze has fossilized to become chert and flint.

radiology (rā′dē-ŏl′ə-jē) The branch of medicine that deals with diagnostic images of anatomic structures through the use of electromagnetic radiation or sound waves and that treats disease through the use of radioactive compounds. Radiologic imaging techniques include x-rays, CAT scans, PET scans, MRIs, and ultrasonograms.

radiometer (rā′dē-ŏm′ĭ-tər) A device used to detect or measure radiation. Radiometers generally consist of a glass bulb containing a rarefied gas in which four diamond-shaped paddles are mounted on a central axis. Each paddle is black on one side and silvery on the other. When radiation such as sunlight strikes them, the black side absorbs radiation and the silvery side reflects it, resulting in a temperature difference between the two sides and causing motion of gas molecules around the edges of the paddles. This motion of the surrounding gas molecules causes the paddles to spin. Precision radiometers, which use a complete vacuum rather than a gas, exploit the difference in **radiation pressure** on either side of the paddles to cause them to spin. Radiometers measure the intensity of radiation by measuring the rate of spin of the paddles. Also called *light mill.*

radiometric dating (rā′dē-ō-mĕt′rĭk) A method for determining the age of an object based on the concentration of a particular radioactive isotope contained within it. For inorganic materials, such as rocks containing the radioactive isotope rubidium, the amount of the isotope in the object is compared to the amount of the isotope's decay products (in this case strontium). The object's approximate age can then be figured out using the known rate of decay of the isotope. For organic materials, the comparison is between the current ratio of a radioactive isotope to a stable isotope of the same element and the known ratio of the two isotopes in living organisms. Radiocarbon dating is one such type of radiometric dating.

radionuclide (rā′dē-ō-no͞o′klīd′) A nuclide that exhibits radioactivity.

radio telescope An instrument that consists of a radio receiver and antenna system mounted on a wide, bowl-shaped reflector, used to detect radio-frequency emissions from astronomical objects. The reflector and receiver form a **parabolic antenna**; incoming radio waves are focused by the reflector onto the receiver, where the radio signals are translated into electrical signals for further processing or electronic display. Due to the long wavelengths of radio waves, the reflectors of radio telescopes must be very large to focus the waves at a good resolution. Separate reflectors are sometimes linked in fixed arrays to act as a single collector.

radiotoxic (rā′dē-ō-tŏk′sĭk) Relating to or being a radioactive substance that is toxic to living cells or tissues.

radio wave A very low frequency electromagnetic wave (from roughly 30 kilohertz to 100 gigahertz). Radio waves are used for the transmission of radio and television signals; the microwaves used in radar and microwave ovens are also radio waves. Many celestial objects, such as pulsars, emit radio waves. See more at **electromagnetic spectrum.**

radium (rā′dē-əm) *Symbol* **Ra** A rare, bright-white, highly radioactive element of the alkaline-earth group. It occurs naturally in very small amounts in ores and minerals containing uranium, and it is naturally luminescent. Radium is used as a source of radon gas for the treatment of disease and as a neutron source for scientific research. Its most stable isotope is Ra 226 with a half-life of 1,622 years. Atomic number 88; melting point 700°C; boiling point 1,737°C; valence 2. See **Periodic Table.**

radius (rā′dē-əs) Plural **radii** (rā′dē-ī′) or **radiuses. 1.** A line segment that joins the center of a circle or sphere with any point on the circumference of the circle or the surface of the sphere. It is half the length of the diameter. **2.** The shorter and thicker of the two bones of the forearm or the lower portion of the foreleg. See more at **skeleton.**

radius vector 1. A line segment that joins the origin and a variable point in a system of polar or spherical coordinates. **2.** The imaginary straight line that connects the center of the Sun or another body with the center of a planet, comet, or other body that orbits it.

radix (rā′dĭks) Plural **radices** (răd′ĭ-sēz′, rā′dĭ-) or **radixes. 1.** *Biology.* The primary or beginning portion of a part or organ, as of a nerve at its origin from the brainstem or spinal cord. **2.** *Mathematics.* The base of a system of numbers, such as 2 in the binary system and 10 in the decimal system.

radon (rā′dŏn) *Symbol* **Rn** A colorless, odorless, radioactive element in the noble gas group. It is produced by the radioactive decay of radium and occurs in minute amounts in soil, rocks, and the air near the ground. Radon is used as a source of radiation for the treatment of cancer and other diseases. Its most stable isotope is Rn 222 with a half-life of 3.82 days. Atomic number 86; melting point –71°C; boiling point –61.8°C; specific gravity (solid) 4. See **Periodic Table.**

raffinate (răf′ə-nāt′) The portion of an original liquid that remains after other components have been dissolved by a solvent. The term is often used to refer to the oil that is not dissolved in petroleum refining operations.

raffinose (răf′ə-nōs′) A white crystalline sugar obtained from cottonseed meal, sugar beets, and molasses. Raffinose is an oligosaccharide, consisting of three simple sugars (fructose, galactose, and glucose) linked together. *Chemical formula:* $C_{18}H_{32}O_{16}$.

rain (rān) Water that condenses from water vapor in the atmosphere and falls to Earth as separate drops from clouds. Rain forms primarily in three ways: at weather fronts, when the water vapor in the warmer mass of air cools and condenses; along mountain ranges, when a warm mass of air is forced to rise over a mountain and its water vapor cools and condenses; and by convection in hot climates, when the water vapor in suddenly rising masses of warm air cools and condenses. See also **hydrologic cycle.**

rainbow (rān′bō′) An arc-shaped spectrum of color seen in the sky opposite the Sun, especially after rain, caused by the refraction and reflection of sunlight by droplets of water suspended in the air. Secondary rainbows that are larger and paler sometimes appear within the primary arc with the colors reversed (red being inside). These result from two reflections and refractions of a light ray inside a droplet.

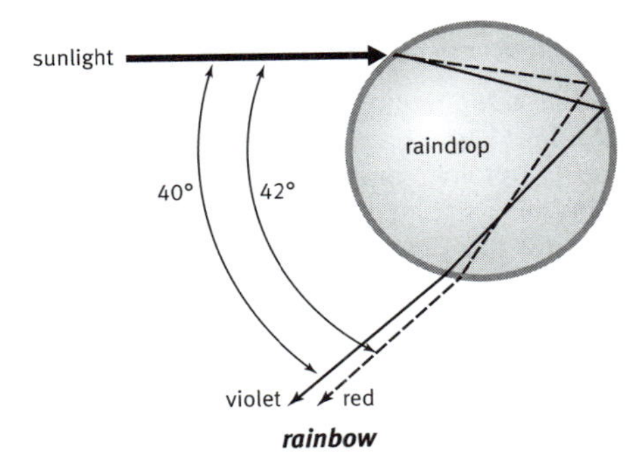

rainbow

In a primary rainbow, light is refracted upon entering the raindrop, reflected off the back of the raindrop, then refracted as it exits. The angle between the ray of sunlight that strikes the drop and the rays of varying wavelengths that leave the drop ranges from 40° for violet to 42° for red. (Angles in this diagram are approximate.)

rainfall (rān′fôl′) The quantity of water, usually expressed in millimeters or inches, that is precipitated in liquid form in a specified area and time interval. Rainfall is often considered to include solid precipitation such as snow, hail, and sleet as well.

rainforest (rān′fôr′ĭst) A dense evergreen forest with an annual rainfall of at least 406 cm (160 inches).

rain shadow An area having relatively little precipitation due to the effect of a topographic barrier, especially a mountain range, that causes the prevailing winds to lose their moisture on the windward side, causing the leeward side to be dry.

A CLOSER LOOK **rainforest**

Most of the world's *rainforests* lie near the equator and have tropical climates. However, cooler rainforests exist in the Pacific Northwest region of the United States and Canada. The world's largest rainforest is located in the Amazon River basin. The Amazon rainforest has been described as the "lungs of our planet" because it continuously recycles carbon dioxide into oxygen, with a significant percentage of the world's atmospheric oxygen being produced in this region. Besides helping to regulate the world's climate, rainforests host an extraordinary diversity of life. Scientists believe that as many as half of the Earth's different species of plants and animals are found only in the rainforests, which take up a mere 7 percent of the world's landmass. By some estimates, more than half of the Earth's original rainforests have already been burned or cut down for timber or grazing land, and more than 130 plant, animal, and insect species are thought to be going extinct daily as a result of the lost habitat. Currently 25 percent of Western pharmaceuticals are derived from tropical rainforest ingredients, and 70 percent of the plants with anticancer properties are found only in this shrinking biome.

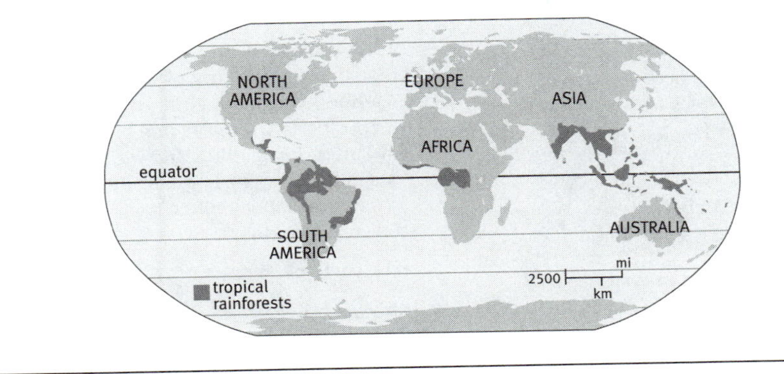

RAM (răm) Short for *random access memory.* The main memory of a computer, in which data can be stored or retrieved from all locations at the same (usually very high) speed. See also **dynamic RAM, static RAM.**

Raman (rä′mən), Sir **Chandrasekhara Venkata** 1888–1970. Indian physicist who in 1928 demonstrated that when light traverses a transparent material, some of the light that is deflected changes in frequency. For the discovery of this effect, which is now named after him, Raman received the 1930 Nobel Prize for physics. He also conducted research in the physiology of vision.

Raman effect The alteration of the frequency and the phase of light as it passes through a transparent medium. The Raman effect is caused by small differences between the energy of photons absorbed by the molecules that make up the medium and the energy of photons re-emitted.

ramjet (răm′jĕt′) A cylindrical jet propulsion engine consisting of air intake and combustion chambers into which burning fuel is injected, forcing hot air out of the rear of the engine at high pressures to provide forward thrust. See Note at **turbojet.**

Ramsay (răm′zē), Sir **William** 1852–1916. British chemist who discovered the noble gases argon (with Lord Rayleigh), helium, neon, xenon, and krypton. For this work he was awarded the 1904 Nobel Prize for chemistry. In 1908 his research showed that radon was also a noble gas.

William Ramsay

random (răn′dəm) **1.** Relating to a type of circumstance or event that is described by a probability distribution. **2.** Relating to an event in which all outcomes are equally likely, as in the testing of a blood sample for the presence of a substance.

random-access memory See **RAM.**

random coil A sequence of amino acids that has neither alpha-helical nor beta-sheet structure. Proteins consisting of alpha helices or beta sheets are reduced to random coils upon denaturing. Compare **alpha helix, beta sheet.**

range (rānj) **1.** The set of all values that a given function may have. Compare **domain. 2.** The difference between the smallest and largest values in a set of data. If the lowest test score of a group of students is 54 and the highest is 94, the range is 40.

Rankine scale (răng′kĭn) A scale of absolute temperature having the same degree increments as those of the Fahrenheit scale, in which the freezing point of water is 491.69° and the boiling point is 671.69°. A temperature in degrees Fahrenheit is converted to a temperature in degrees Rankine by the addition of 459.67°.

raphe (rā′fē′) Plural **raphae** (rā′fē′). **1.** A seamlike line or ridge between two similar parts of a body organ, as in the scrotum. **2.** The portion of the funiculus that is united to the ovule wall, commonly visible as a line or ridge on the seed coat. **3.** A groove in the frustule of some diatoms.

raphide (rā′fīd) Plural **raphides** (răf′ĭ-dēz′). One of a bundle of needlelike crystals of calcium oxalate occurring in many plant cells. The crystals discourage animals from eating the plant by irritating their tissues.

rapid eye movement sleep (răp′ĭd) See **REM sleep.**

raptor (răp′tər) **1.** A bird of prey, such as a hawk, eagle, or owl. **2.** Any of various mostly small, slender, carnivorous dinosaurs of the Cretaceous Period. Raptors had hind legs that were adapted for leaping and large, curved claws used for grasping and tearing at prey. Raptors were probably related to birds, and some even had feathers.

rare-earth element (râr′ûrth′) See **lanthanide.**

rarefaction (râr′ə-făk′shən) **1.** A decrease in density and pressure in a medium, such as air, especially when caused by the passage of a wave, such as a sound wave. **2.** The region in which this occurs.

ratio (rā′shō, rā′shē-ō′) A relationship between two quantities, normally expressed as the quotient of one divided by the other. For example, if a box contains six red marbles and four blue marbles, the ratio of red marbles to blue marbles is 6 to 4, also written 6:4. A ratio can also be expressed as a decimal or percentage.

rational horizon (răsh′ə-nəl) See **celestial horizon.**

rational number A number that can be expressed as an integer or a quotient of integers. For example, 2, –5, and $\frac{1}{2}$ are rational numbers.

ray (rā) **1.** A thin line or narrow beam of light or other radiation. **2.** A geometric figure consisting of the part of a line that is on one side of a point on the line. **3.** See **ray flower.**

ray-finned fish Any of various bony fishes belonging to the subclass Actinopterygii, having fins supported by thin bony rays. Ray-finned fish evolved in the early Devonian Period and include most species of fish today. Also called *actinopterygian.* Compare **lobe-finned fish.**

ray flower One of the narrow flowers, resembling single petals, that surround the central disk in the capitulum or flower cluster of a plant of the composite family (Asteraceae or Compositae), such as the daisy or sunflower. Rays are often male flowers with sterile stamens. Also called *ray, ray floret.* See illustration at **disk flower.**

Rayleigh (rā′lē), Third Baron. Title of **John William Strutt.** 1842–1919. British physicist whose investigation of the densities of gases led to his discovery (with Sir William Ramsay) of the noble gas argon in 1894. For this work he won the 1904 Nobel Prize for physics.

Rayleigh scattering The scattering of electromagnetic radiation by particles with dimensions much smaller than the wavelength of the radiation. The frequency of the radiation is not altered by this form of scattering, though the phase of the light is usually changed. Because the amount of Rayleigh scattering is greater at shorter frequencies, more scattering of the sun's rays by the Earth's atmosphere occurs on the blue end of the spectrum than at the red end, thus more blue light reaches the Earth, and the sky generally appears blue. Compare **Raman effect.** See also **Compton effect.**

Rayleigh wave A type of seismic surface wave that moves with a rolling motion that consists of a combination of particle motion perpendicular and parallel to the main direction of wave propagation. The amplitude of this motion decreases with depth. Like primary waves, Rayleigh waves are alternatingly compressional and extensional (they cause changes in the volume of the rocks they pass through). Rayleigh waves travel slower than **Love waves**.

Rb The symbol for **rubidium.**

rDNA Abbreviation of **recombinant DNA.**

Re The symbol for **rhenium.**

reactant (rē-ăk′tənt) A substance participating in a chemical reaction, especially one present at the start of the reaction.

reaction (rē-ăk′shən) **1.** A rearrangement of the atoms or molecules of two or more substances that come into contact with each other, resulting in the formation of one or more new substances. Chemical reactions are caused by electrons of one substance interacting with those of another. The reaction of an acid with a base, for example, results in the creation of a salt and water. Some, but not all, reactions can be reversed. **2.** See **nuclear reaction. 3.** An action that results directly from or counteracts another action, especially the change in a body's motion as a result of a force applied to it. Some reactions counteract forces and are not readily apparent. When an object rests on a surface, such as a table, for example, the downward force it applies to the surface is counteracted by an equal but upwards force, or reaction, applied by the surface. See more at **Newton's laws of motion. 4.** A response to a stimulus, such as a reflex. **5.** The response of cells or tissues to an antigen, as in a test for immunization.

reactive thrust (rē-ăk′tĭv) Thrust generated by an engine through the rapid expulsion of propellant or its exhaust in a particular direction. By Newton's third law, the force required to expel the propellant or exhaust is counterbalanced by an equal and opposite force, driving the engine (and its vehicle) in the opposite direction. See also **ion engine, Newton's laws of motion.**

read-only memory (rēd′ōn′lē) See **ROM.**

reagent (rē-ā′jənt) A substance participating in a chemical reaction, especially one used to detect, measure, or produce another substance. Compare **agent.**

real number (rē′əl) A number that can be written as a terminating or nonterminating decimal; a rational or irrational number. The numbers 2, –12.5, $\frac{3}{7}$, and pi (π) are all real numbers.

receiver (rĭ-sē′vər) A device, as in a radio or telephone, that converts incoming radio, microwave, or electrical signals to a form, such as sound or light, that can be perceived by humans. Compare **transmitter.**

Recent (rē′sənt) See **Holocene.**

receptacle (rĭ-sĕp′tə-kəl) The enlarged upper end of a flower stalk that bears the flower or group of flowers. The fleshy edible part of an apple is actually a modified receptacle. See more at **flower.**

receptor (rĭ-sĕp′tər) **1.** A nerve ending or other structure in the body, such as a photoreceptor, specialized to sense or receive stimuli. Skin receptors respond to stimuli such as touch and pressure and signal the brain by activating portions of the nervous system. Receptors in the nose detect the presence of certain chemicals, leading to the perception of odor. **2.** A structure or site, found on the surface of a cell or within a cell, that can bind to a hormone, antigen, or other chemical substance and thereby begin a change in the cell. For example, when a mast cell within the body encounters an allergen, specialized receptors on the mast cell bind to the allergen, resulting in the release of histamine by the mast cell. The histamine then binds to histamine receptors in other cells of the body, which initiate the response known as inflammation as well as other responses. In this way, the symptoms of an allergic reaction are produced. Antihistamine drugs work by preventing the binding of histamine to histamine receptors.

recessive (rĭ-sĕs′ĭv) Relating to the form of a gene that is not expressed as a trait in an individual unless two such genes are inherited, one from each parent. In an organism having two different genes for a trait, the recessive form is overpowered by its counterpart, or dominant, form located on the other of a pair of chromosomes. In humans, lack of dimples is a recessive trait, while the presence of dimples is dominant. See more at **carrier, inheritance.** Compare **dominant.**

reciprocal (rĭ-sĭp′rə-kəl) Either of a pair of numbers whose product is 1. For example, the number 3 is the reciprocal of $\frac{1}{3}$.

recombinant DNA (rē-kŏm′bə-nənt) A form of DNA produced by combining genetic material from two or more different sources by means of genetic engineering. Recombinant DNA can be used to change the genetic makeup of a cell, as in adding a gene to make a bacterial cell produce insulin.

rectangle (rĕk′tăng′gəl) A four-sided plane figure with four right angles.

rectifier (rĕk′tə-fī′ər) An electrical device that converts alternating current to direct current. Rectifiers are most often made of a combination of diodes, which allow current to pass in one direction only. Compare **converter, transformer.**

rectilinear (rĕk′tə-lĭn′ē-ər) Relating to, consisting of, or moving in a straight line or lines.

rectum (rĕk′təm) Plural **rectums** or **recta.** The last section of the digestive tract, extending from the colon to the anus, in which feces is stored for elimination from the body. *—Adjective* **rectal.**

recurring decimal (rĭ-kûr′ĭng) See **repeating decimal.**

recycling (rē-sī′klĭng) The collection and often reprocessing of discarded materials for reuse. Recycled materials include those used in manufacturing processes and those used in consumer products. The recycled material is often degraded somewhat by use or processing and therefore must be converted to another purpose. For example, the processing of recycled newspaper and other paper wastes usually shortens their fibers, and the material cannot be used to make high-grade paper. Instead, it can be reprocessed to make cardboard or insulation. Recycling helps reduce pollution, prolong the usefulness of landfills, and conserve natural resources.

red alga (rĕd) Any of various photosynthetic protists belonging to the phylum Rhodophyta.

Most red algae are marine seaweeds, with bladelike or filamentous bodies. Others deposit calcium carbonate in their cell walls and grow as crusty layers on rocks, and these species are essential reef-building organisms. Unicellular forms are also known, and a few red algae live in fresh water. Red algae have chlorophyll a as well as pigments called phycobilins, which impart the reddish colors characteristic of the group. Phycobilins absorb the blue and green wavelengths of light that reach deeper coastal waters where red algae are abundant. Red algae often show complicated life cycles involving an alternation of a haploid generation with two distinct diploid generations, both of which produce a different kind of spore. Their cell walls are made of cellulose and gelatinous polysaccharide compounds, some of which have commercial value. Agar and carrageenan are extracted from red algae, and the genus *Porphyra* is cultivated and dried to make nori, the edible seaweed used in sushi. See more at **alga.**

red blood cell Any of the oval or disc-shaped cells that circulate in the blood of vertebrate animals, contain hemoglobin, and give blood its red color. The hemoglobin in red blood cells binds to oxygen for transport and delivery to body tissues, and it transports carbon dioxide, excreted as a metabolic waste product, out of the tissues. The red blood cells of mammals have no nucleus, while those of other vertebrates do contain nuclei. Red blood cells are formed in the bone marrow. Also called *erythrocyte.*

A CLOSER LOOK **red blood cells**

While 60 percent of the US population is eligible to donate blood, only about 5 percent does. There is no substitute for human blood, which is used for numerous medical situations, including surgery for trauma, cancer treatment, organ transplants, burns, open heart surgeries, anemia, clotting disorders, and treating premature babies. The average red blood cell transfusion is 3.4 pints. Blood, which is made in the bone marrow, has four main components—*red cells, platelets, plasma,* and *white cells.* Red blood cells, or *erythrocytes,* transport oxygen from the lungs to the rest of the body. These disk-shaped cells contain *hemoglobin,* an iron-containing protein that picks up oxygen molecules as the blood exchanges gases in the lungs. The red blood cells, which can live about 120 days in the circulatory system, deliver oxygen to the far reaches of the body, where it is released for use by other cells, such as those of the brain and muscles. Red blood cells also pick up carbon dioxide and return it to the lungs to be exhaled. All animals have some form of oxygen distribution system, but only vertebrates use red blood cells. In some invertebrates, such as the earthworm, oxygen is transported using hemoglobin that is freely dissolved in the blood. Other invertebrates don't use hemoglobin at all. The horseshoe crab, for instance, uses copper instead of iron, making its blood blue instead of red.

red dwarf A small, dim star with relatively cool surface temperatures that is positioned to the lower right on the **main sequence** in the Hertzsprung-Russell diagram. Red dwarfs, at about 0.1 to 0.5 solar mass, consume their nuclear fuel very slowly and live for about 100 billion years. Although they are difficult to see, they are so long-lived that they are likely the most abundant type of star; of the 30 nearest stars to Earth, 21 are red dwarfs, including the closest star, Proxima Centauri. See Note at **dwarf star.**

red giant A giant star that has a relatively low surface temperature, giving it a reddish or orange hue. Red giants are non-main-sequence stars positioned in the upper right of the **Hertzsprung-Russell diagram.** They are not massive stars but rather late, expanded stages of lower-mass main-sequence stars that have exhausted the hydrogen in their core and are fusing their remaining hydrogen into helium in a luminous outer shell. The Sun is expected to become a red giant in about 5 billion years, expanding to 70 times its current size and bringing its surface extremely close to Earth's present orbit. See more at **star.** See Note at **dwarf star.**

red shift See under **Doppler effect.**

red supergiant An extremely large red giant star with a minimum of 15 solar masses. The best known red supergiant is Betelgeuse, with a luminosity about 10,000 times that of the Sun. When a supergiant collapses into a supernova, it may result in either a **neutron star** or a **black hole.**

red tide A population explosion of certain species of dinoflagellates, a kind of protozoan found in plankton. The dinoflagellates color the water red or reddish-brown and secrete a toxin that kills fish. Red tide usually occurs in warm coastal waters.

reduction (rĭ-dŭk′shən) **1.** The changing of a fraction into a simpler form, especially by dividing the numerator and denominator by a common factor. For example, the fraction $\frac{8}{12}$ can be reduced to $\frac{4}{6}$, which can be further reduced to $\frac{2}{3}$, in each case by dividing both the numerator and denominator by 2. **2.** A chemical reaction in which an atom or ion gains electrons, thus undergoing a decrease in valence. If an iron atom having a valence of +3 gains an electron, the valence decreases to +2. Compare **oxidation.**

USAGE **reduction/oxidation**

Beginning students of chemistry are understandably puzzled by the term *reduction:* shouldn't a reduced atom or ion be one that *loses* electrons rather than gains them? The reason for the apparent contradiction comes from the early days of chemistry, where reduction and its counterpart, oxidation, were terms invented to describe reactions in which one substance lost an oxygen atom and the other substance gained it. In a reaction such as that between two molecules of hydrogen ($2H_2$) and one of oxygen (O_2) combining to produce two molecules of water ($2H_2O$), the hydrogen atoms have gained oxygen atoms and were said to have become "oxidized," while the oxygen atoms have (as it were) lost them by attaching themselves to the hydrogens, and were said to have become "reduced." Importantly, though, in the process of gaining an oxygen atom, the hydrogen atoms have had to give up their electrons and share them with the oxygen atoms, while the oxygen atoms have gained electrons. Thus comes the apparent paradox that the "reduced" oxygen has in fact *gained* something, namely electrons. Today the terms *oxidation* and *reduction* are used of any reaction, not just one involving oxygen, where electrons are (respectively) lost or gained.

Reed (rēd), **Walter** 1851–1902. American physician and army surgeon who proved in 1900 that yellow fever was transmitted by the *Aedes aegypti* mosquito. His research led to the mosquito eradication programs carried out by William Gorgas that virtually eradicated yellow fever from Havana, Cuba, and from the Panama Canal Zone.

reef (rēf) A strip or ridge of rocks, sand, or coral that rises to or near the surface of a body of water. See more at **coral reef.**

reference frame (rĕf′ər-əns) A basis of a four-dimensional coordinate system in which the first coordinate is understood as a time coordinate, while the other three coordinates represent spatial dimensions. Inertial frames and non-inertial frames are both examples of reference frames. Also called *frame of reference.* See also **General Relativity, space-time, Special Relativity.**

refinery (rĭ-fī′nə-rē) An industrial plant that uses mechanical and chemical means to purify a substance, such as petroleum or sugar, or to convert it to a form that is more useful.

reflecting telescope (rĭ-flĕk′tĭng) See under **telescope.**

reflection (rĭ-flĕk′shən) **1.** The change in direction of a wave, such as a light or sound wave, away from a boundary the wave encounters. Reflected waves remain in their original medium rather than entering the medium they encounter. ▸ According to the **law of reflection,** the angle of reflection of a reflected wave is equal to its angle of incidence. Compare **refraction.** See more at **wave. 2.** Something, such as sound, light, or heat, that is reflected.

reflection nebula A nebula that reflects the light of a nearby star or stars toward Earth rather than producing its own. See more at **nebula.**

reflex (rē′flĕks′) **1.** An involuntary physiological response to a stimulus, as the withdrawal of a body part from burning heat. **2.** An unlearned or instinctive response to a stimulus. Also called *unconditioned response.* See more at **classical conditioning.**

reflexive (rĭ-flĕk′sĭv) Of or relating to a mathematical or logical relation such that, for any given element, that element has the given relation to itself. Equality in mathematics is a reflexive relation, since $a = a$ for all a, whereas the relation of being 'less than' is not, since it is not true that $a < a$ for any a.

refracting telescope (rĭ-frăk′tĭng) See under **telescope.**

refraction (rĭ-frăk′shən) **1.** The bending of a wave, such as a light or sound wave, as it passes from one medium to another medium of different density. The change in the angle of propagation depends on the difference between the **index of refraction** of the original medium and the medium entered by the wave, as well as on the frequency of the wave. Compare **reflection.** See also **lens, wave. 2.** The apparent change in position of a celestial body caused by the bending of light as it enters the Earth's atmosphere.

refractory (rĭ-frăk′tə-rē) **1.** Having a high melting point. Ceramics that are made from clay and minerals are often refractory, as are metal oxides and carbides. Refractory materials are often used as liners in furnaces. **2.** Resistant to heat. **3.** Of or relating to a refractory period.

refractory period The period immediately following the transmission of an impulse in nerve or muscle, in which a neuron or muscle cell regains its ability to transmit another impulse. See more at **action potential.**

A CLOSER LOOK **refraction/reflection**

The terms *refraction* and *reflection* describe two ways that waves, as of sound or light, change course upon encountering a boundary between two media. The media might consist of two different substances, such as glass and air, or a single substance in different states in different regions, such as air at different temperatures or densities in different layers. Reflection occurs, as in a mirror, when a wave encounters the boundary but does not pass into the second medium, instead immediately changing course and returning to the original medium, typically reflecting from the surface at the same angle at which it contacted it. Refraction occurs, as in a lens, when a wave passes from one medium into the second, deviating from the straight path it otherwise would have taken. The amount of deviation or "bending" depends on the *indexes of refraction* of each medium, determined by the relative speed of the wave in the two media. Waves entering a medium with a higher index of refraction are slowed, leaving the boundary and entering the second medium at a greater angle than the incident wave. Waves entering a medium with a lower index are accelerated and leave the boundary and enter the second medium at a lesser angle. Incident light waves tend to be fully reflected from a boundary met at a shallow angle; at a certain *critical angle* and at greater angles, some of the light is also refracted; looking at the surface of water from a boat, for instance, one can see down into the water only out to where the sight line reaches the critical angle with the surface. Light passing through a prism is mostly refracted, or bent, both when it enters the prism and again when it leaves the prism. Since the index of refraction in most substances depends on the frequency of the wave, light of different colors is refracted by different amounts—hence the colorful rainbow effect of prisms. The boundary between media does not have to be abrupt for reflection or refraction to occur. On a hot day, the air directly over the surface of an asphalt road is warmer than the air higher up. Light travels more quickly in the lower region, so light coming down from the sky (from not too steep an angle) is refracted back up again, giving a "blue puddle" appearance to the asphalt—a mirage.

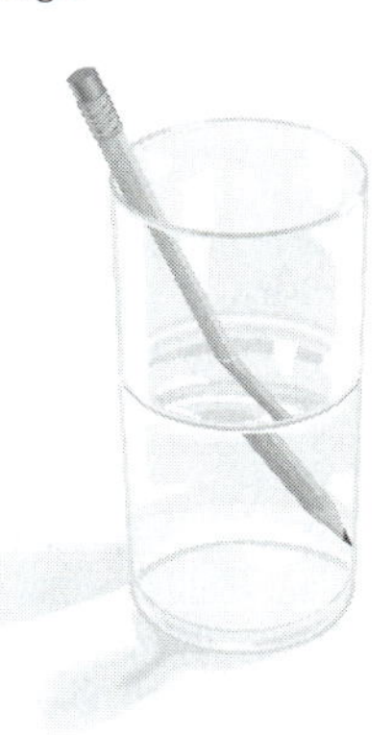

Light waves bend as they pass from one substance to another. This pencil appears to be bent at various angles as the light passes through air only; through air and glass; through water, air, and glass; and through water and glass.

refrigerant (rĭ-frĭj′ər-ənt) A substance, such as ice or ammonia, used to cool something by absorbing heat from it. Refrigerants are usually substances that evaporate quickly. In the process of evaporation they draw heat from surrounding substances.

regeneration (rĭ-jĕn′ə-rā′shən) The regrowth of lost or destroyed parts or organs.

A CLOSER LOOK **regeneration**

Regeneration of parts or, in some cases, nearly the entire body of an organism from a part, is more common than one might think. Many protists like the amoeba that have been cut in half can grow back into a complete organism so long as enough of the nuclear material is undamaged. Severed cell parts, such as flagella, can also be regrown in protists. New plants can be grown from cuttings, and plants can often be regenerated from a mass of fully differentiated cells (such as a section of a carrot root), which, if isolated in a suitable environment, turn into a mass of undifferentiated cells that develop into a fully differentiated organism. The capacity for regeneration varies widely in animals, with some able to regenerate whole limbs and others not, but the capacity is reduced significantly in more complex animals. Certain simple invertebrates like the hydra are always regenerating themselves. If cut into tiny pieces that are then mixed up, the pieces can reorganize themselves and grow back into a complete organism. Flatworms have the capacity to regenerate themselves from only a small mass of cells. If they are chopped up into fine pieces, each piece has the capacity to develop into an entire organism. Starfish, which are echinoderms, can regenerate their entire body from their central section and a single arm. Newts and salamanders can regenerate lost legs and parts of eyes, but many other amphibians such as frogs and toads cannot. Certain lizards can regenerate their tails. In many animals, these regenerated body parts are not as large as the originals but are usually sufficient to be functional. Many higher animals such as mammals regularly regenerate certain tissues such as hair and skin and portions of others such as bone, but most tissues cannot be regenerated. About 75 percent of the human liver can be removed, and it will regenerate into a functional organ. The physiological reasons for this are still not understood. Regeneration in this case takes the form of the enlargement of the remaining structures rather than the re-creation of the lost ones. Thus, there are four mechanisms for tissue regeneration in animals: the reorganization of existing cells (as in the hydra), the differentiation of stored stem cells into the specific tissues needed (as in the salamander), the dedifferentiation of neighboring tissue cells and their subsequent regrowth as cells of the needed type (as in plants as well as certain animals like the salamander), and the compensatory growth of the surviving cells of the specific tissue (as in the human liver). There is a great interest in stem cells because of their potential use in regenerating body tissues, such as nerve cells and heart muscle. The biochemical mechanisms for dedifferentiation are also the subject of intense study.

regional metamorphism (rē′jə-nəl) A type of metamorphism in which the mineralogy and texture of rocks are changed over a wide area by deep burial and heating associated with the large-scale forces of plate tectonics. In regional metamorphism, rocks that form closer to the margin of the tectonic plates, where the heat and pressure are greatest, often differ in their minerals and texture from those that form farther away. Compare **contact metamorphism.**

regolith (rĕg′ə-lĭth′) The layer of rock and mineral fragments that rests on bedrock and is produced by the weathering of rocks. Regolith constitutes the surface of most land.

regression (rĭ-grĕsh′ən) **1.** A subsiding of the symptoms or process of a disease. **2.** The return of a population to an earlier or less complex physical type in successive generations. **3.** The relationship between the mean value of a random variable and the corresponding values of one or more independent variables. **4.** A relative fall in sea level resulting in deposition of terrestrial strata over marine strata. Compare **transgression. 5.** Retrograde motion of a celestial body.

regular (rĕg′yə-lər) Having all sides or faces equal. For example, a square is a regular polygon, and a cube is a regular polyhedron.

regular flower A radially symmetric flower. In a regular flower, all of the members of a single whorl, such as the petals, are similar in shape and size. Lilies and the apple tree, for example, bear regular flowers. Compare **irregular flower.**

Regulus (rĕg′yə-ləs) A bright triple star in the constellation Leo, with an apparent magnitude of 1.35. *Scientific name:* Alpha Leonis.

regurgitation (rē-gûr′jĭ-tā′shən) The return of partially digested food from the stomach to the mouth.

relational database (rĭ-lā′shə-nəl) A database system in which any field can be a component of more than one of the database's tables.

relative atomic mass (rĕl′ə-tĭv) See **atomic weight.**

relative density The ratio of the density of one substance to that of a reference substance, typically water. See more at **specific gravity.**

relative humidity The ratio of the actual amount of water vapor present in a volume of air at a given temperature to the maximum amount that the air could hold at that temperature, expressed as a percentage. Warm air can hold more water vapor than cool air, so a particular amount of water vapor will yield a lower relative humidity in warm air than it does in cool air. Compare **absolute humidity.**

relative molecular mass See **molecular weight.**

relative permittivity The ratio of the magnetic **permittivity** of a substance to the permittivity of a vacuum.

relative temperature scale A temperature scale in which measurements are amounts that are more or less than a reference amount. In the Celsius scale, for example, the reference amount is set as the freezing point of water, or zero. Other measurements are made relative to this point. Relative temperature scales have both positive and negative numbers. The Celsius scale and the Fahrenheit scale are relative temperature scales. Compare **absolute temperature scale.**

relativistic mass (rĕl′ə-tə-vĭs′tĭk) In Special and General Relativity, the observed mass of an object moving with respect to the observer. The relativistic mass is a function of the rest mass and the velocity of the object. Compare **rest mass.**

relativity (rĕl′ə-tĭv′ĭ-tē) Either of two theories in physics developed by Albert Einstein, **General Relativity** or **Special Relativity.** See Notes at **Einstein, gravity, space-time.** —*Adjective* **relativistic.**

A CLOSER LOOK **relativity**

Albert Einstein's two theories of *relativity* were the first successful revisions of Newtonian mechanics—a mechanics so simple and intuitive that it was held to be a permanent fixture of physics. Uniting the theories is the idea that two observers traveling relative to each other may have different perceptions of time and space, yet the laws of nature are still uniform, and certain properties always remain *invariant.* Einstein developed the first theory, the theory of Special Relativity (1905), to explain and extend certain consequences of Maxwell's equations describing electromagnetism, in particular, addressing a puzzle surrounding the speed of light in a vacuum, which was predicted always to be the same, whether the light source is stationary or moving. Special Relativity considers the laws of nature from the point of view of frames of reference upon which no forces are acting, and describes the way time, distance, mass, and energy must be perceived by observers who are in uniform motion relative to each other if the speed of light must always turn out the same for all observers. Two implications of Special Relativity are space and time dilation. As speed increases, space is compressed in the direction of the motion, and time slows down. A famous example is the space traveler who returns to Earth younger than his Earth-dwelling twin, his biological processes proceeding more slowly due to his relative speed. These effects are very small at the speeds we normally experience but become significant at speeds approaching the speed of light (known as *relativistic* speeds). Perhaps the best-known implication of Special Relativity is the equation $E=mc^2$, which expresses a close relation between energy and mass. The speed of light is a large number (about 300,000 km per second, or 186,000 mi per second), so the equation suggests that even small amounts of mass can be converted into enormous amounts of energy, a fact exploited by atomic power and weaponry. Einstein's General Theory of relativity extended his Special Theory to include *non-inertial* reference frames, frames acted on by forces and undergoing acceleration, as in cases involving *gravity.* The General Theory revolutionized the way gravity, too, was understood. Since Einstein, gravity is seen as a curvature in space-time itself.

---

relay (rē′lā) An electrical switch that is operated by an electromagnet, such as a solenoid. When a small current passes through the electromagnet's coiled wire, it produces a magnetic field that attracts a movable iron bar, causing it to pivot and open or close the switch.

relief map (rĭ-lēf′) A map that depicts land configuration, often with contour lines. Some relief maps are three-dimensional.

rem (rĕm) The amount of ionizing radiation required to produce the same biological effect as one rad of high-penetration x-rays. The rem has been replaced in most scientific contexts by the sievert.

remainder (rĭ-mān′dər) In division, the difference between the dividend and the product of the quotient and divisor. Dividing 14 by 3 gives 4 and a remainder of 2.

remanence (rĕm′ə-nəns) The magnetic flux density remaining in a material, especially a **ferromagnetic** material, after removal of the magnetizing field. Good permanent magnets have a high degree of remanence. Remanence is measured in teslas. Also called *retentivity.* Compare **coercivity.**

remission (rĭ-mĭsh′ən) Abatement or subsiding of the symptoms of a disease.

REM sleep (rĕm) Short for *rapid eye movement sleep.* A period of sleep characterized by rapid periodic twitching movements of the eye muscles and other physiological changes, such as accelerated respiration and heart rate, increased brain activity, and muscle relaxation. REM sleep is associated with activity in the pons of the brainstem; when the pons is eliminated, REM sleep does not occur. REM sleep is the stage of sleep in which most dreaming takes place and is thought to allow for the organization of memories and the retention of learning. REM sleep is the fifth and last stage of sleep that occurs in the sleep cycle, which repeats itself about five times throughout a period of sleep. It is preceded by four stages of non-REM sleep. REM stages become longer with each cycle and account for about 20–25 percent of total sleep in adult humans. In infants, roughly 50 percent of sleep is REM sleep, which is believed to be necessary for the maturation of the central nervous system. Compare **non-REM sleep.** See more at **sleep.**

renal (rē′nəl) Relating to or involving the kidneys.

renewable (rĭ-no͞o′ə-bəl) Relating to a natural resource, such as solar energy, water, or wood, that is never used up or that can be replaced by new growth. Resources that are dependent on regrowth can sometimes be depleted beyond the point of renewability, as when the deforestation of land leads to desertification or when a commercially valuable species is harvested to extinction. Pollution can also make a renewable resource such as water unusable in a particular location. Compare **nonrenewable.**

reniform (rĕn′ə-fôrm′, rē′nə-) Shaped like a kidney. Used to describe certain minerals and certain leaves.

renin (rē′nĭn, rĕn′ĭn) A proteinase enzyme of high specificity that is released by the kidney and acts to raise blood pressure by activating angiotensin. See also **angiotensin.**

rennin (rĕn′ĭn) An enzyme that catalyzes the coagulation of milk. Rennin is found in the gastric juice of the fourth stomach of young ruminants and is used in making cheese. Also called *chymosin.*

repeating decimal (rĭ-pē′tĭng) A decimal in which a pattern of one or more digits is repeated indefinitely, such as 0.353535...

repetend (rĕp′ĭ-tĕnd′) The digit or group of digits that repeats infinitely in a repeating decimal. The repetend of the decimal form of $\frac{5}{12}$ (0.4166666...) is 6.

repetitive strain injury (rĭ-pĕt′ĭ-tĭv) Any of various musculoskeletal disorders, such as carpal tunnel syndrome, characterized by numbness, pain, or weakness of muscles, tendons, nerves, and other soft tissues, and caused by the persistent repetition of certain physical movements.

repressor (rĭ-prĕs′ər) A protein that binds to an operator, blocking transcription of an operon and the enzymes for which the operon codes.

reproduction (rē′prə-dŭk′shən) The process by which cells and organisms produce other cells and organisms of the same kind. ▸ The reproduction of organisms by the union of male and female reproductive cells (gametes) is called **sexual reproduction.** Many unicellular and most multicellular organisms reproduce sexually. ▸ Reproduction in which offspring are produced by a single parent, without the union of reproductive cells, is called **asexual reproduction.** The fission (splitting) of bacterial cells and the cells of multicellular organisms by mitosis is a form of asexual reproduction, as is the budding of yeast cells and the generation of clones by runners in plants. Many plants and fungi are capable of reproducing both sexually and asexually, as are some animals, such as sponges and aphids.

reproductive cell (rē′prə-dŭk′tĭv) See **gamete.**

reproductive system 1. The system of organs involved with animal reproduction, especially sexual reproduction. The structure of animal reproductive systems depends on the type of fertilization (internal or external) and whether the animal lays eggs or bears live offspring. In mammals, the reproductive system consists mainly of the ovaries, fallopian tubes, uterus, and vagina in females and the testes, sperm ducts, and penis in males. **2.** The system of organs involved with the reproduction, especially the sexual reproduction, of plants and other complex multicellular organisms that are not animals. In flowering plants, the reproductive system consists of pistils and stamens. In conifers and most other gymnosperms, the reproductive system consists of male and female cones. The male gametes are produced directly from a cell in the pollen grain of these gymnosperms rather than in a distinct reproductive structure, while the female cones have ovules containing archegonia. In seedless plants (bryophytes and ferns), the reproductive system consists of archegonia and antheridia.

reptile (rĕp′tīl′) Any of various cold-blooded vertebrates of the class Reptilia, having skin covered with scales or horny plates, breathing air with lungs, and usually having a three-chambered heart. Unlike amphibians, whose eggs are fertilized outside the female body, reptiles reproduce by eggs that are fertilized inside the female. Though once varied, widespread, and numerous, reptilian lineages, including the pterosaurs, ichthyosaurs, plesiosaurs, and dinosaurs, have mostly become extinct (though birds are living descendants of dinosaurs). The earliest reptiles were the cotylosaurs (or stem reptiles) of the late Mississippian or early Pennsylvanian Period, from which mammals evolved. Modern reptiles include crocodiles, snakes, turtles, and lizards.

reptilian (rĕp-tĭl′ē-ən) Relating to or characteristic of reptiles.

resection (rĭ-sĕk′shən) Surgical removal of all or part of an organ, tissue, or structure. A *wedge resection* is removal of a piece of tissue that is triangularly shaped.

reservoir (rĕz′ər-vwär′) **1.** A natural or artificial pond or lake used for the storage of water. **2.** An underground mass of rock or sediment that is porous and permeable enough to allow oil or natural gas to accumulate in it. **3.** An organism that is the host for a parasitic pathogen or that directly or indirectly transmits a pathogen to which it is immune.

resin (rĕz′ĭn) **1.** Any of numerous clear or translucent, yellowish or brownish substances that ooze from certain trees and plants. Resins are used in products such as varnishes, lacquers, adhesives, plastics, and drugs. Balsam is a resin. **2.** Any of various artificial substances, such as polyurethane, that have similar properties to natural resins and are used to make plastics.

resistance (rĭ-zĭs′təns) **1.** A force, such as friction, that operates opposite the direction of motion of a body and tends to prevent or slow down the body's motion. **2.** A measure of the degree to which a substance impedes the flow of electric current induced by a voltage. Resistance is measured in ohms. Good conductors, such as copper, have low resistance. Good insulators, such as rubber, have high resistance. Resistance causes electrical energy to be dissipated as heat. See also **Ohm's law. 3.** The capacity of an organism, tissue, or cell to withstand the effects of a harmful physical or environmental agent, such as a microorganism or pollutant.

resistivity (rē′zĭs-tĭv′ĭ-tē) A measure of the potential electrical resistance of a conductive material. It is determined experimentally using the equation $\rho = RA/l$, where R is the measured resistance of some length of the material, A is its cross-sectional area (which must be uniform), and l is its length. It is measured in ohm-meters.

resistor (rĭ-zĭs′tər) A device used in electrical circuits to maintain a constant relation between current flow and voltage. Resistors are used to step up or lower the voltage at different points in a circuit and to transform a current signal into a voltage signal or vice versa, among other uses. The electrical behavior of a resistor obeys Ohm's law for a constant resistance; however, some resistors are sensitive to heat, light, or other variables. ▸ **Variable resistors,** or **rheostats,** have a resistance that may be varied across a certain range, usually by means of a mechanical device that alters the position of one terminal of the resistor along a strip of resistant material. The length of the intervening material determines the resistance. Mechanical variable resistors are also called *potentiometers,* and are used in the volume knobs of audio equipment and in many other devices. Compare **capacitor.** See more at **Ohm's law.**

resonance (rĕz′ə-nəns) Oscillation induced in a physical system when it is affected by another system that is itself oscillating at the right frequency. For example, a swing will swing to greater heights if each consecutive push on it is timed to be in rhythm with the initial swing. Radios are tuned to pick up one radio frequency rather than another using a **resonant circuit** that resonates strongly with the incoming signal at only a narrow band of frequencies. The soundboards of musical instruments, contrastingly, are designed to resonate with a large range of frequencies produced by the instrument. See also **harmonic motion.**

resonant circuit (rĕz′ə-nənt) An electric circuit which has very low impedance at a certain frequency. Resonant circuits are often built using an inductor, such as a coil, connected in parallel to a capacitor. The response of the circuit to signals of different frequencies is a function of the inductance and capacitance of the circuit and peaks at one frequency value, at which the current flow resonates most strongly with the input signal. Resonant circuits are used in radio and television tuners to pick out broadcast signals of specific frequencies.

respiration (rĕs′pə-rā′shən) **1.** The process by which organisms exchange gases, especially oxygen and carbon dioxide, with the environment. In air-breathing vertebrates, respiration takes place in the lungs. In fish and many invertebrates, respiration takes place through the gills. Respiration in green plants occurs during photosynthesis. **2.** See **cellular respiration.**

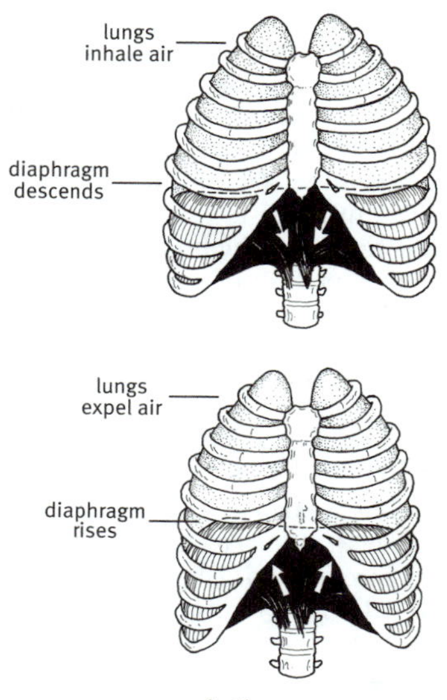

respiration

During inhalation (top), *the diaphragm descends and air fills the lungs. During exhalation* (bottom), *the diaphragm rises and the lungs expel air.*

respiratory syncytial virus (rĕs′pər-ə-tôr′ē) A virus of the family Paromyxoviridae and the genus *Pneumovirus* that causes severe childhood respiratory infections, especially bronchiolitis in infants.

respiratory system The system of organs and structures in which gas exchange takes place,

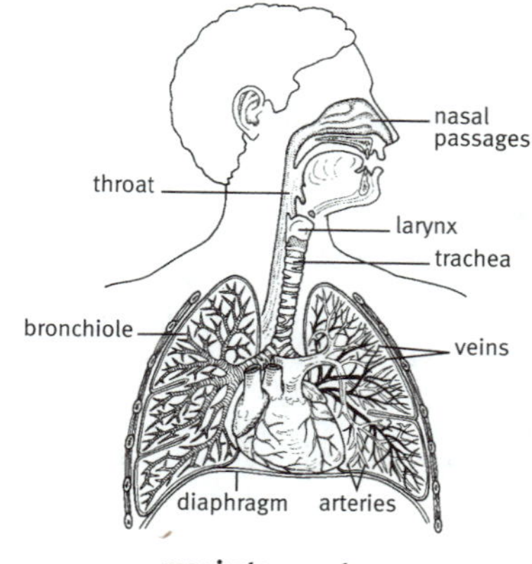

respiratory system

consisting of the lungs and airways in air-breathing vertebrates, gills in fish and many invertebrates, the outer covering of the body in worms, and specialized air ducts in insects.

response (rĭ-spŏns′) A reaction, as that of an organism or any of its parts, to a specific stimulus. See more at **classical conditioning.**

rest energy (rĕst) In Special Relativity, the energy corresponding to the **rest mass** of a body, equal to the rest mass multiplied by the speed of light squared. See more at **mass-energy equivalence.**

rest mass In Special and General Relativity, the observed mass of a body that is not in motion with respect to the observer. Also called *invariant mass.* Compare **relativistic mass.**

resultant (rĭ-zŭl′tənt) A single vector that is the equivalent of a set of vectors.

retentivity (rē′tĕn-tĭv′ĭ-tē) See **remanence.**

reticular formation (rĭ-tĭk′yə-lər) A complex network of neurons and axons that is located throughout the brainstem. The reticular formation regulates consciousness, sleep, and wakefulness.

reticulate (rĭ-tĭk′yə-lĭt) Resembling or forming a net or network, as the veins of some leaves.

reticulum (rĭ-tĭk′yə-ləm) Plural **reticula.** The second division of the stomach in ruminant animals, which together with the rumen contains microorganisms that digest fiber. The reticulum's contents are regurgitated for further chewing as part of the cud. See more at **ruminant.**

retina (rĕt′n-ə) Plural **retinas** or **retinae** (rĕt′n-ē′). The light-sensitive membrane that lines the inside of the back of the eyeball and connects to the brain by the optic nerve. The retina of vertebrate animals contains rods and cones, specialized cells that absorb light.

retinitis pigmentosa (rĕt′n-ī′tĭs pĭg′mĕn-tō′sə) A hereditary degenerative disease of the retina, characterized by difficulty seeing at night, pigmentary changes within the retina, and eventual loss of vision.

retinol (rĕt′n-ôl′) See **vitamin A.**

retort (rĭ-tôrt′, rē′-) A glass laboratory vessel in the shape of a bulb with a long, downward-pointing outlet tube. It is used for distillation or decomposition by heat.

retrograde (rĕt′rə-grād′) **1.** Having a rotational or orbital movement that is opposite to the movement of most bodies within a celestial system. In the solar system, retrograde bodies are those that rotate or orbit in a clockwise direction (east to west) when viewed from a vantage point above the Earth's north pole. Venus, Uranus, and Pluto have retrograde rotational movements. No planets in the solar system have retrograde orbital movements, but four of Jupiter's moons exhibit such movement. **2.** Having a brief, regularly occurring, apparently backward movement in the sky as viewed from Earth against the background of fixed stars. Retrograde movement of the planets is caused by the differing orbital velocities of Earth and the body observed. For example, the outer planets normally appear to drift gradually eastward in the sky in relation to the fixed stars; that is, they appear night after night to fall a little farther behind the neighboring stars in their westward passage across the sky. However, at certain times a particular planet appears briefly to speed up and move westward a bit more quickly than the neighboring stars. This happens as Earth, in its faster inner orbit, overtakes and passes the planet in its slower outer orbit; the appearance of moving counter to its usual eastward drift is thus simply the result of perspective as seen from Earth. Compare **prograde.**

retrovirus (rĕt′rō-vī′rəs) Any of a group of RNA viruses whose RNA is used as a template inside a host cell for the formation of DNA by means of the enzyme reverse transcriptase. The DNA thus formed is inserted into the host cell's genome. Most retroviruses can cause cancer. Retroviruses also include HIV. —*Adjective* **retroviral.**

return stroke (rĭ-tûrn′) See **reverse lightning.**

reverse fault (rĭ-vûrs′) A geologic fault in which the hanging wall has moved upward relative to the footwall. Reverse faults occur where two blocks of rock are forced together by compression. Compare **normal fault.** See Note and illustration at **fault.**

reverse lightning A flash of light in the sky caused by an electrical discharge from the Earth's surface to a cloud. Reverse lightning is so called because lightning typically occurs between clouds or travels from a cloud to the Earth's surface. Also called *return stroke.* See note at **lightning.**

reverse osmosis A method of producing pure water by forcing saline or impure water through a semipermeable membrane across which salts or impurities cannot pass. Reverse osmosis is used for water filtration, for desalinization of seawater, and in kidney dialysis machines.

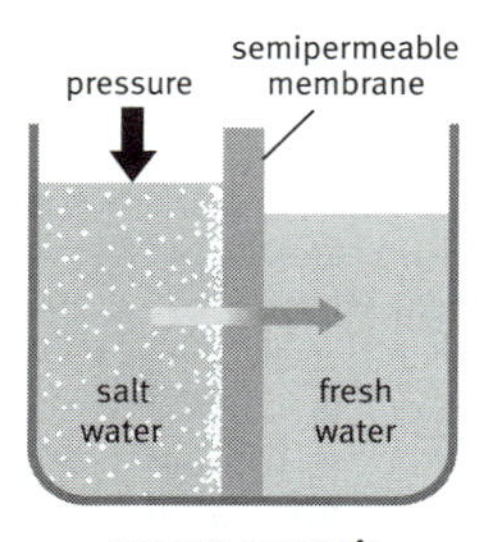

reverse osmosis

reverse transcriptase Any of a class of enzymes that catalyze the formation of DNA from an RNA template and are found in retroviruses, and also in certain body cells (such as stem cells) as the enzyme **telomerase**. The action of reverse transcriptase runs in the opposite direction from normal genetic transcription in the cell, in which RNA is copied from DNA. Drugs that inhibit the action of viral reverse transcriptase have been used to treat retroviral infections such as AIDS, and those that inhibit telomerase are potential anticancer agents.

reverse transcription The process by which DNA is synthesized from an RNA template by means of the enzyme reverse transcriptase.

revolution (rĕv′ə-lo͞o′shən) **1.** The motion of an object around a point, especially around another object or a center of mass. **2.** A single complete cycle of such motion.

USAGE **revolution/rotation**

In everyday speech *revolution* and *rotation* are often used as synonyms, but in science they are not synonyms and have distinct meanings. The difference between the two terms lies in the location of the central axis that the object turns about. If the axis is outside the body itself—that is, if the object is orbiting about another object—then one complete orbit is called a *revolution*. But if the object is turning about an axis that passes through itself, then one complete cycle is called a *rotation*. This difference is often summed up in the statement "Earth *rotates* on its axis and *revolves* around the Sun."

Reye's syndrome (rīz) A rare, acute encephalopathy characterized by fever, vomiting, fatty infiltration of the liver, disorientation, and coma, occurring mainly in children and usually following a viral infection, such as chickenpox or influenza.

Rf The symbol for **rutherfordium.**

Rh The symbol for **rhodium.**

rhenium (rē′nē-əm) *Symbol* **Re** A very rare, dense, silvery-white metallic element with a very high melting point. It is used to make catalysts and electrical contacts. Atomic number 75; atomic weight 186.2; melting point 3,180°C; boiling point 5,627°C; specific gravity 21.02; valence 1, 2, 3, 4, 5, 6, 7. See **Periodic Table.**

rheostat (rē′ə-stăt′) See under **resistor.**

rhesus monkey (rē′səs) A small, yellowish-brown monkey (*Macaca mulatta*) of India, widely used in biological and medical research. The Rh (Rhesus) factor was first discovered in rhesus monkeys.

rheumatic fever (ro͞o-măt′ĭk) An acute inflammatory disease resulting from infections that are caused by a certain strain of bacteria of the genus *Streptococcus*, such as strep throat, usually in the absence of antibiotic treatment. It is marked by fever and inflammation of the joints, nerves, and heart, where it can progress to scarring and permanent dysfunction of the valves.

rheumatoid arthritis (ro͞o′mə-toid′) A chronic autoimmune disease characterized by progressive arthritis of several small or medium-sized joints, especially in the hands. Symptoms can include morning stiffness, joint swelling and weakness, and deformity and disability.

Rh factor (är′āch′) Any of several antigens present on the surface of red blood cells in most humans. People with Rh factors are classified as having a blood type that is *Rh positive*, while people who lack the antigen are said to be *Rh negative* and can produce powerful antibodies that destroy red blood cells if given a blood transfusion from an Rh-positive donor. A woman who is Rh negative and is pregnant with an Rh–positive fetus can produce antibodies that are life threatening to the fetus. See Note at **blood type.**

rhinovirus (rī′nō-vī′rəs) Any of a group of viruses of the family Picornaviridae and the genus *Rhinovirus* that cause the common cold.

rhizoid (rī′zoid′) **1.** A slender, rootlike filament by which mosses, liverworts, and the gametophytes of ferns attach themselves to the material in which they grow. **2.** A branching, rootlike extension by which algae and fungi absorb water and nutrients.

rhizome (rī′zōm′) A plant stem that grows horizontally under or along the ground and often sends out roots and shoots. New plants develop from the shoots. Ginger, iris, and violets have rhizomes. Also called *rootstock.* Compare **bulb, corm, runner, tuber.**

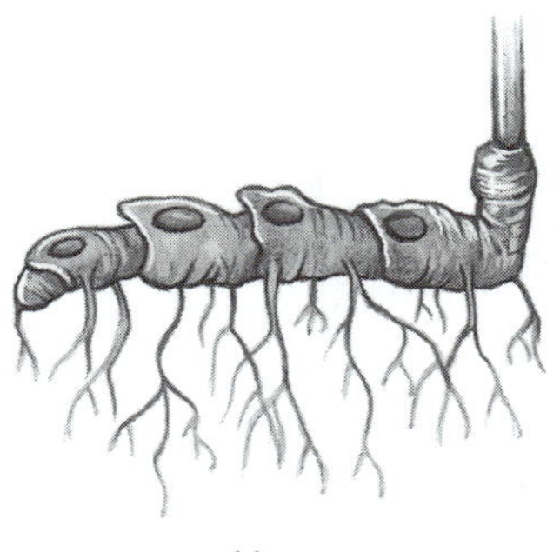

rhizome

of a Solomon's seal plant

rhizopod (rī′zə-pŏd′) Any of various protozoans of the group Rhizopoda, characteristically moving and taking in food by means of pseudopods. Rhizopods include amoebas and radiolarians.

rhizosphere (rī′zə-sfîr′) The soil zone that surrounds and is influenced by the roots of plants. Within the rhizosphere, roots secrete a slimy lubricating substance, called mucigel, that cause the particles of soil to adhere to the roots, assisting in the uptake of water, and encourages the growth of nitrogen-fixing bacteria and other beneficial microorganisms.

Rhodesian man (rō-dē′zhən) A fossil hominid specimen found in south-central Africa, previously classed as a distinct species (*Homo rhodesiensis*) but now generally regarded as an archaic example of *Homo sapiens.*

rhodium (rō′dē-əm) *Symbol* **Rh** A rare, silvery-white metallic element that is hard, durable, and resistant to acids. It is used as a permanent plating for jewelry and is added to platinum to make high-temperature alloys. Atomic number 45; atomic weight 102.905; melting point 1,966°C; boiling point 3,727°C; specific gravity 12.41; valence 2, 3, 4, 5, 6. See **Periodic Table.**

rhombohedral (rŏm′bō-hē′drəl) See **trigonal.**

rhombohedron (rŏm′bō-hē′drən) Plural **rhombohedrons** or **rhombohedra.** A prism with six faces, each a rhombus.

rhomboid (rŏm′boid′) A parallelogram with unequal adjacent sides. —*Adjective* **rhomboid, rhomboidal.**

rhombus (rŏm′bəs) Plural **rhombuses** or **rhombi** (rŏm′bī). A parallelogram with four equal sides; an equilateral parallelogram. —*Adjective* **rhomboid, rhomboidal.**

rhynchocephalian (rĭng′kō-sə-fāl′yən) Any of various mostly extinct lizardlike reptiles of the order Rhynchocephalia, whose only living representative is the tuatara (*Sphenodon punctatus* and *S. guntheri*). Rhynchocephalians have several primitive skeletal features, including two temporal arches on each side of the skull rather than one, and teeth attached to the edge of the jaw rather than set in sockets. Most rhynchocephalians died out by the end of the Jurassic Period.

rhyolite (rī′ə-līt′) A usually light-colored, fine-grained extrusive igneous rock that is compositionally similar to granite. It often includes flow lines formed during the extrusion.

rib (rĭb) **1.** Any of a series of long, curved bones extending from the spine and enclosing the chest cavity. In mammals, reptiles, and birds, the ribs curve toward the center of the chest and in most cases attach to the sternum (breastbone). There are 12 pairs of ribs in humans. See more at **skeleton. 2.** One of the main veins of a leaf.

rib cage The bony structure in the chest formed by the ribs and sternum (breastbone) that encloses and protects the heart and lungs.

riboflavin (rī′bō-flā′vĭn) A water-soluble compound belonging to the vitamin B complex that is important in carbohydrate metabolism and the maintenance of mucous membranes. Riboflavin is found in milk, leafy vegetables, meat, and egg yolks. Also called *vitamin B_2*. *Chemical formula:* $C_{17}H_{20}N_4O_6$.

ribonucleic acid (rī′bō-no͞o-klē′ĭk) See **RNA.**

ribose (rī′bōs′) A pentose sugar with a furanose structure that occurs as a component of riboflavin and RNA. *Chemical formula:* $C_5H_{10}O_5$.

ribosomal RNA (rī′bə-sō′məl) The RNA that is a structural component of a ribosome. Comparison of gene sequences of the ribosomal RNA of different organisms has been used to determine evolutionary relationships among the organisms. See more at **RNA.**

ribosome (rī′bə-sōm′) A sphere-shaped structure within the cytoplasm of a cell that is composed of RNA and protein and is the site of protein synthesis. Ribosomes are free in the cytoplasm and often attached to the membrane of the **endoplasmic reticulum**. Ribosomes exist in both eukaryotic and prokaryotic cells. Plastids and mitochondria in eukaryotic cells have smaller ribosomes similar to those of prokaryotes. See more at **cell.**

Richards (rĭch′ərdz), **Ellen Swallow** 1842–1911. American chemist and educator whose survey of water quality in Massachusetts led to the establishment of the first water quality standards in the United States and the first modern sewage treatment plant.

Ellen Swallow Richards

Richter scale (rĭk′tər) A logarithmic scale used to rate the strength or total energy of earthquakes. The scale has no upper limit but usually ranges from 1 to 9. Because it is logarithmic, an earthquake rated as 5 is ten times as powerful as one rated as 4. An earthquake with a magnitude of 1 is detectable only by seismographs; one with a magnitude of 7 is a major earthquake. The Richter scale is named after the American seismologist Charles Francis Richter (1900–1985). See Note at **earthquake.**

ricin (rī′sĭn, rĭs′ĭn) An extremely poisonous protein extracted from the castor bean. Ricin inhibits protein synthesis in cells, and is used as a biochemical reagent and in cancer research.

rickets (rĭk′ĭts) A bone disease seen mostly in children, caused by a deficiency of vitamin D, usually as a result of inadequate dietary intake or lack of exposure to sunlight. This deficiency causes decreased calcium absorption from the intestine and abnormalities in formation and mineralization of skeletal bone, resulting in defective bone growth and deformity.

ridge (rĭj) **1.** A long narrow chain of hills or mountains. **2.** See **mid-ocean ridge. 3.** A narrow, elongated zone of relatively high atmospheric pressure associated with an area of peak anticyclonic circulation. Compare **trough.**

Riemann (rē′mən, -män′), **(Georg Friedrich) Bernhard** 1826–1866. German mathematician who originated the non-Euclidean system of geometry that is now named after him. Riemann also studied optics and electromagnetic theory, and his work influenced Albert Einstein's theory of General Relativity.

Riemannian geometry (rē-män′ē-ən) A non-Euclidean system of geometry based on the postulate that within a plane every pair of lines intersects.

rift (rĭft) **1.** A continental rift. **2.** A narrow break, crack, or other opening in a rock, usually made by cracking or splitting.

rift valley 1. A long, narrow valley lying between two normal geologic faults. Rift valleys usually form where the Earth's lithosphere has become thin through extension associated with plate-tectonic processes. Unlike river valleys and glacial valleys that form primarily through erosional processes, rift valleys form by the subsidence of the intermediate land as the faults are pulled apart. They are on the order of thousands of kilometers long and wide. See more at **normal fault.** **2.** The deep undersea valley located along the center of the mid-ocean ridge. It is associated with the crustal thinning and extensional forces that cause magma to upwell onto the ocean floor.

Rigel (rī′jəl) A very bright, bluish-white supergiant star in the constellation Orion. It is a binary star, with an average apparent magnitude of 0.12. *Scientific name:* Beta Orionis.

WORD HISTORY **Rigel and star names**

The history of astronomy owes much to Arabic scientists of the Middle Ages, who preserved the astronomical learning of ancient Greece and made improvements on it. The English names of many of the brightest stars in the heavens are Arabic in origin. The name of the supergiant star *Rigel,* for example, comes from the Arabic word for "foot" (the foot of the constellation Orion, that is). Some other important stars whose names are Arabic include *Aldebaran,* "the one following (the Pleiades)"; *Betelgeuse,* "hand of Orion"; *Deneb,* "tail" (of the constellation Cygnus, the swan); and *Altair,* "the flying eagle" (in the constellation Aquila, the eagle). The names of other stars are usually Greek or Latin, such as Antares or Sirius, as are the names of the constellations.

right angle (rīt) An angle having a measure of 90°.

right ascension The position of a celestial object east of the vernal equinox along the **celestial equator**. Right ascension is measured in hours, minutes, and seconds from the vernal equinox (0 hours) to the point where a great circle drawn through the object and the north and south celestial poles intersects the celestial equator. Each hour corresponds to 15° of angular distance along the celestial equator for a total of 24 hours. See more at **equatorial coordinate system.**

right brain The cerebral hemisphere located on the right side of the corpus callosum. The right brain controls activities on the left side of the body and, in humans, usually controls perception of spatial relationships and the ability to recognize common shapes and objects. The thought processes involved in creativity and imagination are generally associated with the right brain.

right-hand rule A rule that uses the shape the right hand to established the standard orientation of vector quantities normal to a plane, especially when calculating a vector product or the helicity of particle spin. In the case of the vector product $C = A \times B$, the direction of C is obtained by pointing the right hand with fingers straight in the direction of A, and then bending the fingers in the direction of B; the extended thumb now roughly points in the direction of C. For spin, the fingers of the right hand should curl in the direction of motion, and the thumb shows the direction of the spin vector.

right triangle A triangle having a right angle.

rigor mortis (rĭg′ər môr′tĭs) Muscular stiffening following death, resulting from the unavailability of energy needed to interrupt contraction of the muscle fibers.

ring (rĭng) **1.** A set of elements subject to the operations of addition and multiplication, in which the set is an abelian group under addition and associative under multiplication and in which the two operations are related by distributive laws. **2.** A group of atoms linked by bonds that may be represented graphically in circular or triangular form. Benzene, for example, contains a ring of six carbon atoms. All cyclic compounds contain one or more rings. **3.** See **annulus.** **4.** See **growth ring.**

ringworm (rĭng′wûrm′) Any of a number of contagious fungal infections of the skin, hair, or nails caused chiefly by species of the genera *Microsporum, Trichophyton,* and *Epidermophyton.* Ringworm often causes scaly, itching ring-shaped patches, especially on the skin. Also called *tinea.*

rip (rĭp) **1.** A stretch of water in a river, estuary, or tidal channel made rough by waves meeting an opposing current. **2.** A rip current.

riparian (rĭ-pâr′ē-ən) Relating to or inhabiting the banks of a natural course of water. Riparian zones are ecologically diverse and contribute to the health of other aquatic ecosystems by filtering out pollutants and preventing erosion. Salmon in the Pacific Northwest feed off riparian insects; trees such as the black walnut, the American sycamore, and the cottonwood thrive in riparian environments.

rip current A strong, narrow surface current that flows rapidly away from the shore. Rip currents form when excess water that has accumulated along a shore due to wind and waves rushes back suddenly to deeper waters. Also called *rip tide*.

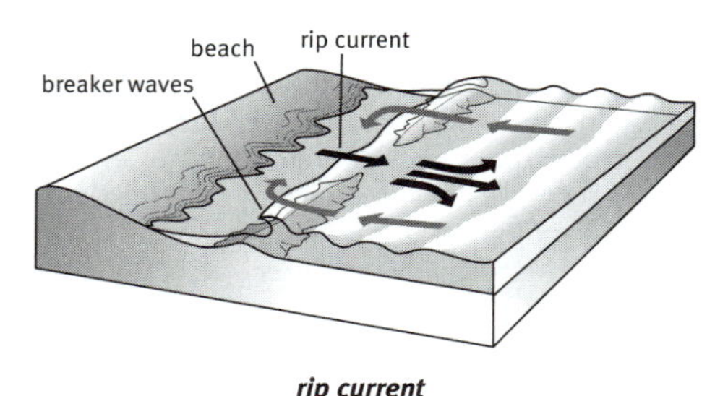

rip current

When water that comes ashore is channeled back out to sea through a narrow passage, such as a break in a sandbar, the strength of the flow of water is concentrated and forms a rip current.

river (rĭv′ər) A wide, natural stream of fresh water that flows into an ocean or other large body of water and is usually fed by smaller streams, called tributaries, that enter it along its course. A river and its tributaries form a drainage basin, or watershed, that collects the runoff throughout the region and channels it along with erosional sediments toward the river. The sediments are typically deposited most heavily along the river's lower course, forming floodplains along its banks and a delta at its mouth.

river basin The land area that is drained by a river and its tributaries. The Mississippi River basin, for example, is a vast area that covers much of the central United States from the central ranges of the Appalachian Mountains in the east to the eastern ranges of the Rocky Mountains in the west, funneling toward its delta in southern Louisiana and emptying into the Gulf of Mexico.

riverine (rĭv′ə-rīn′, -rēn′) **1.** Relating to, formed by, or resembling a river. **2.** Relating to a system of inland wetlands and deep-water habitats associated with nontidal flowing water, characterized by the absence of trees, shrubs, or emergent vegetation. Compare **lacustrine, marine, palustrine.**

Rn The symbol for **radon.**

RNA (är′ĕn-ā′) Short for *ribonucleic acid.* The nucleic acid that is used in key metabolic processes for all steps of protein synthesis in all living cells and carries the genetic information of many viruses. Unlike double-stranded DNA, RNA consists of a single strand of nucleotides, and it occurs in a variety of lengths and shapes. RNA also differs from DNA in having the pyrimidine base uracil instead of thymine and in having ribose instead of deoxyribose in its sugar-phosphate backbone. In eukaryotes, RNA is produced in the cell nucleus. ▸ **Messenger RNA** is RNA that carries genetic information from the cell nucleus to the structures in the cytoplasm (known as ribosomes) where protein synthesis takes place. ▸ **Ribosomal RNA** is the main structural component of the ribosome. ▸ **Transfer RNA** is RNA that delivers the amino acids necessary for protein synthesis to the ribosomes. Compare **DNA.**

RNA

A. *adenine*, C. *cytosine*, G. *guanine*, U. *uracil*

RNA polymerase A polymerase that catalyzes the synthesis of a complementary strand of RNA from a DNA template, or, in some viruses, from an RNA template.

RNA virus A virus whose genome is composed of RNA. Retroviruses are RNA viruses. Compare **DNA virus.**

robot (rō′bŏt′) A machine designed to replace human beings in performing a variety of tasks, either on command or by being programmed in advance.

Roche limit (rōsh) The shortest distance at which a satellite not held together by any force other than its own gravity can orbit another celestial body without being torn apart by the **tidal force** between them. The distance depends on the densities of the two bodies and the orbit of the satellite. If the satellite and the object are of similar densities, the Roche limit is about two and a half times the radius of the larger object. Since most natural satellites are rigid bodies, their tensile strength allows them to orbit much closer than their Roche limit; however, rigid bodies too may be broken up by tidal forces. The rings surrounding Saturn and the other gas giants in the outer solar system may be the orbiting debris of moons that approached much closer than the Roche limit and were fragmented by tidal forces. The limit is named after the French mathematician Edouard Roche (1820–83).

roche moutonnée (rôsh′ mo͞ot′n-ā′, mo͞o′tô-nā′) An elongate mound of bedrock worn smooth and rounded by glacial abrasion. A roche moutonnée has a long axis parallel to the direction of glacial movement, a gently sloping, striated side facing the direction from which the glacier originated, and a steeper side facing the direction of glacial movement. The height, length, and width of roche moutonnées are on the order of a few meters (tens of feet).

rock (rŏk) **1.** A relatively hard, naturally occurring mineral material. Rock can consist of a single mineral or of several minerals that are either tightly compacted or held together by a cementlike mineral matrix. The three main types of rock are **igneous, sedimentary,** and **metamorphic. 2.** A piece of such material; a stone.

rocket (rŏk′ĭt) A vehicle or device propelled by one or more rocket engines, especially such a vehicle designed to travel through space.

rocket engine An engine used to produce a jet of hot gases to propel a rocket. The jet is produced by combustion of a fuel with other chemicals stored in the rocket. Since they do not rely on the oxygen in the atmosphere for combustion, rocket engines can operate in space. Compare **turbojet.**

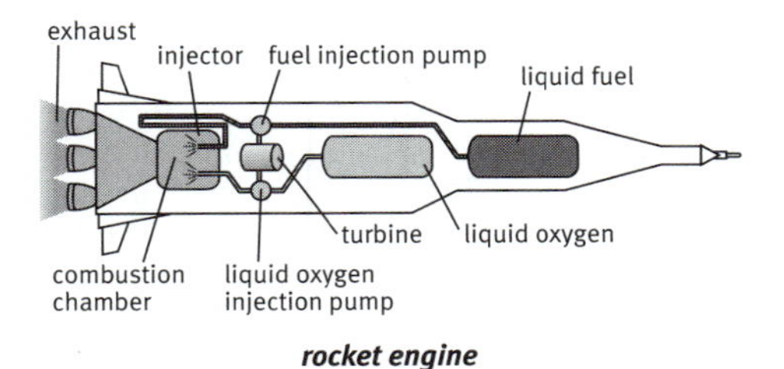

rocket engine

The combination of a mixture of liquid fuel and oxygen in the combustion chamber generates the thrust that propels the rocket.

rocketsonde (rŏk′ĭt-sŏnd′) An instrument transported to the upper atmosphere by rocket, used to study meteorological conditions. A rocketsonde is capable of measuring and transmitting information to altitudes of about 76,000 m (249,280 ft).

Rocky Mountain spotted fever (rŏk′ē) An acute infection caused by the bacteria *Rickettssia Rickettii* and transmitted by ticks of the family *Ixodidae*. It is characterized by fever, exhaustion, muscle pains, and skin rash.

rod (rŏd) One of the rod-shaped cells in the retina of the eye of many vertebrate animals. Rods are more sensitive to light than cones and are responsible for the ability to see in dim light. However, rods are insensitive to red wavelengths of light and do not contribute greatly to the perception of color. Compare **cone.**

rodent (rōd′nt) Any of various very numerous, mostly small mammals of the order Rodentia, having large front teeth used for gnawing. The teeth grow throughout the animal's life, and are kept from getting too long by gnawing. Rodents make up about half the living species of mammals, and include rats, mice, beavers, squirrels, lemmings, shrews, and hamsters.

rodenticide (rō-dĕn′tĭ-sīd′) A pesticide used to kill rodents. Warfarin is a rodenticide. Compare **fungicide, herbicide, insecticide.**

roe (rō) The eggs of a fish, often together with the membrane of the ovary they are held in.

Roentgen (rĕnt′gən, rĕnt′jən), **Wilhelm Konrad** 1845–1923. German physicist who discovered x-rays in 1895 and went on to develop x-ray photography, which revolutionized medical diagnosis. In 1901 he was awarded the Nobel Prize for physics.

Wilhelm Roentgen

roentgenium (rĕnt-gĕn′ē-əm, -jĕn′-, rŭnt-, rœnt-gĕn′-) *Symbol* **Rg** An artificially produced radioactive element with a mass number of 280. Its most stable known isotope has a half-life of 3.6 seconds. Atomic number 111. See **Periodic Table.**

ROM (rŏm) Short for *read-only memory.* Computer hardware that stores programs or data that cannot be added to, modified, or deleted. ROM does not require power to maintain its contents and is often used to save instructions that enable the computer's operating system to communicate with other hardware.

root (ro͞ot, ro͝ot) **1.** A plant part that usually grows underground, secures the plant in place, absorbs minerals and water, and stores food manufactured by leaves and other plant parts. Roots grow in a root system. Eudicots and magnoliids have a central, longer, and larger **taproot** with many narrower lateral roots branching off, while monocots have a mass of threadlike **fibrous roots**, which are roughly the same length and remain close to the surface of the soil. In vascular plants, roots usually consist of a central cylinder of vascular tissue, surrounded by the pericycle and endodermis, then a thick layer of cortex, and finally an outer epidermis or (in woody plants) periderm. Only finer roots (known as feeder roots) actively take up water and minerals, generally in the uppermost meter of soil. These roots absorb minerals primarily through small epidermal structures known as root hairs. In certain plants, **adventitious roots** grow out from the stem above ground as aerial roots or prop roots, bending down into the soil, to facilitate the exchange of gases or increase support. Certain plants (such as the carrot and beet) have fleshy storage roots with abundant parenchyma in their vascular tissues. See also **fibrous root, taproot. 2.** Any of various other plant parts that grow underground, especially an underground stem such as a corm, rhizome, or tuber. **3.** The part of a tooth that is embedded in the jaw and not covered by enamel. **4a.** A number that, when multiplied by itself a given number of times, produces a specified number. Since $2 \times 2 \times 2 \times 2 = 16$, 2 is a fourth root of 16. **b.** A solution to an equation. For example, a root of the equation $x^2 - 4 = 0$ is 2, since $2^2 - 4 = 0$.

root cap A thimble-shaped mass of cells that covers and protects the root tip of plants. Also called *calyptra.*

root hair A hairlike outgrowth of a plant root that absorbs water and minerals from the soil. Root hairs are tubular extensions of the epidermis that greatly increase the surface area of the root. They are constantly dying off and being replaced by new ones as the root grows and extends itself into the soil.

root knot A disease of plants characterized by protuberant swellings on the roots caused by infestation with nematodes.

root pressure Pressure produced in the roots of plants, causing exudation of sap from cut stems and guttation of water from leaves. The pressure is generated by the concentration of solutes in the xylem of the root and stem, which in turn causes water to move into the xylem by osmosis.

rootstock (ro͞ot′stŏk′, ro͝ot′-) See **rhizome.**

root system The configuration of a plant's various roots. See more at **fibrous root, taproot.**

Rorschach test (rôr′shäk′, -shäkh′) A psychological test in which a subject's interpretations of a series of standard inkblots are analyzed as an indication of personality traits, preoccupations, and conflicts. The test is named after its inventor, Swiss psychiatrist and neurologist Hermann Rorschach (1884–1922).

rot (rŏt) *Verb.* **1.** To undergo decomposition, especially organic decomposition; decay. —*Noun.* **2.** Any of several plant diseases characterized by the breakdown of tissue and caused by various bacteria or fungi.

rotation (rō-tā′shən) **1.** The motion of an object around an internal axis. **2.** A single complete cycle of such motion. See Note at **revolution. 3.** A transformation of a coordinate system in which the new axes have a specified angular displacement from their original position while the origin remains fixed.

rotation vector A vector quantity whose magnitude is proportional to the amount or speed of a rotation, and whose direction is perpendicular to the plane of that rotation (following the **right-hand rule**). Spin vectors, for example, are rotation vectors.

rotator cuff (rō′tā′tər) A group of muscles and tendons attaching the shoulder to the scapula (shoulder blade) that provide stability to the shoulder joint and act to rotate the arm.

Injuries to the rotator cuff often happen when the arm is repeatedly moved over the head with great force, as when pitching a baseball.

rotifer (rō′tə-fər) Any of various tiny, multicellular aquatic animals of the phylum Rotifera, having a wheel-like ring of cilia at their front ends. The cilia trap small organisms for food. Rotifers are grouped by some scientists together with nematodes and some other invertebrates as aschelminths.

router (rou′tər) A device in a network that handles message transfers between computers. A router receives information and forwards it based on what the router determines to be the most efficient route at the time of transfer.

Roux (ro͞o), **(Pierre Paul) Émile** 1853–1933. French bacteriologist who assisted Louis Pasteur on most of his major discoveries. Later, working with Alexandre Yersin, he showed that the symptoms of diphtheria are caused by a lethal toxin produced by the diphtheria bacillus. Roux carried out early work on the rabies vaccine and directed the first tests of the diphtheria antitoxin.

rRNA Abbreviation of **ribosomal RNA.**

Ru The symbol for **ruthenium.**

RU 486 (är′yo͞o fôr′ā-tē-sĭks′) An oral drug that terminates early pregnancy by interfering with the action of progesterone, thereby preventing the attachment of a fertilized ovum to the uterine wall.

rubber (rŭb′ər) **1.** An elastic material prepared from the milky sap of certain tropical plants, especially the tree *Hevea brasiliensis.* Rubber is a polymer that is used, after processing, in a great variety of products, including electric insulation and tires. In its pure form, it is white and consists of repeating units of C_5H_8. **2.** Any of various synthetic materials having properties that are similar to those of this substance.

Rubbia (ro͞o′bē-ə), **Carlo** Born 1934. Italian physicist who discovered the W and Z bosons that carry the weak nuclear force. The existence of these subatomic particles strongly confirmed the validity of the theory of the electroweak force. For this work Rubbia shared with Dutch physicist Simon van der Meer the 1984 Nobel Prize for physics.

rubella (ro͞o-bĕl′ə) See **German measles.**

rubeola (ro͞o-bē′ə-lə, ro͞o′bē-ō′lə) See **measles.**

rubidium (ro͞o-bĭd′ē-əm) *Symbol* **Rb** A soft, silvery-white metallic element of the alkali group. It ignites spontaneously in air and reacts violently with water. Rubidium is used in photoelectric cells, in making vacuum tubes, and in radiometric dating. Atomic number 37; atomic weight 85.47; melting point 38.89°C; boiling point 688°C; specific gravity (solid) 1.532; valence 1, 2, 3, 4. See **Periodic Table.**

ruby (ro͞o′bē) A deep-red, translucent variety of the mineral corundum, containing small amounts of chromium and valued as a gem. Compare **sapphire.**

rumen (ro͞o′mən) The first and largest division of the stomach in ruminant animals, in which the food is fermented by microorganisms. See more at **ruminant.**

ruminant (ro͞o′mə-nənt) Any of various even-toed hoofed mammals of the suborder Ruminantia. Ruminants usually have a stomach divided into four compartments (called the rumen, reticulum, omasum, and abomasum), and chew a cud consisting of regurgitated, partially digested food. Ruminants include cattle, sheep, goats, deer, giraffes, antelopes, and camels.

runner (rŭn′ər) A slender stem that grows horizontally and puts down roots to form new plants. Strawberries spread by runners. Also called *stolon.* Compare **bulb, corm, rhizome, tuber.**

Ruska (rŭs′kə), **Ernst August Friedrich** 1906–1988. German electrical engineer who in 1931 developed the world's first electron microscope, which he continued to improve in subsequent work. For this work he shared with German physicist Gerd Binnig and Swiss physicist Heinrich Rohrer the 1986 Nobel Prize for physics.

Russell (rŭs′əl), **Henry Norris** 1877–1957. American astronomer who studied binary stars and developed methods to calculate their mass and distances. Working independently of Ejnar Hertzsprung, Russell also demonstrated the relationship between types of stars and their absolute magnitude. This correlation is now known as the **Hertzsprung-Russell diagram.**

rust (rŭst) *Noun.* **1.** Any of the various reddish-brown oxides of iron that form on iron and many of its alloys when they are exposed to oxygen in the presence of moisture. **2a.** Any of various basidiomycete fungi that are parasitic on plants and produce reddish or brownish spots on leaves. Rusts attack a wide variety of plants and can cause enormous damage to crops. **2b.** Any of the various plant diseases caused by these fungi. —*Verb.* **3.** To become corroded or oxidized.

ruthenium (ro͞o-thē′nē-əm) *Symbol* **Ru** A rare, silvery-gray metallic element that is hard, brittle, and very resistant to corrosion. It is used to harden alloys of platinum and palladium for jewelry and electrical contacts. Atomic number 44; atomic weight 101.07; melting point 2,310°C; boiling point 3,900°C; specific gravity 12.41; valence 0, 1, 2, 3, 4, 5, 6, 7, 8. See **Periodic Table.**

Rutherford (rŭ*th*′ər-fərd), **Ernest** First Baron Rutherford of Nelson. 1871–1937. New Zealand-born British physicist who was a pioneer of subatomic physics. He discovered the atomic nucleus and named the proton. Rutherford demonstrated that radioactive elements give off three types of rays, which he named alpha, beta, and gamma, and invented the term *half-life* to measure the rate of radioactive decay. For this work he was awarded the Nobel Prize for chemistry in 1908.

BIOGRAPHY Ernest Rutherford

Current theories of nuclear fission and fusion reactions are well accepted; these reactions now drive nuclear power plants and atomic bombs. But when the notion that some atoms could spontaneously disintegrate into other atoms was first advanced in 1902 by Ernest Rutherford, it found resistance among his colleagues, who believed that the chemical elements of which known matter was composed were indestructible and immutable. Undaunted, this New Zealand–born physicist then made a large number of discoveries in rapid succession, including the discovery of three kinds of radioactivity (alpha, beta, and gamma rays), and his brilliance and prodigious output soon won over his critics. By the time he garnered the Nobel Prize for chemistry six years later, he had written 80 more scientific papers. His explanation in 1903 of the radioactive decay of uranium—that pieces of uranium atoms were literally breaking off and being emitted, thereby transforming the uranium into a new element—was compelling and soon well accepted. Astonishingly, what are arguably his greatest discoveries came three years after he won the Prize. In 1911, he showed that atoms were composed of smaller constituents: electrons orbiting around a positively charged nucleus. While the rudiments of this idea had already been proposed by others, Rutherford's experimental research conclusively demonstrated its correctness. Rutherford later identified the proton, one of the particles found in the nucleus. The *Rutherford atom,* as it came to be known, is the model of atomic structure from which today's well-established quantum mechanical theories of atomic structure derive. Rutherford also succeeded in inducing the first artificial fusion, fusing deuterium atoms together into radioactive tritium and a light isotope of helium.

rutherfordium (rŭ*th*′ər-fôr′dē-əm) *Symbol* **Rf** A synthetic, radioactive element that is produced by bombarding plutonium with carbon or neon ions. Its most stable isotope is Rf 261 with a half-life of 62 seconds. Atomic number 104. See **Periodic Table.**

Rutherford scattering The scattering undergone by a stream of heavy charged particles, generally **alpha particles** fired at a sample of a heavy metal, caused by exposure to **coulomb forces** exerted by the atomic nuclei of the sample. The patterns produced by such scattering off an extremely thin sheet of gold were early evidence that atoms contain a tiny, positively charged core, what is now called the atomic **nucleus.**

rutile (ro͞o′tēl′, -tīl′) A lustrous red, reddish-brown, or black tetragonal mineral that is an ore of titanium. Rutile usually occurs as prismatic crystals in other minerals, especially as dark needlelike crystals in quartz. *Chemical formula:* TiO_2.

s 1. Abbreviation of **second** (of time), **second** (of an arc). **2.** The symbol for **strangeness.**

S The symbol for **sulfur.**

sabin (sā′bĭn) A unit of acoustic absorption such that one square meter of material of one sabin absorbs 100 percent of the sound energy that strikes it.

Sabin, Albert Bruce 1906–1993. American microbiologist and physician who developed a vaccine against polio that contained an active form of the polio virus (1957). This replaced a less effective vaccine, invented by Jonas Salk, that contained an inactivated form of the virus.

Albert Sabin

sabkha (săb′kə) A flat area between a desert and an ocean, characterized by a crusty surface consisting of evaporite deposits (including salt, gypsum, and calcium carbonate), windblown sediments, and tidal deposits. Sabkhas form primarily through the evaporation of sea water that seeps upward from a shallow water table and through the drying of windblown sea spray.

sac (săk) A pouch or pouch-shaped structure in an animal or plant, often containing liquids. The human bladder is a sac.

saccharide (săk′ə-rīd′) Any of a series of sweet-tasting, crystalline carbohydrates, especially a simple sugar (a monosaccharide) or a chain of two or more simple sugars (a disaccharide, oligosaccharide, or polysaccharide). Glucose, lactose, and cellulose are saccharides.

saccharin (săk′ər-ĭn) A white, crystalline powder used as a calorie-free sweetener. It tastes about 500 times sweeter than sugar. Saccharin is made from a compound of toluene, which is derived from petroleum. *Chemical formula:* $C_7H_5NO_3S$.

Saccharomyces (săk′ə-rō-mī′sēz′) See under **yeast.**

sacrum (sā′krəm, săk′rəm) Plural **sacra.** A triangular bone at the base of the spine, above the coccyx (tailbone), that forms the rear section of the pelvis. In humans it is made up of five vertebrae that fuse together by adulthood. See more at **skeleton.**

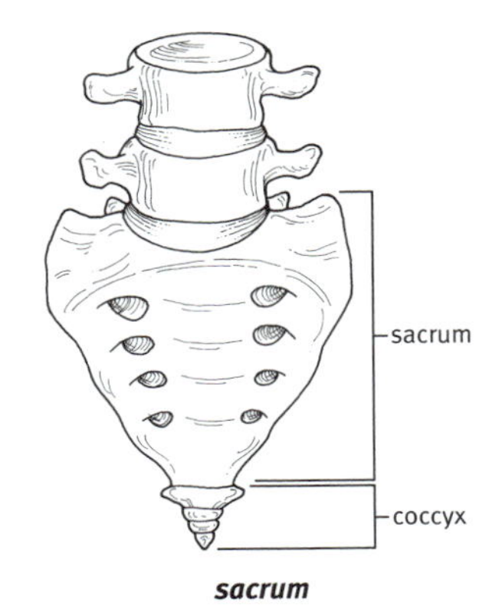

sacrum

Sagittarius (săj′ĭ-târ′ē-əs) A constellation in the Southern Hemisphere near Scorpius and Capricornus. Sagittarius (the Archer) is the ninth sign of the zodiac.

Saint Elmo's fire (sānt ĕl′mōz) A visible and sometimes audible electric discharge projecting from a pointed object such as the mast of a ship or the wing of an airplane, during an electrical storm. First identified as an electrical phenomenon by Benjamin Franklin in 1749, St. Elmo's fire is a bluish-white plasma caused by the release of electrons in a strong electric field (200 or more volts per centimeter); the electrons have enough energy to ionize atoms in the air and cause them to glow. The phenomenon appears near pointed objects because electrical fields generated by charged surfaces are strongest where curves are sharpest. It is named after St. Elmo, the patron saint of mariners, as the phenomenon was often observed by sailors during thunderstorms at sea. See also **lightning rod.**

Salam (sä-läm′), **Abdus** 1926–1996. Pakistani theoretical physicist who helped the develop the theory of the electroweak force, explaining the relationship between two of the four fundamental forces of nature, the electromagnetic force and the weak force. For this work he shared with Sheldon Glashow and Steven Weinberg the 1979 Nobel Prize for physics.

sal ammoniac (săl ə-mō′nē-ăk′) See **ammonium chloride.**

salicylate (sə-lĭs′ə-lāt′, -lĭt, săl′ə-sĭl′ĭt) A salt or ester of salicylic acid, containing the group $C_7H_5O_3$.

salicylic acid (săl′ĭ-sĭl′ĭk) A white, crystalline acid used to make aspirin, to treat certain skin conditions, and to preserve and flavor foods. Salicylic acid is benzoic acid with a hydroxyl group (OH) attached to the carboxyl group (COOH). *Chemical formula:* $\mathbf{C_7H_6O_3}$.

salina (sə-lī′nə, -lē′-) **1.** An area of land encrusted with crystalline salt, especially a salt pan or a salt-encrusted playa. **2.** A body of water, such as a salt marsh, spring, pond, or lake, having a high saline content.

saline (sā′lēn′) Of or containing salt.

saliva (sə-lī′və) The watery fluid that is secreted into the mouth by the salivary glands. In many animals, including humans, it contains the enzyme amylase, which breaks down carbohydrates. Saliva also contains mucus, which lubricates food for swallowing, and various proteins and mineral salts. Some special chemicals occur in the saliva of other animals, such as anticoagulants in the saliva of mosquitoes.

salivary gland (săl′ə-vĕr′ē) A gland in terrestrial animals that secretes saliva. In humans, three pairs of large glands, which include the parotid glands, secrete saliva into the mouth.

Salk (sôlk), **Jonas Edward** 1914–1995. American microbiologist who in 1954 developed the first effective vaccine against polio, using an inactivated form of the virus. Salk's vaccine, which was administered by injection, was widely used until 1959 when Albert Sabin introduced an orally administered vaccine derived from a live form of the virus.

Jonas Salk

salmonella (săl′mə-nĕl′ə) Plural **salmonellae** (săl′mə-nĕl′ē) or **salmonellas** or **salmonella.** Any of various gram-negative, rod-shaped bacteria of the genus *Salmonella* that cause food poisoning and typhoid fever in humans and other mammals.

salt (sôlt) **1.** Any of a large class of chemical compounds formed when a positively charged ion (a cation) bonds with a negatively charged ion (an anion), as when a halogen bonds with a metal. Salts are water soluble; when dissolved, the ions are freed from each other, and the electrical conductivity of the water is increased. See more at **complex salt, double salt, simple salt. 2.** A colorless or white crystalline salt in which a sodium atom (the cation) is bonded to a chlorine atom (the anion). This salt is found naturally in all animal fluids, in seawater, and in underground deposits (when it is often called *halite*). It is used widely as a food seasoning and preservative. Also called *common salt, sodium chloride, table salt. Chemical formula:* **NaCl.**

saltation (săl-tā′shən, sôl-) A single mutation that drastically alters the phenotype.

salt marsh A marsh in which the water is saline, especially a coastal wetland that has halophyte vegetation and is regularly flooded at high tide. Coastal salt marshes help to preserve the shoreline by accommodating storm tides.

salt pan **1.** A small, undrained, shallow depression in which water accumulates, evaporates, and deposits salt. **2.** A small lake of brackish water occupying such a depression.

saltpeter (sôlt′pē′tər) See **potassium nitrate.**

saltwater (sôlt′wô′tər) Consisting of or living in salty water, especially seawater.

samara (săm′ər-ə) An achene (a dry, one-seeded fruit) in which the pericarp is modified into a winglike structure adapted for airborne dispersal. The seeds of the ash, elm, and maple are contained in samaras.

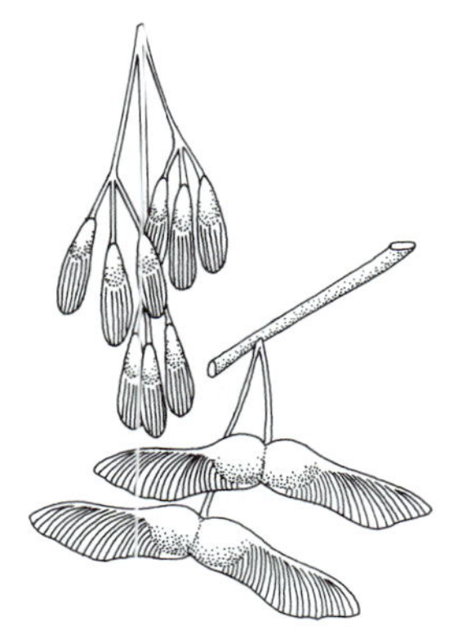

samara

top: *Flowering ash* (Fraxinus ornus); bottom: *Field maple* (Acer campestre)

samarium (sə-mâr′ē-əm) *Symbol* **Sm** A silvery-white metallic element of the lanthanide series that exists in several forms and has seven naturally occurring isotopes. It is used to make glass that absorbs infrared light and to absorb neutrons in the fuel rods of nuclear reactors. Atomic number 62; atomic weight 150.36; melting point 1,072°C; boiling point 1,791°C; specific gravity approximately 7.50; valence 2, 3. See **Periodic Table.**

sampling distribution (săm′plĭng) The distribution of a statistic, such as occurs when a number of sample means are calculated for a given population.

sand (sănd) A sedimentary material consisting of small, often rounded grains or particles of disintegrated rock, smaller than granules and larger than silt. The diameter of the particles ranges from 0.0625 to 2 mm. Although sand often consists of quartz, it can consist of any other mineral or rock fragment as well. Coral sand, for example, consists of limestone fragments.

sandbar (sănd′bär′) A long mass or low ridge of submerged or partially exposed sand built up in the water along a shore or beach by the action of waves or currents.

sandstone (sănd′stōn′) A medium-grained sedimentary rock consisting of fine to coarse sand-sized grains that have been either compacted or cemented together by a material such as silica, iron oxide, or calcium carbonate. Although sandstone usually consists primarily of quartz, it can also consist of other minerals, and it can vary in color from yellow or red to gray or brown.

sandstorm (sănd′stôrm′) A strong wind that carries clouds of sand and dust through the air. Most of the particles in a sandstorm are between 0.08 and 1 mm (0.0032 and 0.04 inches) in size. Sandstorms usually are limited to within 3 m (10 ft) of the ground, rarely getting more than 15 m (49 ft) high. They develop in desert areas where loose sand can be stirred up by wind. Most sandstorms occur during the day when the Earth's surface heats up and dissipate at night as it cools.

Sanger (săng′ər), **Frederick** Born 1918. British biochemist who determined the order of amino acids in the insulin molecule, thereby making it possible to manufacture synthetic insulin. For this work, he received the Nobel Prize for chemistry in 1958. In 1980 Sanger received another Nobel Prize for chemistry (jointly with American molecular biologists Paul Berg and Walter Gilbert) for his development of methods for mapping the structure and function of DNA.

sanitary landfill (săn′ĭ-tĕr′ē) See **landfill.**

Santa Ana (săn′tə ăn′ə) A strong, dry, hot wind that blows from the desert regions of southern California toward the Pacific coast, usually in winter. The Santa Ana wind occurs when a region of high pressure forms over the Great Basin (the high plateau east of the Sierra mountains and west of the Rocky mountains). The clockwise rotation of air around this high-pressure region forces air down from the plateau. As it descends, the air warms up and picks up speed. Because the wind is hot and dry, it contributes to the hazardous fire conditions of southern California.

sap (săp) **1.** The watery fluid that circulates through a plant that has vascular tissues. Sap moving up the xylem carries water and minerals, while sap moving down the phloem carries water and food. **2.** See **cell sap.**

saphenous vein (sə-fē′nəs) Either of two main superficial veins of the leg, one larger than the other, that begin at the foot. A portion of the larger saphenous vein is often used in surgery for coronary artery bypass graft.

saponin (săp′ə-nĭn, sə-pō′-) Any of various plant glucosides that form soapy lathers when mixed and agitated with water. They are used in detergents, foaming agents, and emulsifiers. Some saponins, such as digitalis, affect the heart and have been used as medicines and arrow poisons by indigenous peoples of Africa and South America.

sapphire (săf′īr′) A clear, fairly pure form of the mineral corundum that is usually blue but may be any color except red. It often contains small amounts of oxides of cobalt, chromium, and titanium and is valued as a gem. Compare **ruby.**

saprolite (săp′rə-līt′) Soft, thoroughly decomposed and porous rock, often rich in clay, formed by the in-place chemical weathering of igneous, metamorphic, or sedimentary rocks. Saprolite is especially common in humid and tropical climates. It is usually reddish brown or grayish white and contains those structures (such as cross-stratification) that were present in the original rock from which it formed.

sapropel (săp′rə-pĕl′) An unconsolidated ooze consisting mainly of putrefied plant remains and found in anaerobic areas at the bottom of swamps, lakes, and shallow seas.

saprophyte (săp′rə-fīt′) An organism, especially a fungus or bacterium, that lives on and gets its nourishment from dead organisms or decaying organic material. Saprophytes recycle organic material in the soil, breaking it down into in simpler compounds that can be taken up by other organisms. *—Adjective* **saprophytic** (săp′rə-fĭt′ĭk).

sapwood (săp′wo͝od′) The younger layers of new wood produced by the interior side of the vascular cambium within a tree trunk. Sapwood is active in the conduction of water and is usually lighter in color than heartwood.

sarcoma (sär-kō′mə) Plural **sarcomas** or **sarcomata** (sär-kō′mə-tə). A malignant tumor originating from **mesodermal** tissue, such as fat, muscle, or bone. Compare **carcinoma.**

sarcomere (sär′kə-mîr′) The contractile unit of a skeletal muscle fiber. Sarcomeres are divided into bands of filaments made of actin or myosin. During muscle contraction, the filaments slide over each other to cause shortening of the sarcomere.

sarcopterygian (sär-kŏp′tə-rĭj′ē-ən) See **lobe-finned fish.**

sarin (sâr′ĭn) A poisonous liquid that inhibits the body's ability to catalyze acetylcholine. It is used as a nerve gas in chemical warfare. *Chemical formula:* $C_4H_{10}FO_2P$.

sastrugi (să-stro͞o′gə, sä′strə-) also **zastrugi** (ză-stro͞o′gə, zä′strə-) Long, wavelike ridges of snow, formed by the wind and found on the polar plains. Sastrugi are usually up to several meters high and are often parallel to the prevailing wind direction.

satellite (săt′l-īt′) **1.** A small body in **orbit** around a larger body. See Note at **moon. 2.** An object launched to orbit Earth or another celestial body. Satellites are used for research, communications, weather information, and navigation. The first artificial Earth satellite was Sputnik 1, launched by the Soviet Union in October 1957; the first successful American satellite was launched in January 1958.

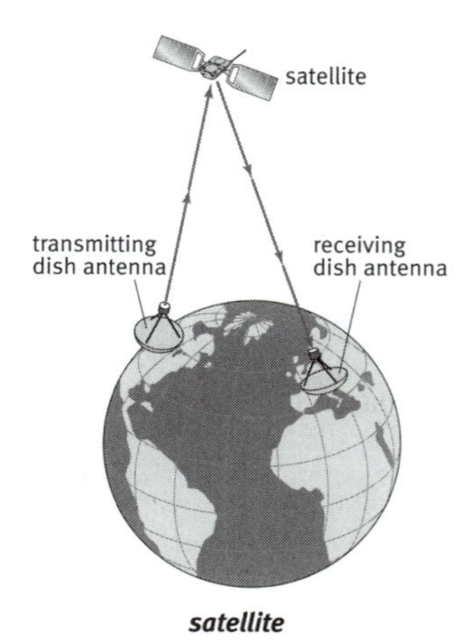

satellite

Communications satellites receive, amplify, and transmit radio signals between dish antennas that may be hundreds of miles apart.

satellite dish A parabolic antenna used to receive signals relayed by satellite.

saturated (săch′ə-rā′tĭd) **1.** Relating to an organic compound in which all the carbon atoms are joined by single bonds and therefore cannot be combined with any additional atoms or radicals. Propane and cyclopentane are examples of saturated hydrocarbons. Compare **unsaturated. 2.** Relating to a solution that is unable to dissolve more of a solute. **3.** Containing as much water vapor as is possible at a given temperature. Air that is saturated has a relative humidity of 100 percent.

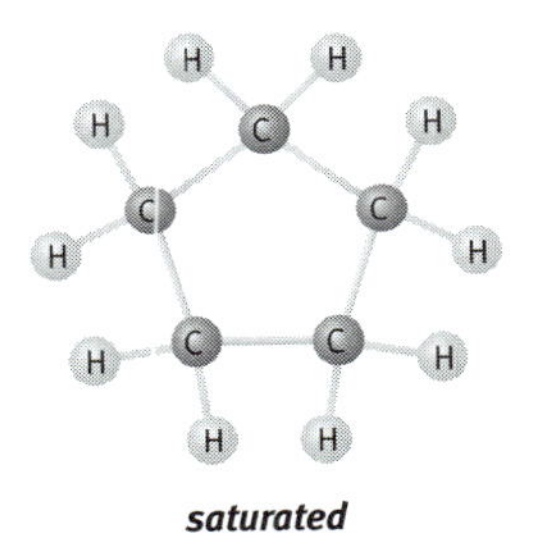

saturated

a molecule of cyclopentane

saturated adiabatic lapse rate See under **lapse rate.**

saturated fat A fat with a **triglyceride** molecule containing three saturated fatty acids. All carbon atoms in the fatty acid chains of saturated fats are connected by single bonds. Most fats derived from animal sources are saturated fats. Eating foods high in saturated fats can lead to elevated cholesterol levels in the blood. Compare **unsaturated fat.**

saturation (săch′ə-rā′shən) **1.** The state of a physical system, such as a solution, containing as much of another substance, such as a solute, as is possible at a given temperature or pressure. **2.** The vividness of a color's hue. Saturation measures the degree to which a color differs from a gray of the same darkness or lightness. Compare **hue, value. 3.** The state of being a saturated organic compound. See more at **saturated.**

saturation point The point at which a substance, under given conditions, can receive no more of another substance in solution.

Saturn (săt′ərn) The sixth planet from the Sun and the second largest, with a diameter about ten times that of Earth. Saturn is a **gas giant** that is almost as large as Jupiter in diameter but with only about 30 percent of Jupiter's mass. Its mainly gaseous composition together with its rapid axial rotation (it rotates once every 10.7 hours) cause a noticeable flattening at the poles and a prominent equatorial bulge. Saturn is encircled by a large, flat system of rings made up of rock fragments and tiny ice crystals, first observed by Galileo in 1610. The rings are believed to be unstable and therefore likely of recent origin; they may have been formed from bodies such as asteroids or moons that were shattered as they approached closer than the **Roche limit**. Saturn has numerous moons, of which the largest is Titan, the second largest moon in the solar system after Jupiter's Ganymede and larger than both Mercury and Pluto. See Table at **solar system.**

saurian (sôr′ē-ən) A lizard or similar reptile.

saurischian (sô-rĭs′kē-ən) Any of various dinosaurs belonging to the group Saurischia, one of the two main divisions of dinosaurs. Saurischians had a pelvis similar to that of modern reptiles, in which the pubis pointed forward. Saurischians include all the carnivorous dinosaurs (**theropods**) as well as the herbivorous **sauropods,** and often grew to great size. Birds are descended from one group of saurischians. Compare **ornithischian.**

sauropod (sôr′ə-pŏd′) One of the two types of saurischian dinosaurs, widespread during the Jurassic and Cretaceous Periods. Sauropods were plant-eaters and often grew to tremendous size, having a stout body with thick legs, long slender necks with a small head, and long tails. Sauropods included the apatosaurus (brontosaurus) and brachiosaurus. Compare **theropod.**

sauropterygian (sô-rŏp′tə-rĭj′ē-ən) Any of various extinct aquatic reptiles of the group Sauropterygia of the Mesozoic Era. Sauropterygians were carnivorous, had paddlelike limbs, and included the plesiosaurs. They were distantly related to the dinosaurs.

savanna or **savannah** (sə-văn′ə) A flat, grass-covered area of tropical or subtropical regions, nearly treeless in some places but generally having a mix of widely spaced trees and bushes. Savannas have distinct wet and dry seasons, with the mix of vegetation dependent primarily on the relative length of the two seasons.

Sb The symbol for **antimony.**

Sc The symbol for **scandium.**

scab (skăb) A crust that forms over a healing wound, consisting of dried blood, plasma, and other secretions.

scalar (skā′lər) A quantity, such as mass, length, or speed, whose only property is magnitude; a number. Compare **vector.**

scalar product The numerical product of the lengths of two vectors and the cosine of the angle between them.

scale[1] (skāl) **1.** One of the small thin plates forming the outer covering of fish, reptiles, and certain other animals. **2.** A similar part, such as one of the minute structures overlapping to form the covering on the wings of butterflies and moths. **3.** A small, thin, usually dry plant part, such as one of the protective leaves that cover a tree bud or one of the structures that contain the reproductive organs on the cones of a conifer. **4.** A plant disease caused by scale insects.

scale[2] (skāl) **1.** An ordered system of numbering or indexing that is used as a reference standard in measurement, in which each number corresponds to some physical quantity. Some scales, such as temperature scales, have equal intervals; other scales, such as the Richter scale, are arranged as a geometric progression. **2.** An instrument or a machine for weighing.

scale insect Any of various small homopterous insects of the superfamily Coccoidea that suck the juices of plants, the females of which secrete and remain under waxy scales on plant tissue.

scalene (skā′lēn′) Having three unequal sides, as a triangle that is neither equilateral nor isosceles.

scandium (skăn′dē-əm) *Symbol* **Sc** A soft, silvery, very lightweight metallic element that is found in various rare minerals and is a byproduct in the processing of certain uranium ores. It has a high melting point and is used to make high-intensity lights. Atomic number 21; atomic weight 44.956; melting point 1,540°C; boiling point 2,850°C; specific gravity 2.99; valence 3. See **Periodic Table.**

scanning electron microscope (skăn′ĭng) An electron microscope that moves a narrowly focused beam of electrons across an object and detects the patterns made by the electrons scattered by the object and the electrons knocked loose from the object. From these patterns a three-dimensional image of the object is created.

scanning force microscope See **atomic force microscope.**

scanning tunneling microscope A microscope used to make images of individual atoms on the surface of a metal. The microscope has a probe with a small voltage applied to it ending in a tiny sharp tip (ideally consisting of one atom) that is moved close the material's surface. **Quantum tunneling** of electrons between tip and the metal provides a small current, and that current is held constant by varying the distance between the tip and the material's surface atoms. As the probe is moved across the surface, a three-dimension image of the surface is formed, based on the continual adjustments made to the height of the tip to keep the electron flow constant.

scapula (skăp′yə-lə) Plural **scapulae** (skăp′yə-lē′) or **scapulas.** Either of two flat, triangular bones forming part of the shoulder. In humans and other primates, the scapulae lie on the upper part of the back on either side of the spine. Also called *shoulder blade.* See more at **skeleton.**

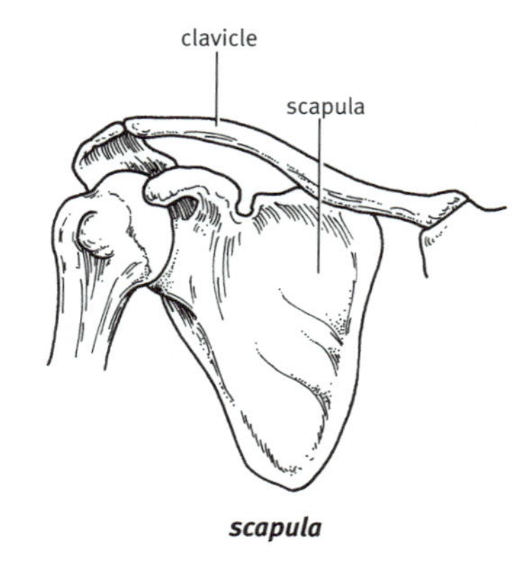

scapula

scarlet fever (skär′lĭt) A severe acute infectious disease caused by the bacterium *Streptococcus pyogenes,* occurring mainly in children, and marked by high fever, sore throat and a red skin rash.

scarp (skärp) A continuous line of cliffs produced by vertical movement of the Earth's crust along a fault or by erosion. The term is often used interchangeably with *escarpment* but is more accurately associated with cliffs produced by faulting rather than those produced by erosional processes.

scattering (skăt′ər-ĭng) The spreading of a stream of particles or a beam of rays, as of light, over a range of directions as a result of collisions with other particles. The sky appears blue due to the tendency of air molecules to scatter blue and violet light more than light of other frequencies. The scattering probabilities and patterns of subatomic particles, accelerated by particle accelerators and aimed at a target, is a major component of experimental particle physics. See also **diffusion, cross section.**

scattering layer A concentrated layer of marine organisms found in most oceanic waters that reflects and scatters sound waves, as from sonar. The layer is of varying composition and can include both plankton and nekton (free-swimming organisms such as copepods, krill, and small fish). Scattering layers, which may occur at more than one depth in the same location, typically move upward at night to feed on phytoplankton and downward during the day, as deep as 1,000 m (3,280 ft), probably to escape predators. Also called *deep scattering layer.*

scavenger (skăv′ən-jər) An animal that feeds on dead organisms, especially a carnivorous animal that eats dead animals rather than or in addition to hunting live prey. Vultures, hyenas, and wolves are scavengers.

Scheele (shā′lə), **Karl Wilhelm** 1742–1786. Swedish chemist who discovered a number of compounds and elements. He discovered oxygen around 1771, but because the results of his experiments were not published until 1777, Joseph Priestley is usually credited with the discovery. Scheele made extensive investigations of plant and animal materials, and his work was fundamental to the development of organic chemistry.

schist (shĭst) A highly foliated, medium-grained metamorphic rock that splits easily into flakes or slabs along well-defined planes of mica. The mineral composition of schist is varied and is often reflected in the name given to the rock. For example, a schist that contains garnet is called a *garnet schist.* A schist containing chlorite is called a *chlorite schist. —Adjective* **schistose** (shĭs′tōs′).

schistosome (shĭs′tə-sōm′) Any of several chiefly tropical trematodes of the genus *Schistosoma,* many of which are parasitic in the blood of birds and mammals, including three species that cause infection in humans, as in schistosomiasis.

schistosomiasis (shĭs′tə-sə-mī′ə-sĭs) Any of a group of diseases caused by flatworm parasites of the genus *Schistosoma* that infest the blood of humans and other mammals, characterized by severe diarrhea and damage to vital organs, including the intestine and bladder. Schistosomiasis is seen in rural areas of Africa, Latin America, and Asia, and it is transmitted through contact with contaminated water.

schizocarp (skĭz′ə-kärp′, skĭt′sə-) A dry fruit that splits at maturity into two or more closed, one-seeded parts, as in the carrot or mallow.

schizophrenia (skĭt′sə-frē′nē-ə, skĭt′sə-) Any of a group of psychiatric disorders characterized by withdrawal from reality, illogical patterns of thinking, delusions, hallucinations, and psychotic behavior. Schizophrenia is associated with an imbalance of the neurotransmitter dopamine in the brain and may have an underlying genetic cause.

schlieren (shlîr′ən) **1.** Irregular dark or light streaks in plutonic igneous rock. Schlieren have the same general mineral composition as the rocks in which they are found, but they are usually slightly darker or lighter than the rest of the rock because of differences in the ratios of the mineral types they include. They are typically a few centimeters to tens of meters long and can form in various ways, including by sorting of minerals during magma flow and through the gravitational settling of minerals during magma cooling and solidification. **2.** Regions of a transparent medium, as of a flowing gas, that are visible as light or dark areas because their densities are different from that of the bulk of the medium.

schott (shŏt) See **shott.**

Schrödinger (shrō′dĭng-ər, shrā′-), **Erwin** 1887–1961. Austrian physicist who founded the study of wave mechanics when he developed a mathematical equation that describes the wavelike behavior of subatomic particles.

Schrödinger's equation was fundamental to Paul Dirac's development of quantum mechanics, and he and Dirac shared the Nobel Prize for physics in 1933.

A CLOSER LOOK **Schrödinger's cat**

Schrödinger's cat is a thought experiment proposed by the physicist and philosopher Erwin Schrödinger. It shows how quantum-mechanical indeterminacy at a microscopic level can cause indeterminacy at a macroscopic level. The indeterminacy can be resolved by observation but entails a paradox. Schrödinger would have us imagine a cat inside a closed box with a tiny bit of a radioactive substance and an apparatus consisting of a Geiger counter, hammer, and flask of cyanide. Over the course of an hour, there is a chance that the radioactive substance might emit an alpha particle, and an equal chance that it might not. If an alpha particle is emitted and hits the Geiger counter, a relay is set in motion whereby the hammer shatters the flask, releasing the gas and killing the cat. Thus, after an hour, there is equal probability of the cat being alive or dead, and an observer can open the box and see which state the cat is in. But until the observation is made, the two possible states, which are mutually exclusive, coexist. In quantum-mechanical terms they are in a condition of *superposition*. The act of observation changes that, and one state becomes established to the exclusion of the other. In other words, without being observed, the cat does not exist in a particular state at all. It is neither alive nor dead, or it is both alive and dead, depending on how you want to look at it. This thought experiment provides an extreme case of the condition in which small-scale objects always exist. Under the laws of quantum mechanics, electrons, photons, and all other particles are each in a number of superposed states that interact with each other, forming interference patterns, and giving rise to an overall behavior that often seems wavelike. But a measurement made to determine the state that the particle is in destroys the interference patterns and gives rise to different behavior that seems more like that of tiny particles; hence the phenomenon of *wave-particle duality*. Whether the cat in the box can also be considered an observer in quantum mechanics currently remains unclear.

Schrödinger's equation (shrō′dĭng-ərz, shrā′-) An equation describing the state and evolution of a quantum mechanical system, given boundary conditions. Different solutions to the equation are associated with different wave functions, usually associated with different energy levels. This equation is fundamental to the study of wave mechanics. See also **wave function, wave mechanics.**

Schwann cell (shwän, shvän) Any of the cells that cover the axons in the peripheral nervous system and form the myelin sheath. See more at **myelin.**

Schwarzschild radius (shwôrts′chīld′, shvärts′-shĭld) A radius defined for a body of a given mass and proportional to that mass, such that if the body is smaller than that radius, the force of gravity is strong enough to prevent matter and energy to escape from within that radius. The Earth is much larger than its Schwarzschild radius, which is approximately 7 mm (0.28 inches). **Black holes** are examples of objects smaller than their Schwarzschild radius, which defines the radius of their **event horizon**. The Schwarzschild radius is a consequence of Einstein's General Relativity theory. It is named after the German astronomer Karl Schwarzschild (1873–1916).

sciatic nerve (sī-ăt′ĭk) A thick nerve that arises in the lower part of the spine and passes through the pelvis on its way to the back of the leg. It carries sensory information from the leg to the central nervous system and controls the action of many muscles. The sciatic nerve is the largest nerve in the body.

science (sī′əns) The investigation of natural phenomena through observation, theoretical explanation, and experimentation, or the knowledge produced by such investigation. ► Science makes use of the **scientific method**, which includes the careful observation of natural phenomena, the formulation of a hypothesis, the conducting of one or more experiments to test the hypothesis, and the drawing of a conclusion that confirms or modifies the hypothesis. See Note at **hypothesis.**

scientific name (sī′ən-tĭf′ĭk) A name used by scientists, especially the taxonomic name of an organism that consists of the genus and species. Scientific names usually come from Latin or Greek. An example is *Homo sapiens*, the scientific name for humans.

scientific notation A method of expressing numbers in terms of a decimal number between 1 and 10 multiplied by a power of 10. The scientific notation for 10,492, for example, is 1.0492×10^4.

scion (sī′ən) A detached shoot or twig containing buds from a woody plant, used in grafting.

sclera (sklîr′ə) The tough, white, fibrous tissue that covers all of the eyeball except the cornea.

sclereid (sklĕr′ē-ĭd) A thick-walled lignified plant cell, often branched in shape. Sclereids form many hard structures such as seed coats and nut shells. They are a type of sclerenchyma cell but are usually shorter than fibers.

sclerenchyma (sklə-rĕng′kə-mə) A supportive tissue of vascular plants, consisting of thick-walled, usually lignified cells. Sclerenchyma cells normally die upon reaching maturity but continue to fulfill their structural purpose in the plant. There are two types of sclerenchyma cells: fiber cells and sclereids. Compare **collenchyma, parenchyma.**

scleroderma (sklîr′ə-dûr′mə) A connective tissue disease characterized by the deposition of fibrous tissue into the skin and often other organs, causing tissue hardening and thickening.

scoliosis (skō′lē-ō′sĭs) A lateral curvature of the spine, usually having no known cause and occurring most commonly in preteen and adolescent girls.

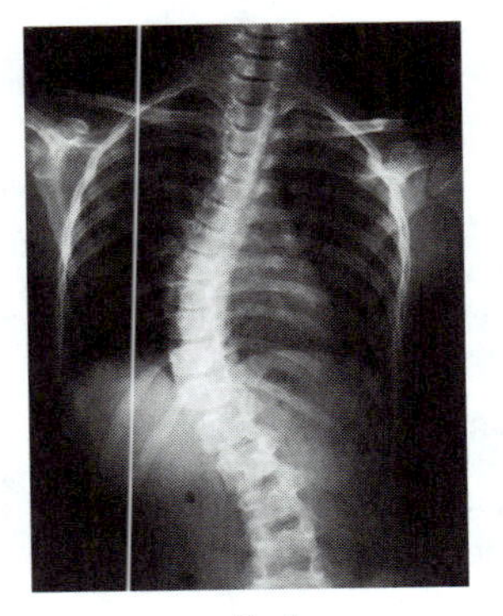

scoliosis

scoliosis of a human spine

scopolamine (skə-pŏl′ə-mēn′, -mĭn) A poisonous, syrupy, colorless alkaloid extracted from plants such as deadly nightshade and henbane. Scopolamine depresses the central nervous system and is used primarily as a sedative and to dilate the pupils, treat nausea, and prevent motion sickness. *Chemical formula:* $C_{17}H_{21}NO_4$.

scoria (skôr′ē-ə) Plural **scoriae** (skôr′ē-ē′). Rough, crusty, solidified lava containing numerous vesicles that originated as gas bubbles in the lava while it was still in a molten state.

Scorpius (skôr′pē-əs) or **Scorpio** (skôr′pē-ō′) A constellation in the Southern Hemisphere near Libra and Sagittarius. Scorpius (the Scorpion) contains the bright star Antares and is the eighth sign of the zodiac.

scrambler (skrăm′blər) An electronic device that scrambles telecommunication signals to make them unintelligible to anyone without a special receiver.

scrapie (skrā′pē, skrăp′ē) A usually fatal, infectious disease of sheep and goats that is marked by chronic itching, loss of muscular coordination, and progressive deterioration of the central nervous system, thought to be caused by a prion.

screen (skrēn) **1.** The surface on which an image is displayed, as on a television, computer monitor, or radar receiver. **2.** An electrode placed between the plate (anode) and the control grid in a tetrode valve, used to reduce the capacitance between the grid and the plate, increasing its ability to respond to high frequencies, especially radio frequencies.

scrotum (skrō′təm) Plural **scrota** or **scrotums.** The external sac of skin that encloses the testes in most mammals. The scrotum keeps the testes at the optimal temperature (slightly below body temperature) for producing sperm.

SCSI (skŭz′ē) Short for *small computer system interface.* A computer interface used for connecting peripheral devices, such as external disk drives and scanners, to personal computers and each other, consisting of 25–50 individual signal paths (usually wires) bundled together and sharing a single connector plug.

scurvy (skûr′vē) A disease caused by vitamin C deficiency, characterized by bleeding of the gums, rupture of capillaries under the skin, loose teeth, and generalized weakness.

scutellum (skyo͞o-tĕl′əm) Plural **scutella. 1.** A shieldlike bony plate or scale, as on the thorax of some insects. **2.** The large, shield-shaped cotyledon of the embryo of a grass plant, specialized for the absorption of food from the endosperm.

Se The symbol for **selenium.**

sea (sē) **1.** The continuous body of salt water that covers most of the Earth's surface. **2.** A region of water within an ocean and partly enclosed by land, such as the North Sea. See Note at **ocean. 3.** A large body of either fresh or salt water that is completely enclosed by land, such as the Caspian Sea. **4.** *Astronomy.* A mare.

sea anemone Any of numerous, often brightly colored cnidarians of the class Anthozoa, having flexible cylindrical bodies with tentacles surrounding a central mouth. Sea anemones are related to jellyfish and corals, but have no free-swimming (medusoid) stage, and resemble flowers.

Seaborg (sē′bôrg′), **Glenn Theodore** 1912–1999. American chemist who led the team that discovered the element plutonium in 1941. In 1944 they discovered americium and curium, and by bombarding these two elements with alpha rays, Seaborg produced the elements berkelium and californium. In 1951 Seaborg shared the Nobel Prize for chemistry with American atomic scientist Edwin McMillan, who had predicted the existence of plutonium in 1939.

seaborgium (sē-bôr′gē-əm) *Symbol* **Sg** A synthetic, radioactive element that is produced by bombarding californium with oxygen ions or bombarding lead with chromium ions. Its most long-lived isotopes have mass numbers 259, 261, 263, 265, and 266 with half-lives of 0.9, 0.23, 0.8, 16, and 20 seconds, respectively. Atomic number 106. See **Periodic Table.**

sea cucumber See **holothurian.**

sea-floor spreading In the theory of plate tectonics, the process by which new oceanic crust is formed by the convective upwelling of magma at mid-ocean ridges, resulting in the continuous lateral displacement of existing oceanic crust. See more at **magentic reversal.**

seal (sēl) Any of various aquatic carnivorous mammals of the families Phocidae and Otariidae, having a sleek, torpedo-shaped body and limbs that are modified into paddlelike flippers. Seals live chiefly in the Northern Hemisphere and, like walruses, are pinnipeds.

sea level The level of the ocean's surface. Sea level at a particular location changes regularly with the tides and irregularly due to conditions such as wind and currents. Other factors that contribute to such fluctuation include water temperature and salinity, air pressure, seasonal changes, the amount of stream runoff, and the amount of water that is stored as ice or snow. ► The reference point used as a standard for determining terrestrial and atmospheric elevation or ocean depths is called the **mean sea level** and is calculated as the average of hourly tide levels measured by mechanical tide gauges over extended periods of time.

seam (sēm) A thin layer or stratum, as of coal or rock.

seamount (sē′mount′) A large underwater mountain, usually conical in shape and at least 1,000 m (3,280 ft) above the ocean floor. Seamounts are usually isolated and are volcanic in origin.

seaquake (sē′kwāk′) An earthquake originating under the ocean floor. Seaquakes are caused by shifting of the tectonic plates at the bottom of the ocean. The seabed pushes and pulls on the water above it, sending violent pressure waves toward the surface and often creating **tsunamis.**

sea salt Salt that is produced by the evaporation of sea water and that contains sodium chloride and trace elements such as sulfur, magnesium, zinc, potassium, calcium, and iron.

sea slug Any of various colorful marine gastropods of the suborder Nudibranchia, lacking a shell and gills but having fringelike projections that serve as respiratory organs. Also called *nudibranch.*

season (sē′zən) **1.** One of four natural divisions of the year—spring, summer, autumn, and winter—in temperate zones. Each season has its own characteristic weather and lasts approximately three months. The change in the seasons is brought about by the shift in the angle at which the Sun's rays strike the Earth. This angle changes as the Earth orbits in its yearly cycle around the Sun due to the tilt of the Earth's axis. For example, when the northern or southern hemisphere of the Earth is at an angle predominantly facing the Sun and has more daylight hours of direct, overhead sunlight than nighttime hours, it is in its summer season; the opposite hemisphere is in then opposite condition and is in its winter season. See also **equinox, solstice. 2.** In some tropical climates, either of the two divisions—rainy and dry—into which the year is divided. These divisions are defined on the basis of levels of precipitation.

seasonal affective disorder (sē′zə-nəl) A mood disorder that occurs during seasons when exposure to sunlight is limited, characterized by symptoms of depression.

sea squirt Any of various tunicates of the class Ascidiacea, having a transparent sac-shaped body with two siphons. One of the siphons is used to draw water (carrying oxygen and food particles) into the body, while the other expels it. Sea squirts are free-swimming as larvae but sessile as adults. Like other tunicates, sea squirts are chordates, but they have a notochord only in the larval stage.

seawater (sē′wô′tər) Salt water, normally with a salinity of 35 parts per thousand (3.5%), in or coming from the sea or ocean. Although seawater contains more than 70 elements, most seawater salts are ions of six major elements: chloride, sodium, sulfate, magnesium, calcium, and potassium. The major sources of these salts are underwater volcanic eruptions, chemical reactions involving volcanic matter, and chemical weathering of rocks on the coasts. Seawater is believed to have had the same salinity for billions of years.

seaweed (sē′wēd′) Any of various red, green, or brown algae that live in ocean waters. Some species of seaweed are free-floating, while others are attached to the ocean bottom. Seaweed range from the size of a pinhead to having large fronds (such as those of many kelps) that can be as much as 30.5 m (100 ft) in length. Certain species are used for food (such as nori) and fertilizer, and others are harvested for carrageenan and other substances used as thickening, stabilizing, emulsifying, or suspending agents in industrial, pharmaceutical, and food products. Seaweed is also a natural source of the element iodine, which is otherwise found only in very small amounts. See more at **brown alga, green alga, red alga.**

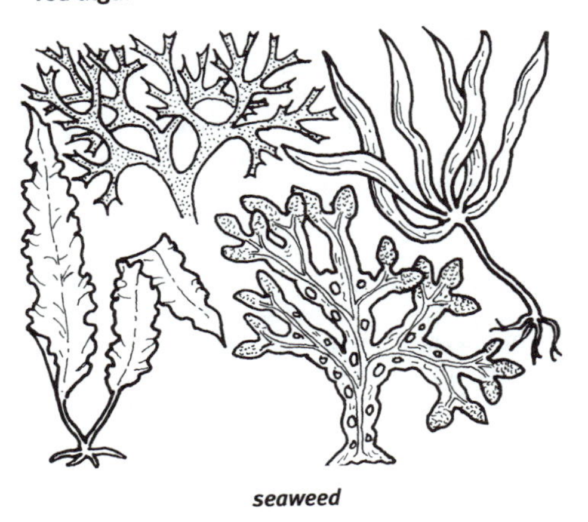

seaweed

top left to right: *Irish moss and horsetail kelp;*
bottom left to right: *sugar kelp and bladder wrack*

sebaceous gland (sĭ-bā′shəs) Any of the glands that are found in the skin of mammals and secrete sebum into the hair follicles.

sebum (sē′bəm) The fatty substance secreted by the sebaceous glands of mammals that protects and lubricates the skin and hair.

sec Abbreviation of **secant.**

secant (sē′kănt′) **1.** A straight line or ray that intersects a curve, especially a circle, at two or more points. **2.** The ratio of the length of the hypotenuse in a right triangle to the side adjacent to an acute angle. The secant is the inverse of the cosine. **3.** The reciprocal of the abscissa of the endpoint of an arc of a unit circle centered at the origin of a Cartesian coordinate system, the arc being of length x and measured counterclockwise from the point (1, 0) if x is positive or clockwise if x is negative. **4.** A function of a number x, equal to the secant of an angle whose measure in radians is equal to x.

Secchi depth (sĕk′ē) A measure of the clarity of water, especially seawater. Secchi depth is measured using a circular plate, known as a Secchi disk, which is lowered into the water until it is no longer visible. High Secchi depths indicate clear water; whereas low Secchi depths indicate cloudy or turbid water. See also **Secchi disk.**

Secchi disk An instrument used for measuring the clarity of water, especially seawater. It consists of a circular plate divided into alternating black and white quadrants and attached to a long measuring tape. The plate is lowered into the water, and the depth at which it is no longer visible from the surface is recorded. The Secchi disk is named after its inventor, Italian astronomer Angelo Secchi (1818–1878). See also **Secchi depth.**

sech Abbreviation of **hyperbolic secant.**

second (sĕk′ənd) **1.** A unit of time equal to $\frac{1}{60}$ of a minute. ▸ A **sidereal second** is $\frac{1}{60}$ of a sidereal minute, and a **mean solar second** is $\frac{1}{60}$ of a mean solar minute. See more at **sidereal time, solar time. 2.** A unit of angular measurement, such as longitude or right ascension, equal to $\frac{1}{60}$ of a minute of arc.

secondary (sĕk′ən-dĕr′ē) **1.** Relating to a secondary color. **2.** Relating to or derived from either of the lateral meristems (the **cork cambium** or the **vascular cambium**) of vascular plants. For example, secondary xylem in a stem is produced by the vascular cambium, as opposed to primary xylem produced by the apical meristem during the original growth of the stem from a seedling. See also **secondary growth. 3a.** Relating to or having a carbon atom that is attached to two other carbon atoms in a molecule. **3b.** Relating to an organic molecule, such as an alcohol, in which the functional group is attached to a secondary carbon. **4.** Relating to a medical condition that arises as a result of another disorder, disease process, or injury. Compare **primary, tertiary.**

secondary color A color produced by mixing two additive primary colors in equal proportions. The secondary colors are cyan (a mixture of blue and green), magenta (a mixture of blue and red), and yellow (a mixture of green and red). Each secondary color is also the **complementary color** (or **complement**) of the primary color whose wavelength it does not contain. Thus cyan is the complement of red, magenta is the complement of green, and yellow is the complement of blue. See Note at **color.**

secondary consumer See under **consumer.**

secondary electron An electron produced by **secondary emission.**

secondary emission The emission of electrons or ions from a material, caused by collisions of electrons or other charged particles with the material. Electron multipliers, photomultipliers, and Geiger counters all work by means of secondary emission.

secondary growth Growth in vascular plants resulting from the production of layers of secondary tissue by a lateral meristem (the **cork cambium** or the **vascular cambium**). The new tissue accumulates and results in thicker branches and stems. Secondary growth occurs in gymnosperms, most eudicots, and woody magnoliids (such as the magnolia). Most monocots and herbaceous plants undergo little or no secondary growth but simply stop growing when their primary tissues mature. Compare **primary growth.**

secondary meristem See **lateral meristem.**

second quantum number The quantum number corresponding to the angular momentum of an electron in an orbital around an atom, determining the shape of its orbit. See Note at **quantum number.**

secondary sex characteristic Any of the physical traits in a sexually mature animal that are specific to one sex but are not directly involved in the act of reproducing. Secondary sex characteristics are thought to have evolved to give an individual an advantage in mating by making the individual more attractive to mates or by allowing the individual to defeat rivals in competition for mates. Some secondary sex characteristics include the facial hair of the human male, the relatively prominent breasts of the human female, the antlers found only in the male of most species of deer, and the colorful plumage of the males of many species of birds. The appearance of secondary sex characteristics is determined by the sex hormones. See more at **sexual selection.**

secondary structure 1. The protein structure characterized by folding of the peptide chain into an alpha helix, beta sheet, or random coil. See also **primary structure, quaternary structure, tertiary structure. 2.** The folded, helical structure of double-stranded DNA.

secondary succession See under **succession.**

secondary wave A type of seismic body wave in which rock particles vibrate at right angles to the direction of wave travel. Secondary waves cause the rocks they pass through to change in shape. These waves are the second fastest traveling seismic waves (after primary waves) and can travel through solids but not through liquids or gases. Also called *shear wave, S wave.* See Note at **earthquake.**

secrete (sĭ-krēt′) To produce and discharge a substance, especially from the cells of specialized glands. For example, the islets of Langerhans in the pancreas secrete the hormone insulin.

secretion (sĭ-krē′shən) **1.** The process of secreting a substance from a cell or gland. **2.** A substance, such as saliva, mucus, tears, bile, or a hormone, that is secreted.

sector (sĕk′tər) The part of a circle bounded by two radii and the arc between them.

sedative (sĕd′ə-tĭv) A drug having a calming or quieting effect, often given to reduce anxiety or to promote relaxation.

sediment (sĕd′ə-mənt) **1.** *Geology.* Solid fragmented material, such as silt, sand, gravel, chemical precipitates, and fossil fragments, that is transported and deposited by water, ice, or wind or that accumulates through chemical precipitation or secretion by organisms, and that forms layers on the Earth's surface. Sedimentary rocks consist of consolidated sediment. **2.** *Chemistry.* Particles of solid matter that settle out of a suspension to the bottom of the liquid.

sedimentary (sĕd′ə-mĕn′tə-rē) Relating to rocks formed when sediment is deposited and becomes tightly compacted. Depending on the origin of the sediments they contain, sedimentary rocks are classified as **clastic sedimentary rocks, chemical sedimentary rocks,** or **evaporites.** Sandstone and conglomerate, for example, consist of fragments of broken preexisting rocks or minerals and are classified as clastic sedimentary rocks. Limestone forms from the precipitation of calcium carbonate through water and is classified as a chemical sedimentary rock. Gypsum and halite deposits form through the evaporation of mineral-rich water and are classified as evaporites.

sedimentology (sĕd′ə-mən-tŏl′ə-jē) The science that deals with the description, classification, and origin of sedimentary rock.

Seebeck effect (zā′bĕk) The creation of an electrical potential across points in a metal that are at different temperatures. The effect is caused by the thermal energy of the valence electrons in the warmer part of the metal; the kinetic energy of these electrons, which are very free in metals, allows them to migrate toward the colder part more readily than the colder electrons migrate to the warmer part. The colder part of the metal is therefore more negatively charged than the warmer part, resulting in electric potential. The Seebeck effect is used in **thermocouples**. It was discovered by the German physicist Thomas Seebeck (1770–1831).

seed (sēd) *Noun.* **1.** A mature fertilized ovule of angiosperms and gymnosperms that contains an embryo and the food it will need to grow into a new plant. Seeds provide a great reproductive advantage in being able to survive for extended periods until conditions are favorable for germination and growth. The seeds of gymnosperms (such as the conifers) develop on scales of cones or similar structures, while the seeds of angiosperms are enclosed in an ovary that develops into a fruit, such as a pome or nut. The structure of seeds varies somewhat. All seeds are enclosed in a protective seed coat. In certain angiosperms the embryo is enclosed in or attached to an **endosperm**, a tissue that it uses as a food source either before or during germination. All angiosperm embryos also have at least one **cotyledon**. The first seed-bearing plants emerged at least 365 million years ago in the late Devonian Period. Many angiosperms have evolved specific fruits for dispersal of seeds by the wind, water, or animals. See more at **germination, ovule.** —*Verb.* **2.** To plant seeds in soil. **3.** To initiate rainfall or to generate additional rainfall by artificially increasing the precipitation efficiency of clouds. See more at **cloud seeding.**

seed-bearing plant A plant that produces seeds. The gymnosperms and the angiosperms together form the seed-bearing plants. The seed-bearing plants have been an enormously successful group in the history of life, owing to the evolution of seeds and pollen. The seed is a superior unit of dispersal to the naked spore, since it includes a food reserve and, among angiosperms, a protective layer. Also, seedless plants are dependent upon the presence of liquid water for sperm dispersal. Pollen makes water unnecessary for sperm transport. Instead, eggs are fertilized after male nuclei have been transported within the protective pollen grain to the female reproductive parts, usually by pollinating agents such as insects or the wind. Also called *seed plant, spermatophyte.*

seed coat The outer protective covering of a seed. The seed coat develops from the integument of the ovule. Also called *testa.*

seed leaf See **cotyledon.**

seedless plant (sēd′lĭs) A plant that does not produce seeds. Ferns, horsetails, and all the bryophytes are seedless plants. See more at **bryophyte.**

seedling (sēd′lĭng) A young plant, especially one that grows from a seed rather than from a cutting. See Note at **germination.**

seed plant See **seed-bearing plant.**

segment (sĕg′mənt) **1.** The portion of a line between any two of its points. **2.** The region bounded by an arc of a circle and the chord that connects the endpoints of the arc. **3.** The portion of a sphere included between a pair of parallel planes that intersect it or are tangent to it.

seiche (sāsh, sēch) An oscillating wave in an enclosed body of water. A seiche may have a period from a few minutes to a few hours and is usually a result of seismic or atmospheric disturbances.

seif dune (sāf, sīf) A sharp-crested longitudinal sand dune or chain of sand dunes common to the Sahara and ranging up to 300 m (984 ft) in height and 300 km (186 mi) in length.

seismic (sīz′mĭk) Relating to an earthquake or to other tremors of the Earth, such as those caused by large explosions.

seismicity (sīz-mĭs′ĭ-tē) The frequency or magnitude of earthquake activity in a given area. Global seismicity maps show that the regions where seismicity is the highest correspond with the edges of the tectonic plates.

seismogram (sīz′mə-grăm′) The record that is produced by a seismograph.

seismograph (sīz′mə-grăf′) An instrument that detects and records vibrations and movements in the Earth, especially during an earthquake. Most seismographs employ a pendulum mounted within a rigid framework and connected to a mechanical, optical, or electromagnetic recording device. When the Earth vibrates or shakes, inertia keeps the pendulum steady with respect to the movements of the frame, producing a graphic record of the duration and intensity of the Earth's movements. Separate instruments are needed to record the north-south horizontal, east-west horizontal, and vertical components of a tremor. By comparing the records produced by seismographs located in three or more locations across the Earth, the location and strength of an earthquake can be determined.

seismology (sīz-mŏl′ə-jē) The scientific study of earthquakes and of the internal structure of the Earth. It includes the study of the origin, geographic distribution, effects, and possible prediction of earthquakes.

seismometer (sīz-mŏm′ĭ-tər) A detecting device that receives seismic impulses. It is the detecting component of a seismograph. See more at **seismograph.**

seizure (sē′zhər) A sudden episode of transient neurologic symptoms such as involuntary muscle movements, sensory disturbances and altered consciousness. A seizure is caused by abnormal electrical activity in the brain, which is often diagnosed on an **electroencephalogram.** See also **epilepsy.**

selective serotonin reuptake inhibitor (sĭ-lĕk′tĭv) See **SSRI.**

selenium (sĭ-lē′nē-əm) *Symbol* **Se** A nonmetallic element that occurs in a gray crystalline form, as a red powder, or as a black glassy material. It is highly photosensitive and can be used to convert light into electricity. Its ability to conduct electricity also increases with higher exposure to light. For these reasons selenium is used in photocopying technology, photography, and solar cells. Atomic number 34; atomic weight 78.96; melting point 217°C; boiling point 684.9°C; specific gravity (gray) 4.79; (red) 4.5; (black) 4.28; valence 2, 4, or 6. See **Periodic Table.**

self-fertilization (sĕlf′fûr′tl-ĭ-zā′shən) Fertilization that occurs when male and female gametes produced by the same organism unite. Self-fertilization occurs in many protozoans and invertebrate animals. It results from **self-pollination** in plants. Self-fertilization allows an isolated individual organism to reproduce but restricts the genetic diversity of a community. Compare **cross-fertilization.**

self-pollination (sĕlf′pŏl′ə-nā′shən) The transfer of pollen from a male reproductive structure (an anther or male cone) to a female reproductive structure (a stigma or female cone) of the same plant or of the same flower. Self-pollination tends to decrease the genetic diversity (increase the number of homozygous individuals) in a population, and is much less common than cross-fertilization. Many species of plants have evolved mechanisms to promote cross-pollination and avoid self-pollination, though certain plants, such as the pea, regularly self-pollinate. Compare **cross-pollination.**

self-similarity (sĕlf′sĭm′ə-lăr′ĭ-tē) The property of having a substructure analogous or identical to an overall structure. For example, a part of a line segment is itself a line segment, and thus a line segment exhibits self-similarity. By contrast, no part of a circle is a circle, and thus a circle does not exhibit self-similarity. Fractals such the Sierpinski triangle are self-similar to an arbitrary level of magnification; many natural phenomena, such as clouds and plants, are self-similar to some degree. See more at **fractal.**

semen (sē′mən) A thick, whitish fluid that is produced during ejaculation by male mammals and carries sperm cells.

semi– A prefix that means "half," (as in *semicircle,* half a circle) or "partly, somewhat, less than fully," (as in *semiconscious,* partly conscious).

semiaquatic (sĕm′ē-ə-kwăt′ĭk) Adapted for living or growing in or near water, but not entirely aquatic.

semiarid (sĕm′ē-ăr′ĭd) Having low precipitation but able to support grassland and scrubby vegetation. Steppes have semiarid climates.

semicircular canal (sĕm′ĭ-sûr′kyə-lər) Any of the three looped tubes of the inner ear that together with the vestibules makes up the organs that maintain equilibrium in vertebrates.

semiconductor (sĕm′ē-kən-dŭk′tər) Any of various solid substances, such as silicon or germanium, that conduct electricity more easily than insulators but less easily than conductors. In semiconductors, thermal energy is enough to cause a small number of electrons to escape from the valence bonds between the atoms (the **valence band**); they orbit instead in the higher-energy **conduction band,** in which they are relatively free. The resulting gaps in the valence band are called **holes.** Semiconductors are vital to the design of electronic components and circuitry, including transistors, laser diodes, and memory and computer processing circuits.

semiconductor diode A diode made of semiconductor components, usually silicon. The cathode, which is negatively charged and has an excess of electrons, is placed adjacent to the anode, which has an inherently positive charge, carrying an excess of holes. At this junction a **depletion region** forms, with neither holes nor electrons. A positive voltage at the anode makes the depletion region small, and current flows; a negative voltage at the anode makes the depletion region large, preventing current flow.

semiconductor laser A solid-state laser constructed like a semiconductor diode. At the junction of anode and cathode, electrons and holes come together; when an electron falls into a hole, bandgap radiation is emitted, as in a **light emitting diode.** Semiconductor lasers are constructed so that the resulting light tends to stimulate other nearby pairs of electrons and holes to undergo the same process, emitting more light. Also called *diode laser, laser diode.* See more at **laser.**

semiheavy water (sĕm′ē-hĕv′ē) See under **heavy water.**

Semmelweis (zĕm′əl-vīs′), **Ignaz Phillipp** 1818–1865. Hungarian physician who was a pioneer of sterile surgical practices. He proved that infectious disease and death in the obstetric clinic were caused by medical students going directly from performing autopsies to treating patients. Semmelweis instituted strict rules governing hygiene that dramatically reduced the mortality rate.

sense organ (sĕns) In animals, an organ or part that is sensitive to a stimulus, as of sound, touch, or light. Sense organs include the eye, ear, and nose, as well as the taste buds on the tongue.

sensible horizon (sĕn′sə-bəl) The plane of an observer's position lying at a right angle to the line formed by the observer's zenith and nadir. The plane of the sensible horizon is parallel to the plane of the observer's celestial horizon but is tangential to the Earth's surface rather than passing through the Earth's center. Both the celestial and sensible horizons change with the observer's position. Compare **celestial horizon.**

sensory (sĕn′sə-rē) Involving the sense organs or the nerves that relay messages from them. Compare **motor.**

sepal (sē′pəl) One of the usually separate, green parts that surround and protect the flower bud and extend from the base of a flower after it has opened. Sepals tend to occur in the same number as the petals and to be centered over the petal divisions. In some species sepals are colored like petals, and they can even be indistinguishable from petals, as in the lilies (in what are called tepals). In some groups, such as the poppies, the sepals fall off after the flower bud opens. See more at **flower.**

sepsis (sĕp′sĭs) A severe infection caused by pathogenic organisms, especially bacteria, in the blood or tissues. If untreated, a localized infection, as in the respiratory or urinary tracts, can lead to infection in the bloodstream and widespread inflammation, characterized initially by fever, chills, and other symptoms and later by septic **shock.** —*Adjective* **septic.**

septum (sĕp′təm) Plural **septa.** A thin wall or membrane that separates two parts or structures in an organism. Septae separate the chambers of the heart and subdivide the hyphae of some fungi.

sequence (sē′kwəns) *Noun.* **1.** A set of quantities ordered in the same manner as the positive integers, in which there is always the same relation between each quantity and the one succeeding it. A sequence can be finite, such as {1, 3, 5, 7, 9}, or it can be infinite, such as $\{1, \frac{1}{2}, \frac{1}{3}, \frac{1}{4}, \ldots \frac{1}{n}\}$. Also called *progression.* **2.** The order of subunits that make up a polymer, especially the order of nucleotides in a nucleic acid or of the amino acids in a protein. —*Verb.* **3.** To determine the order of subunits of a polymer.

sere (sîr) The entire sequence of ecological communities successively occupying an area from the initial stage to the climax community. See more at **succession.**

series (sîr′ēz) **1.** The sum of a sequence of terms, for example $2 + 2^2 + 2^3 + 2^4 + 2^5 + \ldots$ **2.** A group of rock formations closely related in time of origin and distinct as a group from other formations.

series circuit See under **circuit.**

serine (sĕr′ēn′) A nonessential amino acid. *Chemical formula:* $C_3H_7NO_3$. See more at **amino acid.**

serotinous (sĭ-rŏt′n-əs, sĕr′ə-tī′nəs) Late in developing, opening, or blooming. For example, serotinous pine cones may persist unopened on the tree for years and only burst open during a forest fire. Serotinous flowers on trees develop only after the tree has produced leaves.

serotonin (sĕr′ə-tō′nĭn, sîr′-) A monoamine substance that is formed from tryptophan and found in many animal tissues, including the intestine and central nervous system. In the brain, serotonin acts as a neurotransmitter that is involved in the control of pain perception, the sleep-wake cycle, and mood. Serotonin is also produced in some bacteria and plants.

serpentine (sûr′pən-tēn′, -tīn′) Any of a group of greenish, brownish, or yellowish monoclinic minerals, occurring in igneous or metamorphic rocks. They are used as a source of magnesium and asbestos. *Chemical formula:* $(Mg,Fe)_3Si_2O_5(OH)_4$.

serpentinite (sər-pĕn′tə-nīt′) A metamorphic rock consisting almost entirely of minerals in the serpentine group. Serpentinite forms from the alteration of ferromagnesian silicate materials, such as olivine and pyroxene, during metamorphism.

serum (sîr′əm) Plural **serums** or **sera. 1.** See **blood serum. 2.** Blood serum extracted from an animal that has immunity to a particular disease. The serum contains antibodies to one or more specific disease antigens, and when injected into humans or other animals, it can transfer immunity to those diseases.

server (sûr′vər) A computer that manages centralized data storage or network communications resources. A server provides and organizes access to these resources for other computers linked to it.

sessile (sĕs′īl′) **1.** Permanently attached or fixed and not free-moving, as corals and mussels. **2.** Stalkless and attached directly at the base, as certain kinds of leaves and fruit.

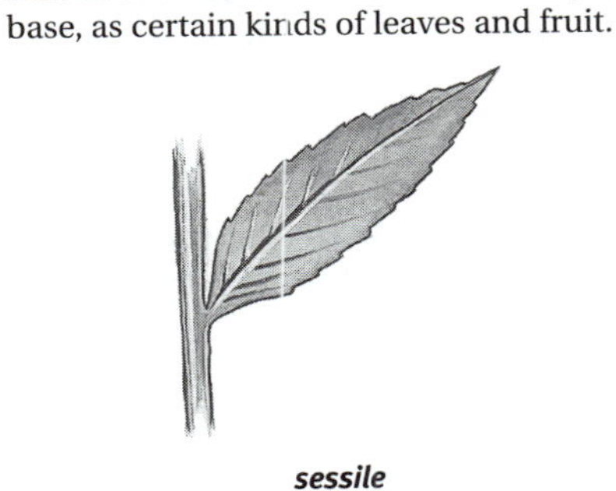

sessile

set (sĕt) A collection of distinct elements that have something in common. In mathematics, sets are commonly represented by enclosing the members of a set in curly braces, as {1, 2, 3, 4, 5}, the set of all positive integers from 1 to 5.

seta (sē′tə) Plural **setae** (sē′tē). A stiff hair, bristle, or bristlelike process or part on an organism. Setae on the bodies of spiders are used as sensory organs, while setae on the bodies of many polychaete worms, such as earthworms, are used for locomotion. Microscopic setae on the feet of geckos allow adhesion to vertical surfaces.

sex (sĕks) Either of two divisions, male and female, into which most sexually reproducing organisms are grouped. Sex is usually determined by anatomy, the makeup of the sex chromosomes, and the type and amount of hormones produced. When the sex of an organism is determined by the sex chromosomes, males and females are generally produced in equal numbers. In other organisms, such as bees and wasps, in which females develop from fertilized eggs and males develop from unfertilized eggs, distribution of the sexes is unequal.

A CLOSER LOOK **sex**

Thanks to high school biology, we are accustomed to thinking of the sex of an organism as being determined by the chromosomes, notably the sex chromosome in humans (designated X or Y). But this is not the whole story, and it applies universally only to mammals and birds. In other animals sex is often determined by environmental factors and can be a variable phenomenon. In a species of slipper limpet (*Crepidula fornicata*), a kind of mollusk, all individuals begin life as females. Clinging to rocks and to each other, they form piles. The limpet on top of the pile changes into a male. If another limpet attaches itself on top of the male limpet, the newcomer becomes male, and the male limpet beneath it reverts to being female. These slipper limpets show the evolutionarily advanced feature of internal fertilization, and the male on top extends his reproductive organ down the pile of females below him to fertilize their eggs. For some fish, the number of males in the population determines the sex of the fish. If there are not enough males, some females become males. In these examples, the same animal can make fertile eggs and fertile sperm at different times in its life. These animals are not hermaphrodites, like some worms, but literally change sex. Some animals have only one sex. For instance, some species of lizards reproduce only by parthenogenesis—that is, their unfertilized eggs grow into adults, and these species no longer have males. Sometimes the external temperature determines the sex of an animal during its early development. If the eggs of the American alligator (*Alligator mississippiensis*) are incubated at above 34 degrees Celsius (93°F), all of the offspring become males. If they are incubated below 30 degrees Celsius (86°F), they become females. The midrange of temperatures results in both male and female offspring.

---

sex cell See **gamete.**

sex chromosome Either of a pair of chromosomes, usually called X and Y, that in combination determine the sex of an individual in many animals and in some plants. In mammals, XX results in a female and XY in a male, while the opposite is true in birds (where the designations ZW for female and ZZ for male are often used). Sex chromosomes carry the genes that control the development of reproductive organs and secondary sex characteristics. In some organisms, sex is determined by environmental influences. See also **X-chromosome, Y-chromosome.**

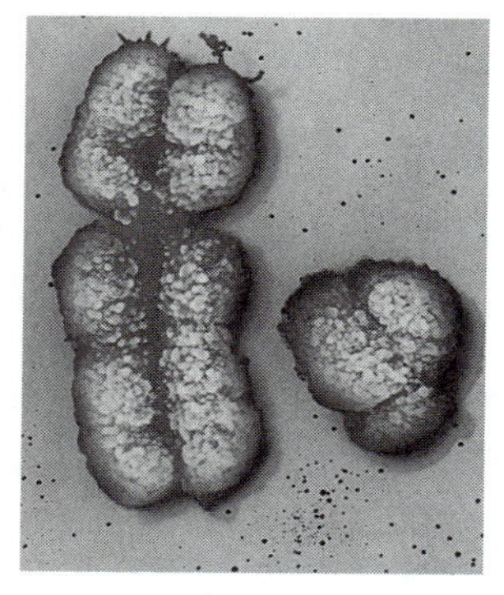

sex chromosome

human sex chromosomes X (left) *and Y* (right)

sex hormone Any of various steroid hormones that regulate the sexual development of an organism and are needed for reproduction. Testosterone and estrogen are sex hormones.

sex linkage The condition in which a gene responsible for a trait is located on a sex chromosome, resulting in sexually dependent inheritance of the trait.

sex-linked Relating to a gene carried on, or a trait transmitted by, a sex chromosome. Color blindness and hemophilia in humans are sex-linked traits.

sextant (sĕk′stənt) An instrument containing a graduated 60° arc and a movable pivoted arm corresponding to the radius of the arc's circle, used in **celestial navigation** to measure the altitude of a celestial body in order to determine the observer's latitude and longitude. A horizontally mounted telescope and two small mirrors are arranged so that the observer can, by moving the pivoted arm, sight the horizon and the reflected image of the celestial body in the same line, giving a reading along the arc that is used to look up the observer's position in a published table.

sexually transmitted disease (sĕk′sho͞o-ə-lē) Any of various infectious diseases, such as chlamydia, gonorrhea, and syphilis, that are transmitted through sexual intercourse or other intimate sexual contact. Also called *venereal disease.*

sexual reproduction See under **reproduction.**

sexual selection The process by which certain organisms produce more offspring by mating more frequently than other organisms of the same sex and thereby ensure the survival of more of their genetic traits. Sexual selection is a form of natural selection in which organisms are competing not for food or other resources in the environment but for mates. The development of size difference between males and females in mammals and birds, in which the greater strength (and often aggressiveness) of larger males allows them to have greater success mating, is seen as a consequence of sexual selection. The development of secondary sex characteristics, such as colored feathers in male birds or large antlers in male deer, which are attractive to the opposite sex as signs of fitness but are not directly involved in reproduction is also attributed to sexual selection. These features are often disadvantageous to the organism's survival—the colored feathers make the male bird more visible to predators, for instance—but can provide the organism with a competitive advantage over rivals in mating. The theory of sexual selection was first proposed by Charles Darwin in *The Origin of Species.*

Seyfert galaxy (sē′fərt, sī′-) A spiral galaxy with a small, compact, bright nucleus that exhibits variable light intensity and radio-wave emission. Seyfert galaxies are **active galaxies** and are thought to contain a black hole in their **galactic nucleus**. The nuclei of Seyfert galaxies generate an emission spectrum characteristic of hot, ionized clouds of gas, shooting out from the accretion disk around the black hole. The observations during the 1940s of American astronomer Carl Seyfert (1911–60), after whom the galaxies are named, demonstrated that these jets of gas are expelled from the nucleus at speeds up to millions of miles per hour. See also **blazar, quasar.**

sferic also **spheric** (sfîr′ĭk, sfĕr′-) A radio emission from a stroke of lightning.

sferics also **spherics** (sfîr′ĭks, sfĕr′-) The study of **atmospherics**, especially through the use of electronic detectors.

Sg The symbol for **seaborgium.**

SGML (ĕs′jē-ĕm-ĕl′) Short for *Standard Generalized Markup Language.* A standardized markup language for describing the structure and formatting of a computer document. Sections of the document are set off by embedded tags. The tags and the relationships among the groups they represent are described in a DTD (Document Type Definition). The tags do not directly specify what the display of the document will look like, so different applications can display the information differently.

shaken baby syndrome (shā′kən) A condition in infants in which brain injury is caused by violent shaking that causes the child's brain to rebound against the skull. This results in bruising, swelling, and bleeding of the brain, often leading to permanent, severe brain damage or death.

shale (shāl) A fine-grained sedimentary rock consisting of compacted and hardened clay, silt, or mud. Shale forms in many distinct layers and splits easily into thin sheets or slabs. It varies in color from black or gray to brown or red.

shatter cone (shăt′ər) A fractured, conical fragment of rock that has striations radiating outward from the apex. Shatter cones are believed to form when rocks are subjected to shock waves associated with a meteoric impact. They are typically between a few centimeters and a few meters long.

shear (shîr) **1.** A force, movement or pressure applied to an object perpendicular to a given axis, with greater value on one side of the axis than the other. See more at **shear force, stress, strain. 2.** See **skew.**

shear force A force acting in a direction parallel to a surface or to a planar cross section of a body, as for example the pressure of air along the front of an airplane wing. Shear forces often result in shear strain. Resistance to such forces in a fluid is linked to its **viscosity.** Also called *shearing force.*

shear modulus See under **modulus of elasticity.**

shear strain or **shearing strain** See under **strain.**

shear stress A form of **stress** that subjects an object to which force is applied to **skew**, tending to cause shear strain. For example, shear stress on a block of wood would arise by fixing one end and applying force to this other; this would tend to change the block's shape from a rectangle to a parallelogram. See also **strain.**

shear wave See **secondary wave.**

sheath (shēth) An enveloping tubular structure, such as the base of a grass leaf that surrounds the stem or the tissue that encloses a muscle or nerve fiber.

sheet lightning (shēt) A broad, sheetlike illumination of a portion of a thundercloud, caused by an unseen, often internal or reflected lightning flash. See also **heat lightning.** See more at **lightning.**

shelf (shĕlf) See **continental shelf.**

shelf fungus Any of various basidiomycete fungi that form shelflike carpophores (spore-bearing structures) on tree trunks or decaying wood. Also called *bracket fungus.*

shelf ice An extension of glacial ice into coastal waters that is in contact with the bottom near the shore but not toward the outer edge of the shelf.

shell (shĕl) **1a.** The usually hard outer covering of certain animals, such as mollusks, insects, and turtles. **b.** The hard outer covering of a bird's egg. **c.** The hard outer covering of a seed, nut, or fruit. **2a.** A set of electron orbitals that have nearly the same energy. Electrons in outer shells have greater energy than those in shells closer to the nucleus. Elements in the Periodic Table range from the lightest elements with electrons normally occupying one shell (hydrogen and helium) to the heaviest, with electrons in seven shells (radium and uranium, for instance). See more at **atomic spectrum, orbital, subshell.** See Note at **metal. 2b.** Any of the stable states of other particles or collections of particles (such as the nucleons in an atomic nucleus) at a given energy or small range of energies.

shepherd satellite (shĕp′ərd) A moon that orbits near the edge of a planetary ring, stabilizing the ring's particles through gravitational pull and confining the ring to a sharply defined band. For example, the moons of Uranus known as Cordelia and Ophelia are shepherd satellites that constrain Uranus's rings to a narrow band. It is believed that the gravity of the fast-moving inner satellite Cordelia causes the particles of the inner ring to speed up and move to outer orbits, while at the same time the gravity of the slower, outer satellite Ophelia decelerates outer particles, pushing them inward. The result is a compressed, well-defined ring. Also called *shepherd moon.*

shield (shēld) **1.** A wall or housing of an absorbing material, such as concrete or lead, built around a nuclear reactor to prevent the escape of radiation. **2.** A structure or arrangement of metal plates or mesh designed to protect a piece of electronic equipment from electrostatic or magnetic interference. **3.** A large geographic area where rocks of a continent's craton (the ancient, relatively undisturbed portion of a continental plate) are visible at the surface. A shield is often surrounded by platforms covered with sediment.

shield volcano See under **volcano.**

shingles (shĭng′gəlz) See under **herpes.**

shoal (shōl) A submerged mound or ridge of sediment in a body of shallow water.

shock (shŏk) **1.** An instance of the passage of an electric current through the body. The amount of injury caused by electric shock depends on the type and strength of the current, the length of time the current is applied, and the route the current takes once it enters the body. **2.** A life-threatening condition marked by a severe drop in blood pressure, resulting from serious injury or illness.

Shockley (shŏk′lē), **William Bradford** 1910–1989. American physicist who, with John Bardeen and Walter Brattain, invented the transistor in 1947. For this work, all three shared the Nobel Prize for physics in 1956. Shockley went on to make improvements to the transistor that made it easier to manufacture.

shock wave A large-amplitude wave formed by the sudden compression of the medium through which the wave moves. Shock waves can be caused by explosions or by objects moving through a fluid at a speed greater than the speed of sound.

shoot (sho͞ot) The part of a vascular plant that is above ground, including the stem and leaves. The tips of shoots contain the **apical meristem**.

shooting star (sho͞o′tĭng) See **meteor** (sense 1).

short circuit (shôrt) An electrical path in a circuit that causes most of the current to flow around or away from some other path in the circuit. Accidental short circuits, especially between the high and low voltages of a power supply, can cause very strong current to flow, possibly damaging or overheating the circuit.

short-day plant A plant that flowers only after being exposed to light periods shorter than a certain critical length, as in early spring or fall. Chrysanthemums and strawberries are short-day plants. Compare **day-neutral plant, long-day plant.** See more at **photoperiodism.**

short ton See **ton** (sense 1).

shortwave (shôrt′wāv′) A radio wave with a frequency between 5.9 megahertz and 26.1 megahertz. Shortwaves broadcast from the Earth's surface are reflected by the upper atmosphere and can travel great distances around the planet. The shortwave band of the electromagnetic spectrum is used for amateur radio communications.

shott or **chott** (shŏt) A shallow lake or marsh with brackish or saline water, especially in northern Africa. Shotts are dry during the summer, at which time they are also characterized by salt deposits and a lack of vegetation.

shoulder blade (shōl′dər) See **scapula.**

shrub (shrŭb) A woody plant that is smaller than a tree, usually having several stems rather than a single trunk; a bush.

Si The symbol for **silicon.**

sibling species (sĭb′lĭng) Any of two or more related species that are morphologically nearly identical but are incapable of producing fertile hybrids. Sibling species can only be identified by genetic, biochemical, behavioral, or ecological factors, and are thought to have become divergent very recently.

sick building syndrome (sĭk) An illness affecting workers in office buildings, characterized by skin irritation, headache, and respiratory problems, and thought to be caused by indoor pollutants, microorganisms, or inadequate ventilation.

sickle cell anemia (sĭk′əl) A hereditary disease characterized by red blood cells that are sickle-shaped instead of round because of an abnormality in their hemoglobin, the protein that carries oxygen in the blood. Because of their shape, the cells can cause blockage of small blood vessels in the organs and bones, reducing the amount of available oxygen.

side chain (sīd) A chain or ring in an organic molecule that branches off from a central structure or from a part that is a point of reference.

A CLOSER LOOK **sickle cell anemia**

Sickle cell anemia is a genetic mutation that can be either detrimental or beneficial depending on the number of copies of the mutated gene a person inherits. While it is harmful if a person inherits two copies of the mutated gene (one from each parent), a person could actually benefit if only one copy of the gene is inherited. The defective gene causes red blood cells to be distorted into a sickle shape, which makes it hard for them to pass through the tiny blood vessels where they give oxygen to body tissues. Inheriting two copies of the mutated gene results in a lifelong disease that causes anemia, pain, and other complications. With just one copy of the gene, though, only mild sickling occurs, and the disease does not manifest itself. This mild sickling, however, is also harmful to the parasites that cause malaria and can protect a person from that disease. In a region like tropical Africa where malaria is common, people who have the mutation in one gene are more likely to ward off a malaria infection and to live long enough to have children, who then inherit the gene. And because a person is less likely to inherit two copies of the gene instead of just one, the benefits of the gene outweigh its risks for most people in these regions. About one in 500 African-American newborns and one out of every 1,000 to 1,400 Hispanic babies are diagnosed with sickle cell anemia each year in the United States. Almost ten percent of African Americans carry the sickle cell gene. There is no cure for the disease, but treatment can reduce pain and prolong life.

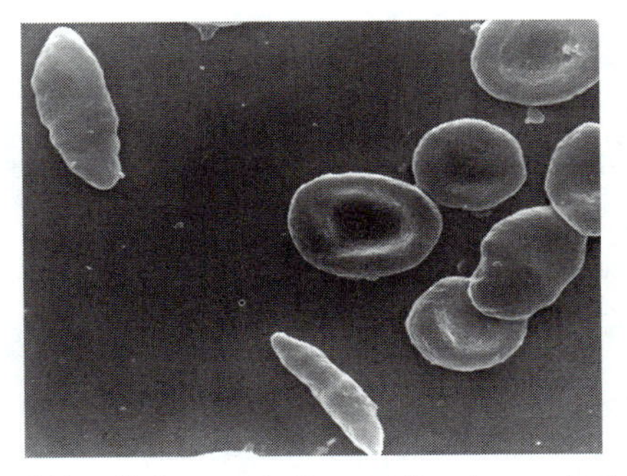

sickle cells (top and bottom right) *and normal red blood cells*

sidereal (sī-dîr′ē-əl) **1.** Relating to the stars or constellations. **2.** Measured with respect to the background of fixed stars instead of the Sun.

sidereal time Time based on the rotation of the Earth with respect to the background of fixed stars. Astronomers generally use sidereal time rather than solar time because it is better suited to observations beyond the solar system. ► A **sidereal day** is the time required for one complete rotation of the Earth on its axis with respect to a fixed star. It is an unvarying unit equal to 23 hours, 56 minutes, 4.09 seconds of solar time. ► A **sidereal month** is the average period of revolution of the Moon around the Earth with respect to a fixed star, equal to 27 days, 7 hours, 43 minutes of solar time. ► A **sidereal year** is the time required for one complete revolution of the Earth around the Sun with respect to a fixed star, equal to 365 days, 6 hours, 9 minutes, 9.54 seconds of solar time. Compare **solar time.**

SI derived unit (ĕs′ī′) Any of the units of measure derived from combinations of the seven base units and two supplementary units of the International System. See Table at **measurement.**

SIDS (sĭdz) See **sudden infant death syndrome.**

siemens (sē′mənz) Plural **siemens.** See **mho.**

sierra (sē-ĕr′ə) A high, rugged range of mountains having an irregular outline somewhat like the teeth of a saw.

sieve cell (sĭv) An elongated, food-conducting cell in phloem characteristic of gymnosperms. Sieve cells have pores through which nutrients flow from cell to cell, but they have no sieve plates like the more specialized sieve-tube elements of angiosperms. Compare **sieve-tube element.**

sievert (sē′vərt) The SI derived unit used to measure the amount of radiation necessary to produce the same effect on living tissue as one gray of high-penetration x-rays. The sievert is named after Swedish physicist Rolf Sievert (1896–1966).

sieve-tube element An elongated, food-conducting cell in phloem in angiosperms.

Unlike the tracheary elements of xylem, sieve elements have living protoplasts when mature, but they lack a nucleus and are dependent upon **companion cells** for certain functions. At the ends of each sieve-tube element are **sieve plates,** areas of the cell wall containing numerous large pores. The sieve-tube elements are aligned end to end and form continuous tubes, along which nutrients flow relatively unimpeded from cell to cell through the sieve plates. Sieve-tube elements are more specialized than the sieve cells found in gymnosperms. Compare **sieve cell.**

sigma (sĭg′mə) A sigma baryon.

sigma baryon Any of three unstable subatomic particles in the baryon family, having a mass 2,328 to 2,343 times that of the electron and a positive, neutral, or negative electron charge. See Table at **subatomic particle.**

sign (sīn) **1.** A body manifestation, usually detected on physical examination or through laboratory tests or xrays, that indicates the presence of abnormality or disease. Compare **symptom. 2.** See **symbol.** See Table at **symbol.**

signal (sĭg′nəl) A fluctuating quantity or impulse whose variations represent information. The amplitude or frequency of voltage, current, electric field strength, light, and sound can be varied as signals representing information.

significant digits (sĭg-nĭf′ĭ-kənt) The digits in a decimal number that are warranted by the accuracy of the means of measurement. Significant digits are all the numbers beginning with the leftmost nonzero digit, or beginning with the first digit after the decimal point if there are no nonzero digits to the left of the decimal point, and extending to the right. For example, 302, 3.20, and 0.023 all have three significant digits.

silent (sī′lənt) **1.** Relating to a mutation that changes a nucleotide in a codon without a difference in the amino acid for which it is coded. See more at **point mutation. 2.** Producing no detectable signs or symptoms, as a medical condition such as heart attack.

silica (sĭl′ĭ-kə) A chemical compound that is the main constituent of most of the Earth's rocks. Silica occurs naturally in five crystalline forms (quartz, tridymite, cristobalite, coesite, and stishovite), in a cryptocrystalline form (chalcedony), and in an amorphous form (opal). It is also the main chemical compound in sand. Silica is used to make glass, concrete, and other materials. Also called *silicon dioxide. Chemical formula:* **SiO_2.**

silicate (sĭl′ĭ-kāt′) **1.** Any of a large class of chemical compounds composed of silicon, oxygen, and at least one metal. Most rocks and minerals are silicates. **2.** Any mineral containing the group SiO_4, either isolated, or joined to other groups in chains, sheets, or three-dimensional groups with metal elements. Micas and feldspars are silicate minerals.

siliceous (sĭ-lĭsh′əs) Resembling or containing silica.

silicon (sĭl′ĭ-kŏn′) *Symbol* **Si** A metalloid element that occurs in both gray crystalline and brown noncrystalline forms. It is the second most abundant element in the Earth's crust and can be found only in silica and silicates. Silicon is used in glass, semiconductors, concrete, and ceramics. Atomic number 14; atomic weight 28.086; melting point 1,410°C; boiling point 2,355°C; specific gravity 2.33; valence 4. See **Periodic Table.**

silicon dioxide See **silica.**

silicone (sĭl′ĭ-kōn′) Any of a class of chemical compounds consisting of long chains of alternating silicon and oxygen atoms, with two organic radicals, typically a methyl (CH_3) and a phenyl (C_6H_5) group, attached to each silicon atom. Silicones are very stable and resist the effects of water, heat, and oxidizing agents. They are used to make adhesives, lubricants and synthetic rubber.

silique (sĭ-lēk′) An elongated dry dehiscent seed pod that is the characteristic fruit of the mustard family. The two sides split off at maturity and leave a central partition to which the seeds are attached.

silk (sĭlk) **1.** A fiber produced by silkworms to form cocoons. Silk is strong, flexible, and fibrous, and is essentially a long continuous strand of protein. It is widely used to make thread and fabric. **2.** A substance similar to the silk of the silkworm but produced by other insect larvae or by spiders to spin webs.

sill (sĭl) A sheet of igneous rock intruded between layers of older rock. See illustration at **batholith.**

sillimanite (sĭl**′**ə-mə-nīt′) A usually white, pale-green or brown orthorhombic mineral occurring as long, slender, fibrous crystals in metamorphic rocks. It is a polymorph of andalusite and kyanite, but can form at higher temperatures than either of these. *Chemical formula:* Al_2SiO_5.

silt (sĭlt) A sedimentary material consisting of grains or particles of disintegrated rock, smaller than sand and larger than clay. The diameter of the particles ranges from 0.0039 to 0.0625 mm. Silt is often found at the bottom of bodies of water where it accumulates slowly by settling through the water.

siltstone (sĭlt**′**stōn′) A fine-grained sedimentary rock consisting primarily of compacted and hardened silt. Siltstones are similar to shale but without laminations. They vary in color from black or gray to brown or red.

Silurian (sĭ-lo͝or**′**ē-ən) The third period of the Paleozoic Era, from about 438 to 408 million years ago. During this time glaciers that formed during the late Ordovician melted, causing sea levels to rise. The first coral reefs, fish with jaws, and freshwater fish appeared, and jawless fish continued to spread rapidly. The first vascular plants also appeared, as did land invertebrates including relatives of spiders and centipedes. See Chart at **geologic time.**

silver (sĭl**′**vər) *Symbol* **Ag** A soft, shiny, white metallic element that is found in many ores, especially together with copper, lead, and zinc. It conducts heat and electricity better than any other metal. Silver is used in photography and in making electrical circuits and conductors. Atomic number 47; atomic weight 107.868; melting point 960.8°C; boiling point 2,212°C; specific gravity 10.50; valence 1, 2. See also **sterling silver.** See **Periodic Table.** See Note at **element.**

silver iodide A pale-yellow, odorless powder that darkens when it is exposed to light. It is used in photography, as an antiseptic in medicine, and in cloud seeding. *Chemical formula:* AgI.

silver nitrate A poisonous, clear, crystalline compound that darkens when exposed to light. It is used in photography and silver plating, and as an external antiseptic. *Chemical formula:* $AgNO_3$.

simian (sĭm**′**ē-ən) *Adjective.* **1.** Resembling or characteristic of apes or monkeys. —*Noun.* **2.** An ape or monkey. Ther term *simian* is no longer used in scientific contexts. Compare **prosimian.**

simple fraction (sĭm**′**pəl) A fraction in which both the numerator and denominator are whole numbers, such as $\frac{5}{7}$.

simple fruit A fruit that develops from a single ovary in a single flower. Simple fruits may be fleshy or dry. There are three main kinds of fleshy simple fruit: the berry, the drupe, and the pome. Dry simple fruits are classified as dehiscent (splitting open to free the seeds, as in the milkweed and pea pod) and indehiscent (enclosing the seeds until after the fruit has left the plant, as in grains, nuts, and winged fruits like the maple). Compare **accessory fruit, aggregate fruit, multiple fruit.**

simple harmonic motion See **harmonic motion.**

simple interest Interest computed only on the original principal and not on the sum of the principal plus accrued interest. The amount of simple interest remains constant. Compare **compound interest.**

simple machine See under **machine.**

Simple Mail Transfer Protocol See **SMTP.**

simple salt A salt in which no hydrogen or hydroxyl (OH) ion is replaced by a metallic ion. Sodium chloride (NaCl) is a simple salt. Compare **complex salt, double salt.**

simple sugar See **monosaccharide.**

sin Abbreviation of **sine.**

sinanthropus (sī-năn**′**thrə-pəs, sĭ-, sī′năn-thrō**′**-pəs, sĭn′ăn-) An extinct hominid postulated from bones found in China in the late 1920s and originally designated *Sinanthropus pekinensis* in the belief that it represented a species evolutionarily preceding humans. Sinanthropus is now classified as *Homo erectus.* Also called *Peking man.* See more at **Homo erectus.**

sine (sīn) **1.** The ratio of the length of the side opposite an acute angle in a right triangle to the length of the hypotenuse. **2.** The ordinate of the endpoint of an arc of a unit circle centered at the origin of a Cartesian coordinate system, the arc being of length x and measured counterclockwise from the point (1, 0) if x is positive or clockwise if x is negative. **3.** A function of a number x, equal to the sine of an angle whose measure in radians is equal to x.

singularity (sĭng′gyə-lăr**′**ĭ-tē) **1.** A point of infinite density and infinitesimal volume, at which space and time become infinitely distorted according to the theory of General Relativity. According to the big bang theory, a gravitational singularity existed at the beginning of the universe. Singularities are also believed to exist at the center of black holes. **2.** *Mathematics.* A point at which the derivative does not exist for a given function but every neighborhood of which contains points for which the derivative exists.

sinh Abbreviation of **hyperbolic sine.**

sink (sĭngk) **1.** A part of the physical environment, or more generally any physical system, that absorbs some form of matter or energy. For example, a forest acts as a sink for carbon dioxide because it absorbs more of the gas in photosynthesis than it releases in respiration. Coral reefs are a long-lasting sink for carbon, which they sequester in their skeletons in the form of calcium carbonate. **2.** *Geology.* **a.** See **playa. b.** See **sinkhole. c.** A circular depression on the flank of a volcano, caused by the collapse of a volcanic wall.

sinkhole (sĭngk**′**hōl′) A natural depression in a land surface formed by the dissolution and collapse of a cavern roof. Sinkholes are roughly funnel-shaped and on the order of tens of meters in size. They generally occur in limestone regions and are connected to subteranean passages. Also called *sink.* See more at **karst topography.**

sinoatrial node (sī′nō-ā**′**trē-əl) A small mass of specialized cardiac muscle fibers that acts as a pacemaker by initiating and maintaining normal heartbeat and cardiac rhythm. It is located in the right atrium.

sinus (sī**′**nəs) **1.** A cavity or hollow space in a bone of the skull, especially one that connects with the nose. **2.** A channel for the passage of a body fluid, such as blood.

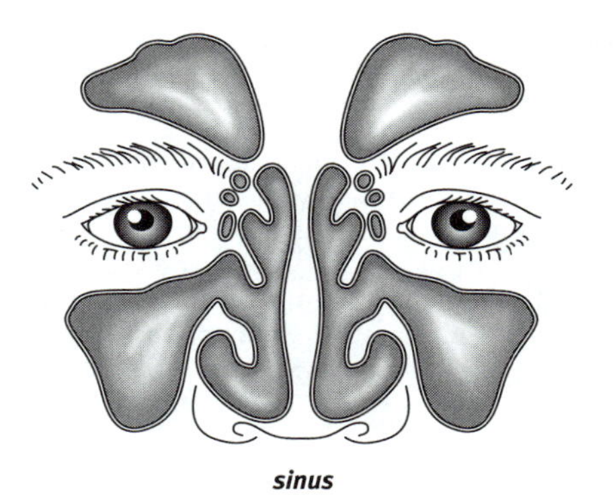

sinus

healthy sinus cavities

sinusoidal projection (sī′nə-soid**′**l) A map projection in which the parallels and a central meridian, usually the prime meridian, are straight lines and the other meridians are curved outward from the central meridian. Sinusoidal projection maps present accurate area and distance at every parallel and at the central meridian; distortion increases at the outer meridians and at high latitudes. It is often used in atlases to map Africa and South America. Compare **conic projection, homolosine projection, Mercator projection.**

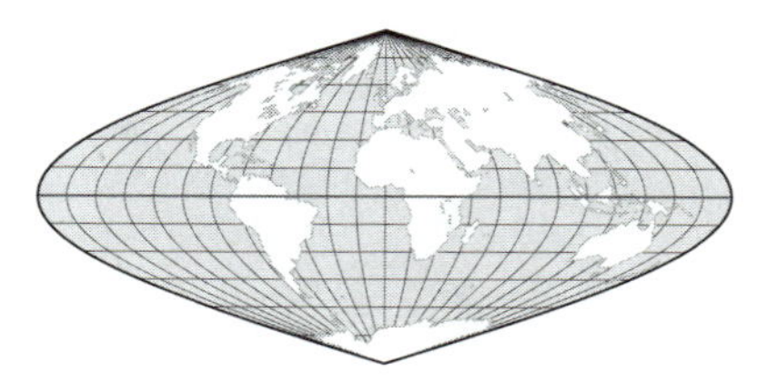

sinusoidal projection

siphon (sī**′**fən) **1.** A pipe or tube in the form of an upside-down U, filled with liquid and arranged so that the pressure of the atmosphere forces liquid to flow upward from a container through the tube, over a barrier, and into a lower container. **2.** A tubular animal part, as of a clam, through which water is taken in or expelled.

siphonophore (sī-fŏn**′**ə-fôr′, sī**′**fə-nə-) Any of various transparent, often subtly colored marine cnidarians of the order Siphonophora, consisting of a delicate floating or swimming colony of polyplike and medusalike individuals. Siphonophores can reach lengths of 40 m (131 ft), making them longer than blue whales. The Portuguese man-of-war is a siphonophore.

Sirius (sĭr′ē-əs) The brightest star seen in the night sky. It is in the constellation Canis Major. It is a white main-sequence star on the Hertzsprung-Russell diagram, with an apparent magnitude of −1.5. Sirius is a binary star, and its companion is a white dwarf star referred to as the Pup. Sirius is also known by the name *Dog Star. Scientific name:* Alpha Canis.

SI unit Any of the units of measure in the International System. See more at **International System.**

skarn (skärn) A coarse-grained metamorphic rock formed by the contact metamorphism of carbonate rocks. Skarn typically contains garnet, pyroxene, epidote, and wollastonite.

skeleton (skĕl′ĭ-tn) **1.** The internal structure of vertebrate animals, composed of bone or cartilage, that supports the body, serves as a framework for the attachment of muscles, and protects the vital organs and associated structures. **2.** A hard protective covering or supporting structure of invertebrate animals. See also **endoskeleton, exoskeleton.** *—Adjective* **skeletal.**

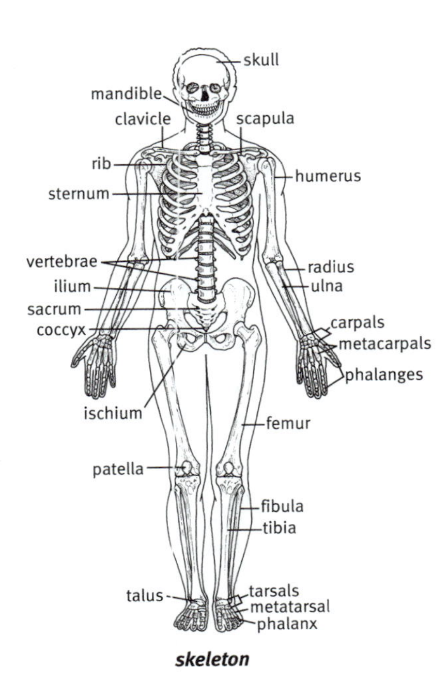

skeleton

skew (skyo͞o) A transformation of coordinates in which one coordinate is displaced in one direction in proportion to its distance from a coordinate plane or axis. A rectangle, for example, that undergoes skew is transformed into a parallelogram. Also called *shear.*

skin (skĭn) The outer covering of a vertebrate animal, consisting of two layers of cells, a thick inner layer (the dermis) and a thin outer layer (the epidermis). Structures such as hair, scales, or feathers are contained in the skin, as are fat cells, sweat glands, and sensory receptors. Skin provides a protective barrier against disease-causing microorganisms and against the sun's ultraviolet rays. In warm-blooded animals, it aids in temperature regulation, as by insulating against the cold.

skin drag See under **drag.**

skin friction See under **drag.**

skull (skŭl) The part of the skeleton that forms the framework of the head, consisting of the bones of the cranium, which protect the brain, and the bones of the face. See more at **skeleton.**

sky (skī) The atmosphere, as seen from a given point on the Earth's surface. The sky appears to be blue because the wavelengths associated with blue light are scattered more easily than those that are associated with the other colors.

slag (slăg) The vitreous mass left as a residue by the smelting of metallic ore. It consists mostly of the siliceous and aluminous impurities from the iron ore.

slate (slāt) A fine-grained metamorphic rock that forms when shale undergoes metamorphosis. Slate splits into thin layers with smooth surfaces. It ranges in color from gray to black or from red to green, depending on the minerals contained in the shale from which it formed.

sleep (slēp) A natural, reversible state of rest in most vertebrate animals, occurring at regular intervals and necessary for the maintenance of health. During sleep, the eyes usually close, the muscles relax, and responsiveness to external stimuli decreases. Growth and repair of the tissues of the body are thought to occur, and energy is conserved and stored. In humans and certain other animals, sleep occurs in five stages, the first four consisting of **non-REM sleep** and the last stage consisting of **REM sleep**. These stages constitute a sleep cycle that repeats itself about five times during a normal episode of sleep. Each cycle is longer that the one preceding it because the length of the REM stage increases with every cycle until waking occurs. Stage I is characterized by drowsiness, Stage II by light sleep, and Stages III and IV by deep sleep. Stages II and III repeat themselves before REM sleep (Stage V), which occurs about 90 minutes after the onset of sleep. During REM sleep, dreams occur, and memory is thought to be organized. In the stages of non-REM sleep, there are no dreams, and brain activity decreases while the body recovers from wakeful activity. The amount and periodicity of sleep in humans vary with age, with infants sleeping frequently for shorter periods, and mature adults sleeping for longer uninterrupted periods. See also **non-REM sleep, REM sleep.**

sleep apnea Apnea caused by upper airway obstruction during sleep, associated with frequent awakening and often with daytime sleepiness. It occurs most often in people who are obese or who have an obstructed respiratory tract or neurological abnormalities.

sleeping sickness (slē′pĭng) **1.** An often fatal protozoan infection prevalent in tropical Africa, caused by either of two **trypanosomes** spread by the bite of the tsetse fly and characterized by fever and extreme lethargy. **2.** Encephalitis that is caused by any of various viruses and is characterized by lethargy and extreme muscular weakness.

sleet (slēt) Precipitation that falls to earth in the form of frozen or partially frozen raindrops, often when the temperature is near the freezing point. Sleet usually leaves the cloud in the form of snow that melts as it passes through warm layers of air during its descent. The raindrops and partially melted snowflakes then freeze in the colder layers nearer the earth before striking the ground as pellets of ice, which usually bounce. By contrast, **hail** forms by the accumulation of layers of ice on the hailstone as it moves up and down in the cloud, and hailstones can become much larger than sleet pellets. The word *sleet* is also used informally to describe a mixture of snow, sleet, and rain.

slickenside (slĭk′ən-sīd′) A polished, striated rock surface formed by one rock mass sliding over another in a fault plane.

slide (slīd) **1.** A mass movement of earth, rocks, snow, or ice down a slope. Slides can be caused by an accumulation of new matter or of moisture in the overlying material, or by erosion within or below the material. They are often triggered by an earthquake or other disturbance such as an explosion. **2.** The mass of material resulting from such a process.

slime (slīm) A slippery or sticky mucous substance secreted by certain animals, such as slugs or snails.

slime mold Any of various organisms that exist as slimy masses and are commonly found on decaying plant matter. They are classified among the protists as two distinct phyla, *Dictyolsteliomycota* (the cellular slime molds) and *Myxomycota* (the plasmodial slime molds). The two phyla are not directly related to each other. ▸ **Cellular slime molds** live as single, amoeba-like cells moving about feeding on bacteria. When food becomes scarce, they combine into a large, slug-like, mobile colony. This migrates to a new area before developing into a multicellular stalked structure that produces and releases spores. Each spore then develops into a new amoeba-like cell. ▸ **Plasmodial slime molds** exist as a mass of amoeba-like protoplasm (called a plasmodium) that contains many nuclei within a single cell membrane. A single organism can spread out thinly and cover up to several square meters. The slimy mass moves along ingesting bacteria, fungi, and other organic matter. When food grows scarce, they stop moving and grow multicellular, spore-producing stalks. The plasmodial slime molds are also called *myxomycetes.*

slough (slŭf) *Noun.* **1.** The dead outer skin shed by a reptile or an amphibian. —*Verb.* **2.** To shed an outer layer of skin.

slow neutron (slō) A neutron that is in thermal equilibrium with the surrounding medium, especially one produced by fission in a nuclear reactor, and slowed by a **moderator**. The uranium 235 used in thermal nuclear reactors undergoes fission more readily when struck by slow neutrons than when struck by the fast neutrons released by fission. Also called *thermal neutron.* Compare **fast neutron.**

sluiceway (slo͞os′wā′) An artificial channel, especially one for carrying off a portion of the current of a stream, canal, or other larger body of water.

Sm The symbol for **samarium.**

small calorie (smôl) See **calorie** (sense 1).

small intestine The long, narrow, coiled section of the intestine that extends from the stomach to the beginning of the large intestine. Nutrients from food are absorbed into the bloodstream from the small intestine. In mammals, it is made up of the duodenum, jejunum, and ileum.

smallpox (smôl′pŏks′) A highly infectious and often fatal disease caused by the variola virus of the genus *Orthopoxvirus* and characterized by fever, headache, and severely inflamed skin sores that result in extensive scarring. Once a dreaded killer of children that caused the deaths of millions of Native Americans after the arrival of European settlers in the Americas, smallpox was declared eradicated in 1980 following a global vaccination campaign. Samples of the virus have been preserved in laboratories in the US and Russia. Also called *variola.* See Note at **Jenner.**

small solar system body Any of various celestial bodies that orbit the Sun and that are neither planets nor dwarf planets. This category includes asteroids and comets.

smelt (smĕlt) To melt ores in order to extract the metals they contain. Oxide ores, such as iron ore, are smelted with carbon, which serves as a fuel and changes the ore into a reduced metal.

Smith (smĭth), **Hamilton Othanel** Born 1931. American microbiologist who isolated bacterial enzymes that could split genetic DNA into fragments large enough to retain genetic information but small enough to permit chemical analysis. The existence of these compounds (called *restriction enzymes*) was earlier predicted by Werner Arber, and their discovery revolutionized genetic engineering. For this work Smith shared the 1978 Nobel Prize for physiology or medicine with Arber and Daniel Nathans.

Smith, Michael 1932–2000. British-born Canadian biochemist who developed a method for making a specific genetic mutation at any spot on a DNA molecule. He shared with American biochemist Kary B. Mullis the 1993 Nobel Prize for chemistry.

smog (smŏg) **1.** A form of air pollution produced by the reaction of sunlight with hydrocarbons, nitrogen compounds, and other gases primarily released in automobile exhaust. Smog is common in large urban areas, especially during hot, sunny weather, where it appears as a brownish haze that can irritate the eyes and lungs. Ozone, a toxic gas that is not normally produced at lower atmospheric levels, is one of the primary pollutants created in this kind of smog. Also called *photochemical smog.* **2.** Fog that has become polluted with smoke and particulates, especially from burning coal.

smoke (smōk) A mixture of carbon dioxide, water vapor, and other gases, usually containing particles of soot or other solids, produced by the burning of carbon-containing materials such as wood and coal.

smut (smŭt) **1.** Any of various bacidiomycete fungi that are parasitic on plants and are distinguished by the black, powdery masses of spores that appear as sooty smudges on the affected plant parts. Smuts are parasitic chiefly on cereal grasses like corn and wheat and can cause enormous damage to crops. **2.** Any of the various plant diseases caused by smuts, such as corn smut.

Sn The symbol for **tin.**

snow (snō) Precipitation that falls to earth in the form of ice crystals that have complex branched hexagonal patterns. Snow usually falls from stratus and stratocumulus clouds, but it can also fall from cumulus and cumulonimbus clouds.

snowfield (snō′fēld′) A large expanse of snow, usually with a smooth and uniform surface, and especially at the head of a glacier.

snow line **1.** The boundary marking the lowest altitude at which a given area, such as the top of a mountain, is always covered with snow. **2.** The boundary marking the furthest extent around the polar regions at which there is snow cover. The polar snow lines vary with the seasons.

snowpack (snō′păk′) **1.** An area of naturally formed, packed snow that usually melts during the warmer months. **2.** The amount of snow that accumulates annually in a mountainous area.

snow pellet See **graupel.**

soap (sōp) A substance used for washing or cleaning, consisting of a mixture of sodium or potassium salts of naturally occurring fatty acids. Like detergents, soaps work by surrounding particles of grease or dirt with their molecules, thereby allowing them to be carried away. Unlike detergents, soaps react with the minerals common in most water, forming an insoluble film that remains on fabrics. For this reason soap is not as efficient a cleaner as most detergents. Additionally, the film is what causes rings to form in bathtubs. Compare **detergent.**

soapstone (sōp′stōn′) A soft metamorphic rock composed mostly of the mineral talc, but also including chlorite, pyroxene, and amphibole. It has a schistose texture and is greasy to the touch. Soapstone forms through the alteration of ferromagnesian silicate minerals during metamorphism.

social science (sō′shəl) Any of various disciplines that study human society and social relationships, including sociology, psychology, anthropology, economics, political science, and history.

sociobiology (sō′sē-ō-bī-ŏl′ə-jē) The scientific study of the biological basis of the social behavior of animals, based on the theory that such behavior is often genetically determined and that the genes governing this behavior are subject to the usual mechanisms of evolution. Sociobiology posits that an animal will normally behave in ways that will increase the survival of its genes in the gene pool, either by increasing its own reproductive success or the reproductive success of an individual or group that is closely related and thus shares some of the genes of the organism. Sociobiologists seek to find explanations for animal behaviors such as **aggression** and **altruism** in relation to the survival of the animal's genes.

sociology (sō′sē-ŏl′ə-jē) The scientific study of human social behavior and its origins, development, organizations, and institutions.

Soddy (sŏd′ē), **Frederick** 1877–1956. British chemist who was a pioneer in the study of radioactivity. With Ernest Rutherford, he explained the atomic disintegration of radioactive elements. Soddy also coined the word *isotope* to describe elements that were chemically identical but had different atomic weights. He was awarded the Nobel Prize for chemistry in 1921.

sodic (sō′dĭk) Relating to or containing sodium.

sodium (sō′dē-əm) *Symbol* **Na** A soft, lightweight, silvery-white metallic element of the alkali group that reacts explosively with water. It is the most abundant alkali metal on Earth, occurring especially in common salt. Sodium is very malleable, and its compounds have many important uses in industry. Atomic number 11; atomic weight 22.99; melting point 97.8°C; boiling point 892°C; specific gravity 0.971; valence 1. See **Periodic Table.**

sodium bicarbonate A white crystalline compound used in beverages and as a leavening agent to make baked goods. In medicine, sodium bicarbonate is used as an antacid and cleanser and to replenish electrolytes. Also called *baking soda, bicarbonate of soda. Chemical formula:* **$NaHCO_3$.**

A CLOSER LOOK **sodium bicarbonate**

A white, chalky powder, sodium bicarbonate also goes by its household name, *baking soda.* Sodium bicarbonate is a base and reacts with acids in what is called *neutralization,* because both the acid and the base are converted into more neutral substances on the pH scale. Neutralization with sodium bicarbonate usually produces carbon dioxide gas, which bubbles forth whenever vinegar (an acid) and baking soda are mixed (as they frequently are in "kitchen science" experiments). Such reactions are an important factor in baking, where the production of the gas is what causes cakes to rise. Sodium bicarbonate has long been used in small amounts (about a half teaspoon) mixed with water to neutralize excess stomach acid. Sodium bicarbonate also has the unique ability to neutralize substances that are more basic than it is. It can do this because in water, sodium bicarbonate breaks down ultimately into carbonic acid (H_2CO_3), an unstable acid, which can then react with a base to neutralize it. This ability to neutralize both acids and many bases is why baking soda is so effective at reducing odors.

sodium chloride See **salt** (sense 2).

sodium fluoride A colorless, crystalline salt used to fluoridate water and treat tooth decay.

It is also used as an insecticide and a disinfectant. *Chemical formula:* **NaF.**

sodium hydroxide A white, corrosive, solid compound that absorbs water and carbon dioxide from the atmosphere and forms lye when in solution. Sodium hydroxide is toxic and strongly alkaline and is used to make chemicals and soaps and to refine petroleum. *Chemical formula:* **NaOH.**

sodium hypochlorite An unstable salt that is usually stored in solution and used as a fungicide and an oxidizing bleach. *Chemical formula:* **$NaOCl \cdot 5H_2O$.**

sodium nitrate A poisonous, white, crystalline compound used in solid rocket propellants, in matches, as a fertilizer, and in curing meat. *Chemical formula:* **$NaNO_3$.**

soft hail (sôft) See **graupel.**

soft palate See under **palate.**

software (sôft′wâr′) The programs, programming languages, and data that direct the operations of a computer system. Word processing programs and Internet browsers are examples of software. Compare **hardware.**

softwood (sôft′wo͝od′) **1.** A coniferous tree, especially as distinguished from an angiosperm, or hardwood, tree. **2.** The wood of a coniferous tree. Softwoods are in general softer than hardwoods. However, some softwoods, such as yew, are comparatively hard, while some hardwoods, such as basswood, are comparatively soft.

soil (soil) The loose top layer of the Earth's surface, consisting of rock and mineral particles mixed with decayed organic matter (humus), and capable of retaining water, providing nutrients for plants, and supporting a wide range of biotic communities. Soil is formed by a combination of depositional, chemical, and biological processes and plays an important role in the carbon, nitrogen, and hydrologic cycles. Soil types vary widely from one region to another, depending on the type of bedrock they overlie and the climate in which they form. In wet and humid regions, for example, soils tend to be thicker than they do in dry regions. See more at **A horizon, B horizon, C horizon.** See illustration at **ABC soil.**

solar (sō′lər) **1.** Relating to the Sun. **2.** Using or operated by energy from the Sun. **3.** Measured in reference to the Sun.

solar cell A photoelectric cell designed to convert sunlight into electrical energy, typically consisting of layers or sheets of specially prepared silicon. Electrons, displaced through the **photoelectric effect** by the Sun's radiant energy in one layer, flow across a junction to the other layer, creating a voltage across the layers that can provide power to an external circuit. Solar cells are used as power supplies in calculators, satellites, and other devices, and as a primary source of electricity in remote locations.

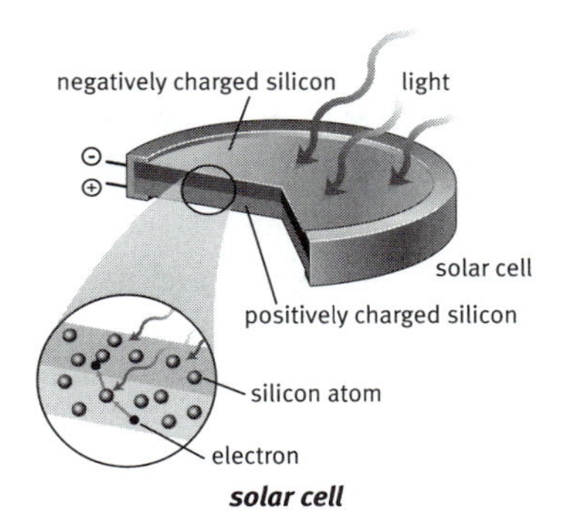

solar cell

When light penetrates a solar cell and reaches the lower charged layer, its energy causes atoms there to release electrons, which drift to the upper layer, giving the upper layer a net negative charge and the lower layer a net positive charge. This voltage difference can be used as a source of electrical energy.

solar constant The average amount of solar radiation received by the Earth's atmosphere, per unit area, when the Earth is at its mean distance from the Sun. This constant is equal to 1370 watts per square meter. Solar radiation varies with the distance of the Earthfrom the Sun and with the appearance or decay of sunspots.

solar day See under **solar time.**

solar eclipse See under **eclipse.**

solar-electric propulsion A form of **ionic propulsion** in which the power for the ion engine is provided by solar cells.

solar energy 1. The radiant energy emitted by the Sun. **2.** Energy derived from the Sun's radiation. Passive solar energy can be exploited through architectural design, as by positioning windows to allow sunlight to enter and help heat a space. Active solar energy involves the conversion of sunlight to electrical energy, especially in solar (photovoltaic) cells. See also **solar cell.**

solar flare A sudden eruption of hydrogen gas in the chromosphere of the Sun, usually associated with sunspots. Solar flares may last between several hours and several days, and have temperatures ranging from 20 to 100 million degrees K. The energy of a solar flare, which consists primarily of charged particles and x-rays, is comparable to tens of millions of hydrogen bombs, but is less than one-tenth the total energy emitted by the Sun every second. First observed in 1859, solar flares dramatically affect the Sun's weather and the **solar wind,** and are correlated with the appearance of **auroras** on the Earth. See also **prominence.**

solar month See under **solar time.**

solar sail A saillike device that is made of lightweight and highly reflective material and attached to a spacecraft to harness the radiation pressure of the solar wind and light for propulsion. Also called *light sail.*

solar system **1.** Often **Solar System.** The Sun together with the eight planets, their moons, and all other bodies that orbit it, including dwarf planets, asteroids, comets, meteoroids, and Kuiper belt objects. The outer limit of the solar system is formed by the **heliopause**. See more at **nebular hypothesis.** **2.** A similar system surrounding another star. Over two dozen stars are known to have planets in orbit around them, though none is known to have as extensive or diverse a group of orbiting bodies as the Sun's system.

A CLOSER LOOK **solar system**

The *solar system* consists of much more than just the Sun and planets. It contains billions of other objects and extends far beyond the outermost planets. Several hundred thousand asteroids revolve around the Sun in orbits mainly between Mars and Jupiter. Countless smaller meteoroids, including cometary debris and fragments from the collision of larger bodies, are also present, some of which approach Earth's orbit closely enough to be known as *near Earth objects.* In addition, as many as a billion objects, most the size of a speck of dust, cross through our atmosphere as meteors or micrometeoroids each day, though the vast majority are invisible to observers on the ground. Astronomers have recorded more than 800 comets passing through the inner part of the solar system. Billions more lie in the area surrounding the solar system, in the disk of debris known as the Kuiper belt and in the swarm of comets known as the Oort cloud. All of these objects orbit the Sun at high speeds. Some orbits, like those of the planets near the Sun, are almost circular. Other orbits, like those of comets that make their way in among the planets, are stretched out into long ellipses. As in most scientific fields, new discoveries are constantly changing our understanding and definitions. The objects in the Kuiper belt, for example, were discovered in the 1990s. When the new planetarium at the American Museum of Natural History opened in 2000, many visitors were shocked to find that Pluto had been demoted. In 2006, the International Astronomical Union classified Pluto as a dwarf planet.

---

solar time Time based on the rotation of the Earth with respect to the Sun. Solar time units are slightly longer than sidereal units due to the continuous movement of the Earth along its orbital path. For example, by the time the Earth has completed one full rotation on its axis with respect to the fixed stars, it has also moved a short distance in its orbit and is oriented slightly differently to the Sun, so that it must turn slightly more on its axis to complete a full rotation with respect to the Sun. ▸ The time it takes the Earth to rotate fully with respect to the Sun is called a **solar day.** The length of a solar day varies throughout the year due to variations in the Earth's orbital speed and other factors. ▸ The average value of all solar days in the solar year is called a **mean solar day**; it is 24 hours long and by convention is measured from midnight to midnight. ▸ A **solar year** is the period of time required for the Earth to make a complete orbit with respect to the Sun as measured from one vernal equinox to the next; it is equal to 365 days, 5 hours, 48 minutes, 45.51 seconds. A solar year is also called an *astronomical year* and a *tropical year.* ▸ A **solar month** is one twelfth of a solar year, totaling 30 days, 10 hours, 29 minutes, 3.8 seconds. Compare **sidereal time.**

Solar System

PHYSICAL PROPERTIES OF THE PLANETS

Listed below are the eight planets that have been identified in our solar system. The **sidereal period** is the amount of time it takes for a planet to make one complete revolution around the Sun. This is measured in relation to the fixed stars, in order to observe the orbit from a place outside our solar system. The sidereal period of Earth is exactly one year. The **rotational period** is the amount of time it takes for a body to make one complete rotation about its own axis. Earth rotates about its axis in one day. Earth rotates from west to east, counterclockwise when seen from above the North Pole. If one of the other planets rotates about its axis clockwise from east to west (the manner opposite to that of Earth), the rotation is called **retrograde** and a minus sign (–) appears in front of the number of days in the planet's rotational period.

	Equatorial Diameter		Mass	Sidereal Period	Rotational Period	Average Surface Temperature		Average Distance from the Sun	
	kilometers	*miles*	*x 10^{24} kilograms*	*years*	*days*	*degrees Celsius*	*degrees Fahrenheit*	*(in millions) kilometers*	*miles*
Mercury	4,880	3,032	0.33	0.24	58.65	179	354	57.9	36.0
Venus	12,104	7,521	4.87	0.62	–243.02	453	847	108.2	67.2
Earth	12,756	7,926	5.97	1.00	0.99727	8	46	149.6	92.96
Mars	6,794	4,222	0.64	1.88	1.026	–63	–81	227.9	141.6
Jupiter	142,984	88,846	1,898.6	11.86	0.414	–153	–243	778.6	483.8
Saturn	120,536	74,897	568.46	29.46	0.444	–185	–301	1,433.5	890.8
Uranus	51,118	31,763	86.83	84.01	–0.718	–215	–355	2,872.4	1,784.9
Neptune	49,528	30,775	102.43	164.79	0.671	–225	–373	4,495.1	2,793.1

PHYSICAL PROPERTIES OF THE SUN

The **corona,** an irregular envelope of gas that surrounds the Sun, extends about 13,000,000 kilometers (8,000,000 miles) past the Sun's visible surface. The low-density gas in the corona ranges in temperature from 1 to 2 million degrees Celsius (1.8 to 3.6 million degrees Fahrenheit). The Sun is more than 330,000 times more massive than Earth and contains more than 99.8 percent of the mass of the entire solar system. Because the Sun is not a solid body like Earth, its rotational period is not the same everywhere on its surface. Its outer layers rotate at different rates in different places, taking a longer time at the poles. The Sun's core, on the other hand, does rotate as a solid body.

	Diameter		Mass	Rotational Period (outer layers)	Temperature		
	kilometers	*miles*	*x 10^{24} kilograms*	*days*		*degrees Celsius*	*degrees Fahrenheit*
Sun	1,392,000	865,000	1,989,100	*Equator:* 25.4 *Poles:* 36	*Surface:* *Core:*	6,051 15,710,000	10,924 28,278,500

solar wind A continuous stream of plasma ejected by the Sun, flowing outward from the corona. This plasma, which consists mostly of protons and electrons, has enough energy to escape the Sun's gravitational field at speeds ranging from about 300 to 800 km (186 to 496 mi) per second and averaging 1,610,000 km (1,000,000 mi) per hour, which allows the solar wind to reach Earth in about 3.9 days. The speed and intensity of the solar wind depends on magnetic activity at different regions of the Sun. The solar wind spreads out from the Sun in a pinwheel pattern as a result of the Sun's rotation, pushing back the interstellar medium to the boundary known as the **heliopause.** The tails of comets, which always extend away from the Sun regardless of the direction of the comet's motion, are a result of the impact of solar wind, which dislodges ice and other particles from the comet's surface. Similar winds flowing from other stars are called **stellar winds**. See also **aurora.**

solar year See under **solar time.**

solenoid (sō′lə-noid′) A coil of wire that acts as an **electromagnet** when electric current is passed through it, often used to control the motion of metal objects, such as the switch of a relay.

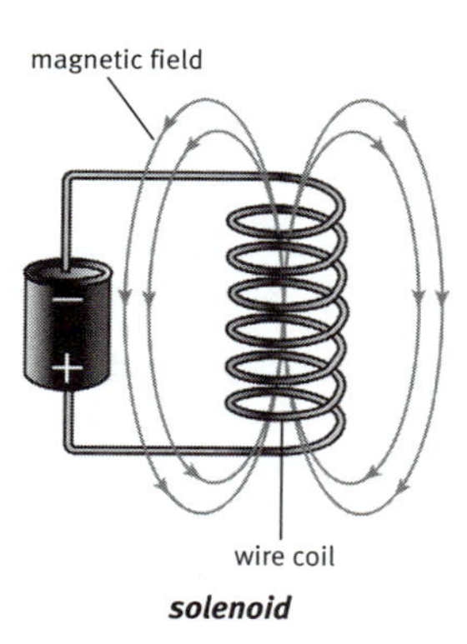

solenoid

solid (sŏl′ĭd) **1.** *Physics.* One of four main **states of matter**, in which the molecules vibrate about fixed positions and cannot migrate to other positions in the substance. Unlike a gas or liquid, a solid has a fixed shape, and unlike a gas, a solid has a fixed volume. In most solids (with exceptions such as glass), the molecules are arranged in crystal lattices of various sizes. **2.** *Mathematics.* A geometric figure that has three dimensions.

solid angle A three-dimensional angle, formed by three or more planes intersecting at a common point. Its magnitude is measured in steradians, a unitless measure. The corner of a room forms a solid angle, as does the apex of a cone; one can imagine an indefinite number of planes forming the smooth round surface of the cone all intersecting at the apex. Solid angles are commonly used in photometry.

solid geometry The branch of mathematics that deals with three-dimensional figures and surfaces.

solid solution A uniform mixture of substances in solid form. Solid solutions often consist of two or more types of atoms or molecules that share a crystal lattice, as in certain metal alloys. Much of the steel used in construction, for example, is actually a solid solution of iron and carbon. The carbon atoms, which fit neatly within the iron's crystal lattice, add strength to its structure.

solid-state physics The branch of condensed matter physics that specializes in the study of solids, especially in the electric and magnetic properties of solid crystalline materials, such as semiconductors.

solidus (sŏl′ĭ-dəs) Plural **solidi** (sŏl′ĭ-dī′). The maximum temperature at which all components of a mixture (such as an alloy) can be in a solid state. Above the solidus some or all of the mixture will be in a liquid state. See illustration at **eutectic.** Compare **liquidus.**

Solo man (sō′lō) A fossil hominid specimen previously classed as a distinct species (*Homo soloensis*) but now generally regarded as an archaic example of *Homo sapiens.*

solstice (sŏl′stĭs, sōl′-) **1.** Either of the two points on the celestial sphere where the **ecliptic** (the apparent path of the Sun) reaches its greatest distance north or south of the **celestial equator.** ▸ The northernmost point of the Sun's path, called the **summer solstice**, lies on the Tropic of Cancer at 23°27′ north latitude. ▸ The southernmost point of the Sun's path, called the **winter solstice**, lies on the Tropic of Capricorn at 23°27′ south latitude. **2.** Either of the two corresponding moments of the year when the Sun is directly above either the Tropic of Cancer or the Tropic of Capricorn. The summer solstice occurs on June 20 or 21 and the winter solstice on December 21 or 22, marking the beginning of summer and winter in the Northern Hemisphere (and the reverse in the Southern Hemisphere). The days on which a solstice falls have the greatest difference of the year between the hours of daylight and darkness, with the most daylight hours at the beginning of summer and the most darkness at the beginning of winter. Compare **equinox.**

soluble (sŏl′yə-bəl) Capable of being dissolved. Salt, for example, is soluble in water.

solum (sō′ləm) Plural **sola** or **solums.** The upper layers of a soil profile in which soil formation occurs. The A and B horizons in an ABC soil are part of the solum.

solute (sŏl′yōōt) A substance that is dissolved in another substance (a solvent), forming a solution.

solution (sə-lōō′shən) **1.** *Chemistry.* A mixture in which particles of one or more substances (the solute) are distributed uniformly throughout another substance (the solvent), so that the mixture is homogeneous at the molecular or ionic level. The particles in a solution are smaller than those in either a colloid or a suspension. Compare **colloid, suspension. 2.** *Mathematics.* A value or values which, when substituted for a variable in an equation, make the equation true. For example, the solutions to the equation $x^2 = 4$ are 2 and –2.

Solutrean also **Solutrian** (sə-lōō′trē-ən) Relating to an Upper Paleolithic culture in southwestern Europe between the Aurignacian and Magdalenian cultures, dating from around 21,000 to 17,000 years ago. The short-lived Solutrean culture was characterized by finely crafted tools, such as slender, leaf-shaped blades and shouldered points, as well as ornaments, carvings, and cave paintings.

Solvay process (sŏl′vā, sôl-vā′) A process used to produce large quantities of sodium carbonate. In the Solvay process, salt (sodium chloride) is treated with ammonia and then carbon dioxide, producing sodium bicarbonate and ammonium chloride. The ammonium chloride is usually combined with lime to produce ammonia (recycled for reuse) and calcium chloride.

solvent (sŏl′vənt) A substance that can dissolve another substance, or in which another substance is dissolved, forming a solution. Water is the most common solvent.

somatic (sō-măt′ĭk) Relating to the body. ► The cells of the body with the exception of the reproductive cells (gametes) are known as **somatic cells.** See Note at **mitosis.**

somatomedin (sō-măt′ə-mēd′n, sō′mə-tə-) Any of several peptides that are synthesized in the liver and other tissues and are capable of stimulating growth processes, especially in bone, cartilage, and muscle. Somatomedins are secreted and activated in response to **growth hormone**. Somatomedin levels rise progressively during childhood, peak at puberty, and then stabilize at lower levels in adulthood. Somatomedins influence calcium, phosphate, carbohydrate, and lipid metabolism and have also been associated with the growth of certain cancers. Also called *insulinlike growth factor.*

somatostatin (sō-măt′ə-stăt′n, sō′mə-tə-) A polypeptide produced by the hypothalamus and the pancreas. Somatostatin produced by the hypothalamus acts as a neurohormone that inhibits the secretion of other hormones, especially growth hormone and thyrotropin. Somatostatin secreted by the pancreas acts as a hormone that inhibits the secretion of the other pancreatic hormones, insulin and glucagon, and reduces the activity of the digestive system.

somatotropin (sə-măt′ə-trō′pĭn, sō′mə-tə-) See **growth hormone** (sense 1).

Somerville (sŭm′ər-vĭl′), **Mary Fairfax Greig** 1780–1872. Scottish astronomer and mathematician who wrote expository works on mathematics, physical geography, microscopic science, and astronomy. Her writings explained complex scientific ideas to the general public through simple illustrations and experiments that the average reader could easily understand.

sonar (sō′när′) **1.** Short for *sound navigation and ranging.* A method of detecting, locating, and determining the speed of objects through the use of reflected sound waves. A sound signal is produced, and the time it takes for the signal to reach an object and for its echo to return is used to calculate the object's distance. The **Doppler effect** can also be used to determine the object's relative velocity. Electronic sonar systems are used for submarine navigation and for detecting schools of fish. Some mammals, especially bats, use biological sonar to navigate and detect prey in dark conditions, commonly called *echolocation.* **2.** The equipment or physiology used in doing this. See also **Doppler effect, lidar, radar.**

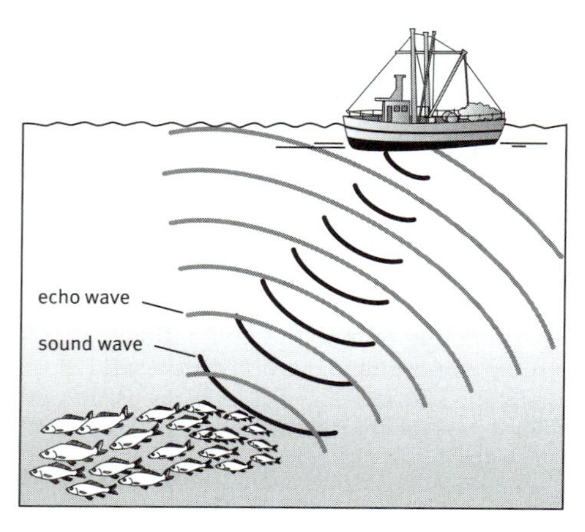

sonar

An electrical impulse is converted into sound waves that are transmitted underwater. The sound waves are reflected off objects in their paths, creating echoes that return to the vessel and are picked up by the sonar equipment.

sonic barrier (sŏn′ĭk) See **sound barrier.**

sonic boom A shock wave of compressed air caused by an aircraft traveling faster than the speed of sound. It is often audible as a loud, explosive sound, and it sometimes causes damage to structures on the ground.

soot (so͝ot) A black, powdery substance that consists mainly of carbon and is formed through the incomplete combustion of wood, coal, diesel oil, or other materials. Because it absorbs energy from sunlight rather than reflecting it, soot is believed to be a cause of global warming, especially when it settles on snow and ice, reducing their reflectivity. Soot particles in the air are a contributing factor in respiratory diseases.

sorbic acid (sôr′bĭk) A white crystalline solid found in the berries of the mountain ash or prepared synthetically and used as a food preservative and fungicide. *Chemical formula:* $C_6H_8O_2$.

sorbitol (sôr′bĭ-tôl′, -tōl′) A white, sweetish, crystalline alcohol found in various berries and fruits or prepared synthetically. It is used as a flavoring agent, a sugar substitute for people with diabetes, and a moisturizer in cosmetics and other products. *Chemical formula:* $C_6H_{14}O_6$.

sorption (sôrp′shən) The taking up and holding of one substance by another. Sorption includes the processes of absorption and adsorption.

sorting (sôr′tĭng) **1.** The process by which sediment particles that have a certain characteristic, such as a given shape or grain size, are separated from other associated particles by an active agent of transportation, such as wind, a stream, or a glacier. **2.** A measure of the degree to which this process has occurred within a body of sediment. Wind-blown sediments are usually well-sorted because only a small range of grain sizes can be lifted by a particular wind velocity. Glacially derived sediments are usually poorly sorted because of the great range of particle sizes that are picked up by a moving glacier.

sorus (sôr′əs) Plural **sori** (sôr′ī). **1.** A cluster of sporangia borne on the underside of a fern frond. A sorus is sometimes covered by an indusium. Also called *fruitdot.* **2.** A reproductive structure consisting of masses of spores, characteristic of certain fungi such as rusts and smuts.

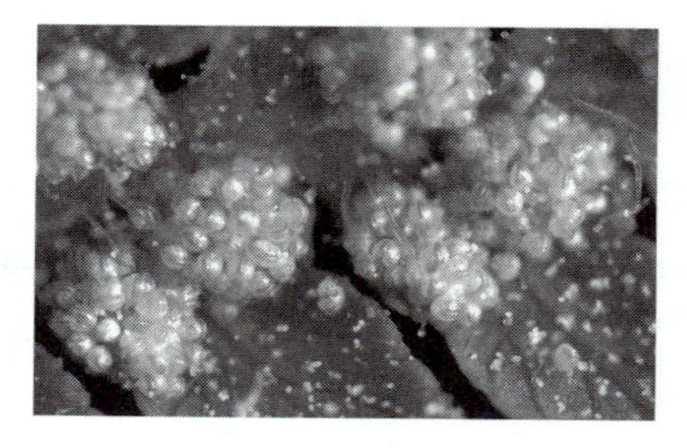

sorus

magnified view of sporangia on the underside of fern leaves

sound[1] (sound) **1.** A type of longitudinal wave that originates as the vibration of a medium (such as a person's vocal cords or a guitar string) and travels through gases, liquids, and elastic solids as variations of pressure and density. The loudness of a sound perceived by the ear depends on the amplitude of the sound wave and is measured in decibels, while its pitch depends on its frequency, measured in hertz. **2.** The sensation produced in the organs of hearing by waves of this type. See Note at **ultrasound.**

sound² (sound) **1.** A long, wide inlet of the ocean, often parallel to the coast. Long Island Sound, between Long Island and the coast of New England, is an example. **2.** A long body of water, wider than a strait, that connects larger bodies of water.

sound barrier The sharp increase in **drag** experienced by aircraft approaching the speed of sound. Also called *sonic barrier.*

source code (sôrs) Code written by a programmer in a high-level language and readable by people but not computers. Source code must be converted to object code or machine language by a compiler before a computer can read or execute the program. Compare **object code.**

Southern Cross (sŭth′ərn) A constellation in the polar region of the Southern Hemisphere near Centaurus.

Southern Hemisphere 1. The half of the Earth south of the equator. **2.** The half of the celestial sphere south of the celestial equator.

southern lights See under **aurora.**

southern oscillation The cyclic variation of atmospheric pressure at sea level between the eastern and western regions of the southern Pacific Ocean as a result of the periodic cooling and warming of surface water known as **La Niña** and **El Niño**. It has an average period of 2.33 years.

South Frigid Zone (south) See **Frigid Zone.**

South Pole The southern end of the Earth's axis of rotation, located at 90° south latitude at a point in Antarctica. See more at **axis.**

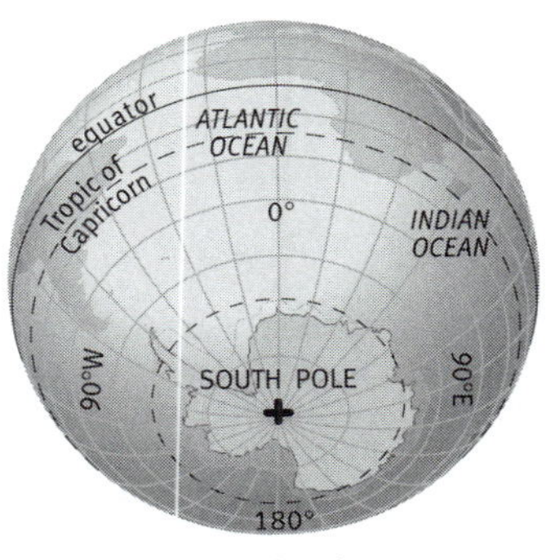

South Pole

South Temperate Zone See **Temperate Zone.**

space (spās) **1.** The region of the universe beyond Earth's atmosphere. ► The part of this region within the solar system is known as **interplanetary space.** ► The part of this region beyond the solar system but within the Milky Way or within another galaxy is known as **interstellar space.** ► The part of this region between the Milky Way and other galaxies is known as **intergalactic space. 2.** The familiar three-dimensional region or field of everyday experience. **3.** *Mathematics.* A mathematical object, typically a set of sets, that is usually structured to define a range across which variables or other objects (such as a coordinate system) can be defined.

space-time A four-dimensional reference frame, consisting of three dimensions in space and one dimension in time, used especially in Relativity Theory as a basis for coordinate systems for identifying the location and timing of objects and events. In **General Relativity**, space-time is thought to be curved by the presence of mass, much as the space defined by the surface of a piece of paper can be curved by bending the paper. See more at **relativity.**

A CLOSER LOOK **space-time**

Albert Einstein's theory of *General Relativity,* published in 1915, extended his theory of *Special Relativity* to systems that are accelerating. One of the primary causes of acceleration in the universe is gravity, and Einstein showed that the effects of acceleration are actually the same as those of the force of gravity; in fact, they are locally indistinguishable. For instance, both in an accelerating rocket in space and in a rocket standing on its launch pad on Earth, the astronauts are pushed back into their seats. Unlike Newtonian physics, which views gravity as an attractive force between all bodies in the universe, General Relativity describes the universe in terms of a continuous *space-time* fabric that is curved by masses located within it. In the space-time continuum of General Relativity, events are defined in terms of four dimensions: three of space, and one of time, with one coordinate for each dimension; we continuously "move" along the time dimension. What does it mean, though, for space-time to be curved? One way of conceptualizing this is to imagine just a two-dimensional space-time, with one spatial dimension and one time dimension. But instead of an infinite plane, imagine a tube, with an object's position in time defined by a coordinate of length along the tube, and position in space by a coordinate around the circumference of the tube. An object traveling uniformly through space then describes a helix along this tube, eventually returning to its starting space-coordinate position, but at a different time. (It is an open question in cosmology as to whether our universe has a similar curvature in three dimensions; if so, traveling in one direction long enough would bring you back to where you began.) An important consequence of the notion of curved space-time is that the curvature should affect all motion; thus, even light, which has no mass, should follow a curved path wherever gravity has warped space-time. An important verification of this—which made headlines around the world—took place during a solar eclipse on May 29, 1919, when it was observed that light from stars near the Sun was bent by an angle exactly predicted by the expected curvature of space-time near the massive Sun. Space-time can in principle be warped so strongly by a huge mass that any radiation emitted from the mass curves back in again and cannot escape. These huge masses are thought to exist as *black holes*.

---

spadix (spā′dĭks) Plural **spadices** (spā′dĭ-sēz′). A fleshy spike of minute flowers, usually enclosed within a spathe, as in the arums.

spark arrester (spärk) **1.** A device designed to keep sparks from escaping, as at a chimney opening or where materials are being welded. **2.** A device used to control or curtail electric sparking at a point where a circuit is made or broken.

spark coil An **induction coil** used to induce sparks in the spark plugs of an internal combustion engine. See more at **induction coil.**

spark gap A gap in an otherwise closed electric circuit across which a discharge occurs at a prescribed voltage. The discharge often causes ionization of the surrounding gas and thereby very high temperatures and can be exploited, as in spark plugs, to ignite engine fuel in a combustion chamber.

spark plug A device in the cylinder head of an internal-combustion engine that ignites the fuel mixture by means of an electric spark. Spark plugs consist of two electrodes that are separated by an insulator except at their tips. When high voltage electricity is fed to one electrode, it jumps across the gap from one tip to the other as a spark. The voltage required to generate a spark is directly proportional to the size of the gap; in a car engine it is about 18,000 volts.

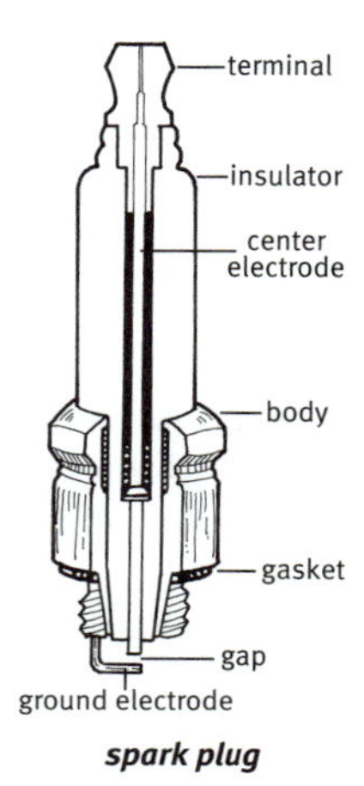

spark plug

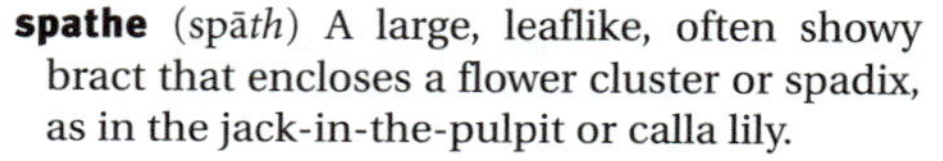

spathe (spā*th*) A large, leaflike, often showy bract that encloses a flower cluster or spadix, as in the jack-in-the-pulpit or calla lily.

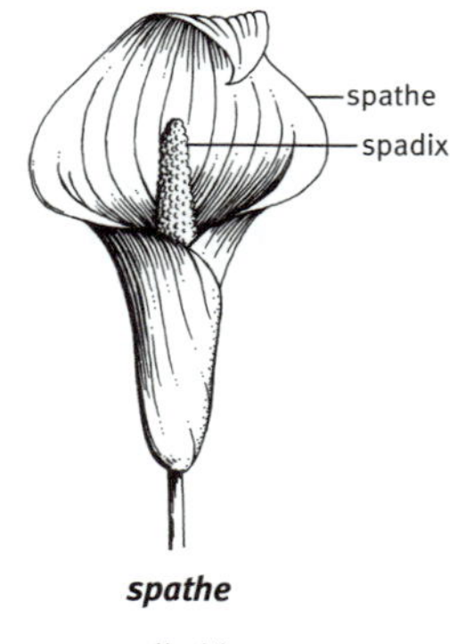

spathe

calla lily

spawn (spôn) *Noun.* **1.** The eggs of water animals such as fish, amphibians, and mollusks. **2.** Offspring that are produced in large numbers. —*Verb.* **3.** To lay eggs; produce spawn.

Special Relativity (spĕsh′əl) The theory of space and time developed by Albert Einstein, based on the postulates that all the laws of physics are equally valid in all reference frames moving at a constant speed relative to each other (that is, in all **inertial frames**), and that the speed of light is observed to be the same in all reference frames. To compare measurements of length or time made in different inertial frames, their values must be regarded as components of vectors in a 4-dimensional space called **Minkowski spacetime**. Different frames are related to each other mathematically by **Lorentz transformations**. Special Relativity predicts that all massless particles, such as photons, are always necessarily moving at the speed of light, while all particles with mass are always necessarily moving slower than the speed of light, and only particles with mass can be at rest. The theory also states that velocities are not simply additive. For example, if an object (such as a spacecraft) moving with velocity v_1 observes another faster object (such as a rocket fired from the spacecraft) moving ahead of it in the same direction with velocity v_2, an observer at rest observes the faster object moving at a velocity that is not just the sum of v_1 and v_2, but at the slower speed

$$\frac{v_1 + v_2}{1 + \frac{v_1 v_2}{c^2}}$$

See also **General Relativity, time dilation.**

speciation (spē′shē-ā′shən) The formation of new biological species by the development or branching of one species into two or more genetically distinct ones. The divergence of species is thought to result primarily from the geographic isolation of a population, especially when confronted with environmental conditions that vary from those experienced by the rest of the species, and from the random change in the frequency of certain alleles (known as **genetic drift**). According to the theory of evolution, all life on Earth has resulted from the speciation of earlier organisms. See also **adaptive radiation.**

species (spē′shēz, spē′sēz) A group of organisms having many characteristics in common and ranking below a genus. Organisms that reproduce sexually and belong to the same species interbreed and produce fertile offspring. Species names are usually written lower case and in italics, as *rex* in *Tyrannosaurus rex*. See Table at **taxonomy.**

specific gravity (spĭ-sĭf′ĭk) The **relative density** of a solid or liquid, usually when measured at a temperature of 20°C, compared with the maximum density of water (at 4°C). For example, the specific gravity of carbon steel is 7.8, that of lead is 11.34, and that of pure gold is 19.32.

specific heat See under **heat capacity.**

spectral line (spĕk′trəl) An isolated bright or dark line in a spectrograph produced by emission or absorption of light of a single wavelength, generally corresponding to a specific shift in the energy of an electron moving from one orbital to another.

spectral type A classification system for stars based on the strength of their spectral lines, using the letters O, B, A, F, G, K, M, L, and T to denote a range from blue (as in blue giant stars) to dim red (as in brown dwarfs). The spectrum of a star correlates with its surface temperature, ranging from over 60,000°K (O type) to less than 3,500°K (L and T types). See also **Hertzsprung-Russell diagram.**

spectrograph (spĕk′trə-grăf′) **1.** A photograph or computer image of a spectrum produced by a spectroscope. Also called *spectrogram.* **2.** See **spectroscope.**

spectrography (spĕk-trŏg′rə-fē) See **spectroscopy.**

spectrometer (spĕk-trŏm′ĭ-tər) A spectroscope equipped with devices for measuring the frequencies of the radiation observed by it.

spectroscope (spĕk′trə-skōp′) Any of various instruments used to analyze the component parts of a sample by separating its parts into a spectrum. ► In a **light spectroscope,** light is focused into a thin beam of parallel rays by a lens, and then passed through a prism or diffraction grating that separates the light into a frequency spectrum. The intensity of light at different frequencies in the spectrum can be analyzed to determine certain properties of the source of the light, such as its chemical composition or how quickly it is moving. ► In a **mass spectroscope,** sample ions are beamed through an electric or magnetic field that deflects the ions; the amount of deflection depends on the ratio of their mass to their electric charge. The ion beam is thus split into separate bands; the collection of bands is called the **mass spectrum** of the sample, and can be analyzed to determine the distribution of ions in the sample. Spectroscopes are also called *spectrographs.*

spectroscopic binary (spĕk′trə-skŏp′ĭk) See under **binary star.**

spectroscopy (spĕk-trŏs′kə-pē) The analysis of spectra, especially light or mass spectra, to determine properties of their source. ► In **light**

or **optical spectroscopy,** the spectrum of a light source is analyzed through a spectroscope to determine atomic composition of a substance. In astronomy, phenomena such as **red shift** can also be analyzed. ▸ In **mass spectroscopy,** a spectroscope is used to determine the composition of ions or charged molecules in a sample. Spectroscopy is also called *spectrography.* See also **atomic spectrum, spectroscope.**

spectrum (spĕk′trəm) Plural **spectra** (spĕk′trə) or **spectrums. 1.** A range over which some measurable property of a physical phenomenon, such as the frequency of sound or electromagnetic radiation, or the mass of specific kinds of particles, can vary. For example, the spectrum of visible light is the range of electromagnetic radiation with frequencies between between 4.7×10^{14} and 7.5×10^{14} hertz. **2.** The observed distribution of a phenomenon across a range of measurement. See more at **atomic spectrum, spectroscopy.**

speed (spēd) The ratio of the distance traveled by an object (regardless of its direction) to the time required to travel that distance. Compare **velocity.**

speedometer (spĭ-dŏm′ĭ-tər) An instrument for indicating the speed of a vehicle, typically by measuring the rate of rotation of a wheel or fan whose rate of rotation depends on the speed of the vehicle. Compare **odometer.**

speleology (spē′lē-ŏl′ə-jē) The exploration and scientific study of the geological and ecological features of caves.

sperm (spûrm) The smaller, usually motile male reproductive cell of most organisms that reproduce sexually. Sperm cells are haploid (they have half the number of chromosomes as the other cells in the organism's body). Sperm often have at least one flagellum. During fertilization, the nucleus of a sperm fuses with the nucleus of the much larger egg cell (the female reproductive cell) to form a new organism. In male animals, sperm are normally produced by the testes in extremely large numbers in order to increase the chances of fertilizing an egg. Motile sperm cells produced by some multicellular protist groups (such as the algae), the bryophyte plants, and the seedless vascular plants, require water to swim to the egg cell. In gymnosperms and angiosperms, sperm do not need water for mobility but are carried to the female reproductive organs in the pollen grain. In the cycads and the gingko (both gymnosperms), the sperm are motile and propel themselves down the pollen tube to reach the egg cell. In the conifers and angiosperms, the sperm are not themselves motile but are conveyed to the ovule by the growing pollen tube.

A CLOSER LOOK sperm

The human sperm cell is divided into a head that contains the nucleus, a mid-section that contains mitochondria to provide energy for the sperm, and a flagellum that allows the sperm to move. When fertilization occurs, the nucleus and other contents from the sperm cells are drawn into the cytoplasm of the egg, but the mitochondria in the sperm are destroyed and do not survive in the zygote. Since mitochondria contain their own DNA (thought to be a relic from an existence as separate symbiotic organisms), all of the mitochrondrial DNA in humans is thus inherited from the female. The semen produced by the male reproductive tract as a medium for sperm typically contains over 100 million sperm cells, all of which have but one purpose: to fertilize the single available egg.

spermaceti (spûr′mə-sē′tē) A white, waxy substance that is obtained from the head of the sperm whale and sometimes other whales, porpoises, and dolphins, and was once widely used to make candles, ointments, and cosmetics. Spermaceti is a liquid at body temperature and consists primarily of fatty alcohols and esters of fatty acids. Spermaceti is produced by the spermaceti organ, a huge melon-shaped sac that may function as a battering ram in aggressive behavior between males.

spermatheca (spûr′mə-thē′kə) Plural **spermathecae** (spûr′mə-thē′sē). A receptacle in the reproductive tracts of certain female invertebrates, especially insects, in which spermatozoa are received and stored until needed to fertilize the ova.

spermatium (spər-mā′shē-əm) Plural **spermatia.** A nonmotile cell in red algae and certain ascomycete fungi that functions as a male gamete.

spermatocyte (spər-măt′ə-sīt′) A diploid cell that undergoes meiosis to form four spermatids, cells which then develop into sperm. See more at **spermatogenesis.**

spermatogenesis (spər-măt′ə-jĕn′ĭ-sĭs, spûr′-mə-tə-) The formation and development of spermatozoa. Spermatogenesis in humans begins with the spermatogonium, the diploid cell that undergoes mitosis to form new spermatogonia as well as cells called primary spermatocytes. Each primary spermatocyte then undergoes the first meiotic division to produce two secondary spermatocytes. Each secondary spermatocyte undergoes the second meiotic division to produce two nonmotile cells called spermatids. The four spermatids then develop flagella and become sperm. Since some of the original spermatogonia replace themselves, the males are able to produce large numbers of sperm continuously after sexual maturity.

spermatogonium (spər-măt′ə-gō′nē-əm) Plural **spermatogonia.** Any of the cells of the gonads in male organisms that undergo mitosis to form spermatocytes as well as new spermatogonia. See more at **spermatogenesis.**

spermatophore (spər-măt′ə-fôr′) A capsule or compact mass of spermatozoa extruded by the males of certain invertebrates and primitive vertebrates and directly transferred to the reproductive parts of the female.

spermatophyte (spər-măt′ə-fīt′) See **seed-bearing plant.**

spermatozoid (spər-măt′ə-zō′ĭd) A male reproductive cell produced in an antheridium, as in algae, fungi, and nonflowering plants. Each spermatozoid has cilia that propel it toward the archegonium. Also called *antherozoid.*

spermatozoon (spər-măt′ə-zō′ŏn′) Plural **spermatozoa.** A sperm cell produced in the testis of an animal.

spermicide (spûr′mĭ-sīd′) An agent that kills spermatozoa, especially one that is used as a contraceptive.

Sperry (spĕr′ē), **Roger Wolcott** 1913–1994. American neurobiologist who pioneered the behavioral investigation of "split-brain" animals and humans, establishing that each hemisphere of the brain controls specific higher functions. He shared with American neurophysiologist David H. Hubel and Swedish neurophysiologist Torsten N. Wiesel the 1981 Nobel Prize for physiology or medicine.

BIOGRAPHY **Roger Wolcott Sperry**

Ever wondered what it's like to see the world upside-down and backwards? Some salamanders found out in the 1930s. They were experimental subjects in the lab of Roger Wolcott Sperry, who had made his first big splash on the scientific community by showing that the functions of specific motor nerves in mammals were hardwired and unchangeable. Salamanders, unlike mammals, can regenerate nerves, so Sperry cut through their optic nerves and rotated their eyeballs 180 degrees. When the nerve grew back, it was somehow "guided back" to its original termination sites, resulting in the salamanders' visual field being radically altered. While this work was path-breaking, Sperry's most famous experiments involved work with the brain in which the corpus callosum, the thick network of nerves that connects the left and right cerebral hemispheres, had been severed (resulting in a "split brain"). Sperry showed first that the hemispheres of split-brain cats learned tasks separately, and with equal facility, and were essentially independent cognitive organs. He then turned to humans, using patients whose corpora callosa had been severed as treatment for epilepsy (widely done at the time). Using these patients Sperry was able to demonstrate that the two hemispheres are functionally distinct: the left hemisphere is dominant in verbal and analytical tasks, while the right hemisphere is dominant in music and spatial tasks. The results of Sperry's and his colleagues' research led to the construction of a map of the brain and also to his sharing the Nobel Prize for physiology or medicine in 1981.

sphagnum (sfăg′nəm) See **peat moss.**

sphalerite (sfăl′ə-rīt′) A usually yellow-brown or brownish-black cubic mineral occurring either as single dodecahedral crystals or as granular masses. It often contains cadmium, arsenic, or manganese and has a distinct rotten egg odor. It is often found in association with the mineral galena and is the primary ore of zinc. *Chemical formula:* **ZnS.**

sphene (sfēn) A brown or yellow monoclinic mineral occurring as an accessory mineral in igneous and metamorphic rocks. It usually occurs as wedge or lozenge-shaped crystals. *Chemical formula:* $CaTiSiO_5$.

sphere (sfîr) A three-dimensional geometric surface having all of its points the same distance from a given point.

spherical aberration (sfîr′ĭ-kəl) See under **aberration.**

spherical coordinate Any of a set of coordinates in a three-dimensional system for locating points in space by means of a radius vector and two angles measured from the center of a sphere with respect to two arbitrary, fixed, perpendicular directions.

spherical geometry The geometry of circles, angles, and figures on the surface of a sphere.

spheroid (sfîr′oid′) A three-dimensional geometric surface generated by rotating an ellipse on or about one of its axes.

sphincter (sfĭngk′tər) A ring-shaped muscle that encircles an opening or passage in the body. The opening and closing of the anus is controlled by contraction and relaxation of a sphincter, as is the opening that leads to the stomach from the esophagus.

Spica (spī′kə) A bright bluish-white binary star in the constellation Virgo, with an apparent magnitude of 0.96. *Scientific name:* Alpha Virginis.

spicule (spĭk′yo͞ol) **1.** A needlelike structure or part, such as one of the mineral structures supporting the soft tissue of certain invertebrates, especially sponges. **2.** Any of numerous short-lived vertical jets of hot gas rising from the solar chromosphere and extending into the corona. Spicules, which only last for about five to ten minutes, are usually several hundred kilometers wide and several thousand kilometers high.

spike (spīk) An elongated indeterminate inflorescence in which the flowers are attached directly to a common stem, rather than borne on individual stalks arising from the stem. The gladiolus produces spikes. The distinctive spikes of grasses such as wheat or barley are known as **spikelets.** See illustration at **inflorescence.**

spikelet (spīk′lĭt) A small spike, especially one that is part of the characteristic inflorescence of grasses and sedges. A grass spikelet consists of one or more florets (reduced flowers). Each floret contains a pistil and stamens and is enclosed by two bracts, the lemma and the palea. At the base of the entire spikelet are two additional scalelike bracts, the glumes.

spin (spĭn) **1.** The intrinsic angular momentum of a rigid body or particle, especially a subatomic particle. Also called *spin angular momentum.* **2.** The total angular momentum of a physical system, such as an electron orbital or an atomic nucleus. **3.** A quantum number expressing spin angular momentum; the actual angular momentum is a quantum number multiplied by Dirac's constant. Fermions have spin values that are integer multiples of $\frac{1}{2}$, while bosons have spin values that are integer multiples of 1.

spina bifida (spī′nə bĭf′ĭ-də) A congenital defect caused by incomplete formation of the neural tube, in which the vertebral column is not fully closed, resulting in protrusion of the meninges and sometimes the spinal cord. Damage to the exposed spinal cord can cause neurological abnormalites, including paralysis.

spinal column (spī′nəl) See **vertebral column.**

spinal cord The long, cordlike part of the central nervous system that is enclosed within the vertrbral column (spine) and descends from the base of the brain, with which it is continuous. the spinal cord branches to form the nerves that convey motor and sensory impulses to and from the tissues of the body.

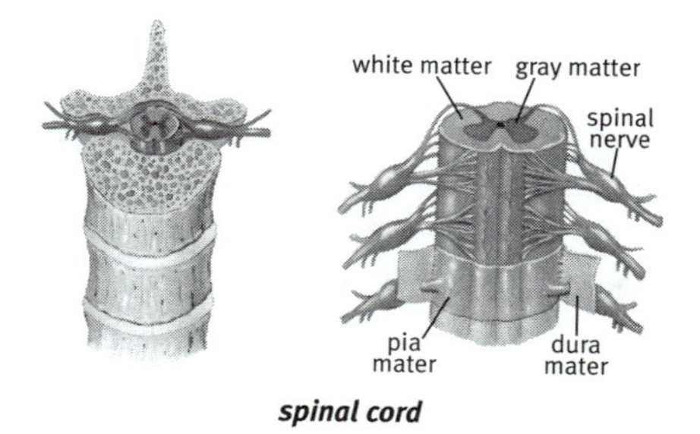

spinal cord

left: *section of vertebral column showing the spinal cord within the column;* right: *segment of spinal cord with associated nerve fibers*

spinal nerve Any of the nerves that arise in pairs from the spinal cord and form an important part of the peripheral nervous system. The spinal nerves contain both sensory and motor nerve fibers. There are 31 pairs of spinal nerves in the human body.

spinal tap See **lumbar puncture.**

spin angular momentum See **spin** (sense 1).

spindle (spĭn′dl) A network of protein fibers that forms in the cytoplasm of a cell during cell division. The spindle grows forth from the centrosomes and attaches to the chromosomes after the latter have been duplicated, and the nuclear membrane dissolves. Once attached, the spindle fibers contract, pulling the duplicate chromosomes apart to opposite poles of the dividing cell. See more at **meiosis, mitosis.**

spine (spīn) **1.** See **vertebral column. 2.** Any of various pointed projections, processes, or appendages of animals. **3.** A sharp-pointed projection on a plant, especially a hard, narrow modified leaf, as on a cactus, that is adapted to reduce water loss. Compare **thorn.** See more at **leaf.** —*Adjective* **spinal.**

spinel also **spinelle** (spĭ-nĕl′) **1.** A hard, variously colored cubic mineral, having usually octahedral crystals and occurring in igneous and metamorphosed carbonate rocks. The red variety is valued as a gem and is sometimes confused with the ruby. *Chemical formula:* $\mathbf{MgAl_2O_4}$. **2.** Any of a group of minerals that are oxides of magnesium, iron, zinc, manganese, or aluminum.

spinneret (spĭn′ə-rĕt′) One of the small openings in the back part of a spider or silk-producing insect larva, through which the sticky fluid that dries into silk is released.

spin quantum number See under **quantum number.**

spin vector *Physics.* A vector quantity used to represent the spin angular momentum of a particle. Its magnitude represents the magnitude of the spin, while its direction lies along the axis of rotation of the particle, following the **right hand rule**.

spin wave A wave propagated through a crystal lattice as a result of shifts in atomic magnetic fields associated with the spin angular momentum of electrons in the lattice.

spiny-headed worm See **acanthocephalan.**

spiracle (spĭr′ə-kəl, spī′rə-) An opening through which certain animals breathe, such as the blowhole of a whale.

spiral galaxy (spī′rəl) A galaxy consisting of a rotating flattened disk with an ellipsoidal central bulge from which extend a pattern of two or more luminous spiral arms. Spiral galaxies range from large bulges with tightly wound arms (classified as Sa) to small bulges with loosely wound arms (classified as Sc and in some cases Sd). The majority of the mass of a spiral galaxy is contained in its bulge, made up mostly of old stars, while the arms are composed mostly of younger stars and large amounts of interstellar gas and dust. A spherical, relatively dust-free region known as a **galactic halo** surrounding a spiral galaxy may contain large amounts of dark matter. The Milky Way is a spiral galaxy. See illustration at **galaxy.** ► A spiral galaxy whose central bulge has the shape of a bar from whose ends the spiral arms emanate is called a **barred spiral galaxy**. About a third of spiral galaxies have this straight or lozenge-shaped bar of stars, gas, and dust extending out from the nucleus. Barred spiral galaxies are classified similarly to regular spirals, from SBa (large bulge, tightly wound arms) to SBc (smaller bulge, looser arms). Astronomers believe that some elliptical galaxies containing hints of a bar and spiral might once have been barred spiral galaxies. Compare **elliptical galaxy, irregular galaxy, lenticular galaxy.** See more at **Hubble classification system.**

spirillum (spī-rĭl′əm) Plural **spirilla.** Any of various bacteria that are shaped like a spiral, such as the spirochete *Treponema pallidum,* which causes syphilis.

spirochete (spī′rə-kēt′) Any of various bacteria of the order Spirochaetales that are shaped like a spiral, such as *Treponema pallidum,* the pathogen that causes syphilis.

spirogyra (spī′rə-jī′rə) Any of a genus of filamentous freshwater green algae having cylindrically shaped cells with spiral-shaped bands of chloroplasts. Species of *Spirogyra* reproduce asexually by cell division and fragmentation and sexually by conjugation. They form green scum on ponds, floated by the oxygen produced by photosynthesis.

spleen (splēn) An organ in vertebrate animals that in humans is located on the left side of the abdomen near the stomach. The spleen is mainly composed of lymph nodes and blood vessels. It filters the blood, stores red blood cells (erythrocytes) and destroys old ones, and produces white blood cells (lymphocytes).

splice (splīs) To join together genes or gene fragments or insert them into a cell or other structure, such as a virus, by means of enzymes. In genetic engineering, scientists splice together genetic material to produce new genes or to alter a genetic structure. In messenger RNA, the introns are removed, and exons are spliced together to yield the final messenger RNA that is translated. See also **exon, intron.**

sponge (spŭnj) **1.** Any of numerous aquatic, chiefly marine invertebrate animals of the phylum Porifera. Sponges characteristically have a porous skeleton, usually containing an intricate system of canals, that is composed of fibrous material or siliceous or calcareous spicules. Water passing through the pores brings food to the organism. Sponges live in all depths of the sea, are sessile, and often form irregularly shaped colonies attached to an underwater surface. Sponges are considered the most primitive members of the animal kingdom, since they lack a nervous system and differentiated body tissues or organs. Adults do not have moving parts, but the larvae are free-swimming. Sponges have great regenerative capacities, with some species able to regenerate a complete adult organism from fragments as small as a single cell. Sponges first appear during the early Cambrian Period and may have evolved from protozoa. Also called *poriferan.* See Note at **regeneration. 2.** The light, fibrous, flexible, absorbent skeleton of certain of these organisms, used for bathing, cleaning, and other purposes. **3.** A piece of porous plastic, rubber, cellulose, or other material, similar in absorbency to this skeleton and used for the same purposes.

spongy parenchyma (spŭn′jē) A layer of cells in the interior of leaves, consisting of loosely arranged, irregularly shaped cells that have chloroplasts. The spongy parenchyma has many spaces between cells to facilitate the circulation of air and the exchange of gases. It lies just below the **palisade layer**. Also called *spongy mesophyll.* See more at **photosynthesis.**

spontaneous abortion (spŏn-tā′nē-əs) See **miscarriage.**

spontaneous combustion The bursting into flame of a mass of material as a result of chemical reactions within the substance, without the addition of heat from an external source. Oily rags and damp hay, for example, are subject to spontaneous combustion.

spontaneous generation The supposed development of living organisms from nonliving matter, as maggots from rotting meat. The theory of spontaneous generation for larger organisms was easily shown to be false, but the theory was not fully discredited until the mid-19th century with the demonstration of the existence and reproduction of microorganisms, most notably by Louis Pasteur. Also called *abiogenesis.*

spool (spo͞ol) To store data that is sent to a device, such as a printer, in a buffer that the device reads. This procedure allows the program that sent the data to the device to resume its normal operation without waiting for the device to process the data.

sporangiophore (spə-răn′jē-ə-fôr′) A stalk or branch that bears sporangia.

sporangium (spə-răn′jē-əm) Plural **sporangia.** A cell or structure in which spores are produced. Ferns, fungi, mosses, and algae release spores from sporangia. Also called *spore case.*

spore (spôr) **1.** A usually one-celled reproductive body that can grow into a new organism without uniting with another cell. Spores are haploid (having only a single set of chromosomes). Fungi, algae, seedless plants, and certain protozoans reproduce asexually by spores. Plant spores that are dispersed by the wind have walls containing sporopollenin. See more at **alternation of generations. 2.** A similar one-celled body in seed-bearing plants; the macrospore or microspore. The **macrospore** of seed-bearing plants develops into a female gametophyte or **megagametophyte,** which is contained within the ovule and eventually produces the egg cells. (The megagametophyte is also called the embryo sac in angiosperms.) The **microspore** of seed-bearing plants develops into the male **microgametophyte** or pollen grain. **3.** See **endospore.**

spore case See **sporangium.**

sporogenesis (spôr′ə-jĕn′ĭ-sĭs) The formation or production of spores. Sporogenesis may result from mitosis or meiosis. Plant spores result from meiosis, whereas fungal spores may result from either mitosis or meiosis depending on the species.

sporogony (spə-rŏg′ə-nē) Reproduction by multiple fission of a spore or zygote, characteristic of many apicomplexans. Sporogony results in the production of sporozoites.

sporophyll (spôr′ə-fĭl′) A leaf or leaflike organ that bears sporangia.

sporophyte (spôr′ə-fīt′) Among organisms which display an **alternation of generations** (such as plants, fungi, and certain algae), the individual diploid organism that produces spores. A sporophyte develops from an embryo resulting from the union of two gametes. Each of its cells has two sets of chromosomes, as opposed to the haploid **gametophyte** generation. See more at **alternation of generations, gametophyte.**

sporopollenin (spôr′ə-pŏl′ə-nĭn) An organic polymer that is extremely resistant to degradation. Pollen grains and many kinds of spores have a protective outer coating of sporopollenin, which is so durable that microfossils of sporopollenin are found in rocks over 500 million years old.

sporozoan (spôr′ə-zō′ən) Any of numerous parasitic protozoans now classified as **apicomplexans**.

sporozoite (spôr′ə-zō′īt′) Any of the minute, undeveloped apicomplexans produced by multiple fission of a zygote or spore, especially at the stage just before infection of a new host cell.

spreading zone (sprĕd′ĭng) See **divergent plate boundary.**

spring (sprĭng) **1.** A device, such as a coil of wire, that returns to its original shape after being compressed or stretched. Because of their ability to return to their original shape, springs are used to store energy, as in mechanical clocks, and to absorb or lessen energy, as in the suspension system of vehicles. **2.** A small stream of water flowing naturally from the Earth.

spring tide A tide in which the difference between high and low tide is the greatest. Spring tides occur when the Moon is either new or full, and the Sun, the Moon, and the Earth are aligned. When this is the case, their collective gravitational pull on the Earth's water is strengthened. Compare **neap tide.** See more at **tide.**

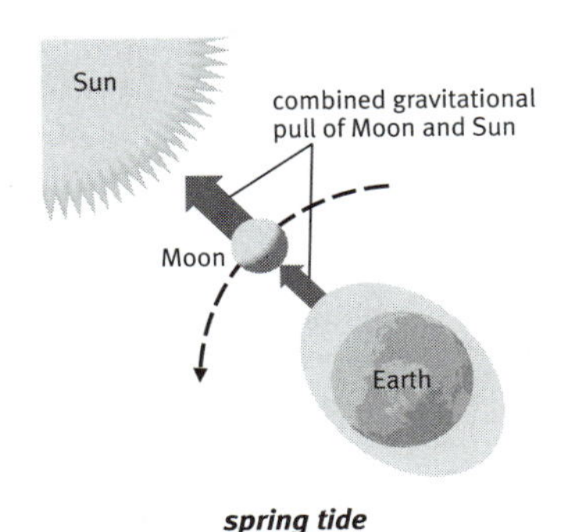

spring tide

Spring tides occur when the Sun and Moon are in line with the Earth and their gravitational pulls reinforce each other.

spur (spûr) **1.** A small ridge that projects sharply from the side of a larger hill or mountain. **2.** A projection from a bone, as on the heel of the foot.

squalene (skwā′lēn′) A colorless, unsaturated hydrocarbon found especially in the liver oil of sharks and in human sebum. It is an intermediate compound in the body's synthesis of cholesterol. *Chemical formula:* $C_{30}H_{50}$.

squall (skwôl) A brief, sudden, violent windstorm, often accompanied by rain or snow. A squall is said to occur if a wind having a sustained speed of 40 km (25 mi) per hour lasts at least 1 minute and then decreases rapidly. See also **squall line.**

squall line A line of sudden, sometimes violent thunderstorms that develop on the leading edge of a cold front. Squall lines can form up to 80 to 240 km (50 to 149 mi) in front of an advancing cold front and can be more than 160 km (99 mi) long. The thunderstorms of a squall line can produce severe weather conditions, such as hail and rain accompanied by winds of over 96 km (60 mi) per hour; they are also associated with tornadoes, especially in spring and early summer.

squamous cell carcinoma (skwā′məs, skwä′-) Any of various carcinomas that arise from a kind of flat, scaly epithelial cell, found in organs such as the skin, cervix, oral cavity, larynx, and vulva. Squamous cell carcinoma of the skin is associated with sun exposure.

square (skwâr) *Noun.* **1.** A rectangle having four equal sides. **2.** The product that results when a number or quantity is multiplied by itself. The square of 8, for example, is 64. —*Adjective.* **3.** Of, being, or using units that express the measure of area. —*Verb.* **4.** To multiply a number, quantity, or expression by itself.

square root A number that, when squared, yields a given number. For example, since $5 \times 5 = 25$, the square root of 25 (written $\sqrt{25}$) is 5.

Sr The symbol for **strontium.**

SSRI (ĕs′ĕs-är-ī′) Short for *selective serotonin reuptake inhibitor.* Any of a class of drugs that inhibit the uptake of serotonin in the central nervous system and are used to treat depression and other psychiatric disorders.

stable (stā′bəl) **1.** Not susceptible to a process of decay, such as radioactivity. For example, the most common isotope of carbon, carbon 12, is stable. Protons and photons are examples of stable subatomic particles. See more at **decay. 2.** Relating to a chemical compound that does not easily decompose or change into other compounds. Water is an example of a stable compound. **3.** Relating to an atom or chemical element that is unlikely to share electrons with another atom or element. **4.** Not likely to change significantly or to deteriorate suddenly, as an individual's medical condition.

stack (stăk) An isolated, columnar mass or island of rock along a coastal cliff. Stacks are formed by the erosion of cliffs through wave action and are larger than chimneys.

stainless steel (stān′lĭs) Any of various alloys of iron that contain chromium, nickel, and small amounts of carbon. They may also contain minor amounts of other elements, such as molybdenum. Stainless steel is resistant to rusting and corrosion.

stalactite (stə-lăk′tīt′) A cylindrical or conical mineral deposit projecting downward from the roof of a cave or cavern, formed by the dripping of water saturated with minerals. Stalactites form gradually as the minerals precipitate out of the saturated water. They usually consist of calcite but can also consist of other minerals. Compare **stalagmite.**

stalagmite (stə-lăg′mīt′) A cylindrical or conical mineral deposit, similar to a stalactite but built up from the floor of a cave or cavern. Stalagmites are typically broader than stalactites. The two formations are often, but not always, paired, and they sometimes join at a midpoint to form a pillar. Compare **stalactite.**

stalk (stôk) **1a.** The main stem of a plant. **b.** A slender structure that supports a plant part, such as a flower or leaf. **2.** A slender supporting structure in certain other organisms, such as the reproductive structure in plasmodial slime molds or the part of a mushroom below the cap. **3.** A slender supporting or connecting part of an animal, such as the eyestalk of a lobster.

stamen (stā′mən) Plural **stamens** or **stamina** (stā′mə-nə, stăm′ə-). The male reproductive organ of a flower, consisting of a filament and a pollen-bearing anther at its tip. See more at **anther, flower.**

staminate (stā′mə-nĭt) Having stamens but no carpels. Male flowers are staminate.

standard deviation (stăn′dərd) A statistic used as a measure of the dispersion or variation in a distribution, equal to the square root of the arithmetic mean of the squares of the deviations from the arithmetic mean.

standard error The standard deviations of the sample in a frequency distribution, obtained by dividing the standard deviation by the total number of cases in the frequency distribution.

standard model A theory of subatomic particles and their interactions. The standard model is a kind of **quantum field theory** and states that all matter consists of three types of particles: leptons, quarks, and the gauge bosons (gluons, intermediate vector bosons, and photons), which are responsible for the strong, weak, and electromagnetic forces, along with the more controversial Higgs boson. The standard model was first formulated in the 1970s and was tested repeatedly in the 1980s. It correctly predicted the existence and properties of W and Z bosons. It does not, however, unify all forces, since it does not include an explanation of gravity; a boson called the **graviton** that might be the mediator of gravity has not been found and cannot be accounted for in the model.

standard time The time in any of the 24 time zones into which the Earth's surface is divided, usually the mean time at the central meridian of the given zone. There are four standard time zones in the contiguous continental United States: Eastern, using the 75th meridian; Central, using the 90th meridian; Mountain, using the 105th meridian; and Pacific, using the 120th meridian. Alaska Standard Time, centered on the 135th meridian, is one hour behind Pacific time, and Hawaii Standard Time, centered on the 150th meridian, is one hour behind Alaska time. See more at **daylight-saving time, time zone.**

standing wave (stăn′dĭng) A wave that oscillates in place, without transmitting energy along its extent. Standing waves tend to have stable points, called nodes, where there is no oscillation. Examples of standing waves include the vibration of a violin string and electron orbitals in an atom. Also called *stationary wave.* See also **harmonic oscillator.**

stannic (stăn′ĭk) Containing tin, especially tin with a valence of 4. Compare **stannous.**

stannous (stăn′əs) Containing tin, especially tin with a valence of 2. Compare **stannic.**

stapes (stā′pēz) Plural **stapes** or **stapedes** (stā′pĭ-dēz′). The roughly stirrup-shaped bone that is the innermost of the three small bones (ossicles) of the middle ear.

staphylococcus (stăf′ə-lō-kŏk′əs) Plural **staphylococci** (stăf′ə-lō-kŏk′sī, -kŏk′ī). Any of various bacteria of the genus *Staphylococcus* that are gram-positive cocci and are normally found on the skin and mucous membranes of warm-blooded animals. Pathogenic strains such as *S. aureus* commonly cause infections of the skin, bones, lungs and other organs. Some staphylococcal disease, such as food poisoning, is caused by a toxin produced by the bacteria.

star (stär) **1.** A large, spherical celestial body consisting of a mass of gas that is hot enough to sustain nuclear fusion and thus produce radiant energy. Stars begin their life cycle as clouds of gas and dust called **nebulae** and develop, through gravitation and accretion, into increasingly hot and dense **protostars.** In order to reach the temperature at which nuclear reactions are ignited (about 5 million degrees K), a protostar must have at least 80 times the mass of Jupiter. For most of its life a star fuses hydrogen into helium in its core, during which period it is known as a **dwarf star** and is classed according to its surface temperature and luminosity (or spectral type) on a continuum called the **main sequence** in the Hertzsprung-Russell diagram. When a star exhausts the hydrogen in its core, it typically develops into one of several non-main-sequence forms depending on how massive it is. Smaller stars, with masses less than eight times that of the Sun, become **red giants** and end their lives, after blowing away their outer layers, as **white dwarfs.** More massive stars become **supergiants** and end their lives, after exploding in a supernova, as either a **neutron star** or a **black hole. 2.** Any of the celestial bodies visible to the naked eye at night as fixed, usually twinkling points of light, including binary and multiple star systems.

starburst (stär′bûrst′) The rapid formation of large numbers of new stars in a galaxy at a rate high enough to alter the structure of the galaxy significantly. Starburst galaxies, which are very luminous, form stars at rates that are between tens and hundreds of times faster than those of ordinary galaxies. Initial star formation is believed to be brought on by violent events such as collisions, or near collisions, with other galaxies, in which shockwaves cause the gases in the interstellar medium to collapse into **protostars.** The resulting stars are generally massive and short-lived. These stars become **supernovae** that create further shock waves, triggering yet more star formation.

starch (stärch) **1.** A carbohydrate that is the chief form of stored energy in plants, especially wheat, corn, rice, and potatoes. Starch is a mixture of two different polysaccharides built out of glucose units, and forms a white, tasteless powder when purified. It is an important source of nutrition and is also used to make adhesives, paper, and textiles. **2.** Any of various substances, including natural starch, used to stiffen fabrics.

starfish (stär′fĭsh′) Any of various marine echinoderms of the class Asteroidea, having a star-shaped body usually with five arms. The arms have rows of little suckers on the undersides, called tube feet, with which the animal moves around and grasps prey. Many species extrude their stomach onto prey and digest it externally. Starfish can grow new arms if any are lost, and in one species, a whole individual can be regenerated from a single piece of arm. Starfish are related to sea urchins and sea cucumbers.

Stark effect (stärk) The splitting of single spectral lines of an emission or absorption spectrum of a substance into several components when the substance is placed in an electric field. The effect occurs when several electron orbitals in the same shell, which normally have the same energy level, have different energies due to their different orientations in the electric field. Quantum mechanical predictions of this effect are extremely accurate, a fact that provided compelling early evidence for quantum mechanics. The Stark effect is named after its discoverer, German physicist Johannes Stark (1874–1957). Compare **Zeeman effect.**

state of matter (stāt) One of the four principal conditions in which matter exists—solid, liquid, gas, and plasma. See also **phase, phase transition.**

static (stăt′ĭk) *Adjective.* **1.** Having no motion; being at rest. Compare **dynamic. 2.** Relating to or producing static electricity. —*Noun.* **3.** Distortion or interruption of a broadcast signal, such as crackling or noise in a receiver or specks on a television screen, often produced when background electromagnetic radiation in the atmosphere disturbs signal reception or when there are loose connections in the transmission or reception circuits.

static electricity Electric charge that has accumulated on an object. Static electricity is often created when two objects that are not good electrical conductors are rubbed together, and electrons from one of the objects rub off onto the other. This happens, for example, when combing one's hair or taking off a sweater. Sudden releases of built-up static electricity can take the form of an **electric arc**. See Note at **electric charge.**

static friction See under **friction.**

static memory 1. Computer memory used by a piece of software that contains fixed information that is not altered while the software is running. **2.** See **static RAM.**

static pressure The pressure exerted by a liquid or gas, especially water or air, when the bodies on which the pressure is exerted are not in motion.

static RAM A type of RAM that stores data in transistor circuits. Static RAM is faster than dynamic RAM and does not need to be continuously refreshed. Because it is more expensive and holds less data than dynamic RAM, it is used primarily for cache memory.

statics (stăt′ĭks) The branch of physics that deals with physical systems in equilibrium, in which no bodies are in motion, and all forces are offset or counterbalanced by other forces.

statin (stăt′n) Any of a class of drugs that inhibit a key enzyme involved in the synthesis of cholesterol and promote receptor binding of LDL-cholesterol, resulting in decreased levels of serum cholesterol and LDL-cholesterol and increased levels of HDL-cholesterol.

stationary front (stā′shə-nĕr′ē) A transition zone between two nearly stationary air masses of different density. See more at **front.**

stationary wave See **standing wave.**

statistical mechanics (stə-tĭs′tĭ-kəl) The branch of physics that applies statistical principles to the mechanical behavior of large numbers of small particles (such as molecules, atoms, or subatomic particles) in order to explain the overall properties of the matter composed of such particles. The **kinetic theory** of heat is an example of statistical mechanics; the laws of thermodynamics can all be explained using statistical mechanics. Both classical physics and quantum mechanics have been used in the development of statistical mechanical theories. ▸ **Bose-Einstein statistics** explains the behavior of large numbers of bosons, which are particles that can simultaneously occupy the same quantum state (such as photons in a laser beam). ▸ **Fermi-Dirac statistics** explains the behavior of large numbers of particles that obey the Pauli exclusion principle (such as electrons) and cannot simultaneously occupy the same quantum state.

statistics (stə-tĭs′tĭks) **1.** *Used with a singular verb.* The branch of mathematics that deals with the collection, organization, analysis, and interpretation of numerical data. Statistics is especially useful in drawing general conclusions about a set of data from a sample of the data. **2.** *Used with a plural verb.* Numerical data.

statoscope (stăt′ə-skōp′) A barometer for measuring or recording very small variations in atmospheric pressure, often used as an instrument for indicating changes in the altitude of an aircraft.

statute mile (stăch′o͞ot) See **mile** (sense 1).

staurolite (stou′rə-līt′) A brownish to black orthorhombic mineral, often having crossed intergrown crystals. Staurolite is found in mica schists and gneisses. *Chemical formula:* $\mathbf{(FeMg)_2Al_9Si_4O_{23}(OH)}$.

STD Abbreviation of **sexually transmitted disease.**

steady state (stĕd′ē) A condition of a physical system or device that does not change over time, or in which any one change is continually balanced by another, such as the stable condition of a system in equilibrium.

steady state theory A cosmological theory stating that the universe has always expanded at a uniform rate with no beginning or end, that it will continue to expand and have constant density, and that the distribution of old and new objects in the universe is basically even. The theory has been largely abandoned in favor of the **big bang** theory, largely due to the discovery of **quasars** and other entities that appear only at very great distances, suggesting an absolute relationship between the age of objects and their distance. Steady state theory was also discredited by the discovery of **cosmic background radiation**, which was predicted by the big bang theory but not by the steady state theory.

steam (stēm) **1.** Water in its gaseous state, especially at a temperature above the boiling point of water (above 100°C, or 212°F, at sea level). See Note at **vapor. 2.** A mist of condensed water vapor.

steam engine An engine in which the energy of hot steam is converted into mechanical power, especially an engine in which the force of expanding steam is used to drive one or more pistons. The source of the steam is typically external to the part of the machine that converts the steam energy into mechanical energy. Compare **internal-combustion engine.**

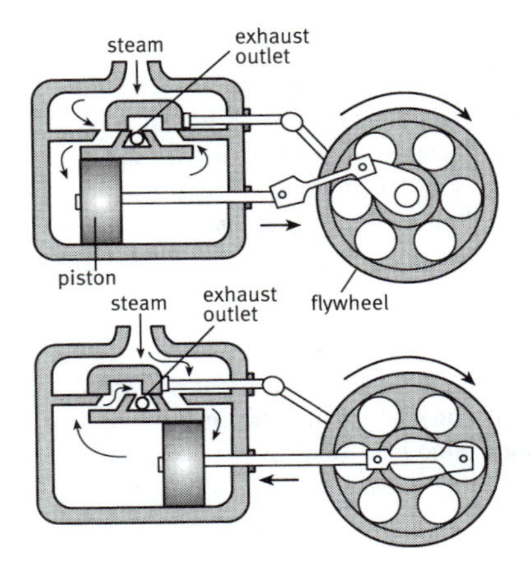

steam engine

Rightward (top) *and leftward* (bottom) *movements of a slide valve steam engine.*

stearate (stē′ə-rāt′, stîr′āt′) A salt or ester of stearic acid, containing the group $C_{17}H_{35}COO$.

stearic acid (stē-ăr′ĭk, stîr′ĭk) A colorless, odorless, waxlike fatty acid occurring in animal and vegetable fats and used in making soaps, candles, lubricants, and other products. *Chemical formula:* $\mathbf{C_{18}H_{36}O_2}$.

stearin (stē′ər-ĭn, stîr′ĭn) also **stearine** (stē′ər-ĭn, -ə-rēn′, stîr′ĭn) **1.** A colorless, odorless, tasteless ester of glycerol and stearic acid found in most animal and vegetable fats and used in the manufacture of soaps, candles, metal polishes, and adhesives. *Chemical formula:* $\mathbf{C_{57}H_{110}O_6}$. **2.** The solid form of fat.

steel (stēl) Any of various hard, strong, flexible alloys of iron and carbon. Often, other metals are added to give steel a particular property, such as chromium and nickel to make it stainless. Steel is widely used in many kinds of tools and as a structural material in building.

stegosaur (stĕg′ə-sôr′) Any of several large herbivorous ornithischian dinosaurs of the group Stegosauria of the Jurassic and early Cretaceous Periods. The largest genus, *Stegosaurus,* had a tail with two horizontal spikes for defense, and an arched back with an alternating double row of large, triangular, upright bony plates. Stegosaurs grew over 6 m (20 ft) long, but had extremely small heads with brains the size of a walnut. The hindquarters were controlled by a neural ganglion in the hip region that was larger than the brain.

stele (stēl, stē′lē) The central core of primary vascular tissues in the stem or root of a vascular plant, consisting of xylem and phloem together with pith.

stellar (stĕl′ər) Relating to or consisting of stars.

stellar wind (wĭnd) The flow of plasma (extremely hot gas composed of ions) ejected from the surface of a star into space. The Sun's **solar wind** is a stellar wind.

stem (stĕm) **1.** The main, often long or slender part of a plant that usually grows upward above the ground and supports other parts, such as branches and leaves. Plants have evolved a number of tissue arrangements in the stem. Seedless vascular plants (such as mosses and ferns) have primary vascular tissue in an inner core, a cylindrical ring, or individual strands scattered amid the ground tissue. In eudicots, magnoliids, and conifers, the stem develops a continuous cylindrical layer or a ring of separate bundles of vascular tissue (including secondary vascular tissue) embedded in the ground tissue. In monocots and some herbaceous eudicots, individual strands of primary vascular tissue are scattered in the ground tissue. **2.** A slender stalk supporting or connecting another plant part, such as a leaf or flower.

stem cell An unspecialized cell found in fetuses, embryos, and some adult body tissues that has the potential to develop into specialized cells or divide into other stem cells. Stem cells from fetuses or embryos can develop into any type of differentiated cells, while those found in mature tissues develop only into specific cells. Stem cells can potentially be used to replace tissue damaged or destroyed by disease or injury, but the use of embryonic stem cells for this purpose is controversial. Also called *progenitor cell.*

stem reptile See **cotylosaur.**

steppe (stĕp) A vast, semiarid grassland, as found in southeast Europe, Siberia, and central North America.

steradian (stĭ-rā′dē-ən) A unitless measure of solid angles. A solid angle projecting from the center of a sphere and cutting its surface has a measure of $\frac{s}{r^2}$ steradians, where s is the surface area of the sphere cut out by the solid angle, and r is the radius of the sphere. See also **radian.**

stereochemistry (stĕr′ē-ō-kĕm′ĭ-strē) The branch of chemistry that deals with the spatial arrangements of atoms in molecules and with the chemical and physical effects of these arrangements.

stereoisomer (stĕr′ē-ō-ī′sə-mər) Any of two or more isomers having the same linkages between the atoms but differing in the way these atoms are arranged in space. There are two types of stereoisomers, geometric isomers and enantiomers. Compare **structural isomer.**

stereoscope (stĕr′ē-ə-skōp′) An optical instrument through which two slightly different images (typically photographs) of the same scene are presented, one to each eye, providing an illusion of three dimensions. Modern **virtual reality** equipment often uses a stereoscope that presents animated, computer-generated images to the eyes, rather than photographic images. ► A **stereogram** is a single pair of photographic images used in a stereograph. See also **stereoscopic vision.**

stereoscopic vision (stĕr′ē-ə-skŏp′ĭk) See **binocular vision.**

sterile (stĕr′əl, stĕr′īl′) **1.** Not able to produce offspring, seeds, or fruit; unable to reproduce. **2.** Free from disease-causing microorganisms. —*Noun* **sterility** (stə-rĭl′ĭ-tē).

sterilization (stĕr′ə-lĭ-zā′shən) **1.** The procedure of destroying all microorganisms in or on a given environment, such as a surgical instrument, in order to prevent the spread of infection. This is usually done by using heat, radiation, or chemical agents. **2.** Any of various surgical procedures intended to eliminate the capacity to reproduce in humans or animals.

sterling silver (stûr′lĭng) An alloy that contains at least 92.5 percent silver and up to 7.5 percent copper or another metal, used to make jewelry and silverware.

sternum (stûr′nəm) A long, flat bone located in the center of the chest, serving as a support for the collarbone and ribs. Also called *breastbone.* See more at **skeleton.**

steroid (stĕr′oid′) **1.** Any of a large class of organic compounds having as a basis 17 carbon atoms arranged in four rings fused together. Steroids include many biologically important compounds, including cholesterol and other sterols, the sex hormones (such as testosterone and estrogen), bile acids, adrenal hormones, plant alkaloids, and certain forms of vitamins. **2.** Any of various hormones having the structure of a steroid that are made synthetically, especially for use in medicine. **3.** An anabolic steroid. *—Adjective* **steroidal.**

sterol (stîr′ôl′) Any of various alcohols having the structure of a steroid, usually with a hydroxyl group (OH) attached to the third carbon atom. Sterols are found in the tissues of animals, plants, fungi, and yeasts and include cholesterol.

Stevens (stē′vənz), **Nettie Maria** 1861–1912. American biologist who identified the role of X and Y chromosomes in determining the sex of an organism. Stevens studied the chromosomes of mealworm beetles, first establishing that chromosomes are inherited in pairs. She later showed that eggs fertilized by X-carrying sperm produced female offspring, while Y-carrying sperm produced male offspring. She extended this work to studies of sex determination in various plants and insects.

Nettie Stevens

stigma (stĭg′mə) The sticky tip of a flower pistil, on which pollen is deposited at the beginning of pollination. See more at **flower.**

stilb (stĭlb) The unit of illuminance in the centimeter-gram-second system, equal to one candela per square centimeter.

stilbene (stĭl′bēn′) A colorless or yellowish crystalline compound used in the manufacture of dyes and as a fabric brightener. *Chemical formula:* $C_{14}H_{12}$.

stimulant (stĭm′yə-lənt) An agent, especially a drug, that causes increased activity, especially of the nervous or cardiovascular systems. Caffeine is a commonly used stimulant.

stimulated emission (stĭm′yə-lā′tĭd) The emission of electromagnetic radiation in the form of photons of a given frequency, triggered by photons of the same frequency. For example, an excited atom, with an electron in an energy orbit higher than normal, releases a photon of a specific frequency when the electron drops back to a lower energy orbit; if this photon strikes another electron in the same high-energy orbit in another atom, another photon of the same frequency is released. The emitted photons and the triggering photons are always in phase, have the same polarization, and travel in the same direction. Also called *induced emission.* See also **population inversion.**

stimulus (stĭm′yə-ləs) Plural **stimuli** (stĭm′yə-lī′). *Physiology.* **1.** Something that can elicit or evoke a physiological response in a cell, a tissue, or an organism. A stimulus can be internal or external. Sense organs, such as the ear, and sensory receptors, such as those in the skin, are sensitive to external stimuli such as sound and touch. **2.** Something that has an impact or an effect on an organism so that its behavior is modified in a detectable way. See more at **classical conditioning**

stinger (stĭng′ər) A sharp stinging organ, such as that of a bee, scorpion, or stingray. Stingers usually inject venom.

stipe (stīp) A supporting stalk or stemlike structure, especially the stalk of a pistil, the petiole of a fern frond, or the stalk that supports the cap of a mushroom.

stipule (stĭp′yo͞ol) One of the usually small, paired parts resembling leaves at the base of a leafstalk in certain plants, such as roses and beans.

stochastic (stō-kăs′tĭk) **1.** Involving or containing a random variable or variables. **2.** Involving chance or probability.

stock (stŏk) **1.** The trunk or main stem of a tree or another plant. **2.** A plant or stem onto which a graft is made. **3.** A plant or tree from which cuttings and scions are taken.

stokes (stōks) Plural **stokes.** The unit of kinematic viscosity in the centimeter-gram-second system, measured in square centimeters per second. See more at **viscosity.**

Stokes, Sir **George Gabriel** 1819–1903. Irish mathematician and physicist who investigated the wave theory of light and described the phenomena of diffraction (1849) and fluorescence (1852) and the nature of x-rays. He also investigated fluid dynamics, developing the modern theory of motion of viscous fluids. A unit of kinematic viscosity is named for him.

stolon (stō′lŏn′) **1.** *Botany.* See **runner. 2.** *Zoology.* A stemlike structure of certain colonial organisms, such as hydroids, from which new individuals arise by budding.

stoma (stō′mə) Plural **stomata** (stō′mə-tə). **1.** *Botany.* One of the tiny openings in the epidermis of a plant, through which gases and water vapor pass. Stomata permit the absorption of carbon dioxide necessary for photosynthesis from the air, as well as the removal of excess oxygen. Stomata occur on all living plant parts that have contact with the air; they are especially abundant on leaves. A single leaf may have many thousands of stomata. Each stoma is generally between 10 to 30 microns in length and is surrounded by a pair of crescent-shaped cells, called **guard cells.** The guard cells can change shape and close the stoma in order to prevent the loss of water vapor. See Note at **transpiration. 2.** *Zoology.* A mouthlike opening, such as the oral cavity of a nematode. **3.** *Medicine.* A temporary or permanent opening in a body surface, especially the abdomen or throat, that is created by a surgical procedure, such as a colostomy or tracheostomy.

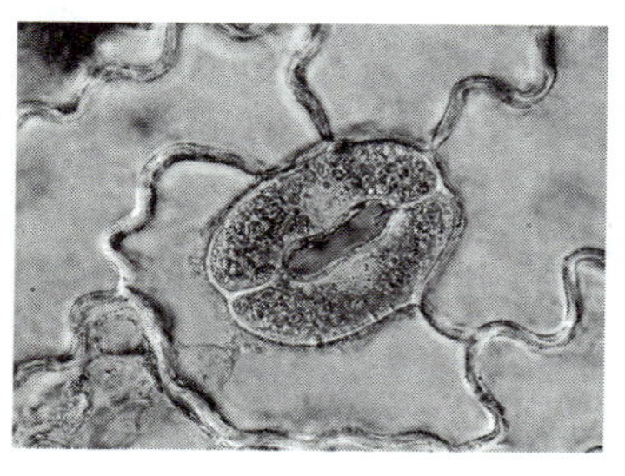

stoma

photomicrograph of a stoma on the frond of a polypod fern (Polypodium sp.)

stomach (stŭm′ək) **1.** A saclike muscular organ in vertebrate animals that stores and breaks down ingested food. Food enters the stomach from the esophagus and passes to the small intestine through the pylorus. Glands in the stomach secrete hydrochloric acid and the digestive enzyme pepsin. **2.** A similar digestive structure of many invertebrates. **3.** Any of the four compartments into which the stomach of a ruminant is divided (the rumen, reticulum, omasum, or abomasum).

stone (stōn) **1.** Rock, especially when used in construction. **2.** The hard, woody inner layer (the endocarp) of a drupe such as a cherry or peach. Not in scientific use. **3.** See **calculus** (sense 2).

Stone Age The earliest known period of human culture, marked by the use of stone tools. See **Mesolithic, Neolithic, Paleolithic.** See Note at **Three Age system.**

stony-iron meteorite (stō′nē-ī′ərn) A relatively rare type of meteorite consisting of approximately equal amounts by weight of silicate minerals and an alloy of nickel and iron. See more at **meteorite.**

stony meteorite See under **meteorite.**

stop codon (stŏp) Any of three codons in a molecule of messenger RNA that do not code for an amino acid and thereby signal the termination of the synthesis of a protein. The three stop codons are UAA, UAG, and UGA, where U is uracil, A is adenine, and G is guanine. Also called *termination codon.*

storage device (stôr′ĭj) A hardware device, such as a CD-ROM or hard disk in a computer, used to record and store data.

storm (stôrm) **1.** A low-pressure atmospheric disturbance resulting in strong winds accompanied by rain, snow, or other precipitation and often by thunder and lightning. **2.** A wind with a speed from 103 to 117 km (64 to 73 mi) per hour, rating 11 on the Beaufort scale.

storm cell An air mass that contains up and down drafts in convective loops, moves and reacts as a single entity, and functions as the smallest unit of a storm-producing system.

storm center The point of lowest barometric pressure within the central area of a storm.

storm surge See **tidal wave** (sense 2).

straight angle (strāt) An angle having a measure of 180°.

straight chain An arrangement of atoms of the same type that forms an unbranched open chain. Propane is a straight-chain compound with a chain of three carbon atoms.

strain (strān) **1.** A group of organisms of the same species, sharing certain hereditary characteristics not typical of the entire species but minor enough not to warrant classification as a separate breed or variety. Resistance to specific antibiotics is a feature of certain strains of bacteria. **2.** The extent to which a body is distorted when it is subjected to a deforming force, as when under stress. The distortion can involve a change both in shape and in size. All measures of strain are dimensionless (they have no unit of measure). ▸ **Axial strain** is equal to the ratio between the change in length of an object and its original length. ▸ **Volume strain** is equal to the ratio between the change in volume of an object and its original volume. It is also called *bulk strain.* ▸ **Shear strain** is equal to the ratio between the amount by which an object is skewed and its length. Compare **stress.** See more at **Hooke's law.**

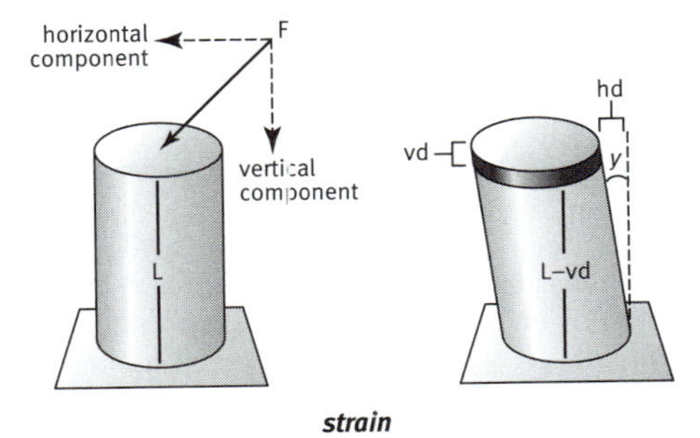

strain

The vertical component of the force F *applied to the metal rod causes a vertical deformation of the rod, or* axial strain, *shown as* vd *in the diagram on the right. The horizontal component of* F *causes a horizontal skewing, or* shear strain, *shown by* hd *in the diagram on the right. The axial strain is measured as* vd/L. *The shear strain is measured as* hd/L, *which is equal to the tangent of the angle y.*

strait (strāt) A narrow waterway joining two larger bodies of water. The Strait of Gibraltar, for example, connects the Mediterranean Sea with the Atlantic Ocean.

strange attractor (strānj) See under **attractor.**

strange matter A hypothetical, extremely dense form of matter composed entirely of **strange quarks**. These quarks are not grouped into separate particles (like the quarks bundled into protons and neutrons in atomic nuclei) but form a single undifferentiated mass. The material of which hypothetical **quark stars** are composed is thought to be strange matter; some evidence for strange matter has also been found in collisions in particle accelerators and in seismic events possibly triggered by the passing of strange matter through the Earth.

strangeness (strānj′nĭs) The property of containing a strange quark or antiquark. Strangeness is expressed in terms of an integer quantum number, –1 for each strange quark and +1 for each strange antiquark. Hadrons that possess strangeness are called **strange**. The total strangeness of a quantum system is unchanged by decay processes involving the strong or electromagnetic forces; however, decay through the weak force can change the total strangeness of the system. See also **baryon number, isospin.**

strange particle Any of various unstable elementary particles, having a short half-life and a nonzero **strangeness** quantum number. Sigma and xi baryons, for example, are strange particles.

strange quark A quark with an electric charge of $-\frac{1}{3}$ and a mass of 391 electron masses, greater than that of the up quark and down quark, but smaller than that of other quarks. The presence of a strange quark contributes strangeness of –1 to the physical system containing it, while a strange antiquark contributes strangeness of +1. See Table at **subatomic particle.**

strange star See **quark star.**

stratification (străt′ə-fĭ-kā′shən) Formation or deposition of layers, as of rock, sediments, or atmospheric regions.

stratigraphy (strə-tĭg′rə-fē) The scientific study of rock strata, especially the distribution, deposition, correlation, and age of sedimentary rocks.

stratocumulus (străt′ō-kyo͞om′yə-ləs) Plural **stratocumuli** (străt′ō-kyo͞om′yə-lī′). A low-altitude, often patchy cloud occurring in extensive horizontal layers with distinct, rounded tops. See illustration at **cloud.**

stratopause (străt′ə-pôz′) The boundary between the stratosphere and the mesosphere, located at an altitude of about 50 km (31 mi) above the Earth's surface.

stratosphere (străt′ə-sfîr′) The region of the Earth's atmosphere extending from the tropopause to about 50 km (31 mi) above the Earth's surface. The stratosphere is characterized by the presence of ozone gas (in the **ozone layer**) and by temperatures which rise slightly with altitude, due to the absorption of ultraviolet radiation. See also **exosphere, mesosphere, thermosphere, troposphere.** See illustration at **atmosphere.**

stratovolcano (străt′ō-vŏl-kā′nō) See under **volcano.**

stratum (strā′təm, străt′əm) Plural **strata** or **stratums. 1.** A layer of sedimentary rock whose composition is more or less the same throughout and that is visibly different from the rock layers above and below it. **2.** A layer of tissue, as of the skin or another organ.

stratus (străt′əs, strā′təs) Plural **strati** (străt′ī, strā′tī). A diffuse, grayish cloud that often produces drizzle and is formed primarily in altitudes no higher than 2,000 m (6,560 ft). A stratus cloud close to the ground or water is called fog. See illustration at **cloud.**

streak (strēk) **1.** The characteristic color of a mineral after it has been ground into a powder. Because the streak of a mineral is not always the same as its natural color, it is a useful tool in mineral identification. **2.** A bacterial culture inoculated by drawing a bacteria-laden needle across the surface of a solid culture medium. Also called *streak plate.* **3.** Any of various viral diseases of plants characterized by the appearance of discolored stripes on the leaves or stems.

streak plate 1. An unglazed piece of porcelain, such as a tile, used to test the characteristic streak of minerals by rubbing the mineral across the tile. Streak plates have a hardness of about 6.5 on the Mohs scale and cannot be used for testing harder minerals. **2.** See **streak** (sense 2).

stream (strēm) **1.** A flow of water in a channel or bed, as a brook, rivulet, or small river. **2.** A flow of a watery substance, such as blood in blood vessels or cytoplasm in fungal hyphae, in an organism or in part of an organism.

streaming (strē′mĭng) Relating to information that is transmitted in real time over the Internet, instead of being sent first as a file and then opened after it has been downloaded.

streamline (strēm′līn′) To construct or reconstruct an object to reduce the amount of **drag** it undergoes as it moves through a fluid, especially air or water.

strep throat (strĕp) Infection of the throat caused by the bacterium *Streptococcus pyogenes.* Symptoms usually include fever, redness of the throat, lymph node enlargement, and inflammation of the tonsils.

streptococcus (strĕp′tə-kŏk′əs) Plural **streptococci** (strĕp′tə-kŏk′sī, -kŏk′ī). Any of various bacteria of the genus *Streptococcus* that are gram-positive cocci and are normally found on the skin and mucous membranes and in the digestive tract of mammals. One type of streptococcus, Group A, is a common pathogen in humans and causes various infections, including strep throat, scarlet fever, pneumonia, and some types of impetigo.

streptomycin (strĕp′tə-mī′sĭn) An aminoglycoside antibiotic, $C_{21}H_{39}O_{12}N_7$, produced by the actinomycete *Streptomyces griseus,* given as an intramuscular injection to treat tuberculosis and other bacterial infections.

stress (strĕs) **1.** The force per unit area applied to an object. Objects subject to stress tend to become distorted or deformed. Compare **strain.** See also **axial stress, shear stress.** See more at **Hooke's law. 2a.** A physiologic reaction by an organism to an uncomfortable or unfamiliar physical or psychological stimulus. Biological changes result from stimulation of the sympathetic nervous system, including a heightened state of alertness, anxiety, increased heart rate, and sweating. **b.** The stimulus or circumstance causing such a reaction.

stress fracture See under **fracture.**

striation (strī-ā′shən) One of multiple, usually parallel grooves or scratches on a rock surface, produced by abrasion associated with glacial movement, stream flow, a geologic fault, or meteoric impact. See more at **glacial striation, shatter cone, slickenside.**

strike (strīk) The course or bearing of a structural surface, such as an inclined bed or a fault plane, as it intersects a horizontal plane. See illustration at **dip.**

strike-slip fault A geologic fault in which the blocks of rock on either side of the fault slide horizontally, parallel to the strike of the fault. See Note and illustration at **fault.**

string theory (strĭng) Any of various theories in physics hypothesizing that **space-time** has more than four dimensions, and that some of the dimensions are exceedingly small and stringlike in shape. Elementary particles in string theory are understood as standing waves in such space-time strings, rather than as pointlike objects. String theories attempt to unify gravity with the other **fundamental forces**.

strip mine (strĭp) An open mine, especially a coal mine, whose seams or outcrops run close to ground level and are exposed by the removal of overlying soil and rock. Strip mining can extract coal covered by as much as 60 m (200 ft) of rock and soil. If the seams roll through hills instead of lying flat, a series of tiers or contours are used to extract the coal or ore. Strip mining has been criticized for being ecologically destructive and for causing pollution of water resources, as the removed soil and rock are often dumped in lower-lying areas. Mining operators are sometimes required to restore soil and vegetation and to clean up the mining site.

strobe (strōb) **1.** A strobe light. **2.** A stroboscope. **3.** A spot of higher than normal intensity in the sweep of an indicator on a scanning device, as on a radar screen, used as a reference mark for determining the position or distance of the object scanned or detected.

strobe light A lamp that produces very short, intense flashes of light by means of an electric discharge in a gas. The ability of strobe lights to "freeze" the motion of rapidly moving objects by making them visible for only a fraction of a second makes them very useful in photography and in measuring vibration and other types of high-speed motion.

strobilus (strō-bī′ləs) Plural **strobili** (strō-bī′lī). **1.** A reproductive structure that consists of sporophylls or scales arranged spirally or in an overlapping fashion along a central stem, as in horsetails, some lycophytes, and many kinds of gymnosperms. For example, the **cones** of pine trees are strobili. **2.** An inflorescence or ripened multiple fruit in which many floral bracts are arranged in an overlapping fashion, as in the hop plant.

stroboscope (strō′bə-skōp′) Any of various instruments used to observe moving objects by making them appear stationary, especially with pulsed illumination or mechanical devices that intermittently interrupt observation.

stroke (strōk) A sudden loss of brain function caused by a blockage or rupture of a blood vessel of the brain, resulting in necrosis of brain tissue (called a cerebral **infarct**) and characterized by loss of muscular control, weakening or loss of sensation or consciousness, dizziness, slurred speech, or other symptoms that vary with the extent and severity of brain damage. Also called *cerebrovascular accident.*

stromatolite (strō-măt′l-īt′) A dome-shaped structure consisting of alternating layers of carbonate or silicate sediment and fossilized algal mats. Stromatolites are produced over geologic time by the trapping, binding, or precipitating of sediment by groups of microorganisms, primarily cyanobacteria. They are widely distributed in the fossil record and contain some of the oldest recorded forms of life, from over three billion years ago. They continue to form today especially in western Australia.

strong force (strông) The fundamental force that mediates interactions between particles with **color charge**, such as quarks and gluons. The strong force binds quarks together to form baryons such as protons and neutrons, maintains the binding of protons and neutrons together in atomic nuclei, and is responsible for many particle decay processes. Particles that interact through the strong force exchange gluons, much as particles involved in electromagnetic interactions exchange photons. Quark color, but not flavor, is changed by the exchange of gluons. The strong force is stronger than the weak force, the electromagnetic force, and gravity, but has been known to apply only across distances the size of atomic nuclei or smaller. Also called *color force, strong interaction, strong nuclear force.*

strong interaction See **strong nuclear force.**

strong nuclear force See **strong force.**

strontium (strŏn′chē-əm, -tē-əm) *Symbol* **Sr** A soft, silvery metallic element of the alkaline-earth group that occurs naturally only as a sulfate or carbonate. One of its isotopes is used in the radiometric dating of rocks. Because strontium salts burn with a red flame, they are used to make fireworks and signal flares. Atomic number 38; atomic weight 87.62; melting point 769°C; boiling point 1,384°C; specific gravity 2.54; valence 2. See **Periodic Table.**

strontium 90 A radioactive isotope of strontium having a mass number of 90 and a half-life of 28 years. Strontium 90 is the most dangerous component of the fallout from nuclear explosions because it is easily absorbed by the body. It is also used in medicine to treat cancer.

strophoid (strō′foid′) The curve traced out by points *P* and *P′* which lie on lines through a fixed point *A* where the midpoint *M* of *PP′* is on a fixed line and the absolute value |*PM*| equals the absolute value |*P′M*| equals the absolute value |*MO*|, where *O* is the projection of *A* onto that fixed line.

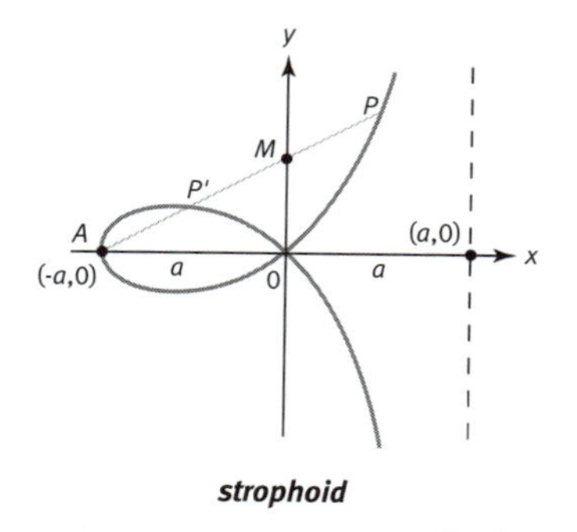

strophoid

graph of the equation $y^2 = \frac{x^2(a+x)}{(a-x)}$

structural formula (strŭk′chər-əl) A chemical formula that shows how the atoms making up a compound are arranged within the molecule. For example, the structural formula of aspirin is $CH_3COOC_6H_4COOH$, indicating that it consists of an acetyl group (CH_3COO) attached to the carboxylic acid (COOH) of a phenyl group (C_6H_4). Compare **empirical formula, molecular formula.**

structural isomer Any of two or more compounds with identical chemical formulas, such as propyl alcohol and isopropyl alcohol, that differ structurally in the sequence in which the atoms are linked. Structural isomers do not exhibit the same chemical behavior. Compare **stereoisomer.**

strychnine (strĭk′nīn′) An extremely poisonous, white crystalline compound derived from the seeds of the nux vomica tree. Strychnine is an alkaloid and was formerly used in medicine to stimulate the nervous system. It is currently used as a rat poison. *Chemical formula:* $C_{21}H_{22}O_2N_2$.

stutter (stŭt′ər) A speech disorder characterized by spasmodic repetition of the initial consonant or syllable of words and frequent pauses or prolongation of sounds.

style (stīl) The slender part of a flower pistil, extending from the ovary to the stigma. The pollen tube grows through the style delivering the pollen nuclei to the ovary. See more at **flower, pollination.**

stylolite (stī′lə-līt′) A contact surface usually found between two calcareous rock layers, marked by a series of jagged interlocking up-and-down projections that resemble a suture or the tracing of a stylus. Stylolites are found in rocks that have been subjected to pressure and are formed by a process, called pressure dissolution, whereby insoluble material, such as clay and hydrocarbons, becomes concentrated along a surface while other more soluble material, such as calcite, is carried away.

styrene (stī′rēn′) A colorless, oily aromatic hydrocarbon that readily undergoes polymerization. It is used in making polystyrene, polyesters, synthetic rubber, and other products. *Chemical formula:* C_8H_8.

sub– A prefix that means "underneath or lower" (as in *subsoil*), "a subordinate or secondary part of something else" (as in *subphylum.*), or "less than completely" (as in *subtropical.*)

subantarctic (sŭb′ănt-ärk′tĭk, -är′tĭk) Relating to the geographic area just north of the Antarctic Circle. The subantarctic region is the coldest part of the South Temperate Zone.

subaqueous (sŭb-ā′kwē-əs) **1.** Found or occurring underwater. **2.** Formed or adapted for underwater use or operation.

subarctic (sŭb′-ärk′tĭk, -är′tĭk) Relating to the geographic area just south of the Arctic Circle.

The subarctic region is the coldest part of the North Temperate Zone, characterized by warm but very brief summers, and bitterly cold winters. Little vegetation exists in this climate, as temperatures are extreme, ranging from below –30°C (–22°F) in the winter and as warm as 30°C (86°F) in the summer.

subarid (sŭb-ăr′ĭd) Somewhat arid; moderately dry.

subatomic particle (sŭb′ə-tŏm′ĭk) Any of various particles of matter that are smaller than a hydrogen atom. Protons, neutrons, and electrons are subatomic particles, as are all hadrons and leptons. See Table on page 602. See also **composite particle, elementary particle.**

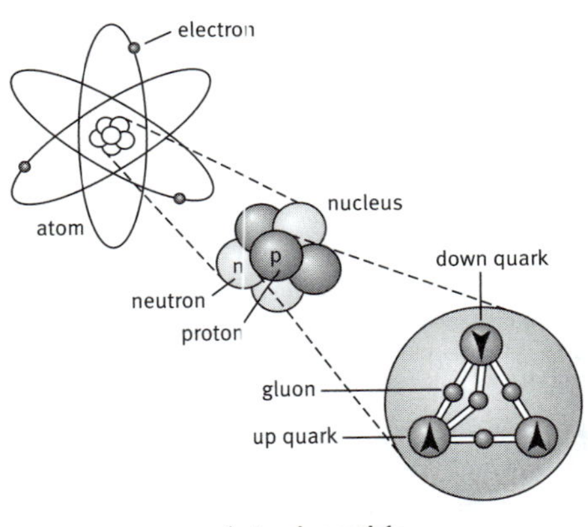

subatomic particle

anatomy of a lithium atom

subclass (sŭb′klăs′) A taxonomic category of related organisms ranking below a class and containing one or more orders.

subclimax (sŭb-klī′măks′) A stage in the ecological succession of a plant or animal community immediately preceding a climax, often persisting because of the repeated effects of fire, flood, or other conditions. See also **climax community, succession.**

subcutaneous (sŭb′kyo͞o-tā′nē-əs) Located or placed just beneath the skin.

subduction (səb-dŭk′shən) A geologic process in which one edge of one lithospheric plate is forced below the edge of another. The denser of the two plates sinks beneath the other. As it descends, the plate often generates seismic and volcanic activity (from melting and upward migration of magma) in the overriding plate. Compare **obduction.** —*Verb* **subduct.**

subduction zone A convergent plate boundary where one plate subducts beneath the other, usually because it is denser. The western coast of South America is roughly coincident with a subduction zone in which a plate consisting of ocean floor is subducting beneath the continental mass of South America.

suberin (so͞o′bər-ĭn) A polyester composed of fatty acids and aromatic compounds that occurs naturally in the cell walls of cork tissue in plants. Suberin acts together with waxes to protect plant surfaces from water loss and microbial attack, and helps to close tears.

subfamily (sŭb′făm′ə-lē) A subdivision of a family of organisms, containing one or more genera. The names of subfamilies in the animal kingdom typically end in *–inae,* as in Cerambycinae, a subfamily of the longhorn beetle family Cerambycidae.

subkingdom (sŭb′kĭng′dəm) A subdivision of a kingdom of organisms. A subkingdom contains one or more phyla.

sublimation (sŭb′lə-mā′shən) The process of changing from a solid to a gas without passing through an intermediate liquid phase. Carbon dioxide, at a pressure of one atmosphere, sublimates at about –78 degrees Celsius. Ice and snow on the Earth's surface also sublimate at temperatures below the freezing point of water. Compare **deposition.**

sublittoral (sŭb-lĭt′ər-əl) **1.** Relating to the region of the ocean bottom between the low tide line and the edge of the continental shelf, ranging in depth to about 200 m (656 ft). Unlike areas of the littoral zone, the sublittoral zone is always submerged. Compare **littoral. 2.** Relating to the deeper part of a lake below the area in which rooted plants grow.

suborder (sŭb′ôr′dər) A taxonomic category of related organisms ranking below an order and containing one or more families.

subphylum (sŭb′fī′ləm) Plural **subphyla.** A subdivision of a phylum of organisms. A subphylum contains one or more classes.

subpixel (sŭb′pĭk′səl) See under **pixel.**

subset (sŭb′sĕt′) A set whose members are all contained in another set. The set of positive

Subatomic Particles

ELEMENTARY PARTICLES

The elementary particles are those subatomic particles that are not made up of smaller units. They include the quarks, the leptons, and the gauge bosons. The antiparticles of the quarks and leptons are also elementary particles. Each of these elementary particles interacts with other elementary particles through one or more forces. These forces are mediated (carried) by the gauge bosons. The photon mediates the electromagnetic force, the gluon mediates the strong nuclear force, the intermediate vector bosons mediate the weak nuclear force, and the graviton (if it exists) mediates gravity.

Family Name	Particle Name	Approximate Mass*	Electric Charge
Quarks	up (symbol u**)	10	$+\frac{2}{3}$
	down (symbol d)	20	$-\frac{1}{3}$
	charm (symbol c)	2,935	$+\frac{2}{3}$
	strange (symbol s)	391	$-\frac{1}{3}$
	top (symbol t)	352,000	$+\frac{2}{3}$
	bottom (symbol b)	9,200	$-\frac{1}{3}$
Leptons	electron	1	–1
	electron neutrino	$<3 \times 10^{-5}$	0
	muon	207	–1
	muon neutrino	<0.33	0
	tau	3,480	–1
	tau neutrino	<59	0
Gauge Bosons	photon	0	0
	graviton	0	0
	gluon	0	0
	W boson	160,000	–1 *or* +1
	Z boson	182,000	0

*Masses are expressed as multiples of an electron's mass, 9.1066×10^{-28} g, and are approximate.

**The antiparticles of the quarks are denoted by the same symbols as the quarks but with a bar over the symbols, as in $\bar{u}$.

SUBATOMIC PARTICLES THAT ARE NOT ELEMENTARY PARTICLES

Family Name	Particle Name	Composition	Approximate Mass*	Electric Charge
Baryons	proton	uud	1,836	+1
	neutron	udd	1,839	0
	lambda	uds	2,183	0
	lambda-b	udb	11,000	0
	lambda-c	udc	4,471	+1
	sigma	uus *or* $(ud \pm du)s/\sqrt{2}$ *or* dds	2,328 *or* 2,334 *or* 2,343	+1 *or* 0 *or* –1
	xi	uss *or* dss	2,572 *or* 2,585	0 *or* –1
	xi-c	dsc *or* usc	4,834 *or* 4,826	0 *or* +1
	omega	sss	3,272	–1
	omega-c	ssc	5,292	0
Mesons	pion	$u\bar{d}$ *or* $(u\bar{u}-d\bar{d})/\sqrt{2}$	273 *or* 264	+1 *or* 0
	kaon	$u\bar{s}$ *or* $d\bar{s}$	966 *or* 974	+1 *or* 0
	J/psi	$c\bar{c}$	6,060	0
	omega	$(u\bar{u}+d\bar{d})/\sqrt{2}$	1,532	0
	eta	$(u\bar{u}+d\bar{d})/\sqrt{2}$	1,071	0
	eta-c	$c\bar{c}$	5,832	0
	B	$d\bar{b}$ *or* $u\bar{b}$	10,331 *or* 10,331	0 *or* +1
	B-s	$s\bar{b}$	10,507	0
	D	$c\bar{u}$ *or* $c\bar{d}$	3,649 *or* 3,658	0 *or* +1
	D-s	$c\bar{s}$	3,852	+1
	chi	$c\bar{c}$	6,687	0
	psi′	$c\bar{c}$	7,213	0
	upsilon	$b\bar{b}$	18,513	0

integers, for example, is a subset of the set of integers.

subshell (sŭb′shĕl′) One or more orbitals in the electron shell of an atom with the same energy level. Subshells have different shapes and are distinguished by their magnetic quantum number. See more at **orbital, quantum number.**

subsidiary cell (səb-sĭd′ē-ĕr′ē) A plant epidermal cell that is located next to a guard cell in the stoma of a leaf and differs in structure from other epidermal cells. Also called *accessory cell.*

subsoil (sŭb′soil′) In an ABC soil, the B horizon. The term was formerly used to mean the layer of earth below the humus or surface soil.

subsonic (sŭb-sŏn′ĭk) Having a speed less than that of sound in a designated medium, usually air; having a velocity less than Mach 1. Compare **hypersonic, supersonic, transonic.**

subspecies (sŭb′spē′shēz, -sēz) A subdivision of a species of organisms, usually based on geographic distribution. The subspecies name is written in lowercase italics following the species name. For example, *Gorilla gorilla gorilla* is the western lowland gorilla, and *Gorilla gorilla graueri* is the eastern lowland gorilla.

substrate (sŭb′strāt′) **1.** The material or substance on which an enzyme acts. See more at **enzyme. 2.** The surface on or in which plants, algae, or certain animals, such as barnacles or clams, live or grow. A substrate may serve as a source of food for an organism or simply provide support.

substratum (sŭb′strā′təm, -străt′əm) Plural **substrata** or **substratums. 1.** An underlying layer or stratum. **2.** A surface on which an organism grows or is attached; a substrate.

subtemperate (sŭb-tĕm′pər-ĭt) Relating to the colder regions of the Temperate Zones.

subtorrid (sŭb-tôr′ĭd) Subtropical.

subtraction (səb-trăk′shən) The operation of finding the difference between two numbers or quantities.

subtractive (səb-trăk′tĭv) **1.** Relating to the production of color by the blocking or removal of varying wavelengths, as with colored filters, or by the mixing of pigments that absorb certain wavelengths and reflect others. ▸ The **subtractive primaries** cyan, magenta, and yellow are those colors whose wavelengths can be filtered or absorbed in different proportions to produce all other colors. Compare **additive.** See Note at **color. 2.** Marked by or involving subtraction.

subtrahend (sŭb′trə-hĕnd′) A number subtracted from another. For example, in the expression $4 - 3$, 3 is the subtrahend.

subtropical (sŭb-trŏp′ĭ-kəl) Relating to the regions of the Earth bordering on the tropics, just north of the Tropic of Cancer or just south of the Tropic of Capricorn. Subtropical regions are the warmest parts of the two Temperate Zones.

succession (sək-sĕsh′ən) The gradual replacement of one type of ecological community by another in the same area, involving a series of orderly changes, especially in the dominant vegetation. Succession is usually initiated by a significant disturbance of an existing community. Each succeeding community modifies the physical environment, as by introducing shade or changing the fertility or acidity of the soil, creating new conditions that benefit certain species and inhibit others until a climax community is established. ▸ The sequential development of plant and animal communities in an area in which no topsoil exists, as on a new lava flow, is called **primary succession.** ▸ The development of such communities in an area that has been disturbed but still retains its topsoil, as in a burned-over area, is called **secondary succession.** See more at **climax community.**

succinate (sŭk′sə-nāt′) A salt or ester of succinic acid.

succinic acid (sək-sĭn′ĭk) A colorless, crystalline organic acid that is important in the Krebs cycle and occurs naturally in amber. It is synthesized for use in pharmaceuticals and perfumes. *Chemical formula:* $C_4H_6O_4$.

succulent (sŭk′yə-lənt) Any of various plants having fleshy leaves or stems that store water. Cacti and the jade plant are succulents. Succulents are usually adapted to drier environments and display other characteristics that reduce water loss, such as waxy coatings on leaves and stems, fewer stomata than occur on other plants, and stout, rounded stems that minimize surface area.

sucker (sŭk′ər) **1.** A part by which an animal sucks blood from or uses suction to cling to another animal. Leeches and remoras have suckers. **2.** A shoot growing from the base or root of a tree or shrub and giving rise to a new plant, a clone of the plant from which it comes. The growth of suckers is a form of asexual reproduction.

sucrose (so͞o′krōs′) A crystalline sugar found in many plants, especially sugar cane, sugar beets, and sugar maple. It is used widely as a sweetener. Sucrose is a disaccharide composed of fructose and glucose. Also called *table sugar. Chemical formula:* $C_{12}H_{22}O_{11}$.

suction (sŭk′shən) **1.** A force acting on a fluid caused by difference in pressure between two regions, tending to make the fluid flow from the region of higher pressure to the region of lower pressure. **2.** The act of reducing pressure to create such a force, as by the use of a pump or fan.

sudden infant death syndrome (sŭd′n) A fatal condition that affects sleeping infants that are less than one year old and appear to be healthy. It is characterized by a sudden cessation of breathing and is thought to be caused by a defect in the central nervous system. Also called *crib death, SIDS.*

Suess (zo͞os), **Eduard** 1831–1914. Austrian geologist who studied the evolution of the features of the Earth's surface, particularly the formation of mountains. His concept of the early supercontinent Gondwanaland was a forerunner of modern theories of continental drift.

sugar (sho͝og′ər) Any of a class of crystalline carbohydrates that are water-soluble, have a characteristic sweet taste, and are universally present in animals and plants. They are characterized by the many OH groups they contain. Sugars are monosaccharides or small oligosaccharides, and include sucrose, glucose, and lactose.

sulfa drug (sŭl′fə) See under **sulfonamide.**

sulfate (sŭl′fāt′) A salt or ester of sulfuric acid, containing the group SO_4.

sulfide (sŭl′fīd′) A chemical compound of sulfur and another element or radical, such as hydrogen sulfide.

sulfite (sŭl′fīt′) A salt or ester of sulfurous acid, containing the group SO_3.

sulfonamide (sŭl-fŏn′ə-mīd′, -mĭd) Any of a group of organic sulfur compounds containing the radical SO_2NH_2. ► Certain sulfonamides known as **sulfa drugs** are used as antibiotics to treat bacterial infections.

sulfonate (sŭl′fə-nāt′) A salt or ester of sulfonic acid, containing the group SO_3H.

sulfone (sŭl′fōn′) Any of various organic sulfur compounds having a sulfonyl group (SO_2) attached to two carbon atoms. Sulfones were formerly used to treat leprosy and tuberculosis, and are now used in making synthetic resins for wire coatings, automotive parts, and plumbing items.

sulfonic acid (sŭl-fŏn′ĭk) Any of several acids containing the group SO_3H attached to an organic radical. Sulfonic acids and their salts are important in industry as catalysts in organic synthesis and in making sulfonamide drugs, detergents, and dyes.

sulfonyl (sŭl′fə-nĭl′) The radical SO_2. Also called *sulfuryl.*

sulfur also **sulphur** (sŭl′fər) *Symbol* **S** A pale-yellow, brittle nonmetallic element that occurs widely in nature, especially in volcanic deposits, minerals, natural gas, and petroleum. It is used to make gunpowder and fertilizer, to vulcanize rubber, and to produce sulfuric acid. Atomic number 16; atomic weight 32.066; melting point (rhombic) 112.8°C; (monoclinic) 119.0°C; boiling point 444.6°C; specific gravity (rhombic) 2.07; (monoclinic) 1.957; valence 2, 4, 6. See **Periodic Table.**

sulfur dioxide A colorless, poisonous gas or liquid with a strong odor. It is formed naturally by volcanic activity, and is a waste gas produced by burning coal and oil and by many industrial processes, such as smelting. It is also a hazardous air pollutant and a major contributor to acid rain. *Chemical formula:* SO_2.

sulfuric (sŭl-fyo͝or′ĭk) Containing sulfur, especially sulfur with a valence of 6. Compare **sulfurous.**

sulfuric acid A strong corrosive acid. It combines very easily with water, making it a good drying agent. Sulfuric acid is the most widely used acid in industry. It is used to make detergents, dyes, drugs, explosives, pigments, fertilizers, and many other products. It is also the acid in lead-acid electric batteries. *Chemical formula:* H_2SO_4.

sulfurous (sŭl′fər-əs, sŭl-fyo͝or′əs) **1.** Containing sulfur, especially sulfur with a valence of 4 or 3. Compare **sulfuric. 2.** Characteristic of or emanating from burning sulfur.

sulfurous acid A colorless solution of sulfur dioxide in water, characterized by a suffocating sulfurous odor. It is used as a bleaching agent, preservative, and disinfectant. *Chemical formula:* H_2SO_3.

sulfur trioxide A colorless, highly reactive, corrosive compound that exists either as a liquid or as a solid. As a solid, it exists in three forms that may coexist in a given sample. Sulfur trioxide is used as an intermediate in manufacturing sulfuric acid, and in making sulfonates from organic compounds. *Chemical formula:* SO_3.

sum (sŭm) The result of adding numbers or quantities. The sum of 6 and 9, for example, is 15, and the sum of $4x$ and $5x$ is $9x$.

summer solstice (sŭm′ər) See under **solstice.**

sun (sŭn) Often **Sun.** A medium-sized, main-sequence star located in a spiral arm of the Milky Way galaxy, orbited by all of the planets and other bodies in our solar system and supplying the heat and light that sustain life on Earth. Its diameter is approximately 1,392,000 km (865,000 mi), and its mass, about 330,000 times that of Earth, comprises more than 99 percent of the matter in the solar system. It has a temperature of some 15.7 million degrees C (28.3 million degrees F) at its core, where nuclear fusion produces tremendous amounts of energy, mainly through the series of reactions known as the **proton-proton chain**. The energy generated in the core radiates through a radiation zone to an opaque convection zone, where it rises to the surface through convection currents of the Sun's plasma. The Sun's surface temperature (at its **photosphere**) is approximately 6,200 degrees C (11,200 degrees F). Turbulent surface phenomena analogous to the Earth's weather are prevalent, including magnetic storms, sunspots, and solar flares. The Sun was formed along with the rest of the solar system about 4.5 billion years ago and is expected to run out of its current hydrogen fuel in another 5 billion years, at which point it will develop into a red giant and ultimately into a white dwarf. See Table at **solar system.** See Note at **dwarf star.**

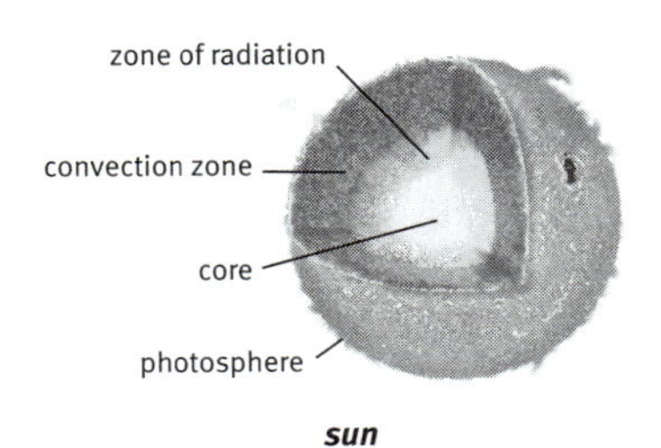

sun

cutaway of the Sun

sunbow (sŭn′bō′) A prismatic arc of colors, similar to a rainbow, resulting from the refraction of sunlight through a mist or spray of water.

sun protection factor A measure, expressed numerically, of the degree to which a preparation containing sunscreen protects the skin from ultraviolet rays. The higher the value, the greater the level of protection from sun damage to the skin.

sunspot (sŭn′spŏt′) Any of the dark, irregular spots that usually appear in groups on the surface of the Sun (its **photosphere**), lasting from a few days to several weeks or more. Sunspots appear dark because they are cooler, by up to 1,500°K, than the surrounding photosphere. They are associated with strong magnetic fields and solar magnetic storms moving in a vortex pattern, similar to a tornado on Earth. The number of sunspots waxes and wanes over an 11-year period; at maximum activity there are often increased numbers of **solar flares**.

superclass (so͞o′pər-klăs′) A taxonomic category of related organisms ranking below a phylum and containing one or more classes.

supercluster (so͞o′pər-klŭs′tər) A large group of neighboring clusters of galaxies, along with isolated galaxies scattered between them, the entire collection ranging in size from roughly 10 million to 1 billion light-years in diameter. The distribution of clusters is often close to that of a flattened sphere or disk. The Milky Way belongs to a galactic cluster known as the Local Group, which in turn is a member, along with about a hundred other such clusters, of the Virgo or Local Supercluster. There is some question as to whether gravity holds superclusters together and whether any larger structures exist within which superclusters are organized.

superconductivity (so͞o′pər-kŏn′dŭk-tĭv**′**ĭ-tē) The ability of certain metals or alloys to conduct an electric current with almost no resistance. Superconductivity usually occurs close to absolute zero, at temperatures approaching –459.67°F (–273.15°C), but has also been observed at temperatures as high as –200°F (–128.88°C). —*Noun* **superconductor.**

supercontinent (so͞o**′**pər-kŏn′tə-nənt) A large continent that, according to the theory of plate tectonics, is thought to have split into smaller continents in the geologic past. The supercontinent **Pangaea** is believed to have formed when earlier continental landmasses came together sometime before the Permian Period, staying together until after the Triassic Period, when it broke into the smaller supercontinents **Laurasia** and **Gondwanaland**. These supercontinents are believed to have later separated into the landmasses that correspond to the current continents. Other supercontinents are hypothesized to have formed and broken apart earlier in geologic time.

supercool (so͞o′pər-ko͞ol**′**) To cool a substance below a phase-transition temperature without the transition occurring. For example, water can be cooled well below the freezing point without freezing (as often happens in the upper atmosphere); the introduction of an impurity or surface can then trigger freezing. Supercooling is an example of **hysteresis.** Compare **superheat.**

superfamily (so͞o**′**pər-făm′ə-lē) A taxonomic category of related organisms ranking below an order or its subdivisions and containing one or more families. The names of superfamilies in the animal kingdom end in *–oidea,* such as Hominoidea, the primate superfamily containing the family of apes (Pongidae) and the family of humans (Hominidae).

superfluid (so͞o′pər-flo͞o**′**ĭd) A fluid, such as liquid helium, that flows with little or no friction at temperatures close to absolute zero.

supergiant (so͞o**′**pər-jī′ənt) A star that is larger, brighter, and more massive than a giant star, being thousands of times brighter than the Sun and having a relatively short lifespan—only about 10 to 50 million years as opposed to around 5 billion years for the Sun. Supergiants, such as **Betelgeuse** and **Rigel** in Orion, are only found in young cosmic structures such as the arms of spiral galaxies. Red supergiants such as Betelgeuse are late-stage stars, having burned most of their hydrogen in an earlier stage as **main-sequence** stars, and now fuse helium into heavier elements through the triple alpha process. Blue supergiants such as Rigel are thought to have evolved from red giants, though some are considered main-sequence stars. Supergiants are thought to eventually undergo a supernova, ending up as neutron stars or black holes.

supergravity (so͞o**′**pər-grăv′ĭ-tē) A quantum field theory that combines general relativity with supersymmetry in order to unify gravity with the other fundamental forces of nature. Supergravity predicts the existence of the graviton as a carrier for the force of gravity, as well as a corresponding particle called the gravitino, neither of which have been observed experimentally.

superheat (so͞o′pər-hēt**′**) To heat a substance above a phase-transition temperature without the transition occurring. For example, water can be heated above its boiling point without boiling; the introduction of an impurity or physical disturbance can then trigger boiling. Superheating is an example of **hysteresis.** Compare **supercool.**

superior conjunction (so͝o-pîr**′**ē-ər) See under **conjunction.**

superior planet Any of the planets Mars, Jupiter, Saturn, Uranus, and Neptune, whose orbits lie beyond that of Earth. Because these planets never come between the Earth and Sun, they do not exhibit crescent phases, only full and gibbous. Unlike the inferior planets Mercury and Venus, superior planets rise in the east and set in the west in the normal pattern of celestial objects. Compare **inferior planet.** See also **inner planet.**

supernova (so͞o′pər-nō**′**və) Plural **supernovae** (so͞o′pər-nō**′**vē) or **supernovas.** A massive star that undergoes a sudden, extreme increase in brightness across the electromagnetic spectrum, followed by a more gradual decrease lasting from several days to several months. Supernovae occur when a supergiant star collapses suddenly at the end of its life, condensing its core material into an extremely compact mass that then undergoes a slight rebound. The resulting shock wave sends all matter surrounding the core flying into space, leaving a **neutron star** or **black hole** at the site of the core's collapse. Supernovae may also occur when a white dwarf accretes material from a companion red giant star, resulting in an increase in mass that eventually triggers carbon fusion in the core of the white dwarf; the sudden increase in available fuel causes energy to be released in a violent explosion. In both cases the shock waves induce further fusion in the matter surrounding the collapsed core; the many elements resulting from this fusion and from the various other stages of **nucleosynthesis** over the lifetime of the star are scattered into space. These elements serve as the material from which new stellar and planetary systems are formed; in fact, every heavy element found on Earth is thought to have been the product of supernovae explosions. The last supernova to be observed in the Milky Way was seen in 1604 by Johannes Kepler and was used by Galileo, at his trial, as evidence against the presupposition that the universe never changes. Compare **nova.**

superorder (sōō′pər-ôr′dər) A taxonomic category of related organisms ranking below a class or subclass and containing one or more orders.

superposition (sōō′pər-pə-zĭsh′ən) **1.** The principle that in a group of stratified sedimentary rocks the lowest were the earliest to be deposited. **2.** The principle by which the description of the state of a physical system can be broken down into descriptions that are themselves possible states of the system. For example, harmonic motion, as of a violin string, can be analyzed as the sum of harmonic frequencies or harmonics, each of which is itself a kind of harmonic motion; harmonic motion is therefore a superposition of individual harmonics. **3.** The combination of two or more physical states, such as waves, to form a new physical state in accordance with this principle. See also **wave.** See Note at **Schrödinger.**

supersonic (sōō′pər-sŏn′ĭk) Having a speed greater than that of sound in a designated medium, usually air; having a speed greater than Mach 1. Compare **hypersonic, subsonic, transonic.**

superstring theory (sōō′pər-strĭng′) A type of string theory that includes the theoretical assumptions of **supersymmetry.**

supersymmetry (sōō′pər-sĭm′ĭ-trē) A theory of physics that states that for each boson (a subatomic particle that carries a fundamental force, such as the photon, which carries the electromagnetic force) there is a corresponding fermion with the same mass. The theory is an attempt to unify the fundamental forces of matter under one theory. Supersymmetry has not been shown to hold in the real world, though some scientists suspect that evidence for it may be found only at extremely high energies; some also believe that certain particles predicted by the theory may make up **dark matter.**

supplementary angles (sŭp′lə-mĕn′tə-rē) Two angles whose sum is 180°.

suppressor (sə-prĕs′ər) **1.** A mutant gene that suppresses the phenotypic expression of another usually mutant gene. **2.** A device, such as a resistor or grid, that is used in an electrical or electronic system to reduce unwanted currents. ► A **suppressor grid** in a vacuum tube such as a **pentode** is designed to prevent the secondary emission of electrons from the **plate.** When electrons emitted by the tube's cathode strike the plate, their energies can be high enough to cause secondary emission of low-energy electrons from the plate, and these electrons can drift away into other positively charged electrodes in the tube (like the **screen** or the **control grid**), drawing current from the plate. A negatively charged suppressor grid near the plate repels these low-energy electrons and pushes them back toward the plate so that no current is lost, increasing the efficiency of the tube.

surf (sûrf) The waves of the sea as they break upon a shore or a reef.

surface tension (sûr′fəs) A property of liquids such that their surfaces behave like a thin, elastic film. Surface tension is an effect of intermolecular attraction, in which molecules at or near the surface undergo a net attraction to the rest of the fluid, while molecules not near the surface are attracted to other molecules equally in all directions and undergo no net attraction. Because of surface tension, the surface of a liquid can support light objects (such as water beetles on the surface of a pond). Surface tension is responsible for the spherical shape of drops of liquid; spheres minimize the surface area of the drop and thus minimize surface tension. See also **capillary action, meniscus.**

surface wave A seismic wave that travels across the surface of the Earth as opposed to through it. Surface waves usually have larger amplitudes and longer wavelengths than body waves, and they travel more slowly than body waves do. Love waves and Rayleigh waves are kinds of surface waves. Compare **body wave.** See Note at **earthquake.**

surfactant (sər-făk′tənt) **1.** A substance that, when dissolved in water, lowers the surface tension of the water and increases the solubility of organic compounds. Surfactants are used in inks to increase the effects of capillary action; detergents are surfactants that help remove organic compounds from a substance by making them dissolve more readily in the water in which the substance is washed. **2.** A substance composed of lipoprotein that is secreted by the alveolar cells of the lung and maintains the stability of pulmonary tissue by reducing the surface tension of fluids that coat the lung.

surf zone See **breaker zone.**

surge (sûrj) A coastal rise in water level caused by wind.

surge protector A device, typically portable and containing its own electrical outlets, used to provide regulated electric power for other electric devices. Surge protectors get their power from standard household or industrial electrical sources and contain circuitry that prevents damaging or disruptive spikes or surges in the source voltage from reaching the devices plugged into them.

surjection (sər-jĕk′shən) A function such that each member of its range is mapped onto by a member of the domain. Compare **bijection, injection.**

suspension (sə-spĕn′shən) A mixture in which small particles of a substance are dispersed throughout a gas or liquid. If a suspension is left undisturbed, the particles are likely to settle to the bottom. The particles in a suspension are larger than those in either a colloid or a solution. Muddy water is an example of a suspension. Compare **colloid, solution.**

swamp (swŏmp) An area of low-lying wet or seasonally flooded land, often having trees and dense shrubs or thickets.

S wave See **secondary wave.**

sweat (swĕt) The salty liquid given off by sweat glands in the skin of mammals. As sweat evaporates, the skin cools, causing a reduction in body heat.

sweat gland Any of the numerous small, tubular glands that are found in the skin of many mammals and that secrete sweat externally through pores to help regulate body temperature. In humans and some other mammals, sweat glands are distributed over most of the body surface.

swim bladder (swĭm) See **air bladder** (sense 1).

swimmeret (swĭm′ə-rĕt′) One of the paired abdominal appendages of certain aquatic crustaceans, such as shrimp, lobsters, and isopods. Swimmerets are normally found on the first five abdominal segments and typically terminate in paired oarlike branches. They function primarily for carrying the eggs in females and are usually adapted for swimming. Also called *pleopod.*

swine flu (swīn) A highly contagious form of influenza seen in swine, caused by a virus of the family Orthomyxoviridae. The infection is communicable to humans and caused a worldwide epidemic in 1918.

syenite (sī′ə-nīt′) A light-colored, coarse-grained igneous rock consisting primarily of alkali feldspar together with some mafic minerals, especially hornblende. Unlike most igneous rocks, syenite has little or no quartz. It is believed to form from the cooling of magma that forms at very high temperatures and at great depths. It is the coarse-grained equivalent of trachyte.

symbiont (sĭm′bē-ŏnt′, -bī-) or **symbiote** (sĭm′bē-ōt′, -bī-) An organism in a symbiotic relationship. In cases in which a distinction is made between two interacting organisms, the symbiont is the smaller of the two and is always a beneficiary in the relationship, while the larger organism is the host and may or may not derive a benefit. See also **host, parasite.**

symbiosis (sĭm′bē-ō′sĭs) The close association between two or more organisms of different species, often but not necessarily benefiting each member. The association of algae and fungi in lichens and of bacteria living in the intestines or on the skin of animals are forms of symbiosis. Some scientists believe that many multicellular organisms evolved from symbiotic relationships between unicellular ones and that the DNA-containing organelles within certain eukaryotic cells (such as mitochondria and chloroplasts) are the product of symbiotic relationships in which the participants became interdependent. There are four forms of symbiosis: **amensalism, commensalism, mutualism,** and **parasitism**. —*Adjective* **symbiotic.**

symbol (sĭm′bəl) A conventional, printed or written figure used to represent an operation, element, quantity, relation, unit of measurement, phenomenon, or descriptor. Also called *sign*. See Table on page 610.

symmetric (sĭ-mĕt′rĭk) or **symmetrical** (sĭ-mĕt′rĭ-kəl) **1.** Relating to or exhibiting symmetry. **2.** Relating to a logical or mathematical relation between two elements such that if the first element is related to the second element, the second element is related in like manner to the first. The relation $a = b$ is symmetric, whereas the relation $a > b$ is not.

symmetry (sĭm′ĭ-trē) **1.** An exact matching of form and arrangement of parts on opposite sides of a boundary, such as a plane or line, or around a central point or axis. **2.** *Physics.* See **invariance.**

sympathetic nervous system (sĭm′pə-thĕt′ĭk) The part of the autonomic nervous system that tends to act in opposition to the parasympathetic nervous system, as by speeding up the heartbeat and causing contraction of the blood vessels. It regulates the function of the sweat glands and stimulates the secretion of glucose in the liver. The sympathetic nervous system is activated especially under conditions of stress. Compare **parasympathetic nervous system.**

sympatric (sĭm-păt′rĭk) *Ecology.* Occupying the same or overlapping geographic areas without interbreeding. Although they share the same geographic range, sympatric populations of related organisms become isolated from each other reproductively. This can happen by the development of subpopulations that become dependent on distinct food sources or that evolve distinct seasonal mating behavior. Flowering plants frequently become reproductively isolated through the development of polyploid hybrids (hybrids with three or more sets of chromosomes) that cannot backcross with either parent. ► The development of new species as a result of the reproductive isolation of populations that share the same geographic range is called **sympatric speciation**. Compare **allopatric.**

symptom (sĭm′təm) A subjective indication of a disorder or disease, such as pain, nausea or weakness. Symptoms may be accompanied by objective signs of disease such as abnormal laboratory test results or findings during a physical examination. Compare **sign.**

synapse (sĭn′ăps′) The small junction across which a nerve impulse passes from one nerve cell to another nerve cell, a muscle cell, or a gland cell. The synapse consists of the *synaptic terminal*, or presynaptic ending, of a sending neuron, a postsynaptic ending of the receiving cell that contains receptor sites, and the space between them (the *synaptic cleft*). The synaptic terminal contains **neurotransmitters** and cell organelles including mitochondria. An electrical impulse in the sending neuron triggers the migration of vesicles containing neurotransmitters toward the membrane of the synaptic terminal. The vesicle membrane fuses with the presynaptic membrane, and the neurotransmitters are released into the synaptic cleft and bind to receptors

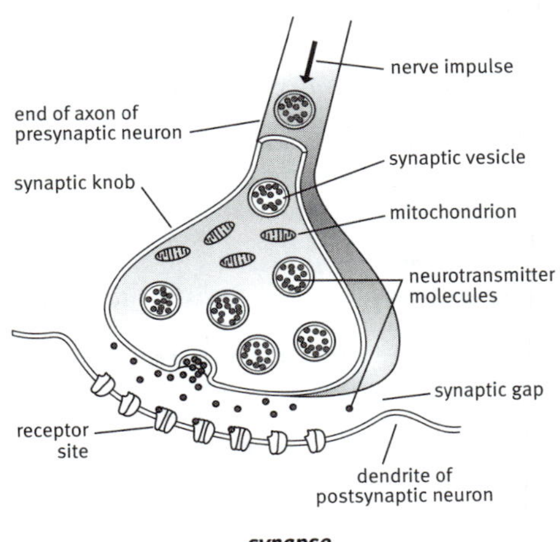

synapse

Symbols and Signs

The following symbols and signs and their designations are among those most commonly used in the sciences. Symbols consisting of letters of the alphabet, such as chemical elements or the mathematical abbreviations *sin* and *cos,* are entered in the regular alphabetical sequence of entries. See also symbols in tables at **measurement, Periodic Table,** and **subatomic particle.**

ASTRONOMY

Symbol	Meaning
☉ or ☼	Sun
● or ☻	new moon
☽	first quarter
○ or ☺	full moon
☾	last quarter
☿	Mercury
♀	Venus
⊖ or ⊕	Earth
♂	Mars
♃	Jupiter
♄	Saturn
♅	Uranus
♆	Neptune
♇	Pluto
♈	Aries
♉	Taurus
♊	Gemini
♋	Cancer
♌	Leo
♍	Virgo
♎	Libra
♏	Scorpio
♐	Sagittarius
♑	Capricorn
♒	Aquarius
♓	Pisces

BIOLOGY

Symbol	Meaning
☉ or ①	annual
⚇ or ②	biennial
♃	perennial
♂ or ♂	male
♀	female
□	male (in charts)
○	female (in charts)

CHEMISTRY

Symbol	Meaning
⌬	benzene ring
→	reaction direction
⇌	reversible reaction
↓	precipitate
↑	gas

MATHEMATICS

Symbol	Meaning
+	plus
−	minus
±	plus or minus
∓	minus or plus
×	multiplied by
÷	divided by
=	equal to
≠	not equal to
≈	approximately equal to
≡	identical with
≢	not identical with
≎	equivalent
≅	congruent to
>	greater than
≯	not greater than
<	less than
≮	not less than
≧ or ≥	greater than or equal to
≦ or ≤	less than or equal to
| |	absolute value
≐ or →	approaches
∝	proportional to, varies as
∥	parallel
⊥	perpendicular
∠	angle
∟	right angle
△	triangle
□	square
▭	rectangle
▱	parallelogram
○	circle
⌒	arc of circle
⊥	equilateral
≜	equiangular
√	radical; root; square root
∛	cube root
∜	fourth root
Σ	sum
! or ⌊	factorial product
∞	infinity
∫	integral
f	function
∂ or δ	differential; variation
π	pi
∪	logical sum or union
∩	logical product or intersection
⊂	is contained in
⊃	implication
∊	is a member of; mean error
:	is to; ratio
::	as; proportion
∴	therefore
∵	because

METEOROLOGY

Symbol	Meaning
◍	rain
*	snow
⊠	snow on ground
←	ice crystals
◬	hail
△	sleet
∨	frostwork
⊔	hoarfrost
≡	fog
∞	haze; dust haze
T	thunder
<	lightning
⦶	solar corona
⊕	solar halo
ꓘ	thunderstorm

PHARMACOLOGY

Symbol	Meaning
℞	take
ĀĀ	of each
℔	pound
℥	ounce
ʒ	dram
℈	scruple
f℥	fluid ounce
*f*ʒ	fluid dram
♏	minim

PHYSICS

Symbol	Meaning
°	degree
′	minute; foot
″	second; inch
Δ	increment, change
ω	angular velocity; solid angle
Ω	ohm
μΩ	microhm
MΩ	megohm
Φ	magnetic flux
→	direction of flow
⇄	electric current

of the connecting cell where they excite or inhibit electrical impulses. See also **neurotransmitter.**

synapsid (sĭ-năp′sĭd) A reptile with one temporal opening on each side of the skull. Synapsids evolved in the late Permian Period and were characterized by carrying their limbs under their body and developing front teeth that were different from their back teeth. One group of synapsids, the therapsids, gave rise to the mammals. Compare **anapsid, diapsid, therapsid.**

syncarp (sĭn′kärp′) A fleshy compound fruit composed either of the fruits of several flowers, as in the pineapple and mulberry, or of several carpels of a single flower, as in the magnolia and raspberry.

synchrocyclotron (sĭng′krō-sī′klə-trŏn′) A type of cyclotron that modulates the frequency of the electric fields that accelerate the particles, thereby keeping the accelerating forces synchronized with the particle as its mass increases with velocity due to the effects of relativity, and providing greater energies for the accelerated particles than an unsynchronized cyclotron. Also called *frequency modulated cyclotron.* See also **synchrotron.** See Note at **particle accelerator.**

synchronous orbit An orbit of a satellite around a rotating body, such that one orbit is completed in the time it takes for the body to make one revolution on its own axis. From the viewpoint of the rotating body (such as Earth), a satellite in synchronous orbit appears to hover over a single position or else to sweep back and forth once per revolution along a single line in the sky. *Synchronous orbit* is sometimes used synonymously with **geostationary orbit,** although the two terms are identical only if the synchronous orbit is in the equatorial plane (when the satellite appears to hover motionless over the rotating body). Artificial satellites in synchronous orbits are commonly used in telecommunications, due to their relatively stable overhead position with respect to the Earth's surface. Compare **synchronous rotation.**

synchronous rotation The rotation of an orbiting body on its axis in the same amount of time as it takes to complete a full orbit, with the result that the same face is always turned toward the body it is orbiting. Synchronous rotation is the result of **tidal forces** that over time slow the rotation of the smaller body until it is synchronized with its period of revolution around the larger body. The Earth's Moon exhibits synchronous rotation, as do a majority of moons in the solar system. Also called *captured rotation*. Compare **synchronous orbit.**

synchrotron (sĭng′krə-trŏn′) A type of particle accelerator that accelerates charged subatomic particles (generally protons) in a circular path. Unlike cyclotrons and synchrocyclotrons, in which particles follow a spiral path, synchrotrons consist of a single ring-shaped tube through which the particles loop numerous times, guided by precisely synchronized magnetic fields and accelerated at various points in the loop by electric field bursts. Synchrotrons are currently the most powerful particle accelerators, and the study of high-energy collisions driven by synchrotrons has lead to the discovery of many subatomic particles. See also **cyclotron, synchrocyclotron.** See Note at **particle accelerator.**

synchrotron radiation Electromagnetic radiation emitted by high-energy particles that are moving in a circular path, as in a synchrotron particle accelerator. Energy lost by synchrotron radiation increases rapidly as particles approach the speed of light.

syncline (sĭn′klīn′) A fold of rock layers that slope upward on both sides of a common low point. Synclines form when rocks are compressed by plate-tectonic forces. They can be as small as the side of a cliff or as large as an entire valley. Compare **anticline.**

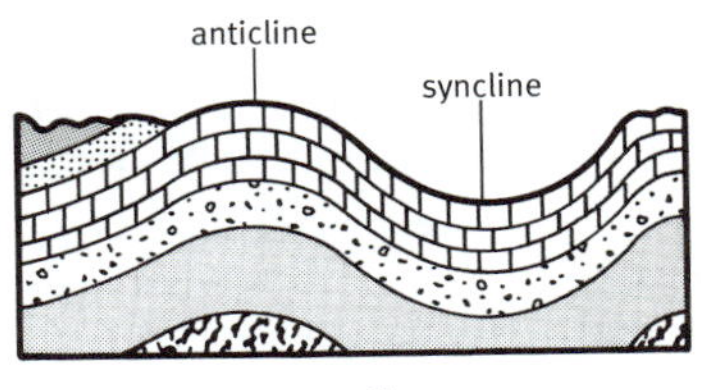

syncline

syndrome (sĭn′drōm′) An abnormal condition or disease that is identified by an established group of signs and symptoms.

synecology (sĭn′ĭ-kŏl′ə-jē) The branch of ecology that deals with the ecological interrelationships among communities of organisms. Compare **autecology.**

synergid (sĭ-nûr′jĭd, sĭn′ər-) One of two small, short-lived nuclei lying near the egg in the mature embryo sac of a flowering plant. The synergids are part of the **egg apparatus** and are thought to help the pollen nucleus reach the egg cell for fertilization. See more at **egg apparatus, embryo sac.**

syngamy (sĭng′gə-mē) The fusion of two gametes in fertilization.

synodic (sĭ-nŏd′ĭk) Relating to the conjunction of celestial bodies, especially to the interval between two successive conjunctions of a planet or the Moon with the Sun as viewed from Earth. For example, the new moon comes at the conjunction of the Moon with the Sun; the interval between successive new moons (the time it takes for the Moon to orbit the Earth and return to conjunction with the Sun) is the Moon's synodic period, also called a **lunar month.** Synodic time differs from **sidereal time,** which is measured in relation to the stars and is generally more appropriate to astronomical observation.

synodic month See **lunar month.**

synovial fluid (sĭ-nō′vē-əl) A clear fluid secreted by membranes in joint cavities, tendon sheaths, and bursae, and functioning as a lubricant. When a joint disorder is present, the synovial fluid that is removed and examined can contain indicators of disease, such as white blood cells or crystals.

synovial membrane The connective-tissue membrane that lines the cavity of a joint and produces the synovial fluid.

synthesis (sĭn′thĭ-sĭs) Plural **syntheses** (sĭn′thĭ-sēz′). The formation of a chemical compound through the combination of simpler compounds or elements. —*Verb* **synthesize.**

synthetic (sĭn-thĕt′ĭk) Produced artificially, especially in a laboratory or other man-made environment. Nylon is a synthetic chemical compound.

synthetic speech Speech that is produced by an electronic synthesizer activated by a keyboard. People who are incapable of speech can communicate by means of synthetic speech.

syphilis (sĭf′ə-lĭs) A sexually transmitted disease caused by the spirochete *Treponema pallidum* that is characterized in its primary stage by genital sores. If untreated, skin ulcers develop in the next stage, called *secondary syphilis.* As the disease progresses to potentially fatal *tertiary syphilis,* neurologic involvement with weakness and skeletal or cardiovascular damage can occur.

syringe (sə-rĭnj′) A medical instrument used to inject fluids into the body or draw them from it. Syringes have several different forms. Bulb syringes are usually made of rubber and work by squeezing the bulb to expel a fluid from it, as in ear irrigation. Needle syringes have hypodermic needles attached to plastic or glass tubes that contain plungers to create force or suction.

systemic lupus erythematosus (ĕr′ə-thē′mə-tō′sĭs) An inflammatory autoimmune disease of the connective tissue, characterized by fever, skin rash, joint pain or arthritis, and anemia. It occurs most commonly in women and often affects the kidneys and nervous system, as well as other organs.

systole (sĭs′tə-lē) The period during the normal beating of the heart in which the chambers of the heart, especially the ventricles, contract to force blood into the aorta and pulmonary artery. Compare **diastole.** —*Adjective* **systolic** (sĭ-stŏl′ĭk).

syzygy (sĭz′ə-jē) **1.** Either of two points in the orbit of a celestial body where the body is in opposition to or in conjunction with the Sun. **2a.** Either of the two points in the orbit of the Moon when it lies in a straight line with the Sun and Earth. A new moon syzygy occurs when the Moon is between the Sun and the Earth; a full moon syzygy occurs when the Earth is between the Moon and the Sun. **b.** The configuration of the Sun, Moon, and Earth when lying in a straight line.

Szilard (zĭl′ərd), **Leo** 1898–1964. Hungarian-born American physicist who introduced the concept of the nuclear chain reaction. With Enrico Fermi, he built the world's first nuclear reactor. Szilard was instrumental in the development of the atomic bomb.

T 1. Abbreviation of **temperature, tesla, thymine. 2.** The symbol for the isotope **tritium.**

Ta The symbol for **tantalum.**

tableland (tā′bəl-lănd′) A flat, elevated region, such as a plateau or mesa.

table salt (tā′bəl) See **salt** (sense 2).

table sugar See **sucrose.**

tachyon (tăk′ē-ŏn′) A hypothetical subatomic particle that travels faster than the speed of light. Although the principles of **relativity** forbid the acceleration of a particle past the speed of light, they do not prohibit the existence of particles that have always moved faster than that speed. No such particle has ever been detected.

tactile (tăk′təl, tăk′tīl′) Used for or sensitive to touch.

tactile hair A hair or hairlike structure that is highly sensitive to pressure or touch. Tactile hairs help organisms sense their environment. They cover the bodies of most insects and are found in many mammals, especially as whiskers.

tag (tăg) A sequence of characters in a markup language used to provide information, such as formatting specifications, about a document. Tags are enclosed in a pair of angle brackets that indicate to the browser how the text is to be displayed.

taiga (tī′gə) A forest located in the Earth's far northern regions, consisting mainly of cone-bearing evergreens, such as firs, pines, and spruces, and some deciduous trees, such as larches, birches, and aspens. The taiga is found just south of the tundra.

tail (tāl) **1.** The rear, elongated part of many animals, extending beyond the trunk or main part of the body. Tails are used variously for balance, combat, communication, mating displays, fat storage, propulsion and course correction in water, and course correction in air. **2.** A long, stream of gas or dust forced from the head of a comet when it is close to the Sun. Tails can be up to 150 million km (93 million miles) long, and they always point away from the Sun because of the force of the solar wind. ► **Plasma tails,** or **ion tails,** appear bluish and straight and narrow, and are formed when solar wind forces ionized gas to stream off the coma. **Dust tails** are wide and curved, and are formed when solar heat forces trails of dust off the coma; solid particles reflecting the Sun's light create their bright yellow color.

tailbone (tāl′bōn′) See **coccyx.**

talc (tălk) A very soft white, greenish, or gray monoclinic mineral usually occurring as massive micalike flakes in igneous or metamorphic rocks. It has a soapy texture and is used in face powder and talcum powder, for coating paper, and as a filler in paints and plastics. *Chemical formula:* $\mathbf{Mg_3Si_4O_{10}(OH)_2}$.

talon (tăl′ən) One of the sharp, curved claws on a limb of a bird or other animal such as a lizard, used for seizing and tearing prey. Most talons are situated at the ends of digits.

talus[1] (tā′ləs) Plural **tali** (tā′lī′). The bone of the ankle that articulates with the tibia and fibula to form the ankle joint.

talus[2] (tā′ləs) Plural **taluses.** Rock fragments that have accumulated at the base of a cliff or slope. ► The concave slope formed by such an accumulation of rock fragments is called a **talus slope**.

tamoxifen (tə-mŏk′sə-fĕn) An estrogen antagonist drug used to treat advanced breast cancer in women whose tumors are dependent on estrogen and to prevent breast cancer in some women who are at high risk.

tan Abbreviation of **tangent.**

tangent (tăn′jənt) **1.** A line, curve, or surface touching but not intersecting another. **2.** The ratio of the length of the side opposite an acute angle in a right triangle to the side adjacent to the angle. The tangent of an angle is equal to the sine of the angle divided by the cosine of the angle. **3.** The ratio of the ordinate to the abscissa of the endpoint of an arc of a unit circle centered at the origin of a Cartesian coordinate system, the arc being of length x and measured counterclockwise from the point (1, 0) if *x* is positive or clockwise if *x* is negative. **4.** A function of a number *x*, equal to the tangent of an angle whose measure in radians is equal to *x*.

tanh Abbreviation of **hyperbolic tangent.**

tannic acid (tăn**′**ĭk) A lustrous, yellow-brown, amorphous **tannin**, having the chemical composition $C_{76}H_{52}O_{46}$. It is derived from the bark and fruit of many plants and used in photography, as a mordant, and as a clarifying agent for wine and beer.

tannin (tăn**′**ĭn) Any of various compounds, including tannic acid, that occur naturally in the bark and fruit of various plants, especially the nutgalls, certain oaks, and sumac. Tannins are polyphenols, and form yellowish to light brown amorphous masses that can be powdery, flaky, or spongy. They are used in photography, dyeing, in tanning leather, in clarifying wine and beer, and as an astringent in medicine. Tannins also give color and flavor to black tea.

tantalum (tăn**′**tə-ləm) *Symbol* **Ta** A hard, heavy, gray metallic element that is highly resistant to corrosion at low temperatures. It is used to make light-bulb filaments, surgical instruments, and glass for camera lenses. Atomic number 73; atomic weight 180.948; melting point 2,996°C; boiling point 5,425°C; specific gravity 16.6; valence 2, 3, 4, 5. See **Periodic Table.**

tapeworm (tāp**′**wûrm′) See **cestode.**

taproot (tăp**′**ro͞ot′, -ro͝ot′) The main root in gymnosperms, eudicotyledons, and magnoliids, usually stouter than the lateral roots and growing straight downward from the stem. The taproot develops from the **primary root.** The taproot and its lateral roots penetrate deeper into the soil than the fibrous roots characteristic of monocotyledons. Compare **fibrous root.**

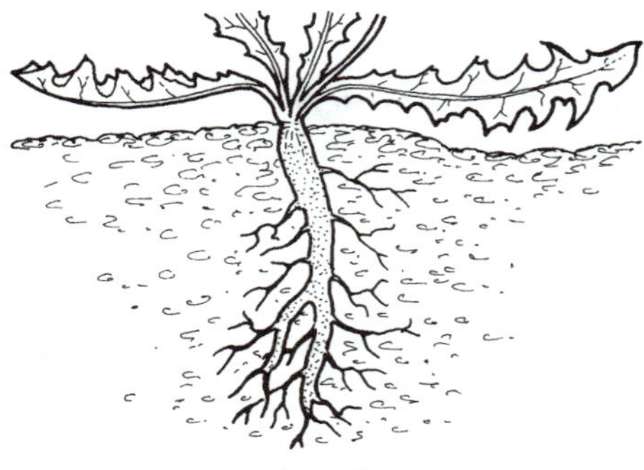

taproot

of a dandelion plant

tar (tär) **1.** A dark, oily, viscous material, consisting mainly of hydrocarbons, produced by the destructive distillation of organic substances such as wood, coal, or peat. **2.** See **coal tar. 3.** A solid, sticky substance that remains when tobacco is burned. It accumulates in the lungs of smokers and is considered carcinogenic.

tardigrade (tär**′**dĭ-grād′) Any of various slow-moving, minute invertebrates of the phylum Tardigrada. Tardigrades have a head and four fused body segments, each of which has a pair of stubby legs ending in claws. They live in water, damp moss, flower petals, or sand, and are usually 1 mm (0.04 inches) or less in size. Tardigrades are able to resist extremely low temperature, pressure, and humidity, and go into dormant states for months or years. They are believed to be intermediate in evolutionary development between annelids and arthropods. Also called *water bear.*

tarn (tärn) A small mountain lake, especially one formed as a glacier melts, filling a cirque with water.

tar pit An accumulation of natural tar or asphalt at the Earth's surface, especially one that traps animals and preserves their hard parts. Tar pits form when petroleum in subterranean petroleum-bearing rocks oozes up to the surface. As it rises, the petroleum loses its volatile components, forming a thick tar or asphalt deposit.

tarsal (tär**′**səl) *Adjective.* **1.** Relating to or involving the ankle. —*Noun.* **2.** Any of the seven bones of the tarsus.

tarsus (tär**′**səs) Plural **tarsi** (tär**′**sī, -sē). **1.** The group of seven bones lying between the leg and the metatarsals and forming part of the ankle. **2.** The group of bones lying between the leg and metatarsals in the hind feet in some vertebrates, such as dinosaurs and birds. **3.** A fibrous plate that supports and shapes the edge of the eyelid. **4.** The lower part of the leg of an arthropod, usually divided into segments.

tartar (tär**′**tər) **1.** A hard yellowish deposit on the teeth, consisting of organic secretions and food particles deposited in various salts, such as calcium carbonate. **2.** A reddish acid compound consisting of a tartrate of potassium, found in the juice of grapes and deposited on the sides of wine casks.

tartaric acid (tär-tăr′ĭk) A crystalline organic acid that exists in four isomeric forms and occurs widely in plants. It is found in byproducts of wine fermentation and has a wide variety of uses, including to make cream of tartar and baking powder, to add effervescence to beverages, to polish metal, in printing and dyeing, and to make photographic chemicals. *Chemical formula:* $\mathbf{C_4H_6O_6}$.

tartrate (tär′trāt′) A salt or ester of tartaric acid, containing the group $C_4H_4O_6$.

taskbar (tăsk′bär′) A row of buttons or graphical controls on a computer screen that represent open programs, among which the user can switch back and forth by clicking on the appropriate one.

taste bud (tāst) Any of numerous sense organs in most vertebrate animals that are specialized to detect taste. Taste buds are sensitive to four types of taste: sweet, sour, salty, or bitter. In land vertebrates, the taste buds are found on the surface of the tongue. Fish have taste buds all over their bodies.

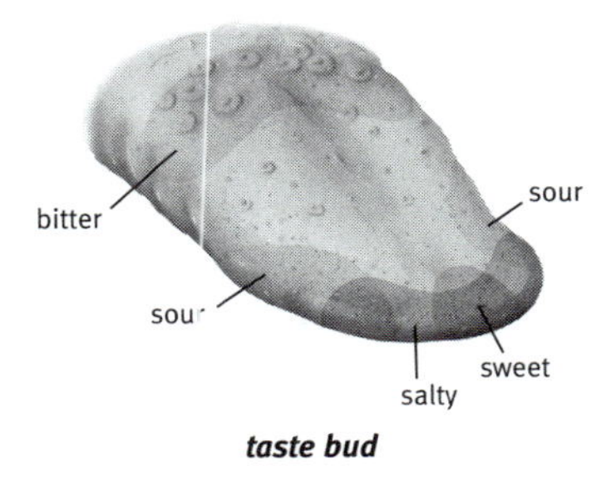

taste bud

The human tongue can detect four distinct tastes.

tau neutrino (tou, tô) A type of neutrino associated with the tau particle, often created in particle interactions involving tau particles (such as the decay of these particles). The tau neutrino has a mass of zero or near zero, and has no electric charge. See more at **neutrino.** See Table at **subatomic particle.**

tau particle An elementary particle of the lepton family, having a mass about 3,550 times that of the electron, a negative electric charge, and a mean lifetime of 3×10^{-13} seconds. See Table at **subatomic particle.**

Taurus (tôr′əs) A constellation in the Northern Hemisphere near Orion and Aries. Taurus (the Bull) contains the bright star Aldebaran and the grouping known as the Pleiades. It is the second sign of the zodiac.

taxon (tăk′sŏn′) Plural **taxa.** A taxonomic category or group, such as a phylum, order, family, genus, or species.

taxonomy (tăk-sŏn′ə-mē) The scientific classification of organisms into specially named groups based either on shared characteristics or on evolutionary relationships as inferred from the fossil record or established by genetic analysis. See Table on page 616.

Tay-Sachs disease (tā′săks′) A genetic disease in which the products of fat metabolism accumulate in the nervous system, causing retardation, paralysis, and death by preschool age. Individuals of eastern European Jewish descent have a higher risk of inheriting Tay-Sachs disease. The disease is named after its describers, British ophthalmologist Warren Tay (1843–1927) and American neurologist Bernard Sachs (1858–1944).

Tb The symbol for **terbium.**

TB **1.** Abbreviation of **terabyte.** **2.** Abbreviation of **tuberculosis.**

Tc The symbol for **technetium.**

T cell Any of the lymphocytes that develop in the thymus gland and that act in the immune system by binding antigens to receptors on the surface of their cells in what is called the **cell-mediated immune response**. T cells are also involved in the regulation of the function of B cells. Also called *T lymphocyte*. See more at **cell-mediated immune response.** Compare **B cell.**

TCP/IP Short for *Transmission Control Protocol/Internet Protocol.* A protocol for communication between computers, used as a standard for transmitting data over networks and as the basis for standard Internet protocols.

Te The symbol for **tellurium.**

tear (tîr) A drop of the clear salty liquid secreted by glands (lacrimal glands) in the eyes. Tears wet the membrane covering the eye and help rid the eye of irritating substances.

teat (tēt, tĭt) A small projection near the center of the mammary gland of many female mam-

Taxonomy

The scientific classification of life is called taxonomy, and was originally developed in the 18th century by the Swedish botanist Carolus Linnaeus. Linnaeus divided all forms of life into two large groups—the animal and plant kingdoms. Modern biologists have expanded his system to more kingdoms. One popular system is shown below:

Kingdom	Type of Organisms
Prokaryotes *or* Monerans	bacteria and blue-green algae
Protists *or* Protoctists	single-celled organisms such as the amoeba, euglena, and paramecium; also dinoflagellates, slime molds, and most algae
Fungi	mushrooms, yeasts, molds
Plantae	plants
Animalia	multicellular animals

In the traditional Linnean system, organisms are classified according to shared physical characteristics into a fixed number of hierarchical levels, which (in descending order) are **kingdom, phylum** (or **division,** in the plant kingdom), **class, order, family, genus,** and **species.** Intermediate levels, such as *subphylum* and *infraclass,* are also used as needed. (Modern taxonomy seeks to group organisms strictly by evolutionary descent rather than by shared characteristics; see **cladistics.**)

By convention, taxonomic categories are capitalized except for species and subspecies names. Genus and species names are written in italic, as in *Tyrannosaurus rex* or the bacillus *Escherichia coli.* The chart below compares the taxonomy of three organisms—a human, a horse, and a goldfish. As the chart shows, the three species share important taxonomic characteristics despite their many differences.

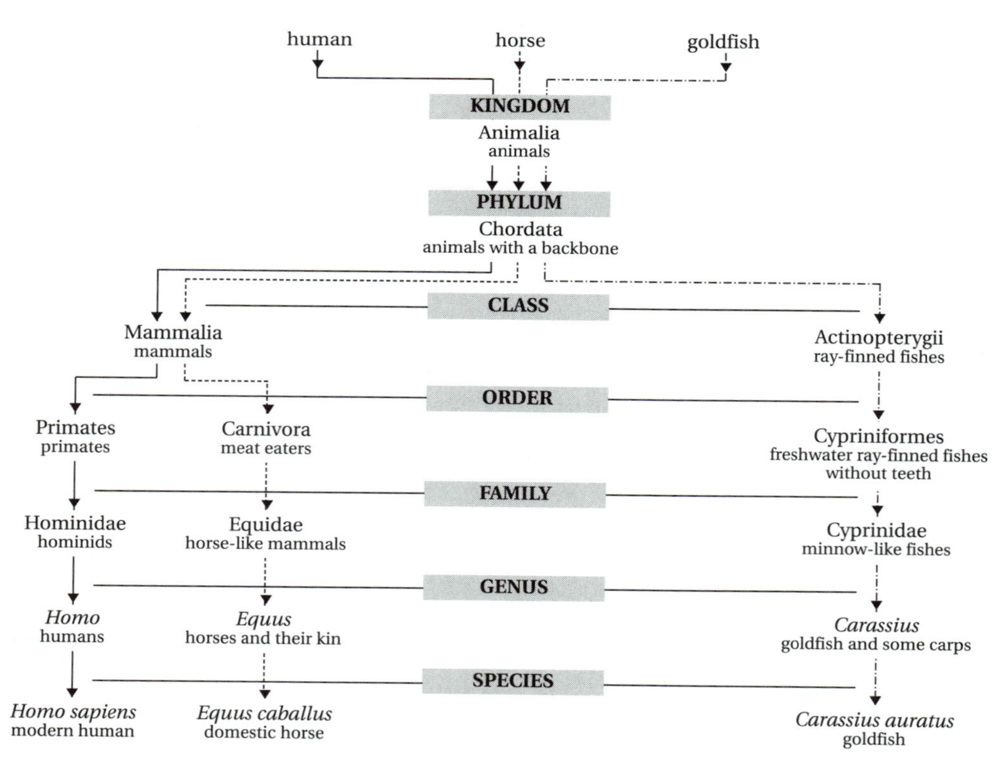

mals that contains the outlet of the milk ducts. Each teat contains a single milk duct, while nipples each contain more than one.

technetium (tĕk-nē′shē-əm) *Symbol* **Tc** A silvery-gray, radioactive metallic element. It was the first element to be artificially made, and it is produced naturally in extremely small amounts during the radioactive decay of uranium. Technetium is used to remove corrosion from steel. Its longest-lived isotope is Tc 98 with a half-life of 4,200,000 years. Atomic number 43; melting point 2,200°C; specific gravity 11.50; valence 0, 2, 4, 5, 6, 7. See **Periodic Table.**

technology (tĕk-nŏl′ə-jē) **1.** The use of scientific knowledge to solve practical problems, especially in industry and commerce. **2.** The specific methods, materials, and devices used to solve practical problems.

tectonic (tĕk-tŏn′ĭk) Relating to the forces involved in plate tectonics or the structural features resulting from them.

tectonic boundary In the theory of plate tectonics, a boundary between two or more plates. The plates can be moving toward each other (at convergent plate boundaries), away from each other (at divergent plate boundaries), or past each other (at transform faults). Maps of seismic and volcanic activity across the Earth indicate that most earthquakes and volcanic eruptions occur along or near tectonic boundaries. This is because earthquakes and volcanic eruptions are caused by the scraping, pushing, pulling, and melting of the Earth's lithosphere along these boundaries.

tectonics (tĕk-tŏn′ĭks) The branch of geology that deals with the broad structural and deformational features of the outer part of the Earth, their origins, and the relationships between them. See more at **plate tectonics.**

tegmen (tĕg′mən) A covering or integument, such as the tough leathery forewing of certain insects or the inner coat of a seed.

Teisserenc de Bort (tĕs-rä′ də bôr′), **Léon Philippe** 1855–1913. French physicist and meteorologist who pioneered the use of unmanned balloons outfitted with weather instruments for meteorological studies. Using these balloons he discovered and named the stratosphere in 1899.

tektite (tĕk′tīt′) Any of numerous dark-brown to green glassy objects, usually small (about the size of a walnut) and round with pitted surfaces. Tektites consist primarily (65% to 90%) of silica and have a very low water content. They are found in groups in several widely separated parts of the world and bear no relation to surrounding geologic formations. Some have shapes that show the kind of melting and deformation typical of objects that fall through the Earth's atmosphere. Tektites are believed to be of extraterrestrial origin or to have formed during high-velocity impacts on terrestrial rocks.

tele– A prefix that means "at a distance," as in *telemetry.*

telecommunication (tĕl′ĭ-kə-myo͞o′nĭ-kā′shən) The science and technology of sending and receiving information such as sound, visual images, or computer data over long distances through the use of electrical, radio, or light signals, using electronic devices to encode the information as signals and to decode the signals as information.

telegraph (tĕl′ĭ-grăf′) A communications system in which a message in the form of short, rapid electric impulses is sent, either by wire or radio, to a receiving station. **Morse code** is often used to encode messages in a form that is easily transmitted through electric impulses.

telemanipulator (tĕl′ə-mə-nĭp′yə-lā′tər) A device for transmitting hand and finger movements to a remote robotic device, allowing the manipulation of objects that are too heavy, dangerous, small, or otherwise difficult to handle directly.

telemetry (tə-lĕm′ĭ-trē) The measurement of data at a remote source and transmission of the data (typically by radio) to a monitoring station. Telemetry is used, for example, to track the movements of wild animals that have been tagged with radio transmitters, and to transmit meteorological data from weather balloons to weather stations.

teleost (tĕl′ē-ŏst′, tē′lē-) See **bony fish.**

telescope (tĕl′ĭ-skōp′) **1.** An arrangement of lenses, mirrors, or both that collects visible light, allowing direct observation or photographic recording of distant objects. ► A **refracting telescope** uses lenses to focus light to produce a magnified image. **Compound lenses** are used to avoid distortions such as spherical and chromatic aberrations. ► A **reflecting telescope** uses mirrors to view celestial objects at high levels of magnification. Most large optical telescopes are reflecting telescopes because very large mirrors, which are necessary to maximize the amount of light received by the telescope, are easier to build than very large lenses. **2.** Any of various devices, such as a radio telescope, used to detect and observe distant objects by collecting radiation other than visible light.

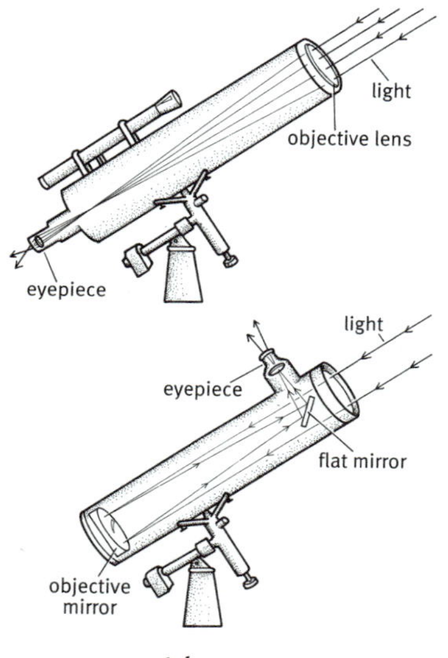

telescope

top: *refracting telescope;* bottom: *reflecting telescope*

telluride (tĕl′yə-rīd′) A chemical compound of tellurium and another element, including gold and silver. Telluride is found in mineral ores.

tellurium (tĕ-lo͝or′ē-əm) *Symbol* **Te** A metalloid element that occurs as either a brittle, shiny, silvery-white crystal or a gray or brown powder. Small amounts of tellurium are used to improve the alloys of various metals. Atomic number 52; atomic weight 127.60; melting point 449.5°C; boiling point 989.8°C; specific gravity 6.24; valence 2, 4, 6. See **Periodic Table.**

telomerase (tə-lŏm′ə-rās′, -rāz′) An enzyme that preserves the length of telomeres across cell divisions in germ cells, stem cells, and most cancer cells. A kind of **reverse transcriptase**, telomerase is an RNA-containing enzyme that synthesizes the DNA of telomeres by reverse transcription. It is active during DNA replication and is thought to play a role in the proliferation and apparent immortality of cells in which it is present. In cells that lack telomerase (that is, in most somatic cells of the body), the telomeres of chromosomes shorten and eventually disappear over repeated cell divisions. The inhibition of telomerase is being investigated as a method of killing cancerous cells. See more at **telomere.**

telomere (tĕl′ə-mîr′, tē′lə-) Either of the sections of DNA occurring at the extreme ends of each chromosome in a eukaryotic cell. Telomeres consist of highly repetitive sequences of DNA that do not code for proteins, but function as caps to keep chromosomes from fusing together. The length of the telomere influences the stability of genetic information just interior of the telomere, since the nucleotide sequences at the ends of a chromosome are not copied by DNA polymerase. Successive copying can thus shorten telomeres, sometimes to the point that functional genes near the telomeres are lost, and this may play a role in cellular senescence and age-related diseases. In germ cells, stem cells, and some cancer cells, shortened telomeres can be extended by the enzyme *telomerase*, thus keeping both the telomeres and the genes near them functioning. Most somatic cells do not express telomerase, and the shortening of telomeres during each round of cell division may be part of the natural aging of cells.

telophase (tĕl′ə-fāz′) The final phase of cell division, in which membranes form around the two groups of chromosomes, each at opposite ends of the cell, to produce the two nuclei of the daughter cells. The spindle disappears, and the cytoplasm usually divides (in the process called cytokinesis). In mitosis, telophase is preceded by anaphase. In meiosis, telophase occurs twice, once as part of the first meiotic division (when it is usually called telophase I) and once during the second meiotic division (when it is usually called telophase II). During telophase I, the members of pairs of homologous chromosomes which have separated during anaphase I (anaphase of the first meiotic division) regroup at the two ends of the cell. During telophase II, the individual chromatids that separated during anaphase II (anaphase of the second meiotic division) regroup at the ends of the cell. See more at **meiosis, mitosis.**

temperate (tĕm′pər-ĭt) Marked by moderate temperatures, weather, or climate.

Temperate Zone Either of two regions of the Earth of intermediate latitude, the **North Temperate Zone,** between the Arctic Circle and the Tropic of Cancer, or the **South Temperate Zone,** between the Antarctic Circle and the Tropic of Capricorn.

temperature (tĕm′pər-ə-cho͝or′) **1.** A measure of the ability of a substance, or more generally of any physical system, to transfer heat energy to another physical system. The temperature of a substance is closely related to the average kinetic energy of its molecules. See also **Boyle's law. 2.** Any of various standardized numerical measures of this ability, such as the Kelvin, Fahrenheit, and Celsius scales. **3.** An abnormally high body temperature; a fever.

USAGE **temperature/heat**

Heat and *temperature* are closely related but distinct and sometimes subtle ideas. Heat is simply transferred thermal energy—most commonly, the kinetic energy of molecules making up substance, vibrating and bouncing against each other. A substance's temperature, on the other hand, is a measure of its ability to transfer heat, rather than the amount of heat transferred. For example, a match lit under a pot of boiling water reaches a much higher temperature than the water, but it is able to give off much less heat, since only a small amount of thermal energy is created and released by it. When any two substances of different temperatures are in thermal contact, the *laws of thermodynamics* state that heat flows from the higher-temperature substance into the lower-temperature substance, raising the temperature of the heated body and lowering the temperature of the body releasing heat until *thermal equilibrium* is reached, and the temperatures are the same. Thus temperature describes a characteristic of matter that determines the direction and extent of heat transfer, so the match with little heat but high temperature still adds energy to the water when placed under the pot. Providing a closed physical system with heat generally raises its temperature but not necessarily; for example, ice at zero degrees Celsius requires considerable additional heat in order to melt into water at zero degrees Celsius. Temperature can be related to the average kinetic energy of the molecules of gases, though this relation breaks down in most real cases involving liquids, solids, substances with larger molecules, and radiation with no mass, such as light. The two most common temperature scales, Celsius (C) and Fahrenheit (F), are based on the freezing and boiling points of water. On the Celsius scale there are 100 increments between the two points, and on the Fahrenheit scale there are 180. Scientists also use the International System units called Kelvins (K). A difference in temperature of one degree is equivalent in the Celsius and Kelvin scales, but their absolute scales are different: while zero degrees C is the temperature at which water freezes (at a pressure of one atmosphere), zero degrees K (–273.72 degrees C), also called *absolute zero,* is the least possible temperature for a system, representing a theoretical state from which no heat can be extracted.

temperature gradient The rate of change of temperature with displacement in a given direction from a given reference point.

template (tĕm′plĭt) A molecule of a nucleic acid, such as DNA, that serves as a pattern for the synthesis of another molecule of a nucleic acid.

temporal (tĕm′pər-əl) Relating to or near the bones that form the sides and part of the base of the skull.

temporal lobe The lobe of each cerebral hemisphere lying to the side and rear of the frontal lobe. The temporal lobe controls hearing and some aspects of language perception, emotion, and memory.

temporomandibular joint (tĕm′pə-rō-măn-dĭb′yə-lər) The joint that facilitates mandibular movement, located between the head of the mandible and the temporal bone.

tendon (tĕn′dən) A band of tough, fibrous, inelastic tissue that connects a muscle to a bone. Tendons are made chiefly of collagen.

tendril (tĕn′drəl) A slender, coiling plant part, often a modified leaf or leaf part, that helps support the stem of some climbing angiosperms by clinging to or winding around an object. Peas, squash, and grapes produce tendrils.

tensile strength (tĕn′səl, tĕn′sīl′) A measure of the ability of material to resist a force that tends to pull it apart. It is expressed as the minimum tensile **stress** (force per unit area) needed to split the material apart.

tensile stress See under **axial stress.**

tension (tĕn′shən) **1.** A force that tends to stretch or elongate something. **2.** An electrical potential (voltage), especially as measured in electrical components such as transformers or power lines involved in the transmission of electrical power.

tensor (tĕn′sər, -sôr′) **1.** A structure of quantities arranged by zero or more indices, such as a scalar (zero indices), a vector (one index), or a matrix (two indices), which is invariant under transformations of coordinates. **2.** Any of various muscles that stretch or tighten a body part, as the muscle that acts to tense the soft palate, called the *tensor palati.*

tentacle (tĕn′tə-kəl) A narrow, flexible, unjointed part extending from the body of certain animals, such as an octopus, jellyfish, or sea anemone. Tentacles are used for feeling, grasping, or moving.

tepal (tē′pəl, tĕp′əl) A division of the perianth of a flower in which the sepals and petals are indistinguishable, as in tulips and lilies.

tephra (tĕf′rə) Solid matter, such as ash, dust, and cinders, that is ejected into the air by an erupting volcano. Tephra is a general term for all pyroclastic materials ejected from a volcano.

tera– A prefix that means: **1.** One trillion (10^{12}), as in *terahertz,* one trillion hertz. **2.** 2^{40} (that is, 1,099,511,627,776), which is the power of two closest to a trillion, as in *terabyte.*

terabyte (tĕr′ə-bīt′) **1.** A unit of computer memory or data storage capacity equal to 1,024 gigabytes (2^{40} bytes). **2.** One trillion bytes. See Note at **megabyte.**

teraflop (tĕr′ə-flŏp′) A measure of computing speed equal to one trillion floating-point operations per second.

teratogen (tə-răt′ə-jən, tĕr′ə-tə-) An agent, such as a virus, a drug, or radiation, that can cause malformations in an embryo or fetus.

teratology (tĕr′ə-tŏl′ə-jē) The scientific study of birth defects.

terbium (tûr′bē-əm) *Symbol* **Tb** A soft, silvery-gray metallic element of the lanthanide series. It is used in color television tubes, x-ray machines, and lasers. Atomic number 65; atomic weight 158.925; melting point 1,356°C; boiling point 3,123°C; specific gravity 8.229; valence 3, 4. See **Periodic Table.**

term (tûrm) **1.** Each of the quantities or expressions that form the parts of a ratio or the numerator and denominator of a fraction. **2.** Any of the quantities in an equation that are connected to other quantities by a plus sign or a minus sign.

terminal (tûr′mə-nəl) **1.** *Electricity.* A position in a circuit or device at which a connection can be made or broken. See Note at **battery. 2.** *Computer Science.* A device, often equipped with a keyboard and a video display, by which one can read, enter, or manipulate information in a computer system.

terminal bud See under **bud.**

terminal moraine See under **moraine.**

termination codon (tûr′mə-nā′shən) See **stop codon.**

termite (tûr′mīt′) Any of various pale-colored insects of the order Isoptera that live in large colonies and feed on wood. Termites resemble ants in their appearance, manner of living, and social organization, but are not closely related. Termites can be very destructive to wooden buildings and structures. Also called *isopteran.*

terpene (tûr′pēn′) Any of a class of hydrocarbons consisting of two or more isoprene (C_5H_8) units joined together. Simple terpenes are found in the essential oils and resins of plants such as conifers. Turpentine, for example, is such an oil. More complex terpenes include vitamin A, carotenoid pigments (such as lycopene), squalene, and rubber. Terpenes are used in organic synthesis.

terpenoid (tûr′pə-noid′) Any of a class of hydrocarbons that consist of terpenes attached to an oxygen-containing group. Terpenoids are widely found in plants, and can form cyclic structures such as sterols.

terra (tĕr′ə) **1.** A rough highland or mountainous region of the moon with a relatively high albedo. Compare **mare. 2.** A vast highland region on a planet.

terrestrial (tə-rĕs′trē-əl) **1.** Relating to Earth or its inhabitants. **2.** Relating to, living on, or growing on land.

terrestrial planet See **inner planet.**

terrestrial radiation Long-wave electromagnetic radiation originating from Earth and its atmosphere. It is the radiation emitted by naturally radioactive materials on Earth including uranium, thorium, and radon.

terrigenous (tĕ-rĭj′ə-nəs) Derived from the land, especially by erosive action. Used primarily of shallow marine sediments.

territoriality (tĕr′ĭ-tôr′ē-ăl′ĭ-tē) A behavior pattern in animals consisting of the occupation and defense of a territory.

territory (tĕr′ĭ-tôr′ē) A geographic area occupied by a single animal, mating pair, or group. Animals usually defend their territory vigorously against intruders, especially of the same species, but the defense often takes the form of prominent, threatening displays rather than out-and-out fighting. Different animals mark off territory in different ways, as by leaving traces of their scent along the boundaries or, in the case of birds, modifying their calls to keep out intruders.

tertiary (tûr′shē-ĕr′ē) *Noun.* **1. Tertiary.** The first period of the Cenozoic Era, from about 65 to 2 million years ago. During this time the continents took on their present form, and the climate changed from being warmer and wetter, in the early part of the period, to being drier and cooler in the later part. Mammals replaced dinosaurs as the dominant form of terrestrial animal life, and many modern types of flowering plants, insects, mollusks, fish, amphibians, reptiles, and birds appeared. The Tertiary is subdivided into the Paleogene and the Neogene, although these terms are not as widely used as are the names of the epochs that constitute them. See Chart at **geologic time.** *—Adjective.* **2a.** Relating to or having a carbon atom that is attached to three other carbon atoms in a molecule. **2b.** Relating to an organic molecule, such as an alcohol, in which the functional group is attached to a tertiary carbon. **3.** Relating to an advanced level of medical care, usually provided by subspecialists after the delivery of primary medical care. Compare **primary, secondary.**

tertiary consumer See under **consumer.**

tertiary structure The three-dimensional structure of a protein or nucleic acid. Amino acids form secondary structures such as alpha helices, beta sheets, and random coils, which in turn fold on themselves to form the tertiary structure of the protein. Only if a protein is correctly folded will it have its intended biological activity. Several diseases, such as cystic fibrosis and mad cow disease, are caused by alterations in the tertiary structure of one or more proteins. See also **primary structure, quaternary structure, secondary structure.**

tesla (tĕs′lə) The SI derived unit of magnetic flux density, equal to the magnitude of the magnetic field vector necessary to produce a force of one newton on a charge of one coulomb moving perpendicular to the direction of the magnetic field vector with a velocity of one meter per second. It is equivalent to one weber per square meter.

Tesla, Nikola 1856–1943. Serbian-born American electrical engineer and physicist who in 1881 discovered the principles of alternating current. He went on to invent numerous devices and procedures that were essential to the harnessing of electricity and the development of radio.

Nikola Tesla

BIOGRAPHY **Nikola Tesla**

The Serbian-born inventor Nikola Tesla came to America when he was 28 years old. After working briefly for Thomas Edison, Tesla set up his own laboratory and immediately launched a succession of discoveries and inventions. At the time, most commercially generated electricity was distributed over a direct current (DC) system invented by Edison's lab. This system was very expensive and inefficient for a variety of reasons. To be practical and safe, everyday use of electricity generally required low voltages, but transmission of low-voltage power is very inefficient. Generators at the time easily generated alternating current (AC), but not steady DC, and conversion was difficult. Finally, converting high-voltage DC power required for efficient transmission to low-voltage power presented yet another set of technical difficulties. Tesla was a staunch proponent of using AC throughout the power supply chain. He demonstrated that AC power could be transmitted efficiently at high voltages over very long distances, and it could be brought down to safe voltages easily with the use of transformers. After Tesla sold the patents to his AC system to George Westinghouse in 1885, there ensued a competition for dominance between Edison's DC system and Westinghouse's AC. Tesla gave public demonstrations of electricity to ease people's fears about the safety of the AC system, even to the point of having currents passed through his body to ignite flames. Tesla's approach won out; the first power plants at Niagara Falls used the AC system to power the city of Buffalo, New York. Tesla's invention of motors and generators using the AC system helped to ensure its success at replacing direct current throughout the country. Beyond his pioneering work in the production and transmission of electromagnetic energy (including what we now know as radio transmission), Tesla's inventions include the Tesla coil (an induction coil used in radio and television technology), a kind of bladeless turbine, remote control systems, and dozens of other devices—over 700 patents in all.

---

tesla coil An electrical device that generates extremely high voltages, usually for the purpose of creating dramatic electric arcs and lightning effects or for producing x-rays. Tesla coils use step-up transformers to boost the voltage of a power supply and build up large charges in a capacitor. A spark gap periodically shorts out the capacitor, releasing its charge in huge current flows that generate extremely high voltages (up to ten million volts) through an open-air transformer.

tesseract (tĕs′ə-răkt′) A four-dimensional hypercube, having sixteen corners. See more at **hypercube.**

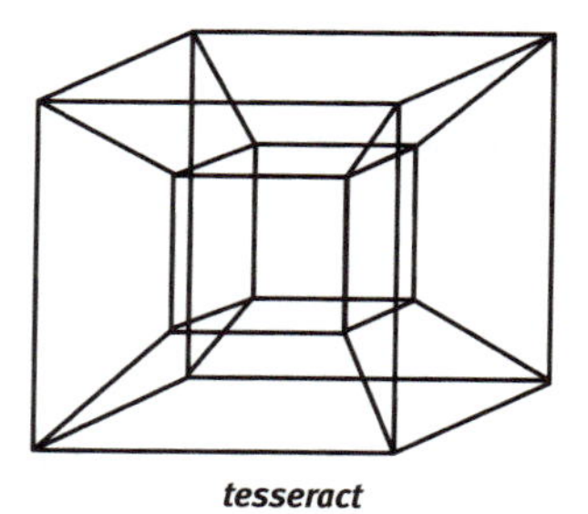

tesseract

A two-dimensional depiction of a tesseract. In a real tesseract, which cannot exist in our three-dimensional space, each edge would be the same length.

testa (tĕs′tə) Plural **testae** (tĕs′tē′). See **seed coat.**

testacean (tĕ-stā′shən) Any of various protozoans of the order Arcellinida (or Testacida) that are encased in a shell. Testaceans are rhizopods and are related to amoebas. Their shell usually has one chamber and may be secreted by the organism as a chitinous or siliceous covering, or may be glued together out of sand particles.

testicle (tĕs′tĭ-kəl) Either of the testes of a male mammal together with the scrotum that contains it.

testis (tĕs′tĭs) Plural **testes** (tĕs′tēz). The primary reproductive organ of male animals, in which sperm and the male sex hormones (androgens) are produced. In most vertebrates, the testes are contained inside the body. In many mammals, however, the testes are enclosed in an external scrotum.

testosterone (tĕs-tŏs′tə-rōn′) A steroid hormone that is the most potent naturally occurring androgen and that regulates the development of the male reproductive system and male secondary sex characteristics. Testosterone is produced mainly in the testes and is also used as a drug in the treatment of certain medical disorders. *Chemical formula:* $C_{19}H_{28}O_2$.

test tube (tĕst) A cylindrical tube of clear glass, usually open at one end and rounded at the other, used as a container for small amounts of a substance in laboratory tests and experiments.

tetanus (tĕt′n-əs) An acute, often fatal infectious disease caused by the bacterium *Clostridium tetani,* which usually enters the body through a wound and produces a toxin that affects nerve conduction. Tetanus is characterized by painful, spasmodic contractions of voluntary muscles, especially of the jaw.

tetrachloroethylene (tĕt′rə-klôr′ō-ĕth′ə-lēn′) A colorless, nonflammable organic liquid used in dry-cleaning solutions, as an industrial solvent, and as an agent for expelling or destroying parasitic intestinal worms. Also called *perchloroethylene. Chemical formula:* **C_2Cl_4.**

tetracycline (tĕt′rə-sī′klēn′, -klĭn) A yellow crystalline compound, $C_{22}H_{24}N_2O_8$, synthesized or derived from several bacterial species of the genus *Streptomyces* and used as an antibiotic in bacterial infections. Other drugs of the tetracycline class have a similar chemical structure.

tetrad (tĕt′răd′) **1.** A four-part structure that forms during prophase I of meiosis and consists of two homologous chromosomes, each composed of two identical chromatids. During prophase I of meiosis, one chromosome exchanges corresponding segments of genetic material with the other chromosome in the tetrad in the process called **crossing over.** See more at **meiosis. 2.** A group of four cells, as of spores or pollen grains, formed from a parent cell by meiosis. As part of the process of spermatogenesis, a spermatocyte divides into a tetrad of four spermatids, cells which go on to develop into sperm. See more at **spermatogenesis.**

tetraethyl lead (tĕt′rə-ĕth′əl) A colorless, poisonous, oily liquid, formerly in wide use as an antiknock agent in gasoline for internal-combustion engines. *Chemical formula:* **$C_8H_{20}Pb$.**

tetragonal (tĕ-trăg′ə-nəl) Relating to a crystal having three axes, two of which are of the same length and are at right angles to each other. The third axis is perpendicular to these. The mineral zircon has tetragonal crystals. See illustration at **crystal.**

tetrahedron (tĕt′rə-hē′drən) Plural **tetrahedrons** or **tetrahedra.** A polyhedron having four faces.

tetrahydrofuran (tĕt′rə-hī′drə-fyo͝or′ăn′) A clear, highly flammable liquid used as a solvent for natural and synthetic resins and in the production of nylon. Tetrahydrofuran is a furan ring without double bonds between any of the carbon atoms. It is the building block of the sugars ribose, deoxyribose, and fructose.*Chemical formula:* **C_4H_8O.**

tetralogy of Fallot (tĕ-trăl′ə-jē, -trŏl′-; fă-lō′) A congenital malformation of the heart characterized by a defect in the ventricular septum, misplacement of the origin of the aorta, narrowing of the pulmonary artery, and enlargement of the right ventricle. It is named after its describer, French physician Etienne-Louis Arthur Fallot (1850–1911).

tetramerous (tĕ-trăm′ər-əs) **1.** Having four similar segments or parts. **2.** Having flower parts, such as petals, sepals, and stamens, in sets of four.

tetrapod (tĕt′rə-pŏd′) Having four feet, legs, or leglike appendages. Any of various mostly terrestrial vertebrates that breathe air with lungs. Most tetrapods have two pairs of limbs, though some, such as whales and snakes, have lost one or both pairs. Tetrapods include the amphibians, reptiles, birds, mammals, and various extinct groups, and evolved from lobe-finned fish during the late Devonian Period. Tetrapods are classified according to the structure of their skull into anapsids, diapsids, and synapsids.

tetrode (tĕt′rōd′) A four-element electron tube containing an **anode**, **cathode**, **control grid**, and additional electrode called the **screen**. They function in the same manner as **triode**, but are more effective at higher frequencies due to the effect of the screen. See more at **screen.** See also **pentode.**

texture (tĕks′chər) The general physical appearance of a rock, especially with respect to the size, shape, size variability, and geometric arrangement of its mineral crystals (for igneous and metamorphic rocks) and of its constituent elements (for sedimentary rocks). A sandstone that forms as part of an eolian (wind-blown) deposit, for example, has a texture that reflects its small, rounded sand grains of uniform size, while a sandstone that formed as part of a fluvial deposit has a texture reflecting the presence of grains of varying sizes, with some more rounded than others.

Th The symbol for **thorium.**

thalamus (thăl′ə-məs) Plural **thalami** (thăl′ə-mī′). The part of the vertebrate brain that lies at the rear of the forebrain. It relays sensory information to the cerebral cortex and regulates the perception of touch, pain, and temperature.

thalassemia (thăl′ə-sē′mē-ə) Any of a group of inherited forms of anemia occurring chiefly among people of Mediterranean descent, caused by faulty synthesis of part of the hemoglobin molecule

thalassic (thə-lăs′ĭk) **1.** Relating to seas or oceans, especially smaller or inland seas. **2.** Relating to rocks formed from sediment that has been deposited on the ocean floor.

Thales (thā′lēz), 624?–546? BCE. Greek philosopher who was considered by later Greek writers to be a founder of geometry and abstract astronomy. He is said to have accurately predicted a solar eclipse in 585 BCE, although this and certain other stories associated with Thales have been questioned by modern scholars.

thalidomide (thə-lĭd′ə-mīd′) A drug used to treat leprosy. It was previously prescribed to treat nausea during early pregnancy, but was found to cause severe birth defects, including stunting or absence of the limbs. *Chemical formula:* $\mathbf{C_{13}H_{10}N_2O_4}$.

thallium (thăl′ē-əm) *Symbol* **Tl** A soft, malleable, very poisonous metallic element that is used in photography, in making low-melting and highly refractive glass, and in treating skin infections. Atomic number 81; atomic weight 204.38; melting point 303.5°C; boiling point 1,457°C; specific gravity 11.85; valence 1, 3. See **Periodic Table.**

thallophyte (thăl′ə-fīt′) Any of a former group of plantlike organisms showing no differentiation into stem, root, or leaf. Thallophytes were regarded as constituting a major division of the plant kingdom and included the algae, fungi, and lichens. No longer in scientific use.

thallus (thăl′əs) Plural **thalli** (thăl′ī). A type of body found among plants and fungi that is not differentiated into roots, stems, or leaves. Thalli are found among lichens, mosses, liverworts, and many algae, as well as the gametophyte generations of horsetails and ferns, which have rhizoids but not true roots.

thalweg (täl′vĕg′) **1.** The line defining the lowest points along the length of a river bed or valley, whether underwater or not. **2.** A subterranean stream following a course similar to that of an overlying surficial stream.

theca (thē′kə) Plural **thecae** (thē′sē′, -kē′). A case, covering, or sheath, such as the pollen sac of an anther, the spore case of a moss, or the outer covering of the pupa of certain insects.

thecodont (thē′kə-dŏnt′) Any of various extinct primitive archosaurs of the order Thecodontia of the late Permian and Triassic Periods. Thecodonts had teeth in sockets and were probably ancestral to the dinosaurs, pterosaurs, and crocodilians.

Thénard (tā-när′), **Louis Jacques** 1777–1857. French chemist who is best known for his 1818 discovery of hydrogen peroxide. Earlier, working with Joseph Gay-Lussac, Thénard also discovered boron (1808).

theobromine (thē′ō-brō′mēn′) A bitter, colorless alkaloid that occurs in the cacao bean, cola nuts, and tea. It is found in chocolate products and used in medicine as a diuretic, vasodilator, and myocardial stimulant. Theobromine is a xanthine and similar in structure to caffeine and theophylline. *Chemical formula:* $\mathbf{C_7H_8N_4O_2}$.

theodolite (thē-ŏd′l-īt′) An optical instrument used to measure angles in surveying, meteorology, and navigation. In meteorology, it is used to track the motion of a weather balloon by measuring its elevation and azimuth angle. The earliest theodolite consisted of a small mounted telescope that rotated horizontally and vertically; modern versions are sophisticated computerized devices, capable of tracking weather balloons, airplanes, and other moving objects, at distances of up to 20,000 m (65,600 ft).

theophylline (thē-ŏf′ə-lĭn) A colorless, crystalline alkaloid derived from tea leaves or made synthetically. It is used in medicine especially as a bronchial dilator. Theophylline is a xanthine that is similar in structure to caffeine and is a structural isomer of theobromine. *Chemical formula:* $\mathbf{C_7H_8N_4O_2}$.

theorem (thē′ər-əm, thîr′əm) A mathematical statement whose truth can be proved on the basis of a given set of axioms or assumptions.

theory (thē′ə-rē, thîr′ē) A set of statements or principles devised to explain a group of facts or phenomena. Most theories that are accepted by scientists have been repeatedly tested by experiments and can be used to make predictions about natural phenomena. See Note at **hypothesis.**

theory of everything A physical theory, such as a grand unified theory, that attempts to explain all physical matter and interactions under a single unified set of principles.

therapeutic cloning (thĕr′ə-pyo͞o′tĭk) The production of embryonic stem cells for use in replacing or repairing damaged tissues or organs, achieved by transferring a diploid nucleus from a body cell into an egg whose nucleus has been removed. The stem cells are harvested from the blastocyst that develops from the egg, which, if implanted into a uterus, could produce a clone of the nucleus donor.

therapsid (thə-răp′sĭd) An advanced type of synapsid reptile that evolved in the Permian Period. Therapsids further differentiated their dentition into nipping, biting, and crushing teeth, and (unlike diapsids) had forelimbs that were more greatly developed than hindlimbs. Therapsids include the so-called mammallike reptiles of the Permian and Triassic Periods, as well as mammals. Compare **anapsid, diapsid, synapsid.**

thermal (thûr′məl) *Adjective.* **1.** Relating to heat. —*Noun.* **2.** A usually columnar mass of warm air that rises in the lower atmosphere because it is less dense than the air around it. Thermals form because the ground surface is heated unevenly by the Sun. The air usually rises until it is in equilibrium with the air surrounding it.

thermal capacity See **heat capacity.**

thermal conductance A measure of the ability of a material to transfer heat per unit time, given one unit area of the material and a temperature gradient through the thickness of the material. It is measured in watts per meter per degree Kelvin.

thermal conductivity A measure of the ability of a material to transfer heat. Given two surfaces on either side of the material with a temperature difference between them, the thermal conductivity is the heat energy transferred per unit time and per unit surface area, divided by the temperature difference. It is measured in watts per degree Kelvin.

thermal equilibrium The condition under which two substances in physical contact with each other exchange no heat energy. Two substances in thermal equilibrium are said to be at the same **temperature.** See also **thermodynamics.**

thermal neutron See **slow neutron.**

thermal vent An opening in the Earth, especially on the ocean floor, that emits hot water and dissolved minerals. See more at **hydrothermal vent.**

thermion (thûr′mī′ən) An electrically charged particle or ion that is emitted by a heated conducting material. The electrons emitted from the cathodes of electron tubes (such as **cathode ray tubes**) are thermions.

thermionic tube (thûr′mī-ŏn′ĭk) An electrical component, typically an electron tube, through which electric current flows via the emission of thermions from a heated cathode.

thermistor (thûr′mĭs′tər) A resistor whose resistance varies as a function of temperature. Thermistors are used in electrical devices such as thermometers and thermostats that measure, monitor, or regulate temperature.

thermo– or **therm–** A prefix that means "heat," as in *thermometer.*

thermochemistry (thûr′mō-kĕm′ĭ-strē) The branch of chemistry that is concerned with the heat generated by chemical reactions and by changes of state.

thermocline (thûr′mə-klīn′) A distinct layer in a large body of water, such as an ocean or lake, in which temperature changes more rapidly with depth than it does in the layers above or below. Thermoclines may be a permanent feature of the body of water in which they occur, or they may form temporarily in response to phenomena such as the solar heating of surface water during the day. Factors that affect the depth and thickness of a thermocline include seasonal weather variations, latitude and longitude, and local environmental conditions.

thermocouple (thûr′mə-kŭp′əl) A thermoelectric device used to make accurate measurements of temperatures, especially high temperatures. It consists of a circuit having two wires of different metals or metal alloys welded together. A temperature gradient across the junction of the wires gives rise to an electric potential by the Seebeck effect. This potential varies with the strength of the temperature gradient and can be measured by a voltmeter. Thermocouples can also be used to generate small amounts of electricity for powering other devices.

thermodynamics (thûr′mō-dī-năm′ĭks) The branch of physics that deals with the relationships between heat and other forms of energy. Four basic laws have been established. ► The **first law** states that the amount of energy added to a system is equal to the sum of its increase in heat energy and the work done on the system. The first law is an example of the principle of conservation of energy. ► The **second law** states that heat energy cannot be transferred from a body at a lower temperature to a body with a higher one without the addition of energy. Thus, warm air outside can transfer its energy to a cold room, but transferring energy out of a cold room to the air outside requires extra energy (as with an air conditioner). ► The **third law** states that the **entropy** of a pure crystal at absolute zero is zero. Since there can be no physical system with lower entropy, all entropy is thus defined to have a positive value. ► The **zeroth law** states that if two bodies are in thermal equilibrium with some third body, then they are also in equilibrium with each other. This law has its name because it was implicitly assumed in the development of the other laws, and is in fact more fundamental than the others, but was only later established as a law itself.

thermoelectric (thûr′mō-ĭ-lĕk′trĭk) Relating to electric potential or power produced by heat, or to heat produced by electric energy. The thermoelectric energy of a nuclear power plant is produced by the heat generated from nuclear fission. The thermoelectric properties of materials such as selenium and semiconductors are exploited in devices such as thermistors and thermocouples used in temperature gauges. See also **Seebeck effect.**

thermometer (thər-mŏm′ĭ-tər) An instrument used to measure temperature. There are many types of thermometers; the most common consist of a closed, graduated glass tube in which a liquid expands or contracts as the temperature increases or decreases. Other types of thermometers work by detecting changes in the volume or pressure of an enclosed gas or by registering thermoelectric changes in a conductor (such as a thermistor or thermocouple).

thermonuclear (thûr′mō-no͞o′klē-ər) **1.** Relating to the fusion of atomic nuclei at high temperatures or to the energy produced in this way. **2.** Relating to weapons based on nuclear fusion, especially as distinguished from those based on nuclear fission.

thermopile (thûr′mə-pīl′) A device consisting of a number of thermocouples connected in series or parallel, used for measuring temperature or generating current.

thermoplastic (thûr′mə-plăs′tĭk) Of or relating to a compound that can be repeatedly made soft and hard through heating and cooling. Polyethylene and polystyrene are thermoplastic resins. Compare **thermosetting.**

thermoregulation (thûr′mō-rĕg′yə-lā′shən) Maintenance of a constant internal body temperature independent of the environmental temperature. Thermoregulation in humans is effected through metabolic activity and sweating.

thermoremanent magnetization (thûr′mō-rĕm′ə-nənt) The magnetization that an igneous rock acquires as the temperature of the magma or lava from which it forms falls below the **Curie point** during the cooling and solidification process. The thermoremanent magnetization bears the signature of the magnetic field (usually the Earth's magnetic field) in which the rock is located as it forms. See Note at **magnetic reversal.**

thermosetting (thûr′mō-sĕt′ĭng) Relating to a compound that softens when initially heated, but hardens permanently once it has cooled. Thermosetting materials are made of long-chain polymers that cross-link with each other after they have been heated, rendering the substance permanently hard. Compare **thermoplastic.**

thermosphere (thûr′mə-sfîr′) The region of the Earth's upper atmosphere lying above the mesosphere and extending from a height of approximately 80 km (50 mi) to between 550 and 700 km (341 and 434 mi) above the Earth's surface. In the thermosphere temperatures increase steadily with altitude, reaching as high as 1,727°C (3,140°F) at the highest elevations. Chemical reactions occur much faster here than on the surface of the Earth. See also **exosphere, mesosphere, stratosphere, troposphere.** See illustration at **atmosphere.**

thermostat (thûr′mə-stăt′) A device that automatically controls heating or cooling equipment in such a way as to maintain a temperature at a constant level or within a specified range, generally using a thermometer capable of triggering electrical switches that activate or deactivate the equipment.

theropod (thîr′ə-pŏd′) Any of various carnivorous saurischian dinosaurs of the group Theropoda. Theropods walked on two legs and had small forelimbs and a large skull with long jaws and sharp teeth. Most theropods were of small or medium size, but some grew very large, like *Tyrannosaurus.* Theropods lived throughout the Mesozoic Era. Compare **sauropod.**

thiamine (thī′ə-mĭn) A water-soluble pyrimidine derivative belonging to the vitamin B complex that is important in carbohydrate metabolism and normal activity of the nervous system. It is found in pork, organ meats, whole grain cereals, legumes, and nuts. Deficiency of thiamine in the diet results in beriberi. Also called *vitamin* B_1. *Chemical formula:* $C_{12}H_{17}ClN_4OS$.

thiazide (thī′ə-zīd′, -zĭd) Any of a group of drugs that block reabsorption of sodium in the kidneys, used as diuretics primarily in the treatment of hypertension.

thiazole (thī′ə-zōl′) **1.** Any of a class of organic compounds containing a ring that consists of three carbon atoms, one nitrogen atom, and one sulfur atom. Thiamine, penicillin and its derivatives, various other drugs, and numerous dyes are thiazoles. **2.** A colorless or pale yellow liquid used in making dyes and fungicides. *Chemical formula:* C_3H_3NS.

thio- or **thi-** A prefix that means "containing sulfur," used especially of a compound in which an oxygen atom has been replaced by a sulfur atom, as in *thiourea.*

thiol (thī′ôl′, -ōl′) A sulfur-containing organic compound having the general formula RSH, where R is another element or radical. Thiols are typically very volatile and strong-smelling, and are responsible for the odor of onions, garlic, rotting flesh, and skunk musk. Also called *mercaptan.*

thiophene (thī′ə-fēn′) A colorless liquid used as a solvent. The chemical properties of thiophene resemble those of benzene, which occurs with it in coal tar. Thiophene has a ring of four carbon atoms and one sulfur atom, and among its derivatives are biotin, various plant pigments, and some pharmaceuticals. *Chemical formula:* C_4H_4S.

thiosulfate (thī′ō-sŭl′fāt′) A compound containing the group S_2O_3.

thiourea (thī′ō-yo͝o-rē′ə) A lustrous white crystalline compound used as a developer in photography and photocopying and in various organic syntheses. Thiourea has the same structure as urea, but with a sulfur atom in place of the oxygen atom. *Chemical formula:* CH_4N_2S.

third quantum number (thûrd) See **magnetic quantum number.**

tholeiite (tō′lə-īt′) Any of a series of igneous rocks that are similar in composition to basalt, but are richer in silica and iron and poorer in aluminum than basalt is. Tholeiites form especially at mid-ocean ridges and in continental rift areas.

Thompson (tŏmp′sən, tŏm′-), **Benjamin.** Count Rumford. 1753–1814. American-born British physicist who conducted numerous experiments on heat and friction, which led him to discover that heat is produced by moving particles.

Thomsen (tŏm′sən), **Christian Jürgensen** 1788–1865. Danish archaeologist who introduced the three-part system for the chronological classification of prehistoric artifacts, divided into the Stone, Bronze, and Iron ages. See more at **Three-Age system.**

Thomson (tŏm′sən), Sir **J(oseph) J(ohn)** 1856–1940. British physicist who discovered the electron in 1897. While experimenting with cathode rays, he deduced that the particles he observed were smaller than an atom. Thomson also made noteworthy studies of the conduction of electricity through gases.

He was awarded the Nobel Prize for physics in 1906.

BIOGRAPHY **J. J. Thomson**

Nowadays we take for granted the existence of electrons, but this was not true just over 100 years ago, when the atom was thought to be a single unit that had no parts. The breakthroughs came in the late 1890s, when the British physicist J. J. Thomson was studying what we now call cathode-ray tubes. As an electric current passed from the cathode at one end of the tube to the anode at the other, raylike emanations were seen to proceed from the cathode to the anode. Thomson examined the nature of the rays' charge by bringing a positively charged and a negatively charged plate near the path of the rays, and observed that the rays were deflected toward the positive plate, suggesting they had negative charge. A series of experiments in which various objects were placed in the path of the rays showed that they also had momentum (they would cause a small paddle wheel to turn, for example). If they had momentum, that meant (in the physics of the time) that they had mass, suggesting that the rays were composed of tiny particles. Other experimental results, some by other scientists, suggested that the ratio of the charge to the mass of these particles had to be less than one-thousandth the ratio for charged hydrogen atoms. By examining both the energy of the rays and the amount by which an electric charge deflected them, Thomson was able to calculate that these particles had one two-thousandth the mass of a hydrogen atom. The particles, first named *corpuscles,* were later called *electrons.* (The term *electron* was not completely new; it had been invented in 1891 for the rays themselves.) Thomson was thus the first to discover that particles smaller than atoms existed, and for his pioneering work he was awarded the 1906 Nobel Prize for physics.

---

thoracic (thə-răs′ĭk) Relating to or located in or near the thorax.

thorax (thôr′ăks′) Plural **thoraxes** or **thoraces** (thôr′ə-sēz′). **1.** The upper part of the trunk in vertebrate animals. The thorax includes the rib cage, which encloses the heart and lungs. In mammals, the thorax lies above the abdomen and below the neck. **2.** The middle division of the body of an insect, to which the wings and legs are attached. The thorax lies between the head and the abdomen.

thorium (thôr′ē-əm) *Symbol* **Th** A silvery-white, radioactive metallic element of the actinide series. It is used for fuel in some nuclear reactors and for improving the high-temperature strength of magnesium alloys. The only naturally occurring isotope of thorium, Th 232, is also its most stable, having a half-life of 14.1 billion years. Atomic number 90; atomic weight 232.038; approximate melting point 1,750°C; approximate boiling point 4,500°C; approximate specific gravity 11.7; valence 4. See **Periodic Table.**

thorn (thôrn) A short, hard, pointed part of a stem or branch of a woody plant. Compare **spine.**

Three Age system (thrē) A system for classifying prehistoric artifacts according to successive stages of technological development, divided into the Stone, Bronze, and Iron ages.

A CLOSER LOOK **Three Age system**

In organizing the extensive collection of artifacts at the National Museum of Denmark, the 19th-century Danish archaeologist Christian Thomsen proposed an innovative system based on the assumption of a progression in human technology from stone to bronze to iron. His insight that early technology had developed in chronological stages, rather than concurrently at different levels of society, proved essentially correct, though ultimately of limited use in describing the various progressions in parts of the world other than Europe, western Asia, and northern Africa. Once empirical study of archaeological collections began, Thomsen's Three Age system was rapidly modifed into four ages by the subdivision of the Stone Age into the Old Stone (now Paleolithic) and New Stone (Neolithic) ages. Subsequent refinement has added Mesolithic (Middle Stone) and Chalcolithic (Copper and Stone) to the original terms, which are now known as periods rather than ages. Use of the full terminology—Paleolithic, Mesolithic, Neolithic, Chalcolithic, Bronze, and Iron—is appropriate only for Europe, the Middle East, and Egypt, and even there it is not uniformly accepted among archaeologists today.

---

threonine (thrē′ə-nēn′) An essential amino acid. *Chemical formula:* $C_4H_9NO_3$. See more at **amino acid.**

thrombin (thrŏm′bĭn) An enzyme in blood that catalyzes the conversion of fibrinogen to fibrin and is essential to the coagulation of blood.

thrombosis (thrŏm-bō′sĭs) The formation or presence of a thrombus.

thrombus (thrŏm′bəs) Plural **thrombi** (thrŏm′bī′). A clot consisting of fibrin, platelets, red blood cells, and white blood cells that forms in a blood vessel or in a chamber of the heart and can obstruct blood flow. The rupture of atherosclerotic plaques can cause arterial **thrombosis** (the formation of thrombi), while tissue injury, decreased movement, oral contraceptives, prosthetic heart valves, and various metabolic disorders increase the risk for venous thrombosis. A thrombus in a coronary artery can cause a heart attack. Compare **embolus.**

thrush (thrŭsh) **1.** An infectious disease, caused by the fungus *Candida albicans,* characterized by small whitish eruptions on the mouth, throat, and tongue, and sometimes accompanied by fever, colic, and diarrhea. Thrush is most often seen in infants, children, and people with impaired immune systems. **2.** A degenerative condition of a horse's foot, usually caused by unhygienic management.

thrust (thrŭst) The force that propels an object in a given direction, especially when generated by the object itself, as by an engine or rocket.

thrust fault A reverse fault in which the fault plane is inclined at an angle equal to or less than 45°. See Note at **fault.**

thulium (tho͞o′lē-əm) *Symbol* **Tm** A soft, silver-gray metallic element of the lanthanide series. One of its artificial radioactive isotopes is used as a radiation source in small, portable x-ray machines. Atomic number 69; atomic weight 168.934; melting point 1,545°C; boiling point 1,727°C; specific gravity 9.3; valence 2, 3. See **Periodic Table.**

thunder (thŭn′dər) The explosive noise that accompanies a stroke of lightning. Thunder is a series of sound waves produced by the rapid expansion of the air through which the lightning passes. Sound travels about 1 km in 3 seconds (about 1 mi in 5 seconds). The distance between an observer and a lightning flash can be calculated by counting the number of seconds between the flash and the thunder. See Note at **lightning.**

thundercloud (thŭn′dər-kloud′) See **cumulonimbus.**

thunderstorm (thŭn′dər-stôrm′) A storm of heavy rain accompanied by lightning, thunder, wind, and sometimes hail. Thunderstorms occur when moist air near the ground becomes heated, especially in the summer, and rises, forming cumulonimbus clouds that produce precipitation. Electrical charges accumulate at the bases of the clouds until lightning is discharged. Air in the path of the lightning expands as a result of being heated, causing thunder. Thunderstorms can also be caused by temperature changes triggered by volcanic eruptions and forest fires, and they occur with much greater frequency at the equatorial regions than in polar regions.

thylakoid (thī′lə-koid′) A saclike membrane that contains the chlorophyll in cyanobacteria and in the chloroplasts of plant cells and green algae. In chloroplasts, thylakoids are arranged in stacks called **grana.**

thymidine (thī′mĭ-dēn′) A nucleoside composed of thymine and deoxyribose that (with the addition of phosphate to form the nucleotide thymine) occurs in DNA. Radioactively tagged thymidine has been used in genetics research to study DNA synthesis in the nuclei of cells undergoing cell division. *Chemical formula:* $C_{10}H_{14}N_2O_5$.

thymine (thī′mēn′) A pyrimidine base that is a component of DNA. It forms a base pair with adenine. *Chemical formula:* $C_5H_6N_2O_2$.

thymus (thī′məs) An organ of the lymphatic system located behind the upper sternum (breastbone). T cells (T lymphocytes) develop and mature in the thymus before entering the circulation. In humans, the thymus stops growing in early childhood and gradually shrinks in size through adulthood, resulting in a gradual decline in immune system function.

thyroid gland (thī′roid′) A two-lobed gland that wraps around the trachea and is located at the base of the neck in vertebrate animals. The thyroid gland secretes two important hormones: thyroxine, which regulates the cell metabolism necessary for normal growth and development, and calcitonin, which stimulates the formation of bone and helps regulate the amount of calcium in the blood.

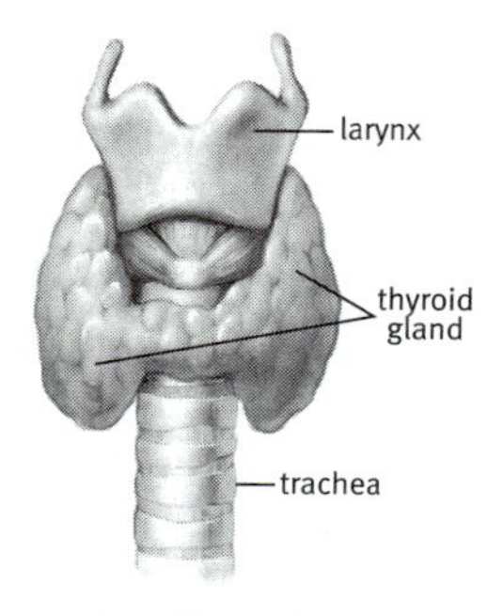

thyroid gland

thyroid hormone A hormone, such as thyroxine, produced by the thyroid gland.

thyroid stimulating hormone A glycoprotein hormone secreted by the anterior portion of the pituitary gland that stimulates and regulates the activity of the thyroid gland. Also called *thyrotropin.*

thyrotropin (thī′rə-trō′pĭn, thī-rŏt′rə-) See **thyroid stimulating hormone.**

thyroxine (thī-rŏk′sēn′, -sĭn) An iodine-containing hormone secreted by the thyroid gland that increases the rate of cell metabolism and regulates growth. Thyroxine can also be made synthetically for treatment of hypothyroidism.

thyrse (thûrs) A dense inflorescence in which the side branches end in cymes, as in the lilac. Also called *thyrsus.* See more at **inflorescence.**

Ti The symbol for **titanium.**

tibia (tĭb′ē-ə) The larger of the two bones of the lower leg or lower portion of the hind leg. See more at **skeleton.**

tick (tĭk) Any of numerous small, parasitic arachnids of the suborder Ixodida that feed on the blood of animals. Like their close relatives the mites and unlike spiders, ticks have no division between cephalothorax and abdomen. Ticks differ from mites by being generally larger and having a sensory pit at the end of their first pair of legs. Many ticks transmit febrile diseases, such as Rocky Mountain spotted fever and Lyme disease.

tidal air (tīd′l) See **tidal volume.**

tidal basin An area that holds water during high tide, especially a body of water in an area subject to tides whose water level is maintained at a desired level by artificial means.

tidal bulge See **tidal wave** (sense 1).

tidal flat A nearly flat coastal area that is alternately covered and exposed by tides and consisting of unconsolidated sediments and precipitated salts. A tidal flat is the middle part of a tidal basin, below the vegetation-supporting salt marsh and the low-tide mark.

tidal force A secondary effect of the gravitational forces between two objects orbiting each other, such as the Earth and the Moon, that tends to elongate each body along the axis of a line connecting their centers. Tidal forces are responsible for the fluctuation of the tides as well as for the **synchronous rotation** of certain moons as they orbit their planets.

tidal pool A pool of water remaining after a tide has retreated. The physical conditions of the tidal pool can change significantly throughout the day as the temperature of its water fluctuates with solar heating and its salinity increases with evaporation. As a result, tidal pools are harsh environments for the organisms that inhabit them. Also called *tide pool.*

tidal volume The volume of air inhaled and exhaled with each breath. Also called *tidal air.*

tidal wave 1. Either of the two swells or crests of surface ocean water created by the gravitational effects of the Moon and Sun and circling the globe on opposite sides to create the daily periods of high and low tides. Also called *tidal bulge.* **2.** An unusual rise in the level of water along a seacoast, as from a storm or a combination of wind and tide. Also called *storm surge.* **3.** A tsunami.

USAGE **tidal wave/tsunami**

The term *tidal wave* is used in everyday speech to refer to a gigantic and enormously destructive wave caused by an underwater earthquake or volcanic eruption—what scientists would properly call a *tsunami.* When scientists use the word *tidal wave,* they normally are referring to an unusually large wave or bulge of water that sometimes occurs around a high tide. These tidal waves are certainly big and powerful, but they are tiny in comparison with tsunamis.

---

tide (tīd) The regular rise and fall in the surface level of the Earth's oceans, seas, and bays caused by the gravitational attraction of the Moon and to a lesser extent of the Sun. The maximum high tides (or spring tides) occur when the Moon and Sun are directly aligned with Earth, so that their gravitational pull on Earth's waters is along the same line and is reinforced. The lowest high tides (or neap tides) occur when the Moon and Sun are at right angles to each other, so that their gravitational pull on Earth's waters originates from two different directions and is mitigated. Tides vary greatly by region and are influenced by sea-floor topography, storms, and water currents. See also **ebb tide, flood tide, neap tide, spring tide.**

tidemark (tīd′märk′) A line or mark on a shore indicating the highest or lowest level reached by the tide.

tide pool See **tidal pool.**

tidewater (tīd′wô′tər) **1.** Water that inundates land at flood tide. **2.** Water affected by the tides, especially tidal streams. **3.** Low coastal land drained by tidal streams.

till (tĭl) An unstratified, unconsolidated mass of boulders, pebbles, sand, and mud deposited by the movement or melting of a glacier. The size and shape of the sediments that constitute till vary widely.

timberline (tĭm′bər-līn′) A geographic boundary beyond which trees cannot grow. On the Earth as a whole, the timberline is the northernmost or southernmost latitude at which trees can survive; in a mountainous region, it is the highest elevation at which trees can survive. Also called *tree line.*

timberline

time (tīm) **1.** A continuous, measurable quantity in which events occur in a sequence proceeding from the past through the present to the future. See Note at **space-time. 2a.** An interval separating two points of this quantity; a duration. **b.** A system or reference frame in which such intervals are measured or such quantities are calculated.

time dilation The relativistic effect of the slowing of a clock with respect to an observer. In Special Relativity, a clock moving with respect to an observer appears to run more slowly than to an observer moving with the clock. In General Relativity, time dilation is also caused by gravity; clocks on the earth's surface, for example, run more slowly than clocks at high altitudes, where gravitational forces are weaker.

time reversal A mathematical operation on the description of a physical system in which the signs of variables representing time are reversed. Also called *time conjugation.* See more at **invariance.**

time zone Any of the 24 divisions of the Earth's surface used to determine the local time for any given locality. Each zone is roughly 15° of longitude in width, with local variations for economic and political convenience. Local time is one hour ahead for each time zone as one travels east and one hour behind for each time zone as one travels west. The International Meridian Conference in 1884 established the **prime meridian** as the starting point for the 24 zones. See more at **International Date Line, standard time.**

tin (tĭn) *Symbol* **Sn** A malleable, silvery metallic element that occurs in igneous rocks. It has a crystalline structure and crackles when bent. Tin is used as an anticorrosion agent and is a part of numerous alloys, including bronze. Atomic number 50; atomic weight 118.71; melting point 231.89°C; boiling point 2,270°C; specific gravity 7.31; valence 2, 4. See **Periodic Table.** See Note at **element.**

tinea (tĭn′ē-ə) See **ringworm.**

tinnitus (tĭn′ĭ-təs, tĭ-nī′-) A buzzing, ringing, or whistling sound in one or both ears occurring without an external stimulus. Its causes include ear infection or blockage, certain drugs, head injury, and neurologic disease.

tissue (tĭsh′o͞o) A large mass of similar cells that make up a part of an organism and perform a specific function. The internal organs and connective structures (including bone and cartilage) of vertebrates, and cambium, xylem, and phloem in plants are made up of different types of tissue.

tissue plasminogen activator An enzyme that dissolves blood clots. It can be produced naturally by cells in the walls of blood vessels, or prepared through the use of genetic engineering. Tissue plasminogen activator is used in the coronary arteries during heart attacks and in the cranial arteries in certain types of strokes.

titanium (tī-tā′nē-əm) *Symbol* **Ti** A shiny, white metallic element that occurs in all kinds of rocks and soils. It is lightweight, strong, and highly resistant to corrosion. Titanium alloys are used especially to make parts for aircraft and ships. Atomic number 22; atomic weight 47.87; melting point 1,660°C; boiling point 3,287°C; specific gravity 4.54; valence 2, 3, 4. See **Periodic Table.**

titanium dioxide A white powder used as an opaque white pigment. It occurs naturally as the mineral rutile. *Chemical formula:* TiO_2.

titanosaur (tī-tăn′ə-sôr′, tīt′n-) Any of various very large sauropod dinosaurs of the group Titanosauria of the Cretaceous Period. The titanosaurs were the last group of sauropods to evolve and probably include the largest sauropods ever (belonging to the genus *Argentinosaurus*). Some specimens were armored with bony plates.

titanothere (tī-tăn′ə-thîr′) Any of various extinct herbivorous hoofed mammals of the family Brontotheriidae of the Eocene and Oligocene Epochs. Titanotheres were mostly large animals resembling rhinoceroses and had massive skulls with horns and stout bodies.

titer (tī′tər) **1.** The concentration of a substance in solution or the strength of such a substance as determined by titration. **2.** The minimum volume of a solution needed to cause a particular result in titration. **3.** The concentration of antibodies present in the highest dilution of a serum sample at which visible clumps with an appropriate antigen are formed.

titration (tī-trā′shən) The process or operation of determining the concentration of a substance in solution. Titration is performed by adding to a known volume of the solution a standard reagent of known concentration in carefully measured amounts until a reaction of definite and known proportion is completed (as shown by a color change or by electrical measurement) and then calculating the unknown concentration.

Tl The symbol for **thallium.**

T lymphocyte also **T-lymphocyte** See **T cell.**

Tm The symbol for **thulium.**

TNT (tē′ĕn-tē′) Short for *trinitrotoluene.* A yellow, crystalline compound used mainly as an explosive. As it can only explode by means of a detonator and is not affected by shock, it is safe to handle and is used especially in munitions and for demolitions. *Chemical formula:* $C_7H_5N_3O_6$.

tocopherol (tō-kŏf′ə-rôl′, -rōl′) Any of a group of closely related, fat-soluble alcohols that are types of vitamin E, especially alpha-tocopherol.

toluene (tŏl′yo͞o-ēn′) A clear, toxic, flammable liquid that is used in fuels, explosives, dyes, medicines, and many industrial chemicals. Toluene consists of a methyl group attached to benzene. Also called *methylbenzene. Chemical formula:* C_7H_8.

toluidine (tə-lo͞o′ĭ-dēn′) Any of three isomeric compounds containing a benzene ring with a methyl (CH_3) and amino (NH_2) group attached to it. Toluidine is used to make dyes. *Chemical formula:* C_7H_9N.

tolyl (tŏl′əl) The group C_7H_7, derived from toluene.

Tombaugh (tŏm′bô′), **Clyde William** 1906–1997. American astronomer who discovered Pluto in 1930. The existence of a planet whose orbit is beyond Neptune, known as Planet X, had been predicted some years earlier by the American astronomer Percival Lowell based on apparent perturbations in the orbits of Neptune and Uranus. Tombaugh's careful analysis of celestial photographs did reveal a new celestial body, although it was too small to have caused the perturbations. Planet X itself was never found.

tombolo (tŏm′bə-lō′) A sand or gravel bar that connects an island to the mainland or to another island.

tombolo

St. Ninian's Island, Orkney Islands, Scotland

tomography (tō-mŏg′rə-fē) Any of several radiologic techniques for making detailed three-dimensional images of a plane section of a solid object, such as the body, while blurring out the images of other planes. See also **computerized axial tomography, positron emission tomography.**

ton (tŭn) **1.** A unit of weight in the US Customary System equal to 2,000 pounds (900 kilograms). Also called *short ton.* See Table at **measurement. 2.** A unit of weight in the US Customary System equal to 2,240 pounds (1,008 kilograms). Also called *long ton.* See Table at **measurement. 3.** See **metric ton.**

tongue (tŭng) **1.** A muscular organ in most vertebrates that is usually attached to the bottom of the mouth. In snakes, the tongue is used as a sense organ. In frogs, the tongue is chiefly used to capture prey. In mammals, the tongue is the main organ of taste and is an important organ of digestion. In humans, the tongue is used to produce speech. **2.** A similar organ in certain invertebrate animals.

tongue worm Any of various small, colorless, tongue-shaped, wormlike invertebrates of the group Pentastoma, which is considered by many zoologists to be a phylum intermediate in evolutionary development between the annelids and arthropods. Tongue worms are parasites that live embedded in the respiratory systems of vertebrates, especially reptiles and also humans, in tropical regions. They have simple nervous and digestive systems but lack circulatory and respiratory systems. Also called *pentastome.*

tonsillectomy (tŏn′sə-lĕk′tə-mē) Surgical removal of the tonsils.

tonsillitis (tŏn′sə-lī′tĭs) Inflammation of the tonsils, usually caused by bacterial infection.

tonsils (tŏn′səlz) The two oval-shaped masses of tissue at the back of the throat that lie between the mouth and the pharynx. The tonsils are thought to prevent infections of the breathing passages but often become infected themselves.

tooth (to͞oth) Plural **teeth** (tēth). **1.** Any of the hard bony structures in the mouth used to grasp and chew food and as weapons of attack and defense. In mammals and many other vertebrates, the teeth are set in sockets in the jaw. In fish and amphibians, they grow in and around the palate. See also **dentition. 2.** A similar structure in certain invertebrate animals.

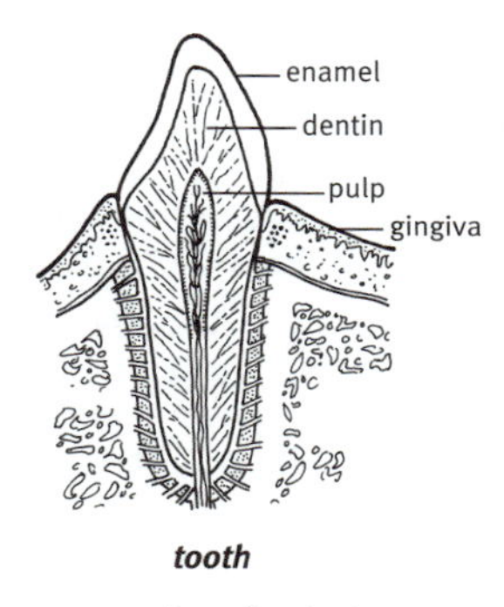

tooth

cross section of an incisor

toothed whale (to͞otht, to͞o*th*d) Any of various whales of the suborder Odontoceti, having an asymmetrical skull with one blowhole and numerous cone-shaped teeth. Toothed whales include the sperm, beluga, pilot, and beaked whales, and the narwhal, orca (killer whale), dolphins, and porpoises. Compare **baleen whale.**

topaz (tō′păz′) **1.** A colorless, blue, yellow, brown, or pink orthorhombic mineral valued as a gem. Topaz occurs as transparent or translucent prisms in silica-rich igneous rocks, such as pegmatite, and in tin-bearing rock veins. *Chemical formula:* $Al_2SiO_4(F,OH)_2$. **2.** Any of various yellow gemstones, especially a yellow variety of sapphire or corundum.

topography (tə-pŏg′rə-fē) **1.** The three-dimensional arrangement of physical attributes (such as shape, height, and depth) of a land surface in a place or region. Physical features that make up the topography of an area include mountains, valleys, plains, and bodies of water. Human-made features such as roads, railroads, and landfills are also often considered part of a region's topography. **2.** The detailed description or drawing of the physical features of a place or region, especially in the form of **contour maps**.

topology (tə-pŏl′ə-jē) The mathematical study of the geometric properties that are not normally affected by changes in the size or shape of geometric figures. In topology, a donut and a coffee cup with a handle are equivalent shapes, because each has a single hole.

top quark (tŏp) A quark with a charge of $+\frac{2}{3}$. Its mass is larger than that of all the other quarks and is about 360,000 times that of the electron. See Table at **subatomic particle.**

topsoil (tŏp′soil′) The upper portion of a soil, usually dark colored and rich in organic material. It is more or less equivalent to the upper portion of an A horizon in an ABC soil.

tornado (tôr-nā′dō) A violently rotating column of air extending from a cumulonimbus cloud to the Earth, ranging in width from a few meters to more than a kilometer and whirling at speeds between 64 km (40 mi) and 509 km (316 mi) per hour or higher with comparable updrafts in the center of the vortex. The vortex may contain several smaller vortices rotating within it. Tornadoes typically take the form of a twisting, funnel-shaped cloud extending downward from storm clouds, often reaching the ground, and dissolving into thin, ropelike clouds as the tornado dissipates. Tornadoes may travel from a few dozen meters to hundreds of kilometers along the ground. Tornadoes usually form in the tail end of violent thunderstorms, with weaker funnels sometimes forming in groups along a leading **squall line** of an advancing cold front or in areas near a **hurricane.** The strongest tornadoes, which may last several hours and travel hundreds of kilometers, can cause massive destruction in a relatively narrow strip along their path. The causes of tornado formation are not well understood.

tornado

toroid (tôr′oid′) A surface generated by rotating a closed curve about an axis that is in the same plane as the curve but does not intersect it.

torque (tôrk) The tendency of a force applied to an object to make it rotate about an axis. For a force applied at a single point, the magnitude of the torque is equal to the magnitude of the force multiplied by the distance from its point of application to an axis of rotation. Torque is also a vector quantity, equal to the vector product of the vector pointing from the axis to the point of application of force and the vector of force; torque thus points upward from a counterclockwise rotation. See also **angular momentum, lever.**

Torricelli (tō′rə-chĕl′ē), **Evangelista** 1608–1647. Italian mathematician and physicist noted for discovering that the atmosphere exerts pressure. He demonstrated that this pressure affected the level of mercury in a tube, thereby inventing the mercury barometer (1643).

torrid (tôr′ĭd) Parched with the heat of the sun.

Torrid Zone See **tropics** at **tropic** (sense 2).

torsion (tôr′shən) **1.** The stress on an object when torque is applied to it. **2.** A mathematical operation in geometry measuring how tightly a plane is twisted.

torus (tôr′əs) Plural **tori** (tôr′ī). **1.** A surface generated by rotating a circle about an axis that is in the same plane as the circle but does not intersect it. A torus resembles a donut and is a subtype of toroid. **2.** The torus-shaped apparatus that contains plasma in nuclear fusion reactors.

total eclipse (tōt′l) An eclipse in which the entire surface of a celestial object is obscured. See more at **eclipse.**

touchscreen (tŭch′skrēn′) A monitor screen that can detect and respond to something, such as a finger or stylus, pressing on it.

Tourette's syndrome (to͝o-rĕts′) or **Tourette syndrome** (to͝o-rĕt′) A neurological disorder characterized by multiple facial and other body tics, usually beginning in childhood or adolescence and often accompanied by grunts and compulsive utterances, such as interjections or obscenities. It is named for its discoverer, French neurologist Georges Gilles de la Tourette (1857–1904).

tourmaline (to͝or′mə-lĭn, -lēn′) Any of several minerals having the general chemical formula **$(Na,Ca)(Mg,Fe,Al,Li)_3Al_6(BO_3)_3Si_6O_{18}(OH)_4$**. Tourmaline occurs in many different translucent colors, usually in crystals shaped like 3-, 6-, or 9-sided prisms. It occurs in igneous and metamorphic rocks, especially in pegmatites.

Townes (tounz), **Charles Hard** Born 1915. American physicist who invented the maser, laying the foundation for the development of laser technology. In 1964 he shared with Russian physicists Nicolay Gennadiyevich Basov and Aleksandr Mikhailovich Prokhorov the Nobel Prize for physics.

Charles Townes

toxemia (tŏk-sē′mē-ə) A condition in which the blood contains bacterial toxins disseminated from a local source of infection or metabolic toxins resulting from organ failure or other disease. Also called *blood poisoning.*

toxic (tŏk′sĭk) **1.** Relating to or caused by a toxin. **2.** Capable of causing injury or death, especially by chemical means; poisonous.

toxicology (tŏk′sĭ-kŏl′ə-jē) The scientific study of poisons, of their effects and detection, and of the treatment of poisoning.

toxic shock syndrome (tŏk′sĭk) An acute infection characterized by high fever, a sunburnlike rash, vomiting, and diarrhea, followed in severe cases by shock, that is caused by a toxin-producing strain of the common bacterium *Staphylococcus aureus.* It occurs chiefly among menstruating women who use tampons.

toxin (tŏk′sĭn) A poisonous substance, especially one produced by a living organism. Toxins can be products or byproducts of ordinary metabolism, such as lactic acid, and they must be broken down or excreted before building up to dangerous levels. Toxins can facilitate survival, as with snake venom that kills or immobilizes prey, or cyanide produced by some plants as a defense against being eaten. Bacterial toxins can sometimes be neutralized with antitoxins. Compare **antitoxin.**

toxoid (tŏk′soid′) A substance that is normally toxic but has been treated to destroy its toxic properties without eliminating its capacity to stimulate the production of antitoxins by the immune system.

toxoplasmosis (tŏk′sō-plăz-mō′sĭs) An infectious disease caused by the protozoan *Toxoplasma gondii* that can be transmitted by infected humans and animals, especially cats, often by contact with feces. Toxoplasmosis can be a mild illness with fever and swollen lymph nodes, or progress to severe damage to the liver, heart, lungs, and brain. Fetuses that become infected during pregnancy may have congenital blindness and brain damage.

trace element (trās) An element present in an organism in only very small amounts but essential for normal metabolism. Iodine and cobalt are trace elements required by humans.

trace fossil A fossil consisting of an imprint of or a mark left by an organism, as opposed to physical remains. Trace fossils are produced in soft sediments and include surface tracks, molded impressions left by organisms or tissues that later decayed, and subsurface burrows or tunnelings.

tracer (trā′sər) An identifiable substance, such as a dye or radioactive isotope, that can be followed through the course of a mechanical, chemical, or biological process. Tracers are used in **radioimmunoassays** and other laboratory testing. The use of radioactive iodine, for example, can give information about thyroid gland metabolism. Also called *label.*

trachea (trā′kē-ə) Plural **tracheae** (trā′kē-ē′) or **tracheas. 1.** The tube in vertebrate animals that leads from the larynx to the bronchial tubes and carries air to the lungs. In mammals the trachea is strengthened by rings of cartilage. Also called *windpipe.* **2.** Any of the tiny tubes originating from the spiracles of many terrestrial arthropods and forming a branching network that brings air directly to body cells.

tracheary element (trā′kē-ĕr′ē) Either of two types of elongated cells, tracheids and vessel elements, found in xylem in vascular plants. Tracheids are found in all vascular plants, but vessel elements are unique to angiosperms. Both kinds of cells die at maturity, but their lignified cell walls remain as the conduits through which water is carried in the xylem. See more at **tracheid, vessel element.**

tracheid (trā′kē-ĭd, -kēd′) An elongated, water-conducting cell in xylem, one of the two kinds of tracheary elements. Tracheids have pits where the cell wall is modified into a thin membrane, across which water flows from tracheid to tracheid. The cells die when mature, leaving only their lignified cell walls. Tracheids are found in all vascular plants. Compare **vessel element.**

tracheophyte (trā′kē-ə-fīt′) See **vascular plant.**

tracheostomy (trā′kē-ŏs′tə-mē) Surgical construction of an opening in the trachea, usually by making an incision in the front of the neck, for the insertion of a catheter or tube to facilitate breathing.

trachoma (trə-kō′mə) A contagious disease of the conjunctiva and cornea, caused by the bacteria *Chlamydia trachomatis* and characterized by granules of inflammatory tissue. It is a major cause of blindness in Asia and Africa.

trachyte (trā′kīt′, trăk′īt′) A light-colored, fine-grained igneous rock consisting primarily of alkali feldspar together with some mafic minerals, especially hornblende. Unlike most igneous rocks, trachyte has little or no quartz. Trachyte is the fine-grained equivalent of syenite.

tract (trăkt) **1.** A series of body organs that work together to perform a specialized function, such as digestion. **2.** A bundle of nerve fibers, especially in the central nervous system, that begin and end in the same place and share a common function.

traction (trăk′shən) **1.** Static friction, as of a wheel on a track or a tire on a road. See more at **friction. 2.** A sustained pulling force applied mechanically to a part of the body by means of a weighted apparatus in order to correct the position of fractured or dislocated bones, especially of the arm, leg, or neck.

trade winds (trād) Winds that blow steadily from east to west and toward the equator over most of the Torrid Zone. The trade winds are caused by hot air rising at the equator, with cool air moving in to take its place from the north and from the south. The winds are deflected westward because of the Earth's west-to-east rotation. Compare **antitrades.**

trait (trāt) A genetically determined characteristic or condition. Traits may be physical, such as hair color or leaf shape, or they may be behavioral, such as nesting in birds and burrowing in rodents. Traits typically result from the combined action of several genes, though some traits are expressed by a single gene.

trajectory (trə-jĕk′tə-rē) **1.** *Physics.* The line or curve described by an object moving through space. **2.** *Mathematics.* A curve or surface that passes through a given set of points or intersects a given series of curves or surfaces at a constant angle.

transcription (trăn-skrĭp′shən) The process in a cell by which genetic material is copied from a strand of DNA to a complementary strand of RNA (called **messenger RNA**). In eukaryotes, transcription takes place in the nucleus before messenger RNA is transported to the ribosomes for protein synthesis. Compare **translation.**

transducer (trăns-do͞o′sər) A device that converts one type of energy or signal into another. For example, a microphone is a transducer that converts sound waves into electric impulses; an electric motor is a transducer that converts electricity into mechanical energy.

trans fat (trăns) A fat containing trans fatty acids.

trans fatty acid A fatty acid that is commonly produced by the partial hydrogenation of the unsaturated fatty acid vegetable oils. Trans fatty acids are present in hardened vegetable oils, most margarines, commercial baked foods, and many fried foods. An excess of these fats in the diet raises lipid levels in the blood. The term *trans* refers to the opposed positioning of hydrogen atoms when unsaturated fats are partially hydrogenated.

transfer RNA (trăns′fər) See under **RNA.**

transfinite number (trăns-fī′nīt′) A number that is greater than any finite number.

transformation (trăns′fər-mā′shən) **1.** The genetic alteration of a bacteria cell by the introduction of DNA from another cell or from a virus. **Plasmids,** which contain extrachromosomal DNA, are used to transform bacteria in recombinant DNA research. **2.** The change undergone by an animal cell upon infection by a cancer-causing virus.

transformer (trăns-fôr′mər) A device used to change the voltage of an alternating current in one circuit to a different voltage in a second circuit, or to partially isolate two circuits from each other. Transformers consist of two or more coils of conducting material, such as wire, wrapped around a core (often made of iron). The magnetic field produced by an alternating current in one coil induces a similar current in the other coils. ► If there are fewer turns on the coil that carries the source of the power than there are on a second coil, the second coil will provide the same power but at a higher voltage. This is called a **step-up transformer.** ► If there are fewer turns on the second coil than on the source coil, the outgoing power will have a lower voltage. This is called a **step-down transformer.** Compare **converter, rectifier.**

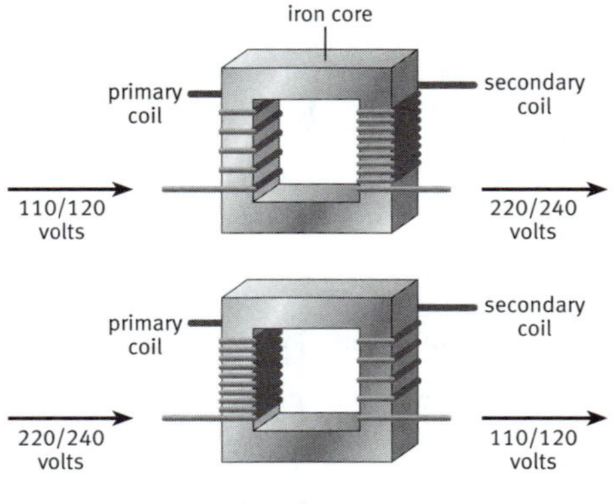

transformer

The increase or decrease in the voltage of a signal is related to the ratio of the number of turns of wire in the coils. In the upper transformer, the outgoing voltage is doubled because the secondary coil has twice as many turns as the first. In the lower transformer, the voltage is halved.

transform fault (trăns′fôrm′) A type of strike-slip fault that accommodates the relative horizontal slip between other tectonic elements, such as tectonic plates, and is common along the edges of plates in mid-ocean ridge regions. The lateral displacement along transform faults often ends or changes form abruptly. See Note at **fault.**

transfusion (trăns-fyo͞o′zhən) The transfer of blood or a component of blood, such as red blood cells, plasma, or platelets, from one person to another to replace losses caused by injury, surgery, or disease. Donated blood products are tested for blood type and certain infectious diseases and stored in blood banks until they are used. The blood of the donor is shown to be histologically compatible, or *crossmatched,* with that of the recipient before transfusion. See more at **Rh factor.** See Note at **blood type.**

transgenic (trăns-jĕn′ĭk) Relating to an organism whose genome has been altered by the transfer of a gene or genes from another species or breed. Transgenic organisms are used in research to help determine the function of the inserted gene, while in industry they are used to produce a desired substance.

transgression (trăns-grĕsh′ən) A relative rise in sea level resulting in deposition of marine strata over terrestrial strata. The sequence of sedimentary strata formed by transgressions and regressions provides information about the changes in sea level during a particular geologic time. Compare **regression** (sense 4).

transistor (trăn-zĭs′tər) An electronic device that controls the flow of an electric current, most often used as an amplifier or switch. Transistors usually consist of three layers of semiconductor material, in which the flow of electric current across the outer layer is regulated by the voltage or current applied at the middle layer. Having replaced the vacuum tube, transistors are the basis of much modern electronic technology, including the microprocessor. See also **logic circuit, logic gate.**

transit (trăn′sĭt) **1.** The passage of a smaller celestial body or its shadow across the disk of a larger celestial body. As observed from Earth, Mercury and Venus are the only planets of the solar system that make transits of the Sun, because they are the only planets with orbits that lie between Earth and the Sun. Mercury makes an average of 13 transits of the Sun each century. Transits of Venus across the Sun are much rarer, with only 7 of them having occurred between 1639 and 2004. In contrast, transits of Jupiter's moons across its disk are common occurrences. Compare **occultation. 2.** The passage of a celestial body across the celestial meridian (the great circle on the celestial sphere passing through the celestial poles and an observer's zenith). For any observer, the object is at its highest in the sky at its transit of the observer's meridian. See more at **celestial meridian.**

transition element (trăn-zĭsh′ən) Any of the metallic elements within Groups 3 through 12 in the Periodic Table. All the transition metals have two electrons in their outermost shell, and all but zinc, cadmium, and mercury have an incompletely filled inner shell (just inside the outermost shell). Transition elements form alloys easily, have high melting points, and have more than one valence because of their incomplete inner shells. See **Periodic Table.** See Note at **metal.**

transition region 1. The area in a seed-bearing plant where the vascular tissue of the root changes into the vascular tissue of the stem. In eudicotyledons, the roots have a solid cylinder of vascular tissue surrounded by cortex. The vascular tissues gradually branch apart and reorient themselves in the stem around a central pith. **2.** The thin layer of the solar atmosphere that separates the chromosphere from the corona. In the transition region, temperatures increase from a relatively cool 20,000°K in the chromosphere to the 1,000,000°K and higher temperatures of the corona. Transition region chemistry involves the ionization of hydrogen and other elements; the light produced by this region comes from ionized forms of carbon, oxygen, and silicon.

transitive (trăn′sĭ-tĭv) Of or relating to a mathematical or logical relation between three elements such that if the relation holds between the first and second elements and between the second and third elements, it necessarily holds between the first and third elements. The relation of being greater than in mathematics is transitive, since if $a > b$ and $b > c$, then $a > c$.

translation (trăns-lā′shən) **1.** *Biochemistry.* The process in the ribosomes of a cell by which a strand of messenger RNA directs the assembly of a sequence of amino acids to make a protein. Compare **transcription. 2.** *Physics.* Motion of a body in which every point of the body moves parallel to and the same distance as every other point of the body. **3.** *Mathematics.* The changing of the coordinates of points to coordinates that are referred to new axes that are parallel to the old axes.

translocation (trăns′lō-kā′shən) **1.** A chromosomal aberration in which a chromosomal segment changes position, usually moving from one chromosome to a different, nonhomologous chromosome. In one type of Down Syndrome, for example, translocation of a large segment of chromosome 21 to another chromosome results in an individual who has the genetic equivalent of three chromosomes 21 and thus has the phenotype of Down syndrome but who has a normal total number of chromosomes. A translocation within a given chromosome is called a *shift.* **2.** A chromosomal segment that is translocated.

translucent (trăns-lo͞o′sənt) Allowing radiation (most commonly light) to pass through, but causing diffusion. Frosted glass, for example, is translucent to visible light. Compare **transparent.**

transmitter (trăns′mĭt-ər) A device that converts sound, light, or electrical signals into radio, microwave, or other electrical signals of sufficient strength for the purpose of telecommunication. Compare **receiver.**

transmutation (trăns′myo͞o-tā′shən) The changing of one chemical element into another. Transmutations occur naturally through radioactive decay, or artificially by bombarding the nucleus of a substance with subatomic particles.

transonic (trăn-sŏn′ĭk) Relating to or capable of speeds at or near the speed of sound (at or approaching Mach 1) or to aerodynamic conditions for bodies travelling at such speeds. Compare **hypersonic, subsonic, supersonic.**

transparent (trăns-pâr′ənt) Allowing radiation or matter to pass through with little or no resistance or diffusion. Compare **opaque, translucent.** See Note at **glass.**

transpiration (trăn′spə-rā′shən) The process of giving off vapor containing water and waste products, especially through the stomata on leaves or the pores of the skin.

transplant (trăns′plănt′) **1.** A plant that has been uprooted and replanted. **2.** A surgical procedure in a human or animal in which a body tissue or organ is transferred from a donor to a recipient or from one part of the body to another. Heart, lung, liver, kidney, corneal, and bone-marrow transplants are performed to treat life-threatening illness. Donated tissue must be histocompatible with that of the recipient to prevent immunological rejection. See also **graft.**

transponder (trăn-spŏn′dər) A radio or radar transmitter and receiver that responds to an incoming signal either by broadcasting its own predetermined signal (as in aircraft identification systems) or by relaying the incoming signal at a different frequency (as in satellite communications).

transpose (trăns-pōz′) To move a term or quantity from one side of an algebraic equation to the other by adding or subtracting that term to or from both sides. By subtracting 2 from both sides of the equation $2 + x = 4$, one can transpose the 2 to the other side, yielding $x = 4 - 2$, and thus determine that x equals 2.

transposon (trăns-pō′zŏn) A segment of DNA that is capable of independently replicating itself and inserting the copy into a new position within the same or another chromosome or plasmid. Transposons act somewhat similarly to viruses and in humans are an underlying cause of hemophilia, certain cancers, and other diseases. In other organisms, they can become a permanent and even beneficial part of the genome, as in maize corn, where transposons account for half the genome, and certain bacteria, where genes for antibiotic resistance can spread by means of transposons. Also called *jumping gene.*

transuranic (trăns′yo͝o-răn′ĭk, -rā′nĭk) Having an atomic number greater than 92.

transverse dune (trăns-vûrs′, trăns′vûrs′) A large, strongly asymmetrical, elongated dune lying at right angles to the prevailing wind direction. Transverse dunes have a gently sloping windward side and a steeply sloping leeward side. They generally form in areas of sparse vegetation and abundant sand. Most beach dunes are transverse dunes. See illustration at **dune.** Compare **longitudinal dune.**

A CLOSER LOOK **transpiration**

Plants need much more water than animals do. But why? Plants use water not only to carry nutrients throughout their tissues, but also to exchange gases with the air in the process known as *transpiration.* Air, which contains the carbon dioxide that plant cells need for photosynthesis, enters the plant mainly through the stomata (tiny holes under its leaves). The air travels through tiny spaces in the leaf tissue to the cells that conduct photosynthesis. These cells are coated with a thin layer of water. The cell walls do not permit gases to pass through them, but the carbon dioxide can move across the cell walls by dissolving in the water on their surface. The cells remove the carbon dioxide from the water and use the same water to carry out oxygen, the main waste product of photosynthesis. All this mixing of water and air in transpiration, though, has one drawback: more than 90 percent of the water that a plant's roots suck up is lost by evaporation through the stomata. This is why a plant always needs water and why plants that live in dry climates, such as cacti, have reduced leaf surfaces from which less water can escape.

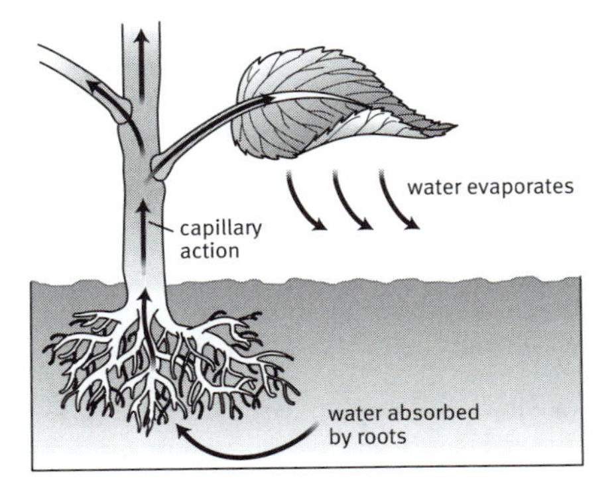

transverse wave A wave that oscillates perpendicular to the axis along which the wave travels. Electromagnetic waves are transverse waves, since the electric and magnetic fields oscillate at a right angle to the direction of motion. Waves in bodies of water are also transverse waves, since the molecules of water oscillate up and down perpendicular to the direction of the wave's motion. Compare **longitudinal wave.** See more at **wave.**

trapezium (trə-pē′zē-əm) Plural **trapeziums** or **trapezia.** A four-sided plane figure having no parallel sides.

trapezoid (trăp′ĭ-zoid′) A four-sided plane figure having two parallel sides.

trauma (trô′mə, trou′-) **1.** Severe bodily injury, as from a gunshot wound or a motor vehicle accident. **2.** Psychological or emotional injury caused by a deeply disturbing experience.

traumatology (trô′mə-tŏl′ə-jē, trou′-) The branch of medicine that deals with the treatment of serious wounds and injuries.

travertine (trăv′ər-tēn′, -tĭn) A white, tan, or cream-colored form of limestone, often having a fibrous or concentric appearance. Travertine is formed through the rapid precipitation of calcium carbonate, especially at the mouth of a hot spring or in limestone caves, where it forms stalactites and stalagmites. It is similar to, but harder, than **tufa.**

tree (trē) Any of a wide variety of perennial plants typically having a single woody stem, and usually branches and leaves. Many species of both gymnosperms (notably the conifers) and angiosperms grow in the form of trees. The ancient forests of the Devonian, Mississippian, and Pennsylvanian periods of the Paleozoic Era were dominated by trees belonging to groups of seedless plants such as the lycophytes. The strength and height of trees are made possible by the supportive conductive tissue known as **vascular tissue**.

tree line See **timberline.**

trematode (trĕm′ə-tōd′) Any of numerous parasitic flatworms of the class Trematoda, having a thick outer cuticle and one or more suckers or hooks for attaching to host tissue. Flatworms include both external and internal parasites of animal hosts, and some cause diseases of humans in tropical regions, such as schistosomiasis. Liver flukes, blood flukes, and planarians are flatworms. Also called *fluke.*

tremor (trĕm′ər) **1.** A relatively minor seismic shaking or vibrating movement. Tremors often precede larger earthquakes or volcanic eruptions. **2.** An involuntary shaking or trembling of the head or extremities that can be idiopathic or associated with any of various medical conditions, such as Parkinson's disease.

trench (trĕnch) A long, steep-sided valley on the ocean floor. Trenches form when one tectonic plate slides beneath another plate at a subduction zone. The Marianas Trench, located in the western Pacific east of the Philippines, is the deepest known trench (10,924 m or 35,831 ft) and the deepest area in the ocean.

triangle (trī′ăng′gəl) A closed geometric figure consisting of three sides.

triangulation (trī-ăng′gyə-lā′shən) A method of determining the relative positions of points in space by measuring the distances, and sometimes angles, between those points and other reference points whose positions are known. Triangulation often involves the use of **trigonometry**. It is commonly used in the navigation of aircraft and boats, and is the method used in the **Global Positioning System**, in which the reference points are satellites.

Triassic (trī-ăs′ĭk) The earliest period of the Mesozoic Era, from about 245 to 208 million years ago. During the early part of the Triassic Period the supercontinent Pangaea was located along the equator; by the end of the Triassic it had started to split up. Land life diversified in the Triassic in response to the mass extinctions of the end of the Paleozoic. Conifers, cycads, marine reptiles, dinosaurs, and the earliest mammals first appeared. See Chart at **geologic time.**

triboelectricity (trī′bō-ĭ-lĕk-trĭs′ĭ-tē, trĭb′ō-) An electrical charge produced by friction between two objects that are nonconductive. Rubbing glass with fur, or a comb through the hair, can built up triboelectricity. Most everyday static electricity is triboelectric.

tributary (trĭb′yə-tĕr′ē) A stream that flows into a river, a larger stream, or a lake.

tributary

a system of tributaries flowing into the Mispillion River, Delaware

triceps (trī′sĕps′) The muscle at the back of the upper arm that raises and lowers the forearm. The triceps has three points of attachment to bone at its origin.

triceratops (trī-sĕr′ə-tŏps′) A large herbivorous dinosaur of the genus *Triceratops* of the late Cretaceous Period, measuring up to 7.6 m (25 ft) in length. Triceratops had a squat, tanklike body, a beaklike mouth with a short horn over it, and a long horn over each eye. The back of its neck was covered with a wide, bony plate.

trichinosis (trĭk′ə-nō′sĭs) A disease caused by the parasitic nematode *Trichinella spiralis* that is ingested as larvae found in the muscle tissue of undercooked meat, especially pork. Once digested, the larvae develop into adult worms in the intestinal tract. Trichinosis is characterized by fever, intestinal pain, nausea, muscular pain, and edema.

trichloroacetic acid (trī-klôr′ō-ə-sē′tĭk) A colorless, deliquescent, corrosive, crystalline compound used as a herbicide and topically as an astringent and antiseptic. *Chemical formula:* $C_2Cl_3O_2H$.

trichloroethane (trī-klôr′ō-ĕth′ān′) A colorless, nonflammable compound having a sweet odor, existing in two isomers. It is used as a solvent for adhesives, pesticides, and lubricants, and in industrial cleaning solutions. *Chemical formula:* $C_2H_3Cl_3$.

trichloroethylene (trī-klôr′ō-ĕth′ə-lēn′) A heavy, colorless, toxic liquid. It is used to degrease metals, to extract oil from nuts and fruit, as a refrigerant, in dry cleaning, and as a fumigant. *Chemical formula:* C_2HCl_3.

trichogyne (trĭk′ə-jīn′, -gīn′) A hairlike terminal process forming the receptive part of the female reproductive structure (called the gametangium) in red algae and certain ascomycete and basidiomycete fungi. Male gametes attach themselves to the trichogyne.

trichome (trĭk′ōm′, trī′kōm′) One of the hairlike or bristlelike outgrowths on the epidermis of a plant. Trichomes serve a variety of functions, depending on their location. As root hairs (and as leaf hairs in epiphytes), trichomes absorb water and minerals. As leaf hairs, they reflect radiation, lower plant temperature, and reduce water loss. They also provide defense against insects.

trichomonas (trĭk′ə-mō′năs′) Any of various flagellated protozoans of the genus *Trichomonas* that are parasitic and inhabit the digestive and genitourinary tract of the host. *Trichomonas vaginalis* is a common cause of vaginal infections in humans.

trichopteran (trĭ-kŏp′tər-ən) Any of various small, freshwater insects of the order Trichoptera. Trichopterans have two pairs of wings covered with hairs, and often have hair on the head and thorax. The caddisflies are trichopterans.

triclinic (trī-klĭn′ĭk) Relating to a crystal having three axes of different lengths intersecting at oblique angles. The mineral microcline (a type of feldspar) has triclinic crystals. See illustration at **crystal.**

tricuspid (trī-kŭs′pĭd) *Adjective.* **1.** Having three points or cusps. —*Noun.* **2.** A tooth having three points or cusps, especially a molar. **3.** See **tricuspid valve.**

tricuspid valve A valve of the heart, usually having three cusps, that is located between the right atrium and right ventricle. The tricuspid valve prevents the backflow of blood to the right atrium.

tricyclic (trī-sī′klĭk, -sĭk′lĭk) **1.** Relating to a chemical compound having three closed rings. Anthracene is a tricyclic hydrocarbon. **2.** Relating to a class of drugs used to treat depression and having a tricyclic chemical structure consisting of two benzene rings fused to opposite sides of a seven-member ring. The seven-member ring consists of six carbon atoms and one nitrogen atom. Tricyclic antidepressants enhance the activity of monoamine neurotransmitters in the brain by inhibiting their reuptake by the cells that secrete them. **3.** Composed of or arranged in three distinct whorls, as the petals of a flower.

trigeminal nerve (trī-jĕm′ə-nəl) Either of the fifth pair of cranial nerves, having sensory and motor functions in the face, teeth, mouth, and nasal cavity.

trigeneration (trī′jĕn′ə-rā′shən) A process in which an industrial facility uses its waste energy to produce heat or electricity as well as cooling. Compare **cogeneration.**

triglyceride (trī-glĭs′ə-rīd′) Any of a class of organic compounds that are esters consisting of three fatty acids joined to glycerol. The fatty acids may be the same or may be different. Triglycerides are the chief lipids constituting fats and oils and function to store chemical energy in plants and animals.

trigonal (trī-gō′nəl) Relating to a crystal having three axes of equal length intersecting at oblique angles. This crystal system is considered a subset of the hexagonal system. The mineral quartz has trigonal crystals. Also called *rhombohedral.* See illustration at **crystal.**

trigonometric function (trĭg′ə-nə-mĕt′rĭk) A function of an angle, as the sine, cosine, or tangent, whose value is expressed as a ratio of two of the sides of the right triangle that contains the angle.

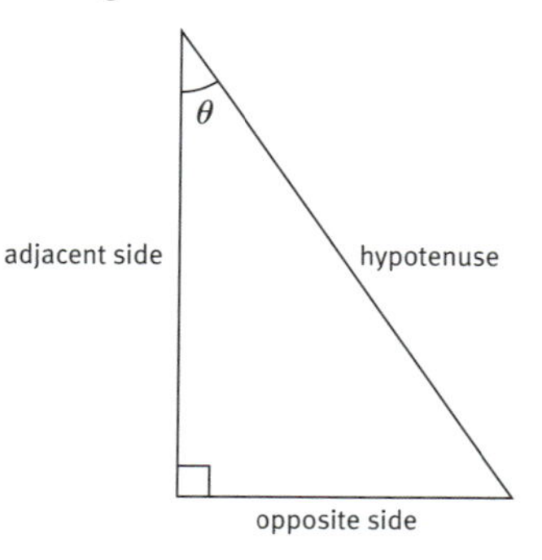

trigonometric function

In a right triangle, the trigonometric functions are:

$$\textit{sine}\ \theta = \frac{\textit{opposite}}{\textit{hypotenuse}} \quad \textit{cosine}\ \theta = \frac{\textit{adjacent}}{\textit{hypotenuse}}$$

$$\textit{tangent}\ \theta = \frac{\textit{opposite}}{\textit{adjacent}}$$

trigonometry (trĭg′ə-nŏm′ĭ-trē) The study of the properties and uses of trigonometric functions.

trihalomethane (trī′hăl-ə-mĕth′ān′) Any of various organic compounds containing three halogen atoms substituted for three of the hydrogen atoms in a methane molecule. It can occur in chlorinated water as a result of reaction between organic materials in the water and chlorine added as a disinfectant. Chloroform is a trihalomethane.

trilobite (trī′lə-bīt′) Any of numerous extinct and mostly small arthropods of the subphylum Trilobita that lived during the Paleozoic Era and are extremely common as fossils. Trilobites had a hard outer covering divided into three lengthwise and three widthwise sections. Their heads had two prominent compound eyes similar in structure to those of modern insects.

trilobite

trimer (trī′mər) Any of various chemical compounds made of three smaller identical or similar molecules (called monomers) that are linked together. Trimers are linked by hydrogen bonds, coordinate bonds, or covalent bonds. Raffinose is a trimer composed of the monomers glucose, fructose, and galactose.

trimerous (trĭm′ər-əs) **1.** Having three similar parts or segments. **2.** Having flower parts, such as petals, sepals, and stamens, in sets of three.

trinitrotoluene (trī-nī′trō-tŏl′yo͞o-ēn′) See **TNT.**

trinomial (trī-nō′mē-əl) **1.** A mathematical expression that is the sum of three monomials, such as $ax^2 + bx - c$. **2.** A taxonomic designation indicating genus, species, and subspecies or variety, as in *Brassica oleracea botrytis,* the cauliflower.

triode (trī′ōd′) An electron tube used mostly for signal amplification, consisting of a cathode and anode (or **plate**) as in a **diode**, and an intervening wire mesh called the **control grid.** With little voltage on the grid, large currents can flow between the cathode and plate, but small variations in the voltage on the grid cause large variations in this current, allowing for large amplification of a signal applied to the control grid. See also **pentode, tetrode.**

triol (trī′ôl′, -ōl′) An alcohol containing three hydroxyl groups (OH) per molecule.

triple alpha process (trĭp′əl) A nuclear fusion reaction in which three helium nuclei (alpha particles) fuse to form a carbon nucleus, thereby releasing energy. Two helium nuclei fuse to form an unstable isotope of beryllium, which under conditions of sufficient temperature and pressure fuses with a third helium nucleus to form carbon before it decays. Triple alpha processes take place in stars in which large quantities of helium have accumulated as the product of **proton-proton chain** and **carbon cycle** reactions.

triple point The temperature and pressure at which a substance can exist in equilibrium in the liquid, solid, and gaseous states. The triple point of pure water is at 0.01°C (273.16K, 32.01°F) and 4.58 mm (611.2Pa) of mercury and is used to calibrate thermometers. Compare **critical point.**

trisomy (trī-sō′mē, trī′sō′-) The condition of having three copies of a given chromosome in each body cell rather than the normal number of two.

trisomy 21 See **Down syndrome.**

tritium (trĭt′ē-əm, trĭsh′ē-əm) A radioactive isotope of hydrogen whose nucleus has one proton and two neutrons with atomic mass of about 3 and a half life of 12.5 years. Tritium is rare in nature but can be made artificially in nuclear reactions. It is used in thermonuclear weapons and luminescent paints, and sometimes as a tracer. See more at **hydrogen.**

trivalent (trī-vā′lənt) *Chemistry.* Having a valence of 3.

trivial name (trĭv′ē-əl) **1.** A common or vernacular name as distinguished from a scientific name, as *chimpanzee* for *Pan troglodytes.* **2.** A common, historic, or convenient name for a substance. The trivial name is often derived from the source in which the substance was discovered. It is not systematic and is not used in modern official nomenclature. **Sucrose** is the trivial name for β-*D-fructofuranosyl*-α-*D-glucopyranoside.* Compare **chemical name.**

tRNA Abbreviation of **transfer RNA.**

trophic (trŏf′ĭk) Relating to the feeding habits of different organisms in a food chain or web.

trophic level Any of the sequential stages in a food chain, occupied by producers at the bottom and in turn by primary, secondary, and tertiary consumers. Decomposers (detritivores) are sometimes considered to occupy their own trophic level. ► The rate at which energy is transferred from one trophic level to the next is called the **ecological efficiency.** Consumers at each level convert an average of only about 10 percent of the chemical energy in their food to their own organic tissue. Since plants can only convert approximately 1 percent of incident sunlight into chemical energy at the lowest trophic level (the bottom of the food chain), the percentage of the energy in incident sunlight that reaches a tertiary consumer is about 0.0001.

trophoblast (trō′fə-blăst′) The outermost layer of cells of the blastocyst, which attaches the fertilized ovum to the uterine wall and serves as a nutritive pathway for the embryo. The trophoblast eventually differentiates into such tissues as the amnion, the placenta, and the umbilical cord.

tropic (trŏp′ĭk) **1.** Either of the two parallels of latitude representing the points farthest north and south at which the Sun can shine directly overhead. The northern tropic is the **Tropic of Cancer** and the southern one is the **Tropic of Capricorn. 2. tropics.** The region of the Earth lying between these latitudes. The tropics are generally the warmest and most humid region of the Earth. Also called *Torrid Zone.*

tropical cyclone (trŏp′ĭ-kəl) See under **cyclone.**

tropical depression A tropical cyclone having sustained surface winds less than 39 mi (63 km) per hour. See Note at **cyclone.**

tropical storm A tropical cyclone having sustained surface winds between 39 and 73 mi (63 and 118 km) per hour. See Note at **cyclone.**

tropical year A solar year. See under **solar time.**

Tropic of Cancer The parallel of latitude approximately 23°27' north. It forms the boundary between the Torrid and North Temperate zones.

Tropic of Capricorn The parallel of latitude approximately 23°27' south. It forms the boundary between the Torrid and South Temperate zones.

tropism (trō′pĭz′əm) The growth or movement of a living organism or anatomical structure toward or away from an external stimulus, such as light, heat, or gravity. See also **geotropism, hydrotropism, phototropism.** —*Adjective* **tropistic.**

troponin (trō′pə-nĭn, trŏp′ə-) One of the proteins that make up the thin filaments of muscle tissue and that regulate muscle contraction and relaxation. Troponin occurs in three forms bound together in a complex. One of the three forms is a receptor of calcium ions that induces structural changes that allow the actin in the thin filaments to interact with myosin, causing contraction.

tropopause (trō′pə-pôz′, trŏp′ə-) The boundary between the upper troposphere and the lower stratosphere, varying in altitude from about 8 km (5 mi) at the poles to 18 km (11 mi) at the equator.

tropophyte (trō′pə-fīt′, trŏp′ə-) A plant adapted to climatic conditions in which periods of abundant precipitation alternate with periods of dryness.

troposphere (trō′pə-sfîr′, trŏp′ə-) The lowest and densest region of the Earth's atmosphere, extending from the Earth's surface to the tropopause. The troposphere is characterized by temperatures that decrease with increasing altitude. At the top of this region, temperatures are close to –55°C (–67°F). The weather, major wind systems, and cloud formations occur mostly in the troposphere. See also **exosphere, mesosphere, stratosphere, thermosphere.** See illustration at **atmosphere.**

trough (trôf) **1.** The part of a wave with the least magnitude; the lowest part of a wave. Compare **crest.** See more at **wave. 2.** A narrow, elongated region of relatively low atmospheric pressure occurring at the ground surface or in the upper atmosphere, and often associated with a front. Compare **ridge.**

troy weight (troi) A system of weights and measures in which the grain is the same as in the avoirdupois system, and a pound contains 12 ounces, or 5,760 grains. Troy weight is used primarily by miners and gold dealers. Compare **avoirdupois weight.**

true bug (tro͞o) Any of various insects of the group Heteroptera. True bugs usually have soft flat bodies, well-developed antennae, and stink glands. They include the water bugs, water striders, bedbugs, cinch bugs, lace bugs, and assassin bugs. Some scientists classify the true bugs as a suborder of the order Hemiptera rather than as a separate insect order. See Note at **bug.**

true fruit A fruit in which all tissues are derived from a ripened ovary and its contents. See more at **fruit.** Compare **accessory fruit.**

true north See **geographic north.**

truth-value (tro͞oth′văl′yo͞o) The truth or falsity of a logical proposition.

trypanosome (trĭ-păn′ə-sōm′) Any of various parasitic flagellate protozoans of the genus *Trypanosoma* that can cause serious diseases, such as sleeping sickness. They are transmitted by the bite of certain insects, such as tsetse flies.

trypanosomiasis (trĭ-păn′ə-sō-mī′ə-sĭs) A disease or infection caused by a trypanosome.

trypsin (trĭp′sĭn) An enzyme that aids digestion by breaking down proteins. It is produced by the pancreas and secreted into the small intestine, where it catalyzes the cleavage of peptide bonds connecting arginine or lysine to other amino acids.

tryptophan (trĭp′tə-făn′) An essential amino acid. *Chemical formula:* $C_{11}H_{12}N_2O_2$. See more at **amino acid.**

tsetse fly (tsĕt′sē) Any of several bloodsucking African flies of the genus *Glossina,* two species of which (*G. palpalis* and *G. morsitans*) often carry and transmit trypanosomes, the protozoans that cause sleeping sickness.

TSH Abbreviation of **thyroid-stimulating hormone.**

tsunami (tso͞o-nä′mē) A very large ocean wave that is caused by an underwater earthquake or volcanic eruption and often causes extreme destruction when it strikes land. Tsunamis can have heights of up to 30 m (98 ft) and reach speeds of 950 km (589 mi) per hour. They are characterized by long wavelengths of up to 200 km (124 mi) and long periods, usually between 10 and 60 minutes. See Note at **tidal wave.**

tubal ligation (to͞o′bəl) A method of sterilization in which the fallopian tubes are surgically tied.

tube cell (to͞ob) The cell in the pollen grain that develops into the pollen tube (the tube which conveys the male gametes of seed-bearing plants to the ovule).

tube foot One of the numerous external, fluid-filled muscular tubes of echinoderms, such as the starfish or sea urchin, used for locomotion, respiration, and grasping food or prey.

tuber (to͞o′bər) The thickened part of an underground stem of a plant, such as the potato, bearing buds from which new plant shoots arise. Compare **bulb, corm, rhizome, runner.**

tubercle (to͞o′bər-kəl) A small rounded projection, swelling, or lump, as on the roots of legumes or on bodily tissue, especially the cluster of inflammatory cells that form in the lungs in tuberculosis.

tuberculosis (to͝o-bûr′kyə-lō′sĭs) An infectious disease caused by the bacterium *Mycobacterium tuberculosis* that is transmitted through inhalation and is characterized by cough, fever, shortness of breath, weight loss, and the appearance of inflammatory substances and tubercles in the lungs. Tuberculosis is highly contagious and can spread to other parts of the body, especially in people with weakened immune systems. Although the incidence of the disease has declined since the introduction of antibiotic treatment in the 1950's, it is still a major public-health problem throughout the world, especially in Asia and Africa.

tufa (to͞o′fə) A soft, friable, and porous sedimentary rock consisting of calcium carbonate and formed by the evaporation of water, especially at the mouth of a hot spring or on a drying lakebed. It is similar to, but harder than, **travertine.**

tuff (tŭf) A rock made up of particles of volcanic ash, varying in size from fine sand to coarse gravel.

tularemia (to͞o′lə-rē′mē-ə) An infectious disease characterized by intermittent fever and swelling of the lymph nodes, caused by the bacterium *Francisella tularensis.* It chiefly affects wild rabbits and rodents but can also be transmitted to humans through the bite of various insects or through contact with infected animals.

tumor (to͞o′mər) See **neoplasm.** See Note at **cancer.**

tumor necrosis factor A protein that is produced by monocytes and macrophages and is able to attack and destroy tumor cells. It can also exacerbate certain chronic inflammatory diseases.

tumor suppressor gene A gene that suppresses cellular proliferation. When inherited in a mutated state in which it is no longer suppressive, it is associated with the development of various cancers, including most familial cancers.

tundra (tŭn′drə) A cold, treeless, usually lowland area of far northern regions. The lower strata of soil of tundras are permanently frozen, but in summer the top layer of soil thaws and can support low-growing mosses, lichens, grasses, and small shrubs.

tungsten (tŭng′stən) *Symbol* **W** A hard, gray to white metallic element that is very resistant to corrosion. It has the highest melting point of all elements, and it retains its strength at high temperatures. It is used to make light-bulb filaments and to increase the hardness and strength of steel. Atomic number 74; atomic weight 183.84; melting point 3,410°C; boiling point 5,900°C; specific gravity 19.3 (20°C); valence 2, 3, 4, 5, 6. Also called *wolfram.* See **Periodic Table.**

tungsten carbide An inorganic carbon compound that forms a fine gray powder whose grains are dense and extremely hard. Tungsten carbide is used in tools, dies, wear-resistant machine parts, and abrasives. *Chemical formula:* **WC.**

tungsten lamp An incandescent electric lamp having a tungsten filament.

tunicate (to͞o′nĭ-kĭt) Any of various primitive marine chordate animals of the subphylum Tunicata, having a rounded or cylindrical body that is enclosed in a tough outer covering. Tunicates start out life as free-swimming, tadpolelike animals with a notochord (a primitive backbone), but many, such as the sea squirts, lose the notochord and most of their nervous system as adults and become fixed to rocks or other objects. Tunicates often form colonies.

tunneling (tŭn′ə-lĭng) See **quantum tunneling.**

turbidite (tûr′bĭ-dīt′) A sedimentary deposit formed by a turbidity current. Turbidites usually consist of a sequence of sediments in which the bottom layers contain the coarsest grains and the upper layers the finest, such as a sequence of sand that is overlain by silt, which in turn is overlain by clay.

turbidity current (tər-bĭd′ĭ-tē) A swift downhill current in water, air, or other fluid, triggered by the weight of suspended material such as silt in a current flowing down a continental shelf or snow in an avalanche.

turbine (tûr′bĭn, -bīn′) Any of various machines in which the kinetic energy of a moving fluid, such as water, steam, or gas, is converted to rotary motion. Turbines are used in boat propulsion systems, hydroelectric power generators, and jet aircraft engines. See also **gas turbine.**

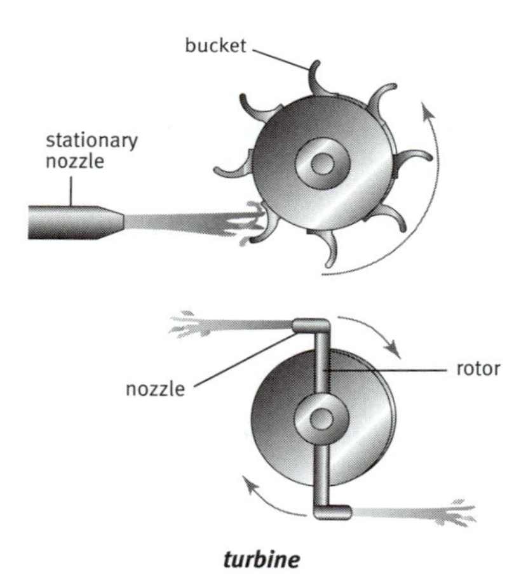

turbine

top: *impulse turbine;* bottom: *reaction turbine*

turbofan (tûr′bō-făn′) A type of gas turbine in which the fan driving air into a **turbojet** also forces additional air around the outside of the turbine, combining it with the exhaust of the turbojet to provide thrust. Turbofans are quieter than simple turbojets and somewhat more fuel efficient, and are widely used in commercial aircraft.

turbojet (tûr′bō-jĕt′) **1.** A type of gas turbine in which air, drawn into a combustion chamber by fans, is rapidly heated by combusted fuel, creating air pressure that drives turbines and provides jet propulsion. The turbines drive the fans, while the air and exhaust expelled from the rear of the engine provide the forward thrust. Compare **rocket engine, turbofan. 2.** An aircraft powered by an engine or engines of this type.

turbulence (tûr′byə-ləns) Chaotic or unstable eddying motion in a fluid. Avoiding excessive turbulence generated around moving objects (such as airplanes), which can make their motion inefficient and difficult to control, is a major factor in aerodynamic design.

turbulent flow (tûr′byə-lənt) Movement of a fluid in which subcurrents in the fluid display turbulence, moving in irregular patterns, while the overall flow is in one direction. Turbulent flow is common in nonviscous fluids moving at high velocities. Compare **laminar flow.**

turgor (tûr′gər, -gôr′) The normal fullness or tension produced by the fluid content of blood vessels, capillaries, and plant or animal cells.

Turing (to͝or′ĭng), **Alan Mathison** 1912–1954. British mathematician who in 1937 formulated a precise mathematical concept for a theoretical computing machine, a key step in the development of the first computer. After the war he designed computers for the British government and helped in developing the concept of artificial intelligence. See Note at **artificial intelligence.**

BIOGRAPHY **Alan Turing**

Alan Turing—father of computer science, codebreaker, cognitive scientist, theoretician in artificial intelligence—achieved fame in 1936 at the age of 24, with a paper in which he showed that no universal algorithm exists that can determine whether a proposition in a given mathematical system is true or false. In the process of his proof he invented what has been called the *Turing machine,* an imaginary idealized computer that can compute any calculable mathematical function. The essentials of this machine (an input/output device, a memory, and a central processing unit) formed the basis for the design of all digital computers. After World War II broke out, he worked in England as a cryptanalyst, where he put his extraordinary talents to work on breaking the famous Enigma code used by the

German military. By 1940, Turing was instrumental in designing a machine that broke the German code, allowing the Allies to secretly decipher intercepted German messages for the rest of the war. At war's end, Turing was hired to help develop the world's first electronic computer and ultimately designed the programming system of the Ferranti Mark 1, the first commercially available digital computer, in 1948. His guiding principle that the brain is simply a computer was an important founding assumption for the new fields of cognitive science and artificial intelligence. He was making advances in modeling the chemical mechanisms by which genes control the structural development of organisms when he suddenly died, just before his forty-second birthday.

Turing machine An abstract model of a computing device, used in mathematical studies of computability. A Turing machine takes a tape with a string of symbols on it as an input, and can respond to a given symbol by changing its internal state, writing a new symbol on the tape, shifting the tape right or left to the next symbol, or halting. The inner state of the Tur-

A CLOSER LOOK **turbojet**

Fully loaded, a jumbo-sized aircraft weighs nearly 800,000 pounds. Yet its jet propulsion engines are so powerful that it hurtles down the runway fast enough to lift it into the air and climb to 35,000 feet. The *turbofan* engines now so common on commercial aircraft are descendants of the *turbojet,* itself a descendant of the *ramjet,* and all three are remarkably simple devices. In each case, thrust comes from the backward propulsion of hot pressurized air, which—according to Newton's third law of motion—produces an equal and opposite reaction in the body from which it was propelled. The ramjet is a simple hollow cylinder with a central combustion chamber in which incoming air is sprayed with burning fuel and blasts out the rear of the cylinder at high pressure. In a turbojet, greater efficiency and control is obtained by having the escaping exhaust drive a turbine before leaving the engine and by having that turbine drive intake fans that compress the incoming air before it is ignited. Though only a small amount of air is accelerated by these engines, it is accelerated by a huge amount, and extremely high thrust, on the order of thousands of horsepower, can be achieved. The turbojet engine is used in high-performance fighter jets, but if speeds are kept down to a few hundred miles per hour, the aerodynamics of the propeller or fan offers certain advantages; for this reason, the turbofan was developed. The turbofan engine encloses the combustion chamber in a larger cylinder and uses larger intake fans to blow some of the incoming air in an outer path around the combustion chamber, straight out the rear of the engine, without the addition of burning fuel. This way, a good part of the thrust is achieved by accelerating a large quantity of "clean" air by a small amount. This makes the turbofan both more fuel-efficient and considerably quieter than the turbojet design. For the extra thrust needed at takeoff, fuel can even be injected into the outer air path of the turbofan so that all of the air takes part in combustion as in turbojet operation; this fuel flow can then be turned off when the extra power is no longer needed.

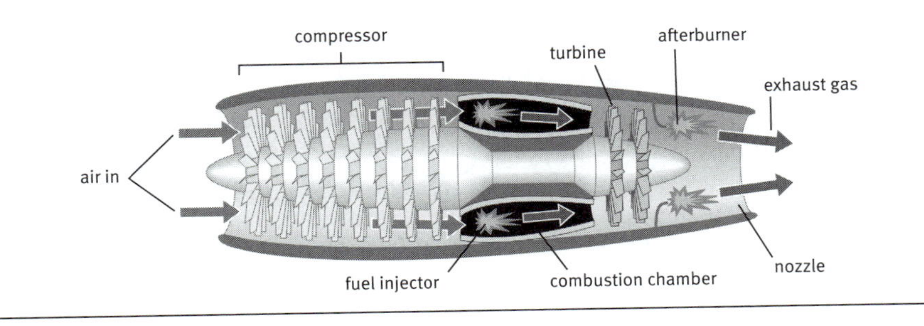

ing machine is described by a **finite state machine.** It has been shown that if the answer to a computational problem can be computed in a finite amount of time, then there exists an abstract Turing machine that can compute it.

turpentine (tûr′pən-tīn′) **1.** A thin, easily vaporized oil that is distilled from the wood or resin of certain pine trees. It is used as a paint thinner and solvent. *Chemical formula:* $C_{10}H_{16}$. **2.** The sticky mixture of resin and oil from which this oil is distilled.

turquoise (tûr′kwoiz′, -koiz′) A blue to bluish-green or yellowish-green triclinic mineral that occurs in reniform (kidney-shaped) masses with surfaces shaped like a bunch of grapes, especially in aluminum-rich igneous rocks such as trachyte. In its polished blue form, turquoise is prized as a gem. *Chemical formula:* $CuAl_6(PO_4)_4(OH)_8 \cdot 5H_2O$

tusk (tŭsk) A long, pointed tooth, usually one of a pair, projecting from the mouth of certain animals, such as elephants, walruses, and wild pigs. Tusks are used for procuring food and as weapons.

twenty-twenty or **20/20** (twĕn′tē-twĕn′tē) Having normal visual acuity. A person with twenty-twenty vision can see type as small as one-third of an inch at a distance of twenty feet.

twin (twĭn) **1.** One of two offspring born of a single gestation. Identical twins result from the division of a fertilized egg. Fraternal twins result from the fertilization of two separate eggs at the same time. **2.** A crystal structure consisting of two intergrown crystals that are mirror images of each other. Mineral twins can form as result of defective crystal growth in response to stress from rock deformation or during magma cooling.

tympanic membrane (tĭm-păn′ĭk) See **eardrum.**

type genus (tīp) The taxonomic genus that is designated as representative of the family to which it belongs. The type genus is usually the first genus in the family to be named, or the genus that is considered most important. Families are named after type genera. For example, the dinosaur *Tyrannosaurus* is the type genus for the family Tyrannosauridae.

type locality 1. The place or source where a holotype or type species was first found and recognized. **2.** The place or region chosen as a standard of reference when describing a type of mineral, rock, series of rocks, or formation. The type locality is the place where that particular type occurs in its most classic form.

type site An archaeological or anthropological site regarded as definitively characteristic of a particular culture and whose name is often applied to the culture. For example, the Olduvai Gorge in northern Tanzania is the type site for the Oldowan culture.

type species See **holotype.**

typhoid fever (tī′foid′) A life-threatening infectious disease caused by the bacterium *Salmonella typhi* and transmitted through contaminated food and water. It is characterized by high fever, intestinal bleeding, diarrhea, and skin rash.

typhoon (tī-fo͞on′) A violent cyclonic storm occurring in the western Pacific Ocean. See Note at **cyclone.**

typhus (tī′fəs) Any of several forms of an infectious disease caused by bacteria of the genus *Rickettsia* transmitted by fleas, mites, or especially lice, and characterized by severe headache, high fever, and skin rash. Louse-born bacteria that cause typhus are especially virulent and can cause epidemics of the disease, which can be fatal in people with weakened immune systems.

tyrannosaur (tĭ-răn′ə-sôr′) Any of various very large carnivorous dinosaurs of the genus *Tyrannosaurus* and related genera of the Cretaceous Period. Tyrannosaurs had very small forelimbs and a large head with sharp teeth. They walked on two legs, probably bent forward with their long tail stretched out as a counterbalance. Tyrannosaurs were theropods and probably distantly related to birds. The largest species, *T. rex,* grew to lengths of 14.3 m (47 ft) or more and may have been the largest land predator that ever lived.

tyrosine (tī′rə-sēn′) A nonessential amino acid. *Chemical formula:* $C_9H_{11}NO_3$. See more at **amino acid.**

U

U **1.** Abbreviation of **uracil.** **2.** The symbol for **uranium.**

ubiquinone (yo͞o′bĭ-kwĭ-nōn′) Any of various fat-soluble quinone compounds found in most aerobic organisms and serving as electron carriers in cellular respiration. Also called *coenzyme Q*.

udder (ŭd′ər) A bag-shaped part of a cow and the females of other ruminants in which milk is formed and stored and from which it is taken in suckling or milking.

udometer (yo͞o-dŏm′ĭ-tər) See **pluviometer.**

UHF Abbreviation of **ultrahigh frequency.**

ulcer (ŭl′sər) A break in the skin or a mucous membrane, such as the one lining the stomach or duodenum, accompanied by inflammation, pus, and loss of tissue.

ulexite (yo͞o′lĭk-sīt′, yo͞o-lĕk′-) A white triclinic mineral that forms rounded, reniform (kidney-shaped) masses of very fine needle-shaped crystals. Ulexite has the unusual optical property of projecting an image of an object placed against a transverse section of it to the opposite surface. It is usually associated with borax deposits and is found in arid environments. *Chemical formula:* $NaCaB_5O_9 \cdot 8H_2O$.

ulna (ŭl′nə) Plural **ulnas** or **ulnae** (ŭl′nē). The longer of the two bones of the forearm or lower portion of the foreleg. See more at **skeleton.**

ultrabasic (ŭl′trə-bā′sĭk) Containing magnesium and iron and only a very small amount of silica. Used of igneous rock, and often used interchangeably with **ultramafic.** Peridotite is an ultrabasic rock.

ultracentrifuge (ŭl′trə-sĕn′trə-fyo͞oj′) A high-velocity centrifuge used in the separation of colloidal or submicroscopic particles.

ultramafic (ŭl′trə-măf′ĭk) Containing mainly mafic minerals. Used of igneous rocks and often used interchangeably with **ultrabasic.** Dunite is an ultramafic rock.

ultrasonography (ŭl′trə-sə-nŏg′rə-fē) **1.** Diagnostic imaging in which ultrasound is used to image an internal body structure or a developing fetus. See Note at **ultrasound.** **2.** An imaging technology that uses high-frequency sound waves to visualize underwater objects, topography, boundaries between layers, and currents. It is often used to locate underwater vehicles on the ocean floor. The sound waves are broadcast, and the timing and frequency shift of their echoes are analyzed in much the same manner as in **sonar** to produce an image or map of the phenomena or objects under investigation. Also called *ultrasound.*

ultrasound (ŭl′trə-sound′) **1.** Sound whose frequency is above the upper limit of the range of human hearing (approximately 20 kilohertz). **2.** See **ultrasonography.** **3.** An image produced by ultrasonography. *—Adjective* **ultrasonic** (ŭl′trə-sŏn′ĭk).

A CLOSER LOOK **ultrasound**

Many people use simple ultrasound generators. Dog whistles, for example, produce tones that dogs can hear but that are too high to be heard by humans. Sound whose frequency is higher than the upper end of the normal range of human hearing (higher than about 20,000 hertz) is called *ultrasound.* (Sound at frequencies too low to be audible—about 20 hertz or lower—is called *infrasound.*) Medical ultrasound images, such as those of a fetus in the womb, are made by directing ultrasonic waves into the body, where they bounce off internal organs and other objects and are reflected back to a detector. Ultrasound imaging, also known as *ultrasonography,* is particularly useful in conditions such as pregnancy, when x-rays can be harmful. Because ultrasonic waves have very short wavelengths, they interact with very small objects and thus provide images with high resolution. For this reason ultrasound is also used in some microscopes. Ultrasound can also be used to focus large amounts of energy into very small spaces by aiming multiple ultrasonic beams in such a way that the waves are in phase at one precise location, making it possible, for example, to break up kidney stones without surgical incision and without disturbing surrounding tissue. Ultrasound's industrial uses include measuring thicknesses of materials, testing for structural defects, welding, and aquatic sonar.

---

ultraviolet (ŭl′trə-vī′ə-lĭt) *Adjective.* **1.** Relating to electromagnetic radiation having frequencies higher than those of visible light but lower than those of x-rays, approximately 10^{15}–10^{16} hertz. Some animals, such as bees, are capable of seeing ultraviolet radiation invisible to the human eye. See more at **electromagnetic spectrum.** —*Noun.* **2.** Ultraviolet light or the ultraviolet part of the spectrum. See Note at **infrared.**

ultraviolet astronomy The study of celestial objects by measurement of the ultraviolet radiation they emit. Celestial objects with strong emissions in the ultraviolet range are usually significantly hotter than those whose emissions are stronger in the visible range. Because the ozone layer absorbs most ultraviolet radiation before it reaches the Earth's surface, most ultraviolet observations are made from rocket-borne or space-based instruments.

umbel (ŭm′bəl) A flat or rounded indeterminate inflorescence in which the individual flower stalks (called pedicels) arise from about the same point on the stem at the tip of the peduncle. The geranium, milkweed, and onion have umbels. Umbels usually show centripetal inflorescence, with the lower or outer flowers blooming first.

umbilical cord (ŭm-bĭl′ĭ-kəl) The flexible cord that attaches an embryo or fetus to the placenta. The umbilical cord contains blood vessels that supply nutrients and oxygen to the fetus and remove its wastes, including carbon dioxide.

umbra (ŭm′brə) Plural **umbras** or **umbrae** (ŭm′brē). **1.** The darkest part of a shadow, especially the cone-shaped region of full shadow cast by Earth, the Moon, or another body during an eclipse. In a full lunar eclipse, which generally lasts for one or two hours, the entire disk of the Moon is darkened as it passes through the umbra. During this period the Moon takes on a faint reddish glow due to illumination by a small amount of sunlight that is refracted through the Earth's atmosphere and bent toward the darkened Moon; the reddish tint is caused by the filtering out of blue wavelengths as the sunlight passes through the Earth's atmosphere, leaving only the longer wavelengths on the red end of the spectrum. See Note at **eclipse. 2.** The dark central region of a sunspot. Compare **penumbra.**

uncertainty principle (ŭn-sûr′tn-tē) A principle, especially as formulated in quantum mechanics, that greater accuracy of measurement for one observable entails less accuracy of measurement for another. For example, it is in principle impossible to measure both the momentum and the position of a particle at the same time with perfect accuracy. Any pair of observables whose operators do not **commute** have this property. As defined in quantum mechanics, it is also called *Heisenberg's uncertainty principle.* Similar uncertainty principles hold for non-quantum mechanical systems involving waves as well.

unconformity (ŭn′kən-fôr′mĭ-tē) A surface between successive strata representing a missing interval in the geologic record of time, produced either by an interruption in deposition or by the erosion of depositionally continuous strata followed by renewed deposition. An unconformity is a type of discontinuity.

undertow (ŭn′dər-tō′) An underwater current flowing strongly away from shore. Undertows are generally caused by the seaward return of water from waves that have broken against the shore.

ungulate (ŭng′gyə-lĭt) A hoofed mammal. Ungulates belong to two orders, Artiodactyla (those having an even number of toes) and Perissodactyla (those having an odd number of toes). See more at **artiodactyl, perissodactyl.**

unicellular (yo͞o′nĭ-sĕl′yə-lər) Having or consisting of a single cell. Compare **multicellular.**

Unicode (yo͞o′nĭ-kōd′) A computer standard for encoding characters. Each character is represented by sixteen bits. Whereas ASCII, being an 8-bit encoding scheme, can only represent 256 characters, Unicode has 65,536 combinations, enabling it to encode the letters of all written languages as well as thousands of characters in languages such as Japanese and Chinese.

unifacial (yo͞o′nə-fā′shəl) Flaked in such a way as to produce a cutting edge that is sharp on one side only. Used of a stone tool. ► Unifacial tools are known as a **unifaces** and include early **core tools** such as choppers as well as later **flake tools** such as scrapers and adzes. Compare **bifacial.**

unified field theory (yo͞o′nə-fīd′) A theory that unites and explains the basic forces of nature (strong, electroweak, and gravitational forces) as manifestations of a single physical principle. No unified field theory that has been proposed so far has gained broad acceptance.

uniformitarianism (yo͞o′nə-fôr′mĭ-târ′ē-ə-nĭz′-əm) The theory that all geologic phenomena may be explained as the result of existing forces having operated uniformly from the origin of the Earth to the present time. See Note at **Lyell.**

uniform resource locator (yo͞o′nə-fôrm′) See **URL.**

union (yo͞on′yən) A set whose members belong to at least one of a group of two or more given sets. The union of the sets {1,2,3} and {3,4,5} is the set {1,2,3,4,5}, and the union of the sets {6,7,8} and {11,12,13} is the set {6,7,8,11,12,13}. The symbol for *union* is ∪. Compare **intersection.**

unisexual (yo͞o′nĭ-sĕk′sho͞o-əl) **1.** Having the sex organs of only one sex; not hermaphroditic. Used of organisms. **2.** Having stamens and pistils in separate flowers borne on the same plant; diclinous. Used of flowers.

United States Customary System (yo͞o-nī′tĭd) *See* **US Customary System.**

univalent (yo͞o′nĭ-vā′lənt) Having a valence of 1.

univalve (yo͞o′nĭ-vălv′) A gastropod, especially one with a single shell, such as a snail, cone, whelk, abalone, or limpet. Univalves belong to the subclass Prosobranchia. Their shells are usually spiral and can hold the whole animal inside. Compare **bivalve.**

universal donor (yo͞o′nə-vûr′səl) A person who has group O blood and is therefore able to serve as a donor to a person with any other blood type. Group O blood contains red blood cells that do not have the antigen A or B and thus do not react with antibodies to these antigens, which are found in the plasma of the other blood types. See also **blood type.**

universal gas law See **ideal gas law.**

universal recipient A person who has group AB blood and is therefore able to receive blood from a person with any other blood type. Group AB blood contains red blood cells that have both antigens A and B and thus does not have reactive antibodies in its plasma to these antigens, which are found in some other blood types. See also **blood type.**

universal time The mean time for the meridian at Greenwich, England (0° longitude), which runs through the former site of the Royal Observatory. It is based on the sidereal period of Earth's rotation and is used as a basis for calculating standard clock time throughout most of the world. Also called *Greenwich Mean Time.* Compare **coordinated universal time.**

universe (yo͞o′nə-vûrs′) The totality of matter, energy, and space, including the **Solar System,** the **galaxies,** and the contents of the space between the galaxies. Current theories of **cosmology** suggest that the universe is constantly expanding.

unsaturated (ŭn-săch′ə-rā′tĭd) **1.** Relating to an organic compound in which two or more of the carbon atoms are joined by a double or triple bond and therefore can be combined with additional atoms or radicals. Benzene and acetylene are examples of unsaturated compounds. Compare **saturated.** See also **monounsaturated, polyunsaturated. 2.** Relating to a solution that is capable of dissolving more solute than it already contains.

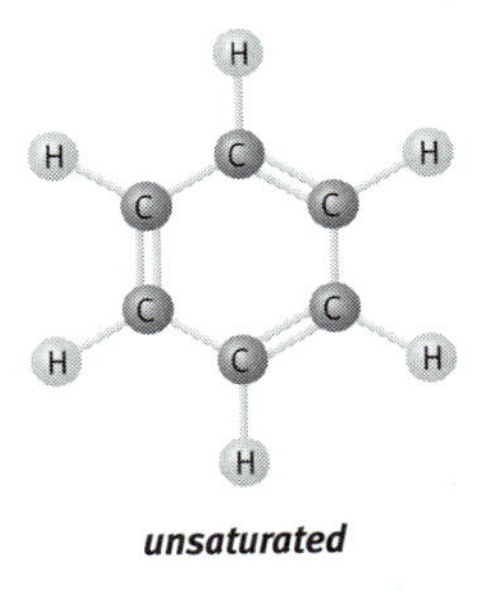

unsaturated

a molecule of benzene

unsaturated fat A triglyceride fat containing at least one unsaturated fatty acid. Fats derived from plants are often unsaturated fats. Eating foods high in unsaturated fats can reduce the amount of cholesterol in the blood. Compare **saturated fat.**

unstable (ŭn-stā′bəl) **1.** Liable to undergo spontaneous decay into some other form. For example, the nucleus of uranium 238 atom is unstable and changes by radioactive decay into the nucleus of thorium 234, a lighter element. Many subatomic particles, such as muons and neutrons, are unstable and decay quickly into other particles. See more at **decay. 2.** Relating to a chemical compound that readily decomposes or changes into other compounds or into elements. **3.** Relating to an atom or chemical element that is likely to share electrons; reactive. **4.** Characterized by uncertain or inadequate response to treatment and the potential for unfavorable outcome, as the status of a medical condition or disease.

ununbium (ə-nŭn′bē-əm) *Symbol* **Uub.** See **element 112.**

ununhexium (ə-nŭn′hĕk′sē-əm) *Symbol* **Uuh.** See **element 116.**

ununoctium (ə-nŭn′ŏk′tē-əm) *Symbol* **Uuo.** See **element 118.**

ununpentium (ə-nŭn′pĕn′tē-əm) *Symbol* **Uup.** See **element 115.**

ununquadium (ə-nŭn′kwŏd′ē-əm) *Symbol* **Uuq.** See **element 114.**

ununtrium (ə-nŭn′trē-əm) *Symbol* **Uut.** See **element 113.**

updraft (ŭp′drăft′) An upward current of warm, moist air. With enough moisture, the current may visibly condense into a cumulus or cumulonimbus cloud. Compare **downdraft.**

upload (ŭp′lōd′) To transfer data or programs from one's own computer or digital device to a server or host computer. Compare **download.**

upper (ŭp′ər) Being a later or more recent division of the geological or archaeological period named. Compare **lower.**

upper atmosphere The part of the atmosphere above the troposphere. The upper atmosphere includes the mesosphere, the ionosphere, the thermosphere, and the exosphere. Temperatures in the upper atmosphere decrease with increasing altitude except in the thermosphere, where temperatures increase significantly with altitude, because the few molecules present receive intense solar radiation.

upper bound A number that is greater than or equal to every number in a given set.

up quark (ŭp) A quark with a charge of $+\frac{2}{3}$. Its mass is approximately 10 times the electron mass and is the least mass of all quarks. See Table at **subatomic particle.**

upsilon particle (ŭp′sə-lŏn′, yo͞op′-) An electrically neutral meson having a mass 18,513 times that of the electron and a mean lifetime of approximately 8.0×10^{-20} seconds. See Table at **subatomic particle.**

upwelling (ŭp-wĕl′ĭng) The rising of cold, usually nutrient-rich waters from the ocean depths to the warmer, sunlit zone at the surface. Upwelling usually occurs in the subtropics along the western continental coasts, where prevailing trade winds drive the surface water away from shore, drawing deeper water upward to take its place. Because of the abundance of krill and other nutrients in the colder waters, these regions are rich feeding grounds for a variety of marine and avian species. Upwelling can also occur in the middle of oceans where cyclonic circulation is relatively permanent or where southern trade winds cross the Equator.

uracil (yo͝or′ə-sĭl) A pyrimidine base that is a component of RNA. It forms a base pair with adenine during transcription. Uracil is therefore structurally analogous to thymine in molecules of DNA. *Chemical formula:* $C_4H_4N_2O_2$.

uranium (yo͝o-rā′nē-əm) *Symbol* **U** A heavy, silvery-white, highly toxic, radioactive metallic element of the actinide series. It has 14 known isotopes, of which U 238 is the most naturally abundant, occurring in several minerals. Fissionable isotopes, especially U 235, are used in nuclear reactors and nuclear weapons. Atomic number 92; atomic weight 238.03; melting point 1,132°C; boiling point 3,818°C; specific gravity 18.95; valence 2, 3, 4, 5, 6. See **Periodic Table.**

Uranus (yo͝or′ə-nəs, yo͝o-rā′-) The seventh planet from the Sun and the third largest, with a diameter about four times that of Earth. Though slightly larger than Nepture, Uranus is the least massive of the four **gas**

giants and is the only one with no internal heat source. A cloud layer of frozen methane gives it a faint bluish-green color, and it is encircled by a thin system of 11 rings and 27 moons. Uranus's axis is tilted 98° from the vertical—the greatest such tilt in the solar system—with the result that its poles are in continuous darkness or continuous sunlight for nearly half of its 84-year orbital period. See Table at **solar system.**

urban wind (ûr′bən wĭnd) A strong wind generated as a result of temperature gradients near or around a group of high-rise buildings. It creates areas of intense air turbulence, especially at street level.

urea (yo͝o-rē′ə) The chief nitrogen-containing waste product excreted in the urine of mammals and some fish. It is the final nitrogenous product in the breakdown of proteins by the body, during which amino groups (NH_2) are removed from amino acids and converted into ammonium ions (NH_4), which are toxic at high concentrations. The liver then converts the ammonium ions into urea. Urea is also made artificially for use in fertilizers and medicine. *Chemical formula:* $\mathbf{CON_2H_4}$.

ureter (yo͝o-rē′tər, yo͝or′ĭ-tər) Either of two long, narrow ducts that in vertebrates carry urine from each kidney to the urinary bladder.

urethane (yo͝or′ĭ-thān′) also **urethan** (yo͝or′ĭ-thăn′) A colorless or white crystalline compound used in organic synthesis. Formerly it was also used to relieve symptoms associated with leukemia. Also called *ethyl carbamate. Chemical formula:* $\mathbf{C_3H_7NO_2}$.

urethra (yo͝o-rē′thrə) Plural **urethras** or **urethrae** (yo͝o-rē′thrē). The duct through which urine passes from the bladder to the outside of the body in most mammals and some fish and birds. In males, the urethra passes through the penis and also serves as the duct for the release of sperm, which enter the urethra from the vas deferens.

Urey (yo͝or′ē), **Harold Clayton** 1893–1981. American chemist who is best known for his discovery of deuterium (or heavy hydrogen) in 1932, for which he was awarded the 1934 Nobel Prize for chemistry. He also developed theories on the formation of the planets and on the synthesis of organic compounds in the Earth's primitive atmosphere.

Harold Urey

uric acid (yo͝or′ĭk) The chief nitrogen-containing waste product excreted in the urine of birds, insects, and most reptiles. It is produced by the breakdown of amino acids in the liver. Uric acid is also produced in small quantities in humans by the breakdown of purines, and elevated levels in the blood can lead to gout. *Chemical formula:* $\mathbf{C_5H_4N_4O_3}$.

urinalysis (yo͝or′ə-năl′ĭ-sĭs) Laboratory analysis of urine, used to aid in the diagnosis of disease or to detect the presence of a specific substance, such as an illegal drug.

urinary system (yo͝or′ə-nĕr′ē) The system of organs and tissues involved with regulation of water content and salt concentration in the body and with the excretion of metabolic wastes and excess water and salt in the form of urine. In mammals, the urinary system consists of the urinary tract and associated tissues.

urinary tract The series of organs in the urinary system in which urine is formed and excreted.

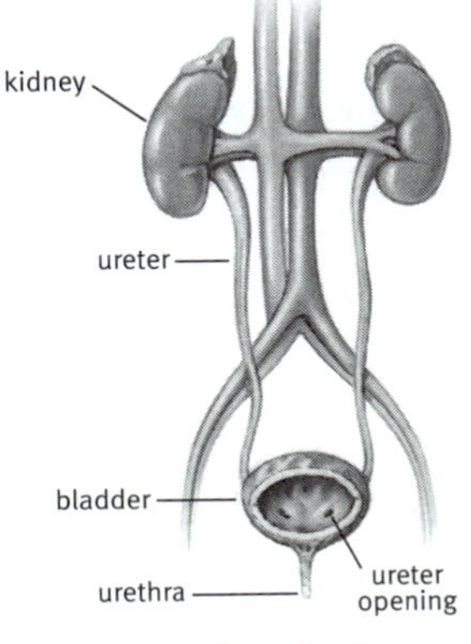

urinary tract

In most vertebrates, the urinary tract consists mainly of the kidneys, ureters, and bladder. In most mammals, and some fish and birds, the urinary tract also includes the urethra.

urine (yo͝or′ĭn) A liquid containing multiple waste products of metabolism, especially urea and other nitrogenous compounds, that are filtered from the blood by the kidneys. Urine is stored in the urinary bladder and is excreted from the body through the urethra.

URL (yo͞o′är-ĕl′) Short for *Uniform Resource Locator*. An Internet address (for example, *http://www.hmco.com/trade/*), usually consisting of the access protocol (*http*), the domain name (*www.hmco.com*), and optionally the path to a file or resource residing on that server (*trade*).

urology (yo͝o-rŏl′ə-jē) The branch of medicine that deals with the diagnosis and treatment of diseases of the urinary tract and urogenital system.

Ursa Major (ûr′sə) A constellation in the polar region of the Northern Hemisphere near Draco and Leo. Ursa Major (the Great Bear) contains the seven stars that form the Big Dipper.

Ursa Minor A ladle-shaped constellation very near the north celestial pole. Ursa Minor (the Lesser Bear) contains the seven stars that form the Little Dipper. Polaris, the North Star, is at the end of the dipper's handle.

ursine (ûr′sīn′) Resembling or characteristic of bears.

urticaria (ûr′tĭ-kâr′ē-ə) See **hives.**

urushiol (o͝o-ro͞o′shē-ôl′, -ōl′) A toxic substance present in the resin or on the surface of plants of the genus *Rhus* (syn. *Toxicodendron*). Urushiol is a mixture of several derivatives of catechol and is the irritating substance in poison ivy.

US Customary System Short for *United States Customary System.* The main system of weights and measures used in the United States and a few other countries. The system is based on the yard as a unit of length, the pound as a unit of weight, the gallon as a unit of liquid volume, and the bushel as a unit of dry volume. See Table at **measurement.** Compare **metric system.**

uterus (yo͞o′tər-əs) Plural **uteri** (yo͞o′tə-rī′) or **uteruses.** The hollow, muscular organ of female mammals in which the embryo develops. In most mammals the uterus is divided into two saclike parts, whereas in primates it is a single structure. It lies between the bladder and rectum and is attached to the vagina and the fallopian tubes. During the menstrual cycle (estrus), the lining of the uterus (endometrium) undergoes changes that permit the implantation of a fertilized egg. Also called *womb.* See more at **menstrual cycle.**

Uub The symbol for **element 112.**

Uuh The symbol for **element 116.**

Uuo The symbol for **element 118.**

Uup The symbol for **element 115.**

Uuq The symbol for **element 114.**

Uut The symbol for **element 113.**

uvarovite (yo͞o-vär′ə-vīt′, o͞o-) An emerald-green variety of garnet found in chromium deposits. *Chemical formula:* $\mathbf{Ca_3Cr_2(SiO_4)_3}$.

UV index (yo͞o′vē′) A scale ranging from 0 to 10, used to estimate the risk for sunburn in midday sunlight, with 0 indicating no risk and 10 indicating maximal risk. The UV index is effectively a rough measure of the amount of harmful ultraviolet radiation in the sunlight reaching the Earth's surface at a given location, given the time of year and current atmospheric conditions, expressed in terms of the risks that are associated with exposure to that amount of radiation.

uvula (yo͞o′vyə-lə) A small mass of fleshy tissue that hangs from the back of the soft palate.

V **1.** The symbol for **vanadium.** **2.** The symbol for **voltage.** **3.** Abbreviation of **volume.**

vaccination (văk′sə-nā′shən) **1.** Inoculation with a vaccine in order to protect against a particular disease. **2.** A scar left on the skin by vaccinating.

vaccine (văk-sēn′) A preparation of a weakened or killed pathogen, such as a bacterium or virus, or of a portion of the pathogen's structure, that stimulates immune cells to recognize and attack it, especially through antibody production. Most vaccines are given orally or by intramuscular or subcutaneous injection. See Note at **Jenner.**

A CLOSER LOOK **vaccine**

In the 1950s, polio epidemics left thousands of children with permanent physical disabilities. Today, infants are given a *vaccine* to prevent infection with the polio virus. That vaccine, like most others, works by stimulating the body's immune system to produce antibodies that destroy pathogens. Scientists usually prepare vaccines by taking a sample of the pathogen and destroying or weakening it with heat or chemicals. The inactivated or attenuated pathogen loses its ability to cause serious illness but is still able to stimulate antibody production, thereby conferring immunity. The Salk polio vaccine contains "killed" virus, while the Sabin polio vaccine contains weakened "live" poliovirus. (Many scientists no longer consider viruses to be living organisms.) Scientists are also able to change the structure of viruses and bacteria at the molecular level, altering DNA so that the potential of the vaccine to cause disease is decreased. New vaccines containing harmless bits of DNA have also been developed.

vaccinology (văk′sə-nŏl′ə-jē) The scientific study of vaccine development.

vacuole (văk′yo͞o-ōl′) A cavity within the cytoplasm of a cell, surrounded by a single membrane and containing fluid, food, or metabolic waste. Vacuoles are found in the cells of plants, protists, and some primitive animals. In mature plant cells, there is usually one large vacuole which occupies a large part of the cell's volume and is filled with a liquid called cell sap. The cell sap stores food reserves, pigments, defensive toxins, and waste products to be expelled or broken down. In the cells of protists, however, there may be many small specialized vacuoles, such as digestive vacuoles for the absorption of captured food and contractile vacuoles for the expulsion of excess water or wastes. See more at **cell.**

vacuum (văk′yo͞om) Plural **vacuums** or **vacuua.** **1.** A region of space in which there is no matter. **2.** A region of space having extremely low gas pressure relative to surrounding pressure. The air pump of a vacuum cleaner, for example, drastically reduces the air pressure inside the device, creating a vacuum; the pressure difference causes air to rush into it, carrying dust and debris along with it.

vacuum bottle A container with a double wall and a partial vacuum in the space between the two walls. Vacuum bottles are used to minimize the transfer of heat between the inside and the outside and thus keep the contents at a desired temperature.

vacuum energy Background energy in a vacuum associated with constant **vacuum fluctuations.** Some astronomers believe vacuum energy to be the source of energy for the apparent acceleration of the expansion of the universe.

vacuum fluctuation A spontaneous, short-lived fluctuation in the energy level of a vacuum, as described by quantum field theory. Although these variations are violations of the law of conservation of energy, they are tolerated in quantum mechanics by virtue of the uncertainty principle. Such fluctuations are associated with **virtual particles.**

vacuum tube An electron tube from which all air has been removed. The vacuum ensures transparency inside the tube for electric fields and moving electrons. Most electron tubes are vacuum tubes; cathode-ray tubes, which include television picture tubes and other video display tubes, are the most widely used vacuum tubes. In other electronic applications, vacuum tubes have largely been replaced by transistors.

vadose zone (vā′dōs′) A subsurface zone of soil or rock containing fluid under pressure that is less than that of the atmosphere. Pore spaces in the vadose zone are partly filled with water and partly filled with air. The vadose zone is limited by the land surface above and by the water table below. Compare **phreatic zone.**

vagina (və-jī′nə) The tube-shaped part of the reproductive tract in female mammals that is connected to the uterus at one end and opens to the outside of the body on the other end. The fully developed fetus passes through the vagina during birth.

vagus nerve (vā′gəs) Either of the tenth pair of cranial nerves that carries motor impulses from the brain to many major organs. The vagus nerve controls the muscles of the larynx (voice box), stimulates digestion, and regulates the heartbeat.

valence (vā′ləns) A whole number that represents the ability of an atom or a group of atoms to combine with other atoms or groups of atoms. The valence is determined by the number of electrons that an atom can lose, add, or share. An atom's valence is positive if its own electrons are used in forming the bond, or negative if another atom's electrons are used. For example, a carbon atom can share four of its electrons with other atoms and therefore has a valence of +4. A sodium atom can receive an electron from another atom and therefore has a valence of −1. (In this book the distinction between positive and negative valences is ignored unless it is relevant.) The valence of an atom generally indicates how many chemical bonds it is capable of forming with other atoms. Also called *valence number, oxidation state.*

valence band The outermost electron shell of atoms in an insulator or semiconductor, in which the electrons are too tightly bound to the atom to carry electric current. Compare **conduction band.** See also **bandgap.**

valence electron An electron in one of the outer shells of an atom that can participate in forming chemical bonds with other atoms.

valeric acid (və-lîr′ĭk, -lĕr′-) A colorless, liquid organic acid that occurs in four isomeric forms and has a disagreeable odor. It occurs naturally in oils from certain marine animals and plants, and is used in flavorings, perfumes, plasticizers, and pharmaceuticals. *Chemical formula:* $\mathbf{C_5H_{10}O_2}$.

valine (văl′ēn′) An essential amino acid. *Chemical formula:* $\mathbf{C_5H_{11}NO_2}$. See more at **amino acid.**

valley (văl′ē) A long, narrow region of low land between ranges of mountains, hills, or other high areas, often having a river or stream running along the bottom. Valleys are most commonly formed through the erosion of land by rivers or glaciers. They also form where large regions of land are lowered because of geological faults.

value (văl′yo͞o) **1.** *Mathematics.* An assigned or calculated numerical quantity. **2.** The relative darkness or lightness of a color. Value measures where a color falls on an achromatic scale from white to black. Compare **hue, saturation.**

valve (vălv) **1a.** Any of various mechanical devices that control the flow of liquids, gases, or loose material through pipes or channels by blocking and uncovering openings. **b.** The movable part or element of such a device. **2.** Any of various structures that prevent the backward flow of a body fluid, such as blood or lymph. Valves in the heart, veins, and lymphatic vessels contain flaps (known as cusps) that close in response to pressure created by the backflow of fluid. **3.** One of the paired hinged shells of certain mollusks, such as clams and oysters. **4.** See **electron tube.**

vanadium (və-nā′dē-əm) *Symbol* **V** A soft, bright-white metallic element that occurs naturally in several minerals. It has good structural strength and is used especially to make strong varieties of steel. Atomic number 23; atomic weight 50.942; melting point 1,890°C; boiling point 3,000°C; specific gravity 6.11; valence 2, 3, 4, 5. See **Periodic Table.**

Van Allen belt (văn ăl′ən) Either of the two zones surrounding the Earth in which a thin distribution of atomic particles with very high energies are trapped by Earth's magnetic field. The inner belt lies between about 500 and 6,440 km (310 to 4,000 mi) above Earth's equator and consists mostly of a mix of protons and electrons. The upper belt lies between about 15,000 and 30,000 km (9,300 to 18,600 mi) and consists mostly of protons. Both belts are thickest at the equator and have irregular densities of particles. The outer radiation belt is much larger and the number of particles varies, increasing in the aftermath of solar flares. The polar **auroras** are caused when some of the charged particles from the outer Van Allen belt hit the upper atmosphere. The Van Allen belt is named after its discoverer, American astrophysicist James Van Allen (1914–2006).

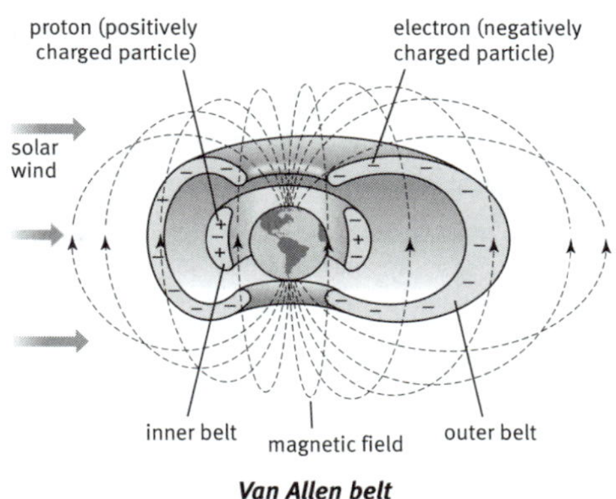

Van Allen belt

Positively and negatively charged particles (mostly protons and electrons) are trapped in the inner Van Allen belt, and negatively charged particles are trapped in the outer Van Allen belt.

Van de Graaff (văn′ də grăf′), **Robert Jemison** 1901–1967. American physicist who in 1929 invented an improved electrostatic generator (later called the Van de Graaff generator) that was adapted for use as a particle accelerator. The generator became an important research tool for atomic physicists and was also used to produce high-energy x-rays helpful in treating cancer.

Van de Graaff generator A type of electrostatic generator used to build up static electrical charge of very high voltages by transferring electric charge from a power supply to a spherical metal terminal. A high-voltage source transfers charge to a nonconducting conveyor belt, usually made of silk or rubber, which continuously redeposits the charge on the insulated metal terminal, where it accumulates. Even small Van de Graaff generators can accumulate a static charge of 100,000 volts; the largest, up to 10 million volts.

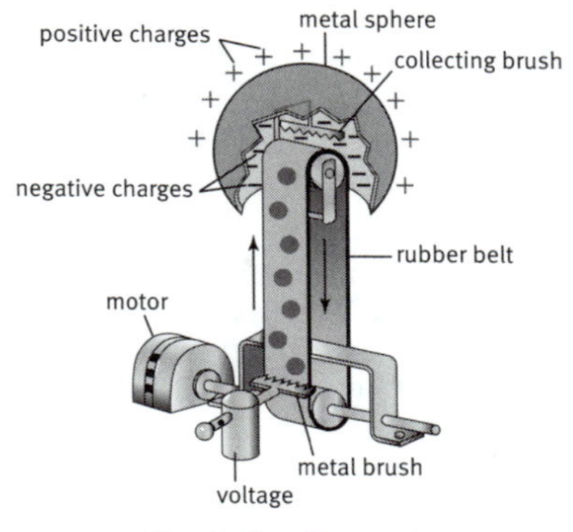

Van de Graaff generator

Positive charge is transferred by a metal brush to a rotating belt, which carries the charge up to the spherical electrode where it accumulates.

van der Waals (văn′ dər wôlz′), **Johannes Diderik** 1837–1923. Dutch physicist who accounted for many phenomena concerning gases and liquids by postulating the existence of intermolecular forces and a finite molecular volume. He derived a new equation of state for gases and liquids (now named for him), and for this work he received the 1910 Nobel Prize for physics.

van der Waals equation An equation that relates the pressure, volume, and absolute temperature of a gas taking into account the finite size of molecules and their intermolecular attraction, having the form $RT = (P + av^{-2})(v - b)$, where R is the gas constant, T is the absolute temperature, P is the pressure, v is the volume of fluid per molecule, a is a measure of the attraction of the molecules for each other (due to **van der Waals forces**), and b is the volume occupied by a single molecule. The equation accurately captures phase transitions between liquid and gas phases of substances. See also **ideal gas law.**

van der Waals force A weak force of attraction between electrically neutral molecules that collide with or pass very close to each other. The van der Waals force is caused by the attraction between electron-rich regions of one molecule and electron-poor regions of another (the attraction between the molecules seen as electric dipoles). The attraction is much weaker than a chemical bond. Van der Waals forces are the intermolecular forces that cause molecules to cohere in liquid and solid states of matter, and are responsible for surface tension and capillary action.

vane (vān) The flattened, weblike part of a feather, consisting of a series of barbs on either side of the rachis.

vapor (vā′pər) **1.** The gaseous state of a substance that is normally liquid or solid at room temperature, such as water that has evaporated into the air. See more at **vapor pressure.** See also **water vapor. 2.** A faintly visible suspension of fine particles of matter in the air, as mist, fumes, or smoke. **3.** A mixture of fine droplets of a substance and air, as the fuel mixture of an internal-combustion engine. —*Verb* **vaporize.**

USAGE **vapor/steam**

The words *vapor* and *steam* usually call to mind a fine mist, such as that in the jet of water droplets near the spout of a boiling teakettle or in a bathroom after a shower. Vapor and steam, however, refer to the gaseous state of a substance. The fumes that arise when volatile substances such as alcohol and gasoline evaporate, for example, are vapors. The visible stream of water droplets rushing out of a teakettle spout is not steam. As the gaseous state of water heated past its boiling point, steam is invisible. Usually, there is a space of an inch or two between the spout and the beginning of the stream of droplets. This space contains steam. The steam loses its heat to the surrounding air, then falls below the boiling point and condenses in the air as water droplets. All liquids and solids give off vapors consisting of molecules that have evaporated from the substance. In a closed system, the *vapor pressure* of these molecules reaches an equilibrium at which the substance evaporates from the liquid (or solid) and recondenses on it in equal amounts.

vapor pressure 1. The pressure exerted by a vapor on the solid or liquid phase with which it is in equilibrium. At pressures lower than the vapor pressure, more atoms or molecules of the liquid or solid vaporize and escape from the surface of the liquid or solid than are absorbed from the vapor, resulting in evaporation. At the vapor pressure the exchange is equal and there is no net evaporation. Also called *evaporation pressure.* **2.** The pressure exerted by water vapor in the atmosphere.

variable (vâr′ē-ə-bəl) **1.** A mathematical quantity capable of assuming any of a set of values, such as x in the expression $3x + 2$. **2.** A factor or condition that is subject to change, especially one that is allowed to change in a scientific experiment to test a hypothesis. See more at **control.**

variable region The portion of the amino (NH_2) terminal of an antibody's heavy and light chains having a variable amino acid sequence. The structure of the variable region determines the antigenic specificity of the antibody. Compare **constant region.**

variable resistor See under **resistor.**

variable star A star whose actual or observed brightness varies periodically. These changes can occur with varying degrees of regularity and intensity, over times ranging from a fraction of a second to many years. Intrinsic variation occurs because of changes of the star itself, often due to internal vibration or eruptions, or to influx of nearby material. **Cepheids** and **novae** are examples of intrinsically variable stars. Extrinsic variation in a star's observed brightness that does not reflect physical changes in the star also occur, as when a darker star periodically eclipses a brighter star in an **eclipsing binary** system, or with the rotation of a star.

varicella (văr′ĭ-sĕl′ə) See **chickenpox.**

varicose veins (văr′ĭ-kōs′) Abnormally prominent and swollen veins, especially in the legs.

variola (və-rī′ə-lə, vâr′ē-ō′lə) See **smallpox.**

Varmus (vär′məs), **Harold Elliot** Born 1939. American molecular biologist who, working with Michael Bishop, discovered oncogenes. For this work, Varmus and Bishop shared the 1989 Nobel Prize for physiology or medicine.

varve (värv) A layer or series of layers of sediment deposited in a body of still water in one year. Varves are typically associated with glacial lake deposits and consist of two layers—a lower, light-colored layer that consists primarily of sand and silt, and a darker upper layer that consists primarily of clay and organic matter. The lower layer is typically deposited in the summer by the rapid melt of glacial ice, and the upper layer is usually deposited in the winter by the slower settling of sediment through calm water. The thickness of the layers in a varve varies depending on the proximity to the margin of the glacier, with thicker layers forming closer to the glacial margin and thinner layers forming farther away from it. Varves have been used, like tree rings, to measure the ages of glacial deposits from the Pleistocene.

vascular (văs′kyə-lər) **1.** Relating to the vessels of the body, especially the arteries and veins, that carry blood and lymph. **2.** Relating to or having xylem and phloem, plant tissues highly specialized for carrying water, dissolved nutrients, and food from one part of a plant to another. Ferns and all seed-bearing plants have vascular tissues; bryophytes, such as mosses, do not. See more at **phloem, xylem.**

vascular bundle A strand of primary tissues found within the stem of a plant and consisting of xylem and phloem, along with cambium. The vascular bundles develop from the procambium of the growing stem. In the monocotyledons, vascular bundles are found in complex arrangements dispersed throughout the stem. In eudicotyledons, they are arranged in a ring, with the xylem of each bundle on the inside and the phloem on the outside. Also called *fibrovascular bundle.* See also **fascicular cambium, interfascicular cambium.**

vascular cambium A cylindrical layer of cambium that runs through the stem of a plant that undergoes **secondary growth.** The vascular cambium produces vascular tissues, new xylem on its interior side and new phloem on its exterior side. All woody plants and most eudicots have vascular cambium. See more at **cambium.**

vascular plant Any of various plants that have the vascular tissues xylem and phloem. The vascular plants include all seed-bearing plants (the gymnosperms and angiosperms) and the pteridophytes (including the ferns, lycophytes, and horsetails). Also called *tracheophyte.* Compare **nonvascular plant.**

vascular tissue The tissue in vascular plants that circulates fluid and nutrients. There are two kinds of vascular tissue: **xylem,** which conducts water and nutrients up from the roots, and **phloem,** which distributes food from the leaves to other parts of the plant. Vascular tissue can be primary (growing from the apical meristem and elongating the plant body) or secondary (growing from the cambium and increasing stem girth). Seedless plants, and nearly all monocotyledons and herbaceous eudicotyledons, have only primary vascular tissue. The evolution of vascular tissue, especially xylem with its rigid water-conducting cells known as **tracheids,** provided the plant stem with greater support and allowed plants to grow upright to great heights. See also **cambium, ground tissue, procambium.** See more at **phloem, xylem.**

vas deferens (văs′ dĕf′ə-rĕnz′) Plural **vasa deferentia** (vā′zə dĕf′ə-rĕn′shē-ə). Either of two ducts through which sperm passes from a testis to the outside of the body. In mammals, the vas deferens connects the testis to the urethra.

vasectomy (və-sĕk′tə-mē) Surgical removal of all or part of the vas deferens, resulting in sterility.

vasoconstriction (vā′zō-kən-strĭk′shən) Constriction of a blood vessel, as by a nerve or drug.

vasodilation (vā′zō-dī-lā′shən, -dĭ-) also **vasodilatation** (vā′zō-dĭl′ə-tā′shən, -dī′lə-) Dilation of a blood vessel, as by the action of a nerve or drug.

vasopressin (vā′zō-prĕs′ĭn) See **antidiuretic hormone.**

vector (vĕk′tər) **1.** A quantity, such as the velocity of an object or the force acting on an object, that has both magnitude and direction. Compare **scalar. 2.** An organism, such as a mosquito or tick, that spreads pathogens from one host to another. **3.** A bacteriophage, plasmid, or other agent that transfers genetic material from one cell to another.

vector boson One of the four subatomic particles that mediate the four fundamental forces. The particles and their corresponding forces are: the photon and the electromagnetic force, the graviton and the gravitational force, the intermediate vector boson and the weak nuclear force, and the gluon and the strong nuclear force. The graviton has yet to be isolated in an experiment.

vector product A vector $\mathbf{c}$, depending on two other vectors $\mathbf{a}$ and $\mathbf{b}$, whose magnitude is the product of the magnitude of $\mathbf{a}$, the magnitude of $\mathbf{b}$, and the sine of the angle between $\mathbf{a}$ and $\mathbf{b}$. Its direction is perpendicular to the plane through $\mathbf{a}$ and $\mathbf{b}$ and oriented so that a right-handed rotation about it carries ***a*** into ***b*** through an angle not greater than 180°. The notation for ***c*** is ***c*** = ***a*** × ***b***. Also called *cross product.* See also **right-hand rule.**

vector space A set of generalized vectors and a field of scalars, together with rules for their addition and multiplication (the same rules used for ordinary vectors and scalars).

Vega (vē′gə, vā′gə) A star in the constellation Lyra and one of the five brightest stars in the night sky. It is a white main-sequence star in the Hertzsprung-Russell diagram, with an apparent magnitude of 0.04. Vega, along with Altair and Deneb, form the Summer Triangle asterism. *Scientific name:* Alpha Lyra.

vegetable (vĕj′tə-bəl) **1.** A plant that is cultivated for an edible part, such as the leaf of spinach, the root of the carrot, or the stem of celery. **2.** An edible part of one of these plants. See Note at **fruit.**

vegetation (vĕj′ĭ-tā′shən) **1.** The plants of an area or a region; plant life. **2.** An abnormal bodily accretion, especially a clot composed largely of fused blood platelets, fibrin, and sometimes bacteria, that adheres to a diseased heart valve.

vegetative (vĕj′ĭ-tā′tĭv) **1.** Relating to or characteristic of plants or their growth. **2.** Relating to vegetative reproduction. **3.** Relating to feeding and growth rather than reproduction, as in the mobile phase of plasmodial slime molds. **4.** Relating to an impaired level of brain function in which a person responds reflexively to certain sensory stimuli but demonstrates no cognitive function.

vegetative reproduction A form of asexual reproduction in plants, in which multicellular structures become detached from the parent plant and develop into new individuals that are genetically identical to the parent plant. For example, liverworts and mosses form small clumps of tissue (called gemmae) that are dispersed by splashing raindrops to form new plants. Bulbs, corms, offsets, rhizomes, runners, suckers, and tubers are all important means of vegetative reproduction and propagation in cultivated plants.

veil (vāl) A membranous covering or part, especially a membrane surrounding the young mushrooms of certain basidiomycete fungi. In some species the membrane (called a *partial veil*) extends only from the stalk to the cap. As the cap expands, the veil breaks, leaving a ring called an **annulus** on the stalk and often scalelike pieces on the cap. These veil remnants are important for identifying species of mushrooms.

vein (vān) **1.** Any of the blood vessels that carry blood toward the heart from the body's cells, tissues, and organs. Veins are thin-walled and contain valves that prevent the backflow of blood. All veins except the pulmonary vein carry blood with low levels of oxygen. **2.** One of the narrow, usually branching tubes or supporting parts forming the framework of an insect's wing or a leaf. Veins in insect wings carry hemolymph and contain a nerve. Veins in leaves contain vascular tissue, with the xylem usually occurring on the upper side of the vein (bringing in water and nutrients) and the phloem on the lower side (carrying away food). See more at **leaf, venation. 3.** A long, narrow deposit of mineral or rock that fills the void formed by a fracture or fault in another rock. The mineralogy of the host rock surrounding the vein is often altered where it is in contact with the vein because of chemical reactions between the two rock types. —*Adjective* **venous** (vē′nəs).

veldt also **veld** (vĕlt, fĕlt) An extensive, treeless grassland of southern Africa.

velocimeter (vē′lō-sĭm′ĭ-tər, vĕl′ō-) A device that utilizes the Doppler effect to measure the speed of sound in water. Developed by the US Army Corps of Engineers, a velocimeter transmits a pulse of sound through water and measures its echo for changes in either pitch or frequency. Oceanographers employ velocimeters to study the interior structure of breaking waves; they are also used in hydraulic engineering and to study fluid flow.

velociraptor (və-lŏs′ə-răp′tər) A small, fast, carnivorous dinosaur of the genus *Velociraptor* of the Cretaceous Period that was about 2 m (6.5 ft) in length. It had long curved claws for grasping and tearing at prey, walked on two legs that were adapted for leaping, and had a long stiff tail used as a counterweight. Velociraptors were a kind of raptor.

velocity (və-lŏs′ĭ-tē) The speed and direction of motion of a moving body. Velocity is a vector quantity. Compare **acceleration, speed.**

vena cava (vē′nə kā′və) Plural **venae cavae** (vē′nē kā′vē). Either of two large veins that carry blood with low levels of oxygen to the right atrium of the heart. ► The **superior vena cava** receives blood from the brain and upper limbs or forelimbs. ► The **inferior vena cava** drains blood from the trunk and lower limbs or hindlimbs and is the largest vein in the body.

venation (vē-nā′shən) **1.** The distribution or arrangement of a system of veins, as in an insect's wing or a leaf blade. Patterns of venation in insect wings are often used to identify and differentiate species. In angiosperm plants, the venation of eudicot and magnoliid leaves is generally **netted** or **reticulate**, with smaller veins branching out from larger ones in a pinnate or palmate pattern, while that of monocots is **parallel,** with many veins of similar size running parallel to each other along the length of the plant part. These parallel veins are connected to each other by much smaller cross veins. **2.** The veins of such a system considered as a group.

venereal disease (və-nîr′ē-əl) See **sexually transmitted disease.**

Venn diagram (vĕn) A diagram that uses circles to represent sets, in which the relations between the sets are indicated by the arrangement of the circles. For example, drawing one circle within another indicates that the set represented by the first circle is a subset of the second set. The Venn diagram is named after its inventor, British mathematician John Venn (1834–1923).

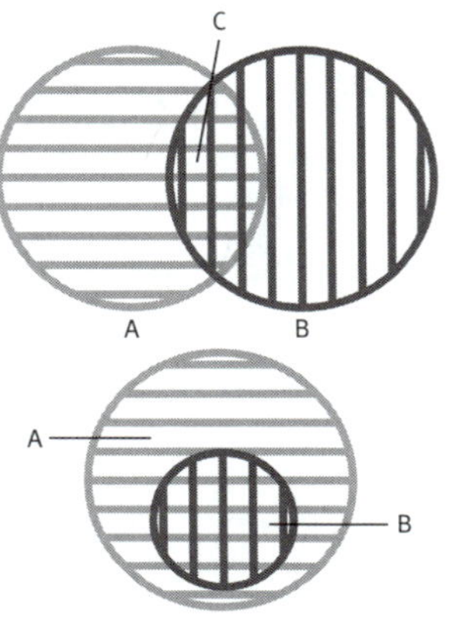

Venn diagram

top: *Sets* A *and* B *intersect to form set* C. *All members of* C *are also members of* A *and* B. bottom: *Set* B *is a subset of* A. *All members of* B *are also members of* A.

venom (vĕn′əm) Any of various poisonous substances secreted by certain snakes, spiders, scorpions, and insects and transmitted to a victim by a bite or sting. Venoms are highly concentrated fluids that typically consist of dozens or hundreds of powerful enzymes, peptides, and smaller organic compounds. These compounds target and disable specific chemicals in the victim, damaging cellular and organ system function. Snake venoms, for example, contain substances that block platelet aggregation (causing bleeding) and that prevent the release of acetylcholine by nerve endings (causing muscle paralysis). Many substances contained in venoms are under investigation for use as pharmaceuticals.

vent (vĕnt) **1.** An opening, and the conduit leading to it, in the side or at the top of a volcano, permitting the escape of fumes, a liquid, a gas, or steam. **2a.** The excretory opening of the digestive tract in animals such as birds, reptiles, amphibians, and fish. **2b.** See **cloaca** (sense 1).

ventifact (vĕn′tə-făkt′) A rock or pebble that has been shaped, polished, or faceted by wind-driven sand.

ventral (vĕn′trəl) Relating to or on the front or lower surface of an animal.

ventricle (vĕn′trĭ-kəl) **1.** A chamber of the heart that receives blood from one or more atria and pumps it by muscular contraction into the arteries. Mammals, birds, and reptiles have two ventricles; amphibians and fish have one. **2.** Any of four fluid-filled cavities in the brain of vertebrate animals. The ventricles are filled with cerebrospinal fluid. —*Adjective* **ventricular** (vĕn-trĭk′yə-lər).

venule (vĕn′yo͞ol) Any of the smaller veins that connect the capillaries to larger veins.

Venus (vē′nəs) The second planet from the Sun, with a diameter about 400 miles less than that of Earth. Venus is a **terrestrial** or **inner planet** and at inferior conjunction comes nearer to Earth than any other planet; depending on its phase, it is also the brightest object in the night sky aside from Earth's moon. Because Venus is an inferior planet (located between Earth and the Sun), it is only visible relatively near the horizon in the first few hours before sunrise or after sunset. It has a dense atmosphere consisting primarily of carbon dioxide, which, together with its proximity to the Sun, creates an intense greenhouse effect, making it the hottest planet in the solar system with an average surface temperature of 464°C (867°F). Venus is completely shrouded by a thick layer of clouds made up mainly of droplets of sulfuric acid with other clouds of vaporous and particulate sulfur dioxide below it. Radar mapping of the Venutian surface shows rolling hills, plains, and numerous volcanoes as well as large impact craters and extensive lava flows. See Table at **solar system.**

vermiform appendix (vûr′mə-fôrm′) See **appendix.**

vernal equinox (vûr′nəl) See under **equinox.**

vernalization (vûr′nə-lĭ-zā′shən) The subjection of seeds or seedlings to low temperature in order to hasten plant development and flowering. Vernalization is commonly used for crop plants such as winter rye and is possible because the seeds and buds of many plants require cold in order to break dormancy.

vernal pool A seasonal body of standing water that typically forms in the spring from melting snow and other runoff, dries out completely in the hotter months of summer, and often refills in the autumn. Vernal pools range from broad, heavily vegetated lowland bodies to smaller, isolated upland bodies with little permanent vegetation. They are free of fish and provide important breeding habitat for many terrestrial or semiaquatic species such as frogs, salamanders, and turtles.

vernation (vûr-nā′shən) The arrangement of young foliage leaves within a bud. The coiled arrangement of young leaves in fern fiddleheads is known as **circinate vernation** and protects the delicate leaf tips as they develop.

Vernier (vĕr-nyā′), **Pierre** 1580–1637. French mathematician and maker of scientific instruments, known especially for his invention of an auxiliary scale (named after him) used for obtaining a highly precise reading of a subdivision of an ordinary scale.

vernier caliper (vûr′nē-ər) A measuring instrument consisting of an L-shaped frame with a linear scale along its longer arm and an L-shaped sliding attachment with a vernier scale, used to read directly the dimension of an object represented by the separation between the inner or outer edges of the two shorter arms.

vernier scale A small, movable auxiliary graduated scale attached parallel to a main graduated scale and calibrated to indicate fractional parts of the subdivisions of the larger scale. Vernier scales are used on certain precision instruments to increase accuracy in measurement.

vertebra (vûr′tə-brə) Plural **vertebrae** (vûr′tə-brā′, -brē′) or **vertebras.** Any of the bones that make up the vertebral column. Each vertebra contains an arched, hollow section through which the spinal cord passes. In humans, the vertebrae are divided into cervical, thoracic, and lumbar sections, and the sacrum and coccyx are both made up of a series of fused vertebrae. The vertebrae are separated by cartilaginous intervertebral disks. See more at **skeleton.**

vertebral column (vûr′tə-brəl) The series of vertebrae extending from the base of the skull to the tip of the tail that forms the supporting axis of the body in vertebrate animals. In humans and tailless apes, the vertebral column ends with the coccyx (tailbone). It encloses and protects the spinal cord and provides a stable attachment for the muscles of the trunk. Also called *backbone, spinal column, spine.*

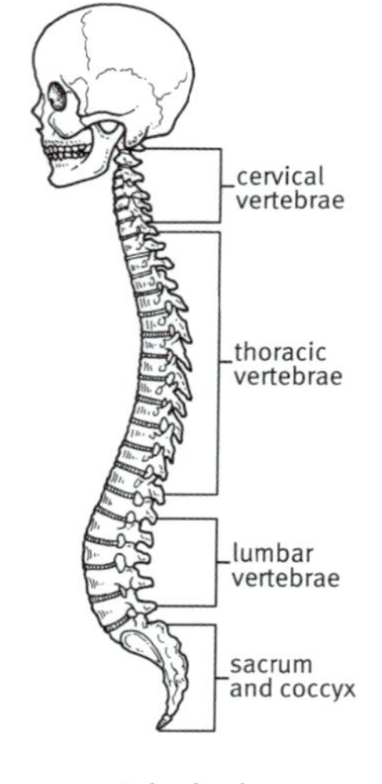

vertebral column

vertebrate (vûr′tə-brĭt, -brāt′) Any of a large group of chordates of the subphylum Vertebrata (or Craniata), characterized by having a backbone. Vertebrates are bilaterally symmetrical and have an internal skeleton of bone or cartilage, a nervous system divided into brain and spinal cord, and not more than two pairs of limbs. Vertebrates have a well-developed body cavity (called a coelom) containing a chambered heart, large digestive organs, liver, pancreas, and paired kidneys, and their blood contains both red and white corpuscles. Vertebrates include fish, amphibians, reptiles, birds, and mammals.

vertex (vûr′tĕks′) Plural **vertices** (vûr′tĭ-sēz′) or **vertexes. 1.** The point at which the sides of an angle intersect. **2.** The point of a triangle, cone, or pyramid that is opposite to and farthest away from its base. **3.** A point of a polyhedron at which three or more of the edges intersect.

vertical angles (vûr′tĭ-kəl) Two angles formed by two intersecting lines and lying on opposite sides of the point of intersection.

vertical circle A great circle on the celestial sphere that passes through the zenith and the nadir and thus is perpendicular to the horizon.

vertigo (vûr′tĭ-gō′) Dizziness characterized by a sensation of whirling motion, either of oneself or of external objects. Vertigo is often caused by damage or disease in the inner ear.

Vesalius (vĭ-sā′lē-əs), **Andreas** 1514–1564. Flemish anatomist and surgeon who is considered the father of modern anatomy. His rigorous descriptions of the structure of the human body, based on his own personal dissections of cadavers, established a new level of clarity and accuracy in the study of human anatomy.

Andreas Vesalius

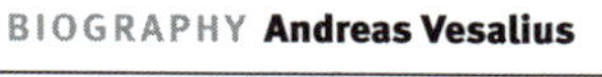

BIOGRAPHY **Andreas Vesalius**

After receiving his medical degree in 1537, Andreas Vesalius began lecturing on surgery and anatomy at the University of Padua. To further his knowledge, he personally dissected cadavers, a task that others in his position would have delegated to an assistant. Through this work Vesalius became convinced that the anatomical theories of the Greek physician Galen, whose ideas had been accepted as authoritative for more than 1,000 years, were not correct. Although Vesalius had begun his career as a Galenist, his hands-on experience led him to believe that Galen's descriptions of the human body were based on dissections of pigs, dogs, and other animals rather than humans, a procedure that was prohibited during Galen's time. Vesalius compared Galen's anatomical texts with his own observations made during dissections. After five years spent compiling his findings, in 1543 he published *De humani corporis fabrica (On the Structure of the Human Body)*, which was the most accurate and comprehensive anatomy textbook to date and included artists' engravings based on Vesalius's own drawings. By relying on careful observation instead of received wisdom, Vesalius transformed the field of anatomy, as well as medicine and biology.

vesicle (vĕs′ĭ-kəl) **1.** A small fluid-filled sac in the body. **2.** A membrane-bound sac in eukaryotic cells that stores or transports the products of metabolism in the cell and is sometimes the site for the breaking down of metabolic wastes. Vesicles bulge out and break off from the endoplasmic reticulum and the Golgi apparatus. Vesicles get their energy for mobility from ATP. Lysosomes and peroxisomes are vesicles. **3.** A small cavity formed in volcanic rock by entrapment of a gas bubble during solidification. *—Adjective* **vesicular** (vĕ-sĭk′yə-lər, və-).

vessel (vĕs′əl) **1.** A blood vessel. **2.** A long, continuous column made of the lignified walls of dead vessel elements, along which water flows in the xylem of angiosperms.

vessel element An elongated, water-conducting cell in xylem, one of the two kinds of tracheary elements. The cells die when mature, leaving only their lignified cell walls. They are stacked one on top of another in long columns, called vessels. Water flows unimpeded from cell to cell along these columns through perforations in the cell walls. Vessel elements are found only in angiosperms. Compare **tracheid.**

Vesta (vĕs′tə) The brightest of all the asteroids and the fourth to be discovered, in 1807. It is the third largest, with a diameter of about 530 km (329 mi), and the only asteroid that can be seen without a telescope. Evidence of lava flows on its surface suggests that it once had a molten interior. See more at **asteroid.**

vestibule (vĕs′tə-byo͞ol′) An oval cavity in the inner ear that together with the semicircular canals makes up the organ that maintains equilibrium in vertebrates.

vestigial (vĕ-stĭj′ē-əl) Relating to a body part that has become small and lost its use because of evolutionary change. Whales, for example, have small bones located in the muscles of their body walls that are vestigial bones of hips and hind limbs.

veterinary medicine (vĕt′ər-ə-nĕr′ē) The branch of medicine that deals with the diseases or injuries of animals and their treatment.

VHF Abbreviation of **very high frequency.**

vibration (vī-brā′shən) A rapid oscillation of a particle, particles, or elastic solid or surface, back and forth across a central position.

vibrissa (vī-brĭs′ə, və-) Plural **vibrissae** (vī-brĭs′ē, və-). **1.** Any of the long, stiff, bristlelike hairs that project from the snout or brow of most mammals, as the whiskers of a cat or rat. Vibrissae often serve as tactile organs, especially in nocturnal animals and marine mammals such as seals and manatees. **2.** Any of several long modified feathers that grow along the sides of the beak of certain birds and help trap insects caught in flight.

vicariance (vī-kâr′ē-əns, vĭ-) The separation or division of a group of organisms by a geographic barrier, such as a mountain or a body of water, resulting in differentiation of the original group into new varieties or species. See also **speciation.**

video display terminal (vĭd′ē-ō′) A computer terminal having a video display that uses a cathode-ray tube.

villus (vĭl′əs) Plural **villi** (vĭl′ī). A small projection on the surface of a mucous membrane, such as that of the small intestine.

vinyl (vī′nəl) **1.** The group C_2H_3, derived from ethylene. **2.** Any of various chemical compounds, typically highly reactive, that contain this group and are used in making plastics. **3.** Any of various plastics made of vinyl, typically tough, flexible, and shiny, often used in upholstery and clothing.

vinyl chloride A flammable gas used to make polyvinyl chloride. *Chemical formula:* C_2H_3Cl.

viral load (vī′rəl) The concentration of a virus, such as HIV, in the blood.

virga (vûr′gə) Light wisps of precipitation streaming from a cloud but evaporating before reaching the ground, especially when the air below is low in humidity.

Virgo (vûr′gō) A constellation in the region of the celestial equator near Leo and Libra. Virgo (the Virgin) contains the bright star Spica and is the sixth sign of the zodiac.

virion (vī′rē-ŏn′, vîr′ē-) A complete viral particle, consisting of RNA or DNA surrounded by a protein shell and constituting the infective form of a virus. The shell, called a **capsid**, protects the interior core that includes the genome and other proteins. After the virion binds to the surface of a specific host cell, its DNA or RNA is injected into the host cell and viral replication occurs with eventual spread of the infection to other host cells.

viroid (vī′roid′) An infectious agent that consists solely of a single strand of RNA and causes disease in certain plants. Viroids lack the protein coat (known as a capsid) of viruses and are the smallest known infectious agents. Containing only about 250 to 375 base pairs, they are much smaller than the smallest genomes of viruses and have no genes for encoding proteins. After invading a host cell, viroids are thought to mimic the cell's DNA, so that the cell's RNA polymerase replicates them in the nucleus. Viroids are believed to cause disease by interfering with the host cell's gene regulation. They are destructive to many important commercial plants, including potatoes, tomatoes, cucumbers, coconuts, and chrysanthemums.

virology (vī-rŏl′ə-jē) The scientific study of viruses and viral diseases.

virtual memory (vûr′cho͞o-əl) A memory management system in a computer that temporarily stores inactive parts of the content

RAM on a disk, restoring it to RAM when quick access to it is needed. This allows software to operate as though the computer has more RAM than it actually does.

virtual particle A short-lived subatomic particle whose existence briefly violates the principle of conservation of energy. The **uncertainty principle** of quantum mechanics allows violations of conservation of energy for short periods, meaning that even a physical system with zero energy can spontaneously produce energetic particles. The more energy a virtual particle has, the shorter its existence. Interactions between normal particles and virtual particles play a crucial role in **quantum field theory** analyses of interactions between real particles. See also **Casimir effect, Feynman diagram, vacuum fluctuation.**

virtual reality A computer simulation of a real or imaginary world or scenario, in which a user may interact with simulated objects or living things in real time. More sophisticated virtual reality systems place sensors on the user's body to sense movements that are then interpreted by the system as movements in the simulated world; binocular goggles are sometimes used to simulate the appearance of objects in three dimensions.

virus (vīʹrəs) Plural **viruses. 1.** Any of various extremely small, often disease-causing agents consisting of a particle (the **virion**), containing a segment of RNA or DNA within a protein coat known as a **capsid.** Viruses are not technically considered living organisms because they are devoid of biological processes (such as metabolism and respiration) and cannot reproduce on their own but require a living cell (of a plant, animal, or bacterium) to make more viruses. Viruses reproduce first either by injecting their genetic material into the host cell or by fully entering the cell and shedding their protein coat. The genetic material may then be incorporated into the cell's own genome or remain in the cytoplasm. Eventually the viral genes instruct the cell to produce new viruses, which often cause the cell to die upon their exit. Rather than being primordial forms of life, viruses probably evolved from rogue pieces of cellular nucleic acids. The common cold, influenza, chickenpox, smallpox, measles, mumps, yellow fever, hemorrhagic fevers, and some cancers are among the diseases caused by viruses. **2.** *Computer Science.* A computer program that duplicates itself in a manner that is harmful to normal computer use. Most viruses work by attaching themselves to another program. The amount of damage varies; viruses may erase all data or do nothing but reproduce themselves. *—Adjective* **viral.**

viscera (vĭsʹər-ə) The soft internal organs of the body, especially those contained within the abdominal and thoracic cavities. *—Adjective* **visceral.**

viscosity (vĭ-skŏsʹĭ-tē) The resistance of a substance to flow. For example, water has a lower viscosity than molasses and flows more easily. Viscosity is related to the concept of **shear force**; it can be understood as the effect of different layers of the fluid exerting shearing force on each other, or on other surfaces, as they move against each other. Viscosity lies behind the **skin friction** component of drag. ► **Kinematic viscosity** is a measure of the rate at which momentum is transferred through a fluid. It is measured in **stokes.** ► **Dynamic viscosity** is a measure of the ratio of the **stress** on a region of a fluid to the rate of change of **strain** it undergoes. It is equal to the kinematic viscosity times the density of the fluid. It is measured in **pascal-seconds** or **poises**.

viscous (vĭsʹkəs) Having relatively high resistance to flow (high **viscosity**).

visual binary (vĭzhʹo͞o-əl) See under **binary star.**

visual field The area that is visible to an immobile eye at a given time.

vital capacity (vītʹl) The maximum amount of air that can be expelled from the lungs after breathing in as deeply as possible.

vital signs The pulse rate, temperature, respiratory rate, and usually blood pressure of an individual.

vitamin (vīʹtə-mĭn) Any of various organic compounds that are needed in small amounts for normal growth and activity of the body. Most vitamins cannot be synthesized by the body, but are found naturally in foods obtained from plants and animals. Vitamins are either water-soluble or fat-soluble. Most water-soluble vitamins, such as the vitamin B complex, act as catalysts and coenzymes in metabolic processes and energy transfer and are excreted fairly rapidly. Fat-soluble vitamins, such as vitamins A, D, and E are necessary for the function or structural integrity of specific body tissues and membranes and are retained in the body.

A CLOSER LOOK **vitamins**

Although it has been known for thousands of years that certain diseases can be treated with specific foods, the scientific link between vitamins and good health wasn't made until the early 1900s by Polish-born American biochemist Casimir Funk. While studying beriberi, a disease that causes depression, fatigue, and nerve damage, Funk discovered an organic compound in rice husks that prevents the illness. He named the compound *vitamine,* derived from the chemical name *amine* and the Latin word *vita,* "life," because vitamins are required for life and were originally thought to be amines. Funk's compound is now known as vitamin B_1, or thiamine. His research and discovery led him, along with English biochemist Sir Frederick Gowland Hopkins, to propose the *vitamin hypothesis of deficiency,* which stated that certain diseases, such as scurvy or rickets, are caused by dietary deficiencies and can be avoided by taking vitamins. Further research allowed scientists to isolate and identify the vitamins that we know today to be essential for human health. Vitamins include A, C, D, E, K, thiamine, riboflavin, niacin, B_6, B_{12}, folic acid, biotin, and pantothenic acid. Vitamins are distinguished from minerals, such as calcium, iron, and magnesium, which are also essential for optimum health.

vitamin A A fat-soluble vitamin important for normal vision, tissue growth, and healthy skin. It is found in fish-liver oils, milk, green leafy vegetables, and red, orange, and yellow vegetables and fruits. A deficiency of vitamin A in humans causes poor vision at night and damage to the skin and mucous membranes. Also called *retinol. Chemical formula:* $C_{20}H_{30}O$.

vitamin B **1.** Vitamin B complex. **2.** A member of the vitamin B complex, especially thiamine (vitamin B_1).

vitamin B_1 See **thiamine.**

vitamin B_2 See **riboflavin.**

vitamin B_6 Any of several water-soluble pyridine derivatives, especially pyridoxine, that are coenzymes in amino acid synthesis and are important in protein and fat metabolism and in healthy nerve function. Vitamin B_6 is found in liver, whole-grain foods, fish, yeast, and many vegetables.

vitamin B_{12} A water-soluble, complex organic compound containing cobalt, found especially in meat, liver, eggs, milk, and milk products. Vitamin B_{12} is necessary for the synthesis of DNA by the body, for the production of blood cells, and for maintaining the health of nerves. A deficiency of vitamin B_{12} in the diet results in pernicious anemia. Also called *cobalamin, cyanocobalamin.*

vitamin B complex Any of a group of water-soluble vitamins important for normal cell growth and metabolism. The vitamins of the vitamin B complex include thiamine (vitamin B_1), riboflavin (vitamin B_2), pyridoxine (vitamin B_6), cobalamin (vitamin B_{12}), pantothenic acid, niacin, biotin, and folic acid.

vitamin C A water-soluble vitamin important for healthy skin, teeth, bones, and blood vessels. It is found especially in citrus fruits, tomatoes, potatoes, and green leafy vegetables. A deficiency of vitamin C in the diet causes **scurvy.** Also called *ascorbic acid. Chemical formula:* $C_6H_8O_6$.

vitamin D Any of a group of fat-soluble sterols necessary for normal bone growth, especially vitamin D_2 (*ergocalciferol*) and vitamin D_3 (*cholecalciferol*). Vitamin D is found in milk, fish, and eggs and can be produced in the skin on exposure to sunlight. A deficiency of vitamin D in the diet causes **rickets** in children.

vitamin D_2 A white crystalline sterol produced by ultraviolet irradiation of ergosterol and also occurring naturally in fungi and some fish oils. Vitamin D_2 is the form of vitamin D generally used as a dietary supplement. Also called *calciferol, ergocalciferol. Chemical formula:* $C_{28}H_{44}O$.

vitamin D_3 A colorless, crystalline steroid hormone that the body synthesizes in the skin when its precursor, a derivative of cholesterol, is irradiated by sunlight. Vitamin D_3 is also found in fish-liver oils, irradiated milk, and all irradiated animal foodstuffs. Also called *calciferol, cholecalciferol. Chemical formula:* $C_{27}H_{44}O$.

vitamin E **1.** A fat-soluble vitamin important for normal cell growth and function. It is found in vegetable oils, wheat germ, green leafy vegetables, and egg yolks. Vitamin E has at least eight different forms, the most prevalent of which is **alpha-tocopherol**.

vitamin K Any of a group of fat-soluble vitamins that are involved in the formation of prothrombin and other clotting factors in the liver and are essential for normal clotting of the blood. (The *K* is derived from the German word *koagulation*.) Vitamin K is also involved in bone formation and repair. Two forms occur naturally: vitamin K_1, which is synthesized by plants, and vitamin K_2, which is mainly synthesized by intestinal bacteria. The other forms are synthetic substances with similar chemical structures.

vitamin K_1 The major dietary form of vitamin K that is synthesized in plants and found primarily in green, leafy vegetables such as alfalfa and in vegetable oils. It can be made synthetically and is given orally to treat prothrombin deficiency that results from heparin and other anticoagulant drugs. Also called *phylloquinone. Chemical formula:* $C_{31}H_{46}O_2$.

vitamin K_2 A form of vitamin K that is synthesized by bacteria in the intestine and is also found in fish and other foods. Also called *menaquinone. Chemical formula:* $C_{41}H_{56}O_2$.

vitreous (vĭt′rē-əs) Relating to or resembling glass.

vitreous humor The gelatinous substance that fills the chamber of the eye between the retina and the lens.

vitriol (vĭt′rē-ōl′, -əl) **1.** A former name for sulfuric acid. **2.** Any of various sulfates of metals, such as ferrous sulfate (green vitriol), zinc sulfate, or copper sulfate (blue vitriol). See also **blue vitriol.**

viverrid (vī-vĕr′ĭd) Any of various mostly small carnivorous mammals of the family Viverridae, having long, slender bodies with short legs and bushy tails. Viverrids include the civets, mongooses, meerkats, and binturong.

viverrine (vī-vĕr′ĭn, -īn′) Characteristic of or resembling civets or their kin.

viviparous (vī-vĭp′ər-əs) Giving birth to living young that develop within the mother's body rather than hatching from eggs. All mammals except the monotremes are viviparous. Compare **oviparous, ovoviviparous.**

vivisection (vĭv′ĭ-sĕk′shən) The practice of examining internal organs and tissues by cutting into or dissecting a living animal, especially for the purpose of scientific research.

vocal cords (vō′kəl) The two folded pairs of membranes in the larynx (voice box) that vibrate when air that is exhaled passes through them, producing sound.

Vogelstein (vō′gəl-stēn′), **Bert** Born 1949. American medical researcher who led the team that identified the specific mutations responsible for cancer of the colon, a discovery that greatly advanced the understanding of the genetic basis of cancer.

voice box (vois) See **larynx.**

VoIP (voip, vē′ō′ī′pē′) Short for *Voice over Internet Protocol.* A protocol for transmitting the human voice in digital form over the Internet or other networks as an audio stream, instead of using traditional telephone lines. VoIP uses the Internet Protocol (IP), but is not limited to communication by computer—even phone-to-phone communication can be conducted using this technology.

volatile (vŏl′ə-tl) Changing easily from liquid to vapor at normal temperatures and pressures. Essential oils used in perfumes are highly volatile.

volcanic arc (vŏl-kăn′ĭk, vôl-) A curved chain of volcanoes in the overriding tectonic plate of a subduction zone. Volcanic arcs form as the result of rising magma formed by the melting of the downgoing plate. They are curved because of the curvature of the Earth. The Andes mountains form a volcanic arc. See also **island arc.**

volcanic glass Natural glass produced by the cooling of molten lava too quickly to permit crystallization. Obsidian is a type of volcanic glass.

volcanic pipe A vertical conduit below a volcano through which magma has passed and that has become filled with solidified magma, volcanic breccia, and fragments of older rock.

volcano (vŏl-kā′nō) **1.** An opening in the Earth's crust from which lava, ash, and hot gases flow or are ejected during an eruption. **2.** A usually cone-shaped mountain formed by the materials issuing from such an opening. Volcanoes are usually associated with plate boundaries but can also occur within the interior areas of a tectonic plate. Their shape is directly related to the type of magma that flows from them—the more viscous the magma, the steeper the sides of the volcano. ▸ A volcano composed of gently sloping sheets of basaltic lava from successive volcanic eruptions is called a **shield volcano.** The lava flows associated with shield volcanos, such as Mauna Loa, on Hawaii, are very fluid. ▸ A volcano composed of steep, alternating layers of lava and pyroclastic materials, including ash, is called a **stratovolcano.** Stratovolcanos are associated with relatively viscous lava and with explosive eruptions. They are the most common form of large continental volcanos. Mount Vesuvius, Mount Fuji, and Mount St. Helens are stratovolcanos. Also called *composite volcano.* See more at **hot spot, island arc, tectonic boundary, volcanic arc.**

volcanology (vŏl′kə-nŏl**′**ə-jē, vôl′-) The scientific study of volcanoes and volcanic phenomena.

volt (vōlt) The SI derived unit used to measure electric potential at a given point, usually a point in an electric circuit. A voltage difference of one volt drives one ampere of current through a conductor that has a resistance of one ohm. One joule of work is required to move an electric charge of one coulomb across a potential difference of one volt. One volt is equivalent to one joule per coulomb. See also **Ohm's law.**

Volta (vōl**′**tə), Count **Alessandro** 1745–1827. Italian physicist who in 1800 invented the voltaic pile, which was the first source of continuous electric current. The volt unit of electromotive force is named for him.

voltage (vōl**′**tĭj) A measure of the difference in electric potential between two points in space, a material, or an electric circuit, expressed in volts.

voltage divider A set of two or more resistors connected in series between an applied voltage, so that the voltage at points between the resistors is a fixed fraction of the applied voltage. Voltage dividers are common in the power supplies of devices such as amplifiers, in which different subcomponents require different voltage levels for their own power supplies.

voltaic cell (vŏl-tā**′**ĭk) See **electric cell.**

voltaic pile A source of electricity consisting of a number of disks that alternate between two different metals and are separated by acid-moistened pads, forming a set of galvanic cells connected in series. See more at **galvanic.** See Note at **battery.**

voltammeter (vōl-tăm**′**mē′tər, vōlt**′**ăm′-) **1.** A device consisting of a voltmeter combined with an ammeter. **2.** An instrument for measuring and sometimes analyzing electrical current and potential in voltammetry.

voltammetry (vōl-tăm**′**mē′tər, vōlt**′**ăm′-) A method of determining the chemical makeup of a sample substance by measuring electrical activity, or the accumulation of chemicals, on electrodes placed in the substance. Voltammetry is often used to determine the amount of trace metals and toxins in water or other solutions.

volt-ampere A unit of electric power equal to the product of one volt and one ampere, equivalent to one watt.

voltmeter (vōlt**′**mē′tər) An instrument used for measuring the difference in voltage between two points in an electric circuit. Voltmeters typically make use of an ammeter that measures current flow across a known resistance inside the voltmeter; direct-current voltages can then be determined by Ohm's law. Digital voltmeters employ A/D converters to provide the numerical value of the voltage displayed. Compare **ohmmeter.**

volume (vŏl**′**yo͞om) **1.** The amount of space occupied by a three-dimensional object or region of space. Volumes are expressed in cubic units. **2.** A measure of the loudness or intensity of a sound.

volumetric (vŏl′yo͞o-mĕt**′**rĭk) Relating to measurement by volume or to a unit that is used to measure volume.

volva (vŏl**′**və, vôl**′**-) A cuplike structure around the base of the stalk of certain basidiomycete fungi.

vomit (vŏm**′**ĭt) Matter ejected from the stomach through the mouth, usually as a result of involuntary muscle contractions.

vortex (vôr**′**tĕks′) Plural **vortexes** or **vortices** (vôr**′**tĭ-sēz′). A circular, spiral, or helical motion in a fluid (such as a gas) or the fluid in such a motion. A vortex often forms around areas of low pressure and attracts the fluid (and the objects moving within it) toward its center. Tornados are examples of vortexes;

vortexes that form around flying objects are a source of turbulence and drag. See also **eddy.**

vulcanize (vŭl′kə-nīz′) To harden rubber by combining it with sulfur or other substances in the presence of heat and pressure. Vulcanization gives rubber strength, resistance, and elasticity.

vulva (vŭl′və) Plural **vulvas** or **vulvae** (vŭl′vē′). The external genitals of female mammals, including the labia and clitoris.

W

w Abbreviation of **width.**

W 1. The symbol for **tungsten. 2.** Abbreviation of **watt.**

wadi (wä′dē) A gully or streambed in northern Africa and southwest Asia that remains dry except during the rainy season.

Waksman (wăks′mən), **Selman Abraham** 1888–1973. Ukrainian-born American microbiologist who pioneered the development of antibiotics. His discoveries included streptomycin (1944), the first drug effective against tuberculosis. He received the Nobel Prize for physiology or medicine in 1952.

Wallace (wŏl′ĭs), **Alfred Russel** 1823–1913. British naturalist who formulated a theory of evolution by natural selection independently of Charles Darwin. Wallace spent eight years (1854–62) traveling in Malaysia and assembling evidence for his theories, which he sent to Darwin in England. Their findings were first presented to the public in 1858.

Walton (wôl′tən), **Ernest Thomas Sinton** 1903–1995. Irish physicist who, with John Cockcroft, was the first to successfully split an atom using a particle accelerator in 1932. For this work they shared the 1951 Nobel Prize for physics.

WAN (wăn) Short for *wide area network.* A communications network that uses such devices as telephone lines, satellite dishes, or radio waves to span a larger geographic area than can be covered by a LAN. The Internet is a WAN.

-ware A suffix that means "software," as in *shareware.*

warfarin (wôr′fər-ĭn) A white crystalline compound used as a rodenticide and as an anticoagulant in medicine. Warfarin is a derivative of coumarin. *Chemical formula:* $C_{19}H_{16}O_4$.

warm-blooded (wôrm′blŭd′ĭd) Having a relatively warm body temperature that stays about the same regardless of changes in the temperature of the surroundings. Birds and mammals are warm-blooded.

warm front The forward edge of an advancing mass of warm air that rises over and replaces a retreating mass of cooler air. As it rises, the warm air cools and the water vapor in it condenses, usually forming steady rain, sleet, or snow. On a weather map, a warm front is depicted as a red line with half circles whose curved sides point in the direction in which the warm air is moving. Compare **cold front, occluded front.** See illustration at **front.**

warning coloration (wôr′nĭng) Conspicuously recognizable markings of an animal that serve to warn potential predators of the nuisance or harm that would come from attacking or eating it. The bold patterns of skunks and the bright colors of poison arrow frogs are examples of warning coloration. Also called *aposematic coloration.* Compare **camouflage.**

wart (wôrt) **1.** A small growth on the skin caused by a virus, occurring typically on the hands or feet. **2.** A similar growth on a plant, especially one caused by a fungal disease.

waste (wāst) *Noun.* **1.** An unusable or unwanted substance or material, such as a waste product. See also **hazardous waste, landfill.** — *Verb.* **2.** To lose or cause to lose energy, strength, weight, or vigor, as by the progressive effects of a disease such as metastatic cancer.

waste product An unusable or unwanted substance or material produced during or as a result of a process, such as metabolism or manufacturing.

water (wô′tər) A colorless, odorless compound of hydrogen and oxygen. Water covers about three-quarters of the Earth's surface in solid form (ice) and liquid form, and is prevalent in the lower atmosphere in its gaseous form, water vapor. Water is an unusually good solvent for a large variety of substances, and is an essential component of all organisms, being necessary for most biological processes. Unlike most substances, water is less dense as ice than in liquid form; thus, ice floats on liquid water. Water freezes at 0°C (32°F) and boils at 100°C (212°F). *Chemical formula:* H_2O.

water bear See **tardigrade.**

water cycle See **hydrologic cycle.**

water gap A transverse cleft in a mountain ridge through which a stream flows.

water of crystallization Water in a crystal structure that is chemically combined with the other elements in the crystal. The water of crystallization is necessary for the maintenance of crystalline properties, but it can be removed by heat.

water of hydration Water that is chemically combined with other molecules to form a hydrate mineral and that is easily removed through heating.

watershed (wô′tər-shĕd′) **1.** A continuous ridge of high ground forming a divide between two different drainage basins or river systems. **2.** The region enclosed by such a divide and draining into a river, river system, or other body of water.

water table The upper surface of an area filled with groundwater, separating the **zone of aeration** (the subsurface region of soil and rocks in which the pores are filled with air and usually some water) from the **zone of saturation** (the subsurface region in which the pores are filled only with water). Water tables rise and fall with seasonal moisture, water absorption by vegetation, and the withdrawal of groundwater from wells, among other factors. The water table is not flat but has peaks and valleys that generally conform to the overlying land surface. Compare **potentiometric surface.**

water vapor Water in its gaseous state, especially in the atmosphere and at a temperature below the boiling point. Water vapor in the atmosphere serves as the raw material for cloud and rain formation. It also helps regulate the Earth's temperature by reflecting and scattering radiation from the Sun and by absorbing the Earth's infrared radiation. See also **vapor.**

Watson (wŏt′sən), **James Dewey** Born 1928. American biologist who, working with Francis Crick, identified the structure of DNA in 1953. By analyzing the patterns cast by x-rays striking DNA molecules, they discovered that DNA has the structure of a double helix, two spirals linked together by bases in ladderlike rungs. For this work Watson and Crick shared with Maurice Wilkins the 1962 Nobel Prize for physiology or medicine. See Note at **Rosalind Franklin.**

Watson-Watt (wŏt′sən-wŏt′), Sir **Robert Alexander** 1892–1973. British physicist who pioneered the development of radar. In 1919 he produced a system for locating thunderstorms by tracking their radio emissions. In the 1930s Watson-Watt led the team that developed radar into a practical system for locating aircraft.

watt (wŏt) The SI derived unit used to measure power, equal to one joule per second. In electricity, a watt is equal to current (in amperes) multiplied by voltage (in volts).

Watt, James 1736–1819. British engineer and inventor who patented a much improved version of the steam engine (1769) and devised the unit of horsepower. The watt unit of power is named for him.

wattage (wŏt′ĭj) An amount of power, especially electrical power, expressed in watts or kilowatts.

watt-hour A unit of energy, especially electrical energy, equal to the work done by one watt acting for one hour. It is equivalent to 3,600 joules.

watt-second A unit of energy equal to the work done by one watt acting for one second, equivalent to 1 joule.

wave (wāv) A disturbance, oscillation, or vibration, either of a medium and moving through that medium (such as water and sound waves), or of some quantity with different values at different points in space, moving through space (such as electromagnetic waves or a quantum mechanical wave described by the wave function). See also **longitudinal wave, transverse wave, wave function.** See Note at **refraction.**

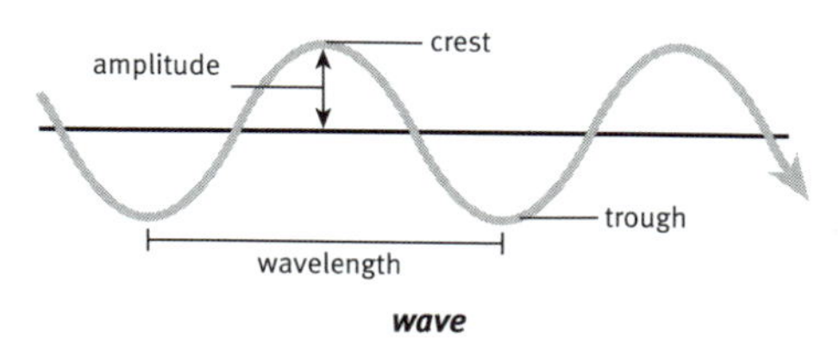

wave

structure of a wave

wave equation 1. A partial differential equation that describes the shape and movement of waves, given a set of boundary conditions (such as the initial shape of the wave, or the evolution of a force affecting the wave). **2.** The fundamental equation of wave mechanics. See also **Schrödinger's equation.**

wave front The set of points in space reached by a wave or vibration at the same instant as the wave travels through a medium. Wave fronts generally form a continuous line or surface. The lines formed by crests of ripples on a pond, for example, correspond to curved wave fronts.

wave function A mathematical function used in quantum mechanics to describe the propagation of the wave associated with a particle or group of particles. The wave function is a solution to Schrödinger's equation, given the boundary conditions that describe the physical system in which the particle is found. The square of the function evaluated at a given point in space is proportional to the probability of finding the particle in the immediate vicinity of that position.

wavelength (wāv′lĕngkth′) The distance between one peak or crest of a wave and the next peak or crest. It is equal to the speed of the wave divided by its frequency, and to the speed of a wave times its period.

wavelet (wāv′lĭt) A small wave; a ripple. See more at **wave.**

wave mechanics A theory that interprets the behavior of matter (especially subatomic or other small particles) in terms of the properties of waves. A broad range of physical phenomena, from the propagation of earthquakes to the structures of electron orbitals in atoms, have been understood using wave mechanics. Quantum mechanics uses a form of wave mechanics and involves wave equations such as Schrödinger's equation to capture both the wavelike and particlelike properties of matter.

wave number The number of wave cycles per unit distance for a wave of a given wavelength.

wave-particle duality The exhibition of both wavelike and particlelike properties by a single entity. For example, electrons undergo diffraction and can interfere with each other as waves, but they also act as pointlike masses and electric charges. The theory of quantum mechanics is a attempt to explain these apparently contradictory properties exhibited by matter. See also **complementarity.**

wave train A succession of wave cycles moving at the same speed and typically having the same wavelength.

wax (wăks) Any of various solid, usually yellow substances that melt or soften easily when heated. They are similar to fats, but are less greasy and more brittle. Naturally occurring animal and plant waxes are esters of saturated fatty acids and alcohols of high molecular weight, including sterols. Waxes are also manufactured synthetically from petroleum, and are used to make polishers, lubricants, coatings, waterproofing, crayons, candles, and many other products.

W boson A subatomic particle with positive electric charge that mediates the weak nuclear force. The W boson has a mass 160,000 times that of the electron. Unlike the other weak force mediator (the Z boson), the W boson changes particles it interacts with into other kinds of particles; for example, in **beta-plus decay**(a kind of **beta decay**), an up quark in a proton decays into a down quark by emitting a W boson, changing the proton into a neutron. The W boson itself decays into a positron and an electron neutrino. See Table at **subatomic particle.**

weak force (wēk) The fundamental force that acts between leptons and is involved in the decay of hadrons. The weak nuclear force is responsible for nuclear beta decay (by changing the flavor of quarks) and for neutrino absorption and emission. It is mediated by the intermediate vector bosons (the W boson and the Z boson), and is weaker than the strong nuclear force and the electromagnetic force but stronger than gravity. Some scientists believe that the weak nuclear force and the electromagnetic force are both aspects of a single force called the **electroweak force.** Also called *weak nuclear force* or *weak interaction.* Compare **electromagnetic force.**

weak interaction See **weak force.**

weak nuclear force See **weak force.**

weather (wĕ*th*′ər) The state of the atmosphere at a particular time and place. Weather is described in terms of variable conditions such as temperature, humidity, wind velocity, precipitation, and barometric pressure.

Weather on Earth occurs primarily in the troposphere, or lower atmosphere, and is driven by energy from the Sun and the rotation of the Earth. The average weather conditions of a region over time are used to define a region's climate.

weather balloon A balloon used to carry instruments aloft to gather meteorological data in the atmosphere. Weather balloons are tracked by radar, Global Positioning System, or **theodolites.** The balloon itself usually contains devices to measure temperature, pressure and humidity, although it can contain specialized devices to measure the ozone in the ozone layer. Filled with hydrogen or helium gas, weather balloons can reach 35 km (22 mi) in altitude before disintegrating due to the decreased pressure of the lower stratosphere.

weather balloon

weathering (wĕ*th*′ər-ĭng) Any of the chemical or mechanical processes by which rocks exposed to the weather undergo chemical decomposition and physical disintegration. Although weathering usually occurs at the Earth's surface, it can also occur at significant depths, for example through the percolation of groundwater through fractures in bedrock. It usually results in changes in the color, texture, composition, or hardness of the affected rocks.

web (wĕb) **1.** A structure of fine, elastic, threadlike filaments characteristically spun by spiders to catch insect prey. The larvae of certain insects also weave webs that serve as protective shelters for feeding and may include leaves or other plant parts. **2.** A membrane or fold of skin connecting the toes in certain animals, especially ones that swim, such as water birds and otters. The web improves the ability of the foot to push against water. **3.** also **Web** The World Wide Web.

weber (wĕb′ər, vā′bər) The SI derived unit of magnetic flux. A magnetic flux of one weber, passing through a conducting loop and reduced to zero at a uniform rate in one second, induces an electric potential of one volt in the loop. One weber is equal to one volt per second, or 10^8 maxwells. The weber is named after German scientist Wilhelm Eduard Weber (1804–1891).

Weber (vā′bər), **Ernst Heinrich** 1795–1878. German physiologist who is noted for his study of sensory response, particularly in the ear and the skin. He also demonstrated that the digestive juices are the specific products of glands.

weblog (wĕb′lôg′) A website that displays in chronological order the postings by one or more individuals and usually has links to comments on specific postings.

webpage or **Web page** (wĕb′pāj′) A document on the World Wide Web, consisting of an HTML file and any related files for scripts and graphics, and often hyperlinked to other documents on the Web. The content of webpages is normally accessed by using a browser.

website or **Web site** (wĕb′sīt′) A set of interconnected webpages, usually including a homepage, generally located on the same server, and prepared and maintained as a collection of information by a person, group, or organization.

Wegener (vā′gə-nər), **Alfred Lothar** 1880–1930. German physicist, meteorologist, and explorer who introduced the theory of continental

Alfred Wegener

drift in 1915. His hypothesis was controversial and remained so until the 1960s, when new scientific understanding of the structure of the ocean floors provided evidence that his theory was correct. See more at plate tectonics.

BIOGRAPHY Alfred Wegener

A look at a map shows that the eastern coast of South America and the western coast of Africa have roughly the same outline. By the early twentieth century scientists observing this similarity speculated that the two continents must have once been joined together with the other continents as one large landmass. But they were unable to come up with a satisfactory model to explain how this giant landmass could become the modern continents positioned around the globe. In 1912 Alfred Wegener proposed a radical new idea, called *continental drift,* claiming that the primeval supercontinent (which he named Pangaea) had slowly broken apart, its pieces separating over millions of years into the locations observed today. In the geological record he found corroborating evidence in the form of identical rock strata and fossil remains in eastern South America and western Africa and in other locations that would align if the continents fit together. Wegener's theory was soon rejected because he could not propose a suitable mechanism to account for the supposed drifting of continents. It was not until the 1960s that a mechanism would appear—the theory of *plate tectonics,* stating that the Earth's lithosphere is divided into plates on top of which the continents ride, slowly carried along by huge convection currents within the Earth's mantle. Sadly, Wegener did not live to see the vindication of his theory, having died on an expedition to Greenland in 1930.

weight (wāt) **1.** The force with which an object near the Earth or another celestial body is attracted toward the center of the body by gravity. An object's weight depends on its mass and the strength of the gravitational pull. The weight of an object in an aircraft flying at high altitude is less than its weight at sea level, since the strength of gravity decreases with increasing distance from the Earth's surface. The SI unit of weight is the newton, though units of mass such as grams or kilograms are used more informally to denote the weight of some mass, understood as the force acting on it in a gravitational field with a strength of one G. The pound is also still used as a unit of weight. **2.** A system of such measures, such as avoirdupois weight or troy weight.

USAGE weight/mass

Although most hand-held calculators can translate pounds into kilograms, an absolute conversion factor between these two units is not technically sound. A pound is a unit of force, and a kilogram is a unit of mass. When the unit pound is used to indicate the force that a gravitational field exerts on a mass, the pound is a unit of weight. Mistaking weight for mass is tantamount to confusing the electric charges on two objects with the forces of attraction (or repulsion) between them. Like charge, the mass of an object is an intrinsic property of that object: electrons have a unique mass, protons have a unique mass, and some particles, such as photons, have no mass. Weight, on the other hand, is a force due to the gravitational attraction between two bodies. For example, one's weight on the Moon is $\frac{1}{6}$ of one's weight on Earth. Nevertheless, one's mass on the Moon is identical to one's mass on Earth. The reason that hand-held calculators can translate between units of weight and units of mass is that the majority of us use calculators on the planet Earth at sea level, where the conversion factor is constant for all practical purposes.

Weinberg (wīn′bûrg′), **Steven** Born 1933. American nuclear physicist who helped develop the theory of the electroweak force, explaining the relationship between two of the four fundamental forces of nature, the electromagnetic force and the weak force. For this work he shared with Sheldon Glashow and Abdus Salam the 1979 Nobel Prize for physics.

well (wĕl) A deep hole or shaft sunk into the Earth to tap a liquid or gaseous substance such as water, oil, gas, or brine. If the substance is not under sufficient pressure to flow freely from the well, it must be pumped or raised mechanically to the surface. Water or pressurized gas is sometimes pumped into a nonproducing oil well to push petroleum resources out of underground reservoirs. See also **artesian well.**

Wernicke's area (vĕr′nĭ-kēz, -kəz) An area located in the rear of the left temporal lobe of the brain. It is associated with the ability to recognize and understand spoken language. It is named for its discoverer, German neurologist and psychologist Carl Wernicke (1848–1905).

westerly (wĕs′tər-lē) A wind, especially a prevailing wind, that blows from the west. The prevailing winds in the middle latitudes are westerlies. See illustration at **wind.**

Western Hemisphere (wĕs′tərn) The half of the Earth that includes North America, Central America, and South America, as divided roughly by the 0° and 180° meridians. See more at **prime meridian.**

Westinghouse (wĕs′tĭng-hous′), **George** 1846–1914. American engineer and manufacturer who introduced the high-voltage alternating current system for the transmission of electricity in the United States. A prolific inventor, Westinghouse received hundreds of patents in his lifetime, including the air brake (1869), automated train-switching signals, and devices for the transmission of natural gas. His inventions made an important contribution to the growth of railroads.

West Nile virus (wĕst nīl) An infectious disease caused by a virus of the genus *Flavivirus,* occurring in Africa, Asia, the Mediterranean, and parts of North America. The infection is transmitted by a mosquito and can cause mild illness as well as severe, sometimes fatal encephalitis and meningitis.

wet cell (wĕt) An electric cell in which the chemicals producing the current are in the form of a liquid rather than in the form of a paste (as in a **dry cell**). Car batteries consist of a series of wet cells. Compare **dry cell.**

wet deposition See under **acid deposition.**

wetland (wĕt′lănd′) A low-lying area of land that is saturated with moisture, especially when regarded as the natural habitat of wildlife. Marshes, swamps, and bogs are examples of wetlands. See more at **lacustrine, marine, palustrine, riverine.**

A CLOSER LOOK **wetland**

Wetlands are areas such as swamps, bogs, and marshes where water either covers the soil or is present at or near the surface, particularly in the root zone, at least a good portion of the year, including the growing season. In the past, wetlands were generally considered unproductive or undesirable lands—smelly and unhealthful, a breeding ground for mosquitoes and other pests—and many were filled in to create farmland or to develop land for housing and industrial use. More than half of the original wetlands in the continental United States have disappeared in the name of reclamation, disease prevention, and flood control. Scientists now realize that, far from being noxious barrens, wetlands play a key role in the ecosystem. They act as filters, removing pollutants, including metals, from waters. They serve as reservoirs, and they aid flood and erosion control by absorbing excess water. Wetlands are home to a great variety of plant and animal species, some endangered, that have evolved to live in the wetland's unique conditions. The preservation and, where possible, restoration of these vital habitats has become a primary goal of environmentalists around the world.

whirlpool (wûrl′po͞ol′) A rapidly rotating current of water or other liquid that sucks everything near it toward its center. The meeting of two tides can create a whirlpool.

white blood cell (wīt) Any of various white or colorless cells in the blood of vertebrate animals, many of which participate in the inflammatory and immune responses to protect the body against infection and to repair injuries to tissues. White blood cells are formed mainly in the bone marrow, and unlike red blood cells, have a cell nucleus. The major types of white blood cells are **granulocytes, lymphocytes,** and **monocytes.** White blood cells are far less numerous in the blood than red blood cells, but their amount usually increases in response to infection and can be monitored as part of a clinical assessment. Also called *leukocyte.*

white dwarf A small, extremely dense star characterized by high temperature and luminosity. A white dwarf is believed to be in its final stage of evolution, having either used up most of its nuclear fuel in its main-sequence stage, or else moved through a giant stage and shed any remaining fuel in its outer layer as a **planetary nebula**, leaving only a glowing core. Some 10 percent of all stars in the **Milky Way** are white dwarfs, but despite their intrinsic luminosity, they are so small that none are visible to the naked eye. See Note at **dwarf.**

white light 1. Electromagnetic radiation composed of a fairly even distribution of all of the frequencies in the visible range of the spectrum, appearing white to the eye. Light from the Sun is nearly perfect white light, although the Sun does not itself appear white when viewed on Earth due to the scattering of light with frequencies in the blue range by the atmosphere, leaving the Sun with a yellow color. **2.** Light that appears white to the eye, composed of some combination of light with frequencies in the red, blue, and green parts of the spectrum. See also **color.**

white matter The whitish tissue of the vertebrate brain and spinal cord, made up chiefly of nerve fibers (axons) covered in myelin sheaths. Compare **gray matter.**

white room See **clean room.**

white supergiant A rare type of extremely bright supergiant with surface temperature of around 10,000°K. Deneb, one of the brightest stars in the **Milky Way,** is a white supergiant; it has luminosity approximately 60,000 times that of the Sun.

Whittle (wĭt′l), Sir **Frank** 1907–1996. British aeronautical engineer and inventor who developed the first aircraft engine powered by jet propulsion in 1937.

whole blood (hōl) Blood from which no constituent, such as red blood cells, white blood cells, plasma, or platelets, has been removed. Whole blood is commonly obtained through blood donation and can be transfused directly or broken down into blood components that can be transfused separately.

whole number 1. A member of the set of positive integers and zero. **2.** A positive integer. **3.** An integer.

whooping cough (ho͞o′pĭng, ho͝op′ĭng, wo͞o′pĭng, wo͝op′ĭng) An infectious disease caused by the bacterium *Bordatella pertussis,* seen most commonly in children and characterized by coughing spasms often ending in loud gasps. Vaccinations usually given during infancy confer immunity to the disease. Also called *pertussis.*

whorl (hwôrl, wôrl, hwûrl, wûrl) **1.** An arrangement of three or more appendages radiating in a circular or spiral arrangement from a point on a plant, as leaves around the node of a stem. The sepals, petals, stamens, and carpels of angiosperms form four separate whorls within a complete flower. **2.** A single turn of a spiral shell of a mollusk.

wide area network (wīd) See **WAN.**

Wilkins (wĭl′kĭnz), **Maurice Hugh Frederick** 1916–2004. British biophysicist who contributed to the discovery of the structure of DNA. He worked with Rosalind Franklin to produce x-ray studies of DNA that helped Francis Crick and James Watson establish its structure as a double helix. For this work Wilkins shared with Crick and Watson the 1962 Nobel Prize for physiology or medicine. See Note at **Rosalind Franklin.**

Wilson (wĭl′sən), **Charles Thomson Rees** 1869–1959. British physicist noted for his research on atmospheric electricity. He developed the Wilson cloud chamber, a device that makes it possible to study and photograph the movement and interaction of electrically charged particles. He shared the 1927 Nobel Prize for physics with Arthur Compton. See more at **cloud chamber.**

Wilson, Edmund Beecher 1856–1939. American zoologist who was one of the founders of modern genetics. He researched the function, structure, and organization of cells, emphasizing their importance as the building blocks of life. He also demonstrated the significance of chromosomes, especially sex chromosomes, in heredity.

Wilson cycle The cyclical opening and closing of ocean basins caused by movement of the Earth's plates. The Wilson cycle begins with a rising plume of magma and the thinning of the overlying crust. As the crust continues to thin due to extensional tectonic forces, an ocean basin forms and sediments accumulate along its margins. Subsequently subduction is initiated on one of the ocean basin's margins and the ocean basin closes up.

When the crust begins to thin again, another cycle begins. The Wilson cycle is named after the Canadian geophysicist J. Tuzo Wilson (1908–1993).

WIMP (wĭmp) Short for *weakly interacting massive particle.* Any of various hypothetical particles, some predicted by certain theories such as supersymmetry, which interact with other particles by the force of gravity alone. WIMPs are considered by some scientists to be candidates for the **dark matter** that makes up much of the mass of the universe.

Wimshurst machine (wĭmz′hûrst′) An electrostatic generator used to generate static electricity at high voltages, consisting of mica or glass disks covered with metal sectors of plates that are rotated in opposite directions. The rotational energy is used to build up static charge across the plates, while each plate periodically makes electrical contact with another as it rotates, building up a voltage by induction. The Wimshurst machine is named after its inventor, British engineer James Wimshurst (1832–1903).

wind (wĭnd) A current of air, especially a natural one that moves along or parallel to the ground, moving from an area of high pressure to an area of low pressure. Surface wind is measured by anemometers or its effect on objects, such as trees. The large-scale pattern of winds on Earth is governed primarily by differences in the net solar radiation received at the Earth's surface, but it is also influenced by the Earth's rotation, by the distribution of continents and oceans, by ocean currents, and by topography. On a local scale, the differences in rate of heating and cooling of land versus bodies of water greatly affect wind formation. Prevailing global winds are classified into three major belts in the Northern Hemisphere and three corresponding belts in the Southern Hemisphere. The **trade winds** blow generally east to west toward a low-pressure zone at the equator throughout the region from 30° north to 30° south of the equator. The **westerlies** blow from west to east in the temperate mid-latitude regions (from 30° to 60° north and south of the equator), and the polar **easterlies** blow from east to west out of high-pressure areas in the polar regions. See also **Beaufort scale, chinook, foehn, monsoon, Santa Ana.**

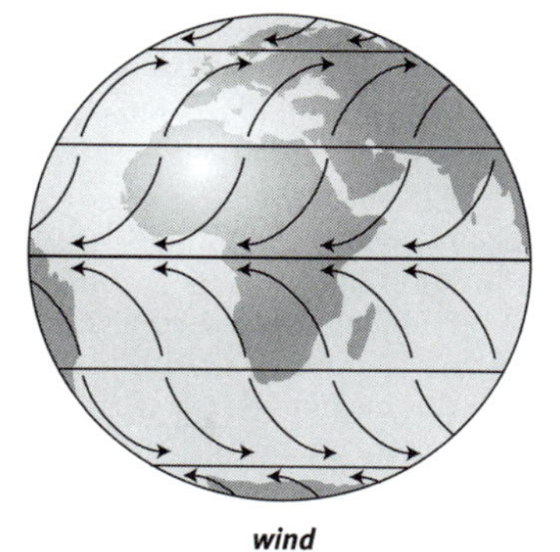

wind

Global wind patterns are determined by differences in atmospheric pressure resulting from the uneven heating of the Earth's surface by the Sun. As air is heated along the equator it rises, creating a zone of low pressure that draws air toward it throughout the tropics and produces the surface flow known as the trade winds. After it rises, the warm equatorial air flows north and south, cooling in the upper atmosphere until it is dense enough to descend near the Tropics of Cancer and Capricorn. Part of this cooler air flows at surface level toward the poles, where it meets the colder, drier air flowing away from the poles. The Coriolis effect deflects these broad surface flows, turning winds that blow toward the equator into easterlies and those that blow toward the poles into westerlies.

wind-chill factor The temperature of windless air that would have the same effect on exposed human skin as a particular combination of wind speed and air temperature. As the wind blows faster, heat is lost more quickly from exposed skin, making a person feel colder even though the air temperature remains the same. Also called *chill factor*.

wind farm A power plant that uses windmills or wind turbines to generate electricity.

wind gap A shallow notch in the crest of a mountain ridge.

windpipe (wĭnd′pīp′) See **trachea** (sense 1).

wind shear A change in wind direction and speed between slightly different altitudes, especially a sudden downdraft. Depending on its scale, wind shear can cause a variety of effects, from minor turbulence to tornadoes. Wind shear caused by interactions between oceanic and atmospheric winds can be so strong that it can dissipate hurricanes. See also **shear force.**

windstorm (wĭnd′stôrm′) A storm with high winds or violent gusts but little or no rain. Winds with speeds up to 241 km (149 mi) per hour have been recorded during windstorms. See also **tornado.**

wind tunnel A chamber through which air is blown at controlled speeds to simulate the motion of objects placed in the chamber through the air, used to study the aerodynamic properties of objects such as automobiles, airplanes, and missiles.

wing (wĭng) **1.** One of a pair of specialized parts used for flying, as in birds, bats, or insects. **2.** A thin, papery projection on certain fruits that are dispersed by the wind, such as the fruits of ash, elm, and maple trees. See also **samara. 3.** A part extending from the side of an aircraft, such as an airplane, having a curved upper surface that causes the pressure of air rushing over it to decrease, thereby providing lift.

winter solstice (wĭn′tər) See under **solstice.**

wisdom tooth (wĭz′dəm) One of four molars, the last on each side of both jaws in humans, usually appearing in young adulthood.

wishbone (wĭsh′bōn′) The forked bone in front of the breastbone in most birds, consisting of the two collarbones partly fused together. It serves as a spring, capturing some of the energy during the downward stroke of the wings for release on the upward stroke.

witch of Agnesi (wĭch əv än-yā′zē) A planar cubic curve that is symmetric about the y-axis and that approaches the x-axis as an asymptote. Its equation is $x^2y = 4a^2(2a - y)$, where a is a constant.

withdrawal (wĭ*th*-drô′əl, wĭth-) Discontinuation of the use of an addictive substance. The symptoms of withdrawal include headache, diarrhea, and tremors and can range from mild to life-threatening, depending on the extent of the body's reliance on the addictive substance.

wolfram (wo͝ol′frəm) See **tungsten.**

Wollaston (wo͝ol′ə-stən), **William Hyde** 1766–1828. British chemist and physicist who discovered the elements palladium (1803) and rhodium (1804). In 1805 he devised a process for producing malleable platinum that could be used to make various utensils and apparatus. Wollaston was also one of the first scientists to realize that the arrangement of atoms in a molecule must be three-dimensional.

wollastonite (wo͝ol′ə-stə-nīt′) A white to gray triclinic or monoclinic mineral found in metamorphic rocks and used in ceramics, paints, plastics, and cements. *Chemical formula:* $\mathbf{CaSiO_3}$.

womb (wo͞om) See **uterus.**

wood (wo͝od) The thick xylem of trees and shrubs, resulting from secondary growth by the vascular cambium, which produces new layers of living xylem. The accumulated living xylem is the **sapwood.** The older, dead xylem in the interior of the tree forms the **heartwood.** Often each cycle of growth of new wood is evident as a **growth ring.** The main components of wood are cellulose and lignin. —*Adjective* **woody.**

wood alcohol See **methanol.**

work (wûrk) The transfer of energy from one object to another, especially in order to make the second object move in a certain direction. Work is equal to the amount of force multiplied by the distance over which it is applied. If a force of 10 newtons, for example, is applied over a distance of 3 meters, the work is equal to 30 newtons per meter, or 30 joules. The unit for measuring work is the same as that for energy in any system of units, since work is simply a transfer of energy. Compare **energy, power.**

World Wide Web (wûrld) The complete set of electronic documents stored on computers that are connected over the Internet and are made available by the protocol known as HTTP. The World Wide Web makes up a large part of the Internet. See more at **Internet.**

worm (wûrm) **1.** Any of various invertebrate animals having a soft, long body that is round or flattened and usually lacks limbs. The term *worm* is used variously to refer to the segmented worms (or annelids, such as the earthworm), roundworms (or nematodes), flatworms (or platyhelminths), and various other groups. **2.** A destructive computer program that copies itself over and over until it fills all of the storage space on a computer's hard drive or on a network.

A CLOSER LOOK **worm**

Earthworms are one of many types of *worms,* including those of the flat and round species. Over a century ago, Charles Darwin spent 39 years studying earthworms and wrote *The Formation of Vegetable Mould Through the*

Action of Worms with Observations on Their Habits, an entire book that described his research on earthworm behavior and intelligence and further explained how important earthworms are to agriculture. "Long before [the plow] existed," he wrote, "the land was, in fact, regularly plowed and still continues to be thus plowed by earthworms. It may be doubted whether there are many other animals which have played so important a part in the history of the world." Darwin was referring to the way that earthworms naturally mix and till soil, while both improving its structure and increasing its nutrients. As they tunnel in the soil, earthworms open channels that allow in air and water, improving drainage and easing the way for plants to send down roots; they also carry nutrients from deep soils to the surface. Earthworms eat plant material in the soil, decaying leaves, and leaf litter, and their own waste provides nourishment for plants and other organisms. Slime, a secretion of earthworms, contains nitrogen, an important plant nutrient. It is estimated that each year earthworms in one acre of land move 18 or more tons of soil.

wormhole (wûrm′hōl′) A hole made by a burrowing worm. A theoretical distortion of space-time that would link points in space through a second set of paths, some of which could be shorter than the shortest path without the wormhole. It is not known whether workholes are possible. See more at **space-time.**

W particle See **W boson.**

Wu (wo͞o), **Chien-Shiung** 1912–1997. Chinese-born American physicist. Research with her colleagues on electron emission in the decay of radioactive elements showed that parity symmetry, long thought to hold for all physical laws, is in fact violated; the decay processes displayed odd parity, essentially entailing that nature distinguishes between right-handed and left-handed processes.

WWW Abbreviation of **World Wide Web.**

WYSIWYG (wĭz′ē-wĭg′) Short for *what you see is what you get.* A word-processing or desktop publishing system in which text is displayed on a monitor exactly as it will be printed.

xanthan gum (zăn′thən) A natural gum of high molecular weight produced by fermentation of glucose (usually in the form of corn syrup) with bacteria. Xanthan gum is used as a stabilizer in commercial food preparation.

xanthine (zăn′thēn′, -thĭn) **1.** Any of various purines having two oxygen atoms attached to the six-member ring of carbon and nitrogen atoms. Xanthines include caffeine, theophylline (a toxic alkaloid found in tea leaves), and theobromine (a toxic alkaloid found in cocoa). **2.** The simplest of this class of compounds, forming yellowish-white crystals. It is produced in the body as an intermediate stage in the breakdown of purines to uric acid. It is also found in blood and in certain plants. *Chemical formula:* $C_5H_4N_4O_2$. **3.** Any of several derivatives of this compound.

xanthophyll (zăn′thə-fĭl′) Any of various yellow pigments occurring in the leaves of plants and giving young shoots and late autumn leaves their characteristic color. This color is masked by chlorophyll when the leaf is mature. Xanthophylls aid in the absorption of light by capturing certain wavelengths not captured by chlorophyll and rapidly transferring the energy to chlorophyll by boosting one of its electrons to a higher energy level. Xanthophylls are carotenoids, differing from carotenes in having one or more oxygen-containing groups attached. See also **carotene.**

x-axis (ĕks′ăk′sĭs) **1.** The horizontal axis of a two-dimensional Cartesian coordinate system. **2.** One of the three axes of a three-dimensional Cartesian coordinate system.

X-chromosome (ĕks′krō′mə-sōm′) The sex chromosome that in female mammals is paired with another X-chromosome and in males is paired with a Y-chromosome. Very few genes on the X-chromosome have counterparts on the Y-chromosome, and since males have only one X-chromosome, any gene present on it (even if the gene is recessive in females) is expressed in males. In females, one of the two X-chromosomes in each cell is deactivated. See more at **sex chromosome.** See note at **sex.**

Xe The symbol for **xenon.**

xenoblast (zĕn′ə-blăst′, zē′nə-) A mineral that has not developed its characteristic crystalline faces and that has formed in a rock that has undergone metamorphism.

xenocryst (zĕn′ə-krĭst′, zē′nə-) A crystal foreign to the igneous rock in which it occurs. Xenocrysts usually form during metamorphism. Compare **xenolith.**

xenograft (zĕn′ə-grăft′, zē′nə-) A graft in which the donor and recipient are of different species. Compare **allograft, autograft.**

xenolith (zĕn′ə-lĭth′, zē′nə-) A rock fragment foreign to the igneous mass in which it occurs. Xenoliths usually become incorporated into a cooling magma body when pieces of the rock into which the magma was injected break off and fall into it. Compare **xenocryst.**

xenon (zē′nŏn′) *Symbol* **Xe** A colorless, odorless element in the noble gas group occurring in extremely small amounts in the atmosphere. It was the first noble gas found to form compounds with other elements. Xenon is used in lamps that make intense flashes, such as strobe lights and flashbulbs for photography. Atomic number 54; atomic weight 131.29; melting point –111.9°C; boiling point –107.1°C; density (gas) 5.887 grams per liter; specific gravity (liquid) 3.52 (–109°C). See **Periodic Table.**

xenotransplantation (zĕn′ə-trăns′plăn-tā′shən, zē′nə-) The surgical transfer of cells, tissues, or especially whole organs from one species to another.

xeric (zĕr′ĭk, zîr′-) Relating or adapted to an extremely dry habitat. Succulents such as cacti, aloes, and agaves are xeric plants. Compare **hydric, mesic.**

xerophyte (zîr′ə-fīt′) A plant that is adapted to an arid environment. Many xerophytes have specialized tissues (usually nonphotosynthetic parenchyma cells) for storing water, as in the stems of cacti and the leaves of succulents. Others have thin, narrow leaves, or even spines, for minimizing water loss. Xerophyte leaves often have abundant stomata to maximize gas exchange during periods in which water is available, and the stomata are recessed in depressions, which are covered with fine hairs to help trap moisture in the air. Compare **hydrophyte, mesophyte.**

xi baryon (zī, sī, ksē) Either of two subatomic particles in the baryon family, one neutral and one negatively charged, with masses of 2,572 and 2,585 times that of the electron and average lifetimes of 2.9×10^{-10} and 1.6×10^{-10} seconds. See Table at **subatomic particle.**

xi-c baryon Either of two subatomic particles in the baryon family, one neutral and one positively charged, with masses of 4,826 and 4,834 times that of an electron and average lifetimes of 3.5×10^{-13} and 9.8×10^{-14} seconds, respectively. See Table at **subatomic particle.**

X-linked (ĕks′lĭngkt′) Relating to an inherited trait controlled by a gene on an X-chromosome. Orange fur color is an X-linked trait in cats.

XML (ĕks′ĕm-ĕl′) Short for *extensible markup language.* A version of SGML that allows one to design a customized markup language, used to allow for the easy interchange of documents and data on the World Wide Web or between software components.

x-ray also **X-ray** (ĕks′rā′) **1.** A high-energy stream of electromagnetic radiation having a frequency higher than that of ultraviolet light but less than that of a gamma ray (in the range of approximately 10^{16} to 10^{19} hertz). X-rays are absorbed by many forms of matter, including body tissues, and are used in medicine and industry to produce images of internal structures. See more at **electromagnetic spectrum. 2.** An image of an internal structure, such as a body part, taken with x-rays.

x-ray astronomy The study of celestial objects by measurement of the x-rays they emit. Because the Earth's atmosphere absorbs x-rays, x-ray detectors are usually carried into space on satellites. X-rays are emitted by high-energy objects such as active galactic nuclei, supernova remnants, x-ray binary stars, neutron stars, and the regions around black holes. Objects which do not produce their own x-rays can reflect radiation from nearby stars, making it possible to study the objects using x-ray astronomy; one notable example of this is the analysis of solar x-rays reflected by the Moon.

x-ray binary star A binary star system that produces a significant amount of x-ray radiation. When one of the components of a binary system is an extremely dense body such as a white dwarf or a neutron star, its enormous gravitational attraction can pull material from the other star into its own orbit, causing the accelerated matter to release high-energy photons in the x-ray spectrum. See more at **accretion disk.**

x-ray crystallography The study of molecular structure by examining diffraction patterns made by x-rays beamed through a crystalline form of the molecules. X-ray crystallography is used extensively in biochemistry to examine the molecular structure of such molecules as proteins and DNA.

x-ray diffraction The scattering of x-rays by crystal atoms, producing a diffraction pattern that yields information about the structure of the crystal. X-ray diffraction is used in **x-ray crystallography**.

xylem (zī′ləm) A tissue in vascular plants that carries water and dissolved minerals from the roots and provides support for softer tissues. Xylem consists of several different types of cells: fibers for support, parenchyma for storage, and tracheary elements for the transport of water. The tracheary elements are arranged as long tubes through which columns of water are raised. In a tree trunk, the inner-

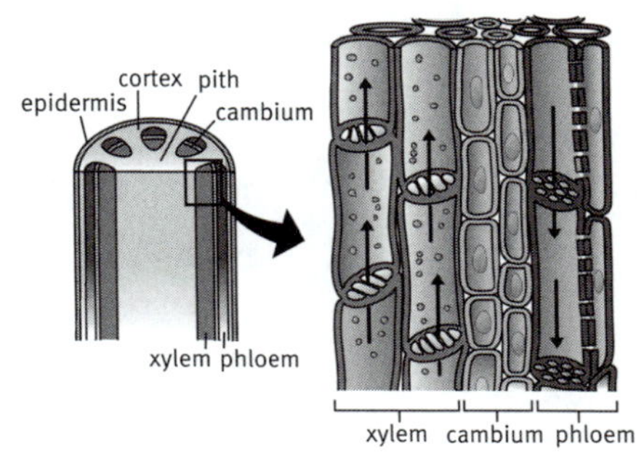

xylem

Xylem cells in a stem carry water from a plant's roots to its leaves. Phloem distributes food that is made in the plant's leaves to other parts of the plant. Cambium cells divide into either xylem or phloem cells. The cortex and pith, seen in the stem cross section (left), *provide structural support.*

most part of the wood is dead but structurally strong xylem, while the outer part consists of living xylem, and beyond it, layers of cambium and phloem. See more at **cambium, capillary action.** Compare **phloem.**

xylene (zī-lēn′, zī′lēn′) also **xylol** (zī′lôl′, -lōl′) A flammable hydrocarbon obtained from wood and coal tar. Xylene consists of a benzene ring with two methyl (CH_3) groups attached, and occurs in three isomeric forms. It is used as a solvent, in jet fuel, and in the manufacture of dyes, fibers, perfumes, and films. *Chemical formula:* $\mathbf{C_8H_{10}}$. A mixture of xylene isomers used as a solvent in making lacquers and rubber cement and as an aviation fuel.

xylitol (zī′lĭ-tôl′, -tōl′) A sweet white crystalline alcohol derived from xylose and used as a sugar substitute. *Chemical formula:* $\mathbf{C_5H_{12}O_5}$.

xylose (zī′lōs′) A white crystalline sugar extracted from wood, straw, and corn. It is used in dyeing and tanning and as a substitute for sucrose in diabetic diets. *Chemical formula:* $\mathbf{C_5H_{10}O_5}$.

Y

Y The symbol for **yttrium.**

Yalow (yăl′ō), **Rosalyn Sussman** Born 1921. American physicist who, working with the biophysicist Solomon A. Berson, developed the radioimmunoassay (RIA), an extremely sensitive technique for measuring very small quantities of substances such as hormones, enzymes, and drugs in the blood. For this work, she won a 1977 Nobel Prize for physiology or medicine.

Rosalyn Yalow

yard (yärd) A unit of length in the US Customary System equal to 3 feet or 36 inches (0.91 meter). See Table at **measurement.**

yaws (yôz) A highly contagious tropical disease that chiefly affects children, caused by the spirochete *Treponema pertenue* and characterized by raspberrylike sores, especially on the hands, feet, and face.

y-axis (wī′ăk′sĭs) **1.** The vertical axis of a two-dimensional Cartesian coordinate system. **2.** One of the three axes of a three-dimensional Cartesian coordinate system.

Yb The symbol for **ytterbium.**

Y-chromosome The sex chromosome that is paired with an X-chromosome in males. Most of the genes on the X-chromosome determine physical attributes associated with males, such as the development of testes. See more at **sex chromosome.** See Note at **sex.**

yd Abbreviation of **yard.**

yeast (yēst) Any of various one-celled fungi that reproduce by budding and can cause the fermentation of carbohydrates, producing carbon dioxide and ethanol. There are some 600 known species of yeast, though they do not form a natural phylogenic group. Most yeasts are ascomycetes, but there are also yeast species among the basidiomycetes and zygomycetes. The budding processes in yeasts show a wide range of variations. In many yeasts, for example, the buds break away as diploid cells. Other yeasts reproduce asexually only after meiosis, and their haploid buds act as gametes that can combine to form a diploid cell, which functions as an ascus and undergoes meiosis to produce haploid spores. Still other yeasts form buds in both haploid and diploid phases. The ascomycete yeast *Saccharomyces cerevisiae* is used in baking to produce the carbon dioxide that leavens dough and batter. It has been the subject of extensive research in cell biology, and its genome was the first to be sequenced among eukaryotes. A variety of yeasts of the genus *Saccharomyces* are used in making beer and wine to provide alcohol content and flavor. Certain other yeasts, such as *Candida albicans*, are pathogenic in humans.

yellow fever (yĕl′ō) A life-threatening infectious disease caused by a virus of the genus *Flavivirus* and characterized by fever, jaundice, and internal bleeding. Yellow fever occurs mainly in tropical regions of Africa and Latin America and is transmitted by mosquitoes.

yellows (yĕl′ōz) Any of various plant diseases characterized by yellowish discoloration and often by wilting, deformation, and stunted growth. Yellows may be caused by **phytoplasmas,** by ascomycete fungi of the genus *Fusarium,* or by a virus, especially of the genus *Chlorogenus.*

Yersin (yĕr-sä′), **Alexandre Émile John** 1863–1943. French bacteriologist who, working with Émile Roux, isolated the toxin that causes the symptoms of diphtheria. Yersin later discovered the bacillus that causes bubonic plague and developed a serum to protect against it.

-yl A suffix used to form the chemical names of organic compounds when they are radicals (parts of larger compounds), such as *ethyl* and *phenyl.*

-yne A suffix used to form the names of hydrocarbons having one or more triple bonds, as in *ethyne.*

yolk (yōk) The yellow internal part of the egg of a bird or reptile. The yolk is surrounded by the albumen and supplies food to the developing young.

yolk sac A sac that is attached to the gut of an embryo and encloses the yolk in bony fish, sharks, reptiles, mammals, and birds. In most mammals, the yolk sac functions as part of the embryo's circulatory system before the placenta develops.

Young (yŭng), **Thomas** 1773–1829. British physicist and physician who is best known for his contributions to the wave theory of light and his discovery of how the lens of the human eye changes shape to focus on objects of different distances. He also studied surface tension and elasticity, and Young's modulus (a measure of the rigidity of materials) is named for him. He is also credited with the first scientific definition of the word *energy.*

Young's modulus (yŭngz) See under **modulus of elasticity.**

young stellar object A protostar in which the dust and gas have collapsed into a sufficiently massive body for fusion to begin. The characteristic trait of young stellar objects is the presence of jet outflows—either of molecular gas or of radioactive emissions—that interact with the interstellar medium out of which the star has just been born. See more at **protostar.**

ytterbium (ĭ-tûr′bē-əm) *Symbol* **Yb** A soft, silvery-white metallic element of the lanthanide series that occurs as seven stable isotopes. It is used as a radiation source for portable x-ray machines. Atomic number 70; atomic weight 173.04; melting point 824°C; boiling point 1,196°C; specific gravity 6.972 or 6.54 (25°C) depending on allotropic form; valence 2, 3. See **Periodic Table.**

yttrium (ĭt′rē-əm) *Symbol* **Y** A silvery metallic element found in the same ores as elements of the lanthanide series. Yttrium is used to strengthen magnesium and aluminum alloys, to provide the red color in color televisions, and as a component of various optical and electronic devices. Atomic number 39; atomic weight 88.906; melting point 1,522°C; boiling point 3,338°C; specific gravity 4.45 (25°C); valence 3. See **Periodic Table.**

Yukawa (yo͞o-kä′wä), **Hideki** 1907–1981. Japanese physicist who in 1935 mathematically predicted the existence of the meson, for which he was awarded the 1949 Nobel Prize for physics.

zastrugi (ză-stro͞o′gə, ză′strə-) Another spelling of **sastrugi.**

z-axis (zē′ăk′sĭs) One of the three axes of a three-dimensional Cartesian coordinate system.

Z boson An electrically neutral subatomic particle that mediates the weak nuclear force. The Z boson has a mass 182,000 times that of the electron. Unlike the other weak force mediator (the W boson), it does not change particles it interacts with into other types of particles. Like the **photon**, the Z boson is its own antiparticle. See Table at **subatomic particle.**

Zeeman (zā′măn′), **Pieter** 1865–1943. Dutch physicist whose study of light sources in a magnetic field resulted in his discovery and explanation of the Zeeman effect. For this work he shared with Hendrik Lorentz the 1902 Nobel Prize for physics. He also studied the absorption and motion of electricity in fluids and magnetic fields on the surface of the Sun.

Zeeman effect The splitting of single spectral lines of an emission or absorption spectrum of a substance into three or more components when the substance is placed in a magnetic field. The effect occurs when several electron orbitals in the same shell, which normally have the same energy level, have different energies due to their different orientations in the magnetic field. A *normal Zeeman effect* is observed when a spectral line of an atom splits into three lines under a magnetic field. An *anomalous Zeeman effect* is observed if the spectral line splits into more than three lines. Astronomers can use the Zeeman effect to measure magnetic fields of stars. Compare **Stark effect.**

zenith (zē′nĭth) The point on the **celestial sphere** that is directly above the observer (90 degrees above the **celestial horizon**). Compare **nadir.**

zeolite (zē′ə-līt′) Any of a family of hydrous aluminum silicate minerals, whose molecules enclose cations of sodium, potassium, calcium, strontium, or barium. Zeolites are usually white or colorless, but they can also be red or yellow. They are characterized by their easy and reversible loss of water of hydration. They usually occur within cavities in basalt.

zero (zîr′ō) The numerical symbol 0, representing a number that when added to another number leaves the original number unchanged.

A CLOSER LOOK zero

Although the origin of *zero* is controversial, some historians believe that it was invented by the Babylonians in about 500 BCE. In the sixth century, it was discovered by the Hindus and Chinese, and 700 years later, it reached the Western world via the Arabs. Zero is the only integer (whole number) that is neither positive nor negative. In a sense, zero makes negative numbers possible, as a negative number added to its positive counterpart always equals zero. When zero is added to or subtracted from a number, it leaves the number at its original value. Zero is essential as a position holder in the system known as *positional notation.* In the number 203, for example, there are two hundreds, zero tens, and three ones. Zero indicates that the value of the tens place is zero. In the number 1024, zero indicates that the value of the hundreds place is zero. Scientists use the term *absolute zero* (0° Kelvin) to refer to the (unattainable) theoretically lowest possible temperature, at which the kinetic energy of molecules is zero.

zero gravity The condition of real or apparent weightlessness occurring when any gravitational forces acting on a body meet with no resistance so the body is allowed to accelerate freely. Bodies in free fall (including trajectories like orbits) experience zero gravity; bodies at rest on the Earth's surface do not, since they are subject to the counterforce of the surface supporting them.

zero-point energy The irreducible minimum energy possessed by a physical system, as of a substance at a temperature of absolute zero. The zero point energy of a vacuum is not zero due to **vacuum fluctuations.** See Note at **absolute zero.**

zero-sum game A game in which the sum of the winnings by all the players is zero. In a zero-sum game, a gain by one player must be matched by a loss by another player. Poker is a zero-sum game if the house does not take a cut as a charge for playing.

zinc (zĭngk) *Symbol* **Zn** A shiny, bluish-white metallic element that is brittle at room temperature but is malleable when heated. It is used in alloys such as brass and bronze, as a coating for iron and steel, and in various household objects. Zinc is essential to human and animal growth. Atomic number 30; atomic weight 65.39; melting point 419.4°C; boiling point 907°C; specific gravity 7.133 (25°C); valence 2. See **Periodic Table.**

zincate (zĭng′kāt′) A chemical compound containing the group ZnO_2.

zinc chloride A white, water-soluble crystalline compound used as a wood preservative, as a soldering flux, and for a variety of industrial purposes, including the manufacture of cements and paper parchment. *Chemical formula:* $ZnCl_2$.

zinc oxide A white or yellowish powdery compound used in paints and in various medicines and skin cosmetics. *Chemical formula:* **ZnO.**

zinc sulfate A colorless crystalline compound used especially in hydrated form as an emetic and astringent, as a fungicide, and in wood and skin preservatives. *Chemical formula:* $ZnSO_4$.

zinjanthropus (zĭn-jăn′thrə-pəs, zĭn′jăn-thrō′-) An extinct hominid postulated from bones found in Tanzania in 1959 and originally designated *Zinjanthropus boisei* by Louis S.B. Leakey. It was later shown to be an australopithecine and renamed *Australopithecus boisei.*

zircon (zûr′kŏn′) A brown, reddish to bluish, gray, green, or colorless tetragonal mineral that occurs in igneous, metamorphic, and sedimentary rocks, and especially in sand. The colorless varieties are valued as gems. *Chemical formula:* $ZrSiO_4$.

zirconium (zûr-kō′nē-əm) *Symbol* **Zr** A shiny, grayish-white metallic element that occurs primarily in zircon. It is used to build nuclear reactors because of its ability to withstand bombardment by neutrons even at high temperatures. Zirconium is also highly resistant to corrosion, making it a useful component of pumps, valves, and alloys. Atomic number 40; atomic weight 91.22; melting point 1,852°C; boiling point 4,377°C; specific gravity 6.56 (20°C); valence 2, 3, 4. See **Periodic Table.**

Zn The symbol for **zinc.**

zodiac (zō′dē-ăk′) A band of the **celestial sphere** extending about eight degrees north and south of the **ecliptic**, representing the portion of the sky within which the paths of the Sun, the Moon, and the planets are found. In astrology, the zodiac is divided into 12 equal segments, each of which is named after a constellation through which the ecliptic passes in that region of the sky. The traditional beginning point of constellations is Aries, followed in calendrical order by Taurus, Gemini, Cancer, Leo, Virgo, Libra, Scorpius, Sagittarius, Capricorn, Aquarius, and Pisces. See also **equinox.**

zodiacal light (zō-dī′ə-kəl) A faint hazy cone of light, often visible in the west just after sunset or in the east just before sunrise, and elongated in the direction of the ecliptic on each side of the Sun. It is apparently caused by the reflection of sunlight from meteoric particles in the plane of the ecliptic.

zone (zōn) **1.** Any of the five regions of the surface of the Earth that are loosely divided according to prevailing climate and latitude, including the Torrid Zone, the North and South Temperate zones, and the North and

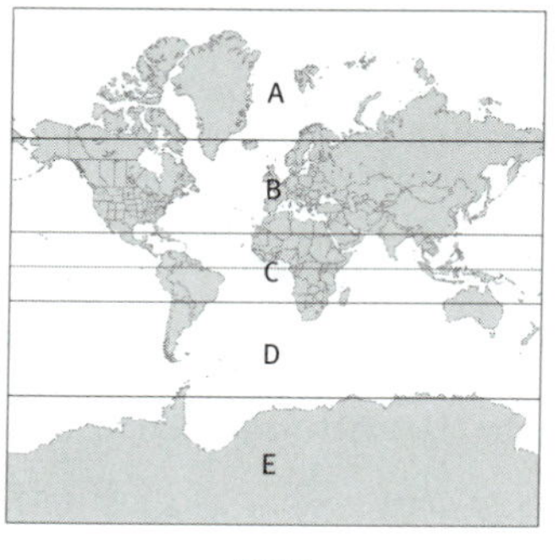

zone

climatic zones: A. *North Frigid Zone,* B. *North Temperate Zone,* C. *Torrid Zone,* D. *South Temperate Zone,* E. *South Frigid Zone*

South Frigid zones. **2.** *Ecology.* An area characterized by distinct physical conditions and populated by communities of certain kinds of organisms. **3.** *Mathematics.* A portion of a sphere bounded by the intersections of two parallel planes with the sphere. **4.** *Anatomy.* An area or a region distinguished from adjacent parts by a distinctive feature or characteristic. **5.** *Geology.* A region or stratum distinguished by composition or content.

zone of accumulation See **B horizon.**

zone of illuviation See **B horizon.**

zone of leaching See **A horizon.**

zoogeography (zō′ə-jē-ŏg′rə-fē) The scientific study of the geographic distribution of animals. See more at **biogeography.**

zoology (zō-ŏl′ə-jē, zo͞o-ŏl′-) The scientific study of organisms in the kingdom Animalia, including their growth and structure.

zooplankton (zō′ə-plăngk′tən) Plankton that consists of tiny animals, such as rotifers, copepods, and krill, and of microorganisms once classified as animals, such as dinoflagellates and other protozoans.

zoosporangium (zō′ə-spə-răn′jē-əm) Plural **zoosporangia.** A sporangium in which zoospores develop.

zoospore (zō′ə-spôr) A motile flagellated spore that serves as a means of asexual reproduction among certain algae, fungi, and protoctists.

Z particle See **Z boson**

Zr The symbol for **zirconium.**

zwitterion (zwĭt′ər-ī′ən, swĭt′-, tsvĭt′-) A molecule, especially an amino acid, containing a positively charged ion at one end and a negatively charged ion at the other.

zygomorphic (zī′gə-môr′fĭk, zĭg′ə-) Relating to a flower that can be divided into equal halves along only one line; bilaterally symmetrical. The flowers of the iris and the snapdragon are zygomorphic. Compare **actinomorphic.**

zygomycete (zī′gə-mī′sēt′, zĭg′ə-) Any of various fungi belonging to the phylum Zygomycota, characterized by the absence of cross walls (called septa) in all of their hyphae except reproductive hyphae. The absence of septa allows cytoplasm to stream along the hyphae, and most species produce abundant, fast-growing hyphae. Many species of zygomycetes live on decaying plant and animal matter in soil, though some are parasites on plants, insects, and certain soil animals, and a few cause disease in domestic animals and humans. Zygomycetes reproduce both by producing asexual haploid spores in conidia at the end of their hyphae and by producing sexual haploid spores by meiosis after hyphae of different mating types conjugate and their nuclei fuse.

zygospore (zī′gə-spôr′, zĭg′ə-) A large, multinucleate spore formed by union of isogametes (gametes that are not distinguished by size and structure), as in certain algae or fungi. A zygospore develops thick, resistant walls and enters a period of dormancy before germinating.

zygote (zī′gōt′) The cell formed by the union of the nuclei of two reproductive cells (gametes), especially a fertilized egg cell.

–zygous A suffix meaning "having zygotes of a specified kind," as in *heterozygous.*

TIMELINE OF SCIENTIFIC DISCOVERY

before 2000 BCE	Egyptian priests develop the world's first sophisticated medical practice
1300 BCE	Chinese astronomers establish the length of the year as 365 $\frac{1}{4}$ days
585 BCE	Thales proves the first theorems of deductive geometry
520 BCE	Pythagoras recognizes that the Earth is a sphere and proposes that the Sun, Moon, planets, and stars move in paths around the Earth, which is suspended in space
450 BCE	The Greeks develop a method of writing numbers based on letters of the alphabet
430 BCE	Democritus develops the concept of atoms as minute, indivisible bodies, of which all matter is made
400 BCE	Hippocrates takes steps to make medicine scientific by basing treatment on observation and logic rather than received knowledge
350 BCE	Aristotle dissects plants and animals and recognizes the need for classification, categorizing more than 500 types
335 BCE	Aristotle founds the Lyceum, where he teaches the importance of observation, logic, and theory
300 BCE	Euclid writes his *Elements*, 13 books containing all that is known about geometry in his time. The books will be used until the 19th century
250 BCE	Archimedes works out the principle of the lever and other simple machines
240 BCE	Eratosthenes estimates the circumference of the Earth and the distance to the Sun and Moon
129 BCE	Hipparchus creates the first chart of stars and planets
140 CE	Ptolemy develops a model of the universe that has the Earth as the center of the solar system
170 CE	Galen describes the parts and functions of the human body, based on his dissection and studies of animals. His theories will form the basis of European anatomical knowledge until Vesalius disproves them in the 16th century
200 CE	Chinese mathematicians use powers of 10 to express numbers
300 CE	The Chinese begin to develop the abacus
400 CE	The term *chemistry* is used for the first time by Alexandrian scholars to describe the activity of changing matter
683 CE	The first uses of a "goose-egg" symbol for zero appear in Cambodia and Sumatra
800 CE	The decimal system is introduced by the Arabs into Spain
829 CE	Arab and Persian astronomers build observatories that use very large instruments to measure the positions of stars more accurately
1041–1048	Movable type is invented in China
1127	Ibn Sina's *The Canon of Medicine* is translated into Latin. It will remain a standard work in European medical studies until the 17th century
1202	Fibonacci introduces zero to Europe and popularizes the decimal system that is still in use today
1266–1267	Bacon writes *Opus majus*, an extensive scientific encyclopedia that emphasizes the need for mathematics, observation, experimentation, and rational thinking in effective scientific method
c. 1440	Movable type is introduced in Europe
1506–1513	Leonardo da Vinci creates detailed drawings in astronomy, mechanics, and anatomy and draws up plans for machines that are not built until centuries later

1543 —Copernicus proposes that the planets revolve around the Sun
—Vesalius publishes books on human anatomy based on dissection of human cadavers

1572 Brahe observes a supernova (later called Tycho's star) before the invention of the telescope

1595 Mercator's atlas of Europe is published

1600 Gilbert introduces the terms *electricity* and *magnetic pole* in the first major scientific book published in England

1609 —Galileo constructs the first astronomical telescope
—Kepler publishes his discovery that planets move around the Sun in elliptical orbits

1628 Harvey describes the circulation of blood through the body

1637 Descartes publishes a group of writings in which he uses algebra to solve geometry problems. This method of combining geometry with algebra was independently developed in 1636 or earlier by Pierre de Fermat, but Fermat's work will not be published until 1670

1642 Pascal develops a mechanical calculator that can add and subtract

1643 Torricelli invents the barometer

1659 Huygens becomes the first to observe the surface features of Mars

1662 Boyle and Hooke publish findings on the behavior of gases

1665 Hooke publishes *Micrographia*, his microscopic observations of plants, and coins the term *cell*

1666–1675 Newton and Leibniz, working independently, develop calculus

1668 Newton invents the first reflecting telescope

1674–1677 Leeuwenhoek studies bacteria and protozoans under a microscope, and confirms the existence of sperm cells, discovering that they are one of the two cells that allow reproduction

1679 Leibniz introduces binary arithmetic, demonstrating that every number can be represented using only the symbols 0 and 1

1687 Newton publishes works on mathematics and explains gravitation

1714 Fahrenheit invents the mercury thermometer

1724 Fahrenheit develops the Fahrenheit scale

1735 Linnaeus introduces the system of classification for plants and animals that is still in use today

1742 Celsius develops the Celsius scale

1756 Black discovers carbon dioxide

1766 Cavendish discovers hydrogen, which he calls "inflammable air"

1769 Watt invents the modern steam engine

1771 Scheele discovers oxygen, but the results of his experiments are not published until 1777

1774 Priestley independently discovers oxygen

1778 Lavoisier discovers that air is a mixture of gases

1781 Herschel discovers the planet Uranus

1785 Hutton proposes that the arrangement and structure of the Earth's rocks suggests that they were formed through a very slow process over the course of millions of years

1789 Lavoisier publishes the *Elementary Treatise of Chemistry*, which firmly establishes the oxygen theory of combustion, introduces new chemical nomenclature, and states the law of conservation of mass

1796 Jenner performs the first smallpox vaccination

1800 Volta invents the first electric battery

1805 Dalton describes how matter is made up of small particles called *atoms*

1807 Fulton invents the steamboat

1808 Gay-Lussac describes the chemical behavior of combining gases

1811 Berzelius introduces the modern system of chemical symbols

1812 —Mohs develops a scale for determining hardness in rocks

—Cuvier introduces the theory of the extinction of animal groups during catastrophes

1816 Fresnel demonstrates the wave nature of light and explains polarization

1819 Oersted observes that electric current produces a magnetic force and founds the study of electromagnetism

1831 Faraday develops the first electric generator and transformer

1833 Lyell finishes his book *Principles of Geology*, which identifies the Recent, Pliocene, Miocene, and Eocene periods of Earth's history and popularizes the field of geology

1837 Babbage draws up plans for the analytical engine, a machine that performs mathematical calculations and is the forerunner of the modern computer

1840 Agassiz introduces the idea of the ice age

1844 Morse develops the telegraph

1847 Helmholtz independently proposes the law of conservation of energy. The law was previously stated in a little-read 1842 paper by German physician Julius Robert von Mayer

1848 Kelvin introduces the Kelvin scale

1850 Foucault measures the velocity of light

1851 Foucault uses a pendulum to demonstrate that the Earth rotates on its axis

1858 Darwin and Wallace together present the theory of evolution by natural selection to the public

1859 Lenoir designs the first practical internal-combustion engine

1862–1877 Pasteur's study of microorganisms leads to the theory that germs cause disease

1865 —Maxwell shows that light is a form of electromagnetic radiation

—Lister introduces antiseptics to the practice of surgery

—Mendel proposes theories about heredity in plants

1869 Mendeleev develops the Periodic Table

1876 —Bell invents the telephone

—Kelvin designs a mechanical computer for solving differential equations, demonstrating that machines can be programmed to solve all sorts of mathematical problems

1879 Edison invents the incandescent lamp

1880 Hollerith invents a system of recording and retrieving information using punched cards, an important step in the development of computing

1884 Tesla immigrates to the US where he develops numerous inventions, including alternating current power supplies, electric motors, and radio transmission

1888 Roux and Yersin discover the toxin that causes diphtheria

1892–1898 Ramsay discovers the noble gases

1895 Roentgen discovers x-rays

1896 Becquerel demonstrates that uranium is radioactive

1897 Thomson discovers the electron

1898 Marie and Pierre Curie discover the elements polonium and radium

1899 Teisserenc de Bort discovers and names the stratosphere

1900 —Ehrlich describes how antibodies form

—Planck discovers units of energy called *quanta*

—Edison invents the nickel-alkaline battery

1901 —Marconi sends radio waves across the Atlantic

	—Landsteiner discovers three human blood groups (A, B, and O)
	—Hollerith develops the first numerical keyboard for punching cards in tabulating machines
1902	Landsteiner's colleagues discover the fourth major human blood group, AB
1903	The first successful airplane is launched by Wilbur and Orville Wright
1904	John Fleming invents the electron tube
1905	—Einstein introduces Special Relativity, the first part of his theory of relativity
	—Nettie Stevens identifies the role of X and Y chromosomes in determining the sex of organisms
1906	Thomson demonstrates that the hydrogen atom has only a single electron
1911	Rutherford discovers the nucleus of the atom and describes the basic structure of the atom
1912	Funk postulates the existence of vitamins
1913	Bohr develops a new theory of the structure of the atom, based on the work of Rutherford and Planck
1915	Wegener introduces the theory of continental drift
1916	Einstein introduces General Relativity, the second part of his theory of relativity
1921	Banting and Best discover the hormone insulin
1924	Hubble presents evidence for the existence of other galaxies
1926	Goddard builds the first successful liquid-fueled rocket
1927	—Alexanderson builds the first practical television system
	—Landsteiner discovers human blood groups M and N
1928	Fleming discovers penicillin
1929	—Lawrence builds the first cyclotron
	—Van de Graaf builds the Van de Graaf generator
1931	Ruska invents the electron microscope
1932	—Chadwick discovers the neutron
	—Cockcroft and Walton perform the first successful splitting of an atom with a particle accelerator
1935	Watson-Watt invents radar
1937	—Krebs shows how cells store energy and make food
	—Turing develops the idea for a theoretical computing machine, laying the foundation for modern digital computers
	—Whittle develops the first jet-propulsion aircraft engine
1938	Work by Hahn and colleagues on uranium leads to the discovery of nuclear fission
1939	Dubos discovers tyrothricin, the first commercially produced antibiotic
1941	Seaborg leads the team that creates plutonium
1942	Fermi creates the first man-made nuclear chain reaction
1942–1945	The Manhattan Project in the US develops the first atomic bomb
1942–1949	Hodgkin discovers the structure of penicillin and vitamin B_{12}
1944	Waksman and colleagues discover the first antibiotic for treating tuberculosis
1945	ENIAC, the first electronic computer, is invented
1947	Bardeen, Brattain, and Shockley invent the transistor
1949	Forrester invents the first magnetic core memory for an electronic digital computer
1951	—Rosalind Franklin and Maurice Wilkins use x-rays to study the structure of DNA
	—Linus Pauling elucidates the alpha-helical structure of proteins
1953	Watson and Crick explain the structure of DNA, based in part on studies by Franklin and Wilkins
1954	—Salk introduces the first polio vaccine
	—Sanger explains the structure of the hormone insulin
1957	—Sabin introduces an improved polio vaccine

—The Soviet Union launches *Sputnik I*, the first artificial satellite, and *Sputnik 2*, which carries the live dog Laika for 10 days, showing that life can survive aboard spacecraft

1959–1960 Louis and Mary Leakey discover early human fossils in Africa

1960 —Maiman invents the first working laser

—Goodall begins her study of chimpanzees in Tanzania

—Perutz determines the structure of hemoglobin

1961 The Soviet Union's *Vostok I* makes the first manned spaceflight

1962 —Sperry discovers that the two sides of the brain control different body functions

—Carson publishes a book that details the effects of pesticides on the environment

—The US space probe *Mariner 2* lands on Venus, becoming the first artificial satellite to reach the vicinity of another planet and return scientific information

1964 —Gell-Mann introduces the idea of quarks

—The US space probe *Ranger 7* takes the first good close-range photographs of the Moon

—The US launches *Mariner 4*, which takes 21 pictures of Mars' surface and successfully transmits them back to Earth

1967 —Barnard performs the first successful human heart transplant

—The concept of plate tectonics is introduced at a meeting of the American Geophysical Union

1969 —The US *Apollo 11* mission successfully lands humans on the Moon

—ARPANET, the US Defense Department's precursor to the Internet, is born

1971 —The US launches *Mariner 9*, which becomes the first man-made object to orbit another planet when it enters orbit around Mars

—The first microprocessor ("chip") introduced by Intel in the US

1972 Lexitron puts the first word processing system on the market

1973 The US launches *Pioneer 10* and *Pioneer 11*, the first space probes to reach the vicinities of Jupiter and Saturn, respectively

1974 Hawking describes the energy in a black hole using mathematics

1975 —Köhler and Milstein develop a technique to produce monoclonal antibodies

—The Altair 8800, the first successful personal computer, is introduced in the US

1979 Human insulin is synthesized by genetic engineering methods

1980 Luis and Walter Alvarez introduce the theory that an asteroid caused extinction of the dinosaurs

1985 Toshiba introduces the first true laptop computer, the T1100

1990 The Hubble Space Telescope is launched

1991 The World Wide Web becomes available on the Internet

1992 The first websites appear on the World Wide Web

1997 —Handheld computers become popular

—"Dolly," cloned from a mature body cell of another sheep, is born in Edinburgh, Scotland. She will give birth to a daughter, "Bonnie," the following year, demonstrating that clones can survive, mature to adulthood, and reproduce

2003 The Human Genome Project is completed

PICTURE CREDITS

The editorial and production staff wishes to thank the many individuals, organizations, and agencies that have contributed to the art program of this dictionary.

Credits on the following pages are arranged alphabetically by boldface entry word. At entries for which there are two or more picture sources, the sources follow the order of the illustrations.

The following source abbreviations are used throughout the credits: AA\Academy Artworks; AA-ES\Animals Animals-Earth Scenes; AP-WWP\Associated Press-Wide World Photos; AR\Art Resource, New York; CC\Chris Costello; CI\Carlyn Iverson; CB\Corbis-Bettmann; COR\Corbis; CRF\Corbis Royalty-Free; EM\Elizabeth Morales; GH\Grant Heilman Photography, Inc.; HM\© School Division, Houghton Mifflin Company; JM\Jerry Malone; LCL\Laurel Cook Lhowe; NASA\National Aeronautics and Space Administration; PDI-GI\© 2004 PhotoDisc, Inc.-Getty Images; PG\Precision Graphics; PR\Photo Researchers, Inc.; PT\Phototake; RS\Robin Storesund; and UG-GGS\UG-GGS Information Services.

A-Z Text

ABC soil PG **abscissa** AA **acid deposition** COR/Massimo Listri **acoustics** UG-GGS **acute angle** AA **adrenal gland** CI **adsorption** AA **aerial root** COR/Theo Allofs **aerodynamic** AA **aerosol** Tech-Graphics **Louis Agassiz** CB **Maria Agnesi** CB **aliphatic** RS **Luis Walter Alvarez** COR/Roger Ressmeyer **amber** HM **ammonite** COR/Layne Kennedy **anemometer** GH/Barry Runk **aneroid barometer** PG **angioplasty** LCL **angle** RS **antibody** CI **anticline** Gail Piazza **aphelion** AA **aromatic** RS **artesian well** EM **arthritis** LCL **astigmatism** Cecile Duray-Bito **asymptote** AA **atherosclerosis** EM **atmosphere** AA **atoll** AA **awn** LCL **bacteria** PR/Scimat; PT/Dennis Kunkel; PR/Scimat **ball bearing** EM **Benjamin Banneker** Photographs and Prints Division, Schomburg Center for Research in Black Culture, The New York Public Library, Astor, Lenox and Tilden Foundations, neg. no. SC-CN-79-0053 **batholith** EM **battery** UG-GGS **beard** PDI-GI/Siede Preis **bell curve** PG **benzene ring** RS **biceps** LCL **binocular** PG **bipinnate** LCL **black hole** PG **Elizabeth Blackwell** Courtesy of the National Library of Medicine, #B030136 **blood cell** COR/Science Pictures Limited **Niels Bohr** CB **bond** RS **braided stream** COR/Lowell Georgia **brain** LCL **bridge** EM **buckminsterfullerene** PT/Michael Freeman **burl** Margaret Anne Miles **bypass** LCL **caliper** Gail Piazza **capacitor** UG-GGS **caries** LCL **Rachel Carson** COR/Underwood & Underwood **George Washington Carver** AP-WWP **caste** LCL **cathode-ray tube** PG **celestial sphere** JM **cell** EM **centripetal force** PG **chromosome** LCL **circle** AA **circuit** EM **circulatory system** LCL **cladogram** EM **cloud** PG **cloud forest** COR/Michael and Patricia Fogden **coaxial cable** EM **coelacanth** Gail Piazza **comet** PDI-GI/StockTrek **complementary angles** AA **compound eye** AA-ES/OSF, John Forsdyke **computerized axial tomography** PT/© Collection CNRI **cone** RS **conic projection** AA **constellation** Tech-Graphics **continental rise** RS **contour line** U.S. Geological Survey map **Coriolis effect** EM **Jacques Cousteau** CB **covalent bond** RS **crater** PDI-GI/StockTrek **crystal** RS **Marie Curie** AP-WWP **cycloid** Tech-Graphics **cyclotron** PG **cylinder** RS **Charles Darwin** CB **dehiscence** GH/Runk & Schoenberger **desalinization** PG **diatom** COR/Lester V. Bergman **diesel engine** Tech-Graphics **digestion** EM **digestive system** LCL **dimorphism** COR/Paul A. Souders **diode** PG **dip** EM **disc brake** PG **disk flower** COR/George D. Lepp **diverticulum** CI **DNA** LCL **Doppler effect** PG **drumlin** EM **dune** EM **ear[1]** LCL **earthquake** RS **eclipse** EM **Paul Ehrlich** CB **Albert Einstein** CB **electromagnetic spectrum** Tech-Graphics **El Niño** EM **elytron** EM **enzyme** EM **epiphyte** PR/B.G. Thomson **equator** JM **erosion** COR/Chinch Gryniewicz **eudicotyledon** LCL **euglena** EM **eustasy** EM **eutectic** EM **exocarp** LCL **exterior angle** AA **eye** LCL **Fahrenheit** LCL **fault** LCL **feather** Cecile Duray-Bito **Enrico Fermi** AP-WWP **Fibonacci sequence** EM **fish** EM **fission** PG **flower** EM **fluorescent lamp** LCL **fold** EM **food chain** EM **fossil** COR/Wolfgang Kaehler; COR/Layne Kennedy **fractal** EM **Rosalind Franklin** Courtesy of the James D. Watson Collection, Cold Spring Harbor Laboratory Archives; photograph by Ray Gosling **front** UG-GGS **fulcrum** LCL **fuse** PG **fusion** PG **galaxy** NASA, The Hubble Heritage Team, STScl, AURA ODRing Arc; PR/John Chumack **Galileo Galilei** AR/Réunion des Musées Nationaux **gear** UG-GGS **germination** EM **gill** EM **Global Positioning System** EM **golden section** EM **Jane Goodall** AP-

WWP/Jean-Marc Bouju **gradient** EM **graft** LCL **gravimeter** photograph courtesy of T.M. Niebauer of Micro-g Solutions, Inc. **greenhouse effect** PG **growth ring** LCL **Gulf Stream** EM **hair** CI **harmonic** EM **heart** LCL **heat exchanger** CC **William and Caroline Herschel** CB **hertz** PG **Dorothy Crowfoot Hodgkin** © The Nobel Foundation **homolosine projection** JM **horst** Gail Piazza **hurricane** UG-GGS **James Hutton** U.S. Geological Survey **hydroelectric** Tech-Graphics **hydroid** PT/Dennis Kunkel **hydrologic cycle** EM **hydroponics** PDI-GI **hyperbola** AA **hyperopia** Cecile Duray-Bito **hypocycloid** Tech-Graphics **iceberg** CI **icosahedron** Tech-Graphics **incandescent lamp** PG **induction** PG **inflorescence** EM **interior angle** AA **internal-combustion engine** LCL **International Date Line** JM **intestine** LCL **involucre** CRF **Edward Jenner** The Wellcome Library, London **jet stream** EM **joint** CI **Irène and Frédéric Joliot-Curie** CB **karst topography** EM; Photo by William K. Jones, courtesy of the Cave Conservancy of the Virginias **kidney** CI **Klein bottle** AA **Robert Koch** CB **lagoon** COR/B.S.P.I. **laser** PG **Antoine Lavoisier** COR/Michael Nicholson **leaf** EM; PR/John Kaprielian; CRF; COR/Craig Tuttle; AA-ES/Robert Comport **leaf scar** GH **Louis and Mary Leakey** CB **lens** PG **lenticel** PT/Ken Wagner; PT/Ken Wagner **Leonardo da Vinci** AR/Alinari **lightning** EM **Linnaeus** CB **Joseph Lister** COR **lithosphere** EM **liver** LCL **liverwort** GH/Runk & Schoenberger **longitude** JM **lungfish** CC **magic square** PG **magnetic field** PT/Yoav Levy **magnetic reversal** EM **manometer** PG **Guglielmo Marconi** COR/Hulton-Deutsch Collection **Barbara McClintock** CB **meiosis** Cecile Duray-Bito **Gregor Mendel** CB **Dmitri Mendeleev** CB **meniscus** EM **Mercator projection** JM **mesa** COR/Macduff Everton **metamorphic** COR/Kevin Schafer **metamorphosis** GH/Runk & Schoenberger; GH/Runk & Schoenberger; PR/Robert J. Erwin **mirage** EM **Maria Mitchell** Vassar College Libraries, Department of Archives and Special Collections **mitosis** Cecile Duray-Bito **Möbius strip** PG **molecule** RS **monocotyledon** LCL **moon** David Mackay Ballard **moraine** COR/Tom Bean **moss** GH/Runk & Schoenberger **MRI** PT/© Collection CNRI **mushroom** LCL **mutualism** PR/Nigel J. Dennis **myopia** Cecile Duray-Bito **nanotube** Dr. Jiří Kolafa, Prague Institute of Chemical Technology (MACSIMUS software) **neap tide** PG **nephron** EM **neuron** AA **Sir Isaac Newton** COR/Leonard de Selva **North Pole** Thom Gillis **nuclear reactor** CC **oasis** COR/Phil Banko **ocellus** COR/D. Robert & Lorri Franz **ophiolite suite** EM **J. Robert Oppenheimer** AP-WWP **opposite** LCL **orbital** EM **osmosis** RS **osteoporosis** PT/CNRI; PT/CNRI **oxbow** COR/Yann Arthus-Bertrand **pahoehoe** PDI-GI/InterNetwork Media **pancreas** LCL **Pangaea** Tech-Graphics **parabola** RS **parallax** PG **Blaise Pascal** COR **Louis Pasteur** COR/Hulton-Deutsch Collection **Linus Pauling** AP-WWP **Marguerite Perey** American Institute of Physics, Emilio Segrè Visual Archives **perihelion** AA **photosynthesis** PG **pinnate** LCL **piston** Matthew Pippin **pixel** Library of Congress; Tech-Graphics **Max Planck** CB **pollen** PR/SPL, R.E. Litchfield; PR/Scimat; PR/Eye of Science **polymorphism** College Division, Houghton Mifflin Company and James Scherer **prop root** GH/Grant Heilman **pterodactyl** Tech-Graphics **Ptolemaic system** RS **pump** Gail Piazza **pupa** CC **radar** UG-GGS **rainbow** EM **rain forest** UG-GGS **William Ramsay** COR **refraction** JM **respiration** LCL **respiratory system** LCL **reverse osmosis** PG **rhizome** EM **Ellen Swallow Richards** MIT Museum **rip current** EM **RNA** CI **rocket engine** EM **Wilhelm Roentgen** AP-WWP **Albert Sabin** CB **sacrum** LCL **Jonas Salk** AP-WWP **samara** Cecile Duray-Bito **satellite** PG **saturated** RS **scapula** LCL **scoliosis** COR/Lester V. Bergman **seaweed** LCL **sessile** EM **sex chromosome** PR/Biophoto Associates, Science Source **sickle cell anemia** Custom Medical Stock Photo, Inc./© Hossler **sinus** EM **sinusoidal projection** JM **skeleton** LCL **solar cell** PG **solenoid** AA **sonar** EM **sorus** PT/Charles Kingery **South Pole** Thom Gillis **spark plug** LCL **spathe** CC **spinal cord** CI **spring tide** PG **steam engine** Tech-Graphics **Nettie Stevens** Carnegie Institution of Washington **stoma** PT/ISM **strain** EM **strophoid** AA **subatomic particle** EM **sun** AA **synapse** EM **syncline** Gail Piazza **taproot** LCL **taste bud** CI **telescope** LCL **Nikola Tesla** PR/SPL **tesseract** Tech-Graphics **thyroid gland** CI **timberline** CRF **tombolo** Image courtesy C. Palmer, www.buyimage.co.uk, 01279 757917 **tooth** LCL **tornado** PDI-GI/Don Farrall **Charles Townes** AP-WWP **transformer** RS **transpiration** EM **tributary** PR/Jeffrey Lepore **trigonometric function** RS **trilobite** AA-ES/OSF, Sinclair Stammers **turbine** AA **turbojet** PG **unsaturated** RS **Harold Urey** CB **urinary tract** CI **Van Allen belt** EM **Van de Graaff generator** AA **Venn diagram** EM **vertebral column** LCL **Andreas Vesalius** Courtesy of the National Library of Medicine, #B025721 **wave** RS **weather balloon** NASA/JPL/Caltech **Alfred Wegener** CB **wind** EM **xylem** EM **Rosalyn Yalow** CB **zone** JM

Charts and Tables

Geologic Time and **Taxonomy:** Precision Graphics **Measurement Table**, **Organic Compounds**, **Periodic Table of the Elements**, **Solar System**, **Subatomic Particles**, and **Symbols and Signs**: Catherine Hawkes, Cat & Mouse